ANIMMA 2019 – Advancements in Nuclear Instrumentation Measurement Methods and Their Applications

EPJ Web of Conferences Volume 225

Portoroz, Slovenia
17 – 21 June 2019

Editors:

A. Lyoussi
M. Giot
M. Carette
I. Jenčič
C. Reynard-Carette
L. Vermeeren
L. Snoj
P. Le Dû

ISBN: 978-1-7138-0288-4

Printed from e-media with permission by:

Curran Associates, Inc.
57 Morehouse Lane
Red Hook, NY 12571

CURRAN ASSOCIATES INC. proceedings .com

Some format issues inherent in the e-media version may also appear in this print version.

For additional information, please contact EDP Sciences – Web of Conferences at the address below.

EDP Sciences – Web of Conferences
17, Avenue du Hoggar
Parc d'Activité de Courtabœuf
BP 112
F-91944 Les Ulis Cedex A
France

Phone: +33 (0) 1 69 18 75 75
Fax: +33 (0) 1 69 28 84 91

contact-edps@webofconferences.org

Additional copies of this publication are available from:

Curran Associates, Inc.
57 Morehouse Lane
Red Hook, NY 12571 USA
Phone: 845-758-0400
Fax: 845-758-2633
Email: curran@proceedings.com
Web: www.proceedings.com

TABLE OF CONTENTS

Dear Colleagues and Friends,

Instrumentation and measurement in nuclear media as well as radiation detection, measurement, diagnostic and monitoring are key aspects that contribute to the quality of scientific programmes in the fields of physics, energy, fuel cycle, radioactive waste management, safeguards and homeland security. Furthermore, measurements relying on nuclear physics now play an important role in various fields of application such as biology, medicine and the environment. Since the first edition ANIMMA* international conference becomes now a real opportunity to get together with colleagues, partners and friends to exchange ideas and share knowledge and experience in the nuclear instrumentation, measurement methods and nuclear experimental sciences in general.

The ANIMMA Conference has always strived to create a special meeting place shared by all those working in nuclear instrumentation and its applications, as we strongly believe that cross-border exchanges between scientists, engineers and industrialists can only lead to the most developed ideas, the best solutions and the most efficient collaborations and partnerships. ANIMMA continues to maintain a high level of scientific and technical quality by presenting not only the latest advances but also the state of the art in each field through the participation of international specialists and experts. It is an ideal meeting for scientists and engineers in the field of nuclear measurement, instrumentation in specific media, radiation instrumentation, software engineering, data acquisition analysis and treatment, and related applications to present their work and network with their colleagues from around the world.

ANIMMA conference was founded in 2009, thus, we celebrated all together the 10th anniversary last June in Portoroz. This achievement was made possible thanks to your contributions, your participations and above all thanks to your continued confidence.

ANIMMA 2019 was the sixth issue of a series of conferences with a scientific program during 5 days dealing with:

- A full-day short-courses
- 4 international workshops
- At least a plenary lecture each day followed by parallel sessions.
- Intensive oral sessions for poster presentations everyday followed by plenary poster session
- Technical visit to JSI Triga research reactor.

ANIMMA 2019 was a great success thanks to your contributions and to your participation and also thanks to ANIMMA committees; partners and sponsors. 230 presentations represented by 31 nationalities have been presented and discussed during parallel oral sessions and intensive oral sessions. In addition, a permanent exhibition from industry and research institutes took place during the whole conference week.

Short-courses and workshops were also successful.

Finally, on behalf of ANIMMA committees I should like to express my heartfelt thanks to all attendees for having made this scientific event a great success.

We look forward to welcoming the community to the seventh edition of ANIMMA international conference which will be held from June 21 to 25, 2021 at the wonderful city of Prague (Czech Republic).

Prof. Dr. Abdallah LYOUSSI
ANIMMA General Chairman

(*) Advancements in Nuclear Instrumentation Measurement Methods and their Applications

edp sciences

EPJ Web of Conferences

Statement of Peer review

In submitting conference proceedings to EPJ Web of Conferences, I certify to the Publisher that I adhere to the **Policy on Publishing Integrity** of the journal in order to safeguard good scientific practice in publishing.

1. All articles have been subjected to peer review administered by the proceedings editors.
2. Reviews have been conducted by expert referees, who have been requested to provide unbiased and constructive comments aimed, whenever possible, at improving the work.
3. Proceedings editors have taken all reasonable steps to ensure the quality of the materials they publish and their decision to accept or reject a paper for publication has been based only on the merits of the work and the relevance to the journal.

Title, date and place of the conference

ANIMMA 2019
Advancements in Nuclear Instrumentation Measurement Methods and their Applications"
PORTOROŽ, SLOVENIA. JUNE 17-21, 2019

Proceedings editor(s):

Prof. Dr. Abdallah Lyoussi; Prof. Dr. Michel Giot
Dr. Michel Carette; Dr. Igor Jenčič;
Dr. Christel-Reynard-Carette, Dr. Ludo Vermeeren
Dr. Luka Snoj; Dr. Patrick Le Dû.

Date and editor's signature

December, 2, 2019

EPJ Web of Conferences **225**, 01001 (2020)
ANIMMA 2019
https://doi.org/10.1051/epjconf/202022501001

SCALP: a detector for (n,α) cross-section measurements

G. Lehaut*, M. Bourgeot*, B. Galhaut† D. Goupillière*, M. Henri*, F.R. Lecolley*, X. Ledoux‡, J. Lory*, L. Manduci†, N. Marie*, J. Perronnel* and Ch. Vandamme*
*Laboratoire de Physique Corpusculaire de Caen, ENSICAEN/Univ. de Caen/CNRS-IN2P3, France.
†Ecole des Applications Militaires de l'Energie Atomique, BP. 19, 50115 Cherbourg Armées, France
‡GANIL, Grand Accélérateur National d'Ions Lourds, CEA-DSM/CNRS-IN2P3, BP 55027, 14076 Caen Cedex 5, France

***Abstract*—Neutron induced reactions on oxygen have been studied with strong interest because of the uncertainties generated on the helium production in fuel and on the neutron multiplication factor in nuclear reactors [1], [2]. Still large discrepancies exist and new measurements are welcome in order to acquire new data aiming at the uncertainty reduction [3].**

SCALP (Scintillating ionization Chamber for ALPha particle production in neutron induced reaction) is a new scintillating ionization chamber [4] used as an active target to measure the cross section of (n, alpha) reactions on various gaseous targets such as ^{19}F or ^{16}O, from the reaction threshold up to 15 MeV. It consists of an ionization chamber filled with CF_4 (for fluorine measurements) or CF_4+CO_2(for oxygen measurements) allowing the detection of the energy deposed by the light charged particles emitted in the (n, alpha) reaction. In addition, four Photo-Multiplier Tubes detect the scintillation light produced by the interaction of the particles in the gas active volume. Taking advantage of the fast response of the scintillation, the neutron kinetic energy can be inferred by time-of-flight measurements. SCALP is then well adapted to mono-energetic neutron beams or to white neutron beams that will be delivered at the NFS Facility [5].

Because of its good resolution, SCALP discriminates different channel outputs, enabling to disentangle the different reactions [6]. We will present the performances of the SCALP detector in terms of temporal and energetic features. We will also present tests made at the GENESIS plate-form at the LPSC Grenoble [7].

I. Introduction

In the context of global warming, nuclear reactors are considered as an important supply of nearly carbon free energy production. However, great efforts are still deployed to improve their safety as well as the reduction of the ^{235}U consumption in view of its exhaustion in the next century.

In this context, reactor simulations are recommended to make future plants more safe and efficient. Some codes (Monte-Carlo or deterministic) allow specific simulations in neutron transport (based on the Boltzman equation) and/or in the thermohydraulics. These simulations give access with a very high precision to the neutron flux in the reactor core or in the surrounding structure. The drawback is that, since these codes need measured or evaluated nuclear data inputs (such as nuclear cross sections, decay constants...), the uncertainties of the simulated results depend on those of the nuclear data.

For reactor calculations, all possible reactions will not have individually the same impact on integral observables.

The ^{16}O(n,α)^{13}C figures among these key reactions because it affects the knowledge of helium production in the fuel since 25% of this helium in produced by ^{16}O(n,α) and induces an uncertainty of a few hundred of pcm in the neutron multiplication factor (k_{eff}) which is a fundamental parameter for the reactor control [3]. Therefore this reaction belongs to the High Priority List of the NEA who demands more measurements for this reaction from its threshold up to 10 MeV.

The SCALP (Scintillating ionization Chamber for ALPpha particle) detector was designed and built to satisfy this request. The section 2 presents a description of the detector and its electronic acquistion. Then in section 3, some results obtain with a tri-α-source and also with a 15 MeV neutron source are shown.

II. SCALP detector

A. Description

SCALP is an active gaseous target designed to estimate the cross-section of the ^{16}O(n,α)^{13}C reaction. It consists of a Ionization Chamber IC coupled to four Photo-Multiplier Tubes PMTs which read the scintillating light produced by the neutron interaction in the CF_4 scintillating gas. Taking advantage of the fast response of the scintillation, the neutron kinetic energy can then be inferred by time-of-flight measurements.

Fig. 1 shows the SCALP detector. It is made of a sealed gas vessel with four quartz portholes to allow the scintillation from the gas to be detected by Hammatsu R2059 PMTs having an extended wavelength photo cathode. In the gas volume there is also a IC delimited by two electrodes (surface 123×123 mm^2) spaced by 125 mm and between which a potential difference of 4 kV is applied. A constant drift electric field is ensured in the active volume by field-shaping wires. The induction of the ionization signal takes place in the last 5 mm of the chamber thanks to a Frisch grid. The electrons produced during the ionizations are collected and this signal is then amplified by a charge preamplifier.

B. Detector acquisition system

The signals from PMTs are sent to a FASTER [8] Caras Module in QDC/TDC (charge/time digital converter) mode, where a Constant Fraction Discriminator is implemented to allow a good resolution of the time of PMT signal t_{PMT}

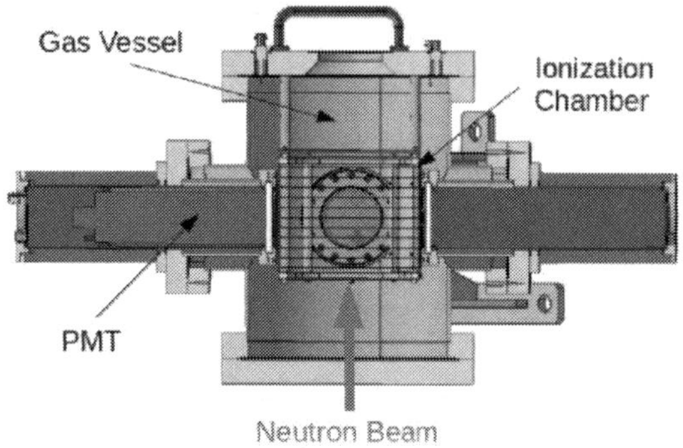

Fig. 1. Mechanical sketch of SCALP showing the principal component of the detector: a gas vessel, the PMTs and the ionization chamber.

(for a gate shape signal, the TDC achieves a resolution $\sigma_{TDC} \approx 10$ ps). The charge signal of each PMT is integrated over a 100 ns windows.

Also the IC signal is sent to a Caras Module in a ADC (amplitude digital converter) mode. The threshold trigger for the IC is carried out on a fast signal ($\tau_{shapping}$ = 60 ns) at a time t_{IC}, the pulse height (PH) being measured after a spectroscopic filter ($\tau_{shapping} \sim \mu$s).

During the acquisition time, each signal is independent. PMT signals are merged in offline analysis with a temporal cut $\Delta t_{PMT} < 10$ ns. PMT and IC signals are also merged with the condition $\Delta t_{IC-PMT} < 10$ μs. In this way coincidences are selected and the event-by-event analysis can be done.

III. Detector Test

A. α-source response

In order to test the SCALP response in laboratory, an α-source was used. It was composed by 3 isotopes ^{239}Pu, ^{241}Am and ^{244}Cm, leading to α-particle mean energies of, respectively E_{α} = 5.14; 5.44; 5.8 MeV.

The source was placed in the active volume in such a way as to emit in the half solid angle parallel to the electric-field. It was collimated.

The first tests were done with a pure CF_4 gas at 1 bar.

Fig. 2 shows the light output obtained for the α-particle source. The mean value is around 15 photo electrons (p.e.). With the help of coincidences among the photomultipliers, it is possible to reduce the background noise. The resulting curve is in red.

The correlation between the lights measured by two PMTs are shown on fig. 3. The similar response of the two PMTs are expected since their positions are symmetric relative to the source position.

The difference between PMT-time signals was also studied, see fig. 4. The dispersion around the average value is σ_{CRT} = 1.15 ns (corresponding to the coincidence resolving

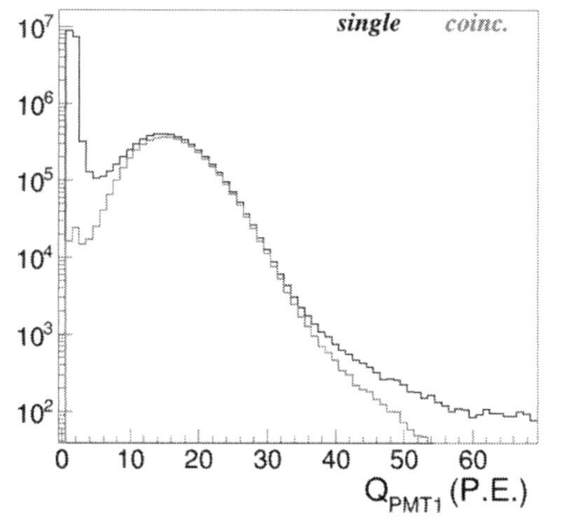

Fig. 2. Distribution of the light detected by a PMT for 5.5 MeV α-particles. The black curve shows all events and the red curve shows event in coincidence with at least one another PMT.

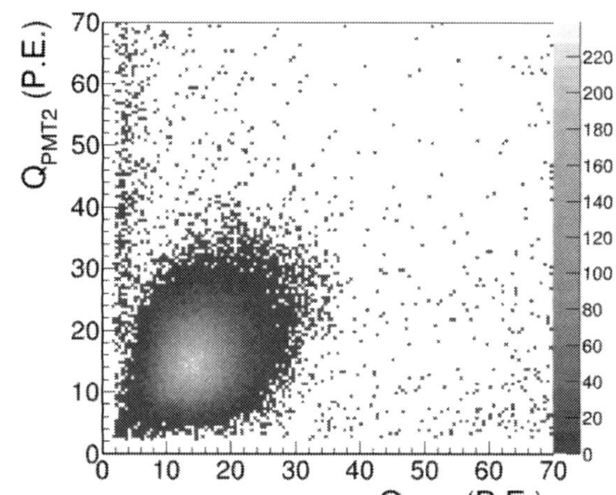

Fig. 3. Correlation between the light detected by two PMTs for the same set of events.

https://doi.org/10.1051/epjconf/202022501001

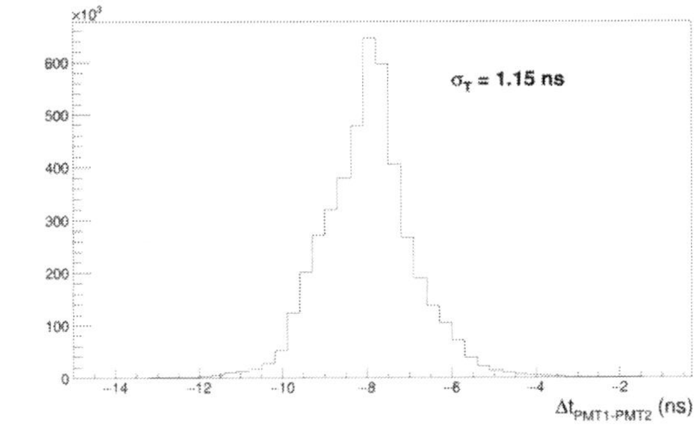

Fig. 4. Distribution of the time difference between two PMTs for the same set of events.

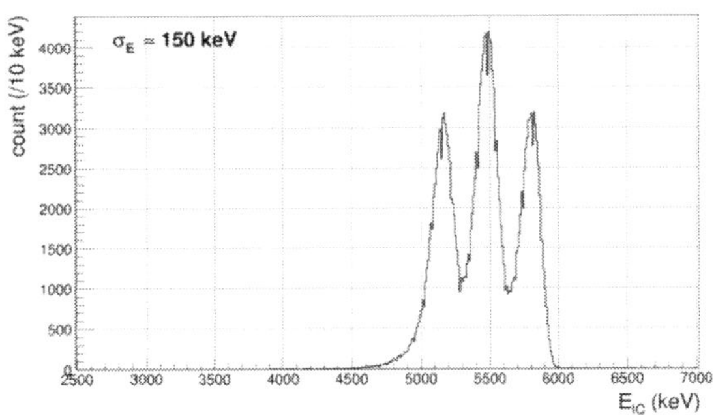

Fig. 5. Energy spectrum of the tri-α source, after calibration.

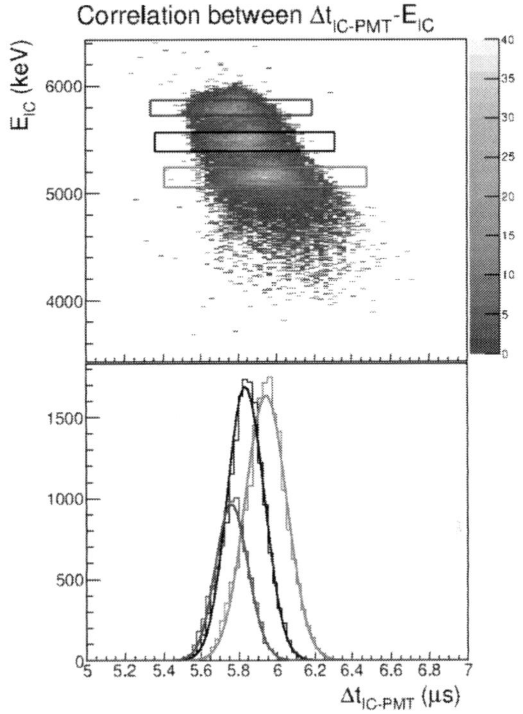

Fig. 6. Upper panel: Correlation betwenn the difference between PMTs and IC time signals, Δt_{IC-PMT} and the α-particle energy, E_α, detected in the IC. Lower panel: Δt_{IC-PMT} distributions obtained for the three different α-particle energy selections associated to each of the isotope of the tri-α source and shwn as rectangmes in the upper panel.

time of our set-up), which constitue a good result for our purpose.

With a linear energy calibration inferred from the tri-α source measurements, the α energy distribution obtained from the IC response is displayed in fig. 5. The energy resolution is about σ_{IC} = 150 keV, for the 5-6 MeV energy range.

SCALP allows also the vertex position determination, i.e., the distance of the interaction point from the Frisch Grid. This can be obtained using the time difference between the IC signal and the involved PMT signal: Δt_{IC-PMT}. In fact, using the correlation between the α-particle energy E_α and the time difference Δt_{IC-PMT} (as shown in the upper panel of fig. 6), one can observe different time values induced by the pulse height (due to the threshold trigger) and the difference of fews millimeters walked by each α. In the lower panel of fig. 6 are presented the Δt_{IC-PMT} distributions obtained for three different α-particle energy selections associated to each of the isotope of the tri-α-source and shown as rectangles in the upper panel.

Measurements of the time difference Δt_{IC-PMT} were done for different source positions along the electric field axis. The results inferred for each of the three α-particle energy selections are shown in fig. 7. A linear correlation between the difference Δt_{IC-PMT} and the source position is observed. The slope extracted, which correspond to the electron drift velocity, equals to v^e_{drift} = 5.2 ± 0.4 cm/μs. This velocity value is in agreement with the one given in reference [9].

A second set of test was performed using a mixture of a few percent of CO_2 with the CF_4 gas. The CO_2 gas constitutes the target and its concentration in the active volume must be accurately determined. As expected, the introduction of even a small quantity of CO_2 changes the intensity of the output light. Fig. 8 displays the loss in light output as a function of the CO_2 concentration fraction. This loss amounts at nearly 40% for 3% CO_2 fraction. As expected, the resolution of the energy IC signal increases with the CO_2 concentration (see fig. 9). So, in order to maintain good specifications of our detector for physics measurements, the fraction of CO_2 in the active volume must not exceed 2-3%.

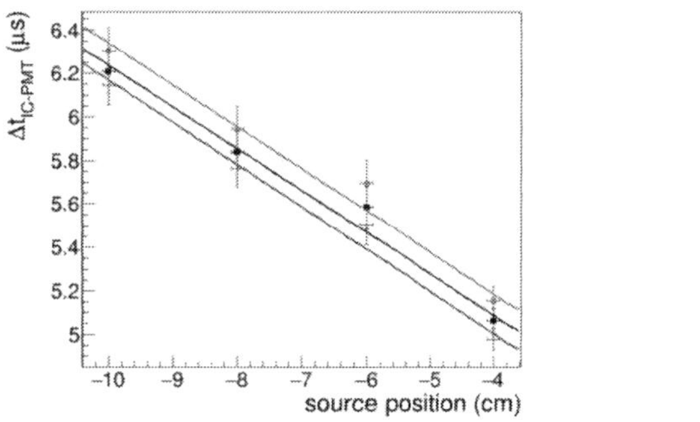

Fig. 7. Time difference Δt_{IC-PMT} for different source positions along the electric field axis and for each of the three α-particle energy selections.

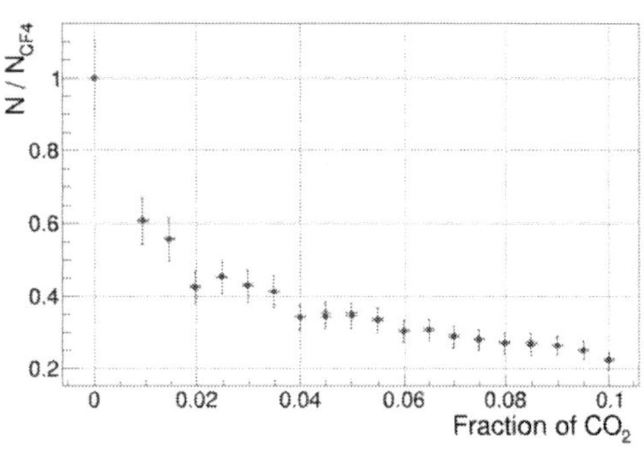

Fig. 8. Evolution of the light fraction detected by the PMTs as a function of the fraction of CO_2 gas added in the active volume. The evolutions obtained for two different PMTs are presented with red and blue points.

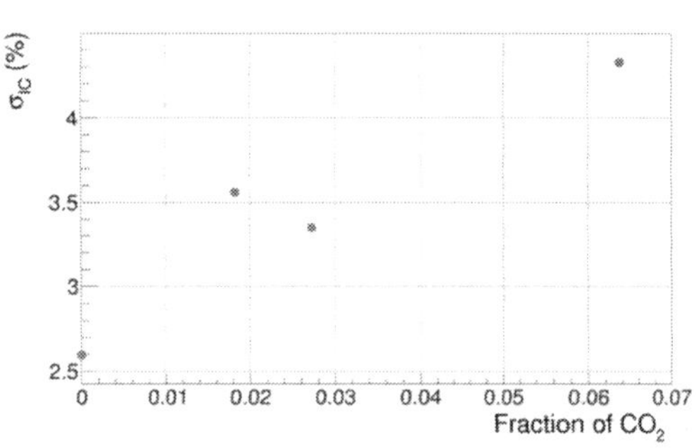

Fig. 9. Evolution of the IC energy resolution as function of the fraction of CO_2 gas added in the active volume.

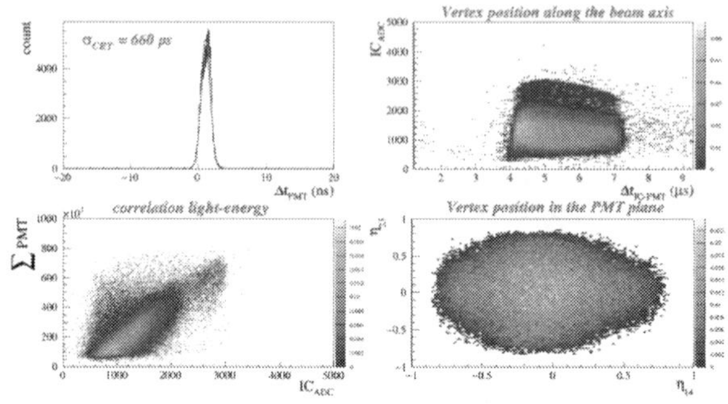

Fig. 10. Panel a): differences between the PMT time signals. Panel b): correlation between the time difference Δt_{IC-PMT} and the IC pulse height, ADC. Panel c): correlation between the ADC signal and the sum of lights detected by the four PMTs. Panel d): Correlation between the contrast variable of PMTs (see text for definition) along two orthogonal directions.

B. Test at GENESIS

To complete our tests done first with α-sources, the detector SCALP was installed at the GENESIS facility of the LPSC of Grenoble. This facility can deliver non-collimated neutron produced by $d(d,n)$ or $d(t,n)$ reactions using 220 keV deuton electrostatic accelerator GENEPI2 [7]. For the SCALP test, a tritium target was used and neutrons were produced with an incident energy of E_n = 15.19 MeV.

Fig. 10 presents the response of the detector filled with pure CF_4 gas at a 1 bar pressure. In panel a), the time differences between the four PMT signals show that the detection of the neutron interaction time in the gas is achieved with a resolution of a $\sigma_{CRT} \approx$ 600 ps. The event distribution in the whole active volume is illustrated in panel b) where the IC pulse height (ADC) is presented as a function of the time difference Δt_{IC-PMT}. Compare to the α source measurement at fixed position, the distribution of the time difference Δt_{IC-PMT} the neutron source distribuion is larger corresponding to events distributed all along the electric field axis. In panel c), the correlation between the ADC signal and the sum of the lights detected by the four PMTs is presented. As expected, the correlation is linear (due to common origin of the scintillation and of the ionisation).

Defining the η_{ij} observable as it follows:

$$\eta_{ij} = \frac{QDC_i - QDC_j}{QDC_i + QDC_j}$$

and using these observable for face-to-face PMTs in the two orthogonal directions, one can obtains an image of the energy deposit in the active volume in a plane perpendicular to the PMT window (see panel d)). This combined with Δt_{IC-PMT} allows a 3D image of the interactions points.

All these results are quite good despite the fact that the detector was not operating at the nominal electric field because of a technical problem encountered.

IV. Conclusion and outlooks

There is a need for new measurements of the $^{16}O(n,\alpha)$ cross-section. For this purpose, we developed a new detector, SCALP, allowing a simultaneous measurement of the time-of-flight of the incoming neutron and of the energy deposited by the charged particles produced in the reaction. This set-up makes use of both properties: scintillation and ionization of the CF_4 gas filling the detector active volume. First, laboratory tests were done with a tri-α-source of mean energy $E_\alpha \approx 5.5$ MeV. At this energy, the light detected by each PMT was estimated to be around 15 p.e.E for the pure CF_4 gas and 7 p.e. for a mixture of CF_4 and CO_2 (97/3). The coincidence resolving time of our detector was also extracted. It is of the order of one ns. Moreover, the energy resolution of the IC is 150 keV and 220 keV for respectively, pure CF_4 and a mixture CF_4/CO_2. The time and energy resolutions are suitable for our experimental aim.

A test with a neutron source was also performed at the GENESIS platform at LPSC in Grenoble. The SCALP device is able to detect the neutron interaction in his entire active volume.

The next step of our program would consit in making the measurements at different neutron facilities (NFS@SPIRAL2, IRMM, Amande). The first campaign will be devoted to a pure CF_4 target in order to estimate the reaction rate of neutrons on carbon and fluorine nuclei. A follow-up campaign would be performed with a CF_4/CO_2 target. The data analysis and collaboration with evaluators will take place in frame of the SANDA (Supplying Accurate Nuclear Data for energy and non-energy Applications) project from H2020 program.

Acknowledgment

We would like to thank, the GENESIS team for providing the neutron source at the LPSC Grenoble: E. Labussière, S. Rey, M. Baylac, B. Cheymol and A. Billebaud. We also would like to thank colleagues, J.M. Fontbonne and L. Leterrier from LPC Caen for their technical support, especially for the IC instrumentation. This work was partially supported by the NACRE project of IN2P3 institute and by the NEEDS french program.

References

[1] M. Salvatores and R. Jacqmin, *Uncertainty and Target Accuracy assessment for innovative systems using recent covariance data evaluations* (International Evaluation Co-operation, NEA/WPEC-26, 2008), 464
[2] M. B. Chadwick et al., *The CIELO collaboration*, Nuclear Data Sheets 118 (2014).
[3] Courcelle, Nuclear Science NEA/WPEC-22, 2006
[4] G. Lehaut, et al. *Scintillation properties of N 2 and CF 4 and performances of a scintillating ionization chamber*, Nucl. Instrum. Meth. A 797 (2015), 57-63.
[5] X. Ledoux, et al., *The Neutrons for Science Facility at SPIRAL-2*, Nucl.Data Sheets 119 (2014) 353-356.
[6] B. Galhaut, et al., *SCALP: Scintillating ionization chamber for ALPha particle production in neutron induced reactions*, EPJ Web Conf. 146 (2017) 03014.
[7] http://lpsc.in2p3.fr/index.php/fr/groupes-de-physique/enjeux-societaux/physique-des-reacteurs/plateforme-genesis
[8] http://faster.in2p3.fr/
[9] J. Va'vra, P. Coyle, J. Kadyk, J. Wise, Nuclear Instruments and Methods in Physics Research Section A: Accelerators, Spectrometers, Detectors and Associated Equipment 324 (12) (1993) 113.

Timepix3 Luminosity Determination of 13 TeV Proton-Proton Collisions at the ATLAS Experiment

André Sopczak
ATLAS Collaboration
andre.sopczak@cern.ch

***Abstract*—Medipix and Timepix devices, installed in the ATLAS cavern at the LHC, have provided valuable complementary luminosity information. Results are presented from measurements with Timepix3 (TPX3) detectors. In contrast with previously employed frame-based data acquisition, the TPX3 detector remains active continuously, sending information on pixel hits as they occur. Hit- and cluster-counting methods were used for the luminosity determination of the LHC proton-proton collisions. The LHC luminosity versus time is determined using these two methods and fitted to a simple model, which incorporates luminosity reduction from single bunch and beam-beam interactions. The precision of the luminosity determination could be improved by counting the number of clusters, instead of just pixel hits. The internal precision and long-term stability of the TPX3 luminosity measurement are below 0.5%. TPX3, owing to its 1.56 ns time-tagging, is able to resolve the time structure of the luminosity due to the collisions of individual proton bunches when integrated over an LHC fill.**

I. Introduction

Precision luminosity measurements are of particular importance for many analyses in high-energy physics. Networks of hybrid active pixel detectors from the Medipix/Timepix family [1] installed in ATLAS cavern [2] successfully demonstrated their potential to determine luminosity [3], [4]. Recently, the latest generation Timepix detectors, namely Timepix3 (TPX3) [5], were installed in the ATLAS cavern at CERN [6]. The TPX3 device closest to the LHC interaction point is used in this analysis to measure the primary and secondary particle fluxes originating from 13 TeV proton-proton collisions. The data were taken between April and October 2018 during the LHC Run-2 operation.

The use of the TPX3 device for luminosity measurements has several advantages compared to the previous luminosity measurements at the LHC during Run-1 (MPX devices) [3] and Run-2 (TPX devices) [4].

1) The possibility of (quasi-) continuous operation: The dead-time caused by the readout of the frames is about 6 s for MPX and 0.12 s for TPX devices, requiring a compromise between sensor occupancy and relative dead-time minimisation. Cluster-counting with MPX and TPX devices is possible with high particle fluxes only when the exposure time is much shorter than the dead-time. Otherwise, overlapping clusters could not be separated. Therefore, hit-counting is necessarily used for previous high rate luminosity analyses. Furthermore, the TPX devices can be operated in either of three modes: time-over-threshold (energy deposits, ToT), time-of-arrival (ToA), and Medipix (counting) mode; in contrast, the TPX3 device measures ToT and ToA information simultaneously. A novelty of TPX3 is the pixel-comparator-driven (event-driven) data transmission which replaces the frame-based transmission of the TPX devices.

2) An improved cluster separation: In the previous MPX and TPX luminosity analyses, no cluster (particle) counting with high statistics was possible. Thus, to achieve statistically significant short-term luminosity measurements, a hit-counting method was applied. The devices were used in the Medipix mode, where each pixel counts the number of pixel hits per frame. A source of statistical uncertainty arose from the large variance of hit counts per interacting particle. Owing to the precise pixel time-stamping of TPX3, cluster identification and separation could also be done in high flux conditions. This leads to a higher precision of the luminosity measurements. Furthermore in TPX3 data, clusters were defined by temporally and spatially concurrent hits on the sensor.

3) A higher time resolution: The previous MPX and TPX luminosity measurements were LHC bunch-integrated as the exposure time was much longer than the bunch separation of 25 ns. As already demonstrated, with the time resolution of TPX3 and proper triggering, it was possible to resolve the bunch structure of the LHC beam [6]. This article provides a proof of principle, as it demonstrates the luminosity determination of individual colliding proton bunches when the data are integrated over an LHC fill.

In summary, an important innovation is that the TPX3 device allows for cluster counting up to high particle rates, since pixel data are digitized and read out as a stream rather than accumulated on-pixel until a full frame is transmitted. Additionally, if clusters partially overlap, they can now be reconstructed separately using the time stamps in the pixels.

Moreover, the two-layer structure of the TPX3 device doubles the measurement statistics and, in this analysis, allows one to determine the internal precision and long-term time stability of individual TPX3 devices. Here, 'internal' refers to the luminosity precision being determined only based on the TPX3 device and not with respect to other luminosity determination possibilities.

This extended abstract outlines a future article which is structured as follows. First, the TPX3 device is described, along with the concept of LHC luminosity monitoring by hit-counting and cluster-counting. Further, the LHC luminosity versus time is determined and the internal TPX3 measurement precision is evaluated. The internal precision of luminosity determination is obtained by computing the difference between two layers of the TPX3 device. The internal long-

A. Sopczak is with the Institute of Experimental and Applied Physics, Czech Technical University in Prague, Husova 240/5, CZ-11000, Czech Republic. (e-mail: andre.sopczak@cern.ch).

term luminosity precision is given from the comparison of layer-1 and layer-2 luminosity measurements. Additionally, the relative long-term stability is also given with respect to other ATLAS luminosity monitors (Section II). Finally, as a proof of principle, the luminosity determination of individual LHC bunch crossings is described, followed by conclusions (Section III).

II. Long-Term Stability with Respect to Other ATLAS Luminometers

The relative long-term stability of TPX3 luminosity measurements was already compared to that of other ATLAS luminometers for 2017 data [7]. Figure 1 shows the fractional differences between the run-integrated luminosities measured from TPX3, track-counting, EMEC, FCal, and TILE with respect to the baseline LUCID algorithm. The luminosity measurements from the other detectors have been normalised to that of LUCID for a reference run. The values for each run are plotted as a function of the cumulative delivered luminosity fraction, ranging from zero at the start of the year to one at the end of the year. This results in an axis monotonically increasing in time. Short runs with less than about two hours of data-taking are not shown. The run-to-run agreement between the various luminosity measurements is generally at the percent level or better for the bulk of the data, with various short- and long-term trends being visible.

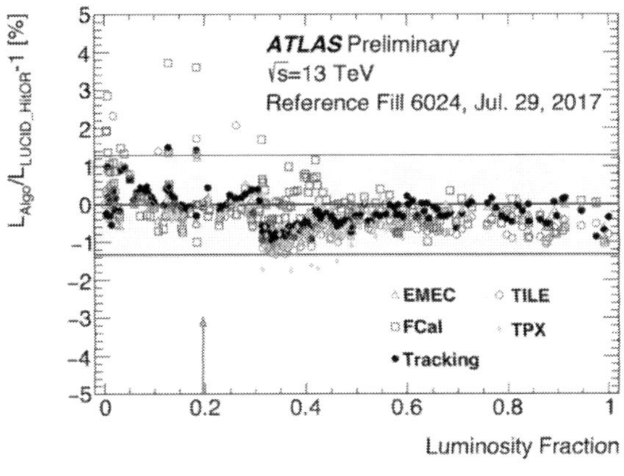

Fig. 1. Fractional differences in run-integrated luminosity between the LUCID BiHitOR algorithm and the TPX3, track-counting, EMEC, FCal, and TILE measurements, plotted as a function of the cumulative delivered luminosity normalised to the 2017 total. The luminosity measurements from the other detectors have been normalised to that of LUCID in the reference run indicated by the arrow. The assigned ±1.3% long-term stability uncertainty is shown by the yellow band [7].

III. Conclusions

A double layer TPX3 device installed in the ATLAS detector cavern has successfully taken data at the LHC during Run-2 with 13 TeV proton-proton collisions. The relative internal short-term precision of the TPX3 luminosity measurements was determined from the LHC luminosity curve to be less than 0.5% for 60 s time intervals. It is observed that cluster-counting improves the luminosity precision compared to hit-counting, since one particle typically corresponds to one cluster of hits. The internal short-term precision was determined in a complementary evaluation by studying the relative difference in luminosities measured by the two TPX3 sensor layers using cluster-counting and a relative precision of 0.4% was observed. The deviation from the internal long-term time stability of the TPX3 device for luminosity measurements was below 0.4%. When integrated over an entire LHC fill, it is demonstrated that the TPX3 device used in this study even has the capability to determine the time structure of the luminosity due to the collisions of individual LHC proton bunches.

Acknowledgment

The author would like to thank warmly the ATLAS Luminosity Group for useful discussions and interactions, and the Medipix Collaboration for providing the Timepix3 assemblies. The project is supported by the Natural Sciences and Engineering Research Council of Canada (NSERC), and by the European Regional Development Fund-Project "Van de Graaff Accelerator - a Tunable Source of Monoenergetic Neutrons and Light Ions" (No. CZ.02.1.01/0.0/0.0/16_013/0001785).

References

[1] C. Leroy, S. Pospisil, M. Suk, and Z. Vykydal, "Proposal to Measure Radiation Field Characteristics, Luminosity and Induced Radioactivity in ATLAS with TIMEPIX Devices," Project Proposal [Online]. Available: http://cds.cern.ch/record/1646970, 2014.

[2] ATLAS Collaboration, "The ATLAS Experiment at the CERN Large Hadron Collider," *JINST*, vol. 3, p. S08003, 2008.

[3] A. Sopczak, B. Ali, N. Asbah, B. Bergmann, K. Bekhouche, D. Caforio *et al.*, "MPX Detectors as LHC Luminosity Monitor," *IEEE Trans. Nucl. Sci.*, vol. 62, pp. 3225–3241, 2015.

[4] A. Sopczak, B. Ali, T. Asawatavonvanich, J. Begera, B. Bergmann, T. Billoud *et al.*, "Precision Luminosity of LHC Proton-Proton Collisions at 13 TeV Using Hit Counting With TPX Pixel Devices," *IEEE Trans. Nucl. Sci.*, vol. 64, pp. 915–924, 2017.

[5] T. Poikela, J. Plosila, T. Westerlund, M. Campbell, M. De Gaspari, X. Llopart, V. Gromov, R. Kluit, M. van Beuzekom, F. Zappon, V. Zivkovic, B. C. K. Desch, Y. Fu, and A. Kruth, "Timepix3: a 65k channel hybrid pixel readout chip with simultaneous ToA/ToT and sparse readout," *JINST*, vol. 9, no. 05, p. C05013–C05013, May 2014.

[6] P. Burian, P. Broulím, B. Bergmann, V. Georgiev, S. Pospíšil, L. Pušman, and J. Zich, "Timepix3 detector network at ATLAS experiment," *JINST*, vol. 13, no. 11, p. C11024, 2018.

[7] ATLAS Collaboration, "Luminosity determination in pp collisions at $\sqrt{s}$ = 13 TeV using the ATLAS detector," 2019, ATLAS Conference Note, ATLAS-CONF-2019-021.

Energy calibration via correlation using an adaptive mesh refinement

Daniel Maier*, Olivier Limousin*, Geoffrey Daniel*†
*AIM, CEA, CNRS, Université Paris-Saclay – Université Paris Diderot
Sorbonne Paris Cité – F-91191 Gif-sur-Yvette Cedex, France – Email: daniel.maier@cea.fr
†École des Ponts ParisTech – 77420 Champs-sur-Marne, France

***Abstract*—Energy calibration via correlation has proven to be a powerful calibration method especially in the case of calibration data sets with low counting statistics. This work reports on an evolved version of this approach that uses an adaptive mesh refinement to sample the parameter space which results in the same calibration performance but in a considerably reduced computation time.**

I. Introduction

A precise energy calibration is a critical step for many experiments using X-ray and gamma-ray detectors, like spectrometry in general, X-ray fluorescence analysis, Compton imaging, source identification... Non-linear effects introduced by the detector response as well as non-linear effects within the signal processing chain, like in amplifier ASICs or ADCs, often require to use non-linear models for the energy calibration, i.e. a parameter space with a dimensionality bigger than two. In addition, repetitive re-calibration that is used to correct time-dependent response instabilities, or a strong pixelation of the detector result in low counting statistics per detector channel implying a reduced accuracy of the energy calibration. Section II shortly summarizes the main idea behind the classical energy calibration via correlation (ECC) approach. Section III presents an evolved parameter sampling and optimization method used for ECC, the adaptive mesh refinement (AMR). Section IV gives an outlook for the implementation of an adaptive reduction of dimensionality, a feature foreseen for the next version of ECC. Section V summarizes the results.

II. A first solution: ECC

A method that uses an Energy Calibration via Correlation approach was presented [1] to address the problems of low counting statistics in the calibration data set. The idea of ECC is not to use a peak fitting approach, which examines only specific line positions, but to use the complete spectral response of the calibration measurement. The procedure of ECC can be summarized as follows:

1) Make a synthetic line spectrum based on prior knowledge of the used calibration source (Fig. 1 (a)).
2) Modify the line spectrum to a synthetic energy spectrum by applying the detector response model. This model corrects for a non-ideal detector response and includes typically the energy resolution and the quantum efficiency of the detector (Fig. 1 (b)).

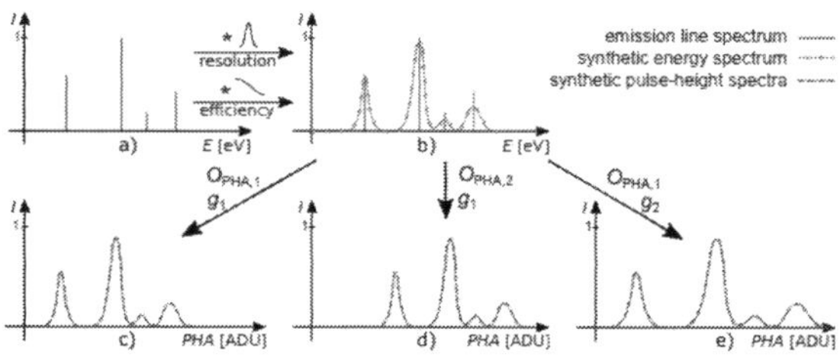

Fig. 1. Concept of ECC: creating synthetic energy spectra (a-b) and transforming them into synthetic pulse-height spectra (c-e) that are correlated with the observed pulse-height spectrum.

3) Discretize the parameter space and calculate for every combination of parameters the corresponding synthetic pulse-height spectrum (Fig. 1 (c) to (e)).
4) Compute the correlation value of all synthetic pulse-height spectra with the observed pulse-height spectrum.
5) The best calibration parameters are found by searching for the highest correlation (Fig. 2).

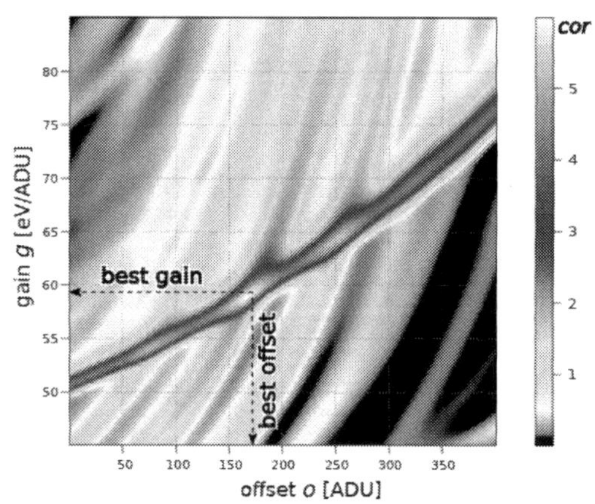

Fig. 2. Correlation map: the correlation value cor indicates the goodness of the parameter set which is here defined by the offset o and the gain g.

The correlation value is defined as

$$cor = \sum_{i=0}^{N-1} S_{\mathrm{syn},\,i} \cdot S_{\mathrm{obs},\,i} \tag{1}$$

where i is the binning index of the synthetic pulse-height spectrum S_{syn} and the observed pulse-height spectrum S_{obs} and N is the total number of bins. The higher the correlation value the better is the match between the two spectra. Therefore, the correlation value can be used as a measure of the goodness of the parameter set.

Even though the presented examples show a linear calibration, the concept of ECC is not limited to any number of parameters. An example of a non-linear ECC using three parameters can be found in [1].

Compared to a peak fitting approach that is based on finding PHA-energy pairs and fitting these pairs to the proposed calibration model, ECC performs considerably better, especially for calibration sets with a small number of photons. Another benefit of ECC is that it is not restricted to line peaks but can use the full spectral response like Comptonization structures. See reference [1] for more details on ECC and a quantitative comparison between ECC and a Gaussian peak fitting approach.

III. Adaptive mesh refinement

The equally spaced homogeneous grid that was used for the previous presented implementation of ECC samples each region of the parameter space equally fine. If a high parameter resolution is required and the parameter space is large because of high dimensionality or large pixel-to-pixel variations, the computation time for the calibration becomes an issue. This is especially true for applications which aim for near real-time calibration or re-calibration or applications which have many detection channels, i.e. strong pixelation. Adaptive mesh refinement is a numerical analysis method that samples the parameter space dynamically during the computation time with an increasing resolution for interesting regions. It is frequently used in simulations of turbulent regions of fluid dynamics [2].

A. *Adaptive mesh zooming*

Compared to the previous, classical ECC procedure described in Sect. II, only the 3rd step is changing for $\mathrm{ECC}_{\mathrm{AMR}}$. The adaptive mesh approach samples the parameter space in two steps. First, an initial homogeneous mesh, similar to the one used in the classical approach but with a much lower parameter resolution, is applied and the corresponding correlation map is calculated. Based on the obtained correlation values $cor_{0,j}$, regions are classified as interesting if

$$cor_{0,j} > t_0 \quad \text{for all mesh elements } j \tag{2}$$

$$t_0 = f_0 \cdot cor_0^* \tag{3}$$

where f_0 defines a fraction of the best obtained initial correlation value cor_0^*. The initial zooming threshold t_0 triggers a zoom into an interesting region with a three times increased resolution for each parameter; i.e. for two parameters an interesting region is sampled with a sub-resolution of 3×3. The correlation values of each of these 9 sub-regions are calculated and new interesting regions are defined via

$$cor_{i,j} > t_i \tag{4}$$

$$t_i = t_{i-1} + f \cdot (cor_i^* - t_{i-1}) \quad \text{for } i = 1..N_z \tag{5}$$

where i is the iteration index and N_z defines the maximal number of iterative zoomings. The zooming-threshold definition via two independent parameters f_0 and f is motivated by the idea to have a less strict selection for the first zooming but an increasing harder selection for the following iterations. Figure 3 shows the described zooming method schematically.

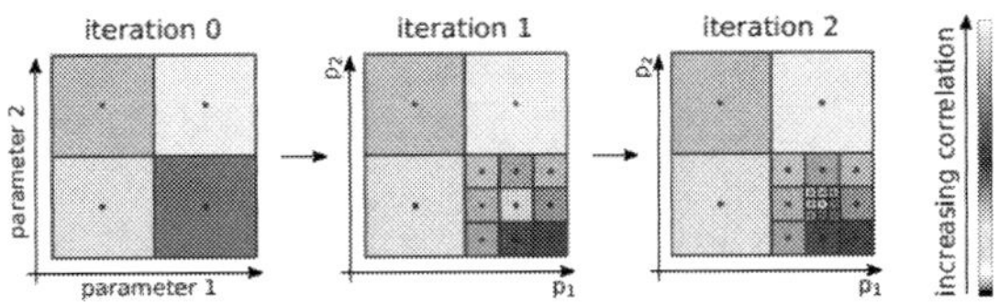

Fig. 3. Schematic zooming into interesting regions. The parameter set for each region is indicated with a dot. For iteration 2 the total number of calculated correlations is 4+8+8=20.

B. *Adaptive mesh expansion*

Because the center of a parameter region is used to calculate the correlation value for this region, the region might appear as not interesting even if the optimal parameter set is positioned within the region but close to the edge of the region. In this case, the rough sampling prevents from finding the optimal parameter set.

To minimize this problem, a mesh expansion method is developed that allows to step backwards iteratively and zoom into a region that was initially classified as uninteresting if the difference of the iteration index i between neighboring regions is bigger than two. For an element with index i_e and its neighboring elements with index $i_{e,k}$, the expansion condition becomes

$$|i_e - i_{e,k}| > 2 \quad \text{for all } e \text{ and } k. \tag{6}$$

In order to keep the number of neighbors low, only neighbors to the edges of a region are considered. This is especially important for a high dimensional parameter space. Given the dimensionality of the parameter space as d, the number of edge neighbors that differ only in one parameter value is $2 \cdot d$ while it is maximally $3^d - 1$ for all possible neighbors.

After the expansion, the zooming criterion is applied for the new region according to the saved thresholds t_i. Finally the expansion condition is checked for the new region. Due to this method, the algorithm expands into interesting parameter regions even if they seemed to be uninteresting within the previous iterations. In this way, the expansion method works as a local gradient ascent optimizer.

Figure 4 shows schematically the described expansion method followed by a re-zoom. After the zooming iteration $i = 2$ ends, the zooming-threshold t_2 is calculated. The region indicated in red exceeds this threshold and a zoom will be conducted for $i = 3$. This would violate the condition of Eq. (6) so that an expansion is conducted before iteration 3 begins. After the neighboring region is expanded to $i = 1$, it is re-zoomed according to threshold t_1. Finally, the expanding condition is checked again and another expansion might occur. If no more expansion is to be done, iteration $i = 3$ starts.

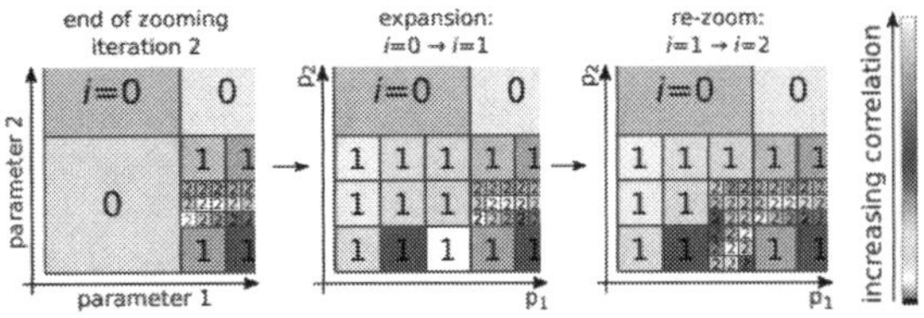

Fig. 4. Schematic expansion into an interesting region followed by a re-zoom. After the re-zoom, another expansion to the bottom region (not shown) might occur or the next iteration $i = 3$ starts.

C. *Calibration example*

Figure 5 shows a correlation map for the calibration of a camera system [3] that is based on a CdTe Caliste-HD detector [4]. The calibration uses an Am-241 source and calculates the best calibration parameters via classical ECC which was conducted by using ECC_{AMR} with no iteration ($i = 0$) and with a fine resolved initial mesh of $\Delta o = 5\,\mathrm{ADU}$ and $\Delta g = 0.25\,\mathrm{ADU/keV}$. The parameter space is equally sampled with 1000×720 sampling points and the calculation of the $7.2 \cdot 10^5$ correlations takes approximately 100 seconds on an ordinary computer with single core use.

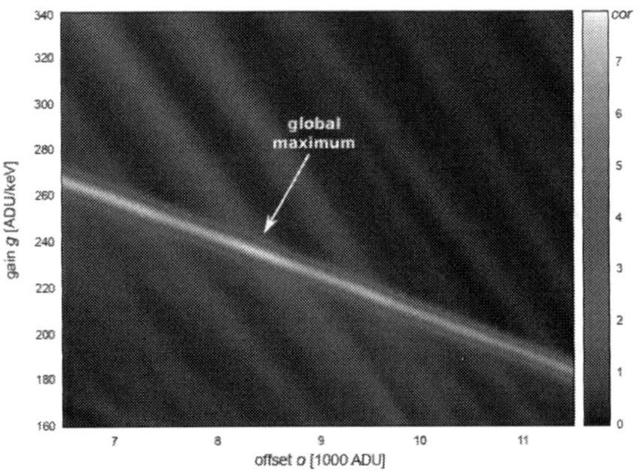

Fig. 5. Correlation map for an Am-241 calibration obtained via a classical fine mesh ECC with a resolution of $\Delta o = 5\,\mathrm{ADU}$ and $\Delta g = 0.25\,\mathrm{ADU/keV}$.

This realistic data set shows that the optimization problem is not convex but has multiple local maxima. Furthermore, due to absorption of low energy photons in the camera housing (0.5 mm Al), the source spectrum is dominated by the emission line around 60 keV. Consequently, all parameter combinations that match the synthetic 60 keV line with the observed line result in a large correlation which can be seen by the yellow line in Fig. 5. Nevertheless, the obtained parameter set for the best correlation fits the measurement well, see Fig. 6.

Figure 7 shows for the same calibration measurement the result of ECC_{AMR} after 5 iterations. The initial grid was chosen to $\Delta o = 500\,\mathrm{ADU}$ and $\Delta g = 10\,\mathrm{ADU/keV}$, leading to 400 initial correlations and a final resolution of $\Delta o \approx 2\,\mathrm{ADU}$

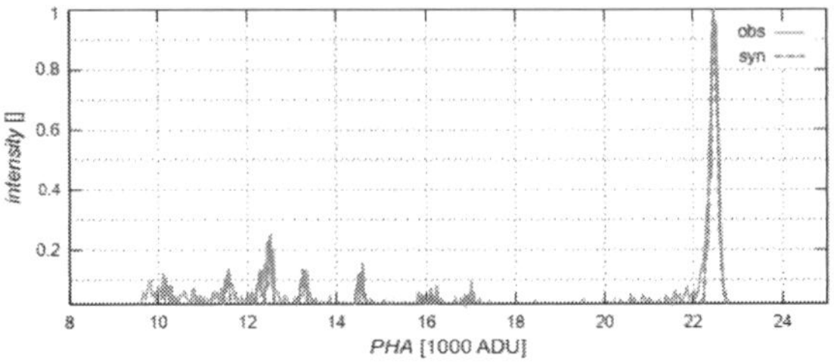

Fig. 6. Observed (obs) and best synthetic (syn) spectra for an Am-241 calibration for a single detector channel. The observation has 2590 events. Both spectra are normalized to a maximal intensity of 1.

and $\Delta g \approx 0.05\,\mathrm{ADU/keV}$. The threshold parameters were set to $f_0 = 0.6$ and $f = 0.7$. The numbers of computed correlations for each iteration are shown in Fig. 8. In total, 1602 correlations are calculated which is 450 times less compared to classical ECC. The number of correlations directly measures the computation time within ECC. The total computation time is about 2 seconds, which is, compared to classical ECC, a reduction by a factor of 50, including also reading the input data, binning the spectra, and writing the results.

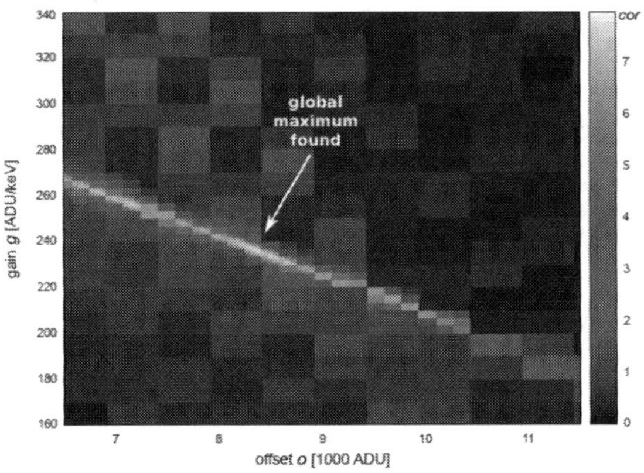

Fig. 7. ECC_{AMR} correlation map after 5 iterations. The final resolution at the global maximum is $\Delta o \approx 2\,\mathrm{ADU}$ and $\Delta g \approx 0.05\,\mathrm{ADU/keV}$.

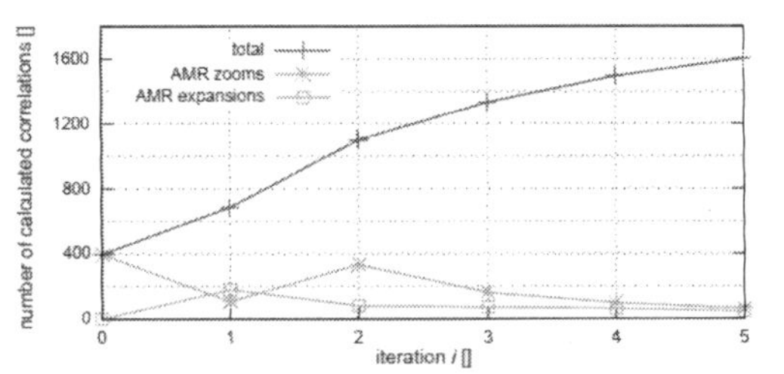

Fig. 8. Number of calculated correlations as a function of the iteration i. The total number is the cumulative sum of the number of zooms and the number of expansions. In total, 1602 correlations are calculated.

Figure 9 shows the correlation map for the same dataset but using ECC_{AMR} with disabled adaptive mesh expansion. The global maximum cannot be found in this case because the initial grid shows near the global maximum a relatively low correlation value of $cor \approx 3.2$ which is only 44 % of the maximal initial correlation value. A finer initial grid or a lower threshold factor f_0 could resolve this problem but only at the expense of an increased computation time.

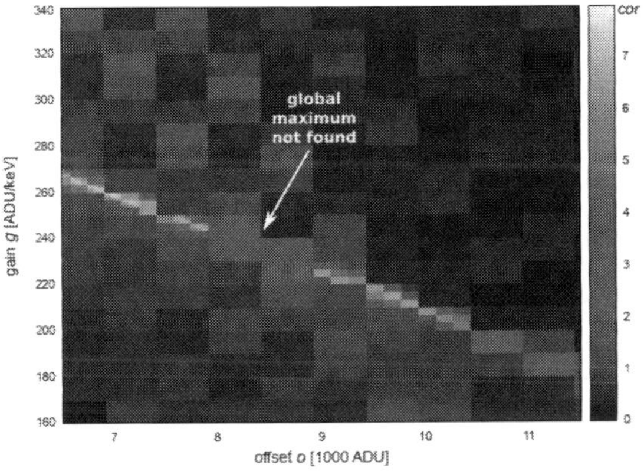

Fig. 9. ECC_{AMR} correlation map after 5 iterations with disabled adaptive mesh expansion. The global maximum is not found.

IV. Outlook

In the current version, the resolution of each parameter is increased by a factor of 3 for each iteration. If the required resolutions of the parameters are considerably different, it would be beneficial if the zooming factor could change during the computation for each parameter individually. In particular, an implementation is foreseen where the zooming of a parameter stops if a given resolution is obtained. This would result in an adaptive reduction of the dimensionality of the parameter space — a consistent further development in the sense of an adaptive mesh. Figure 10 shows this idea schematically for the case of two parameters. The stopping criterion, which is in the current version defined by the number of iterations, would be reached if all required parameter resolutions are obtained.

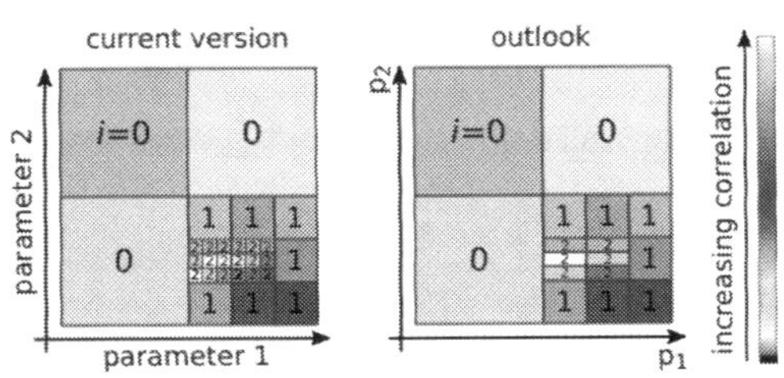

Fig. 10. After iteration $i = 1$, the zooming is reduced from 3×3 (current version) to 1×3 (next version), i.e. only 3 and not 9 zooms are calculated for each interesting region. The stopping condition is reached when the resolution requirements of all parameters are achieved and no more zooming is required.

V. Conclusions

ECC is compatible with an adaptive mesh refinement that reduces the computation time dramatically: in the given example, the computation time is reduced by a factor of ~ 450 for an even slightly better parameter resolution compared to the classical ECC approach. Both methods, classical ECC and ECC_{AMR}, result in a maximal calibration error of $\Delta E = 275\,\text{eV}$ for the shown example of an Am-241 calibration.

The application of AMR requires to use an adaptive expansion method in order to explore regions that are interesting but that seemed to be uninteresting during previous iterations. The expansion works recursively at the end of each iteration, so that the algorithm always looks for the best correlation value for a given parameter resolution.

Even though the algorithm is shown for only two parameters and for the example of energy calibration, it can be applied for any multi-dimensional optimization problem. Like ECC, ECC_{AMR} is highly parallelizable with a total memory usage of less than 100 MB for the given examples.

References

[1] D. Maier, O. Limousin, Energy calibration via correlation, Nuclear Instruments and Methods in Physics Research A 812 (2016) 43–49. arXiv:1512.04584, doi:10.1016/j.nima.2015.11.149.

[2] M. Berger, P. Colella, Local adaptive mesh refinement for shock hydrodynamics, Journal of Computational Physics 82 (1) (1989) 64 – 84. doi:https://doi.org/10.1016/0021-9991(89)90035-1.

[3] D. Maier, C. Blondel, C. Delisle, O. Limousin, J. Martignac, A. Meuris, F. Visticot, G. Daniel, P.-A. Bausson, O. Gevin, G. Amoyal, F. Carrel, V. Schoepff, C. Mahé, F. Soufflet, M.-C. Vassal, Second generation of portable gamma camera based on Caliste CdTe hybrid technology, Nuclear Instruments and Methods in Physics Research Section A: Accelerators, Spectrometers, Detectors and Associated Equipment 912 (2018) 338 – 342, new Developments In Photodetection 2017. doi:https://doi.org/10.1016/j.nima.2017.12.027.

[4] A. Meuris, O. Limousin, O. Gevin, F. Lugiez, I. L. Mer, F. Pinsard, M. Donati, C. Blondel, A. Michalowska, E. Delagnes, M.-C. Vassal, F. Soufflet, Caliste HD: A new fine pitch Cd(Zn)Te imaging spectrometer from 2 keV up to 1 MeV, 2011 IEEE Nuclear Science Symposium Conference Record (2011) 4485–4488.

Deep Learning Methods On Neutron Scattering Data

Guanghan SONG*, Lionel PORCAR, Martin BOEHM, Franck CECILLON, Charles DEWHURST
Yannick LE GOC, Jérome LOCATELLI, Paolo MUTTI† and Tobias WEBER
Institut Laue-Langevin
Grenoble, France
Email: *songg@ill.fr, †mutti@ill.fr

***Abstract*—Recently, by using deep learning methods, a computer is able to surpass or come close to matching human performance on image analysis and recognition. This advanced methods could also help extracting features from neutron scattering experimental data. Those data contain rich scientific information about structure and dynamics of materials under investigation. Deep learning could help researchers better understand the link between experimental data and materials properties. Moreover,it could also help to optimize neutron scattering experiment by predicting the best possible instrument configuration. Among all possible experimental methods, we begin our study on the small-angle neutron scattering (SANS) data and by predicting the structure geometry of the sample material at an early stage. This step is a keystone to predict the experimental parameters to properly setup the instrument as well as the best measurement strategy. In this paper, we propose to use transfer learning to retrain a convolutional neural networks (CNNs) based pre-trained model to adapt the scattering images classification, which could predict the structure of the materials at an early stage in the SANS experiment. This deep neural network is trained and validated on simulated database, and tested on real scattering images.**

***Index Terms*—artificial intelligence, deep learning, convolutional neural network, transfer learning, small-angle neutron scattering**

I. Introduction

The fundamental properties of the neutron make it a powerful tool to investigate atomic-scale structure and dynamics of materials. Compared to other scattering techniques such as x-ray or light, neutron has advantages such as negligible radiation damage, exceptional penetration power and ability to selectively highlight specific parts of materials via isotope labelling [1]. However, because the flux of neutrons is lower than the flux of the photons at modern light sources, long measurement time (even tens of hours) is required for a high-quality data collection. In this case, early decision becomes a critical component in the experimental work flow. On the other hand, the instrument settings of neutron scattering experiments are manually decided by scientists according to their experiences and knowledges. Due to large differences in the level of experience of the different users, high quality data could become difficult to obtain and quite time consuming. Hence, we propose to use recent advanced machine learning methods to help interpreting experimental data in order to optimize effective use of limited beam time and ence increasing science productivity.

Small-angle neutron scattering (SANS) is an increasingly popular method in structural biology that usefully complements high-resolution structural techniques such as x-ray crystallography or nuclear magnetic resonance spectroscopy [2]. Over time, many methods have been developed to extract relevant information directly from the experimental scattering intensities, working exclusively with the experimentally obtained data. Inthi work, we consider the application of machine learning methods to extract sample structural information from SANS data.

Deep learning has become one of the most active research field in the machine learning community since it was presented in 2006 [3]. Recently, it impacted deeply data science, and its popularity has grown exponentially, especially in computer vision related tasks [4]. Convolutional neural networks (CNNs) is a well-known deep learning architecture inspired by the natural visual perception mechanism of the living creatures. In recent years, CNNs have been widely adopted for a variety of applications in image classification [5] [6], video analysis [7] and natural language processing [8]. CNNs are widely used in image related tasks since they present several advantages. Firstly, CNNs decompose images into multiple patches that are partially overlapped. In this case, each cortex neuron only corresponds to a single patch, which enables the network to classify images with little data pre-processing [9]. Secondly, CNNs are weight sharing, which enable CNNs share learned features from a single sample via back-propagation with other samples. It efficiently decreases the necessary required input sample size and quantity. [10].

In practices, large amount of data is required to train the CNNs from scratch. Due to limited calculation power and lack of labeled big data set for relevant problems, the concept of transfer learning has been introduced. When applying transfer learning, a network is trained on a new dataset of samples using the stored weights for some or all layers from a pre-trained network as an initialization, rather than initializing weights randomly. The selection of the pre-trained model in transfer learning is influenced by two factors : problem size and similarity [11]. The selection of the pre-trained model is made on the basis of the associated problem which is relevant to the target problem. If the size of the target data set is smaller and similar to the source training data set then the chance of over fitting is high. Similarly, if the size of the target data is larger and similar to the source data sets then the chance of over fitting is low and it requires only the fine tuning of the pre-trained model [12].

In this paper, we applied transfer learning method on a CNNs architecture based pre-trained model, to analyze and classify neutron scattering image from SANS experiment. This model named Inception-v3 is pre-trained on ImageNet database. With fine-tuning, this model is first retrained and validated by simulated scattering images, then verified on real SANS experimental data collected at the Institut Laue-Langevin (ILL).

II. METHODS

A. *Data simulation*

To generate two-dimensional Small-Angle Neutron Scattering (SANS) images we use the data reduction and simulation code GRASP [13] . GRASP is a MatlabTM script application designed for the graphical inspection, analysis and reduction of position sensitive detector(PSD) data produced by the 3 SANS instruments of the ILL. GRASP is using a mixed analitical and Monte-Carlo approach for the neutron event generetion and can include the real intrument resolution function, the measured backgrounds from a variety of different samples and the influence of the sample environment. By using GRASP, we could generate simulated SANS 2D detector images with variable sets of parameters. We could select different materials' structures and also freely change the instrument configuration. In this work we have choosen D22 SANS instrument and its schematic layout is presented in Fig. 1

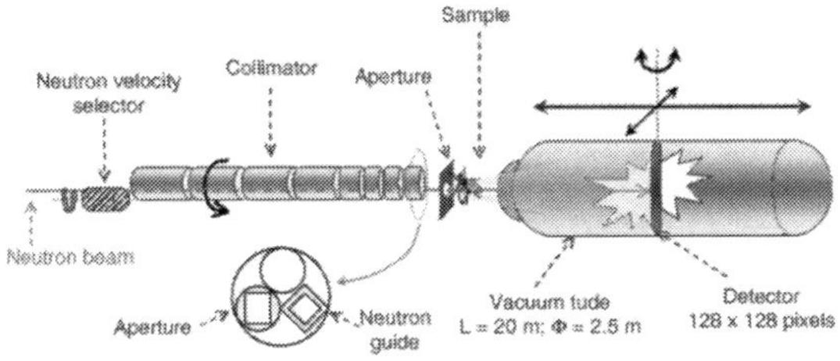

Fig. 1: Schematic representation the steady-state instrument D22 at the Institut Laue-Langevin [2].

Starting from a cold neutron beam arising from the H512 neutron guide, the desired neutron *wavelength* can be selected through a mechanical velocity selector in a range between 4.6 and 40 Å. The wavelength spread $\Delta\lambda/\lambda$ is 10% (FWHM). The virtual source-to-sample distance is chosen by a collimation system consisting of eight sections. Each section comprises three tubes, any one of which can be positioned on the beam axis. One tube contains a neutron guide of 40 x 55 mm; the second is equipped with an antiparasitic aperture; the third tube bears fixations for future neutron-optical equipment. D22 possesses a large ^{3}He multidetector with an active area of 1 m^2. The detector is subdivided in pixels with a size of 8 x 8 mm. The detector consists of 128 linear sensitive Reuter-Stokes tubes arranged vertically with a spacing of 8 mm. Each tube has a dead time of about 1 s roughly corresponding to the analog pulse shaping time. It moves inside a 2.5 m wide and 20 m long vacuum tube providing sample-to-detector distances from 1.1 m to 17.6 m. D22 thus covers a total q range of 4 x 10^{-4} to 0.44 Å^{-1} (no detector offset) or 0.85 Å^{-1} (with detector offset) in standard conditions.

A large amounts of SANS diffusion patterns were obtained from samples with geometrical structure and uniform scattering length density such as spheres, cylinders, core-shell spheres, ellipsoid and flux line bragg peak. The corresponding dimensions of the geometric bodies, i.e., inner and outer radius, length, width, height etc., were respectively randomly and independently sampled in ranges from 10 to 800 Å. Each instrument configuration of D22 such as sample to detector distance, collimation distance, wavelength bands introduces unique features in the resulting detector images. In our simulation, the sample is inserted in an empty cell filled with background liquid: water (H_2O) or heavy water (D_2O).

B. *Neural Network Model*

In this paper, we have adopted transfer learning using the pre-trained CNNs based Inception-v3 [14] architecture to classify the 2-D neutron scattering images. Instead of creating a model from scratch, we have started from a previously trained model able to solve a similar classification problem. This model was known to classify an image across 1000 categories supplied by the ImageNet academic competition with an error rate that approached human performance. The Inception deep convolutional architecture was introduced as GoogLeNet (Inception-v1) [15]. Since Inception networks are fully convolutional, each weight corresponds to one multiplication per activation. Therefore, by introducing factorizing convolutions in this model, Inception-v3 is able to reduce the number of connections/parameters (42-layer deep learning network) and similar complexity as VGGNet [16], without decreasing the network efficiency. Inception-v3 is a commonly used architecture for deep learning tasks. This model is a state-of-the-art CNNs and it is widly use in the field of computer vision. It consists of two parts:

- Feature extraction part with a CNN.
- Classification part with fully-connected and softmax layers.

The pre-trained Inception-v3 model achieves state-of-the-art accuracy for recognizing general objects with 1000 classes. The model extracts general features from input images in the first part and classifies them based on those features in the second part. The model itself is made up of symmetric and asymmetric building blocks, including convolutions, average pooling, max pooling, concats, dropouts, and fully connected layers. BatchNorm is used extensively throughout the model and applied to activation inputs. Loss is computed via Softmax.

In transfer learning, when we build a new model to classify our original dataset, we reuse the feature extraction part and re-train the classification part with our dataset. Since we don't have to train the feature extraction part, which is the most complex part of the model, we can train the model with less

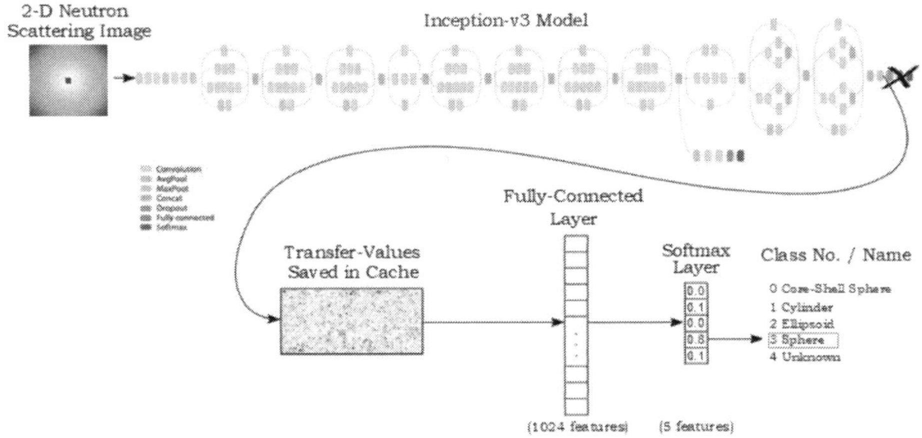

Fig. 2: Schematic diagram of Inception-v3 with transfer learning.

computational resources and training time. The work flow of Inception-v3 model with transfer learning is shown in Fig.2.

III. Experimental results

The small angle neutron scattering instrument D22 is used to investigate a wide range of samples, to identify their atomic structure and composition. Hence, we classify the samples with a 2-D scattering image at an early stage and then, we tried to predict its structure. We have started to consider four different geometrical structures: sphere, core-shell sphere, cylinder and ellipsoid, which procuce a quite similar scattering patterns. All the simulations are performed on a Linux Server (Ubuntu 16.04.5) with two NVIDIA Tesla P100 GPUs.

A. Experiment I: sphere vs. flux line bragg peak

Spheres and flux-line bragg peak produce quite different images on the detector. Unfortunately, in real experiments, the instrument parameters are set according to a "guess" of the structure type and sample parameters, which sometimes are far away from the real structure. Hence,we first try to recognize the scattering image at an early stage without any priori knowledge.

1) Dataset: By using GRASP, we have generated 2000 sample images are generated respectively for the sphere and flux line bragg peak model. The corresponding dimensions of the geometric bodies, i.e., radius for sphere, rocking angle for flux line bragg peak etc. are generated in a certain range randomly and independently. Fig.3 shows some examples in this database, and a square beam stop is applied in the center of detector. As shown in Fig.3, because of data collection time is not long enough or the parameters setting of the instrument are not always perfect, to obtain a high quality scattering images is not easy for further analysis.

2) Result: We begin the training trail with the pre-trained model Inception-v3. The initial weights are taken from the pre-trained weights based on ImageNet. Only the weights of the attached dense layers are updated. It means all layers are set to frozen except the last dense layer. The model is compiled with binary cross-entropy loss function for the training and validation sets. The initial learning rate is randomly set to 0.01 and the model is trained for 300 epochs, which means the network examined each image 300 times. For each training trail, 80% data are selected randomly for training, 10% data are used for testing and 10% data are used for validation.

(a) sphere: centralized

(b) sphere: low resolution

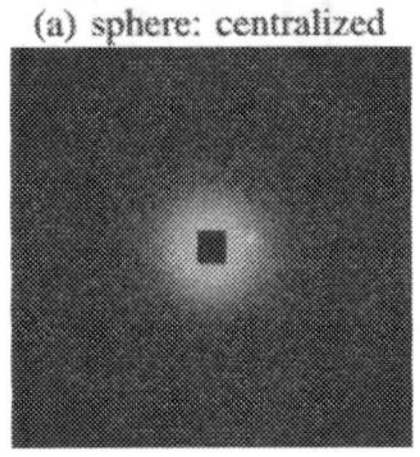
(c) flux line bragg peak: centralized

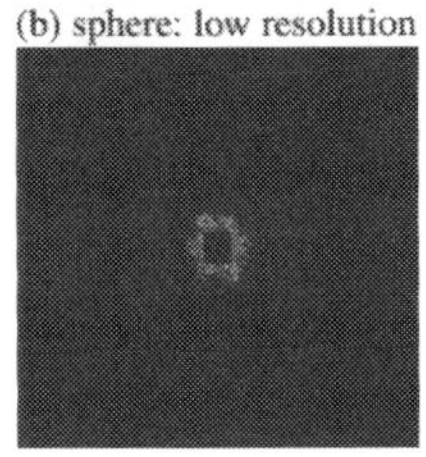
(d) flux line bragg peak: low resolution

Fig. 3: Simulated database examples of experiment I

Fig.4 shows the average accuracy of classification and loss of cross-entropy. After 200 epochs, the average accuracy is stable to 100%, which indicates all the simulated samples in the dataset are corrected classified by current classifier.

Then this classifier is tested on nine scattering image of

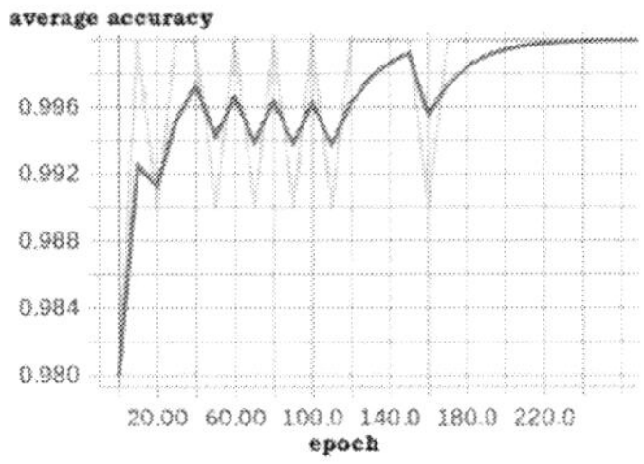

(a) average accuracy

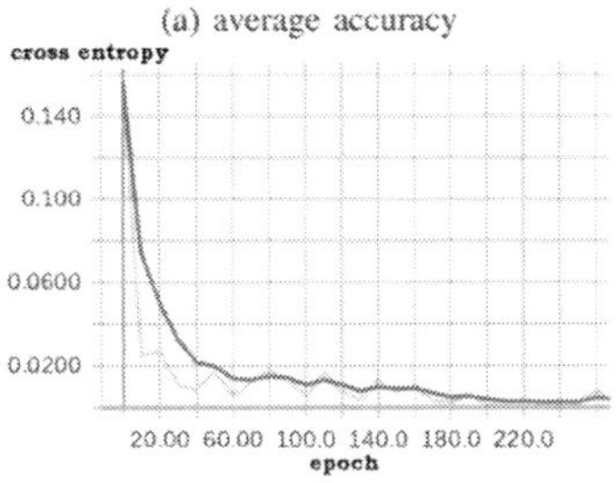

(b) cross entropy

Fig. 4: Validation accuracy of experiment I.

material silica sphere from real experiment shown in Fig.5. All the images are correctly recognized with 99% confidence.

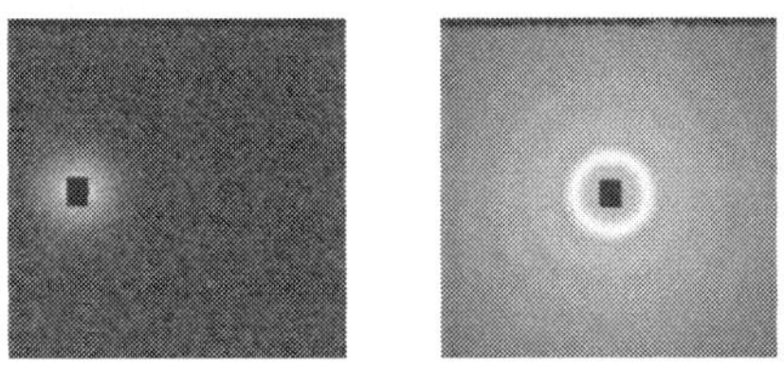

Fig. 5: Scattering image from real experiment of a silica sphere structure.

B. Experiment II: sphere vs. core-shell sphere vs. cylinder vs. ellipsoid

The four models: sphere, core-shell sphere, cylinder and ellipsoid geometrically are similar to each other under some particular parameters in the structure. For example, if the model cylinder with length 0, it becomes a sphere model. Recognizing these four structures only from scattering image is a challenge task.

1) Dataset: 10000 simulated scattering images are generated by GRASP,for each individual sample structure. The corresponding dimensions of the geometric bodies, i.e., length, inner and outer radius etc. are generated randomly and independently. The dataset contains complex situation of the models, including monodispersity, polydispersity, low resolution, centralized scattering images etc...

2) Result: The same pre-trained model as experiment I is applied. Fig.6 shows the results of average accuracy and loss of cross entropy. As a result of an average accuracy of 73%, the most incorrect and correct classification images are shown in Fig.7.

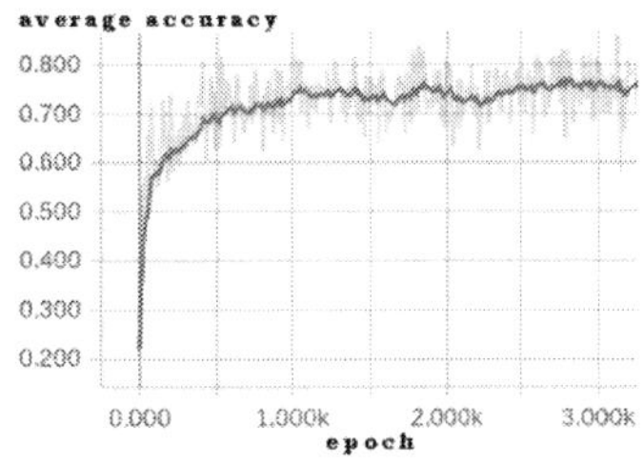

(a) average accuracy

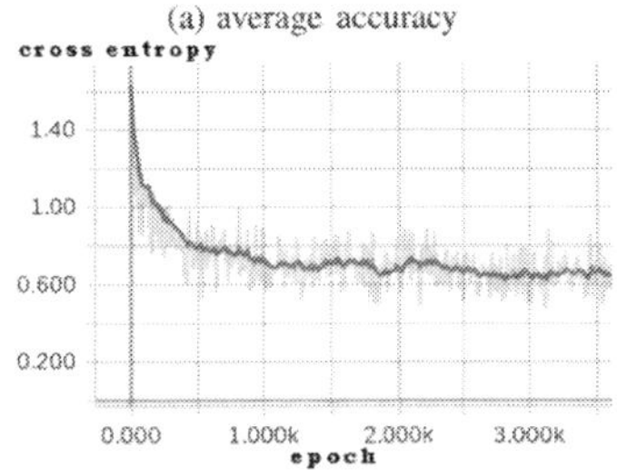

(b) cross entropy

Fig. 6: Validation accuracy of experiment II.

We have employed data augmentation and fine-tuning methods to improve the recognition accuracy. The model is optimized by stochastic gradient descent optimizer [17], which restarts techniques to find an optimal learning rate. Optimal learning rate finder is basically boosts the learning rate until the validation loss stops to decrease. Because of limited field of view of the instrument, in real experiment, sometimes we choose to make the neutron beam arrival at the detector with a certain bias from the center, to obtain a larger field of view of the material structure. 2000 simulated scattering images with beam stream arrival with certain bias from the center are added respectively to each class. Fig.8 shows two scattering samples in this kind of scattering data.

Table. I shows the average precision, sensitivity (recall), and F1-Score of the model with fine-tuning and augmented database. The average accuracy of this model is 79.8%, which is better than the performance of the model without fine-tuning. Except class unknown, the other four classes have proximity F1-score (between 75% and 80%). This unknown class is consisted of scattering images generated from simulated program, without or less recording useful scattering information by improper instrument parameters setting. For cylinder, the precision is high 90%, which is the ratio of correctly predicted positive observations to the total predicted positive observations, meanwhile the sensitivity is low 65%, which is the ratio of correctly predicted positive observations

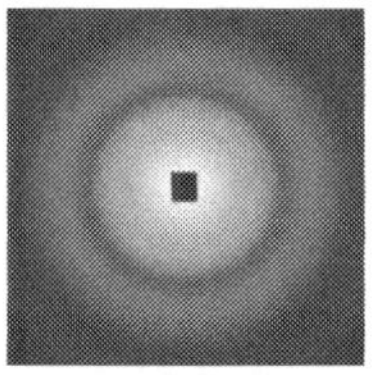

(a) sphere: correct recognition

(b) sphere: incorrect recognition

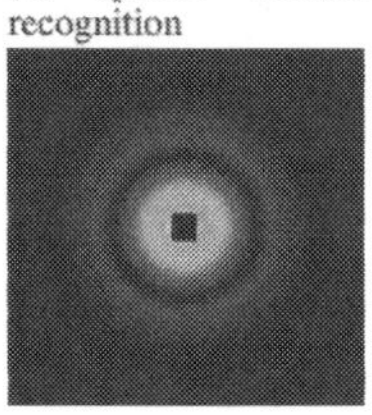

(c) core-shell sphere: correct recognition

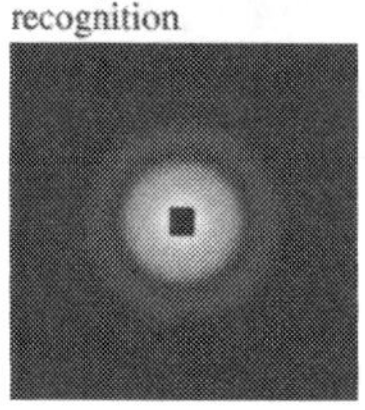

(d) core-shell sphere: incorrect recognition

(e) cylinder: correct recognition

(f) cylinder: incorrect recognition

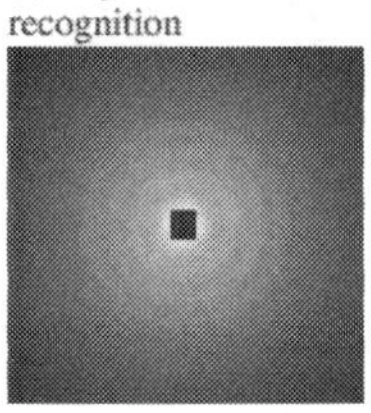

(g) ellipsoid: correct recognition

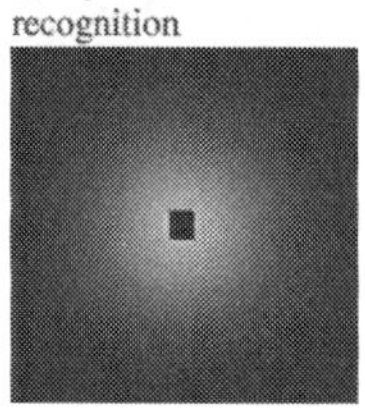

(h) ellipsoid: incorrect recognition

Fig. 7: Display of correctly and incorrectly classified images.

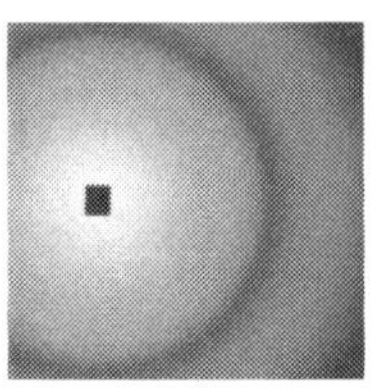

(a) sphere

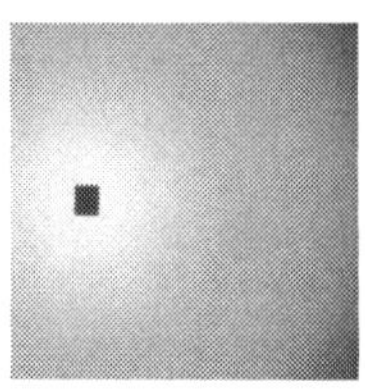

(b) cylinder

Fig. 8: Example of simulated scattering images with center bias of the neutron beam arrival.

to the all observations in actual class, turns out similar F1-score with other classes.

TABLE I: The average precision, sensitivity, F1-Score on the model with fine-tuning.

Classes	Precision(%)	Sensitivity(%)	F1-Score(%)
Core-Shell Sphere	75	75	75
Cylinder	90	65	75
Ellipsoid	76	83	79
Sphere	70	85	77
Unknown	92	91	92

Then, this classifier is tested on 9 silica sphere scattering image, 9 cylinder scattering image, 2 ellipsoid scattering image and 1 core-shell sphere scattering image from real experiment. The scattering images from real experiment are raw data, without any pre-process. The most incorrect and correct classification images are shown in Fig.9. In 9 silica sphere real scattering images, 8 image are recognized as sphere with an average confidence value of 92%. As shown in Fig.9e, the incorrect classification happened on the image with less scattering information comparing to correct classifications. For cylinder class, only 3 images are correctly recognized with an average confidence 63%. In the rest 6 images with misclassification, 5 images are similar to Fig.9f. It seems with scattering images which contain the scattering information from the whole directions, that means the neutron beam arrived in the center of detector obtain higher recognition rate than scattering images with center bias of neutron beam arrivals. And for the images which contain only little information around beam stop, it seems difficult to recognize among these four structures.

IV. DISCUSSION & CONCLUSIONS

In this article, we introduced a deep learning based method to recognize material's structure category by its 2-D neutron scattering image. The performance of accuracy is quite depending on the similarity among the material's structure. As shown in experiment I (secton. III-A), the classifier is able to separate structure model with high accuracy between sphere and flux line bragg peaks even in low resolution, less scattering information, and low neutron beam intensity scattering images. However to predict the scattering images from close geometrical structures, such as sphere, core-shell sphere, cylinder and ellipsoid, the accuracy of our classifier decreased. From observation, the accuracy is influenced by the quality and type of scattering images. For example, the classifier could get higher accuracy on high resolution rather than low resolution scattering images. Also it performs better on scattering images with neutron beam arrival at the center of detector which contain scattering information in all directions rather than with center bias scattering images for which part of the information is missing.

The training trail of deep CNNs is done on simulated database. And the trained classifier is tested on several raw scattering images from real experiments. With this small

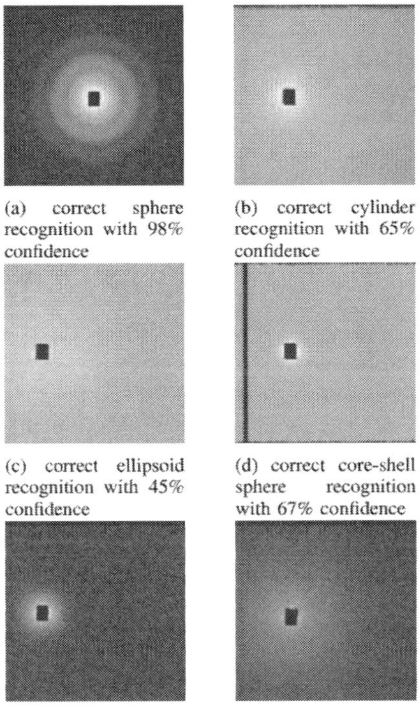

(a) correct sphere recognition with 98% confidence

(b) correct cylinder recognition with 65% confidence

(c) correct ellipsoid recognition with 45% confidence

(d) correct core-shell sphere recognition with 67% confidence

(e) incorrect sphere recognition

(f) incorrect cylinder recognition

Fig. 9: Display of scattering images from real experiment with correct and incorrect recognitions.

among of scattering images from real experiment, we couldn't get solid conclusion about the performance of our classifier. However, the result indicates the tendency that our classifier is able to correctly recognize real scattering images without any training samples from real scattering data, it could work equally on simulated and real images.

REFERENCES

[1] T. Zemb and P. Lindner, *Neutrons, X-rays and light: scattering methods applied to soft condensed matter*. North-Holland, 2002.

[2] I. Grillo, "Small-angle neutron scattering and applications in soft condensed matter," *Soft matter characterization*, pp. 723–782, 2008.

[3] J. Schmidhuber, "Deep learning in neural networks: An overview," *Neural networks*, vol. 61, pp. 85–117, 2015.

[4] S. Lathuilière, P. Mesejo, X. Alameda-Pineda, and R. Horaud, "A comprehensive analysis of deep regression," *IEEE transactions on pattern analysis and machine intelligence*, 2019.

[5] A. Krizhevsky, I. Sutskever, and G. E. Hinton, "Imagenet classification with deep convolutional neural networks," in *Advances in neural information processing systems*, 2012, pp. 1097–1105.

[6] U. R. Acharya, S. L. Oh, Y. Hagiwara, J. H. Tan, and H. Adeli, "Deep convolutional neural network for the automated detection and diagnosis of seizure using eeg signals," *Computers in biology and medicine*, vol. 100, pp. 270–278, 2018.

[7] A. Karpathy, G. Toderici, S. Shetty, T. Leung, R. Sukthankar, and L. Fei-Fei, "Large-scale video classification with convolutional neural networks," in *Proceedings of the IEEE conference on Computer Vision and Pattern Recognition*, 2014, pp. 1725–1732.

[8] Y. Kim, "Convolutional neural networks for sentence classification," *arXiv preprint arXiv:1408.5882*, 2014.

[9] A. Krizhevsky, I. Sutskever, and G. E. Hinton, "Advances in neural information processing systems," *Curran Associates, Inc*, pp. 1097–1105, 2012.

[10] H. Tang, A. M. Scaife, and J. Leahy, "Transfer learning for radio galaxy classification," *arXiv preprint arXiv:1903.11921*, 2019.

[11] S. Khan, N. Islam, Z. Jan, I. U. Din, and J. J. C. Rodrigues, "A novel deep learning based framework for the detection and classification of breast cancer using transfer learning," *Pattern Recognition Letters*, 2019.

[12] A. Kaya, A. S. Keceli, C. Catal, H. Y. Yalic, H. Temucin, and B. Tekinerdogan, "Analysis of transfer learning for deep neural network based plant classification models," *Computers and Electronics in Agriculture*, vol. 158, pp. 20–29, 2019.

[13] "Grasp," https://www.ill.eu/fr/users-en/scientific-groups/large-scale-structures/grasp/, institut Laue-Langevin.

[14] C. Szegedy, V. Vanhoucke, S. Ioffe, J. Shlens, and Z. Wojna, "Rethinking the inception architecture for computer vision," in *Proceedings of the IEEE conference on computer vision and pattern recognition*, 2016, pp. 2818–2826.

[15] C. Szegedy, W. Liu, Y. Jia, P. Sermanet, S. Reed, D. Anguelov, D. Erhan, V. Vanhoucke, and A. Rabinovich, "Going deeper with convolutions," in *Proceedings of the IEEE conference on computer vision and pattern recognition*, 2015, pp. 1–9.

[16] K. Simonyan and A. Zisserman, "Very deep convolutional networks for large-scale image recognition," *arXiv preprint arXiv:1409.1556*, 2014.

[17] M. Talo, U. B. Baloglu, Ö. Yıldırım, and U. R. Acharya, "Application of deep transfer learning for automated brain abnormality classification using mr images," *Cognitive Systems Research*, vol. 54, pp. 176–188, 2019.

Characteristics of 3D Printed Plastic Scintillator

Dong-geon Kim, Sangmin Lee, Junesic Park, Jaebum Son, Yong Hyun Kim, and Yong Kyun Kim
Hanyang University, Nuclear Engineering, Seoul, South Korea

***Abstract*— Digital Light Processing (DLP) 3D printing technique can be a powerful tool to fabricate plastic scintillator with a geometrically desired shape in innovatively fast time. Plastic scintillator with the size of 30 mm × 30 mm × 10 mm was fabricated by using the plastic resin and the DLP 3D printer (ASIGA, Pico2^{HD}). The characteristics of decay time, energy resolution, intrinsic detection efficiency were analyzed and compared between the fabricated 3D printing plastic scintillator and a commercial plastic scintillator BC408 (Saint-Gobain Crystal). Decay time profile of the tested plastic scintillators was measured for ^{137}Cs Compton maximum electron 477 keV by using a modified time correlated single photon counting (TCSPC) setup. The time profile was fitted by reconvolution function, and each decay time component and contribution was analyzed. For energy resolution of plastic scintillator, the Gaussian spectrum for ^{137}Cs Compton maximum electron 477 keV was selectively measured by using the γ-γ coincidence experimental setup. As a result, it was confirmed that the 3D printing plastic scintillator showed average decay time 15.6 ns and energy resolution 15.4%. These characteristics demonstrates the feasibility of 3D printing plastic scintillator as a radiation detector.**

***Index Terms*— Digital Light Processing (DLP), 3D Printer, UV polymerization, Plastic Scintillator, Scintillation Performance**

I. Introduction

PLASTIC scintillator has been developed with various designs to expand its applications in radiation detection field since the 1950s when plastic scintillator was first invented. These plastic scintillators can be commonly fabricated by thermal polymerization technique [1]. A liquid solution with scintillation components is polymerized in heater, depending on a temperature profile with time. However, this technique has some limitations: (1) its geometrically confined and simple shapes of rods, cylinder, flat sheets and (2) long fabrication time of one week or more.

In an effort to overcome these limitations of this thermal polymerization, 3D printing plastic scintillator has been studied [2], [3]. In our group, for the application to Digital Light Processing (DLP) 3D printing, the UV-cured plastic scintillator was developed using UV LED curing machine. It was confirmed that the characteristics of this UV-cured plastic scintillator were relative light output 34% to BC408 (Saint-Gobain Crystal), transmittance 49%, and decay time 2.46 ns [4]. Recently, a novel plastic scintillator with a new wavelength shifter and intermediate solvent was developed by DLP 3D printing technique, and its characteristics were relative light output 67% to BC408 and transmittance 56% [3]. Although this 3D printing plastic scintillator showed excellent light output and transmission over the previous results, the demonstration for other main characteristics is required for practical use in radiation detection field.

In this work, for practical use, the characteristics of the 3D printing plastic scintillator were demonstrated. The characteristics of decay time and energy resolution were analyzed and compared with a commercial plastic scintillator BC408 (Saint-Gobain Crystal).

II. Materials and Methods

3D printing plastic scintillator with the size of 30 mm × 30 mm × 10 mm was fabricated by the Digital Light Processing (DLP) 3D printer (ASIGA, Pico2^{HD}). The material composition of plastic resin developed in the previous study [3] was used. Each experimental setup was constructed to measure and compare the characteristics of decay time and energy resolution for the 3D printing plastic scintillator (Sample) and commercial plastic scintillator (BC408). A modified time correlated single photon counting (TCSPC) setup was constructed to measure scintillation time profile. The measured time profile as shown in Figure 1 was fitted with reconvolution function to analyze each decay time constant and contribution.

In the case of energy resolution, it is difficult to measure the energy resolution since the photopeak is not found in the γ-ray energy spectrum due to low photoelectric cross section of plastic scintillators. In other to overcome this problem, we constructed the γ-γ coincidence experimental setup and measured the Gaussian energy spectrum for the maximum Compton electron reaction (Compton edge) among several Compton electrons reacted with the scintillator as shown in Figure 2.

This work was supported by the National Research Foundation of Korea (NRF) grant funded by the Ministry of Science and ICT (No. NRF-2016M2A2A6A03912636).

Dong-geon Kim is with the Nuclear Engineering Department, University of Hanyang, Seoul 04763, Korea (e-mail:kdgeon79@hanyang.ac.kr).

Sangmin Lee is with the Nuclear Engineering Department, University of Hanyang, Seoul 04763, Korea (e-mail: s6245299@hanyang.ac.kr).

Junesic Park is with the Nuclear Engineering Department, University of Hanyang, Seoul 04763, Korea (e-mail: ergosphere84@gmail.com).

Jaebum Son is with the Nuclear Engineering Department, University of Hanyang, Seoul 04763, Korea (e-mail: jbs1000@hanyang.ac.kr).

Yong Hyun Kim is with the Nuclear Engineering Department, University of Hanyang, Seoul 04763, Korea (e-mail:kimyh04@naver.com).

Yong Kyun Kim is with the Nuclear Engineering Department, University of Hanyang, Seoul 04763, Korea (e-mail:ykkim4@hanyang.ac.kr).

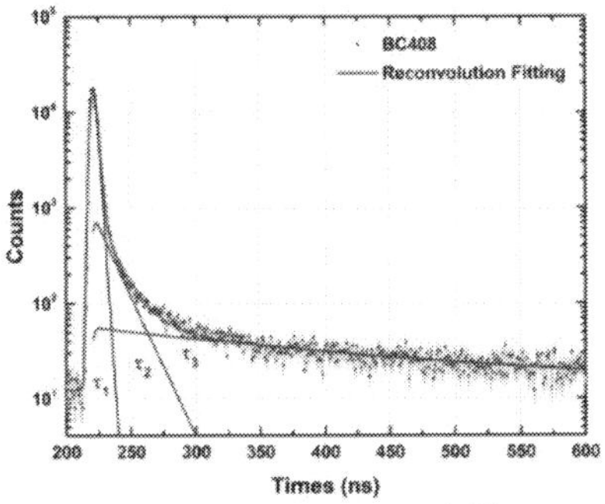

Fig. 1. Scintillation decay time profile of plastic scintillator BC408 for ^{137}Cs Compton maximum electron 477 keV.

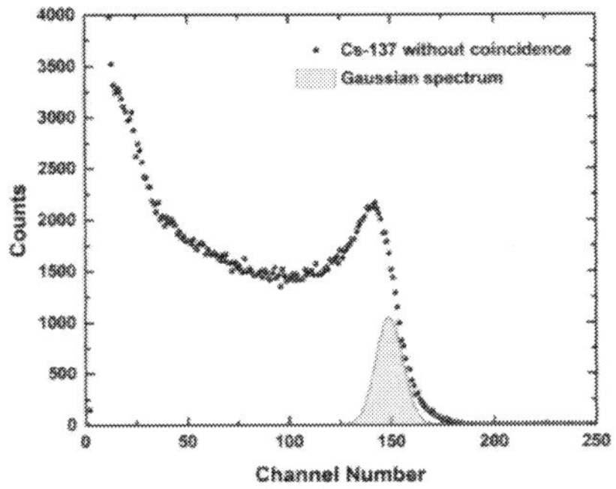

Fig. 2. ^{137}Cs γ-ray pulse height spectrum (without coincidence spectrum) and Gaussian spectrum (γ-γ coincidence spectrum) for plastic scintillator BC408.

III. Results

Table 1 shows the characteristics of plastic scintillators BC408 and Sample. It was confirmed that the average decay time 15.6 ns for Sample were compatible to BC408, which verifies that this 3D printing plastic scintillator can be used in high radiation environment of γ-ray. The energy resolution 15.4% of Sample was somewhat lower than 9.7% of BC408.

TABLE I
CHARACTERISTICS OF PLASTIC SCINTILLATORS

Properties	Plastic scintillator	
	BC408	Sample
Peak wavelength [a]	425 nm	470 nm
Light output [a]	10,470 ph/MeV	7,000 ph/MeV
Transmittance [a]	76%	56%
Average decay time	18.3 ns	15.6 ns
Energy Resolution [b]	9.7%	15.4%

[a]These measurements are referred to J. Son *et al.* [3].
[b]This values represent the measurements at 477 keV for Compton electron produced by ^{137}Cs γ-ray source.

IV. Conclusion

In this work, the characteristics of 3D printing plastic scintillator were analyzed for decay time, energy resolution and intrinsic detection efficiency. It was found that the average faster decay time 15.6 ns of 3D printed plastic scintillator than 18.3 ns of BC408. These characteristics demonstrates the feasibility of 3D printing plastic scintillator as a radiation detector.

References

[1] C. H. Lee, J. Son, T. H. Kim, and Y. K. Kim, "Characteristics of Plastic Scintillators Fabricated by a Polymerization Reaction," *Nucl. Eng. Technol.*, vol. 49, no. 3, pp. 592–597, 2017.
[2] Y. Mishnayot, M. Layani, I. Cooperstein, S. Magdassi, and G. Ron, "3D Printing of Scintillating Materials," no. June, pp. 1–4, 2014.
[3] J. Son, D. G. Kim, S. Lee, J. Park, Y. Kim, T. Schaarschmidt, and Y. K. Kim, "Improved 3D Printing Plastic Scintillator Fabrication," *J. Korean Phys. Soc.*, vol. 73, no. 7, pp. 887–892, 2018.
[4] S. Lee, J. Son, D. G. Kim, J. Choi, and Y. K. Kim, "Characterization of plastic scintillator fabricated by UV LED curing machine," *Nucl. Instruments Methods Phys. Res. Sect. A Accel. Spectrometers, Detect. Assoc. Equip.*, vol. 929, no. July 2018, pp. 23–28, 2019.

EPJ Web of Conferences **225**, 01006 (2020) https://doi.org/10.1051/epjconf/202022501006
ANIMMA 2019

Read-out Electronics and Signal Processing in GERDA and Future Prospects

Valerio D'Andrea* on behalf of the GERDA Collaboration
*Dipartimento di Scienze Fisiche e Chimiche, Università degli Studi dell'Aquila
& INFN Laboratori Nazionali del Gran Sasso, Assergi, Italy
Email: valerio.dandrea@lngs.infn.it

***Abstract*—The GERDA experiment searches for the neutrinoless double beta decay of ^{76}Ge. The experiment is using 36 kg of high-purity germanium detectors, simultaneously as source and detector, deployed into ultra-pure cryogenic liquid argon. GERDA is one the leading experiment in the field, reporting the highest sensitivity on the half-life of $0\nu\beta\beta$ decay with $1.1{\cdot}10^{26}$ yr, the lowest background index with $6{\cdot}10^{-4}$ cts/(keV·kg·yr) and an excellent energy resolution of 0.12% (FWHM).**

The search for the $0\nu\beta\beta$ decay of the isotope ^{76}Ge will be continued in the next years by the LEGEND-200 experiment, that aims to reach a sensitivity up to 10^{27} yr using 200 kg of enriched HPGe detectors. The preparation of this experiment already started.

The basic concepts of the GERDA read-out electronics, obeying both the severe requirements of ultra high radio-purity and cryogenic operation, are summarized. For LEGEND-200 a new electronics design, including a separation of the preamplifier in two stages, has been already designed and realized: results from tests are presented.

Additionally, we will introduce the digital signal processing adopted for the energy reconstruction in GERDA and a new implementation of an optimum digital filter by means of the DPLMS method. This method are discussed and the first application to GERDA data are presented.

I. GERDA EXPERIMENT

The GErmanium Detector Array (GERDA) experiment [1] is searching for neutrinoless double beta ($0\nu\beta\beta$) decay of the isotope ^{76}Ge. It is located at the underground Laboratori Nazionali del Gran Sasso (LNGS) of INFN in Italy. A rock overburden of about 3500 m water equivalent removes the hadronic components of cosmic ray showers and reduces the muon flux at the experiment.

The GERDA setup, illustrated in Fig. 1, has been designed following a multi-layer approach. High-Purity Germanium (HPGe) detectors enriched to about 87% in ^{76}Ge are operated bare in liquid argon (LAr), being both source and detector of $0\nu\beta\beta$ decay. The LAr cryostat is complemented by a water tank with 10 m diameter which further shields from neutron and γ backgrounds and also works as muon veto.

After a first physics data taking campaign [2] carried out from 2011 to 2013, the GERDA setup has been upgraded to perform the next step [3]. The major upgrade is the introduction of 30 new BEGe detectors from Canberra [4] with an optimal energy resolution, due to the low input capacitance (~ pF), and a powerful pulse shape discrimination (PSD), thanks to the configuration of the p$^+$ and n$^+$ contacts that produce a highly non-uniform electrical field. In addition, an active suppression of the background by detecting the LAr scintillation light, using PMTs and wavelength shifting fibers coupled to SiPMs, has been introduced [5]. The core of the GERDA setup is shown in Fig. 2: the Ge detector array (30 BEGe, 7 enriched coaxial and 3 natural coaxial detectors) is at the center of the instrumented LAr volume.

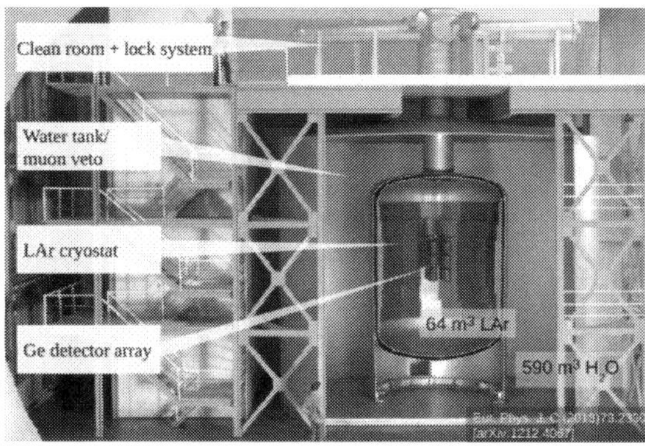

Fig. 1. GERDA schematic view [1].

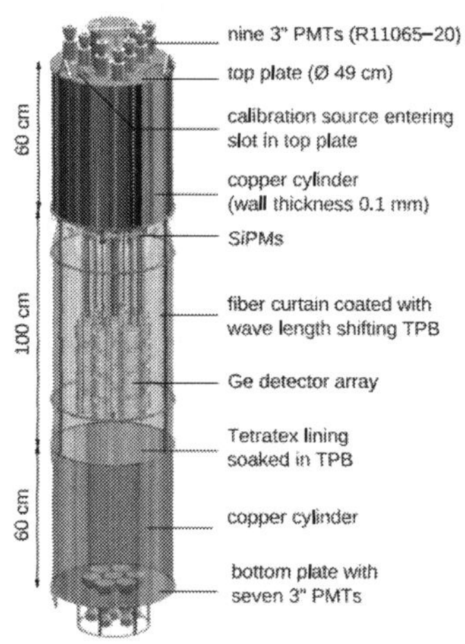

Fig. 2. Assembly of detector array and LAr veto system [3].

Initial results from the second phase, with about 10 kg·yr exposure, indicate that the target background of 10^{-3} cts/(keV·kg·yr) is achieved, thus making GERDA the first experiment in the field which will be "background-free" up to the design exposure of 100 kg·yr [6]. This condition has been confirmed also in the following data release [7].

After more than two years of data taking after the upgrade, the total available enriched Ge exposure at the last data release of June 2018 is 82.4 kg·yr [8]. The $0\nu\beta\beta$ decay analysis yielded no signal, setting a new limit on the ^{76}Ge $0\nu\beta\beta$ decay half-life of $T_{1/2}^{0\nu} > 0.9 \cdot 10^{26}$ yr (90% C.L.) with a median sensitivity of $1.1 \cdot 10^{26}$ yr (90% C.L.), thus making GERDA the first $0\nu\beta\beta$ decay experiment to surpass 10^{26} yr sensitivity. The probability to obtain a limit stronger than the actual one in an ensemble of repeated experiments with null signal is 63%.

II. LEGEND EXPERIMENT

Based on the success of GERDA and MAJORANA [9], the search for $0\nu\beta\beta$ decay in ^{76}Ge will be continued in the next years by LEGEND [10] (Large Enriched Germanium Experiment for Neutrinoless $\beta\beta$ Decay). LEGEND will proceed in phases towards a $0\nu\beta\beta$ decay discovery potential at a half-life beyond 10^{28} yr. The best technologies will be selected based on lessons learned in GERDA and MAJORANA, as well as contributions from other groups.

In its first phase, LEGEND-200, the existing GERDA infrastructure at LNGS will be modified to deploy 200 kg of detectors in the cryostat. LEGEND-200 has a background goal of less than 0.6 cts/(FWHM·ton·yr). Achieving this background rate will allow to reach a sensitivity greater than 10^{27} yr with 1 ton·yr of exposure.

Multiple techniques are already planned to achieve the required background reduction, such as the use of the MAJORANA electroformed copper, the upgrade of the GERDA liquid argon veto and the improvement of the front-end electronics, described in next sections. A crucial point is the choice of a new detector geometry, the Inverted Coaxial Point Contact (ICPC) detector [11], with similar performance to the BEGe and PPC detectors and a mass as large as a coaxial detector. Five enriched ICPC detectors have been already produced and deployed in GERDA in May 2018.

The second stage of LEGEND will occur in a new infrastructure, with 1000 kg of detectors deployed. The background goal for LEGEND-1000 is less than 0.1 cts/(FWHM·ton·yr). This background reduction is necessary to achieve a $0\nu\beta\beta$ decay discovery potential at a half-life greater than 10^{28} yr on a reasonable timescale. The required depth to keep cosmogenic activation backgrounds (e.g. ^{77m}Ge) within the background budget is currently under investigation and will be a contributing factor in the choice of site.

III. READ-OUT ELECTRONICS IN GERDA

The peculiar GERDA situation, with Ge detector working directly in LAr, and the requirement of very low radioactive background, introduces several constraints on the design of the Ge detector readout.

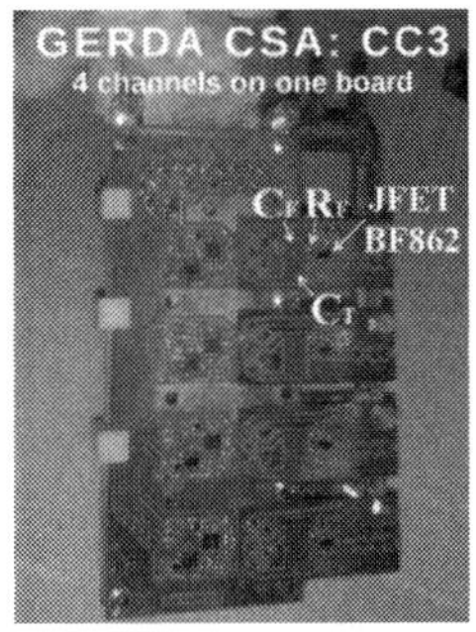

Fig. 3. CC3: GERDA read-out electronics.

The front-end electronics selected for GERDA is the CC3 [12], a hybrid and cryogenic Charge Sensitive Amplifier (CSA), designed to preserve the energy resolution of Ge detectors and to provide higher bandwidth for PSD. A picture of the CC3 board, placing 4 channels, is shown in Fig. 3.

The CC3 is manufactured with highly radio-pure materials on a 0.8 mm thick Cuflon substrate with two low-noise amplifiers with CMOS and Si Bipolar technology and the very front-end (VFE) section with the JFET BF862 from NXP Semiconductors and the feedback components: a resistor of 500 MΩ and a capacitor of 0.4 pF.

During GERDA data taking a performance degradation related to the Ge detector position in the strings has been observed [13]. In best cases an energy resolution of 0.1% (FWHM at 2.6 MeV) is observed in BEGe detectors, in agreement with previous results. Going down in the string the resolution deteriorate until 0.15%. A suitable explanation for this behavior is that the signal cable (from detector to CC3) introduces a stray capacitance, increasing the noise and reducing energy resolution and PSD power. This observation motivated to put effort in the realization of a new read-out electronics for the LEGEND-200 experiment, including an amplifier stage close to each detector.

IV. NEW READ-OUT ELECTRONICS FOR LEGEND-200

A new read-out with the CSA separated in 2 stages, has been already designed, realized and tested. The first stage, namely the Low Mass Front End (LMFE), to be placed in the vicinity of the detector, and the second stage, called CC4, to be located 40-50 cm far from detector array.

The LMFE has been designed and assembled in the Lawrence Berkeley National Laboratory, a picture is shown in Fig. 4. It is realized with ultra-pure materials to fit the stringent condition of LEGEND-200: the circuit is made on a Suprasil substrate, a high purity synthetic fused silica material, with sputtered Ti/Au traces. The LMFE is placing a bare die JFET (two options are tested: the Moxtek MX11 and the SF291 from Semefab), the feedback resistor (in amorphous germanium), feedback and test capacitors (made between traces).

The second stage of the new read-out has been designed and assembled in Milano and is based on GERDA CSA with two

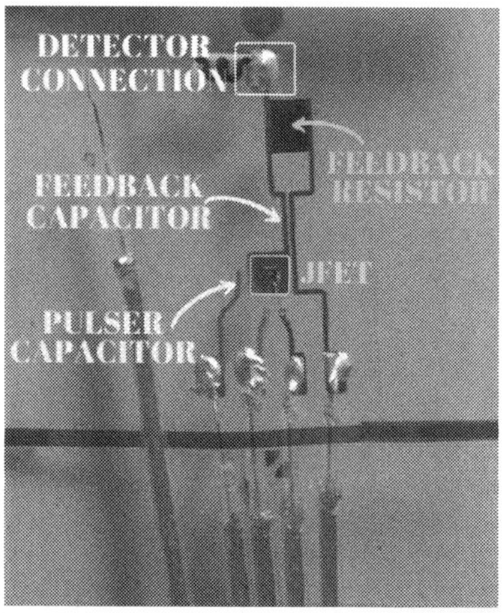

Fig. 4. Low Mass Front End (LMFE) realized in the Lawrence Berkeley National Laboratory [14].

Fig. 5. Prototype of the CC4 (in this version the LMH6655 operational amplifier from Texas Instruments is used).

LMH6654 operational amplifier from Texas Instruments and the addition of a differential output to increase the dynamic range. The picture of the prototype is shown in Fig. 5: connectors with "honeycomb" shape are introduced in order to increase the mechanical stability of the system. The final version will be realized on a Cuflon or Kapton substrate.

First tests of the integration between LMFE and CC4 reported an energy resolution with a test pulse signal of 0.7 keV (FWHM), a dynamic range larger than 9 MeV with a 0.48 pF feedback capacitor, a signal rise time of $\lesssim 50$ ns with 3 pF of input capacitor and a power consumption of ~ 200 mW/chn.

V. Signal Processing in Gerda

It has been observed that GERDA detectors are sometimes affected by low frequency noise because of detector to front-end distance, microphonism induced by liquid argon and picked-up 50 Hz disturbances. In order to preserve their

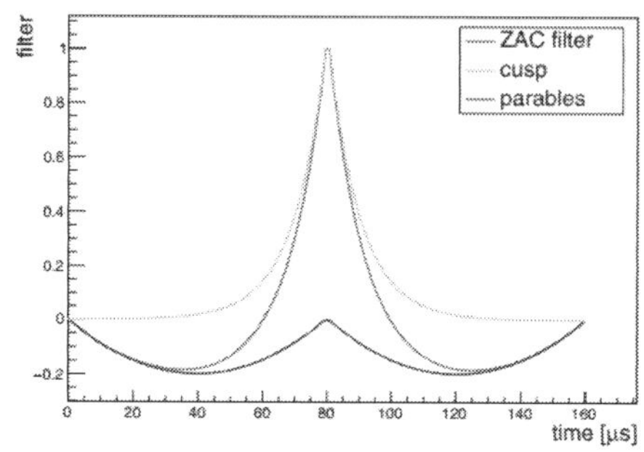

Fig. 6. ZAC filter: finite-length cusp (in green) and two negative parabolas (in blue) are summed to obtain the ZAC filter (in red) [13].

intrinsic energy resolution and PSD capability, a non-standard digital processing technique had to be developed.

From theory it can be shown that in case of series and parallel noise, the optimum shaping filter for energy estimation is an infinite cusp. Then, using a filter with total area equal to zero the low-frequency noise can be removed. In addition, the low-frequency baseline fluctuations are well subtracted by filters with parabolic shape.

Following these arguments a good shaping filter for GERDA is a cusp-like filter with zero total area, obtained by subtracting two parabolas. Also, to allow the full charge collection of the detector a flat-top must be added. The resulting filter is a Zero-Area finite-length Cusp-like (ZAC) filter with central flat-top [15], currently used for energy estimation in GERDA [13]. Fig. 6 shows an example of the ZAC filter construction: a cusp-like filter (in green) and two negative parabolas (in blue) are summed up to obtain the final filter (in red).

This filter improves the energy resolution with respect to a standard gaussian filter. The FWHM at $Q_{\beta\beta} = 2039$ keV obtained in GERDA is 3.0 (3.6) keV for BEGe (coaxial) detectors.

However, the ZAC filter is not able to treat sinusoidal disturbances which are also observed in GERDA detectors, as shown in the FFT power spectrum of Fig. 7 obtained with a BEGe detector. These disturbances reduce the effectiveness of the ZAC filter.

VI. Implementation of the DPLMS filter

An optimum filter, based on the individual detector, is needed in GERDA in order to treat a general noise situation and remove sinusoidal and other disturbances. In addition, the use of an optimum filter is more suitable with a high number of detectors (e.g. LEGEND) since the filter is adapted to the specific noise situation and no further adjustment is needed.

An optimum filter suitable for the GERDA situation is the Digital Penalized Least Mean Square (DPLMS) filter [16]. The method is based on the minimization of a figure of merit composed of experimental noise variance at the filter output and additional weighted constraints that impose some

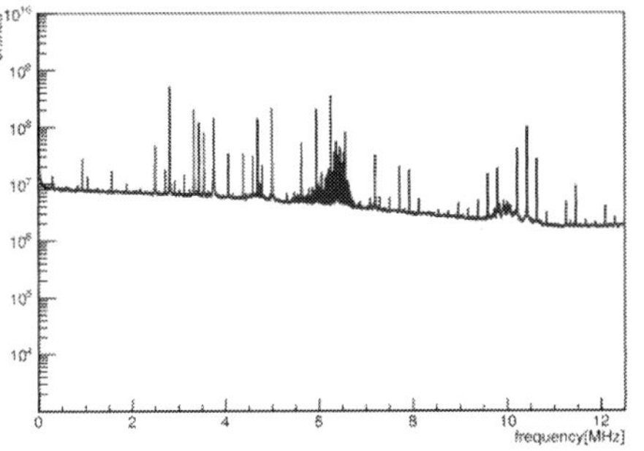

Fig. 7. FFT power spectrum obtained in GERDA with a BEGe detector.

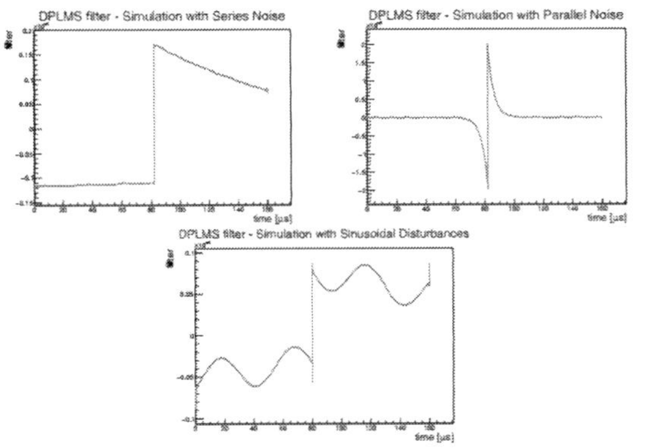

Fig. 8. DPLMS filter obtained with different simulated signals as indicated in the top legend.

appropriate deterministic behavior: pile-up removal, addition of a proper flat-top and zero-area condition. This filter works directly in the time domain, no deconvolution from frequency domain is needed (as the case of other optimum filters).

A C++ based software has been developed to create DPLMS filters starting from experimental signals.

A. DPLMS filter with simulated waveforms

Simulated waveforms that reproduce a signal coming from a GERDA detector have been used to characterize the DPLMS filter. Different types of experimental situations can be added to signals: series and parallel noise, sinusoidal disturbances at various frequencies, baseline movements and pile-up. Few examples of the DPLMS filter produced in different simulations are reported in Fig. 8: the case of only series noise (left upper panel), with the addition of parallel noise (right upper panel) and with a sinusoidal disturbance at frequency of 20 kHz (lower panel).

The resolutions obtained with the reconstruction using the DPLMS filter have been compared with standard gaussian and ZAC filters. As shown in Fig. 9, with increasing series and parallel noise, the DPLMS filter gives a larger improvement with respect to other filters.

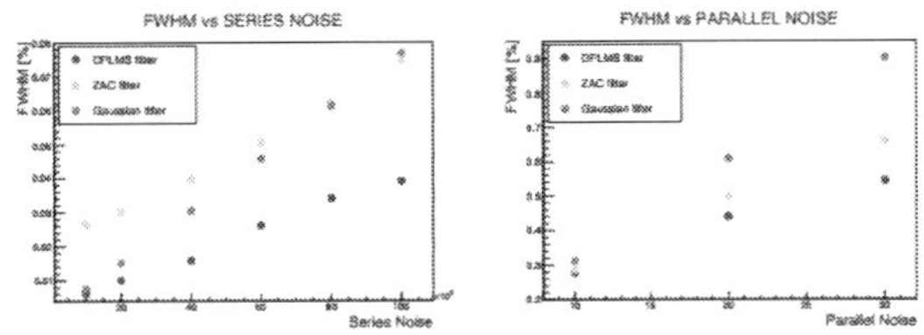

Fig. 9. Comparison of resolutions (in FWHM) obtained using DPLMS (in blue), ZAC (in green) and gaussian (in red) filters as function of the series (left panel) and parallel (right panel) noises.

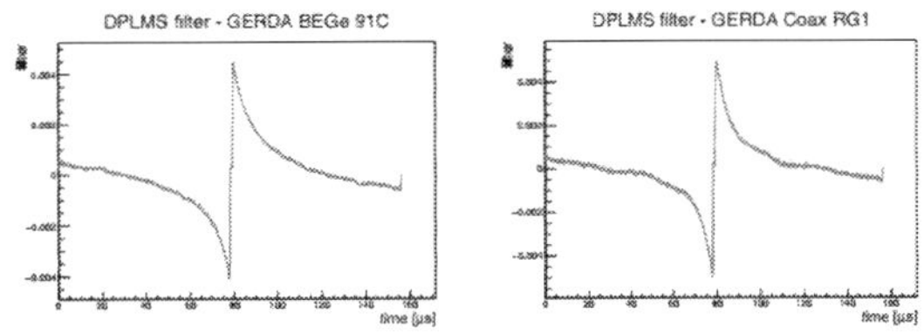

Fig. 10. DPLMS filters obtained in GERDA with a BEGe (left) and a coaxial (right) detector.

TABLE I
COMPARISON OF FWHM AT THE 2.6 MEV PEAK OBTAINED WITH GAUSSIAN, ZAC AND DPLMS FILTERS IN THREE GERDA DETECTORS.

detector	Gaussian	ZAC filter	DPLMS filter
BEGe-91A	2.77 ± 0.06	2.67 ± 0.06	2.61 ± 0.05
BEGe-91C	3.52 ± 0.06	3.38 ± 0.06	3.24 ± 0.05
Coax-RG1	4.77 ± 0.04	4.34 ± 0.04	4.22 ± 0.03

Sinusoidal disturbances are well removed by the DPLMS filter, while other filters are not able to treat this situation and very bad resolution is achieved.

B. DPLMS filter with GERDA *data*

In a second step, the DPLMS filter method has been applied to signals coming from GERDA detectors. Three detectors have been selected: the 91A BEGe detector, that is reporting a very good energy resolution, the 91C BEGe detector, that shows a bad resolution, and the RG1 coaxial detectors.

To create DPLMS filters, pulses from ^{228}Th calibration runs are used, the noise is evaluated during physics runs. The flat-top condition and the pile-up removal are needed for proper energy reconstruction. DPLMS filters obtained for the 91C BEGe and the RG1 coaxial detectors are reported in Fig. 10.

The comparison with other filters is performed by fitting the 2.6 MeV peak of the calibration spectra. The results are reported in Tab. I for gaussian, ZAC and DPLMS filters in the three selected GERDA detectors.

The energy resolution improves using the ZAC filter in comparison with the results of the gaussian filter. A further improvement is produced with the DPLMS filter: the FWHM improvement is in the range of 6-12% and 2-4% with respect to gaussian and ZAC filter respectively.

VII. Conclusions and Outlook

The adopted solution for the GERDA read-out electronics has been presented. A new read-out has been realized for the LEGEND-200 experiment, that will replace GERDA. The new design includes a separation in two stages: the goal is not to degrade the excellent performance of Ge detectors. First tests reported good results in terms of energy resolution, dynamic range, bandwidth and power consumption.

The particular GERDA condition requires a non-standard shaping filter, currently a modified cusp-like filter is used. To take into account a more general noise situation, we started the implementation of an optimum filter: our choice is the DPLMS filter. Promising preliminary results has been obtained in a first application to signal coming from GERDA detectors and a further improvement is expected with more specialization. In addition a future implementation of the DPLMS filter is planned to improve the PSD power, that is much more affected by high-frequency sinusoidal disturbances.

Acknowledgment

The GERDA experiment is supported by the German Federal Ministry for Education and Research (BMBF), the German Research Foundation (DFG) via the Excellence Cluster Universe, the Italian Istituto Nazionale di Fisica Nucleare (INFN), the Max Planck Society (MPG), the Polish National Science Centre (NCN), the Foundation for Polish Science (TEAM/2016-2/17), the Russian Foundation for Basic Research (RFBR) and the Swiss National Science Foundation (SNF). These research institutions acknowledge internal financial support. This project has received funding/support from the European Unions Horizon 2020 research and innovation programme under the Marie Sklodowska-Curie grant agreements No 690575 and No 674896. The GERDA Collaboration thanks the directors and the staff of the LNGS for their support of the GERDA experiment.

References

[1] GERDA Collaboration, Eur. Phys. J. C 73 2330 (2013)
[2] GERDA Collaboration, Phys. Rev. Lett. 111, 122503 (2013)
[3] GERDA Collaboration, Eur. Phys. J 78 388 (2018)
[4] GERDA Collaboration, Eur. Phys. J. C 75 39 (2015)
[5] Janicskò-Csàthy J. et al., (arXiv:1606.04254)
[6] GERDA Collaboration, Nature 544, 7648 (2017)
[7] GERDA Collaboration, Phys. Rev. Lett. 120 132503 (2018)
[8] Zsigmond A.J. for the GERDA Collaboration, Neutrino 2018, Heidelberg
[9] MAJORANA Collaboration, (arXiv:1902.02299)
[10] LEGEND Collaboration, AIP Conf. Proc. 1894, 020027 (2017)
[11] Kermaidic Y. et al., Nucl. Instrum. Meth. A 891, 106 (2018)
[12] Riboldi S. et al., ANIMMA 2015, Lisbon (Portugal), pp. 1-6
[13] D'Andrea V., PhD Thesis (2017)
[14] M. Willers, Private communication
[15] GERDA Collaboration, Eur. Phys. J. C 75 255 (2015)
[16] Gatti E. et al., Nucl. Instr. Meth. Phys. Res. A 523 (2004) 167-185

Theoretical Approach to the Energy Resolution of a Scintillation Spectrometer with Several Photodetectors

Victor V. Samedov
National Research Nuclear University MEPhI (Moscow Engineering Physics Institute)
v-samedov@yandex.ru

Abstract **— The exact mathematical description of processes at registration of incident particles by a scintillation spectrometer with several photodetectors allows receiving correct formulae for the mean value and the variance of the amplitude at the output of photodetectors, and for the covariance between signals of photodetectors. It was shown that using the relative covariance of signals allows the experimental determination of the covariance in the light yield and the Fano factor in a scintillator. The important advantage of the proposed method is its independence on the gains and the noises of photodetectors electronics.**

Index Terms— **scintillation spectrometer, scintillation detector, photodetector, energy resolution, covariance between signals, Fano factor, light yield.**

I. Introduction

THERE are a number of processes in scintillation detectors that lead to output signal broadening in response to impinging monochromatic radiation. There are semiempirical formulae that account for the contribution of different factors to the energy resolution of scintillation detectors [1], [2], [3]. However, these formulae do not contain any information on relationship of these contributions with scintillator characteristics and characteristics of the processes that occur in a scintillation detector at registration of incident particles.

The correct formula for the variance of the output signal of a scintillation detector cannot be semiempirical. It must result strictly from the exact mathematical description of the processes that take place at registration of the incident particle by a scintillation detector. Only such formula has the predictive ability. Only in this case dependence of the energy resolution on characteristics of scintillation crystal and other parameters of the detector can be revealed, and the conditions at which the characteristics of the processes can be derived from the output signal. As the process of signal formation at the output of detector represents a random branching cascade process, the formalism of probability generating functions (PGF) is the most adequate for its formulation [4].

Our approach to formulation of the mathematical model of an incident particle registration by a scintillation detector is based on the approach developed in formulation of the theory of shower spectrometers in [5], and published in [6] and [7].

2. Mathematical Model of a Scintillation Detector

Let us formulate the mathematical model of an incident particle registration by a scintillation detector with N photodetectors, which we number with the index n, $(n=\overline{1,N})$. The process of an incident particle energy transformation into the output signals includes the following successive stages.

1. <u>The stage of interaction of an incident particle with a scintillation crystal</u>. Let divide the coordinate space into volume elements ΔV_i, $(i=\overline{1,I})$ that belong to the detector volume V, and ΔV_0 the exterior to the detector volume. An incident particle at interacting with the detector volume produces in the volume elements ΔV_i secondary particles of the type α, $(\alpha=\overline{1,A})$ where index α determines secondary particles such as photons, electrons, positrons, phonons, and may include protons, if incident particles are protons or neutrons, and alpha-particles, if incident particles are alpha-particles. Let us divide the range of possible secondary-particle energies into the energy intervals ΔE_j, $(j=\overline{1,J})$ and the space of possible directions of secondary particles into solid-angle elements $\Delta\Omega_k$, $(k=\overline{1,K})$. Let introduce a $A\times I\times J\times K$-dimensional random vector $\{N_{\alpha ijk}\}$, whose components $N_{\alpha ijk}$ are the number of secondary particles of the type α in the element of the phase space $\Delta\Gamma_{ijk}=\Delta V_i\Delta E_j\Delta\Omega_k$. Let introduce the probabilities $p^c_{\alpha ijk}$ of all possible combinations of components of the random vector $\{N^c_{\alpha ijk}\}$, which characterize all possible distributions of secondary particles in the phase space elements after finishing all processes of downconversion of the energy of the incident particle into the energy of secondary particles that cannot produce electron-hole pairs in the scintillation crystal.

2. <u>The stage of electron-hole pairs generation</u>. The scintillation properties are strongly related to the crystal structure. The scintillation crystals have the property of semiconductor material, and generation of light photons is

Manuscript received May 29, 2019.
Victor V. Samedov is with the National Research Nuclear University MEPhI (Moscow Engineering Physics Institute), 31, Kashirskoye Shosse, 115409, Moscow, Russian Federation (telephone: 007-495-323-9038, e-mail: v-samedov@yandex.ru)

going via the generation of electron-hole pairs. Let $f_{v_{\alpha ijk}}[s]$ be the PGF of the process of electron-hole pairs generation by a secondary particle of the type α from the element of the phase space $\Delta\Gamma_{ijk}$. Hereafter s is the auxiliary variable of the probability generating function. In a semiconductor crystal, only secondary particles with energies greater than the threshold energy $E_{\alpha ij_{\min}k}$, which may depend on the point in the detector volume, and on the energy and moving direction of the secondary particle, can produce electron-hole pairs. This fact can be accounted for by assuming that the mean value $\langle v_{\alpha ijk}\rangle$ and the variance $\sigma^2_{v_{\alpha ijk}}$ of the number of electron-hole pairs are equal to zero for secondary particles with energies less than the threshold energy $j \le j_{\alpha\min}$.

Despite the diversity of mechanisms in various inorganic scintillators, the process of light photon emission in general consists of the following successive stages.

3. The stage of electron-hole pairs recombination. This stage is characterized by the probability of an electron-hole pair generated in the volume element ΔV_i to survive in recombination or to form an exciton $\varepsilon_{r\,\alpha ijk}$. This probability strongly depends on the stopping power of the secondary particle of the type α from the element of the phase space $\Delta\Gamma_{ijk}$. This stage is binomial with the PGF

$$f_r\,[s] = 1-\varepsilon_{r\,\alpha ijk} + \varepsilon_{r\,\alpha ijk}\cdot s,\quad (1)$$

4. The stage of diffusion of carriers (electrons, holes and excitons) in a scintillator. This process is also binomial with the PGF

$$f_D\,[s] = 1-p_{D\,ii'} + p_{D\,ii'}\cdot s,(2)$$

where $p_{D\,ii'}$ is the probability of diffusion of an electron-hole pair produced in the volume element ΔV_i to the volume element $\Delta V_{i'}$, $(i' = \overline{1,I})$.

5. The stage of luminescent centers activation. This stage obeys the binomial distribution with the PGF

$$f_a\,[s] = 1-\varepsilon_{a\,i'} + \varepsilon_{a\,i'}\cdot s,\ (3)$$

where $\varepsilon_{a\,i'}$ is the probability of luminescent center activation in the volume element $\Delta V_{i'}$ by an electron-hole pair.

6. The stage of a light photon emission by the luminescent center. Because of a random spatial distribution of luminescent centers in the host matrix, the energy levels of each luminescent center experience different Stark shifts arising from the electric field of the surrounding ions. Therefore, the number of light photons per elementary mode of radiation is less than unity, and we can consider that photons are emitted by the luminescent centers independently. In this case, the process of a photon emission by the luminescent center belonging to the volume element $\Delta V_{i'}$ is binomial with the PGF

$$f_e\,[s] = 1-\sum_{l=1}^{L} p_l \sum_{k'=1}^{K'} \varepsilon_{e\,i'lk'} + \sum_{l=1}^{L} p_l \sum_{k'=1}^{K'} \varepsilon_{e\,i'lk'}\cdot s,\ (4)$$

where $\varepsilon_{e\,i'lk'}$ is the probability of the radiative transition with emission of a photon with the wavelength λ_l, $(l = \overline{1,L})$ and the direction in the solid-angle element $\Delta\Omega_{k'}$, $(k' = \overline{1,K'})$. In (4) p_l is the probability of the l radiative transition mode $\left(\sum_{l=1}^{L} p_l = 1\right)$.

It is necessary to notice that this model is also applicable to organic scintillators if the stage of electron-hole pairs generation is related to the stage of ionization and excitation of organic molecules; the stage of electron-hole pairs recombination to quenching processes; the stage of carriers diffusion to the process of excitation energy migration to other molecules; the stage of luminescent centers activation to the stage of internal conversion of the excitation energy into the energy of fluorescent transition; the stage of a light photon emission by the luminescent center to the stage of a fluorescent transition with emission of a light photon.

7. The stage of a photon collection on the entrance window of one of the photodetectors. The fundamental nature of indivisibility of a light photon requires that it must be counted by only one of photodetectors. Let $\tau_{n\,i'lk'mk''}$ be the probability of a photon with the wavelength λ_l emitted in the solid-angle element $\Delta\Omega_{k'}$ by the luminescent center in the $\Delta V_{i'}$ to get the surface element $\Delta S_{n\,m}$, $(m = \overline{1,M})$ of the entrance window of n-th photodetector in the direction belonging to the solid-angle element $\Delta\Omega_{n\,k''}$, $(k'' = \overline{1,K''})$ with respect to the surface normal. The PGF of this process has the form

$$f_\tau\,[\{s_n\}] = 1-\sum_{n=1}^{N}\sum_{m=1}^{M}\sum_{k''=1}^{K''}\tau_{n\,i'lk'mk''} + \sum_{n=1}^{N}\sum_{m=1}^{M}\sum_{k''=1}^{K''}\tau_{n\,i'lk'mk''}s_n$$

, (5)

where $\{s_n\}$ are the auxiliary variables of photodetectors.

8. The stage of a photon conversion to a photoelectron. The process of a photon conversion to a photoelectron in n-th photodetector is binomial with the PGF

$$f_\eta\,[s] = 1-\eta_{n\,lmk''} + \eta_{n\,lmk''}\cdot s,\ (6)$$

where $\eta_{n\,lmk''}$ is the quantum efficiency of the surface element $\Delta S_{n\,m}$ of the n-th photodetector to the photon with the wavelength λ_l crossing the entrance window in the direction belonging to the solid-angle element $\Delta\Omega_{n\,k''}$.

As the processes of successive stages 3 – 8 are binomial, so the PGF of this compound cascade process is also binomial

$$f_\chi\,[\{s_n\}] = 1-\sum_{n=1}^{N}\chi_{n\,\alpha ijk} + \sum_{n=1}^{N}\chi_{n\,\alpha ijk}\cdot s_n,\ (7)$$

where

$$\chi_{n\ \alpha ijk} = \varepsilon_{r\ \alpha ijk} \sum_{i'=1}^{I} p_{D\ ii'} \sum_{l=1}^{L} p_l \sum_{k'=1}^{K'} \sum_{m=1}^{M} \sum_{k''=1}^{K''} \varepsilon_{a\ i'} \varepsilon_{e\ i'lk'} \tau_{n\ i'lk'mk''} \eta_{n\ lmk''} \qquad (8)$$

is the probability of a photoelectron creation in the n-th photodetector by a secondary particle of the type α from the element of the phase space $\Delta\Gamma_{ijk}$.

9. <u>The stage of signal amplification in the n-th photodetector</u>. Let $f_{g\,n}[s_n]$, $(n = \overline{1, N})$, be the PGF of the gain in the n-th photodetector with the mean value $\langle g_n \rangle$ and the variance σ_{gn}^2. Let $f_{\text{noise}\,n}[s_n]$ be the PGF of the electronic noise at the output of n-th photodetector with the zero mean value and the variance $\sigma_{\text{noise}\,n}^2$.

For the model described above, the PGF of incident particles registration by a scintillation detector with several photodetectors is

$$f_{\{Q_n\}}\left[\{s_n\}\right]=\lim_{IJK\to\infty}\sum_c p^c_{\alpha ijk}\prod_{\alpha=1}^{A}\prod_{i=1}^{I}\prod_{j=j_{\min}}^{J}\prod_{k=1}^{K}\left(f_{\nu_{\alpha ijk}}\left[1-\sum_{n=1}^{N}\chi_{n\ \alpha ijk}+\sum_{n=1}^{N}\chi_{n\ \alpha ijk}f_{g\,n}\left[s_n\right]\right]\right)^{N^c_{\alpha ijk}}\prod_{n=1}^{N}f_{\text{noise}\,n}\left[s_n\right],\quad(9)$$

This PGF will be used for determination of the covariance of photodetectors signals.

If we set $\{s_n=s\}$, $(n=\overline{1,N})$, then we have the PGF for the sum signal of scintillation spectrometer photodetectors

$$f_Q\left[s\right]=\lim_{IJK\to\infty}\sum_c p^c_{\alpha ijk}\prod_{\alpha=1}^{A}\prod_{i=1}^{I}\prod_{j=j_{\min}}^{J}\prod_{k=1}^{K}\left(f_{\nu_{\alpha ijk}}\left[1-\sum_{n=1}^{N}\chi_{n\ \alpha ijk}+\sum_{n=1}^{N}\chi_{n\ \alpha ijk}f_{g\,n}\left[s\right]\right]\right)^{N^c_{\alpha ijk}}\prod_{n=1}^{N}f_{\text{noise}\,n}\left[s\right],\quad(10)$$

From the probability generating function (10), the mean value and the variance of the sum signal take forms:

$$\langle Q\rangle=\left.\frac{\partial f_Q\left[s\right]}{\partial s}\right|_{s=1}=\lim_{IJK\to\infty}\sum_c p^c_{\alpha ijk}\sum_{\alpha=1}^{A}\sum_{i=1}^{I}\sum_{j=1}^{J}\sum_{k=1}^{K}N^c_{\alpha ijk}\left\langle\nu_{\alpha ijk}\right\rangle\sum_{n=1}^{N}\chi_{n\ \alpha ijk}\left\langle g_n\right\rangle,\quad(11)$$

$$\sigma_Q^2=\left.\frac{\partial f_Q^2\left[s\right]}{\partial s^2}\right|_{s=1}+\left.\frac{\partial f_Q\left[s\right]}{\partial s}\right|_{s=1}-\left(\left.\frac{\partial f_Q\left[s\right]}{\partial s}\right|_{s=1}\right)^2=\sigma^2_{\text{cov}}+\sigma^2_{\text{pair}}+\sigma^2_{\text{ph}}+\sigma^2_{gain}+\sigma^2_{\text{noise}},\quad(12)$$

In (12),

$$\begin{aligned}\sigma^2_{\text{cov}}=&\lim_{IJK\to\infty}\sum_c p^c_{\alpha ijk\alpha'i'j'k'}\sum_{\alpha=1}^{A}\sum_{i=1}^{I}\sum_{j=1}^{J}\sum_{k=1}^{K}N^c_{\alpha ijk}\left\langle\nu_{\alpha ijk}\right\rangle\sum_{n=1}^{N}\left\langle\chi_{n\ \alpha ijk}\right\rangle\left\langle g_n\right\rangle\sum_{\alpha'=1}^{A}\sum_{i'=1}^{I}\sum_{j'=1}^{J}\sum_{k'=1}^{K}N^c_{\alpha'i'j'k'}\left\langle\nu_{\alpha'i'j'k'}\right\rangle\sum_{n=1}^{N}\left\langle\chi_{n\ \alpha'i'j'k'}\right\rangle\left\langle g_n\right\rangle\\&-\left(\lim_{IJK\to\infty}\sum_c p^c_{\alpha ijk}\sum_{\alpha=1}^{A}\sum_{i=1}^{I}\sum_{j=1}^{J}\sum_{k=1}^{K}N^c_{\alpha ijk}\left\langle\nu_{\alpha ijk}\right\rangle\sum_{n=1}^{N}\left\langle\chi_{n\ \alpha ijk}\right\rangle\left\langle g_n\right\rangle\right)\times\\&\left(\lim_{IJK\to\infty}\sum_c p^c_{\alpha'i'j'k'}\sum_{\alpha'=1}^{A}\sum_{i'=1}^{I}\sum_{j'=1}^{J}\sum_{k'=1}^{K}N^c_{\alpha'i'j'k'}\left\langle\nu_{\alpha'i'j'k'}\right\rangle\sum_{n=1}^{N}\left\langle\chi_{n\ \alpha'i'j'k'}\right\rangle\left\langle g_n\right\rangle\right)\end{aligned},\quad(13)$$

is the variance of the output signal due to covariance of secondary particles in the phase space, where $p^c_{\alpha ijk\alpha'i'j'k'}$ is the joint distribution function of all possible combinations of secondary particles in the joint phase space.

$$\begin{aligned}\sigma^2_{\text{pair}}&=\lim_{IJK\to\infty}\sum_c p^c_{\alpha ijk}\sum_{\alpha=1}^{A}\sum_{i=1}^{I}\sum_{j=1}^{J}\sum_{k=1}^{K}N^c_{\alpha ijk}\sigma^2_{\nu_{\alpha ijk}}\left(\sum_{n=1}^{N}\left\langle\chi_{n\ \alpha ijk}\right\rangle\left\langle g_n\right\rangle\right)^2\\&=\lim_{IJK\to\infty}\sum_c p^c_{\alpha ijk}\sum_{\alpha=1}^{A}\sum_{i=1}^{I}\sum_{j=1}^{J}\sum_{k=1}^{K}N^c_{\alpha ijk}F_{\alpha ijk}\left\langle\nu_{\alpha ijk}\right\rangle\left(\sum_{n=1}^{N}\left\langle\chi_{n\ \alpha ijk}\right\rangle\left\langle g_n\right\rangle\right)^2\end{aligned},\quad(14)$$

is the variance of the output signal due to fluctuations in the number of electron–hole pair. In (14), according to the Fano model [8], we assume that fluctuations in the number of electron–hole pairs are proportional to their average

$$\sigma^2_{\nu_{\alpha ijk}}=F_{\alpha ijk}\left\langle\nu_{\alpha ijk}\right\rangle,\quad(15)$$

where $F_{\alpha ijk}$ is the Fano factor.

$$\sigma^2_{\text{ph}}=\lim_{IJK\to\infty}\sum_c p^c_{\alpha ijk}\sum_{\alpha=1}^{A}\sum_{i=1}^{I}\sum_{j=1}^{J}\sum_{k=1}^{K}N^c_{\alpha ijk}\left\langle\nu_{\alpha ijk}\right\rangle\left(\sum_{n=1}^{N}\left\langle\chi_{n\ \alpha'i'j'k'}\right\rangle\left\langle g_n\right\rangle^2-\left(\sum_{n=1}^{N}\left\langle\chi_{n\ \alpha ijk}\right\rangle\left\langle g_n\right\rangle\right)^2\right)\quad(16)$$

is the variance of the output signal due to fluctuations in the number of photoelectrons in photodetectors.

$$\sigma^2_{gain}=\lim_{IJK\to\infty}\sum_c p^c_{\alpha ijk}\sum_{\alpha=1}^{A}\sum_{i=1}^{I}\sum_{j=1}^{J}\sum_{k=1}^{K}N^c_{\alpha ijk}\left\langle\nu_{\alpha ijk}\right\rangle\sum_{n=1}^{N}\left\langle\chi_{n\ \alpha'i'j'k'}\right\rangle\sigma^2_{gn}\quad(17)$$

is the variance of the output signal due to fluctuations in the electronic gains of photodetectors.

$$\sigma^2_{\text{noise}}=\sum_{n=1}^{N}\sigma^2_{\text{noise}\,n}\quad(18)$$

is the variance due to electronic noise.

As the average number of electron–hole pairs is determined by the energy absorbed in the detector volume element, it is convenient to move from the particle distribution to the absorbed energy distribution as follows:

$$N^{c}_{\alpha ijk}\left\langle v_{\alpha ijk}\right\rangle = W^{c}_{\alpha ijk} / \varepsilon_{\text{e-h}\,\alpha ijk} , \quad (19)$$

where $W^{c}_{\alpha ijk}$ is the energy absorbed in the volume element ΔV_i given $N^{c}_{\alpha ijk}$ secondary particles of type α are in the phase space element $\Delta\Gamma_{ijk}$;and $\varepsilon_{\text{e-h}\,\alpha ijk}$ is the average energy of the electron–hole pair production.

Any detector is usually characterized by the energy resolution of the total absorption peak, when the total energy of the primary monoenergetic particle E_0 is absorbed in the detector volume

$$\lim_{IJK\to\infty}\sum_{\alpha=1}^{A}\sum_{i=1}^{I}\sum_{j=1}^{J}\sum_{k=1}^{K}W^{c}_{\alpha ijk} = \sum_{\alpha=1}^{A}\int_V\int_E\int_\Omega dVdEd\vec{\Omega}\frac{\partial^3 W^{c}_{\alpha}(E_0,\vec{r},E,\vec{\Omega})}{\partial V\partial E\partial\vec{\Omega}} = \sum_{\alpha=1}^{A}\int_V\int_E\int_\Omega dVdEd\vec{\Omega}w^{c}_{\alpha}(E_0,\vec{r},E,\vec{\Omega}) = E_0 . \quad (20)$$

where $w^{c}_{\alpha}(E_0,\vec{r},E,\vec{\Omega})$ is the differential density of absorbed energy in the detector volume given $N^{c}_{\alpha ijk}$ secondary particles of the type α are in the phase space element $\Delta\Gamma_{ijk}$.

For the case when primary monoenergetic particles are totally absorbed in the detector volume, the mean value of signal takes the form

$$\langle Q\rangle = \sum_{\alpha=1}^{A}\int_V\int_E\int_\Omega dVdEd\vec{\Omega}u(E-E_{\alpha\ \min}(\vec{r},E,\vec{\Omega}))\left\langle\frac{w^{c}_{\alpha}(E_0,\vec{r},E,\vec{\Omega})}{\varepsilon_{\text{e-h}\,\alpha}(\vec{r},E,\vec{\Omega})}\left(\sum_{n=1}^{N}\left\langle\chi_{n\,\alpha}(\vec{r},E,\vec{\Omega})\right\rangle_D\langle g_n\rangle\right)\right\rangle_c \quad (21)$$

where the Heaviside unit function, ($u(x)=0$ for $x<0$ and $u(x)=1$ for $x\geq 0$) accounts for the threshold energy of electron–hole pair production by secondary particles. The subscript c of the angle brackets denotes averaging over all possible distributions of secondary particles in the phase space elements;

$$\begin{aligned}\left\langle\chi_{n\,\alpha}(\vec{r},E,\vec{\Omega})\right\rangle_D &= \int_V dV'\rho_D(\vec{r},\vec{r}')\sum_{l=1}^{L}p_l(\lambda_l)\int_{\Omega'}d\vec{\Omega}'\varepsilon_{r\,\alpha}(\vec{r},E,\vec{\Omega},\vec{r}')\varepsilon_a(\vec{r}')\varepsilon_e(\vec{r}',\lambda_l,\vec{\Omega}')\times \\ &\int_{S_n}\int_{\Omega''_n}dS_n d\vec{\Omega}''_n\tau_n(\vec{r}',\lambda_l,\vec{\Omega}',S_n,\vec{\Omega}''_n)\eta_n(\lambda_l,S_n,\vec{\Omega}''_n)\end{aligned} . \quad (22)$$

The subscript D of the angle brackets denotes averaging over the diffusion probability density.

In a uniform scintillator the spreading of carriers produced at the point $\vec{r}$ is symmetrical about the origin, and the probability density of finding a carrier at the point $\vec{r}'$ of the scintillator depends only on the relative distance between the points and obeys the normal distribution

$$\rho_D(\vec{r}',\vec{r}) = \rho_D(|\vec{r}'-\vec{r}|) = (\sqrt{4\pi}\lambda_D)^{-3}\exp\left(-(\vec{r}'-\vec{r})^2/4\lambda_D^2\right) \quad (23)$$

with characteristic diffusion length

$$\lambda_D = \sqrt{D\tau} , \quad (24)$$

where D is the effective diffusion coefficient, and τ is the mean lifetime of carriers.

The Taylor expansion of the coefficient (22) up to the second order on displacement $\vec{r}'-\vec{r}$ has the form

$$\chi_{n\,\alpha}(\vec{r}',E,\vec{\Omega}) = \chi_{n\,\alpha}(\vec{r},E,\vec{\Omega}) + \left((\vec{r}'-\vec{r})\cdot\nabla\right)\chi_{n\,\alpha}(\vec{r},E,\vec{\Omega}) + \frac{1}{2!}\left((\vec{r}'-\vec{r})\cdot\nabla\right)^2\chi_{n\,\alpha}(\vec{r},E,\vec{\Omega}) . \quad (25)$$

At averaging (25) over the probability density of diffusion (23), owing to isotropy the linear term vanishes, and the second order term comes to

$$\begin{aligned}&\left\langle\frac{1}{2!}\left((\vec{r}'-\vec{r})\cdot\nabla\right)^2\chi_{n\,\alpha}(\vec{r},E,\vec{\Omega})\right\rangle_D = \frac{1}{2}\left\langle\sum_{i,i'=1}^{3}(\xi'_i-\xi_i)\frac{\partial^2\chi_{n\,\alpha}(\vec{r},E,\vec{\Omega})}{\partial\xi_i\partial\xi'_i}\Big|_{\xi_i,\xi'_i=x,y,z}(\xi'_{i'}-\xi_{i'})\right\rangle_D = \\ &\frac{1}{2}\sum_{i=1}^{3}\frac{\partial^2\chi_{n\,\alpha}(\vec{r},E,\vec{\Omega})}{\partial\xi_i^2}\Big|_{\xi_i=x,y,z}\left\langle(\xi'_i-\xi_i)^2\right\rangle_D = \frac{1}{6}\Delta\chi_{n\,\alpha}(\vec{r},E,\vec{\Omega})\left\langle(\vec{r}'-\vec{r})^2\right\rangle_D = \Delta\chi_{n\,\alpha}(\vec{r},E,\vec{\Omega})\lambda_D^2\end{aligned} , \quad (26)$$

As a result,

$$\left\langle \chi_{n\,\alpha}(\vec{r},E,\vec{\Omega})\right\rangle_D = \chi_{n\,\alpha}(\vec{r},E,\vec{\Omega}) + \Delta\chi_{n\,\alpha}(\vec{r},E,\vec{\Omega})\lambda_D^2 = (1+\lambda_D^2\Delta)\chi_{n\,\alpha}(\vec{r},E,\vec{\Omega}) =$$

$$= \sum_{l=1}^{L} p_l(\lambda_l)(1+\lambda_D^2\Delta)\int_{\Omega'} d\vec{\Omega}'\varepsilon_{r\,\alpha}(\vec{r},E,\vec{\Omega})\varepsilon_a(\vec{r})\varepsilon_e(\vec{r},\lambda_l,\vec{\Omega}')\int_{S_n}\int_{\vec{\Omega}''_n} dS_n d\vec{\Omega}''_n \tau_n(\vec{r},\lambda_l,\vec{\Omega}',S_n,\vec{\Omega}''_n)\eta_n(\lambda_l,S_n,\vec{\Omega}''_n), \quad (27)$$

As we move to the absorbed energy distribution in the detector volume, the contributions to the output signal variance are

$$\sigma^2_{\text{cov}} = \sum_{\alpha=1}^{A}\sum_{\alpha'=1}^{A\cdot}\int_V\int_E\int_\Omega\int_{V'}\int_{E'}\int_{\Omega'} dVdEd\vec{\Omega}dV'dE'd\vec{\Omega}'u(E-E_{\alpha\ \min}(\vec{r},E,\vec{\Omega}))u(E'-E_{\alpha'\ \min}(\vec{r}',E',\vec{\Omega}'))\times$$

$$\left\langle \frac{w_\alpha^c(E_0,\vec{r},E,\vec{\Omega})}{\varepsilon_{\text{e-h}\,\alpha}(\vec{r},E,\vec{\Omega})}\left(\sum_{n=1}^{N}\left\langle\chi_{n\,\alpha}(\vec{r},E,\vec{\Omega})\right\rangle_D\langle g_n\rangle\right)\frac{w_{\alpha'}^c(E_0,\vec{r}',E',\vec{\Omega}')}{\varepsilon_{\text{e-h}\,\alpha'}(\vec{r}',E',\vec{\Omega}')}\left(\sum_{n'=1}^{N}\left\langle\chi_{n'\,\alpha'}(\vec{r}',E',\vec{\Omega}')\right\rangle_D\langle g_{n'}\rangle\right)\right\rangle_c, \quad (28)$$

$$-\left(\sum_{\alpha=1}^{A}\int_V\int_E\int_\Omega dVdEd\vec{\Omega}u(E-E_{\alpha\ \min}(\vec{r},E,\vec{\Omega}))\left\langle\frac{w_\alpha^c(E_0,\vec{r},E,\vec{\Omega})}{\varepsilon_{\text{e-h}\,\alpha}(\vec{r},E,\vec{\Omega})}\left(\sum_{n=1}^{N}\left\langle\chi_{n\,\alpha}(\vec{r},E,\vec{\Omega})\right\rangle_D\langle g_n\rangle\right)\right\rangle_c\right)^2$$

$$\sigma^2_{\text{pair}} = \sum_{\alpha=1}^{A}\int_V\int_E\int_\Omega dVdEd\vec{\Omega}u(E-E_{\alpha\ \min}(\vec{r},E,\vec{\Omega}))\times$$

$$\left\langle F_\alpha(\vec{r},E,\vec{\Omega})\frac{w_\alpha^c(E_0,\vec{r},E,\vec{\Omega})}{\varepsilon_{\text{e-h}\,\alpha}(\vec{r},E,\vec{\Omega})}\left(\sum_{n=1}^{N}\left\langle\chi_{n\,\alpha}(\vec{r},E,\vec{\Omega})\right\rangle_D\langle g_n\rangle\right)^2\right\rangle_c, \quad (29)$$

$$\sigma^2_{\text{ph}} = \sum_{\alpha=1}^{A}\int_V\int_E\int_\Omega dVdEd\vec{\Omega}u(E-E_{\alpha\ \min}(\vec{r},E,\vec{\Omega}))\times$$

$$\left\langle\frac{w_\alpha^c(E_0,\vec{r},E,\vec{\Omega})}{\varepsilon_{\text{e-h}\,\alpha}(\vec{r},E,\vec{\Omega})}\left(\sum_{n=1}^{N}\left\langle\chi_{n\,\alpha}(\vec{r},E,\vec{\Omega})\right\rangle_D\langle g_n\rangle^2-\left(\sum_{n=1}^{N}\left\langle\chi_{n\,\alpha}(\vec{r},E,\vec{\Omega})\right\rangle_D\langle g_n\rangle\right)^2\right)\right\rangle_c \quad (30)$$

$$\sigma^2_{gain} = \sum_{\alpha=1}^{A}\int_V\int_E\int_\Omega dVdEd\vec{\Omega}u(E-E_{\alpha\ \min}(\vec{r},E,\vec{\Omega}))\left\langle\frac{w_\alpha^c(E_0,\vec{r},E,\vec{\Omega})}{\varepsilon_{\text{e-h}\,\alpha}(\vec{r},E,\vec{\Omega})}\sum_{n=1}^{N}\left\langle\chi_{n\,\alpha}(\vec{r},E,\vec{\Omega})\right\rangle_D\sigma^2_{g\,n}\right\rangle_c \quad (31)$$

$$\sigma^2_{\text{noise}} = \sum_{n=1}^{N}\sigma^2_{\text{noise}\,n} \quad (32)$$

The expressions (21) (27) – (32) are the most general formulae for the average and the variance of a scintillation detector with several photodetectors and are the basis for various approximations. These expressions include the formulae for the average and the variance of any photodetector signal if we leave in the sum only one term for the given photodetector.

3. APPROXIMATION FOR LOW-ENERGY X-RAYS

For the following, we need approximation for low-energy X-rays, when we can ignore the process of multiple scattering of X-quantum and can consider that its energy is transmitted to the kinetic energy of a photoelectron at the point of interaction. If the volume of downconversion of the photoelectron energy into the energy of electron-hole pairs is small compared with the volume of the detector, we can consider that all electron-hole pairs are generated at the point of X-ray quantum interaction with the detector volume.

For uniform and isotropic scintillator with the only one radiative transition mode the expression (27) is simplified to

$$\left\langle\chi_n(\vec{r},E,\vec{\Omega})\right\rangle_D = S(E)QT_n(\vec{r}), \quad (33)$$

where $S(E)=\varepsilon_r(E)\varepsilon_a$ is the standard notation of the probability of the luminescent center activation by one electron-hole pair that can depend only on the electron energy; Q is the standard notation of the quantum efficiency of the luminescence process defined as $Q/4\pi=\varepsilon_e(\vec{r},\lambda_l,\vec{\Omega}')$;

$$T_n(\vec{r}) = (1+\lambda_D^2\Delta)\int_{\Omega'}\frac{d\vec{\Omega}'}{4\pi}\int_{S_n}\int_{\vec{\Omega}''_n} dS_n d\vec{\Omega}''_n\tau_n(\vec{r},\vec{\Omega}',S_n,\vec{\Omega}''_n)\eta_n(S_n,\vec{\Omega}''_n) \quad (34)$$

is the probability of a photoelectron creation in the n -th photodetector by a photon emitted by the luminescent center at the point $\vec{r}$ of the scintillator.

In a uniform and isotropic scintillator $\varepsilon_{\text{e-h}}(\vec{r},E,\vec{\Omega})=\varepsilon_{\text{e-h}}(E)$. For the case of the local absorption of the energy of an incident particle at the point $\vec{r}_c$ of the scintillator, the differential density of absorbed energy in the detector volume is

$$w^c(E_0,\vec{r},E)=\int_\Omega d\vec{\Omega}w^c(E_0,\vec{r},E,\vec{\Omega})=w^c(E_0,E)\delta(\vec{r}-\vec{r}_c)\,. \quad (35)$$

The averaging over of all possible combinations of secondary particles in the phase space is factorized, and the averaging over the spatial coordinate is due to variation of the X-ray quantum interaction point.
In this case the mean value and the variance of the sum signal take forms

$$\langle Q\rangle=\int_E dEu(E-E_{\min})\left\langle\frac{w^c(E_0,E)}{\varepsilon_{\text{e-h}}(E)}S(E)Q\right\rangle_c\left\langle\sum_{n=1}^N T_n(\vec{r}_c)\langle g_n\rangle\right\rangle_c ,(36)$$

$$\sigma^2_{\text{cov}}=\int_E\int_{E'}dEdE'u(E-E_{\min})u(E'-E_{\min})\left\langle\frac{w^c(E_0,E)}{\varepsilon_{\text{e-h}}(E)}S(E)Q\frac{w^c(E_0,E')}{\varepsilon_{\text{e-h}}(E')}S(E')Q\right\rangle_c\left\langle\left(\sum_{n=1}^N T_n(\vec{r}_c)\langle g_n\rangle\right)^2\right\rangle_c$$
$$-\left(\int_E dEu(E-E_{\min})\left\langle\frac{w^c(E_0,E)}{\varepsilon_{\text{e-h}}(E)}S(E)Q\right\rangle_c\left\langle\sum_{n=1}^N T_n(\vec{r}_c)\langle g_n\rangle\right\rangle_c\right)^2 , (37)$$

$$\sigma^2_{\text{pair}}=\int_E dEu(E-E_{\min})\left\langle F(E)\frac{w^c(E_0,E)}{\varepsilon_{\text{e-h}}(E)}S^2(E)Q^2\right\rangle_c\left\langle\left(\sum_{n=1}^N T_n(\vec{r}_c)\langle g_n\rangle\right)^2\right\rangle_c , (38)$$

$$\sigma^2_{\text{ph}}=\int_E dEu(E-E_{\min})\left\langle\frac{w^c(E_0,E)}{\varepsilon_{\text{e-h}}(E)}S(E)Q\right\rangle_c\left\langle\sum_{n=1}^N T_n(\vec{r}_c)\langle g_n\rangle^2\right\rangle_c$$
$$-\int_E dEu(E-E_{\min})\left\langle\frac{w^c(E_0,E)}{\varepsilon_{\text{e-h}}(E)}S^2(E)Q^2\right\rangle_c\left\langle\left(\sum_{n=1}^N T_n(\vec{r}_c)\langle g_n\rangle\right)^2\right\rangle_c , (39)$$

$$\sigma^2_{gain}=\int_E dEu(E-E_{\min})\left\langle\frac{w^c(E_0,E)}{\varepsilon_{\text{e-h}}(E)}S(E)Q\right\rangle_c\left\langle\sum_{n=1}^N T_n(\vec{r}_c)\sigma^2_{g\,n}\right\rangle_c , (40)$$

For the scintillation spectrometer with the one photodetector, for comparing with existing in literature formulae, it is more convenient to use the relative variance related to the energy resolution of the spectrometer by the expression

$$\frac{\Delta E}{E}=2.36\eta_Q\,, (41)$$

where ΔE is the full width at half maximum (FWHM) of the line with the energy E.
The relative variance for a scintillation spectrometer with the one photodetector is

$$\eta_Q^2=\frac{\sigma_Q^2(E_0)}{\langle Q(E_0)\rangle^2}=\eta^2_{\text{cov}}+\eta^2_{\text{pair}}+\eta^2_{ph}+\eta^2_{gain}+\eta^2_{\text{noise}}\,, (42)$$

where

$$\langle Q(E_0)\rangle=\int_E dEu(E-E_{\min})\left\langle\frac{w^c(E_0,E)}{\varepsilon_{\text{e-h}}(E)}S(E)Q\right\rangle_c\langle T(\vec{r}_c)\rangle_c\langle g\rangle , (43)$$

$$\eta^2_{\text{cov}}=\frac{\int_E\int_{E'}dEdE'u(E-E_{\min})u(E'-E_{\min})\left\langle\frac{w^c(E_0,E)}{\varepsilon_{\text{e-h}}(E)}S(E)\frac{w^c(E_0,E')}{\varepsilon_{\text{e-h}}(E')}S(E')\right\rangle_c\langle T^2(\vec{r}_c)\rangle_c}{\left(\int_E dEu(E-E_{\min})\left\langle\frac{w^c(E_0,E)}{\varepsilon_{\text{e-h}}(E)}S(E)\right\rangle_c\right)^2\langle T(\vec{r}_c)\rangle_c^2}-1,\quad (44)$$

$$\eta^2_{\mathrm{pair}} = \frac{\int_E dEu(E-E_{\min})\left\langle F(E)\frac{w^c(E_0,E)}{\varepsilon_{\mathrm{e\text{-}h}}(E)}S^2(E)\right\rangle_c \left\langle T^2(\vec{r}_c)\right\rangle_c}{\left(\int_E dEu(E-E_{\min})\left\langle \frac{w^c(E_0,E)}{\varepsilon_{\mathrm{e\text{-}h}}(E)}S(E)\right\rangle_c\right)^2 \left\langle T(\vec{r}_c)\right\rangle_c^2}, \quad (45)$$

$$\eta^2_{\mathrm{ph}} = \frac{1}{\int_E dEu(E-E_{\min})\left\langle \frac{w^c(E_0,E)}{\varepsilon_{\mathrm{e\text{-}h}}(E)}S(E)Q\right\rangle_c \left\langle T(\vec{r}_c)\right\rangle_c} - \frac{\int_E dEu(E-E_{\min})\left\langle \frac{w^c(E_0,E)}{\varepsilon_{\mathrm{e\text{-}h}}(E)}S^2(E)\right\rangle_c \left\langle T^2(\vec{r}_c)\right\rangle_c}{\left(\int_E dEu(E-E_{\min})\left\langle \frac{w^c(E_0,E)}{\varepsilon_{\mathrm{e\text{-}h}}(E)}S(E)\right\rangle_c\right)^2 \left\langle T(\vec{r}_c)\right\rangle_c^2}, \quad (46)$$

$$\eta^2_{gain} = \frac{1}{\int_E dEu(E-E_{\min})\left\langle \frac{w^c(E_0,E)}{\varepsilon_{\mathrm{e\text{-}h}}(E)}S(E)Q\right\rangle_c \left\langle T(\vec{r}_c)\right\rangle_c}\frac{\sigma_g^2}{\langle g\rangle^2}, \quad (47)$$

$$\eta^2_{\mathrm{noise}} = \frac{\sigma^2_{\mathrm{noise}}}{\langle Q(E_0)\rangle^2}. \quad (48)$$

As opposed to semiempirical formulae, the formulae (42)-(48) contain all information of dependence of all contributions to the energy resolution on scintillator characteristics and characteristics of the processes that occur in a scintillation detector at registration of incident particles.

For our consideration it is more convenient to introduce the differential light yield

$$\frac{\partial Y^c(E_0,E)}{\partial E} = \frac{w^c(E_0,E)}{\varepsilon_{\mathrm{e\text{-}h}}(E)}S(E)Q, \quad (49)$$

and the total light yield from a scintillator for incident particles with the energy E_0

$$Y^c(E_0) = \int_E dEu(E-E_{\min})\frac{\partial Y^c(E_0,E)}{\partial E} = E_0 L, \quad (50)$$

where L is the standard notation for specific light yield.

In this case, the average value and the relative variance are

$$\langle Q(E_0)\rangle = \langle Y^c(E_0)\rangle_c \langle T(\vec{r}_c)\rangle_c \langle g\rangle, (51)$$

$$\eta_Q^2 = \eta^2_{\mathrm{cov}} + \eta^2_{\mathrm{pair}} + \eta^2_{phe} + \eta^2_{gain} + \eta^2_{\mathrm{noise}}, (52)$$

$$\eta^2_{\mathrm{cov}} = \frac{\int_E\int_{E'} dEdE'u(E-E_{\min})u(E'-E_{\min})\left\langle \frac{\partial Y^c(E_0,E)}{\partial E}\frac{\partial Y^c(E_0,E')}{\partial E'}\right\rangle_c \left\langle T^2(\vec{r}_c)\right\rangle_c}{\langle Y^c(E_0)\rangle_c^2 \langle T(\vec{r}_c)\rangle_c^2} - 1, (53)$$

$$\eta^2_{\mathrm{pair}} = \frac{\int_E dEu(E-E_{\min})\left\langle F(E)\frac{\partial Y^c(E_0,E)}{\partial E}S(E)Q\right\rangle_c \left\langle T^2(\vec{r}_c)\right\rangle_c}{\langle Y^c(E_0)\rangle_c^2 \langle T(\vec{r}_c)\rangle_c^2}, (54)$$

$$\eta^2_{\mathrm{ph}} = \frac{1}{\langle Y^c(E_0)\rangle_c \langle T(\vec{r}_c)\rangle_c} - \frac{\int_E dEu(E-E_{\min})\left\langle \frac{\partial Y^c(E_0,E)}{\partial E}S(E)Q\right\rangle_c \left\langle T^2(\vec{r}_c)\right\rangle_c}{\langle Y^c(E_0)\rangle_c^2 \langle T(\vec{r}_c)\rangle_c^2}, (55)$$

$$\eta^2_{gain} = \frac{1}{\langle Y^c(E_0)\rangle_c \langle T(\vec{r}_c)\rangle_c}\frac{\sigma_g^2}{\langle g\rangle^2}, (56)$$

$$\eta^2_{\text{noise}} = \frac{\sigma^2_{\text{noise}}}{\langle Q(E_0)\rangle^2} \quad (57)$$

The average value and the relative variance become the most simple in the case when all incident particles interact with scintillator in the vicinity of the point $\vec{r}_0$, and the volume element in which the energy of incident particle converts to the light photons, is small enough that the coefficients $\langle T(\vec{r}_c)\rangle_c = T(\vec{r}_0)$ and $\langle T^2(\vec{r}_c)\rangle_c = T^2(\vec{r}_0)$ are almost constant, then

$$\langle Q(E_0)\rangle = \langle Y^c(E_0)\rangle_c T(\vec{r}_0)\langle g\rangle, \ (58)$$

$$\eta_Q^2 = \eta_Y^2 + \eta^2_{\text{pair}} + \eta^2_{phe} + \eta^2_{gain} + \eta^2_{\text{noise}}, \ (59)$$

$$\eta_Y^2 = \frac{\int\limits_E \int\limits_{E'} dE dE' u(E-E_{\min})u(E'-E_{\min})\left\langle \frac{\partial Y^c(E_0,E)}{\partial E}\frac{\partial Y^c(E_0,E')}{\partial E'}\right\rangle_c}{\langle Y^c(E_0)\rangle_c^2} - 1, (60)$$

is the relative variance of the photodetector signal caused by the covariance of the differential light yield. This term is due to dependence of the differential light yield on the energy of secondary particles in the course of downconversion of the energy of an incident particle. If the differential light yield does not depend on the energy of the secondary particles, then the total light yield from a scintillator linearly depends on energy of an incident particle, and this term is equal to zero.

If we define the average value of the probability of the luminescent center activation, the average value of the Fano factor, and the average energy of the electron–hole pair production by expressions

$$\langle Y^c(E_0)\rangle_c = \int\limits_E dEu(E-E_{\min})\left\langle \frac{\partial Y^c(E_0,E)}{\partial E}\right\rangle_c = \int\limits_E dEu(E-E_{\min})\left\langle \frac{w^c(E_0,E)}{\varepsilon_{\text{e-h}}(E)}S(E)Q\right\rangle_c$$
$$= \frac{\langle W^c(E_0)\rangle_c}{\langle \varepsilon_{\text{e-h}}\rangle}\langle S\rangle Q \cong \frac{E_0}{\langle \varepsilon_{\text{e-h}}\rangle}\langle S\rangle Q, \ (61)$$

$$\int\limits_E dEu(E-E_{\min})\left\langle \frac{\partial Y^c(E_0,E)}{\partial E}S(E)Q\right\rangle_c = \int\limits_E dEu(E-E_{\min})\left\langle \frac{w^c(E_0,E)}{\varepsilon_{\text{e-h}}(E)}S^2(E)Q^2\right\rangle_c$$
$$= \frac{\langle W^c(E_0)\rangle_c}{\langle \varepsilon_{\text{e-h}}\rangle}\langle S^2\rangle Q^2 \cong \frac{E_0}{\langle \varepsilon_{\text{e-h}}\rangle}\langle S^2\rangle Q^2, \ (62)$$

$$\int\limits_E dEu(E-E_{\min})\left\langle F(E)\frac{\partial Y^c(E_0,E)}{\partial E}S(E)Q\right\rangle_c = \int\limits_E dEu(E-E_{\min})\left\langle F(E)\frac{w^c(E_0,E)}{\varepsilon_{\text{e-h}}(E)}S^2(E)Q^2\right\rangle_c$$
$$= \frac{\langle W^c(E_0)\rangle_c}{\langle \varepsilon_{\text{e-h}}\rangle}\langle F\rangle\langle S^2\rangle Q^2 \cong \frac{E_0}{\langle \varepsilon_{\text{e-h}}\rangle}\langle F\rangle\langle S^2\rangle Q^2, \ (63)$$

then the contributions of the last three terms to the relative variance, can be rewritten as follows

$$\eta^2_{\text{pair}} + \eta^2_{ph} + \eta^2_{gain} + \eta^2_{\text{noise}} = \frac{\langle \varepsilon_{\text{e-h}}\rangle}{E_0}\left(\langle F\rangle - 1\right)\frac{\langle S^2\rangle}{\langle S\rangle^2} + \frac{\langle \varepsilon_{\text{e-h}}\rangle}{E_0}\frac{1+\eta_g^2}{\langle S\rangle Q} + \frac{\sigma^2_{\text{noise}}}{\langle Q(E_0)\rangle^2}, \ (64)$$

where $\eta_g^2 = \sigma_g^2 / \langle g\rangle^2$ is the relative variance of the gain.

4. COVARIANCE BETWEEN SIGNALS OF PHOTODETECTORS OF A SCINTILLATION DETECTOR

On the importance of the covariance between signals of photodetectors was emphasized in [9], [10]. From the PGF (9), the covariance between signals of photodetectors $Q_n(E)$ and $Q_{n'}(E)$ of a scintillation spectrometer has the form

$$\mathrm{cov}(Q_n,Q_{n'}) = \left.\frac{\partial^2 f_{\{Q_n\}}\left[\{s_n\}\right]}{\partial s_n \partial s_{n'}}\right|_{\{s_n=1\}} - \left.\frac{\partial f_{\{Q_n\}}\left[\{s_n\}\right]}{\partial s_n}\right|_{\{s_n=1\}} \cdot \left.\frac{\partial f_{\{Q_n\}}\left[\{s_n\}\right]}{\partial s_{n'}}\right|_{\{s_n=1\}} =$$

$$= \lim_{IJK\to\infty} \sum_c p^c_{\alpha ijk\alpha' i'j'k'} \sum_{\alpha=1}^{A}\sum_{i=1}^{I}\sum_{j=1}^{J}\sum_{k=1}^{K} N^c_{\alpha ijk} \left\langle v_{\alpha ijk}\right\rangle \chi_{n\ \alpha ijk} \left\langle g_n\right\rangle \sum_{\alpha'=1}^{A}\sum_{i'=1}^{I}\sum_{j'=1}^{J}\sum_{k'=1}^{K} N^c_{\alpha' i'j'k'} \left\langle v_{\alpha' i'j'k'}\right\rangle \chi_{n'\ \alpha' i'j'k'} \left\langle g_{n'}\right\rangle$$

$$-\left(\lim_{IJK\to\infty} \sum_c p^c_{\alpha ijk} \sum_{\alpha=1}^{A}\sum_{i=1}^{I}\sum_{j=1}^{J}\sum_{k=1}^{K} N^c_{\alpha ijk} \left\langle v_{\alpha ijk}\right\rangle \chi_{n\ \alpha ijk} \left\langle g_n\right\rangle\right)\left(\lim_{IJK\to\infty} \sum_c p^c_{\alpha' i'j'k'} \sum_{\alpha'=1}^{A}\sum_{i'=1}^{I}\sum_{j'=1}^{J}\sum_{k'=1}^{K} N^c_{\alpha' i'j'k'} \left\langle v_{\alpha' i'j'k'}\right\rangle \chi_{n'\ \alpha' i'j'k'} \left\langle g_{n'}\right\rangle\right)$$

$$+ \lim_{IJK\to\infty} \sum_c p^c_{\alpha ijk} \sum_{\alpha=1}^{A}\sum_{i=1}^{I}\sum_{j=1}^{J}\sum_{k=1}^{K} N^c_{\alpha ijk} \chi_{n\ \alpha ijk} \left(\sigma^2_{v_{\alpha ijk}} - \left\langle v_{\alpha ijk}\right\rangle\right)\left\langle g_n\right\rangle \chi_{n'\ \alpha ijk} \left\langle g_{n'}\right\rangle$$

(65)

As we move to the absorbed energy distribution in the detector volume, the covariance between signals of photodetectors turns to

$$\mathrm{cov}(Q_n,Q_{n'}) = \sum_{\alpha=1}^{A}\sum_{\alpha'=1}^{A} \int_V\int_E\int_\Omega\int_{V'}\int_{E'}\int_{\Omega'} dVdEd\vec{\Omega}dV'dE'd\vec{\Omega}' u(E-E_{\alpha\ \min}(\vec{r},E,\vec{\Omega}))u(E'-E_{\alpha'\ \min}(\vec{r}',E',\vec{\Omega}'))\times$$
$$\left\langle \frac{w^c_\alpha(E_0,\vec{r},E,\vec{\Omega})}{\varepsilon_{\text{e-h}\ \alpha}(\vec{r},E,\vec{\Omega})} \left\langle \chi_{n\ \alpha}(\vec{r},E,\vec{\Omega})\right\rangle_D \left\langle g_n\right\rangle \frac{w^c_{\alpha'}(E_0,\vec{r}',E',\vec{\Omega}')}{\varepsilon_{\text{e-h}\ \alpha'}(\vec{r}',E',\vec{\Omega}')} \left\langle \chi_{n'\ \alpha'}(\vec{r}',E',\vec{\Omega}')\right\rangle_D \left\langle g_{n'}\right\rangle \right\rangle_c$$
$$-\sum_{\alpha=1}^{A}\int_V\int_E\int_\Omega dVdEd\vec{\Omega} u(E-E_{\alpha\ \min}(\vec{r},E,\vec{\Omega}))\left\langle \frac{w^c_\alpha(E_0,\vec{r},E,\vec{\Omega})}{\varepsilon_{\text{e-h}\ \alpha}(\vec{r},E,\vec{\Omega})} \left\langle \chi_{n\ \alpha}(\vec{r},E,\vec{\Omega})\right\rangle_D \left\langle g_n\right\rangle \right\rangle_c \times \qquad (66)$$
$$\sum_{\alpha'=1}^{A}\int_{V'}\int_{E'}\int_{\Omega'} dV'dE'd\vec{\Omega}' u(E'-E_{\alpha'\ \min}(\vec{r}',E',\vec{\Omega}'))\left\langle \frac{w^c_{\alpha'}(E_0,\vec{r}',E',\vec{\Omega}')}{\varepsilon_{\text{e-h}\ \alpha'}(\vec{r}',E',\vec{\Omega}')} \left\langle \chi_{n'\ \alpha'}(\vec{r}',E',\vec{\Omega}')\right\rangle_D \left\langle g_{n'}\right\rangle \right\rangle_c$$
$$+\sum_{\alpha=1}^{A}\int_V\int_E\int_\Omega dVdEd\vec{\Omega} u(E-E_{\alpha\ \min}(\vec{r},E,\vec{\Omega}))\times$$
$$\left\langle \frac{w^c_\alpha(E_0,\vec{r},E,\vec{\Omega})}{\varepsilon_{\text{e-h}\ \alpha}(\vec{r},E,\vec{\Omega})} \left(F_\alpha(\vec{r},E,\vec{\Omega})-1\right) \left\langle \chi_{n\ \alpha}(\vec{r},E,\vec{\Omega})\right\rangle_D \left\langle g_n\right\rangle \left\langle \chi_{n'\ \alpha}(\vec{r},E,\vec{\Omega})\right\rangle_D \left\langle g_{n'}\right\rangle \right\rangle_c$$

The expressions (65) and (66) are the most general for the covariance between signals of photodetectors of a scintillation spectrometer.

If the secondary particles are only electrons, and the scintillator is uniform and isotropic with only one radiative transition mode, then for the local absorption of the energy of an incident particle the covariance between signals of photodetectors turns to

$$\mathrm{cov}(Q_n,Q_{n'}) = \int_E\int_{E'} dEdE' u(E-E_{\min})u(E'-E_{\min})\times$$
$$\left\langle \frac{w^c(E_0,E)}{\varepsilon_{\text{e-h}}(E)} S(E)Q \frac{w^c(E_0,E')}{\varepsilon_{\text{e-h}}(E')} S(E')Q \right\rangle_c \left\langle T_n(\vec{r}_c)T_{n'}(\vec{r}_c)\right\rangle_c \left\langle g_n\right\rangle\left\langle g_{n'}\right\rangle$$
$$-\int_E dEu(E-E_{\min})\left\langle \frac{w^c(E_0,E)}{\varepsilon_{\text{e-h}}(E)} S(E)Q \right\rangle_c \left\langle T_n(\vec{r}_c)\right\rangle_c \left\langle g_n\right\rangle \int_{E'} dE'u(E'-E_{\min})\left\langle \frac{w^c(E_0,E')}{\varepsilon_{\text{e-h}}(E')} S(E')Q \right\rangle_c \left\langle T_{n'}(\vec{r}_c)\right\rangle_c \left\langle g_{n'}\right\rangle$$
$$+\int_E dEu(E-E_{\min})\left\langle \frac{w^c(E_0,E)}{\varepsilon_{\text{e-h}}(E)} (F(E)-1)S^2(E)Q^2 \right\rangle_c \left\langle T_n(\vec{r}_c)T_{n'}(\vec{r}_c)\right\rangle_c \left\langle g_n\right\rangle\left\langle g_{n'}\right\rangle$$

(67)

The covariance between signals of photodetectors of a spectrometer underlies an experimental method of the Fano factor determination for the first time proposed in [11]. For determination of the Fano factor the use of the relative covariance between signals of photodetectors was proposed

$$\frac{\operatorname{cov}(Q_n,Q_{n'})}{\langle Q_n\rangle\langle Q_{n'}\rangle}=\frac{\int\limits_E\int\limits_{E'} dEdE'u(E-E_{\min})u(E'-E_{\min})\left\langle \frac{w^c(E_0,E)}{\varepsilon_{\text{e-h}}(E)}S(E)\frac{w^c(E_0,E')}{\varepsilon_{\text{e-h}}(E')}S(E')\right\rangle_c}{\left(\int\limits_E dEu(E-E_{\min})\left\langle \frac{w^c(E_0,E)}{\varepsilon_{\text{e-h}}(E)}S(E)\right\rangle_c\right)^2}\frac{\langle T_n(\vec r_c)T_{n'}(\vec r_c)\rangle_c}{\langle T_n(\vec r_c)\rangle_c\langle T_{n'}(\vec r_c)\rangle_c}-1$$
$$+\frac{\int\limits_E dEu(E-E_{\min})\left\langle \frac{w^c(E_0,E)}{\varepsilon_{\text{e-h}}(E)}(F(E)-1)S^2(E)\right\rangle_c}{\left(\int\limits_E dEu(E-E_{\min})\left\langle \frac{w^c(E_0,E)}{\varepsilon_{\text{e-h}}(E)}S(E)\right\rangle_c\right)^2}\frac{\langle T_n(\vec r_c)T_{n'}(\vec r_c)\rangle_c}{\langle T_n(\vec r_c)\rangle_c\langle T_{n'}(\vec r_c)\rangle_c}$$

(68)

It is necessary to emphasize that unlike a correlation coefficient

$$\rho(Q_n,Q_{n'})=\frac{\operatorname{cov}(Q_n,Q_{n'})}{\sigma_{Q_n}\sigma_{Q_{n'}}},\ (69)$$

which depends on all the variances of signals of the photodetectors, the relative covariance (68) does not depend on the gains and the noises of photodetectors electronics. It is the important advantage of the proposed experimental method, in comparison with existing methods of the Fano factor determination based on the subtraction of the electronic noise from the variance of the signal of the photodetector [12].

The expression (68) becomes simpler in the case if incident particles interact with a scintillator in the vicinity of the point $\vec r_0$, and the volume element in which the energy of incident particle converts to the light photons, is small enough that the coefficients $\langle T_n(\vec r_0)\rangle_c=T_n(\vec r_0)$ $\langle T_{n'}(\vec r_0)\rangle_c=T_{n'}(\vec r_0)$ are almost constant. In this case $\langle T_n(\vec r_0)T_{n'}(\vec r_0)\rangle_c=T_n(\vec r_0)T_{n'}(\vec r_0)$ and the expression (68) reduces to

$$\frac{\operatorname{cov}(Q_n,Q_{n'})}{\langle Q_n\rangle\langle Q_{n'}\rangle}=\frac{\int\limits_E\int\limits_{E'} dEdE'u(E-E_{\min})u(E'-E_{\min})\left\langle \frac{w^c(E_0,E)}{\varepsilon_{\text{e-h}}(E)}S(E)\frac{w^c(E_0,E')}{\varepsilon_{\text{e-h}}(E')}S(E')\right\rangle_c}{\left(\int\limits_E dEu(E-E_{\min})\left\langle \frac{w^c(E_0,E)}{\varepsilon_{\text{e-h}}(E)}S(E)\right\rangle_c\right)^2}-1$$
$$+\frac{\int\limits_E dEu(E-E_{\min})\left\langle \frac{w^c(E_0,E)}{\varepsilon_{\text{e-h}}(E)}(F(E)-1)S^2(E)\right\rangle_c}{\left(\int\limits_E dEu(E-E_{\min})\left\langle \frac{w^c(E_0,E)}{\varepsilon_{\text{e-h}}(E)}S(E)\right\rangle_c\right)^2}\ .\ (70)$$

As we move to the differential light yield and the total light yield

$$\frac{\operatorname{cov}(Q_n,Q_{n'})}{\langle Q_n\rangle\langle Q_{n'}\rangle}=\frac{\int\limits_E\int\limits_{E'} dEdE'u(E-E_{\min})u(E'-E_{\min})\left\langle \frac{\partial Y^c(E_0,E)}{\partial E}\frac{\partial Y^c(E_0,E')}{\partial E'}\right\rangle_c}{\langle Y^c(E_0)\rangle_c^2}-1$$
$$+\frac{\int\limits_E dEu(E-E_{\min})\left\langle \frac{\partial Y^c(E_0,E)}{\partial E}(F(E)-1)S(E)\right\rangle_c}{\langle Y^c(E_0)\rangle_c^2}\ ,\ (71)$$

With expressions (61)-(63), the relative covariance between signals of photodetectors reduces to

$$\frac{\operatorname{cov}(Q_n,Q_{n'})}{\langle Q_n\rangle\langle Q_{n'}\rangle}=\eta_Y^2(E_0)+\frac{\langle\varepsilon_{\text{e-h}}\rangle}{E_0}(\langle F\rangle-1)\frac{\langle S^2\rangle}{\langle S\rangle^2}\ .\ (72)$$

The dependence of the last term on the inverse energy of incident particles allows separation of the contributions of the covariance in the light yield and the Fano factor, and its experimental determination from the dependence of the relative covariance on the energy of incident particles. As the contribution of the second term in (72) is inversely proportional to the number of electron–hole pairs, then for high energy of incident particles the relative covariance between signals of photodetectors defines the covariance of the differential light yield in the course of downconversion of the energy of an incident particle to energies of the secondary particles.

5. CONCLUSION

The exact mathematical description of process at registration of incident particles by a scintillation spectrometer with several photodetectors allows receiving correct formulas for the mean value and the variance of the amplitude at the output of photodetectors, and the covariance between signals of photodetectors. It was shown that using the relative covariance of signals allows the experimental determination of the covariance in the light yield and the Fano factor in a scintillator.

REFERENCES

[1] M. Moszyński, A. Syntfeld-Każuch, L. Swiderski, M. Grodzicka, J. Iwanowska, P. Sibczyński, T. Szczęśniak "Energy resolution of scintillation detectors", *Nucl. Instr. Meth. in Phys. Res.* A, vol. 805, pp. 25–35, 2016.
[2] P. Lecoq, A. Annenkov, A. Gektin, M. Korzhik, C. Pedrini, *Inorganic Scintillators for Detector Systems*, Springer-Verlag Berlin Heidelberg, 2006.
[3] Qi Li, J. Q. Grim, R. T. Williams, G. A. Bizarri, W. W. Moses "The role of hole mobility in scintillator proportionality", *Nucl. Instr. Meth. in Phys. Res.* A, vol. 652, pp. 288–291, 2011.
[4] W. Feller, *An Introduction to Probability Theory and Its Applications*, John Wiley and Sons, New York - London - Sydney, 1968.
[5] V.V. Samedov, "Accounting for fluctuations of electron-photon showers in the theory of shower spectrometers", Ph.D. dissertation, Moscow Engineering Physics Institute, Moscow,1972.
[6] V.V. Samedov, "Energy Resolution of Chernkov and Scintillation Shower Spectrometers for High-Energy Particles", *Instruments and Experimental Techniques*, vol. 28, pp. 1037-1039, 1985.
[7] V.V. Samedov, "Correct method for calculating energy resolution and instrument functions of scintillation spectrometers", *Measurement Techniques*, vol. 28, pp. 265-271, 1985.
[8] U. Fano, "Ionization yield of radiation II, The fluctuation of the number of ions", *Phys. Rev.*, vol. 72, pp. 26-29, 1947.
[9] V.V. Samedov, "Position resolution in DROIDs", *J. Low Temp. Phys.*, vol. 151, pp. 333-336, 2008.
[10] V.V. Samedov, "Once More on the Position Resolution of DROIDs", *AIP Conference Proceedings*, vol. 1185, pp. 397-400, 2009.
[11] V.V. Samedov, "How to Determine Experimentally the Fano Factor in DROIDs", *AIP Conference Proceedings*, vol. 1185, pp. 462-464, 2009.
[12] R. Devanathan, L.R. Corrales, F. Gao, W.J. Weber, "Signal variance in gamma-ray detectors-A review", *Nucl. Instr. Meth. in Phys. Res.* A, vol. 565, pp. 637–649, 2006.

EPJ Web of Conferences **225**, 01008 (2020) https://doi.org/10.1051/epjconf/202022501008
ANIMMA 2019

METU-DEFOCUSING BEAMLINE : A 15- 30 MeV PROTON IRRADIATION FACILITY and BEAM MEASUREMENT SYSTEM

M.Bilge Demirkoz
Middle East Technical University
Ankara, Turkey
demirkoz@metu.edu.tr

Caner Seckin
Middle East Technical University
Ankara, Turkey
cseckin@metu.edu.tr

Akanay Avaroglu
Middle East Technical University
Ankara, Turkey
akanay@metu.edu.tr

Besna Bulbul
Middle East Technical University
Ankara, Turkey
besna.bulbul@metu.edu.tr

Pelin Uslu
Middle East Technical University
Ankara, Turkey
puslu@metu.edu.tr

Erinc Kılıc
Middle East Technical University
Ankara, Turkey
erinck@metu.edu.tr

Yusuf Orhan
Ministry of Defense
Ankara, Turkey
yusuf.orhan@msb.gov.tr

Selen Akcelik
Middle East Technical University
Ankara, Turkey
sakcelik@metu.edu.tr

Merve Yigitoglu
Middle East Technical University
Ankara, Turkey
merve.yigitoglu@metu.edu.tr

Caglar Saral
Middle East Technical University
Ankara, Turkey
csaral@metu.edu.tr

Selcen Uzun Duran
Karadeniz Technical University
Trabzon, Turkey
selcenduran@ktu.edu.tr

Ugur Kılıc
Middle East Technical University
Ankara, Turkey
ukilic@metu.edu.tr

Ilias Efthymiopoulos
CERN
Geneva, Switzerland
ilias.efthymiopoulos@cern.ch

Anıl Berkay Poyrazoglu
Middle East Technical University
Ankara, Turkey
apoyraz@metu.edu.tr

Abdulrahman Albarodi
Middle East Technical University
Ankara, Turkey
abdul.albaroudi@metu.edu.tr

Nazire Celik
Middle East Technical University
Ankara, Turkey
nazcelik@metu.edu.tr

***Abstract*— Middle East Technical University – Defocusing Beam Line (METU-DBL) project is an irradiation facility providing 15 MeV to 30 MeV kinetic energy protons for testing various high radiation level applications, ranging from Hi-Lumi LHC upgrade, space electronic components to nuclear material research. The project located inside the premises of the TAEA (Turkish Atomic Energy Agency) SANAEM (Saraykoy Nuclear Education and Research Center) close to Ankara, provides users a wide selectable flux menu (10^5–10^{10} p/cm²/s). The facility is now being commissioned and the facility will be providing a large test area (20 cm x 15 cm) for material, detector and electronics tests. The proton beam is monitored along the beamline using aluminum oxide screens and the flux and uniformity is measured using three detectors attached to the robotic system for cross- checks. A fiber scintillator detector scans the large irradiation area while small area diamond detector and Timepix3 detector are used for spot checks for calibration. Several samples can be radiated simultaneously inside the irradiation area and the robotic system provides 5 separate holders for samples which can be moved in or out, providing users flexibility for the desired fluence. This talk will first introduce METU- DBL as a radiation test facility, then discuss the radiation monitoring of the beam area and the radiation room, while highlighting how this facility can be used for future testing of materials for radiation tolerance.**

Keywords*— *Single Event Effect, Defocusing beamline, Space Radiation, Proton Accelerator Facility, Radiation tests

I. INTRODUCTION

Electronic devices are exposed to high dose radiation in the Hi-Lumi LHC [1] and space environment. Radiation exposure to these devices affects their performance and the reliability so they must be tested. In the scope of METU-DBL (Defocusing Beam Line) project, a beamline has been constructed at Turkish Atomic Energy Authority (TAEA)'s proton accelerator facility's (PAF) R&D room. The schematic design of the proton accelerator facility is given in Fig. 1, where a central cyclotron provides proton beams to two simultaneous irradiation rooms. METU-DBL's goal is to study single event effects (SEEs), are caused by single, energetic particles, namely protons. In the space environment, protons which are trapped in radiation belts around the Earth, or solar flares, can cause both direct ionization and indirect ionization SEEs in sensitive electronic devices [2].

TAEA SANAEM PAF has a selectable energy range 15-30 MeV cyclotron proton accelerator in the range of 0.1 μA to 1.2 mA. This facility has four irradiation rooms. Three of them are used in order to produce radiopharmaceuticals for medical purposes and the other room is reserved for R&D purposes [3].

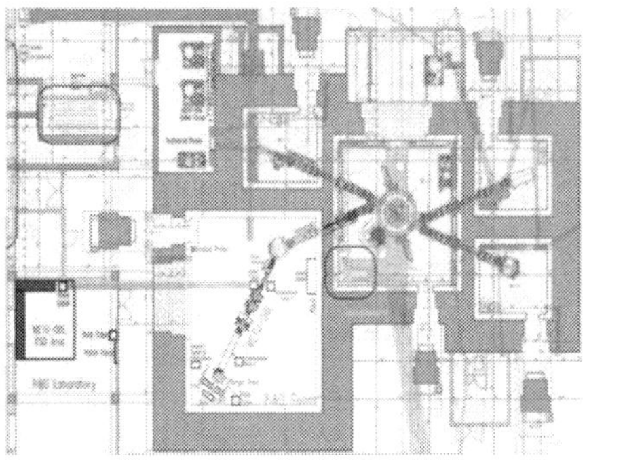

Fig. 1. The schematic design of TAEA SANAEM Proton Accelerator Facility. The cyclotron is seen in red on the right, where 4 arms extend out to 4 different irradiation rooms. The one on the lower left is the R&D room in which METU-DBL is shown in black. The connection between this R&D room and the control system in the R&D laboratory is provided by the underground trench shown here in red.

Proton tests are crucial for space applications because protons are the dominant component of cosmic rays that penetrate through the shielding of a satellite. For radiation tests, METU-DBL follows the ESA/ESCC No: 25100 standard [4]. This standard includes the parameters that must be provided at the end of the beamline, and these parameters are shown in TABLE I.

TABLE I. The requirements of ESA/ESCC No: 25100 Standard for SEE tests.

Parameters	Value
Energy range	20 MeV- 200 MeV
Flux range	10^5 p/cm^2/s to at least 10^8 p/cm^2/s
Beam area	15.40 x 21.55 cm
Uniformity	± 10%

The 2D design of the METU-DBL is shown in Fig. 2. There is a 5-port switching magnet at the beginning of the beamline, which can guide the beam into 5 different arms to perform different experiments.

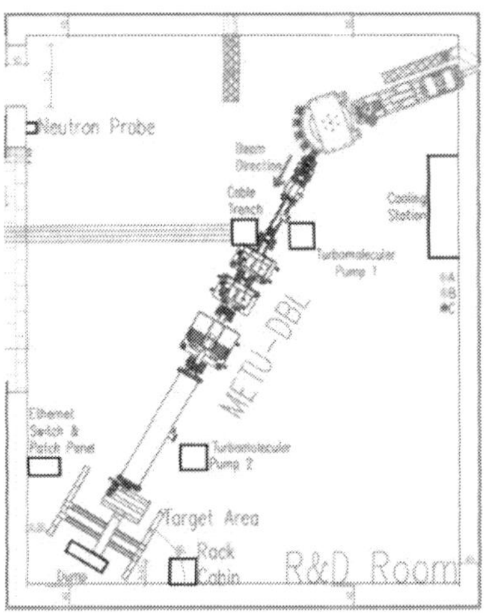

Fig. 2. A 2D drawing of METU-DBL and the R&D room. A 5- port switching magnet is shown in red color at the beginning point of METU-DBL. The length of the beamline is approximately 7.5 m starting from the 5-port switching magnet.

The METU-DBL is installed on the right-most port of the switching magnet.

Fig. 3. Final design of METU-DBL. The beamline starts at the exit of the 5-port magnet. A large custom-design quadrupole magnet follows two smaller ones. The beam pipe gets wider as the beam blows up towards the end of the beamline. 5-port switching magnet and radiation shields are not shown.

The final setup of METU-DBL includes a beam stopper and a vacuum shutter to protect the beamline and the vacuum of METU-DBL. A helium section improves the uniformity at the target area by scattering the beam between the titanium foils. An adjustable conical collimator is used to provide the desired flux and also to protect the magnets of METU-DBL. 3 quadrupole magnets is employed to enlarge the beam. A beam measurement system and a beam dump towards the end of the beamline are shown in Fig. 3 [5].

METU-DBL provides users a wide flux range to choose from (10^5-10^{11}p/cm^2/s) and an energy range of 15-30 MeV. Reaching fluences of 10^{14}p/cm^2 only takes about 20 minutes.

METU-DBL has six subsystems: beam optics, vacuum, cooling, robotic, control, test and measurement subsystems.

Beam optics studies of METU-DBL follow the ESA/ESCC No. 25100 standard. This subsystem consists of quadrupole magnets to enlarge the proton beam, collimators to reduce the proton flux and a beam screen to observe the beam profile. C++ based MAD-X [6] and FORTRAN based TRANSPORT were used for the beam optics calculations [7]. A movable Aluminum Oxide screen, at 45° to the beamline and ImageJ program [8] is used to analyze the beam profile.

In METU-DBL, protons must reach the target area with minimum energy loss to achieve the ESA/ESCC No: 25100 standard. A vacuum subsystem consisting of one mechanical pump and two turbomolecular pumps decreases the pressure inside the beam pipe to 10^{-6} mbar is which is the desired vacuum level.

Quadrupole magnets and collimators of METU- DBL get warm due to high currents and secondary particles during irradiation tests. The cooling subsystem of METU-DBL consists of a water tank and 2 circulation pumps providing 50 kW of cooling power. The water in the tank is sent to the beam elements and quadrupole magnets by circulation pumps. Also, thermal analyses of the beam elements directly exposed to the proton beam are simulated using ANSYS Fluent [9] to make sure no element will be deformed by thermal loads. An overhead of 15 kW is included in the cooling budget for the cooling requirements of future users of METU- DBL.

The motions of the electromechanical components are controlled by the METU-DBL robotic subsystem. This subsystem is in charge of the management of the test table components (detectors as well as the devices under tests), the control of the adjustable conical collimator and a mobile radiation measurement robot. These tasks are performed either autonomously or by user commands.

METU-DBL control subsystem provides the communication and the control of the electronics and electromechanics. Control subsystem consists of control software, an industrial computer, a control computer, PLCs, motor drivers and Ethernet switches. During radiation tests, each component of the beamline is monitored from the TAEA SANAEM PAF control room.

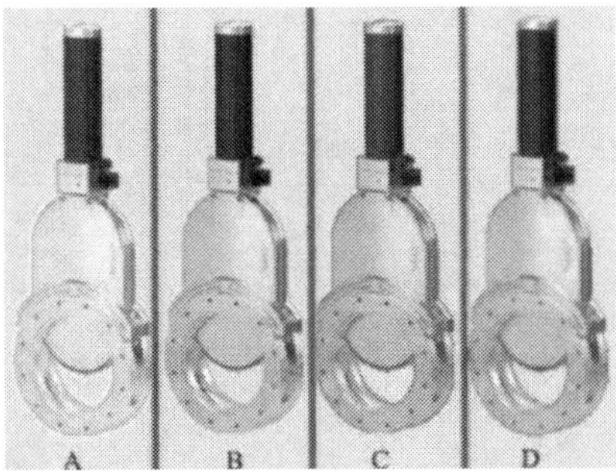

Fig. 4. Color codes of the beam elements of METU-DBL. While green color shows the ideal status, yellow color indicates the warning status. The red color shows the error status and the blue color indicates the safe status.

In the Fig. 4, the instant states of each beam element are indicated by color codes. Green is an ideal condition; yellow in case of a warning, red in case of error, and the blue is the safe condition. This visual aid allows for a fast intervention in case of an emergency.

The test and measurement system has three different detectors: TimePix3 pixel, fiber scintillators and diamond detector which are shown in Fig. 5. These are necessary to measure the proton beam flux and the uniformity of METU-DBL target area.

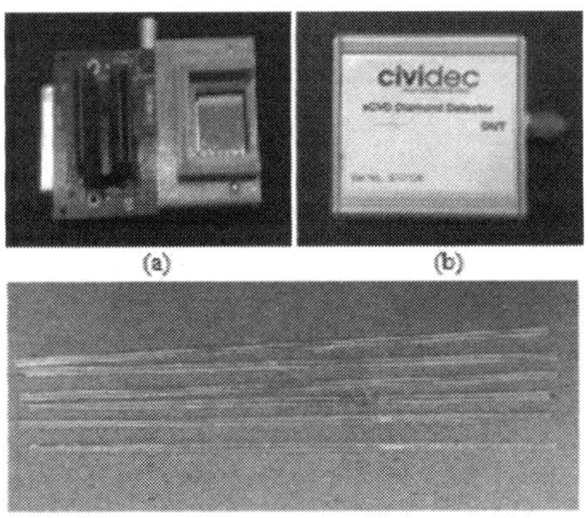

(c)

Fig. 5. Detectors of METU-DBL (a) Timepix3 pixel detector, (b) diamond detector (c) fiber scintillator.

The Timepix3 is a pixel detector which was developed within the Medipix collaboration at CERN. The critical part of this detector is the sensor layer, and it is segmented with a pixel size of 55 µm into a square matrix of 256 x 256 pixels [10]. These pixels have a threshold for particle measurement, and they provide information about particles having more energy than the pre-defined threshold energy. Proton flux may also be calculated with fiber scintillators. The number of photons that are created is proportional to the number of protons reaching the scintillator. The proton flux can be inferred by measuring the photon flux in the fiber scintillator [11]. Due to its radiation endurance, diamond detectors [12] can also be used to measure proton flux. These features enable them to work efficiently for a long time under irradiation. Other detectors' performance might suffer because of radiation damage sustained to their crystal structure. The detector has a single diamond crystal between two gold electrodes.

When a proton passes the crystal, a potential difference occurs between the electrodes that can be used for proton detection. Under this proton irradiation, activation of the materials that exist around the beamline might occur. In high radiation environment dose calculations (for example Hi Lumi LHC), radioisotope and shielding studies are important to protect the beam elements, the sensitive electronic devices that are exposed to radiation directly or indirectly, and the health of the radiation workers. In METU-DBL, the instantaneous dose rate in the R&D room is recorded with Geiger-Müller, gamma and neutron detectors. During radiation tests, the maximum dose rates and cooling times are also recorded.

In addition, in the scope of dose calculations and shielding studies, METU-DBL system was built into the FLUKA simulation program [13]. FLUKA is a tool for the calculations of particle transportation and interactions with matter. FLUKA packages include proton and electron accelerators, calorimetry, activation, dosimetry, detector design, accelerator driven systems, etc. The radioisotope and dose rates of the beam elements can be analyzed with FLUKA.

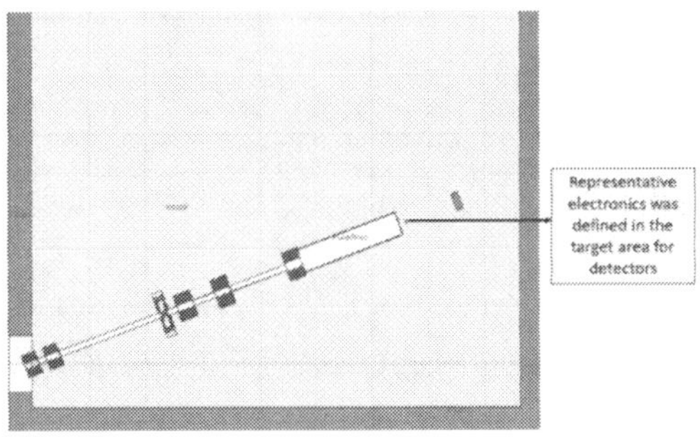

Fig. 6. Modeling of METU-DBL as seen in the graphical interface of FLUKA called Flair. Beam elements such as magnets are seen as brown rectangles around the beam pipe. An electronic component in the target area was placed to represent detectors and their readout electronics to determine the necessary shielding designs.

For the shielding designs, various material thicknesses and sequences were simulated. In order to stop protons, secondary neutrons, and gammas; aluminum, polyethylene, and lead were used respectively. Numerous designs were studied for shielding the detectors, and the best results depending on availability and manufacturability were chosen for each. The shielded

TimePix3 pixel detector, one of the fiber scintillators, and the diamond detector is shown in Fig. 7.

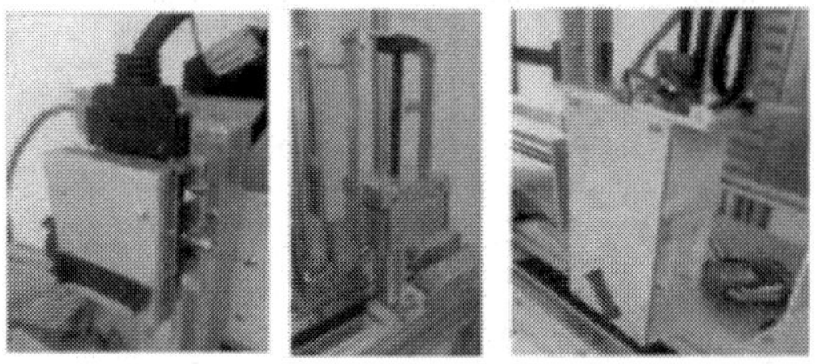

(a) (b) (c)

Fig. 7. From left to right, (a) the shielded TimePix3 pixel detector, (b) one shielded fiber scintillator, and (c) the shielded diamond detector can be seen. Numerous designs were studied for these detectors and the best results were chosen for each.

In the next section, pretest of METU-DBL will be explained.

II. PRETESTS OF METU-DBL

METU-DBL installation was performed by considering the shielding studies and dose calculations. During the pretests, a lot of experience has been gained, and the final installation started with the experiences from the pretests. Radiation tests of nine electronic cards and materials of eight different institutions were tested with the pretest setup between January and March 2018 [14].

Fig. 8. Panoramic photo of the preliminary test design. The beam comes from the accelerator from beamline in the left wall. Initial beam diagnostics are made with CCD camera which looks at the image of the beam on an Al_2O_3 screen. The test and measurement system at the target area measures the flux and uniformity of the beam with 3 movable detectors. There is a beam dump before the beam hits the right wall. The total length of pretest line of METU- DBL is 6.97 m.

In the pretest setup, the beam size could be as large as 4 cm x 6 cm and the flux was between 1×10^8 and 1×10^{10} protons/cm^2/s. The preliminary setup of the METU-DBL at TAEA R&D room is seen in the photograph shown in Fig. 8. The beam comes from the accelerator to beamline connection situated on the left wall. Initial beam diagnostics are made with a CCD camera which looks at the image of the beam on an Al_2O_3 screen. The test and measurement system at the target area measures the flux and uniformity of the beam using the diamond detector, TimePix3 pixel detector and the fiber scintillator, all shielded and positioned on a movable setup. After passing through the measurement system, the beam encounters a beam dump at the right wall to stop the protons. The total length of METU-DBL pretest setup is 6.97 m.

During the pretests stage, the performed radiation tests are shown in TABLE II. Each test was performed using 30 MeV energy protons at different irradiation times. Only the pin diode test was performed at 2.0 μA, while the other tests were performed at 0.1 μA as specified by the users.

TABLE II. Radiation tests which performed during the pretest of METU-DBL.

Institution	Material Type	Irradiation Time
METU- Department of Physics	PIN Diode	30 min. 51 sec.
TÜBİTAK MAM Materials Institute	Solar cells for IMECE Satellite	20 min. 1 sec.
TÜBİTAK MAM Materials Institute	Shielding glass for IMECE Satellite	49 min. 56 sec.
TÜBİTAK MAM Energy Institute	Battery materials for IMECE Satellite	25 sec.
Gazi University	Solar cells	9 min. 57 sec.
TÜBİTAK Uzay	GaN-FET	30 min. 42 sec.
Sabancı University	Coating substance	55 sec.
METU– Department of Metallurgical and Materials Engineering	Bulk metallic glass	30 min
TÜBİTAK BİLGEM UEKAE YİTAL	DDA3	10 min.

During the Pin Diode radiation tests, the arrangement of the 3 detectors on the test table with pretest setup is shown in Fig. 9. The TimePix3 pixel detector and the fiber scintillators were attached to pin diodes for readout and a diamond detector. This system has movable X-Y stages onto which the detectors are attached. The target area is scanned with 3 mm accuracy and flux measurement are taken at regular intervals. The device under test is positioned based on these measurements so that the required homogeneity is met.

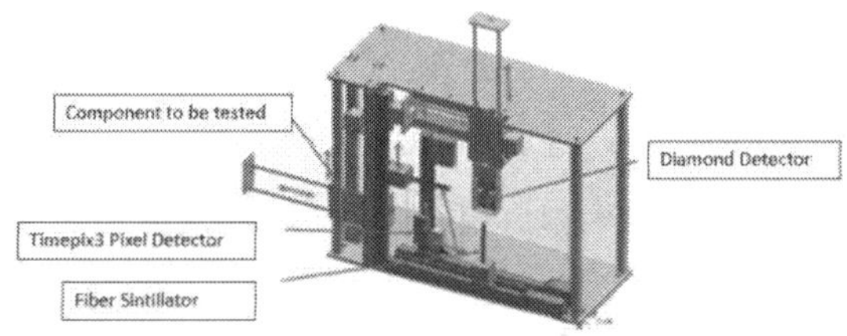

Fig. 9. Pretest test table of METU-DBL. Three detectors which are TimePix3 pixel detector, fiber scintillators are seen. The target area is scanned with 3 mm accuracy and flux measurement are taken at regular intervals. Device under test is positioned based on these measurements so that the required homogeneity is met.

Before the tests, detectors take measurement to ascertain the integrity of the beam at the target area. The components to be tested then are placed on the scanned area with detectors. First proton irradiation test was performed on six pin-diodes. The position of the pin diodes on the test table (left) and measurement positions and fluxes for each measurement taken by the diamond detector (right) are shown Fig. 10. The maximum flux (1.88 x 10^{11} p/cm^2/s) and minimum flux (4.82 x 10^{10} p/cm^2/s) are seen in the test area (red rectangular). Characterization of the pin diodes before and after testing have been performed and are being prepared for publication.

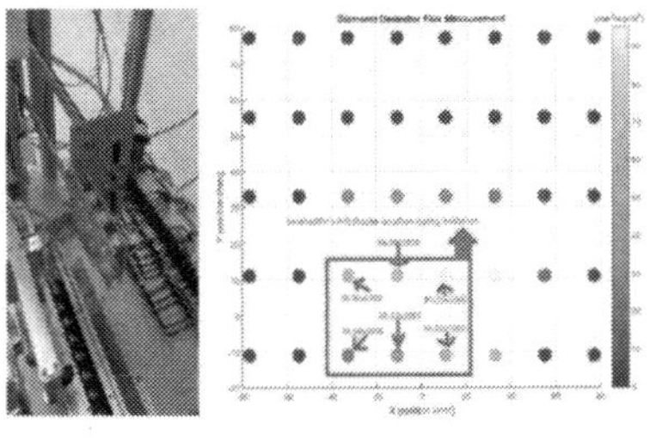

Fig. 10. In (a), the pin diodes under test are seen in the movable holder of the test desk. In (b), measurement positions and fluxes for each measurement taken by the diamond detector (right) is shown with a color scale. The maximum flux (1.88 x 10^{11} p/cm^2/s) and minimum flux (4.82 x 10^{10} p/cm^2/s) seen in the test area shown in red.

After the pretests had been completed, the pretest line was disassembled, and the assembly of the beamline started according to the final design of METU-DBL.

III. INSTALLATION OF THE FINAL SETUP OF METU-DBL

The main differences between the beamline setup in the pretests stage and the final configuration are the employing of a 5-port switching magnet, movable beam screen, custom design quadrupole magnet, new conical collimator and helium section. The differences are summarized in TABLE III.

TABLE III. Differences between the pretest and final design of METU-DBL.

Pretest Line Design	Final Design of METU-DBL
-	5-Port Switching Magnet
2 Quadrupole Magnets	2 Quadrupole Magnets and Custom Design Quadrupole Magnet
Conical Collimator	New Conical Collimator
Fixed Beam Screen	Movable Beam Screen
-	Helium Chamber
Test Table	New Test Table

As of May 2019, production and assembly of 5-port switching magnet, custom design quadrupole magnet, and new test mechanism have been completed and added to the beamline. The helium section, movable beam screen and new conical collimator are in the design and production stages.

The final setup installation started with the assembly of a 3 tons 5-port switching magnet, and it is located at the starting point of the beamline as shown in Fig. 11. This magnet was purchased from IBA and mounting process were performed with the IBA technical team. The METU-DBL installation started on the right-most port of the switching magnet.

Fig. 11. 5-port magnet installed in the R&D room of the TAEA SANAEM PAF. METU-DBL beamline follows from the right- most exit of this magnet.

The other critical elements of beam optics are three quadrupole magnets in the METU-DBL which are shown in Fig. 12. Quadrupole magnets focus the beam in one direction, while defocusing in the other direction. The incoming proton beam is enlarged by quadrupole magnets to satisfy the ESA/ESCC No: 25100 standard.

Fig. 12. Quadrupole magnets of METU-DBL. Custom design quadrupole magnet can be seen at front and first two quadrupole magnets which purchased from Scanditronix can be seen at behind the custom design magnet.

The first two quadrupole magnets of METU-DBL were purchased from Scanditronix Company, and the third quadrupole magnet was designed by the METU-DBL team and manufactured in Turkey, since it was not produced commercially.

METU-DBL pretests were performed with only the first two quadrupole magnets. The properties of these quadrupole magnets in comparison to the new custom design quadrupole magnet are given in the TABLE IV.

TABLE IV. Properties of Quadrupole Magnets of METU-DBL.

	First Two Quadrupole Magnets	Custom Design Quadrupole Magnet
Flow Rate	3.3 l/min	3.4 l/min
Current	70 A (nom.)	160 A
Aperture	110 mm	160 mm
Gradient Strength	5.0 T/m	7.5 T/m
Effective Length	300 mm	300 mm
Number of turns per coils	86	122

In TABLE IV, gradient strength is the most critical parameter for quadrupole magnets. Because ESA/ESCC No: 25100 standard can only be achieved with the given gradient strength values.

Electrical and mechanical acceptance tests of first and second quadrupole magnets were performed at METU Physics Department and the magnetic field measurements were made by TUBITAK UME. The custom design quadrupole magnet underwent mechanical, electronic and magnetic tests, first in Turkey, and then in CERN. The magnet has successfully passed all acceptance tests and was therefore certified by CERN.

In accelerator systems, another essential beam elements are collimators to cut the tails of the beam. Thus, the size of the beam in the x-y axes is reduced, and the number of particles is decreased.

The function of the conical collimator in METU- DBL is to reduce the flux. After the pretests, a collimator of 40 x 40 x 28 cm^3 dimensions with an aperture of 1 x 1 mm^2 to 10 x 10 mm^2 was designed to precisely center the gap where the proton beam passes and to provide sufficient cooling. Different flux options can be offered to users in the target area, with the adjustable aperture of the conical collimator. The design of the conical collimator is given in Fig. 13.

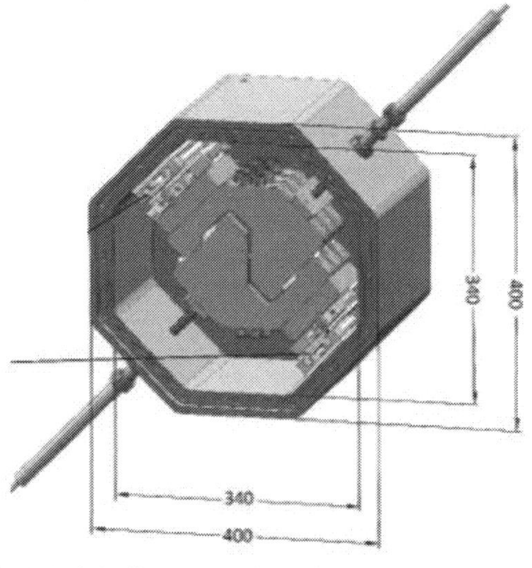

Fig. 13. Conical Collimator design of METU-DBL. Linear drives control leaves which cut the incoming beam. (All dimensions are given in millimeters).

During the pretests, mounting and dismantling the fixed beam screen into the beamline was time consuming. A movable beam screen was designed to increase the efficiency of the final setup of METU-DBL.

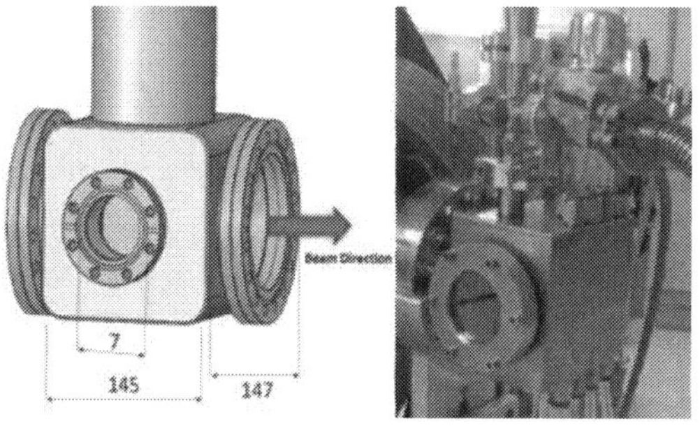

Fig. 14. A 3-D drawing of the movable beam screen designed to observe the proton beam profile. The image of the beam can be viewed from the window which has 7 cm radius (Left). Fixed Beam screen of METU-DBL which is used for pretest. Also the radius of fixed beam screen is 7 cm. (All dimensions are given in millimeters).

On the movable beam screen, an Al_2O_3 plate of 7 x 7 cm^2 size was placed at an angle of 45° to the beam axis and was used to find the center of the beam. Also, sapphire glass was preferred for the glass of the moving beam screen due to its radiation endurance. The setup is shown in Fig. 14. This screen is essential for alignment of the mechanical installation of the accelerators to observe the beam profile at different stages. In the METU-DBL pretest setup, the size of the proton beam can be measured as a function of a magnetic field by using a quadrupole scan technique [15] with a fixed beam screen. The quadrupole scan technique is a widely used method for measuring transverse emittance at particle accelerators.

In METU-DBL, two scattering foils of 50 µm thickness provide a homogeneous radiation field in the target area by enlarging the beam and reducing the proton flux. When the particles pass through the scattering foils, the angle of particles change due to multiple Coulomb scattering. The protons lose their energy while passing through scattering foils, heating the foils up in the process, thus cooling is required. The protons at a current of 100 µA of 30 MeV have a thermal load of 3 kW. A helium section which has 150 x 100 x 140 mm^3 dimensions is placed between the scattering foils to provide the cooling requirements. The cross-sectional area of the helium cooling section is shown in Fig. 15. In this design, stainless steel 316 was chosen as the material of the helium section.

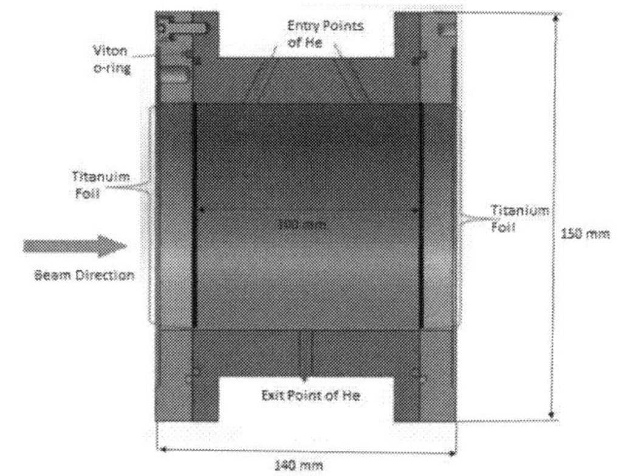

Fig. 15. Helium cooling chamber design of METU-DBL. The titanium foils were placed between the flanges shown in green and purple colors.

In this design, the titanium foils were placed in contact with the surface of the viton o-rings between the flanges shown in green and purple colors. It is aimed to provide more efficient cooling for titanium foils by giving 45° angles to helium gas ducts. The helium gas entering the chamber with two channels is intended to exit from a single point and go to the heat exchanger that cools the helium gas.

The last improvement in METU-DBL is the new test table. The first design of the test table with detectors and DUTs was used throughout the pretests. During the pretest, the risk of collision of detectors with each other was high due to the small size of the previous test table. After the completion of the pretests, the test table was redesigned in the light of the acquired experience, and the final design was completed. The previous test table could scan the area which has limit of ±300 mm in X-axis and ±1200 mm in Y-axis while the new test system could scan the larger area, which has the limit of ±760 mm in X-axis and ±670 mm in Y-axis. It can be seen in Fig. 17.

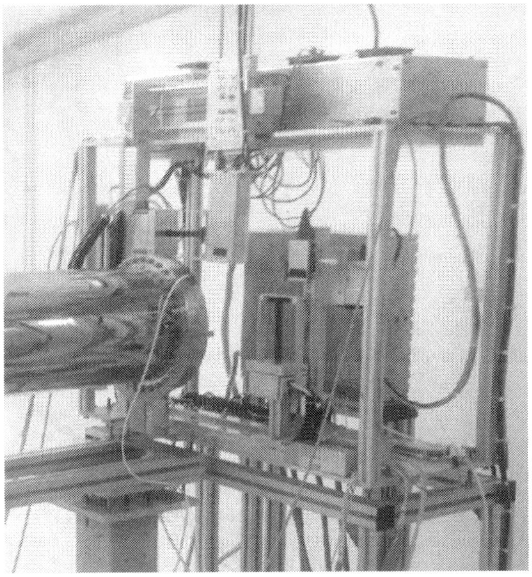

Fig. 16. Test table of METU-DBL pretest.

The detectors and DUTs are attached in the cross-shaped holders and are moved by the servo motors in the new design of METU-DBL test mechanism. The speed, acceleration and position control devices of the servo motors can be set independently of the PLC or can be controlled manually using a software.

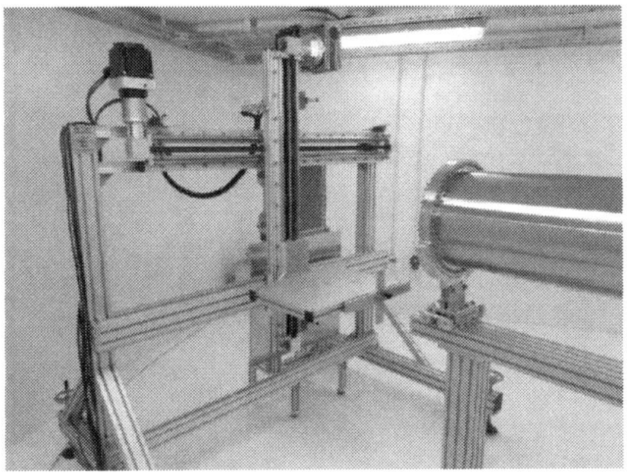

Fig. 17. Final design of the test table of METU-DBL, followed by the beam dump.

CONCLUSION

METU-DBL pretest installation was completed, and nine radiation tests were performed between January and March 2018. After the pretests, the METU-DBL system was dismantled and re-established in accordance with the final design. In contrast to the pretest system, 5-portswitching magnet, custom design quadrupole magnet and helium chamber were added to the beamline. Also, the design of the conical collimator and beam screen has been updated. Radiation tests will be carried out when the METU-DBL installation is completed. Commissioning and tests of electronics will commence in 2019. METU-DBL will serve for testing components intended for high radiation environments such as the Hi Lumi LHC as well as medical and space applications.

ACKNOWLEDGMENT

METU-DBL is funded by Strategy and Budget Presidency (Grant no: 2015K121190) and also supported by the Turkish Academy of Sciences. We want to thank our colleagues from TAEA SANAEM Proton Accelerator Facility.

REFERENCES

[1] Bruce, R., A. Lechner, and S. Redaelli. Energy deposition simulations for upgraded collimation layouts. No. CERN-ACC-2014-0009. 2014

[2] Underwood, Craig I. The single-event-effect behaviour of commercial-off-the-shelf memory devices. A decade in low-Earth orbit. In: RADECS 97. Fourth European Conference on Radiation and its Effects on Components and Systems (Cat. No. 97TH8294). IEEE, 1997. p. 251-258.

[3] TAEA, "Proton Accelerator Facility Booklet", TAEA SANAEM, 2012

[4] ESA, "Single Event Effects Test Method and Guidelines, ESCC Basic Specification No. 25100", 2014

[5] Demirkoz, M. Bilge, et al. Pretest setup installation of the METU-DBL project to perform space radiation tests. In: 2017 8th International Conference on RAST 2017. IEEE, p. 311-316.

[6] Schmidt, F., & Grote, H. (2005). MAD-X User's guide.

[7] Rohrer, U. (2007). PSI graphic transport framework. based on a CERN-SLAC-FERMILAB version by KL Brown et al.

[8] Abràmoff, M. D., Magalhães, P. J., & Ram, S. J. (2004). Image processing with ImageJ. Biophotonics international, 11(7), 36-42.

[9] Fluent, A. N. S. Y. S. "Ansys fluent." Academic Research. Release 14 (2015).

[10] Kraus, V., Holik, M., Jakubek, J., Kroupa, M., Soukup, P., & Vykydal, Z. (2011). FITPix—fast interface for Timepix pixel detectors. Journal of Instrumentation, 6(01), C01079.

[11] Beischer, B., Gast, H., Greim, R., Karpinski, W., Kirn, T., Nakada, T., & Wlochal, M. (2010). A high-resolution scintillating fiber tracker with silicon photomultiplier array readout. Nuclear Instruments and Methods in Physics Research Section A: Accelerators, Spectrometers, Detectors and Associated Equipment, 622(3), 542-554.

[12] Rebai, M., Cazzaniga, C., Croci, G., Tardocchi, M., Cippo, E. P., Calvani, P., & Gorini, G. (2015). Pixelated Single- crystal Diamond Detector for fast neutron measurements. Journal of Instrumentation, 10(03), C03016.

[13] Battistoni, Giuseppe, et al. (2015) Overview of the FLUKA code, Annals of Nuclear Energy 82, p: 10-18.

[14] Demirkoz, B., Nigdelioglu, S., Yigitoglu, M., Aydın, S., Efthymiopoulos, I. (2018). METU defocusing beamline project for the first SEE tests in Turkey and the test results from the METU-DBL preliminary setup. Nucl. Instrum. Meth. A., in press

[15] Green, A., & Shin, Y. M. (2015). Implementation of Quadrupole-scan Emittance Measurement at Fermilab's Advanced Superconducting Test Accelerator (ASTA)

EPJ Web of Conferences **225**, 01009 (2020)
ANIMMA 2019

https://doi.org/10.1051/epjconf/202022501009

Design and development of Ionization Profile Monitor for the Cryogenic sections of the ESS Linac

F. Benedetti, P. Abbon, F. Belloni, G. Coulloux, F. Gougnaud, C. Lahonde-Hamdoun, P. Le Bourlout, Y. Mariette, J. Marroncle, J.P. Mols, V. Nadot, L. Scola and G. Tauzin
CEA Saclay, Institut de Recherche sur les lois Fondamentales de l'Univers
Gif-sur-Yvette, France
R. Tarkeshian and C. Thomas
European Spallation Source ERIC
Lund, Sweden

***Abstract*—Saclay CEA/IRFU is working for the delivery of five Non-Invasive Profile Monitors in the frame of the in-kind contribution agreement signed with the European Spallation Source. Neutrons will be produced by spallation reactions of 2 GeV proton beam impinging on a Tungsten target. To accelerate protons a powerful linear accelerator of 5 MW is under construction. Diagnostic devices are mandatory tools for the tuning and protection of the machine. The non-invasive profile monitors provide a measurement of the beam profile in transverse directions to the beam propagation. This project raises several physical and technical challenges including low signal detection of ions or electrons, profile distortions induced by the beam Space Charge effect and non-uniformities of electric field. Simulation and model of the critical aspects of the detector have been performed in order to prove the performance and the feasibility of the detector. A series of prototypes has been built with different readout types, and tested in real conditions at the 3 MeV proton accelerator IPHI. All of them show some advantages and drawbacks revealed by the tests in real beam conditions. In this paper we present the results of the tests for the various configuration readout systems to agree with the model and simulation of the detector. In concluding remarks, we will discuss the performance of the prototypes and point out the camera-based one to be the more suitable for the final design.**

***Index Terms*—Beam diagnostic, Linear proton accelerators, MCP, Strip detector, Particle beam measurements**

I. Introduction

The European Spallation Source (ESS) will be a European research infrastructure dedicated to neutronic science. The source is currently under commissioning at the ESS site in Lund, Sweden, and will be the brightest pulsed neutron source. The production of neutrons is ensured by the spallation process: high energy protons will impinge on a tungsten target. To accelerate the protons a powerful 2 GeV linear accelerator is being built. The accelerator can be schematized in two parts. The first part accelerates the beam up to 90 MeV by mean of conventional room temperature RF cavities. Then a cold part using superconducting cavities cooled with liquid helium is used to reach the highest energies. An overview of the ESS linac is shown in Fig. 1, and Table I gives its main characteristics.

The project was supported by the French in-kind contribution to ESS (AIK 7.3).

F. Benedetti is with the Département d'Électronique des Détecteurs et d'Informatique pour la Physique, Commissariat à l'Énergie Atomique et aux énergies alternatives, Saclay, 91400 France, (email: florian.benedetti@cea.fr).

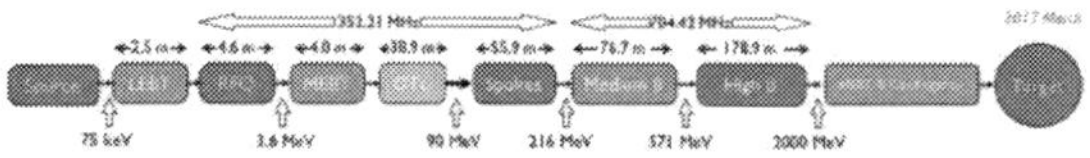

Fig. 1. A simplified representation of the ESS linac. The IPMs will be installed in the cryogenic part (blue).

At ESS, both invasive and non-invasive profilers will be installed along the accelerator. The beam profile will be also recorded at the target location and upstream of the beam dump [1]. The invasive measurements are mainly done with wire scanners. These devices cannot handle the huge beam peak power of ESS at nominal conditions (125 MW), and will be only used at low beam duty cycle [2]. Therefore, Non invasive Profile Monitors (NPM) will take over for higher beam power. In fact, the NPM refers to two types of devices depending on the detection principle. Fluorescence Profile Monitors (FPM) [3] are foreseen in the "warm" parts of the accelerator whereas Ionization Profile Monitors (IPM) will be implemented exclusively in the cryogenic part of the accelerator. Five pairs of IPMs are foreseen at several locations.

TABLE I
ESS LINAC CHARACTERISTICS

Characteristic	Value
Energy	2 GeV
Pulse current	62.5 mA
Pulse duration	2.86 ms
Repetition rate	14 Hz
Duty cycle	4 %
Power (peak)	5 MW (125 MW)
Radio Frequencies	352.21 MHz 704.42 MHz

II. Ionization Profile Monitors

A. State of art

The first use of IPMs dates back to the 1960s [4]. In the 90s, the raise of the MicroChannel Plates have permitted to measure profiles [5], [6] in more critical working conditions. The IPM method is now mature and used in several installations [7]–[10]. Recently, the interest in semiconductor detectors has grown [11] and first results looks promising [12].

The different technologies of detection, presented just before, have been reviewed in order to select the most efficient one with respect to the ESS requirements.

B. Principle of operation

An Ionization Profile Monitor (IPM) is a non-invasive detector that measures the transverse profile of a beam. Its principle of operation is shown in Fig. 2 and can be summarized in 3 main steps:

1) Beam protons pass through the vacuum, inducing ionizations of the residual gas molecules: electron/ion pairs are created.
2) Inside the IPM, a strong electric field drives electrons or ions towards a segmented readout system.
3) The profile is reconstructed in one transverse direction. For a complete profile, a pair of IPMs, rotated by 90° with respect to each other, is mandatory.

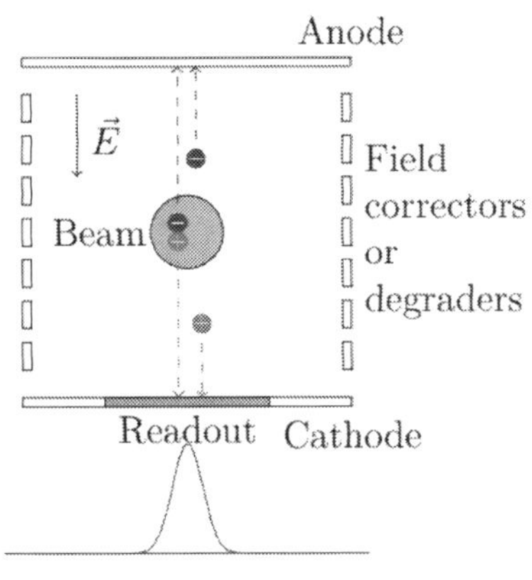

Fig. 2. Visual explanation of how an IPM works. The electric field between the electrodes can be reverted by inverting the polarity, giving the possibility to detect ions or electrons. Field correctors or degraders, left and right, improve the field uniformity.

One can note that an IPM forces the charge collection with an electrical field, whereas a FPM is passive and detects within a certain solid angle. On the other hand, the implantation of IPMs is more complex since it operates in vacuum.

Simulations were done in order to check the feasibility of the IPM design for the given ESS conditions. Each simulation focused on the following hot topics:

- Quantification of the ionization signal in terms of number of produced electron/ion pairs for ensuring that the signal is sufficiently high for reconstructing a profile per pulse beam, despite of the low gas pressure and relatively small ionization cross section at high proton energy.
- The extraction field must be as uniform as possible in order to lead efficiently and correctly the ionization by-products toward the readout.
- The space charge effect induced by the beam and the initial momentum of ionization electrons/ions, which may distort the profile, must be evaluated.
- The choice of an efficient readout technology which must match the ESS working conditions.

All these points will be presented in the next sections.

III. Electron/ion rates

The IPMs rely on the by-product collection of the ionized residuals gas. The number of ionized particles gives the signal strength which has to be compliant to the readout sensitivity. Therefore, the number of ionization particles, that are created by the beam itself along the residual gas, must be evaluated.

A. Calculation using Bethe model

For heavy charged particles like protons, the main interaction is due to electromagnetic interactions of the incident particle with the orbiting electrons of the medium. The Bethe equation describes the mean rate of energy losses per distance unit by a heavy charged particle as follows [13, p. 446]:

$$-\left\langle \frac{dE}{dx}\right\rangle = K\rho\frac{Z}{A}\frac{z^2}{\beta^2}\left[\frac{1}{2}ln\left(\frac{2m_e\beta^2\gamma^2T_{max}c^2}{I^2}\right)-\beta^2\right] \quad (1)$$

where K is a constant factor defined by $K = 4\pi N_a r_e^2 m_e c^2$, r_e is the classical electron radius, m_e is the electron mass, N_a the Avogadro constant and c the speed of light in vacuum. Z, A and ρ are respectively the atomic number, the mass number and the density of the given medium. The mean excitation energy I is the only non-trivial variable in the Bethe equation [14]. The maximum transfer energy for one collision T_{max} is given by the following equation:

$$T_{max} = \frac{2m_e\beta^2\gamma^2c^2}{1+\frac{2\gamma m_e}{M}+\left(\frac{m_e}{M}\right)^2} \quad (2)$$

where, M and m_e are respectively the incident particle and electron masses. The β and γ variables have their normal significance as Lorentz factors.

When the medium is a mixture of several compounds, its mean stopping power needs to be calculated as the sum of the mean stopping power of its components weighted by their mass proportion. By introducing W, the average energy for producing an ion/electron pair in a medium [15], [16], one can estimate the number of ion/electron pairs created in a given readout length Δx of materials as:

$$N_{total} = \sum_{n=First}^{Last}\frac{\left\langle\frac{dE}{dx}\left(\rho_n, I_n, A_n, Z_n\right)\right\rangle}{W_n}\Delta x \quad (3)$$

The Bethe formula can be implemented in a C++ code once the composition of the medium and the I value of each compound is known. The expected pressure in the cryogenic part at ESS is around 10^{-9} mbar, and the gas composition is given by the following: $79\,\%\ H_2$, $10\,\%\ CO$, $10\,\%\ CO_2$ and $1\,\%\ N_2$

B. Calculation using PAI model

The Garfield++ software [17] also has been used to compute the number of primary ionizations. This software is normally intended for the modelization of gaseous detectors. It allows to simulate the creation of electron/ion pairs due to the ionization of gas by an incident particle, the transport and amplification of these electrons in the gas and the signal induced on a readout plane. In our case, we simulated only the pair creation in the residual gas. For this step Garfield++ uses Heed ++ [18], a C ++ code that implements the photoabsorption and ionization (PAI) model [19].

Table II summarizes the results of the two previous calculations done for different IPM positions along the ESS linac, and the factor ratio between the methods.

TABLE II
COMPARISON OF EXPECTED NUMBER OF ELECTRON/ION PAIRS PER cm BETWEEN CALCULATION USING BETHE EQUATION AND RESULTS FROM GARFIELD++.

Energy	N_{Bethe}	$N_{garfield}$	Factor
97.2	100210	52537	0.52
231.4	54970	27463	0.50
278.9	49160	26124	0.53
315.8	45850	23769	0.52
628.3	33600	17522	0.52

C. Limitations

The estimation strongly depends on various vacuum parameters that have not been measured yet. In consequence, the calculated values give only a coarse approximation. Also, the measurement of the W parameter includes the delta ray (or secondary ionization) electrons, hence the W value is biased [4], [13] for the case at hand, since the IPMs work at very low pressure. The real W values may be higher than the ones given in references, therefore the real number of ionization pairs may be 2 to 4 times lower than the expected number.

IV. PROFILE DISTORTIONS

A. Extraction field

The IPMs can be seen as parallel plate detectors. In an ideal case, these plates are infinite sized. The extraction field is then completely oriented in a single direction, normal to the detection plane and the projection of the profile on this plane is perfect. In reality, the plates have finite dimensions, comparable to the gap between the two electrodes. In these conditions the field is no more uniform. In addition, the cross-interaction between the electric fields of two close by IPMs is very strong. The non-uniformity of the electric field is very problematic because it creates mirage effects and prevents the correct measurement of the beam profile. To overcome this effect, two solutions have been considered:

- Using field correctors or field degraders [10]. This is done by placing conductors on each side. Each corrector is set to a certain potential in order to constrain the field.
- Putting grounded conductors between the two IPMs [10] to protect against the IPM cross-interaction.

COMSOL has been mainly used for the simulations of extraction fields, and a particle tracking algorithm has been implemented in C++. The code integrates the motion of particle due to the Lorentz force:

$$\vec{F} = m \cdot \vec{a} = q \cdot (\vec{E}(\vec{r}, t) + \vec{v} \times \vec{B}(\vec{r}, t)) \tag{4}$$

Fig. 3 shows an example of results given by the particle tracking. Without the corrections, the reconstructed profile is 35 % wider than reality. The shift in position is due to the cross interaction between the two IPMs. When corrections are enabled, the obtained transverse profile is much better: the error on the profile width is only 0.4 % and there is no more shift in position.

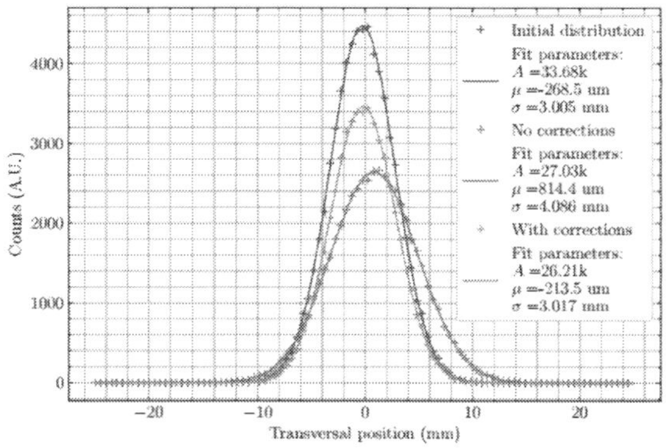

Fig. 3. A simulation of profile measurement with particle tracking for two real field configurations: without correction and with corrections. The initial proton (blue) beam is assumed to be Gaussian with $\sigma_{beam} = 3$ mm (blue). Without correction, a shift of 1 mm and a broadening of 35 % are observed (red). The corrections allows to negate the two previous distortions (green).

B. Space Charge effect

In the context of IPMs, the Space Charge effect refers to the influence of the beam itself on the profile measurement. An ESS bunch is about 10^9 protons, therefore bunches generate a strong electromagnetic field while moving along the accelerator, and the electrons or ions are affected by this electromagnetic field while drifting to the read-outs in a IPM.

The estimation of this electromagnetic field is necessary to ensure that the profile measurement is not too strongly affected by the space charge effects. For this reason, a simulation code has been developed conjointly by ESS and CEA/Saclay. The code computes the electric field of a 3D Gaussian distributed bunch in the moving frame $\bar{K}$ (rest frame for bunches) [20]. The electric field is transformed into an electromagnetic field in the laboratory frame K (rest frame for IPMs) by the Lorentz transformations of electromagnetic fields:

$$\boldsymbol{E} = \begin{pmatrix} \gamma \bar{E}_x \\ \gamma \bar{E}_y \\ \bar{E}_z \end{pmatrix}, \ \boldsymbol{B} = \begin{pmatrix} -\gamma v_b \frac{\bar{E}_y}{c^2} \\ \gamma v_b \frac{\bar{E}_x}{c^2} \\ 0 \end{pmatrix} \tag{5}$$

Then, the motion of the electrons or ions is integrated thanks to the equation (4).

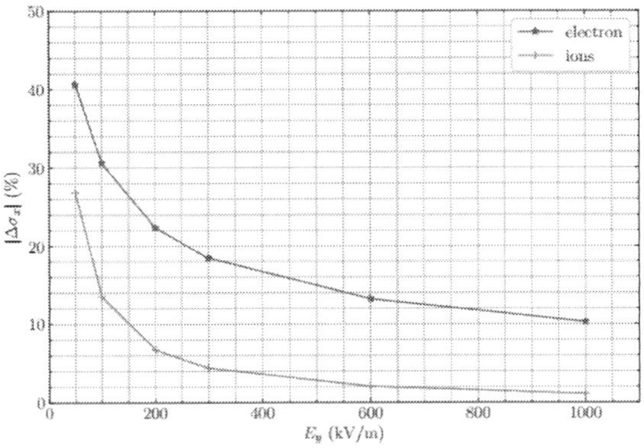

Fig. 4. Error on the profile measurement, for electrons and H_2^+ ions, due to space charge effects versus the extraction field. The beam is assumed to be a Gaussian with $\sigma_x = \sigma_x = \sigma_x = 2\,\mathrm{mm}$

Using the code, it has been shown that space charge effect can be partially compensated by a high extraction field value. It has been seen that profiles from electrons seem to be more affected by space charge effect than from ions. In the Fig. 4, a general example of the ESS beam size measured from profiles obtained with ions and electron is presented, and for various values of the electric field strength. The beam size has been chosen to be 3% smaller than the nominal value. The values of the field strength used in the simulation are chosen in a range that is technically achievable. Within that range it can be seen that the minimization of the space charge effect is limited. It is also clear that the effect is stronger for electrons than for ions.

In the case of ESS beam, the ions and electrons feel a permanent electric field from the HV cage of the IPM, and an variable field due to the bunched structure of the beam. The duty cycle of the variable beam plays a role in the distortion of the profiles. In the case of electrons, due to their small mass, they will reach the detector within the passage of one bunch. Therefore, they feel the combined forces most of the time along their path. In the case of the ions, it takes the time of many bunches for them to reach the detector. Hence, the forces from the bunch are weighted by the duty cycle of the varying electric field. For the case of ESS, the bunches are 3 ps, and the period between them is 2.84 ns. So the integrated force seen by the ions has less strength than in the case with electrons.

The expected uncertainty for the beam size measurement is 10%. The presented case shows the impact of the space charge can be strong, but yet the retrieved values are still within the requirements for the measurement uncertainty. Larger beam sizes are associated with weaker space charge effect, hence, the designed instrument is expected to deliver reliable information on the ESS beam profile and size.

A workaround to totally counterbalance even more the space charge effects consists in adding a magnetic field parallel to the extraction field with a magnet [8]. However, this solution is not possible at ESS since the available space is too limited. Consequently, the profile measurements will be performed with ions.

More detailed reviews of the simulations of space charge effects for the ESS IPM project are available here [21], [22].

C. *Initial momentum*

A first approximation of the initial speed of ions can be done thanks to the distribution of Maxwell-Boltzmann.The speed of the fastest ion H_2^+ is below $5000\,\mathrm{m/s}$. A field of few hundred volts per centimeter is more than enough to compensate this effect, therefore the thermal motion for ions can be completely neglected.

Garfield++ can be used again since it also gives the energy distribution of the ionization electrons as well as the angular distribution. The energy distribution spread up to keV range but most probable energies are below $50\,\mathrm{eV}$. The ions has supposed to be emitted with the opposite momentum. In consequence, the beam measurement with electrons is also more affected by the initial momentum than the measurement with ions. This enforces even more the necessity to measure the profile with ions at ESS.

V. Readout Technologies

A. *Strips*

Conductive strips is the simplest method to implement. Electrodes are etched on a PCB with a thin layer of copper. When electrons or ions move towards the electrodes, they induce a current that can be calculated by mean of Ramo-Schockley theorem [23], [24]. The induced current is then integrated by a front end electronic.

A COTS (Component Of The Shelf) front end electronics has been selected for this purpose: the CARAMEL card from the FASTER platform. This solution is developed by the LPC laboratory at Caen in France. CARAMEL is a charge digitizer card based on the DDC316 chip from Texas Instruments. Each card provides 32 input channels and two cards can be plugged in an AMC mother board. The mother board is compatible with μTCA crates.

This method is very robust and radiation hard since strips are passive components. However, strips alone can not be used when the signal is too low, and it is necessary to find a way to amplify the signal.

B. *MCPs*

A MicroChannel Plate (MCP) generates electrons from incident particles [25]. When a particle hits the MCP hole entrance, secondary electrons are emitted. Due to difference of potential, secondaries are drawn towards the channel output and strike hole walls again, creating more and more electrons. Then, electrons are collected on a detection plane. The detection plane can be a phosphorus screen which converts electrons to visible photons or conductive strips. Both solutions have been implemented.

A vision system is necessary to record light from the phosphorus screen. A camera with a lens should be sufficient in the case at hand. High resolution is not mandatory, so pixels could be relatively big in order to increase light collection and dynamic range. Sony IMX249 fits well with these prerequisites, so the FLIR Blackfly-PGE-23S6M-C camera has been selected.

Unfortunately MCPs suffer of aging effect, the gain is not stable and decreases over the time. Therefore, a calibration must be performed to correct the gain from time to time.

C. Silicon detectors

The last solution consists of a pixelated silicon detector. When a charged particle passes through the silicon it deposits its energy and electron/hole pairs are created. Then, the charge carriers drift in the semiconductor due to the bias voltage, and the induced signal is read by an embedded electronic. A single particle may create thousands of secondaries, therefore silicon detectors are very sensitive and interesting for IPMs.

The use of silicon detectors seems very promising but detection is not assured for ions at low energies since the stopping range is comparable to the deadlayer of silicon detectors. The feasibility of silicon detection for IPM with ions had to be checked. To this end, a small test bench has been developed and installed in a ion-implant facility: IRMA [26]. The test has been done with a tailored TimePix [27], [28] detector, kindly provided by the CERN-BI team.

The most crucial result of the test at IRMA is shown in Fig. 5. The detection seems possible at 15 keV, however at slightly lower energy 12 keV, the signal completely vanishes. Therefore, this solution has been discarded since it give almost no margin for error.

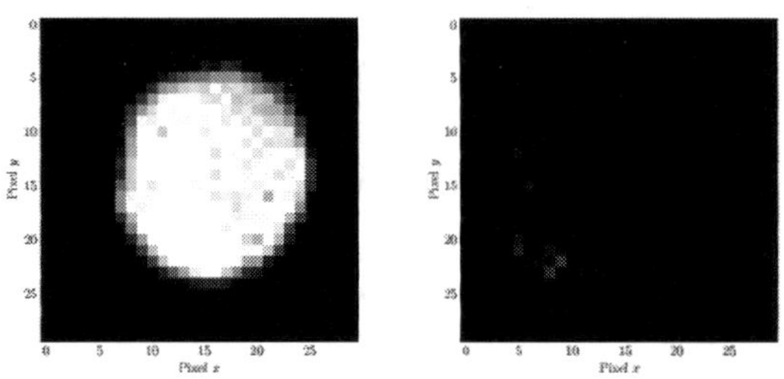

Fig. 5. Images of Time over Threshold signal from a TimePix chip for a incident H_2^+ beam at 15 keV (left) and 12 keV (right).

VI. Prototypes design and test setup

A. IPM and test bench design

The simulations presented in the previous sections show that the profile measurements with IPMs may match the ESS requirements. However, some critical points, particularly the choice of the readout, were not fully clear, so the feasibility had to be proven experimentally. For this purpose three prototypes have been manufactured. The design of the prototypes is independent of the readout technology, so the readout can be changed with only few operations. A prototype using a MCP is shown in Fig. 6

A test bench has been developed in order to validate the prototypes. The bench can be split into two different independent parts. The first part (upstream) tries to mimic the ESS LWU chamber inside which two IPMs can be inserted. The idea is to be close to the ESS conditions in term of high voltages and electrical fields. The second part (downstream) offers one more IPM slot and two viewports for reference measurements in order to compare with the IPMs ones. Two solutions have been considered for the reference measurements: fluorescence profile monitors (FPMs) and scintillating screens. The IPMs can be mounted independently in Y or X direction thanks to their design, thus it is even possible to measure the same profile direction with all three IPMs.

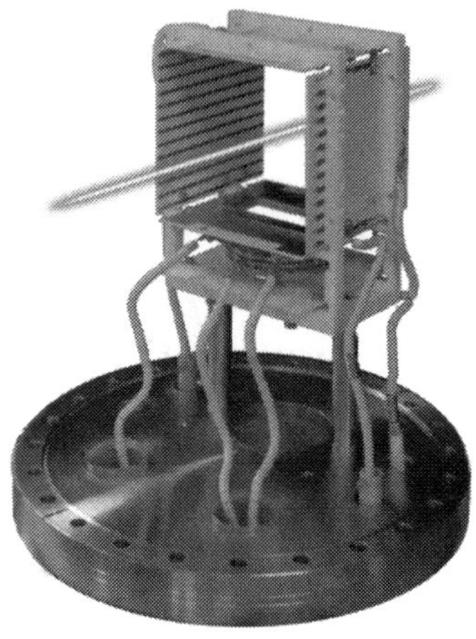

Fig. 6. One of the IPM prototypes. Here the readout is a MCP, visible through the rectangular slit. The copper lines on each side are the field correctors.

B. Control System

The whole system is almost fully compatible with the version 3.16 of EPICS base. The PointGrey GigE cameras are well supported by the AreaDetector module. A custom plugin, developed by ESS, performs a gaussian fit on every image of the profile. Raw images are saved into HDF5 files. This format allows to combine various datasets together, for instance the raw IPM images with some beam information. Since all high voltage power supplies have their own SPCI Ethernet interface, a simple softIOC with StreamDevice is enough to control and monitor them. Three OPIs have been developed in order to control cameras, power supplies and a motor (geobrick) to move scintillating screens vertically to intercept the beam or to be safely moved far from it. These OPIs run under the BOY module of the ESS Control System Studio version 4.5. An Archiver Appliance records and saves slow process variables from the power supplies, the vacuum systems and the accelerator.

C. IPHI accelerator

IPHI is a high intensity linear proton accelerator located at CEA/Saclay. This project started in the late of 90's [29] but protons were accelerated up to 3 MeV in April 2016 [30].

https://doi.org/10.1051/epjconf/202022501009

Proton plasma is created by an electron cyclotron resonance source, and transported toward a radio frequency quadruple (RFQ) by a low energy beam transport line. An iris ensures a fine tuning of the current, and two solenoids focus and filter the plasma before the injection in the RFQ. Then, the protons are accelerated up to 3 MeV and bunched with a frequency of 352 MHz by the RFQ. A medium energy beam transport line, downstream from the RFQ, contains focusing elements, steerers, dipole magnet and beam diagnostics. The dipole magnet can distribute the protons over two beam lines.

The main line has a dedicated beam stop of 300 kW, allowing the commissioning of the accelerator at high intensity and duty cycle. The secondary line is more modular but restricted to lower intensity and duty cycle (few hundred Watt). In 2018, this line was open for external user experiments [31].

Fig. 7. Picture of the IPHI installation. The IPM test bench is also visible in the foreground.

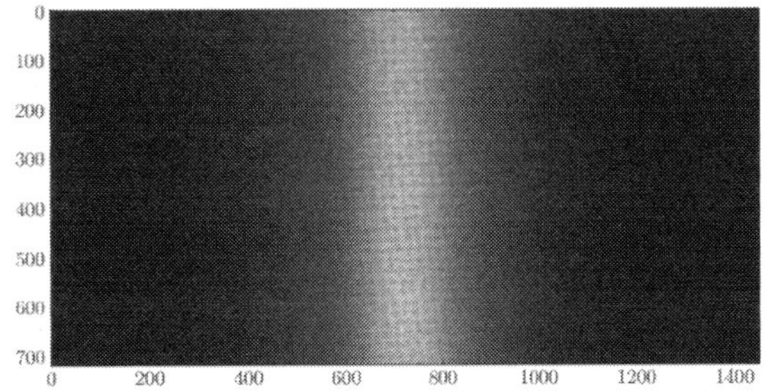

Fig. 8. An example of a beam projection image recorded by the camera. The image covers an area of 38 mm by 18 mm.

VII. Results from IPHI tests

A. First measurements

Two test campaigns were accomplished at IPHI. In the first campaign, the correct operation of the IPMs was checked, the beam was finely tune and first measurements were performed. In the second campaign, further studies were carried out and the first results were confirmed. During both campaigns, the beam was tune to minimize its divergence in the dispersive plane. Both types of IPM have been working in both campaigns.

The optical IPM gives directly an image of the beam in longitudinal and transverse direction. The processing of data is done as follow. First, dead pixels are removed and the image is cropped to a region of interest. Then, a FFT filter is applied for reducing the noise due to sensor if necessary. The profile is reconstructed by summing all pixels in the longitudinal direction. Fig. 8 shows an image acquired from the optical IPM. One can see the shadow of a grid that is placed right in front of the MCP input. The grid has a pitch of $450\,\mu m$ and a wire size of $50\,\mu m$. This gives an idea of the resolution that can be achieved with the optical IPM.

The strips measure the profile in only one transverse direction. The data processing is different because the acquisition is done continuously. First, the pedestals are removed and a search is performed in the whole dataset to find events corresponding to a pulse. Fig. 9 shows the same profile measured by both IPMs.

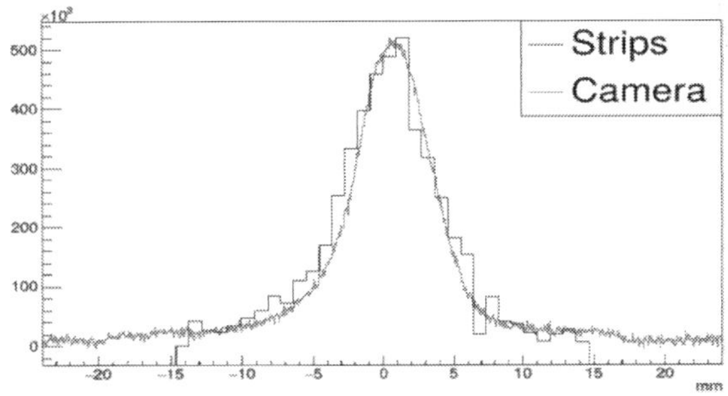

Fig. 9. Superposition of the same beam profile recorded with the strips and optical IPMs.

B. Comparison with IPHI diagnostics

Unfortunately, there is no profile measurement on the IPHI beam line, therefore the measurement of the beam profile can not be compared to an existing diagnostic. However, a complete system of Beam Profile Monitors (BPM) and AC Current Transformer (ACCT) are installed on IPHI. The beam can be also scanned with electrostatic steerers and the current is tuned by an iris aperture close to the source. An IPM is able to measure position and current of a beam, even if it is firstly intended for profile measurement.

An example of beam scanning is illustrated in Fig. 10. The sharp steps are due to steerers whereas the small transitions between two steps are the variations of beam position from pulse to pulse. This is not a desirable effect and it has limited the measurement of the position resolution. The reason of this effect is not clear and may be due to the short pulse duration time used during the tests.

The same can be done for the beam current. The signal is measured at several beam currents for a fixed IPM gain. IPMs response is very linear over an important current range. At IPHI, the number of particles collected by the MCP is an order of magnitude higher than the one expected at ESS, and no

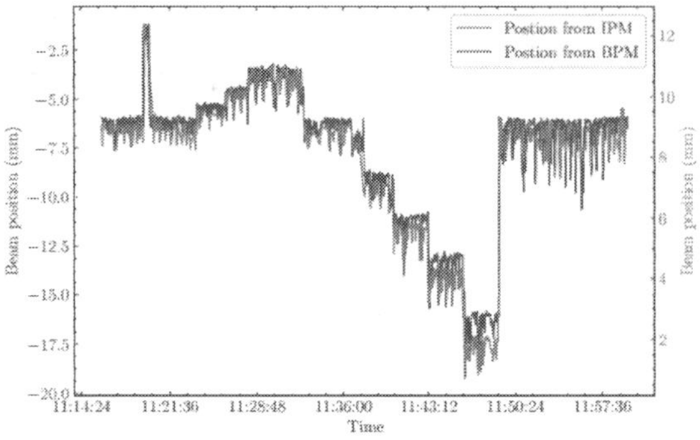

Fig. 10. Beam position over the time, measured with the BPM and the optical IPM. A steerer has been used to move the beam (step transitions). However, small variations between two steerer steps were not expected.

channel saturation has been observed on the MCP signal. This means that a single stage MCP will not suffer of saturation effect at ESS.

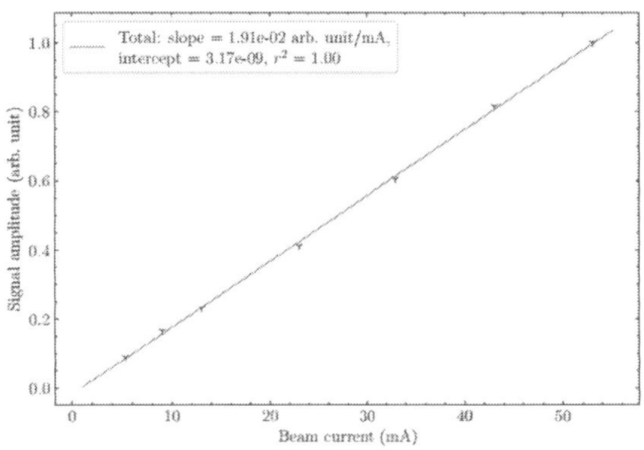

Fig. 11. Evolution of the signal with the beam current. The evolution has a good linearly over an important current range, and no saturation has been observed.

C. *Extrapolation to ESS condition*

One of the important studies conducted at IPHI is the estimation of the detection limit. The objective is to determine which readout can detect the signal in ESS conditions. However the conditions at IPHI are different from the ones of ESS. The proton energy at IPHI is 3 MeV, and according to Bethe equation, the cross section is 60 times higher compared to 2 GeV proton. The vacuum level is also higher by one or two decades. To compensate for these two factors it is possible to reduce the beam current and pulse duration.

For bare strips readout, the number of charges collected was measured for several beam currents, and the intersection of the sensitivity curve with the noise level gives a rough idea of the limit of detection. However, this method greatly underestimates the integration noise of the electronics because the IPHI pulses are very short. According to the first results, the detection limit is just above the most favorable case of ESS. Measuring with bare strips seems difficult in ESS conditions.

Unlike the strips IPM, it is almost impossible to quantify the number of primary particles without a complete calibration of the MCP. Therefore, the extrapolation is done by calculating the Bethe formula with respect to the beam parameters and vacuum conditions measured at IPHI. The beam current and the pulse duration were set to their lowest values, respectively $0.5\,\mathrm{mA}$ and $50\,\mu s$. The pressure was about $4 \cdot 10^{-8}\,\mathrm{mbar}$ In these condition the Bethe scaling is equal to one for the worse ESS case ($628.3\,\mathrm{MeV}$), and the resulting profile is shown in Fig. 12. The measurement may be possible at ESS even with a single stage MCP.

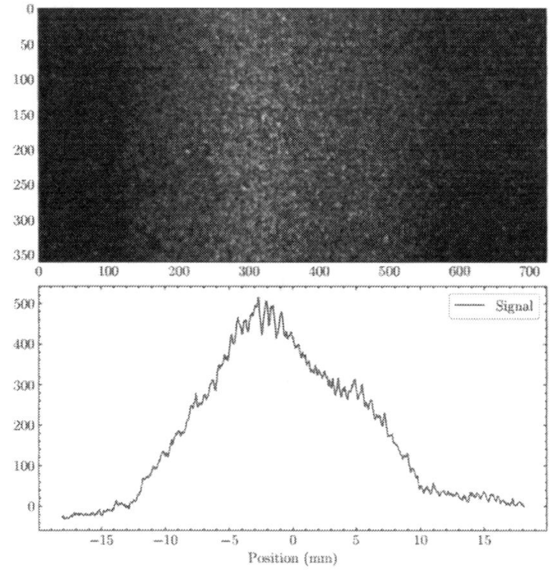

Fig. 12. Image of the beam profile at the lowest current and pulse duration possible at IPHI ($0.5\,\mathrm{mA}$, $50\,\mu s$). In these conditions, the signal is close to the one expected at ESS for the worst case. Note that IPHI was not designed to work a such low current.

VIII. Conclusion

A complete study of all the critical key points concerning the feasibility of this detector has been performed. The expected number of ionization products at the given ESS condition may be enough if the readout is sufficiently sensitive. The non uniformities of the extraction field can be corrected by hardware corrections, and the distortions of the profile due to space charge effects can be reduced if ions are used and sufficient extraction field is provided to the IPMs. Three readout technologies have been considered and a complete platform has been developed in order to test the different readouts. Two types of IPM have been tested in real beam conditions at IPHI, a $3\,\mathrm{MeV}$ proton accelerator. Both strips and MCP were able to measure the beam profile. But approximations to ESS conditions show that the use of MCP is mandatory, therefore the optical IPM is preferred.

The tests at IPHI were a great opportunity to earn experiences and feedback. From this, an improved IPM design is under development. The IPM is now in two part allowing a quick replacement of the MCP without unmounting the cage. A MCP calibration system is also foreseen to monitor and correct the ageing of MCPs.

The production of the IPMs will be done in an ISO-5 particle free environment to insure the compliance with cryogenic cavities. All IPMs will be qualified at Saclay, according to ESS requirements. First detectors will be delivered at ESS in the beginning of 2020 and will be ready for the first protons on dump.

ACKNOWLEDGMENT

The authors are grateful to James Storey and his colleagues (CERN, Geneva, Switzerland) for their help with the TimePix detector. We would like to thank Cyril Bachelet and his team (CSNSM, Orsay, France) for their support during the tests at IRMA. We are thankful to our colleagues at IPHI (CEA, Saclay, France) for their assistance and availability during two beam test campaigns. Finally, we thank all people from the ESS collaboration for their fruitful exchanges and support that help us to design the monitors.

REFERENCES

[1] T. Shea, "Proton beam measurement strategy for the 5 MW European Spallation Source target," in *Proc. IBIC2013*, 2013, pp. 349–352. [Online]. Available: http://accelconf.web.cern.ch/AccelConf/IBIC2013/papers/tupc02.pdf

[2] B. Cheymol, A. Jansson, and T. J. Shea, "Wire scanner design for the European Spallation Source," in *Proc. IBIC2013*, 2013, pp. 830–832. [Online]. Available: http://accelconf.web.cern.ch/AccelConf/IBIC2013/papers/wepf10.pdf

[3] C. Thomas *et al.*, "Design and implementation of non-invasive profile monitors for the ESS LEBT," in *Proc. IBIC'16*, 2016, pp. 551–554. [Online]. Available: http://accelconf.web.cern.ch/AccelConf/ibic2016/papers/tupg80.pdf

[4] W. DeLuca, "Beam Detection using Residual Gas Ionization," in *Proc. PAC1969*, 1969, pp. 813–822. [Online]. Available: http://accelconf.web.cern.ch/AccelConf/p69/PDF/PAC1969_0813.PDF

[5] J. Krider, "Residual gas beam profile monitor," *Nuclear Instruments and Methods in Physics Research Section A: Accelerators, Spectrometers, Detectors and Associated Equipment*, vol. 278, no. 3, pp. 660 – 663, 1989. [Online]. Available: https://doi.org/10.1016/0168-9002(89)91188-1

[6] K. Wittenburg, "Experience with the Residual Gas Ionisation Beam Profile Monitors at the DESY Proton Accelerators," in *Proc. EPAC1992*, 1992, pp. 1133–1135. [Online]. Available: http://accelconf.web.cern.ch/AccelConf/e92/PDF/EPAC1992_1133.PDF

[7] K. Satou, N. Hayashi, S. Lee, and T. Toyama, "A prototype of Residual Gas Ionization Profile Monitor for J-PARC RCS," in *Proc. EPAC2006*, 2006, pp. 1163–1165. [Online]. Available: http://accelconf.web.cern.ch/AccelConf/e06/PAPERS/TUPCH065.PDF

[8] T. Giacomini, P. Forck, J. D. Villiers, J. Dietrich, and D. Liakin, "Ionization Profile monitors - IPM @ GSI," in *Proc. DIPAC2011*, 2011, pp. 419–421. [Online]. Available: http://accelconf.web.cern.ch/AccelConf/DIPAC2011/papers/tupd51.pdf

[9] D. K. Morris *et al.*, "Operational use of ionization profile monitors in the Fermilab Main Injector," in *Proc. NA-PAC2011*, 2011, pp. 519–521. [Online]. Available: http://accelconf.web.cern.ch/AccelConf/PAC2011/papers/mop222.pdf

[10] J. Egberts, "IFMIF-LIPAc Beam Diagnostics. Profiling and Loss Monitoring Systems," Theses, Université Paris Sud - Paris XI, Sep. 2012. [Online]. Available: https://tel.archives-ouvertes.fr/tel-00772158

[11] J. Storey *et al.*, "Development of an Ionization Profile Monitor Based on a Pixel Detector for the CERN Proton Synchrotron," in *Proc. IBIC2015*, 2015, pp. 470–473. [Online]. Available: http://accelconf.web.cern.ch/AccelConf/IBIC2015/papers/tupb059.pdf

[12] ——, "First Results From the Operation of a Rest Gas Ionisation Profile Monitor Based on a Hybrid Pixel Detector," in *Proc. IBIC'17*, 2017, pp. 318–322. [Online]. Available: http://accelconf.web.cern.ch/AccelConf/ibic2017/papers/we2ab5.pdf

[13] Tanabashi *et al.*, "Review of particle physics," *Phys. Rev. D*, vol. 98, p. 030001, Aug 2018. [Online]. Available: https://www.doi.org/10.1103/PhysRevD.98.030001

[14] S. Kamakura, N. Sakamoto, H. Ogawa, H. Tsuchida, and M. Inokuti, "Mean excitation energies for the stopping power of atoms and molecules evaluated from oscillator-strength spectra," *Journal of Applied Physics*, vol. 100, no. 6, p. 064905, sep 2006. [Online]. Available: https://www.doi.org/10.1063/1.2345478

[15] J. Weiss and W. Bernstein, "Energy required to produce one ion pair for several gases," *Physical Review*, vol. 98, no. 6, pp. 1828–1831, jun 1955. [Online]. Available: https://www.doi.org/10.1103/physrev.98.1828

[16] H. Bichsel, D. H. Peirson, J. W. Boring, A. Green, M. Inokuti, and G. Hurst, "Average energy required to produce an ion pair, ICRU Report 31," *Journal of the International Commission on Radiation Units and Measurements*, vol. 16, no. 2, pp. 18–32, may 1979.

[17] H. Schindler, R. Veenhof *et al.* (2019) Garfield++ - simulation of tracking detectors. [Online]. Available: http://garfieldpp.web.cern.ch/garfieldpp/

[18] I. Smirnov, "Modeling of ionization produced by fast charged particles in gases," *Nuclear Instruments and Methods in Physics Research Section A: Accelerators, Spectrometers, Detectors and Associated Equipment*, vol. 554, no. 1-3, pp. 474–493, dec 2005. [Online]. Available: https://www.doi.org/10.1016/j.nima.2005.08.064

[19] W. W. M. Allison and J. H. Cobb, "Relativistic Charged Particle Identification by Energy Loss," *Annual Review of Nuclear and Particle Science*, vol. 30, no. 1, pp. 253–298, dec 1980. [Online]. Available: https://www.doi.org/10.1146/annurev.ns.30.120180.001345

[20] R. Wanzenberg. (2010, May) Nonlinear Motion of a Point Charge in the 3D Space Charge Field of a Gaussian Bunch. Hamburg, Germany. [Online]. Available: http://www.desy.de/~mpywar/paper/2010/Internal_Report_M_10-01.pdf

[21] C. Thomas, F. Belloni, and J. Marroncle, "Space Charge Studies for the Ionisation Profile Monitors for the ESS Cold Linac," in *Proc. IBIC'16*, 2016, pp. 556–559. [Online]. Available: http://accelconf.web.cern.ch/AccelConf/ibic2016/papers/tupg81.pdf

[22] F. Belloni *et al.*, "Space Charge Effects Studies for the ESS Cold Linac Beam Profiler," in *Proc. IBIC'18*, 2018, pp. 371–374. [Online]. Available: http://accelconf.web.cern.ch/AccelConf/ibic2018/papers/weoc04.pdf

[23] S. Ramo, "Currents induced by electron motion," *Proceedings of the IRE*, vol. 27, no. 9, pp. 584–585, sep 1939. [Online]. Available: http://doi.org/10.1109/JRPROC.1939.228757

[24] W. Shockley, "Currents to conductors induced by a moving point charge," *J. Appl. Phys.*, vol. 9, no. 10, p. 635, 1938. [Online]. Available: http://doi.org/10.1063/1.1710367

[25] J. L. Wiza, "Microchannel plate detectors," *Nuclear Instruments and Methods*, vol. 162, no. 1-3, pp. 587–601, jun 1979.

[26] J. Chaumont, F. Lalu, M. Salome, A.-M. Lamoise, and H. Bernas, "A medium energy facility for variable temperature implantation and analysis," *Nuclear Instruments and Methods in Physics Research*, vol. 189, no. 1, pp. 193 – 198, 1981. [Online]. Available: https://doi.org/10.1016/0029-554X(81)90145-2

[27] X. Llopart, R. Ballabriga, M. Campbell, L. Tlustos, and W. Wong, "Timepix, a 65k programmable pixel readout chip for arrival time, energy and/or photon counting measurements," *Nuclear Instruments and Methods in Physics Research Section A: Accelerators, Spectrometers, Detectors and Associated Equipment*, vol. 581, no. 1, pp. 485 – 494, 2007. [Online]. Available: https://doi.org/10.1016/j.nima.2007.08.079

[28] V. Kraus, M. Holik, J. Jakubek, M. Kroupa, P. Soukup, and Z. Vykydal, "FITPix — fast interface for timepix pixel detectors," *Journal of Instrumentation*, vol. 6, no. 01, pp. C01079–C01079, jan 2011. [Online]. Available: https://doi.org/10.1088/1748-0221/6/01/c01079

[29] P.-Y. Beauvais *et al.*, "Status report on the saclay high-intensity proton injector project (IPHI)," in *Proc. EPAC2000*, 2000, pp. 283–285. [Online]. Available: http://accelconf.web.cern.ch/AccelConf/e00/PAPERS/THOAF202.pdf

[30] R. Gobin *et al.*, "High intensity beam production at CEA/Saclay for the IPHI project," in *Proc. ECRIS2016*, 2016, pp. 83–85. [Online]. Available: http://accelconf.web.cern.ch/AccelConf/ecris2016/doi/JACoW-ECRIS2016-WEPP01.html

[31] F. Sene *et al.*, "Increase of IPHI Beam Power at CEA Saclay," in *Proc. IPAC'18*, 2018, pp. 694–696. [Online]. Available: http://accelconf.web.cern.ch/AccelConf/ipac2018/doi/JACoW-IPAC2018-TUPAF016.html

Investigating Artefacts and Repair Mechanisms Associated with α-particle Interactions in Charge Coupled Devices

Rosie Newton[1], Mike J. Scott[2], and Malcolm J. Joyce[1]
[1] Lancaster University, United Kingdom
[2] BIC Technology Ltd., United Kingdom

***Abstract*—Charge coupled devices (CCDs) have been shown to have potential for detecting charged particles and other forms of ionising radiation. In particular, the clusters in the pixel images produced are distinctive for α and β radiations, with α particles causing large, symmetrical clusters or long, vertical tracks often referred to as *blooming*, and β particles causing long, curved tracks. The size and shape of these tracks are also related to the energy of the incident radiation, giving the potential for spectroscopy of these types of radiation. This could be used, for example, to realise a hand-held, portable device for in-situ detection and identification of radioactive contamination. Images have been taken of exposures to a ^{210}Po α-particle source using Sony ICX825AL interline transfer CCDs, covered with a thin, aluminised Mylar film to prevent inadvertent exposure to light. Both vertical streaks and round clusters have been observed in each of the three CCDs used. Increased levels of noise have been seen after several hours of exposure to the ^{210}Po source, found to be caused by damage from α radiation. Three methods have been investigated to reduce this noise: 1) Annealing at 100°C for 24 hours repairs a significant proportion of the damage caused by the α radiation, though the effectiveness is reduced with each subsequent exposure and annealing cycle; 2) Cooling is performed during exposures by using a heat pipe to move heat away from the back of the CCD and a fan to cool the heat pipe and electronics associated with the CCD. This reduces the dark current during CCD measurements, and the method of cooling used allows the device to retain its small, hand-held size; 3) Any remaining noise can be removed later using image processing. These techniques are combined to extend the lifetime of the device. The rate of damage from the ^{210}Po source is similar whether the device is cooled or not, however the noise evident in the images produced when cooled is significantly less, allowing the CCD to be used for a longer time between annealing operations.**

***Index Terms*—Charge coupled device, alpha radiation, radiation damage.**

I. Introduction

THE detection and identification of radioactive contamination is essential in the nuclear industry but may be difficult to achieve in-situ for α radiation, owing to its short range in air. Because α-emitting radioactivity is particularly hazardous if inhaled or ingested, identifying α contamination is critical to managing this risk. Usually, α-particle spectroscopy is performed in a laboratory by processing the sample to isolate the radioisotope and detecting the α radiation under a vacuum. This procedure is not always practical because of the processing and equipment involved.

CCDs are small, semiconductor pixel-imaging devices commonly used in digital cameras. Different types of radiation leave distinctive tracks in CCDs, allowing them to be used for identification of radiation. CCDs are small in size making them suitable for use in a hand-held instrument, and potentially usable in-situ.

As radiation enters a CCD, charge carriers are created via ionization in the active region of the device, as shown in Fig. 1. The evolved charge is then collected by gate electrodes, passed through adjacent pixels, and read out sequentially to form a pixelated image based on the location of the charge carrier distribution in the CCD, and hence that of the associated interactions. Different types of artifacts may constitute these images, depending on the interactions of the radiation in the CCD. As the radiation scatters through several pixels, this will leave a trail of charge carriers, resulting in a track of pixels. If a large amount of energy is deposited in a given pixel, it diffuses through the substrate to create large, symmetrical clusters of pixels. If the amount of created charge reaches the capacity of the pixel, it may overflow vertically. This "blooming" would be expected to leave a vertical streak of pixels in the absence of any blooming correction facilities.

Manuscript received July 11, 2019. This work was supported by the Next Generation Nuclear Centre for Doctoral Training, the Engineering and Physical Sciences Research Council (EPSRC), and the Lloyds Register Foundation. M. J. Joyce acknowledges the support of the Royal Society as a recipient of a Wolfson Research Merit award.

Rosie Newton and Malcolm J. Joyce are with the Engineering Department, Lancaster University, LA1 4YW, UK. (e-mail: r.newton1@lancaster.ac.uk).
Mike J. Scott is with BIC Technology Ltd., Gostin House, 32/36 Hanover St., Liverpool, L1 4LN, UK (e-mail: info@bictechnology.co.uk)

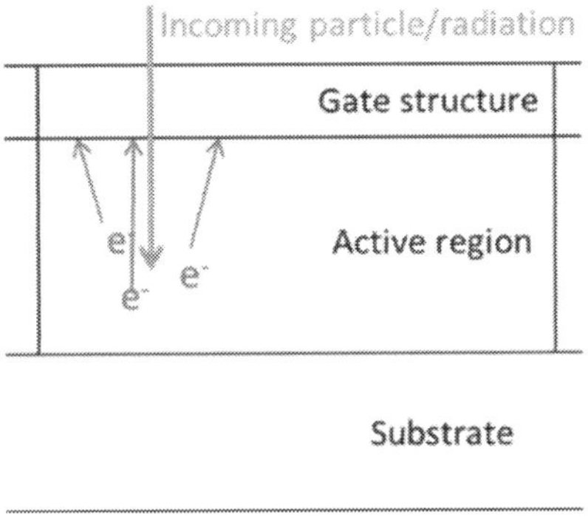

Fig. 1. A schematic cross section of a single pixel within a CCD.

II. Prior Art

It has been reported that β particles often cause long, curved tracks as they scatter through the CCD, and that α particles typically form large, round clusters [1], [2]. This is because the latter interact more strongly, creating a larger number of charge carriers in a single pixel which diffuses into the surrounding pixels. Occasionally, α particles will also bloom into a vertical streak.

In research prior to that presented here, similar differences have been seen, as shown in Fig. 2. Here, the larger round clusters and vertical streaks from α radiation can be seen as a result of exposure to a ^{210}Po source. In the exposures to ^{60}Co and ^{137}Cs, smaller clusters and curved tracks can be seen from the β and γ radiation. These can be used to identify α and β sources, respectively, through the analysis of the sizes and shapes of the clusters produced [3], [4].

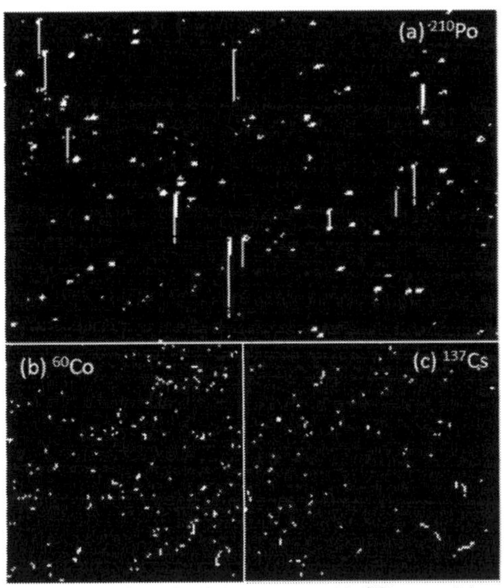

Fig. 2. Example sections of typical exposures to: (a) a ^{210}Po α-particle source for 30 s; (b) a ^{60}Co β/γ source for 5 s; (c) a ^{137}Cs β/γ source for 1 s.

III. CCD Details

The CCD used for this research is a Sony ICX825AL. This is an interline CCD with 1392 × 1040 pixels and a pixel size of 6.45 × 6.45 μm, for a total imaging area of 6.71 × 8.98 mm. The device can be seen in Fig. 3, with the CCD itself covered with an aluminised Mylar film to prevent the interaction of light which would otherwise dominate the image. The CCD can be seen uncovered in the inset image.

Fig. 3. α-particle detection device used in this research showing the CCD covered with Mylar film and the associated electronics. Inset: Uncovered CCD.

IV. Alpha Damage

Over continued exposure of the CCD to α radiation, an increase in noise was observed in the images produced. This noise consists of bright pixels where no interactions with radiation have occurred during the current exposure. At low levels of noise this does not cause a significant problem, however a high level of noise can impact the accuracy of α-particle detection. To confirm that the noise was caused by damage from the α radiation, half of the CCD imaging area was covered with card sufficiently thick to stop the α particles. The CCD was then exposed to a ^{210}Po source with an activity of 21 Bq for five hours. At each hour interval, a background measurement was taken with no source present. Consistent sections of each side of the images produced were then analysed to identify a difference in the number of signals in the background between the covered and uncovered sides of the CCD. The data can be seen in Fig. 4.

It is clear that the uncovered section which has been exposed to the α radiation has increased in background signals with each additional hour of exposure. In contrast, the covered side which prevented α particles from interacting with the CCD showed no change over the 5-hour period. Similar experiments were performed with the 320 kBq ^{137}Cs source and, over a period of 9 hours, no change was seen in the background. This indicates that α radiation is the most significant cause of the increase in background noise.

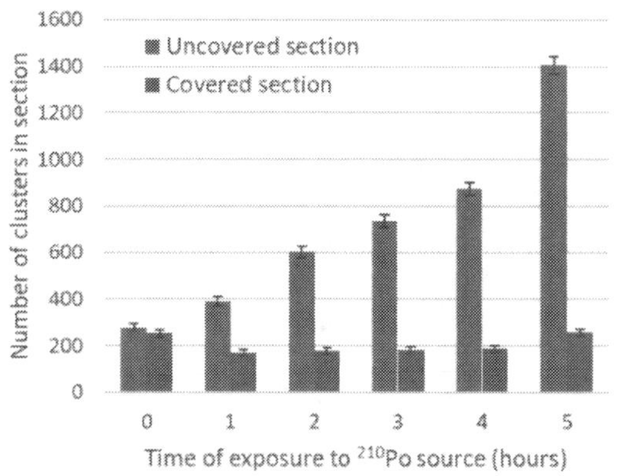

Fig. 4. The number of bright pixels in a 30-s background measurement for equivalent-size sections on the covered and uncovered side of the CCD after each additional hour of exposure to a ^{210}Po source.

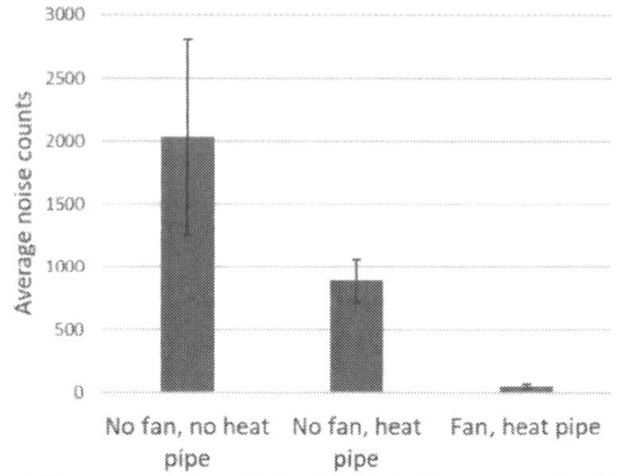

Fig. 5. The average number of bright pixels (noise) in images produced by the CCD in 5-minute exposures with no source present at three different levels of cooling: no cooling, just the heat pipe, and heat pipe with a fan.

V. Repair Mechanisms

Multiple techniques have been considered to repair the damage produced by interactions with α radiation and to mitigate the effects of the damage during exposure. The techniques used have consisted of annealing, cooling and image processing. These may collectively be used to improve the accuracy of α detection and extend the useful lifetime of the device.

A. Annealing

Annealing was performed to repair some of the damage caused by the α radiation. The entire device was heated to a temperature of 100°C for 24 hours, reducing the amount of noise by 89% in one anneal. It has been found that the annealing is slightly less effective with each subsequent cycle, as in each case not all of the damage is repaired.

Heating to a higher temperature has been shown to be effective at repairing a greater amount of damage in other CCDs [5], however this would require dismantling the device due to the problems higher temperatures might cause, such as damaging electronic components. This would not be practical for standard use and so the 100°C temperature was used as this allows the entire device to be heated, while still resulting in significant reduction to the damage.

B. Cooling

To minimize the noise created by the damage during operation, the CCD is cooled. A copper heat pipe is used to draw heat away from the CCD and is then cooled by a fan. The noise reduction from cooling with different combinations of the fan and heat pipe can be seen in Fig. 5. The heat pipe alone is successful at reducing the noise by 56%, which is useful for situations where it may not be practical to be operating a fan, such as where contamination may be in the form of dust. However, if a fan is also able to be used, the noise reduces by 97.5%. This allows the CCD to be used for a much longer time between annealing, as although the rate of damage remains the same, it will not be as evident in the images produced.

C. Image processing

Any remaining noise can be removed by image processing. As the noise consists of pixels with lower intensity values than the pixels where α radiation interacted, and typically occurs in single pixels, the images can be processed to remove this low intensity noise. Images were taken with no CCD present to obtain pure noise images, and the average intensity of the pixels considered. This was compared with a typical α signal, to determine a cut-off intensity of 100. For any cluster of pixels in which the brightest pixel has an intensity of less than 100, the entire cluster is removed. If any single pixel is over this threshold, the entire cluster is kept. This ensures that no information is lost from the α-particle streaks, which are bright where the α particle interacted but tend to low intensities. An example section of an image before and after processing can be seen in Fig. 6. This shows the significant reduction in noise possible with image processing.

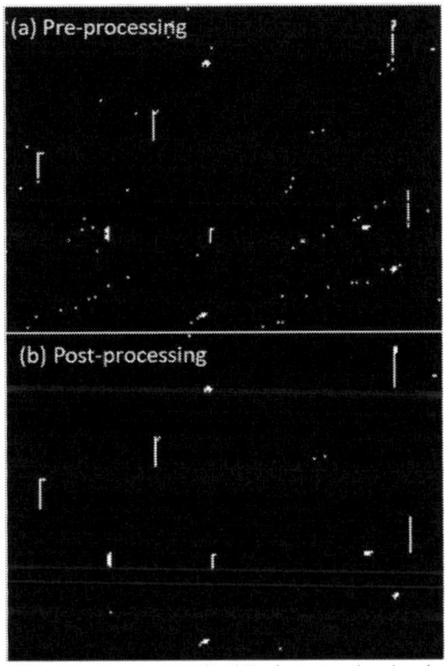

Fig. 6. A representative section of a CCD image produced under a 30-s exposure to a ^{210}Po source (a) before and (b) after processing to remove noise.

VI. CONCLUSION

CCDs can be used to image and identify α radiation, however they are subject to damage from the α particles which produce increasing levels of noise over continued exposure. This noise can be reduced through cooling the device and image processing. As the noise builds to a problematic level, a significant amount of the damage can be repaired through annealing. By combining these techniques, the lifetime of the device as an α-particle detector is greatly extended and the accuracy of detection improved by reducing false measurements or masked signals due to noise.

REFERENCES

[1] A. Aguilar-Arevalo *et al.*, "Measurement of radioactive contamination in the high-resistivity silicon CCDs of the DAMIC experiment," *J. Instrum.*, vol. 10, no. 08, pp. P08014–P08014, Aug. 2015.

[2] J. Estrada, J. Molina, J. J. Blostein, and G. Fernández, "Plasma effect in silicon charge coupled devices (CCDs)," *Nucl. Instruments Methods Phys. Res. Sect. A Accel. Spectrometers, Detect. Assoc. Equip.*, vol. 665, pp. 90–93, 2011.

[3] R. Newton, J. Dykes, M. J. Scott, and M. J. Joyce, "Alpha contamination assay, dosimetry and spectrometry using charge coupled devices," *2016 IEEE Nucl. Sci. Symp. Med. Imaging Conf. Room-Temperature Semicond. Detect. Work. NSS/MIC/RTSD 2016*, vol. 2017–Janua, 2017.

[4] R. Newton, M. J. Scott, and M. J. Joyce, "Investigating Artifacts Associated with beta-Particle Interactions in Charge-Coupled Devices," ... *J. Mod. Phys. Conf. Ser.*, vol. 1, no. 1, pp. 1–5, 2010.

[5] G. R. Hopkinson and A. Mohammadzadeh, "Comparison of CCD Damage Due to 10- and 60-MeV Protons," *IEEE Trans. Nucl. Sci.*, vol. 50, no. 6 I, pp. 1960–1967, 2003.

EPJ Web of Conferences **225**, 01011 (2020) https://doi.org/10.1051/epjconf/202022501011
ANIMMA 2019

Report of High Temperature Measurements with a Fabry-Perot Extensometer

G. Cheymol[1], A. Verneuil[1], P. Grange[2], H. Maskrot[1], C. Destouches[3]
[1] Service d'Etudes Analytiques et de Réactivité des Surfaces (SEARS), CEA, Université Paris-Saclay
[2] Service d'Etude des Matériaux Irradiés, CEA, Université Paris-Saclay
[3] Service de Physique Expérimentale d'essais en Sûreté et d'Instrumentation (SPESI), CEA

***Abstract*— Fabry-Perot (FP) sensors like other Fiber Optic (FO) sensors may be of particular interest for in pile experiments in MTR with little room available thanks to their compact size. Light weight also reduces gamma heating hence limiting the thermal effect. Different physical parameters such as temperature, strain, displacement, vibration, pressure, or refractive index may be sensed through the measurement of the optical path length difference in the cavity. We have developed a Fabry-Perot extensometer able to operate at high temperature (up to 400°C), under a high level of radiation (neutron and gamma flux). The measurement based on interferometry is largely insensitive to radiation induced attenuation (RIA) thanks to the wavelength encoding of the useful signal, but for such high fluence as encountered in a reactor core, a special rad-hard fiber is needed. Operating in the wavelength domain around 1µm remains preferable to minimize the impact of irradiation. Moreover, fast neutron radiation is expected to change the structure of the fiber and possibly others materials in the transducer. Then, we revised the basic scheme of Extrinsic Fabry-Perot Interferometer (EFPI) so that the effects of compaction remain limited. Tests under mixed neutron and gamma irradiation permitted to verify the general behavior and particularly the low drift with radiation induced compaction (RIC). Also, two types of tests have been conducted to verify the accuracy at high temperature. The first ones are "measurements" of thermal dilatation of materials: the sensor is fixed on a sample and knowing its thermal expansion, it is possible to predict the measurement expected from the optical sensor when the temperature is increased from low to high temperature. The comparison between the predicted and experimental outputs informs on how the sensor is accurate. The second types are tests on a tensile test bench operating at high temperature. The Fabry-Perot measurements are compared, in the elastic domain, with the expected strain given by the Young modulus of the material, and also on a larger strain domain, with the measurements of a high temperature axial extensometer. Both types of tests are presented and commented.**

***Index Terms*— Extensometer, Fabry Perot sensor, Material testing reactor, Optical fiber sensor, White light interferometry**

I. Introduction

VARIOUS R&D programs are implemented in order to improve the performance and the safety of existing and future plants as well as to assess and develop new reactor concepts. Among them, irradiation experiments in Material Testing Reactors (MTR) allow to optimize designs for improved technologies and to qualify innovative fuels and materials. In particular, the Jules Horowitz Reactor (JHR) [1], a new Material Testing Reactor under construction in Cadarache (France) is expected to feature highly instrumented experiments. As well, next generations of power reactor could benefit enhanced in-core online monitoring.

Optical Fiber Sensors (OFS) display some interesting features in order to equip irradiation rigs in MTR. In order to make measurement of elongation, we have developed a Fabry-Perot Sensor (FPS) able to work under high temperature and large neutron fluence [2,3]. This development has first been carried out jointly with SCK•CEN in the framework of the Joint Instrumentation Laboratory (JIL), then CEA has continued the development to improve the operation at high temperature.

In this article, we will first present the principle of the sensor; next, we will focus on two sets of different experiments conducted to evaluate the accuracy of the FPS when operating at high temperatures. They are based on the study of the thermal dilatation of the support and on the tests on a tensile testing machine.

II. The Fabry Perot Sensor

Our Fabry-Perot sensor developed for applications in MTR is based on a classical scheme (Fig. 1). In the selected design each face of the cavity is connected to a localized area of the sample separated by the gage length. Any variation of the gage length produces a similar variation of the Fabry-Perot cavity length.

The input beam reflects on both faces of the cavity and the state of interference generated by the two reflected beams depends on the ratio: $\frac{Lc}{\lambda}$, where Lc is the cavity length and λ is the wavelength. White light interferometry takes advantage of the channeled spectrum - obtained thanks to the broad spectrum of the SLED and to the spectrometer with enough resolution - to retrieve the cavity length Lc [3].

Manuscript received xx. This work was supported by French Alternative Energies and Atomic Energy Commission (CEA), Division of Nuclear Energy. It was performed at the CEA Saclay Center.
G. Cheymol, A. Verneuil, H. Maskrot are with Den – Service d'Etudes Analytiques et de Réactivité des Surfaces (SEARS), CEA, Université Paris-Saclay, F-91191, Gif sur Yvette, France (33(0)1.69.08.62.71; e-mail: guy.cheymol@cea.fr.

P. Grange is with Den – Service d'Etude des Matériaux Irradiés (SEMI), CEA, Université Paris-Saclay, F-91191, Gif sur Yvette, France.
C Destouches is with Den – Service de Physique Expérimentale d'essais en Sûreté et d'Instrumentation (SPESI), CEA, F-13108, St Paul lez Durance, France.

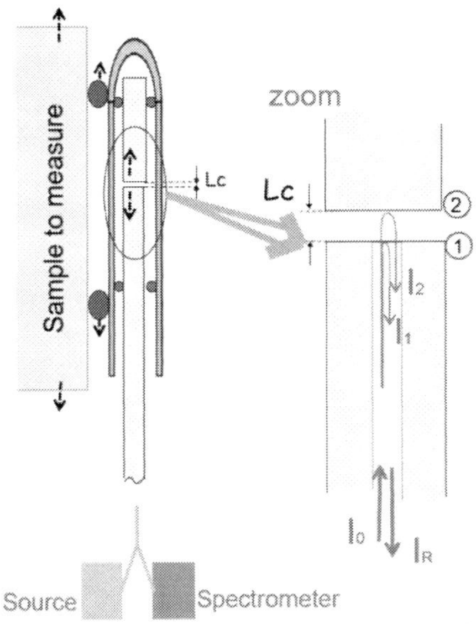

Fig. 1. Principle of the Fabry-Perot sensor: any variation of the gage length produces a similar variation in the Fabry-Perot cavity length, probed through interference of the reflections on both faces of the cavity.

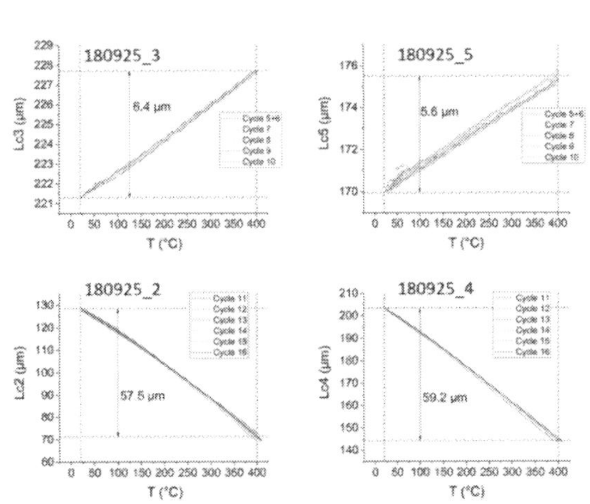

Fig. 2. Graphs of Lc versus the temperature for different cycles, for the four tested sensors.

III. Assessment of the accuracy at high temperature

The following describes some tests performed to evaluate accuracy at a higher temperature.

A. Study of the thermal dilatation of material

Four similar FPS are used for that set of experiments: two sensors, with reference 180925_3 and 180925_5, are fixed on a 316 L stainless steel plate, and two sensors, with reference 180925_3 and 180925_5, are fixed on a tantalum plate. The dimensions of the plate are 10 mm x 50 mm x 2 mm. The four plates are inserted in tubular oven and the temperature rises to 400 °C, then decreases again to room temperature. This cycle is repeated several times, and each time the length of the cavity Lc is plotted as a function of the temperature given by a thermocouple placed near the plates. The graphs resulting from the different cycles are presented in Fig. 2 for the four tested sensors.

The different graphs show the good repeatability of the measurements and make it possible to determine an average value of the variation of the cavity length Lc when the temperature varies from 20 to 400 °C. This value is named ΔLc_{exp} in the following and is compared to the expected theoretical value ΔLc_{th}.

With 18.5 x 10^{-6}/°C [4, 5] and 6.6 x 10^{-6}/°C [6] considered for mean coefficients of thermal expansion (between 20 °C and 400 °C) of 316L stainless-steel and tantalum materials respectively, the associated ΔLc_{th} of a standard FPS are 7.5 µm and -60.3 µm respectively. Small variations in the sensors built compared to the standard FPS implies slight variations in ΔLc_{th}. The revised theoretical values are reported in Table 1, as well as the experimental values.

TABLE 1
THEORETICAL AND EXPERIMENTAL VALUES OF ΔLC WHEN THE TEMPERATURE VARIES FROM 20 TO 400 °C FOR THE FOUR TESTED SENSORS

n°	ΔLc_{exp}	ΔLc_{th}	ΔLc_{exp} - ΔLc_{th}
On 316 L stainless steel plate			
180925_3	6.4µm	7.6 µm	- 1.2 µm
180925_5	5.6µm	7.3 µm	-1.7 µm
On tantalum plate			
180925_2	-57.5 µm	-60.8 µm	3.3µm
180925_4	-59.2 µm	-60.5 µm	1.3 µm

The difference between experimental and theoretical value is ± 1 to 3 µm. This seems acceptable, especially considering that slight uncertainties about the temperatures and the positions of the boundings can lead to uncertainties of about 1 µm for ΔLc_{exp} and ΔLc_{th}. For example, a variation of 0.2 mm in the gage length on the tantalum plates produces a change of nearly 1 µm in ΔLc_{th}.

B. Tensile tests

1) Description

The second set of experiments to assess the accuracy of the FPS at high temperature are tests of elongation with a tensile test machine. Unlike during thermal expansion evaluations, FPS remain at a set temperature when the tests are performed. 316 L stainless steel samples with a useful length of 100 mm are used, with a section of 6 mm x 2 mm.

During the various experiments, we plot stress-strain curves. The strain, given by ΔLc divided by the gage length, is measured by the FPS and is compared first to the expected strain given by the Hooke's law:

$$\sigma = E \cdot \varepsilon, \quad (4)$$

where σ, E, ε are the strain, the Young Modulus, and the stress respectively. The values considered for E vary with the temperature [5]. But the comparison is limited to the elastic domain of the stress strain curve, with a maximum strain of about 0.1 %.

On a larger range of strain, the output of the FPS is also compared to an axial extensometer with ceramic rods (Maytec), used with a gage length of about 20mm (Fig. 3). The error of this measurement is specified to be less than 10µm, error that is expected to be reduced with calibration of the device.

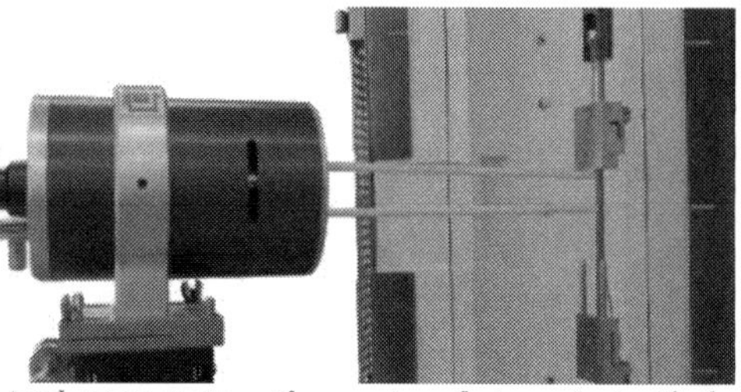

Fig. 3. Axial extensometer with ceramic rods in contact with the sample, on the opposite side of the FPS assembly.

After some statics tests on the elastic domain, a greater non-reversible elongation is achieved during a dynamic test with a constant elongation rate of 2.5 µm/s generated by the tensile test machine.

2) Results of the tests on the elastic domain

Fig. 4. presents the graphs obtained with tests on the elastic domain at high temperature. All tests indicate that FPS is accurate at high temperature on the low range of elongation tested.

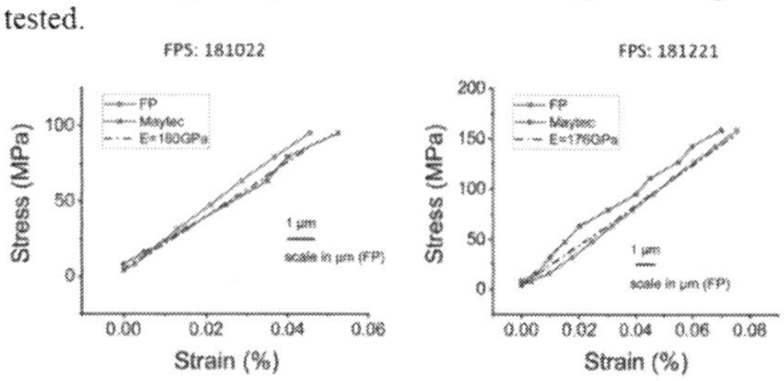

Fig. 4. Stress-strain curves at high temperature: 300°C for FPS 181022 (left) and 350°C for FPS 181221(right) on the elastic domain. Both the FPS and the extensometer with ceramic rods (Maytec) are close to the theoretical values given by Hooke's law.

3) Results of the dynamic tensile tests

Fig. 5. presents the tensile stress-strain curve obtained with the FPS 181022, at 300°C, with a strain up to 1.5%. The plots given by the Maytec extensometer and the FPS are quite close.

In order to get a more precise assessment of the difference between the two sensors, the elongations data given by the Maytec extensometer is normalized to the same gage length (with a rule of three on the elongation measurements in µm) and the difference is calculated (Fig. 6). After filtering, the difference is limited to less than 6µm with an elongation of more than 200µm measured by the FP sensor.

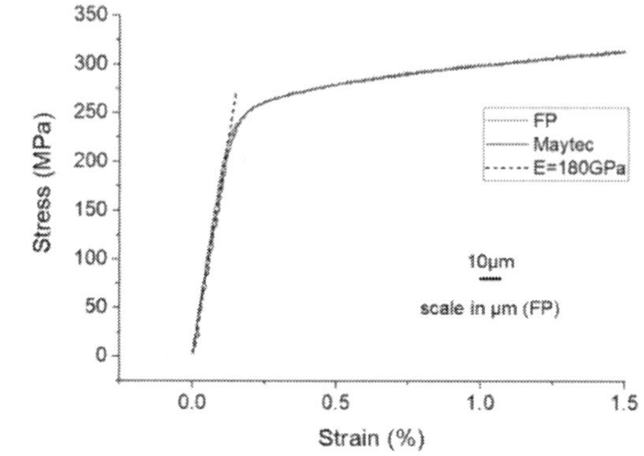

Fig. 5. Tensile stress-strain curve with FPS 181022, up to a strain of 1.5%.

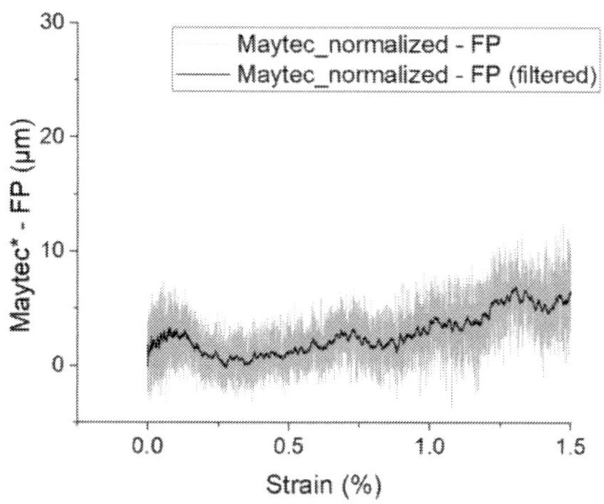

Fig. 6. Difference of elongation measured between FPS and Maytec (normalized to the same gage length) versus the strain on dynamic test with FPS 181022. The curve is noisy (grey); filtering the data given by the Maytec extensometer leads to the dark plot.

IV. Conclusions

The "measurements" of thermal dilatation and also tensiles tests have permitted to assess the accuracy of the FP sensor at high temperature (300°C to 400°C).

By comparison with Hooke's law on the elastic domain (low elongation, about 0.1% at maximum), the error is limited to about 1 µm. On the ranges within the specification of the sensor (1% to 2%), the error can be of some microns. The limited accuracy of the reference extensometer used did not permit to access more precisely the accuracy.

The sensor is designed to be nearly unaffected by gamma + neutron irradiation. This has been partially verified and is expected to be confirmed in a future irradiation (2020-2021).

References

[1] http://www-rjh.cea.fr
[2] G. Cheymol, J.F. Villard, A. Gusarov, B. Brichard, Fibre optic extensometer for high radiation and high temperature nuclear applications, IEEE Trans.Nucl.Sci., vol.60, no.5, pp. 3781-3784, 2013.

[3] G. Cheymol, A. Gusarov, S. Gaillot, C. Destouches and N. Caron, "Dimensional Measurements Under High Radiation With Optical Fibre Sensors Based on White Light Interferometry - Report on Irradiation Tests"; IEEE Trans. Nucl. Sci., vol.61, no.4, pp. 2075-2081, 2014.
[4] https://docplayer.fr/6618826-Selection-de-fiches-techniques.html
[5] L. Depradeux, "Simulation numérique du soudage – acier 316L – validation sur cas tests de complexité croissante'', PhD dissertation, INSA de Lyon, Lyon, France, 2003
[6] P. Hidnert, "Thermal expansion of tantalum" Bureau of Standards Journal of Research, Vol 2, Issue 5, p 887-896, May 1929, http://dx.doi.org/10.6028/jres.002.028.

System noise of a digital pulse processing module for nuclear instrumentation

Y. Ma, W-J. Fischer, J. Henniger, D.Weinberger, T. Kormoll
TU Dresden, Germany
yuzhen.ma@tu-dresden.de

***Abstract*—Digital pulse processing finds a widespeed use in nuclear instrumentation. This work presents a digital pulse processing (DPP) module based on an FPGA and a 16-bit 125 MHz ADC and analyses the system noise distribution by acquired digital data using this system. A moving average filter is utilized to suppress the system noise of the DPP module. Furthermore, digital trapezoidal filter is applied for the use with a charge sensitive preamplifier with High Purity Germanium (HPGe) detector. The energy spectrum and corresponding resolution are demonstrated with a scintillation and semiconductor detector.**

***Index Terms*—digital signal processing; FPGA; noise; Gaussian; nuclear**

I. Introduction

Digital pulse processing (DPP) has been widely used in the ionizing radiation field recently. It has many advantages compared to analog system in the development phase and update period [1]. DPP modules have become more and more popular during the past decades since the development of semiconductor and digital components, such as Field Programmable Gate Arrays (FPGAs) and high-speed analog-to-digital converters (ADCs). The amplifier and ADC are assigned to accurately and reliably amplify or sample the amplitude of a continuous voltage pulse that can be subject to various noise sources [2]. Amplitude depends not only on the system noise, but also on the instantaneous value of the offset, which is very critical for a nuclear instrumentation for it is proportional to the amount of charge.

In the present paper, a DPP module is handled, which contains a free-running sampling ADC and an FPGA for real-time pulse analysis. In this work, the system response to signals from a pulse generator and radiation detectors are investigated.

II. Material and Methods

A digital spectroscopy system contains detector, processor and analyzer, which aims to receive, process and analyze ionizing radiation on event-by-event traces [3], [4]. Most common features of an event scintillation detector are energy, time and pulse shape.

Y. Ma and W. Fischer are with the MST (Institute of Semiconductor and Microsystems Technology), Technische Universität Dresden, Dresden 01069, Germany (e-mail: Yuzhen.Ma@tu-dresden.de, Wolf-Joachim.Fischer@tu-dresden.de).

J. Henniger is with the IKTP (Institute of Nuclear and Particle Physics), Technische Universität Dresden, Dresden 01069, Germany (e-mail: henniger@asp.tu-dresden.de).

D. Weinberger is with the HZDR (Helmholtz-Zentrum Dresden-Rossendorf), Dresden 01328, Germany and Serious Dynamics GbR, Dresden 01139, Germany (e-mail: weinberger@serious-dynamics.de)

T. Kormoll is wih the IKTP (Institute of Nuclear and Particle Physics), Technische Universität Dresden, Dresden 01069, Germany and Serious Dynamics GbR, Dresden 01139, Germany (Thomas.Kormoll@tu-dresden.de)

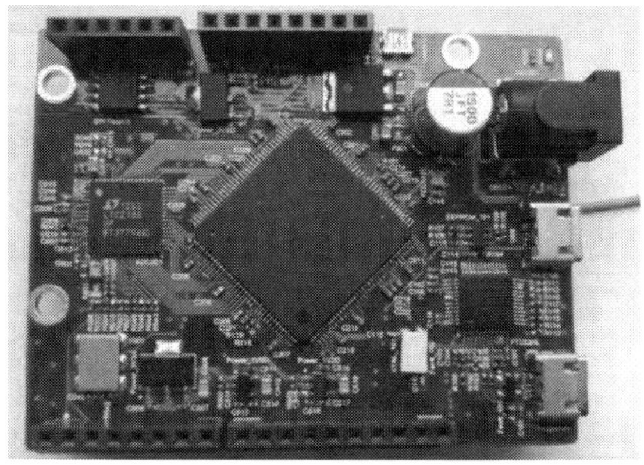

Fig. 1. Hardware design based on a LatticeXP2 FPGA.

A. Design of the digital signal processing system

The DPP architecture described hereafter is called DAQ125 that is a 2-channel, 125 MHz 16-bit digitizer board as shown in figure 1. It has three stages, analogue input, digital processing, and uploading data for analysis. The whole system block is depicted in figure 2, in which these three stages are represented by three different kinds of background colors respectively. In the first stage, an RC lowpass filter acting as an anti-aliasing filter isolates the drive circuitry from the A/D sample-and/hold switching, and limits wideband noise from the drive circuitry. 25 Ω resistor and 12 pF capacitor are chosen as the values of lowpass filter components based on the input radiation signal frequency [5]. The continuous signal from a radiation detector is sampled by the ADC LTC2185 and processed by the FPGA LatticeXP2 in real-time. In the second stage, the acquired data stream is passed to an FPGA firmware which comprises digital modules, i.e., trigger, shift register, waveform memory, integrator and control-memory. The FPGA firmware is self-triggering and allows to readout a portion of waveform around the trigger point or to calculate some features around the trigger and to transfer only certain numbers per event. The trigger mode of this firmware is leading or falling edge and the threshold value is within the input range of ADC. The delaying length of the trigger point is controlled by a shift register that is configured though software. Once input

https://doi.org/10.1051/epjconf/202022501012

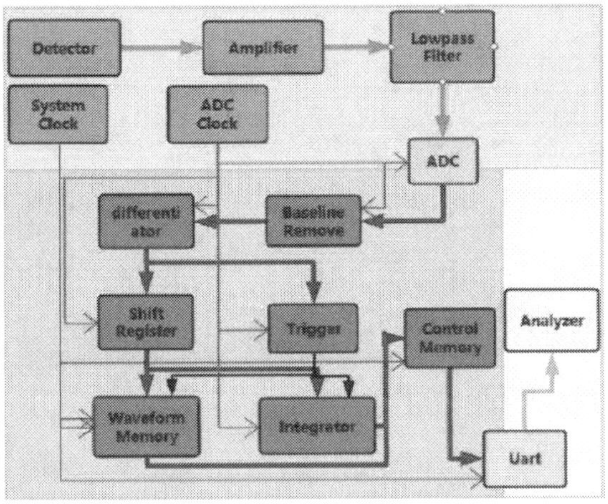

Fig. 2. Architectural block of whole system.

data is triggered and shifted, the transferred data is saved into waveform memory up to a maximum amount of 512 samples.

B. Reference system

The acquired data from an oscilloscope HDO4000A of Teledyne LeCroy is used as a reference comparing to the output of the designed DAQ125 to evaluate and analyze execution of the designed DPP system. Both of these two schemes are connected to a pulse generator or detector by using a BNC-KJK connector simultaneously. The sampling rate of HDO4000A is 10GSPS with 5 μs sampling time per analyzing data stream. Corresponding to the sampling rate and sampling time of LTC2185 ADC, 512 samples are extracted from the raw data of HDO4000A by extracting one sample every 80 times to work as a reference for the acquired data by using DAQ125. In order to get the distribution curve of noise, Probability Distribution Function (PDF) is employed to analyze the characteristic of system noise.

In the primary phase, a pulse generator, TR-0458/B, is utilized. The parameters of the generated signal such as waveform, frequency, amplitude and duty cycle can be manual changed. In this work, a pulse signal with a frequency of 1 MHz, a amplitude of -4 dBV and a duty cycle of 37.2% is generated.

At the last stage, the scintillation detector $CeBr_3$ and semiconductor detector HPGe are employed. A dual digital multichannel analyzer (MCA) of CAEN S.p.A. is utilized as a reference which is comparing to the spectroscopy of the DPP to evaluate its performance. The HPGe detector is biased by the 2000 V high voltage with 300 μA current using one channel of module MCA DT5780. The software MC^2 Analyzer is utilized to acquire data and plot energy spectrum in real time with the energy filter setup that 8 μs rise time, 48.34 μs decay time, 2.5 μs flat top and 80% flat top delay. The sampling rate of DT5780 is 100MS/s with 14-bit for analyzing data. 512 samples are extracted from the raw data of DAQ125

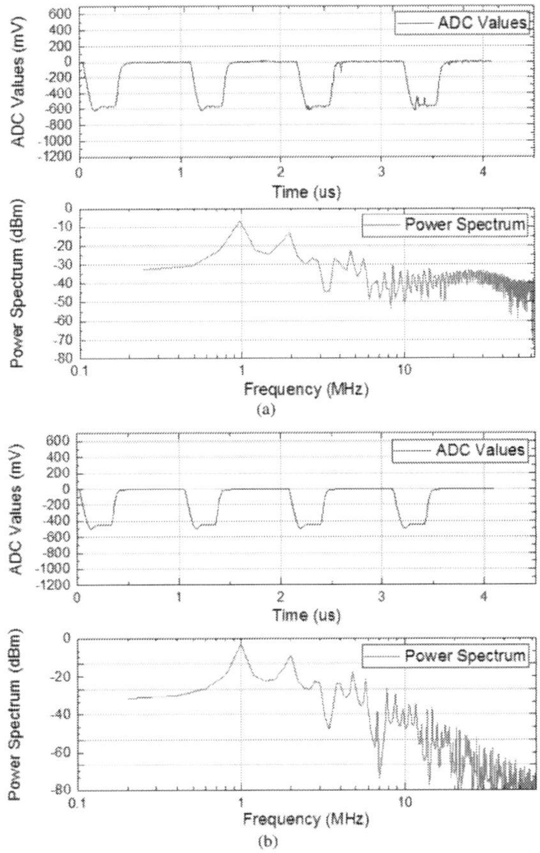

Fig. 3. Acquired data (a) and reference signal (b) shown in time and frequency domain acquired by HDO4000A. The total measurement is 512 samples.

according to the sampling rate of LTC2185 ADC and output signal of HPGe detector.

III. Results and Discussion

According to the material and methods in the above section, two different kinds of schemes were applied to acquire a detector signal to assess the implementation of the developed DPP. Pulse generator is connected to system. The system is trigged at the falling edge of a 1 MHz pulse signal. The blue and red line represent pulse features in the time and frequency domain respectively in figure 3. The experimental result of the obtained data by using the DAQ125 is depicted in figure 3(a). The blue line is the measured pulse time and a timestamp with 8 ns is assigned with 512 samples in the above curve. The red line indicates the accordant power spectrum after applying the Fast Fourier Transform (FFT) algorithm to the processing data. As a comparative outcome, the figure 3(b) represents the time and frequency information of the identical signal under a equivalent status by using HDO4000A.

As a comparison for the different schemes, the blue curve in figure 3(a) is much noisier than the one in figure 3(b) within

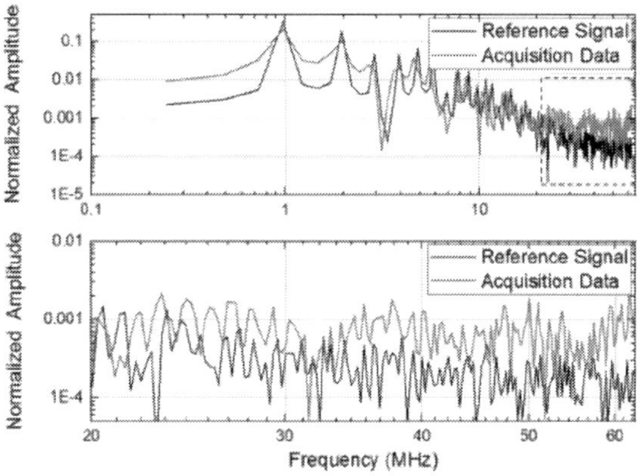

Fig. 4. Amplitude spectrogram comparison of acquired data and reference signal.

4.096 μs sampling time. Consequently, the trend of both power spectrum curves indicate well matching in the lower frequency part, while a higher level is demonstrated in figure 3(a) than figure 3(b) at the high frequency part. For analyzing the system operation of DPP, the normalized amplitude spectrogram is presented in figure 4, which is a direct comparison of the acquired data and the reference signal in the frequency domain. The red plot exhibits data from DAQ125 and the black one displays the signal from the reference system. As can be seen from the amplitude spectrogram, FFT of the 1 MHz periodic signal is composed of a series of discrete spectral lines, all of which are integer multiples of 1 MHz. The amplitude spectrogram envelope of the acquisition data is determined by the pulse duty cycle, whose normalized amplitude is bigger than the one of the reference signal in the frequency section between 20 MHz and 62.5 MHz. Zoom of this piece is shown in the bottom graph of figure 4. It displays a regular harmonic envelope on the frequency-amplitude axis and has an offset noise above the reference signal. This graph indicates that noise suppression is necessary in the high frequency portion which is a harmonic envelope of the pulse. To analyze the system noise level and its distribution, the difference of system noise between DAQ125 and the reference system is measured in the following section.

The acquisition without input signal are extracted by using both DAQ125 and the reference system to measure and process the system noise. The display of signals based on different systems in the time and frequency domain are shown in figure 5. After calculation, the standard deviation and mean of acquisition data is 4.826 mV and 9.093 mV while the values of reference are 1.428 mV and 54.029 mV. From measurements, the system noise amplitude of DAQ125 has a wider range of floats above and below the mean of 9.093 mV which means system noise level of DAQ125 is higher than the reference system. This inference is verified by the normalized amplitude spectrogram of noise in figure 5(b). It indicates that the system noise of the DAQ125 has a higher value than the signal of the reference system in the whole frequency domain.

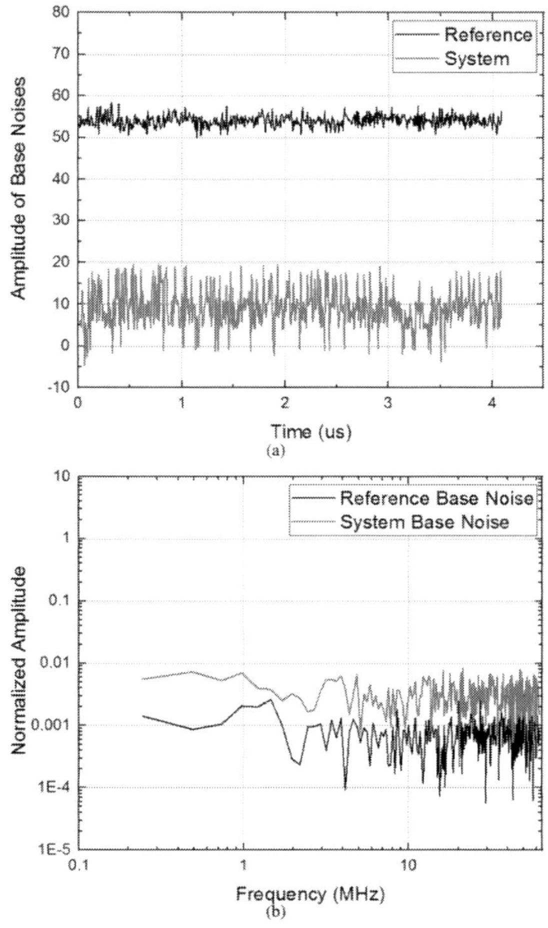

Fig. 5. Acquired data (a) and reference signal (b) shown in time and frequency domain. The total measurement is 512 samples.

A low pass filter is adopted in analog stage while a moving average filter is introduced in digital circuit block to suppress the distribution of system noise. After processing and calculation, the comparison of raw pulses and pulses from the digital filter is depicted in figure 6(a). Their γ spectrum with the scintillation detector $CeBr_3$ is shown in figure 6(b). As predicted, the moving average filter has reduced the amplitude of baseline noise. The resolution at 662 keV γ-rays is 4.7% with radiation source ^{137}Cs.

Simultaneously, radiation pulse and energy spectrum are provided in figure 7 with semiconductor detector HPGe and ^{152}Eu radiation source. A trapezoidal filter is exploited for the use with a charge sensitive preamplifier in HPGe detector. The transformation after different filters is shown in figure 7(a). The figure 7(b) shows low energy part of X-rays acquired by DAQ125 system and MCA DT5780. The orange curve in figure 7(b) is the reference from the MCA DT5780 while the blue one is the acquired data from DAQ125. The first two

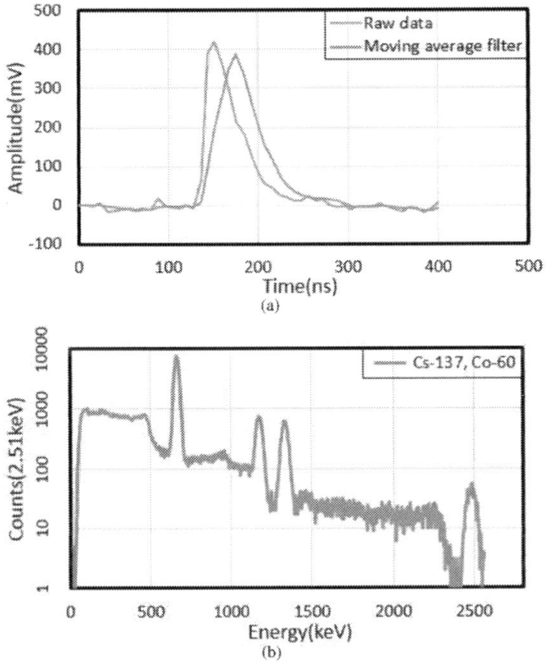

Fig. 6. Radiation pulse(a) and γ-rays spectrum(b) with a CeBr$_3$ detector.

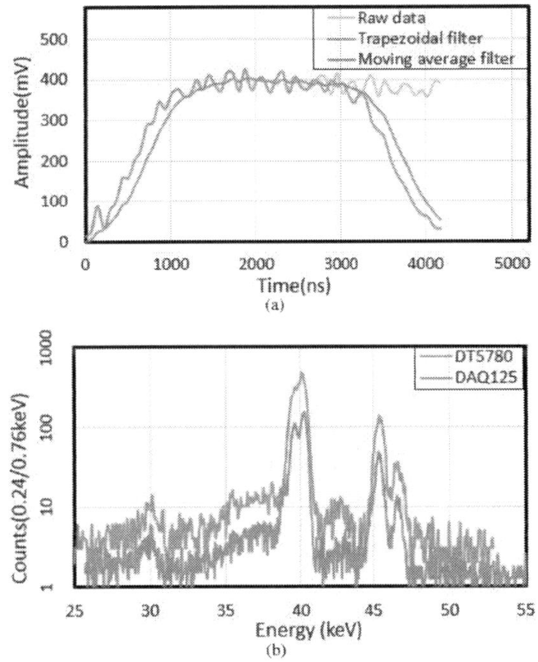

Fig. 7. Radiation pulse(a) and X-rays spectrum(b) with a HPGe detector.

peaks, 39.5 keV and 40.1 keV, are discriminated on the blue line but overlapped in orange line. Meanwhile, the acquired data from DAQ125 shows better performance at peaks 45.4 keV and 46.6 keV since they are separated much deeper than the data from MCA DT5780. As a comparative outcome, the blue plot shows a good energy resolution on the spectrum. The full width at half maximum (FWHM) is calculated at different energy using multiple radioactive source, such as ^{152}Eu, ^{137}Cs, ^{207}Bi, ^{133}Ba and ^{241}Am. Based on measured results and corresponding FWHM, energy resolution of HPGe detectors versus γ-ray energy is plotted in figure 8. It indicates that the range of energy resolution is from 0.23% to 1.00% in the energy of incident γ-ray with HPGe detector.

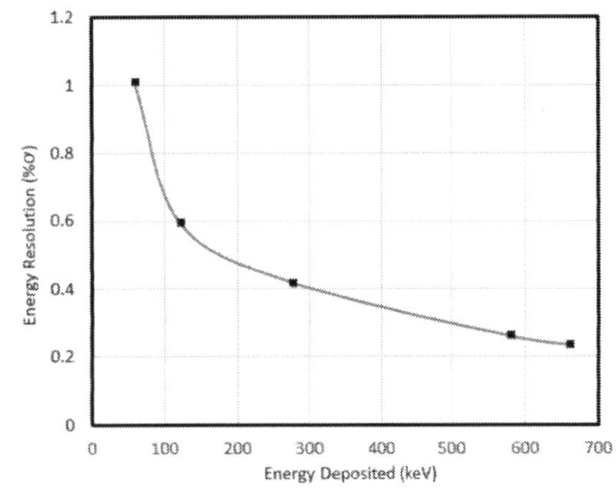

Fig. 8. Resolution range with the HPGe detector in the energy of incident γ-ray.

IV. Conclusion

The experiment and analysis results of DPP module for nuclear instrument are illustrated in this work. The measurement results of system noise is analyzed. Based on this analysis, a corresponding low pass filter and moving average filter are exploited in analog and digital circuit block respectively. Energy spectrum and resolution are presented with scintillating and semiconductor detector applied in nuclear instrumentation.

References

[1] J. B. Simões and C. M. B. A. Correia, "Pulse processing architectures," *Nucl. Instrum. Meth.*, vol. A 422, pp. 405–410, 1999.

[2] W. K. Warburton, M. Momayezi, B. Hubbard-Nelson, and W. Skulski, "Digital pulse processing: New possibilities in nuclear spectroscopy," *Appl. Radiations Isotopes*, vol. 53, pp. 916-920, 2000.

[3] J. M. Cardoso, J. B. Simões, and C. M. B. A. Correia, "A high performance reconfigurable hardware platform for digital pulse processing," *IEEE Trans. Nucl. Sci.*, vol. 51, no. 3, pp. 921–925, Jun. 2004.

[4] Y. Moline, M. Thevenin, G. Corre and M. Paindavoine, "A novel digital pulse processing architecture for nuclear instrumentation," *2015 4th International Conference on Advancements in Nuclear Instrumentation Measurement Methods and their Applications (ANIMMA)*, pp. 1-4, Lisbon, 2015.

[5] Linear Technology, "16-Bit, 125/105/80Msps Low Power Dual ADCs," LTC2185 datasheet.

EPJ Web of Conferences **225**, 01013 (2020) https://doi.org/10.1051/epjconf/202022501013
ANIMMA 2019

Role of chemical substitution in the photoluminescence properties of cerium samarium tungstates $Ce_{(2-x)}Sm_x(WO_4)_3$ $(0 \leq x \leq 0.3)$

K. Derraji [1], C. Favotto [1], J-C, Valmalette [1], S. Villain [1], J-R. Gavarri [1], G. Nolibe [2], A. Lyoussi [3] and F. Guinneton [1*]

[1]*Université de Toulon, Aix-Marseille Univ., CNRS, IM2NP, France*

[2]*CeSigma S.A, Le Pradet, France*

[3]*CEA Cadarache, St Paul lez Durance, France*

**contact:* guinneton@univ-tln.fr

Abstract.

In the general framework of the development of materials with tunable photoluminescence, a series of cerium samarium tungstates $Ce_{(2-x)}Sm_x(WO_4)_3$ with $x \leq 0.3$ was synthesized by a coprecipitation method followed by thermal treatment at 1000 °C. The polycrystalline compounds were characterized by X-ray diffraction, scanning electron microscopy and photoluminescence experiments. In the present work, the objective would be to determine the role of PL emitting centers in the variations of PL intensities. Firstly, Rietveld analysis showed a decrease of cell parameters and confirmed that a solid solution was obtained. Diffraction profile analyses showed that structural distortions increasing with composition x were observed: they were ascribed to difference in cation sizes of Ce^{3+} and Sm^{3+}, and to defects generated during crystal growth. The photoluminescence (PL) spectra were obtained under X-Ray (45 kV-35 mA) and UV (364.5 nm) excitations. Two PL emissions of Ce^{3+} were observed only under UV excitation. Four PL emissions of Sm^{3+} were observed under UV and X-ray excitations, and their intensities decreased with increasing composition x. Two additional transitions were observed under UV and X-ray excitations: they were attributed to oxygen vacancy defects. In the range 800 to 1000 nm, an increasing IR emission is observed: it was
ascribed to emissions due to other oxygen vacancies. The main results are reported in Table 1. The chromaticity diagram (see Figure 1) showed that the colors associated with PL responses vary with Sm composition and excitation energies. This offers the opportunity to develop materials with tunable PL. To better understand this complex behavior, now, we plan to study the solid solution in the composition range x>0.3.

Table 1: Photoluminescence emissions under UV and X-ray excitations observed in the series $Ce_{2-x}Sm_x(WO_4)_3$. Transitions generated by presence of oxygen vacancies Vx , V° and V°°.

UV excitation		**X-ray excitation**		**Attributions**
Wavelength (nm) Color	**Energy (eV)**	**Wavelength (nm) Color**	**Energy (eV)**	**Transitions Ce^{3+} and Sm^{3+}**
497.75 Green	2.49	No emission		Ce: $^2T2g\,/\,^2D_{3/2} \rightarrow {}^2F_{7/2}$
524.61 Green	2.36	No emission		Ce: $^2T2g\,/\,^2D_{5/2} \rightarrow {}^2F_{5/2}$
564.33 Yellow	2.19	563 Yellow	2.20	Sm: $^4G_{5/2}$ -> $^6H_{3/2}$
600.50 Orange	2.064	599 Orange	2.070	Sm: $^4G_{5/2}$ -> $^6H_{7/2}$
616.56 Orange	2.013	617 Orange	2.009	V°* → V°
643.2 Red	1.927	644 Red	1.925	Sm: $^4G_{5/2}$ -> $^6H_{9/2}$
667 Red	1.859	669 Red	1.853	$V^{x*} \rightarrow V^x$
705 Dark red	1.759	705 Dark red	1.759	Sm: $^4G_{5/2}$ -> $^6H_{11/2}$
900 Large IR	1.38	880 Large IR	1.41	V° → holes O2p
950 Large IR	1.30	930 Large IR	1.33	V^x → holes O2p

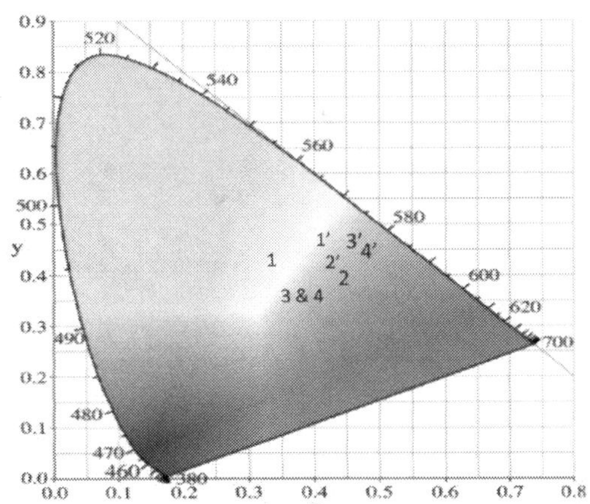

Figure 1: Chromaticity diagram for $Ce_{2-x}Sm_x(WO_4)_3$ system under both excitations.
Under X-ray excitation, points are noted as **1** for x=0, **2** for x=0.1, **3** for x=0.2 and **4** for x=0.3.
Under UV excitation points are noted as **1'** for x=0, **2'** for x=0.1, **3'** for x=0.2, **4'** for x=0.3.

Key words: samarium cerium tungstates, chemical substitution, structural distortions, photoluminescence.

EPJ Web of Conferences **225**, 02001 (2020) https://doi.org/10.1051/epjconf/202022502001
ANIMMA 2019

Self Powered Neutron Detectors with High Energy Sensitivity

S. Fiore[a], O. Aberle[b], M. Angelone[a], M. Calviani[b], F. Di Giambattista[a,c], L. Lepore[d], M. Nyman[e], M. Pillon[a], A. Plompen[e]
[a] ENEA, Department of Fusion and Technology for Nuclear Safety and Security, Frascati, Italy; [b] CERN, Geneva, Switzerland; [c] Università degli studi di Roma "Tor Vergata", Roma, Italy; [d] ENEA, Department of Fusion and Technology for Nuclear Safety and Security, Roma, Italy; [e] EC JRC Geel, Belgium;

THE Roadmap to Fusion Energy poses several technological challenges, many of them related to neutrons generated in the fusion reactions. Fusion reactions between deuterium and tritium will take place in tokamaks releasing a large amount of energy. The generated nuclear power is mainly transferred to the neutrons produced in the reactions, and then converted to thermal power and then to electricity. As a consequence, a high intensity fast neutron flux ($>10^{14}$ n/cm^{2}s) is expected to hit the inner walls of the tokamak. ITER will be the first fusion reactor to be built in which neutron fluxes of the order of 10^{14} n/cm^{2}s will be reached on the first wall surface, decreasing with depth. Monitoring this neutron flux is one of the key diagnostics to understand plasma burning conditions, and with such flux intensity this is a challenging task, also considering the environment conditions such as high temperatures above 200°C. Owing to the high neutron fluxes, radiation damage of structures and devices are one of the main challenges for future fusion reactors design. Several displacements per atoms (DPA) and up to tens of DPA are foreseen on the exposed materials and structures. DEMO is the ITER's successor, and it will be the first prototype of a fusion power plant. In order to evaluate the damage to structural materials in DEMO, the Roadmap to Fusion Energy has foreseen the construction of a dedicated test facility, IFMIF-DONES, which is planned to be built in Escuzar, 30 km far from Granada (Spain).

DONES is based on the acceleration of a deuteron beam up to 40 MeV, 125 mA current in continuous wave mode, with 5 MW beam average power. The rectangular beam (approximately 20 cm$\times$ 5 cm) hits a liquid lithium target flowing at 15 m/s to dissipate the beam power. A neutron flux of 10^{14} n/cm^{2}s, with a broad peak at 14 MeV, is produced by stripping nuclear reactions, reproducing the expected conditions in DEMO [1]. The materials samples to be tested are positioned right behind the lithium target. The neutron flux in the test cell must be monitored. The detectors to be used to this purpose must withstand the high incident neutron flux at high temperatures (between 200°C and 500°C), working consecutively for almost one year without the possibility to access the irradiation cell. Almost any neutron detector currently available would not work in these conditions for such long time, therefore new ad-hoc detectors must be developed.

The DONES collaboration identified two possible candidates for neutron flux monitoring in the test cell: micro fission chambers and Self Powered Neutron Detectors (SPNDs). Although micro fission chambers have the right sensitivity, have small dimensions and are radiation-resistant, they can't work at such high temperatures. On the contrary SPNDs, aside from their radiation hardness and reduced dimensions, are also capable of working at high temperatures, and seem therefore one of the most suitable option for the DONES sample irradiation cell.

SPNDs are electronic devices that exploit the neutron-induced activation of specific materials, which then decay through electron emission: in an SPND, the activated material acts as a first electrode called emitter. A second electrode (collector) placed at close distance from the emitter catches the beta electrons. These two electrodes, connected in a circuit, act as a current generator whose current is proportional to the emitter activation, which is in turn proportional to the incoming neutron flux (Fig.1). SPNDs are used for in-

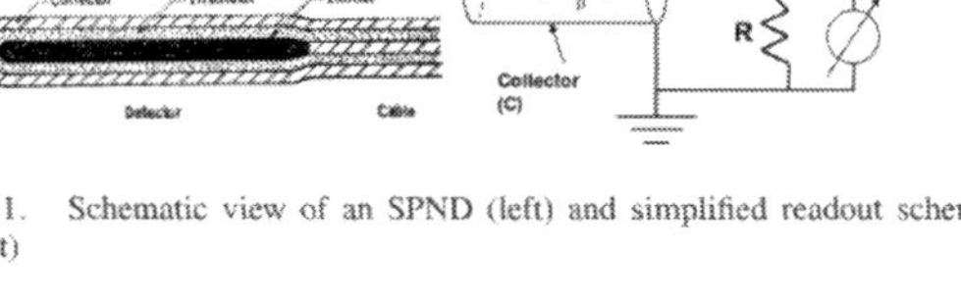

Fig. 1. Schematic view of an SPND (left) and simplified readout scheme (right)

core neutron flux measurements in fission reactors, exploiting thermal neutron induced activation. The typical time response ranges from milliseconds to seconds, and the sensitivity is around 10^{-21} A/(n/cm^{2}s); the assembly is coaxial with an outer diameter of 3 mm and few cm length. Due to the low current over neutron flux ratio, SPNDs are used for the measurement of neutron fluxes above 10^{10} n/cm^{2}s. Their use can be extended over several orders of magnitude above, where other kind of detectors fail due to radiation effects (NIEL, SEE, DDD) especially when exposed for long periods of time. As an SPND does not require bias voltage to work, these detectors are easy to operate.

Different types of reactions can take place in the electrodes and insulator, inducing a current through the emission of

https://doi.org/10.1051/epjconf/202022502001

electrons [2].

- (n,β): after a neutron capture on an emitter nucleus, the nucleus is unstable and beta decays emitting an electron. The decay takes place after a time that depends on the half life of the activated isotope, producing a *delayed* signal. When the signal reaches its saturation value, the current is proportional to the incident neutron flux.
- (n,γ): after a neutron capture the nucleus reaches a stable state through the instantaneous emission of gammas. The gammas interact with the medium through photoelectric effect, Compton effect and pair production, with the consequent emission of secondary electrons. This process produces a *prompt* signal that is proportional to the incident neutron flux.
- (γ,e^-): those elements which are sensitive to radiative capture can also produce a signal through the direct interaction with external gammas, producing a *prompt* signal that is proportional to the incident gamma flux.

The polarity of the current produced by these processes depends on which electrode is involved in the primary interaction. Electrons coming from the emitter that stop in the collector produce a positive current signal, while electrons produced in the collector that stop in the emitter produce a negative current signal. The net current is the algebraic sum of all these contributions. Along with the main signal formation interactions, also charged particles such as electrons can induce a current in the SPND, whose contribution to the detector response must be taken into account.

In order to extend the use of SPNDs for high energy neutron flux measurement, a new set of emitter materials have been studied and proposed. The operation of SPNDs in fusion environments like DONES imposes some constraints on the choice of the materials. Low melting point materials must be avoided, since the temperatures can reach 500°C. There will not be the possibility of accessing the test cell for almost one year, therefore attention must be paid in order to avoid a sensitivity change over time. The value of the neutron capture cross section in the energy region of interest shouldn't be neither too high (to avoid the *burn-up* effect due to transmutation, that reduces the sensitivity) nor too low (so that the sensitivity to neutrons is high enough). The half life of the activated nucleus produced by the neutron capture must be small enough to obtain a fast response, with the signal reaching the saturation value within a maximum of few minutes. Also, the half lives of the daughter nuclei of the activated nuclei must be either very short (similar or less than that of the parent) or very long (at least equal to the duration of the measurement), otherwise the response of the detector would change during the time of the data taking. In light of these considerations two prototypes of SPNDs were produced: one with a rhodium emitter (a common material used in fission reactors neutron flux monitoring) and one with an aluminum alloy emitter. Also one dummy SPND with no emitter material was produced to study the electronic background noise. Rhodium has a neutron radiative capture cross section of $\sim 10^{-1}$-10^{-3} barn between 100 keV and 15 MeV, while the cross section for aluminum in the same energy region is roughly constant, with a value of the order of 10^{-3} barn (Fig. 2) [3]. Both chosen elements would produce a delayed response in the SPNDs due to the beta decay of ^{104}Rh and ^{28}Al.

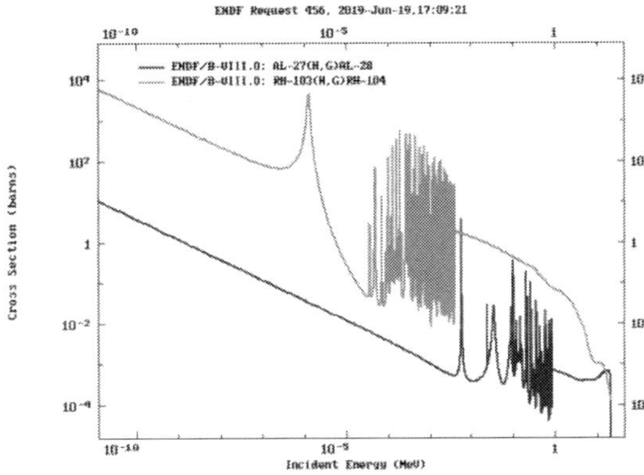

Fig. 2. Radiative capture cross section for ^{103}Rh and ^{27}Al, that constitute 100% of natural rhodium and aluminum, the chosen elements for the emitters of the SPNDs under test [3].

The SPNDs had to be tested in an environment similar to the one expected in DONES, that is a mixed neutron and gamma field, with a neutron spectrum up to 40 MeV, in an accelerator-driven neutron source with a similar background radiation conditions.

The two prototypes were firstly tested at the GELINA neutron time-of-flight facility of the JRC in Geel (Belgium). GELINA is a photoneutron source based on the acceleration of an electron beam on a rotating uranium target. The electron bunches are accelerated by a linac up to 140 MeV. The bunches are then compressed and exit the accelerator with a width of $\sim$1 ns. The repetition rate of the bunches can be varied between 50 Hz and 800 Hz, leading to different average neutron production rate. The electrons interact with uranium producing bremsstrahlung and generate neutrons by photonuclear reactions. Each electron pulse produces 4.3×10^{10} neutrons (with an average production rate of 3.4×10^{13} n/s) with an energy distribution from subthermal up to about 20 MeV [4].

The SPNDs were placed 5 cm away from the target, on a support structure specially designed to hold the detectors as close as possible to the neutron production point (Fig. 3), substituting the moderator that is used for time-of-flight measurements. Two support structures of different materials were used in the test, one in lead and one in polyethylene.

A full MCNP simulation of the interaction between the beam and the target was made, considering the maximum repetition rate of 800 Hz. The resulting expected neutron flux is 6×10^{10} n/cm^{2}s both with the lead and PE support, but with a different energy distribution due to the moderating effect of polyethylene. The expected gamma flux is 9×10^{11} γ/cm^{2}s with the lead support, while it is 6×10^{12} γ/cm^{2}s with the PE support. This difference is due to the higher interaction cross section of gammas in lead with respect to polyethylene. Also the Warren model [6], [7] for SPNDs current production was used to calculate the expected current produced by the

Fig. 3. GELINA experimental setup. The support structure for the SPNDs and the exit of the beam line are visible.

simulated incident neutron and gamma fluxes. The calculated current include the contributions from the prompt and delayed response of the detectors to the incident neutrons and the signal induced by the incident photons. The calculated current is of the order of 10 pA for the aluminum SPND, with a slight difference between the lead and PE case. The rhodium SPND calculated current was found to be of the order of hundreds of pA, more than 10 times greater than the aluminum SPND current, due to the much higher cross section.

Three Keithley picoammeters (models 6485, 6487 and 6517A) were used to measure the current of the three SPNDs. Different runs were made varying the repetition rate of electron pulses on the target, starting from 50 Hz and increasing it by a factor 2 up to 800 Hz, doubling each time the neutron yield. The neutron production was monitored with a calibrated neutron detector located in the target hall. Nickel activation foils were used to measure the flux on the support structure. The flux at the maximum beam intensity turned out to be 5.9×10^{10} n/cm^{2}s, consistent with the simulated value 6×10^{10} n/cm^{2}s.

The current from both the SPNDs was found to be proportional to the neutron rate over the full range of possible values obtained with the variation of the repetition rate.

The time response of the rhodium SPND with the polyethylene support structure was studied in the time window corresponding to the transition between a constant neutron flux and the beam switching off. As expected, the response is delayed, and the signal decays exponentially. The decay of the signal is consistent with the hypothesis of neutron activation of the ^{103}Rh nuclei of the emitter. Note that, while a prompt response is the sum of the contributions from the gammas of the radiative capture, the external gammas and background charged particles, a delayed response can only be due to neutron capture.

The measured signal amplitude differs partially from the expected values. Also the polarity of the current is different from what one would expect. If the dominant process in signal formation is neutron capture, the produced current should be positive under any condition, because it would be induced by the flowing of electrons from the emitter to the collector. The aluminum SPND current is instead always negative without any delay in the response to a sudden change in the neutron flux (Fig. 4). This leads to the conclusion that some processes other than beta activation of the emitter contribute the most in the formation of the signal. The rhodium SPND produced a prompt and negative signal with the lead support, with a much smaller amplitude than the one measured with the polyethylene support. When the support structure was made of polyethylene, the rhodium SPND signal was instead delayed with a positive polarity. This observations may be explained by the presence of external electrons that interact with the detectors inducing a negative current, that in some cases dominates over the other components. This hypothesis could be confirmed by the results of a dedicated MCNP simulation, which gave as a result a current due to external electron fluxes of the order of 10^{-10} A with negative polarity.

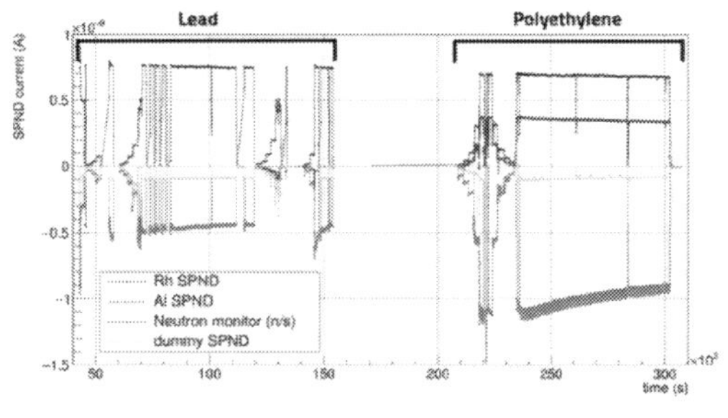

Fig. 4. Data taken at GELINA facility. The rhodium SPND current is shown in blue, the aluminum SPND current is shown in red, and the dummy SPND current is shown in green. The average neutron rate is shown in black. The first part of the data was taken with the lead support, while the last part was taken with the polyethylene support.

The time resolution of the neutron beam at GELINA didn't allow to disentangle the prompt and delayed components of the signal, because of the sampling time (3 s) being much higher than the spacing between neutron pulses (that was a maximum of 20 ms).

The SPNDs were then installed at the n_TOF facility at the European Laboratory for Particle Physics (CERN) for a new test. The n_TOF facility is part of the fixed target experimental program at the CERN accelerator complex. The neutrons are produced by the spallation of 20 GeV/c protons coming from the Proton Synchrotron (PS) on a pure lead target. The high intensity neutron pulses are produced by 7 ns (1 sigma) wide proton bunches of 7×10^{12} p every 1.2 seconds (or multiples of this interval). The neutrons have a wide energy spectrum ranging from 10 meV up to several GeV [5]. There are two experimental areas to perform measurements with the time of flight techniques, EAR1 and EAR2, placed at a distance of about 200 m and 20 m from the target respectively, in which the flux is of the order of 10^4 n/cm^{2}s and 10^6 n/cm^{2}s. In order to get a neutron flux of about $\sim10^{11}$ n/cm^{2}s with the main component in the energy range between 1 MeV and 10 MeV, it was decided to install the SPNDs in the area surrounding the target, inside the concrete shielding (Fig. 5).

The DAQ was installed far from the detectors, in a non-

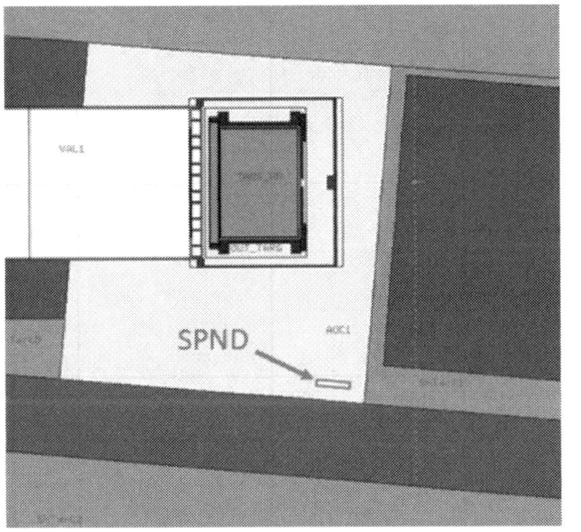

Fig. 5. FLUKA Monte Carlo model of the target area at n_TOF. The SPNDs were installed in the area showed in the red box.

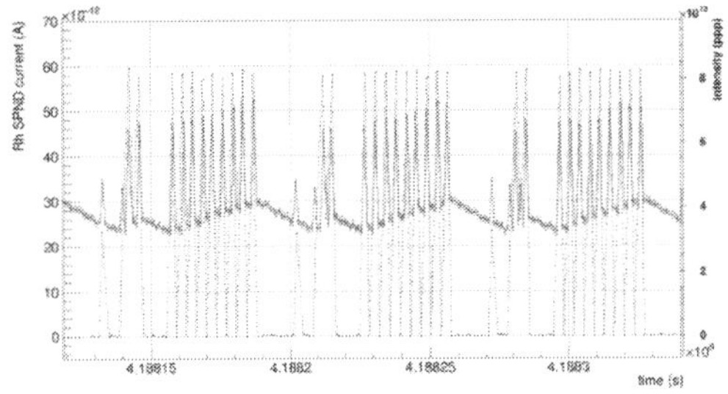

Fig. 6. Rhodium SPND signal shape at n_TOF. The proton pulse intensity is shown with black points connected by a dashed line, the Rh SPND current is shown with blue points. The superposition of a prompt and a delayed signal component is clearly visible.

radiation controlled area at the Intersecting Storage Rings (ISR). The SPNDs were connected to the readout system through three cables 100 m long. Since the target area is not directly accessible, the detectors were pushed through the concrete blocks chamfers, only sustained by the semi rigid mineral cables. Two readout systems were used: one with the Keithley picoammeters employed in the GELINA run, and one with a 4-channel CAENels picoammeter (model AH401D).

The test beam lasted over 50 days, between September and November 2018. During this time the detector response under different beam conditions were tested, including Machine Development stops, technical stops and short beam interruptions, as well as different proton pulse intensities, that have resulted in a wide range of average neutron production rates.

The 10 ms time resolution of the CAENels picoammeter allowed us to resolve the individual current peaks produced by the prompt emission of the target following the proton pulses. In such conditions it was possible to distinguish the prompt and the delayed components of the SPNDs signal.

The rhodium SPND showed a higher sensitivity than the aluminum SPND. The acquired signal is the superposition of a prompt and a delayed component (Fig. 6). The prompt component is represented by the sharp current peaks corresponding to a proton pulse hitting the target, which are produced by the prompt emission of photons and charged particles. The delayed component is visible as a slow variation of the baseline current with the change of the average neutron flux caused by the different sequence of proton pulses on the target. The delayed component is entirely due to neutron activation, and it was therefore isolated from the prompt current peaks in order to study the neutron response of the rhodium SPND.

The time response of the detectors was studied in the time windows in which the beam is switched off. The baseline current decays exponentially, consistently with the data from GELINA and the hypothesis of the beta activation of ^{103}Rh. Because of the particular position of the detectors (currently not employed for measurements), no reference value of the neutron flux was availablee. The only measurable quantity which is proportional to the neutron flux is the average proton current on the target. In order to study the linearity of the signal, the rhodium SPND baseline current was put in relation with the corresponding average proton current, and a linear relation was found. The measured current is also consistent with the analytical calculations based on the Warren model [6], [7], that predict a delayed component of the order of 10^{-11} A in the simulated flux conditions.

According to the Warren model the prompt component of the signal of the aluminum SPND is one order of magnitude lower than the rhodium SPND current, as it was confirmed by the data. Also, the expected delayed component is of the order of $\sim 10^{-15}$ A, much lower than the noise level. This is consistent with the measurements, that show sharp current peaks corresponding to the proton pulses, but no change in the baseline current is visible. This could be explained by the low cross section for neutron capture on aluminum in the energy region of the actual neutron spectrum at n_TOF, which was simulated with FLUKA, resulting softer than expected.

The performances of the SPNDs and the experience of the first on-target experiment at GELINA and n_TOF could open new scenarios. SPNDs are planned to be installed near the third generation n_TOF spallation target to monitor the neutron production independently from the detectors in the experimental areas. This installation could be interesting also for the fusion community, since it will give the chance to perform a long-term test of the detectors before the installation of SPNDs in the DONES test cell.

Before the installation of the SPNDs at n_TOF new Monte Carlo simulations are needed, in order to understand better the SPNDs response in neutron and gamma mixed fields. The development of a new Monte Carlo model of signal formation in the detectors will allow to design new SPND prototypes also for tests at research reactors and other neutron sources. Particular attention will be paid to the materials chosen for the emitter, which must work at high temperatures and produce a signal that is fast enough to monitor the changes in the neutron flux, along with being sensitive to high energy neutrons.

In addition to these considerations, also some methods to reduce the gamma component of the signal will be further investigated.

REFERENCES

[1] Ibarra, A et al., Nucl. Fusion 58 (2018).
[2] Goldstein, N.P. et al., IEEE Transactions on Nuclear Science 1 (1979).
[3] Brown, D.A. et al., Nucl. Data Sheets 148 (2014) 1-142.
[4] Enc, D. et al., NIM A618 1-3 (2010) 54-68.
[5] Guerrero, C. et al., The European Physical Journal A 49 (2013), 1-15
[6] Warren, H.D., Nucl. Science and Engineering 48 (1972), 331-342
[7] Warren, H.D. et al., Nucl. Science and Engineering 54 (1974), 395-415

EPJ Web of Conferences **225**, 02002 (2020) https://doi.org/10.1051/epjconf/202022502002
ANIMMA 2019

Enhancement of irradiation resistance and experimental research of compact electronics for ITER Radial X-ray Camera

Kaiyun CHEN, Hongrui CAO*, Jinlong ZHAO, Xiuli SHENG, Shi LI, Liqun HU

Institute of Plasma Physics, Chinese Academy of Sciences, Hefei 230031, China.

*Corresponding author: caohr@ipp.ac.cn

***Abstract*— International Thermonuclear Experimental Reactor (ITER) Radial X-ray Camera (RXC) is installed in the equatorial port #12 (middle drawer) to measure the poloidal profile of the plasma x-ray emission with high spatial and temporal resolution. According to the irradiation test results with Co-60 gamma ray source, Cf-252 neutron source and the result in Deuterium - Tritium (DT) accelerator neutron irradiation experiment for RXC electronics before, the weak parts of the circuit have been improved by radiation resistance design such as using radiation-hard power chip and optimizing the Field Programmable Gate Array (FPGA) code. Other factors related to radiation, such as easiness of maintenance, compact electronics structure and remote state monitoring are also considered in the design. The performance of the electronics and chassis has been preliminarily verified in Experimental Advanced Superconducting Tokamak (EAST) experiments showing resistance to interference from discharge environment. Besides, the irradiation resistance of electronics was investigated in high fluence neutron irradiation experiment in fission reactor, where the detector, pre-amplifier, mid-amplifier and cables under test received the accumulated neutron fluence over $9.89\times10^{13}n/cm^{-2}$ (1MeV equivalent), and the quality of output signal was monitored in real time. It's found after the experiment that the pre-amplifier worked well, while the detector and mid-amplifier were partly damaged.**

***Index Terms*— Amplifier, Chassis and shielding, Neutron radiation test, X-ray camera.**

I. INTRODUCTION

International Thermonuclear Experimental Reactor (ITER) Radial X-ray Camera (RXC) is installed in equatorial #12 (middle drawer) to measure the poloidal profile of the plasma x-ray emission with high spatial and temporal resolution [1, 2]. To have a whole view of the plasma and keep the detectors from the neutron flux, this camera will be separated into two individual cameras: internal camera and external camera as shown in Fig.1 [1]. Internal camera will be installed inside the port plug and its detectors will be mounted on upper side and bottom side of DSM (Diagnostic Shielding Module) to have the edge views of the plasma. External camera outside of the port plug will have views to cover the core plasma and the detector arrays will be installed in secondary vacuum chamber. In current stage, the detector and electronics used for internal camera are still under development. As a result, only electronics for external camera will be discussed in detail in this paper, although the electronics for internal camera will be quite similar. According to the previously reported irradiation test results with Co-60 gamma ray source, Cf-252 neutron source and the result in Deuterium - Tritium (DT) accelerator neutron irradiation experiment for RXC electronics [2], the weak parts of the circuit are power chip and FPGA, and have been improved by radiation resistance design which is described in the following sections. Meanwhile, considering the nuclear environment (to reduce worker exposure to radiation) and space limitation in port cell (to allow more space to put nuclear shielding), easiness of maintenance, compact electronics structure and remote state monitoring are designed. In addition, design of magnetic shielding, irradiation experiment in reactor and integration test in EAST tokamak are presented.

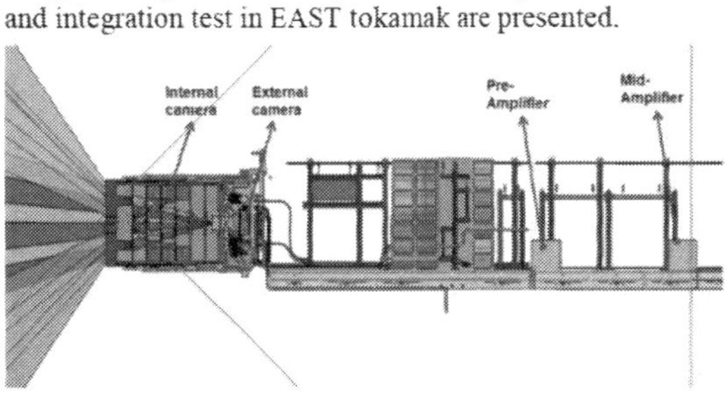

Fig.1 3D model of the RXC components

II. ELECTRONICS DESIGN

As shown in Fig.2, the electronics mainly includes detector, pre-amplifiers, mid-amplifiers and I&C (Instrument and control). The pre-amplifiers will be placed in the front of port cell 12 close to the bio-shield wall and mid-amplifiers in the end of port cell 12 close to the mashing box. Extra-low-voltage power supply will be placed in the corner of diagnostic building. I&C (including DAQ (Data AcQuistion), fast and slow controllers) will be housed in the cubicle area of diagnostic building.

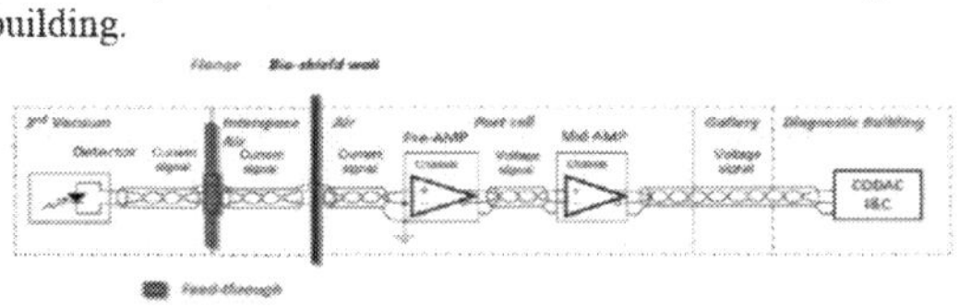

Fig.2 RXC electronics layout

For electronics, the actual measured quantity is chord-integrated soft x-ray intensity with signal current varying from 10nA to 10uA and maximum time resolution 20 kHz. To enhance the flexibility of electronics to both monitor low performance and high performance plasmas, the gain of pre-amplifier should be fixed at the order of magnitude 1×10^{6}V/A and gain of mid-amplifier is designed to be changeable. After a long distance transmission (about 20 meters) from detectors, the current signal from detector is amplified to voltage signal by the trans-impedance pre-amplifier and the output will flow into mid-amplifier which is programmable and is controlled by I&C. The output signal of mid-amplifier will be sent to the DAQ system and the mid-amplifier can drive long distance cable whose length is up to 100 meters.

Fig.3 shows the electronics and I&C architecture. In order to control the gain of mid-amplifier, the radiation resistant fiber to ensure communication is used to connect the I&C cubicle in the diagnostic building with mid-amplifier. When the gain of mid-amplifier needs to be changed, the fast controller in the I&C cubicle will send the digital electronical signal to mid-amplifier chassis with communication mode of RS232.

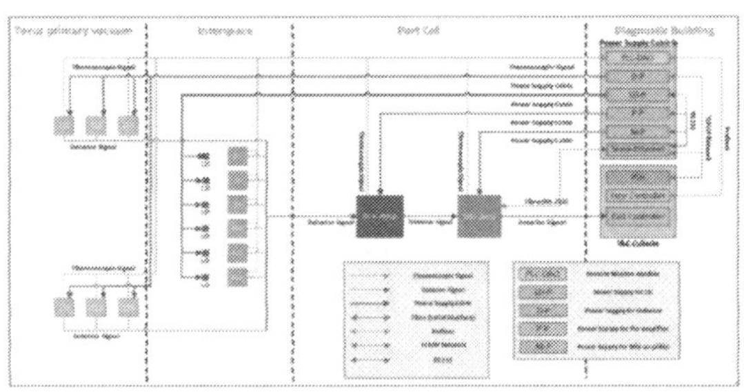

Fig.3 The electronics and I&C architecture

The performance of pre-amplifier for weak current signal will influence the availability of the whole RXC electronics system. Because of the strong electromagnetic interference in ITER environment, the far-front signal transmission, improvement of signal to noise ratio, resistance to electromagnetic interference and other key technologies need to be investigated deeply.

The principle design of pre-amplifier is shown in Fig.4 and Fig.5. The pre-amplifier circuit is divided into two stages. The first stage is transimpedance amplification with single-ended input and differential output and mainly uses OPA2140 chip [3]. The second stage is voltage amplification and differential voltage driving which mainly uses THS4131 [4]. In order to improve the nuclear radiation resistant performance of the power supply module, radiation resistant power chips are used in the pre-amplifier design (RHFL4913A and RHFL7913A) [5, 6].

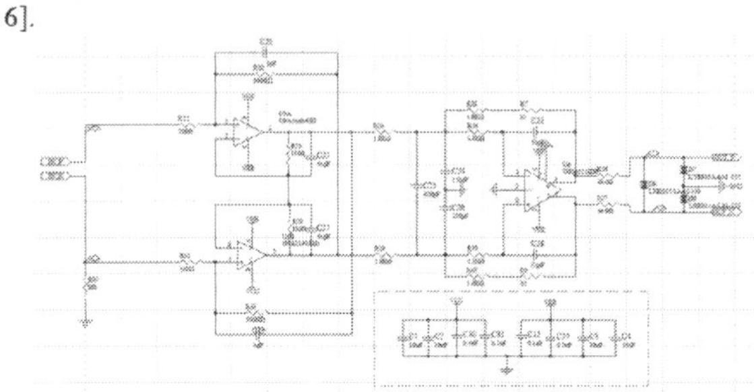

Fig.4 Schematic of differential amplifier part

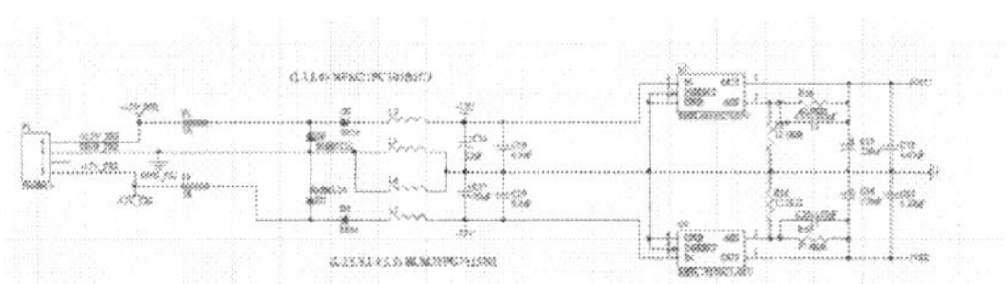

Fig.5 Schematic of power supply part

The principle design of mid-amplifier is shown in Fig.6. One master mid-amplifier and nine slave mid-amplifier boards will be used in the mid-amplifier chassis. The master one includes five parts: 16-channel PGA (programmable gain amplifier), fiber-to-RS232 converting module, FPGA, self-test function part and analog regulated power module. In these five parts, FPGA is a core processing module which can receive gain adjustment command from fiber-to-RS232 module which receive the optical control signal from fast controller in I&C cubicle and control 16-channel PGA to change the amplification. FPGA can also give a monitoring signal such as operating status to I&C for checking. The third function of FPGA is to control the switch for mid-amplifier self-test. Self-test module includes 16 switches and a self-test signal generator which is a reference voltage source. When system operates normally, FPGA controls the switch to connect the original signal from pre-amplifier to the input end of PGA in mid-amplifier. If self-test of mid-amplifier is required, FPGA will change the switch to connect the self-test signal generator to the input end of PGA in mid-amplifier. For final design and electronic test, 10M08SAU169C8 FPGA is used from company of Intel [7].

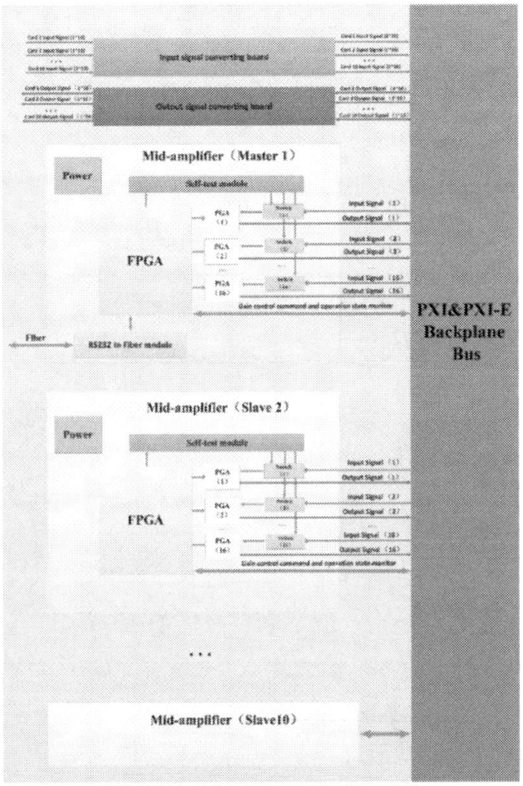

Fig.6 Principle structure of mid-amplifier

In PCB design, pre-amplifier and mid-amplifier are based on PXIE compact architecture as shown in Fig.7 and Fig.8. The multi-channel consistency, heat dissipation and crosstalk have been specially considered in PCB design.

Fig.7 Prototype of pre-amplifier

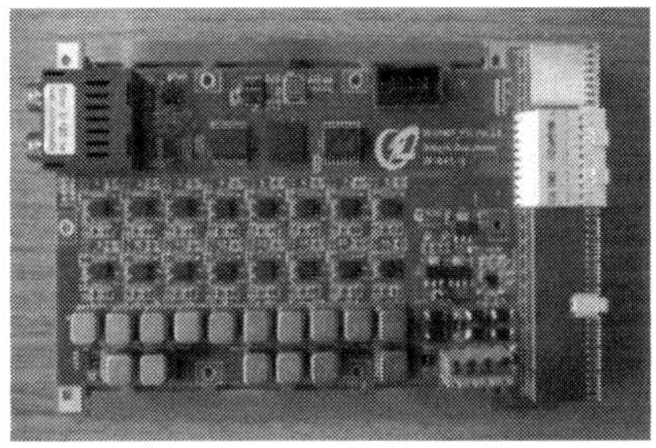

Fig.8 Prototype of mid-amplifier

The state machine is used in the design to remotely control and monitor the working state of mid-amplifier. If a problem is found in mid-amplifier, the mid-amplifier can be restarted by soft reset or hard reset to improve the availability. These designs will reduce human in-situ operations in port cell and consequently occupational exposure. As shown in Fig.9, the state machine has five states: initialization, configuration, self-test, running, and error.

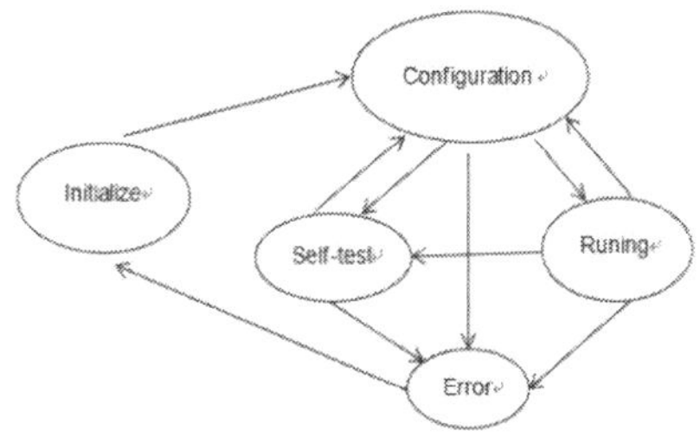

Fig.9 The flow chart of state for FPGA code of mid-amplifier

After the mid-amplifier cards are powered on, the state machine firstly comes to the initialization state, during which all the registers in FPGA are reset, the default gain configuration are restored, the RS232 communication from master mid-amplifier to I&C system are set up, and the communication from master mid-amplifier to the slave mid-amplifier are established. If the initialization state has been prepared, the state machine jumps to the configuration state automatically. In the configuration state, I&C system will send the gain configuration command to master mid-amplifier, and master mid-amplifier will send the command to the other slave mid-amplifier. When the configuration is successful done, the mid-amplifier will send back the information showing configuration is ok. And then I&C will send the command to control the state machine to jump to the self-test state or the running state. In the running state, I&C system can monitor the gain. If the gain of mid-amplifier need change, I&C system will send the command to control the state machine to come back to configuration state. In self-test state, I&C system can control the mid-amplifier to use the internal reference signal source in mid-amplifier circuit board to test signal amplification of the mid-amplifier. When an error occurs in running state or self-teat state, the state machine will come to error state and

https://doi.org/10.1051/epjconf/202022502002

send the information of error to I&C system. When I&C system receives the information of error, it will send the command to control the state machine to change to the initialization state.

Soft and hard reset of the mid-amplifier FPGA system is realized in the design. If there is any error of FPGA generated by SEE (Single Event Effect) or other reasons, I&C control can select soft or hard reset of FPGA. If the hardware reset is selected, I&C will send a shutdown command to turn off the ELV (Extremely Low Voltage) power in power supply cubicle. After stopping for a few seconds, he turn-on command is sent. The mid-amplifier will restart, and the FPGA will automatically jump to the initialization state, and perform the initialized work. If choosing software restart, I&C will directly send restart command to the FPGA in mid-amplifier, and the FPGA will jump back to initialize the state and restart the process.

III. Chassis and shielding design

The pre-amplifiers and mid-amplifiers are housed in chassis. The chassis as shown in Fig.10, which is designed based on PXI/PXIE mixed bus architecture, can separate in space the circuit board and input/output cables so that the circuit boards can be hot pluggable without removing cables in advance, and the efficiency of maintenance can be greatly improved. Considering the standard of IEC 60529, the level of chassis is over IP65 which is expected to prevent from dust and jetting water.

The advantages of the chassis are providing reliable and stable integrated structure, EMC shielding, heat dissipation and efficient maintenance.

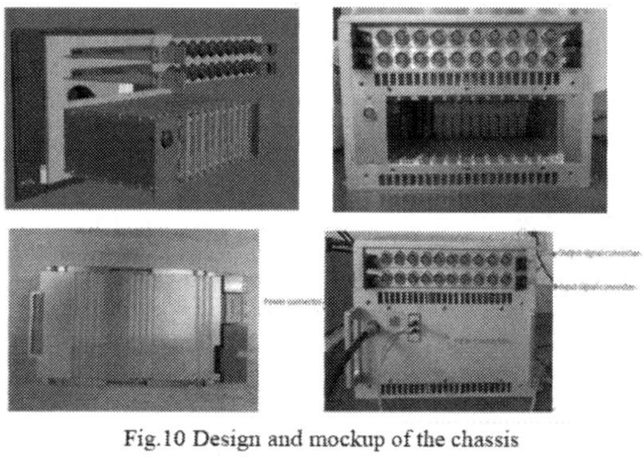

Fig.10 Design and mockup of the chassis

To increase EM-shielding performance and to enhance heat removal, metal plates were used to wrap the pre-amplifier and mid-amplifier.

The power of each card is estimated to be 11.2W by summing the power of 32 chips. The heat removal of the card with air/water was simulated, setting air temperature as 30°C. As shown in Fig.11, air cooling can keep temperature of electronics chips under 41°C, while water cooling will have better performance.

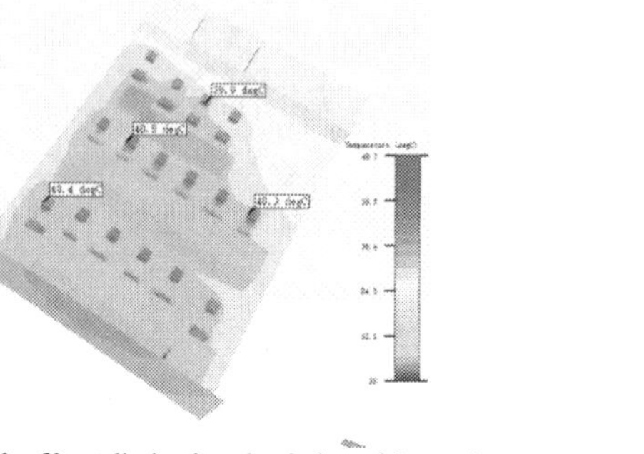

Fig.11 Result of heat dissipation simulation of the card

In addition, magnetic shielding is considered during the design of chassis. Fig.12 indicates the static magnetic field for L1 level in ITER Tokamak [8], and the stray field where the pre-amplifiers and mid-amplifiers will be located could reach respectively 120 mT and 75 mT during 'Full Scenario 17MA'.

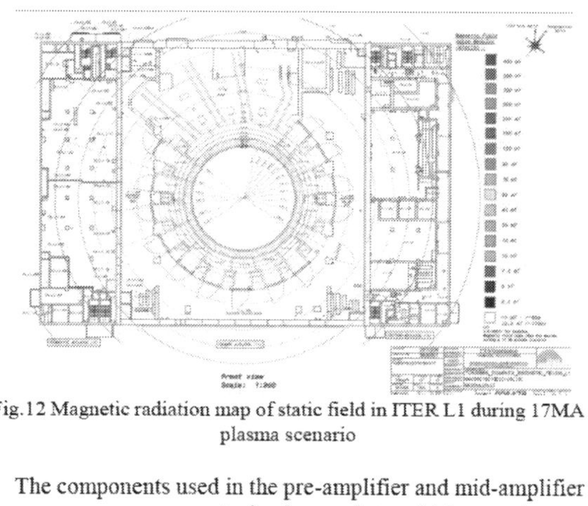

Fig.12 Magnetic radiation map of static field in ITER L1 during 17MA plasma scenario

The components used in the pre-amplifier and mid-amplifier are common components in the market, which may be not resistant to magnetic environment. As shown in Table.1 [9], the magnetic non-qualification threshold for pre-amplifier and mid-amplifier is supposed to be 5 mT. Therefore, magnetic shielding may be needed for these components. On the other hand, the electronics also need to meet the EMC requirements. As a result, customized chassis and dedicated static magnetic shielding enveloping the chassis are designed.

Table.1 EEE thresholds for magnetic field qualification

Static magnetic field	**Non critical**	**Critical**
Less than 5mT	Design with components selected for low susceptibility.	Qualification tests* at 2 times the expected induction (i.e. with a margin of 6 dB)
5mT to 10mT	Check of bill of materials against component susceptibility. In case of doubt, a qualification test will be requested.	Qualification tests* at 2 times the expected induction (i.e. with a margin of 6 dB).
Over 10mT	Qualification tests* at 2 times the expected induction (i.e. with a margin of 6 dB).	Qualification tests* at 2 times the expected induction (i.e. with a margin of 6 dB).

The static magnetic shielding is put outside the chassis. Permalloy will be used as shielding material which has high permeability and saturation, and its magnetic relative permeability is about 4×10^4 which is 10 times higher than commercial type iron. The composition of this permalloy (Model: 1J85) is shown in table.2. As shown in Fig.13, the simulation result is that 1 mm thickness of shielding is enough to attenuate magnetic field from 120mT to below 5mT. Considering the support of the structure, it is recommended to use 1.5-2 mm thickness plates for static magnetic shielding.

Table.2 Composition of permalloy

Composition (%)									
Model	C	Mn	Si	P	S	Ni	Mo	Cu	Fe
1J85	≤0.03	0.3-0.6	0.15-0.3	≤0.02	≤0.02	79-81	4.8-5.2	≤0.2	14-16

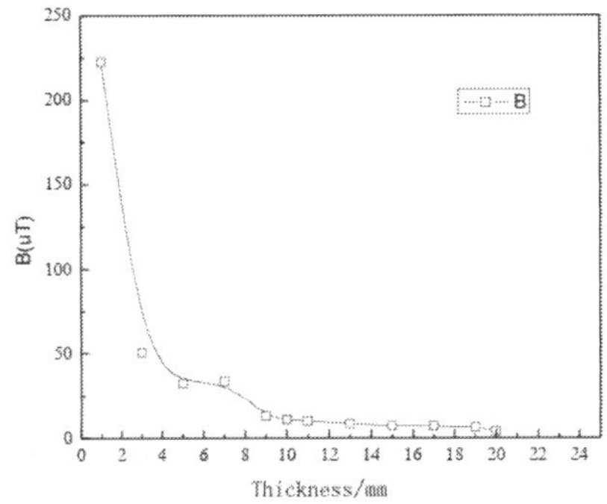

Fig.13 The static magnetic field inside the shielding box with changing thickness

IV. NEUTRON IRRADIATION TEST IN REACTOR

The RXC camera has been tested in strong neutron radiation environment in fission reactor. The reactor power can reach 20MW in full power operation. The purpose of the test is mainly to know electronics performance in strong nuclear environment. The reactor power is 100kW in this test, and the neutron flux is 5.84×10^9 n·cm^{-2}·s^{-1} (1MeV equivalent) in the radiation room where the detector, pre-amplifier, mid-amplifier and cables under test received accumulated neutron fluence over 9.89×10^{13} n·cm^{-2} (1MeV equivalent), and the quality of output signal was monitored in real time. The total radiation time is 16928 s and the ratio of n/γ is 7.7×10^9 n·cm^{-2}·rad^{-1}.

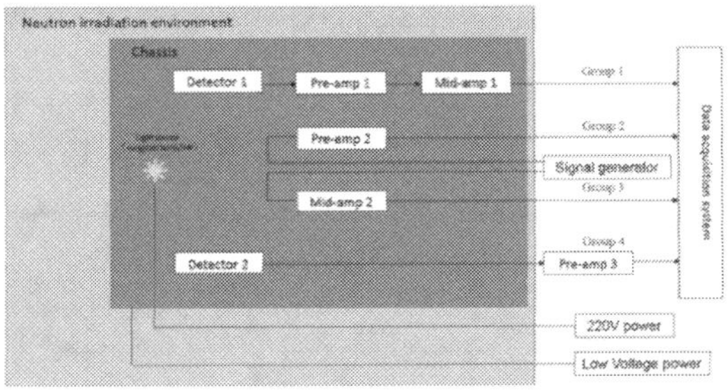

Fig.14 Layout sketch of electronics neutron irradiation test in the reactor

As shown in Fig.14, four groups of detector, pre-amplifier and mid-amplifier are exposed in irradiation environment. In groups 2-4, only pre-amplifier, mid-amplifier and detector is put inside the irradiation room, respectively; in group 1, the detector, pre-amplifier and mid-amplifier are both put inside the irradiation room. It should be noted that, in group 2-3, the input signal comes from signal generator; in group 1 and 4, visible light source (tungsten lamp) is used to illuminate detectors. This experiment arrangement can identify in real time whether the detector, pre-amplifier or mid-amplifier is impaired. To verify the influence of FPGA code on mid-amplifier SEE, the odd channels and even channels of the two amplifier cards are written with different codes. The odd channels are configured only once in the code on receiving command from I&C, while the even channels are automatically configured several times per second on receiving command.

As shown in Fig.15 and Fig.16, a customized PXIE chassis is designed for neutron irradiation test in fission reactor, mainly for support and power supply of tested components. The tungsten lamp and detectors are in left side and right side of the chassis panel respectively. Two pre-amplifier and two detectors are in the same vertical plane where the neutron flux is the same.

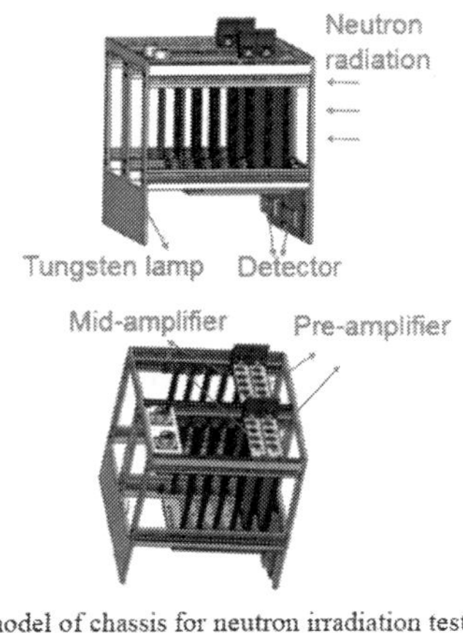

Fig.15 3D model of chassis for neutron irradiation test in reactor

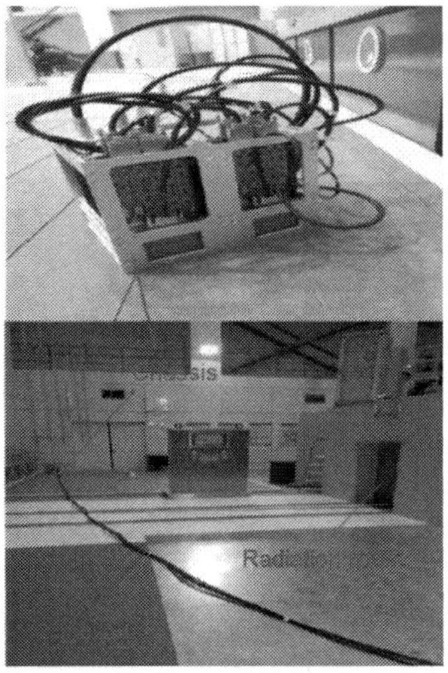

Fig.16 Photo of chassis and the radiation room

As shown in Fig.17, DAQ system put outside the radiation region is used to record and display the output of pre-amplifier and mid-amplifier in real time. The change of signal amplitude is used to assess damage level due to neutron irradiation.

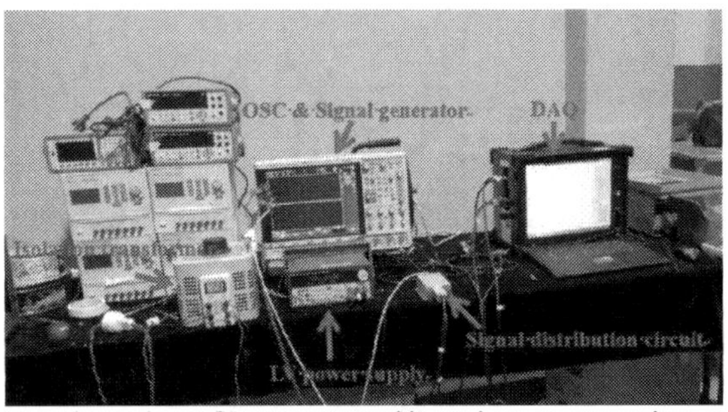

Fig.17 Photo of DAQ system and low voltage power supply

The test results are shown in Fig.18, Fig.19, Fig.20 and Fig.21. For group 2, pre-amplifier works well in the whole test and the amplification of pre-amplifier increased 0.83%. For group 1, only 1 SEE was found in one of the even channel of mid-amplifier. For group 3, 5 SEEs were found in the mid-amplifier which can be recovered by restarting, among which 1 SEE was found in all the channels and these channels saturated, and other SEEs were found only in even channels. When neutron fluence reached 6.4×10^{13} n/cm^2, mid-amplifiers' output began to decrease. By testing after experiment, there were two power supply chips in mid-amplifier damaged (note that no radiation resistant power chip was used in mid-amplifier). For group 4, the detector is not fully damaged, but the responsivity to visible light decreased 40%.

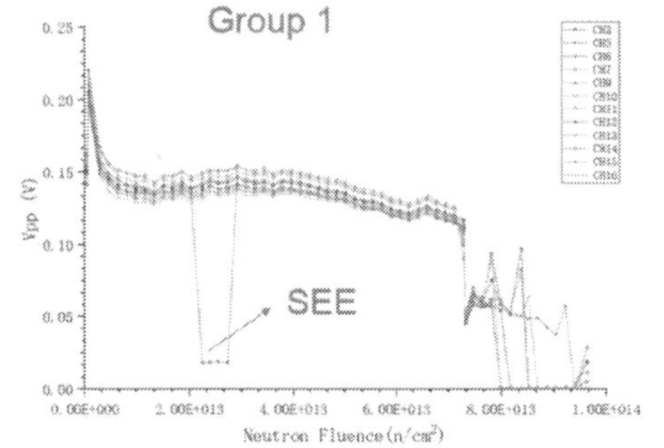

Fig.18 The result of group 1

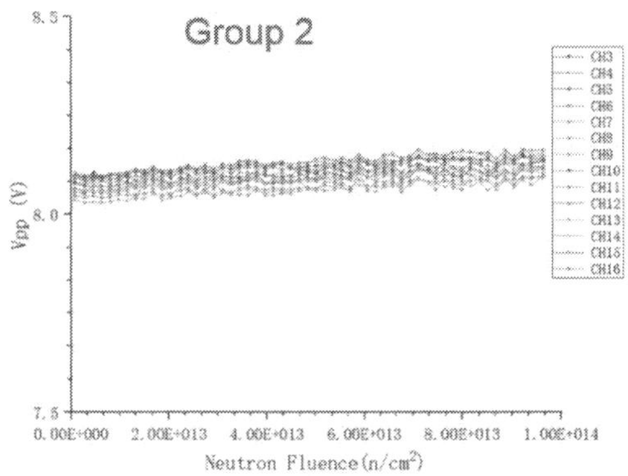

Fig.19 The result of group 2

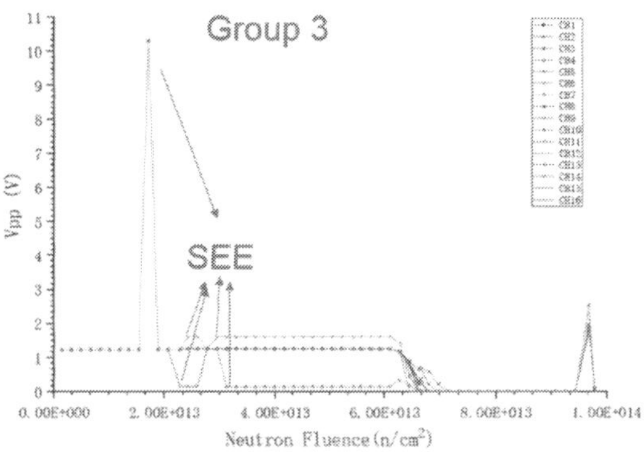

Fig.20 The result of group 3

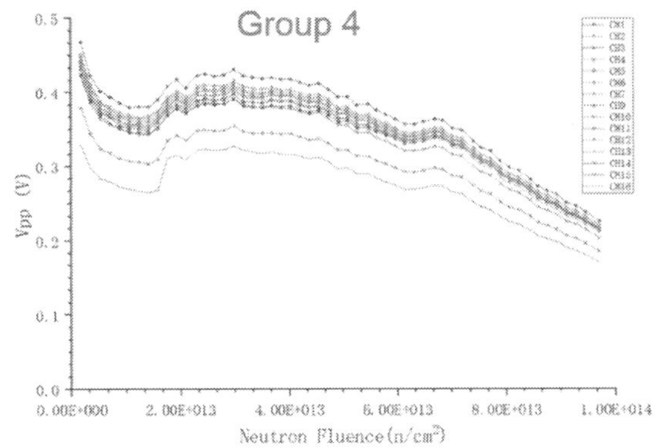

Fig.21 The result of group 4

V. RXC SYSTEM TEST IN EAST TOKAMAK

The diode array detector, pre-amplifier, DAQ system, customized chassis and radiation-hard cables to be used in ITER RXC have been tested in Experimental Advanced Superconducting Tokamak (EAST) experiments.

In the tests, the detector available is Centronic LD35(24)-5T which contains 35 elements each with active area of 12mm×2mm as shown in Fig.22.

Fig.22 Picture of detector

The 0.5 meter Kapton cable is used from detector to feedthrough, 25 meter HELUKABEL (18×2×0.25) cable is used from feedthrough to pre-amplifier, and 100 meter HELUKABEL (18×2×0.25) cable is used from pre-amplifier to DAQ as shown in Fig.23.

Fig.23 Picture of HELUKABEL (18×2×0.25) cable

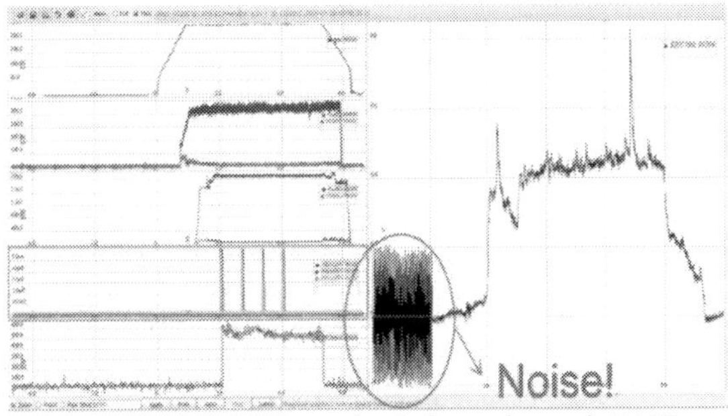

Fig.24 Picture of discharge signals from webscope in EAST

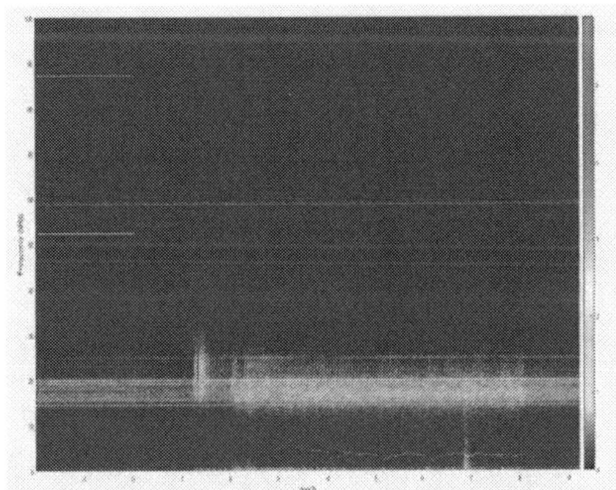

Fig.25 Frequency spectrum with time of the x-ray signal

As shown in Fig.24 and Fig.25, during the plasma discharge, as the ECRH, NBI and lower hybrid waves (LHWs) were jointly injected to the plasma, the whole electronics system worked well. However, there was a lot of noise found in soft x-ray signal, especially before the discharge was started.

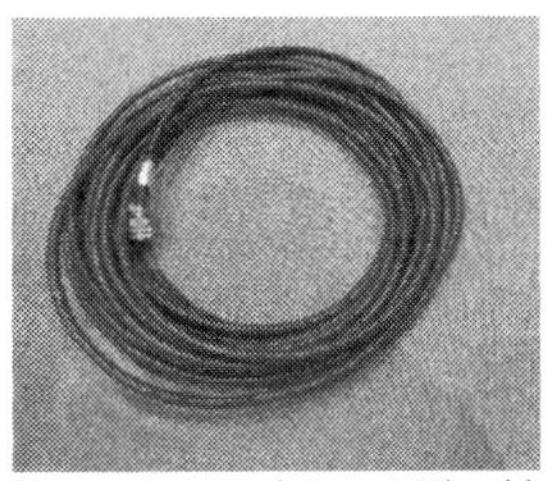

Fig.26 Picture of HELUKABEL (18×2×0.25) cable with aluminum foil shielding added

https://doi.org/10.1051/epjconf/202022502002

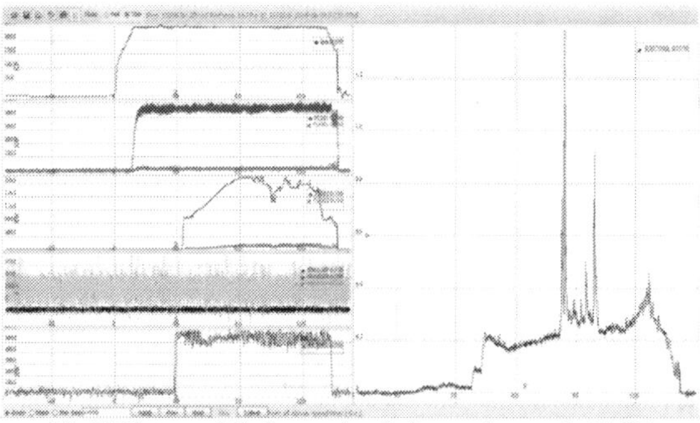

Fig.27 Picture of discharge signals from webscope in EAST

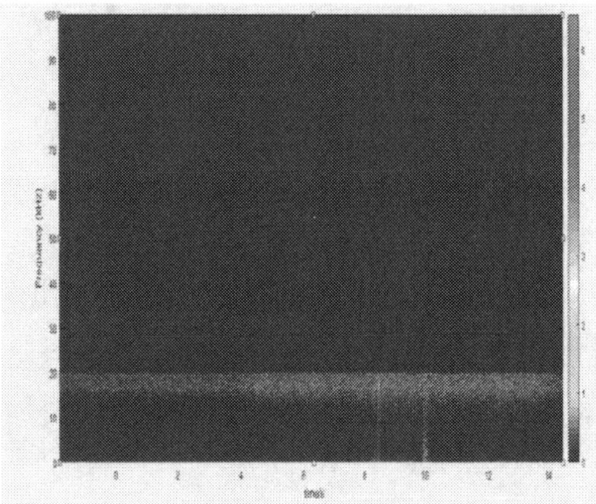

Fig.28 Frequency spectrum with time of the x-ray signal after using Al shielding for cables

As shown in Fig.26, Fig.27 and Fig.28, as aluminum foil shielding was added to the cable, the signal became well immediately (seen from Fig.27 and 28). This indicates the signal quality can be improved by adding aluminium shielding.

VI. Conclusion

In final design phase of ITER Radial X-ray Camera, enhancement of irradiation resistance research of compact electronics has been done. Power chip and FPGA in the circuit which are identified as weak parts have been improved by radiation resistance design. Considering influence of nuclear environment on worker exposure to radiation and optimization of space reservation limited in port cell to allow for more space to put nuclear shielding, easiness of maintenance, compact electronics structure and remote state monitoring are designed, while magnetic shielding was also considered. To exploit and verify the radiation performance, the electronics are tested in irradiation experiment in reactor. No problem was found in pre-amplifier with radiation resistant power chip; the responsivity of detectors decreased 40%; the mid-amplifier with optimized FPGA code was found less SEEs than that in previous experiment. The electronics were also integrated and tested in EAST tokamak, and found work well in EAST experiments with various ancillary heating.

Acknowledgment

ITER is a Nuclear Facility INB-174. The views and opinions expressed herein do not necessarily reflect those of the ITER Organization. Thanks to people in ITER Organization and other organizations or institutes supporting RXC design. Special thanks to Robin Barnsley, Antoine Sirinelli, Artur Malaquias, Barry Alper, and Michael Walsh.

This work was supported by the National Nature Science Foundation of China under Grant NO.11605240 and NO.11605243.

References

[1] L. Hu, et al., "Preliminary design and R&D of ITER diagnostic-radial X-ray camera", Nuclear Inst. and Methods in Physics Research A, vol. 870, pp. 50–54, Jul. 2017.
[2] L. Hu, et al., "Design and Test of Irradiation-Related components in ITER Radial X-Ray Camera", IEEE TRANSACTIONS ON NUCLEAR SCIENCE, vol.65, No.9, pp. 2398-2403, Sep. 2018.
[3] http://www.ti.com/product/OPA2140/technicaldocuments
[4] http://www.ti.com/product/THS4131/technicaldocuments
[5] https://www.st.com/en/space-products/rhfl4913a.html
[6] https://www.st.com/en/space-products/rhfl7913a.html
[7] https://www.intel.co.uk/content/www/uk/en/products/programmable/fpga/max-10.html
[8] R. Roccella, private communication, Oct 2018.
[9] D. Hamilton, private communication, May 2012.

EPJ Web of Conferences **225**, 02003 (2020)
ANIMMA 2019
https://doi.org/10.1051/epjconf/202022502003

Analysis of DD, TT and DT Neutron Streaming Experiments with the ADVANTG Code

Bor Kos
JSI and UL
Ljubljana, Slovenia
bor.kos@ijs.si

Theodora Vasilopoulou
NCSRD
Athens, Greece
dora@ipta.demokritos.gr

Scott W. Mosher
ORNL
Oak Ridge, USA
moshersw@ornl.gov

Ivan A. Kodeli
JSI
Ljubljana, Slovenia
ivan.kodeli@ijs.si

Robert E. Grove
ORNL
Oak Ridge, USA
grovere@ornl.gov

Jonathan Naish
CCFE
Oxford, United Kingdom
jonathan.naish@ukaea.uk

Barbara Obryk
IFJ
Krakow, Poland
barbara.obryk@gmail.com

Rosaria Villari
ENEA
Frascati, Italy
rosaria.villari@enea.it

Paola Batistoni
ENEA
Frascati, Italy
paola.batistoni@enea.it

and JET Contirbutors*

***Abstract*—The paper presents an analysis of DD, TT and DT neutron streaming benchmark experiments with the recently released hybrid transport code ADVANTG (AutomateD VAriaNce reducTion Generator). ADVANTG combines the deterministic neutron transport solver Denovo with the Monte Carlo transport code MCNP via the principle of variance reduction. It automatically produces weight-window and source biasing variance reduction parameters based on the CADIS (Consistent Adjoint Driven Importance Sampling) methodology. Using this novel hybrid methodology Monte Carlo simulations of realistic complex fusion streaming geometries have become possible. In this paper the experimental results from the 2016 DD campaign using measurements with TLDs and activation foils up to 40 m from the plasma source are analyzed. New detailed models of the detector assemblies were incorporated into the JET 360° MCNP model for this analysis. In preparation of the TT and DTE2 campaigns at JET a pre-analysis for these campaigns is also presented.**

***Index Terms*—JET, hybrid transport, MCNP, ADVANTG, TLD, activation foils**

I. Introduction

Neutron transport calculation and estimations of the neutron flux and doses far away from the plasma source in large fusion tokamak devices such as the International Thermonuclear Experimental Reactor (ITER) are challenging not only because of the complexity of the machine but also because of the physical processes involved. Therefore, validation of the state-of-the-art of radiation transport codes and tools and nuclear data against experimental results in real tokamak environment is needed to ensure reliability in ITER predictions. Within JET3 program in the frame of EUROfusion consortium, neutronics experiments are in preparation for the future high performance Deuterium-Tritium campaign (DTE2) at the Joint European Torus (JET) to validate ITER tools, codes and assumptions.

One such neutronics benchmark, the *Neutron Streaming Experiment* [1], [2] is currently being performed. It consists of measurements of neutron flux at several experimental positions throughout the JET Torus Hall with thermo-luminescent detectors (TLDs) and activation foils and simulation of neutron flux and doses at the experimental positions using state-of-the-art deterministic, Monte Carlo and hybrid transport codes alongside the latest evaluations of nuclear data and the calculated data are compared to the experimental quantities

Up until now measurements have been performed only during Deuterium–Deuterium (DD) campaigns since 2012. The last experiment was conducted in 2016 in preparation of future DTE2 (Deuterium-Tritium Experiment 2) campaign [2]. The detectors were located in several positions in the Torus Hall close to the tokamak and in two key locations – the southwest labyrinth and in the southeast chimney. These zones represent an extremely challenging problem for deterministic and Monte Carlo neutron transport simulations. In recent years so called hybrid deterministic/Monte Carlo code systems have been developed that use variance reduction parameters based on an approximate global deterministic adjoint/forward calculation to speed up the convergence of a Monte Carlo simulation.

The **A**utomate**D** **VA**ria**N**ce reduc**T**ion **G**enerator (ADVANTG) [3] code is a Deterministic/Monte Carlo two-step transport code. In the first step it uses the Denovo discrete ordinate deterministic solver to calculate an approximate adjoint and forward global neutron flux field which is used as the basis to determine variance reduction parameters for the second step precise Monte Carlo N-Particle (MCNP) [4] calculation.

In this paper the post-analysis of the DD campaign and the pre-analysis of the Tritium-Tritium (TT) and DTE2 campaigns planed for 2019/2020 using the hybrid simulation methodology is presented. In the first part of the paper the hybrid two-step methodology, specifically the ADVANTG code is described. Details of the experiment are given in the second part of the paper and we concluded with a comparison between the experimental and calculation results for the 2016 DD campaign and pre-analysis computational results for the upcoming TT and DTE2 campaigns.

**see the author list of X. Litaudon et al 2017 Nucl. Fusion 57 102001*

II. Neutral particle transport methodology

Two main approaches to neutral particle transport simulation in fission and fusion applications are usually used - deterministic approach for solving the neutron transport or Boltzmann equation and the Monte Carlo approximation. Each of the two approaches has its benefits and drawbacks. The hybrid methodology tries to combine the benefits of both while diminishing the drawbacks. The deterministic, Monte Carlo and hybrid approaches are discussed in the following subsections.

A. *Deterministic transport simulations*

Analytical solutions of the Boltzmann non-multiplying integro-differential equation 1, given bellow, of real life problems with complex geometries and realistic material composition are difficult or impossible.

$$\hat{\Omega}\cdot\nabla\psi(\vec{r},\hat{\Omega},E)+\Sigma_t(\vec{r},E)\psi(\vec{r},\hat{\Omega},E)= \\ =\int_0^{\infty}\int_{4\pi}\Sigma_s(\vec{r},\hat{\Omega}'\cdot\hat{\Omega},E'\rightarrow E)\psi(\vec{r},\hat{\Omega}',E')d\hat{\Omega}'dE'+ \\ +q(\vec{r},\hat{\Omega},E) \quad (1)$$

Where $\vec{r}$ is the spatial variable, $\hat{\Omega}$ is the unit vector determining the direction of flight, E is the energy and ψ the angular neutron flux. Σ_t is the total cross section and Σ_s the scattering cross section. q represents the external fixed source of neutrons.

Several numerical approximations to solve the Boltzmann equation were derived and one of the most commonly used for fixed source shielding simulations is the discrete ordinates approach (S_N). In the S_N method, used by the deterministic solver Denovo [5], the energy variable is discretized into a number of energy groups and the cross-sections are approximated by histograms in energy. The angular variable is discretized by using the discrete ordinates S_N approximation. The scattering cross sections is approximated using the Legandre polynomial expansion. The spatial variable is discretized using different techniques such as finite-difference, finite-element or nodal methods. The discretization of variables using numerical methods introduces a number of approximations which in turn results in possibly less accurate final results. Deterministic simulations require significant amounts of computer memory which increases with the level of details introduced in the computational model i.e. number of geometrical voxels, angles and energy groups used. On the other hand these simulations are usually inexpensive considering computer processing time and produce global results in one computational run.

B. *Monte Carlo transport simulations*

The Monte Carlo (MC) method for neutron transport simulations is based on the fact that macroscopic cross sections can be interpreted as the probability of interaction per unit distance. By simulating a large enough number of individual particle histories by using random sampling faithful to the underlying physics the average behavior of particles in the physical system can be determined. With MC there is no need to solve the transport equation directly and thusly in principal no error[1] is introduced to the simulation because of the discretization in space, energy and angle. Using MC methods we get an accurate result but with an uncertainty which is related to the number of detected or *tallied* particle histories. The simulation uncertainty decreases with a square-root dependency on the number of tallied particle histories $O(1/\sqrt{N})$. To decrease the statistical uncertainty an ample number of particle histories have to be simulated which in turn requires large amounts of computer processing time. This problem is especially problematic in scenarios where there are highly attenuating materials between the particle source and detector or simply when the detector is far away from the source. Another drawback of MC methods is the lack of global results. In MC simulations one is usually interested only in a few areas of interest rather than in a global result (eg. dose field) because such a calculation would be computationally too expensive.

C. *Hybrid transport simulations*

As we have shown in previous sections, neutron transport can be simulated by fast, less accurate deterministic solvers or by computationally expensive, more accurate Monte Carlo methods. Analog MC simulations, simulations where no variance reduction is used, often do not produce statistically relevant results in a reasonable computer processing time for cases with high attenuating materials between the source and detector. In such cases variance reduction is needed. Several variance reduction techniques can be used in Monte Carlo simulations. The idea behind *hybrid codes* is to combine the best attributes of both families of codes via the principle of variance reduction.

In this analysis the so called weight-window variance reduction technique is used to increase the importance of histories which will contribute to the detector tally and disregard the histories that will not. Weight-windows are a particle population control variance reduction technique, where the particle is split if its weight is above the wight-window upper bound and terminated (Russian roulette is played) if the particle weight is too low. This is done in a statistically consistent way in order not to bias the final results.

One of the current methods used to determine effective weight-windows variance reduction parameters is based on the adjoint flux - a measure of the *importance* of neutrons contributing to the response of the detector. Because calculating the adjoint flux with MC methods would be computationally as expensive as calculating the flux, a computationally inexpensive rough approximation of the adjoint flux is calculated with a deterministic code. The adjoint flux calculated using the rough deterministic approximation is used to construct weight-window variance reduction parameters for a detailed MC calculation.

[1]Inaccuracies in MC simulations usually arise from approximations in computational modeling (geometry, materials, source) and nuclear data uncertainties.

D. ADVANTG

ADVANTG (**A**utomate**D** **VA**ria**N**ce reduc**T**ion **G**enerator) [3] was developed at Oak Ridge National Laboratory and released in 2015. ADVANTG generates space- and energy-dependent mesh-based weight-window bounds and biased source distributions. The variance reduction (VR) parameters are based on a discrete ordinates solution of the adjoint transport equation calculated by Denovo. The variance reduction parameters can be produced for a continuous-energy Monte Carlo simulation of fixed-source neutron, photon, and coupled neutron-photon transport problems. Specifically ADVANTG implements the Consistent Adjoint Driven Importance Sampling (CADIS) method [6] to generate parameters to accelerate the convergence rate of an individual tally in a MCNP5 code version 1.6 [4] simulation. Alternatively, the Forward-Weighted CADIS (FW-CADIS) method [7] can be used to obtain relatively uniform rates of convergence across multiple tallies or the space and energy bins of an arbitrary sized mesh tally in a MCNP defined problem.

The CADIS method was developed to accelerate the estimation of a detector response of the general form (2)

$$R = \int\int q(\vec{r}, E)\Phi^{\dagger}(\vec{r}, E) dV dE \qquad (2)$$

where q is a known source distribution and $\Phi^{\dagger}$ is the scalar adjoint flux corresponding to the adjoint source $q = \sigma_d$. Central values of the weight windows are calculated as (3)

$$w(\vec{r}, E) = \frac{R}{\Phi^{\dagger}} \qquad (3)$$

An important feature of the CADIS method is the use of a consistent biased source distribution (4)

$$\hat{q}(\vec{r}, E) = \frac{1}{R} q(\vec{r}, E)\Phi^{\dagger} \qquad (4)$$

where q is the unbiased source distribution. This source is *consistent* in the sense that source particles will be assigned a weight identical to the value given by equation 3.

The FW-CADIS method was developed to uniformly accelerate the convergence of several tallies in a Monte Carlo simulation which differ in their location and energy response of the form (5)

$$R_i = \int\int \sigma_{d,i}(\vec{r}, E)\Phi(\vec{r}, E) dV dE \qquad (5)$$

where i=1,2,...,N, and both $\sigma_{d,i}$ and N are arbitrary and $\Phi(\vec{r}, E)$ is the scalar flux corresponding to source distribution q. This is accomplished by constructing an adjoint source that consists of appropriately weighted contributions from all tallies of interest. The weights are the inverses of the individual responses (6)

$$q^{\dagger} = \frac{1}{R_1}\sigma_{d,1} + \frac{1}{R_2}\sigma_{d,2} + \cdots + \frac{1}{R_N}\sigma_{d,N} \qquad (6)$$

In the FW-CADIS formalism an initial forward calculation is needed to determine the weighted adjoint sources for the second step adjoint calculation. The importance function calculated using the weighted adjoint sources is then used to construct weight-window bounds and a biased source distribution using the CADIS formalism.

III. Streaming Experiment

During three different experimental campaigns in 2012, 2013-2014 and 2016 measurements were performed with both the very sensitive thermo-luminescence detectors (TLD) and activation foils. Lessons learned in the 2012 and 2013-2014 campaigns were used to repeat the measurements during the 2016 DD campaign [2]. In this paper only the 2016 results are analyzed.

To differentiate between the neutron and non-neutron components of the dose field the team from the Institute of Nuclear Physics (IFJ) in Poland used two different detectors - natural and suppressed ^{6}Li based detectors. The highly sensitive ^{nat}LiF: Mg,Cu,P (MCP-N) and ^{7}LiF:Mg,Cu,P (MCP-7) TLDs. The MCP-N TLDs contained natural lithium with natural abundance 7.59% of ^{6}Li. The MCP-7 TLDs have suppressed ^{6}Li to 0.03% with the rest lithium content composed of to 99.97 % ^{7}Li. The TLD types and the current assembly were optimized on the basis of previous experience (2012–2014 DD campaigns [1]).

Three different kinds of activation foils were used to determine the reaction rate during the 2016 DD campaign. The foils were produced and analyzed by Institute of Nuclear and Radiological Sciences, Energy, Technology & Safety "Demokritos" (NCSRD) in Greece. The three different foils were: cobalt (Co), tantalum (Ta) and silver (Ag). The chemical purity of all foils was above 99.9 %. The foils were used bare and with a 0.5 mm thick Cadmium foil to cut off thermal neutrons. All of the disc-shaped foils had the same dimensions: diameter = 14.9 mm, thickness= 0.5 mm. The measured reaction rates on the foils were for the following reactions: ^{59}Co(n,γ)^{60}Co, ^{181}Ta(n,γ)^{182}Ta, ^{109}Ag(n,γ)^{110m}Ag. The foil activity was measured at NCSRD using a gamma-ray spectrometry system based on a high-purity coaxial germanium semiconductor detector (GEM80) of 85% relative efficiency, 1.37 keV energy resolution (Full Width at Half Maximum) at the 1332 keV and a peak-to-Compton ratio of 93:1. Detailed information on the TLD and activation foil calibration process and composition is provided in references [8] and [9].

To investigate the shadowing effect in the directional neutron field two (vertical and horizontal) high-density polyethylene (HDPE) boxes were used to hold the TLDs and the activation foils. The two boxes were also immersed in a container (diameter = 25.5 cm, height = 25 cm) also made from the HDPE. The configuration of the TLDs in the HPDE boxes and container are shown in Fig. 1. The activations foils were configured in HDPE containers similar to the ones containing TLDs.

The TLD and activation foil HDPE containers were positioned at several locations at different distances from the plasma source in the JET tokamak hall, specifically in the southwest (SW) labyrinth and southeast (SE) chimney. The

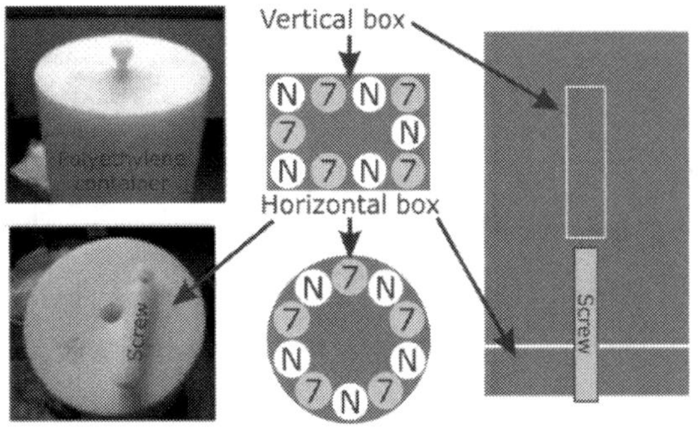

Fig. 1. TLD assemblies of N: MCP-N and 7: MCP-7 TLDs in high density polyethylene holders and container.

labyrinth is a five segment dogleg configuration of borated concrete, and the chimney consists of three large air ducts that connect the basement with the torus hall. A 3D representation of the tokamak, tokamak hall and the detector locations is given in Fig. 2. The TLD detectors were positioned at locations A1 - A8 and B1 - B8. Location B8 was not analyzed because the neutron flux measured there during the 2016 DD campaign with the TLDs was at background levels. Location B5 was also omitted from this analysis because of a positioning mistake during installation.

Assemblies with activation foil were positioned at locations A1, A2, A4 and B2, B3, B5. However, no reliable experimental data were acquired for position B2, due to a displacement of foils within the respective HDPE container. More on the specifics of the experimental data results from the 2016 DD campaign in Section V.

IV. Calculations and Model Information

The full 360° JET MCNP model used in this analysis consists of a detailed plasma containment model and other major streaming relevant components including the SE Chimney and SW labyrinth features. The simplifications done to the model are presumed not to affect the final transport results. The model was developed and verified by the Culham Center for Fusion Energy and other contributors. A 3D representation of the model is presented in Fig.2[2].

The DD and DT neutron source spectrum integrated in MCNP was used for the energy distribution of the plasma source while the spatial distribution is a faithful but simplified representation of a typical thermal plasma shape. The TT energy source spectra was calculated using a source subroutine for MCNP [10]. FENDL 3.1b [11] based libraries were used for both the multigroup deterministic ADVANTG calculation and the continuous energy Monte Carlo MCNP calculation. Some nuclide nuclear data missing from FENDL 3.1b was taken from JEFF-3.2 [12].

For this analysis the complete detailed models of the TLDs, activation foils, holders and containers were produced and incorporated into a single JET MCNP model for the first time. Up to now, two different approaches had been implemented: either a one-step MCNP calculations without detailed detector models (CCFE) or two-step MCNP calculations with detailed detector models (NCSRD). For this paper the lessons learned from previous analyses were used to produce effective variance reduction parameters using ADVANTG. The *optimal* variance reduction parameters produced statistically relevant simulation results in volume averaged track-length estimator tallies positioned in the actual detectors up to 40 m away from the plasma source.

As mentioned before, a hybrid two step workflow combining the deterministic based code ADVANTG version 3.0.3. and stochastic transport code MCNP5 v. 1.6 was used for all calculations in this paper.

In the first step optimal variance reduction parameters were produced using the FENDL 3.1b based 47 neutron energy group library, a dedicated forward peaking quadrature set, 5th order of scattering cross section expansion and, most important, unique non-uniform Cartesian spatial meshes for each detector location. The spatial meshes ranged from a total of 1 million to about 5 million spatial voxels. These input parameters were used by ADVANTG to execute Denovo and produce an approximate importance (adjoint) function using Denovo. The importance function was then used to produce weight window and source biasing VR parameters for each detector (TLD or activation foil) using the CADIS methodology described in Sec. II-D.

Using the VR parameters produced in the first step, the second step MCNP calculations were performed for each detector. The computational results for the DD plasma source are presented and compared to experimental results in Sec. V-A. Preliminary results for the two other plasma sources (TT and DTE2) are presented in Sec. V-B in preparation for the analysis of the TT and DT campaigns currently underway at JET.

V. Results

Two sets of results corresponding to the two different detectors (TLDs and activation foils) are presented in this section. The computational results are presented and compared to the 2016 DD experimental campaign in Sec. V-A. Because of the delays in the TT and DT campaigns only pre-analysis calculation results for the TLDs are presented in Sec. V-B. These calculation results are also compared to the DD results. Such a comparison is beneficial for future experimental configurations.

A. Analysis of DD benchmark experiment

All activation foils and TLDs in positions A1-A4, B2, B3, and B5 were irradiated with a total of $3.52 \cdot 10^{18}$ neutrons. The other TLDs were irradiated up to $6.18 \cdot 10^{18}$. In the case of the TLDs, both the experimental and calculation fluence results were normalized to *per source particle*. Only the 1σ statistical uncertainty of the second step Monte Carlo calculation is included in the uncertainty of the calculation. The

[2]The MCNP model is rotated 90°clockwise relative to true North.

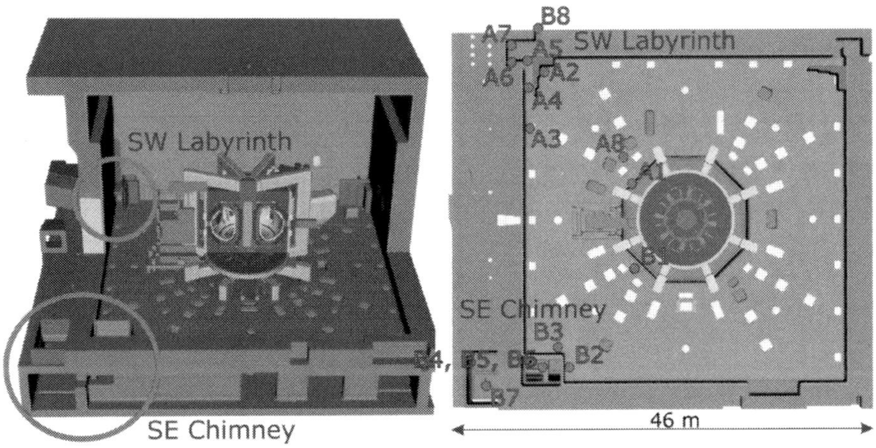

Fig. 2. Southwest labyrinth and southeast chimney (left). TLD and activation foil container locations in the JET tokamak hall (right).

final total experimental uncertainty includes contributions from the uncertainty of: the calibration of TLDs, the actual TLD measurements, neutron yield and the background radiation measurement. Calculation/Experiment (C/E) values, reported in Tab. I, range from about 0.6 for the detector locations closest to the plasma to about 12 for the farthest location at the bottom of the chimney in the basement of the tokamak. The C/E uncertainty is the root sum of the square of experimental and calculation relative uncertainties. These results are slightly better compared to the C/E values calculated using simplified TLD containers mentioned in Sec. IV. One can also observe that the major part of the uncertainty comes from the experimental part, while the statistical uncertainty of the MCNP calculations is low because of efficient ADVANTG produced VR parameters. Calculations and experiments at the farthest positions are extremely difficult and the calculated C/E values are encouraging.

In Tab. II the activation foil results are compared against calculations in terms of reaction rates (RR). The IRDFF [13] activation libraries were used to calculate the activities of the activation foils using volume integrated tally estimators (F4 tally) and tally multiplication cards (FM card) implemented in MCNP. As it can be observed, C/E ratios range from about 0.5 to about 14.5 while, in most positions, the ratios corresponding to cadmium-covered foils are closer to one than those derived by bare foils. The C/E uncertainty is, as in the previous table, the root sum of the square of experimental and calculation relative uncertainties, where the calculation uncertainty is only the 1σ statistical uncertainty of the second step Monte Carlo calculation. The experimental uncertainty of the activation foils was calculated by combining all identified sources of error, namely the counting statistics, detector efficiency and nuclear data (uncertainties related to isotope half-lives and gamma-ray emission probabilities). Among these parameters, the statistical uncertainty of the gamma-ray measurements,

TABLE I
COMPARISON OF TLD CALCULATIONS AND EXPERIMENTAL RESULTS FOR THE DD PLASMA SOURCE.

DD	Experiment [1/cm^2/src. particle]		Calculation [1/cm^2/src. particle]		Calculation / Experiment	
Location	Fluence	Rel. Unc.	Fluence	Rel. Unc.	C/E	Rel. Unc.
A1	4.07E-09	0.17	2.34E-09	0.02	0.58	0.17
A8	1.02E-09	0.16	1.25E-09	0.06	1.22	0.17
A2	1.97E-10	0.17	4.18E-10	0.02	2.12	0.17
A3	2.03E-10	0.17	4.11E-10	0.01	2.02	0.17
A4	7.04E-11	0.18	1.88E-10	0.02	2.67	0.18
A5	3.98E-12	0.17	1.20E-11	0.02	3.01	0.17
A6	2.29E-13	0.25	1.34E-12	0.01	5.85	0.25
A7	2.29E-14	1.59	2.34E-09	0.02	5.20	1.59
B1	6.38E-09	0.16	1.09E-08	0.02	1.72	0.16
B2	8.40E-11	0.17	2.26E-10	0.02	2.68	0.18
B3	1.55E-10	0.17	3.25E-10	0.03	2.10	0.17
B4	1.29E-10	0.16	2.77E-10	0.03	2.15	0.17
B6	7.42E-13	0.20	4.85E-12	0.03	6.54	0.20
B7	5.65E-15	6.30	6.90E-14	0.02	12.20	6.30

which varied from 5 % to approximately 40 %, had the major contribution to the combined uncertainty.

Although the agreement between calculation and experiment is not as good as with the TLDs, the results, especially for the foils covered with cadmium, are very promising, taken into account the complexity of the problem. Moreover, this was the first attempt to analyze the reaction rates in the activation foils using a single Monte Carlo transport calculation sped up with ADVANTG produced VR parameters. In most cases up to now, the results were interpreted in terms of neutron fluence in order to be compared against calculations. Furthermore, most of previous attempts used a two step methodology, where in the first step the source was recorded using the MCNP source-write capability and in the second step the recorded source was used with a simplified model of a segment of the tokamak

 https://doi.org/10.1051/epjconf/202022502003

TABLE II
COMPARISON OF CALCULATED AND EXPERIMENTAL REACTIONS RATES (RR) OF THE ACTIVATION FOILS FOR THE DD PLASMA SOURCE.

DD		Experiment [s^{-1}]		Calculation [s^{-1}]		Calculation/Experiment	
Nuclide	Location	Reaction Rate	Relative Uncertainty	Reaction Rate	Relative Uncertainty	C/E	Relative Uncertainty
$^{109}Ag\ (n,\gamma)\ ^{110m}Ag$	A1	7.58E+03	0.11	1.99E+04	0.02	2.62	0.11
	A1+Cd	1.36E+03	0.17	1.73E+03	0.02	1.27	0.17
$^{59}Co\ (n,\gamma)\ ^{60}Co$	A1	3.80E+05	0.10	4.03E+05	0.02	1.06	0.11
	A1+Cd	5.21E+04	0.11	2.50E+04	0.02	0.45	0.11
	A2	1.36E+04	0.12	8.24E+04	0.02	6.07	0.12
	A2+Cd	2.48E+03	0.19	4.91E+03	0.05	1.98	0.19
	A4	7.02E+03	0.12	3.91E+04	0.02	5.58	0.12
	A4+Cd	1.22E+03	0.26	2.40E+03	0.02	1.97	0.26
	B3	1.66E+04	0.11	8.72E+04	0.03	5.25	0.12
	B3+Cd	7.67E+03	0.22	6.09E+03	0.04	0.79	0.23
	B5	5.54E+03	0.14	9.08E+03	0.02	1.64	0.14
	B5+Cd	1.00E+03	0.34	4.77E+02	0.03	0.48	0.34
$^{181}Ta\ (n,\gamma)\ ^{182}Ta$	A1	1.58E+05	0.10	1.65E+05	0.02	1.04	0.10
	A1+Cd	3.22E+04	0.11	1.68E+04	0.02	0.52	0.11
	A2	5.19E+03	0.13	3.49E+04	0.02	6.73	0.13
	A2+Cd	1.54E+03	0.20	3.40E+03	0.05	2.21	0.21
	A4	1.82E+03	0.14	1.77E+04	0.02	14.48	0.14
	A4+Cd	5.07E+02	0.27	1.45E+03	0.02	2.86	0.27
	B3	3.99E+03	0.12	2.96E+04	0.03	7.42	0.12
	B3+Cd	4.21E+02	0.18	3.06E+03	0.03	7.27	0.18
	B5	1.67E+03	0.16	3.64E+03	0.02	2.18	0.16
	B5+Cd	3.79E+02	0.48	2.67E+02	0.03	0.71	0.48

building to calculate the reaction rates [14]. Nevertheless, the single calculation approach presented here enables a more accurate representation of the studied geometry and thus is able to provide lower uncertainties in the calculated data. The source of the observed discrepancies will be investigated further in the upcoming TT and DTE2 campaigns.

B. Pre-analysis for TT and DTE2 campaign

In preparation for the TT and DTE2 plasma campaigns at JET in 2019/2020 we performed a preliminary simulation of the TLDs at the same experimental positions but with different energy source spectra - the TT and DT source spectrum. The ADVANTG calculations had to be repeated and refined because of the change in the energy spectrum. The results shown in Tab. III predict that the neutron fluence per source neutron measured by the TLDs during the TT campaign will be similar to those recorded during the DD campaign. On the other hand the neutron fluences per source neutron during the DT camping will be, on average, about 20 % higher than the DD campaign. The C/C uncertainty in the last column is the root sum of the square of relative uncertainties given in columns 3 and 5 which represent the 1σ statistical uncertainty of the volume integrated track length estimator tallies (F4 tallies). During the upcoming TT experiment $5.04{\cdot}10^{19}$ TT and $1.04{\cdot}10^{20}$ DT neutrons (due to 1 % Deuterium) will be produced and during the following DTE2 experiment the maximum DT yield will be $1.53{\cdot}10^{21}$ of DT neutrons and $3.06{\cdot}10^{19}$ of TT neutrons [15]. Tab. IV reports the the total neutron fluences predicted in TLDs during TT and DTE2 campaigns considering the expected level of irradiation in future experiments[3]. The relative uncertainties given in columns 3 and 5 are the root sum of the square of uncertainties from contributions of TT and DT sources.

[3]The neutrons from DD reactions were disregarded in these predictions.

TABLE III
COMPARISON OF TLD CALCULATIONS FOR DD, TT AND DT PLASMA SOURCES.

	TT Calculation [$1/cm^2$/src. particle]		DT Calculation [$1/cm^2$/src. particle]		DD/ TT	DD/ DT
Location	Fluence	Rel. Unc.	Fluence	Rel. Unc.	C_{DD}/C_{TT}	C_{DD}/C_{DT}
A1	2.30E-09	0.01	3.28E-09	0.02	1.02	0.71
A8	1.21E-09	0.02	1.71E-09	0.04	1.03	0.73
A2	4.37E-10	0.07	5.58E-10	0.02	0.96	0.75
A3	4.06E-10	0.03	5.53E-10	0.02	1.01	0.74
A4	1.90E-10	0.01	2.60E-10	0.03	0.99	0.72
A5	1.23E-11	0.03	1.59E-11	0.01	0.97	0.75
A6	1.28E-12	0.02	1.64E-12	0.02	1.05	0.82
A7	1.16E-13	0.02	1.53E-13	0.03	1.03	0.78
B1	9.30E-09	0.02	1.15E-08	0.03	0.98	0.79
B2	2.27E-10	0.02	2.90E-10	0.03	0.99	0.78
B3	3.38E-10	0.03	4.06E-10	0.04	0.96	0.80
B4	2.64E-10	0.04	3.45E-10	0.02	1.05	0.80
B6	5.27E-12	0.04	5.99E-12	0.04	0.92	0.81
B7	6.46E-14	0.02	8.53E-14	0.05	1.07	0.81

VI. CONCLUSION

In this paper we presented a post-analysis of the DD neutron streaming experiments performed in 2016 and a pre-analysis of the upcoming TT and DT campaigns using ADVANTG. ADVANTG was used to prepare efficient variance reduction parameters for Monte Carlo MCNP calculations of difficult streaming problems in complex realistic tokamak geometries.

The agreement between the calculation and experimental results for the TLDs ranges from 0.6 close to the plasma to 12 more than 40 m away from the plasma source. The C/E values slightly improved compared to previous analyses because detailed models of the TLD holders and containers were used. Considering the complexity of the problem, model

TABLE IV
PREDICTION OF TOTAL NEUTRON FLUENCE AT TLD LOCATIONS FOR THE UPCOMING TT AND DTE2 CAMPAIGNS.

	TT campaign		DTE2 campaign	
Location	Fluence [1/cm^2]	Rel. Unc.	Fluence [1/cm^2]	Rel. Unc.
A1	4.58E+11	0.01	5.09E+12	0.02
A8	2.39E+11	0.03	2.66E+12	0.04
A2	8.01E+10	0.02	8.67E+11	0.02
A3	7.80E+10	0.02	8.59E+11	0.02
A4	3.67E+10	0.02	4.04E+11	0.03
A5	2.28E+09	0.01	2.47E+10	0.01
A6	2.35E+08	0.01	2.55E+09	0.02
A7	2.17E+07	0.02	2.37E+08	0.03
B1	1.67E+12	0.02	1.79E+13	0.03
B2	4.16E+10	0.02	4.51E+11	0.03
B3	5.93E+10	0.03	6.32E+11	0.04
B4	4.92E+10	0.02	5.36E+11	0.02
B6	8.88E+08	0.03	9.32E+09	0.04
B7	1.21E+07	0.03	1.33E+08	0.05

simplifications and potential uncertainty because of nuclear data uncertainty propagation this is a very encouraging results.

For the first time, activation foil reaction rate measurements were also analyzed using this hybrid ADVANTG/MCNP methodology. The C/E results are not as good as with TLDs but, these are promising preliminary results which need to be analyzed further in the upcoming TT and DTE2 campaigns at JET planed for 2019/2020.

To prepare for the upcoming experimental campaigns a pre-analysis using different (TT and DT) source spectra was performed. The results show that one can expect similar levels of neutron fluence per source particle measured by the TLDs during the TT campaign and on average a 20 % increase of the fluence per source particle during the DT campaign. Total neutron fluences for the upcoming TT and DTE2 campaigns determined using expected neutron yields were also calculated. The expected neutron fluence during the TT campaign ranges from $1.67 \times 10^{12}\,\mathrm{n/cm^2}$ at location B1 to $1.21 \times 10^{7}\,\mathrm{n/cm^2}$ at location B7. On the other hand the expected neutron fluence during the DTE2 campaign is expected to range from $1.79 \times 10^{13}\,\mathrm{n/cm^2}$ at location B1 to $1.33 \times 10^{8}\,\mathrm{n/cm^2}$ at location B7.

ADVANTG has proven to be a reliable tool for analyzing complex fusion streaming problems and reducing the statistical uncertainty of the second-step Monte Carlo calculation. But one has to acknowledge other uncertainties in such a complex simulation, i.e. model simplification and uncertainties because of nuclear data. In the future we plan to investigate the uncertainties due to nuclear data using a newly developed capability of the Sensitivity/Uncertainty code SUSD3D [16], [17].

ACKNOWLEDGMENT

This work has been carried out within the framework of the EUROfusion Consortium and has received funding from the Euratom research and training program 2014–2018 and 2019-2020 under grant agreement No 633053. The views and opinions expressed herein do not necessarily reflect those of the European Commission.

The authors also acknowledge the financial support from the Slovenian Research Agency (research core funding No. P2-0073 and project for training young researchers No. 1000-15-0106).

REFERENCES

[1] P. Batistoni, S. Conroy, B. Obryk, S. Popovichev, I. Stamatelatos, B. Syme, and T. Vasilopoulou and JET Contributors 2015, "Benchmark experiments on neutron streaming through JET Torus Hall penetrations", Nucl. Fusion 55 053028.
[2] R. Villari, P. Batistoni, J.P. Catalan, B. Colling, D. Croft, U. Fischer, D. Flammini, N. Fonnesu, L. Jones, A. Klix, B. Kos, M. Klosowski, I. Kodeli, S. Loreti, F. Moro, J. Naish, B. Obryk, L. Packer, P. Pereslavtsev, R. Pilotti, S. Popovichev, P. Sauvan, I.E. Stamatelatos, T. Vasilopoulou and JET Contributors, "ITER oriented neutronics benchmark experiments on neutron streaming and shutdown dose rate at JET", Fus. Eng. Des., 123, pp 171-176, 2017.
[3] S. W. Mosher, S. R. Johnson, A. M. Bevill, A. M. Ibrahim, C. R. Daily, T. M. Evans, J. C. Wagner, J. O. Johnson, and R. E. Grove, "ADVANTG—An automated variance reduction parameter generator", ORNL/TM-2013/416 Rev. 1, Oak Ridge National Laboratory, Oak Ridge, TN, 2015.
[4] X-5 MONTE CARLO TEAM, "MCNP – A general Monte Carlo N-particle transport code, Version 5 – Volume I: Overview and theory", LA-UR-03-1987, Los Alamos National Laboratory, Apr. 2003, revised Feb. 2008.
[5] T. M. Evans, A. S. Stafford, R. N. Slaybaugh, K. T. Clarno, "Denovo: A new three-dimensional parallel discrete ordinates code in SCALE", Nuclear Technology, vol. 171, pp. 171–200, 2010.
[6] J.C. Wagner, A. Haghighat, "Automated variance reduction of Monte Carlo shielding calculations using the discrete ordinates adjoint function", Nuclear Science and Engineering, 128, 186, 1998.
[7] J. C. Wagner, D. E. Peplow, and S. W. Mosher, "FW-CADIS method for global and semi-global variance reduction of Monte Carlo radiation transport calculations", Nuclear Science and Engineering, 176, 37-572014.
[8] B. Obryk, R. Villari, P. Batistoni, A. Colangeli, P. De Felice, N. Fonnesu, M. Kłosowski, S. Loreti, K. Malik, J. Nash, M. Pillon, M. Pimpinella, L. Quintieri and JET Contributors, "TLD calibration for neutron fluence measurements at JET fusion facility", Nuclear Inst. and Methods in Physics Research, A 904, 202–213, 2018.
[9] T. Vasilopoulou, I. E. Stamatelatos, P. Batistoni, A. Colangeli, D. Flammini, N. Fonnesu, S. Loreti and JET Contributors, "Improved Neutron Activation Dosimetry for Fusion." Fusion Engineering and Design 139: 109–114, 2019.
[10] I. Lengar, A. Čufar, V. Radulović, P. Batistoni, S. Popovichev, L. Packer, Z. Ghani, I. A. Kodeli, S. Conroy, L. Snoj and JET contributors, "Activation Material Selection for Multiple Foil Activation Detectors in JET TT campaign", Fusion Engineering and Design (2018) in press.
[11] A. Koning, A. Trkov, "FENDL-3.1b: Fusion Evaluated Nuclear Data Library Ver.3.1b", https://www-nds.iaea.org/fendl/.
[12] OECD/NEA, "The JEFF-3.2 Nuclear Data Library", NEA Data Bank, Available at https://www.oecd-nea.org/dbforms/data/eva/evatapes/je_32/, 2014.
[13] R. Capote, K.I. Zolotarev, V.G. Pronyaev, and A. Trkov, "Updating and Extending the IRDF-2002 Dosimetry Library" J. ASTM International, Volume 9, Issue 4, 2012.
[14] T. Vasilopoulou, I. E. Stamatelatos, P. Batistoni, N, Fonnesu, R. Villari, J. Naish, S. Popovichev, B. Obryk and JET Contributors, "Activation foil measurements at JET in preparation for D-T plasma operation" Fusion Engineering and Design, In-Press, 2018.
[15] R. Villari and G. Sips, personal communication.
[16] B. Kos, I. A. Kodeli, "Coupling of the SUSD3D S/U Code with the Denovo Deterministic Transport Solver", Proceedings, 27th International Conference Nuclear Energy for New Europe - NENE 2018, Portorož, Slovenia, 2018.
[17] I. A. Kodeli, "Multidimensional deterministic nuclear data sensitivity and uncertainty code system: Method and application." Nuclear Science and Engineering, 138, pp. 45–66, 2001.

Application of fluorine-based threshold activation detector for neutron flux calculation from D-T neutron generator

Paweł Sibczyński*, Andrzej Brosławski*, Szymon Burakowski*, Arkadiusz Chłopik*, Marek Dryll*, Andrzej Dziedzic*, Michał Gierlik*, Krystian Grodzicki*, Łukasz Kaźmierczak*, Tomasz Krakowski*, Kamil Niechcielski*, Jacek Rzadkiewicz*, Ivan Sliphukin*, Krystian Trela*, Arkadiusz Urban*, Kacper Zezuliński*, Alessandro Iovene† and Carlo Tintori†
*National Centre for Nuclear Research (NCBJ), A. Soltana 7, 05-400 Otwock, Poland
Electronics and Detection Systems Division (TJ2)
Email: pawel.sibczynski@ncbj.gov.pl
† CAEN S.p.A., Via Vetraia 11, 55049 Viareggio (LU), Italy

***Abstract*—In this paper we propose a method of fast neutron flux estimation from a pulsed D-T neutron generator with application of single CaF_2 scintillation crystal. The analysis method relies on $^{19}F(n,\alpha)^{16}N$ threshold activation reaction having neutron energy threshold at 1.6 MeV. As a result, the ^{16}N undergo β^- decay with half-life of 7.1 s, emitting β particles with endpoint up to 10.4 MeV in the scintillator medium. Integration of the β distribution curve, preceded by calculation of (n,α) rate on F with Monte Carlo N-Particle Transport Code v6 (MCNP6) for fixed geometry, allows to estimate the neutron flux in 4π per second within few minutes.**

***Index Terms*—neutron detection, Threshold Activation Detection, scintillator, neutron generator**

I. Introduction

One of a technique useful for fast neutron detection is known as a Threshold Activation Detection (TAD) [1], [2]. It relies on registration of characteristic radiation emitted from specific nuclei after threshold reaction with neutrons. A possible approach is to use a scintillation material containing itself a material, which undergoes the activation reaction, such as fluorine. In this work we investigated a CaF_2 scintillator, which mass fraction of Ca and F is about 50 / 50% [3]. The F nuclei, present in the scintillator medium, undergo i.e. $^{19}F(n,\alpha)^{16}N$ reaction, resulting in emission of high energy beta particles (endpoints at 10.4 MeV and 4.3 MeV) and γ rays (6.1 MeV) with the half-life of 7.1 s. The (n,α) reaction cross section reaches about 100 mb at about 5 MeV (see Fig. 1), which is sufficient for such application. In this work, we propose a new method of neutron flux estimation from deuterium-tritium (D-T) neutron generators operated in pulse mode. An undoped CaF_2 with size of ⌀5"×3" was used, however, other scintillators containing fluorine such as EJ-313/BC-509, BaF_2 or fluorocarbon plastic [4] can be also considered for that application. The method usefulness for

This project has received funding from the European Union's Horizon 2020 research and innovation programme under grant agreement No 653323 and CentriX project No RPMA.01.01.00-14-8516/17.

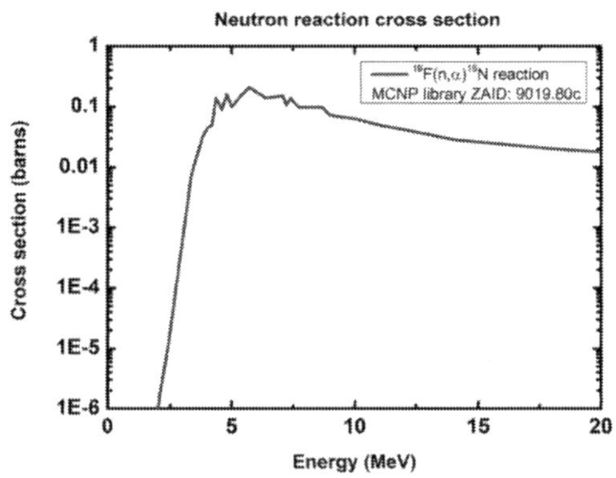

Fig. 1. The $^{19}F(n,\alpha)^{16}N$ neutron cross section used in the MCNP simulations, a ZAID 9019.80c library was chosen.

deuterium-deuterium (D-D) neutron generators flux monitoring will be also discussed. The CaF_2 scintillator was previously used for simultaneous detection of prompt photofission neutrons and delayed γ-rays from nuclear materials, potentially hidden in cargo container. These field trials were performed at Maasvlakte Rotterdam Seaport in the frame of C-BORD project.

II. Methodology

A. Experimental setup

Fig. 2 presents the experimental setup at NCBJ measurement facility, equipped with a Sodern Genie 16D D-T neutron generator. Data were registered with a CAEN DT5730 8 channel 500 MS/s desktop digitizer with DPP-PSD firmware and DigiTes 4.5.10 control software. The 1 ms excitation and 10 ms repetition neutron generator pulse mode was set. The trigger width was shaped by a laboratory-grade digital logic pulse shaper based on Arduino,

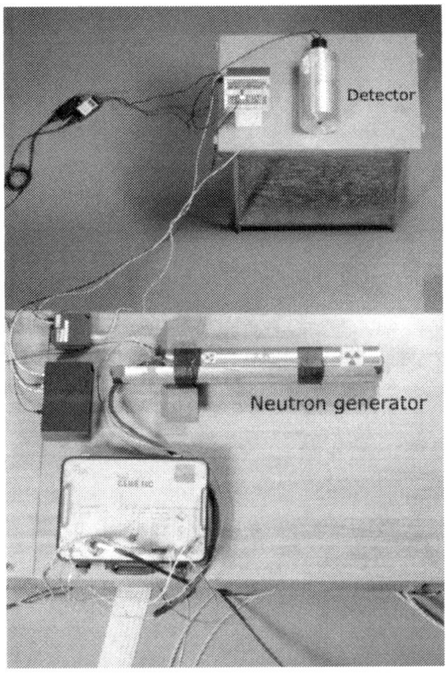

Fig. 2. The experimental setup for measurements of a neutron generator flux.

which blocked acquisition for 5 ms after each pulse in order to prevent detection of neutron capture γ-rays. The detector was placed 1 m from neutron generator tritium target and 83 cm above the concrete floor level. Each acquisition run last 10 minutes.

B. Monte Carlo simulations and data analysis

An MCNP v6.11 software package [5], [6] was used for investigation of the CaF_2 (n, α) reaction rate, expressed per one source neutron. We used average flux tally with FM4 card in order to calculate the reaction rate separately for Ca and F and run 1E8 histories. The obtained (n, α) rates for D-T and D-D neutron generators were presented in Tab. I. The uncertainties (expressed in percentage of the expected value) were taken from the MCNP calculation output, which are related only with the number of launched histories. The MCNP geometry model is shown in Fig. 3.

TABLE I
THE SIMULATED (n, α) REACTION RATES IN THE CaF_2 DETECTOR.

Neutron generator	(n, α) **rate for** ^{19}**F**
D-T ($E_n = 14\ MeV$)	$1.37 \times 10^{-5} \pm 0.0041\%$
D-D ($E_n = 2.5\ MeV$)	$6.46 \times 10^{-9} \pm 0.0037\%$

The (n, α) reaction rate for the simulated setup for D-T neutron generator is over four orders of magnitude greater than that for D-D one. This is in line with expectations, as the total (n, α) cross section for the 2.5 MeV neutrons is significantly lower than that for the D-T neutron generator. A D-D neutron generator would need to provide neutron flux of 2.5×10^{11} n/s to obtain the same statistics as the Sodern D-T generator at 90kV/40μA.

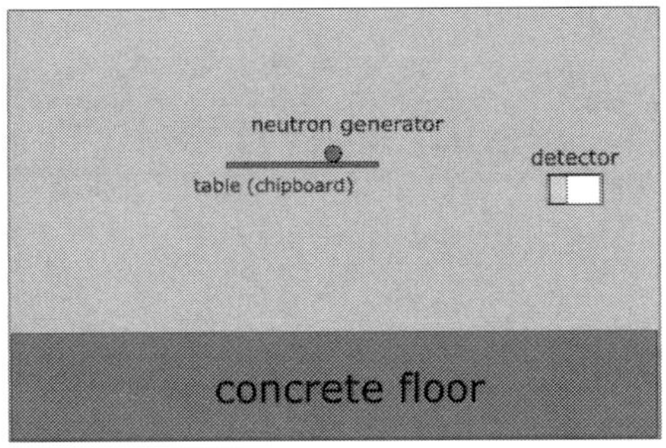

Fig. 3. Visualization of the MCNP geometry model.

After the (n, α) reaction rate calculation, one need to calculate the number of generated β particle with the endpoint at 10.4 MeV. The intensity of this transition is 28%. In contrast, the intensity of the 4.3 MeV transition is 66%. The exemplary energy spectrum for neutron generator HV of 90 kV and current of 40 μA is presented in Fig. 4.

The β continuum with endpoints at 4.3 MeV and 10.4 MeV as well as 6.1 MeV γ-rays are clearly visible. The β continuum energy distribution was fitted with use of the following approximation (1):

$$\mathrm{N(E)} \sim \mathrm{C}/\mathrm{c}^5 \times \sqrt{(\mathrm{E}^2 + 2\mathrm{E}\mathrm{m_e}\mathrm{c}^2)} \times (\mathrm{Q} - \mathrm{E})^2 \times (\mathrm{E} + \mathrm{m_e}\mathrm{c}^2), \tag{1}$$

where Q - beta endpoint, E - electron energy, m_e - electron mass, c - speed of light, C - constant.

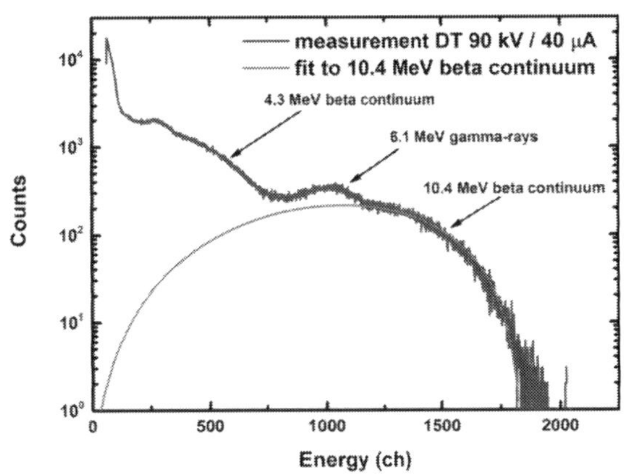

Fig. 4. The CaF_2 energy spectra after 600 s of exposition to 14 MeV neutrons from D-T neutron generator.

C. Results

The calculated neutron flux from the D-T neutron generator is presented in Fig. 5. The neutron flux per second was calculated with the following equation (2):

$$\phi_n = \frac{\int_0^Q N(E)dE}{R(n,\alpha) \times B}, \tag{2}$$

where $\int_0^Q N(E)dE$ is the integral over the beta energy spectrum distribution from 0 to the beta endpoint Q, $R(n,\alpha)$ is the (n,α) reaction rate on ^{19}F, obtained from the MCNP simulations, and B is the branching ratio of (n,α) for 10.4 MeV beta endpoint, equal to 0.28. Results were normalized to obtain the neutron flux per one second. The obtained results are in very good agreement with the neutron flux data provided by the neutron generator manufacturer.

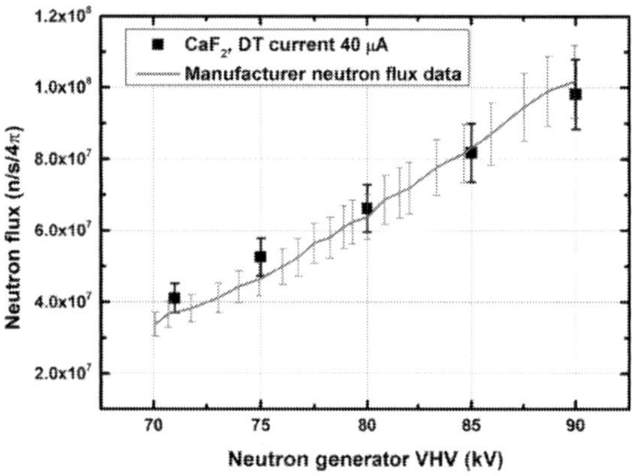

Fig. 5. Calculated neutron flux from Sodern Genie 16D neutron generator with use of CaF_2 detector and comparison with the manufacturer data.

III. CONCLUSIONS

We presented the approach for fast neutron flux estimation from D-T neutron generator with use of a single CaF_2 as the threshold activation detector. Main advantage of the method is the fact that neither activation foils nor High Purity Germanium (HPGe) detector is required to perform the flux calculation. After preceded MCNP simulations, required for (n,α) rate estimation at fixed detector and neutron generator positions, the neutron flux can be calculated in real time even for intense and short neutron pulses. This solution can be exceptionally useful in neutron generator test-bench facility, where at fixed positions one need to calculate the absolute neutron flux for any type of D-T neutron generator.

REFERENCES

[1] T. Gozani, M. J. King, and J. Stevenson, "The investigation of fast neutron Threshold Activation Detectors (TAD)," *J. Inst.*, vol. 7, pp. C02042–C02042, Feb. 2012.

[2] J. Stevenson, T. Gozani, M. Elsalim, C. Condron, and C. Brown, "Linac based photofission inspection system employing novel detection concepts," *Nuclear Instruments and Methods in Physics Research Section A: Accelerators, Spectrometers, Detectors and Associated Equipment*, vol. 652, pp. 124–128, Oct. 2011.

[3] P. Sibczynski, A. Dziedzic, K. Grodzicki, J. Iwanowska-Hanke, M. Moszyński, L. Swiderski, A. Syntfeld-Każuch, D. Wolski, F. Carrel, A. Grabowski, M. Hamel, F. Laine, A. Sari, A. Iovene, C. Tintori, C. Fontana, and F. Pino, "New perspectives for undoped CaF_2 scintillator as a threshold activation neutron detector," *EPJ Web of Conferences*, vol. 170, p. 07012, 2018.

[4] M. Hamel, P. Sibczynski, P. Blanc, J. Iwanowska, F. Carrel, A. Syntfeld-Każuch, and S. Normand, "A fluorocarbon plastic scintillator for neutron detection: Proof of concept," *Nuclear Instruments and Methods in Physics Research Section A: Accelerators, Spectrometers, Detectors and Associated Equipment*, vol. 768, pp. 26–31, Dec. 2014.

[5] T. Goorley, M. James, T. Booth, F. Brown, J. Bull, L. J. Cox, J. Durkee, J. Elson, M. Fensin, R. A. Forster, J. Hendricks, H. G. Hughes, R. Johns, B. Kiedrowski, R. Martz, S. Mashnik, G. McKinney, D. Pelowitz, R. Prael, J. Sweezy, L. Waters, T. Wilcox, and T. Zukaitis, "Initial MCNP6 Release Overview," *Nuclear Technology*, vol. 180, pp. 298–315, Dec. 2012.

[6] J. K. Shultis and R. E. Faw, "An MCNP primer," tech. rep., Dept. of Mechanical and Nuclear Engineering, Kansas State University, Manhattan, KS 66506, 2011.

Development of a driftless Johnson noise thermometer for nuclear applications

Jonathan V. Pearce[1], Paul Bramley[2], and David Cruickshank[2]
[1] National Physical Laboratory, Teddington, UK
[2] Metrosol Ltd, Paulersbury, UK

Abstract—**Existing temperature sensors such as thermocouples and platinum resistance thermometers suffer from calibration drift, especially in harsh environments, due to mechanical and chemical changes (and transmutation in the case of nuclear applications). A solution to the drift problem is to use temperature sensors based on fundamental thermometry (primary thermometers) where the measured property is related to absolute temperature by a fundamental physical law. A Johnson noise thermometer is such a sensor and uses the measurement of the extremely small thermal voltage noise signals generated by any resistive element to determine temperature using the Johnson-Nyquist equation. A Johnson noise thermometer never needs calibration and is insensitive to the condition of the sensor material, which makes it ideally suited to long-term temperature measurement in harsh environments. These can include reactor coolant circuits, in-pile measurements, nuclear waste management and storage, and severe accident monitoring. There have been a number of previous attempts to develop a Johnson noise thermometer for the nuclear industry, but none have achieved commercialization because of technical difficulties. We describe the results of a collaboration between the National Physical Laboratory and Metrosol Limited, which has led to a new technique for measuring Johnson noise that overcomes the previous problems that have prevented commercialization. The results from a proof-of-principle prototype that demonstrates performance commensurate with the needs of nuclear applications is presented, together with details of progress towards the commercialization of the technology. The development partners have effected a step change in the application of primary thermometry to industrial applications and seek partners for field trials and further exploitation.**

Keywords—**Thermometry, Johnson noise, Johnson noise thermometry, traceability, ITS-90, temperature measurement**

I. Introduction

THERMOMETRY in nuclear processes generally relies on conventional sensors such as thermocouples or resistance thermometers. These thermometers need to be calibrated [1], and as the sensor material changes in harsh operating environments, the calibration also changes in an unknown manner [2,3]. In general calibration drift corresponding to between tens and hundreds of °C can be expected [4,5]. Such drift leads to uncertainty about the process temperature, higher operating costs (frequent and costly sensor recalibrations/replacements), lower generating efficiencies (processes are operated at lower temperatures to ensure adequate safety margins) and compromises safety (temperature uncertainty being critical to the safe monitoring and control of reactors and other processes).

Johnson noise thermometry (JNT) has long been considered as a suitable solution to this problem [6]. JNT relies on the measurement of the electrical noise arising from the random thermal motion of charge carriers (electrons) in a sensor [7], and is completely independent of any material degradation. It is in a class of thermometry techniques called 'primary' thermometry [8], in that it can provide a direct measure of the temperature without recourse to calibration. As no calibration is required, it also does not experience calibration drift. JNTs exist mainly in National Measurement Institutes for extremely accurate metrology applications [6] such as measurement of the Boltzmann constant [9,10] for the redefinition of the kelvin [11,12], but these devices are large, slow, impractical, costly, and do not have sufficient immunity to electromagnetic interference (EMI). A JNT has also been applied to nuclear environments [13] which was successful, but slow, and it was not commercialized.

The reason JNT has not become widespread, despite the application of considerable ingenuity [14,15], is because there are a number of significant technical challenges [6], the most demanding of which is the measurement of the very small signals in the harsh electromagnetic environments encountered in industry. The main goal of the current development work is to resolve the technical challenges that have so far prevented successful deployment of this technique in the nuclear power industry.

We describe a collaboration between Metrosol and NPL to design, build, test for electromagnetic compability (EMC) immunity to the conditions expected in the heavy industrial environment found in nuclear power stations, and metrologically validate a practical JNT. The aim is to provide a technology platform to facilitate the development, with industry partners, of products tailored for the specific technical and regulatory requirements of the industry.

Manuscript received 10 April 2019. This work was supported in part by Innovate UK and the UK government's Department for Business, Energy and Industrial Strategy.

Jonathan Pearce is with the National Physical Laboratory, Hampton Road, Teddington, TW11 0LW, United Kingdom (telephone: +44 (0)20 8943 6886, e-mail: jonathan.pearce@npl.co.uk)

Paul Bramley is with Metrosol Limited, Plum Park Estate, Watling Street, Paulerspury, Northamptonshire, NN12 6LQ, United Kingdom (telephone: +44 (0)1327 810284, e-mail: paul.bramley@metrosol.co.uk)

David Cruickshank is with Metrosol Limited, Plum Park Estate, Watling Street, Paulerspury, Northamptonshire, NN12 6LQ, United Kingdom (telephone: +44(0)1327 810284, e-mail: david.cruickshank@metrosol.co.uk)

As newcomers to the industry, Metrosol and NPL seek partners with whom to collaborate and eventually license the core technology, to provide a faster and more effective route to market.

A first description of the practical Johnson noise thermometer was presented at the ANIMMA 2015 conference [16,17] and some more recent developments have been presented in [18]. In this paper we summarise the key developments since 2015.

II. PRACTICAL JNT

The proposed JNT technology makes use of the Nyquist equation [19] to relate the mean noise voltage V_T measured across a resistor to the absolute temperature T:

$$\overline{V_T} = \sqrt{4kTR\Delta f} \qquad (1)$$

where k is Boltzmann's constant, R is its resistance, and Δf is the bandwidth. This allows the temperature to be determined regardless of the physical properties of the sensor (e.g. if sensor damage changes the resistance, there is no effect on the temperature measurement because resistance is continuously measured). The main difficulty is the magnitude of the noise voltage: even with a bandwidth of 1 MHz, a 1 kΩ resistor will generate an RMS voltage of just 1 μV, which is at the limit of current precision measurement capabilities.

The technical challenge is to measure this extremely small signal in an industrial environment where sensors may be exposed to interference signals up to 10 V/m over the entire frequency band from direct current (DC) to several GHz. The current approach employs the established technique of using correlators to extract the required signal while rejecting unwanted noise, but new technologies have been developed for amplification and for providing a reference noise source. To do this, a more modest two sigma uncertainty (about 0.28 °C) is exchanged for much faster response times (about 6.5 s).

A key difficulty with JNT is the fact that the Nyquist equation cannot be used directly because the bandwidth Δf cannot be known. To overcome this, traditional JNTs switch between a reference noise source with Johnson noise voltage V_0, source resistance $R(T_0)$ at a reference temperature T_0 with bandwidth Δf_0, and another Johnson noise source V_T, source resistance $R(T)$ at an unknown temperature T with bandwidth Δf [20]. The switching eliminates spurious effects due to time dependent drifts in the gain and bandwidth of the amplifiers and filters [21]. The unknown temperature is then given by expressing the Nyquist equation in ratio form:

$$T = \frac{\overline{V_T^2}}{\overline{V_0^2}} \frac{R(T_0)}{R(T)} T_0 \qquad (2)$$

There are several key factors which limit the accuracy of the switched-input correlation approach [20]. These are:

- The random nature of the noise signal, which necessitates averaging of the signal for a long enough period with a sufficiently wide bandwidth to reduce statistical fluctuations to the desired level. The time and bandwidth needed to achieve a given fractional uncertainty is given by Rice [22-24] and amounts to about 10 s to achieve an uncertainty of 0.1 % (i.e. about 1 °C) with a bandwidth of 100 kHz. A key development in the new JNT system is the use of higher resistances that therefore generate more Johnson noise and correspondingly reduces the measurement needed to achieve a given uncertainty.
- Electronic component nonlinearity: the tiny signal requires very high amplifier gain and wide bandwidths, and the thermometer must be operated as a comparator, comparing the noise powers from the reference arm at T_0 and the sensing arm at T. These nonlinearities vary with time and ambient temperature, and cannot be calculated or measured. To minimize errors, the same correlated and uncorrelated noise powers are needed in the two arms of the comparator, which means the sensing resistances must be chosen so that $R(T_0)T_0 = R(T)T$.
- Transmission line effects: the preamplifiers must be located some distance from the sensors, which means that the transfer function of the transmission line is significant. The problem is the interaction of the sensor resistance with the transmission line, which results in a dependence of the transmission line frequency response on the sensor resistance. This causes an error that is most simply eliminated by employing a constant sensing resistance $R(T_0) = R(T)$. This requirement conflicts directly with the previous one needed to reject nonlinearity. Limiting the bandwidth can reduce the transmission line error, but in practice it still limits the achievable accuracy – furthermore, limiting the bandwidth increases the measuring time needed for a given accuracy.

Another way of looking at this is to study the assumption $\Delta f = \Delta f_0$ required by Equation (2), since as we switch between the two measurements there will inevitably be a change in the frequency response of the measurement system. The dominant effect is the capacitance of the cables (C for the sense resistor cable connected to R, C_0 for reference resistor cable connected to R_0) connecting the sensor and reference resistors to the input switch [25]. In order for the bandwidths to match, it is required that $RC = R_0C_0$. This can be met for a particular situation by trimming the circuit (e.g. by adding capacitance to one set of cables to balance the system). However, if either resistor changes value with time or temperature, there will be a mismatch in the effective bandwidths, leading to measurement errors. This means that the system would be susceptible to changes in a property of the sensor – the very problem we are trying to solve. Prior JNTs have solved this by ensuring both RC and R_0C_0 are much less than $1/\Delta f$ so that the measurement is made on the flat part of the system frequency response

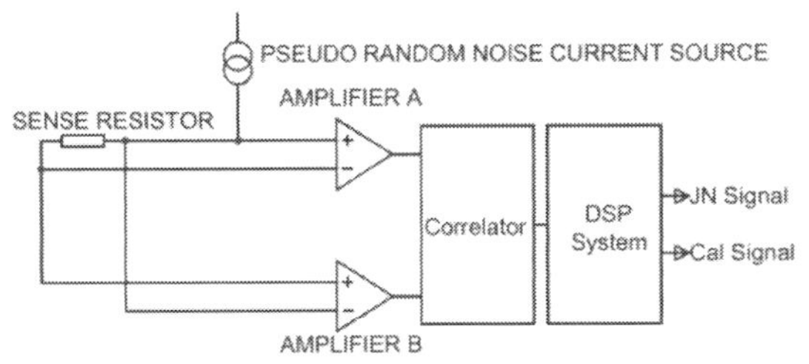

Fig. 1. New JNT topology.

affected by these impedances. This limited sensing resistors to around 100 Ω and bandwidths to around 100 kHz, which in turn limits the size of the Johnson noise signal.

A way around these issues is to replace the reference resistor with a cryogenic quantized voltage noise source based on Josephson junction arrays [26] or, more practically, a white pseudo-random noise (PRN) voltage source [24]. The digitally synthesized PRN signal is the sum of a number of sinewaves (of equal amplitude) that are equally spaced in frequency and with random phase. This generates a Gaussian noise-like waveform that will repeat with a period equal to the reciprocal of the frequency spacing. The waveform can be stored as digital samples and then recreated by streaming the sampled waveform to a digital-to-analogue converter (DAC). The 'comb' of tones comprising the PRN calibration signal then samples the frequency response of the measurement system including tracking changes over time.

III. NEW JNT MEASUREMENT TOPOLOGY

We have seen that a large number of problems stem from switching between a known white noise reference signal and the unknown Johnson noise signal. A new measurement topology was therefore devised to overcome the need to switch, as shown in Fig. 1.

The calibration signal (a PRN waveform) is injected as current into the measurement circuit to generate a calibration voltage signal in the sensor resistor; this is of course overlaid with the Johnson noise signal. The calibration signal and the Johnson noise signal both experience the same frequency response as they pass through the connecting cables, amplifiers and correlator. The Johnson noise and calibration signals are then separated from the composite signal by a digital signal processing (DSP) system. The ratio of the Johnson noise and calibration signal can then be calculated in order to determine the absolute Johnson noise power and hence the temperature of the sense resistor.

Importantly, this topology is tolerant of a 'non-flat' frequency response since the two signals experience the same frequency response. As a result, it is possible to use a much wider bandwidth (1 MHz or more), including frequencies where there is significant attenuation arising from the connecting cables. In the current realization, the PRN signal comprises tones with randomized phase at 152 Hz intervals between 10 kHz and 1.2 MHz. Lower frequency tones are used to determine the resistance of the sense resistor using Ohm's law, since the attenuation by the cable capacitance is negligible at these frequencies. In addition to the current tones, one other piece of information is needed to calculate resistance, namely amplifier gain at low (i.e. flat) frequency, or voltage tones.

For practical reasons, the calibration current is injected at a point adjacent to one of the pre-amplifier outputs. Fig. 2 shows this arrangement after folding out the circuit in Fig. 1.

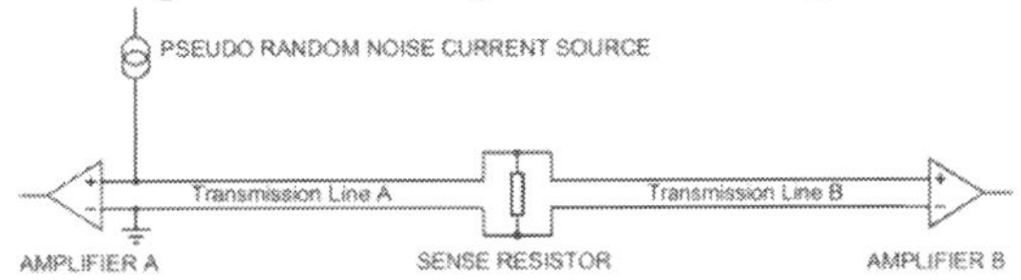

Fig. 2. Transmission lines and current injection for the new Johnson noise measurement topology.

A lumped circuit model of the effect of the transmission lines on the signal attenuation and resulting measurement error was developed, to enable the exploration of parameter space for the system. The errors reduce as the value of the sense resistor is increased. Fig. 3 shows the effect of increasing the sensor resistance for a cable with total series resistance of 1 Ω, transmission lines of capacitance 0 pF to 100 pF and bandwidths of 0 MHz to 1 MHz.

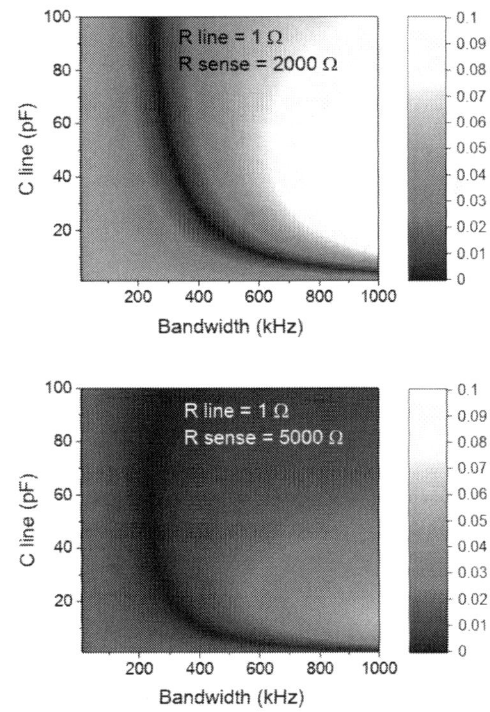

Fig. 3. Effect of sensor resistance on the measurement error, for two different values of the sense resistance [18]. Colour corresponds to the magnitude of the error, ranging from 0 % (blue) to 0.1 % (red).

Fig. 3 shows the benefit of using a higher sense

resistance. Increasing the sensor resistance also increases the magnitude of the Johnson noise signal.

As we can measure the frequency response using the tones, we are also able to use a much higher bandwidth as we are no longer limited to operating on the flat part of the input circuit's frequency response. Overall, the ability to use a high value resistor and the consequent increased Johnson noise signal, the ability to access a higher bandwidth, and the lack of switching mean that the measurement time for a given uncertainty is minimized. In the current implementation this reduces the measurement time by a factor of √20 over prior JNTs (1 MHz c.f. 100 kHz bandwidth and a factor of two for not switching).

The inventions described above are subject to a filed patent, GB2545176 [27].

IV. RESULTS

The first benchtop prototype (JNT1) and the second, more compact prototype (JNT2) are shown in Fig. 4. The device consists of a long probe with a similar external aspect to thermocouples or resistance thermometers, and a 'sensing head' containing the sensing electronics. The probe length is important as longer probes will increase cable capacitance and cause low pass filtering, and this will limit the available bandwidth. For example, a 0.5 m probe may achieve a bandwidth of around 1 MHz, while a 5 m probe of the same construction would only achieve a bandwidth of around 100 kHz. This would increase the measurement time needed to achieve the same uncertainty by a factor of √10.

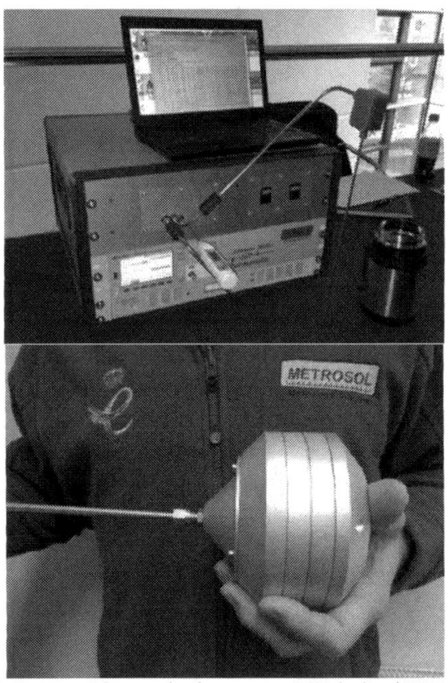

Fig. 4. Prototype Johnson noise thermometers JNT1 (top) and JNT2 (bottom).

Initial tests were conducted on JNT1 to ensure that the measured spectra were clean with no evidence of contamination from external noise sources. Early results with JNT1 showed contaminated spectra and highlighted the sensitivity of the system to external interference (Fig. 5, top panel). This problem was left unresolved in previous work [14]. Considerable effort has been devoted to improving cabling, grounding and shielding. In particular, the sensitive parts of the circuit were modified to use a full tri-axial probe connection. Following these modifications, the spectra became consistently 'clean', as exemplified by the spectrum from JNT2 in Fig. 5, bottom panel.

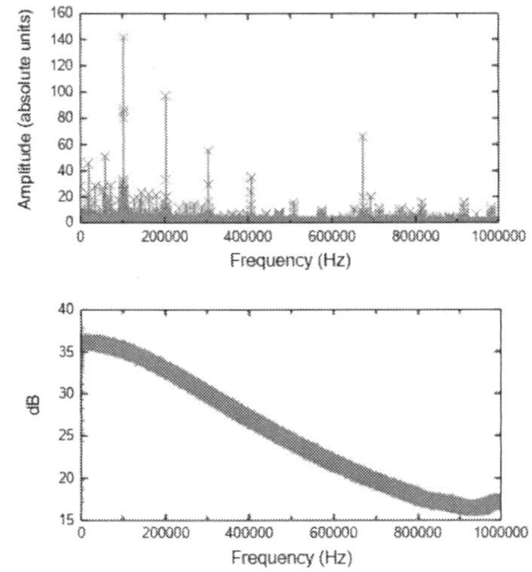

Fig. 5. A very early spectrum from JNT1 showing heavy contamination from EMI (top) and a later spectrum after dramatic circuit improvements in JNT2 (bottom).

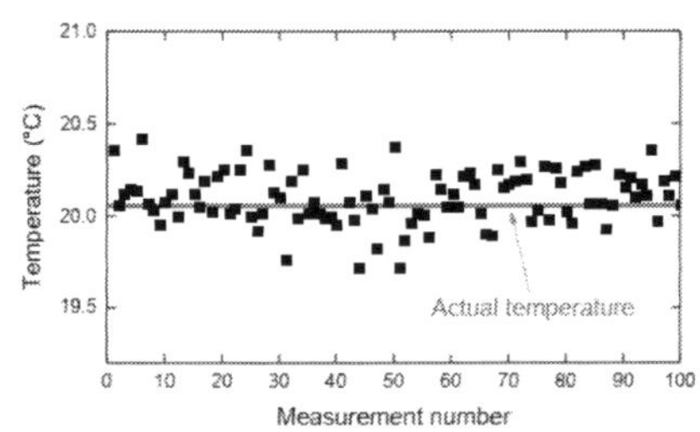

Fig. 6. 100 measurements of temperature made with the JNT over three days (JNT 1 prototype).

The JNT1 system was extended to include the calibration tones allowing the absolute power and hence the temperature of the sensor resistor to be determined. The temperature of a 5 kΩ foil resistor was then measured in an oil bath maintained at 20.06 °C ± 0.01 °C, as measured with a traceably calibrated reference thermometer. A series of 100 readings were taken over a three day period as shown in Fig. 6. The standard deviation of the temperature determined using the Johnson noise measurements was found to be 0.14 °C.

The much improved JNT2 device was submitted for EMC testing applicable to heavy industrial environments. The radiated field immunity was tested to EN 61000-4-3 [28]. Field strengths were 10 V m^{-1} from 80 MHz to 1000 MHz and 3 V m^{-1} from 1 GHz to 2.7 GHz; in both frequency ranges there was no observable effect on the thermometer. It was also tested for radiated emissions to EN 55011 [29] (from 30 MHz to 1000 MHz) and showed no detectable emissions. This represents an extremely important milestone in the development of a practical JNT; insufficient immunity to EMC thwarted previous efforts.

The JNT2 is currently a compact, self-contained thermometer capable of making temperature measurements without any *a-priori* calibration, and has extremely high immunity to EMC. As such it represents a step change improvement over the prototype plans presented at the ANIMMA 2015 conference. In terms of the Technology Readiness Level (TRL) [30], in the last 4 years it has jumped from TRL 2 (Technology concept and/or application formulated) to TRL 5 (Component and/or breadboard critical function verification in relevant environment).

V. CONCLUSION

The continuing development of a practical Johnson noise thermometer (JNT), as part of a joint activity between NPL and Metrosol Limited, has been described. The development, characteristics and performance have been described. The JNT measures the voltage arising from the thermal motion of charge carriers, which are related to temperature through fundamental physics. Importantly, the temperature measurement is unaffected by changes in the property of the sensor (except the resistance which can be measured simultaneously with the temperature) and is independent of any calibration. This opens up the prospect of truly 'driftless' temperature sensing which has substantial benefits for the measurement and control of processes needing reliable temperature measurement. Over the last 3 years the JNT has undergone a step change in size and EMC resilience, and currently measures temperature with an uncertainty (two sigma) of about 0.28 °C with a measuring time of about 6.5 s.

REFERENCES

[1] J.V. Pearce, "Extra points for thermometry", Nat. Phys. Vol. 13, January 2017, p. 104
[2] M.J. Kelly, W.W. Johnston and C.D. Baumann, "The Effects of Nuclear Radiation on Thermocouples", in Temperature: Its Measurement and Control in Science and Industry (TMCSI), Vol. 3 part 2, pp. 265-269, ed. Herzfeld, Reinhold Publishing Corp, NY, 1962
[3] W.E. Browning and C.E. Miller, Calculated Radiation Induced Changes in thermocouple Composition, Temperature: Its Measurement and Control in Science and Industry (TMCSI), Vol. 3 part 2, pp. 271-276, ed. Herzfeld, Reinhold Publishing Corp, NY, 1962
[4] H.M. Hashemian and Jun Jiang, "Nuclear Plant Temperature Instrumentation", Nuclear Engineering and Design 239, 3132-3141 (2009)
[5] J.L. Rempe, D.L. Knudson, K.G. Condie, S. Curtis Wilkins, "Evaluation of Specialized Thermocouples for High-Temperature In-Pile Testing. Proceedings of ICAPP '06 Reno, nV USA, June 4-8, 2006, paper 6086
[6] D.R. White, R. Galleano, A. Actis, H. Brixy, M. De Groot, J. Dubbeldam, A.L. Reesink, F. Edler, H. Sakurai, R.L. Shepard, J.C. Gallop, "The status of Johnson noise thermometry", Metrologia 33, 325-335 (1996)
[7] J.B. Johnson, Thermal agitation of electricity in conductors. Nature 119, 50–51 (1927)
[8] M.R. Moldover, W.L. Tew, H.W. Yoon, "Advances in thermometry", Nat. Phys., 12, 7-11 (2016)
[9] S.P. Benz, A. Pollarolo, J. Qu, H. Rogalla, C. Urano, W.L. Tew, P.D. Dresselhaus, D.R. White, "An electronic measurement of the Boltzmann constant", Metrologia, 48, 142 (2011)
[10] J. Qu, S.P. Benz, K. Coakley, H. Rogalla, W.L. Tew, R. White, K. Zhou, Z. Zhou, "An improved electronic determination of the Boltzmann constant by Johnson noise thermometry", Metrologia, 54, 549 (2017)
[11] M. de Podesta, Rethinking the kelvin, Nat. Phys. 12, 104 (2016)
[12] G. Machin, The kelvin redefined, Meas. Sci. Technol. 29 022001 (2018)
[13] H. Brixy, R. Hecker, J. Oehmen, "Temperature measurements in the high-temperature range (1000 – 2000 °C) by means of noise thermometry", High Temperature High Pressure, 23, 625-631, 1991
[14] Roger A. Kisner et al. 2005. Development of a Johnson Noise Thermometer for Nuclear Power Use, Final Report of the International Nuclear Energy Research Initiative subtask 2.2: Development of Enhanced Reactor Operation Strategy Through Improved Sensing and Control at Nuclear Power Plants
[15] H. Brixy, R. Hecker, J. Oehmen, Temperature measurements in the high-temperature range (1000 – 2000 °C) by means of noise thermometry, High Temperature High Pressure, 23, 625-631, 1991
[16] M. Giot, L. Vermeeren, A. Lyoussi, C. Reynard-Carette, C. Lhuillier, P. Mégret, F. Deconinck, B.S. Gonçalves, "Nuclear instrumentation and measurement: a review based on the ANIMMA conferences", EPJ Nuclear Sci. Technol. 3, 33 (2017)
[17] J.V. Pearce, A. Greenen, P. Bramley, D. Cruickshank, "Towards a practical Johnson noise thermometer for long-term measurements in harsh environments", proc. 2015 4th International Conference on Advancements in Nuclear Instrumentation Measurement Methods and their Applications (ANIMMA) DOI: 10.1109/ANIMMA.2015.7465506 (2016)
[18] P. Bramley, D. Cruickshank, J.V. Pearce, "The development of a practical, drift-free, Johnson noise thermometer for industrial applications", Int. J. Thermophys. 38, 25 (2017)
[19] H. Nyquist, "Thermal Agitation of Electric Charge in Conductors", Phys. Rev. 32, 110-113 (1928)
[20] S.P. Benz, J.M. Martinis, S.W. Nam, W.L. Tew, D.R. White, "A New Approach to Johnson Noise Thermometry using a Josephson Quantized Voltage Source for Calibration", Proceedings of TEMPMEKO 2001: 8th International Symposium on Temperature and Thermal Measurements in Industry and Science, 19-21 June 2001, Berlin, Germany
[21] D.R. White, S.P. Benz, "Constraints on a synthetic-noise source for Johnson noise thermometry", Metrologia 45, 93–101 (2008)
[22] S. O. Rice, "Mathematical Analysis of Random Noise", Bell Syst. Tech. J., 23, 282-332 (1944)
[23] S.O. Rice, "Mathematical analysis of random noise", Bell Syst. Tech. J. 24, 46-156 (1945)
[24] D.R. White et al., "Measurement time and statistics for a noise thermometer with a synthetic-noise reference", Metrologia 45, 395 (2008)
[25] D.R. White and J.-F. Qu, "Frequency-response mismatch effects in Johnson noise thermometry", Metrologia 55(1), 38 (2018)
[26] S.W. Nam et al., "Johnson noise thermometry measurements using a quantized voltage noise source for calibration", IEEE Trans. Instrum. Meas. 52, 550–554 (2003)
[27] https://www.ipo.gov.uk/p-ipsum/Case/PublicationNumber/GB2545176
[28] EN 61000-4-3: Electromagnetic compatibility (EMC) – Part 4-3: Testing and measurement techniques – Radiated, radio-frequency, electromagnetic field immunity test
[29] EN 55011: Industrial, scientific and medical equipment — Radio-frequency disturbance characteristics — Limits and methods of measurement
[30] ISO 16290:2013, Space systems – Definition of the Technology Readiness Levels (TRLs) and their criteria of assessment

APPLICATION OF CDZNTE DETECTORS FOR CONTROL OF INITIAL ENRICHMENT OF FRESH NUCLEAR FUEL AT REFUELING

O.V. Maslov, V.A. Mokritskiy and **A.V. Sokolov**

Odessa National Polytechnic University,
Ukraine, Odessa, 65044, Shevchenko Ave. 1
Emails: *maslov.v.oleg@gmail.com*; maslov@opu.ua;

The task of determining the initial enrichment of nuclear fuel (NF) at nuclear power plants in real time is relevant for both economic indicators of operation of the power unit and for all components of nuclear safety (nuclear safety, security and safeguards). The technology used by nuclear fuel cycle companies is not suitable for use at nuclear power plants. It is proposed to use a measuring system based on CdZnTe semiconductor detectors designed to control the burn-up depth of nuclear fuel for estimating the initial enrichment of fresh nuclear fuel in fuel assemblies (FA). The initial enrichment is estimated with help of the gamma radiation of uranium isotopes of fresh nuclear fuel. A distinctive feature of the system that has been developed is the use of passive tomography technology to take into account the self-absorption of radiation in the FA. The promise of the proposed method for estimating initial enrichment is confirmed by experimental measurements and simulation results. It is shown that the system developed makes it possible to determine the enrichment of nuclear fuel, taking into account its profiling by the volume of FA. Also, the results of measurements confirmed the possibility of using the proposed technical solutions for CdZnTe detectors to create portable monitoring systems for special nuclear material in the field. The solution of this task is of great importance for national security.

Keywords: CdZnTe detectors, gamma-ray imaging, gamma-ray spectroscopy, special nuclear materials monitoring.

Full control over the state of nuclear fuel, at all stages of its life cycle, at nuclear power plants necessarily requires an instrumental assessment of the initial enrichment of fresh fuel before it is loaded into the reactor core. Nuclear fuel (NF) is a mixture of ^{235}U isotopes with a range of content of 1.6-4.4 %, therefore the initial isotope composition of nuclear materials is often the object of measurements. However, there are no such systems for determining the initial enrichment of nuclear fuel at nuclear power plants. Therefore, at present, it is necessary to rely solely on the data declared by the manufacturer. Knowledge of the actual deviation of the initial enrichment is of great importance in order to optimize the fuel permutations, especially when operating at the final stages of the campaign.

The determination of the initial enrichment of NF (i.e., the content of the fissile isotope ^{235}U) in real time can be carried out through passive radiation measurements of the sample's self-radiation. Like all passive measurements, this method of analysis is non-destructive [1]. The determination of uranium enrichment level in samples is a key measurement in technological processes and in the production control on the uranium enrichment and nuclear fuel companies, it acts an important part in the international nuclear safeguards inspections to grantee that uranium fuel is used for peaceful purposes. The principles of enrichment measurement can be used to determine the content of any isotope if their radiation characteristics are known and some special measurement conditions [2,..,5] are met.

The main radiation used in passive non-destructive analysis of uranium samples is gamma radiation. When measuring the enrichment of ^{235}U, the gamma radiation line with a gamma-ray energy of 186 keV is most often used. It is the most distinguished single gamma-emission line for any uranium samples enriched by ^{235}U above the natural level.

The method for determining the enrichment of uranium by gamma radiation was for the first time used to control the gas cylinders with UF_6 [1, 4, 6]. The method of the measurement included recording the radiation of the uranium sample through the channel of the collimator by the detector of the gamma radiation. The enrichment was determined by the intensity of gamma radiation of the ^{235}U isotope with 186 keV energy. If the sample is thick enough, the detector detects gamma rays with an energy of 186 keV only from an insignificant part of the total sample, because of the strong absorption of low-energy quanta in the fuel assembly materials. In this sense, in the terminology of [1], fuel element is a "thick sample". The "visible volume" of the sample is characterized by the configuration of the collimator, the detector geometry and by the length of the free path of radiation with an energy of 186 keV in the sample material [1, 3, 4].

We propose a measuring system based on CdZnTe detectors [7, 8, 9], created for the burn-up analysis of spent nuclear fuel, can be used to estimate the initial enrichment of an overloaded fresh fuel. In this case, the main problem may be the low detection efficiency of the detectors used. To test the possibility of using CdZnTe detectors, fuel pellets with an initial enrichment of 4.4 % were measured. One of the obtained spectra is shown in Fig. 1 and Fig. 2, using different scales of the axes for displaying the count results (i.e. measured intensity) in the channels of the multichannel analyzer. It is done to provide greater visibility of the peak of the total absorption with gamma radiation with energy of ^{235}U 205.3 keV. In the spectrum, well-identifiable lines of 186 keV and 205 keV of self-radiation of ^{235}U are observed. The lines of sample self-radiation of ^{238}U with energies of 742 keV, 766 keV, 786 keV and 1001 keV are visible as well. The 1001 keV line is usually used for analytical purposes.

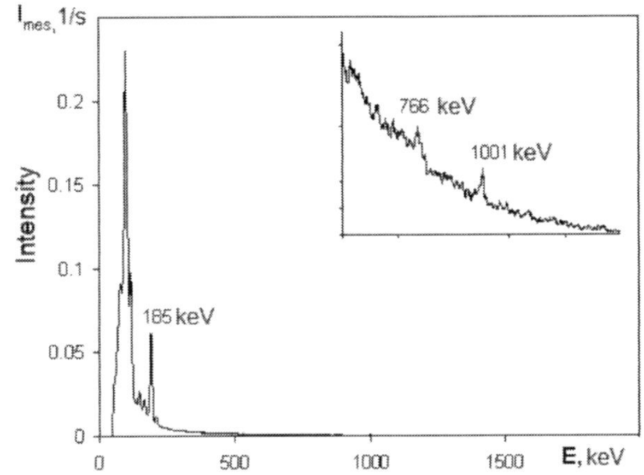

Fig.1 — The spectrum of gamma self-radiation of fuel pellets with initial enrichment of 4.4 %

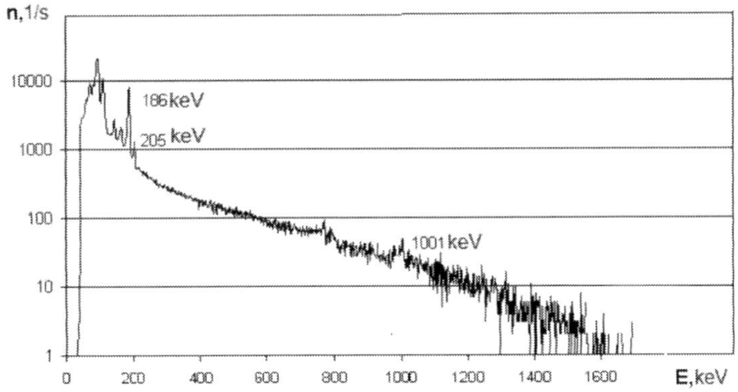

Fig.2 — The spectrum of gamma self-radiation of fuel pellets with an initial enrichment of 4.4 % on a semilogarithmic scale

For comparison, Fig. 3 shows the spectrum of the self-radiation of a metal sample of depleted uranium. The mass of uranium in both series of measurements was approximately the same. It can be seen that the analytical lines of the ^{235}U isotope are absent, and the radiation intensity is much lower. In both cases, the value of the parameter of the thickness of the sample is the "infinite thickness".

The determination of the initial enrichment using the ^{235}U self-radiation intensity measurements with 186 keV energy is carried out according to the following formula [1, 11]:

$$R = a \cdot I_{mes} \cdot F \cdot \exp(\mu_c \rho_c t_c) = F \cdot \exp(\mu_c \rho_c t_c)(a \cdot C_1 - a \cdot f \cdot C_2), \quad (1)$$

where I_{mes} is the net counting rate at the peak of full absorption;
C_1 and C_2 are the total counting rate in the peak of the total absorption and the counting rate of the background and the continuous component of the Compton distribution, respectively;
$F \cdot exp(\mu_c \rho_c t_c)$ are the experimentally determined coefficients for taking into account self-absorption and scattering of radiation in the fuel matrix material and fuel cladding walls.

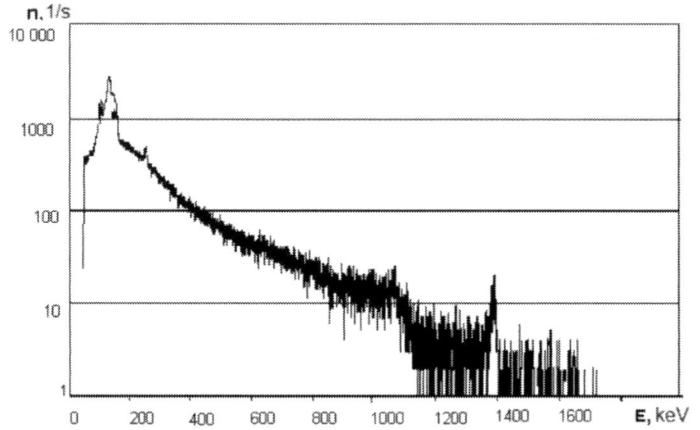

Fig.3 — Spectrum of the self-radiation of a metal sample of depleted uranium

Such a method is fairly simple when carrying out measurements of typical homogeneous samples, for example, in the measurement of barrels containing uranium dioxide. However, FA is an object with a complex geometry and in this case it cannot be considered as a "thick" bung. As a rule, it can be assumed that the main contribution to the counting rate of the full-absorption peak with 186 keV energy is given by one, or at most two rows of external fuel rods. To increase the recording efficiency and reduce the measurement time, the detector should be positioned as close as possible to the fuel assemblies, but in this case, there are observation points where so-called "firing" from fuel rods located in the inner rows are recorded, while the gamma radiation intensity is very uneven depending on the viewing angle [10]. An example of such a phenomenon, obtained by the modeling described below, is shown in Fig. 4. This leads to a strong dependence of the obtained results on the location of the detector relative to the measured FA. This effect is even more pronounced when measuring fuel assemblies containing fuel elements with different enrichment (for example, fuel assemblies marked "ED" contain an external row of fuel rods with an enrichment of 3.6 %, and internal fuel rods have an enrichment of 4.4 %).

To reduce the measurement error, we took into account the effect of self-absorption and scattering of radiation in the material of the fuel matrix and fuel cladding walls, enrichment can be determined on the basis of data on the ratio of the counting rate at the peak of full-energy absorption of 186 keV of the ^{235}U self-radiation and the counting rate at the peak of full absorption of 1001 keV of self-radiation of ^{238}U. Fig. 5 shows the results of measurements of gamma-radiation intensity of one (gray bars), two (black bars), and three (white bars) layers of VVER-1000 fuel pellets. It is clearly seen that an increase in the thickness of the measured layer leads to a significant increase in the emission intensity of the 1001 keV line of self-radiation of ^{238}U and secondary X-ray radiation due to the interaction of the primary radiation with the fuel matrix.

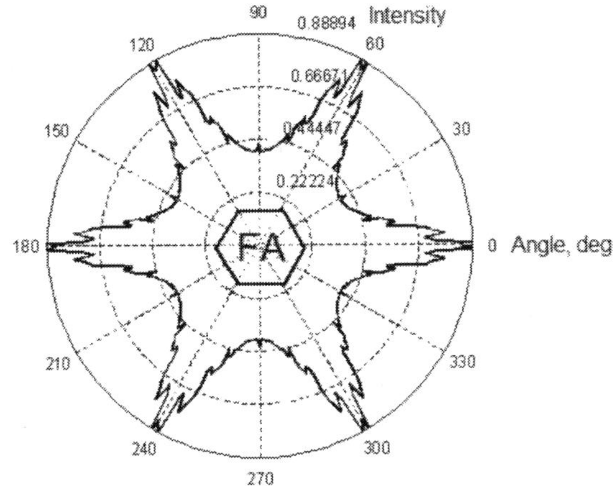

Fig. 4. The intensity of the gamma radiation of ^{235}U 186 keV, depending on the angle of the detector (the distance "FA axis — detector" is 14 cm).

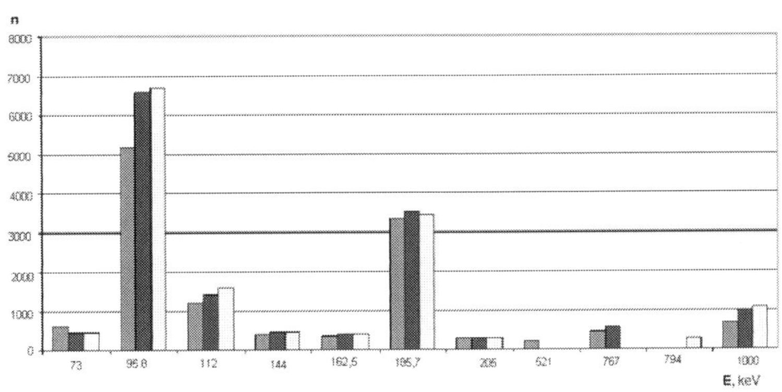

Fig.5 — Intensities of the gamma self-radiation during measurements from one, two and three layers of fuel pellets with an initial enrichment of 4.4%

The dependence of enrichment on the ratio of the self-radiation intensities of ^{235}U and ^{238}U for FA will obviously be nonlinear due to the different absorption of gamma quanta with energies of 186 keV and 1001 keV. For the empirical determination of this dependence, it is necessary to carry out a sufficient number of measurements for a fuel assembly with a different enrichment in a statistical sense, which in principle is not a problem under NPP conditions [10, 12]:

To confirm the possibility of estimating the initial enrichment of nuclear fuel, a two-dimensional plane model of the field of the gamma radiation of uranium isotopes contained in the fuel assemblage was constructed, according to which the location of the detector at the n-th observation point at a distance R_n from the FA axis the gamma-radiation of j-th isotope of uranium with energy E_γ at the location of the detector is:

$$I_n^{jk} = \sum_m A_{mj}\, k_{jk} w_{mn}, \tag{2}$$

where A_{mj} is the activity of the j-th uranium isotope for m-th fuel element taking into account the actual value of enrichment;

k_{jk} is the quantum yield of the k -th gamma line for the j-th isotope;

w_{mn} is the coefficient of the contribution of the m-th fuel element to the intensity of the gamma radiation of the j-th isotope with energy E_k, taking into account the effects of attenuation when the gamma-ray beam propagates from the m-th fuel element to the n-th point of observation, $m = 1, \dots, 312$.

Taking into account the registration efficiency, the measured intensity (counting rate) $\varepsilon(E_k)$ for an isotope with energy E_k the measured gamma radiation intensity of the j -th isotope with energy E_k is expressed as

$$I_n^{jkmes} = \sum_m A_{mj} k_{jk} w_{mn} \, \varepsilon(E_k). \quad (3)$$

In the future, we omit the indices jk , since we are talking about the intensity of gamma radiation at the point of registration of the j-th isotope with energy E_k, i.e. in the simulation it is assumed that the peaks of the gamma spectrum are isolated and processed.

The attenuation in the propagation of gamma radiation is mainly connected with two physical mechanisms [13]:

— geometric attenuation associated with the spatial divergence of the beam

$$P_{geom} = \frac{S_{det}}{4\pi R_{mn}^2}, \quad (4)$$

where S_{det} is the active area of the detector;
R_{mn} is the distance from the center of the m-th fuel rod to the n-th location of the detector;

—physical attenuation of gamma radiation during passage through various media (in particular, when passing through UO_2 in other fuel elements located on the path of gamma-ray run to the detector, in their zirconium shells, and also in the water environment — both inside the FA and outside of it, in the overload pool); the physical attenuation of radiation from the m-th fuel element is expressed as

$$P_{phys} = exp(\sum_l \mu_l \, l_l), \quad (5)$$

where μ_l is the coefficients of linear attenuation of gamma quanta with energy E_k for the corresponding media;
l_l is the length of run of gamma radiation in a medium of type l (uranium dioxide, zirconium, or water).

Thus, the contribution coefficient of the m-th fuel element for the n-th position of the detector, associated with attenuation, is

$$w_{mn} = \frac{S_{det}}{4\pi R_{mn}^2 \, exp(\sum_l \mu_l l_l)}. \quad (6)$$

According to the formulated model, the fields of gamma self-radiation with an energy of 185 keV of the ^{235}U isotope and with an energy of 1001 keV for ^{238}U at different locations of the detector were calculated. All the results presented below are calculated from the relations (2), (3), (6) for the "FA axis — detector" distance of 22.5 cm. This is because of the fact that the developed equipment for monitoring the fuel burn-up depth was specially designed for placement on the working bar of the reloading machine (RM), and the specified size is determined by the diameter of the working rod of the RM. The angular diagram of the intensity distribution of gamma radiation with an energy of 185 keV ^{235}U is shown in Fig. 6. It can be seen that the intensity distribution practically repeats the shape of the fuel assembly. At separate viewing angles, there are irregularities and "sharp increases", the magnitude of which is small and does not exceed 7-8 % of the mean. There are smooth maxima at the angles corresponding to FA edges.

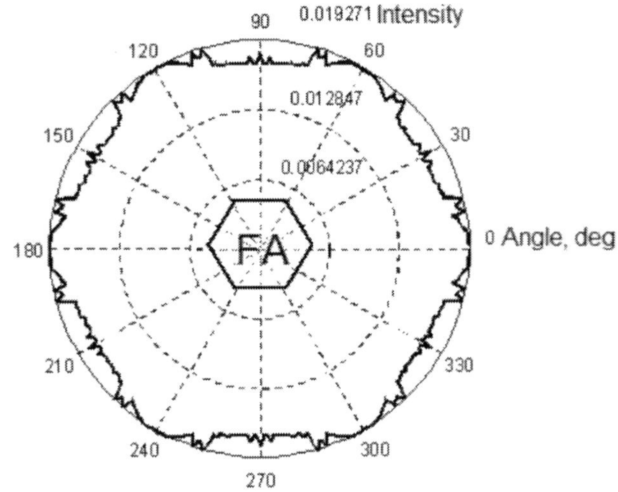

Fig.6 — Angular distribution of intensity of gamma self-radiation for ^{235}U 185 keV (fresh fuel assembly).

The coefficients of the contributions of individual fuel elements w_{mn} are investigated by the relation (6). For the energy 185 keV of ^{235}U isotope in the Fig. 7, the inverse values of the contributions are given in a logarithmic scale. The type of the contribution pattern is typical for low-energy isotopes, in particular, at the observation angles corresponding to the FA edges, there is an acute "ridge", and a pronounced "multi-ridge structure" is observed on the midpoints of the faces.

Fig. 8 shows the angular distribution of the intensity of gamma radiation with an energy of 1001 keV of the ^{238}U isotope. The angular distribution is fairly smooth, the nonuniformity does not exceed 5-7 %. The corresponding energies of 1001 keV ^{238}U of the inverse values which are inverse to fuel element contribution coefficients are shown in Fig. 9. Their appearance is also quite typical for high-energy gamma quanta: in particular, the "ridge structure" is much less pronounced than in Fig. 7.

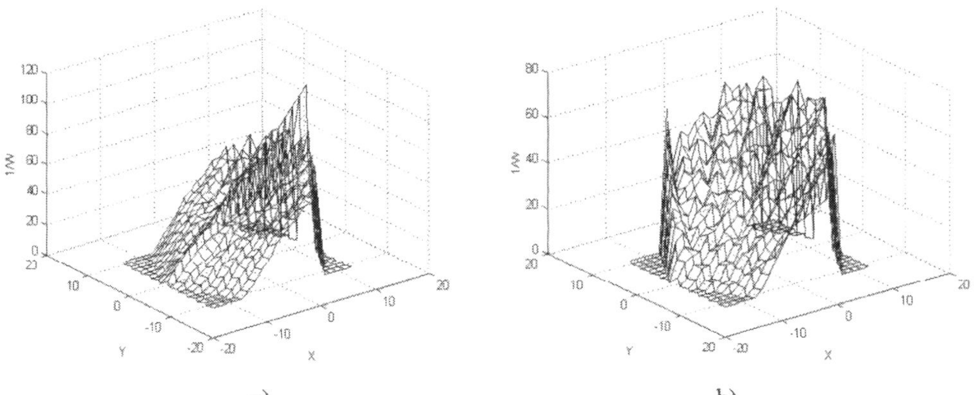

Fig.7 — Coefficients of fuel rod contributions (inverse values) for ^{235}U 185 keV (a – 0^0, b – 30^0).

To research the practical possibility of determining the initial burn-up of fresh nuclear fuel, the ratio of measured intensities of ^{238}U with energy of 1001 keV and ^{235}U with an energy of 185 keV is modelled, taking into account the registration efficiency $\varepsilon(E_k)$ [12] depending on the angle of the detector location (Fig. 10). It can be seen from the figure that the ratio of the measured intensities repeats to a considerable extent the picture in fig. 6: there are irregularities and small "sharp increases" observed in the form of peaks, and besides the peaks being concentrated at the viewing angles corresponding to the midpoints of edges. However, the absolute value of the peaks is small, which allows us to assume that the detectors are arranged exactly opposite to the edges.

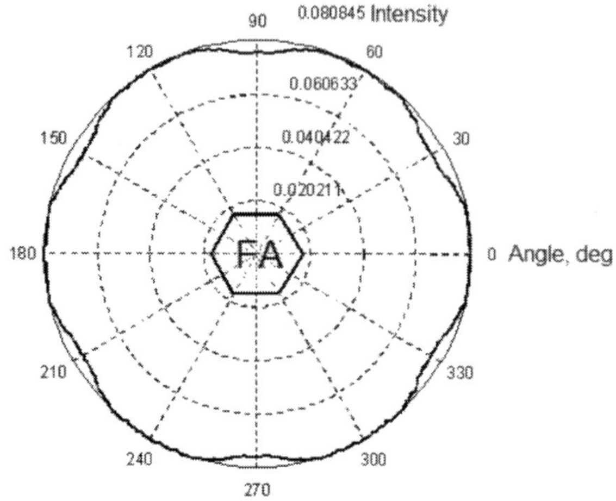

Fig.8 - Angular distribution of the intensity of the gamma self-radiation of ^{238}U with an energy of 1001 keV (fresh FA)

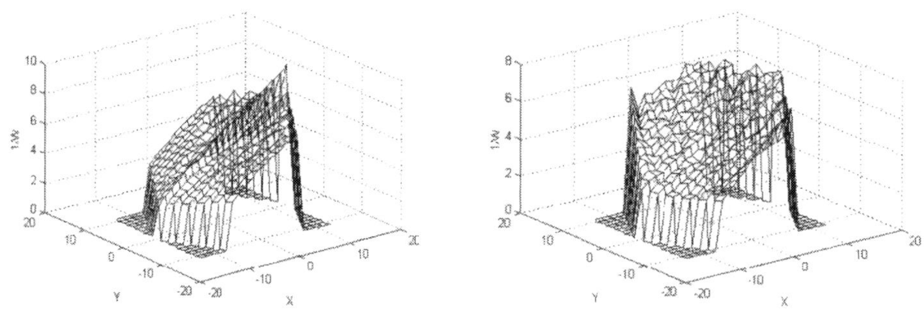

a) b)

Fig. 9 - Coefficients of fuel element contributions (inverse values) for ^{238}U 1001 keV (a –0^0, b – 30^0).

Quantitative analysis of the contribution coefficients w_{mn} for both analytical lines of ^{235}U, ^{238}U isotopes made it possible to determine the following: with an allowable unevenness of contribution coefficient of 40%, placing the detector in opposite to the midpoint of the fuel assembly allows coverage of the entire external row of fuel rods and 4 central fuel elements of the second row for 185 keV ^{235}U energy, and almost completely two fuel rods close to the detector for energy 1001 keV ^{238}U. Considering the homogeneity of fresh nuclear fuel inside the fuel assembly (except for the first row), this is quite sufficient to estimate the initial enrichment. The number of detectors should be determined experimentally, depending on the required accuracy of measurements.

The above experimental spectra and simulation results show that the developed technical facilities to control the burn-out of spent nuclear fuel can be successfully used to determine the initial enrichment of fresh nuclear fuel.

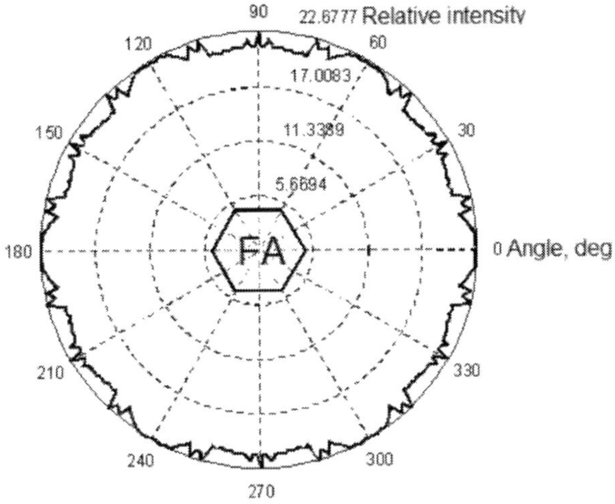

Fig.10 — The ratio of the measured intensities of ^{238}U 1001 keV and ^{235}U 185 keV depending on the detector location angle

The developed equipment makes it possible to determine the enrichment of nuclear fuel when it is transferred by a reloading machine to the reactor core without additional time. The introduction of such a system provides a qualitatively new level of monitoring of the state of nuclear fuel at all stages of its life cycle at nuclear power plants. In particular, there will be an automatic analysis of the initial enrichment of nuclear fuel, a burn-out analysis for each campaign based on the results of measurements during the overload, a control of the fuel cladding tightness during the overload, and an analysis of burn-out during shipment to dry storage. In addition, close integration with the control system of the overload machine ensures automatic maintenance of a database on the history of nuclear fuel, i.e. fulfillment of the IAEA requirements for nuclear safeguards. Continuous accounting of the burn-out of each fuel assembly is also the main source of information to refine the methods of nuclear-physical calculations of the reactor core.

Acknowledgment. This project was supported by Swedish Radiation Safety Authority (SSM) under Partner project agreement P727 between SSM, the Science and Technology Center in Ukraine and Odessa National Polytechnic University.

References

1. Reilly D. Passive Nondestructive Assay of Nuclear Materials / D. Reilly, Norbert Ensslh, Hastings Smith. – Los Alamos National Laboratory, NUREG/CR-5550. 1991. – 700 p. and 2007 Addendum – http://www.lanl.gov/orgs/n/n1/panda/index.shtml

2. Soltys I., Yacout A. M., McKnigh R. D. Application of Control Charts and Isotope Correlations to Spent Fuel Measurements at FCF / 40th Annual Meeting of the Institute of Nuclear Material Management. July 25-29, 1999, Phoenix. – 7 p.

3. Neutron and gamma-ray energy reconstruction for characterization of special nuclear material / Shaun D. Clarke, Michael C. Hamel, Angela Di Fulvio, Sara A. Pozzi // Nuclear Engineering and Technology – 2017 – Vol. 49, No. 6 P. 1354-1357 – DOI:10.1016/j.net.2017.06.005

4. Michael Streicher. A Method to Estimate the Atomic Number and Mass Thickness of Intervening Materials in Uranium and Plutonium Gamma-Ray Spectroscopy Measurements / Michael Streicher, Steven Brown, Yuefeng Zhu, David Goodman, and Zhong He // Nuclear Science IEEE Transactions – 2016 – Vol. 63, No. 5 – P. 2639-2648 – DOI: 10.1109/TNS.2016.2606763

5. Michael Streicher. Special Nuclear Material Characterization Using Digital 3-D Position Sensitive CdZnTe Detectors and High Purity Germanium Spectrometers / Michael Streicher, Steven

Brown, Yuefeng Zhu, David Goodman, Zhong He // Nuclear Science IEEE Transactions – 2016 – Vol. 63, No. 5 – P. 2649-2656 – DOI: 10.1109/TNS.2016.2593631

6. L. Eric Smith. Signatures and Methods for the Automated Nondestructive Assay of UF_6 Cylinders at Uranium Enrichment Plants / L. Eric Smith, Emily K. Mace, Alex C. Misner, and Mark W. Shaver // IEEE Transactions on Nuclear Science – 2010 – Vol. 57, No. 4 – P. 2247 - 2253

7. Fast Neutron Source Localization Using Digital Pixelated 3-D Position-Sensitive CdZnTe Detectors / David Goodman, Michael Streicher, Yuefeng Zhu, Steven Brown and Zhong He // Nuclear Science IEEE Transactions –2017 – Vol. 64, No. 9, – P. 2531 - 2535

8. Oleinik S. G. Determination of the burnup of spent nuclear fuel during reloading / Oleinik S. G., M. V. Maksimov, O V. Maslov // Atomic Energy 2002 – Vol. 92, No. 4, – P. 296-300.

9 Hee Seo. Burnup Measurement of Spent Fuel Assembly by CZT-based Gamma-Ray Spectroscopy for Input Nuclear Material Accountability of Pyroprocessing / Hee Seo, Jong-Myeong Oh, Hee-Sung Shin, Ho-Dong Kim, Seung Kyu Lee, Hyun Su Lee, and Se-Hwan Park // Nuclear Science IEEE Transactions – 2014 –Vol. 61, No. 4 – P. 2169 - 2174

10. Oleinik S. G., V. A. Boltenkov, O V. Maslov, Passive computer γ-tomography of nuclear fuel – Atomic Energy. – 2005 – Vol. 98, No. 3, – P. 227-229.

11. Mokritsky V.A. Automated system for determining the burnup depth of spent nuclear fuel. V.A. Mokritsky, O.V. Maslov, O.V. Banzak // Technology and design in electronic equipment. – 2014. – No. 5 – 6. – P. 63 – 71

12. Maslov O.V. Multiple energies passive computer tomography of nuclear fuel / O.V. Maslov, V.A. Mokritskiy, A.V. Sokolov // International Journal of Tomography & Simulation – 2018 – Issue No. 1, Vo. 31. – P. 1-13.

13. S.J. Svärd. Applicability of a set of tomographic reconstruction algorithms for quantitative SPECT on irradiated nuclear fuel assemblies / Staffan Jacobsson Svärd, Scott Holcombe, Sophie Grape // Nuclear Instruments and Methods in Physics Research Section A: Accelerators, Spectrometers, Detectors and Associated Equipment – 2015 – Vol. 783 – P. 128-141. – DOI: 10.1016/j.nima.2015.02.035

Development of a High Temperature Neutron Flux Detector

Martin Shaw, Bradley Campbell, James Ballantine, Kevin Roberts
Ultra Electronics, Energy, UK

Ultra Electronics, Energy is currently the supplier of neutron flux instrumentation to the UKs Advanced Gas Cooled Reactor (AGR) fleet. Neutron flux instrumentation provides a safety critical function, giving operators the fastest indication of any transient power behaviour in a nuclear reactor.

The operating requirements for these sensors in an AGR reactor are higher than those for equivalent instrumentation in a Pressurised Water Reactor (PWR) or Boiling Water Reactor (BWR). Whilst the underlying physics of these devices is the same, the engineering challenges for AGR instrumentation are different. Design and manufacturing processes have to be more precise due to susceptibility of device performance to a number of factors post installation.

The AGR sensor therefore provides a sound engineering platform for the development of an equivalent device for the harsh environments expected in Generation IV reactors.

This paper discusses the capabilities of the Ultra Electronics neutron flux detector manufacturing facility and how these capabilities are being expanded to cover the anticipated operating conditions for Generation IV reactor designs.

A prototype design has been manufactured and mechanically tested, the sensitive coating process has been developed and the Mineral Insulated (MI) signal cable has been tested at elevated temperature.

I. INTRODUCTION

Ultra Electronics (Ultra) is currently the supplier of neutron flux detectors to the entire UK Advanced Gas Cooled Reactor (AGR) fleet. This follows from its long involvement in the design and development of such devices over a period of 50+ years. The AGR's are unique as commercially operating reactors in that they have an operating temperature much higher than the more prevalent Pressurised Water Reactor (PWR) and Boiling Water Reactor (BWR) designs. As the neutron flux detectors are effectively positioned 'in-core', they and their associated cabling have to operate reliably at temperatures up to 550°C and a pressure of 70 Bar. The plans for the build of future nuclear power platforms now include smaller, more efficient devices that operate at temperatures well beyond even the AGR level of 550°C. The concept designs for these Generation IV High Temperature Reactors utilise a range of heat transfer mediums and in many cases have a requirement for neutron flux detectors to drive control or safety circuits. In anticipation of this requirement, Ultra has been investing in the development and build of a proof of concept fission chamber. The initial concept has the goal of delivering neutron sensitivity comparable to equivalent commercial devices and stability at temperatures of up to 825°C, but with the potential to operate at even higher temperatures.

M. Shaw, B. Campbell, J. Ballantine and K. Roberts are with Ultra Electronics Energy, Unit 7, Lancaster Road, Ferndown Industrial Estate, Dorset, BH21 7SQ.
(e-mail: martin.shaw@ultra-ncs.com).

II. ULTRA ELECTRONICS, ENERGY MANUFACTURING

In order to manufacture neutron flux instrumentation for the UK AGR fleet, it was necessary for Ultra to first establish a manufacturing facility. In 2013, a building was acquired and over the following 3 years fully outfitted with the necessary equipment to perform manufacturing and testing operations.

Due to the lack of available test reactors within the UK, it is necessary to perform characterisation of detector performance at room temperature, 550°C and under the influence of irradiation separately to assess the expected performance in-core.

Consequently, test capabilities include a linear accelerator driven thermal neutron source capable of delivering 1×10^7 $n \cdot cm^{-2} \cdot s^{-1}$ in continuous operation, a 10 TBq ^{137}Cs source for gamma irradiation testing as well as large chamber ovens capable of achieving 650°C.

Failure of these devices during manufacture and service is typically caused by the presence of contamination in the neutron sensitive volume of the detector. To minimise the risk of contamination during build, a clean assembly area has been incorporated into the manufacture facility. Design of the system for providing the gas filling of the sensitive volume has also been designed to minimise the risk of contamination.

Historically, there has been variability within certain manufacturing processes that have contributed to device failures. Wherever possible, automated processes have been implemented to reduce this risk as the time taken to manufacture is of the order of months and the financial cost is significant.

III. AGR NEUTRON FLUX DETECTOR CHALLENGES

Ultra manufactures 4 different designs of detector, all of which are gas-filled ionisation chambers in which the sensitive coating materials are ^{10}B or ^{235}U. The devices are optimised to operate in the following modes: pulse mode, DC mode, logarithmic/linear DC mode and pulse/Campbell mode.

These devices were designed between the 1960s and 1980s and have been used in-core since that date. Ultra has had to demonstrate equivalence of new build devices to these designs to justify their use without having to perform full re-qualification of the devices prior to use. The number of changes permissible are therefore small and have to be justified to the licensee and regulatory body prior to installation.

Legacy designs and specifications include manufacturing information that constrain certain processes, occasionally to the detriment of manufacturing yield or device performance.

Key challenges to successful manufacture are:

- contamination/cleanliness (including purity of the gas in the detection volume)
- Adhesion of the sensitive coating
- Processing under vacuum
- Material selection

As an example of sensitivity of these devices to low levels of contamination, the graph in Fig. 1 shows the effect of gas contamination on the performance of a device over approximately 1 year, at the end of which, the level of contamination is of the order of 50ppm. This leak rate in this case is nitrogen from the containment volume leaking into the neutron sensitive chamber volume.

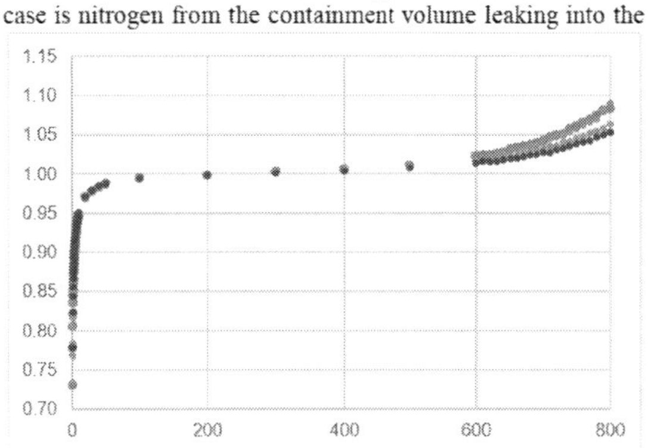

Fig. 1. Normalised current versus applied voltage for a device that is leaking gas into the sensitive volume of the detector, measured over 1 year. The avalanche breakdown region in the region >600V will eventually disappear, the saturation current voltage will increase and the device will become unusable

As a consequence of the failure shown in Fig. 1 and similar problems, processes have evolved to reduce the risk of failure and maximise the yield of useful components.

IV. Generation IV Reactors

Ultra has achieved the ability to satisfy the lifetime requirements of neutron flux instrumentation for the UK AGR fleet. Through this development cycle, a significant level of knowledge and experience has been gained on the impact of high temperature and neutron flux on the performance and behaviour of these devices.

Over the last few years there has been global interest in the development and deployment of new nuclear generating capability, specifically for Small Modular Reactors (SMR) and Generation IV technologies. Ultra is already working in collaboration with NuScale Power in the USA to address the instrumentation and control aspects of their SMR design. These requirements are similar to those for PWR/BWR technology and therefore provide a complex engineering challenge regarding their implementation rather than research and development programmes to identify solutions.

Generation IV reactors are being considered for a number of reasons, most importantly the ability to burn-up waste products, the increased temperature leading to increased efficiency and in certain instances to operate at atmospheric pressure. The conditions expected in a Generation IV reactor type are therefore expected to be very different to that experienced in existing power stations. The neutron fluxes will not only be larger, but also of different energy spectrum due to the make-up of the fuel and the means of operation.

TABLE I
OPERATING CONDITIONS FOR PWR, BWR AND AGR AND EXPECTED OPERATING CONDITIONS FOR SOME GEN IV REACTORS

Reactor	Temperature (°C)	Pressure (bar)	Neutron Flux ($n \cdot cm^{-2} \cdot s^{-1}$)
PWR / BWR	125	1	1×10^{9}
AGR	550	70	1×10^{12}
MFR (e.g. SFR / LFR	700	1	1×10^{15}
VHTR	1000	100	1×10^{15}

All values in the table above are in the location of the sensors, during normal operating conditions.

The Metal cooled Fast Reactor (MFR), including Sodium cooled Fast Reactor and Lead cooled Fast Reactor as well as the VHTR values variants have been estimated. Estimates are based on core outlet temperatures of approximately 800°C X-energy [1] EM2 [2].

Table 1 shows estimates of various Generation IV operating conditions compared to those of PWR/BWR and AGR reactor types. It can be seen from the table that the jump from AGR technology to those of MFR and VHTR designs is significantly less than for PWR/BWR technologies.

The aim of the work presented in this paper is to extend the operational envelope of the core processes used on AGR detector manufacture into the MFR and if possible the VHTR operating regions.

V. Chamber Development

The AGR detectors typically consist of a large number of components, designed to operate successfully in the environment to which they are subjected rather than designed for ease of manufacture.

The chamber design for the high temperature detector has been generated around the critical dimensions of the device, but focussing on design for manufacture. As a result a number of key factors have changed:

- The part count has been significantly reduced
- Design of individual components has been changed to simplify machining
- Components have been designed to accommodate welding and joining techniques
- Vacuum processing time has been addressed in the design

The assembly time for a new device is now estimated to be of the order of 21 days, a reduction of approximately 7 days over the most simple of the existing AGR detector designs.

A chamber has been assembled and welded, then subjected to vacuum processing. It was subsequently filled with gas before temperature cycling to 825°C to represent the mechanical stresses expected in service. The electrical resistivity of the chamber was measured during this temperature excursion to characterise the performance, results of which are shown in Fig. 2, below.

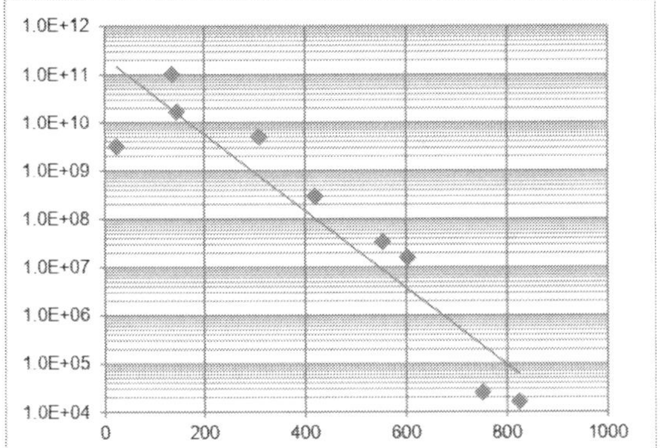

Fig. 2. Resistivity versus temperature for the high temperature detector chamber, tested up to 825°C.

The data from Fig. 2 shows the chamber performance follows the anticipated trend of a 1 decade decrease in resistivity for each increase in temperature of 100°C. This trend is commonly seen in the AGR detectors when tested up to 550°C.

VI. Sensitive Coating Development

Separately to the chamber development, the coating process has been developed based on the technique used for applying ^{235}U to the AGR detector electrodes. This process was considered separately due to the handling risks associated with delamination of coating material during chamber manufacture. To further mitigate the risks, natural U has been used in all trials, which has identical bonding characteristics in the AGR regime.

The change in materials from stainless steel (AGR) to Inconel alloy used here led to modification of the process to ensure that the coating adhered to the electrode surface. Initial trials resulted in some delamination when subjected to temperatures in excess of 650°C.

Changes to the process and surface preparation of the components led to a significant increase in the yield rate of effective coatings. The key contributors to adhesion of the U coating on Inconel have been identified and there are a number of solutions identified to achieve the coating.

The images in Fig. 3 show Inconel electrodes coated with U after completion of trial activities. This process is now ready to be trialled using enriched ^{235}U on electrodes for the next development chamber.

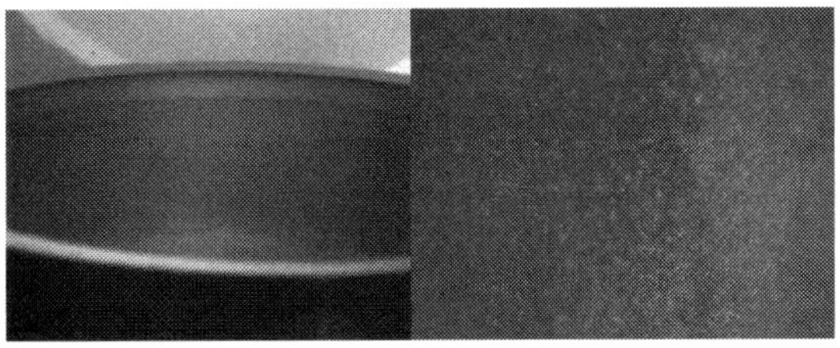

Fig. 3. Examples of uranium coated onto the surface of Inconel after being subjected to 825°C. the left hand image shows the shape of the electrode and the right hand image is a close-up of the coated surface. It can be seen that the coating is well adhered to the surface, there was no evidence of delamination or flaking of the coating post excursion to high temperature.

VII. MI Cable Development

To transmit the signal from the detector to outside the reactor pressure vessel in high temperature power stations, Mineral Insulated (MI) cable is used. In the case of the AGR detectors, this cable is a triaxial copper-copper-stainless steel configuration with a magnesium oxide (MgO) insulating material between each of the metal components. The centre conductor (copper) provides a transmission path for the signal, the inner sheath (copper) acts as a screen for noise rejection and the outer sheath (stainless steel) provides mechanical strength. The stainless steel outer sheath also allows the reactor earth and the instrument earth to be electrically isolated.

The MI cable used in existing detector manufacture was developed by Ultra for use in the AGR fleet. For this study, the cable has been used unchanged but has been subjected to high temperature trials up to 850°C to characterise behaviour at these elevated temperatures.

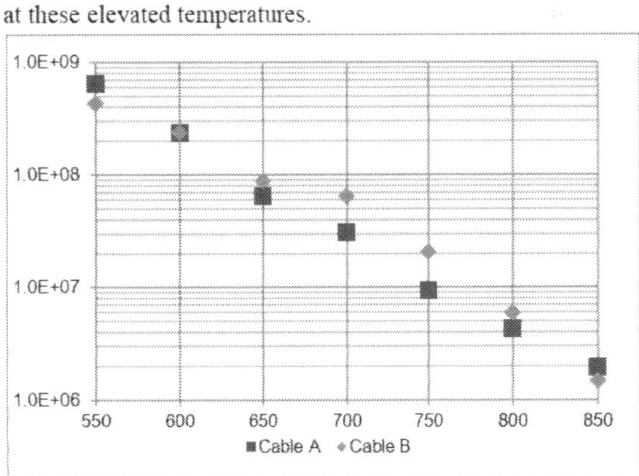

Fig. 4. Resistivity versus temperature for 2 MI cables tested up to 850°C

The graph in Fig. 4 shows the electrical resistivity of two cables as a function of temperature up to 850°C. As was also seen previously in Fig. 2, the data follows the expected trend of a decade decrease in resistivity for every 100°C increase in temperature. This trend is also commonly observed during testing of the MI cables up to 550°C prior to use in manufacture of AGR detectors.

VIII. Further Work

A chamber assembly shall be manufactured with a sensitive coating and employing all processes developed to date. This device shall then be tested under a neutron flux whilst subjected to temperatures up to 825°C to determine operational characteristics.

A containment vessel shall be designed and manufactured to permit connection of the MI cable to the chamber. This will provide both interference immunity for the signal as well as a mechanical pressure boundary for the neutron sensitive volume after installation. This design shall be based on those used for the AGR detectors.

It is believed that the operational envelope of the MI cable can be extended by making material changes to the construction of the cable. It is not yet known whether these changes are required , thus trials and implementation will depend on the performance of the device under neutron flux and high temperature.

Generation of an electronics channel to combine pulse mode, DC mode and Campbell mode shall be performed using elements of the existing instrumentation. A separate study shall also be conducted to determine the feasibility of self-diagnostic electronics to show the health of a device over time.

References

[1] H. Bowers, "X-Energy Xe-100 Reactor", presented at the Initial NRC Meeting, September 2018.

[2] General Atomics, "Advanced Reactors Energy Multiplier Module (EM2)" [Online] http://www.ga.com/advanced-reactors

Method for Prototyping and Testing the Information Minimalism Concept of Nuclear Power Plants

Gwi-sook Jang*, Gee-yong Park
I&C and HFE Research Division, Korea Atomic Energy Research Institute, 989-111,
Daedeok-daero, Yuseong-gu, Daejeon 305-353, Republic of Korea
* gsjang@kaeri.re.kr

***Abstract*—The information structure and visualization design in nuclear power plant is based on careful analysis and understanding of the work domain of the operator. Piping and instrumentation diagram (P&ID) based layouts tend to require more display space. Overuse of mimic layouts can result in visual clutter. P&ID based layouts of controls are usually less easily operated configurations than those provided by other array conventions. Thus, a display method is emerging to minimize P&ID based display. The information minimalism concept is a monitoring and controller display method for each operation mode based on log analyses of operator actions. This method provides the monitoring information and control means necessary for the operator, according to operation mode, to perform a specific operation. This method can reduce the time spent searching for information and the transition between display pages. The aim of this paper was to verify the feasibility of implementing new concepts by establishing an information minimalism prototype. The validity of the function, performance, and operational test methods were verified by testing one such information minimalism prototype.**

In this paper, we describe prototyping and testing methods for the information minimalism concept for NPPs. In the future, this concept is to be added to the concept of operator support with the existing display configuration and navigation, and thereby extend its application range while actually utilizing it.

***Keywords*—Monitoring System, Human Factors Engineering, Human-Machine Interface**

I. Introduction

The information display in a digital based control room of a Korean NPP (Nuclear Power Plant) consists of a hierarchically structured piping and instrumentation (P&ID)-based system, components and function-oriented pages. The operator obtains the monitoring and controller information necessary for particular operation modes by navigating information that is hierarchically distributed over five or more display pages. Due to the distraction from these additional tasks, reaction time is delayed and it is difficult to maintain the cognitive context, which can cause cognitive error in emergencies.

Therefore, it is necessary that the structure used for presentation of information and visualization should be designed to minimize the cognitive burden of recognition, memory, and judgment on the relationship between complex systems when an unexpected event occurs. The information structure and visualization design should be based on careful analysis and understanding of the work domain of the operator.

To solve the above problem, the monitoring system designer has to decide how to reduce the amount of information to be displayed on the screen so that the operators can work more easily. It is also important to provide a means for the operator to move easily among the information display pages and to make a natural cognitive transition to display page with a high degree of information hierarchy.

Two years ago, we proposed a monitoring and controller display method for each operation mode based on log analyses of operator actions to overcome the above problems [1, 2]. This method minimizes the existing information navigation by providing the monitoring and control means necessary for the operator to perform a specific operation mode on one or two display pages. The method provides the monitoring and controller information for each operation mode of an NPP to operators based on understanding of their work domains.

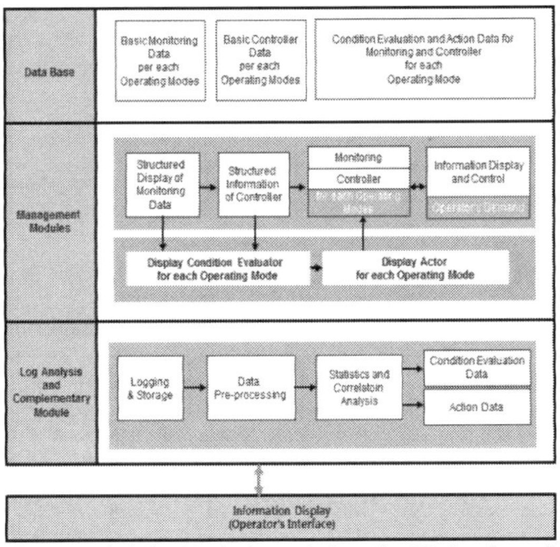

Fig. 1. Method for Monitoring and Controller Display for each Operation Mode based on Log Analyses of Operator Actions

Moreover this method complements the monitoring information and the means of control for each operation mode through analysis of the operation behavior logs obtained during operation, to improve the efficiency of display for each operation mode (see Fig.1).

In this paper, we propose more detailed functions and design issues of the information minimalism concept than in the previous paper (see Fig.2). We also propose methods for prototyping and testing to verify it.

Issues	Description
Information Minimalism	▪ Minimize unnecessary screen components to reduce screen complexity ▪ Visualization scheme to reduce display transition than commercial NPP ▪ Easy display page navigation
Information Structuring and Visualization	▪ Operator task-oriented information display structure ▪ Monitoring and control display by operation mode ▪ Reflecting experience information analyzing operator activity log ▪ Go to the desired screen with information site map ▪ Display bookmarks are provided to collect only the operator selection display
Additional Issue	▪ Preparing the database security in information log ▪ Display sharing scheme for information display device failure

Fig. 2. Design Ideas and Issues

II. PROTOTYPE DEVELOPMENT STRATEGIES

The development principles of the prototype needed to verify the information minimalism concept are as follows:

- Standardization and modularization of the development process by improving development productivity and reuse
- Establishment of a prototype product sharing system
- Establishment of prioritized development items and step-by-step development
- Identification of common risk factors and design security for safe software development
- Performing objective risk analysis and testing of all prototyping outputs

Satisfying the development principles described above, required the following prototype development strategies.

A. Component Based Development (CBD)

The process of developing S/W components and component-based systems differs in many significant ways from the "classical" development process of software systems. The main difference is in the separation of the development process of components from the development process of systems. The primary idea of the component-based approach is the reuse of the existing components instead of adding new ones, thereby building systems from already existing components, whenever possible [3]. Fig. 3 shows a detailed V-shape development process for CBD. Fig. 4 and Fig. 5 show components and a component diagram, respectively, for this prototype according to the CBD method.

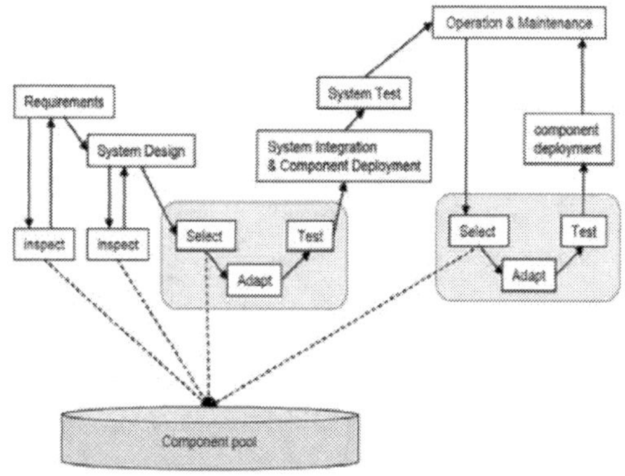

Fig. 3. Detailed V-shape Development Process for CBD

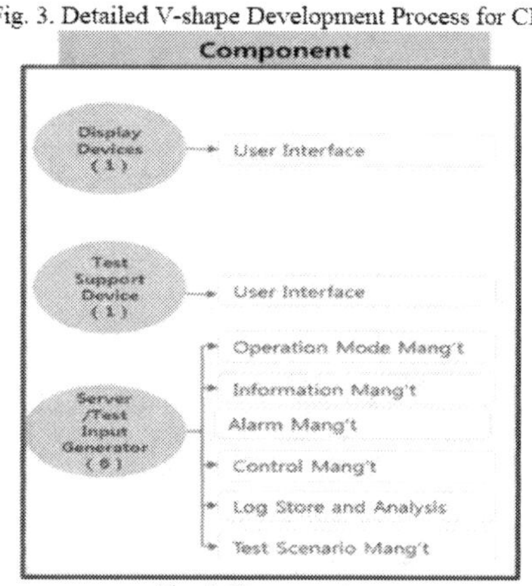

Fig. 4. Components

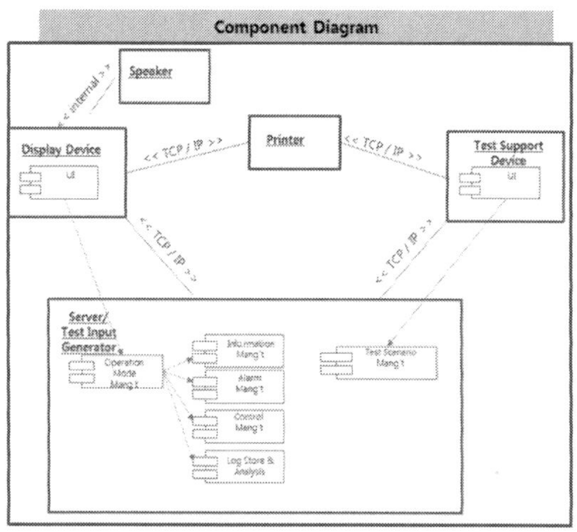

Fig. 5. Prototype Component Design

Because this CBD-based prototyping aims for increased reusability of existing components, the efforts needed for their implementation decreases, and the effort for system verification

https://doi.org/10.1051/epjconf/202022503004

increases. The components of the prototype will be used as inputs for future human factor engineering validation and verification mock-up of a new Korean reactor.

B. Seven Touchpoints of Secure Software

The control and monitoring system of a nuclear power plant is operated in a closed network within the external network. However the non-safety monitoring system is linked to the general business network to provide information about operation of the nuclear power plant to a large number of users. In addition, non-safety monitoring systems use commercial H/W and S/W, so there is a high probability of malicious and non-malicious cyber infestation.

Therefore, in this paper, we propose a secure software development process for the non-safety monitoring system of NPP through this prototyping.

The secure SDLC (Software Development Life Cycle) process ensures that security assurance activities such as penetration testing, code review, and architectural analysis are an integral part pf the development effort. Software security touchpoints are based on good software engineering practice and involve explicitly pondering security throughout the software lifecycle. This means knowing and understanding common risks (including language-based implementation bugs and architectural flaws), designing for security and subjecting all software artifacts to thorough, objective risk analyses and testing. "Software Security Touchpoints" specifies one set of touchpoints and shows how software practitioners can apply them to the various software artifacts produced during software development [4]. This secure method consists of seven major points in the S/W SDLC (see Fig. 6).

1) Code review
2) Risk analysis
3) Penetration testing
4) Risk-based security test
5) Abuse cases
6) Security requirements
7) Security operation

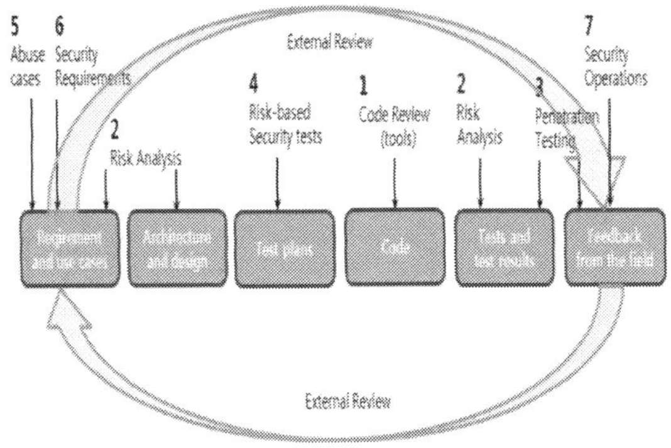

Fig. 6. Seven Touchpoints of Secure Software

TABLE I shows a secure SDLC plan for this prototype based on the seven touchpoints of secure software.

TABLE I
A Secure SDLC Plan for the Prototype

SDLC	Secure Plan	Result
Requirement and use cases	• Converting both overt functional security(e.g., use of applied cryptography) and emergent security propertied (as revealed by abuse case and attack patterns) • Describe the system's behavior when under attack to clarify what areas and components of the software-based system need to be protected, from which threats, and for how long	S/W Req'ts Spec.
Architecture and Design	• Includes documenting assumptions and identifying possible attacks, and uncovering and ranking architectural flaws for mitigation. Recurrent risk tracking, monitoring, and analysis should be ongoing throughout the life cycle	S/W Design Spec.
Test Plans	• Includes testing of security functionality using standard functional testing techniques, and risk-based security of the software as a whole, with test scenario based on attack patterns	Test Plan & Proced ure
Code	• Entails use of static analysis tools to detect common vulnerabilities	
Test and Test Results	• With the architectural risk analysis results driving the selection and implementation of "canned" black-box tests offered by automated application security and penetration testing tools	Test Report
Feedback from the Field	• After development, this includes monitoring the behavior of the fielded software system for indications of attacks and exploits against the software. Knowledge gained through monitoring attacks and exploits should be cycled back into the other touchpoints	S/W Req'ts Spec.

III. PROTOTYPING METHOD

A. Hardware and Software Configurations of the Prototype

The H/W configuration of the prototype follows in Fig. 7 and the S/W configuration of the prototype follows in Fig. 8.

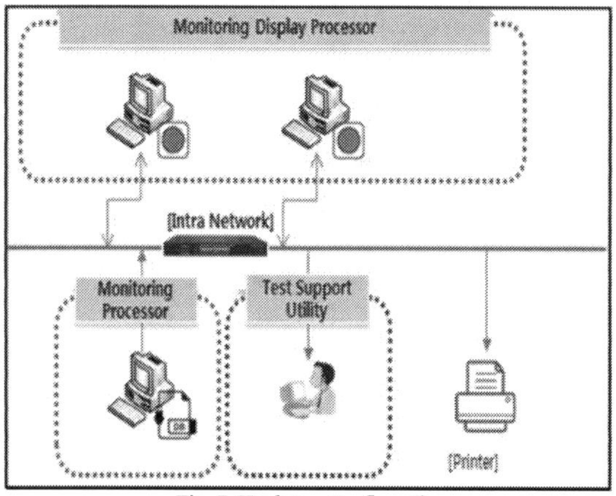

Fig. 7. Hardware Configuration

	Server	Client
H/W	Monitoring Processor	Display Processor Test Support utility
OS	UNIX Ubuntu	Windows 10
S/W	Apache Web Server Database (MySQL)	Web Browser
Development Tools	JSP	JAVASCRIPT HTML SCRIPT

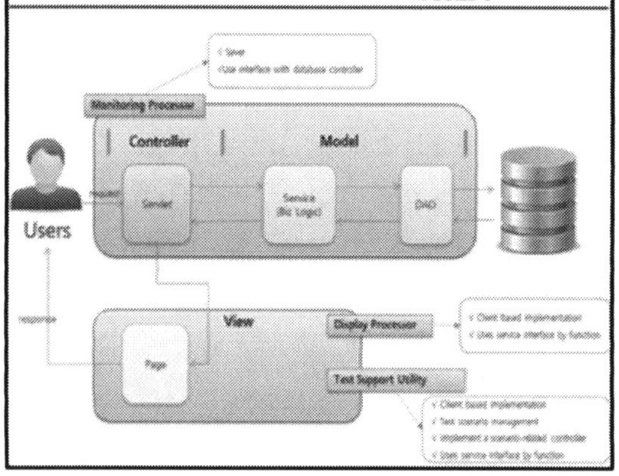

Fig.8. Software Configuration

B. Security Vulnerabilities of the Prototype

The expected security vulnerabilities of the prototype are as follows in Fig. 9. These vulnerabilities are managed according to the plans in TABLE I.

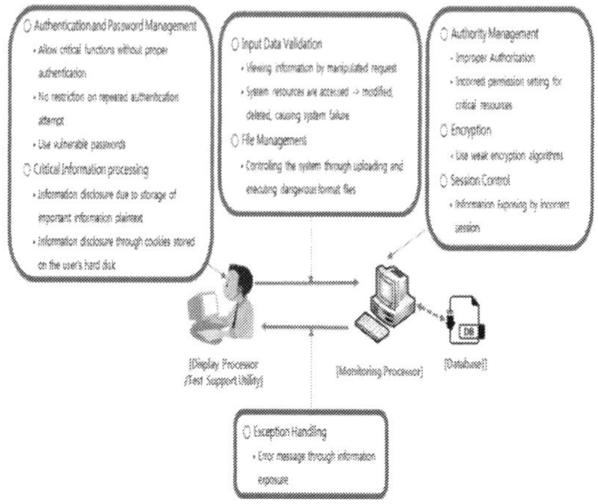

Fig. 9 Expected Security Vulnerabilities of the Prototype

C. Display Design of the Prototype

The monitoring and controller display for each mode based on log analyses of operator actions, is as follows in Fig. 10. The first page is the monitoring and controller display page of a specific operation mode. When an operator clicks a controller name in the controller list, the controller pops-up on the page. The second page is the statistics and correlation analysis page. The operator analyzes the operator activity log information page and adds or deletes monitoring and controller information for a specific operation mode set in the past.

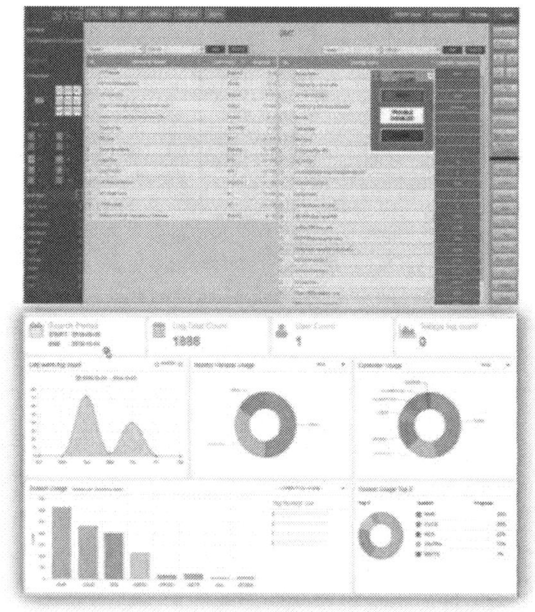

Fig.10. Display Design Examples of the Prototype

https://doi.org/10.1051/epjconf/202022503004

IV. TESTING METHOD

The test steps of this prototype include unit test, functional test, operability test and performance test as shown in Fig. 11. Moreover the detailed test items for each test step according to the CBD based cyber security development process are as shown in Fig. 11.

Step	Test Step	Detailed Test Item
Step 1	Unit Test	Code Vulnerability Test
Step 2	Functional Test	Function and Reliability Test
Step 3	Operability Test	Operability Metric Test
Step 4	Performance Test	Performance Metric Test

	Test Item	Detailed Testing Item
CBD Development	Integration Test	Data Exchangeability
		Data Service Continuity
	Functional Test	Completeness
		Accuracy
7 Touch point Security	Unit Test	Static Test (Code Review)
	Security Test	Access Control (Penetration Test)
		Data Security (Penetration Test)
		Network Security (Penetration Test)
		Database Security (Penetration Test)
General	Operability Test	Message Understandability
		User Interface Understandability
		Message Consistency
		User Interface Consistency
		User Interface Changeability
		Compliance
	Performance Test	Response Time Efficiency
		Work Throughput Efficiency
		Processor Usability
		Memory Usability
		Communication Load Efficiency

Fig. 11. Test Design

The test support utility supports a unit test, performance test, and system utilization analysis test of the prototype, as shown in Fig. 12.

1) *Unit Test*
   - Check the coding rules of the source code
   - Tool : LDRA
2) Performance Test
   - Check the response time
   - Tool : Wireshartk
3) System Utilization Analysis Test
   - Check CPU, memory usage
   - Tool : perfmon.msc

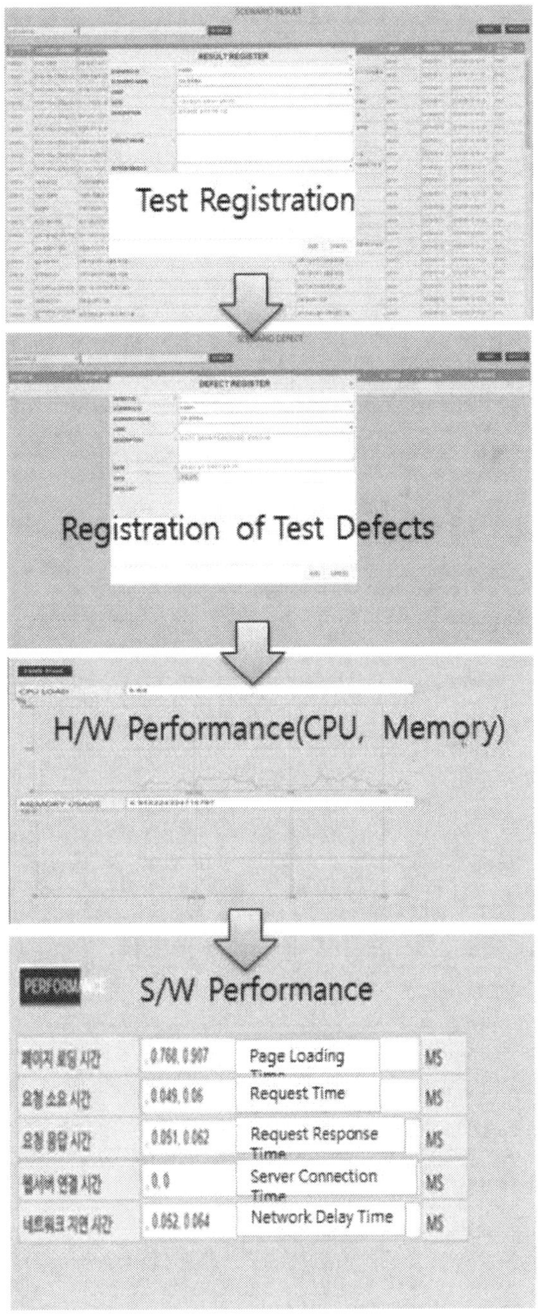

Fig. 12. Test Support Utility

Consideration Factors in Application of Thermocouple Sensors for RCS Temperature Instrumentation

Inkoo Hwang*, Sewoo Cheon, Wonman Park
I&C and HFE Research Division, Korea Atomic Energy Research Institute, 989-111, Daedeok-daero, Yuseong-gu, Daejeon 305-353, Republic of Korea
* ikhwang@kaeri.re.kr

***Abstract*—Because of harsh radiated environmental conditions, it is necessary to use thermocouples (TCs) in the temperature instrumentation channels of a reactor coolant system (RCS) in an integrated pressurized water reactor vessel. Conventionally, resistance temperature detectors (RTDs) have been adopted for RCS temperature measurement.**

Therefore, we have conducted an analysis and review of instrumentation error factors in the measurement circuits of RTD and TC sensors to specify the influence on measurement accuracy for application of TCs instead of RTDs for RCS temperature instrumentation.

From the review and analysis results, it is anticipated that a measurement accuracy deterioration would be an issue and that a drift range should be investigated for the anticipated operational temperature conditions.

***Keywords*—RTD Measurement Circuits, TC Accuracy**

I. Introduction

Resistance temperature detectors (RTDs) and K-type thermocouples (TCs) are the most common devices for measuring the temperature of fluid in nuclear power plants (NPPs). RTDs are typically used for measuring reactor coolant system (RCS) temperature because an accurate and stable sensing is needed for calculating the thermal output power and regulating the reactor during a normal operation. In some instances, however, TCs are favored over RTDs for measuring RCS temperature, as in the case of a core inlet or outlet of an integrated reactor. RTD-based temperature circuits are generally regarded as very accurate and stable in comparison with TC-based circuits, however, there are some considerable factors diminishing the measurement accuracy depending on the measurement methods and calibration circumstances in temperature instrumentation systems.

Thus, this paper introduces the typical factors affecting the measurement errors and specifies the anticipated inaccuracy sizes due to each factor in RTD-based and TC-based temperature instrumentation channels in Section II and III. Based on the investigated inaccuracy numbers, the total sizes of uncertainty deterioration in the application of TCs instead of RTDs are calculated in Section IV. Section V includes the summary of this paper.

II. Factors Affecting The Errors in RTD-based Circuits and Relieving Counter-measures

Although RTDs are commonly used for temperature measurement bellow 500 ℃ in industrial process plants such as NPPs, there are many factors to be considered regarding the measurement accuracy in RTD instrumentation channels. Such factors include lead wire resistance effect, self-heating effect, electromagnetic force error, non-linearity, corrosion, radiation, and insulation loss [1-3]. This section describes the sizes of each effect in the key factors among the above examples, which we have investigated in the previous research [3].

A. *Lead Wire Resistance Effects*

The effect of resistance of the lead wires is not negligible, even in the case where the distance between RTD sensors and instrumentation circuits is a couple of meters. This is because of the low sensitivity characteristic of 100 Ω-RTDs, called Pt-100. The sensitivity (temperature coefficient) is only 0.385 Ω/℃. Accordingly, a 10 m wire of 20 AWG (= 0.33 Ω) contributes an error of 0.84℃.

Therefore, in industrial RTD instrumentation systems, it is necessary to use a compensation method to remove this wire resistance effect for measuring the process temperature correctly.

A simple way to compensate for the wire resistance effect is to adjust the process temperature reading value in the RTD instrumentation system by a proper *in-situ* calibration.

However, if the resistance of the lead wires were to change because of the ambient temperature difference from the *in-situ* calibrated condition, then a temperature reading error dependent on the degree of temperature variation would occur. Should the temperature of the lead wire be increased to 35 ℃ from 10 ℃, the resistance of a 300 m wire of 20 AWG size would increase by 10%; that is, from 10 Ω to 11 Ω. This change would create a reading error of 5.2 ℃ in a 2-wired pt-100 RTD circuit.

Thus, in the case of industrial RTD-based temperature instrumentation systems, it is necessary to use a technical solution for reducing or eliminating the effect of wire resistance change. However, such a solution often requires the use of at least one extra lead wire from the sensor to the instrumentation electronics.

The typical solutions avoiding the wire resistance effect are (*a*) 3-wired bridge connection method, (*b*) 3-wired 1-curent injection method, (*c*) 3-wired 2-current injection method, (*d*) 4-wired 1-current injection method, and (*e5*) local transmitter method.

Usually, a properly designed 3-wired connection circuit can cancel the wire resistance effect completely. In particular, 3-wired with 2-current or 4-wired methods are widely adopted to consistently eliminate the lead wire resistance effect in the industrial RTD-based temperature instrumentation systems.

Accordingly, an appropriate choice of cancellation method for wire resistance can ensure that any error due to a change in lead wire resistance will be negligible.

B. Self-heating Effects

In all RTD-based temperature measurement systems, an excitation current flows though the RTD sensor element. Thus, it is possible to measure the resistance of the RTD and convert the resistance to the medium temperature value. The excitation current flowing through the RTD element generates a Joule's heat energy, which results in a temperature increment beyond the medium temperature to be measured. This additional increment in temperature creates an error with regard to the measurement value.

The error due to this self-heating effect can be expressed as follows:

$$\Delta T = (I^2 \times R_{rtd}) \times h \qquad (1),$$

where, ΔT is the change in temperature due to the power dissipation, I is the excitation current of the RTD sensor element, R_{rtd} is the resistance of the RTD sensor element in Ω, and h is the self-heating coefficient in ℃/mW.

To minimize the error due to self-heating, the excitation current, RTD resistance, and h should be reduced. In this regard, a pt-100 is the most suitable choice because of its low resistance. The size of h is dependent on the surface that is in contact with the medium, the heat transfer coefficient, and the temperature gradient from the RTD sensor element.

Therefore, the size of the excitation current is only the practically manageable parameter to decrease the self-heating effect.

For an example, if we take 0.05 ℃/mW to be a typical self-heating coefficient of an RTD sensor, considering that its range is from 0.01 ℃/mW to 0.1 ℃/mW in water, then the temperature deviation, ΔT, will be +1.05 ℃ when the excitation current is 10 mA and the process temperature is 300 ℃. However, ΔT will decrease to +0.0105 ℃ when 1 mA of current is applied. IEC 60751[4] recommends an excitation current of less than 1 mA.

C. Thermal EMF errors

If there is inhomogeneity in material composition along the RTD connection metal wires, the dissimilar metal constituents will act as noisy TCs in the RTD circuit. These unintended TCs will then generate error voltages related to temperature gradients. This inhomogeneity in RTD connection wire metals can be caused from the manufacturing process or material change due to oxidation, etc. during the long operation period of an RTD system, particularly in high temperature applications.

The materials of the RTD element wires, internal lead wires, and external connection lead wires are usually different from each other. In general, the RTD element wires are made of pure platinum; the internal wires, a platinum or a gold alloy; and the external connection lead wires, copper [5].

Fig. 1 illustrates the potential unintended TC parts in an RTD circuit. If the temperature at one end of an electric conductor differs to that at the other end, then a thermal electromagnetic force (EMF) arises in proportion to the temperature difference because of the Seebeck effect.

In Fig. 1, the total EMF along an RTD instrumentation circuit can be defined as follows:

$$EMF \;\; = E1 + E2 + E3 + E4 + E5 \qquad (2).$$

In Eq. (2), the total thermoelectric EMF is not zero when there is a temperature deviation between T_1 and T_2, or between T_3 and T_4, or T_5 and T_6, or if there is a deviation in homogeneity between the materials of the paired two lead wires. This unnecessary EMF voltage yields a small error in the RTD resistance reading of the voltmeter in this circuit.

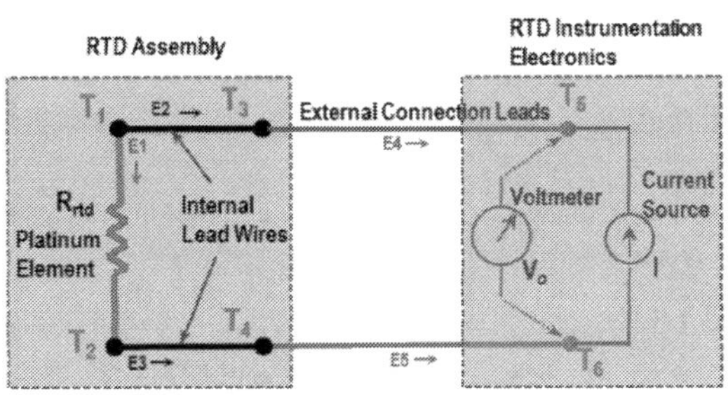

Fig. 1 Unintended TC elements in RTD connection wires

A research has shown that the EMF voltages lie within the range ±0.1~17.9 μV for 25 nuclear grade RTDs at 300 ℃ [7].

The maximum value, ±17.9 μV, corresponds to an error of ±0.047 ℃ when I = 1 mA in the case of Pt-100 RTD. If the excitation current source, i.e. I in Fig. 1, is small, then the error will be smaller.

D. Accuracy of Instrumentation Electronics

To obtain the process temperature signal from the RTD resistance, there must be an electronic or other type of readout instrument capable of converting a resistance signal to a corresponding temperature value. In most industrial process plants, the instrumentation electronics are implemented as a part of a digital control system.

The accuracy ranges of industrial RTD signal conversion modules are within ±0.1 ~ 0.5 % of full scale. One such device currently in use for NPPs has an accuracy of ±0.3 %. If we narrow the full span of measurement range to be 250 ~ 350 ℃ in the signal conversion module to obtain a better accuracy for RCS temperature application, the accuracy limit will be approximately ±0.5 ℃, considering other factors related to accuracy such as power supply effect, stability(drift) effect, etc.

III. Factors Affecting The Errors in TC-based Temperature Measurement Circuits

Although TCs are simple and low-cost sensing elements for measuring fluid or gas temperature at high temperature ranges, having a fast response time performance, instrumentation designers are reluctant to use TCs in control and precise monitoring systems in NPPs, owing to accuracy and stability issues compared to RTDs.

However, because RTDs are not suitable for use inside a nuclear reactor vessel, TCs are the only choice for measuring the temperature of reactor coolant when the temperature sensors need to be installed inside the reactor vessel in an integrated reactor system. Accordingly, we analyzed the extent of accuracy deterioration when TCs are used instead of RTDs for RCS temperature measurement [6].

This section summarizes the key error sources and their associated sizes of that analyzed in Reference [6].

A. Overview of Error Sources in a TC instrument Channel

Fig. 2 illustrates a typical TC instrument channel block diagram, which is generally used for instrumentation and control (I&C) systems in industrial plants such as NPPs. We have classified each error factor present in the measurement loop depicted in the figure with the circled numbers, i.e. ① process medium error, ② error in the TC itself, ③ connector error, ④ extension wire error, ⑤ instrumentation electronic device error, and ⑥ cold junction sensing error. Note that there are other error factors that affect the total uncertainty of measurement; for example, drift error and external noise ingress. However, these are not depicted in Fig. 2 for simplicity in analysis.

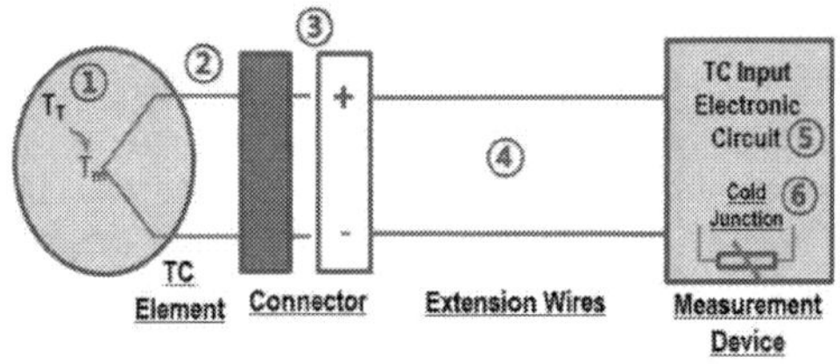

Fig. 2 Error sources in a Thermocouple connection circuit

B. Analysis of Error Terms

The first error factor, ①, indicates the process medium error, which means a deviation between the true process temperature, T_T, and the temperature transferred to the measurement junction of TC, T_m. This type of error is related to the installation and junction types of TCs. It may be large or small depending on the installed situations and locations. However, for our comparison analysis, it is assumed negligible.

The second error source, ②, is an inherent manufactured error in a TC itself and a dominant factor in the aspect of accuracy in a TC-based temperature instrument. For a K-type TC, which is widely used in industrial plants, the error limit of Class 2 is specified as ±2.5 ℃[7]. If we take Class 1 for better accuracy performance, the error limit becomes ±1.5 ℃

The third piece, ③, is an error resulting from temperature difference, ΔT, along the TC connector pins. It is anticipated that the close proximity of the connector pins would not make ΔT more than ±0.2 ℃.

The fourth error factor, ④, is an inherent manufactured error in the extension wires, which must be compatible to the TC element type. For a K-type extension wire, the error limit of Class 2 is specified as ±2.5 ℃ and the error limit of Class 1 is specified ±1.5 ℃ [8].

The fifth error source, ⑤, represents the total uncertainty in the TC signal conversion process, which converts the millivolt signal of the TC to a corresponding temperature value. Assuming that the signal conversion accuracy is the same as the RTD signal conversion module described in Section II-D, the error limit will be ±0.5 ℃.

The sixth part, ⑥, indicates the error of the reference junction temperature, which must be precisely measured for calculating the medium temperature. The reference junction temperature is measured by a high-resolution thermometer such as a thermistor. Therefore, it is reasonable to regard its uncertainty range as ±0.5℃.

Although drift or instability error caused by atomic crystal structure change during a long period of usage is of concern with regard to TC application, we have disregarded the error in our accuracy comparison analysis because the range of drift is highly variable depending on the temperatures in which TCs are exposed, TC structures (bare-wire or mineral-insulated sheath types), and manufacturers [3].

IV. Uncertainty Calculations

With the error sizes included in Sections III and IV, we have calculated the overall uncertainty of the TC-based and RTD-based measurement channels for comparison via the root-sum-square (RSS) method [9, 10]. Excluding the process medium error and drift error, the overall uncertainty of a TC-based system, for Class 2 TCs and extension wires, is as follows:

$$U_{TC-class2} = \pm(2.5^2 + 0.2^2 + 2.5^2 + 0.5^2 + 0.5^2)^{\frac{1}{2}}$$
$$\approx \pm 3.6\ °C \qquad (3).$$

Correspondingly, for Class 1 TCs and extension wires, it is:

$$U_{TC-class1} = \pm(1.5^2 + 0.2^2 + 1.5^2 + 0.5^2 + 0.5^2)^{\frac{1}{2}}$$
$$\approx \pm 2.3\ °C \qquad (4).$$

In the case of RTD systems, if we assume the use of a 3-wired or 4-wired perfect wire resistance compensation circuit, the error factors are the process medium error, RTD element error, self-heating error, thermal EMF effect, drift error, and the conversion error of the signal conversion module. The tolerance of RTD element is specified in a national standard [11]. The tolerance value is ±1.8℃ at 300℃ for Class B and ±0.75℃ at 300℃ for Class A.

Thus, excluding the process medium error and drift error, the overall uncertainty of an RTD-based system, for Class B, is as follows:

$$U_{RTD-classB} = \pm(1.8^2 + 0.047^2 + 0.5^2)^{\frac{1}{2}} + 0.01$$
$$\approx \pm 1.88\ °C \qquad (5).$$

Correspondingly, for Class A, it is:

$$U_{RTD-classA} = \pm(0.75^2 + 0.047^2 + 0.5^2)^{\frac{1}{2}} + 0.01$$
$$\approx \pm 0.91\ °C \qquad (6).$$

In Eqs. (5) and (6), the self-heating error is taken as +0.01 because it is always a positive side error.

With comparison of the uncertainty values from Eq. (3) to Eq. (6), it is concluded that the uncertainty range will increase considerably when TCs are used for measuring RCS temperature instead of RTDs. If we consider the drift factor of TCs, the difference will increase further.

Therefore, the usage of TCs instead of RTDs will bring up some loss of margin in set-point determination for the reactor protection system.

V. Conclusion

This paper addressed the issue of accuracy with regard to the use of K-type TCs for RCS temperature measurement in an integrated reactor. With a comprehensive review and analysis of error factors of both RTD-based and TC-based temperature instrumentation circuits, we have estimated the extent of accuracy loss when TCs are used instead of RTDs.

It has been identified that the measurement uncertainty will increase considerably when TCs are adopted for RCS temperature measurement.

A concern of this paper is that it is not easy to confine drift error tolerance of TC elements, because the drift mechanism is complicated and variable depending on the temperatures in which TCs are exposed and the mechanical and material properties of TCs. In contrast, RTDs generally have a small drift value.

Based on our review and analysis, it can be concluded that the highest accuracy class of TCs and extension wires should be utilized for improving measurement accuracy. Moreover, an experiment should be conducted to investigate the drift error range of the TCs to be used for the normal operating temperature condition during their required operation period.

References

[1] Acromag Inc., Criteria for Temperature Sensor Selection of T/C and RTD Sensor Types: The Basics of Temperature Measurement Using RTDs, Part 1~3 of 3, May 2011.
[2] Rudi Van Nieuwenhove and Ludo Vermeeren, Irradiation effects on temperature sensors, R-3693, ref. INSTR/1248/03-02, March 2003.
[3] Inkoo Hwang, et al., "Considerations regarding Temperature Measurement using RTDs for Industrial Process Plants," KNS Autumn Meeting, Yeosu, Korea, Oct. 25-26, 2018.
[4] IEC 60751, Industrial Platinum Resistance Thermometers and Platinum Temperature Sensors, Ed., 2.0, 2008
[5] Hayashi Denko Inc., Characteristics of RTD Elements, http://hayashidenko.co.jp/en/RTDcharacteristics.html
[6] Inkoo Hwang, et al., "Review of Error Sources for Application of Thermocouples in RCS Temperature Measurement for an Integrated Reactor," to be presented in KNS Spring Meeting, Jeju, Korea, May. 23-24, 2019.
[7] KS C IEC 60584-2, Thermocouples – Part 2: Tolerances, 2004.
[8] KS C IEC 60584-3, Thermocouples – Part 3: Extension and compensation cables – Tolerances and identification system, 2004.
[9] ASME PTC 19.1-1998, Test Uncertainty, Instruments and Apparatus, Supplement to the ASME Performance Test Codes, 1998.
[10] James T. Nakos, Uncertainty Analysis of Thermocouple Measurements Used in Normal and Abnormal Thermal Environment Experiments at Sandia's Radiant heat Facility and Lurance Canyon Burn Site, Sandia Report, Apr. 2004.
[11] KS C 1616, Resistance Temperature Detector, 1992(2017 Reaffirmed), Dec. 1992.

EPJ Web of Conferences **225**, 03006 (2020) https://doi.org/10.1051/epjconf/202022503006
ANIMMA 2019

TEST RESULTS OF AN INNOVATIVE AND MODULAR SENSOR DESIGN FOR FISSION CHAMBER STUDIES

G. de Izarra *, B. Geslot †, S. Bréaud *, A. Pepino *, J-F. Villard *, P. Filliatre*
*CEA, DEN, DER, Instrumentation, Sensors and Dosimetry Laboratory,
Cadarache, F-13108 St-Paul-Lez-Durance, France
†CEA, DEN, DER, LP2E,
Cadarache, F-13108 St-Paul-Lez-Durance, France

***Abstract*—Since many years, the Instrumentation, Sensors and Dosimetry laboratory at CEA Cadarache has been working to propose innovative and robust design of fission chambers. In this framework, few simulation tools like Chester or PyFc were developed to make detector prototyping and optimisation possible. Up to now, the code experimental validations are scarce mainly because of the lack of flexibility of regular fission chambers. In order to set up an experimental database for code validation, an innovative modular twin fission chamber, the CFTM, was designed. It allows to set with precision the inter-electrode gap, the filling gas pressure as well as the fissile deposit used. This detector was irradiated in Minerve zero power reactor in order to gather results for code validation. Thanks to a calibration procedure, a bank of experimental results is available for simulation validation.**

I. INTRODUCTION

Since 2000's, the Instrumentation, Sensor and Dosimetry laboratory is involved in building simulation tools to make nuclear detectors prototyping and optimisation. Few of these tools were dedicated to fission chambers: Chester [1] were able to simulate current pulses at the output of the detector by using Monte-Carlo method. Additional simulation allowed to predict signal in current or in Campbell mode [2]. PyFc [3], another Monte Carlo based simulation was used to investigate fission chamber failure. Yet, most of the simulation tools lacked of experimental qualification. Few reasons may explain this. First, available fission chambers lack of modularity. It is not possible to reconfigure the detector since every part is welded. Second, there is uncertainties on some parameters like the filling pressure of the detector. Third, most of the measurements performed are not absolute since electronic systems are not calibrated in charge or in current.

An experimental work was made in order to produce data valuable for code validation. The major part of this effort consisted in the design of a new fission chamber which can be reconfigured on the fly. Additional work consisted in calibrating properly the measurement system to gather absolute charge and current estimations.

In the first part of the paper, the problem of code validation is discussed and the CFTM, a modular fission chamber is presented. Then, experiments on the Minerve zero power reactor are summarised. In a third part, simplified simulations are performed to link measured quantities to the one derived by simulation. At last, experimental results are discussed.

II. EXPERIMENT DESIGN FOR CODE VALIDATION

A. Code validation

Validation of a simulation tool is not straightforward especially if the range of validity of each model involved in the simulation and its uncertainty is needed. A global picture of the validation process is given in Fig. 1, simulation are made of a set of models and input data describing phenomena like transport of fission fragment in material, generation or drift of charged particles. Those are used to compute intermediate quantities like the amount of charges generated along each fission tracks or charge collection time which are usually not measurable directly. At last, those quantities make possible the estimation of the theoretical current pulses and the charges spectrum related to the detector of interest. Only the current pulses and the charges spectrum can be measured at the output of a fission c hamber. N evertheless, b y c hoosing c arefully the detector parameters which are reduced to its dimensions, the filling g as m ixture a nd t he p olarisation v oltage, i t m ight be possible to retrieve intermediate quantities directly comparable to the simulation results.

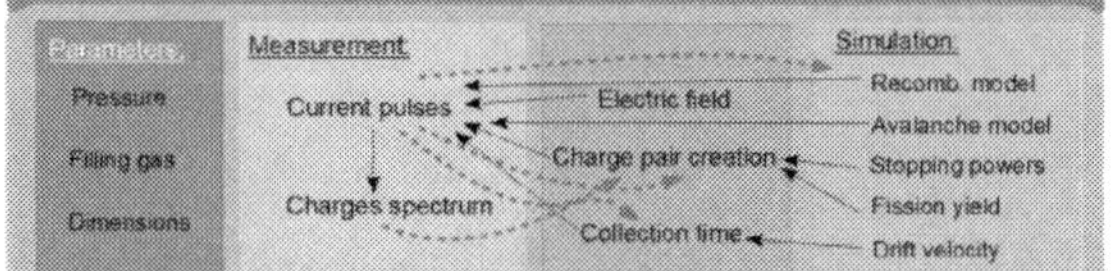

Fig. 1. Schema of the CFTM.

From that procedure, few design constraints might be derived to propose a fission chamber dedicated to experimental code validation:

- Reconfigurable detector: master the dimensions of the detector and their effects on signal.

- Simple electric field: ability to keep the electric field simple regardless of the electrode configuration. Only one working regime during the drift of charged particles.
- Filling pressure between 1 and 5 bar: effect of the filling pressure on the signal.
- Limited effect of impurities on gas transport properties: master the electric charges drift and limit its uncertainty.
- Possibility to lose all the fission fragment energy in the gas: data on heavy ions kinetic energy losses and heavy ion distribution.
- Possibility to detect the two fission fragments emitted: information about fission yield and heavy ions energy losses.
- Possibility to calibrate with a reference reaction: absolute charge and current measurements.
- Possibility to work in saturation or recombination regime: information about recombination/stauration models.

B. CFTM

The design goals were used by the fission c hamber workshop of CEA Cadarache to build a detector dedicated to code validation: the CFTM. This new detector consists of two flat fission chambers set back to back as shown in Fig. 2. It is constituted of a gastight titane body with two electric connectors, a tube for gas filling a nd a p ressure s ensor port. The body is capped with a screwed lid to allow on the fly reconfiguration. Gas tightness is obtained by the use of nitrile O rings. The maximum filling pressure before lid deformation is 6 bars. The detector is modular: ionisation chambers are

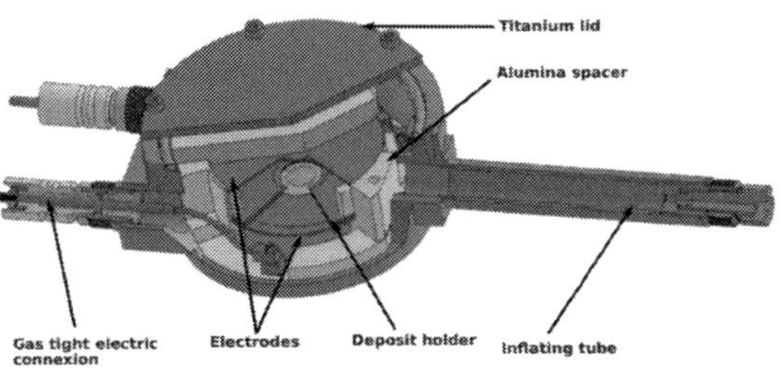

Fig. 2. Schema of the CFTM.

built by setting three electrodes (one central deposit holder and two regular electrodes) and alumina spacers in the body. The inner diameter of the ionisation chamber is 39 mm to allow a complete loss of energy of fission fragments within the 1-5 bar pressure range. The range of average light and heavy fragment in argon at a pressure of 1 bar in normal condition is respectively $R_l = 25$ mm and $R_h = 18$ mm. From this data, it is clear that heavy ions may loose all their kinetic energy at filling pressure above 2 bars.

Thanks to sets of alumina spacers, electrodes can be arranged in three configurations: one with the deposit located in the centre of the detector with two 5 mm inter-electrodes gaps, one with a 9mm/1mm inter-electrodes gaps, and a last one with 7.5mm/2.5mm gaps. The two outer electrodes can be polarised to a different voltage through the two electric connexions while the fissile deposit holder and the detector body are grounded. The electric field is constant in each inter-electrodes gap, it is given by:

$$\vec{E_{plan}} = \frac{\Delta V}{L}\vec{e_v} \tag{1}$$

where ΔV is the polarisation voltage and L is the inter-electrode distance.

Two different deposit holders were built, one is a 0.4 mm thick steel disc for back to back configuration while the other is constituted of two disks with a 9 mm diameter hole in their centre in order to sandwich thin foils of mylar or aluminium with fissile deposit on it.

The back to back configuration has many advantages, spectra ratio can be performed with a well known geometry and similar filling pressure. Alpha emitter deposits like ^{238}Pu or ^{241}Am might be used on one face of the detector to perform an absolute calibration of the charge pair generated by fission fragments in gas.

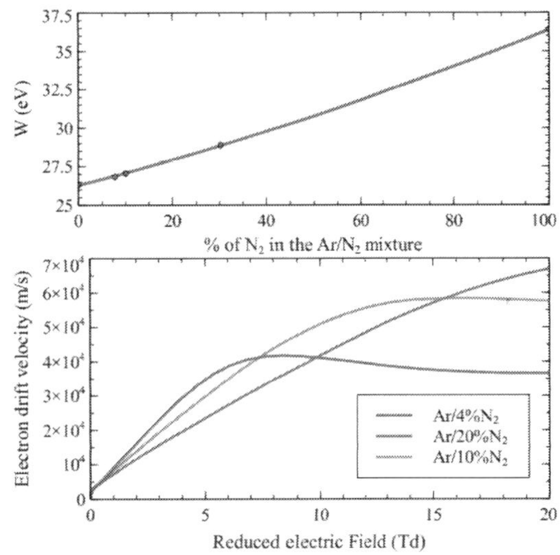

Fig. 3. Up: Mean energy needed to generate an electron/ion pair in an Ar/N_2 mixture as a function of the nitrogen proportion. Down: Electron drift velocty in various Ar/N_2 mixtures. The reference filling gas for fission chamber is Ar+%4 N_2.

Because of on the field reconfiguration, it is needed to fill the detector with a gas mixture almost insensitive to impurities from air. A study was made with different Argon/nitrogen mixtures to keep a high and constant electron drift velocity even with impurities. It was shown that Ar+20% N_2 mixture is the best compromise between electron drift velocity and stability (fig.3). To monitor the filling gas pressure, a Keller M5 transducer is connected to the fission chamber body. It gives a pressure measurement with an accuracy of 0.01 bar.

III. Simulation

The simulation of the detector is mandatory to determine the quantity of interest and how information can be gathered from experiments. A simplified simulation of the detector was built using the COSICAF simulation suite [4]. Because of the specific mixture used in the CFTM, it was necessary to expand the database of COSICAF. W values for argon nitrogen mixture were taken from Tawara & al [5] while the electron drift velocity was computed by using BOLSIG software and the cross section from Phelps & pitchford for nitrogen and Yamabe & Buckman for argon [6]. The Moderation laws used for slowing down fission chamber in the gas were also computed thanks to PRAL algorithm and SRIM stopping powers [7].

Two pressures were investigated as well as two polarisation voltage: Direct polarisation, where the cathode is the deposit holder and inverse polarisation, where the anode correspond to the holder. There are few limitations in this simulation. First only two mean fission fragment are simulated, hence the pulse height spectrum fails to capture the effect of U235 fission yield. Second, COSICAF is only able to handle one heavy ion per fission which limits the simulation to the detector with thick fissile deposit. At last, the code is designed only to simulate current pulse at the output of detector working in saturation regime; It fails to take into account recombination regime.

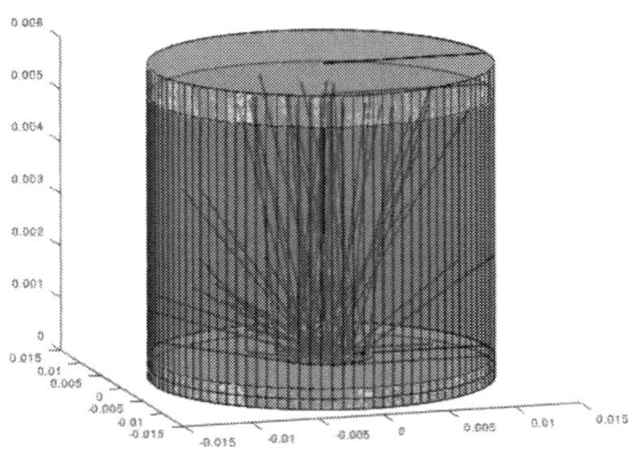

Fig. 4. Few fission fragment trajectories computed in the CFTM geometry with COSICAF.

A picture of the simulated geometry is available in Fig. 4 with few fission fragments tracks.

Few pulses simulated with a direct polarisation are available in Fig. 5. Their amplitude is around 1 μA, which makes them distinguishable by using the typical CEA large bandwidth current preamplifier. Ionic pulses are really long and should not be distinguishable with current amplifier because of amplitudes of the order of nano amps. Nevertheless, electronic charges might be measured easily. Thanks to inverse polarisation, it is possible to estimate the charge carried by the ions during direct polarisation by gathering electrons back

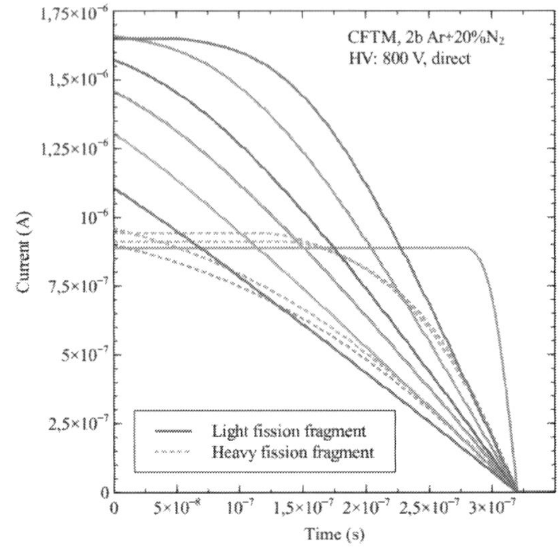

Fig. 5. Rapid simulation of CFTM current pulses in direct polarisation.

on the deposit holder. Useful quantities can be derived from pulses. First, the electron drift time is easily measurable since charge pairs are always generated at the vicinity of the cathode, thus, the pulse duration is linked to the time needed for those electrons to drift up to the anode. Multiple pulse shapes are distinguishable, triangular ones are related to heavy ion trajectories which hit the anode and rectangular pulses are characteristic of trajectories parallel to the electrode. Mixed rectangular/triangular shapes also exist, they are related to trajectories which are oriented toward the alumina wall.

From the plateau of rectangular pulses, it is possible to estimate the amount of charge pair generated along the fission fragment trajectory. It is given by:

$$Q_e = \frac{i_p}{\vec{v_d}(E,N).\vec{E_0}}, \qquad (2)$$

where i_p is the plateau value, Q_e is the total amount of electron generated along the fission fragment trajectory, v_d is the electron drift velocity which depend of the electric field and particle density and E_0 the Shockley Ramo weighting field [8] wich in our case is constant and has similar orientation to the electric field.

Two examples of charge spectrum are depicted in Fig. 6 for a direct and an inverse polarisation. For direct polarisation, only one group of pulses is noticeable because most of the fission fragments are loosing only a fraction of their kinetic energy. In addition, they are producing the same order of magnitude of electrons at the vicinity of the fissile deposit. Since most of the induced charge is due to electrons close to the cathode, measurements give only a limited amount of information about fission yield and difference of stopping powers.

For inverse polarisation, two groups of pulses are clearly noticeable. They correspond to the two groups of fission fragments. This time, electrons generated far away from the fissile deposit are mostly responsible for the recorded charge. Since stopping powers of the fission fragments become more and more different while ions are loosing energy, inverse polarisation tend to exhibit the difference between heavy ions. Hence, inverse polarisation is the optimum configuration to get information on fission yield, and difference on stopping power.

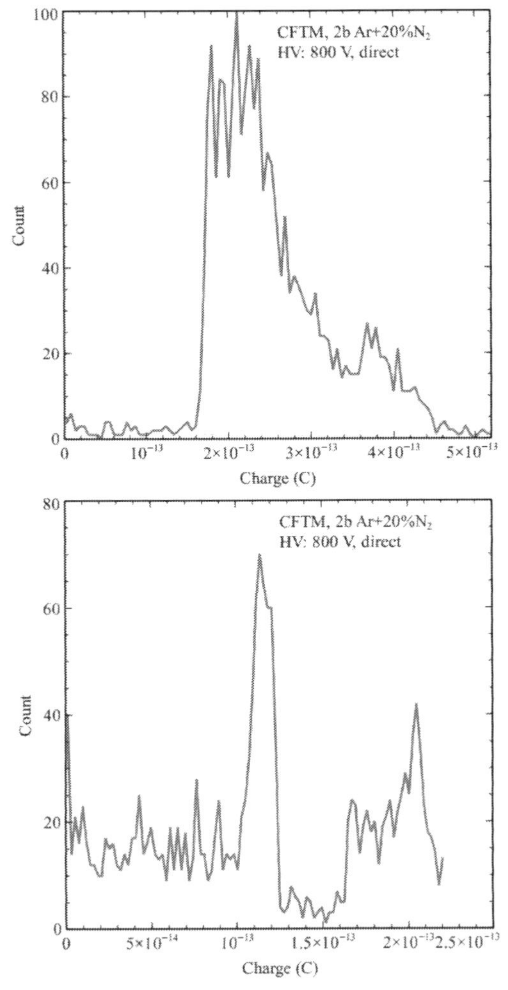

Fig. 6. Up: charge spectrum estimated with a direct polarisation. Down: charge spectrum simulated with an inverse polarisation.

IV. Experiments

An experimental campaign took place in the Minerve zero power reactor with two CFTM manufactured by the fission chamber workshop of CEA cadarache.

Minerve was a french pool type reactor located in CEA Cadarache. The core was constituted of two parts: a driver zone that contains aluminium/uranium plates assembly surrounded by graphite reflector and a central square cavity in the middle of the driver zone which received experimental lattices used to produce a specific neutron spectrum. It was submerged under 3 meter of water. The reactor could be operated to a maximum power of 100 W, which correspond roughly to a neutron flux in the core center of around 10^9 $cm^{-2}.s^{-1}$. Irradiations were performed in a dry tube installed in the reactor reflector for the experiment. Because of its large dimensions, the detector was installed in a dummmy reflector assembly.

Two instrumentations sensitive to charge and to current were used during the experiment. One has two channels with a Canberra PA2006 and a Ortec 142 preamplifiers connected respectively to Canberra amplifier A2024 and A2026. Both channels were connected to a canberra MP2 multichannel analyser. The other electronic instrumentation consisted of two current preamplifiers designed at CEA with a bandwith of $\approx$100 MHz. To keep a reasonable energy resolution, 6 meter of RG58 cable with an estimated capacity of 680 pF were used to connect preamplifiers to the CFTM.

A. *Fission chamber configurations*

Two different detectors were assembled. Each had a symmetric electrode configuration with 5mm/5mm inter-electrode gaps. The first detector, P2 (Fig.7), contains a thick deposit holder with a 0.78 cm^2 U_3O_8 deposit. The deposit contains 10 μg of U5 to reach a count rate of around 1000 cps at reactor full power.

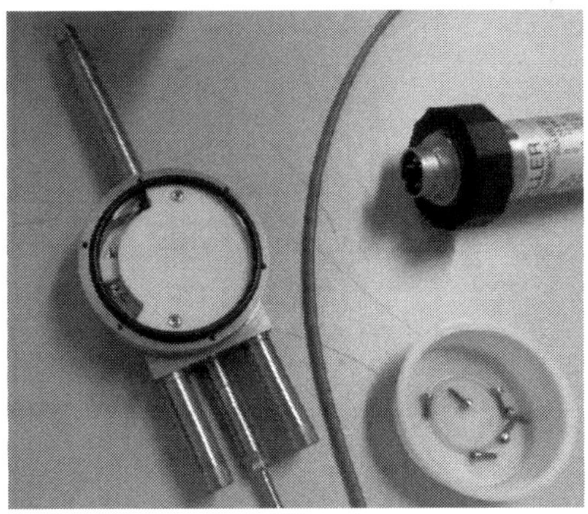

Fig. 7. Picture of a CFTM being configured.

The second detector, P3, was built to detect both fragments produced during a fission. This time the deposit consists of an evaporated droplet of uranyl nitrate sandwiched between two 2 μm thick PET foils.

For preliminary tests, P2 and P3 were filled with a $Ar+4\%N_2$ mixture usually used in regular fission chamber. It was shown that electron drift velocity is less sensitive to gas impurity with a high content of nitrogen in the mixture. Adding more quenching gas also rise the reduced field needed for electron avalanche. As a matter of fact, most of the experiments were conducted with a $Ar+20\%N_2$ mixture as the filling gas.

Numerous pressure and polarisation voltage were investigated to exhibit the recombination and the saturation regime as well as detection of both electronic and ionic charges.

V. Results

Only a few experimental results obtained with current preamplifier and with P2 are presented in the following section. To have absolute measurements, each preamplifier was calibrated. First, a simulation using SPICE software was made to estimate the theoretical transfer function. Even if the transimpedance order of magnitude is good, real preamplifier's gain seem to departure from the computed value because a stage of the preamplification is tunable through potentiometer. We then decided to use a arbitrary wave generator and a well characterized current injection device to measure the current preamplifier transimpedance. Pulses recorded with P2 detector polarised at 700 V with a filling pressure of 2 bar are shown in Fig. 8. Rectangular and triangular pulse shapes are noticeable as predicted by the simulation. Signal to noise ratio, between 2 and 6, is sufficient to compute the induced charge with precision, but make the estimation of charged particles drift time and rectangular pulses plateau values difficult.

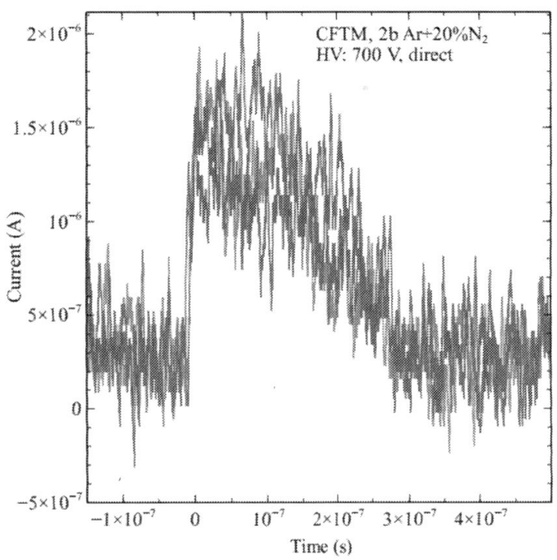

Fig. 8. Example of CFTM current pulses.

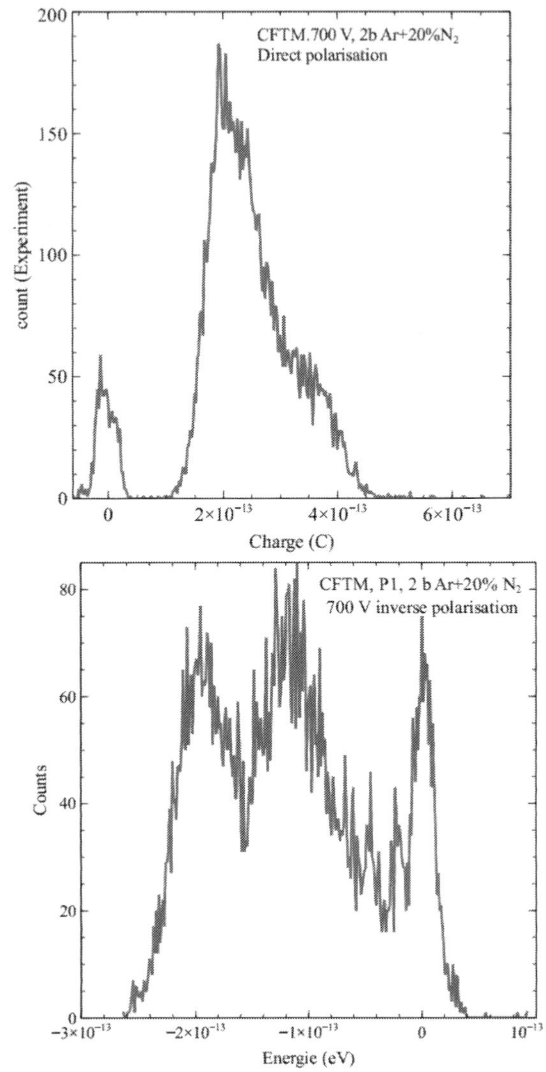

Fig. 9. Up: charge spectrum estimated with a direct polarisation. Down: charge spectrum simulated with an inverse polarisation.

Computed charge spectrum are shown in Fig. 9. The charge has an identical shape as the one from Fig. 6. Moreover, the mean charge value is estimated to $2.53 \pm 0.10\ 10^{-13}$ C while the mean charge estimated by the simulation with the same condition is $2.51\ 10^{-13}$. Other derived quantities like the drift time and the plateau value are available in Table I. Uncertainties on those values don't allow to validate the simulation code since uncertainty on drift velocity and fission fragments energy losses models involved in software are usually lower.

TABLE I
DRIFT TIME AND PLATEAU VALUE OF THE EXPERIMENT CONDUCTED WITH P2 PROTOTYPE FILLED WITH AR+ 20% N_2.

Detector	E/N (Td)	exp. drift time (s)	plateau value (A)
P2 2b, 700 V	2.83	$(3.14 \pm 0.32)\ 10^{-7}$	$1.09\ 10^{-06}$
P2 1b, 400 V	3.23	$(3.09 \pm 0.43)\ 10^{-7}$	$8.03\ 10^{-07}$
P2 4b, 1000 V	2.02	$(4.12 \pm 0.45)\ 10^{-7}$	-

A global validation of the simulation is still possible thanks to the charge spectra. Since it is an integrated quantity, it is immune to high frequency noise. Moreover, the transimpedance of the preamplifier is known with an uncertainty of 4%. Deviation of spectra mean values and other moments give hints on model discrepancies.

VI. CONCLUSION

A new detector, the CFTM, was designed and built for simulation validation purpose. Thanks to its reconfigurability, we showed it is possible to access intermediate quantities generated by simulation like charge pairs created or drift velocity. An experimental campaign took place in Minerve zero power reactor with the CFTM. At this occasion numerous data were gathered at different polarisation voltage and filling gas pressure. The calibration of the electronics measurement system allowed to estimate the absolute charge and current at the output of the detector. Unfortunately, due to our measurement system limitations, the signal to noise ratio is too low to provides most of the quantities awaited with a precision good enough to validate simulation codes. Still, it is possible to assess the global accuracy of a simulation with the help of charge spectra. To enhance the results, new measurements have to be done with a higher polarisation Voltage. An increase of the signal to noise ratio of a factor 2 is awaited while keeping the detector in a saturation regime. An enhanced smoothing procedure based on simulated results and spline approximation is also tried in order to improve the drift time and the plateau value evaluation.

REFERENCES

[1] P. Filliatre, C. Jammes, B. Geslot, and R. Veenhof, "A monte carlo simulation of the fission chambers neutron-induced pulse shape using the garfield suite," *Nuclear Instruments and Methods in Physics Research Section A*, vol. 678, pp. 139–147, 2012.
[2] P. Filliatre and C. Jammes, "A comprehensive numerical modelling of a fission chamber to be operated over a wide dynamics in the vessel of a sodium-cooled reactor," in *Physor 2016 proceedings*, 2016.
[3] Z. Elter, "pyfc: a trim-based fission chamber pulse shape simulator," CHALMERS UNIVERSITY OF TECHNOLOGY, Tech. Rep., 2015.
[4] G. de Izarra, "Cosicaf software," https://github.com/gdeIzarra/cosicaf, 2019.
[5] H. Tawara, J. Kikuchi, and T. Doke, "Energy dependences of W values for MeV alpha particles in Ar-Xe, Ar-N_2 and Ar-CH_4 mixtures," *NIMB*, 1986.
[6] G. Hagelaar and L. Pitchford, "Solving the boltzmann equation to obtain electron transport coefficients and rate coefficients for fluid models," *Plasma Sources Sci. Technol*, vol. 14, pp. 722–733, 2005.
[7] J. Ziegler, M. Ziegler, and J. Biersack, "Srimthe stopping and range of ions in matter (2010). nuclear instruments and methods," *Nuclear Instruments and Methods in Physics Research Section B: Beam Interactions with Materials and Atoms*, vol. 268, pp. 1818–1823, 2010.
[8] Z. He, "Review of the ShockleyRamo theorem and its application in semiconductor gamma-ray detectors," *Nuclear Instruments and Methods in Physics Research A*, vol. 463, pp. 250–267, 2001.

EPJ Web of Conferences **225**, 03007 (2020) https://doi.org/10.1051/epjconf/202022503007
ANIMMA 2019

Predicting Ex-core Detector Response in a PWR with Monte Carlo Neutron Transport Methods

Tanja Goričanec*†, Domen Kotnik*, Žiga Štancar*†, Luka Snoj*† and Marjan Kromar*
*Jožef Stefan Institute, Jamova cesta 39, SI-1000 Ljubljana, Slovenia
†Faculty of Mathematics and Physics, University of Ljubljana, SI-1000 Ljubljana, Slovenia

Abstract—**An approach for calculating ex-core detector response using Monte Carlo code MCNP was developed. As a first step towards ex-core detector response prediction a detailed MCNP model of the reactor core was made. A script called McCord was developed as a link between deterministic program package CORD-2 and Monte Carlo code MCNP. It automatically generates an MCNP input from the CORD-2 data. A detailed MCNP core model was used to calculate 3D power distributions inside the core. Calculated power distributions were verified by comparison to the CORD-2 calculations, which is currently used for core design calculation verification of the Krško nuclear power plant. For the hot zero power configuration, the deviations are within 3 % for majority of fuel assemblies and slightly higher for fuel assemblies located at the core periphery. The computational model was further verified by comparing the calculated control rod worth to the CORD-2 results. The deviations were within 50 pcm and considered acceptable. The research will in future be supplemented with the in-core and ex-core detector signal calculations and neutron transport outside the reactor core.**

I. Introduction

Knowing the reactor power level and its distribution inside the reactor core at any moment is of utmost importance for a safe operation of a nuclear power plant. Neutron detectors positioned outside the reactor core enable continuous power reading. As a first step towards predicting their response, a detailed model of a typical pressurized water reactor (PWR) core was developed with the Monte Carlo neutron transport code MCNP [1]. As a typical pressurized water reactor, a Krško nuclear power plant (NPP) was chosen. The Monte Carlo method was considered because it enables simulation of a more complex and detailed geometry, compared to the deterministic methods, which are more commonly used. However, its downside is its calculation time, which drastically increases for reactor simulations, where a large number of neutron histories and low stochastic errors are needed. Nowadays, with the general increase of the availability of computational resources, also Monte Carlo simulations of a sophisticated reactor became feasible.

To enable safe and continuous operation of the nuclear power plant it is important to accurately control the reactivity. Neutron flux profile inside the reactor core strongly depends on the control rod position [2], [3]. This redistribution alter the reading of ex-core detectors, which can lead to the non-linear power detector response and distort subsequent control rod worth determination from the ex-core detectors reading.

II. Krško Nuclear Power Plant

The Krško NPP is a Westinghouse design and has a two loop pressurized water reactor. Currently, the thermal rating is 1994 MWt with 727 MWe gross electric production. The core is composed out of 121 fuel assemblies (see Fig. 1).

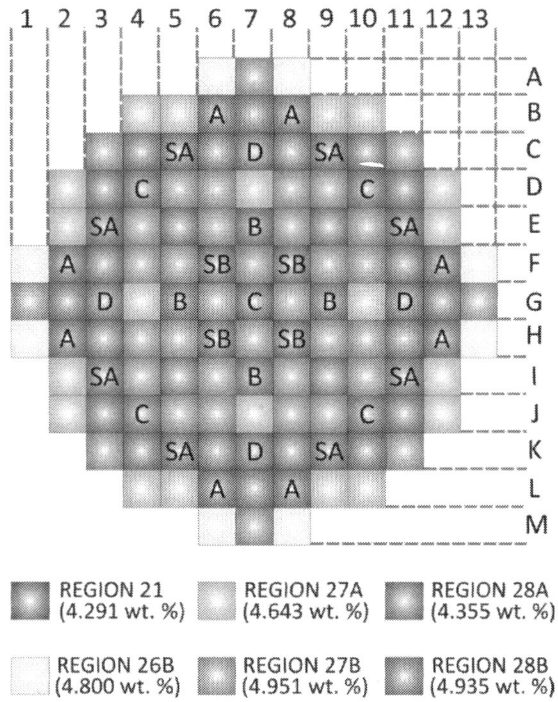

Fig. 1: Core configuration for Cycle 26 with marked fuel regions in different color and control rod cluster positions.

Fuel assembly has a 16×16 lattice filled with 235 fuel rods, 20 guides for control rods and 1 guide for in-core instrumentation as presented in Fig. 2. Integral Fuel Burnable Absorbers (IFBA) are added to some fuel rods to enable long term reactivity control. Typically six different IFBA patterns are used, one of them with 32 IFBA rods is presented in Fig. 2.

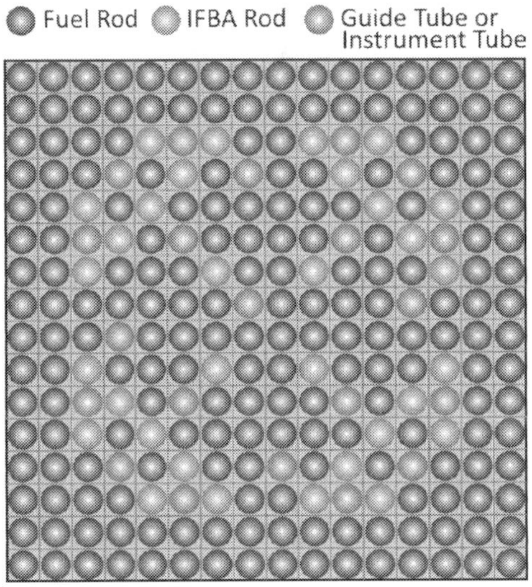

Fig. 2: Schematic drawing of fuel assembly with 32 IFBA rods.

A. Neutron Detectors

The ex-core detector system monitors neutron flux from shutdown conditions to 200 % of full power. This represents neutron flux variations from 10^{-1}-10^{11} n/(cm^2 s), therefore three types of neutron detectors are utilized: BF_3 counter (source range), compensated ionization chamber (intermediate range) and uncompensated ionization chamber (power range). The ex-core detectors are placed in wells, which are located on the cavity wall. Power range detectors are positioned in four equally spaced locations around the core. There are 4 power range channels with 2 vertical detectors per channel.

The in-core neutron instrumentation consist of 4 movable neutron detectors, which can be positioned in 36 in-core positions. The movable detectors are placed in the instrumentation guide thimbles selected so as to monitor a significant number of representative assemblies [4]. Taking into account the reflection of data across the core centrelines, a complete 3D core flux map may be obtained.

B. Control Rod Calibration

Inside the reactor core are 33 rod cluster control assemblies (RCCA) [4] distributed as presented in Fig. 1. A RCCA consist of 20 individual Ag-In-Cd absorber rods fastened at the top to a common spyder assembly. To enable safe operation of the nuclear reactor it is important to accurately know the control rod worth. The control rod worth can be measured using different techniques (i.e. boron dilution and rod swap method). A newer method, called the rod insertion method [5], [6] was developed at the Jožef Stefan Institute (JSI) Reactor Physics Department. It relies on the analysis of the reactor signal, which is recorded during continuous insertion of the control rod bank. Its major advantage is high execution speed (approximately 15 minutes per control rod bank) in contrast to the boron dilution method, which takes about 4 hours. During the insertion of the RCCA the spatial distribution of neutron population is changed [2], [3]. Since the detector measures local neutron flux at the detector location, this can lead to the non-linear power reading. To account for those redistributions, the neutron flux redistribution factors as a function of RCCA axial position are introduced. Neutron flux redistribution factors currently used in the NPP Krško have been determined with a single adjoint flux distribution calculation for the first operational cycle [7]. The calculation was performed using 2D deterministic code DOT [8]. In future research those factors will be updated using Monte Carlo models described in this paper.

III. Calculation Procedure

The calculation procedure designed to obtain ex-core detector response in a typical PWR NPP is schematically presented in Fig. 3. Firstly, the deterministic code package CORD-2 [9] is used to obtain fuel temperature, water temperature, water density, fuel isotopic composition due to the burn-up, burnable absorber layer (IFBA) and boron concentration for each fuel assembly in 10 axial layers. In next step a newly developed subroutine McCord is used to generate MCNP input from the CORD-2 output data. Using MCNP core input power and neutron flux distribution inside the reactor core can be analysed and redistribution due to the control rod movement can be studied. Moreover, in-core detector response can be anlyzed and compared to measurements to validate the MCNP core model. The MCNP core input will be used to calculate the neutron source for the MCNP ex-core model. The MCNP ex-core model includes ex-core structures, such as: reactor vesel, ex-core neutron detectors and all the surrounding concrete structures. Using hybrid code ADVANTG [10] weight windows for neutron transport will be generated and used in the MCNP simulation to study the ex-core detector response and neutron dose field in the containment.

A. CORD-2 calculation

Deterministic CORD-2 system [9] was developed at the JSI Reactor Physics Department and is routinely used for core design verification calculations of the Krško NPP. It consists of two basic reactor physics codes: WIMS-D [11] and GNOMER [12]. WIMS-D is a widely-used lattice code and GNOMER solves the neutron diffusion equasion in three-dimensional Cartesial geometry by Green's function NOdal MEthod [13]. As other deterministic codes, also CORD-2 includes cross-section homogenization. The thermal feedback is taken into account with a simple thermohydraulic module from the CTEMP [14] code. The CORD-2 calculation results were thoroughly verified with core design calculations and its output data (water temperatures, water densities, fuel temperature, fuel isotopic and IFBA layer) will be used as input parameters for MCNP core model.

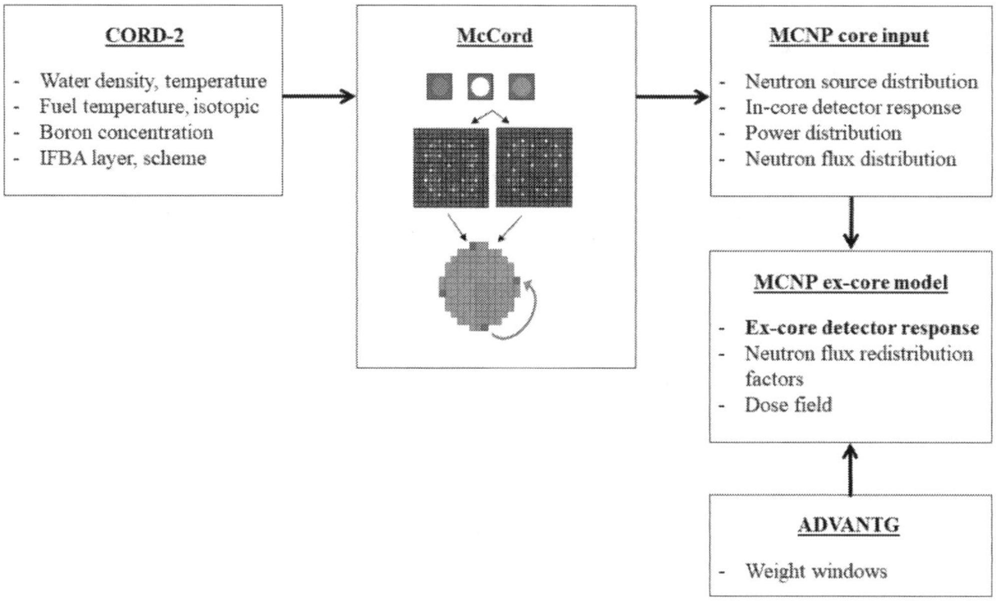

Fig. 3: Calculation procedure scheme to obtain ex-core detector response.

B. Subroutine McCord

The MCNP PWR core model is very complex and detailed (e.g. it includes more than 30.000 cells and approximatelly 1.000 materials, see Figs. 4 and 5) and is currently composed out of approximately 100.000 input lines.

Fig. 4: xz view of the MCNP reactor core model.

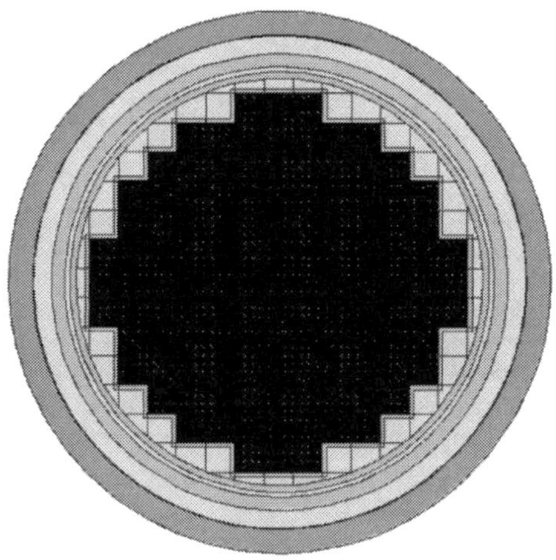

Fig. 5: xy view of the MCNP reactor core model.

To enable prompt generation of MCNP core inputs for different cycles and parts of cycle (e.g. beginning of life or end of life) the process must be completely automated. A subroutine McCord was developed at the JSI Reactor Physics Department and is capable of automated generation of MCNP core input from the CORD-2 output data. It serves as a link between deterministic CORD-2 and Monte Carlo MCNP calculations.

In the McCord code the generation of MCNP input must be very systematic [15]. The basic cell is a fuel pin divided into 10 axial layers and surrounded with water. Different fuel pin cells are described depending on their position inside the fuel assembly and presence of IFBA coating. Each fuel pin layer has dedicated its own water density, fuel density (calculated from fuel isotopic) and IFBA coating, all of which are output from the CORD-2 calculation.

Each axial layer of fuel pin has its own fuel and water material specified taking into account their temperature and fuel isotopic from the CORD-2 output. To describe the materials at desired temperature two neutron cross section libraries at nearest temperatures were selected and atomic fractions of isotopes were calculated using pseudo materials method [16]. Thermal scattering cross sections $S_{\alpha,\beta}$ for fuel (U-UO_2 and O_2-UO_2) were taken at the nearest given temperature. $S_{\alpha,\beta}$ cross sections for water (H-H_2O) have higher impact on the results and were therefore generated in 5 K intervals using MAKXSF code [17] to interpolate pre-calculated thermal scattering libraries, then the $S_{\alpha,\beta}$ cross section at nearest temperature was taken.

The fuel assembly lattice is formed from fuel pin cells and guide tubes for control rods and instrumentation. When forming the fuel assembly lattice the IFBA pattern (number of IFBA coated fuel pins in fuel assembly) must be taken into account. Information about the IFBA pattern of individual fuel assembly is also described in the CORD-2 output file. In total 37 fuel assemblies are generated and translated taking into account quadrant symmetry to form a core lattice. Control rods are added in banks and their axial position can be easily altered. Above the active fuel height top nozzles are placed and below active fuel region bottom nozzles are positioned. Around the active core part the baffle, core barrel, former plates, thermal shield, inner reactor vessel cladding and reactor vessel are modelled as can be observed in Fig. 4 and 5. Between those structures the water with inlet water characteristics is added.

C. MCNP core model calculation and neutron source generation

The MCNP core model produced with McCord subroutine can be used for various purposes, e.g. to calculated 3D power distribution within the core, in-core detector response or to evaluate the control rod worth from the change in the calculated multiplication factor. The MCNP core model will also be used to generate neutron source, which will be implemented in the MCNP containment model and then used for ex-core neutron transport calculations. Fission rate and neutron spectrum will be calculated for each fuel rod in multiple axial layers [18].

D. MCNP containment model calculation

A simplified MCNP model of containment building including reactor pressure vessel and souring structures was already developed [19] and will be supplemented with explicitly modelled ex-core neutron detectors. This model will be used for ex-core neutron transport calculations. To speed up the neutron transport from core to the detectors a hybrid code ADVANTG [10] will be used.

IV. Monte Carlo Computational Model of the Reactor Core

The detailed Monte Carlo model of the reactor core was made using Monte Carlo Neutron Transport Code MCNP [1] and is presented in Fig. 4 and Fig. 5. A great effort was put into designing the core model in a systematic way to enable quick generation of inputs for desired cycle configuration [15] using specially developed subroutine. In the MCNP core model elemental cell is fuel pin and different fuel pins form a fuel assembly lattice. Fig. 6 shows closer view of the MCNP fuel assembly model. For core reconstruction 37 fuel assemblies are modelled and translated taking into account quadrant symmetry. Core is divided into 10 axial layers, which are smaller at the top and bottom of the core and larger in the middle. Individual fuel assembly is specified its own water temperature, water density, fuel temperature, fuel isotopic and IFBA layer for each axial layer.

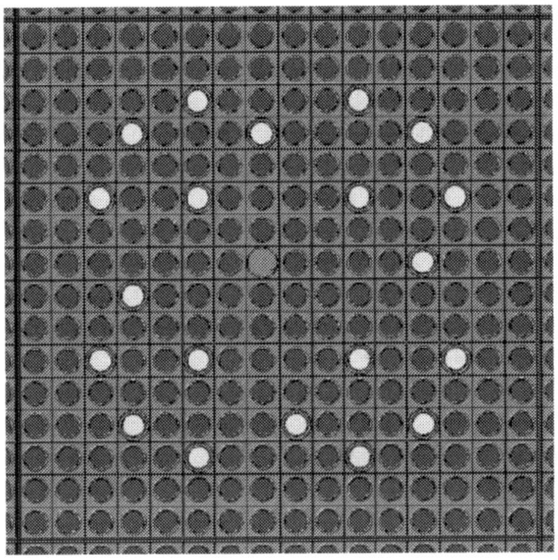

Fig. 6: Closer view of the MCNP fuel assembly model.

The detailed MCNP core model includes several geometric details, such as upper and lower nozzle, simplified model of axial grids, plenum spring, Zircaloy cladding, fuel pin plugs, control rods, etc. Fig. 7 shows closer view of the MCNP top nozzle model and Fig. 8 closer view of the completely inserted control rods. Around the core surrounding structures were added to enable more accurate calculations. In Figs. 4 and 5 baffle, core barrel, former plates, thermal shield ad reactor vessel surrounding the core can be observed.

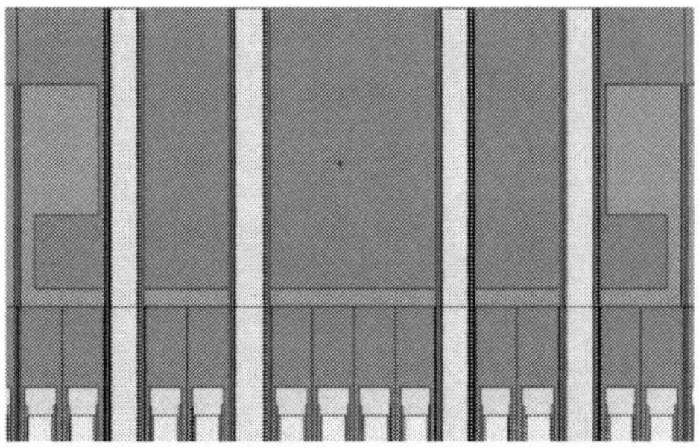

Fig. 7: Closer view of the MCNP top nozzle model.

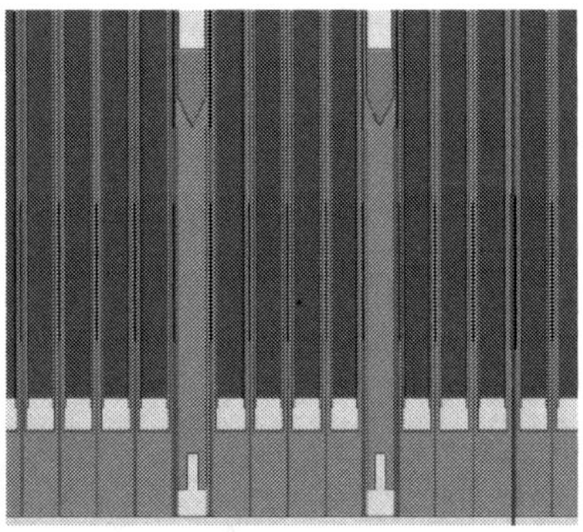

Fig. 8: Closer view of the completely inserted control rods in the MCNP model.

V. RESULTS

A complexity of the MCNP core model makes room for mistakes, which are hard to notice. Therefore, the model has to be thoroughly validated before further use. Its results were validated with extensively verified and validated deterministic code package CORD-2. Specific power (reactor power per unit of volume) distribution inside the reactor core per fuel assembly and axial distributions were analysed. Verification continued with control rod worth determination through the calculation of multiplication factor.

A. Reactor Power Distribution

First step toward validation of the MCNP core model was comparing specific reactor power distribution to the verified and validated CORD-2 results. In Fig. 9 the calculated specific power for cycle 28 hot zero power (HZP) configuration in xy plane normalized to the average value over the entire reactor core is presented. The specific reactor power was calculated through the entire reactor core using fine geometry mesh of 250×250×1 voxels, with one volume being equivalent to approximately one fuel pin. The dips in the specific power within the core are in position of guide tubes for control rod and instrumentation.

To compare results with CORD-2 the calculated specific power over individual fuel assembly was normalized to the total specific power inside the reactor core. The MCNP results for the entire core are presented in Fig. 10, where numbers reported on top stand for normalized specific power to the total specific power and numbers reported on bottom represent stochastic 1σ uncertainty in %. The 1σ statistical uncertainty originating from Monte Carlo method is <0.2 % and therefore considered negligible. Normalized specific power (p_{MCNP}) was compared to the normalized CORD-2 results (p_{CORD}) and deviations for cycle 28 HZP configuration are presented in Fig. 11, where top number represents absolute deviation and bottom number represents relative deviation in term of $p_{CORD}/p_{MCNP} - 1$ in %. It can be observed that the deviations are systematically increasing toward core edge, this can be explained with different treatment of neutron leakage in both codes. The maximum deviation being $\sim$6.5 %, however the majority of fuel assemblies show good agreement between both codes with deviations <3 %.

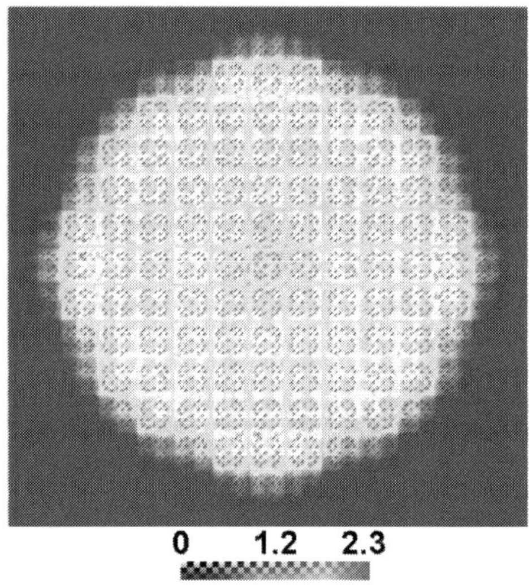

Fig. 9: Specific reactor power in xy plane for cycle 28 HZP configuration calculated using MCNP core model. The specific power values are normalized to the average over the entire core and are presented with colors ranging from low values in blue to high values in red.

To further verify the MCNP core model axial specific power distributions were compared to the CORD-2 results. Specific powers for axial profile analysis were calculated in 10 axial layers, which coincide with geometrical layers in computational model. The MCNP computational relative uncertainty of individual axial layer was below 1 % and considered negligible.

 https://doi.org/10.1051/epjconf/202022503007

	1	2	3	4	5	6	7	8	9	10	11	12	13
A						0.3226 0.1333	0.4486 0.1176	0.3223 0.1334					
B				0.3884 0.1264	0.6145 0.1007	1.1113 0.0819	1.2400 0.0769	1.1133 0.0818	0.6149 0.1007	0.3891 0.1262			
C			0.5101 0.1121	1.0683 0.0828	1.2052 0.0758	1.2017 0.0732	1.2425 0.0720	1.2051 0.0730	1.2066 0.0756	1.0710 0.0826	0.5161 0.1112		
D		0.3836 0.1272	1.0680 0.0828	1.2570 0.0729	1.2423 0.0731	1.1902 0.0739	1.2127 0.0736	1.1971 0.0738	1.2438 0.0730	1.2551 0.0728	1.0718 0.0826	0.3893 0.1263	
E		0.6094 0.1013	1.2074 0.0758	1.2469 0.0730	1.2781 0.0711	1.2591 0.0728	1.3225 0.0712	1.2581 0.0728	1.2777 0.0711	1.2406 0.0731	1.2077 0.0757	0.6145 0.1007	
F	0.3173 0.1343	1.1108 0.0820	1.2093 0.0731	1.2003 0.0737	1.2593 0.0727	1.2134 0.0730	1.1109 0.0764	1.2163 0.0729	1.2599 0.0727	1.1929 0.0738	1.2037 0.0732	1.1062 0.0820	0.3209 0.1338
G	0.4462 0.1179	1.2445 0.0769	1.2520 0.0718	1.2230 0.0733	1.3194 0.0712	1.1049 0.0766	0.8912 0.0812	1.1067 0.0766	1.3151 0.0714	1.2114 0.0736	1.2366 0.0721	1.2335 0.0771	0.4459 0.1180
H	0.3202 0.1339	1.1159 0.0818	1.2129 0.0730	1.2010 0.0738	1.2582 0.0728	1.2118 0.0731	1.1030 0.0766	1.2070 0.0732	1.2480 0.0730	1.1864 0.0740	1.1976 0.0733	1.1039 0.0821	0.3211 0.1338
I		0.6132 0.1010	1.2113 0.0757	1.2488 0.0730	1.2806 0.0711	1.2575 0.0728	1.3170 0.0714	1.2536 0.0729	1.2707 0.0714	1.2359 0.0733	1.2026 0.0759	0.6112 0.1010	
J		0.3833 0.1274	1.0691 0.0829	1.2587 0.0729	1.2475 0.0731	1.1970 0.0739	1.2171 0.0735	1.1956 0.0739	1.2414 0.0732	1.2481 0.0732	1.0663 0.0829	0.3892 0.1265	
K			0.5060 0.1127	1.0692 0.0829	1.2063 0.0759	1.2048 0.0732	1.2481 0.0720	1.2086 0.0732	1.2052 0.0759	1.0632 0.0830	0.5063 0.1125		
L				0.3833 0.1275	0.6061 0.1015	1.1046 0.0823	1.2360 0.0771	1.1102 0.0821	0.6068 0.1015	0.3806 0.1277			
M						0.3160 0.1348	0.4424 0.1185	0.3175 0.1344					

Fig. 10: Specific reactor power per fuel assembly for cycle 28 HZP configuration calculated using MCNP core model. Top number represents specific power normalized to the one over reactor core and bottom number is relative uncertainty in %.

	7	8	9	10	11	12	13
G	0.0342 3.8387	0.0263 2.3757	0.0248 1.8862	0.0177 1.4574	0.0055 0.4443	-0.0340 -2.7432	-0.0233 -5.2294
H	0.0252 2.2724	0.0290 2.3914	0.0269 2.1394	0.0189 1.5828	0.0064 0.5308	-0.0313 -2.8211	-0.0200 -6.2577
I	0.0223 1.6925	0.0254 2.0195	0.0208 1.6302	0.0140 1.1247	-0.0029 -0.2407	-0.0122 -1.9946	
J	0.0200 1.6451	0.0189 1.5781	0.0145 1.1679	0.0077 0.6109	-0.0157 -1.4727	-0.0118 -3.0540	
K	0.0045 0.3626	0.0030 0.2452	-0.0048 -0.3942	-0.0164 -1.5352	-0.0140 -2.7552		
L	-0.0330 -2.6661	-0.0331 -2.9797	-0.0146 -2.3910	-0.0129 -3.3520			
M	-0.0228 -5.1123	-0.0207 -6.4634					

Fig. 11: Deviation in normalized specific reactor power between CORD-2 (p(CORD)) and MCNP (p(MCNP)) for cycle 28 HZP configuration. Deviations are presented per fuel assembly for one quarter of the core. Top number represents absolute difference (p(CORD)-p(MCNP)) and bottom number relative difference (p(CORD)/p(MCNP)-1) in %.

Results for individual axial layer (i) of individual fuel assembly (j) obtained with MCNP ($p_{i,j}$) were normalized as:

$$p_{z,i,j} = \frac{p_{i,j} p_{xy,j} H}{\sum_{i=1}^{10} p_{i,j} H_i}, \tag{1}$$

$$p_{z,i} = \sum_{j=1}^{121} \frac{p_{z,i,j}}{121}, \tag{2}$$

where $p_{xy,j}$ is the normalized specific power for fuel assembly (j) from the xy power profile, H is a total active fuel hight and H_i is a hight of individual axial layer. Averaging is performed through all fuel assemblies within the core to obtain average axial profile (p_z). Results are presented in Fig. 12, where a black line represents MCNP results and a red line CORD-2 results. In Fig. 13 absolute deviations between both calculations are presented. Even though the deviations can be up to 10 %, the shape of the MCNP axial profile is similar to the shape of axial profile calculated with CORD-2 program package, which indicates that there is no sever flaw in the model.

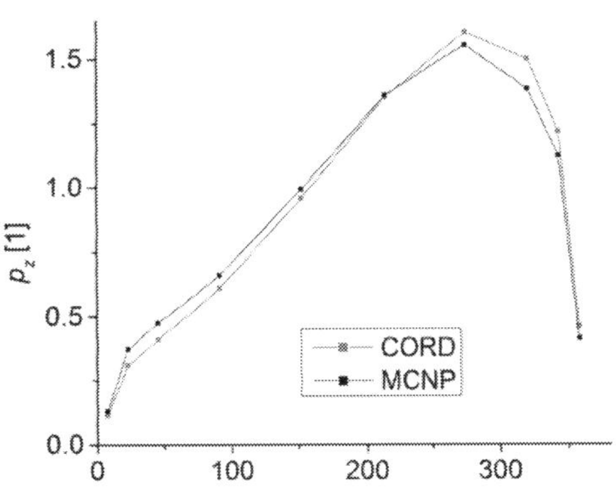

Fig. 12: Normalized specific power average axial profile calculated with CORD-2 (red) and MCNP (black) for fuel cycle 28 HZP configuration.

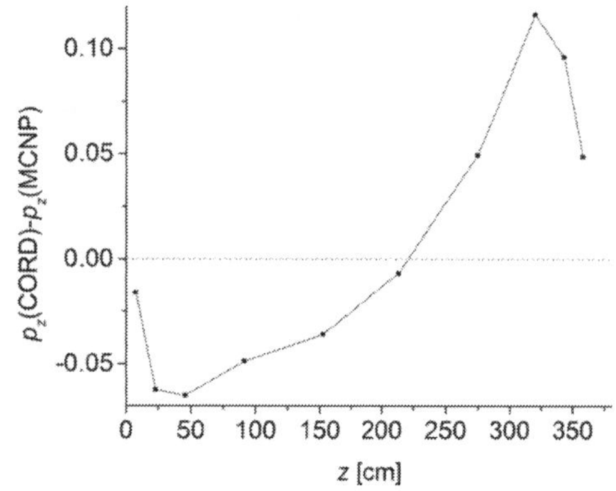

Fig. 13: Absolute deviation between normalized CORD and MCNP average axial profiles for fuel cycle 28 HZP configuration.

B. Control Rod Worth

Control rod worth for different RCCA banks (see Fig. 1) was determined by comparing multiplication factor of calculation with all banks completely withdrawn and calculation with bank of interest fully inserted. The results are presented in Table I, where W_{RI} represents the measured RCCA worth through the rod insertion method and W_M and W_C represent the calculated MCNP and CORD-2 RCCA worth respectively. It can be observed that calculated values between the MCNP and CORD-2 deviate for less then 50 pcm. It can be concluded that the calculations agree well with each other, which additionally confirms the MCNP core model for the HZP configuration. It should be taken into account that the measured values represent the ex-core detector response and cannot be directly compared to the currently calculated values, which take into account only neutron transport within the core. The RCCA worth through calculation of ex-core detector response will be studied in future research and is out of the scope for this paper.

TABLE I: Control Rod Worth for Cycle 28.

Bank	W_{RI} [pcm]	W_M [pcm]	W_C [pcm]	W_M-W_C [pcm]
D	741	703	702	1
C	868	818	833	-15
B	706	721	741	-20
A	929	856	840	16
DCBA	3244	3098	3116	-18
SA	1385	1227	1235	-8
SB	591	638	608	30
SA+SB	1976	1866	1843	23
Total	5220	4964	4959	5

C. Neutron flux redistribution due to the RCCA movement

The neutron flux redistribution due to the RCCA movement was studied and is presented in Fig. 14.

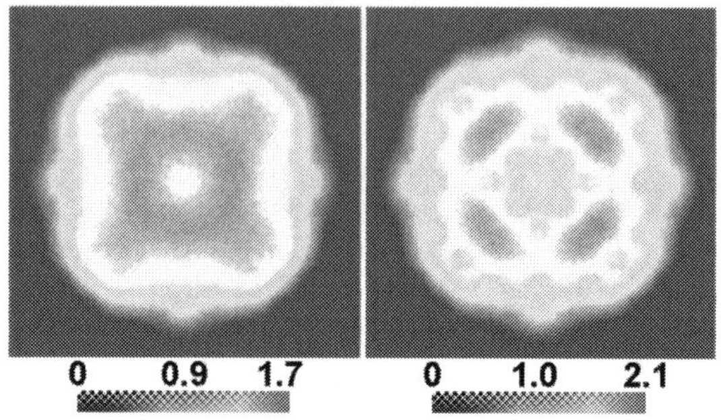

(a) A bank completely inserted. (b) All banks completely inserted.

Fig. 14: Neutron flux in xy plane, normalized to the average value of 1 through the entire core for cycle 28 HZP configuration calculated using MCNP core model. Normalized neutron flux values are represented with colors, ranging from low values in blue to high values in red.

The neutron flux was calculated through the entire reactor core using fine geometry mesh of 250×250×1 voxels, with one volume being equivalent to approximately one fuel pin. Flux was normalized to get 1 as average value through the reactor core. It can be observed that RCCA movement can significantly affect the neutron flux distribution and therefore alters the reading of ex-core neutron detector, which is also used for RCCA worth determination. To account for this effect the neutron flux redistribution factors are introduced when evaluating RCCA worth with rod insertion method. Their determination will be a subject of further studies and is out of the scope for this paper.

VI. Future Research

After validation and verification of the MCNP core model the focus of the future research will be neutron transport outside the reactor core, with the aim to determine ex-core detector response. For those calculations simplified MCNP model of the reactor pressure vessel, surrounding structures including ex-core detectors and containment building presented in Fig. 15 will be used [19]. Firstly, the core model will be used to generate neutron source, which will then be used inside the ex-core model. Fission rate and spectrum will be determined for each fuel rod in multiple axial layers [18]. To speed up neutron transport from core to the e.g. neutron detectors, a hybrid code ADVANTAG [10] will be used. The study will be supplemented with the effect of the control rod movement on the ex-core detector response, which will enable the calculation of neutron flux redistribution factors for each fuel cycle individually, therefore significantly improving the control rod worth determination, which is a safety related parameter.

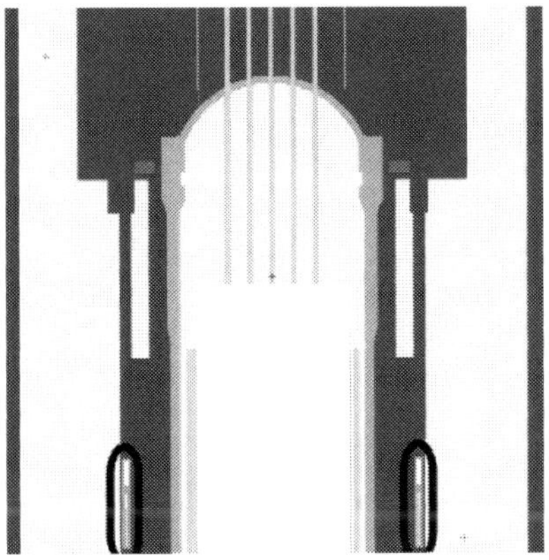

Fig. 15: Simplified reactor vessel and surrounding structure MCNP model with ex-core neutron detectors marked within black ellipse.

VII. Conclusion

The ex-core neutron detectors are in nuclear power plant used to monitor the reactor power during normal operation or when performing different physical tests, e.g. rod insertion for control rod worth determination. Therefore, it is important to know their response as accurately as possible. To predict their response a calculation procedure using Monte Carlo neutron transport was developed. Currently, the majority of core design and control rod worth calculations are performed using deterministic programs, which include geometry simplifications. Using Monte Carlo method would enable simulation of more complex geometry and more accurate prediction of the detector response. The detailed MCNP model of a typical PWR reactor core was developed and automated generation of inputs performed by newly developed subroutine McCord. Hot zero power configuration was verified with deterministic CORD-2 results and the agreement was considered acceptable. In future research the MCNP core model will be validated by in-core measurements and neutron transport outside the reactor core will be performed to study ex-core detector response.

Acknowledgment

The research is funded by the project L2-8163 from the Slovenian Research Agency.

References

[1] J.T. Goorley, et al., *Initial MCNP6 Release Overview - MCNP6 version 1.0*, LA-UR-13-22934, 2013.
[2] T. Kaiba, et al., *Validation of neutron flux redistribution factors in JSI TRIGA reactor due to control rod movements*, Appl. Radiat. Isot., vol. 104, pp. 34-42, 2015. doi:10.1016/j.apradiso.2015.06.026.
[3] T. Goričanec, et al., *Evaluation of neutron flux and fission rate distributions inside the JSI TRIGA Mark II reactor using multiple in-core fission chambers*, Annals of Nuclear Energy, vol. 111, pp. 407-440, 2018.
[4] J. Jensen-Tornehed, et al., *The Nuclear Design and Core Management of the KRSKO Nuclear Power Plant Cycle 26*, revision 1, Westinghouse Electric Company LLC, 2012.
[5] A. Trkov, M. Ravnik, H. Wimmer, B. Glumac, H. Bck, *Application of the rod insertion method for control rod worth measurements in research reactors, Kerntechnik*, vol. 60, pp. 255261, 1995.
[6] V. Merljak, M. Kromar, A. Trkov, *Rod insertion method analysis - A methodology update ad comparison to boron dilution method, Annals of Nuclear Energy*, vol. 113, pp. 96-104, 2018.
[7] M. Kromar, et. al, *The Nuclear Design and Core Management of the Krško NPP Cycle 26,*, Institute Jožef Stefan, Ljubljana, Slovenia, IJS-DP-11027, 2012.
[8] W. A. Rhoades, *DOT IV, Version 4.2: Two Dimensional Discrete Ordinates Radiation Transport Code System*, RSIC Code Package CCC-320, ORNL, Oak Ridge, Tenn., 1979.
[9] M. Kromar, A. Trkov, *Nuclear Design Calculations of the NPP Krško Core*, Journal of Energy Technology, Vol. 2, pp. 41-50, November 2009.
[10] S. W. Mosher, et. al, *ADVANTG An Automated Variance Reduction Parameter Generator*, August 2015.
[11] J. R. Askew, F. J. Fayers, P. B. Kemshell, A General Description of the Code WIMS, *J. Br. Nucl. Energy Soc.*, Vol. 5, p. 564, 1966.
[12] A. Trkov, *GNOMER - Multigroup 3-Dimensional Neutron Diffusion Nodal Code*, Institute Jožef Stefan, Ljubljana, Slovenia, IJS-DP-6688, March 1993.
[13] A. Trkov, M. Najžer, L. Škerget, Variant of Green's function Nodal method for Neutron Diffusion, *J. Nucl. Sci. Technol.*, Vol. 27, pp. 766-777, 1990.
[14] M. Kromar, S. Slavič, A. Trkov, *CTEMP: A Code for Thermohydraulic Calculations,*, Institute Jožef Stefan, Ljubljana, Slovenia, IJS-DP-6143, February 1991.
[15] Ž. Štancar, M. Kromar, B. Kos and L. Snoj, *Construction of a Monte Carlo Benchmark Pressurized Water Reactor Core Model*, In: Proceedings of 25th International Conference Nuclear Energy for New Europe, Slovenia, September 2016.
[16] j. L. Conlin, W. Ji, J. C. Lee, M. R. Martin, *Pseudo Material Construct for Coupled Neutronic-Thermal-Hydraulic Analysis of VHTGR*, Transactions of ANS, Vol. 91, 2005.
[17] F. B. Brown, *The makxsf Code with Doppler Broadening*, LA-UR-06-7002, 2006.
[18] P. Sauvan, et al., *Development of the R2SUNED Code System for Shutdown Dose Rate Calculations*, IEEE Transactions on Nuclear Science, vol. 63, pp. 375-384, 2016.
[19] L. Snoj, B. Kos, Ž. Štancar and M. Kromar, *Neutron Streaming Analysis and Shielding Determination*, IJS work report, IJS-DP-12106, Ljublja

TRITIUM - A Quasi Real-Time Low Activity Tritium Monitor for Water

C.D.R. Azevedo, A. Baeza, M. Brás, T. Cámara, C. Cerna, E. Chauveau, J.M. Gil, J.A. Corbacho,V. Delgado, J. Díaz, J. Domange, C. Marquet, M. Macko, M. Martínez-Roig*, A. Moreno, F. Piquemal, A. Rodríguez, J. Rodríguez, C. Roldán, J.F.C.A. Veloso, N. Yahlali
* Instituto de Física Corpuscular (IFIC), marcos.martinez@ific.uv.es

***Abstract*—Tritium is released abundantly to the environment by nuclear power plants (NPP), as a product of neutron capture by hydrogen and deuterium. In normal running conditions, released cooling waters may contain levels of tritium close to or even larger than the maximum authorised limit for human consumption (drinking and irrigation). The European Council Directive 2013/51/Euratom requires a maximum level of tritium in water for human consumption lower than 100 Bq/L. Current monitoring of tritium activity in water by liquid scintillating method takes about two days and can only be carried out in a dedicated laboratory. This system is not appropriate for real time monitoring. At present, there exists no available detector device with enough sensitivity to monitor waters for human consumption with high enough sensitivity. The goal of the TRITIUM project is to build a tritium monitor capable to measure tritium activities with detection limit close to 100Bq/L, using instrumentation technique developed in recent years for Nuclear and Particle Physics, such as scintillating fibres and silicon photomultipliers (SiPM). In this paper the current status of the TRITIUM project is presented and the results of first prototypes are discussed. A detector system based on scintillating fibers read out either photomultiplier tubes (PMTs) or silicon photomultiplier (SiPM) arrays is under development and will be installed in the vicinity of Almaraz nuclear power plant (Cáceres, Spain) by the fourth term of 2019.**

I. Introduction

A. The SUDOE TRITIUM Project

The TRITIUM project is funded by the INTERREG-SUDOE program of the European Union and composed by a consortium of 5 southwestern European institutions: University of Extremadura (Spain), University of Valencia (Spain), University of Aveiro (Portugal), CENBG Bordeaux (France) and Junta de Extremadura (Spain). The main aim of the project is to develop and build a monitor for low activity tritium in water with near real-time capability for in-situ monitoring of water released by nuclear power plants.

M. Martínez-Roig(corresponding author), T. Cámara, V. Delgado, J. Díaz, and N. Yahlali are with Instituto de Física Corpuscular, Centro mixto CSIC-Universidad de Valencia, Paterna, Spain. (email: marcos.martinez@ific.uv.es)

C.D.R. Azevedo, M. Brás and J.F.C.A. Veloso are with I3N - Departamento de Fisica da Universidade de Aveiro, Aveiro, Portugal

C. Roldán is with Universidad de Valencia, Departamento de Física Aplicada, Burjassot, Spain.

A. Baeza, J.A. Corbacho and A. Rodríguez are with Universidad de Extremadura, Laboratorio de Radioactividad Ambiental. Servicio de apoyo a la investigación, Cáceres, Spain

C. Cerna, E. Chauveau, J. Domange, C. Marquet and F. Piquemal are with Université de Bordeaux and CNRS, CENBG, Gradignan Cedex, France.

J.M. Gil, A. Moreno and J. Rodríguez are with Junta de Extremadura, Plaza del Rastro s/n. 06800, Mérida, Spain.

Fig. 1. Picture of Almaraz NPP and Arrocampo dam (Cáceres, Spain).

The EU Council Directive 2013/51/Euratom regulates the maximum level of tritium in water for human consumption to be lower than 100 Bq/L. This level is considerably higher than that of natural or cosmogenic origin, but may be easily overcome in the cooling water of nuclear power plants even in normal operation. The cooling water is usually released in rivers and may cause direct or indirect contamination of the environment and water for human consumption with tritium, a long-lived hydrogen isotope of more than 12 year half-life. Monitoring the tritium level in the cooling water of nuclear power plants would provide a safety alarm and an indicator of anomalous functioning of the NPP due to an excess of neutrons or a potential water leakage from the primary cooling circuit. Indeed, tritium is the third isotope of hydrogen and is produced by neutron capture of hydrogen and deuterium atoms. Thus, an excess of neutron flux would induce an excess of tritium activity level in the cooling water. In order to measure levels of tritium with the sensitivity required by the European Directive, the consortium is developing a detector system based on scintillating fibers, endowed with active and passive background rejection, both of environmental and cosmic origin. The installation of the detector is planned by the fourth term of 2019, downstream the Arrocampo dam of the Almaraz (Spain) nuclear power plant, in the Tagus river, close to the Portuguese border. A picture of Almaraz NPP is displayed in Fig.1.

B. State of the art

Measurement of tritium activity is one of the routinely controls to be carried out in the surrounding area of NPPs and, in the future, of nuclear fusion facilities which release tritium abundantly. Tritium is an unstable hydrogen isotope with about 12 years half-life, decaying into ^{3}He through β^- decay. The emitted electron has a continuous energy spectrum peaked at 4.7 keV with a maximum energy emission of about 18 keV, resulting in an averaged energy of about 6 keV. Due to such low energy, the detection of tritium decay is not trivial and requires high sensitivity detectors. There are already available devices for the tritium detection: gaseous detectors, semi-conductors and liquid/solid scintillators. The gaseous detection method consists on the introduction of water vapor into an ionisation chamber; this procedure is quite complex for autonomous operation. Solid state detectors, such APDs, can be directly exposed to the beta radiation but the presence of water makes them unusable. Liquid scintillators have been extensively used in laboratory to detect tritium levels in water as low as 1 Bq/L, but they require collection and transportation of water samples and data taking during two or more days, which makes them unsuitable for real-time monitoring. Furthermore, they produce highly toxic residues (as toluene), preventing their use for environmental surveillance. Solid scintillators have been employed by different authors: K.J. Hofstetter in 1993 [1] built a capillary system read out by PMTs, obtaining a detection limit of 22000 Bq/L in 2 min of measurement. Rathnakaran in year 2000 [2] obtained a detection limit of 10000 Bq/L during 10 min of measurement, a number 2 orders of magnitude higher than the detection limit required in this project. An alternative to solid scintillators, which provides higher sensitivity to low levels of tritium in water, is the employment of scintillating optical fibers as a detection medium. This scintillation alternative, which is proposed in the present work, was investigated already in 1998 by J.W. Berthold and L.A. Jeffers [3], using polystyrene optical fibers with fluor-doped cladding read out by photomultiplier tubes (PMTs). These authors aimed at the development of a detector for monitoring tritium in drinking waters to verify compliance with the US EPA drinking water standard of 740 Bq/L, a rather larger limit than that required by the present European Directive.

II. Technological aspects of the TRITIUM detector

The design of TRITIUM detector is based on uncladded scintillating fibers with fluor-doped core, in order to maximize the detection area exposed to water. This is necessary because of the small energy (6 keV on average) of electrons from tritium decay to be measured, which prevent the use of cladded comercial fibers, which clad is too thick to be crossed by the tritium decay electrons. Scintillating fibers of 1 mm and 2 mm diameter have been considered in the R&D phase, while for light detection the options in study are PMTs or silicon photomultiplier arrays(SiPMs). The latter have the advantage of low polarisation voltages ($\sim$ 50 V instead of typically $\sim$ 1kV needed by PMTs), robustness and easy temperature dependence correction that may be implemented automatically. On the other hand, using PMTs has the advantage of a lower number of electronic channels and a larger detection area. In both options, the photodetectors reading out a fiber bundle

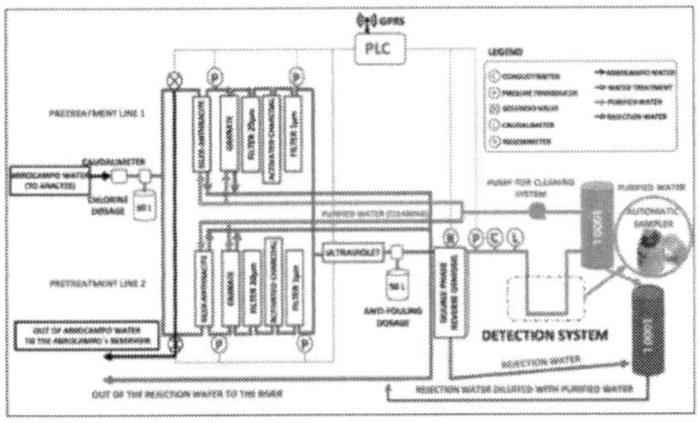

Fig. 2. Scheme of the TRITIUM water purification system delivering ultra pure water to the TRITIUM monitor.

work in coincidence mode in order to reject intrinsic noise (or self-emission in the case of PMTs), and in anti-coincidence mode with a cosmic ray discrimination system made of plastic scintillators. The water supplied to the detector, has to undergo a thorough purification process in order to avoid maintenance of the detector due to organic and mineral depositions on the fiber surface and on the water-vessel inner surfaces. A water purification system was developed to provide a level of water purity specified as hyperpure, characterized by a conductivity of 10 μS/cm. In addition, the purification system has to preserve the in-water tritium level prior to the purification process, and ensure an adequate water flow in the detector, allowing the renewal of its water content every 10 min. The detector system has to be able of providing an alarm signal in case a tritium level in water above 100 Bq/L, after a continuous measurement of about 10 min duration, is found. A scheme of the purification system is displayed in Fig. 2

III. Detector Development

A. Simulations of the scintillation process in the fibers

The proposed detector design using scintillating fibers of 1 mm or 2 mm diameter has been extensively studied through GEANT4 simulations [4]. Due to the low energy of the electrons emitted by tritium, and considering that the measurement medium is water, only a small part of the electrons originated by tritium decay close to the fiber surface are detected, because those electrons produced far away loose their energy in the water. The relationship between the energy deposited by electrons in the fiber and the distance from their emission point to the fiber surface was studied by simulating 2 mm and 1 mm fibers inside a water cylinder of 0.5 mm thickness, containing tritium uniformly distributed in the water volume. The electron emission was considered to be spatially isotropic and its energy was sampled from the tritium emission spectrum (Fig. 3, green distribution). The results are shown in Fig. 4, where it is observed that most (99.7%) of the events reaching the fiber surface are emitted at a distance less than 5 μm from it. Therefore, only are detected those decays happening in a very thin layer of water in the vicinity of the fiber surface. For this reason, scintillating optical fibers cannot be cladded, as the thickness of commercial clads exceed largely 5 μm.

https://doi.org/10.1051/epjconf/202022503008

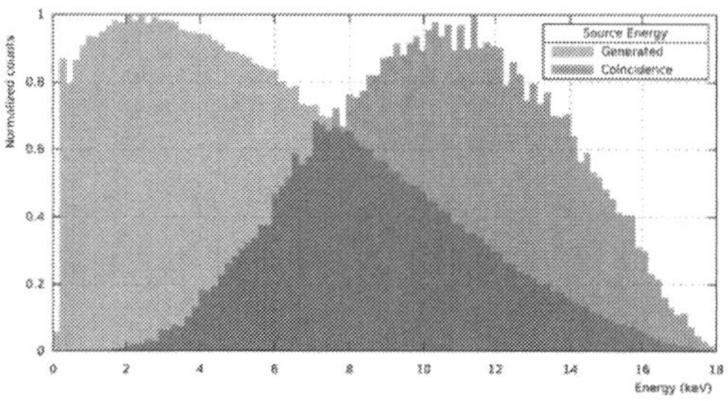

Fig. 3. Energy distribution of electrons from tritium decay (green) and their energy deposited in a scintillation fiber of 2 mm diameter (red) immersed in tritiated water.

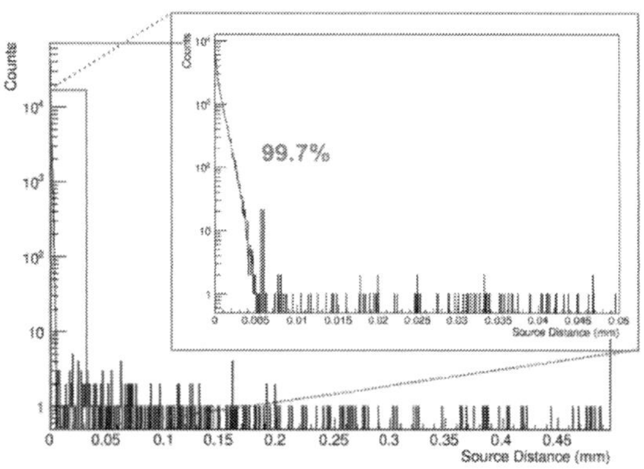

Fig. 4. Distribution of β^- electrons hitting a scintillation fiber immersed in tritiated water as a function of the distance between the beta source and the fiber surface. 99.7% of the events reaching the fiber surface are produced at a distance less then 5 μm from it.

The energy distribution of electrons that reach the fiber surface was studied. The plots displaying the results obtained arepresented in Fig. 3, where, in green, is the tritium emission spectrum, and, in red, the energy distribution of the electrons reaching the fiber, normalized to the emission spectrum in order to be easily compared. The shift of the energy peak of the electrons in the fiber, indicates, that the less energetic electrons can not reach the fibers, as they loose their energy in the water. It must be pointed out that the spectrum of energy deposition in the fiber shows those electrons reaching the fiber surface, which does not necessarily imply that they produce enough photons to be detected.

A crucial issue of the detector design is the diameter of the scintillating optical fibers to be employed. This is determined by a compromise between the detection efficiency of the fibers determined from their light yield collection per unit of energy deposited, and the amount of light produced by energy deposited from the background (cosmic rays and environmental radioactivity essentially), which should be as low as possible. In the actual monitor, a lead shielding will be employed, that should stop most of the environmental radioactive background and soft cosmic rays (essentially electrons and positrons).

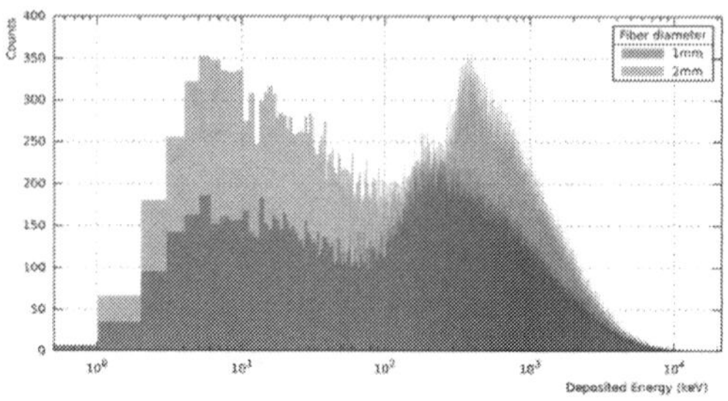

Fig. 5. Distribution of cosmic muon events as a function of the deposited energy in fibers of 1 mm (red) and 2 mm (green) diameter.

Therefore, we have studied the hard component of the cosmic ray spectrum, composed mostly at sea level by muons of energy larger than 200 MeV. In Fig. 5, the energy deposited by cosmic muons in fibers of 1 mm and 2 mm diameter (red and green spectra respectively) are compared. As may have been expected, the counting rate of cosmic muons in a fiber of 1 mm diameter is smaller than in a fiber of 2 mm diameter in the whole energy range, and specially in the region of interest for tritium detection($<$ 8 keV). Hence, scintillating fibers of 1 mm diameter favour background rejection, increasing the detector sensitivity to low tritium activity. Furthermore, it is essential to reduce strongly the cosmic and natural radioactivity backgrounds in the detector. A cosmic ray VETO detector, composed of 1 cm thick plastic scintillators placed both below and above the TRITIUM monitor, read out by photosensors, has been designed. The whole detection system is enclosed inside a lead castle of several cm thickness that will, in addition, reduce cosmic muons and natural radioactivity.

B. Prototypes

Several TRITIUM prototypes were built for the detector proof of concept and efficiency, light yield and sensitivity studies. The first prototype, called TRITIUM-0, contained a bundle of 35 scintillating optical fibers of 1 mm diameter, without cladding, from Saint-Gobain Crystals [5], terminated in the ends by the connectors exhibited in Fig. 6. The fiber-bundle was introduced into a U-shaped PVC tube, that allowed vertical coupling of the fibers to two PMTs, as depicted in Fig. 7. Optical grease BC630 from Saint-Gobain Crystals was used to optimise the light transmission to the photosensors. This prototype was first filled with 40 mL of hyperpure water for background measurements, and afterwards, it was filled with hyperpure highly tritiated water (108.11MBq/L) for tritium activity measurement. This prototype proved the feasibility of the detection principle with scintillating fibers, but also showed important issues related to the light production and collection in the detector.

Two additional prototypes, called TRITIUM-1, were built at IFIC and Aveiro for light collection studies, long-term stability studies of scintillating fibers in water, readout and mechanical studies, including implementation of water flow solutions inside the vessel, water-tightness and optical coupling solutions for the photosensors (PMTs or SiPM arrays).

Fig. 6. One of the two end-connectors of the fiber-bundle of TRITIUM-0 prototype. The fiber ends are polished and coupled to the PMT with optical grease BC600 from Saint-Gobain Crystals [5].

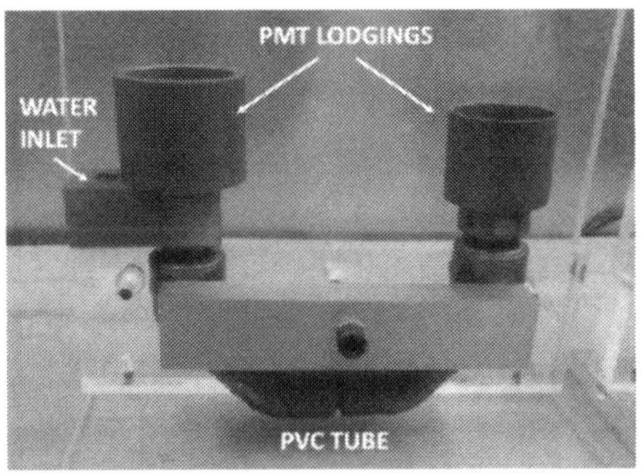

Fig. 7. Proof-of-principle prototype TRITIUM-0 instrumented with two PMTs. The fiber-bundle in the PVC tube was composed of 35 scintillating fibers BCF-12 of 1 mm diameter, 20 *cm* long and without cladding, from Saint-Gobain Crystals.

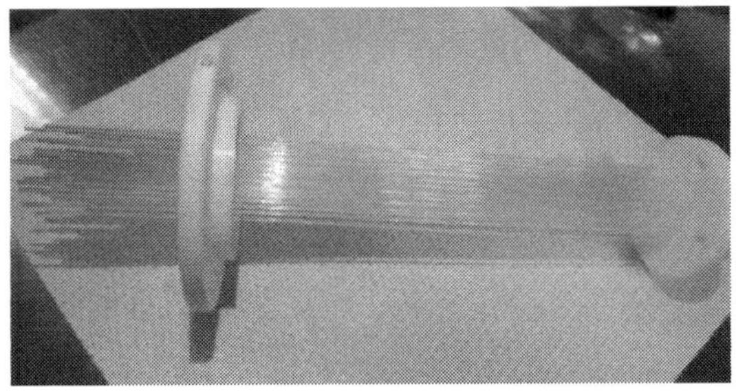

Fig. 8. Scintillating fibers mounted into two PTFE matrices used as top and bottom covers of TRITIUM-1 water vessel. The fibers end were polished and their surface conditioned in a clean room.

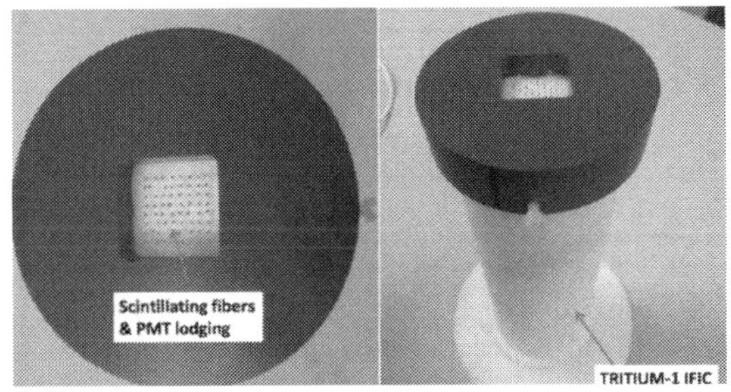

Fig. 9. TRITIUM-1 IFIC prototype with 64 scintillating fibers lodged vertically in a PTFE vessel (right picture), on top of which a PMT was placed in contact with the fibers. The prototype was placed in a dark chamber for data taking.

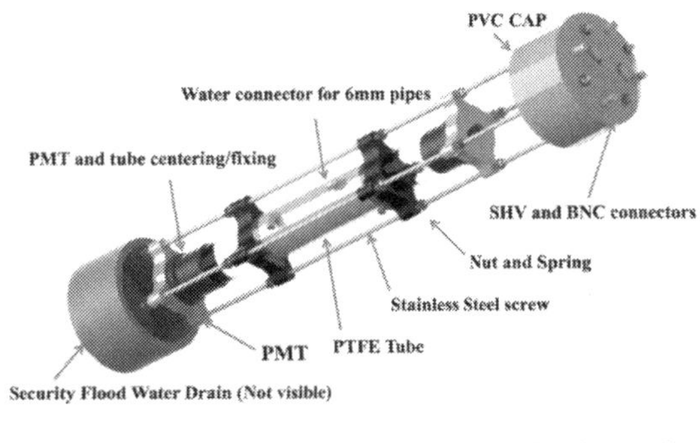

Fig. 10. TRITIUM-1 Aveiro prototype with 400 scintillating fibers, read out by two PMTs in coincidence mode. The fibers are lodged into a PTFE vessel closed by two PMMA windows which transmit the scintillation light to the PMTs. An inlet and an outlet for water flow are foreseen in the vessel body.

IFIC prototype was composed of 64 scintillating fibers BCF-12 from Saint-Gobain Crystals, of 1 mm diameter without cladding, lodged in a highly reflective PTFE vessel. A protocol for conditioning of fiber surface was developed to optimize their in-water detection and light collection capability. This includes cleaving, polishing, and surface conditioning in a clean room, which ensures optimal cleanliness and wetting conditions (optimal contact with water) of the fibers surface. The fibers were then introduced into two PTFE matrices, as shown in Fig. 8, that allowed a separation of 1 mm between neighbour fibers and maintained them in vertical position in the PTFE vessel, as shown in Fig. 9. On top of the vessel, a Hamamatsu R8520-406 PMT was lodged and put into contact with the fibers using optical grease. The PMT signal was preamplified, amplified and acquired in an AmpteK MCA-8000D digital Multichannel Analyzer.

A second version of TRITIUM-1, built in Aveiro, used also a PTFE vessel within which a high number (400) of tightly packed 2 mm diameterscintillating fibers was introduced. These were read out in coincidence mode by two Hamamatsu R2154-02 PMTs, through transparent PMMA windows of the water-vessel. A drawing of this prototype is shown in Fig. 10.

A third monitor TRITIUM-2 was developed for the final monitor to be installed in Arrocampo dam. This monitor makes use of the Aveiro TRITIUM-1 monitor design, and contains 500 1 mm closely packed scintillating fibers, contained in a parallelepiped container with teflon walls to optimise light reflection. This prototype is read-out by SiPM arrays. The details of this prototype are shown in Fig 11

IV. Final Monitor

A final monitor composed of 15 TRITIUM-2 modules will be installed in Arrocampo dam during the third term of 2019. The system is conceived as composed of an arbitrary number

Fig. 11. TRITIUM-2 prototype with 500 scintillating fibers, read out by two photosensors in coincidence mode. The fibers are lodged in a PVC vessel, shown on the right part of the picture, within a parallelepiped container with teflon walls, shown in the picture, to optimise light reflection. An inlet and an outlet for water flow will be placed in the vessel body to allow for water circulation. The prototype is read in coincidence mode by photosensors, through PMMA windows shown on the left part of the picture.

Fig. 12. Metallic structure holding the lead shied in which the TRITIUM-2 modules will be placed.

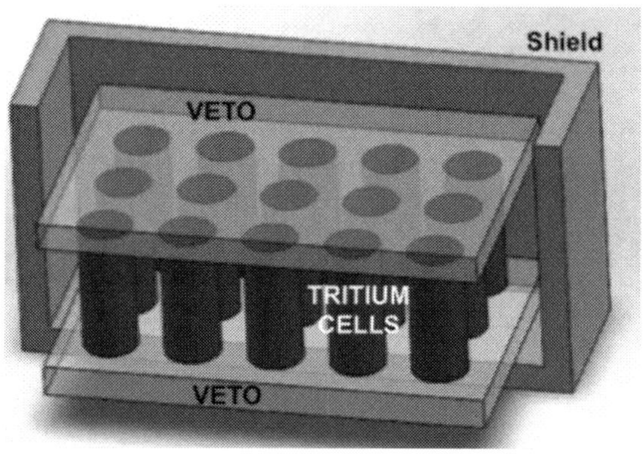

Fig. 13. Scheme of how the TRITIUM-2 modules will be placed within the metallic structure holding the lead shield.

of modules, which allows to improve the sensitivity by just increasing the number of modules. The modules will be placed within the metallic structure holding the lead shield shown in Fig 12. The modules will be read by 4×4 Hamamatsu SiPM arrays. Data will be acquired by a PETSYS system, developed by PETsys Electronics for PET data acquisition. This system has the capability of reading thousands of SiPM channels with good energy and time resolution and the possibility of making custom triggers. The active cosmic veto is made of two 1 cm layers of plastic scintillator povided by EPIC Crystals. Two different veto configuration are being studied at present: two single layers covering the whole set of modules, and partial vetos for groups of three modules as shown in Fig. 14. This latter configuration has mechanical and cabling advantages, and the possibility of putting individual plastics into different trigger configurations and improving the position dependent response. The whole system is depicted in Fig.15.

To verify the lead shielding capability to reject background, an Aveiro module shown in Fig. 8, was installed in Arrocampo dam monitoring station, and measurements were carried out without shielding, with a shielding of 0.5 cm and with a shielding of 1 cm. The count rates obtained are shown in Fig. 16 was installed in the Arrocampo dam. Measurements were carried out without shielding, with a lead shield 2 mm thick, and a lead shield 1 cm thick. The count rate measured is shown in Fig. 17. Curve A corresponds to the measurements carried out without shielding. Curve B corresponds to the measurements acquired with the 2 mm thick shielding and curve C to the 1 cm thick shielding. It is seen from the plots that the shielding reduce strongly the count rate due to low energy background, composed essentially by environmental radioactivity and soft cosmic rays.

V. Preliminary Experimental Results

A clear signal of tritium was observed in the proof-of-concept prototype TRITIUM-0 as shown in Fig. 18, in which the signal from tritiated hyperpure water is observed (blue spectrum) above the background signal (red spectrum) from a similar volume of hyperpure water without tritium. However, the signal observed from highly tritiated water (108.11 MBq/L) was weak, clearly indicating poor light production, collection or transmission to the PMTs. Indeed, among the issues identified to determine the tritium signal magnitude are the geometrical configuration of the fiber-bundle, the condition of the fibers surface, the water circulation pattern in the detector vessel, the reflectivity of the inner surfaces of the water-vessel, and, of crucial importance for low activities, the magnitude of the background signal in the detector.

In TRITIUM-1 prototypes the fibers underwent a thorough conditioning protocol and were set up straight, avoiding light losses due to bending and poor contact with water. In Fig. 19,

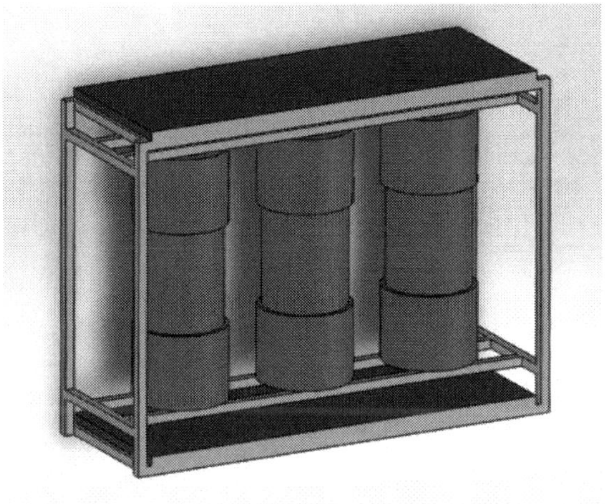

Fig. 14. Scheme of plastic VETO for three TRITIUM-2 modules will be placed within the metallic structure holding the lead shield.

Fig. 15. Scheme of scintillating plastic VETO for 15 TRITIUM-2 modules placed within the metallic structure holding the lead shield.

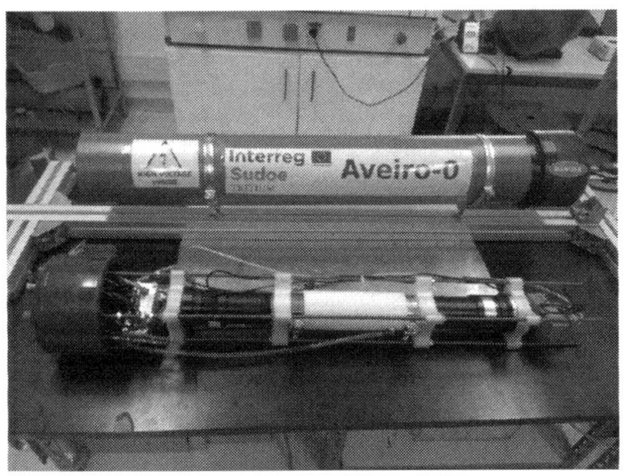

Fig. 16. Aveiro TRITIUM prototype installed in Arrocampo damm

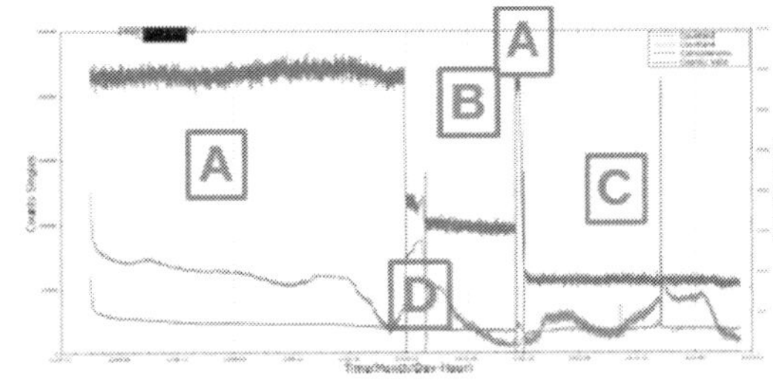

Fig. 17. Results of measurements of the Aveiro TRITIUM prototype installed in Arrocampo damm with different shield conditions.

the signal from tritiated water with similar activity as for TRITIUM-0, shows an improvement of the tritium signal by one order of magnitude compared to TRITIUM-0. TRITIUM-1 IFIC was also used to assess the stability in water of fiber response over time. Through regular measurements in the laboratory, the detector prototype proved a stable response over a period of 9 months. Further measurements are foreseen in the lead shield in Almaraz TRITIUM site, to assess its increase of sensitivity associated to background rejection.

In TRITIUM-1 Aveiro prototype, a substantial increase in tritium level sensitivity compared to the IFIC-0 and IFIC-1 prototypes is expected, due to the higher number of fibers, producing a larger total detection surface, and also, due to the reduction of the PMTs noise through the implementation of a coincidence readout of the fibers. Before the use of tritiated water in this prototype, the capability of low energy detection by the scintillating fibers was first tested. For that, a small bundle of 12 Saint-Gobain BCF-12 fibers without cladding were assembled and wrapped in teflon tape. The fibers were positioned between 2 Hamamatsu R2154-02 PMTs acquiring in a coincidence window of 500 ns through a CAEN V1724 digitizer. The PMTs signals where preamplified by a Cremat Cr-110 module and the fibers where irradiated with 5.9 keV photons provided by a ^{55}Fe source. A study of the number of detected events as a function of the PMTs voltage was carried out and the results are presented in the insert of Fig. 20. The increase of the number of events observed with voltage increase in the PMTs is due to the increase of the PMTs detection efficiency, until a plateau is reached at around 1300 V. In the main plot of Fig. 20 the pulse-height amplitude for both PMTs and the corresponding distribution of their sum can be observed. The different distributions for the PMTs are due to the difference in the intrinsic gain although they were biased with the same voltage (-1250 V). The prototype installed in Arrocampo damm monitoring station shows good stability over time, as shown in Fig. 21, where the count rate over a period of 2 months is displayed. The peaks in the plot correspond to openings of the shielding for maintenance operations. These peaks put in evidence the important role played by the lead shielding.

 https://doi.org/10.1051/epjconf/202022503008

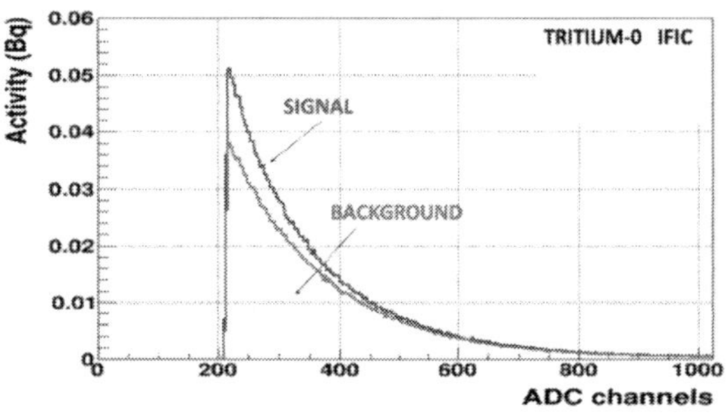

Fig. 18. Data spectra of hyperpure water (red) and of hyperpure tritiated water (108.11 MBq/L, blue) in number of ADC channels, from TRITIUM-0 prototype.

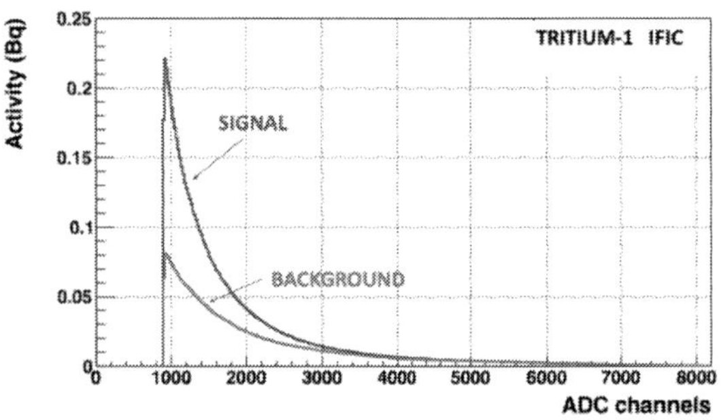

Fig. 19. Data spectra of hyperpure water (red) and of hyperpure tritiated water (108.11 MBq/L, blue) in number of ADC channels, from TRITIUM-1 IFIC prototype.

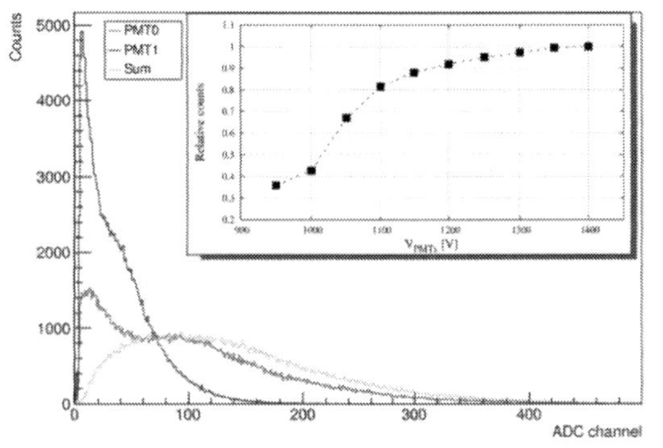

Fig. 20. Pulse-height distribution for each of the PMTs ($V = -1250$ V) of TRITIUM-1 Aveiro, and their sum. In the insert is the normalised count rate as a function of PMTs voltage when operating in coincidence mode. The dashed line is just a guide-to-the-eyes.

VI. Conclusion and Outlook

The TRITIUM project aims at the design, construction and commissioning of a near real-time automatic detector system for monitoring low levels of tritium in water released by nuclear power plants. The tritium monitor has to measure tritium

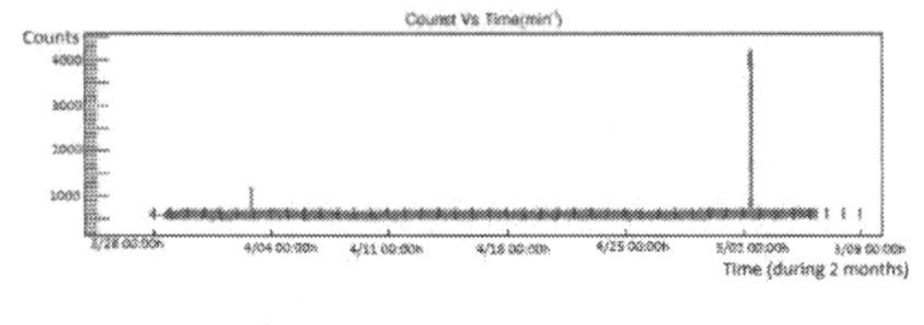

Fig. 21. Count rate of TRITIUM-1 Aveiro prototype during a period of 2 month. The peaks in the plot correspond to openings of the shielding for maintenance operations.

levels below 100 Bq/L, according to the limit required by the European Council Directive 2013/51/EURATOM for water for human consumption. The final design of the TRITIUM detector employs scintillating optical fibers read out by silicon photomultiplier arrays.

GEANT4 simulations were carried out to optimize the detector design (scintillating fiber length, diameter, geometrical configuration) and several prototypes were developed and operated in the laboratories of the consortium, with the aim of providing a proof of concept of the detector, and for studying its sensitivity and mechanical configuration stability. The simulations indicate that the deposited energy in fibers of 2 mm diameter without cladding are peaked around 12 keV, with a distribution ranging from 2 to 18 keV. Moreover the electrons that reach the fiber surface are created at a maximum distance of 5 μm from the fiber surface. The possibility to use scintillating fibers to detect, in coincidence mode, low energies such as the 5.9 keV photons from a ^{55}Fe source, was demonstrated as well as the scaling of the detector sensitivity with the increase of the number of scintillating fibers and the optimization of their surface conditions and geometrical configuration.

The results of this R&D studies allowed the design of a modular TRITIUM detector, composed of 15 cylindrical cells containing each, 500 scintillating optical fibers of 25 cm long, 1 mm diameter, without cladding. Each cell will be instrumented with two SiPM arrays which read out the scintillation signals in coincidence mode to reject the intrinsic noise of the photodetectors. Additionally, active and passive shielding based on scintillating plastics and lead bricks respectively, will be used for the rejection of cosmic and natural radioactivity backgrounds in the tritium detector. This shielding is essential to reach high sensitivity to tritium levels below 100 Bq/L. A water purification system was developed to supply the detector with hyperpure water with conductivity close to 10 μS/cm. This system which has been shown to preserve the tritium level existing prior to the purification process, provides a water flow that enables renewal of the detector water-content each 10 min for tritium measurement.

The construction of the full TRITIUM system is currently in progress. Its commissioning is foreseen in the fourth term of 2019 in the monitoring station of Arrocampo dam at Almaraz (Spain) nuclear power plant.

Acknowledgment

This work was supported by the INTERREG-SUDOE EEC program through the project TRITIUM - SOE1/P4/E0214

References

[1] K. J. Hofstetter, Continuous monitoring for tritium in aqueous effluents at SRS using solid scintillators, Trans. Am. Nucl. Soc. (United States), 1993.
[2] M. Rathnakaran et al., A real-time tritium-in-water monitor for measurement of heavy water leak to the secondary coolant 2000.
[3] J.W. Berthold, and L.A. Jeffers. IN-SITU TRITIUM BETA DETECTOR. United States: N. p., 1998. Web. doi:10.2172/836625.
[4] GEANT4: http://geant4.cern.ch/.
[5] https://www.crystals.saint-gobain.com/.

Impact of the cross section library on ^{93m}Nb activity in VVER-1000 reactor dosimetry

P. Haroková*, M. Lovecký
SKODA JS a.s.
* pavlina.harokova@skoda-js.cz

***Abstract*—One of the objectives of reactor dosimetry is determination of activity of irradiated dosimeters, which are placed on reactor pressure vessel surface, and calculation of neutron flux in their position. The uncertainty of calculation depends mainly on the choice of nuclear data library, especially cross section used for neutron transport and cross section used as the response function for neutron activation. Nowadays, number of libraries already exists and can be still used in some applications. In addition, new nuclear data library was recently released. In this paper, we have investigated the impact of the cross section libraries on activity of niobium, one of the popular materials used as neutron fluence monitor. For this purpose, a MCNP6 model of VVER-1000 was made and we have compared the results between 14 commonly used cross section libraries. A possibility of using IRDFF library in activation calculations was also considered. The results show good agreement between the new libraries, with the exception of the most recent ENDF/B-VIII.0, which should be further validated.**

***Keywords*—Cross-section libraries, MCNP, ^{93m}Nb, Reactor dosimetry, VVER-1000**

I. Introduction

THE radiation effects on the steel material are the main concern in reactor pressure vessel lifetime. These effects can be quantified by neutron fluence for energies above 0.5 MeV for VVER reactor type. Number of activation materials is used for VVER reactor dosimetry. Niobium is the ideal material – due to reasonably low threshold energy and large half-life, almost all neutrons emitted from the reactor and transported to the ex-vessel monitor contribute to the measured niobium activity. Other dosimeters have higher threshold energy and half-life shorter than reactor cycle irradiation time, resulting in the loss of around 90 % of neutron interactions in the ex-vessel area (see Fig. 1).

However, the measurement of niobium activity in practice has several difficulties. Usually, characteristic X-ray peaks of ^{93m}Nb with energies between 16 keV and 19 keV are measured by HPGe detectors. Low energies of emitted photons present the main disadvantage, as they require different approach than typical 1-2 MeV gamma lines emitted by other activation materials commonly used in reactor dosimetry. Moreover, impurities in irradiated niobium (e.g. tantalum) can cause additional issues in the measurement. It is therefore difficult to obtain reliable results with low uncertainties.

Measuring of niobium activity in ex-vessel monitors can be verified (or complemented) by transport calculations. The main uncertainty in transport calculations represents nuclear data, especially the cross sections used in all steps of the calculation. In this paper, impact of the cross section library choice on the ex-vessel niobium monitors activity was investigated.

II. Calculation Model

Transport calculations can be done by either deterministic or statistical calculation codes. Both methods have their advantages and disadvantages. For calculations described in this paper, MCNP6 Monte Carlo code was used. This code allows detailed description of geometry and usage of continuous-energy cross sections. However, the calculation time is influenced by number of output quantities and number of points or cells, for which the desired results are tallied.

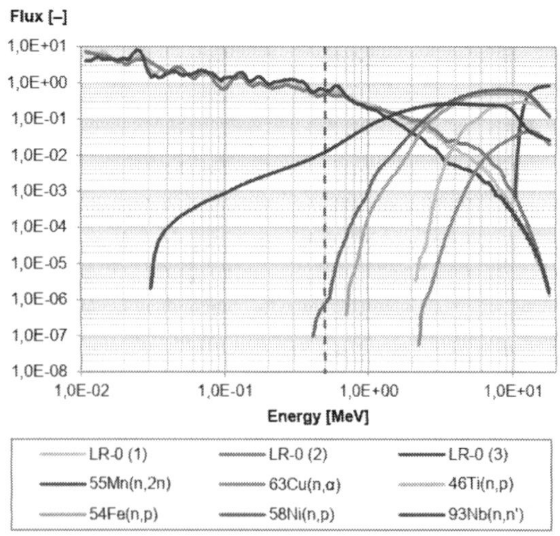

Fig. 1. Neutron spectrum of LR-0 reactor (1 – behind barrel, 2 – before RPV, 3 – behind RPV) and response functions of fluence detectors according to IRDFF-1.05 library. The radiation effects on the steel material can be quantified by neutron fluence for energies above 0.5 MeV (dashed line).

P. Haroková is with Škoda JS a.s., Calculation Department, Orlík 266/15, 316 00 Plzeň, Czech Republic (e-mail: pavlina.harokova@skoda-js.cz).

M. Lovecký is with Škoda JS a.s., Calculation Department, Orlík 266/15, 316 00 Plzeň, Czech Republic (e-mail: martin.lovecky@skoda-js.cz).

In order to compare the impact of library choice and maintain calculation time under reasonable limits, 2-D model of VVER-1000 reactor was used. The model (see Fig. 2) describes all parts of the reactor. The reactor core consists of 163 fuel assemblies of TVSA-T type, which are homogenized at all levels. Inner parts of reactor between the core and reactor pressure vessel (RPV), such as core basket, barrel, etc., correspond with exact geometrical model. Behind RPV is air cavity, homogenized thermal insulation, serpentinite concrete and building concrete (all parts have its own steel coating). The serpentinite concrete contains 27 ionizing chambers and counterweight channels positioned with a step of 12°.

With respect to the typical temperature gradient in the reactor building, mentioned parts of the model were divided into several regions according to temperature, from 578.5 K in reactor core to 563 K in RPV. Between the outer surface of RPV and the air cavity, the temperature decreases linearly down to 303 K. For further statistical accuracy, variance reduction with cell-based weight windows was used in the model to maintain constant neutron population above 0.5 MeV in spatial cells.

III. Transport and Activation Calculations

The neutron source was prepared by a Fortran script using effective source method [2]. Input data are files from MOBY-DICK macrocode [3], which contains information about the number of neutrons generated in the reactor core during the fuel campaign on the selected actinides. Each input file contains about 100 time-points in selected campaign for all axial layers, fuel assemblies and pins (in the case of VVER-1000, it is 48 axial layers, 163 fuel assemblies and 331 pins).

To calculate the neutron fluence, the neutron sources are summed and normalized. The number of neutrons in macrocode is determined for the stationary state by integrating the neutron flux and macroscopic cross-sections for fission. In the activity calculations, an effective source is prepared similarly to the neutron fluence. However, a gradual decrease of source due to radioactive decay of products is also taken into account, and effective neutron source has to be therefore prepared for each activation detector.

To ensure statistical accuracy of the calculation, the simulation was run with 1 billion neutrons for each computational variant. By removing neutrons with energies under 0.4 MeV, the speed of calculation was increased approximately seven times. The limit of 0.4 MeV was chosen sufficiently low under 0.5 MeV due to possible upscattering on coolant's hydrogen nuclei.

The calculations were done for 20th axial layer of fuel height in hot state, which is slightly below half the height of the core. In the first approach, we have used ENDF/B-VII.1 library, which is therefore referred as „reference library" further in this paper. The calculated neutron fluence and Nb detector activity is shown in Fig. 3 and 4. The statistical error is between 2.7 % and 6.4 % for neutron fluence and around 4 % for Nb activity calculation, which is not good enough for RPV-lifetime prediction, but it is sufficient for our intent of nuclear data libraries comparison.

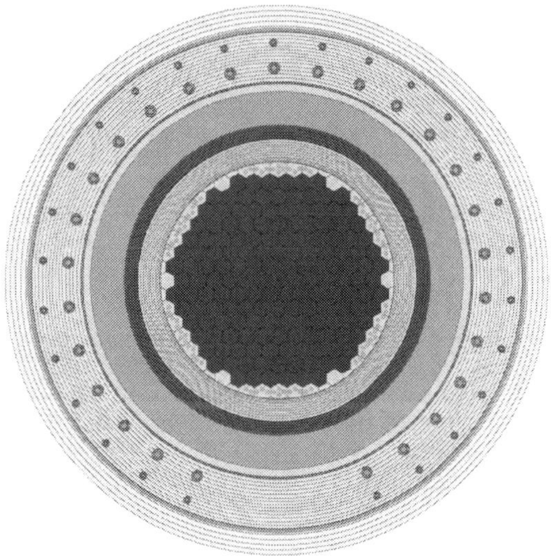

Fig. 2. Calculation model of VVER-1000 in MCNP6 with cell boundaries.

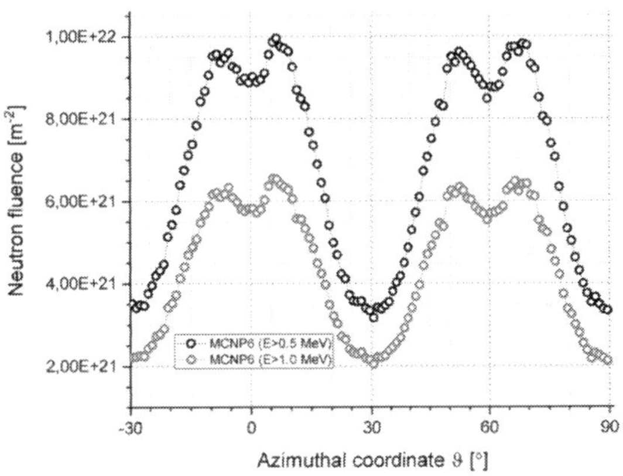

Fig. 3. Neutron fluence on RPV surface (MCNP6, library ENDF/B-VII.1).

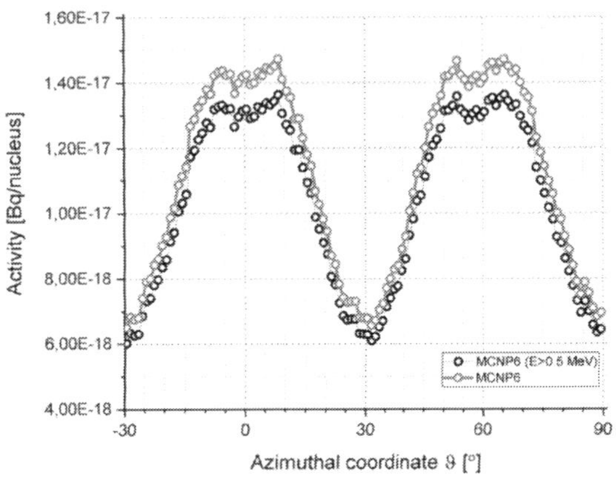

Fig. 4. Activity of niobium detector (MCNP6, library ENDF/B-VII.1).

IV. Nuclear Data

In reactor dosimetry, nuclear data are used in two steps:

1) *Nuclear data for transport*
   Nuclear data for transport include materials cross sections, angle distributions, etc. These data are used for transport of neutrons between the source (reactor core) and the place of detection, i.e. inner surface of RPV for neutron fluence, or outer surface of RPV for activities of ex-vessel monitors.
2) *Nuclear data for activation*
   Nuclear data for activation include materials cross sections. These data describe nuclear reactions on ex-vessel monitors.

Different nuclear data libraries can be used in both of the steps. This approach is sometimes used in combination with dosimetric library IRDFF, which can be used only for an activation part of calculation. In this study, we have also considered this option along with the variant of using the same library for both transport and calculation (i.e. combining the library with itself).

The analysis was focused on comparison of 14 continous-energy libraries from different countries, released between years 1979-2018. This includes various versions of the following libraries:

- Russian Evaluated Neutron Data Library (BROND),
- Russian File of Evaluated Neutron Data (RUSFOND),
- Chinese Evaluated Nuclear Data Library (CENDL),
- US Evaluated Nuclear Data Library (ENDF/B),
- Joint Evaluated Fission and Fusion File (JEF, JEFF),
- Japanese Evaluated Neutron Data Library (JENDL).

The source of all ENDF/B libraries is the official MCNP6 distribution, other libraries were processed by NJOY calculation code.

V. Results

The calculations of neutron fluence and niobium activity were done for all 14 libraries described above. As the libraries ENDF/B-V.0 and CENDL-3.1 does not contain ^{93m}Nb data, we have used cross section values of ^{93m}Nb from reference library. The results were azimuthally centered and compared to ENDF/B-VII.1 (see Fig. 5 and Fig. 6).

With a few exceptions, the influence of nuclear data on

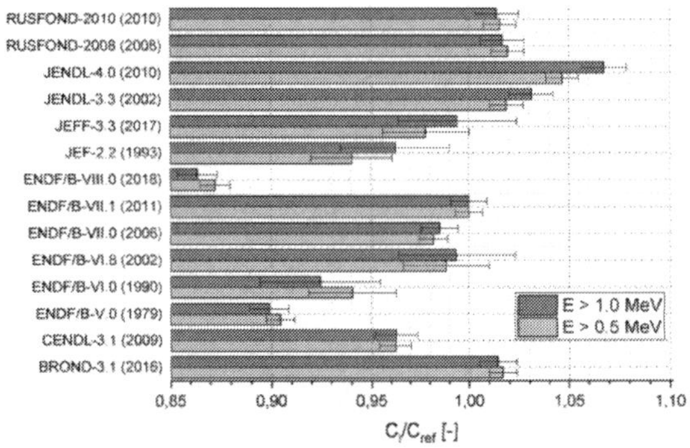

Fig. 5. Relative comparison of selected nuclear data libraries with reference library (ENDF/B-VII.1), showing impact of nuclear data library on calculated neutron fluence on RPV surface.

calculation results is mainly due to the library used for neutron transport – in the first approximation, the conclusions made for neutron fluence can be generalized to activation detectors. In most of the cases, combining the libraries with IRDFF as activation library provides better results than using the same library. The only exception is JENDL-4.0.

The results confirmed some of our expectations about older libraries, e.g. in the case of ENDF/B-V.0, where a clear discrepancy is apparent due to incorrect description of inelastic scattering on iron. As for the more recent libraries, the results show good agreement between most of them. In this case, the impact of library choice on neutron fluence and niobium activity is small, in order of few percent. The exception is the most recent ENDF/B-VIII.0 library, where the results are significantly different.

To explain this difference, the calculations with this library should be further validated. The disagreement could be caused by incorrect format conversion from ENDF-6 to ACE, which is tested by author organization LANL mainly for criticality tasks. Other possibility is error in calculation model or in actual cross-section data, similar to problem presented by [5] with JENDL-3.3 and angle distribution of oxygen-16. Eventually, it may also turn out that the new data are correct. A detailed study on ENDF/B-VIII.0 performance for reactor dosimetry is currently under way.

For the future studies of cross section libraries, there is a possibility of performing nuclear data uncertainty propagation analysis, which uses uncertainties of nuclear data as part of the most recent libraries.

TABLE I
Comparison of cross section libraries

Transport library	Same activation library	IRDFF v.1.05 activation library
BROND-3.1	1.042 (9)	1.013 (14)
CENDL-3.1	0.929 (14)	0.931 (14)
ENDF/B-V.0	0.830 (14)	0.827 (14)
ENDF/B-VI.0	0.999 (30)	1.000 (45)
ENDF/B-VI.8	0.975 (32)	0.974 (46)
ENDF/B-VII.0	0.973 (10)	0.972 (14)
ENDF/B-VII.1	1.000 (10)	1.000 (14)
ENDF/B-VIII.0	0.778 (10)	0.775 (14)
JEF-2.2	0.825 (32)	0.925 (45)
JEFF-3.3	0.910 (32)	1.019 (46)
JENDL-3.3	0.975 (12)	1.014 (31)
JENDL-4.0	1.050 (11)	1.091 (35)
RUSFOND-2008	1.036 (12)	1.010 (32)
RUSFOND-2010	1.097 (11)	1.064 (33)

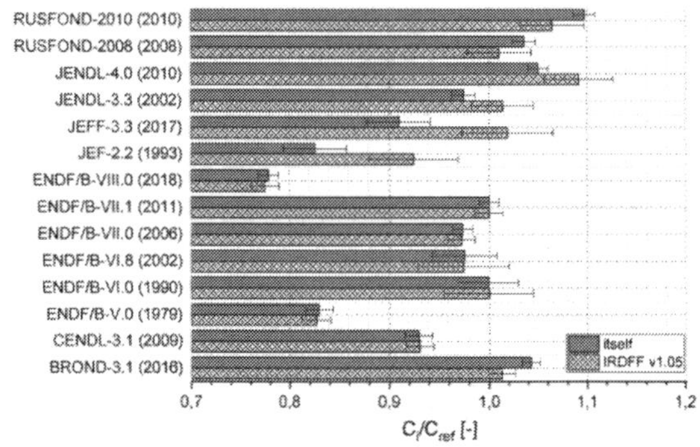

Fig. 6. Relative comparison of selected nuclear data libraries with reference library (ENDF/B-VII.1), showing impact of nuclear data library on calculated activity of Nb detectors using combination of selected library with itself, or IRDFF v.1.05, as activation library.

References

[1] M. Lovecký, V. Smutný, "*Výpočetní model reaktoru VVER-1000 v kódu MCNP pro 2-D výpočty v reaktorové dozimetrii*", Škoda JS report Ae18200, 2018.

[2] V. Smutný, A. Konečná, D. Sprinzl, V. Klupák, M. Vinš, "*The Activation Detector Activity Calculations Using the Effective Source Method and Measurement*", EPJ Web Conf. 153 07041 (2017), DOI: 10.1051/epjconf/201715307041.

[3] V. Krýsl, P. Mikoláš, J. Šůstek, K. Vlachovský, "*Description of program MOBY-DICK with changes for VVER-1000*", (ŠKODA JS, 2015).

[4] M. Lovecký, "*Jaderná data a výpočty v reaktorové dozimetrii reaktoru VVER-440*", Škoda JS report Ae18727, 2018.

[5] A. Vasiliev, H. Ferroukhi, M. Zimmermann, R. Chawla, "*On the effects of oxygen cross-sections in the fast neutron fluence analysis for a reactor pressure vessel scraping test*". Annals of Nuclear Energy 35 (2008) 565-569. DOI: 10.1016/j.anucene.2007.08.018.

Widget-based Alarm Recognition and a Management Support Interface System in Nuclear Power Plants

Seung Min Lee*, Hana Park, Gwi Sook Jang, Inkoo Hwang
Korea Atomic Energy Research Institute, South Korea
* jewellee@kaeri.re.kr

Abstract— **As small modular reactors (SMRs) and various computerized operating support systems have been developed, visual display unit(VDU)-based multifunctional consoles have been proposed. The more a console depends on VDUs, the more human system interface (HSI) designers need to consider the decrease in the human performance caused by the effect of navigation tasks. This paper proposes a widget-based alarm recognition and management support interface system to increase the operator's ability to recognize alarms and to manage alarm-related tasks by reducing additional and excessive navigation tasks. With a small display screen, the system provides operators with quick and accurate access to alarm-related information and allows them to make corrective responses.**

Keywords **—Human-System Interface, Human Performance, Information display, Monitoring system**

I. Introduction

THE change from analogue-based information display panels in the main control rooms (MCRs) in Nuclear Power Plants (NPPs) to digital-based visual display units (VDUs) has brought accompanying changes in the operating environment and operators' tasks.

VDUs and various computerized operating support systems provide an enormous amount of information and facilitate operators' performance of their monitoring and control tasks. However, displaying a large amount of information on the limited VDU screen generates a major secondary task: navigating. Navigation produces a decrease in performance as the operator struggles to access the correct information pages. In a study on the cause of human errors that occur in a digitalized MCR, the task of navigating was pointed out as the second highest human error factor [1-3]. Navigation increases work time, interrupts accurate and rapid operation, and decreases human performance. Thus, the designing of the graphic user interface (GUI) in display screen is an important and necessary consideration in studying the effect of the navigating task.

With the development of small modular reactors (SMRs) and various computerized operating support systems, the concept of local control stations and multifunctional consoles emerged as shown in Figs. 1 and 2.

In such a console, an operator is responsible for one reactor unit or one small plant; thus, he/she needs to access all the monitoring information and to conduct appropriate control activity.

The console is quite dependent on the VDU, and human system interface (HSI) designers must consider the navigation tasks. In other words, they have to minimize the navigation tasks or support performance of navigation tasks. Reducing navigation tasks, especially finding alarm-related information, is necessary because the alarm-related tasks need to be dealt with in real-time.

This paper proposes an alarm recognition and management support interface system. The proposed interface effectively provides the alarm information by efficiently utilizing a small screen space. The system will aid rapid recognition and accurate response, and thus improve the human operators' performance.

Fig. 1. Local control stations of the SMR

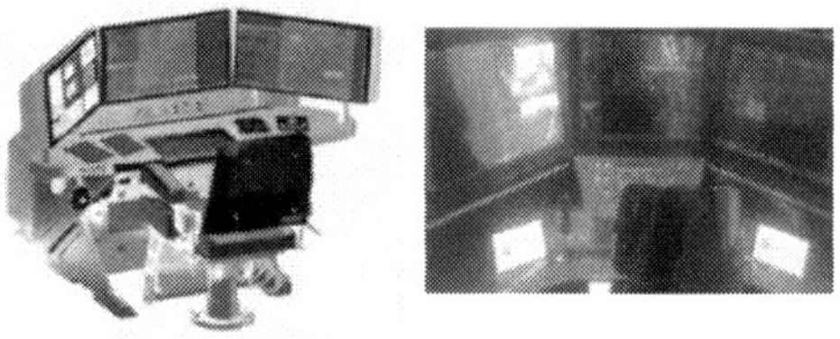

Fig. 2. Multifunctional control stations

II. Alarm Display Systems in NPPs

To design the GUIs of the proposed interface system, we analyzed the characteristics of the existing alarm display screens in conventional and advanced MCRs and considered the pros and cons of existing alarm display screens.

- Characteristics of analogue-based alarm panels (Fig. 3)
  - Alarm tiles for similar subsystems are gathered and located and fixed in the upper space of the panel. It helps operators to observe the real status of plants

S. M. Lee is with the Korea Atomic Energy Research Institute, 34057 Daejeon, Korea (e-mail: jewellee@kaeri.re.kr)
H. N. Park is with the Korea Atomic Energy Research Institute, 34057 Daejeon, Korea (e-mail: hnpark@kaeri.re.kr)
G. S. Jang is with the Korea Atomic Energy Research Institute, 34057 Daejeon, Korea (e-mail: gsjang@kaeri.re.kr)
I. K. Hwang is with the Korea Atomic Energy Research Institute, 34057 Daejeon, Korea (e-mail: ikhwang@kaeir.re.kr)

quickly, accurately and continuously.
  - Alarm tiles generate visual and auditory stimuli in real time when alarms occur. The stimuli also make it possible for the human operators to response quickly and to increase the situation awareness of the status of plants.
  - The space for displaying alarms is limited, and operators do not have enough immediate information on information that is not displayed at the moment.

Fig. 3. Analogue-based alarm display panels in the conventional MCRs

- Characteristics of digitalized alarm screens (Figs.4 and 5)
  - All alarm data and modified information, such as collecting alarm information on similar subsystems, collecting it on the same priority, and listing it in sequence of occurrence, is displayed on the screen through the computerized function.
  - The operators can obtain various kinds of information that the display system provides; however, they are required to conduct navigation tasks to access the right information among the enormous hierarchical information structure, which decreases the efficiency of accessing information rapidly and accurately and understanding the status of plants.

Fig. 4. The advanced MCR

III. Widget-based Alarm Recognition and Management Interface System (WARMI)

Based on our analysis of the existing alarm systems, we isolated the advantages of the existing alarm system. The GUI of the WARMI was designed by reflecting and improving the selected advantages.

- Selected characteristics of the existing alarm system
  - Gathering alarms related to each subsystem and prioritizing alarms effectively improves rapid and accurate recognition.
  - Showing newly generated alarms and deleting acknowledged or processed alarms makes it possible to utilize the small space efficiently and call operators' attention to alarm recognition.
  - Reducing the navigation task load by moving to the right pages is effective eliciting the appropriate operators' action rapidly and accurately.

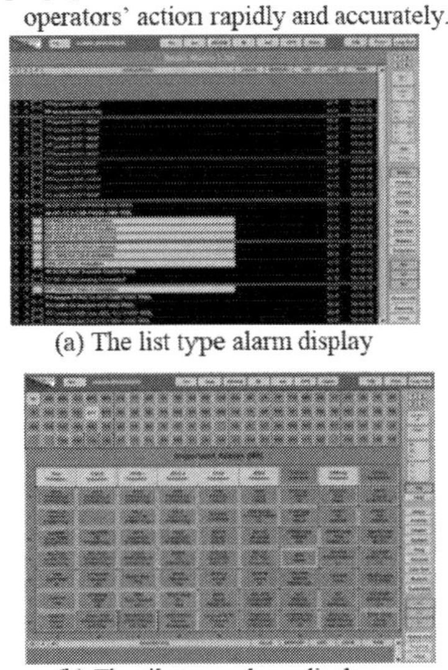

(a) The list type alarm display

(b) The tile type alarm display

Fig. 5. Digitalized alarm display screens

The structure of the WARMI system and how to use it is described as follows.

- The structure of the WARMI system (Fig. 6)
  - The proposed interface was designed using a 10-inch monitor, and operates independently with the existing Information Processing System (IPS) [4] display screen and processors.
  - The information displayed in the WARMI screen is based on the alarm information of the IPS, and the information on alarm generation, acknowledgement, blinking and off is moved from the IPS processor to WARMI in only one direction.
  - The operator can switch the IPS screen to the desired (alarm-related) subsystem by clicking WARMI alarm icons. The screen switching command only moves in one direction from the WARMI to the IPS.
  - The alarm-related operator actions are only available on the IPS screen through the prompt alarm recognition through WARMI and movement to the IPS screen. Human operators' action for alarm through WARMI is not available. The interface provides detailed information about the alarm through each alarm related icon.

- The operation of the WARMI system (Fig. 7)
  - When an alarm occurs, the subsystem related to the generated alarm is expressed by the folder form, and an icon describing the alarm is generated in the folder in WARMI. The intuitive recognition for the information such as the number of generated alarms for each subsystem, and the priority of the alarm is

https://doi.org/10.1051/epjconf/202022503010

availaable.

- By cllicking the folder, the operators get detailed information about the alarms. By clicking the alarm icon, the operators obtain detailed information about the alarm such as the priority, alarm causes, and corrective action to respond the alarm.
- The operators can change the IPS screen page to the related alarm page. It helps operators to directly move to the relevant pages and to perform the relevant operators' actions through the detailed information on the WARMI screen.
- The interface automatically deletes the icon corresponding to an extinguished alarm when the alarm handled by the operator action on the IPS disappears or becomes normal. If all alarm icons are deleted in the folder, the folder is also automatically deleted in WARMI.

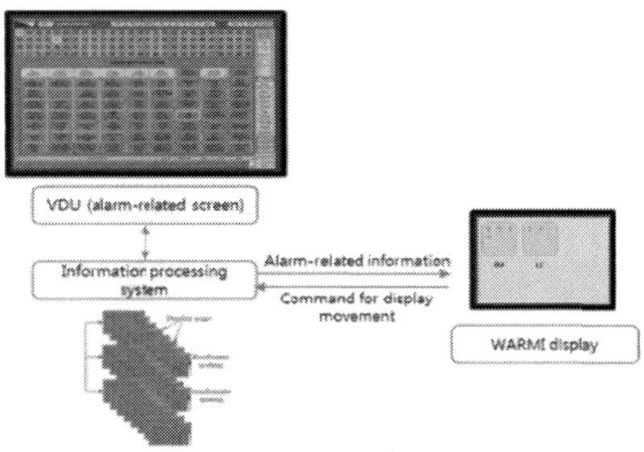

Fig. 6. Structure of WARMI

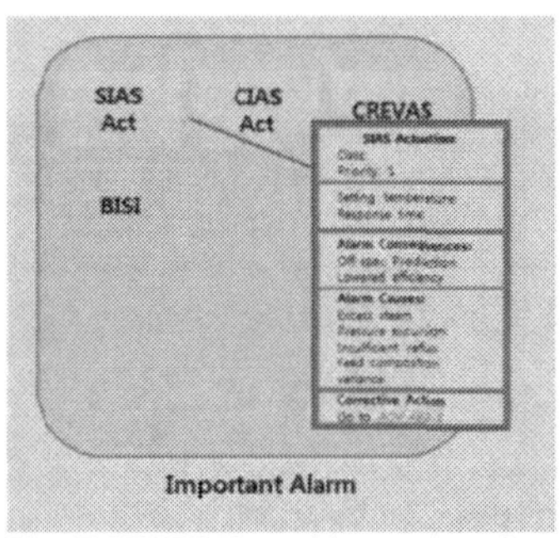

Fig. 7. WARMI graphic user interface

IV. Summary and Future Works

As small modular reactors (SMRs) and various computerized operating support systems were developed, local control stations and multifunctional consoles, which are very dependent on the VDU were introduced.

To reduce the navigation tasks created by using VDUs, the alarm recognition and management support interface system, WARMI, was proposed to aid rapid recognition and accurate response.

We analyzed the characteristics of the existing alarm display screens on conventional and advanced MCRs and designed the GUI of WARMI.

A test will be conducted to verify the effectiveness of the proposed interface system in terms of human performance.

Using WARMI will lighten the navigation task loads of operators using VDUs and improve their performance on alarm recognition and conducting the appropriate responses.

References

[1] I. S. Jang, W. D. Jung, P. H. Seong, "Human Error and the Associated Recovery Probabilities for Soft Control being Used in the Advanced MCRs on NPPs," Annals of Nuclear Energy, Vol. 87, pp. 290-298, 2016.

[2] O'Hara, J., & Brown, W. (2002), "The effects of interface management tasks on crew performance and safety in complex, computer-based systems", NUREG/CR-6633. Washington, D.C.:U.S. NRC.

[3] R. E. Patterson et al., "A Human Cognition Framework for Information Visualization," Computers & Graphics, Vol. 42, pp. 42-58, 2014.

[4] NPX80-IC-SD710, "System Description for the Data Processing System for Nuplex 80+."

EPJ Web of Conferences **225**, 04001 (2020) https://doi.org/10.1051/epjconf/202022504001
ANIMMA 2019

Non-Destructive Examination Development for the JHR Material Testing Reactor

B. Cornu*, C. Roure, D. Moulin, N. Estre, D. Tisseur, M-P. Ferroud-Plattet, P. Kinnunen, P Kotiluoto, A. Revuelta
* bernard.cornu@cea.fr

***Abstract*— The Jules Horowitz Reactor (JHR) is a European material testing reactor (MTR) under construction at the CEA Cadarache centre. It will be dedicated to material and fuel irradiation tests, as well as to the production of medical isotopes. Gamma and X-Ray benches will be implemented in the reactor pool (RER), the irradiated component storage pool (EPI) and in a shielded hot cell for measuring either the whole underwater test device still containing the experimental sample or just the experimental sample before its extraction in the hot cell.**

The CEA/Cadarache Nuclear Measurement Laboratory (LMN) has been working in collaboration with VTT (Technical Research Centre in Finland Ltd.) since 2008 under a Finnish in-kind contribution agreement. This agreement focuses on the development of NDE systems implementing gamma-ray spectroscopy and high-energy X-ray imaging of the sample and irradiation device with the highest definition possible (resolution of 100 µm).

The CEA-VTT technical specifications led to a European call for tenders launched by VTT. The contract was awarded to the Spanish company IDOM for the design, manufacturing, assembly and commissioning of:

- **Underwater gamma and X-ray (UGXR) mechanical benches and their associated gamma and X-ray collimation systems for the RER and EPI pools**
- **Hot cell gamma and X-ray (HGXR) bench in the JHR NDE hot cell.**

The Final Design Reviews (FDR) of the UGXR and HGXR systems were completed in 2016. The design phase has been an iterative process in order to manage interfacing specifications and constraints:

- **Challenging experimental requirements, mainly to cover the wide diversity of sample shapes, sample activity levels and measurement processes, but also to achieve a level of mechanical accuracy to reach the ambitious geometrical resolution target in X-ray imaging,**
- **Environmental constraints (immersion, radiation, compactness, limited accessibility for maintenance),**
- **Nuclear safety constraints (seism, radiation protection).**

The whole design process has produced a number of elaborate and innovative mechatronic systems, which is rather unusual in nuclear applications since the resulting solutions have benefited from IDOM's technological expertise in designing and commissioning large telescopes for the astronomy sector.

Once the manufacturing phase and assembly finalised, the site acceptance tests for the UGXR and HGXR mechanical systems will be performed in 2019-2020 in the TOTEM facility at the CEA Cadarache center. The underwater benches will be tested in the CESARINE pool to check their requirements.

I. Introduction

Gamma and X-Ray examinations systems (NDE) will be implemented in the future Jules Horowitz Reactor (JHR) to perform examinations on JHR experimental samples by high-resolution X-ray imaging and by quantitative measurement of the spatial distribution of gamma emitters.

This article proposes a progress description of the works concerning the whole equipment for this purpose; the current state of the project now allows us to start the first acceptance test programmes for the main components.

The CEA has entrusted its Finnish partner VTT, under an in-kind contribution agreement, with the design and manufacturing of the mechanical part of this equipment project. Three independent European calls for tenders were launched by VTT, with all three having been awarded to the same Spanish company, IDOM ADA.

The purpose of these contracts is to:

- Study and manufacture two identical underwater benches, called UGXR, designed to examine a JHR test device in its entirety,
- Study and manufacture a smaller bench, called HGXR, to be implemented in the hot cell for light experimental samples prior to their extraction from JHR test devices.

In parallel to the major equipment being developed under the Finnish in-kind framework contract, there is a specific project for a 6 MeV linear accelerator whose prototypic design aims at providing a refined spot size X-ray source (<0.5 mm) to meet the high-resolution imaging requirement (geometrical resolution over 100 µm). In addition, the onsite implementation of this accelerator imposes a self-shielding design providing a ratio up to 6 decades in attenuation. The study and manufacturing of this accelerator was entrusted to the French company PMB (ALCEN Group).

X-ray detection will be provided by a linear gamma detector based on solutions provided by ADVACAM s.r.o., a Czech company that develops detection technologies within the framework of the MEDIPIX programme at the CERN research center.

Submitted for review on June 2019. This work was supported by collaboration between CEA and VTT Technical Research Centre of Finland Ldt.

B. Cornu, C. Roure, D. Moulin, N. Estre and D. Tisseur, are with CEA, DEN, Cadarache, DTN, SMTA, Nuclear Measurement Laboratory, F-13108 Saint-Paul-Lez-Durance, France. (e-mail: bernard.cornu@cea.fr).

M.P. Ferroud-Plattet works at the CEA Cadarache centre also, within the DEN/JHR Reactor Service.

P. Kinnunen, P Kotiluoto and A. Revuelta work for the VTT Technical Research Centre of Finland Ldt., P.O. Box 1000 - 02044 VTT, Finland.

https://doi.org/10.1051/epjconf/202022504001

II. Description of the two UGXR benches

The two identical JHR UGXR underwater benches are intended to provide the following high-performance gamma-ray and X-ray imaging examinations, both before and after irradiation: lateral and radial scanning with high spatial resolution suitable for the wide variety of shapes and activities planned for the JHR; emission tomography; radiography; high-resolution (100 µm) 2D X-ray tomography; 3D tomography.

A. Installation of both UGXR benches in the JHR

One of these benches will be installed in the reactor pool. In addition to the fast measurements of short-lived elements afforded by such proximity to the irradiation site, this bench will also be used with particular conditions to conduct measurements on test devices for which the experimental interest is better served by maintaining the connection to the power and/or fluid supply lines during measurements.

The second UGXR bench will be installed in the EPI irradiated components storage pool and used to increase the volume of results produced per cycle. Given that this second bench will also be used outside the inter-cycle periods, it will operate in a less restricted environment, better suited to intense measurement campaigns.

Note also the significant advantage offered by these benches in terms of being able to sequence gamma and X-ray imaging measurements on both benches without the need for intermediate handling of the test device.

B. General design considerations for the underwater benches

The mechanical principle of an UGXR bench is to manoeuvre a whole JHR test device under water (using high-precision movements) in front of a collimation feed-through located at a depth of around 4 m that channels the gamma ray beam and the X-ray beam used for imaging. The UGXR bench must be capable of carrying a JHR test device with a maximum weight of 500 kg and standard overall height of 6.5 m, the shape of which can introduce a dissymmetry of the load given that it sometimes has an off-centred top section.

In addition to the design requirements of such a large-scale, high-precision underwater mechanism, the safety requirements imposed by carrying the test device in close proximity to the core also have to be considered. Implementing all the measures required by the UGXR bench safety classification has substantially increased the workload for the manufacturer, IDOM.

To meet these extremely exacting specifications, IDOM's bench design is based on a mechanism that is almost entirely out of the water, thus essentially avoiding the issue of corrosion in water, which is particularly critical in reactor pools. The principle consists of moving an immersed arm along the X, Y and Z axes and rotating it around the Z-axis by means of a drive system located out of the water (i.e. on the upper part of the bench). The design of this exceedingly high-precision system benefits from IDOM's expertise and comprehensive operating experience in designing mechanisms used for large space observation instruments.

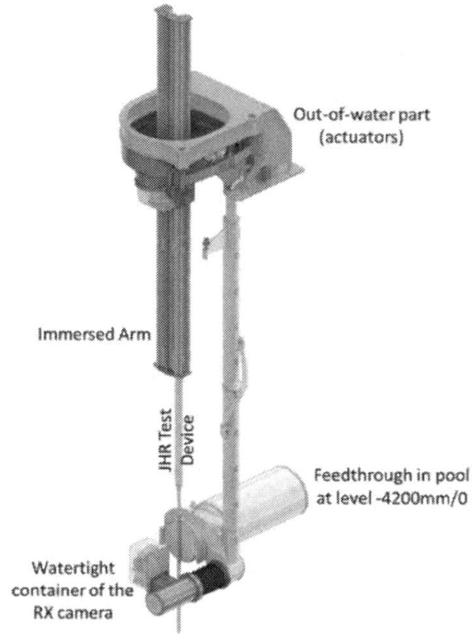

Fig. 1. Overview of the mechanism for moving an immersed JHR test device between the collimation feed-through and the submerged camera

C. Test device gripper

The design optimization and peak performance approach adopted by IDOM led to the test device necessarily being held at a single point, corresponding to its 'mounting plane'. To ensure the most effective means of attachment, the operator will be provided with as many different models of gripping devices as there are geometries of test devices.

Fig. 2. View of an as-built fastening adaptor, here implemented at the factory testing site. The FA locking function is tested using a dummy test device

The specific gripping device concerned (called a fasteningadaptor, or FA) must therefore incorporate all the fundamental safety measures. Note in particular that the mounting plane for each test device pair has a mechanical locating device ensuring the correct correspondence; and lastly that the test device is locked in position by three simultaneous positive logic locks.

D. Specific characteristics of the rotational motion

The X-ray tomography implemented by the UGXR benches imposes a principle of angular stability for the rotational axis, which is a key aspect of the precision sequence used to obtain high-resolution images: the purpose of a geometric resolution better than 100 µm essentially dictates that the angular variation of the rotational axis falls within a an arc deviation of ±2'', which represents an exceptional feat of engineering.

This level of performance can only be achieved when the centre of gravity of the load carried by the bench intercepts the axis of rotation. However, the JHR test devices carried by the bench are often unbalanced by the instrumentation mounted on their top sections. To compensate for an imbalance which is never the same for any two devices, the UGXR bench has a motorised automatic balancing mechanism designed to move two counterweights into corrective positions calculated by an algorithm based on the values supplied by three force sensors mounted on comparative supports.

Another interesting feature of the UGXR bench is the use of a push-pull drive system for the rotational motion, using a pair of opposing synchronised motors, which removes the hysteresis effects of the rotational motion by rendering them undetectable.

E. Vertical stroke

The total vertical stroke of a UGXR bench is 2880 mm. This long stroke is used to examine a test device above the 600 mm height occupied by the experimental sample. For instance, the bench can use gamma spectrometry to examine the lower section of a test device that has been subjected to an accident conditions experimental sequence with gravity displacement of the fuel, immediately after an experiment. Alternatively, it can use X-ray imaging to check the conformity of an assembly of in-vessel equipment placed between the mounting plane and the experimental sample on the test device.

III. Description of UGXR X-ray imaging

The X-ray imaging principle of an UGXR bench consists of placing the test device to be examined between an accelerator placed behind the collimation feed-through and an X-ray camera placed in a built-in watertight container on the underwater bench. The following three components of the UGXR imaging system are described below: the X-ray source comprising the linear electron accelerator; the X-ray beam collimator which is one of the functions of a collimation feed-through in the reactor pool; and the X-ray detector.

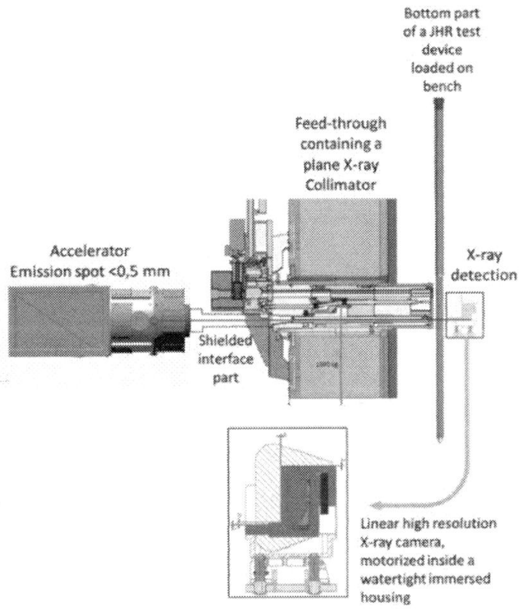

Fig. 3. Principle of X-ray imaging on the UGXR bench. The three main components are the accelerator (which provides a high-energy X-ray source), the pool feed-through (which collimates the X-ray beam toward the object to be examined) and the X-ray detector (comprising a pixelated detection line)

A. Self-shielded accelerator with smaller focal spot

Examining the image of a JHR experimental device contained in a test device requires a high-energy X-ray beam, such as that produced by an electron accelerator.

The JHR has a single accelerator common to both UGXR benches (and the HGXR bench, see section V), which requires it to be capable of adapting to each of the three sites concerned and of being moved between sites to serve successive measurement campaigns as effectively as possible.

PMB, a subsidiary of the ALCEN group, is responsible for designing and manufacturing the accelerator. Assembly is now complete. The accelerator is currently in its testing phase. Work is ongoing to optimise its design in order to achieve the required performance levels.

The device itself is a 6 MeV linear electron accelerator. Development of the prototype was characterised by four major specific design considerations:

- The focal spot needs to be less than 0.5 mm
- The overall dimensions must comply with the highly restrictive constraints imposed by the three different installation sites
- It must be either directly or indirectly connectable to the predefined JHR pool or cell feed-throughs
- The secondary emission levels must allow the accelerator to be used in unshielded facilities

To meet this last specification, the secondary emissions emitted by the docked accelerator on the periphery must be considerably reduced by self-shielding. In order to ensure that the overall weight of the accelerator does not become unmanageable, the additional weight of this shielding has been optimised by incorporating the attenuating structures as close as possible to the X-ray sources that comprise the accelerator line and emission components (electron gun, acceleration line, focus triplet, target) by design.

The reduction of focal spot size required by the first specification is obtained by inserting an assembly comprising a configurable focus coil between the accelerator output and the conversion target.

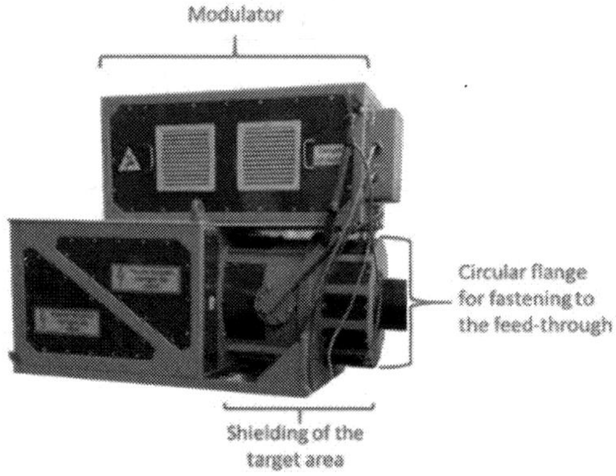

Fig. 4. View of the accelerator in the testing and optimisation phase

B. X-ray beam collimator

The JHR accelerator will emit its main beam toward the object held by the UGXR bench, through the pool wall, via a collimation feed-through to which it is docked.

For the UGXR benches it is intended for specific future contracts that a shielding-collimation interface will be created to allow continuity of the collimated X-ray beam between the accelerator and the feed-through inlet (see the interface shielding on Figure 3).

The pool feed-through X-ray collimator is around 1.5 m long, with a 1 mm planar opening. This generates a planar array beam through the experimental object that impacts the secondary collimator inlet at the submerged linear camera end.

C. X-ray beam detection

The principle of X-ray image formation on the UGXR benches consists of discriminating the useful X-ray flow (the localised emission source of which is targeted by a secondary collimator) from the interfering gamma flow emitted in 4π by the emissive zone on the test device.

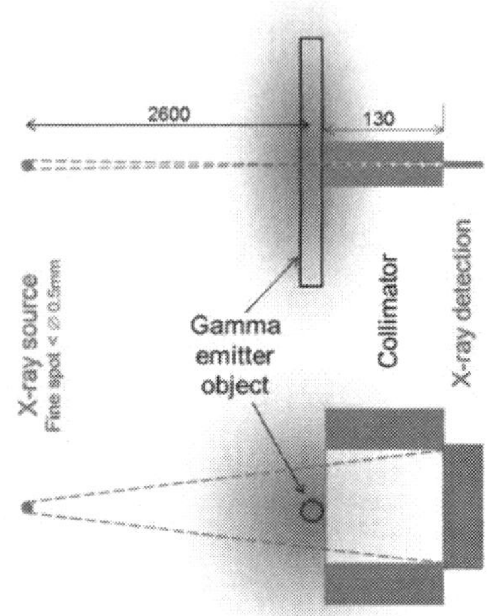

Fig. 5. Principle of linear X-ray detection adapted to a gamma emitting object

This principle is traditionally used in the nuclear industry for X-ray/gamma ray discrimination and imposes line-by-line image acquisition, where each line corresponds to an object displacement increment between the X-ray source and the X-ray detection line. Displacement is axial to form an X-ray; it is rotational to acquire a tomographic sinogram.

For UGXR benches, the linear X-ray camera is aligned in the array beam emitted by the accelerator. It is therefore immersed in the pool at a depth of around 4 m in a watertight container (see Figures 1 and 3). This container is a mechanical UGXR bench component. It is motor-driven to adapt to the varied cross-sectional dimensions of the test objects.

The camera comprises X-ray detection electronics that are shielded from radiation by a tungsten alloy housing. This heavy metal housing incorporates the discriminating secondary collimation function: a 70 mm wide, 130 mm long collimation slit with an adaptable height of 100 µm or 300 µm that can be used to channel the incident beam toward the high-resolution X-ray detection line downstream of the collimator. This secondary collimation function requires very precise alignment of the camera to ensure full lighting of a long collimation slit with just a 100 µm opening. The I&C system for this precision alignment function uses three piezoelectric linear actuators and a dual-axis tilt sensor.

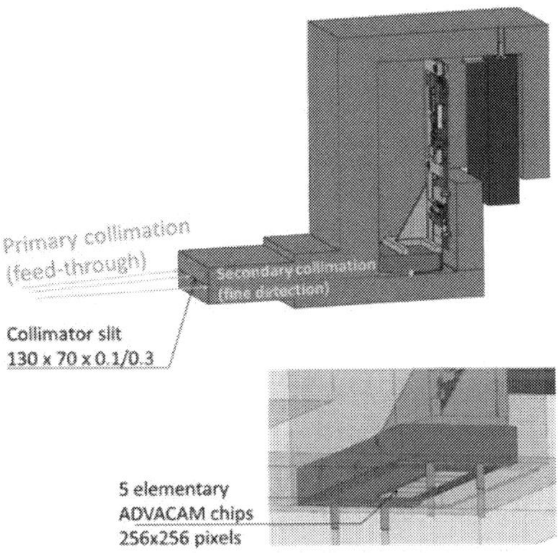

Fig. 6. Cross-section of the X-ray camera comprising a tungsten alloy shielding-collimator housing containing detection electronics based on MEDIPIX technology

The technology that allows each of the X-ray detection lines (projections) to be recorded is based on ADVACAM's WidePIX 1x5 or WidePIX 2x5 system. The principle of this system consists of aligning 5 chips, where each chip is a MEDIPIX 256x256 pixels of 55 µm, component with silicon semiconductor technology. This gives a whole detection made up with 256x1280 pixels, or even 512x1280 pixels if the two-row sensing arrangement is used (WidePIX 2x5).

As shown in Figure 6, the incident flow impacts the thin lateral edge of the 5 components line. The incident photons penetrate thus a layer of semiconducting material corresponding to a trajectory through the whole width of the components, i.e. 14 mm (or 28 mm in case of WidePIX 2x5). This arrangement is used to measure the spatial signal corresponding to every X-ray projection: each signal sample is reconstructed by summing the elementary signals provided by each of the pixels located on the radiation propagation line from the source.

After a series of promising parametric tests conducted by ADVACAM, it is expected that the first prototype camera will be available before the end of 2019. Its test with the accelerator is planned to take place as soon as (heavy) experimental conditions may be set up (see section VII).

IV. Description of the UGXR gamma measuring equipment

The gamma emitters on the test devices containing irradiated fuel can be mapped by the UGXR benches according to three degrees of freedom. This allows axial and radial gamma scanning, as well as emission tomography that combines rotation and radial scanning.

A. Experimental performance of the UGXR gamma measurements

The central requirement for the gammametry function on UGXR benches is that any existing or future gamma-emitting JHR experimental object must be measurable, given that the JHR's MTR vocation implies a wide variety of scenarios. This requirement therefore considerably extends the collimation capabilities to the following:

- Any shape of gamma emitting object: radial strokes adapted to extended shapes, like plate fuels, for instance
- All scanning modes: diversity of pre-collimation slit openings and shapes
- The widest range of gamma activities, including in particular the impact of the widest variety of fuel cooling times: diversity of post-collimation openings and screenings

The UGXR technological response to this requirement consists of:

- Fitting a pre-collimation drum made of tungsten-based material (20 cm thick) inside the pool feed-through to allow the motorised interchange of 5 different types of pre-collimation slits. One of these slits can be interchanged manually, without having to be disassembled mechanically, in order to extend the pre-collimation mode to any other requirement not yet identified at the time the equipment was designed.
- Fitting a drum (15 cm thick) to the gamma-ray germanium detector to allow the motorised interchange of 6 different types of post-collimation apertures; and fitting a second drum allowing the motorised interchange of 5 calibrated copper screens ranging from 0 to 5 cm thick.
- Having a radial stroke of 71 mm on the UGXR bench to cover the majority of widths of gamma emitting objects expected in the JHR.

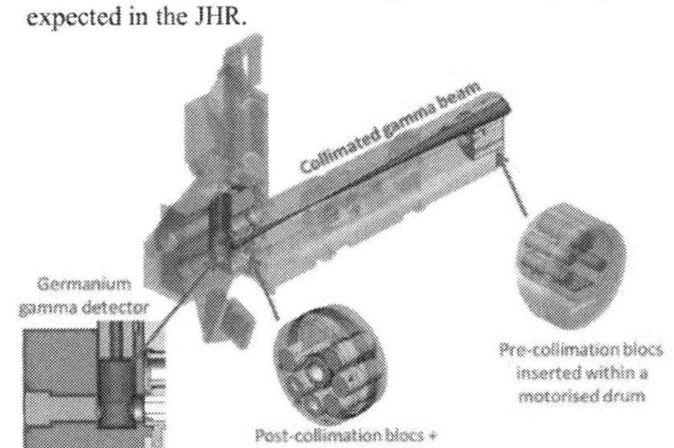

Fig. 7. Collimator-screen assembly fitted in the UGXR pool feed-through

Note that every GXR gamma-ray spectrometers use the Cryo-Pulse 5 Plus provided by Mirion Technologies, with p-type HPGe crystals (30% efficiency), associated to LYNX systems whose Digital Signal Processing technology achieves the performance goals (energy resolution, up to the highest count rates, in particular).

B. Associated radiation protection

The principle of collimation involves an opening which necessarily causes the local rupture of the biological shielding on the out-of-pile side, although the design source term held by the bench from the other side of the feed-through is considerable.

With regard to this specific point, the UGXR safety approach has resulted in a principle which guarantees that a radiation protection seal will be maintained in line with expectations in nominal operation or routine maintenance conditions. To demonstrate this capability, the principle uses mechanical locks and a set of authorisation keys.

The only action then needed from the classified I&C system is the continuous display of the current status of the radiation protection in the work area adjacent to the collimation feed-through.

V. Description of the HGXR bench

As we have seen above, both UGXR benches are intended for examining whole JHR test devices. In addition to these two benches, a smaller bench dedicated to measuring experimental loads extracted from test devices will be installed in one of the JHR's three small shielded hot cells. It offers the same experimental functions: gamma spectrometry and X-ray imaging.

A. Description of the HGXR hot cell bench

The HGXR bench, designed to measure an experimental load that has previously been extracted from its test device in a large hot cell, will support a load weighing up to 30 kg with a maximum height of 1200 mm.

The HGXR equipment comprises a trolley mounted on rails on the floor of the hot cell for a recoil movement according to (0,+2300 mm). It is moved manually along the rails and the electrical connections are established mechanically.

This trolley has a mandrel which holds the experimental object in place. The position of the mandrel is motor-driven in terms of rotation (360°), translational movement on the y-axis (±90 mm), and altimetry on the z-axis (+585 mm/-1150 mm). The trolley also accommodates the shielded camera, identical in every way to that used on the UGXR benches, except that it is not housed in a watertight container.

The design of the HGXR bench means that it may be possible to perform simple dismantling operations in the hot cell by remote handling. The conceptual design has therefore considered the constraints imposed by the clearances, the forces of the remote manipulators in the hot cell and the hanging method imposed by the manipulator grippers.

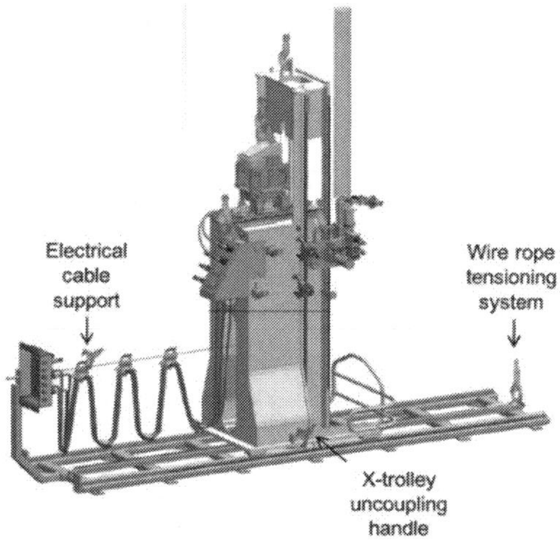

Fig. 8. Overview of the HGXR bench in the JHR ECE hot cell

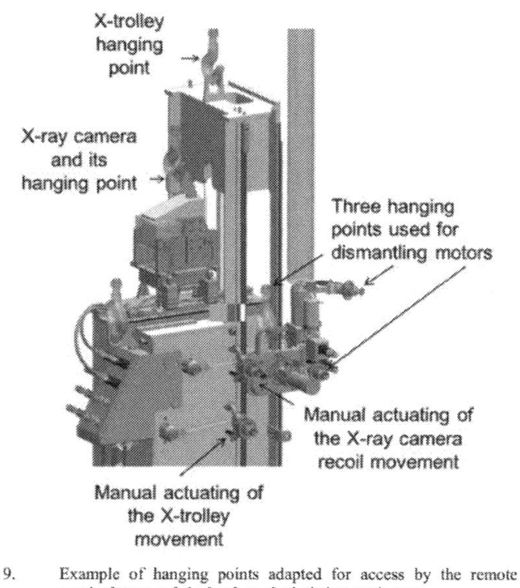

Fig. 9. Example of hanging points adapted for access by the remote manipulators and the hook on the hoisting equipment

B. Description of the HGXR equipment in the airlock adjacent to the hot cell

The principle of the HGXR bench consists of moving an experimental object in the hot cell in front of two equipped feed-throughs in the wall separating the hot cell from the adjacent airlock:

- One of the feed-throughs is dedicated to gamma measurements taken by a gamma-ray detector in the airlock, and via motorised collimation devices similar to those on the UGXR benches
- The other feed-through is dedicated to X-ray imaging via the X-ray camera on the bench, which records the X-ray

https://doi.org/10.1051/epjconf/202022504001

beam generated by the accelerator in the airlock on a projection-by-projection basis.

The design is simplified by having two separate feed-throughs, in contrast to the UGXR bench.

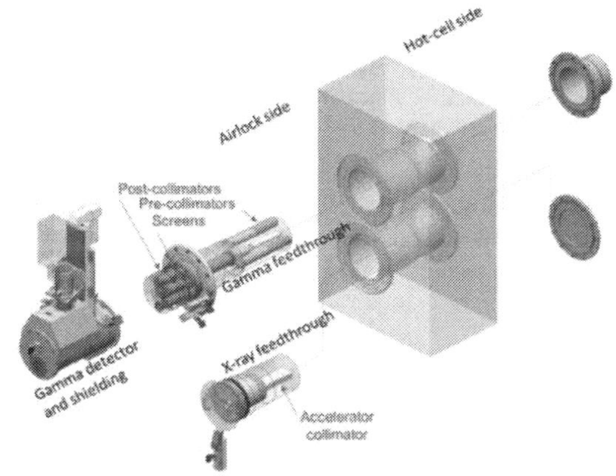

Fig. 10. The HGXR equipment has two separate feed-throughs for gamma measurements and X-ray imaging

The design of the collimator-screen unit for the HGXR bench is similar to that of the UGXR bench, with 3 drums which allow the interchange of 2 different pre-collimators, 4 different post-collimators and 5 different screens using a motorised system.

Similar to the UGXR, the conceptual design of the HGXR implements radiation protection measures based on locking mechanisms combined with authorisation keys.

Fig. 11. Photo of the HGXR bench plate, as built in spring 2019

VI. Generic I&C components for the UGXR and HGXR benches and the accelerator

With two UGXR underwater benches and one HGXR hot cell bench, there are a total of 5 GXR installation sites in the JHR. The I&C racks for each of the two underwater benches are divided across two buildings between level 0 poolside and level -1 near the pool feed-through. The I&C racks for the HGXR hot cell and airlock bench are located in the area in front of the small hot cells.

Each I&C rack is equipped with a SCHNEIDER M340 PLC and associated SCADA station allowing the experimenter to monitor the equipment using specially designed tools such as mimics and alarms. There will also be up to 9 dedicated computers for experimental applications, i.e. the programmes that will be sequencing the gamma measurements (Mirion LYNX system), X-ray measurements (ADVACAM driver), movement of experimental objects (primary PLCs), dynamic positioning of collimator-screen components (secondary PLCs), and the remote controls for the accelerator installed at any one of its three possible sites.

It will be possible to control all this equipment remotely from the JHR NDE control room. However, numerous local actions will also have to be carried out at the actual equipment locations (such as loading, settings, calibrations and maintenance). This flexible, multi-access requirement will be partially or totally done by using the BlackBox Agility KVM-over-IP remote control technology, which uses a matrix switching solution to provide a remote connection between a series of computers and a series of passive monitor-keypad-mouse stations.

VII. Acceptance programme planned at Cadarache

Before being installed on-site at the JHR, all this equipment will pass through the intermediate test facilities at Cadarache. This gradual installation is necessary to adapt the schedules for the tasks being carried out by all the different stakeholders involved (IDOM, VTT, CEA) to the JHR's extremely tight schedule. It will also help in terms of implementing any tasks that need to be completed upstream to ensure the equipment is fully operational in all aspects: operating modes, safety and security, experimental procedures, maintenance, etc.

A. Planned UGXR tests in the TOTEM test hall

The majority of preparatory operations and tests will take place in the TOTEM test hall at Cadarache. This hall is therefore currently being equipped accordingly. The UGXR benches will be included in this equipment, and will be received at two additional workstations:

- The CESARINE pool in the TOTEM hall is equipped with a docking station to receive one of the UGXR benches that is representative of its future installation in the JHR pool. This is a fantastic opportunity to simulate the future installation of the UGXR bench and address the numerous practical issues associated with handling and maintenance operations in the calm surroundings of a dedicated hall. These operations are often difficult given that they involve manoeuvring heavy bulky objects in confined spaces.
- A test tower offers the opportunity to receive a bench that is then fully accessible to operators (full height). This second configuration is particularly suitable for running functional bench tests, including inspectable checks of safety functions, as well as all the measures required to achieve the experimental performance levels necessary for the high spatial resolution capability of these benches.

Fig. 12. Example of the equipment in the TOTEM facility: CESARINE pool equipped with docking station for the UGXR bench

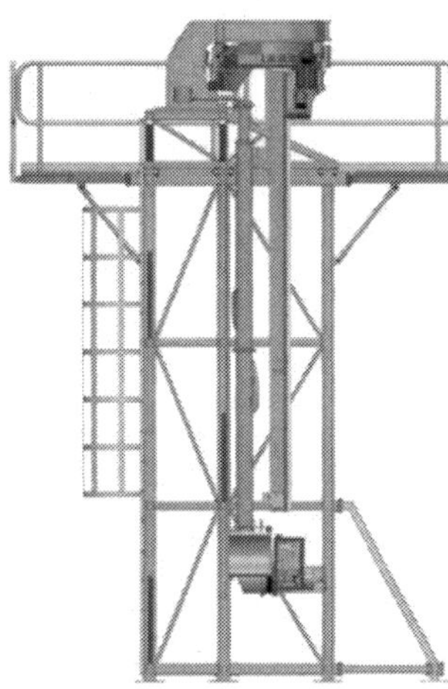

Fig. 13. Example of test equipment in the TOTEM facility: UGXR test tower

The out-of-pile parts of the collimation equipment on the UGXR benches will be placed on test stands representative of their loading in the pool feed-throughs.

To ensure that preparations are exhaustive, all the UGXR I&C functions relating to both the in-pool and out-of-pile parts will be implemented during the tests and all preparatory operations required for the UGXR will be completed in the TOTEM facility.

The UGXR acceptance tests in TOTEM are due to start during 2020.

B. Planned HGXR tests in the TOTEM test hall

The HGXR equipment will also be received, tested and prepared in the TOTEM hall.

The HGXR test programmes are similar to many of those that will be carried out on the UGXR benches. They will be conducted in the TOTEM hall according to significantly less restrictive constraints due to the smaller component size involved.

The specific nature of the remote handling control and maintenance operations required for this hot cell bench remain the priority consideration. Studies conducted by IDOM have taken this into consideration and it has been modelled using 3D tools. However, it is only through testing in an actual hot cell mock-up like the one in the TOTEM hall that details relating to ergonomic aspects and the auxiliary tools which only become apparent at the final preparation stage can be addressed.

Getting started with the HGXR equipment in TOTEM is due to begin in the last quarter of 2019.

C. Planned X-ray imaging tests in CINPHONIE

The CINPHONIE test platform, located in a pit in the CHICADE facility at Cadarache, will be used to test and prepare all the X-ray imaging equipment (the ADVACAM camera and the PMB ORIATRON accelerator) before it is installed at its final destination.

Initial UGXR/HGXR camera tests and next, associated development works for UGXR measurements should be started by the end of 2019. Used with an accelerator with a smaller focal spot size and a motor-driven rotary precision indexing table, it will be tested according to a configuration that will provide a realistic simulation of the future X-ray targets on the UGXR and HGXR benches and generate an image model that will allow the predicted performance levels to be confirmed experimentally.

Regarding the JHR accelerator, it is too early for scheduling preparatory tests in CINPHONIE. A last set of outcomes related to essential optimisation works shall be prior delivered by the contractor PMB. Initial factory tests show that the smaller focal spot size of 0.5 mm is a viable goal. Improving the self-shielding performance, shown to be inadequate during factory tests, represents the bulk of next optimisation works.

VIII. CONCLUSION

The GXR projects, including accelerator and X-ray camera, are starting in the short term their acceptance test phases, thus concluding the industrial partners' roles.

We have seen that these testing programs will be carried out within intermediate test facilities at Cadarache, meaning that for passing from the test sites to the definitive JHR site, numerous implementing tasks, whether electrical, mechanical or dossier type, will remain still necessary for making this equipment fully functional and maintainable on site.

Furthermore, the production of experimental results at the highest level of quality initially aimed by this project will require a demanding development of software tools, whether regarding online processing tools, scientific calculation tools (including for example gamma emission tomography and 3D transmission tomography), and the secured management of any experimental data, preferentially by using a shared relational database.

EPJ Web of Conferences **225**, 04002 (2020) https://doi.org/10.1051/epjconf/202022504002
ANIMMA 2019

Evaluation of Low Dose Silicon Carbide Temperature Monitors

K. L. Davis[1], A. Gusarov[2], T. C. Unruh[1], P. Calderoni[1], B. J. Heidrich[1], K. M. Verner[2], A. Al Rashdan[1], A. A. Lambson[1], S. Van Dyck[2], I. Uytdenhouwen[2]

[1] *Idaho National Laboratory*
[2] *SCK-CEN Belgian Nuclear Research Centre*

***Abstract*— Thermocouples are generally used to provide real-time temperature indications in instrumented tests performed at materials and test reactors. Melt wires or paint spots are often included in such tests as an independent technique of detecting peak temperatures incurred during irradiation. In addition, less expensive static capsule tests, which have no leads attached for real-time data transmission, often rely on melt wires and paint spots as a post-irradiation technique for peak temperature indication. Unfortunately, these techniques are limited in that they can only detect whether a single temperature is or is not exceeded. Silicon carbide (SiC) monitors are advantageous because a single monitor can be used to determine the peak temperature reached within a relatively broad range (200 – 800°C). Although the use of SiC monitors was proposed more than five decades ago, the ultimate performance limits of this technique are not fully understood. The Nuclear Science User Facilities (NSUF) is the United States Department of Energy Office of Nuclear Energy's only designated nuclear energy user facility. Its mission is to provide nuclear energy researchers access to world-class capabilities and to facilitate the advancement of nuclear science and technology. This mission is supported by providing access to state-of-the-art experimental irradiation testing, post irradiation examination facilities, and high performance computing capabilities as well as technical and scientific assistance for the design and execution of projects. As part of an NSUF project, low dose silicon carbide monitors were irradiated in the Belgian Reactor 2 and were then evaluated both at the SCK•CEN and at Idaho National Laboratory's High Temperature Test Laboratory to determine their peak temperature achieved during irradiation. The technical significance of this work was that the monitors were irradiated to a dose that was significantly less than recommended in published literature. This paper will discuss the evaluation process, the irradiation test, and the performance of the low dose silicon carbide temperature monitors.**

***Keywords*—In-pile instrumentation, temperature sensor.**

I. Introduction

SINCE the early 1960s, SiC has been used as a post-irradiation temperature monitor. Several researchers have observed that neutron irradiation induced lattice expansion of SiC annealed out when the post-irradiation annealing temperature exceeds the peak irradiation temperature. Irradiation temperature is determined by measuring a property change after isochronal annealing or during a continuously monitored annealing process [1-8]. There are many properties that may be measured including, electrical resistivity, density, thermal diffusivity, and lattice spacing. In general, electrical resistivity is accepted as a robust measurement technique resulting in accuracies within 20°C [9].

Twelve silicon carbide (SiC) temperature monitors were irradiated in the Belgian Reactor 2 (BR2) as part of a Nuclear Science Users Facilities (NSUF) Project and were delivered to the High Temperature Test Lab (HTTL) for evaluation to determine their peak temperature achieved during irradiation. Each monitor had a sister monitor exposed to identical irradiation test conditions. Monitors with the "A" designation (six in total) were evaluated using an electrical resistivity method [10]. Sister monitors with the "B" designation are to be evaluated using a new method [11]. The evaluation of the "B" temperature monitors will not be described in this report, but will be described in a subsequent report once the evaluations have been completed. TABLE I provides identification for each "A" monitor with its dose and an expected peak irradiation temperature based on thermal analysis.

The quality of material used to manufacture the SiC temperature monitor has a major impact on the radiation-induced swelling and, thus, the ensuing peak irradiation temperature evaluation. Temperature monitors were fabricated from material meeting the Rohm Haas specification SC003. This material was produced via chemical vapor deposition (CVD) process with a high purity (99.9995%) and a density equal to theoretical. No voids or micro cracks were allowed. Independent verification was not performed. Due to its polycrystalline β-cubic structure, the SiC material characteristics are isotropic. Using this characteristic, the SiC monitors were manufactured to exceed a resistivity > 1000 ohm/m. SiC monitors used in the experiment were manufactured as cylinders with a 1 mm diameter and a 12.5 mm length (Fig. 1).

This report discusses the laboratory evaluation method, the irradiation capsule design, thermal analysis, dose calculations, evaluation of the SiC monitors, and subsequent computed tomography evaluations.

II. Method

HTTL uses resistivity measurements to infer peak irradiation temperature [12-14]. SiC monitors may be evaluated for peak irradiation temperatures ranging from 150 – 800°C with a recommended dose ranging from 1 – 8 dpa [10]. For this evaluation, the temperature criterion was met, but it is significant to note SiC monitors with the M1 designation received a dose that was half the minimal recommended dose.

Fig. 2 depicts the equipment at the HTTL used to evaluate the SiC monitors. The SiC monitors are heated in the annealing furnace using isochronal temperature steps. Annealing temperatures are recorded using a National Institute of Standards and Technology (NIST) traceable thermocouple inserted into an alumina tube in the furnace. After each isochronal annealing, the specimens are placed in a resistivity measurement fixture located in the constant temperature chamber (maintained at 40°C) for a minimum of 30 minutes. After the 30 minute wait time, each specimen's resistance is measured. Thanks to the high thermal conductivity of SiC 30 minutes is more than sufficient to achieve thermal equilibrium.

III. IRRADIATION CAPSULE

The reactor exposure was performed using the Basket for Material Irradiation (BAMI) rig of the BR2, using standard non-instrumented capsules. The capsule consists of an external aluminum body which is hermetically sealed in a He atmosphere at two bar pressure using plastic deformation. To provide a required temperature under radiation, samples were inserted into a dedicated holder. In the present experiment, this holder included three sections made as thick-wall stainless steel tubes separated by ceramic spacers. The cross-section of the capsule is schematically shown in Fig. 3 (a) and a photograph of the holders and spacers is shown in Fig. 3 (b). Each 15 mm long stainless steel section had four symmetrically located 1.1 mm diameter channels, in which the SiC monitors were inserted. The 0.1 mm difference in the channel and the sample diameter was made to allow easy retrieval of the SiC samples after irradiation.

The 5 mm thick spacers were made of low thermal conductivity (k = 1.3 W/mK) Alumo-silicate Aremcolox 502-1100 ceramic with a density of 2.70 g/cm^3. This material can sustain long-term temperatures up to 1200°C (melting at ~1600°C). Impurities include Fe_2O_3, MnO, MgO, CaO, TiO_2, K_2O, Na_2O, P_2O_5, C, and S with the total amount less than 3% by mass. The spacers provide thermal insulation between different sections, serve to center the plugs on the body axis, and keep the SiC samples in the channels.

IV. THERMAL ANALYSIS

The capsule has no active heating element; the temperature of the samples during irradiation is defined by the thermal balance between the radiation heat generation in the internal components and heat transfer through the He gas gap and the aluminum body to the reactor cooling water. Therefore, for a given radiation heating level, the inner and outer diameters of the stainless steel holder can be adjusted to obtain a desirable irradiation temperature.

The following phenomena were considered to define the temperature distribution:

1) heat generation inside the holder, spacers, and in the capsule walls,
2) heat exchange through the different materials inside the capsule, including gas gaps,
3) heat exchange on the capsule surface.

The volumetric heat generation is dominated by the gamma heating and can be found using a simple approximation:

$$q = Q\rho \tag{1}$$

where Q [W/g] is gamma heat generation in reference material (aluminum), which is standard material for BR2 gamma-heating evaluation, and ρ is the material density of a component. Heat generation in stainless steel holders is slightly higher. This difference was neglected because it is comparable with the uncertainty of the computational model.

In a closed capsule filled with He gas and small gaps, there is no forced convection, and the influence of the natural convection is small. The maximum temperature considered was 400°C. For this temperature, the radiative heat transfer was negligible when compared to the conduction. Therefore, heat exchange occurs via heat conduction mechanism. For the irradiation, the capsule was placed inside a standard BR2 driver fuel element which was cooled with a water flow at ~10 m/s. It is known that the cooling of the fuel elements occurs in a mono-phase (no boiling) regime. This means that the heat exchange conditions were well defined, and the temperature of the capsule external surface was maintained between 48-52°C throughout the irradiation. A 50°C temperature was used for the capsule design. This conclusion is consistent with the thermal balance calculations below. The design parameters are summarized in TABLE II.

Compared to steel and aluminum, He gas has a relatively low thermal conductivity. As a result, the holder-body gap thickness has a big influence on the holder temperature. This made it possible to obtain three different temperatures in one capsule by fabricating the sections with different He gap thicknesses. The gap between the spacers and the capsule body was minimized to prevent excessive He leakage and to provide accurate centering of the plugs, yet was made large enough to allow for capsule dismantling.

Due to the low thermal conductivity of the spacers and the geometry on the BAMI rig, heat flux in the axial direction is much lower when compared to heat flux in the radial direction. Therefore, the temperature distribution calculation problem is effectively reduced to an axially symmetric 2D problem. This allowed use of the analytical approach for the preliminary capsule design.

For the preliminary design, the system of the heat transfer equations was solved analytically:

$$\frac{1}{r}\frac{d}{dr}\left(k_i r \frac{dT_i}{dr}\right) + q_i = 0, i = 1 \dots n \tag{2}$$

where i is the layer number, n = 4. The following boundary conditions were applied:

$$T_1(0) < \infty \tag{3}$$

$$T_i(r = r_i) = T_{i+1}(r = r_i) \tag{4}$$

$$k_i \frac{dT_i}{dr}(r = r_i) = k_{i+1} \frac{dT_{i+1}}{dr}(r = r_i) \tag{5}$$

$$-k_n \frac{dT_n}{dr}(r = r_n) = h[T_n(r_n) - T_\infty] \tag{6}$$

where T_∞ is the coolant temperature.

The thermal expansion of the capsule components was

taken into account by defining the dimensions at the design temperature. The analytical model was verified using ANSYS numerical simulations. The difference was less than 10°C. The numerical simulations were used for the part fabrication. Calculated peak irradiation temperatures were presented in TABLE I.

V. Dose Calculation

Two identical capsules (M1 and M2) were prepared. The capsule M1 was exposed in the BR2 reactor during two cycles and the capsule M2 during the first cycle only. The irradiation was performed by placing the capsules inside BR2 driver fuel elements with a high burn-up located in the same B120 channel for both cycles. The capsules were dismounted simultaneously following a two month cooling period after the end of the second cycle. The channel selection was done based on the pre-cycle neutronic Monte-Carlo computations. The estimated irradiation conditions are given in TABLE III. The nominal reactor power during two cycles was the same. The gamma-heating levels in the irradiation channel during the two cycles were nearly identical, which means that the temperatures of the samples were also nearly the same. The difference in the neutron fluxes of ~10% is related with a lower fuel burn-up in the second cycle.

DPA calculations were completed using flux outputs from MCNP [15] and displacement cross sections from SPECTER [16]. SPECTER is a computational tool developed at Argonne National Laboratory to assist in material damage calculations. The DPA rate is calculated using,

$$DPA_{rate} = \int_{E_m}^{E_M} \sigma_d(E)\phi(E)dE \qquad (7)$$

with σ_d being the 41 elemental displacement cross sections from SPECTER and ϕ being the neutron flux. Both variables are energy dependent and 100 energy groups from E_m and E_M, 0 to 20 MeV respectively, were used for the calculation. The displacement cross sections are calculated with

$$\sigma_d = \frac{0.8E_a}{2E_d} \qquad (8)$$

where;

E_a is the available energy that is found in the SPECTER manual or can be calculated using NJOY [17],

E_d is the displacement threshold energy, which is also found in the SPECTER manual.

Using the ATR MCNP neutron flux results, the experimental positions with the most similar conditions to those in the BR2 experiment were found [18]. The fast and thermal fluxes of E<0.5eV and E>0.1MeV are much higher in the BR2 reactor than in the ATR. However, the ratio of fast to thermal flux can be used to compare the two differing neutron fluxes. After finding the most similar fast to thermal ratio in the ATR experimental position, the ratio of the BR2 fast flux to the ATR fast flux was found and used as a multiplier for the DPA rate. The DPA rates of silicon and carbon in ATR are 36 days/DPA and 60 days/DPA, respectively. To calculate the DPA rate of the compound, a weighted average is taken based on the atom fraction of silicon and carbon in SiC. This has been shown to be a reasonable method of approximating compound material displacement cross sections when compared to tools like SPECOMP [19] that calculate compound displacement cross sections. Lastly, the time for each experiment is multiplied by the DPA rate, and the actual DPA is found. At 21 days, the DPA is approximately 0.5 and 1.0 at 49 days.

A second irradiation DPA calculation was also performed. The cycle irradiation conditions were calculated using MCNP. The experiment, Motore, was located at the mid-plane of the reactor core within a fuel element in channel B-120. To evaluate the irradiation conditions, F4 tally cards were used, which calculate the neutron flux averaged over a cell in neutrons/cm²s. This calculated value is normalized to obtain the actual neutron flux in the region of interest. The F4 tally normalized result was used for calculating the flux spectrum and the DPA. Fig. 4 presents the neutron flux spectra on the capsules. The DPA is approximately 0.5 and 1.1 at 21 and 49 days, respectively. This updated calculated DPA is in close agreement with the first calculation. The doses for each SiC monitor were presented in TABLE I.

VI. SiC Temperature Monitor Evaluations

This section discusses the evaluation of the SiC monitors and presents the results. This work was conducted in accordance with an approved evaluation plan [10].

An ohmic response curve was generated for each monitor prior to heating. Monitor BR2 M1-High-A exhibited a typical ohmic response and is displayed in Fig. 5. These data were used to check for linearity and to select a target voltage (with corresponding current) that would result in minimal heating of the SiC monitor during resistance testing and remain within the range of the test instrumentation. For this evaluation, voltage ranged from 16 – 20V.

Electrical resistivity is used by HTTL to infer the peak irradiation temperature [1]. Figs. 6 – 11 present the resistivity data taken at each isochronal annealing temperature for each SiC temperature monitor. The peak irradiation temperature, using an electrical resistivity technique, can be taken as the point where the resistivity begins and consistently remains, above the error band. The error band bounds the data and is represented by the dotted lines. For this evaluation, the error band was established as the ±2σ value based on a sample size of the first five data points taken below 150°C.

TABLE IV shows the results for the evaluation. The calculated versus measured peak irradiation temperatures had good agreement comparable to published data [9]. This result is significant considering that M1 SiC monitors received doses that were much less than 1 dpa.

As evident from the presented data (Figs. 6 – 11), all of the monitors responded well with the exception of the BR2 M2-Low-A (see Fig. 10). This temperature monitor received the lowest dose (0.5 dpa) and was exposed to the lowest temperature (255°C). Also the error band was much larger than expected. There are several factors that may be considered as to why BR2 M2-Low-A did not respond to the isochronal heating. Further analysis such as microscopy, computed tomography, and material analysis could be used to determine why M2-Low-A did not respond. Computed tomography was performed and is discussed below.

VII. Computed Tomography Evaluation

In an effort to uncover the cause of the indeterminate response of monitor BR2 M2-Low-A, 3D micro-focus computed tomography (CT) was performed on this monitor and on BR2 M1-LOW-A for comparison. The scan energy was 50 kV at a resolution of 6 µm. The purpose for performing CT was to determine if there were any defects (voids, cracks, porosity, etc.) or foreign material in the BR2 M2-LOW-A monitor that would singularize it from the BR2-M1-LOW-A monitor. Figs. 12 and 13 show CT slices taken from each scan. After evaluating the scans, no apparent anomalous indications were found in the BR2 M2-Low-A monitor. It interesting to note that high density indications were found on or near the surface of the BR2 M1-Low-A monitor, the monitor that performed well.

Both monitors should have been identical in composition. As evident in the scan data, high density material was found on or near the surface of the BR2 M1-Low-A monitor. Visual inspection of the monitor did not find material on the surface; however, it is believed that the material is platinum which transferred from the resistivity measurement apparatus and that it did not influence the annealing behavior. Resistivity measurements were taken after the evaluation but before the CT scan. 3D micro-focus CT did not uncover a plausible cause for the indeterminate response of the BR2 M2-Low-A monitor.

VIII. Conclusion

SiC temperature monitors were irradiated in BR2 as part of an NSUF Project and evaluated at the HTTL to determine their peak temperature achieved during irradiation. Dedicated irradiation capsules were designed and fabricated to allow irradiation at specified temperatures ranging from 240 – 380°C. After the irradiation, the peak irradiation temperature of each monitor was evaluated using the resistance measurement method. This method recommends a minimum dose of 1 dpa. These monitors received doses ranging from 0.5 – 1.1 dpa. Deviations between the calculated temperature and the evaluated temperature were within or near published limits. A significant finding from this evaluation is that it is possible to evaluate SiC temperature monitors at dose levels much less than 1 dpa. SiC monitors were successfully evaluated that were irradiated to 0.5 dpa with temperatures ranging from 240 – 380°C.

References

[1] J. L. Rempe, K. G. Condie, D. L. Knudson, and L.L. Snead, *Silicon Carbide Temperature Monitor Measurements at the High Temperature Test Laboratory*, INL/EXT-10-17608, January 2010.

[2] L. L. Snead, A. M. Williams, and A. L. Qualls, "Revisiting the use of SiC as a Post Irradiation Temperature Monitor," *Effects of Radiation on Materials*, ASTM STP 1447, M L. Grossbeck, Ed, ASTM International, West Conshohocken, PA, 2003.

[3] K. G. Field, J. L. McDuffee, J. W. Geringer, C. M. Petrie, Y. Katoh, "Evaluation of the Continuous Dilatometer Method of Silicon Carbide Thermometry for Passive Irradiation Temperature Determination," Nuclear Instruments and Methods in Physics Research Section B: Beam Interactions with Materials and Atoms, Volume 445, 15 April 2019, Pages 46-56

[4] A. A. Campbell, W. D. Porter, Y. Katoh, and L. L. Snead, "Method for Analyzing Passive Silicon Carbide Thermometry with a Continuous Dilatometer to Determine Irradiation Temperature," Nuclear Instruments and Methods in Physics Research Section B: Beam Interactions with Materials and Atoms, Volume 370, March 2016, Pages 49-58

[5] N.F. Pravdyuk, V.A. Nikolaenko, V.I. Karpuchin, V.N. Kuznetsov, Investigation of Diamond and Silicon Carbide as Indicators of Irradiation Conditions, in: D.J. Littler (Ed.), International Conference on Properties of Reactor Materials and the Effects of Radiation Damage, Butterworths, Gloucestershire, England, 1962, pp. 57–62.

[6] J.I. Bramman, A.S. Fraser, W.H. Martin, Temperature Monitors for Uninstrumented Irradiation Experiments, J. Nucl. Eng. 25 (6) (1971) 223–240.

[7] H. Suzuki, T. Iseki, M. Ito, Annealing Behavior of Neutron Irradiated b-SiC, J. Nucl. Mater. 48 (3) (1973) 247–252.

[8] J.E. Palentine, The Development of Silicon Carbide as a Routine Irradiation Temperature Monitor, and Its Calibration in a Thermal Reactor, J. Nucl. Mater. 61 (3) (1976) 243–253.

[9] Joy L. Rempe, Keith G. Condie, Darrell L. Knudson, and Lance L. Snead, "Comparison Measurements of Silicon Carbide Temperature Monitors", IEEE TRANSACTIONS ON NUCLEAR SCIENCE, VOL. 57, NO. 3, JUNE 2010.

[10] K.L. Davis and T.C. Unruh, "Silicon Carbide Temperature Monitor Evaluation." PLN-3473, Rev. 2, Idaho National Laboratory, February 2018.

[11] A. Al Rashdan, K. Davis, T. Unruh, and J. Daw, "Silicon Carbide Temperature Monitor Online Evaluation," PLN-5465, September 2017.

[12] K. L. Davis, B. M. Chase, T. C. Unruh, D. L. Knudson, J. L. Rempe, *Drexel University Temperature Sensors*, INL/EXT-14-33067, September 2014.

[13] B. M. Chase, J. L. Rempe, K. L. Davis, Evaluation of *Temperature Sensors from the EPRI Zirconium Growth Irradiation Tests*, INL/LTD-14-32210, June 2014.

[14] K. L. Davis, D. L. Knudson, J. L. Rempe, and B. M. Chase, *University of Illinois Temperature Sensors*, INL/EXT-14-33136, September 2014.

[15] C.J. Werner, et al., "MCNP6.2 Release Notes", Los Alamos National Laboratory, report LA-UR-18-20808 (2018).

[16] L.R. Greenwood and R.K. Smither. SPECTER: Neutron damage calculations for materials irradiations. Technical report, Argonne National Lab., ANL-FPP/TM-197, (1985).

[17] U. Ficher, S.P. Simakov, A.Y. Konobeyev. Assessment of survived radiation defects by a modified version of NJOY, report jeff-doc-1223. Technical report, NEA Data Bank, 2007.

[18] J.A. Mascitti and M. Madariaga. Method for the calculation of DPA in the reactor pressure vessel of atucha ii. Science and Technology of Nuclear Installations, 2011.

[19] LR Greenwood. Specomp calculations of radiation damage in compounds. In Reactor Dosimetry: Methods, Applications, and Standardization. ASTM International, 1989.

Tables

TABLE I
BR2 SIC TEMPERATURE MONITORS

Monitor Identification	Dose (dpa)	Calculated Peak Irradiation Temperature (°C)
BR2 M1-Low-A	0.5	255°C
BR2 M2-Low-A	1.1	255°C
BR2 M1-High-A	0.5	310°C
BR2 M2-High-A	1.1	310°C
BR2 M1-Med-A	0.5	410°C
BR2 M2-Med-A	1.1	410°C

TABLE II
MATERIAL CHARACTERISTICS USED IN THERMAL COMPUTATIONS, Q = 10 W/G

Layer n	Material	External radius r, mm	Densityρ, g/cm^3	k, W/m K @ 225°C	q, W/m^3
1	He	4	1.61E-07	0.22	1.61E+00
2	Steel plug	6.45/6.325/5.925	7.85E+00	16.00	7.85E+07
3	He	6.7	1.61E-07	0.22	1.66E+00
4	Al	7.5	2.70E+00	250.00	2.70E+07

TABLE III
BR2 IRRADIATION ENVIRONMENT.

BR2 Cycle Information		Cycle #1 (1/31/17)	Cycle #2 (3/14/17)
Reactor Power [MW]		55	55
Cycle Length [days]		21	28
Fuel Burnup		42%	28%
φ_{th}	n/cm^2/s E<0.5eV	3.27E+14	3.03E+14
φ_{fast}	n/cm^2/s E>0.1MeV	3.93E+14	4.41E+14
Q	W/g (in Al)	10.1	10.3
Φ_{total}	n/cm^2	7.13E+20	1.07E+21

TABLE IV
EVALUATION RESULTS FOR THE BR2 MONITORS.

ID	Measured	Calculated	Deviation (measured - calculated)	% Deviation	Initial resistivity at 40°C
BR2 M1-Low-A	240°C	255°C	-15°C	-6%	20 (Ω-m)
BR2 M2-Low-A	Indeterminate	255°C	n/a	n/a	16 (Ω-m)
BR2 M1-High-A	320°C	310°C	10°C	3%	24 (Ω-m)
BR2 M2-High-A	330°C	310°C	20°C	6%	14 (Ω-m)
BR2 M1-Med-A	390°C	410°C	-20°C	-5%	43 (Ω-m)
BR2 M2-Med-A	380°C	410°C	-30°C	-8%	25 (Ω-m)

Figures

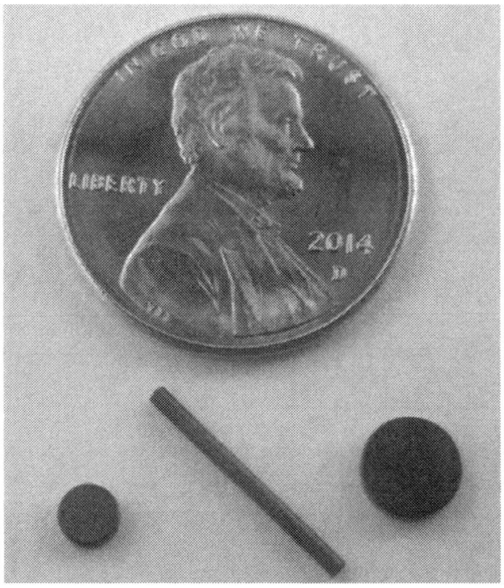

Fig. 1. SiC temperature monitors available for use in irradiation testing include small rods and discs. Only rods were used in this test. Monitors photographed with US cent for size perspective.

Fig. 2. SiC evaluation system components.

https://doi.org/10.1051/epjconf/202022504002

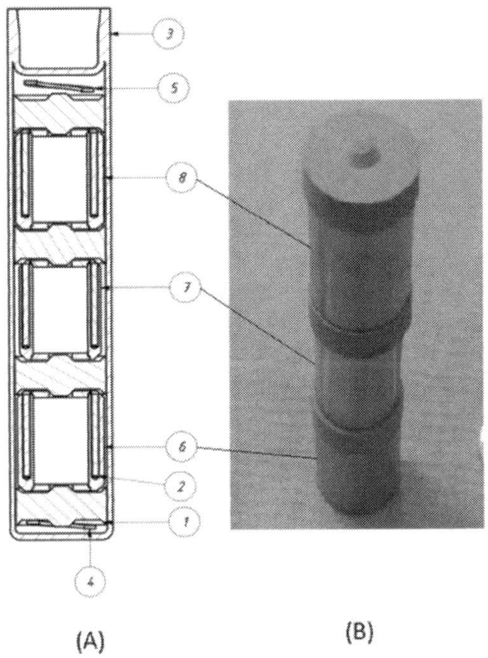

Fig. 3. (a) Cross-section of the aluminum irradiation capsule (3), stainless steel holders (6, 7 and 8) for SiC monitors (2). The holders are separated with ceramic discs (1) which are held in place with springs (4 and 5). (b) Photograph of the three holders and the three ceramic spacers stacked before irradiation.

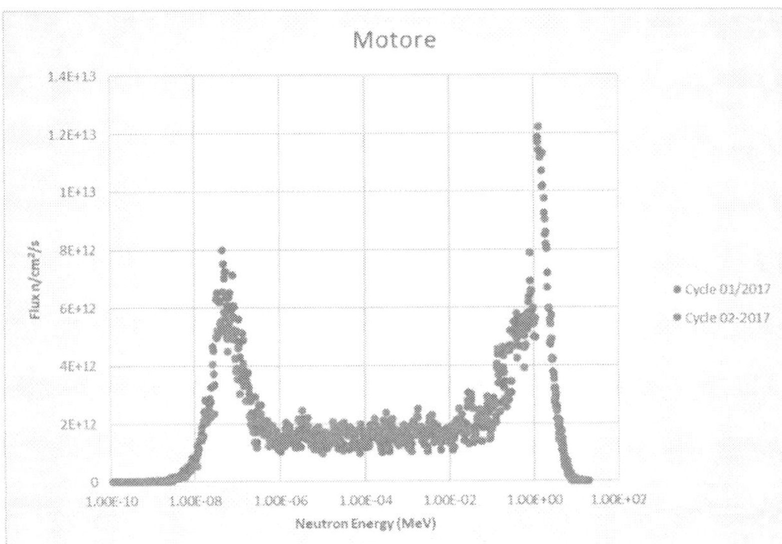

Fig. 4. Neutron flux spectrum on Motore.

https://doi.org/10.1051/epjconf/202022504002

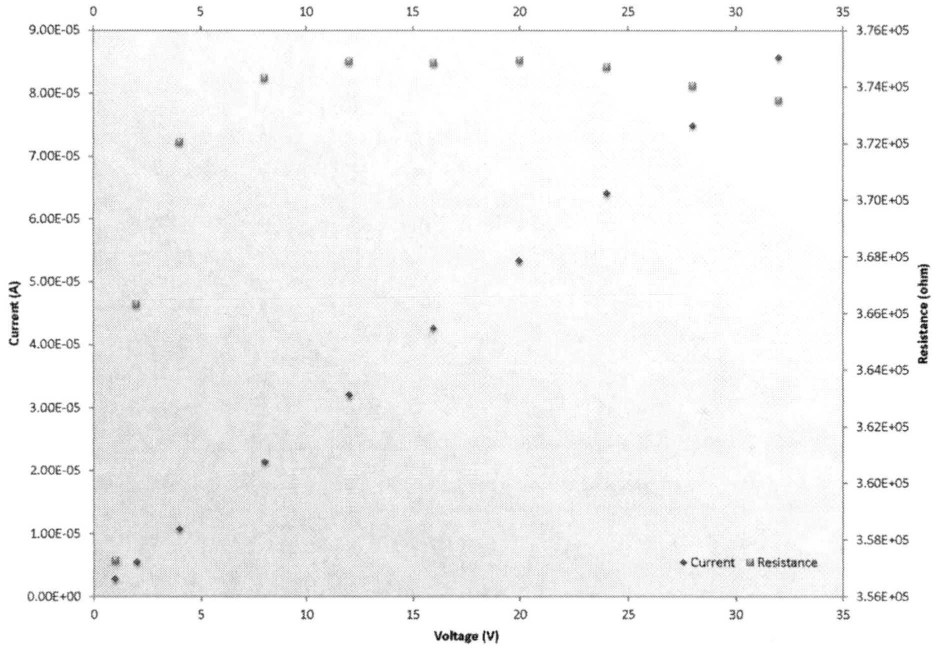

Fig. 5. Typical ohmic response as demonstrated by SiC temperature monitor BR2 M1-High-A.

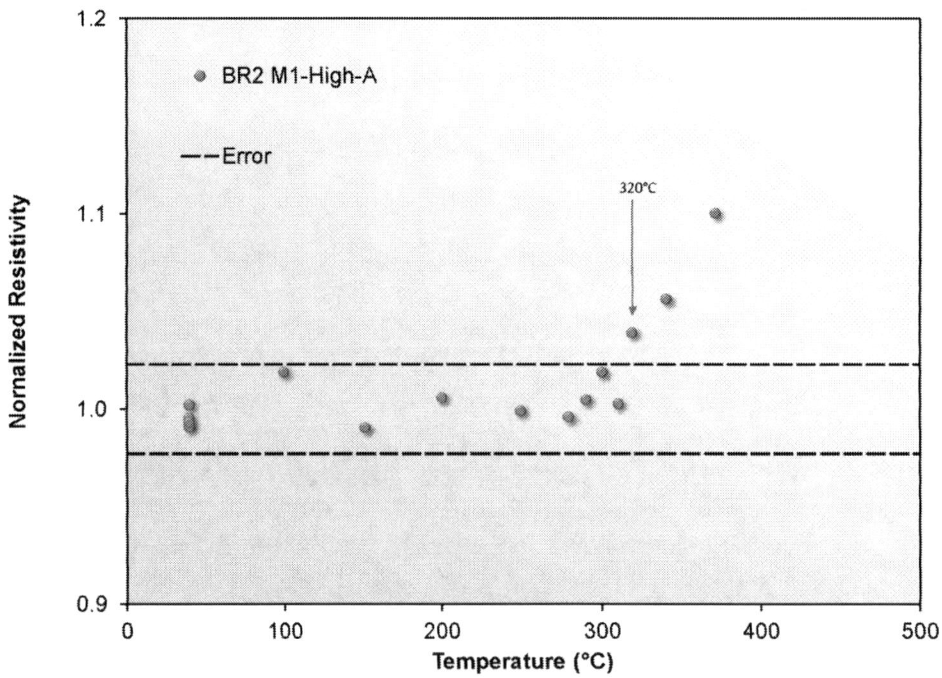

Fig. 6. Resistivity data for BR2 M1-High-A.

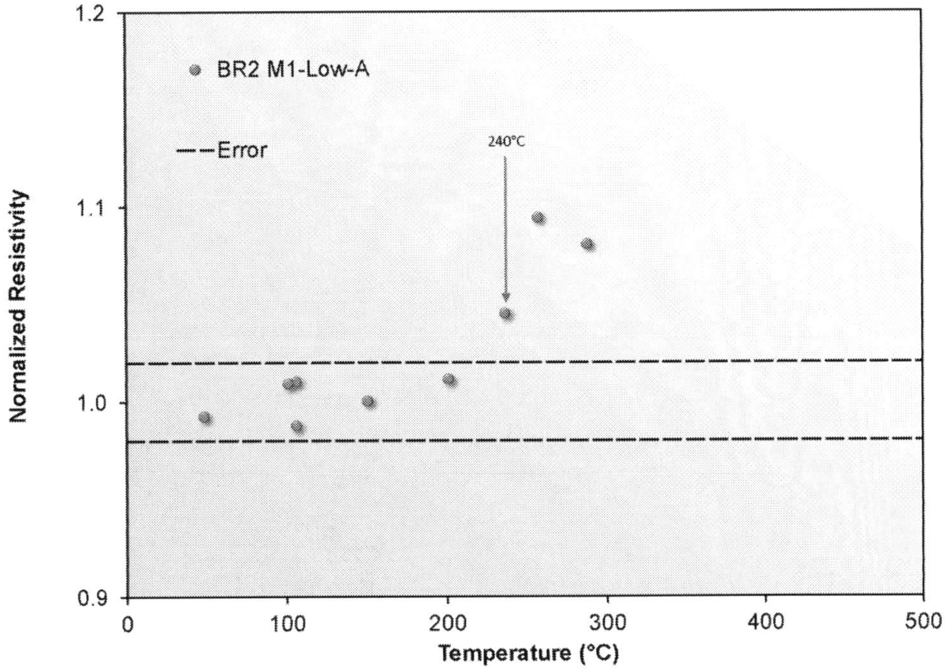

Fig. 7. Resistivity data for BR2 M1-Low-A.

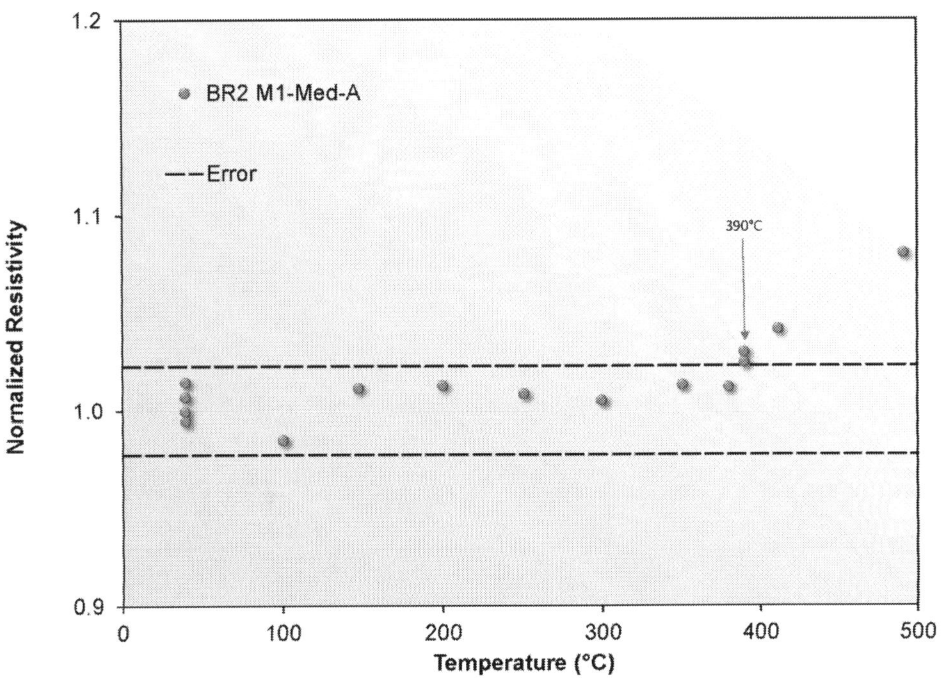

Fig. 8. Resistivity data for BR2 M1-Med-A.

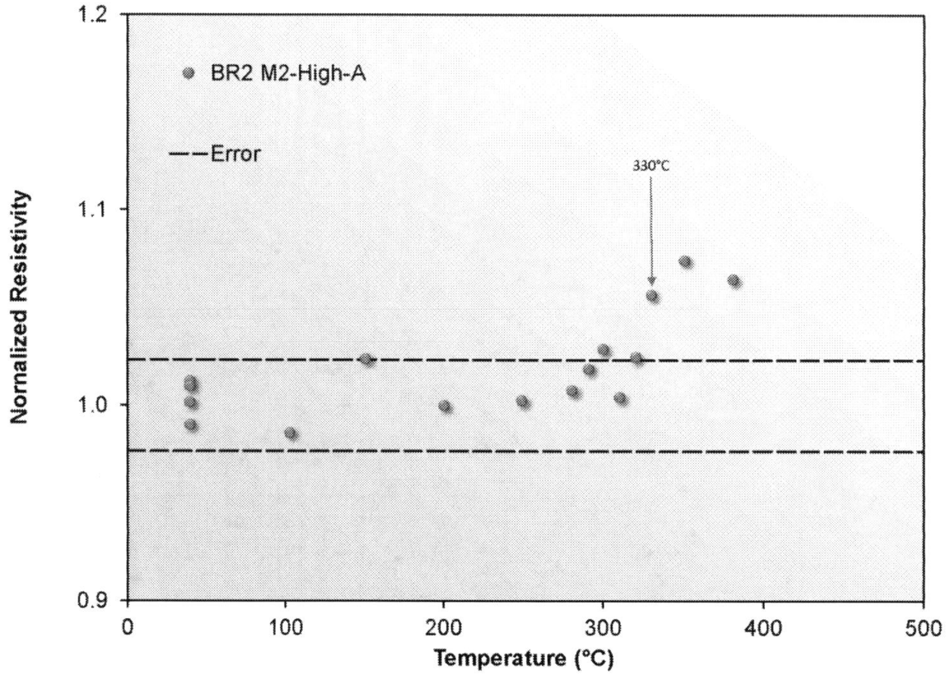

Fig. 9. Resistivity data for BR2 M2-High-A.

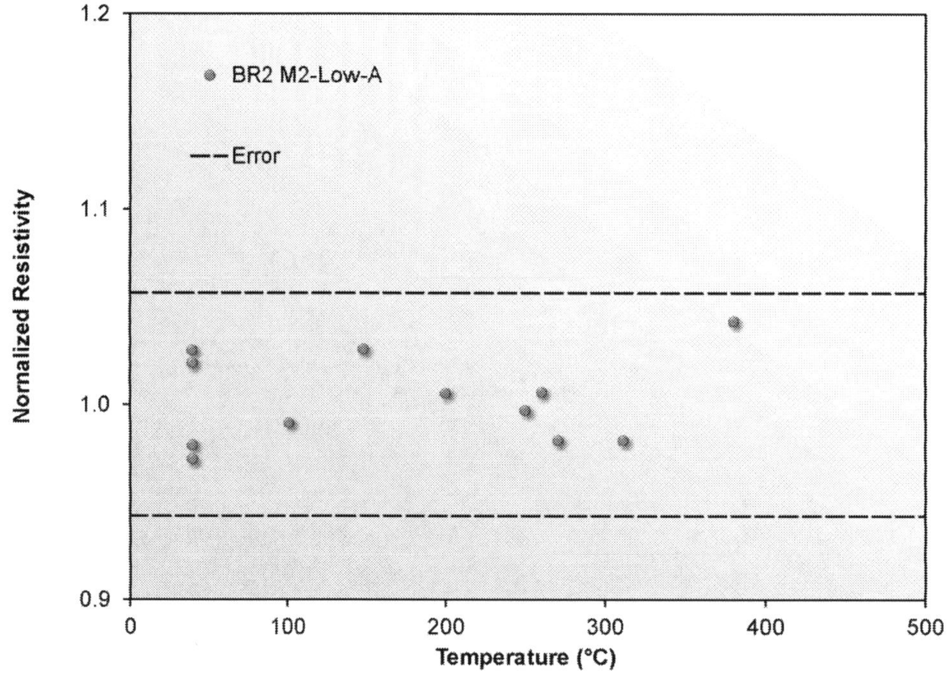

Fig. 10. Resistivity data for BR2 M2-Low-A.

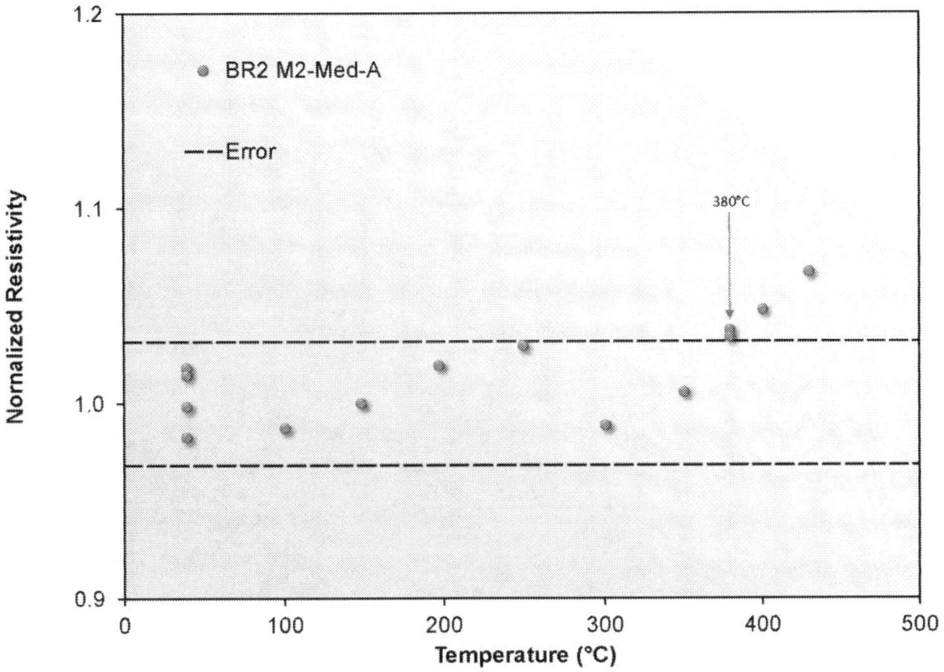

Fig. 11. Resistivity data for BR2 M2-Med-A.

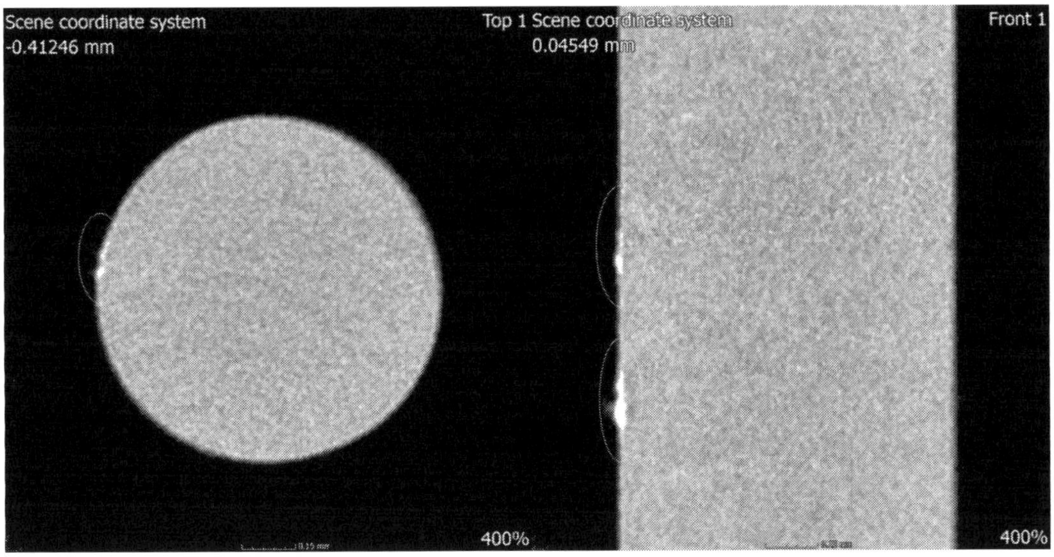

Fig. 12. Computed tomography scan BR2 M1-LOW-A. Areas of high density are highlighted.

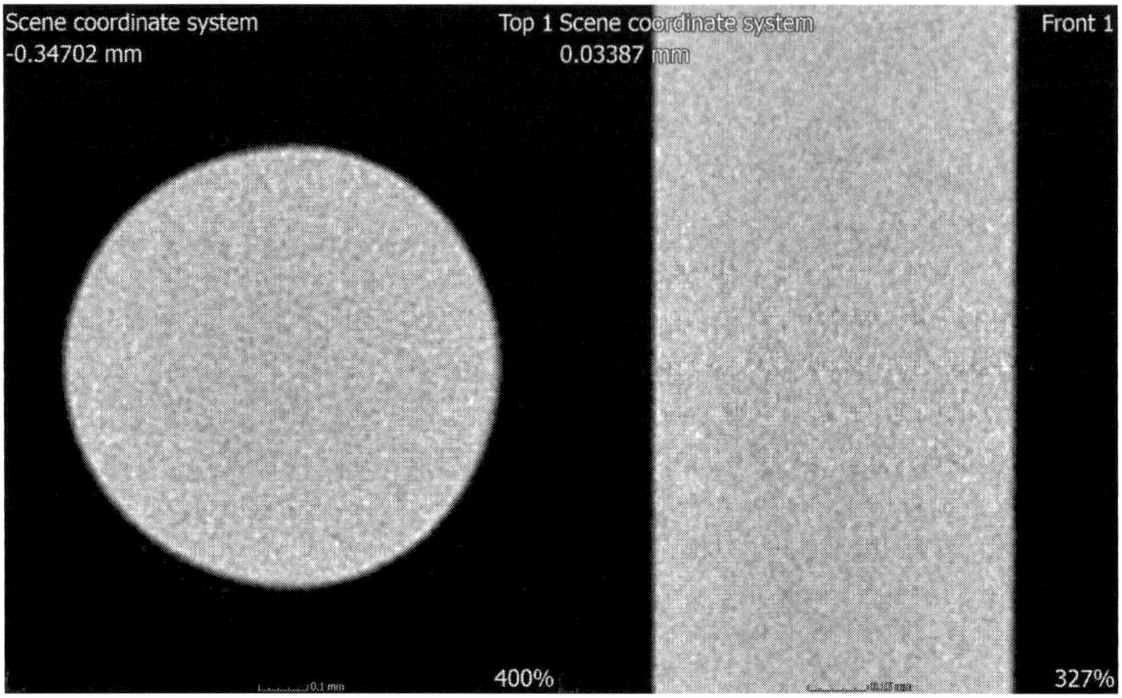

Fig. 13. Computed tomography scan BR2 M2-LOW-A.

https://doi.org/10.1051/epjconf/202022504003

Nuclear heating measurements by gamma and neutron thermometers

R. Van Nieuwenhove*, L. Vermeeren
SCK-CEN
* rvnieuwe@sckcen.be

Abstract— **A gamma thermometer suitable for very high gamma heating levels (up to 20 W/g) has been designed and modelled by means of detailed finite element calculations. Based on a sensitivity analysis, the predicted accuracy of this gamma thermometer is better than 5 %. A novel miniaturized gamma thermometer is proposed in which a single thermocouple is used as the gamma absorption element, allowing a reduction of the sensor diameter down to 3 mm.**
Monte Carlo calculations (by MCNP) have been performed to assess the relative contribution of neutrons to the nuclear heating in a gamma thermometer. Calculations have been performed for gamma thermometers with an inner body made of various materials, such as stainless steel, tungsten, molybdenum and rhodium. By using gamma thermometers made of different materials, it will be possible to deduce the nuclear heating rates in these materials and also to separate out the neutron and gamma heating contributions. The Monte Carlo calculations show that nuclear heating of rhodium is mainly due to neutrons, converting the rhodium gamma thermometer effectively in a neutron thermometer. The sensitivities of the gamma thermometers with W, Mo or Rh as heated materials have been modelled by finite element calculations. It is found that both the Mo and the Rh based sensor have a very linear response up to a nuclear heating of 20 W/g.

***Keywords*—Nuclear measurements, Gamma-ray detectors**

I. Introduction

Nuclear heating measurements in material test reactors are important for various reasons. For the design of experimental rigs, a good knowledge of the gamma heating is required in order to predict the correct irradiation temperature of samples. For power calibrations of test fuels, the knowledge of the contribution of gamma heating is crucial in order to attain the required precision. For nuclear fusion, it is important to know the nuclear heating of various materials such as for tungsten, copper, graphite, chromium, etc. Discrepancies between measured and calculated gamma heating levels (such as by MCNP) have regularly been observed in the past and are partly related to imprecise nuclear data. In order to contribute to a better understanding and predictability, modified gamma thermometers (GT) containing different materials have been designed and modelled.

II. Description of the SCK•CEN Gamma Thermometer

The basic design and principles of the gamma thermometer (GT) are described in [1,2] and have been routinely produced at the Halden Reactor Project since 1962. SCK•CEN designed a slightly different version, as shown in Figure 1.

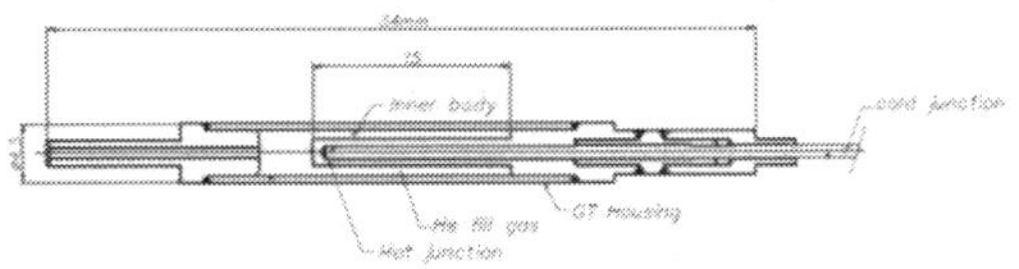

Fig. 1. Schematic drawing of the SCK•CEN gamma thermometer.

The inner body (made of stainless steel) is heated by gamma rays and the temperature difference between the tip of the inner body and the coolant is measured by a differential thermocouple (type K). The inner body (diameter 2.1 mm) is thermally insulated from the outer housing by a gas layer (0.7 mm). The GT is sufficiently strong to be applicable under Pressurized Water Reactor (PWR) conditions. In 1996, the design was modified in order to adapt it to the higher gamma heating (up to 15 W/g) of the BR2 reactor (see Figure 1). The main modifications were i) reduction of the length of the inner body from 30 mm to 15 mm and ii) replacement of the xenon gas by helium. Other minor differences are that the material of the housing and inner body are made of AISI 304 for the SCK•CEN GT (316L in the Halden GT) and that the material of the thermocouple sheath is made in AISI 316L (Inconel 600 in the Halden GT). The reduction of the length of the inner body and the replacement of the gas by He allows to limit the temperature increase of the inner body at the highest gamma heating and to maintain a fairly linear response.

III. Finite Element Modelling

The gamma thermometer was modelled by means of the 2D finite element program QuickField. Temperature dependent material properties were obtained from the database MPDB (JAHM Software). The gas properties were obtained from the NIST website (National Institute of Standards and Technology). Radiation losses from the inner body have an

R. Van Nieuwenhove and L. Vermeeren are with the Nuclear Research Centre SCK•CEN, B-2400 Mol, Belgium (e-mail: rvnieuwe@sckcen.be, lvermeer@sckcen.be).

 https://doi.org/10.1051/epjconf/202022504003

impact on the sensitivity of the GT and are taken into account. The emissivity of the polished inner body is a parameter with considerable uncertainty. Therefore, the emissivity value was adjusted such as to obtain correspondence between the calculated and the measured time constant of the GT. This resulted in an emissivity of 0.3.

The material properties of the thermocouple require also careful consideration because the diameter of the thermocouple (1 mm) is relatively large compared to the diameter of the inner body (2.1 mm). The thickness of the sheath of the thermocouple and the diameter of the two wires (chromel and alumel) are obtained from the manufacturer. Typically, the sheath has a thickness of 0.13 mm while the diameter of the wires is 0.18-0.2 mm. The thermocouple insulation consists of compacted Al_2O_3 powder. By weighing a piece of thermocouple, the density of the Al_2O_3 powder was determined to be 2.56 g/cm3 (considering the theoretical density of the metal), while the average density of the thermocouple was 5.885 g/cm^3. The thermal conductivity of the Al_2O_3 powder was taken to be 0.33 W/mK (based on previous experience at the Halden Project), independent of the temperature. In the finite element calculations, all the other thermal conductivities were described as temperature dependent. For the 2D finite element calculations, the thermocouple (with its 2 inner wires) was modelled as a uniform element with calculated averaged density and averaged thermal conductivity and specific heat.

The temperature of all the outer surfaces of the GT were set to be equal to the coolant temperature.

IV. RESULTS

Finite element calculations have been performed for a helium and an argon filled GT up to 20 W/g at an outer coolant temperature of 100 °C.

The results are shown in Figure 2. It can be seen that the helium filled GT is fairly linear over the whole range while the argon-filled GT exhibits some clear nonlinearity.

In the following, we define the sensitivity of the GT as the temperature increase (as measured by the differential thermocouple) divided by the gamma heating G (expressed in W/g), i.e. $S = \Delta T/G$.

The sensitivity depends on the thermal conductivities of the materials, which are temperature dependent. Hence, the sensitivity depends not only on the outer coolant temperature but also on the gamma heating itself (because a higher gamma heating will result in a higher temperature of the inner body). Because the thermal conductivities generally increase with temperature, the sensitivity of the GT decreases at higher coolant temperature. A calculation has been made for the He-filled GT at 300 °C (see Figure 2). In this case, the sensitivity is 30 % lower as compared to the 100 °C case. It can also be observed that the linearity at 300 °C is even better than the linearity at 100 °C. This is because the relative changes in thermal conductivities are smaller in this case.

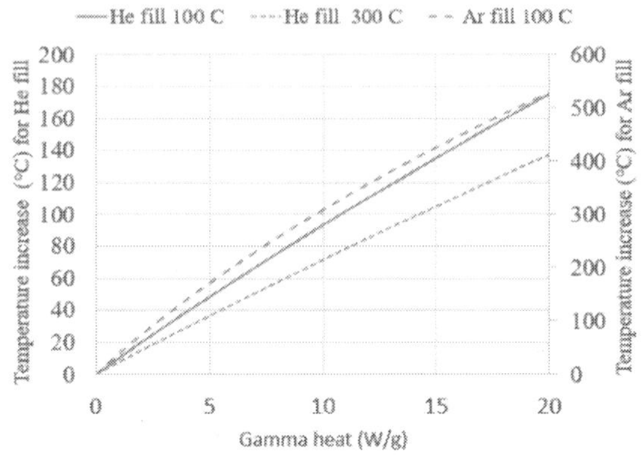

Fig. 2. Calculated temperature increase for a He-filled GT (at 100 °C and 300 °C) and for an Ar filled GT (at 100 °C).

For gamma thermometers filled with xenon or argon, the effect of radiation losses from the inner body is relatively important. However, for a GT filled with helium, the effect of radiation losses on the GT sensitivity is relatively small; 1 % at an environment temperature of 100 °C (at 10 W/g) and 2.6 % at an environment temperature of 300 °C (at 10 W/g).

Using the finite element model, it is also possible to deduce the time constants of the different types of GTs. First, the temperature distribution is calculated for the case in which the thermocouple only is heated (assuming one uses the thermocouple itself as electrical heater), while keeping the outer temperature constant. Thereafter, the heating is suddenly switched off and the time evolution of the tip of the thermocouple is modelled. After some time after switch-off, the decay becomes purely exponential and this part is then fitted by an exponential function to obtain the time constant. Some results are shown in Table III.

V. UNCERTAINTY ANALYSIS

Because of the simple geometry of the GT and the detailed finite element modelling, it is in principle possible to determine the GT response with a very good accuracy. However, because of uncertainties in the thermal conductivities of the materials and the gasses used, as well as due to the unavoidable mechanical tolerances, uncertainties are introduced in the calculated GT sensitivity.

Based on a comparison of various data found in the open literature, we will assume the following uncertainties on the thermal conductivities: 10 % on the thermal conductivity of steel (AISI 304 and AISI 316L) and 5 % on the thermal conductivity of He or Ar.

For the value of the emissivity (radiation from inner body), we assume an uncertainty of 20 %.

For the thermal conductivity of the thermocouple, the main uncertainty is due to the thermal conductivity of the compacted Al_2O_3 powder. In the model, we assume a thermal conductivity of only 0.33 W/mK but here we will allow for a possible increase of a factor 5.

Two different types of uncertainty are considered related to the mechanical tolerances: The first is related to the diameter of the inner body, the second is related to the inner diameter of the outer tube. In the analysis, we assume an uncertainty of 50 µm. Both uncertainties lead to a change in the gas gap width of 7.14 %, as the width of the gas gap is 0.7 mm.

Finally, we also need to consider the uncertainty of the temperature measurement itself. For thermocouples type K, with special tolerance, (ASTM E230) the uncertainty is given as ± 1.1 °C or 0.4 % (whichever is greater). For temperatures up to 275 °C, one has thus to assume an uncertainty given by ± 1.1 °C. The uncertainty on the sensitivity of the GT thus depends on the actual temperature increase (determined by the gamma heating). In what follows, we will calculate the total uncertainty for a He and an Ar filled GT at three different levels of gamma heating, namely at 5, 10 and 15 W/g and for a coolant temperature of 100 °C.

The total uncertainty is calculated as the root of the summed squared errors. The uncertainties for the He and Ar filled GTs are summarized respectively in Tables I and II:

TABLE I
UNCERTAINTIES FOR THE HE-FILLED GT

Variation	Resulting uncertainty in sensitivity (±%)
thermal conductivity of helium by 5 %	4.4
Thermal conductivity of steel by 10 %	1.0
50 µm change of radius of inner body	0.47
50 µm change of inner radius of outer tube	1.40
Thermal conductivity of Al_2O_3 by factor 5	0.14
20 % change in emissivity	0.18
Uncertainty temperature of 1.1 °C	2.3 / 1.18 / 0.81 (for 5, 10, 15 W/g)
Combined uncertainty	*5.28 / 4.89 /4.82* (for 5, 10, 15 W/g)

TABLE II
UNCERTAINTIES FOR THE AR-FILLED GT

Variation	Resulting uncertainty in sensitivity (±%)
thermal conductivity of argon by 5 %	1.08
Thermal conductivity of steel by 10 %	3.67
50 micron change of radius of inner body	0.21
50 µm change of inner radius of outer tube	1.14
Thermal conductivity of Al_2O_3 by factor 5	0.10
20 % change in emissivity	0.30
Uncertainty temperature of 1.1 °C	0.65 / 0.35 / 0.25 (for 5, 10, 15 W/g)
Combined uncertainty	4.06 / 4.03 / 4.02 (for 5, 10, 15 W/g)

VI. CALIBRATION METHODS

When subjecting a GT to transient heating conditions, it is in principle possible to infer its sensitivity. For instance, by preheating a GT to 100 °C (in boiling water) and then dipping it into ice-water (the so-called dip-test), the decay constant can be used to deduce the GT sensitivity by making use of an analytical model. Another transient heating method consists in heating the thermocouple by sending an ac current through it, stop the heating and record the decaying temperature of the inner body (while keeping the outside of the GT at a fixed temperature).

The relation between the time constant (τ) and the sensitivity (S) is given by:

$$S = \frac{\left|1 - \frac{1}{cosh(\sqrt{A}\ L)}\right|}{A \cdot k_b} \cdot \rho \tag{1}$$

in which

$$A = \frac{C_V \cdot \rho}{\tau \cdot k_b} - \left[\frac{\pi}{2 \cdot L}\right]^2 \tag{2}$$

where C_v , ρ, k_b and L are respectively the specific heat, the density, the thermal conductivity and the length of the inner body. In order to arrive at this analytical relation, one has to assume that the thermal material parameters do not vary over the GT, while in general some variations exist because of the non-uniform temperature. In addition, the obtained relation will only be valid at the average temperature during the test.

Since we have however a detailed finite element model of the GT, one can use this model to calculate the time constant and compare it to the experimental value. If both are in good agreement, one can then safely assume that the calculated sensitivity of the GT is correct. In addition, one can then also assume that the calculated sensitivities for other environment temperatures are correct.

VII. MINIATURE GAMMA THERMOMETER

The GT can be miniaturized even further by omitting completely the inner body and by using only the thermocouple as the gamma absorption medium (see Fig. 3).

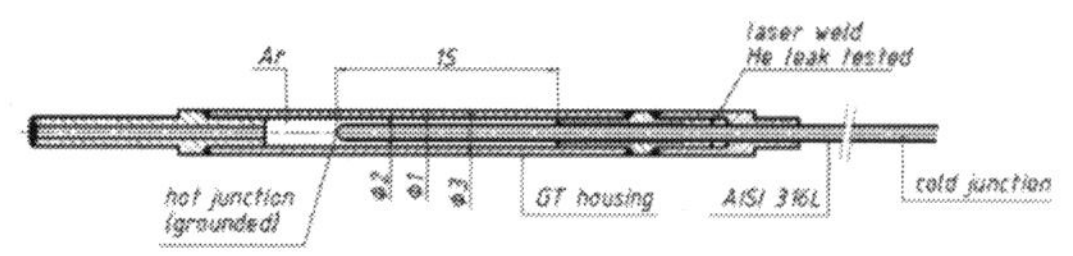

Fig. 3. Miniaturized GT, in which the inner body is reduced to only the thermocouple.

This has the advantages that the outer diameter of the GT can be reduced down to 3 mm and that the GT is easier to fabricate. Such a miniaturized GT, filled with argon has also been modelled by finite element calculations. The resulting sensitivity is about 14.4 °C/(W/g).The calculated time constant is 9.09 s.

A disadvantage of this design is that the uncertainties related to the material properties of the thermocouple (TC) have now a larger impact on the calculated sensitivities. One could remedy the situation by performing the dip-test or the electrical heating transient method, described earlier. For this type of GT one could also use a more direct calibration method: 1) Heat the TC by sending an ac current through it and 2) measure the temperature just after switching off the current. By relating the locally deposited power to the temperature increase, one obtains the sensitivity.

In case one sends a current of 0.5 A through the TC, one obtains an equivalent gamma heat of 1.76 W/g, as compared to only 0.31 W/g for the standard GT. A further improvement can be obtained by using a grounded (differential TC). In this case one could induce a higher current through each TC wire separately towards the sheath. When using a current of 1 A (through both wires), one obtains an equivalent gamma heating of 7.04 W/g. Thus in this case, one obtains a calibration point at a sufficiently high equivalent gamma heat (for the range 0 – 20 W/g). Another advantage of using the grounded TC is that the heating is applied up to the very end of the TC cable. In the case of an ungrounded TC, the hot junction is still 1 to 2 mm away from the end of the TC cable and then this part is not heated and the heat distribution in this case does not correspond to the case of gamma heating which occurs over the full length of the TC and inner body. Also, for the ungrounded TC, an additional fast decay of the temperature is observed just after switching off the current related to the thermal insulation of the TC wires by the Al_2O_3 powder within the TC and this makes the direct electric calibration problematic in this case.

VIII. GAMMA THERMOMETER WITH DIFFERENT ABSORPTION MATERIALS

In order to determine the gamma heating in other materials, the inner body can be made for instance in tungsten, molybdenum or rhodium.

The responses of GTs with an inner body (diameter 2 mm) made of tungsten, molybdenum and rhodium are shown in Fig. 4. It can be seen that the response of the tungsten GT is fairly linear over the range zero to 20 W/g. In fact, the curve is bending slightly upwards. This is a consequence of the fact that the thermal conductivity of tungsten is decreasing with increasing temperature. The responses of the Mo and Rh GTs are almost perfectly linear.

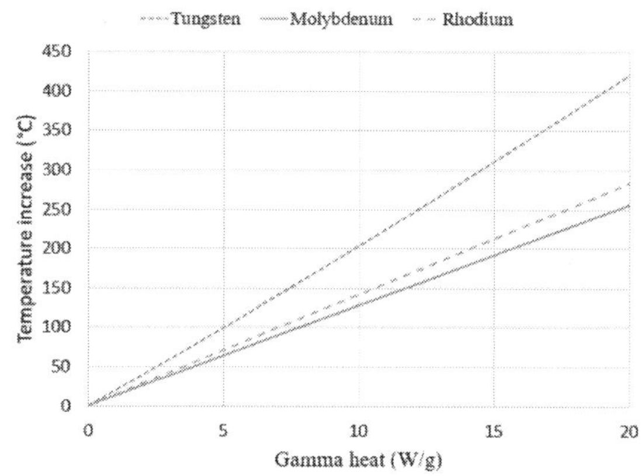

Fig. 4. Calculated temperature increase for an argon filled GT with an inner body made of tungsten, molybdenum and rhodium, at an outer temperature of 100 °C.

The calculated sensitivities and time constants are given in Table III.

TABLE III
OVERVIEW OF SENSITIVITIES (AT 100 °C) AND TIME CONSTANTS FOR THE ARGON FILLED GTS FOR INNER BODIES MADE OF W, MO AND RH.

Material	Fill gas	Sensitivity (°C/(W/g))	Time constant (s)
Tungsten	Ar	20.8	2.84
Molybdenum	Ar	12.8	3.30
Rhodium	Ar	14.2	3.47
AISI 304	Ar	35.8*	18.7
AISI 304	He	10.0*	5.61

* At 1 W/g (nonlinear)

The time constant of the GT with AISI 304 as inner body material and Ar fill gas was determined experimentally by the dip test method (from 100 °C to 0°C) and was found to be 18.3 seconds, in good agreement with the theoretical value (18.7 s).

IX. NEUTRON INDUCED HEATING

When using a GT for gamma heating monitoring in a reactor core, neutrons will also lead to some heating. So in order to interpret the GT signal correctly, the relative heating contribution by neutrons should be assessed. We performed some preliminary MCNP calculations on the gamma and neutron heating in typical reactor conditions, for various materials of the inner body of the GT: graphite, Al, silicon nitride, AISI 304, Mo, Rh, Hf, W, Pb and Bi.

The GT geometry was modeled using a set of infinitely long, concentric cylinders:

- r < 0.55mm: thermocouple (average thermocouple composition, mainly consisting of nickel and alumina, average density 5.885 g/cm)
- 0.55mm < r < 1.05 mm: inner body (various materials, with appropriate densities)
- 1.05 mm < r < 1.75 mm: gas volume; 10 bar nitrogen was taken as a gas in order to assess the maximum neutron heating in the gas (density 0.0125 g/cm)
- 1.75 mm < r < 2.25 mm: AISI304 sheath (7.96 g/cm)

A neutron (or gamma) source was defined on the outer surface of the sheath with a cosine direction distribution simulating an isotropic flux. Typical spectra for reflector channels in the BR2 research reactor were used. The total gamma flux was taken as $3.67 \cdot 10^{14}$ γ/(cm s); the neutron flux calculations were performed for two energy groups: E < 5eV ($1.19 \cdot 10^{14}$ n/(cm s)) and E > 5eV ($0.56 \cdot 10^{14}$ n/(cm s)).

For the pure gamma heating, F6:P tallies were used to get the heating per source gamma in each cell. For the neutron induced heating, coupled neutron-photon calculations were performed and the direct neutron heating rates were obtained from F6:N tallies, while F6:P yielded the heating rates via capture gammas. In the rhodium case, the neutron heating is strongly influenced by the energy deposition by betas emitted upon decay of ^{104}Rh, formed after neutron capture by ^{103}Rh. In this case a separate calculation approach was needed: calculation of the capture rates in rhodium, followed by an electron transport calculation to obtain the deposited beta energy in each cell. For the other materials, this contribution is much smaller and was not investigated in detail.

The results for the heating rates in the thermocouple + the inner body are summarized in Table IV (lower-Z materials) and Table V (higher-Z materials). As expected, the pure gamma heating rates expressed in W/g increase monotonically with the atomic number Z. For most materials, the relative neutron heating contribution is rather small (3-10%). In SiN, (n,p) reactions on ^{14}N lead to a non-negligible neutron heating contribution. For Hf, the neutron induced heating contribution (via capture gammas) is significant. For Rh, the neutron induced heating (via decay betas) is even by far the dominant contribution. So a detector with Rh as inner body material could serve as a (thermal) neutron detector.

TABLE IV
CALCULATED HEATING RATES BY GAMMAS AND NEUTRONS IN THERMOCOUPLES + INNER PARTS FOR VARIOUS LIGHT MATERIALS. THE HEATING RATES ARE EXPRESSED IN W/CM OR IN W/G AS INDICATED. RELATIVE HEATING CONTRIBUTION DUE TO NEUTRONS IS SHOWN IN THE LAST LINE.

	C	Al	SiN	AISI304	Mo
Z	6	13	7/14	26/24/28	42
density (g/cm)	2.23	2.7	3.17	7.96	10.2
G (W/cm)					
pure γ	0.170	0.189	0.207	0.435	0.683
pure n (E<5eV)	0.000	0.000	0.023	0.000	0.000
pure n (E>5eV)	0.000	0.000	0.000	0.000	0.001
n_(E<5eV) → γ	0.008	0.010	0.010	0.035	0.040
n_(E>5eV) → γ	0.007	0.008	0.009	0.016	0.003
G_total (W/cm)	0.186	0.207	0.249	0.486	0.727
G_total (W/g)	1.66	1.68	1.84	1.90	2.33
n-total / total (%)	8.4%	8.6%	16.8%	10.5%	6.1%

TABLE V
CALCULATED HEATING RATES BY GAMMAS AND NEUTRONS IN THERMOCOUPLES + INNER PARTS FOR VARIOUS HEAVY MATERIALS. THE HEATING RATES ARE EXPRESSED IN W/CM OR IN W/G AS INDICATED. RELATIVE HEATING CONTRIBUTION DUE TO NEUTRONS IS SHOWN IN THE LAST LINE.

	Rh	Hf	W	Pb	Bi
Z	45	72	74	82	83
density (g/cm)	12.4	13.31	19.35	11.34	9.8
G (W/cm)					
pure γ	0.823	1.181	1.566	1.163	1.089
pure n (E<5eV)	0.000	0.000	0.000	0.000	0.000
pure n (E>5eV)	0.002	0.002	0.000	0.001	0.001
n_(E<5eV) → γ	0.611	0.477	0.125	0.037	0.033
n_(E>5eV) → γ	0.006	0.030	0.034	0.001	0.001
n_(E<5eV) → β	4.75				
G_total (W/cm)	6.187	1.689	1.725	1.202	1.124
G_total (W/g)	16.85	4.33	3.19	3.53	3.72
n-total / total (%)	86.7%	30.1%	9.2%	3.2%	3.1%

X. CONCLUSION

The sensitivity of the SCK•CEN type gamma thermometer has been calculated over the gamma heating range from zero to 20 W/g. Although the argon filled GT has a higher sensitivity than the helium filled GT, it's response is also more non-linear. The linearity of the helium filled GT was found to be better at 300 °C than at 100 °C.

Based on detailed finite element calculations, an uncertainty analysis has been performed for the SCK•CEN type gamma thermometer, taking into account material and geometrical uncertainties, as well as the uncertainty related to the temperature measurement itself. Although the relative uncertainty depends on the actual gamma heating, one can say that the overall uncertainty remains below 5 % (with the lowest uncertainty obtained at the highest gamma heating).

It is also possible to experimentally determine the time constant of the GT and fine tune the finite element model to obtain agreement between the experimental and the theoretical values. Further, it is also possible to obtain a more direct calibration by establishing a relation between the temperature of the inner

body and the electrical power delivered to the thermocouple (by sending a current through the thermocouple).
A novel type of GT has been proposed in which the inner heated body is reduced to the thermocouple itself. This GT has several advantages, such as reduced inner diameter (down to 3 mm) and even better possibilities for calibration by the electrical heating method.
In order to determine the gamma heating in other materials than stainless steel, different GTs have been considered and modelled in which the inner body is made of W, Mo or Rh. Because the thermal conductivity of these materials decrease with increasing temperature, a compensation for the increasing thermal conductivity of the thermocouple with temperature is obtained resulting in a more linear response than the standard GT. For Mo and Rh, a very linear response was obtained up to a gamma heating of 20 W/g. Because of the high thermal conductivity of W, Mo and Rh, the time constants of these argon filled GTs were found to be rather small, namely below 4 seconds.
MCNP calculations were performed in order to assess the gamma and neutron heating contributions separately. The pure gamma heating rates expressed in W/g increase monotonically with the atomic number Z. For most materials, the relative neutron heating contribution is rather small (3-10%). In SiN and Hf, (n,p) and (n,γ) reactions lead to higher neutron induced heating contributions. For Rh, the neutron induced heating (via decay betas) is even by far the dominant contribution. So a detector with Rh as inner body material could serve as a (thermal) neutron detector.

REFERENCES

1. O. Aarestad (1992, October), In-core Instrumentation for LWRs, Presented at IAEA meeting on "In-core Instrumentation and In-situ Measurements in Connection with Fuel Behaviour", held in Petten, The Netherlands. Available: https://inis.iaea.org/collection/NCLCollectionStore/_Public/28/018/28018794.pdf

2. S. Solstad and R. Van Nieuwenhove (2011) *Nuclear Technology*, vol. 173, no 1, pp. 78-85, Jan. 2011.

Confirmation of Nuclear Heating Rate for Installation of Cold Neutron Source at HANARO

Myong-Seop Kim* and Byung-Gun Park
Korea Atomic Energy Research Institute
* mskim@kaeri.re.kr

***Abstract*—In order to determine the capacity of the cold neutron source refrigerator of HANARO, the nuclear heating rate at CN vertical hole is measured by using the heat-flow calorimetric method and confirmed by the calculation. The heating rate measurement device of HANARO was composed of a calorimeter sensor, an air containing aluminum sleeve for fitting the sensor to the CN hole, aluminum weight and a lead wire assembly. The calorimeter sensor consists of a cylindrical Al sample and container, two thermocouples and the electric heater for the calibration of the calorimeter. The sample is separated by an air gap from the Al container surrounded by an air containing Al sleeve. After installation of the calorimeter at a measurement position of HANARO, the heat transfer inside the calorimeter was simulated by the electric heating for the sample. The nuclear heating rates at the CN hole were determined at three reactor powers of 1, 4 and 8 MW by using the calibration curve and the temperature measurements at each reactor power. The measured nuclear heating rate per unit mass of Al sample at 8 MW reactor power is 0.143 W/g and it is equivalent to the 0.494 W/g at 30 MW. The nuclear heating rate was calculated by using the MCNP code. The calculation model for the whole facility including the reactor core and the reflector tank were established. In the calculation procedure, the heat generations by various radiations were evaluated with considering the prompt, delayed and activation effects. The measured heating rate was reasonably well supported by the calculation using the cold neutron facility design code. It will be very useful for the moderator cell of cold neutron source of HANARO.**

***Keywords*—calculation of heat generation in research reactor, calorimeter, cold neutron source refrigerator, HANARO, nuclear heating rate**

I. Introduction

THE heat removal capacity of the moderator cell is essential information for a cold neutron source (CNS) design in a research reactor. In order to determine the capacity of the refrigerator, the nuclear heating rate at the CNS position should be confirmed. The nuclear heating in a research reactor arises from the nuclear fissions and the interactions with gamma-rays, fast and thermal neutrons. The heating rate can be determined by several methods such as the calorimetry, the ionization chambers and the chemical dosimeters [1].

Calorimetric dosimeters have various advantages for high-dose applications. The operation of the calorimeter is possible by measuring the total amount of energy that is deposited as heat in an isolated mass. The measurement of the temperature rise in a calorimetric dosimeter provides a direct measurement of the full energy imparted to matter by radiation. The thermocouples and the thermistors are sufficiently sensitive and small enough to measure the temperature change with sufficient accuracy and precision [2].

In this work, we measured the nuclear heating rate at the CN hole of HANARO, a 30 MW research reactor, by using the calorimetric method in order to install the cold neutron source. Also, the nuclear heating rate was calculated and compared with the measurements. In the calculation procedure, the heat generation rate caused by the various heat sources such as fission products, neutrons and gamma-rays were considered. Furthermore, the effect of the decay radiations from the activated structural materials in the reactor core and reflector was estimated. It will be very useful for the moderator cell design of the cold neutron source of HANARO.

II. Concept of Calorimeter Operation

The adiabatic calorimeter must be loaded during the reactor power operation with maintaining the vacuum [3,4], and only instantaneous measurements at a low reactor power are possible. Therefore, the concept of a heat flow calorimeter was introduced to measure the heating rate of the CN hole of HANARO instead of the adiabatic calorimeter.

Assuming a good thermal conductivity of the sample and the small heat transfer coefficient, the power integrated over the volume of the sample at a position of the research reactor in equilibrium condition of steady state,

$$P = hs(T_s - T_e) \quad (1)$$

where, T_s, T_e : sample and container temperatures,
h : heat transfer coefficient,
s : sample surface area,
$1/hs$: thermal resistance.

From the above equation, if the surface temperature is uniform, and if the thermal resistance is known, the power can be obtained by measuring the equilibrium temperature

M. S. Kim is with the Korea Atomic Energy Research Institute, Daedeok-daero 989-111, Yuseong, Daejeon, 34057, Korea (e-mail: mskim@kaeri.re.kr).

B. G. Park is with the Korea Atomic Energy Research Institute, Daedeok-daero 989-111, Yuseong, Daejeon, 34057, Korea (e-mail: bgpark@kaeri.re.kr).

difference [3]. It is necessary to know the thermal resistance in order to obtain the heating rate, and it may be determined by the calculation. But, it is preferable to simulate the dose rate by the thermal heating produced in the sample by the Joule effect [5].

III. Experimental Setup

The heating rate measurement facility at the CN hole of HANARO was designed and composed of a calorimeter sensor, an air containing aluminum sleeve for fitting the sensor to the CN hole, aluminum weight and a lead wire assembly as shown in Fig. 1.

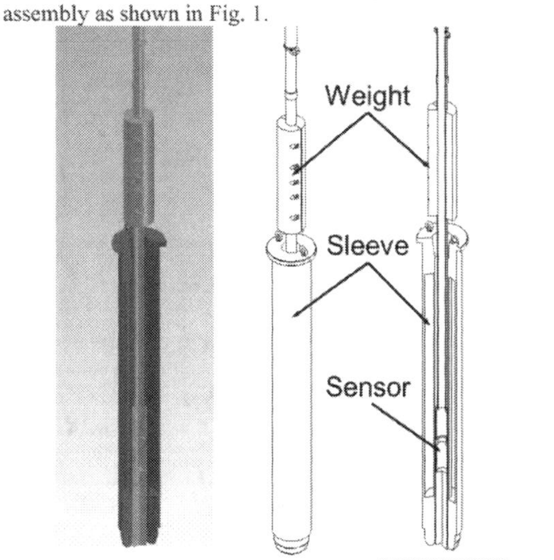

Fig. 1. The heating rate measurement facility at the CN hole of HANARO.

Fig. 2 shows the detailed layout of the sensor part of the calorimeter. The calorimeter sensor consists of a cylindrical aluminum (Al) sample, Al container, two thermocouples and the electric heater for a calibration. The sample is separated by an air gap from the Al container surrounded by an air containing Al sleeve. The longitudinal center of Al sample is equal to the center of reactor core. The calibration heater allows one to determine the thermal resistivity in eq. (1), and this calibration gives a simulation of the heat transfer of the calorimeter over the temperature range and under the irradiation condition. Table. 1 shows the detailed specification of the sample.

TABLE I
DETAILED SPECIFICATION OF THE SENSOR.

Parameters	Figures
Outer diameter of the sensor	56 mm
Al sample	
Diameter	19.3 mm
Length	49.9 mm
Weight	39.345 g
Material	Al6061

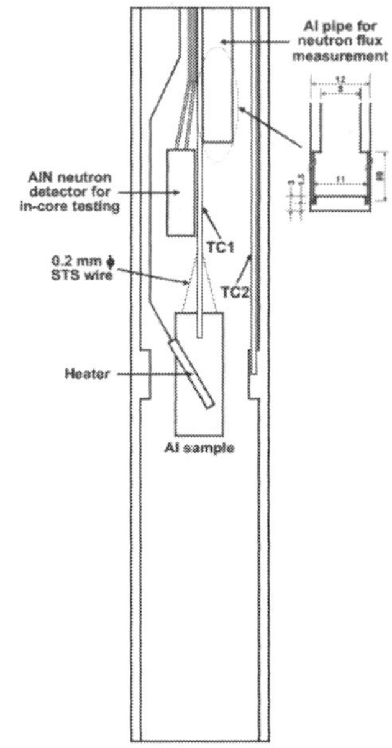

Fig. 2. Detailed layout of the sensor part of calorimeter.

IV. Calibration Experiment

Fabricated facility for the heating rate measurement was installed at the CN hole of HANARO, and the calibration experiments were performed. The temperatures of the sample and container were measured with a variation of the electric power supplied to the heater loaded in the sample. The maximum temperature of the aluminum sample was 385 °C at the electric power of 20 W. The temperature change of the container was several °C in the whole power range.

Fig. 3 shows the whole trends of the temperature changes of the Al sample and the container in the calibration experiment for the high electric power range.

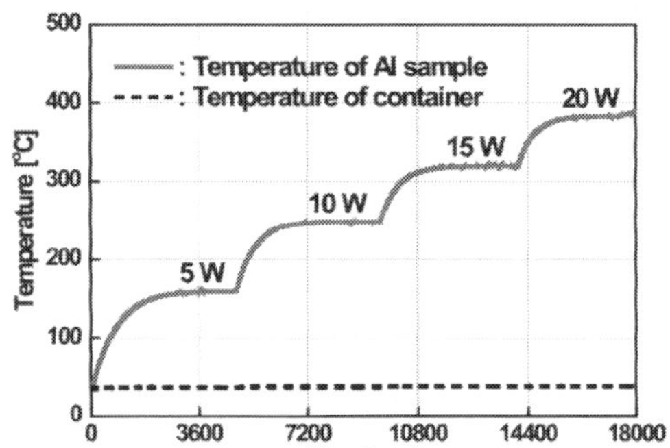

Fig. 3. Whole trends of the temperature changes of the Al sample and the container in the calibration experiment for the high electric power range.

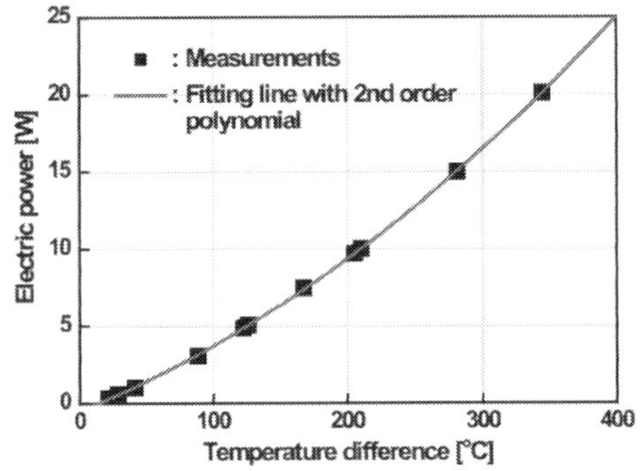

Fig. 4. Result of the calibration experiment.

From this calibration experiment, the relationship between the power supplied to the heater and the temperature difference was determined, and Fig. 4 represents the result of the calibration experiment. In the figure, the solid line is fitting to the measurements with 2nd order polynomial. The calibration curve is not linear because the convective and radiative heat transfers are increased in high sample temperature range [6].

V. Calculation

The nuclear heating rate at CN vertical hole of HANARO was calculated by using the MCNP6.1 code. The calculation model for the whole facility including the reactor core and the reflector tank were established for the reactor operation of 32-2nd beginning-of-cycle of HANARO.

The material composition and source distribution of depleted-core was determined using HANAFMS, HANARO core management program. HANAFMS is based on the WIMS for the lattice calculation and VENTURE for the core calculation. Fig. 5 shows the calculation model for the 32-2nd period of the HANARO operation.

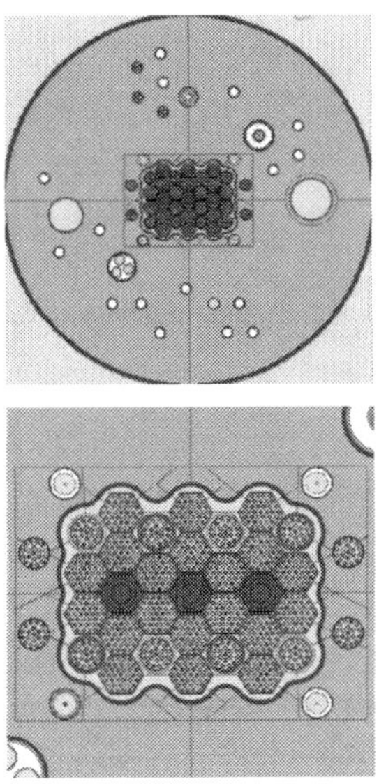

Fig. 5. The reactor model for the calculation of the nuclear heating rate at the CN hole.

The neutron and gamma-ray transport were simulated in almost all materials of the MCNP model including nuclear fuel, heavy water and heavy water tank, reactor flow tube, inner shell, heating rate measurement facility, etc. And each contribution to the nuclear heating of Al sample inside the sensor was evaluated, respectively. The control rod position in the calculation was set on the same as experiment.

The radiation contributions to the nuclear heating rate of the sample through the materials of the nuclear fuel and the structure of the experimental setup are as follow.

- Fission fragments
- Fission neutrons and gamma-rays
- Capture gamma-rays by neutron absorptions
- Beta and gamma-rays from decay of fission fragments
- Beta and gamma-rays from activations of structural materials
- Beta and gamma-rays from activation of aluminum sample

The ACT1 nuclear data library was used for calculation of delayed gamma-rays from $Al^{27}(n,\gamma)Al^{28}$ reaction in aluminum sample. Contribution of delayed gamma-rays from fission fragment in the fuel was deduced from the delayed gamma-ray flux value calculated at the CN hole.

VI. Results

The nuclear heating rates at the CN hole were determined at three reactor powers of 1, 4 and 8 MW by using the calibration curve and the temperature measurements at each reactor power. The measured nuclear heating rate per unit mass of Al sample at 8 MW reactor power is 0.143 W/g as shown in the TABLE 2 and Fig. 6.

TABLE 2
Measurement Data At The Experiment.

Reactor power [MW]	Container temperature [°C]	Sample temperature [°C]	Temperature difference [°C]	Nuclear heating rate at sample [W]	Heating rate per sample unit mass [W/g]
1.18	31.167	63.284	32.118	0.697	0.018±0.00068
4.20	37.051	122.592	85.540	3.000	0.076±0.0023
8.48	40.158	177.220	137.062	5.618	0.143±0.0041
29.30 (30 MW_{th})				19.411	0.494

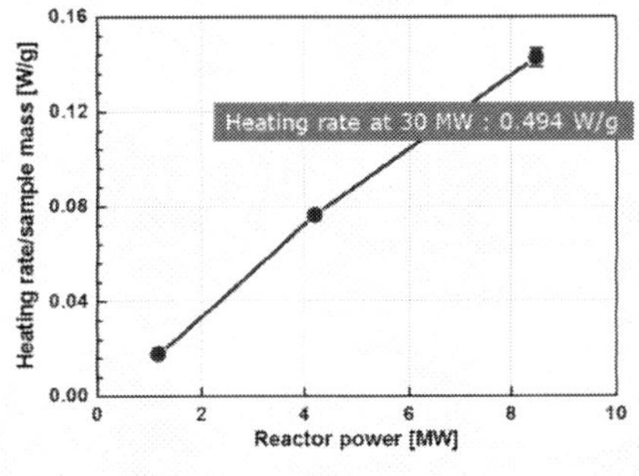

Fig. 6. Nuclear heating rate measurement at the three different reactor power.

TABLE 3
CALCULATION RESULTS OF THE NUCLEAR HEATING RATE AT HANARO.

Contributions			Calculated heating rate per sample unit mass [W/g]
Reactor core	Neutrons		0.006
	Gamma-rays	Prompt, heavy water	0.033
		Prompt, Al sample	0.059
		Prompt, etc.*	0.215
		Delayed, fission fragment	0.050
		Sub-total	0.357
Al sample activation	Beta particles		0.132
	Delayed gamma-rays		0.013
Total			0.508

Table 3 shows the calculation results of nuclear heating rate equivalent to the nominal reactor power of 30 MW. As shown in the Table, it is confirmed that the contributions of the beta particle from aluminum sample activation is pretty big. The contribution of decay gamma-rays from activation of other structure materials such as coolant flow tube is negligible. The calculated heating rate is pretty close to the measured value.

VII. CONCLUSION

The nuclear heating rate at CN vertical hole of HANARO was confirmed by measurement and calculation. The simple heat-flow calorimetric method was applied to the calorimeter design. The calculation model for the whole facility including the reactor core and the reflector tank were developed. The calculation methodology has been established to take into account various factors to contribute the nuclear heating. The measured heating rate was reasonably well supported by the calculation. It will be very useful for the cold neutron source installation of a research reactor..

REFERENCES

[1] A. W. Boyd, "Dosimetry Methods for Structural Materials: Nuclear Heating", AECL-5788, Reprinted from Proceedings of the First ASTM-EURATOM Symposium on Reactor Dosimetry, Petten, Holland, Sept. 22-26, 1975.
[2] F. H. Attix, "Introduction to Radiological Physics and Radiation Dosimetry", New York, John Wiley & Sons, 1986, p.427.
[3] International Atomic Energy Agency, "Determination of Absorbed Dose in Reactors", Technical Reports Series No.127, Vienna, 1971, p.123.
[4] T. Kawai, private communications.
[5] R. Dehn. "Measurement of Gamma Heating in D.F.R.", Nucl. Instr. and Meth., Vol.36, p.141, 1965.
[6] M. J. Todd, J. Bishop and D. K. Cartwright, "A Calorimeter for Graphite Dose Rate in Advanced Gas-cooled Reactors", J. Sci. Instrum, Vol.42, p.576, 1965.

Neutronical aspects of a decay heat measurement: the PRESTO experiment

Francesco Muratori
Cadarache, DER, SPRC, LE2C
CEA, DEN
F-13108 Saint-Paul-lez-Durance, France
francesco.muratori@cea.fr

Frédéric Nguyen
Cadarache , DER, SPRC, LE2C
CEA, DEN
F-13108 Saint-Paul-lez-Durance, France
frederic.nguyen@cea.fr

Christian Gonnier
Cadarache, DER, SRJH
CEA, DEN
F-13108 Saint-Paul-lez-Durance, France
christian.gonnier@cea.fr

Christophe Le Niliot
CNRS, IUSTI
Aix Marseille Univ
Marseille, France
christophe.leniliot@univ-amu.fr

Romain Eschbach
Cadarache , DER, SPRC, LE2C
CEA, DEN
F-13108 Saint-Paul-lez-Durance, France
romain.eschbach@cea.fr

Abstract—**Decay heat is the thermal power released by radioactive decays of unstable isotopes after the nuclear reactor shutdown, and delayed fission reactions. It constitutes a key parameter for the nuclear reactor safety and the nuclear fuel cycle; for this reason, design codes have to be qualified by comparison with experimental measurements. The CEA's package DARWIN2.3 has been qualified for the calculation of PWR decay heat with two integral measurements: the MERCI experience and the CLAB laboratory's experiments; performed respectively on the following cooling time intervals: 40 min – 40 days and 12 years – 25 years. A lack of validation in the first hour of cooling time requires to consider large margins on the calculated decay heat value. As a result, delays in core unloading, intervention of human operators and safety systems dimensioning may occur. The PRESTO experiment, under conception at CEA, deals with a decay heat measurement between 1 and 40 minutes of cooling time for a PWR fuel sample irradiated in the Jules Horowitz Reactor (JHR). A previous thermal study showed that measurements could be sensitive to the decay heat 1 minute after the beginning of the cooling time. Now, a more precise estimation of power sources was performed with the Monte-Carlo code TRIPOLI. In this framework, four different device configurations were considered. Our results show that the irradiation power is not enough elevated in configurations where a tungsten shield is present.**

Keywords **— JHR, PRESTO experiment, TRIPOLI.**

I. Introduction

Decay heat represents the thermal power released by radioactive decays of unstable isotopes after the nuclear reactor shutdown, and delayed fission reactions. Generally, radioactive decay energy is released in the form of beta (electrons) and gamma (photons) radiations; the formers are entirely absorbed in the nuclear fuel, whereas the latter both in the fuel and outside. If N unstable isotopes are present, decay heat time-dependent value is calculated as the summation of the decay heat released by each isotope [1]:

$$P_T(\tau) = \sum_{i=1}^{N} E_i A_i(\tau) \tag{1}$$

where E_i [J] and A_i [s^{-1}] stand respectively for the energy released and the activity, *i.e.* the number of decays per second of the i-th isotope. Decay heat increases with the fuel burn-up until saturation levels of fission products concentrations are reached, and its contributors change according to the cooling time (τ) [2]; as it is shown in [1], they correspond to delayed fissions and fission products in the first minute of cooling time, fission products between one hundred seconds and twenty years and heavy nuclides in the very long term.

The decay heat constitutes a key parameter for the design of nuclear reactor safety systems and the backend of the nuclear fuel cycle. For this reason, design codes have to be validated by comparison with experimental measurements. Nowadays, the CEA's package DARWIN2.3 is qualified in PWR conditions of irradiation with two integral decay heat measurements: the MERCI experiment [3] and the CLAB laboratory's experiments [4]; before, others qualifications were performed concerning the decay heat produced by elementary fission events of fissile materials [5]. The validations cover cooling time intervals going from 40 minutes to 40 days, for the MERCI experiment, and from 12 years to 25 years, for the CLAB's experiments. The lack of a validation in the very short cooling time (and the associated uncertainty), constrains the nuclear industries to take large margins on the calculated decay heat value. As a result, several issues in nuclear reactor safety occur with respect to thermal and radioactivity aspects (*e.g.* delays for core unloading, delays in the intervention of human operators, core cooling dimensioning in case of a Loss Of Coolant Accident, LOCA).

The PRESTO experiment [6] (*PoweR Estimation for Short Time Optimization*), under conception at CEA, deals with a decay heat measurement between 1 and 40 minutes of cooling time for a PWR fuel sample irradiated in the Jules Horowitz Reactor (JHR). Owing to the very short cooling times, the experiment must be performed in the reactor region, which brings on the measurement an important amount of perturbations due to the reactor core decay heat. In a precedent paper [6], we performed a sensitivity study of a determined device configuration by using a preliminary estimation of power sources, thanks to the feedback of SRJH (the CEA division in charge of JHR design), and a numerical scheme simulating heat transfers in the experimental device in 2D cylindrical geometry *(r,z)*. Results showed that measurements were not sensitive enough to estimate the decay heat fraction released out of the fuel sample. As a result, we need a better estimation of power sources and 3D heat transfer modeling considering the non-azimuthal symmetry of irradiation. Thus, we used the TRIPOLI Monte-

Carlo code to estimate nuclear heating during irradiation, at τ=0, and after the reactor shutdown, at τ=2 min and τ=1 h; simulations were performed by considering 4 different device configurations.

In this paper, we compare TRIPOLI results with design parameters: irradiation power, burn-up, irradiation time, required for the conception of the PRESTO experiment. In so doing, we see how the choice of the experimental device influences the experiment conception.

II. The PRESTO experiment

The PRESTO experiment is designed to estimate the decay heat released by a PWR fuel sample, with an aimed confidence interval of 99% for a representative PWR fuel burn-up, between 1 and 40 minutes of cooling time. In this framework, we have to reproduce conditions of irradiation of current PWRs, and keep irradiation times in the JHR no longer than one year.

A. The Jules Horowitz Reactor (JHR) and the experiment development

The Jules Horowitz Reactor [7] (JHR), under construction in the site of Cadarache, was chosen for performing the PRESTO experiment. This reactor is devoted to the study of

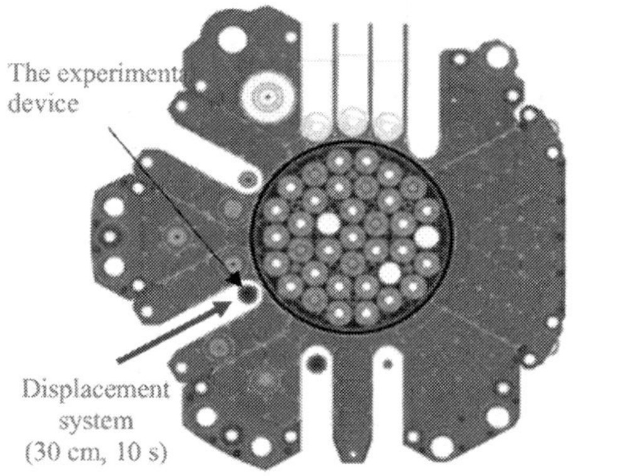

Fig. 1. The Jules Horowitz reactor core and reflector; the position of the experimental device: the JHR reflector on the displacement system.

materials and fuel behavior under irradiation as well as to medical isotopes production. The PRESTO experiment is designed to take place on the displacement system, situated in the JHR reflector (see figure 1); this one allows to master the irradiation power by regulating the distance between the experimental device and the reactor core.

The experiment is foreseen in a sequence of three phases: the irradiation, the stop of the irradiation and the measurement. At first, the experimental device is placed in the vicinity of the reactor core to perform the irradiation. When the desired fuel burn-up is achieved, the stop of the irradiation occurs by carrying out the reactor shutdown (SCRAM), the cooling phase starts, and the experimental device is moved back, by using the displacement system, to preserve it from perturbations coming from the reactor core. In this phase, it is important to remove thermal inertia stored in the device during the irradiation. At 1 minute of cooling time the measurement phase starts.

B. The experimental device

The experimental device scheme is reported in figure 2. The geometry is the cylindrical one, the height and the diameter are respectively around 40 cm and 10 cm. Decay heat power, released by the fuel sample, is deposited in the fuel and the surrounding environnement; particularly, the

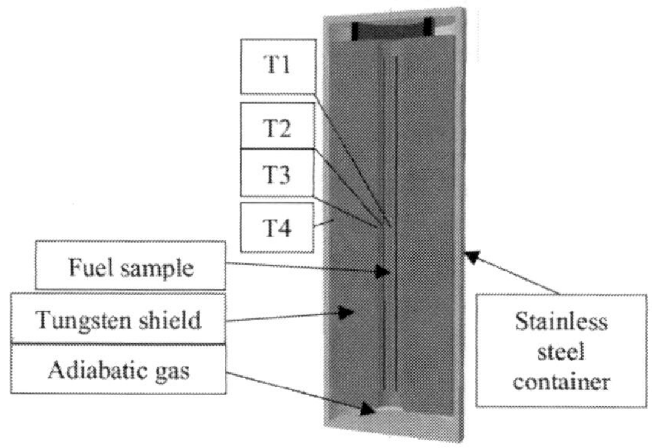

Fig. 2. The experimental device. In red: the fuel sample. In violet: the tungsten shield. In grey: the stainless steel container. T1, T2, T3, T4 are thermocouples placed in the fuel sample and the tungsten shield.

tungsten shield has been designed to collect the decay heat fraction leaving the fuel sample. The measurement principle consists in quantifying decay heat through the structures enthalpy instantaneous variation. In this way, the estimation parameter is the temperature temporal derivative of each structure; particularly we refer to the fuel sample and the tungsten shield. To do this, thermocouples have been located in the experimental device, as it is shown in figure 2, to measure the temperature evolutions. For simplifying the device physics and thus its thermal modeling, heat transfers among structures must be limited. For this reason, an insulating gas is injected in the device chamber (see figure 2).

C. The determination of decay heat

As we are in the very short cooling times, temperature fields are highly unsteady. As a result, to determine decay heat starting from thermocouples measurements, we need for a heat transfer modeling of the experiment. In this way, the decay heat estimation is performed by means of an inverse technique [8]. Figure 3 shows the scheme for the decay heat estimation.

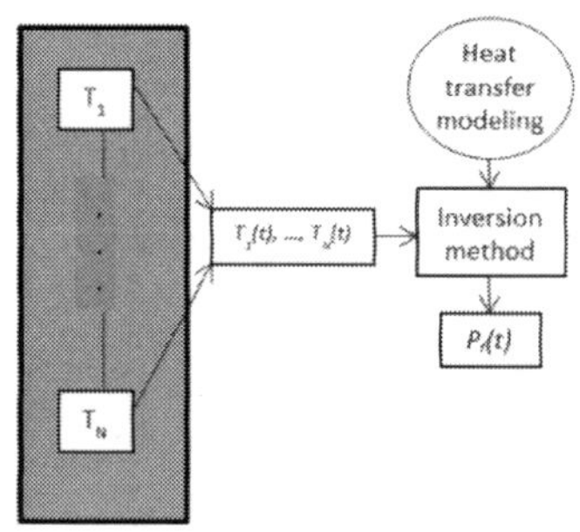

Fig. 3. Decay heat estimation. The *N* thermocouples measure the temperature evolutions. These are used with a heat transfer modeling to estimate decay heat through the linear inversion method.

III. THE ESTIMATION OF NUCLEAR HEATING

As our previous results showed that measurements were not sensitive enough to estimate the decay heat fraction released in the tungsten shield [6], we need for a more accurate sensitivity calculation to evaluate the feasibility of the PRESTO experiment. This involves a better estimation of nuclear heating and a more accurate modeling of heat transfers in the experimental device. To carry out the nuclear heating estimation we determined at first the power sources in the reactor core and the fuel sample through the evolution code MENDEL; results were then used as input in the Monte Carlo code TRIPOLI to evaluate the corresponding nuclear heating of the device.

A. Simulation set-up

Calculations were performed at three instants of cooling time: τ=0 (*i.e.* the reactor shutdown), τ=2 min and τ=1 h, and by considering four different device configurations (see figure 3).

- Configuration 1: the fuel rod (at the center in red) is surrounded by one tungsten shield (in violet), the device is cooled with water (in blue) and enclosed in a stainless steel container (in claret-red).
- Configuration 2: the fuel rod (at the center in red) is surrounded by two tungsten shields (in violet), which are separated by a stainless steel wall (in claret-red). The device is cooled with water (in blue).
- Configuration 3: the fuel rod (at the center in red) is surrounded by an internal shield of tungsten (in violet) and an external shield of zirconium (in green), which are separated by a stainless steel wall (in claret-red). The device is cooled with water (in blue).
- Configuration 4: similar to the number 2, with different thickness for the tungsten shields and the stainless steel wall.

The aim is to see the influence of the device configuration on the experiment conception (*e.g.* technological aspects involved, type of fuel to be used, etc.).

Nuclear heating was calculated by considering neutrons, prompt gammas, and delayed gammas radiation; the latters are produced from fission products decay, and born both in the fuel sample and in the JHR core.

As we deal with spent fuel, we considered a UO_2 fuel sample having an enrichment of 1% in U^{235}.

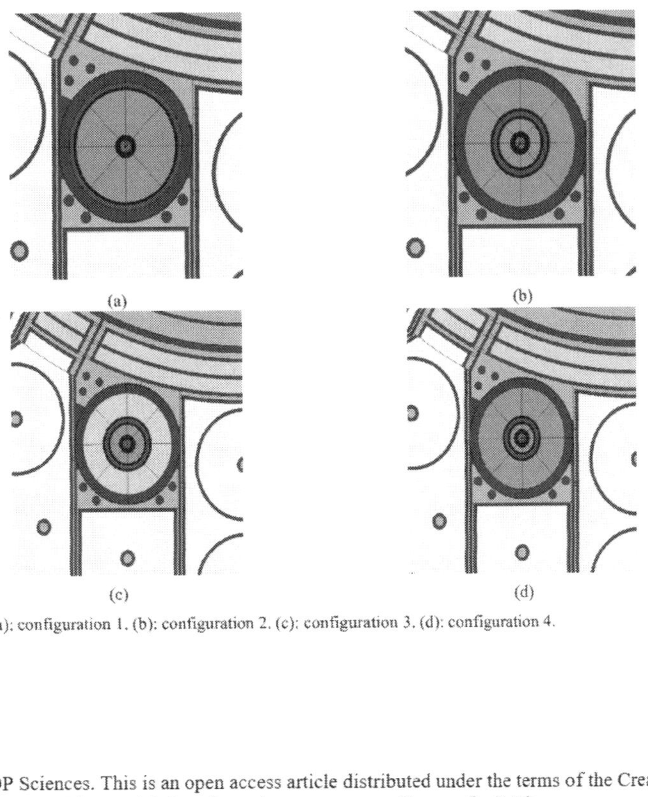

Fig. 4. Device configurations. (a): configuration 1. (b): configuration 2. (c): configuration 3. (d): configuration 4.

B. *Results*

1) *The irradiation phase $\tau = 0$*

To perform the PRESTO experiment, we need for an irradiation power of at least 150 W/cm, in order to obtain a representative PWR decay heat for the fuel sample.

Our results show that the tungsten is a great absorber of thermal neutrons compared to the zirconium (see figure 5). As a result, we do not reach in the fuel sample, for each configuration, the required level of irradiation power; irradiation power is around 10-15 W/cm in configurations 1, 2, and 4, and 60 W/cm in the configuration 3.

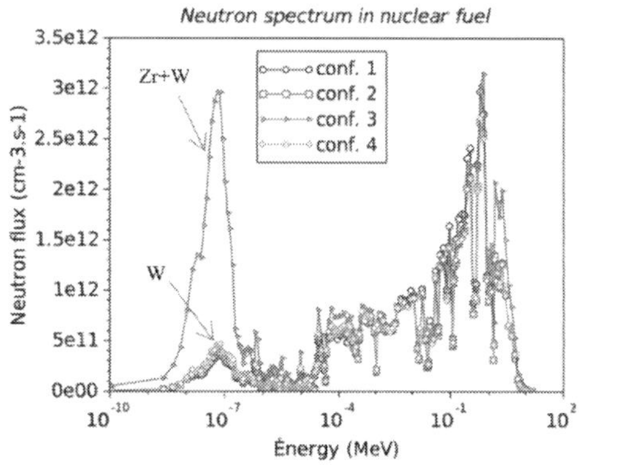

Fig. 5. The neutron spectrum in the fuel sample for each device configuration.

Thus, to achieve a PWR representative decay heat, we need to remove tungsten during the irradiation phase. This makes the experiment design more complicated, and solutions might correspond to design a shielding entirely made of zirconium, or provide vertical displacements for the tungsten shield.

Furthermore, irradiation duration is another important design parameter for the experiment representativity. We need simultaneously to reduce as much as possible the irradiation time, and to improve the experiment representativity.

2) *The measurement phase $\tau = 2$ min*

During the measurement phase, two main power sources were considered to calculate nuclear heating: the decay heat of the fuel sample, and the decay heat of the reactor core. Our results show that they heat the device structures with a different azimuthal form factor. In particular, we found that fuel sample decay heat is released in device structures uniformly according to the azimuth, whereas reactor core decay heat is released mostly in the portion of device facing the reactor vessel (see figure 6).

Estimations were performed by defining, for each configuration, 8 azimuthal regions of equal volume (see figure 3). Such a result, it is important to study a discrimination method between the source to be measured and the perturbation, and improve the estimation of the inverse technique.

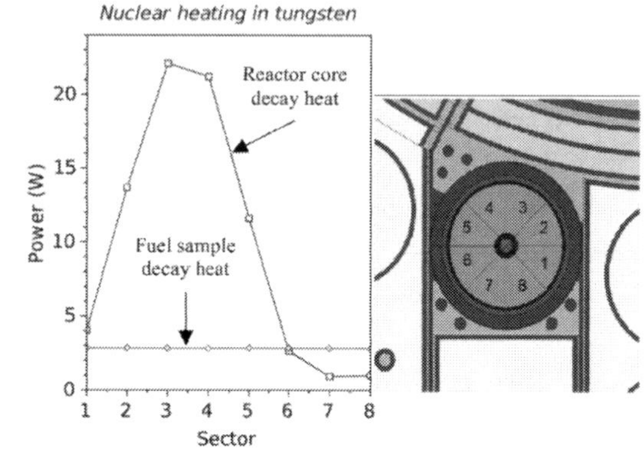

Fig. 6. Nuclear heating in the tungsten shield of the device configuration number 1. Fuel sample and reactor core contributions.

IV. Conclusion and Perspectives

Our results show that the TRIPOLI simulation is necessary to design the experimental device of the PRESTO experiment; in particular, in terms of experiment representativity and gamma shielding design. Furthermore, to maximize the measurement sensitivity, we have to match the neutronic simulation with the technological and metrological aspects of the PRESTO experiment.

Acknowledgment

I would like to thank David Blanchet (CEA, Cadarache), we benefit from his collaboration to perform the TRIPOLI study.

References

[1] J-C. Benoit, "Développement d'un code de propagation des incertitudes des données nucléaires sur la puissance résiduelle dans les réacteurs à neutrons rapides", thèse de doctorat de l'université Paris-Sud (2012).

[2] N. Todreas, M. Kazimi, "Nuclear Systems Vol. I", CRC Press (2012).

[3] S. Bourganel, "Conception d'une experience de mesure de la puissance résiduelle d'un crayon irradié : l'expérience MERCI", thèse de doctorat de l'INPG (2002).

[4] F. Sturek, L. Agrenius, "Measurements of decay heat in spent nuclear fuel at the Swedish interim storage facility", *Clab*, Svensk Karnbranslehantering AB, Rapport R-05-62, (2006).

[5] C. Fiche, "Mesure de la puissance résiduelle totale émise par les produits de fission thermique de 239Pu et 233U", NEACRP/L(1978)212.

[6] F. Muratori, F. Nguyen, C. Gonnier, C. Le Niliot, R. Eschbach, "Modeling and sensitivity study of an experimental device for a decay heat measurement performed on an irradiated nuclear fuel sample in the very short cooling time range: the PRESTO experiment", Congrès Français de Thermique, 3-6 juin 2019, Nantes, France.

[7] J. Estrade, G. Bignan, X. Bravo, "The Jules Horowitz Reactor: a new high performance MTR working as an international user facility in support to nuclear industry, public bodies and research institutes", RRFM (2015).

[8] P.C. Hansen, "Rank deficient and ill-posed problems", siam (1998).

EPJ Web of Conferences **225**, 04006 (2020)
ANIMMA 2019
https://doi.org/10.1051/epjconf/202022504006

Radiation resistance of high frequency ultrasonic devices for distance measurement in the RHF

Meriem Chrifi Alaoui, Emmanuel Le Clézio, Member IEEE,
Gilles Despaux, Member IEEE
University of Montpellier, CNRS, IES UMR 5214, CC 05 003,
860 Rue de Saint Priest, F-34000 Montpellier
Email: emmanuel.le-clezio@umontpellier.fr

Yoann Calzavara
Institute Laue Langevin - ILL
71 Avenue des Martyrs,
F-38000 Grenoble
Email: calzavara@ill.fr

***Abstract*—The fuel element plates of research reactors are subjected to swelling phenomena. These structure modifications impact an initial inter-plate distance of 1.8 mm. An ultrasonic device has been developed to investigate this parameter. With a 1 mm thickness, it relies on two transducers linked to an electronic system. The feasibility of the distance measurement has been proved in a previous study and the irradiation impact on the transducer components is here studied. To do so, a radiation resistance experiment was realized in the Arc-Nucleart Institute of Grenoble. It allowed the study of the influence of radiations on the device active and passive components' characteristics.**

I. Introduction

In a conversion program aiming to the introduction of Low Enriched Uranium in research reactors, investigations are conducted to identify convenient fuel types. Integrated in this program, the Institute Laue Langevin (ILL) studies the conversion of the High Flux Reactor (RHF) of Grenoble. Its fuel element is made of plates separated by a distance of 1.8 mm. During the reactor life cycle, irradiations induce swelling impacting this so-called channel-gap.

To monitor its variations, a non-destructive ultrasonic device was developed to be inserted between two plates. It includes two manufactured high frequency transducers, inserted in a 1 mm blade and related to electronic instruments. The distance measurement feasibility being proved during a RHF experiment [1], the irradiation impact on the transducer components is here studied.

In this view, tests were performed on 10 transducers with central frequencies ranging from 38 MHz to 160 MHz. The device components were irradiated during 4 months in the Arc-Nucleart Institute of Grenoble [2] and the resulting irradiation corresponded to 3500 kGy.

The first section of this paper presents the distance measurement configuration and the proposed ultrasonic device. To analyze the irradiation impact, the second section displays the irradiator and the experimental set-up developed to monitor transducer responses. Results are presented in the last section.

II. Ultrasonic device conception

A. Experimental constraints

As presented in [1], performing distance measurement inside a RHF is quite complicated. First, the fuel element is placed 5 meters below the surface of a water pool. Then, the reactor is made of curved plates with a 1.8 mm inter-plate distance. Finally, the measurements take place in a radiative environment. To integrate these constraints, a device was proposed with a thickness limited to 1 mm.

B. Distance measurement principle

The ultrasonic device structure is presented on figure 1 where the inter-plate distance d depends on the distance h

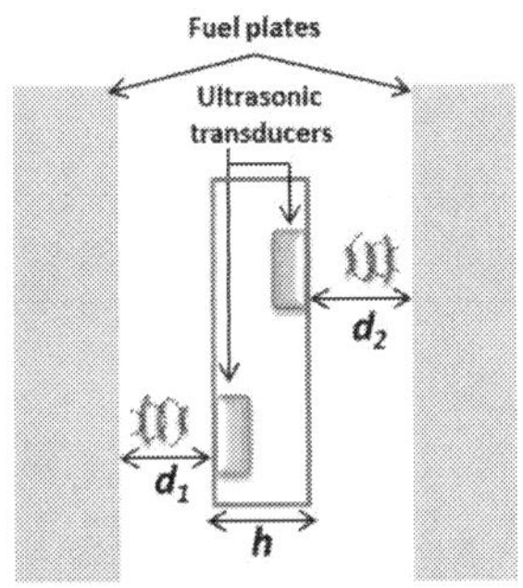

Fig. 1: Inter-plate distance measurement principle.

between the two transducers and distances d_1 and d_2 measured through the estimation of the time of flight [3] of ultrasonic waves propagating in the coupling medium t_1 and t_2:

$$d = h + V * (t_1 + t_2)/2 \quad (1)$$

where V is the sound velocity in water. The feasibility of distance measurement has been proved in a previous study [1] and the irradiation impact on the transducer components is now studied.

III. Irradiation experiment

A. Experimental constraints

10 transducers and several passive components were manufactured. To allow the identification of potentially structural effects, they were distributed in a broad frequency range corresponding to [38 – 160] MHz. They were submitted to irradiation at the Arc-Nucleart Institute of Grenoble [2]. To

https://doi.org/10.1051/epjconf/202022504006

estimate the utrasonic system evolution in a more than 10 hour RHF control, where the device is submitted to a flow of 300 kGy/h, an irradiation was then performed up to a total of 3500 kGy. Two transducers were not stressed to observe environmental impact.

B. Arc-Nucleart Irradiator

The Arc-Nucleart installation includes a water pool and a concrete irradiation cell as presented in figure 2. The gamma sources are of Cobalt ^{60} Co. The transducer elements were installed in the irradiation cell while the sources were in the pool. The irradiator rate ranges from 10 Gy/h to 2 kGy/h.

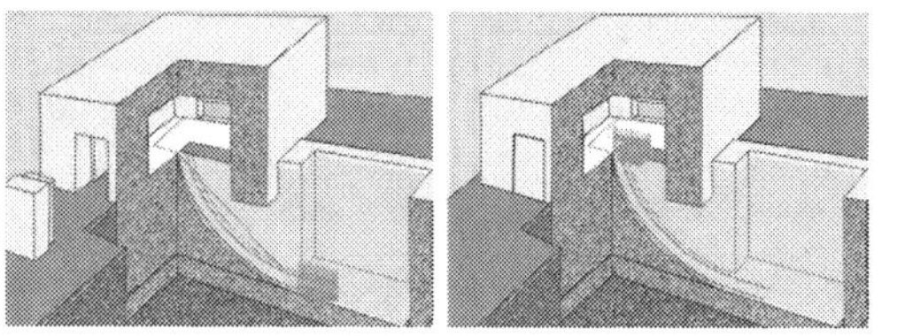

Fig. 2: Nuclear irradiator installation scheme.

For the experiment, a 1.4 kGy/h flow was applied yielding a more than 3 month long experiment. A specific set-up was then realized to allow a continuous monitoring.

C. Experimental set-up

The irradiated devices and passive components positioned in the irradiation cell are presented in figure 3. Communication

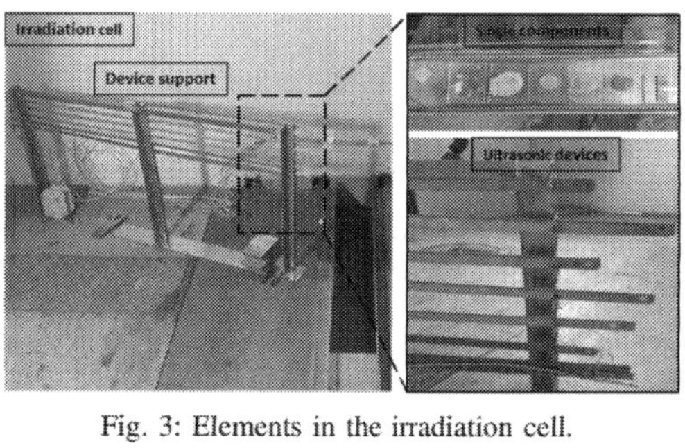

Fig. 3: Elements in the irradiation cell.

paths allow passage of cables linking the transducers to an electronic set-up including:

- A E8356A Agilent network analyzer for fast scanning, wide dynamic range and flexible connectivity traces.
- Acquisition cards inserted in a PXIe−1062 chassis:
  - RF multiplexer switch module PXIe−2747 card, with dual 8 inputs and 2 outputs.
  - PXIe−8135 controller as the interface between electronic components and the acquisition software.

D. Data acquisition and processing

For each transducer, a spectrum acquisition is realized every 6 minutes. To monitor acoustic echoes, an Inverse Fourier Transform and a digital bandpass filtering are applied. Then, a specific echo is extracted and its parameters allow the study of the transducer's behavior during the experiment. To analyze device's behavior, investigated echo parameters are: central frequency, bandwidth and amplitude.

IV. Results and discussion

A. Distance measurement parameters' analyzes

Two transducers' responses are here discussed. They correspond to a 113 MHz reference transducer placed outside the irradiation cell (a) and a 87 MHz irradiated transducer (b). The bandwidth spectrum of the reference transducer measured in the laboratory for 3 days and that obtained outside the irradiation cell are presented in figure 4(a) and (b) respectively. Those of the irradiated transducer are figure 5(a) and (b). These spectra reveal very few variation during irradiation

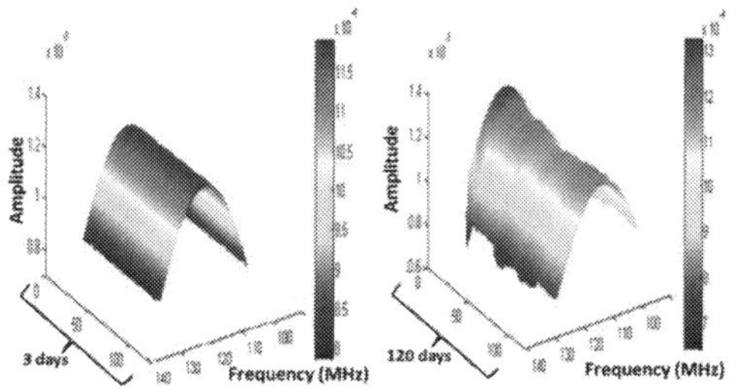

Fig. 4: Bandwidth spectra of the reference transducer (a) in the laboratory (b) outside the irradiation cell.

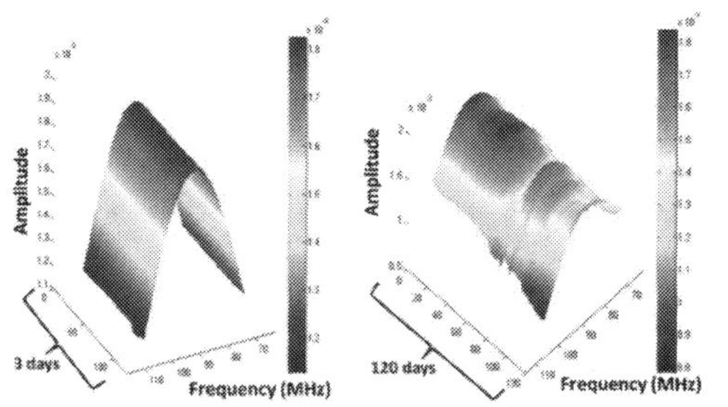

Fig. 5: Bandwidth spectra of the irradiated transducer (a) in the laboratory (b) inside the irradiation cell.

and a very good stability of the transducer characteristics can be stated. Indeed, the frequency, amplitude and bandwidth evolution of the reference transducer are presented in figure 6(a) and (b) and compared to that of the irradiated transducer in figure 7(a) and (b). For the reference transducer, the maximum frequency, amplitude and bandwidth variations are

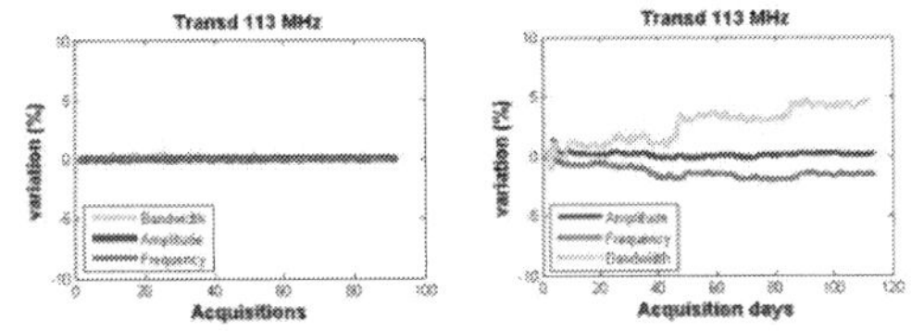

Fig. 6: Reference transducer spectra parameters' variation (a) at the laboratory (b) outside the irradiation cell.

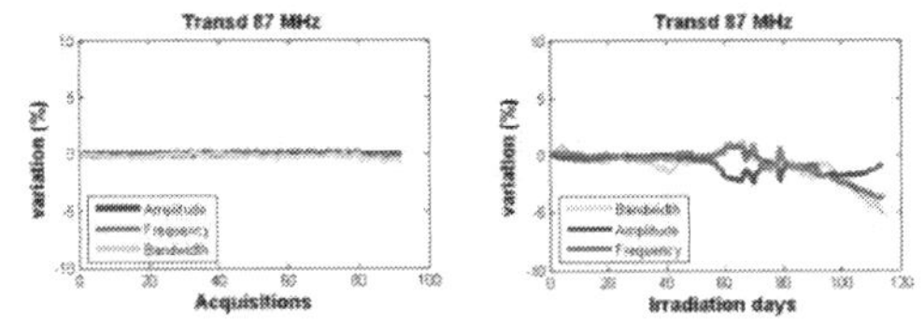

Fig. 7: Irradiated transducer spectra parameters' variation (a) at the laboratory (b) inside the irradiation cell.

of 2%, 1% and 4.8% respectively. For the irradiated transducer, they are equal to 4%, 2.4% and 5.1%. These results clearly indicate a global stability of the transduction characteristics. In agreement with the previous results, the variations of the other transducers' parameters are presented in Table I with

Element	Frequency variation	Amplitude variation	Bandwidth variation
Transducer 38 MHz	8.0 %	5.0 %	4.7 %
Transducer 72 MHz	0.5 %	1.5 %	2.9 %
Transducer 78 MHz	1.5 %	3.2 %	1.3 %
Transducer 87 MHz	4.0 %	2.4 %	5.1 %
Transducer 97 MHz	8.0 %	3.4 %	6.0 %
Transducer 103 MHz	1.0 %	1.0 %	2.7 %
Transducer 116 MHz	3.0 %	4.5 %	3.0 %
Transducer 162 MHz	2.0 %	1.5 %	3.0 %

TABLE I: Transducers parameters' maximum variations.

maximum variations of: 8% for frequency, 5% for amplitude and 6.0% for bandwidth.

B. Passive elements

1) Form and adhesion parameters: 4 elements were manufactured: adhesive, backing and conducting materials deposited on silica supports with a superposed silica, and a conducting material deposited without the second silica layer. Their comparison before and after irradiation is performed thanks to a TePla ultrasonic microscope [4] including a 200 MHz transducer. Initial and after irradiation images are presented in figure 8 first and second lines respectively. They show that, for the 2nd and 3rd elements, disbonding phenomena appear on the contour area of material. The 4th element presents a color modification and an ultrasonic signal attenuation indicating a potential increase in the material porosity. This point will be further studied in future works.

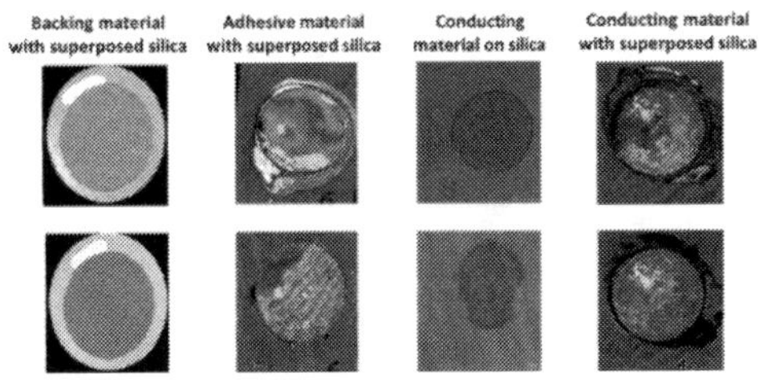

Fig. 8: Passive elements microscopic images.

2) Mechanical properties: Acoustic signatures [5] were realized on the 2 silica elements. Rayleigh velocity and Young modulus estimations are presented in table II, indicating a

Element	Pre-irradiation properties	Post-irradiation properties	Percentage variation
Silica type 1	$V_R = 3395\ m/s$ E = 71.86 GPa	$V_R = 3338\ m/s$ E = 71.13 GPa	1 %
Silica type 2	$V_R = 3110\ m/s$ E = 62.38 GPa	$V_R = 3108\ m/s$ E = 62.30 GPa	0.1 %

TABLE II: Measured mechanical properties of the Silica elements.

negligible maximum variation of 1% after irradiation.

V. CONCLUSION

An irradiation experiment was realized on 10 ultrasonic transducers and passive components. A real-time monitoring system was developed to observe their responses to a 3500 kGy dose. Transducers' frequency, amplitude and bandwidth showed good stability along irradiation. For passive elements, forms and mechanical properties were compared before and after irradiation through acoustic microscopy and revealed high stability. These results clearly indicate that these device components can then be included in systems allowing more than 10 hours of continuous control of RHF fuel elements.

ACKNOWLEDGMENT

The authors would like to thank Y. Launay and O. Guillemard for their contributions to supports and devices manufacturing. Acknowledgments are also dedicated to L. Cortella and Arc-Nucleart members for their great help.

REFERENCES

[1] G. Zaz, E. Le Clézio, A. Dekious, M. Chrifi Alaoui, Y. Calzavara and G. Despaux. In-situ high-resolution measurement of RHF nuclear fuel plates' spacing. *IEEE Transactions on Nuclear Science*, 65, 2018.
[2] www.arc-nucleart.fr.
[3] D. Marioli, E. Sardini, and A. Taroni. Ultrasonic Distance measurement for Linear and Angular Position Control. *IEEE Transactions on Instrumentation and measurement*, 37, 1988.
[4] www.pva-analyticalsystems.com.
[5] M. Marchetti, D. Laux, L. Fongaro, T. Wiss, P. Van Uffelen, et al.. Physical and mechanical characterization of irradiated uranium dioxide with a broad burnup range and different dopants using acoustic microscopy *Journal of Nuclear Materials*, 494, pp.322 – 329, 2017.

Experimental characterization of PCI impact on vibrating fuel rod under axial turbulent flow representative of JHR irradiation device ADELINE: Set-up conception and measurement method

Veronica D'Ambrosi*, Stephane Breaud, Christophe Destouches, Guillaume Ricciardi, Frederic Lebon, Emmanuelle Sarrouy, Jean-Marie Gatt, Jerome Julien and Daniel Parrat
* CEA, DEN, DER, Instrumentation Sensors and Dosimetry Laboratory, Cadarache, France,
veronica.d'ambrosi@cea.fr

***Abstract*—Mechanical loading originating due to Pellet-Cladding Interaction (PCI) in PWR nuclear fuel rod constitutes a first order importance phenomenon when attempting at increasing the flexibility of nuclear reactors operation to follow grid demand. The need to improve the understanding of such complex multi-physics phenomena motivates the realization of particular irradiation sequences carried out in highly instrumented devices located in Material Testing Reactors. Among this kind, ADELINE irradiation device is being developed and will be installed in the future Jules Horowitz Reactor. In the perspective of designing the adapted measurement methodology to detect and characterize PCI phenomenology during irradiation, we present in this paper the experimental bench and its associated measurement program, designed to investigate eventual effects of PCI on the non-linear, flow induced vibrations of nuclear fuel rod. Analytical and numerical models cannot predictively describe the system due to the complexity of phenomena thus the IMPIGRITIA experimental set-up has been developed to reproduce the mechanical interaction between the pellet and the clad at low pressure, room temperature and out of neutron flux. The designed test bench presents a clamped free single short rod, centred in the test section by mean of four centring elements. Different rod configurations are implemented and localized closure of the gap is remotely realized by means of a dilatation system. Laser Doppler Vibrometry is used to measure the transversal displacement of the sample rod in three different conditions: in air, in stagnant water and under turbulent axial flow rate. The experimental program and expected results are presented and discussed.**

Keywords—**Nuclear, Instrumentation, Measurement, Bench, fuel rod, PCI, dynamics, vibrations, LDV**

I. Introduction: Background and Objectives

THROUGH the past years, Pellet-Cladding Interaction (PCI) has constituted an issue in the definition of technical specifications for the operation of nuclear power plants and for the identification of associated safety margins. Consequently, research has attempted to improve the knowledge of this complex phenomenon by numerical simulations, developing multi-physics models, and by experiments in Material Testing Reactors (MTR) [1]. In the latter, environmental constraints have strongly limited the possibility to obtain real time measurements and thus to allow characterisation of the kinetics of the phenomenology. At present days, a need for highly instrumented experiences in MTRs and for on-line measurements to gain real time data during irradiation has been expressed.

In the next future JHR, ADELINE is a specific irradiation device where PWR temperature and pressure conditions are reproduced and which will be partially dedicated at studying this phenomenology.

In this frame we have identified relevant and measurable parameters associated to PCI during normal irradiation and incidental conditions. We also proposed a model to characterize fuel material behaviour undergoing initiating central melting in the attempt at evaluating its macroscopic impact on PCI [2]. We are working to propose innovative and efficient measurement methods for detecting and characterizing this complex multi-physics phenomenology. Among the identified relevant parameters, vibration distinguished as a possible application case.

Flow Induced Vibrations (FIV) of nuclear fuel rods in axial turbulent flow rate have been concerned by a large variety of scientific studies. According to Païdoussis [3] [4] and Axisa [5] flow induced vibrations come from two mechanisms: the drag force in the turbulent boundary layer near the structure (near-field component) and from the wall pressure fluctuations and far field noise. They generally induce transversal vibrations of small amplitudes, typically the vibration amplitude y/D_c does not exceed 10^{-3}, y being the r.m.s. amplitude of vibration and D_c the cylinder diameter. In the nuclear field, works are mainly motivated by fretting concern, phenomenon appearing at the grid-rod interface.

Few researches have instead been carried out on the effects of non linear mechanisms arising from the presence of pellets within the fuel rod. It is generally argued that modeling those effects is rather difficult [6]. Benhamadouche in [7] proposed to evaluate an equivalent mass density for the rod which accounts for the presence of pellets. This density is then used in the evaluation of the dynamic behaviour of the fuel rod taken as a linear Euler Bernoulli beam. Park et al. in [8]

Manuscript received 2019; revised 2019.
V. D'Ambrosi, C. Destouches and S. Breaud were with the CEA, DEN, DER, Instrumentation Sensors and Dosimetry Laboratory, Cadarache, F-13108 St Paul Lez Durance, France. (email veronica.D'AMBROSI@cea.fr)
G. Ricciardi was with CEA Cadarache - DEN/DTN/STCP/LTHC, 13108 Saint Paul ls Durance, France
F. Lebon and E. Sarrouy were with Aix Marseille Univ, CNRS, Centrale Marseille, LMA, Marseille, France
J.M. Gatt, J. Julien and D. Parrat were with DEN/DEC, French Alternative Energies and Atomic Energy Commission (CEA) F- 13108 St-Paul-lez-Durance - France

make the point on the fact that since a gap exists between pellets and gap, the contribution of the pellet mass to the density for dynamic evaluation cannot be clearly determined. They propose to compare the natural frequencies of empty and fulfilled rods to estimate an appropriate density.
The fuel rod vibration is then evaluated, still assuming a linear behaviour.
The non linear vibration of fuel rod has been recently studied by Ferrari et al. [6]: his study concerned a 900 mm zircaloy clad, clamped at both extremities, eventually fulfilled with cylindrical tungsten carbide pellets modelling uranium fuel pellets in terms of dimensions and density.
By comparing the dynamics of empty rod, fulfilled rod freely moving (no spring at one bottom of the fuel column) and fulfilled rod axially constrained in air and in quiescent water, they observed that pellets decrease natural frequencies of the rod and that axial constriction increases natural frequencies and reduced damping with respect to freely moving configuration, both in air and in water. Moreover, they stated that non linear mechanisms were experimentally observed during tests. No studies have been retrieved discussing the effect on the dynamics of the fuel rod of local pellet-clad contact due to gap closure. The impossibility to predictively model the non-linear system justified the design of the experimental set up IMPIGRITIA[1] to state the feasibility of measuring such impact.
This paper is structured as follows: we briefly proceed in section II with the numerical approach of the system, highlighting the empirical parameters to be measured to then present in section III the measurement principles.

Section IV is devoted to the design criteria of the experimental set-up, the definition of rod configurations and associated instrumentation. Section V then briefly describes the experimental program and expected results. Section VI proceeds to conclusions and future work.

II. The numerical approach

A. *Definition of the model*

Figure 1 shows a schematic view of the proposed model. The sample holder and the empty clad are modelled as a cantilevered Euler-Bernoulli beam by a 2D finite element model. Each node of the clad can move along the radial (u_x) and the axial direction (u_y) and can rotate (θ_z). Pellets are considered as discrete masses m_p interacting along the radial direction with corresponding clad nodes through the non linear mechanism of dissipative shock and with each other through Coulomb friction. The equation of motion to be solved is [9]:

$$[M]\{\ddot{x}\} + [C]\{\dot{x}\} + [K]\{x\} + \{fnl\} = \{F_e\} \qquad (1)$$

Where $[M]^{N,N}$ is the total mass matrix, $[C]^{N,N}$ the total damping matrix, $[K]^{N,N}$ the total stiffness one. They are obtained as follows:

$$[M] = \begin{bmatrix} [M_c] & 0 \\ 0 & [M_p] \end{bmatrix} \qquad [C] = \begin{bmatrix} [C_c] & 0 \\ 0 & [C_p] \end{bmatrix}$$

[1]Acronym form "Interaction Mécanique PastIlle Gaine :RéalIsation du Test vIbrAtoire", from *latin* impigritia, impigritiae; means "zeal, dedication"

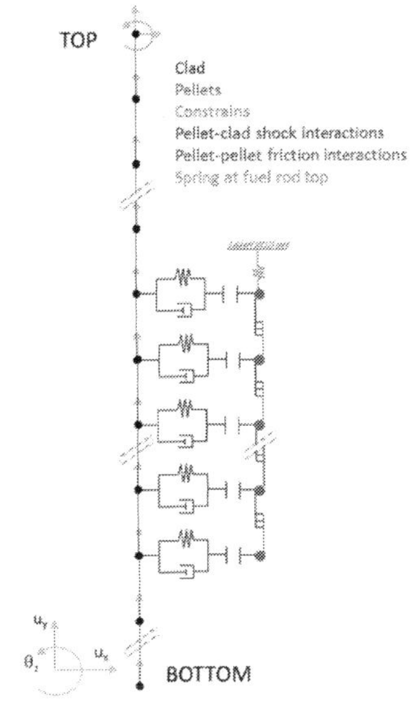

Fig. 1. Numerical model of IMPIGRITIA: schematic view of the configuration

$$[K] = \begin{bmatrix} [K_c] & 0 \\ 0 & [K_p] \end{bmatrix} \qquad (2)$$

where $[M_c]^{n,n}$ is the tridiagonal mass matrix for the clad and $[M_p]^{m,m}$ is the diagonal mass matrix for the pellets, with $N = m + n$. In the same way the total stiffness matrix and the total damping one depends on clad and pellets ones.

Rayleigh damping is used to model the damping matrix [C]:

$$[C] = \alpha[M] + \beta[K] \qquad (3)$$

Assuming proportional damping, the dissipation in the structure can be described by two parameters α and β. Those parameters, which are not a priori known, are experimentally determined measuring the first two eigen-frequencies of the system in air. Indeed it can be shown that [10]:

$$\zeta = \frac{1}{2}\left(\frac{\alpha}{\omega_i} + \beta\omega_i\right) \qquad (4)$$

where ζ is the modal damping, which can be recovered by logarithmic decrement from free damped vibrations of the system and ω_i are the eigen-frequencies.
The non-linear forces acting on the system are expressed by the $\{fnl\}$ term in equation (1). Those forces are the dry friction between discretized pellet nodes and the shock interaction between pellet nodes and clad ones. Gap size is a parameter in the formulation of pellet-clad shock interaction.
The presence of external water plays a role in the external solicitation $\{F_e\}$ but it also induces an additional viscous damping to the structure: according to De Pauw in [11] the effects of flowing fluid on the spectral response of the structure

is described by an added mass m_a, corresponding to the mass of displaced fluid, and a dynamic amplification factor C_m such that:

$$f = \frac{1}{2\pi}\sqrt{\frac{K}{M + C_m m_a}} \quad (5)$$

where K and M are the structural stiffness and mass of the system. The dynamics amplification factor is equal to 1 for quiescent water thus the two parameters can be experimentally obtained by comparing the oscillation frequencies of the system in air, in quiescent water and in fluid flow.
$\{F_e\}$ corresponds to the excitation mechanisms acting on the system, notably the flow solicitation. We decided to apply a statistical approach and model $\{F_e\}$ by taking the envelop PSD defined by Axisa [5] and applying an arbitrary stochastic phasing.

To solve the equation, explicit Newmark scheme has been implemented. As reported in [9] direct time integration is necessary when the problem is non linear in order to avoid strong dependencies of results on integration time step.

B. Investigation of parameters: an empirical model

The displacement $\{x\}$ of the structure, obtained when solving motion equation (1) by mean of explicit Newmark integration scheme, can be expressed as a function of the set of parameters $\mathbb{P}$, resulting from the previous discussion:

$$\{x\} = f\{\mathbb{P}\} =$$
$$f\{\mathbf{M_p}, \mathbf{M_c}, m_a, C_m, \mathbf{K_p}, \mathbf{K_c}, \alpha, \beta, K_c, C_c, \mu, j, c^2\} \quad (6)$$

where $\mathbf{M_i}$, $\mathbf{K_i}$, α and β have already been defined, m_a and C_m are the added mass and the dynamic amplification factor due to the fluid presence and fluid flow. K_c and C_c are the parameters characterising the pellet-clad shock interaction, modelled through a contact stiffness K_c and a viscous dissipation C_c [12]; μ is the friction coefficient to model the pellet-pellet interaction. j is the width of the gap and c is the fluid velocity.
While some of those parameters are completely defined by the geometry of the structure (i.e. the mass matrix of pellets $\mathbf{M_p}$ and sample holder $\mathbf{M_c}$, the stiffness matrix of the sample holder), the others have to be experimentally determined. Identification of model parameters adapted to fit at best experimental data allows calibration of the model. They are resumed in table I.

III. The measurement principle

The presence of pellets inside the clad induces the appearance of non linear mechanical dissipations [6]. Internal attachment between pellets and clad as a consequence of gap closure limits those effects and impacts mode and frequencies of the system. We expect local and progressive mechanical contact between pellets and clad to reduce structural damping as a consequence of reduced dissipative mechanisms and to decrease its eigen-frequencies. Those variations can be retrieved by measuring the displacement of the system $x(t)$. Experimental identification of the frequencies and of modal properties, as the modal damping, can be carried out both in the time domain or in the frequency one through direct or inverse Fourier transform. Indeed:

$$x(t) \quad \leftrightarrow \quad X(\omega)$$

where $X(\omega)$ is the Fourier transform of the time signal of the displacement $x(t)$ of the point characterized by the highest transversal displacement. $X(\omega)$ and $x(t)$ are linked by the relations:

$$X(\omega) = \mathcal{P}[x(t)] = \int_{-\infty}^{\infty} x(t)e^{-i\omega t}dt \quad (7)$$

$$x(t) = \mathcal{P}^{-1}[X(\omega)] = \frac{1}{2\pi}\int_{-\infty}^{\infty} X(\omega)e^{i\omega t}d\omega \quad (8)$$

In both time and frequency domains, system identification shall be carried out on the basis of the model presented in this section, adapting its parameters to fit at best the experimental data. Several methods are available in literature [9] and [11]. The impact of pellet-clad interaction on the transversal vibration $x(t)$ can be experimentally obtained by comparing specifically designed rod configurations under the same excitation conditions. Calibration of parameters previously discussed allows validation of the model capability to simulate the dynamic response of the structure.

TABLE I
Model parameters

Parameter	Description
K_p	Pellet Stiffness
$\zeta(\alpha, \beta)$	Modal damping
m_a	Added mass
C_m	Dynamic amplification factor
K_c	Contact Stiffness
C_c	Contact viscous dissipation
μ	Pellet-Pellet Friction Coefficient
j	Pellet-clad gap
c	Fluid velocity

IV. Design criteria

A. Objectives

We have manufactured the experimental test bench to state the capability to measure whether gap closure impacts fuel rod dynamics. We want to identify if a minimum number of pellets has to interact with the clad to have a measurable effects and to define if FIV are sufficient to have a passive measurement and to study the influence of this axial flow excitation on the feasibility of this measure. Application domain is the ADELINE device.

B. The representativeness issue

ADELINE [13] is designed to be a 4m high irradiation device and constitutes the evolution of ISABELLE 1 system as it was in OSIRIS reactor [14]. As its predecessor, it presents a single rod test section where temperature and pressure conditions characteristics of a PWR are reproduced (155 bar and

275°C inlet water temperature). The tested rod has geometrical characteristics of FABRICE type fuel-rod [14]: a short element (about 60 cm in length) obtained cutting a previously irradiated 4m fuel rod and adapted to MTR geometry. The rod is cooled by an up-streaming turbulent flow rate. It is maintained in the section by a sample holder and centred by four centring elements.
In order to be representative of ADELINE configuration, we accounted for the real environmental conditions, for the geometrical design, material characteristics and hydraulic solicitation.

1) The environment: Thermal and mechanical properties are obviously impacted by the environmental conditions but, through the years, several correlations have been developed to account for these phenomena. Those correlations will be used during post treatment of results to establish the attended behaviour in PWR conditions [15].
The pressurization of the circuit has no impact on the vibration of the component, thus leading to design a law pressure system.

2) The geometry: Mechanical similarity and scale reduction cannot simply be applied and a 1:1 scale system implied non negligible conception, exploitation and cost issues. Still, geometry and fixation conditions strictly impact the dynamics of a structure thus the clamped free conditions of ADELINE sample is maintained in the experimental set-up. Moreover, ADELINE sample holder is made of an upper part and a lower part, the sample rod a centring elements being located in this last one. The two parts are coupled by a mechanical system. The specific configuration and material properties of the real system lead us to decide to limit our study to the lower sample holder part. It results an experimental system of about 1.5 m. This choice is expected to strengthen the impact of a local pellet-clad interaction on the global dynamics of the structure.

3) The hydraulic similarity for FIV: Fuel rod to be installed in the ADELINE device will be submitted to a turbulent flow rate of approximately 1 m/s at about 275°C and 155 bar along the axial direction, from the bottom to the top. Such kind of turbulent flow rate induces the appearance of FIV of the sample rod. Hydraulics is dimensioned in order to reproduce the external solicitation in terms of the axial turbulent flow and pressure fluctuation in the flow. We identified a range of representativeness first by imposing the conservation of wall pressure fluctuations and secondly applying the Burgreen principle. The conservation of wall pressure fluctuation $p_{r.m.s}$ was imposed according to the definition proposed by Axisa in [5] on the basis of the envelop PSD derived from Clinch data [16] and reported in equation (9). The simulation criterion established by Burgreen [17] instead has been recently used by De Paw et al. in [11] to study flow induced vibrations in a prototype representative of the MHYRRA system and it is reported in equation (10).

$$p_{r.m.s} = 0.003(\rho c^2) \tag{9}$$

where $p_{r.m.s}$ is the root mean square of pressure fluctuations, ρ is fluid density and c is average flow velocity.

$$c_{exp} = c\frac{\rho^{0.5}\mu^{-0.33}E^{-0.33}M^{0.165}}{\rho_{exp}^{0.5}\mu_{exp}^{-0.33}E_{exp}^{-0.33}M_{exp}^{0.165}} \tag{10}$$

Subscript 'exp' stands for prototype, μ the fluid dynamic viscosity, E elastic modulus and M the mass of the sample rod at operating temperature.

Applying the described criteria in our test conditions we found a range of representative flow rate varying between 0.9 to 2.6 the fluid velocity at the sample rod in ADELINE.

$$\frac{c_{exp}}{c} \in [0.9 : 2.6] \tag{11}$$

V. The experimental set-up and program

For the experience several representative hollow zircaloy clad were supplied. As previously mentioned, the geometrical configuration of sample rods was defined on the basis of FABRICE type fuel rods used in ADELINE: the zirconium alloy clad has an external diameter of 9.5 mm, its material properties are reported in table II. We defined a length coherent with FABRICE characteristics thus 550 mm long. Each side is welded to a plug which is screwed on a prolongation tube in stainless steel.

The prolongation tube constitutes the sample holder and it is fixed at the top of the experimental section. The other side is instead free to move. A simplified scheme of the sample holder is presented in figure 2.

Due to environmental and realisation issues, the UO_2 constituting the real fuel material was replaced by cylindrical Molybdenum pellets, which is one of the most representative materials according to internal researches carried out at CEA. A comparison of room temperature material properties of UO_2 and Mo are resumed in table II. Pellets diameter was defined according to standard radial dimension of UO_2 fuel pellets, the height instead was deduced by mass conservation of UO_2 fuel pellets.

In order to fulfil the fabrication gap existing between pellets and clad and thus remotely control PCI, a dilatation system capable of local expansion is used. The tool, showed in figure 3, is used in the medical domain for stent implantation: it is made of a balloon which can be externally inflated and a connection tube (the shaft) allowing positioning and inflation of the system.

The component is introduced from the top of the sample holder, through centrally drilled pellets and the balloon is placed in the region where gap closure wants to be induced. There, pellets are replaced by two shells which are displaced by the dilatation of the system, leading to pellet-clad contact.

TABLE II
MATERIAL PROPERTIES OF INTEREST

Material	Density [kg/m^3]	Young's Modulus [GPa]
Zircaloy	6570	95
UO_2	10963	194.3
Mo	10200	320

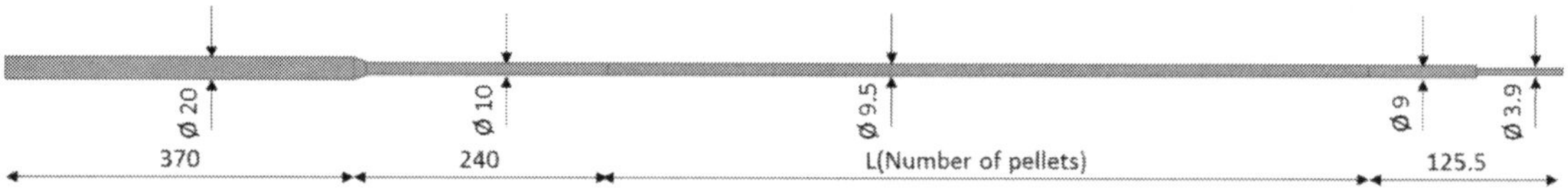

Fig. 2. Sample holder scheme

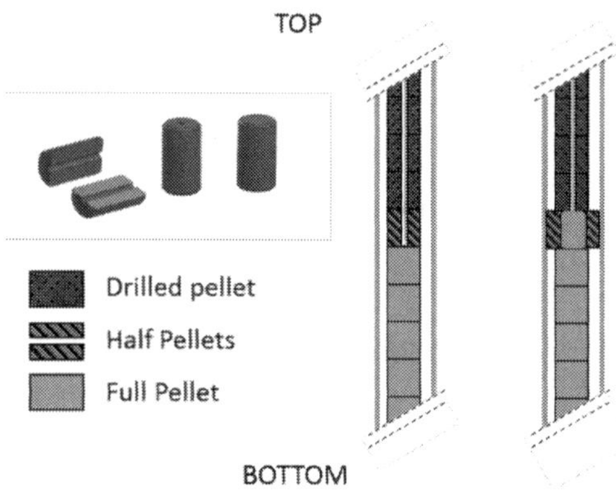

Fig. 3. Schematic view of the functioning of expansion system to induce pellet-clad interaction: inflation of the balloon through the shaft leads to fake-pellets displacement inducing pellet-clad contact in a localized region of the sample rod.

TABLE III
ROD CONFIGURATIONS

Rod name	N° of displaced pellets	Axial extension of gap closure
C2	1	15 mm
C3	3	45 mm
C4	5	75 mm
C5	9	135 mm

We decided to test different rod configurations taking advantage of different length of balloons in order to identify the minimum number of contacting pellets leading to a measurable effects. Rod configurations are resumed in table V. To calibrate the model and characterize system dynamics response, three other rod configurations were manufactured: an empty rod **C0** to characterize modal parameters governing the linear dynamic response of the sample rod; **C1** which corresponds to the fuel rod being fulfilled with Mo pellets, allowing to highlight possible bias due to the presence of dilatation system; and **C6**, where Mo pellets and clad are designed to stick together. This last configuration allows to measure to effect on sample rod dynamics due to complete closure of the gap.

As previously mentioned, the sample rod is maintained by a sample holder, centered in the experimental section by mean of four elements. In order to collect vibration data of the sample rod from Laser Doppler Vibrometer (LDV) system, the experimental section is completely designed in optical quality plexiglass. This type of non-contact measuring system is capable of collecting the vibration of structures even in the presence of surrounding water.

Specifically designed holders allow axial translation of sensible heads. This allows for reconstruction of mode shapes of the rod and experimental modal analysis. The measurement system is showed in figure 4.

To produce and control flow conditions in the experimental section we have designed a dedicated independent water loop. as shown in figures 5 and 6. All conditioning components are located on a separate frame to limit noise effect on LDV measures. To monitor conditions in the circuit we have installed a flow meter at the experimental section inlet, three thermocouples located at different positions to verify temperature evolution in the loop and three pressure gauges to check pressure drops. A summary of instrumentation characteristics and accuracy is reported in table IV.

Measures during the experience are carried out in three different conditions: in air, in quiescent water and under axial, turbulent flow rate. When the system is in air and in stagnant water, the excitation technique is constituted by a magnetic solicitation system attracting the final section of the sample holder, especially made of ferromagnetic material. Logarithmic decrement applied on the collected signal of the under-damped system is used to obtain its damping ratio. Test of the different rod configurations in air and in stagnant water leads

Fig. 4. View of measurement system by LDV: the clad, highlighted in orange, is visible through the plexiglass experimental section; two metallic rods helps supporting the two red structures holding laser heads. The two laser heads located at 90° lead to verify axisymmetric vibrations of the rod

https://doi.org/10.1051/epjconf/202022504007

TABLE IV
INSTRUMENTATION

Parameter	f [Hz]	Instrument	Accuracy
Vibration	1-10^4	Laser Doppler Interfermeter	10^{-7}m/s
Temperature	1	K type thermocouples	+/- 1.5 K
Pressure	1	pressure gauges	+/- 0.3 bars
Flow rate	imposed	Vortex effect flow meter	+/- 2 l/min

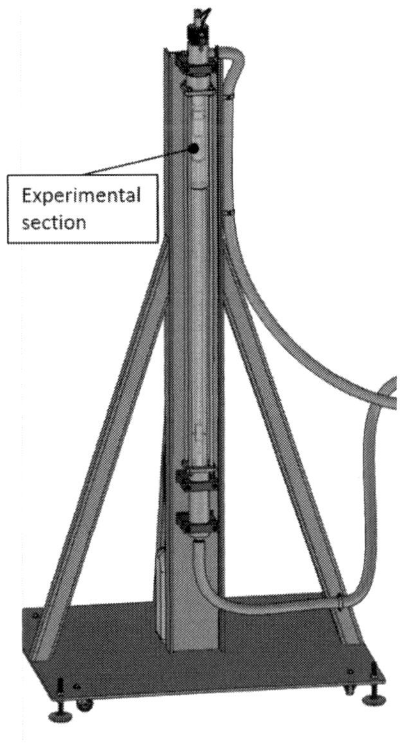

Fig. 5. View of designed experimental test section located on the supporting frame

to the calibration of the unknown parameters defined in section II-B and to verify the possibility of measuring PCI under active solicitation. The test under flow rate excitation then leads first to characterize the amplification factor due to fluid presence and then feasibility of measuring the phenomenon according to a passive approach.

VI. Conclusions and further work

The IMPIGRITIA test bench was designed to investigate PCI and innovative measurement methods. The experiment is conceived to verify the capability of measuring the modification of modal properties and dynamic response as a function of the partial closure of internal pellet-clad gap under turbulent flow excitation representative of PWR environment.
Open literature supports the fact that the non linear mechanisms acting on the dynamics of the fuel rod structure

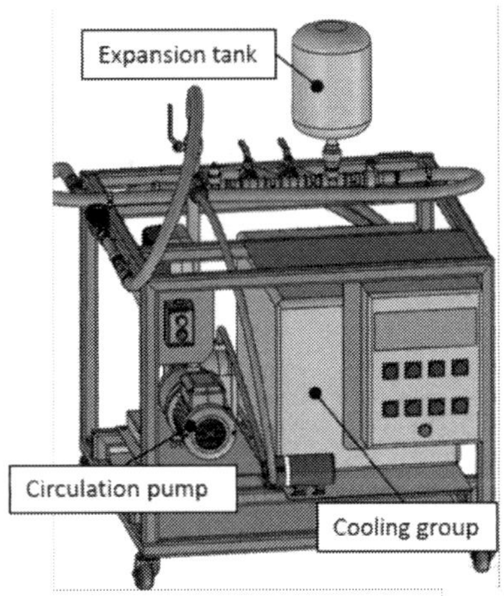

Fig. 6. View of designed water loop conditioning the experimental section

induce measurable variations: the closure of the gap causes a global increase in the isotropy of the dynamic response and a reduction in damping and frequencies of the structure.
The system has been designed though material, mechanical and hydraulic similarity, to reproduce mechanical PCI at room temperature, at low pressure and being representative of the geometry of ADELINE device. Some modifications in the geometry have been carried out which enhance the possibility of measuring the investigated effects.
Due to environmental issues, UO_2 cannot be used in such experiments: the identified representative material is Molybdenum.
According to literature, transversal vibrations in axial turbulent fluid flow are due to pressure fluctuations in the fluid pressure field: hydraulic similarity has been applied and a range of flow velocities varying between [0.9:2.6] ADELINE's one has been identified, using the conservation of pressure fluctuation and the Burgreen principle.
An expansion system has been chosen to locally induce gap closure.
To understand and simulate the physics of the dynamic response of the structure, a model has been developed. It cannot be predictive for the phenomenon under analysis as a large set of parameters has to be first experimentally calibrated. The experimental program has thus been defined to cope with calibration issues, testing in different experimental conditions (air, stagnant water and turbulent axial flow) several rod configurations, notably an empty rod, a fulfilled rod without dilatation system, four rods having PCI axial extension progressively increasing and a full rod with pellets stuck to the clad.
We will characterize vibration mode in terms of damping ratio and eigen-frequencies by measuring the transversal displacement of the system under active controlled solicitation and under passive flow induced vibration. We expect that a

correlation could exists, coupling the axial extension of gap closure with the damping modification of the structure. Whether a measurable modification is clearly identified, we shall proceed to a study to evaluate an adapted instrumentation methodology to cope with ADELINE configuration and nuclear environment, where it would lead to a better understanding of the PCI mechanism under irradiation.

References

[1] B. Michel, J. Sercombe, C. Nonon, F. Michel, and V. Marelle, "Simulation of Pellet-Cladding Interaction with the PLEIADES Fuel Performance Software Environment," *Nuclear Technology*, vol. 182, pp. 124–137, 2013.

[2] V. D'Ambrosi, J. Gatt, F. Lebon, J. Julien, D. Parrat, and C. Destouches, in *Proceedings of MECAMAT 2018*, 2018.

[3] M. P. Païdoussis, *Fluid-Structure Interactions, Slender Structures and Axial Flow*. ELSEVIER, 2004, vol. 2.

[4] M. P. Paidoussis, "Fluidelastic vibration of cylinder arrays in axial and cross flow: State of the art," *Journal of Sound and Vibration*, vol. 76, no. 3, pp. 329–360, 1981. [Online]. Available: http://www.sciencedirect.com/science/article/pii/0022460X81905162

[5] F. Axisa, *Modélisation des systèmes mécaniques : Vibrations sous écoulements*. Hermes, 2001, vol. 4.

[6] G. Ferrari, P. Balasubramanian, and al., "Non-linear vibrations of nuclear fuel rods," *Nuclear Engineering and Design*.

[7] S. Benhamadouche and al., "CFD estimation of the flow induced vibrations of fuel rod downstream a mixing grid," *proceeding of the ASME 2009 Pressure vessels and Piping division conference*.

[8] N. Park, H. Rhee, J.-K. Park, S.-Y. Jeon, and H.-K. Kim, "Indirect estimation method of the turbulence induced fluid force spectrum acting on a fuel rod," vol. 239, no. 7, pp. 1237–1245. [Online]. Available: http://www.sciencedirect.com/science/article/pii/S0029549309001320

[9] M. Gradin and D. J. Rixen, *Mechanical Vibrations, Theory and Application to Structural Dynamics.*, third edition ed. Wiley.

[10] A. Alarcon, "Code aster: Modeling of damping in dynamics linear."

[11] B. De Pauw, W. Weijtjens, S. Vanlanduit, K. Van Tichelen, and F. Berghmans, "Operational modal analysis of flow-induced vibration of nuclear fuel rods in a turbulent axial flow," *Nuclear Engineering and Design*, vol. 284, pp. 19–26, Apr. 2015. [Online]. Available: http://www.sciencedirect.com/science/article/pii/S0029549314006694

[12] P. Wriggers, *Computational Contact Mechanics*, second edition ed. Springer Berlin Heidelberg New York.

[13] C. Gonnier, J. Estrade, G. Bignan, and B. Maugard, in *Proceedings of IGORR 2017*, 2017.

[14] A. Alberman, M. Roche, P. Couffin, S. Bendotti, D. Moulin, and J. Boutfroy, "Technique for power ramp tests in the isabelle 1 loop of osiris reactor," *Nuclear Engineering and Design*, vol. 168, no. 1, pp. 293–303.

[15] J. K. Fink, "Thermophysical properties of uranium dioxide," vol. 279, pp. 1–18.

[16] J. M. Clinch, "Measurements of the wall pressure field at the surface of a smooth-walled pipe containing turbolent water flow," *Journal of Sound and Vibration*, vol. 9, pp. 398–419, 1969.

[17] V. Prakash, M. Thirumalai, M. Anandaraj, P. A. Kumar, D. Ramdasu, G. Pandey, G. Padmakumar, C. Anandbabu, and P. Kalyanasundaram, "Experimental qualification of subassembly design for prototype fast breeder reactor," vol. 241, no. 8, pp. 3325–3332. [Online]. Available: http://linkinghub.elsevier.com/retrieve/pii/S0029549311003487

EPJ Web of Conferences **225**, 04008 (2020) https://doi.org/10.1051/epjconf/202022504008
ANIMMA 2019

Comparison of the Responses of New Reduced-Size Calorimetric Cells Made of Different Structure Materials for High Energy Deposition

A. Volte[1], C. Reynard-Carette[1], A. Lyoussi[2], J. Brun[1], M. Carette[1]
[1]Aix Marseille Univ, Université de Toulon, CNRS, IM2NP, Marseille, France
[2] CEA, DEN, DER, Instrumentation Sensors and Dosimetry Laboratory, Cadarache, France
christelle.carette@univ-amu.fr

***Abstract*—This paper concerns experimental and numerical works on a new differential calorimeter called CALORRE and validated recently under irradiation conditions in MARIA reactor at low nuclear absorbed dose rate level. Works focus on a specific configuration of CALORRE which was designed especially for the measurement of high nuclear energy deposition rates inside Material Testing Reactors. Due to the high level, a new calibration system was fabricated in order to determine the response of the new configuration under laboratory conditions for a very wide range of electrical power never applied in the literature. The response of the new configuration can be considered linear for this very wide range of electrical power. An analytical calculation shows the contribution of each heat transfer in specific zones. 3D thermal simulations performed by means of COMSOL Multiphysics under irradiation conditions give the predictions of the calorimeter behavior under real conditions (up to 20W.g^{-1}).**

***Index Terms*—Calorimeter, Nuclear Dose Rate, On-line Measurements, Calibration, Irradiation campaign.**

I. Introduction

Research in the nuclear field lacks some experimental data essential for numerical simulations in harsh conditions (including intense neutron and photon fluxes and high displacement per atom per year) and thus to better understand the behavior of inert materials or nuclear fuels under irradiations: understanding important for safety challenges, longevity of existing nuclear power plants (NPPs) and development of new concepts of NPPs. Hence, new very efficient instrumentation and devices are required to measure several crucial parameters on-line with a high quality and accuracy, in particular in the framework of the construction of a new material testing reactor (MTR) in Europe (CEA Cadarache, France), the Jules Horowitz Reactor (JHR). Among these parameters, there is the nuclear absorbed dose rate. It is equal to the energy deposition rate per units of mass induced by radiation interactions with matter. It is a key parameter to predict temperature of material samples inside the entire reactor, to design core or reflector experiments and devices, and finally to interpret in-pile experimental results. The JHR with a nominal power of 100MWth will deliver a high fast neutron flux (5.5 10^{14}n.cm^{-2}.s^{-1} from an energy of 1MeV) leading to a high accelerated ageing (up to 16dpa/year) and a high nuclear absorbed dose rate (up to 20W.g^{-1}). This new latter value, not reached in existing European MTRs, implies improvements and innovations of dedicated sensors used to direct measurements such as in-pile calorimeters. Nowadays, two distinct sensors are used for the nuclear absorbed dose rate measurement in MTRs [1-5]: differential calorimeters and single-cell calorimeters including gamma thermometers. In these two cases, the nuclear absorbed dose rate is quantified inside a dedicated reactor channel thanks to temperature measurements (absolute or differential temperatures) and the use of calibration curves achieved preliminary, under laboratory conditions (out-of-reactor) from steady thermal states when heating elements are integrated inside the calorimeter or from non-stationary thermal states for a single-cell calorimeter without heater. At present, one crucial objective corresponds to the measurement of the nuclear absorbed dose rate up to 20W.g^{-1} since no current calorimeter have been designed for such value (up to 15W.g^{-1} for gamma thermometers and 13W.g^{-1} for differential calorimeters). Consequently, Aix-Marseille University and CEA into the framework of a joint laboratory called LIMMEX are involved in research works on instrumentation and measurement methods for online quantification of nuclear and thermal parameters in Material Testing Reactors (MTRs). In particular, studies focus on calorimeters (single-cell and differential calorimeters) used for the quantification of the nuclear absorbed dose rate [2-7]. Research works coupling experiments (under laboratory and real conditions) with thermal numerical works (1D calculations and 3D simulations) allow the design, the characterization, the calibration, the qualification and the miniaturization of various calorimeters owning different metrological advantages (sensitivity, range, size...). This approach led to the design of a new compact calorimetric cell called CALORRE characterized under laboratory conditions and to a first prototype validated under irradiations condition in the MARIA reactor at low nuclear absorbed dose rates (<1W.g^{-1}). Moreover, a complete experimental parametrical study under laboratory conditions allowed the determination of the influence of the sensor geometry (horizontal fin, sensor height) for the same material structure (stainless steel), the influence of the material structure for a specific horizontal fin design, and the influence of coolant fluid conditions on the metrological characteristics of the prototypes (sensitivity, linearity, range) [7].

Thanks to these works, a new CALORRE configuration has been defined to target a nuclear absorbed dose rate up to 20W.g^{-1}. As the preliminary out-of-reactor calibration has never been realized for such a range, a new calibration system has been developed to apply up to 60W (instead of 4.5 or 6W used for previous differential calorimeters, cf. Fig.1). This paper will focus on this new configuration made of Aluminum, studied by coupling experimental works under laboratory conditions by the new calibration device, theoretical calculations with a 1D thermal model and 3D numerical simulations for predicting behavior under irradiated conditions.

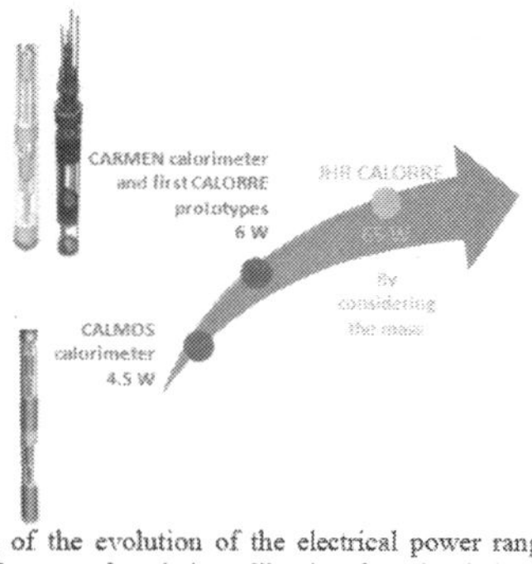

Fig. 1. Diagram of the evolution of the electrical power range applied to differential calorimeters for their calibration by simulating the nuclear absorbed dose rate by means of Joule effect.

The first part of the paper describes the experimental set-up including the new configuration of CALORRE, the new heating system composed of four independent 4-wire heaters used to imitate the nuclear absorbed dose rate by Joule's effect and the new operating protocol adapted for the novel heating system.

The second part is devoted to the experimental characterization of the response of the configuration by imposing an electrical power up to 60W for different conditions. The independent heaters will be used to test the response of the sensor for four identical powers and for asymmetrical depositions. The response and the calibration curves are analyzed by applying a complete 1-direction theoretical thermal model taking into account conductive exchanges (those through the horizontal fin and those inside the gas layers located between the head and the vertical fin) and radiative exchanges between the outer surface of the head and the inner surface of the vertical fin. The contribution of each transfer component is given to explain the sensor sensitivity and the linearity of the calibration curve versus the electrical power.

In the last part, 3D thermal numerical simulations are shown to estimate and validate the whole differential-calorimeter response under irradiation conditions and compare it with a prediction based on a heat balance associated to calibration curves obtained under non-irradiation conditions.

II. Experimental Devices

A. CALORRE Calorimeter

The studied calorimeter CALORRE corresponds to a new heat-flow differential calorimeter (Fig.2 (a)) composed of two superimposed calorimetric cells allowing radial heat transfers (a cell with a sample and the second without to be used as reference cell and remove the nuclear absorbed dose rate due to the cell structure).

Each cell (a head surrounded by an horizontal fin and a vertical fin) is instrumented with two K-type thermocouples (close to the external surface of the sample holder and to the internal surface of the vertical fin respectively) and a heater to perform an out-of-pile calibration (cf. Fig.2 b,c). The studied calorimetric cell (23.1mm in height and 17mm in external diameter) is made of Aluminum (AL5754) and owns a half horizontal fin with 8 metal sectors and 8 empty sectors.

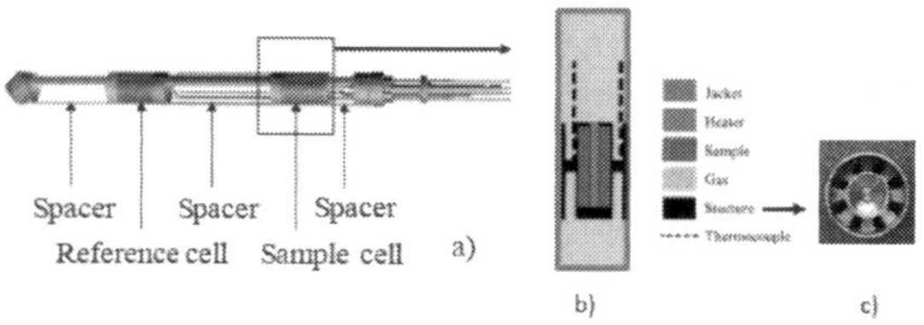

Fig. 2. (a) Diagram of the differential calorimeter assembly composed of two superimposed cells, (b) Diagram of the calorimetric cell, (c) Photograph of the studied calorimetric cell.

The heater for this new configuration has been made especially to be able to inject up to 60W during the out-of-pile calibration step, as said previously.

B. Heating system

The new home-made heating system is composed of four heating elements. Each heating element corresponds to a 4-wire resistance consisting in alumina cylinder drilled with different vertical holes, NiCr wires, an Aluminum sample which holds the 4 heaters (4 connection wires per heater) (cf. Fig.3).

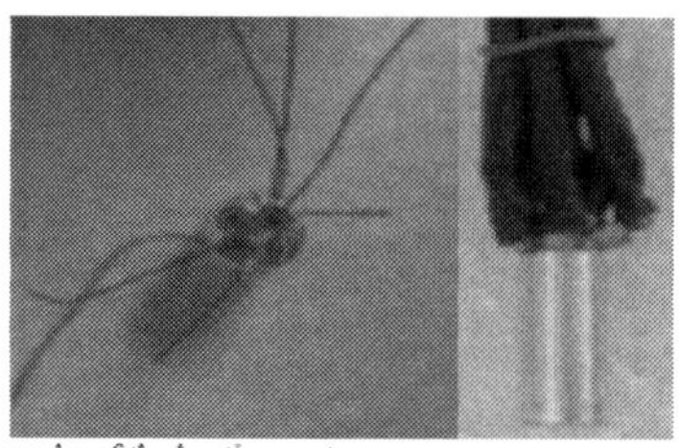

Fig. 3. Photographs of the heating system.

The nature of the material of the resistive wire was changed compared to previous employed heaters (Nickel-chromium instead of Constantan) to have a two times higher electrical resistivity ($1.08\text{x}10^{-6}$ Ω.m against $5.2\text{x}10^{-7}\Omega$.m respectively) leading to a greater resistance for the same length of resistive wire and thus a greater possible injected electrical power range (up to 15W).

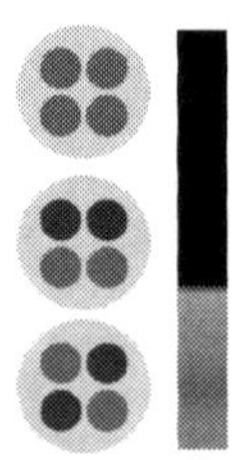

Fig. 4. Diagram for the different operating procedures applied to the heating system.

For the previous works, the power range was only from 0 to 4.5W for CALMOS calorimeter [1] and up to 6W for CALORRE or CARMEN calorimeters [2-7] (cf. Fig.1).

https://doi.org/10.1051/epjconf/202022504008

The four heaters are assembled independently in order to apply different kinds of protocols: the same electrical power in each heater simultaneously, or different values in order to test the influence of a dissymmetric heating (cf. Fig.4).

C. *Experimental Set-Up*

The experimental set-up corresponds to the usual experimental set-up already used to realize the characterization of the temporal response of previous calorimetric cells under laboratory conditions and to do the out-of-pile calibration (cf. Fig.5) [3-4, 7]. It is composed of a fluid circulating bath, 4 accurate resistive-shunts, 4 electrical power suppliers, 1 data acquisition device, 1 computer, 2 K-Type thermocouples and the CALORRE assembly.

Fig.5. Photographs of the experimental set-up.

III. RESPONSE UNDER LABORATORY CONDITIONS

A. *Temporal response and calibration curves*

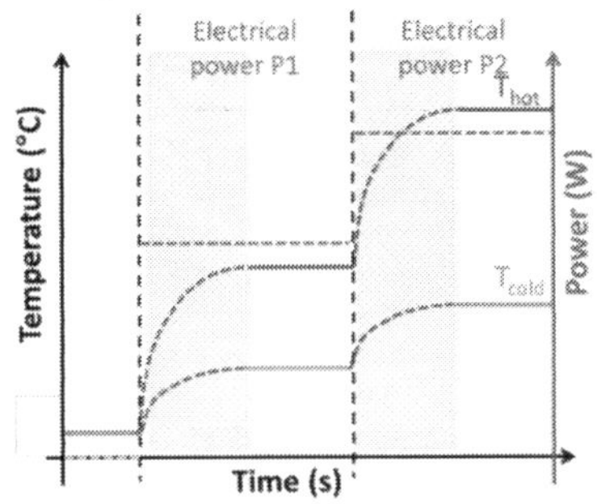

Fig. 6. Diagram of the operating procedure.

For each electrical power applied, the steady state is used to calculate the mean steady cold and hot temperatures (T_{hot} and T_{cold}). Then the calibration curve is obtained by plotting the temperature difference (T_{hot} - T_{cold}) versus the total electrical power applied to all heaters.

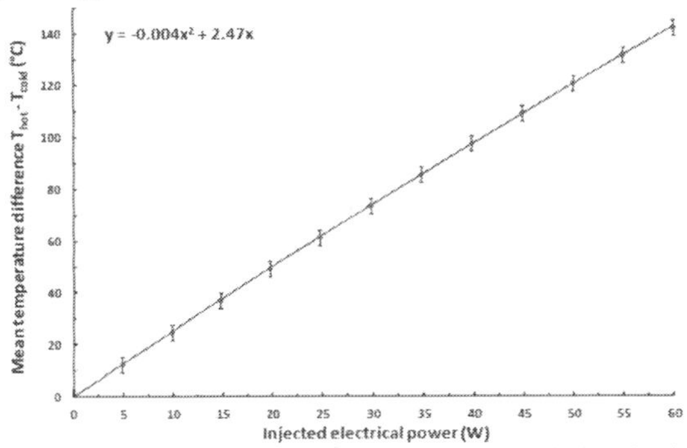

Fig. 7. Calibration curve of the new CALORRE configuration by injecting the same electrical power in each heating elements.

Figure 7 gives the calibration curve for the new configuration, it was obtained by averaging 4 calibration curves effectuated on 4 different days (reproducibility experiments).

B. *Result interpretation*

The calibration curve obtained for the new configuration is almost linear. Its sensitivity is equal to 2.47°C/W. Table 1 shows the results corresponding to other CALORRE configuration. This new CALORRE configuration is the less sensitive configuration but the most linear. By using Aluminum, its sensitivity is divided by 11.5 compared to the same structure geometry made on stainless steel.

TABLE I CALIBRATION FOR FOUR CALORRE CONFIGURATIONS (COOLING FLUID TEMPERATURE EQUAL TO 33°C)

Structure material nature	Horizontal fin geometry	Electrical power range (W)	Coefficient A1 ($°C.W^{-1}$)	Coefficient A2 ($°C.W^{-2}$)
Aluminum	Quarter fin	6	5.04	-0.018
Aluminum	Half fin	60	2.47	-0.004
Stainless steel	Quarter fin	6	22.53	-0.5
Stainless steel	Half fin	6	30.03	-0.816

This is due to the higher value of the thermal conductivity of Aluminum compared to stainless steel (~130$W.°K^{-1}.m^{-1}$ as against ~15$W.°K^{-1}.m^{-1}$).

By using a 1D thermal model including conductive and radiative transfers between the head, the horizontal fin and the vertical fin, the contribution of each transfer can be determined to interpret the behavior of the configurations [7].

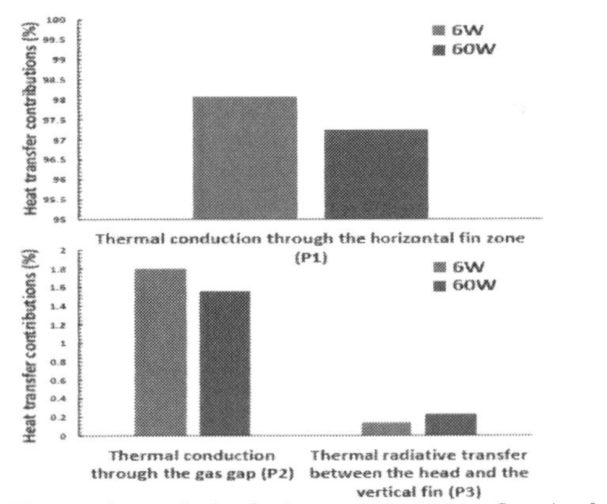

Fig. 8. Heat transfer contribution for the new CALORRE configuration for two applied electrical powers (6W in blue and 60W in red): Thermal conduction through the horizontal fin in the upper figure, thermal conduction through the gas layer localized between the sample holder and the vertical fin and thermal radiative transfer between the sample holder and the vertical fin.

The predominant transfer is due to the thermal conduction through the horizontal fin (metal and gas sectors) (cf. Fig. 8). It corresponds to more than 97% of the total applied electrical power. The contribution of the thermal radiative transfer and that of the thermal conduction through the gas layer are very low (<0.25% and <1.6% at 60W respectively). The thermal radiative transfer is very low because low absolute temperatures are reached (<140°C at 60W). Consequently, the associated thermal distribution leads to a liner response of the sensor on the wide power range (calibration curve with a negligible second-order coefficient).

IV. Prediction for Real Conditions

The thermal behavior of the sensor under real conditions is simulated by using COMSOL Multiphysics 5.3a code. The heat equation is solved for the 3D domain corresponding to the differential calorimeter (two superimposed cells with spacers) and its jacket by neglecting wires (heaters and thermocouples). Thermal conductivities as a function of temperature are considered. Heat sources depending on the material nature are applied to all areas of the considered domain. Convective boundary limits are applied to the external surfaces of the calorimeter jacket by considering a heat transfer coefficient equal to $7325 W.m^{-2}.K^{-1}$. The cooling fluid temperature is equal to 33°C.

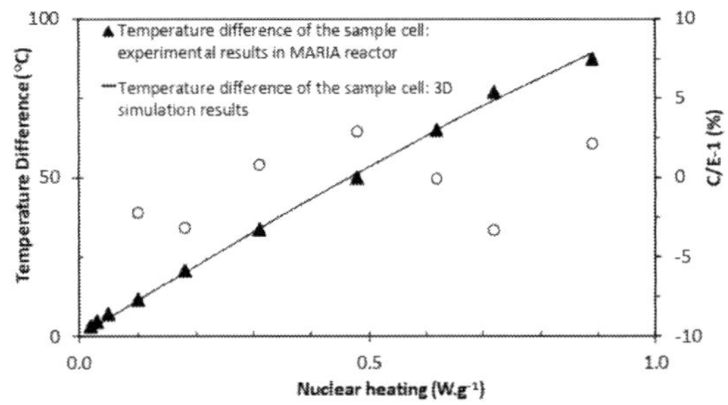

Fig. 9. Temperature differences obtained with the 3D simulations by using COMSOL Multiphysics 5.3a code and experimental temperature differences measured during the MARIA campaign (y-axis on the left) (H-IV-B channel, 24MWth, $h=7325W.°C^{-1}m^{-2}$ and $T_f=33°C$). Relative deviation between numerical and experimental results (y-axis on the right).

Figure 9 gives results obtained with the configuration made of Stainless steel (non-linear response as seen in Table 1) inside the MARIA reactor (H-IV-B channel) at low nuclear heating ($< 1W.g^{-1}$). A good agreement between the calculation and the experiment is observed (discrepancy from -3.3% to 2.9%) (remark: during this campaign, the distance between the two cells remained important because it was kept constant to be equal to the distance between the heads of the CARMEN cells (95mm) in order to do comparison).

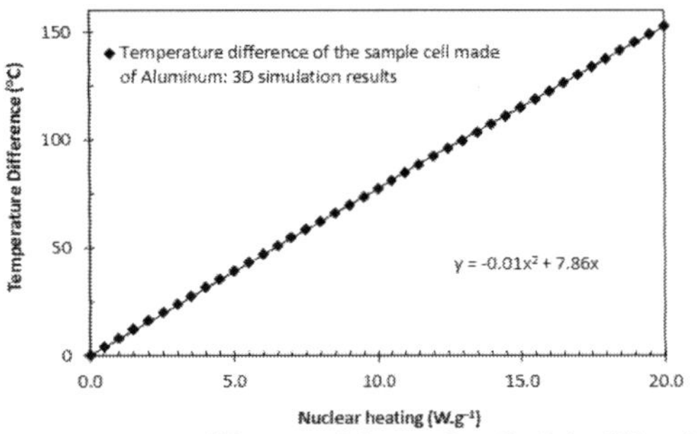

Fig. 10. Temperature differences of the sample cell of the differential calorimeter under irradiation conditions up to 20 $W.g^{-1}$ ($h=7325 W.°C^{-1}m^{-2}$ and $T_f=33°C$).

Considering the validation of the 3D-model, the response of the new compact calorimetric sensor under real conditions and for high nuclear heating rates (up to $20W.g^{-1}$) can be estimated (cf. Fig. 10). The maximal temperature reached inside the calorimeter at $20W.g^{-1}$ (258°C) is much lower than the melting point of Aluminum. This new configuration can be used for the highest nuclear heating rate that will be achieved in the core of the Jules Horowitz reactor at 100MWth by respecting safety margins.

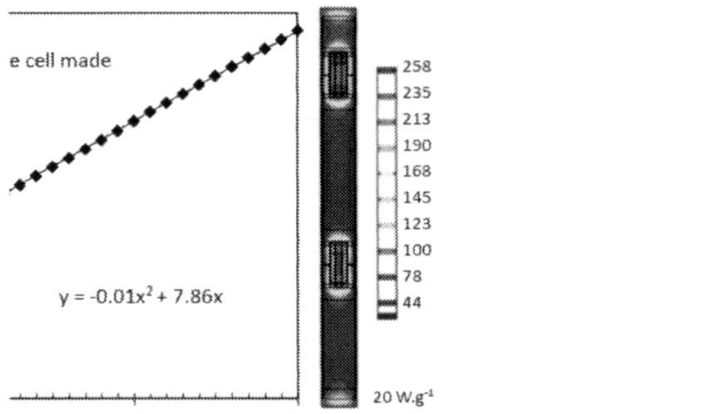

Fig. 11.Temperature field for the new CALORRE configuration for a nuclear heating rate equal to 20 $W.g^{-1}$ ($h=7325 W.°C^{-1}m^{-2}$ and $T_f=33°C$).

V. Conclusions

Thanks to experimental, analytical and numerical works, a new configuration of the CALORRE differential calorimeter was characterized under laboratory conditions and for irradiation conditions (up to $20W.g^{-1}$). A new heating system developed in order to calibrate the sensor for a great electrical power range which represents the total energy deposition in the calorimetric cell head was used successfully. Thanks to it, the power range is multiplied by 10 compared to the literature and previous calibrations (up to 60W). A linear calibration curve for this wide electrical power range was obtained. The linearity is explained by the 1D thermal model. The predominant heat transfer corresponds to thermal conduction through the horizontal fin (area where are localized the two K-type thermocouples). The configuration was validated for high nuclear heating rates (up to $20W.g^{-1}$) thanks to 3D thermal simulations. Parametrical numerical studies will be carried out to optimize the calorimeter response and its size by reducing the inter-cell space in particular.

References

[1] H.Carcreff, "CALMOS: Innovative Device for the Measurement of Nuclear Heating in Material Testing Reactors", *Conference ISRD14*, 2011.

[2] D. Fourmentel *et al.*, "Nuclear Heating Measurements in Material Testing Reactor: A Comparison Between a Differential Calorimeter and a Gamma Thermometer", *IEEE TNS*, Vol. 60, N°1, pp. 328-335, 2013.

[3] J. Brun *et al.*, "Comparison of the thermal calibration response of two calorimetric cells dedicated to nuclear heating measurements", *IEEE TNS*, vol. 61, no. 4, pp. 2254-2261, Aug. 2014.

[4] J. Brun *et al.*, "Responses of Single-Cell and Differential Calorimeters: from Out-of-Pile Calibration to Irradiation Campaigns", *IEEE TNS*, vol. 63, no. 3, pp. 1630-1639, June 2016.

[5] C. Reynard-Carette, *et al.*, " Review Of Nuclear Heating Measurement By Calorimetry In France And USA", ANIMMA 2017 – Advancements in Nuclear Instrumentation Measurement Methods and their Applications, EPJ Web of Conferences **170**, 04019 (2018).

[6] J. Brun *et al.*, “Irradiation Campaign in MARIA Reactor of an Innovative Calorimeter. Comparison of its Nuclear Heating Measurements with Two Kinds of Classical Sensors”, *ConferenceANIMMA)*, 2017.
[7] A. Volte et *al.*, “Study of the Response of a New Compact Calorimetric Cell for Nuclear Heating Rate Measurements”, IEEE Transactions on Nuclear Science, Vol. 65, N°9, pp. 2461-2470, 2018.

EPJ Web of Conferences **225**, 04009 (2020)
ANIMMA 2019
https://doi.org/10.1051/epjconf/202022504009

Calibration of a Single-Cell Calorimeter in a New Transient-state Test Bench

J. Brun*, C. Reynard-Carette, M. Carette, A. Lyoussi, A. Volte, M. Tarchalski, R. Prokopowicz
* Marseille Univ, Université de Toulon, CNRS, IM2NP, Marseille, France,
julie.brun@univ-amu.fr

Abstract– The nuclear radiation energy deposition rate is a key value for the thermal design of experiments, on materials and nuclear fuels, carried out in experimental channels of nuclear research reactors. Studies are led for two kinds of sensor currently dedicated to quantifying this value and corresponding to calorimeter. Development of new sensors but also improvement of their calibration and their associated interpretation methods are necessary. These aims are possible by many ways such as numerical simulations of sensor, characterizations under laboratory conditions and experimental campaign under irradiation conditions. The calibration step under non-irradiation conditions represents a crucial phase. This phase requires the development of specific benches. The present paper focuses on a new thermal-transient bench and its use to perform calibration of a polish single-cell calorimeter. The new bench is detailed. First studies of the influence of external conditions (temperature, velocity) on the calorimeter sensitivity are presented and discussed.

***Keywords*— Calorimetry, MTR, Nuclear Heating Rate, Heat Exchanges, Calibration, Transient State**

I. Introduction

In the field of nuclear energy, the control and measurement of the energy deposition rate per unit mass induced by the interactions of radiation with matter (usually expressed in $W.g^{-1}$) are crucial. This parameter, also called nuclear heating rate, is used in order to carry out accurate studies on the ageing of materials and the behaviour of nuclear fuels under irradiation inside research reactors. In particular, several experiments require this key parameter in order to design thermal device allowing the achievement of specific and stable thermal conditions.

In this context, the IN-CORE program [1-3], run jointly by the CEA and Aix-Marseille University since 2009, was created. Currently, two kinds of in-pile sensors are used to measure nuclear heating rate inside Material Testing Reactors (MTRs): the single-cell calorimeter [4,5], and the differential calorimeter [5-10]. Thus, for each kind of sensor, the IN-CORE program is involved in a scientific axis focusing on the enhancement of sensor design, their out-pile calibration means, methods (protocols) and their in-pile quantification methods, and interpretation analysis. For that, various experimental actions are conducted such as fabrication of specific experimental test benches [10,11] suited to the sensor studied, measurements of thermal properties of materials and developments of multi-sensor mock-ups (cf. Fig.1).

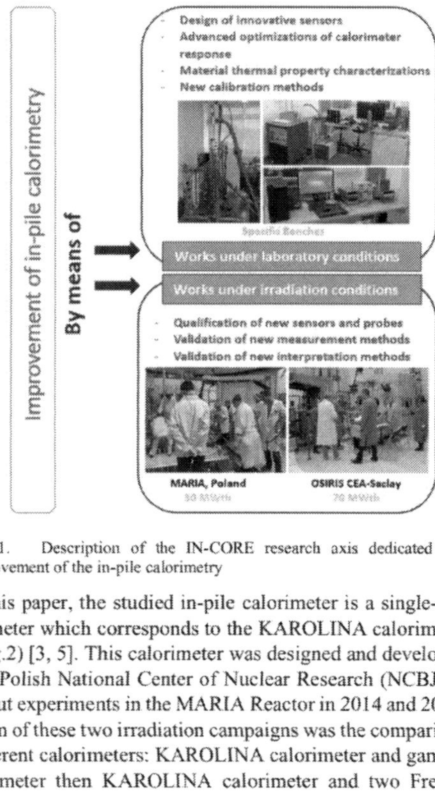

Fig. 1. Description of the IN-CORE research axis dedicated to the improvement of the in-pile calorimetry

In this paper, the studied in-pile calorimeter is a single-cell calorimeter which corresponds to the KAROLINA calorimeter (cf. Fig.2) [3, 5]. This calorimeter was designed and developed by the Polish National Center of Nuclear Research (NCBJ) to carry out experiments in the MARIA Reactor in 2014 and 2015. The aim of these two irradiation campaigns was the comparison of different calorimeters: KAROLINA calorimeter and gamma thermometer then KAROLINA calorimeter and two French differential calorimeters (CARMEN [5] and CALORRE [12,13]).

This single-cell calorimeter designed for a nuclear heating rate range up to $5W.g^{-1}$, is composed of a graphite sample surrounded by a thin layer of Helium gas and a stainless steel jacket. Moreover, this sample is located in the center of the jacket by means of two dedicated sample holders made of porous polymer foam chosen in order to reduce nuclear heating

J. Brun, C. Reynard-Carette, M. Carette and A. Volte are of Aix Marseille Univ, Université de Toulon, CNRS, IM2NP, Marseille, France (e-mail: julie.brun@im2np.fr).
A. Lyoussi is with the CEA, DEN, DER, Instrumentation Sensors and Dosimetry Laboratory, Cadarache, 13108 Saint-Paul-Lez-Durance, France.
M. Tarchalski and R. Prokopowicz are with the National Center of Nuclear Research, Swierk, Poland.

deposition on these two pieces. The gas chosen to fill space between the jacket and the sample is helium due to its high thermal conductivity. Finally, this calorimeter is instrumented by two K-type thermocouples, called Ts and Tf, located at the center of the sample and at the middle height of the external surface of the jacket respectively as shown in Figure 2.

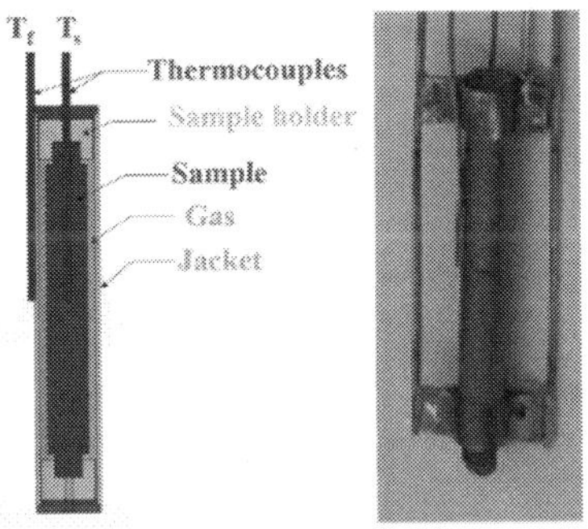

Fig. 2. Scheme of the KAROLINA single-cell calorimeter (in the left–hand section) and its picture integrated inside the INTERCAL probe for the irradiation campaign performed in 2015 (in the right–hand section).

The calibration method of this type of sensor is very different from that of differential calorimeters. In fact, the differential calorimeters integrate heating element in each calorimetric cell, which allows the experimental simulation of an amount of energy, by means of the Joule effect, equivalent to the nuclear heating rate expected in the research reactor and lead to calibration curves by analysing thermal steady states. Therefore, single-cell calorimeters, that do not include heating elements, require other calibration methods. In this case, the calibration methods are based on the analysis of transient states. The processing of the temporal response curve allows the determination of the sensitivity of the sensor from a characteristic value corresponding to thermal time constant.

This paper focuses on the experimental calibration of the KAROLINA calorimeter with a new transient-state test bench. More precisely, this paper begins by the analytical determination of the thermal time constant and the sensitivity of a single-cell calorimeter. Then a detailed presentation of the new transient-state test bench and its experimental protocol are performed. Finally, the experimental temporal response of KAROLINA calorimeter, obtained with this bench, is shown and preliminary studies of the influence of thermal and hydraulic external conditions on this response r and the calorimeter sensitivity are given.

II. THEORETICAL DETERMINATION OF THE SENSITIVITY OF A SINGLE-CELL CALORIMETER

The determination of the sensitivity of a single-cell calorimeter is realized thanks to a transient calibration in two steps. In the first step, the calorimeter is put inside a hot surrounding in order to reach a first thermal steady state at a hot temperature. In the second step, the calorimeter is as fast as possible inserted into a cold surrounding for instance, in order to cool the sensor up to reach a second thermal equilibrium inside it (cf. Fig.3).

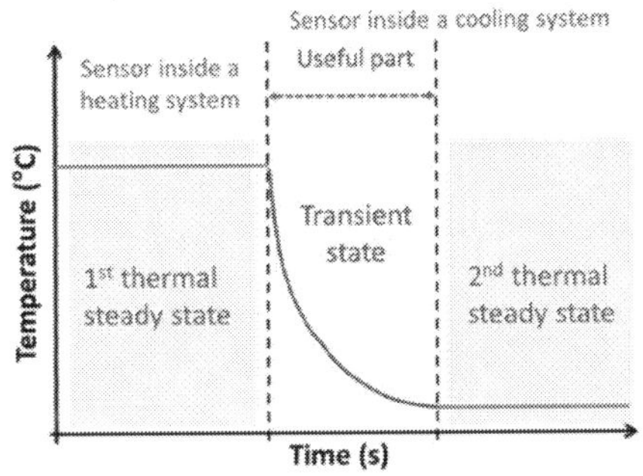

Fig. 3. Diagram of the transient-calibration protocol of a single-cell calorimeter

Thanks to this out-of-pile cooling curve representing temperatures versus time, the thermal time constant (κ in s^{-1}) and the sensitivity of the sensor can be obtained. Indeed, in the case of laboratory conditions by considering conductive transfer only, the heat equation is inside the calorimeter:

$$\rho c_p \frac{\partial T}{\partial t} = \lambda \nabla^2 T \quad (1)$$

where ρ is the density ($kg.m^{-3}$), c_p is the specific heat ($J.kg^{-1}.K^{-1}$), λ is the thermal conductivity ($W.m^{-1}.K^{-1}$).

By applying a heat balance on the calorimeter, the heat accumulation is equal to the heat transfers towards the cooling flow, that leads to this formula:

$$\frac{\partial T}{\partial t} = -\frac{(T_s - T_f)}{\rho c_p V R_{th}} \quad (2)$$

where T_s is the sample temperature (K), T_f is the fluid temperature (K), V is the volume of the area where heat transfers are considered (m^{-3}), and R_{th} is the thermal resistance from the sample to the external surface of the calorimeter jacket.

Thus, with a variable change $(\theta = T_S - T_f)$ and a time integration, the equation (2) becomes:

$$\theta = Ae^{-\kappa t} \quad (3)$$

with

$$\kappa = \frac{1}{\rho V c_p R_{th}} \quad (4)$$

Consequently κ can be determined experimentally by analyzing the transient-state part of the calibration curve and the sensitivity S can be deduced by applying the formula:

$$S = \frac{1}{\kappa c_p} \quad (5)$$

Finally, from this key value, the nuclear heating rate $E_{n+\gamma}$ (in $W.g^{-1}$) can be also calculated with this formula:

$$E_{n+\gamma} = \Delta T_{\infty}.C_p.\kappa \quad (6)$$

To realize this transient-calibration method based on the two successive step, an automatized test bench was developed.

III. Presentation of the BEnch for Calorimeter Calibration under TRANsient conditions (BERTRAN)

To determine accurately the time constant of single-cell calorimeters and their sensitivity, the BERTRAN bench was designed and developed. The main metrological goals for its fabrication were the control and minimization of the transfer time between the hot and the cold surroundings, the capability to perform reproducibility and repeatability and parametrical studies.

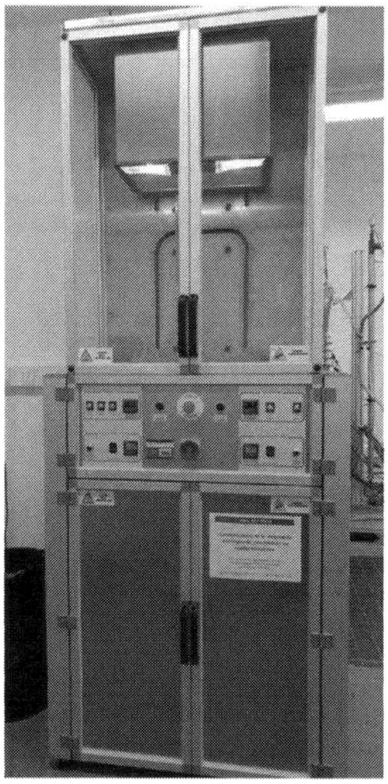

Fig. 4. Picture of the transient-state test bench called BERTRAN

A. Detailed characteristics of this new transient-state test bench

This new experimental setup is composed of:

- Two separate tanks filled with silicon oil in which the single-cell calorimeter is alternatively inserted. These two tanks are used to impose various external thermal and flow boundary conditions to the instrumented cell (a flow in each tank with controlled temperature and velocity) (cf. Fig. 5). The tank located on the right is designed to have a cold fluid temperature from 20°C to 80°C thanks to a heating cartridge of 750W coupled with a cooling device of 300W. The second tank located on the left concerns the hot fluid temperature. A set-up from 120°C to 250°C can be imposed by means of three heating cartridges of 750W each. In each tank, a fluid circulation system is used for the control of the velocity of the flow from 500 rpm to 1250 rpm, which corresponds to a fluid velocity lower than 1.5 m/s.

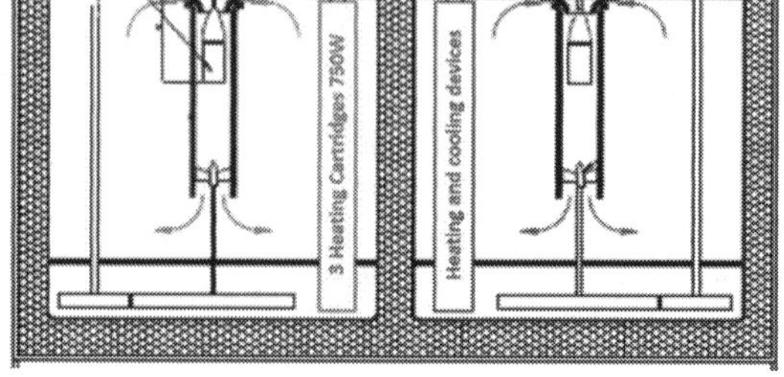

Fig. 5. Diagram of the two tanks: hot and cold surroundings with fluid circulation systems and controlled temperature devices

- A mechanical system composed of two drive chains with two idler sprockets, which move a cart where the calorimeter is fixed thanks to two V-flanges (cf. Fig. 6). This system allows the displacement of the single-cell calorimeter from one tank to the other automatically.

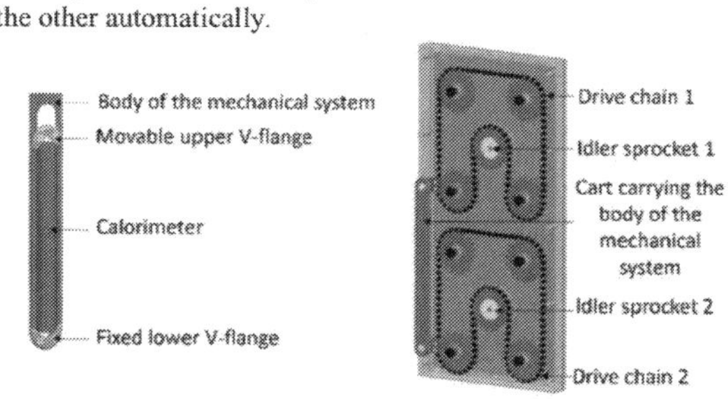

Fig. 6. Diagrams of the system with V-flanges hosting the sensor (in the left–hand section) and the mechanical system ensuring its displacement (in the right–hand section)

The automation of the transfer system of the calorimeter leads to a constant transfer time equal to about 1.3 s.

- A data acquisition system connected to a computer. The device consists of an Agilent data acquisition unit (34970A) containing a 34901A multiplexer of 20 channels used for temperature measurements with the calorimeter K-type thermocouples.

- An extraction system to remove silicon oil vapours.

B. Experimental protocol and temporal response

In order to realise KAROLINA calibration, the temperature of the cold surrounding was fixed to 30°C (a value corresponding to that in the pool of MARIA reactor). The temperature of the warm tank was fixed to 200°C in order to have a significant difference and to reach temperature obtained by the calorimeter sample during irradiation campaigns. An identical velocity in the two baths is also chosen and imposed.

The protocol explained in section II is applied. First of all, the calorimeter is inserted in the hot bath. The temperatures inside the calorimeter increase. When the steady state is reached, the calorimeter is moved and put in the cold bath. The temperatures decrease up to the occurrence of a new steady state as shown on Figure 7. The step 2 is analysed to determine the time constant (cf. Fig. 9). Figure 7 gives results obtained for two experiments (Day 1 and Day 2). A good reproducibility is observed.

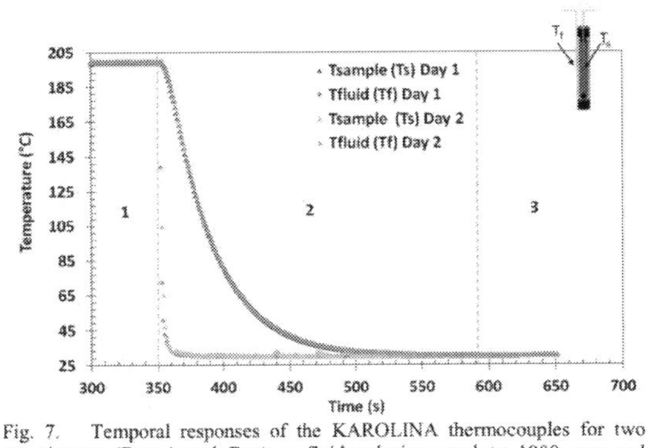

Fig. 7. Temporal responses of the KAROLINA thermocouples for two experiments (Day 1 and Day), a fluid velocity equal to 1000 rpm and temperatures of the hot and cold tanks equal to 200°C and 30°C respectively.

IV. Preliminary Studies with this New Transient-State Test Bench

Preliminary experiments were performed in order to study the influence of the velocity of the fluid flow of the two tanks and the influence of the cold temperature on the sensor sensitivity.

A. Influence of fluid velocity conditions

Various velocities of fluid flow have been tested from 500 to 1250 rpm.

Curves of the sample temperature (Ts) versus time are represented for four velocities in Figure 8. The curve reaches the final cold temperature more quickly when the fluid flow velocity increases. This expected behaviour is due to the intensification of the convective heat exchanges. Nevertheless, a slight influence of the fluid flow velocity can be observed.

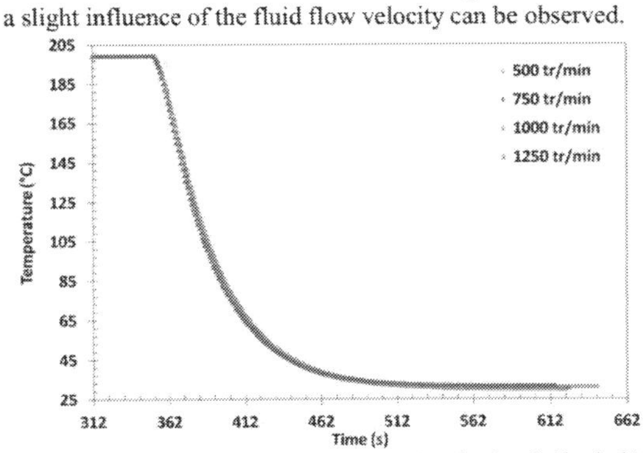

Fig. 8. Calorimeter sample temperature versus time for four fluid velocities from 500 to 1250 rpm and with a hot temperature of 200°C and a cold one of 30°C

Figure 9 shows the cooling part of the temporal curves after a simple mathematical transformation (linearization by considering logarithm functions). Consequently, a direct determination of the constant time κ can be realized as it corresponds to the slope of the linear curve.The four tested velocities lead to this range of time constant: 0.0290 s^{-1}> κ > 0.0283 s^{-1}. This range induces a low variation of the sensitivity (sensitivity calculated by using equation 5) versus the fluid velocity for a same considered temperature of the sample: diminution of about 1°C.g/W from 500 to 1250 rpm (cf. Figure 10).

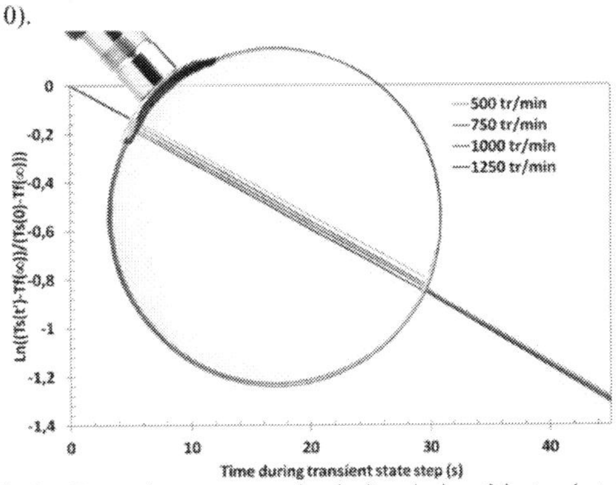

Fig. 9. Temporal curves representing the linearization of the transient state during the cooling step for four fluid velocities from 500 to 1250 rpm and with a hot temperature of 200°C and a cold one of 30°C

But by taking account the variation of the temperature sample leading to a c_p change, a significant variation of the sensitivity versus the sample temperature can be observed (cf. Figure 10). This kind of variation has been also obtained in the case of the differential calorimeters having a high sensitiviy in particular.

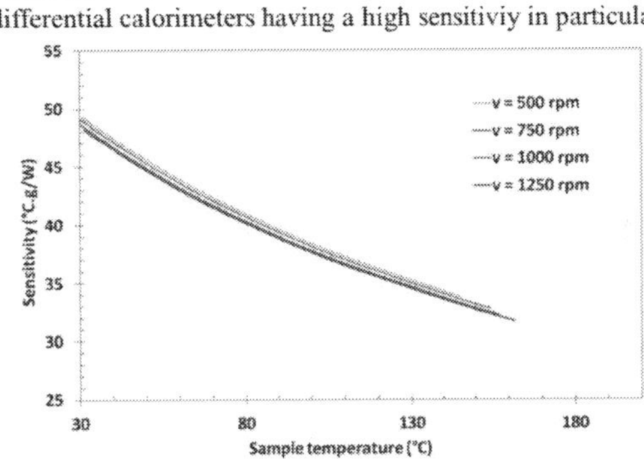

Fig. 10. Sensitivity versus the sample temperature for four fluid velocities from 500 to 1250 rpm and with a hot temperature of 200°C and a cold one of 30°C

The variation in the case of differential calorimeters is linked to the modification of the thermal conductivity of structural materials of the calorimeter and its gas but also to the change of the nature and intensity of heat transfers (increasing of the radiative thermal transfer inside the cell for instance).

B. Influence of fluid temperature conditions

As the fluid temperature inside an experimental channel of a reactor changes versus the axial position, different temperatures of the cold tank were tested.

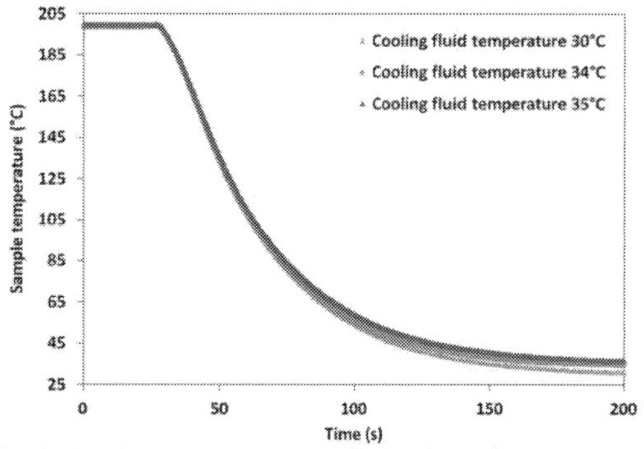

Fig. 11. Sample temperature versus time for three imposed cold temperatures (30°C, 34°C, 35°C), for a hot temperature fixed to 200°C and a fluid velocity of 1000 rpm.

Figure 11 presents the temperature of the sample of the calorimeter versus time for three cold temperatures. It can be also noted a low variation of the temporal response curve. Consequently a low variation of the constant time is observed too.

V. Conclusions and Outlooks

First experiments were performed with a new automated transient-state test bench on a Polish single-cell calorimeter (KAROLINA calorimeter) in order to determine the constant time and the sensitivity of this sensor.

A low influence of the velocity of the fluid flow was shown on the response of the calorimeter, on its time constant and on its sensitivity for a same sample temperature.

By against, a significant influence of the sample temperature on the sensitivity of this sensor was noted. A sensitivity diminution of more than 15°C.g/W has been observed for a temperature increasing from 30°C to 140°C. First experiments of the influence of the cold temperature were carried out.

New parametrical studies by changing hot and cold temperatures and other fluid velocities will be performed to complete these first results of the calibration of single-cell calorimeter. Moreover, a new experimental campaign will be performed for a French compact differential calorimeter called CALORRE. These results will be compared with those obtained by means of a common calibration using heaters (steady state calibration).

Acknowledgment

We acknowledge the IN-CORE program to financially support these works.

References

[1] M.Giot, L. Vermeeren, A. Lyoussi, C. Reynard-Carette, C. Lhuillier, P. Mégret, F. Deconinck, B. Soares Gonçalves. " Nuclear instrumentation and measurement: a review based on the ANIMMA conferences", EPJ Nuclear Sci. Technol. **3**, 33 (2017)

[2] A. Lyoussi, C. Reynard-Carette, "IN-CORE program for on line measurements of neutron, photon and nuclear heating parameters inside Jules Horowitz MTR Reactor, Proc. European Nuclear Conference (ENS), pp. 374-376, May 2014, Marseille, France.

[3] C. Reynard-Carette, J. Brun, M. Carette, A. Volte, A. Lyoussi,"Review of Calorimetry Work dedicated to Absorbed Dose Rate Measurement in MTR and Realized in the Framework of the INCORE and MAHRI-BETHY Programs", Oral presentation Conference IEEE NSS 2019, 26 October - 2 November 2019, Manchester, United Kingdom

[4] M. Tarchalski, K. Pytel, P. Siréta, A. Lyoussi, J. Jagielski, C. Reynard-Carette, C. Gonnier, G. Bignan, "Principle of calibration of the simple calorimeter for nuclear heating measurements in MARIA reactor and transposition to the case of JHR reactor", Proc. ANIMMA 2013, June 2013, Marseille, France.

[5] J. Brun, M. Tarchalski, C. Reynard-Carette ; K. Pytel ; A. Lyoussi ; J. Jagielski ; D. Fourmentel ; J-F. Villard ; M. Carette, "Responses of Single-Cell and Differential Calorimeters: From Out-of-Pile Calibration to Irradiation Campaigns," in IEEE Transactions on Nuclear Science, vol. 63, no. 3, pp. 1630-1639, June 2016.

[6] J. Brun, C. Reynard-Carette, A. Lyoussi, C. De Vita, M. Carette, M. Muraglia, D. Fourmentel, P. Guimbal, J-F. Villard , "Comparison of the thermal response of two calorimetric cells dedicated to nuclear heating measurements during calibration", IEEE TNS, vol. 61, no. 4, pp. 2254 – 2261, August 2014

[7] H. Carcreff, V. Clouté-Cazalaa, L. Salmon , "Development, Calibration, and Experimental Results Obtained With an Innovative Calorimeter (CALMOS) for Nuclear Heating Measurements", IEEE TNS, vol. 59, no. 4, Aug. 2012

[8] H. Amharrak, C? Reynard-Carette, A. Lyoussi, M. Carette, J. Brun, C. De Vita, D. Fourmentel, J-F. Villard, P. Guimbal, "Monte Carlo simulation study of a differential calorimeter measuring the nuclear heating in Material Testing Reactors", Proc. ISRD15, May 2014, Aix-en-Provence, France.

[9] D. Fourmentel, C. Reynard-Carette, A. Lyoussi, J-F. Villard, J-Y. Malo, M. Carette, J. Brun, P. Guimbal, Y. Zerega, "Nuclear Heating Measurements in Material Testing Reactor: A Comparison between a Differential Calorimeter and a Gamma Thermometer," IEEE TNS vol. 60, no. 1, pp. 328 – 335, Feb. 2013

[10] J. Brun, C. Reynard, O. Merroun, A. Lyoussi, M. Carette, A. Janulyte, Y. Zerega, J. Andre, G. Bignan, J-P. Chauvin, D. Fourmentel, C. Gonnier, P. Guimbal, J-Y. Malo, J-F. Villard , "Numerical and experimental calibration of calorimetric sample cell dedicated to nuclear heating measurements", IEEE TNS, Vol. 59, no. 6, Part: 2, pp 3173 – 3179, Dec.2011

[11] C. De Vita, J. Brun, M. Carette, C. Reynard-Carette, A. Lyoussi, D. Fourmentel, J-F. Villard, P. Guimbal, J-Y. Malo, "Qualification of an out-of-pile thermo-hydraulic test bench (BETHY) developed to calibrate calorimetric cells under specific JHR experimental conditions", Proc. ANIMMA 2013, June 2013, Marseille, France.

[12] A. Volte, C. Reynard-Carette, A. Lyoussi, J. Brun, and M. Carette, "Study of the Response of a New Compact Calorimetric Cell for Nuclear Heating Rate Measurements", IEEE TNS, Volume: 65, Issue: 9, Sept. 2018, Pages: 2461-2470.

[13] A. Volte, C. Reynard-Carette, A. Lyoussi, J. Brun, M. Carette., "Comparison of the responses of new reduced-size calorimetric cells made of different structure materials for high energy deposition", Proc. ANIMMA 2017, June 2017, Portoroz, Slovenia.

Performance of Custom-Made Very High Temperature Thermocouples in the Advanced Gas Reactor Experiment AGR-5/6/7 during Irradiation in the Advanced Test Reactor

A. J. Palmer*, R. S. Skifton, M. Scervini, D. C. Haggard, W. D. Swank
* Idaho National Laboratory,
joe.palmer@inl.gov

***Abstract*—The Advanced Gas Reactor-5/6/7 (AGR-5/6/7) experiment is the fourth and final experiment in the AGR experiment series and will serve as the formal fuel qualification test for the TRISO fuels under development by the U.S. Department of Energy. Certain locations in this experiment reach temperatures higher than any of the previous AGR tests, up to 1500°C. Such extreme temperatures create unique challenges for thermocouple-based temperature measurements. High-temperature platinum-rhodium thermocouples (Types S, R, and B) and tungsten-rhenium thermocouples (Type C) suffer rapid decalibration due to transmutation of the thermoelements from neutron absorption. For lower temperature applications, previous experience with Type K thermocouples in nuclear reactors have shown that they are affected by neutron irradiation only to a limited extent. Similarly, Type N thermocouples, which are more stable than Type K at high temperatures, are only slightly affected by neutron fluence. Until recently, the use of these nickel-based thermocouples was limited when the temperature exceeds 1050°C due to drift related to phenomena other than nuclear irradiation. Recognizing the limitations of existing thermometery to measure such high temperatures, the sponsor of the AGR-5/6/7 experiment supported a development and testing program for thermocouples capable of low drift operation at temperatures above 1100°C. High Temperature Irradiation Resistant Thermocouples (HTIR-TCs) based on molybdenum/niobium thermoelements have been under development at Idaho National Laboratory (INL) since circa 2004. A step change in accuracy and long-term stability of this thermocouple type has been achieved as part of the AGR-5/6/7 thermometry development program. Additionally, long-term testing (9000+ hrs) at 1250°C of the Type N thermocouples utilizing a customized sheath developed at the University of Cambridge has been completed with low drift results. Both the improved HTIR and the Cambridge Type N thermocouple types have been incorporated into the AGR-5/6/7 test, which began irradiation in February 2018 in INL's Advanced Test Reactor. Both the improved HTIR and Cambridge Type N thermocouples have exhibited (apparent) low-drift performance during the 170 days of irradiation conducted to date; however, their open circuit failure rates have been quite high (>50%). One of the HTIR thermocouples operated stably at >1450°C for more than 80 days, which we believe is the highest temperature any thermocouple has operated without significant drift in a high neutron flux environment. A description of the performance of the entire AGR-5/6/7 thermocouple set, consisting of 54 thermocouples, is provided in this paper.**

***Keywords*—In-Pile Instrumentation, Material and Test Reactors, High-temperature Irradiation, Thermocouples, Thermocouple Testing.**

I. Introduction

The Advanced Test Reactor (ATR) at Idaho National Laboratory (INL) is being used to conduct an experiment series designated the Advanced Gas Reactor (AGR) tristructural isotopic (TRISO) fuel experiments to qualify this fuel's irradiation performance for High Temperature Gas Reactor (HTGR) applications. The final experiment in this series is denoted AGR-5/6/7. As with the previous AGR experiments, it consists of multiple capsules stacked vertically and with separate temperature-control gas blends for each capsule. It is an instrumented-lead type experiment, with online active temperature control and fission-product monitoring of the effluent gas [1].

Each of the five AGR-5/6/7 capsules consist of a graphite fuel holder with holes drilled axially to accommodate the 12.4 mm diameter × 25 mm long fuel compacts. Metal-sheathed thermocouples and gas lines are inserted into small holes drilled into the graphite holders, as illustrated in Figs. 1 and 2.

Paper submitted 05-15-2019.

This work was supported by the U.S. Department of Energy under DOE Idaho Operations Office Contract DE-AC07-05ID14517. Accordingly, the U.S. Government retains and the publisher, by accepting the article for publication, acknowledges that the U.S. Government retains a nonexclusive, paid-up, irrevocable, worldwide license to publish or reproduce the published form of this manuscript or allow others to do so, for U.S. Government purposes.

A. J. Palmer, R.S. Skifton, D.C. Haggard, and W. D. Swank are with the Idaho National Laboratory, P.O. Box 1625, MS 3840, Idaho Falls, ID, USA 38415-3840. The corresponding author's email address is joe.palmer@inl.gov.

M. Scervini is with the University of Cambridge, Department of Material Science and Metallurgy, 27 Charles Babbage Road, CB3 0FS, Cambridge, UK.

 https://doi.org/10.1051/epjconf/202022504010

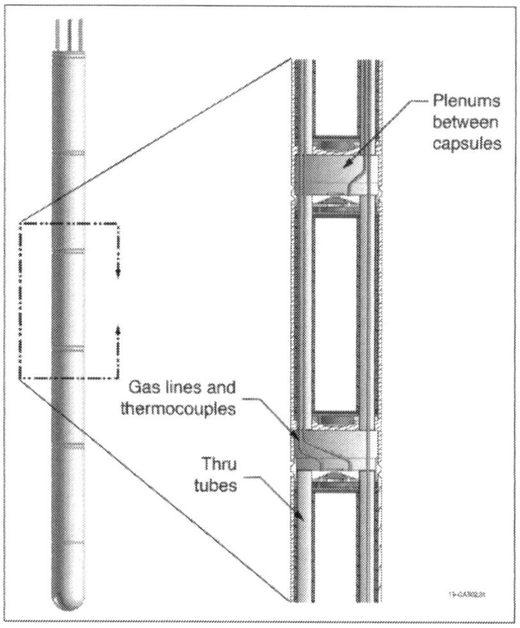

Fig. 1. Elevation view of typical AGR fuel experiment.

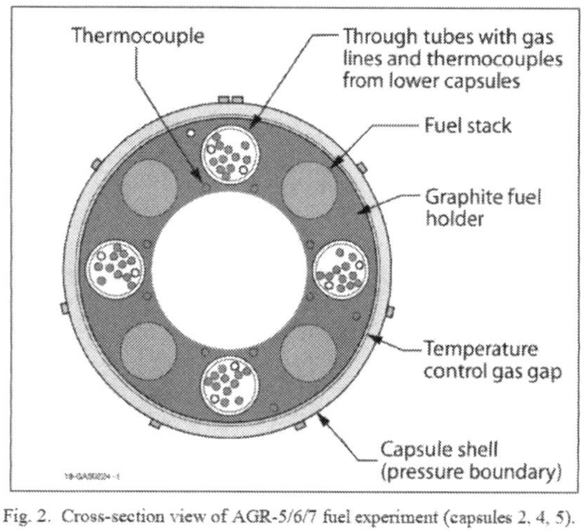

Fig. 2. Cross-section view of AGR-5/6/7 fuel experiment (capsules 2, 4, 5).

AGR-5/6/7 is a combination of three experiments that were envisioned at the beginning of the program. The conditions created in the five capsules making up this test serve as the formal fuel qualification irradiations (AGR-5/6) and the margin test (AGR-7). The purpose of the margin test is to demonstrate that there is a substantial margin between the highest fuel temperature in an operating HTGR and the temperature at which fuel particle failure rate becomes unacceptable.

Temperature measurement has been a critical—but challenging—aspect of all of the AGR experiments because of the high temperatures involved. Commonly used high-temperature commercial thermocouples—such as platinum-rhodium (Types S, R, and B) and tungsten-rhenium (Type C)—suffer dramatic drift as a result of neutron-induced transmutation [2, 3]. These thermocouples types, which are used routinely for high-temperature measurements outside of the reactors, are only used in very special circumstances for reactor experiments. Type K and Type N thermocouples are affected to only a limited extent [4, 5] by neutron irradiation.

However, the use of these nickel-based thermocouples is limited when the temperature exceeds 1050°C due to drift arising from minor alloying elements migrating from the thermocouple's metal sheath to the thermoelements. This change in the composition of the thermolements results in significant decalibration of the signal.

Because it was designed as the fuel qualification and margin test, the AGR-5/6/7 experiment requires measurements in locations with higher operating temperatures than any of the previous experiments. These measurement points experience temperatures up to 1500°C, with many locations above 1200°C. Recognizing the limitations of existing thermometry to measure such high temperatures, the AGR-5/6/7 program sponsor supported a development and testing program for thermocouples capable of low-drift operation at temperatures greater than 1100°C. This program included additional development of high-temperature irradiation-resistant thermocouples (HTIR-TCs) based on molybdenum/niobium thermoelements, which have been studied at INL since circa 2004 [6, 7]. As a result of this development program, a step change in accuracy and long-term stability of this thermocouple type was achieved [8, 9]. Additionally, long-term furnace testing (9000+ hrs) at 1250°C of Type N thermocouples utilizing a customized sheath developed at the University of Cambridge [10] (called herein CAMB Type N) were completed, also with low-drift results [9].

II. AGR-5/6/7 Thermocouples Installation

A. Thermocouple Types and Distribution

The calculated temperature measurement range for the AGR-5/6/7 thermocouples encompassed a range from 700 to 1500°C. It was determined to divide this broad range into three smaller temperature ranges, as follows:

- <1000°C
- 1000–1250°C
- 1250–1500°C.

For the <1000°C range, standard Type N thermocouples are adequate. For the 1000–1250°C range, it was determined that both HTIR-TCs and CAMB Type N would be used. Only HTIR-TCs have the capability of surviving at temperatures above 1300°C, and so only this type was designated for the 1250–1500°C range.

For AGR-5/6/7, the "standard Type N" thermocouples incoporated an Inconel 600 sheath and MgO or Spinel insulation. The CAMB Type N thermocouples incoporated a proprietary high nickel alloy sheath [9, 10] and MgO insulation. The HTIR-TCs incorporated a pure Nb sheath and Al_2O_3 insulation with thermoelements composed of Mo alloyed with La (0.5–1.0%), and Niobium alloyed with P ≤0.1%, Ta ≤0.3%.

Because of the high temperatures involved in AGR-5/6/7,

 https://doi.org/10.1051/epjconf/202022504010

and because two of the three previous AGR tests had ended with a very reduced thermocouple set (for AGR-2, not one thermocouple was operating at the end of the test), it was thought prudent to incorporate as many thermocouples as possible. The number of thermocouples is limited primarily by the number of cables that can be passed through the thru-tubes of the top capsule. Given the space constraints on the size of the thru-tubes, the AGR-5/6/7 project settled on a 54 thermocouple budget. This budget was twice the number of thermocouples installed in any previous AGR experiment.

As discussed below, capsules near the bottom of the train were expected to experience higher failure rates, and higher temperature capsules would also experience higher failure rates. Based on these criteria, the 54 thermocouple budget was distributed as shown in Table I.

In accordance with the ATR numbering convention, capsules in the AGR-5/6/7 test train are numbered from the bottom up; therefore, capsule 1 is the bottom capsule, and capsule 5 is the top one. Capsule 3 represents the margin test and it was designed for the highest fuel temperatures (1300–1500°C); therefore, a large number was allocated—17. Capsule 1 was designed for the second highest temperatures and is the bottom capsule, so it was also given an allocation of 17. Capsules 2, 4, and 5 were designed for temperatures in decreasing order. Since capsule 2 is lower in the test train, and was designed for a marginally higher temperature than capsules 4 and 5, it was assigned 8 thermocouples. The remaining 12 were evenly divided between the two upper capsules.

TABLE I
AGR-5/6/7 THEMOCOUPLE ALLOCATION

Cpsule ID	Std TCs	CAMB TCs	HTIR-TCs	Totals
Cap 5	6	0	0	6
Cap 4	6	0	0	6
Cap 3	4	7	6	17
Cap 2	8	0	0	8
Cap 1	1	7	9	17
Totals	25	14	15	54

B. Protective Sleeves

In the AGR-5/6/7 experiment, each thermocouple was installed inside a protective sleeve. This was done for two reasons: (1) to protect the thermocouple sheaths from carburization and premature failure; and (2) to protect the SiC layer of the TRISO-coated fuel particles from first-row transition elements, such as Cr, Fe, and Ni. The furnace testing program revealed that the selection of protective sleeves could affect drift performance. For example, in the 1200–1250°C range, the CAMB Type N thermocouples drifted about 5°C per 1000 hrs when surrounded by Nb sleeves [11], but only 2.5°C per 1000 hrs with Mo sleeves [9]. This drift is due to solid state diffusion of the sleeve material through the TC sheath and insulation and into the thermoelements, thus changing their composition. Although, Mo was confirmed superior to Nb as a protective sleeve material, it is also very prone to cracking. Mo was not used in the highest temperature regions to surround CAMB Type N TCs to avoid the risk of a crack allowing Ni from the Cambridge TC sheaths to migrate to the SiC-coated fuel particles. For these locations, ZrO_2 sleeves were used instead.

C. Contributors to Thermocouple Failures

A need exists to effect a seal where the thermocouples penetrate the capsule heads; as such, different methods are possible. For the AGR experiment series, brazing was used, primarily because of space constraints. Unfortunately, brazing creates a brittle area (as shown in Fig. 3), where the thermocouple sheath exits the capsule head, and this brittle region is prone to fracture. Fractures in this location have occurred from handling during assembly, and it is assumed that these locations are likely failure points during irradiation.

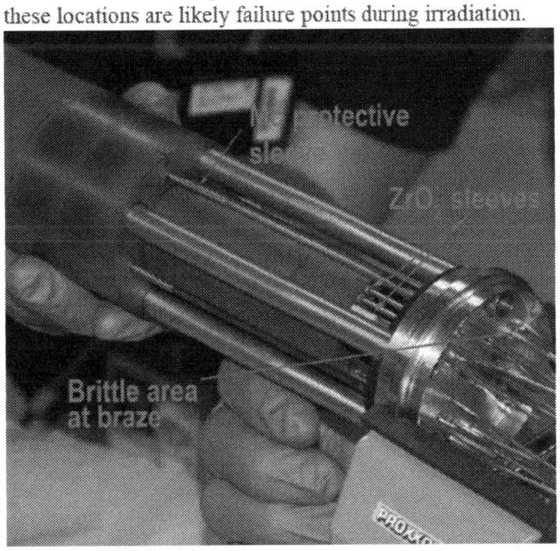

Fig. 3. Capsule 3 protective sleeves and brittle braze area.

All of the test trains in the AGR experiment series incorporated a similar design (i.e., a series of individual capsules stacked on top of each other with thermocouples from lower capsules passing through the bodies of upper capsules by means of small metallic conduits called thru-tubes). This design has the advantage of each capsule being a separate irradiation cell, with a separate gas blend and temperature control system. However, it has the disadvantage of requiring thermocouples from lower capsules to make a long perilous journey through small, hot conduits. Thermocouples from the bottom capsule have a heated length of almost 90 cm and experience thermal expansion-contractions on the order of 6–8 mm each time the reactor stops and starts. An examination of failure data from previous AGR experiments indicated that failures tended to occur from the bottom capsule up, confirming the proposition that thermal expansion was one of the main contributors to thermocouple failure.

The other large contributor to thermocouple failure is peak temperature experienced by the thermocouple. Table II is a temperature limit summary from a major thermocouple distributor [12]. Note that these limits are not the operational limits for Type N thermocouples (Type N thermocouples can operate up to near their melting point for short periods), but rather the recommended diameter for reliable long-term service.

TABLE II
RECOMMENDED SHEATH DIAMETER VS TEMPERATURE FOR MINERAL INSULATED THERMOCOUPLES

Sheath / T/C Dia.	0.032" 0.8 mm	0.040" 1.0 mm	0.062" 1.5 mm	0.093" 2.4 mm	0.125" 3.2 mm
J	260 (500)	260 (500)	440 (825)	480 (900)	520 (970)
K&N	700 (1290)	700 (1290)	920 (1690)	1000 (1830)	1070 (1960)

The sheath diameter of the thermocouples (all three designs) in the AGR-5/6/7 experiment was 1.6 mm. Thermocouple diameters of 1.5 mm or 1.6 mm are very commonly used in high temperature reactor experiments—this size offers a good combination of flexibility and durability. However, according to Table II, temperatures above 920°C are not recommended for this size. The operating temperature of many of the AGR-5/6/7 Type N thermocouples exceed this recommended temperature for a 1.6 mm Type N, so a relatively high failure rate is to be expected. As a practical matter, this is unavoidable for most experiments, as large sheath diameters, such as 3 or 4 mm, are difficult to incorporate and tend to interfere with space allocated for specimens. In addition, large thermocouples are subject to considerable nuclear heating effects, which can raise the junction temperature well above its surroundings.

III. AGR-5/6/7 THERMOCOUPLE IRRADIATION PERFORMANCE

A. Open Circuit Failure

Table III tracks thermocouple failures in the AGR-5/6/7 experiment over the first three standard cycles and one short low-power cycle (163A). This represents approximately 170 days of full power irradiation, but only 155 days of full temperature operation. This is about one third of the planned irradiation period. The failure rate has been higher than expected. To begin with, five thermocouples were lost during assembly. The two samples lost in capsule 3—TC-3-8 and TC-3-16—were HTIR-TCs, which failed because of a manufacturing defect. The sheaths of these two thermocouples

TABLE III
SURVIVING THERMOCOUPLES BY CAPSULE AND IRRADIATION CYCLE

Survivors by Capsule/Type		End of Assem	End of Install	End of 162B	End of 163A	End of 164A	End of 164B
Cap 5	STD	4	3	3	3	3	3
Cap 4	STD	5	4	4	4	4	4
Cap 3	STD	4	4	4	4	4	4
	CAMB	7	5	5	5	2	2
	HTIR	4	4	3	3	3	3
Cap 2	STD	8	8	8	8	6	3
Cap 1	STD	1	1	1	0	0	0
	CAMB	7	7	6	6	3	1
	HTIR	9	9	4	4	4	3
Total Surviving		49	45	38	37	29	23

(pure Nb) were extremely brittle and broke as they were being bent into the thru-tubes of capsule 4. TC-5-6 broke at the braze during assembly, while TC-5-3 and TC-5-4 failed at the transition cup from strain on the cables as the top and bottom sections of the test train were pressed together prior to making the final weld. Unlike previous AGR experiments, AGR-5/6/7 incorporated epoxy only, rather than epoxy-filled metal transition cups, to make the connection between hard cable and soft extension wires. This was apparently the same failure mechanism for the four thermocouples that failed during "installation." It seems that the process of uprighting the test caused a shifting of the thermocouple cables and additional stress on the epoxy transitions. In hindsight, epoxy-filled metal cups would have been a better choice.

The nine thermocouple failures that occurred during assembly and installation were somewhat of an anomaly unique to the AGR-5/6/7 experiment; therefore, they can likely be corrected for subsequent instrumented tests without undue difficulty. The failures during operation are of greater interest. An examination of Table III reveals that of the two variables discussed above (i.e., depth in the test train, and operating temperature), depth has probably been the more dominant factor. For example, the operating temperature of capsule 2 is only a little higher than capsule 4; yet it suffered a 62% failure rate in operation, while none of the capsule 4 thermocouples failed during the first four cycles of operation. Capsule 1 also experienced a high rate of failure, even though all of the Type N thermocouples in this capsule had operating temperatures below 1000°C (all but one were CAMB Type N).

Three CAMB Type N TCs operated at temperatures above 1150°C. Two failed open circuit fairly early in the campaign—TC-3-6 and 3-7. They operated at about 1160°C and 1180°C, respectively, and failed during the second full cycle during an unplanned reactor outage after approximately 62 days of full temperature operation. On the other hand, TC-3-13 has been the control thermocouple for capsule 3 through the first four cycles and has operated successfully at approximately 1160°C for 155 full temperature days.

In comparing the performance between the CAMB Type N and HTIR-TCs installed in the hot locations of capsules 1 and 3, it is clear that the HTIR-TCs survived at a higher rate (even though the HTIR-TCs were installed in higher temperature locations).

An apparent anomaly in Table III is the observation that the four standard Type N thermocouples in capsule 3 all survived the four cycles of irradiation. The graphite holder in capsule 3 was made in two sections (see Fig. 3 where the outer layer is being slid over the inner layer containing the fuel) and the standard Type N were installed in the outer section, which runs at a much cooler temperature (<800°C), and so the survival of these thermocouples was not surprising.

Fig. 4 is a comparison of thermocouple failure trends in AGR-5/6/7 to previous AGR experiments. We note that AGR-2 experienced a modest initial rate of failure, which rapidly accelerated at the end of that test; therefore, the final cycle was run with no functioning thermocouples. In this situation, the thermal model is used to determine an appropriate gas mix for temperature control. The AGR-1 trend line indicates a 50% failure rate. However, this is not a clear picture of the overall thermocouple performance in this experiment. The top capsule had five thermocouples—none of which failed. This represented more than half of the surviving population. The lower three capsules in AGR-1 (e.g., capsules 1, 2, and 3) were irradiated without functioning thermocouples for the last seven cycles (out of a total of 13 cycles). AGR-3/4 had good thermocouple performance. These were standard Type N and only 1 mm diameter, but they operated at lower temperatures—500–700°C for the most part—and the irradiation period was considerably shorter. AGR-5/6/7 has experienced a much higher failure rate during the early phase of irradiation than the previous three experiments. This is partly a function of a much higher percentage of the thermocouple complement being placed in the lower and hotter capsules. However, because of the large initial complement, all capsules had at least three functioning thermocouples at the end of cycle 164B in January 2019, which is about one third of the projected irradiation period.

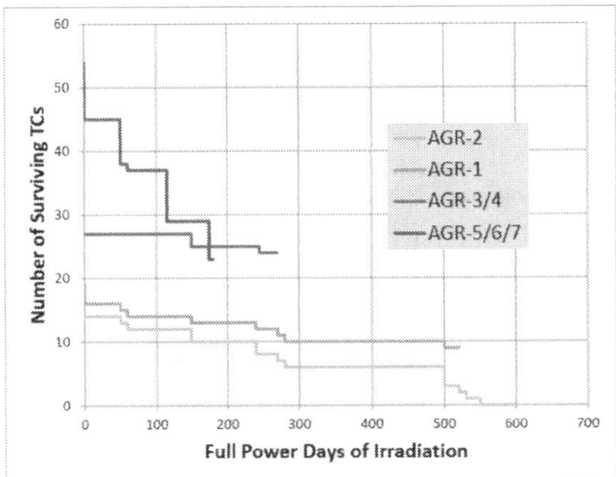

Fig. 4. Thermocouple failure trends for AGR-1, AGR-2, AGR-3/4, and AGR-5/6/7

B. *Initial Accuracy*

All Type N thermocouples for AGR-5/6/7 were fabricated to ASTM E235 specifications. This standard requires calibration checks at 100, 232, and 419°C. ASTM E235 is written around light water reactor applications, and thus, the calibration temperatures are relatively low. Because these Type N thermocouples were to be operated at much higher temperatures than envisioned by E235, samples from all lots of Type N thermocouples that were used in the test were spot-checked against Type B reference thermocouples at 1050°C for standard Type N and 1250°C for CAMB Type N. The results of these spot checks are provided in Table IV.

Table IV provides evidence that the initial lot accuracy of all Type N thermocouples at these elevated temperatures was very good, and substantially exceeded ASTM 230 standard limits of 0.75%. Table IV provides evidence that the initial readings from the Type N thermocouples were likely quite accurate.

Each HTIR-TC was individually calibrated against a Type B reference thermocouple, and a unique fifth order calibration

TABLE IV

LOT CALIBRATION RESULTS FOR TYPE N THERMOCOUPLES IN AGR-5/6/7

TC Type and Lot No.	Check Temp (°C)	Deviation (°C)	Deviation (%)
STD Type N Lot #660	1050	2.6	0.248
STD Type N Lot #565	1050	1.6	0.156
STD Type N Lot #645	1050	1.8	0.178
CAMB N Lot #504	1250	0.2	0.013
CAMB N Lot #613	1250	2.1	0.166

curve was produced for each HTIR-TC. Although all HTIR-TCs were calibrated against Type B reference thermocouples, to date, the EMF output of this thermocouple type has not been as repeatable as Type N. Furnace testing has shown that initial accuracy can be as much as 2% off [9].

Fig. 5 shows calculated temperature contours for capsule 1 during the first few days of the first irradiation cycle, after all capsules were brought to full temperature, and the locations of some of the thermocouples installed in this capsule.

In capsule 1, thermocouples TC-1-1 through TC-1-8 are CAMB Type N, while TC-1-9 through TC-1-17 are HTIR-TCs. In Fig. 5, it can be seen that the calculated temperatures for the Cambridge design were within 1°C of each other (1080°C/1181°C), but the measured temperatures were all more than 100°C lower than the calculated temperatures. Note that there is also considerably more scatter in the measured vs

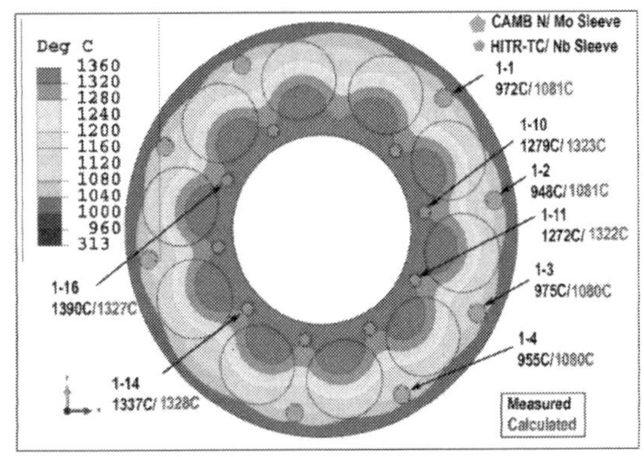

Fig. 5. Measured vs calculated temperatures for capsule 1 early in first irradiation cycle.

calculated temperatures. There is, however, no reason to doubt these measured temperatures this early in their service life, and so this comparison says more about the difficulty of accurately calculating temperatures in a fueled experiment than about the accuracy of the thermocouples (see discussion under "*Drift*" later in this paper). For the HITR-TCs on the right side of the figure, the calculated temperatures exceed the measured values, while the opposite is true on the left side of the figure (measured exceeds calculated). The maximum difference between the measured and calculated values for the HTIR-TCs is 63°C, with the average difference of 43°C, or about 3% of reading. It can be said that the measured temperatures of the HTIR-TCs are in reasonable agreement with the calculated temperatures, but no absolute measure of accuracy can be inferred from this comparison.

Fig. 6 shows the calculated temperature contours of capsule 3 for the first few days of irradiation after achieving full power. Shown in the figure are three Cambridge Type N (3-7, 3-13, and 3-15) and three HTIR-TCs (3-5, 3-12, and 3-14). The three CAMB Type N thermocouples are in good agreement with each other (i.e., within 7°C, as they should be according to the thermal model). Thermocouples 3-13 and 3-14 are worth examining together. This is the only place in the test where a Cambridge Type N and a HTIR-TC are next to each other and the calculated temperatures are essentially the same (i.e., 1185°C vs. 1182°C). In this case, the measured readings are 44°C different, with one being higher and the other lower than

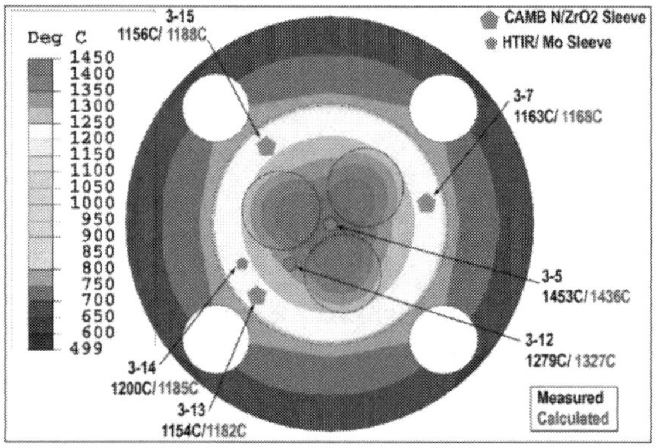

Fig. 6. Measured vs Calculated Temperatures for Capsule 3 Early in First Irradiation Cycle

the calculated temperatures. Again, the overall picture of this situation is that the calculated temperatures offer a rough confirmation of initial accuracy of the developmental HTIR-TCs.

Thermocouple 3-5 measures the highest temperature in the test. The match between the calculated and measured values is quite good, with the measured value coming in at 17°C higher than the calculated one (e.g., an agreement within 1.2%). This may be compared with the next hottest thermocouple in this capsule, 3-12, where the measured value is almost 50°C lower than that calculated. The upshot of this is that the measured temperature gradient between TCs 3-5 and 3-12 is considerably greater than that calculated (e.g., 174°C vs 109°C). The graphite conductivity is possibly less than that estimated in the model. The graphite holder is made from IG-430, and changes in graphite conductivity are known to be dramatic in the first few tenths of a DPA [14].

C. Drift

Long-term thermocouple drift in high-temperature reactor experiments arises from two sources: (1) from thermal effects (as described earlier in this paper); and (2) from nuclear effects (primarily due to transmutation from thermal neutron absorption). Drift due to thermal effects is strongly related to temperature. The standard Type N thermocouples in AGR-5/6/7 have all operated at temperatures less than 950°C; therefore, drift due to thermal effects should be minimal. The maximum operating temperature for CAMB Type N in AGR-5/6/7 has been 1180°C. The test program for this thermocouple type found no drift at 1075°C; at 1250°C, the drift was fairly consistent at 2.5°C/1000 hrs. At temperatures of 1180°C and below, very little drift would be expected for CAMB Type N due to thermal effects, especially over this limited time period. Drift due to nuclear effects is also expected to be quite low for the Type N thermocouple Type, considering the relatively low thermal neutron dose accumulated over the first four cycles of operation (<3.0E+21 n/cm^2).

For the HTIR-TCs, the expected drift discussion is more nuanced. Like the Type N thermocouples, the drift due to nuclear effects at the relatively low dose received in the first four cycles is expected to be minimal. However, unlike the Type N thermocouples, downward drift on the order of 30 or 40°C was expected for HTIR-TCs operating above 1300°C based on results from the furnace testing program [9]. This is because the heat treatment temperature was conducted at 1450°C (to avoid embrittling the Nb sheaths).

Having reviewed the expected drift, we move to the observed drift. Long-term thermocouple drift is difficult to directly quantify in a reactor experiment without resorting to elaborate means, such as incorporating a quantity of pure metal near the thermocouple junction (called a fixed-point cell) [13]. In the absence of such an arrangement, one might assume that drift could be identified by comparing thermocouple measurements to a detailed finite element thermal model. However, as alluded to in the previous section, this approach has its own weaknesses.

An irradiation experiment is a dynamic environment with many variables in play. The overarching difficulty is the large heat flux from the fuel flowing outward to the ultimate heat sink of the reactor coolant. This large heat flux creates steep temperature gradients in the graphite fuel holder (as shown in Fig. 5 and Fig. 6). Second, graphite changes conductivity dramatically under irradiation, which make these steep temperature gradients difficult to calculate accurately. Third, graphite shrinks under irradiation, changing the clearances, and hence, the temperature differences between various components. Fourth, because there are clearances between objects in the capsules, small shifts in relative position are possible: holder relative to shell, fuel relative to its channel, and even thermocouple relative to its hole (as it moves up and down from thermal expansion). Positional shifts during irradiation between the various components can result in significant differences in temperatures between the components. Fifth, the fuel compacts power down over time, which changes the solid body temperature gradients and affects the temperature drops across the clearance gaps between the various components. Sixth, apart from burn up, the volumetric heat generation rate from the fuel is difficult to calculate exactly because each ATR core loading is different (being a mix of used and new assemblies), and the complex interaction of reactor control elements needed to keep power constant during the cycle affects the local neutron flux, and hence, the fuel heat rate. The thermal model attempts to account for most of these factors (e.g., fuel heat rate changes, graphite conductivity change and shrinkage), but there are substantial uncertainties in each of these factors; other factors, such as components shifting relative to each other, are not accounted for at all [14]. Therefore, tracking the temperature difference of two thermocouples as a means of identifying drift, is a useful exercise, but not completely reliable.

As discussed previously, the HTIR thermocouples were only heat-treated and calibrated to 1450°C. The current thinking is that 1600°C fully stabilizes the HTIR-TC. Thus, it was postulated that 1450°C may not have been high enough to fully stabilize the thermoelements [9] and perhaps only a few days of stable operation of TC-3-5 would be achieved because of its high operating temperature. Note that the crystal reordering as a result of heat treatment (or high temperature operation) leads to lower EMF output, and it was expected this thermocouple would drift down after a relatively short period. However, there was no evidence of this phenomenon in TC-3-5 until much later in the irradiation period.

Fig. 7 provides temperature trends for several thermocouples in capsule 3, the margin test capsule. The locations of TC-3-5, TC-3-12, TC-3-13, and TC-3-14 are shown in Fig. 6. The difference between TC-3-5 and TC-3-13 (the control TC) gradually increased from about 300°C to 320°C throughout the first cycle (162B). During the second full cycle, this difference varied between 330°C to 350°C, with TC-3-5 reading approximately 1500°C for most of the cycle. A similar comparison can be made between TC-3-5 and TC-3-12 (i.e., the second highest temperature measurement in capsule 3). At the beginning of the first cycle, the difference between TC-3-5 and TC-3-12 was 175°C and stayed quite stable, gradually increasing to 180°C by the end of the cycle. During the second full cycle, the difference between TC-3-5 and TC-3-12 gradually decreased from 185°C to about 145°C. These comparisons with nearby thermocouples and with the thermal model lead to the conclusion that the true temperature at TC-3-5 was likely between 1430°C and 1500°C for approximately 85 days of high flux irradiation; this likely provided a temperature reading for HTIR-TC within 30 to 50°C of the actual temperature (i.e., 2 to 3% of reading accuracy). We believe this represents the highest temperature any thermocouple has operated without significant drift in a high neutron flux environment.

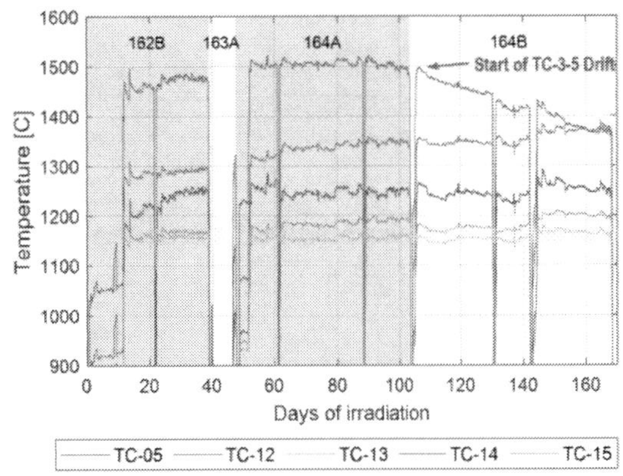

Fig. 7. Capsule 3 selected thermocouple trends.

However, as can be seen in Fig. 7, shortly after the start of cycle 164B, TC-3-5 began drifting downward in a relatively rapid fashion. By the end of cycle 164B, TC-3-5 read less than TC-3-12, which is not physically plausible given that TC-3-5 is located in the center of a three fuel stack array, while TC-3-12 is outboard of the fuel array. The cause of this drift is unknown, but as discussed previously, may have been due to the lower than ideal heat-treatment temperature. Improvements have since been made in the heat-treatment process to allow for heat treatment at 1600°C, which should more fully stabilize the thermoelements and may provide for longer functional operating times at temperatures in the 1400–1600°C range.

Although TC-3-5 was the star performer in terms of peak temperature measured in the AGR-5/6/7 experiment, several other HTIR-TCs have also performed well to date at temperatures above 1250°C. Fig. 8 provides thermocouple trends in capsule 1 for four cycles. Over this period, TC-1-12 and TC-1-14 tracked each other quite well. During the first cycle (162B), TC-1-12 and TC-1-14 showed a constantly increasing temperature difference compared to the control thermocouple operating at a relatively steady 975°C. This increasing temperature difference was probably a result of decreasing conductivity in the IG-430 graphite holder resulting in a larger gradient between the thermocouples outside the fuel ring vs inside (partly evidenced by both thermocouples trending in the same direction). After the first cycle, TC-1-14 has gradually trended downward from 1380°C to about 1340°C, and TC-1-12 has trended downward from 1310°C to about 1255°C. These similar trends provide a degree of assurance that this is likely not due to thermocouple drift, but rather an effect of the fuel compacts burning out (powering down) and resulting in a smaller temperature difference between the control thermocouple (TC-1-4) outside the fuel ring operating at a steady 975°C, and TCs 1-14 and 1-12, which are inside the fuel ring.

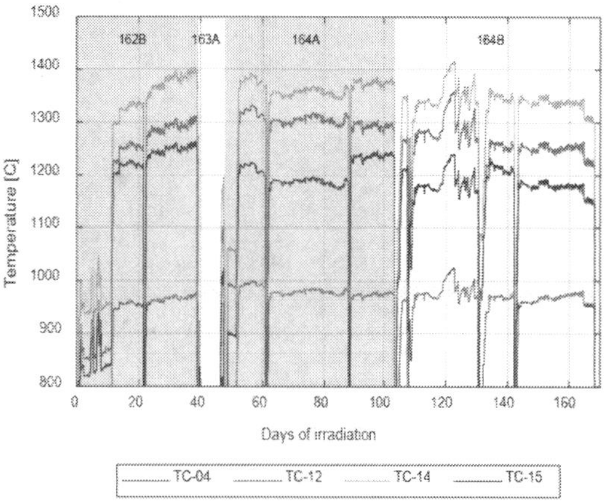

Fig. 8. Capsule 1 selected thermocouple trends.

IV. Conclusions

1. Because of the high operating temperature of the AGR-5/6/7 experiment, a large number of thermocouples were incorporated into the test. In the lower three capsules, the failure rates have been even higher than expected (>60%) during the first four operating cycles. However, all capsules had at least three operating thermocouples at the end of this period.

2. HTIR-TCs survived better than the CAMB Type N in the most severe locations in the experiment. However, one CAMB Type N operated at 1160°C for 155 days without apparent drift.

3. One HTIR-TC operated at 1450–1500°C for 85 days without observable drift, which we believe to be the highest sustained temperature ever recorded in a reactor experiment. Three other HTIR-TCs operated at 1250–1400°C, also without observable drift.

References

[1] A. J. Palmer, D. A. Petti and S. B. Grover, "Advanced Gas Reactor (AGR)-5/6/7 fuel irradiation experiments in the Advanced Test Reactor," in Proceedings of ICAPP 2014, Charlotte, NC, USA, April 6–9, 2014, Paper No. 14017.

[2] M. J. Kelly, W. W. Johnston and C. D. Baumann, "The effects of nuclear irradiation on thermocouples," in Temperature Its Measurement in Science and Industry, 1962, pp. 265–269.

[3] J. D. Heckelman and R. P. Kozar, "Measured drift of irradiated and unirradiated W3%Re/W25%Re thermocouples at nominal 2000K," in Temperature Its Measurement in Science and Industry, 1971, pp. 1935–1949.

[4] B. W. Washburn, "A thermocouple evaluation model and evaluation of Chromel-Alumel thermocouples for high temperature gas cooled reactor applications," LA-NUREG-6768-MS, 1977.

[5] M. Scervini, C. Rae and B. Lindley, "Transmutation of thermocouples in thermal and fast nuclear reactors," in Conference Proceedings of ANIMMA 2013, June 23-27 2013, Marseille, France, pp. 1–8.

[6] J. L. Rempe, et al., "Thermocouples for high-temperature in-pile testing," *Nucl. Technol.*, 156, pp. 320–331, Dec. 2006.

[7] J. L. Rempe, et al., "Long duration performance of high temperature irradiation resistant thermocouples," in Proceedings of ICAPP 2007, Nice, France, May 2008.

[8] R. S. Skifton, A. J. Palmer and P. Calderoni, "Optimization of heat treatment and calibration procedures for high temperature irradiation resistant thermocouples," *Instrum. Sci. Technol.*, 46, pp. 349–363, Dec. 2017.

[9] A. J. Palmer, et al., "Development and testing of thermocouples for the Advanced Gas Reactor fuel experiment AGR-5/6/7," in Proceedings of NPIC and HMIT 2019, Orlando, FL, USA, Feb. 9–14, 2019, Paper No. 25978.

[10] M. Scervini and C. Rae, "Low drift type N thermocouples for nuclear applications," in Conference Proceedings of ANIMMA 2013, Marseille, France, June 23–27, 2013, Paper No. 1054.

[11] A. J. Palmer, et al., "Summary of thermocouple performance during Advanced Gas Reactor fuel irradiation experiments in the Advanced Test Reactor and out-of-pile thermocouple testing in support of such experiments," in Conference Proceeding of ANIMMA 2015, Lisbon, Portugal, April 20–24, 2015, Paper No. 51.

[12] *The Temperature Handbook*, Fifth Ed., Omega Engineering, Inc., p. H-7, 2004.

[13] M. Laurie, et al., "New fixed point mini-cell to investigate thermocouple drift in a high-temperature environment under neutron irradiation," in Conference Proceedings of ANIMMA 2015, Lisbon, Portugal, April 20–24, 2015, Paper No. .

[14] G. L. Hawkes, et al., "Thermal Predictions of the AGR-3/4 Experiment with Post Irradiation Examination Measured Time-Varying Gas Gaps," *ASME J. Nuclear Rad. Sci.* 3(4), 041007 (Jul 31, 2017), Paper No: NERS-17-1008; doi: 10.1115/1.4037095.

EPJ Web of Conferences **225**, 04011 (2020) https://doi.org/10.1051/epjconf/202022504011
ANIMMA 2019

Perspective liquid scintillators for spectrometry of neutron and gamma radiation

Kawa Haji Mahmoud[1], Vaclav Prenosil[1], Zdenek Matej[1], Frantisek Kucera[2], Filip Mravec[3], Michal Kostal[4], Frantisek Cvachovec[5], David Valis[5]

***Abstract*—This paper presents some results of the development of two new scintillation detectors usable for the detection and spectrometry of neutron and photon components of the mixed field. It is a series of liquid organic scintillators containing carbon, hydrogen, nitrogen and oxygen. Elements of individual scintillator designs are arranged in various chemical bonds defining physical and chemical properties of the scintillators. These scintillators have advantageous useful properties. They are non-hygroscopic, not poisonous and are not sensitive to contact with the outside atmosphere and can be stored in conventional laboratory containers.**

***Keywords*—scintillators, scintillation detectors, pulse shape discrimination, Figure of merit**

I. INTRODUCTION

The authors list basic properties of two types of new liquid scintillators LSB205 and LSB206 that are compared with the characteristics of the BC 501 A (NE-213) liquid scintillator, which can be considered the standard used for many years for detection and spectrometry in the mixed field of neutron and photon radiation. Characteristic data of the scintillators developed are focused on the energy range useful, e.g. reactor dosimetry, i.e., neutron energy $\leq$ 15 MeV. This paper mainly contains qualitative and quantitative data on the resolution of neutron and photon particles detected by the scintillator. The resolution of the charged particles is done by pulse shape discrimination, which is based on a variant of the integration method [1], [2]. Quantitatively, the quality of the resolution of charged particles is characterized by the generalized function of Figure of Merit - FOM [2], [3]. If FOM > 1, neutron and gamma radiation responses are perfectly separated. If FOM < 1 then the responses of both particles overlap. Besides, the article contains data on energy resolution and light output of scintillator. The interesting data were observed when investigating the relations of characteristics in a temperature range of 0 – 50 °C. In order to use these scintillators for gamma/neutron mixed field spectrometry, it is necessary to obtain light output function at least for protons. The obtained scintillator temperature dependencies described in this article will enable to refine the measurement of the amplitude resolution of the neutron spectrum measured at the collimated reactor output beam moderated by silicon. This spectrum contains several neutron lines with well-defined energies which could substitute hard to obtain mono-energetic neutron sources [4], [5]. The next step towards the usability for neutron and photon spectrometry in the mixed field is the calculation of the response functions needed for the unfolding of energy spectra.

The properties of newly developed liquid scintillators give them the potential to be used in various areas of nuclear physics, including nuclear safety.

II. DESCRIPTION OF THE MEASURING OBJECT

There are many types of scintillators for mixed neutron and gamma radiation fields. The main requirement for such a scintillator is the ability to distinguish between neutron and gamma radiation interactions. In this paper, we want to focus on comparing the temperature dependencies of the standard BC-501A scintillator and two versions of the newly developed LSB 205 to LSB 206 scintillators containing 88.3 to 91.1% C; 7.6 to 8.7% H; 0.1 to 1.9% N and 0.1 to 2.2% O in the CF^{252} source pool.

III. METHODOLOGY

We compare the thermal properties of the experimental scintillators and the standard liquid scintillator BC-501A. This scintillator has very good separation properties. However, it loses its properties when exposed to air oxygen. The composition of the liquid scintillators tested is not public - it is still protected.

IV. EXPERIMENTAL SETUP

The NGA-01 spectrometric system was used for the measurement. This two-parameter system uses

This work was supported in part by the Ministry of Industry and Trade, within the project TRIO, No. FV20453.
1 Masaryk University, Botanická 15, Brno 612 00, Czech Republic
2 Faculty of Chemistry, Brno University of Technology, Purkyňova 464/118, 612
3 VF, a.s. Svitavská 588, 679 21 Černá Hora, Czech Republic
4 Research Center Rez Ltd, 250 68 Husinec-Rez 130, Czech Republic
5 University of Defence, Kounicova 65, Brno 612 00, Czech Republic

 https://doi.org/10.1051/epjconf/202022504011

2" PMT Hamamatsu R329-02 equipped with an active tapered voltage divider as a detector. This configuration allows the device to be used even with large amplitude ranges of input signals while maintaining excellent output linearity. This configuration ensures excellent results even at high count rates per second. The device can measure up to 700,000 pulses per second.

The NGA-01 measures detector output pulses using two differently amplified branches for two 12-bit ADC inputs with a sample rate of 500 MS/s - see Figure 1. Two channels are used for better signal-to-noise ratio, especially for smaller amplitudes. The data are digitally filtered, summed and evaluated by Field Programmable Gate Array (FPGA).

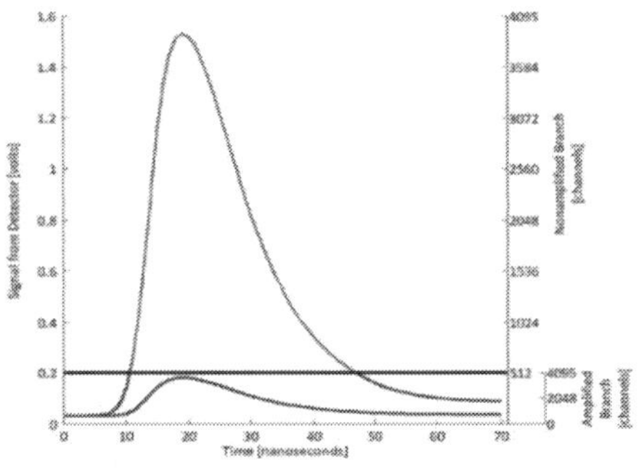

Fig. 1. Amplified and non-amplified impulse selection [3], [6]

The Cs^{137} and Co^{60} emitters were used for calibration of the measurement. The Compton edges in the photon instrument spectrum are for the Cs^{137} a 0.477 MeV energy edge and for the Co^{60} emitter a 1.17 MeV energy edge.

For all scintillators, we used the same PSD (pulse shape discrimination) method based on Charge Comparison [1], [2] implemented directly in the FPGA of the spectrometer NGA-01.

The individual pulse parameters used in the relationship to quantify the PSD are 'energy delta' and 'discrimination delta' values are determined empirically for optimal delimitation of areas with maximum neutron and photon responses in the scintillator being measured.

V. Description of the temperature chamber

The Vötsch Industrietechnik GmbH VT4004 type chamber was used for temperature dependence testing. Temperatures of 0, 5, 10, 20, 30, 40 and 50 degrees Celsius were used for the experiments. A photomultiplier with an active resistor divider was placed in the chamber. The measuring apparatus was outside the air-conditioned space. The connection was made by coaxial cables. The interior of the temperature chamber was shielded from ambient light.

All tested scintillators were air-conditioned at the same time. Calibration sources and test source were mounted on scintillators from their side.

VI. Temperature test results

The results of the experiments were CSV matrices expressing temperature dependence of individual scintillators. The FOM temperature dependencies of each scintillator for the selected test temperature repertoire and concrete energy are shown in the following graphs for each scintillator separately.

We chose a range of 0.5 to 4.5 MeV gamma to display temperature dependencies of separation quality. This area of energy is interesting for our further research.

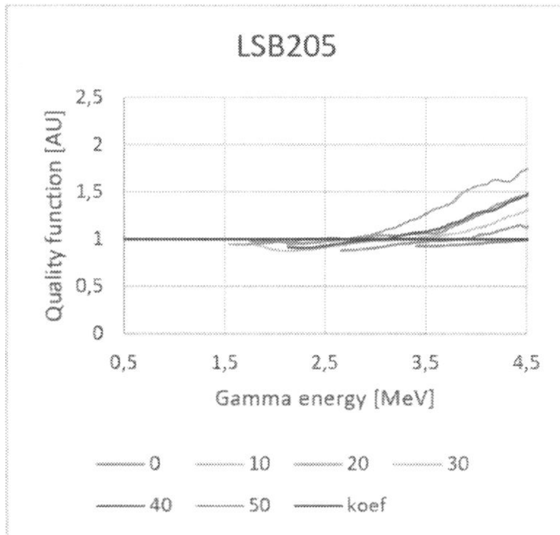

Fig. 2. FOM temperature dependencies of the LSB 205 scintillator

The LSB205 scintillator is not fully satisfactory parameters. It does not register all particles below 1.5 MeV, and therefore it is not possible to evaluate the FOM. It provides reliable separation of particle responses in the order over 3 MeV.

The influence of temperature on the FOM is substantial. It best separates at temperatures around 10 degrees Celsius. For other temperatures, the quality of separation is considerably worse.

The LSB206 scintillator has better parameters. It reliably registers particles with energies below 0.5 MeV, and thus the FOM can be evaluated in the range of selected energies. It provides reliable separation of particle responses in the order of up to 2 MeV.

The influence of temperature on the FOM is again substantial. It best separates at temperatures around 10 degrees Celsius. For other temperatures, the quality of separation is considerably worse.

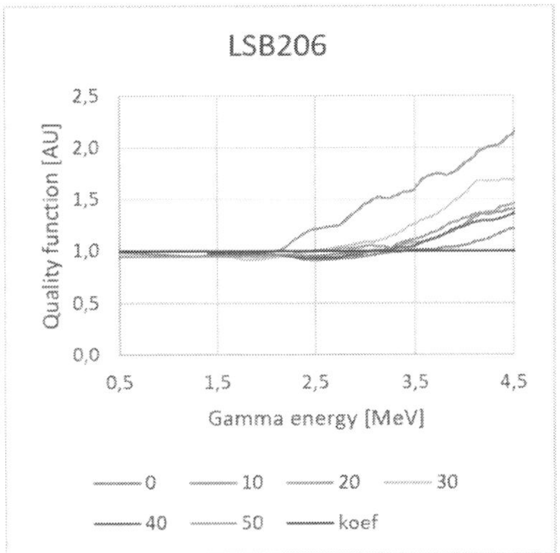

Fig. 3. FOM temperature dependencies of the LSB 206 scintillator

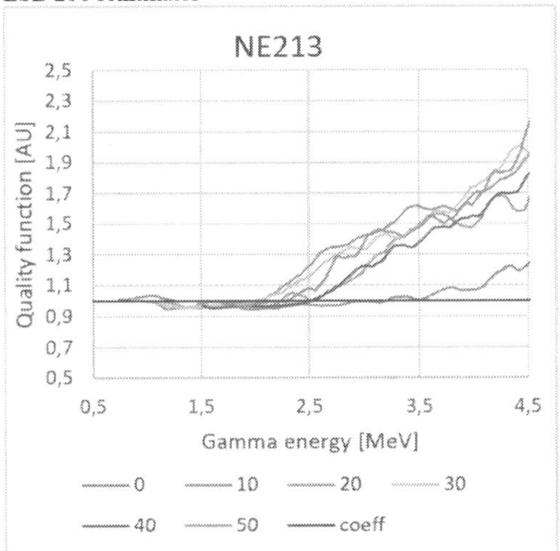

Fig. 4. FOM temperature dependencies of the NE213 scintillator

The NE213 scintillator is basically the liquid scintillator standard for neutron and gamma radiation separation. The dependence of the FOM on the temperature is shown in FIG. 4. Has the best parameters from the studied scintillators. It reliably registers particles with energies below 0.5 MeV, and thus the FOM can be evaluated in the range of selected energies. It provides reliable separation of particle responses in the order of up to 2 MeV.

The influence of temperature on the FOM is again substantial. NE213 scintillator worst separates at temperatures around 0 degrees Celsius. For other temperatures, the quality of separation is essentially the same.

Differences in scintillator behavior will be the subject of further research and experimentation.

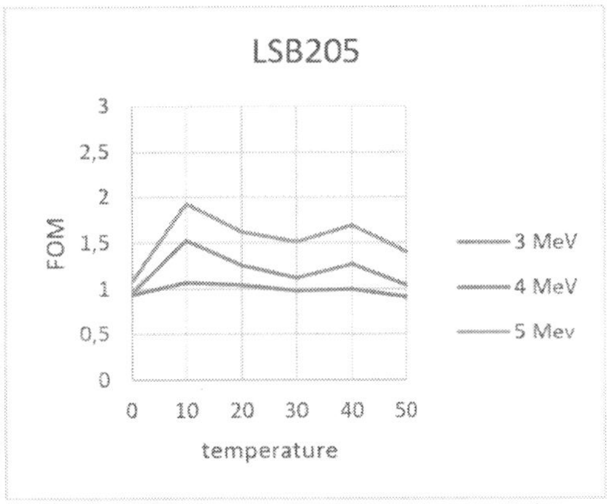

Fig. 5. FOM temperature dependence of the LSB 205 for some energy

The graphs of Figures 5 to 7 show the temperature dependence on the generalized function of Figure of Merit for three particle energies. Specifically for 3, 4 and 5 MeV gamma. It is clear from the graphs that the effect of temperature on FOM has a similar course for all scintillators studied.

The effect of temperature on the FOM of the scintillators LSB205 and LSB206 is most pronounced for 10 degrees Celsius and then decreases. More markedly decreases for the scintillator LSB206.

The temperature effect on the NE213 FOM scintillator is most pronounced for 0 degrees Celsius and then almost constant.

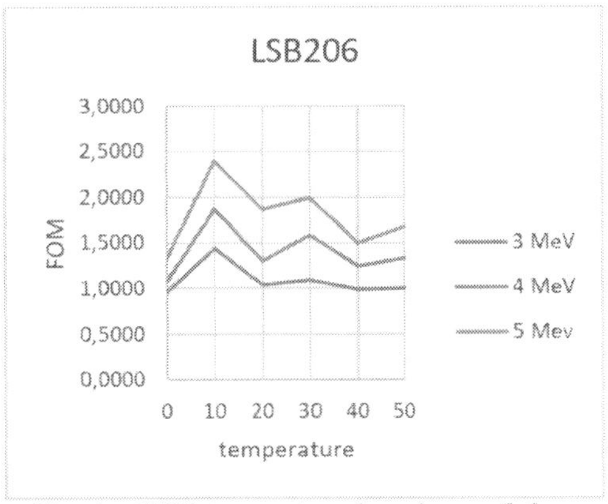

Fig. 6. FOM temperature dependence of the LSB 206 for some energy

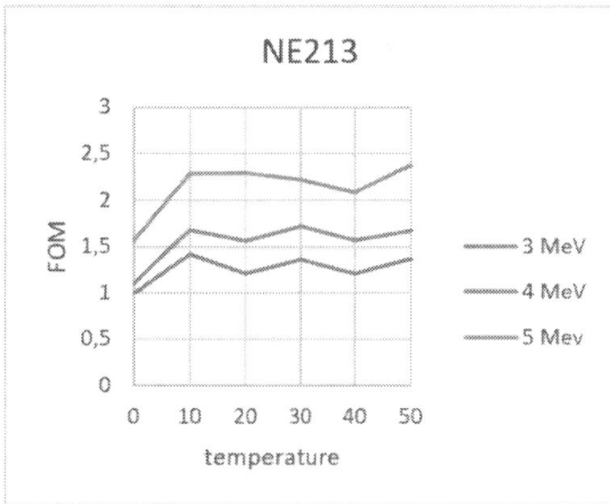

Fig. 7. FOM temperature dependence of the NE213 for some energy

VII. Conclusion

It can be seen from the graphs that the FOM temperature dependence is neither linear nor monotonic. The reason for the non-linearity will be the subject of further research. Knowledge of the FOM temperature dependence of scintillators allows eliminating this dependence in calculations of energy spectra of measured neutron-gamma mixed radiation.

References

[1] M. Amiri, V. Prenosil, F. Cvachovec, Z. Matej, and F. Mravec. "Quick algorithms for real-time discrimination of neutrons and gamma rays," *Journal of Radioanalytical and Nuclear Chemistry*, 2015; 303(1): 583–599. Published online 2014 Aug 24. Doi: 10.1007/s10967-014-3406-5

[2] Z. Matej, M. Amiri, F. Cvachovec, V. Prenosil, F. Mravec, Z. Kopecky, "Digital two-parametric processing of the output data from radiation detectors," *Progress in Nuclear Science and Technology*. Volume 4 (2014) pp. 670-674. DOI:10.15669/pnst.4.670

[3] M. Amiri, F. Cvachovec, Z. Matej, V. Prenosil, F. Mravec, "Digital data processing of stilbene," *3rd International Conference on Advancements in Nuclear Instrumentation, Measurement Methods and their Applications (ANIMMA)*. June 2013. DOI: 10.1109/ANIMMA.2013.6728010

[4] M. Kostal, M. Schulc, J. Soltes, E. Losa, L. Viererbl, Z. Matej, F. Cvachovec, V. Rypar, "Measurements of neutron transport of well-defined silicon filtered beam in lead." *Applied Radiation and Isotopes*, Elsevier, 2018, roč. 2018, č. 142, s. 160-166.

[5] M. Kostal, Z. Matej, E. Losa, O. Huml, M. Stefanik, F. Cvachovec, M. Sulc, B. Jansky, E. Novak, D. Harutyunyan, V. Rypar, "On similarity of various reactor spectra and 235U prompt fission neutron spectrum," *Applied Radiation and Isotopes*, Elsevier, 2018, roč. 135, May, s. 83-91. ISSN 0969-8043 doi: 10.1016/j.apradiso.2018.01.028.

[6] M. Pavelek, Z. Matej, O. Herman, F. Mravec, M. Veskrna, F. Cvachovec, M. Kostal, V. Prenosil, "Fast Digital Spectrometer for Mixed Radiation Fields,". *2017 IEEE SENSORS Proceedings*. IEEE, 2017. s. 436-438, 3 pp. ISBN 978-1-5386-4056-2. doi:10.1109/ICSENS.2017.8234012

EPJ Web of Conferences **225**, 04012 (2020) https://doi.org/10.1051/epjconf/202022504012
ANIMMA 2019

Acoustic instrumentation of the new generation of MTR: effect of nuclear radiation on modified Bismuth Titanate piezoelectric elements

JY. FERRANDIS[1], O. GATSA[1], P. COMBETTE[1], D. FOURMENTEL[2], C. DESTOUCHES[2], V. RADULOVIC[3], L. SNOJ[3]

1. IES- CNRS - Montpellier University, France
2. CEA Cadarache, France
3. Jozef Stefan Institute, Slovenia

Abstract :

In this article we present a first part of the results obtained during an irradiation campaign conducted at the Jozef Stefan Institute to observe the behaviour of piezoelectric materials under gamma and neutron flux. Specific instrumentation has been developed and has enabled the monitoring throughout the irradiation of several materials such as lead zirconate titanate (PZT) or modified Bismuth Titanate (BiT) in either massive or thick film form. Various parameters such as resonance frequency, electromechanical coupling coefficient, electrical capacitance, dielectric losses were measured as a function of the flow and dose received. The results obtained confirm that the samples work up to doses of 10^{18} n°/cm^2 and that the behaviour of the samples varies according to their composition and form.

Keywords : Harsh environment, ultrasonic sensors, Material Test Reactor, piezoelectric materials, radiation effects

I. Introduction

Nuclear fuel rods lifetime optimization helps in increasing combustion rate of nuclear fuel, hence reducing the volume of waste to be recycled. It can be achieved via in-pile control of the fuel rods interior environment. The measurements of the gas mixture (Helium and fission gases) in REMORA-3 experiment [1, 2] prove the possibility of measurements in plenum (upper part of combustive rod, whose role is to collect the fission gases) during operational conditions at up to 200 °C. A representation of REMORA device is given in Fig. 1.

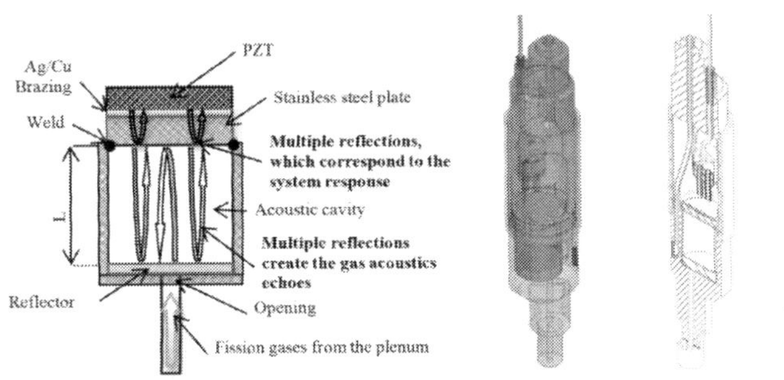

(a) (b)

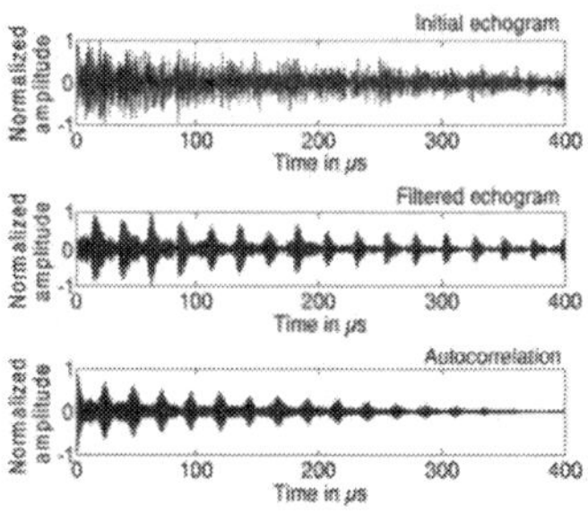

(c)

Fig. 1 : REMORA acoustic gas sensor. (a) Schematic representation [1]. (b) Sensor overview [2]. (c) Experimental results including data processing [2].

The determination of the gas composition is based on the measurement of the sound velocity in a cavity filled with the gas to be analyzed. The emission and detection of the acoustic wave is achieved by coupling an acoustic element and a stainless steel plate that also serves as a confinement barrier [1]. In practice, the difficulties in brazing and ensuring of parallelism within the active element (Pz27 Ferroperm), stainless steel plate and reflector (Fig. 1) caused difficulties in signal measurement. Transducer's limitation in operation temperature renders it impossible for application in a new type material testing reactor (MTR). Since 2011, our researches concern instrumentation for the future RJH reactor (to be used for testing devices for GEN IV reactors) and new requirements appeared such as : working temperature up to nearly 400°C, high level of nuclear radiations or minimally invasive measurement devices. Based on this limitation and the need to operate at higher temperatures, a second approach (which reduces the influence of the parallelism problem) is to use a screen-printing process in order to deposit piezoelectric material as the active element on a dedicated substrate. In previous work we developed a specific sensor based on screen printed lead zirconate titanate (PZT) and we proved the possibility of active element fabrication using material characterization [3] and ultrasonic measurements [4].
Our previous works demonstrated that the technology based on screen-printed piezoelectric layers was viable in a non-radiative environment and that the technology using solid materials was also viable for a radiative environment but for temperatures below 140°C. In order to observe the influence of a radiative environment on the performance of piezoelectric materials of different kinds, we have set up an experimental campaign at the Jozef Stephan Institute (JSI) with two objectives. On the one hand, to observe and compare the behavior of two families of piezoelectric materials: Lead Zirconate Titanate (PZT) and modified Bismuth Titanate (BiT). For the first type, the samples are of type PZ27 (Meggit-Ferroperm) and for the second type they are of type PZ46 (Meggit-Ferroperm). On the other hand, ensure that sensor based on screen-printed process can be used in a radiation environment. For the latter, the piezoelectric screen printing layers will be made from PZ27 type materials as mentioned in [3].

II. Materials and Methods

A. Specimens

Two different types of piezoelectric materials were analyzed. Firstly, bulk materials as PZ27 and PZ46 were purchased from Ferroperm (Meggit) in the form of 10 mm in diameter, 600 um thick, acting as a thickness mode resonator. Secondly, thick films of PZ27 were deposited by screen printing process on alumina substrate. PZT ink has been made by mixing of 74% in weight of PZT powder (PZ27 from

Ferroperm), 6 % binder (CF 7575 from Ferro Electronic company) and 20% of an organic vehicle (ESL400, ESL electronics). These compounds are mixed together using a tri-cylinder mixer until the compound is homogeneous. Similarly, an electrode allowing electrical contact in the form of a thick layer was made using an Ag/Pd type screen printing ink (ESL 9693-G from Ferro Electronic company). Fig. 2 provides a schematic of transducers. The effective area of device is delimited by a mask pattern of 1 cm diameter. The piezoelectric structure is made by a stack of bottom electrode, five layers of PZT forming a final layer of about 100 µm and upper electrode. Two woven mesh are used to build the complete structure including electrodes. After deposition by screen-printing, each layer is dried at 120° C for 15 minutes.

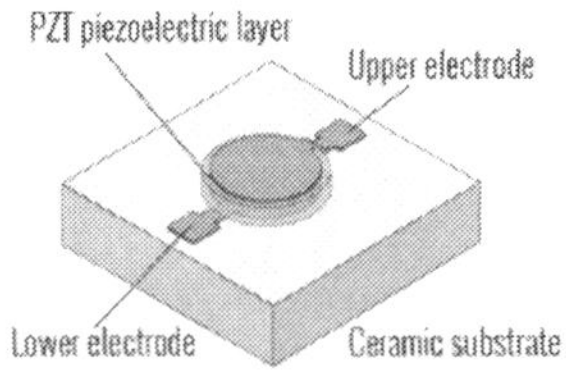

Fig. 2 : Piezoelectric thick film

In this work, for PZT ceramics elaboration, conventional sintering method was used. During this step, specimen were placed inside a Carbolite RWF1100 furnace and sintered under lead oxide-saturated atmosphere at 850 °C. The limitation of sintering temperature by 850 °C was due to the effect of the electrode deposition in sensor production.

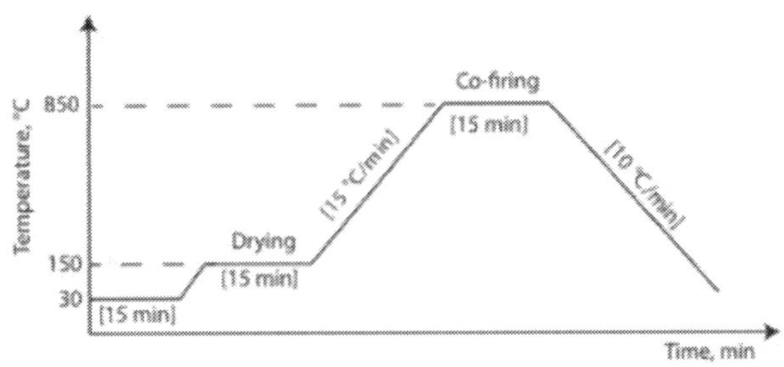

Fig. 3 : Standard sintering protocol used for fabrication of PZT

The sintering protocol used for ceramic fabrication is shown in Fig. 3 with the dwell period at 850 °C of 15 minutes. The time of sintering was adapted according to the protocol of Ag/Pd9693-G conductive ink and was successfully used for PZT components in previous studies.

B. Experimental set-up

1. JSI TRIGA reactor irradiation facilities

JSI TRIGA Mark II reactor is a typical TRIGA type reactor that is to say a pool type reactor, cooled by natural convection, with maximal steady thermal power of 250 kW. The reactor core has an annular configuration with a diameter of 44.2 cm, with 91 in-core positions for fuel elements, 4 control rods, and numerous irradiation positions (Fig. 4).

 https://doi.org/10.1051/epjconf/202022504012

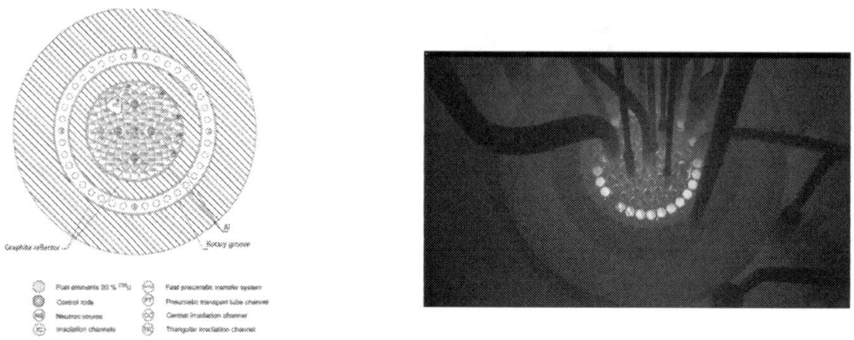

Fig. 4 : JSI TRIGA Mark II reactor core with typical fuel element configuration and irradiation positions.

Nuclear fuel is a homogeneous mixture of fuel and moderator in form of uranium zirconium hydride (U-ZrH). Fuel meat, with central zirconium rod and graphite reflector on either side, encased in stainless steel cylindrical enclosure (radius 1.89 cm and length of 72.5 cm) with bottom and top pin comprise a fuel element. Fuel elements are held in place by top and bottom aluminium grid, each consisting of a plate with holes for fuel elements. Reactor usually operates with central position (A1 or CC), and most of the F-ring positions empty and available for irradiations and two control rods (Transient and Safety) completely withdrawn.It is equipped with numerous irradiation positions, where samples can be irradiated by neutrons and γ-rays. Irradiation position selection is based on its properties, such as physical size and accessibility, as well as neutron and γ-ray spectra, flux and dose intensities.

2. Instrumental Bench

In order to monitoring the electrical behavior of different piezoelectric materials and devices, impedance analyzer (Keysight 4990A) associated with a multiplexer (Agilent 34902) were used. Instruments were connected by specific coaxial cables of 10 m length, purchased from thermocoax (silica core and 50 ohm matching), to a specific specimen holder built in aluminum which contained the different specimens to be analyzed (Fig. 5). Altogether, 10 channels were used allowing for the studies of 6 piezoelectric elements, 3 for impedance calibration and 1 channel for the temperature measurement with type K thermocouple.

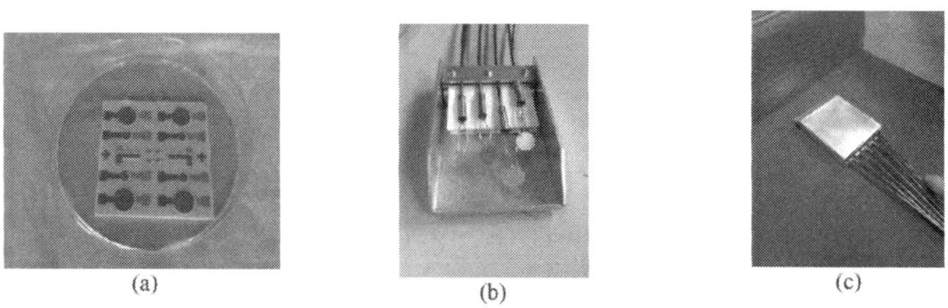

(a) (b) (c)

Fig. 5 : pictures detailing the different piezoelectric elements inserted in the specimen holder. (a) piezoelectric thick films on alumina substrate, (b) all piezoelectric samples and associated cables before closing the cavity, (c) sealed cavity

During 5 days, the samples were placed in air into the triangular irradiation channel where calculated neutron flux at full power (250kW) and total dose during experiment are described on Fig. 6 (uncertainty is about 5%).

Thermal (below 0.625eV	*4.46E12 n/cm².s*
Epithermal (0.625eV to 100keV)	*3.5E12 n/cm².s*
Fast (above 100keV)	*3.8E12 n/cm².s*
Total :	***1.175E13 n/cm².s***
Total dose :	***1.75E17 n/cm²***

Fig. 6 : calculated neutron flux for JSI facilities

Data concerning piezoelectric elements, regardless of the characteristics of the samples, such as real and imaginary parts of electrical impedance (induced data processing), capacitance and dielectric loss obtained at different frequencies and different temperatures (measurement channel, water, ...), were acquired through a specific software during experiment.

3. Data processing

Following a theoretical approach, based on the KLM model [17], the electro-mechanical parameters of piezoelectric elements were calculated by matching experimental measurement with theoretical values. Three complex parameters, parallel frequency (Fp*), electro-mechanical coupling factor (kt*), capacitance (C*) were adjusted using software in order to fit measured data. Theoretical impedance was determined by equation 1 for the bulk elements and by equation 2 for thick films :

$$Z_{elec} = \left(-\frac{i}{C^*\omega}\right)\frac{\left[1-\left(4Fp^*kt^{*2}\tan\left(\frac{\omega}{4Fp^*}\right)\right)\right]}{\omega} \quad (1), \qquad Z_{in} = Z_c\frac{Z_{out} + jZ_c\tan(\varphi)}{Z_c + jZ_{out}\tan(\varphi)} \quad (2)$$

whereby Fp*, kt*, C* are complex values of Fp, kt, capa; and ω is the angular frequency.
Complex values of these parameters can be found from the standard equations of piezoelectricity [18]:

$$Fp^* = Fp\left(1 + \frac{i}{QFp}\right)\ (3), \quad kt^* = kt\sqrt{1 - i/qkt^2}\ (4), \quad C^* = capa\left(1 - \frac{i}{qcapa}\right)(5)$$

whereby QFp – mechanical quality factor, qkt – electro-mechanical loss and qcapa associated to dielectric loss. The parameters are then deduced from the experimental curves fitted with previously described model as shown in Fig. 7.

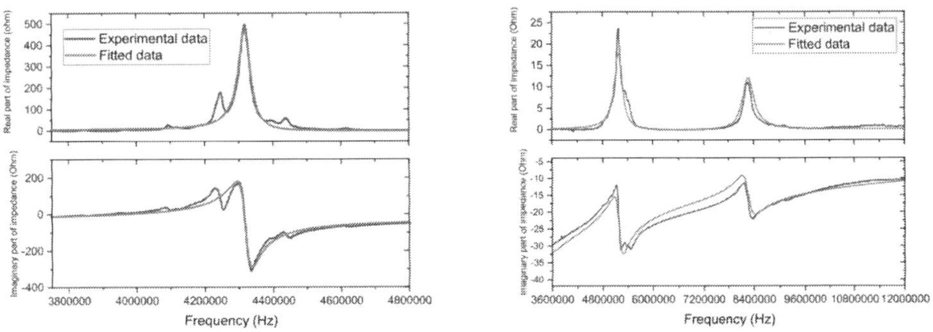

Fig. 7 : real and imaginary parts of electrical impedance : left PZ27, right screen-printed PZ27

Table 1 shows an example of results obtained, but with difficulties in obtaining the Qcapa parameter for PZ46. In view of this, we have chosen not to monitor the evolution of complex parameters and to focus only on the Fp, kt and capa parameters.

TABLE 1
Electro-mechanical parameters of PZ27 and PZ46 samples fitted with model.

Symbol	Quantity	Value PZ27	Value PZ27 thick film
Fp	parallel frequency	4.262 MHz	13.798 MHz
QFp	mechanical quality factor	121	17,57
kt	electro-mechanical coupling factor	0.46	0.31
Qkt	electro-mechanical loss	35.13	NA
capa	capacitance	1.11 nF , (ε_{33} = 958)	1.23 nF, (ε_{33} = 470)
qcapa	dielectric loss	154	NA

These three parameters generally represent the different properties of piezoelectric materials and their monitoring provides important information on their different behaviours : mechanical for F_P, electrical for C_P and electromechanical coupling for K_t.

III. Results

We can observe on Fig. 8 an experimental curve showing the relative variation of the resonance frequency of a PZ27 element as a function of the reactor operating parameters, in particular with the dose rate and the total dose. The temperature curve corresponds to that recorded in the sample holder inserted in the irradiation channel.

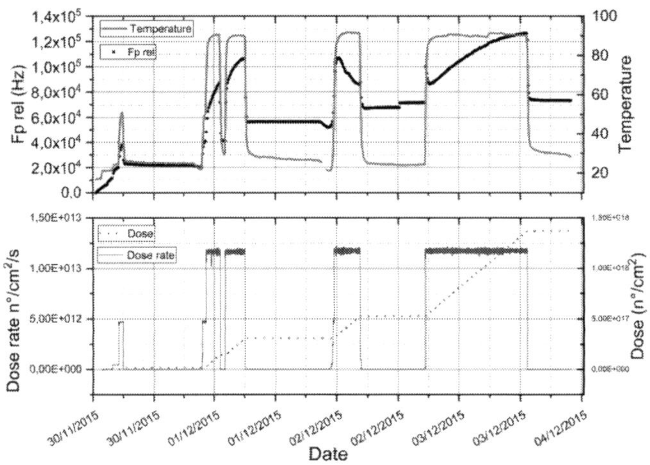

Fig. 8 : example of experimental results obtained during 5 days

We noticed that during the transient phases, particularly during the reactor's ramp-up or during the decrease in its power, some difficulties in adjusting the curves on the theoretical model. We assumed that these observations could be due either to dose rate effects (inducing generations of electrical charges by pyroelectric effect), or to alterations of resonance peaks, or to resonance mode couplings that the 1-dimensional model was not able to take into account. The results presented below correspond only to steady states obtained after returning the irradiation channel to room temperature. We will therefore exclusively consider the influence of the effect of the neutron dose on the measured parameters.

 https://doi.org/10.1051/epjconf/202022504012

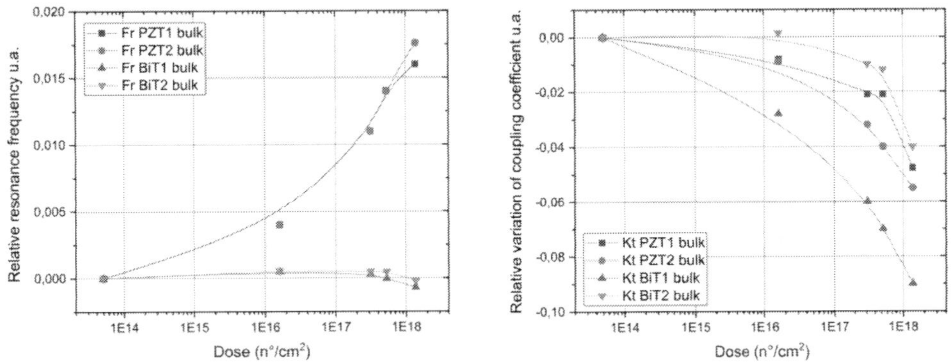

(a) (b)

Fig. 9 : comparison between PT27 and PZ46. (a) relative variation of resonance frequency, (b) relative variation of electromechanical coupling.

Curves shown in Fig. 9 correspond respectively to the relative variation of the resonance frequency Fig. (a) and to the electromechanical coupling coefficient Fig. (b) for 4 piezoelectric disks based on 2 Lead Zironate Titanate and 2 modified Bismuth Titanate. Concerning the variation of the resonance frequency, we observe a significant difference in behaviour between the two types of samples revealing a better performance for PZ46 but this variation does not exceed a maximum of 2% for the PZ27 sample for a dose up to 10^{18} n°/cm^2. As already mentioned by Holber [5] and Sinclair [6] variation of resonance frequency could be induced by modifications of stress constant of piezoelectric materials due to ionization effect. Concerning the evolution of the electromechanical coupling coefficient, the works of Broomfield [7] and on piezoelectric materials reveal a decrease in remnant polarization associated with neutron dose and thus a decrease in electro-mechanical coupling. This effect can be associated with the presence of oxygen deficiencies that actually contribute to an increase in the intrinsic electrical conductivity of the material and a decrease in the electrical-mechanical conversion efficiency. We also noted, specifically for PZ46 samples, a heterogeneity in the results relating to the coupling coefficient. Although these results are not presented in this document, this is also the case when measuring the capacitance of samples of the same material. Although we have taken care to ensure that similar samples are taken before irradiation, with more difficulties for PZ46 materials, their behaviour under irradiation seems more homogeneous for PZ27 samples than for PZ46.

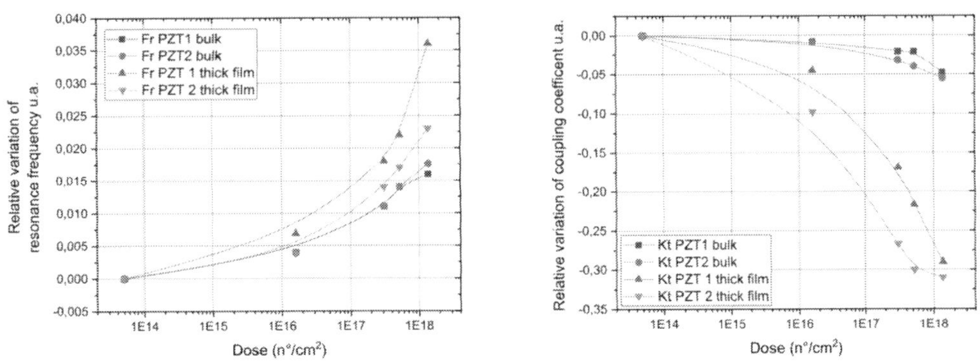

(a) (b)

Figure 10 : comparison between PZ27 bulk and thick film. (a) relative variation of resonance frequency, (b) relative variation of electromechanical coupling.

Through the results presented in Figure 10, it can be observed that the behaviour of the massive zirconate titanate lead and its declination in thick layers evolves in the same way according to the dose received. The same assumptions as above can be made to explain the increase in the relative variation in resonance frequency, but not exceeding 4% for a dose up to 10^{18} n°/cm^2. With regard to the electromechanical coupling coefficient, the difference between thick layers and solid materials is greater. Indeed, the relative variation of this coefficient is 6 times greater for PZ27 layers deposited on alumina and is up to 30%. We can also hypothesize that piezoelectric materials with lower thicknesses may have higher sensitivity to the irradiation dose, as ionization phenomena occurring at lower thicknesses would lead to faster degradation of the material's intrinsic properties and gamma self-heating may have been more important for thick layers deposited on an alumina substrate with a volume 40 times greater than massive piezoelectric elements and in contact with the sample carrier.

IV. Conclusion

An experimental bench has been developed taking into account the specificities associated with a Material Test Reactor in order to observe the radiative effects on specific parameters of piezoelectric materials. The experimental campaign carried out at JSI made it possible to achieve the two initial objectives:

- We achieved the first objective of observing and comparing the performances of materials such as PZ27 and PZ46. The experimental results, in agreement with the state of the art, confirm the fact that modified bismuth titanate (PZ46) has better performances than lead zirconate titanate (PZ27) due to its higher Curie temperature and its behavior on radiative environment.
- For the first time, thick layers of piezoelectric materials deposited on alumina substrates could be tested in a radiative environment. Based on results, we also confirmed that screen-printing technology could be used to produce miniaturized acoustic sensors for *in situ* research reactor instrumentation.

V. References

[1] E. Rosenkrantz, J.Y. Ferrandis, F. Augereau, T. Lambert, D. Fourmentel, X. Tiratay, "An Innovative Acoustic Sensor for In-Pile Fission Gas Composition Measurements," *IEEE Transactions on Nuclear Science*, vol. 6, no. 2, pp. 1346-1353, 2013.

[2] D. Fourmentel, J.F. Villard, J.Y. Ferrandis, F. Augereau, E. Rosenkrantz, M. Dierckx, "Acoustic sensor for in-pile fuel rod fission gas release measurement," *IEEE Transactions on Nuclear Science* vol. 58, no. 1, pp. 151-155, 2011.

[3] F. Very, P. Combette, D. Coudouel, A. Giani, E. Rosenkrantz, J.Y. Ferrandis, D. Fourmentel, "Piezoelectric thick film sensors: Production and characterization," *Applications of Ferroelectric and Workshop on the Piezoresponse Force Microscopy (ISAF/PFM), IEEE International Symposium*, 2013, pp. 287-290.

[4] D. Lapeine, J.Y.Ferrandis, D. Laux, F. Pascal, P. Combette, "Development and integration of screen printed transducers for the detection of pollutants in aquatic environments," *Design, Test, Integration and Packaging of MEMS/MOEMS (DTIP), IEEE*, pp. 1-5, May 2016.

[5] K.E. Holbert, S. Sankaranarayanan, S.S. McCready, D.R. Spearing, A.S. Heger, "Response of piezoelectric acoustic emission sensors to gamma radiation", Proceedings of the 7th European Conference on Radiation and Its Effects on Components and System, RADECS 2003, Noordwijk, Netherlands, September 2003.

[6] A.N. Sinclair, A.M. Chertov, "Radiation endurance of piezoelectric ultrasonic transducers – A review", Ultrasonics, Volume 57, 2015, Pages 1-10,

[7] G.H. Broomfield, "The effects of temperature and irradiation on piezoelectric acoustic transducers and materials", UKAEA Report AERE-R 11 942, Harwell, UK, 1985.

[8] K.E. Holbert, S. Sankaranarayanan, S.S. McCready, D.R. Spearing, Response of lead metaniobate acoustic emission sensors to gamma irradiation, IEEE Trans. Nucl. Sci. 52 (2005) 2583–2590.

 https://doi.org/10.1051/epjconf/202022504013

Investigation of the neutron-gamma ray discrimination performance at low neutron energy of a solution-grown stilbene scintillator

Augusto Di Chicco[1,3,*], Michael Petit[2], Robert Jacqmin[1], Vincent Gressier[2], Brian Stout[3]

[1]*CEA Cadarache, DEN/DER, 13108 Saint-Paul-lez-Durance, France.*
[2]*IRSN Cadarache, SDOS/LMDN, 13108 Saint-Paul-lez-Durance, France.*
[3]*Aix-Marseille Université, Institut Fresnel – UMR 7249, 13397 Marseille, France.*

Abstract **— The results of experiments performed with a Ø25 x 25 mm solution-grown stilbene crystal in mono-energetic neutron fields in the 80-to-230 keV energy range are presented. The goal of the measurements, performed at the AMANDE facility, was to explore the capabilities of this organic scintillator to measure neutrons at the lowest possible energy with good pulse shape discrimination (PSD). The time of flight (TOF) technique was used in order to help with the neutron-gamma discrimination. The data are collected via a programmable digital acquisition (DAQ) system CAEN DT7530 using the software CoMPASS with the charge comparison method (CCM). The data are analysed using post-processing codes developed in the ROOT environment. The results show that the stilbene detector has discrimination capabilities for energies as low as 80 keV.**

Keywords **— PSD, Organic scintillators, Solution-grown stilbene, Mono-energetic neutrons, Time of flight, digital acquisition system, AMANDE facility.**

1. INTRODUCTION

For several nuclear applications such as neutron spectrum measurements in zero-power experimental nuclear reactors, nuclear material detection, fusion diagnostics etc., there is a need for fast neutron detectors that can cover a wide energy range, especially from a few 10 keV's to 10 MeV, and that can operate in mixed neutron-gamma radiation fields. These neutron spectrometers must have a good energy resolution (<10%), high detection efficiency, be compact and easy to handle, while not being too sensitive to thermal neutrons. Organic scintillators have many of these desirable characteristics, their performance being dependent on the type of scintillating material used: liquid, plastic or crystal. However, organic scintillators are sensitive to gamma rays. Consequently, pulse shape discrimination (PSD) techniques have to be applied to the output signals given by the photomultiplier (PMT) to enable neutron-gamma discrimination. For energies below 1 MeV, the discrimination becomes increasingly difficult due to the weakness of the signals given by the light produced by the scintillators. The organic scintillators commonly used as references for neutron spectroscopy in mixed fields are liquid scintillators such as the BC501A and NE213 [1], but these devices have limitations in terms of energy performance (poor discrimination below 1 MeV) and operational security (fire hazard). In recent years, organic scintillator technology has been significantly improved and digital acquisition systems are now able to replace the traditional analogic systems. The main advantages are the reduction of the electronic acquisition chain and the possibility to store each event and its information like time and pulse shapes. In this way, it is possible to perform offline studies. This is much more difficult with analog systems, therefore the signal generally needs to be processed as it is generated. Recently, a new organic scintillator based on a Ø25 x 25 mm solution-grown stilbene single crystal was developed [2] and tested at 5 MeV and 565 keV. However, the available measurements were insufficient for establishing the low neutron-energy threshold below which neutron-gamma discrimination becomes impossible. Our goal is therefore to investigate this threshold in mono-energetic neutrons fields between 80 keV and 230 keV.

2. EXPERIMENTAL DETAILS

A. The AMANDE facility

The AMANDE facility of the Institute for radiological protection and nuclear safety (IRSN) at Cadarache is able to produce continuous or pulsed monoenergetic neutrons from ~2 keV up to 20 MeV [3]. The neutron field for this work was produced using the ^{7}Li(p,n) reaction with a proton beam of 2067 keV energy and 1 μA intensity sent on a LiF target of 107 μg/cm^2 thickness. The neutron energy at 0°, i.e. in the proton beam direction is 304 keV, but more neutron energies are actually accessible by varying the detection angle as shown in Table 1. The experimental area (20 x 20 x 16 m^3), is designed so as to limit neutron returns by scattering on the walls.

*Corresponding author:
E-mail: Augusto.DICHICCO@cea.fr
Tel : +33 6 13 18 92 56

Figure 1 Experimental area with the stilbene detector mounted on the mobile arm

Table 1 Angles relative to proton beam direction and corresponding neutron energies

Angle	Neutron energies	Monoenergetic peak FWHM
45°	223 keV	7 keV
60°	179 keV	6 keV
80°	125 keV	5 keV
100°	84 keV	5 keV

Moreover, the presence of a fully-automated transport system can place the detectors at a distance of up to 6 m from the target, and at any angle between 0° and ±150°. The time-of-flight measurements relies on a pick-up ring placed in the beam at 1 m in front of the target. The ion beam pulsation is performed using a chopper-buncher system [3b] allowing pulse width around 2 ns.

B. Stilbene Organic Scintillator

The solution-grown single crystal of stilbene was produced by InradOptics [4], coupled with a photomultiplier ET 9214B of the ET Enterprise [5]. The whole detector was assembled by SCIONIX [6]. The stilbene is encapsulated in an aluminium canister and is optically coupled to the photomultiplier, which is surrounded by a solid mu-metal shield. When an ionizing particle interacts with the scintillator, light is produced. The signals due to incoming gamma rays are different from those produced by neutrons. Gamma-ray interaction leads to electrons and a low-density charge deposit, which excites atoms principally to singlet states. The relaxation times of excited singlets states are very short (less than 10 ns) and the related light production is called prompt fluorescence. Neutrons, on the other hand, generate protons via elastic scattering on hydrogen, leading to a high-density charge deposit, which excites atoms to single and triplet states. The lifetime of the excited triplet states is longer (more than µs) and the corresponding light emission is called delayed [7]. The output signal is taken from the anode of the photomultiplier and the difference in light emission is used to discriminate neutrons (signal with longer decay time due to delayed fluorescence) from photons (prompt fluorescence mainly), as illustrated in Figure 2. However, as the signal amplitude is directly related to the energy of the incident particle (neutron or photon), when neutron energy decreases, the baseline variation (see Section C below) has a higher impact on the ratio between prompt and delayed fluorescence, and it becomes increasingly difficult to distinguish between the two types of particles.

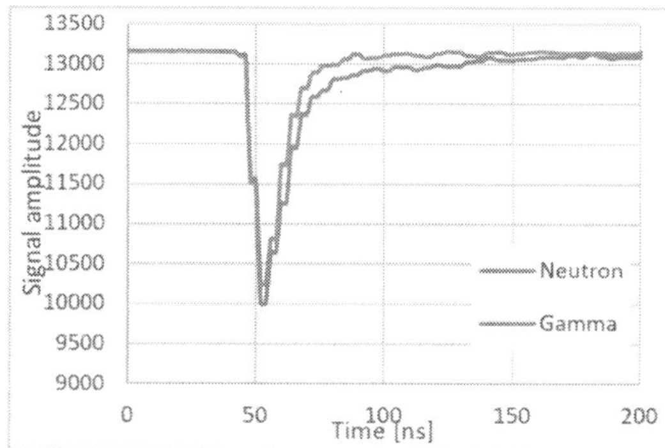

Figure 2 Typical shapes of gamma rays and neutrons

C. Digital Acquisition System and Charge Comparison Method

The anode signal from the PMT is directly connected to a DAQ CAEN 5730 [8]. This device has eight channel, 14-bit resolution, 500 MHz sampling rate and a maximum digitizable voltage of 2 V. The online measurements have been performed with the CAEN software CoMPASS [9] using the Digital Pulse Processing – Pulse Shape Discrimination (DPP-PSD) firmware. In order to perform a good online analysis, the DAQ parameters, shown in Figure 3, have to be optimized according to the type of detector and the experimental situation.

https://doi.org/10.1051/epjconf/202022504013

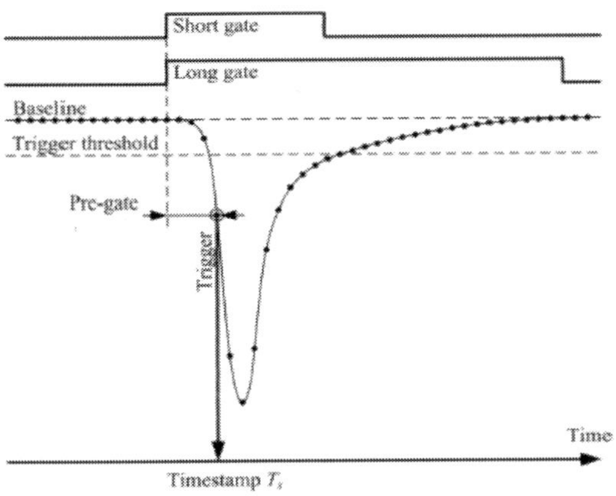

Figure 3 Most relevant parameters for the DPP firmware

Once the recorded signal $i(t)$ exceeds a certain threshold value, the firmware uses the Charge Comparison Method (CCM) to calculate the integrals of the pulses over two different time windows, called "short gate" and "long gate", respectively. These two integral quantities are used in turn to estimate the total light output and the prompt light output. In this way, it is possible to infer a PSD value and discriminate between the detected particles. The equations used to calculate the two integrals are the following:

$$Q_S = \int_{t_0}^{t_S} i(t)dt \quad (1)$$

$$Q_L = \int_{t_0}^{t_L} i(t)dt \quad (2)$$

where t_0 is the start time, called pre-gate, for the integrals calculation, while t_s and t_L correspond respectively to the short and long gates (the optimal values are 34 ns and 414 ns for the stilbene). The baseline is determined by averaging over 256 samples, it is subtracted from the signal during the calculation. The PSD is obtained by the following equation:

$$PSD = \frac{Q_L - Q_S}{Q_L} \quad (3)$$

The two-dimensional (2D) histogram of the PSD for an AmBe source measured with the stilbene is plotted as a function of Q_L (directly related to the energy) in Figure 4 (a). The two areas represent respectively neutrons (high PSD) and gamma rays (low PSD). The PMT was set at -930 V in order to avoid saturation and excessive pile-up counts. The data were handled with ROOT [10].

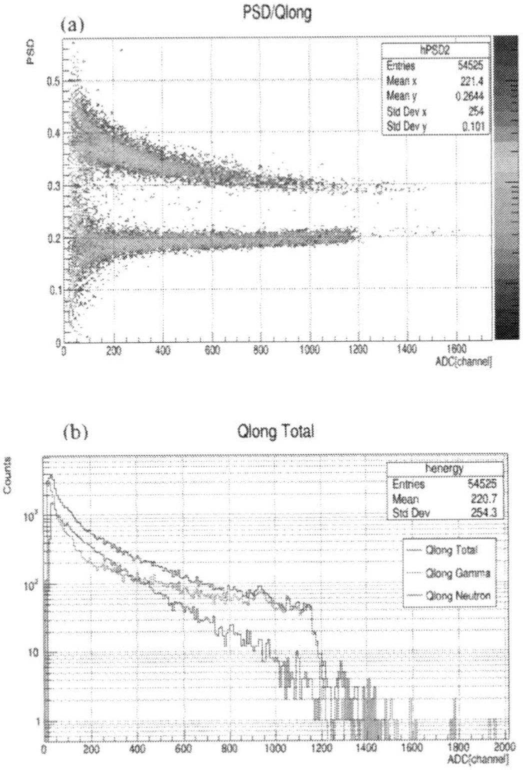

Figure 4 (a) The PSD vs. Q_L of the AmBe source; (b) Light amplitude for various Q_L when applying a cutoff value to the PSD

Figure 4 (b) shows the total light produced (in black) in log scale. The Compton Edge of the 4.438 MeV gamma ray due to the $^{9}Be(\alpha,n)^{12}C^*$ reaction is clearly visible. By considering a cutoff value of 0.26 in the PSD, it is possible to separate the individual contributions of neutrons (0.26 < PSD <0.6, in blue) and gamma-rays (0.06 < PSD <0.26, in red) to the total signal.

D. *Time of Flight technique*

In the low neutron energy range, the weakness of the signals does not always allow to obtain a clear double-peak PSD profile, and hence setting a cutoff value is difficult. At AMANDE, the time-of-flight technique has been used to help distinguish neutrons from gamma-rays, and to improve the resolution of the PSD. To that aim, the proton beam has been pulsed and the pick-up (PR) ring signal has been connected to the second channel of the DAQ and used as stop signal. The stilbene detector, positioned at a fixed distance L = 1,907 m from the target, is connected to the first channel of the digitizer and used as start signal.

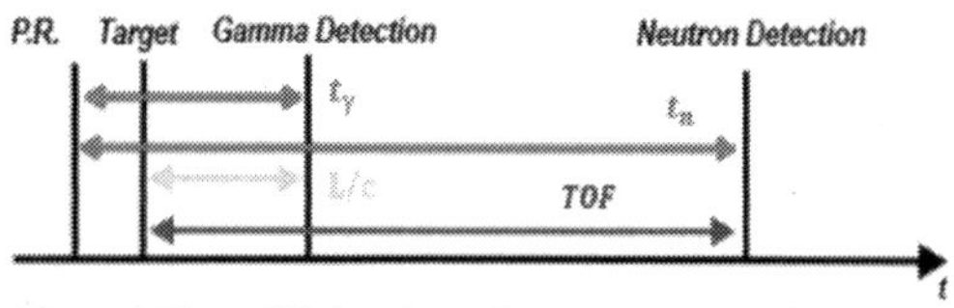

Figure 5 Time of flight scheme for neutron-gamma beams

The DAQ, working in coincidence mode, records the start time of each channel. The difference between these start times is determined with a total resolution of less than 3 ns, for gamma-rays and neutrons simultaneously produced in the target by the pulsed proton beam. The neutron time-of-flight between the target and the detector is simply obtained by writing:

$$ToF_{measured} = t_n - t_\gamma + \frac{L}{c} \quad (4)$$

where t_n and t_γ are the neutron and gamma detection times, respectively, as shown in Figure 5.

3. Experimental Results

For the low energy range measurements, the voltage of the PMT was increased up to -1550 V. However, with this fairly-high voltage setting for the PMT, many signals are lost as their amplitude exceed the 2 V digitizing window, as shown in Figure 6. In addition, there are spurious contributions that give negative values of Q_L and Q_S, or even some where the total light output value is lower than the prompt one. All these contributions had therefore to be removed from the data analysis.

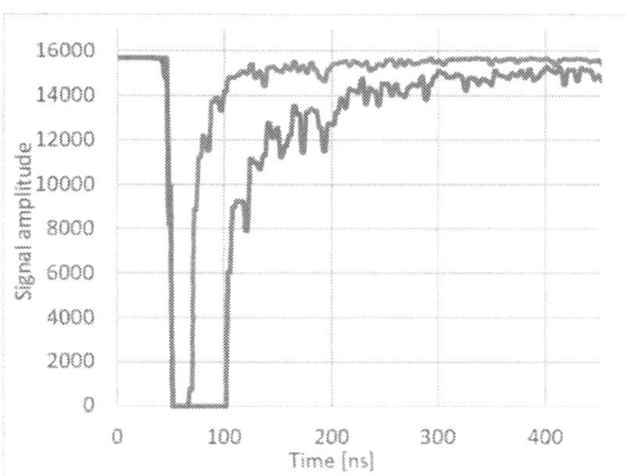

Figure 6 Examples of unusual signals

In Figure 7 (a), the PSD is plotted versus Q_L where it is still possible to discriminate the neutrons from the photons. The upper part of the plot (PSD> 0.7) corresponds to signals with bad pulse shape that must be eliminated. In Figure 7 (b), the PSD is plotted versus the time difference between the start and stop from the DAQ, in pulse mode. The areas corresponding to the gamma-rays at 100 ns and to the neutrons at 404 ± 10 ns are clearly visible. Below the neutron structure, it is also possible to see the gamma rays produced by neutron interactions in the stilbene crystal.

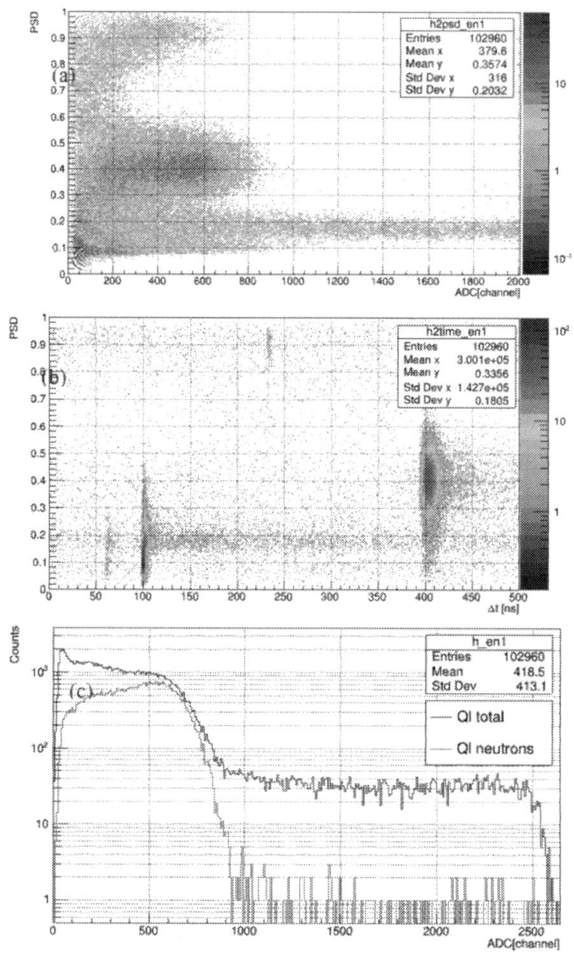

Figure 7 Stilbene data from the DAQ for 230 keV neutrons

In Figure 7 (c), the total pulse height spectra (in blue) is plotted, as well as the derived neutron PHS after selecting the neutron contributions in the time interval 390 < Δt < 425 ns. Figure 8 shows the time distributions recorded in the four neutron fields from Table 1. The monoenergetic peak is clearly visible even at 84 keV, the lowest energy. The corresponding recoil proton energy distributions were obtained by selecting only the events corresponding to the monoenergetic neutrons in the time interval, as shown in Figure 9. These distributions will be used to validate upcoming calculated responses of the stilbene scintillator.

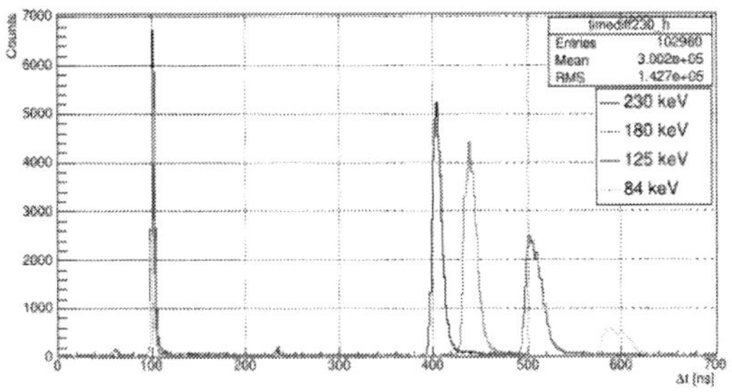

Figure 8 Time of flight distributions for the stilbene

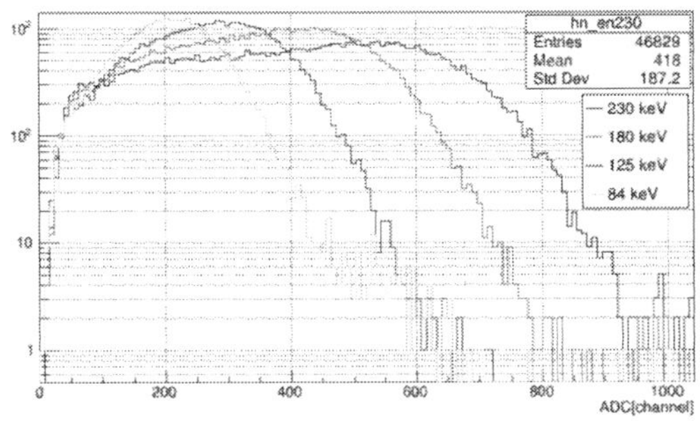

Figure 9 Proton recoil spectra

4. Conclusions

In this article, the capability of an organic scintillator based on a Ø25 x 25 mm solution-grown stilbene single crystal to measure neutrons with energies below 100 keV has been tested at the AMANDE facility, with the help of the time-of-flight technique to improve the PSD analyses. The results show that, even at such "low" energies, the stilbene detector still has some neutron-gamma discrimination capabilities, which is remarkable. Upcoming work will focus on the study of individual pulse shapes, so as to have more precise criteria to distinguish neutrons and gamma rays. In addition, the crystal anisotropy will be studied in this energy range as it can significantly influence the detector response. The next step of these studies will be to calculate the stilbene detector response matrix with Monte Carlo codes and to compare it with the experimental data.

Acknowledgment

The authors thank the entire AMANDE team of IRSN, especially Dr. Richard Babut, for his collaboration and invaluable assistance during the experiments.

References

[1] H. Klein, "Neutron spectrometry in mixed fields: NE213/BC501A liquid scintillation spectrometers," *Radiat. Prot. Dosimetry*, vol. 107, no. 1–3, pp. 95–109, Nov. 2003.

[2] L. Dioni, V. Gressier, G. Nardin, R. Jacqmin, B. Stout, and M. Sumini, "Tests of a solution-grown stilbene scintillator in mono-energetic neutron beams of 565 keV and 5 MeV," *Nucl. Instrum. Methods Phys. Res. Sect. Accel. Spectrometers Detect. Assoc. Equip.*, vol. 880, pp. 210–215, Feb. 2018.

[3] V. Gressier *et al.*, "AMANDE: a new facility for monoenergetic neutron fields production between 2 keV and 20 MeV," *Radiat. Prot. Dosimetry*, vol. 110, no. 1–4, pp. 49–52, Aug. 2004.

[4] Inforad, "Production of Stilbene for Fast Neutron Detection."

[5] "9214B Series - ET-Enterprises Ltd." [Online]. Available: http://et-enterprises.com/products/photomultipliers/product/p9214b-series. [Accessed: 05-May-2019].

[6] "Homepage Scionix.nl ⋆ Scionix," *Scionix*. [Online]. Available: https://scionix.nl/. [Accessed: 05-May-2019].

[7] *The Theory and Practice of Scintillation Counting*. Elsevier, 1964.

[8] "DT5730," *CAEN - Tools for Discovery*. [Online]. Available: https://www.caen.it/products/dt5730/. [Accessed: 05-May-2019].

[9] "CoMPASS," *CAEN - Tools for Discovery*. [Online]. Available: https://www.caen.it/products/compass/. [Accessed: 05-May-2019].

[10] "ROOT a Data analysis Framework | ROOT a Data analysis Framework." [Online]. Available: https://root.cern.ch/. [Accessed: 06-May-2019].

Conceptual Design of Irradiation Facility with 6 MeV and 7 MeV Gamma Rays at the JSI TRIGA Mark II Research Reactor

Andrej Žohar
Reactor Physics Department
Jožef Stefan Institute
Ljubljana, Slovenia
andrej.zohar@ijs.si

Anže Pungerčič
Reactor Physics Department
Jožef Stefan Institute
Ljubljana, Slovenia
anze.pungercic@ijs.si

Klemen Ambrožič
Reactor Physics Department
Jožef Stefan Institute
Ljubljana, Slovenia
klemen.ambrozic@ijs.si

Vladimir Radulović
Reactor Physics Department
Jožef Stefan Institute
Ljubljana, Slovenia
vladimir.radulovic@ijs.si

Anže Jazbec
Reactor Infrastructure Centre
Jožef Stefan Institute
Ljubljana, Slovenia
anze.jazbec@ijs.si

Sebastjan Rupnik
Reactor Infrastructure Centre
Jožef Stefan Institute
Ljubljana, Slovenia
sebastjan.rupnik@ijs.si

Igor Lengar
Reactor Physics Department
Jožef Stefan Institute
Ljubljana, Slovenia
igor.lengar@ijs.si

Luka Snoj
Reactor Physics Department
Jožef Stefan Institute
Ljubljana, Slovenia
luka.snoj@ijs.si

***Abstract*—Activated cooling water in nuclear facilities can present a significant r adiation s ource a round p rimary cooling system causing radiation damage to electrical components, increasing doses to personnel and in the case of fusion facilities additional heating to superconducting coils. As there are only few sources of gamma rays with energies in the range of 6 MeV and 7 MeV an irradiation system using activated cooling water as the source of energetic gamma rays is proposed at the Jožef Stefan Institute (JSI) TRIGA Mark II research reactor. Two different conceptual designs, one utilizing central irradiation channel and one utilizing radial piercing port for water activation, are presented and analysed in the paper. Despite an order of magnitude higher water activation in central channel compared to radial piercing port the ^{16}N decay rate in the irradiation facility is comparable between both design (order of 10^8 decays per second) due to longer transient time from central channel to irradiation facility. In the irradiation facility the expected biological dose rates due to the ^{16}N decay rate are in order of several mSv/h. From the results he conceptual design utilizing the radial piercing port currently presents the best option for the irradiation facility due to the simpler design of the irradiation loop, already present shielding of the loop and comparable number of ^{16}N decay rates to central channel.**

Keywords—**Activated cooling water, Monte Carlo, TRIGA, Irradiation facility.**

I. Introduction

Water is the cooling fluid i n m any fi ssion nu clear power plants, research reactors and it is also considered as a cooling fluid i n f usion r eactors. D uring t he c ooling o f t he reactor core in case of fission r eactors a nd b lankets i n c ase of fusion reactors the water is exposed to neutrons and becomes activated. The activation products in activated water are activated oxygen and hydrogen atoms, activated dissolved gases, corrosion products and water impurities. After irradiation the activated cooling water flows t hrough t he p rimary cooling circuit, commonly outside the primary biological shielding surrounding the reactor vessel, where the activated products decay and emit radiation. The emitted radiation may cause radiation damage to electrical components, increased doses to personnel working close to the cooling circuit and in the case of fusion facilities additional nuclear heating to various cold components such as superconducting coils cooled by liquid helium [1], [2].

There are several water activation products of which the most important are the ^{16}N due to high natural abundance of ^{16}O and the high energy of gamma rays (6.13 MeV and 7.11 MeV) and ^{17}N which emits neutrons with average energy of 0.88 MeV [3]. In case of fission reactors the water contribution to the total radiation field is usually negligible. The situation is however different for fusion devices such as ITER, where the dose-rate due to activated water is estimated to be several 100 Gy/h, exposing workers and sensitive equipment to high energy radiation [4]. Since most of today's gamma irradiations are performed by ^{137}Cs (0.662 MeV) or ^{60}Co (1.1732 MeV and 1.3325 MeV) sources, an irradiation facility with high energy gamma rays would enable rigorous testing of equipment and material samples to assess degradation.

As there are only few sources of gamma rays with energies in the range of 6 MeV and 7 MeV an irradiation system using activated cooling water as the source of energetic gamma rays is proposed at the Jožef Stefan Institute (JSI) TRIGA Mark II research reactor [5]. A conceptual design of such a facility consists of a closed water loop leading the water through one of the reactor's irradiation channels into a shielded irradiation facility outside the reactor core. The analysis of two different water activation loop irradiation systems was performed using the MCNP6 Monte Carlo particle transport code [6] and their usability as a source of high energy gamma rays was evaluated.

The paper is organized as follows. The principle of neutron

https://doi.org/10.1051/epjconf/202022504014

activation of water, the most important activated isotopes, the available nuclear data for neutron water activation and activation in the proposed activation loops at the JSI TRIGA reactor are presented in the first part of the paper. The second part of the paper presents the description of the two proposed water activation loops at the JSI TRIGA reactor. In the last part of the paper the analysis of the calculated results for both activation loops are presented. The proposed activation loops are comped by the water activity in the irradiation facility as the designs of both loop vary significantly. An analysis of the achievable water activity in the irradiation facility due to the water flow rate, reactor power and water retention time in the irradiation facility is also presented.

II. Water Activation

Due to neutron activation of oxygen isotopes in cooling water, the isotopes ^{16}N, ^{17}N and ^{19}O are produced. From the listed isotopes the most important isotope is ^{16}N which is produced via the $^{16}O(n,p)^{16}N$ reaction. Isotopes ^{17}N and ^{19}O are produced via the reactions $^{17}O(n,p)^{17}N$ and $^{18}O(n,\gamma)^{19}O$ [3].

Various high energy decay radiation types are emitted due to decay of activated isotopes in cooling water. As ^{16}N decays via decay path $^{16}N \rightarrow {}^{16}O + \beta^- + \gamma$, high energy gamma rays are emitted (6.13 MeV and 7.12 MeV) with half-life of 7.13 s. ^{17}N decays via decay path $^{17}N \rightarrow {}^{16}O + n + \beta^- + \gamma$ with average energy of 0.88 MeV and a half-life of 4.14 s. The emitted neutrons represent a neutron source outside the primary biological shielding and can activate components outside the primary circuit and produce neutron induced gamma rays. Activated isotope ^{19}O has a half-life of 26.9 s and decays via $^{19}O \rightarrow {}^{19}F + \beta^- + \gamma$ decay path and average gamma ray energy of 0.94 MeV.

The $^{16}O(n,p)^{16}N$ and $^{17}O(n,p)^{17}N$ reactions are threshold reactions with threshold energy at 10 MeV and 8 MeV respectively. The $^{18}O(n,\gamma)^{19}O$ reaction however takes places even at low incidents neutron energies. Fig. 1 displays the cross-section energy dependence for the above mentioned reactions. Due to threshold reactions the activation of water is expected to be higher in fusion reactors like ITER compared to fission reactors due to higher neutron energies (14 MeV neutrons from deuterium-tritium fusion).

The cross-sections the above mentioned water activation reactions vary significantly between the most common evaluated nuclear data libraries as ENDF/B-VIII.0 [3], JEFF-3.3 [7], FENDL-3.1b [8] and TENDL-2017 [9]. For the calculation of water activation presented in this paper only the ENDF/B-VIII.0 librarby was used while a detailed analysis of the cross-sections from different nuclear data libraries is presented in [4].

A closed loop irradiation facility using activated water as a high intensity, high energy gamma ray source is proposed at the JSI TRIGA research reactor. As the proposed system is still in the design phase the activation of water inside closed water loop systems needs to be calculated. The change of specific

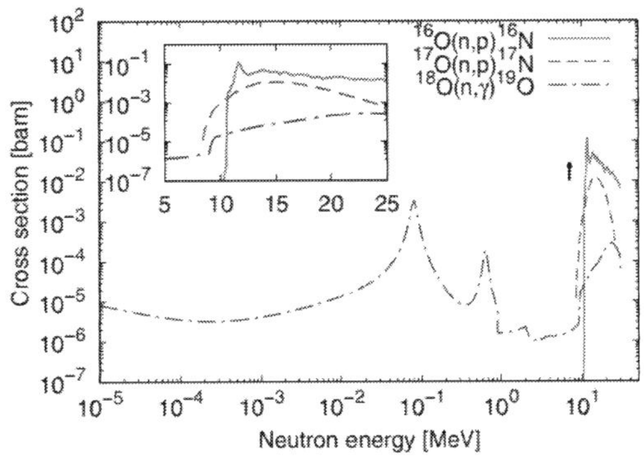

Fig. 1: Cross-section energy dependence for activation of oxygen nuclides taken from the JEFF-3.3 data library [7].

activity of a studied isotope in the water a is described using the equation [10]:

$$a(t) = R(1 - e^{-\lambda t}), \tag{1}$$

where λ is a decay constant and R is an average reaction rate in region of interest of the studied isotope. However, the water in the loop system circulates and is exposed to the neutron flux in the reactor core for a short period of time, in the order of seconds. Thus the activity of the activated water at the inlet to the irradiation facility is described by equation [10]:

$$a_i = R\frac{1 - e^{-\lambda t_i}}{1 - e^{-\lambda T}} e^{-\lambda t_f}, \tag{2}$$

where t_i is the exposure time in the reactor core, T is the total circulation time of the proposed closed water loop system and t_f is the transit time from the reactor core to the irradiation facility [10]. In order to calculate water activity, one most know the water flow rate, irradiation and cooling time, as well as water activation reaction rates inside the reactor core. In this paper we present water activation reaction rates inside the JSI TRIGA irradiation channels as calsulated using the MCNP code. [6].

The goal of the proposed irradition facility is the utilization of energetic gamma rays from the decay of ^{16}N isotope. As the activity and gamma ray emission of ^{17}N and ^{19}O are negligible compared to activity of ^{16}N due to their low natural abundance the analysis of the water activation at the JSI TRIGA research reactor will focus only on the activation of isotope ^{16}O [4].

III. Activation Systems at JSI TRIGA Mark II

The Jožef Stefan Institute TRIGA Mark II research reactor is a light water pool type research reactor with maximum steady state power of 250 kW. The core is submerged into a 6.25 m high and 2 m wide aluminium pool filled with water. The reactor core consists of six concentric rings loaded with stainless steel clad cylindrical fuel rods and aluminium in-core irradiation channels. The fuel material is a homogeneous mixture of U-ZrH with 12 wt% of 20% enriched uranium.

https://doi.org/10.1051/epjconf/202022504014

The fuel rods are placed into two 1.9 cm thick aluminium supporting grids which serve for positioning of fuel elements, control road and addiational in-core experiemnts in concentric rings (A - F) The reactror power is controlled by four individually operated control rods named the safety (S), transient (T), compensating (C) and regulating (R) control rod. The first two are completely withdrawn during normal reactor operation, while the last two are inserted to an approximately equal level. The core is surrounded by a circular graphite reflector [11]. The irradiation channels are normally located in the outermost ring with one exception in the central position of the reactor core. There are also three horizontal channels, one tangential and two radial irradiation channels penetrating the concrete structure of the reactor. One of the radial channels penetrates the graphite reflector to the reactor core while the other ends on the outside of the graphite reflector. A schematic drawing of the reactor core, graphite reflector and irradiation positions is presented in Fig. 2.

Two different conceptual designs for the activated water irradiation facility are currently under consideration. The first concept utilizes the central channel while the second concept utilizes the radial piercing port for water activation (Fig. 2).

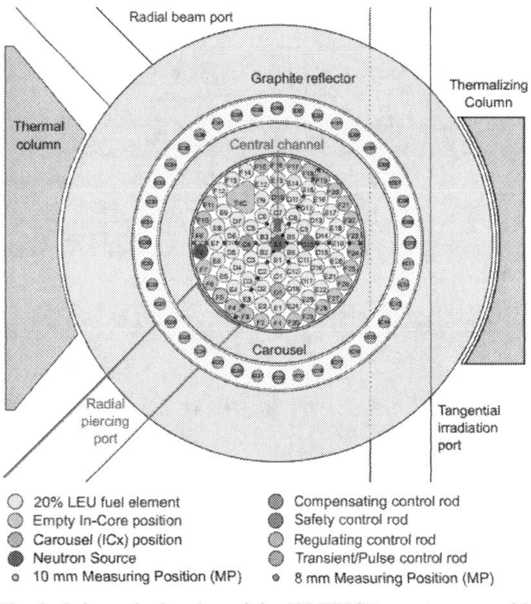

Fig. 2: Schematic drawing of the JSI TRIGA reactor core with marked irradiation channels under consideration for activated water irradiation facility.

In the first conceptual design the higher neutron flux in the central channel compared to the radial piercing port will cause higher activation rate of water. The proposed water activation loop, located in the central irradiation position would consists of a long pipe with the same diameter as the central channel, split into two parts due to limited space in the reactor pool. The water would then be guided trough the reactor tank to the reactor platform, and back down to the ground floor, requiring a 15 m long connection. A schematic model of the proposed conceptual design is presented in Fig. 3. Due to long pipe connection between the reactor core and the irradiation facility the transit time t_f of such a device would be in the order of the ^{16}N half-life ($\sim$ 8 s) depending on the water flow rate. Additional complication due to the increased dose rates at the reactor platform and in the reactor hall during the operation of the loop would require additional shielding. The total length of pipe in the reactor core is estimated to be around 40 cm while the length from the reactor core to the irradiation facility is estimated to be around 15 m (Fig. 3).

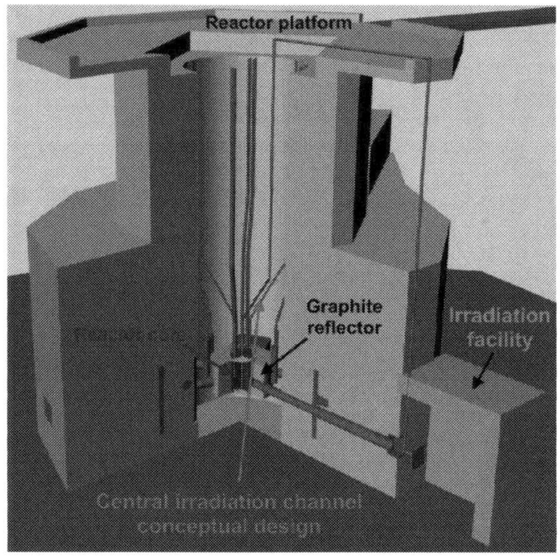

Fig. 3: Model of the proposed conceptual design for utilization of water activation in central irradiation channel.

In the second conceptual design the water is activated in the radial piercing port. Due to the port position adjacent to the reactor core the activation of the water will be lower compared to the central channel. To activate water in radial piercing port a pipe system will be inserted into the port with a large water cask at the end of the port to increase the exposure of the water to neutrons. However, unlike in the first design, the pipes with activated water will be guided outside of the port directly to the irradiation facility next to the port opening (the pipe length being around 3 m) thus reducing the transport time. A schematic model of the proposed conceptual design is presented in Fig. 4. Another benefit of the radial piercing port is its position inside the concrete biological shielding which would provide shielding from activated water.

The summed properties of both proposed conceptual designs are presented in table I.

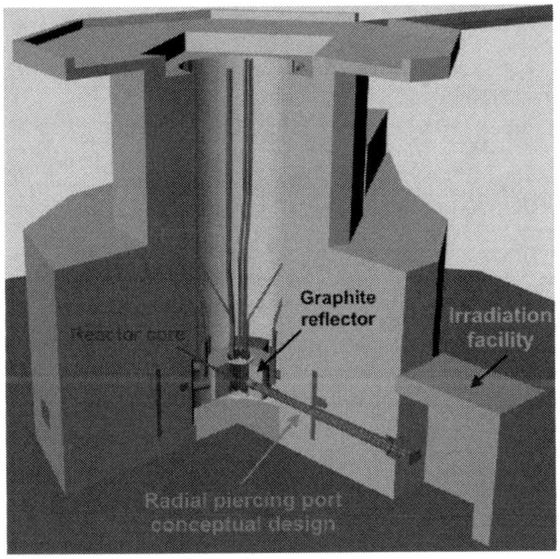

Fig. 4: Model of the proposed conceptual design for utilization of water activation in radial piercing port.

TABLE I: Tabulated data of both proposed conceptual designs for water activation at the JSI TRIGA reactor.

	Central channel	Radial piercing port
Pipe radius [cm]	1.12	1.50
Pipe length [m] (one direction)	15	3
Irradiation time [s] (flow rate = 0.5 l/s)	0.82	2.74
Irradiation time [s] (flow rate = 1 l/s)	0.41	1.37
Transit time [s] (flow rate = 0.5 l/s)	11.76	4.24
Transit time [s] (flow rate = 1 l/s)	5.88	2.12

IV. Conceptual Design Analysis

To calculate the reaction rates for activation of isotope ^{16}O at the JSI TRIGA reactor the water in irradiation channels in the detailed MCNP model was divided in smaller sections to thoroughly analyse both proposed conceptual designs. For the calculation of the reaction rates the ENDF/B-VIII.0 [3] library was used as it is the most up-to-date nuclear data library and there are no significant differences in the cross-section between nuclear data libraries. In the case of water activation in the central channel the highest water activation is in the middle of the active fuel as presented in Fig. 5 due to highest neutron flux. From Fig. 5 it is also visible that the water activation is higher in the bottom half of the reactor core. This is due to position of the control rods as two of the control rods are completely withdrawn while two are withdrawn to approximately same hight at the middle of the reactor core.

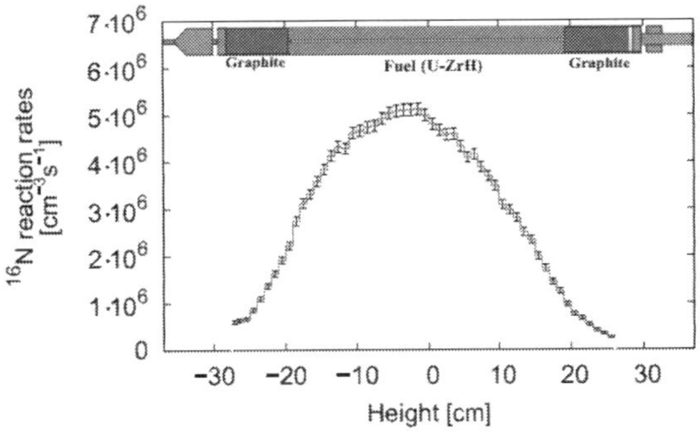

Fig. 5: Reaction rates for water activation in central irradiation channel normalized to reactor full power. The shift of the reaction rates to lower half of the reactor core is due to the position of control rods as two of them were withdrawn to half of the reactor core hight.

In case of water activation in the radial piercing port the reaction rates were calculated at different distanced from the core center and the results are presented in Fig. 6. The highest reaction rates are closest to the reactor core and falls with distance from the reactor core, at 10 cm distance from the core in the port the reaction rates are five times lower compared to the end of the port. By comparison of water reaction rates in central channel (Fig. 5) and radial piercing port (Fig. 6) the reaction rates are on average an order of magnitude higher in the central channel compared to the radial piercing port.

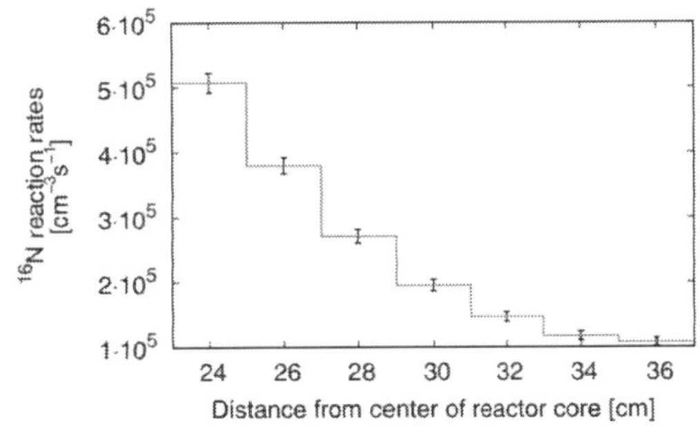

Fig. 6: Reaction rates for water activation in radial piercing port normalized to reactor full power. For the calculations the radial piercing port was filled with water for accurate representation of the proposed design.

To compare viability of both conceptual designs for different experiments utilizing high energy gamma rays the ^{16}N decay rate in the irradiation facility was calculated and compared. The design of the water activation loop allows for the water

 https://doi.org/10.1051/epjconf/202022504014

activity to be controlled by two independent systems - reactor power and water flow rate. However, as the dependency of the ^{16}N decay rate on the reactor power is linear while for the water flow rate it is asymptotic, full reactor power was assumed in all calculations and only water flow rate was varied for analysis of both conceptual designs. The water flow rate was varied for several $0.1\,l/s$ to several l/s as this flow rates are achievable in the proposed conceptual designs. In all cases a $4\,s$ time was assumed in the irradiation facility to produce high energy gamma rays. The results are presented in Fig. 7.

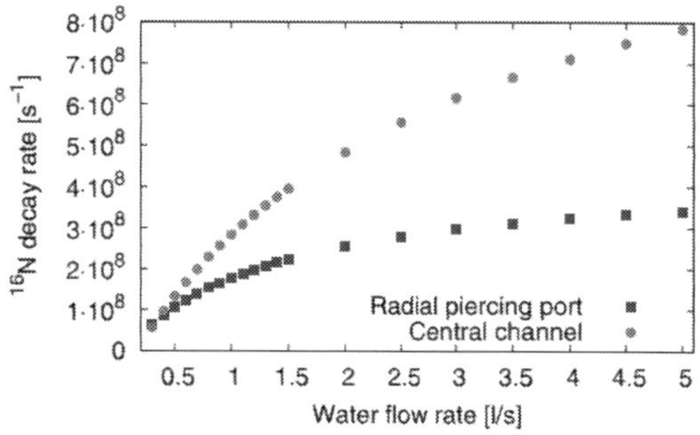

Fig. 7: ^{16}N decay rate dependence on the water flow rate at reactor full power.

For flow rates below $0.5\,l/s$ the ^{16}N decay rate in the irradiation facility is higher for water activated in the radial piercing port despite an order of magnitude higher activation in the central channel. This is due to the long transient times (in the order of ^{16}N half-life) in the conceptual design utilizing the central channel. However, at higher flow rates, above $1\,l/s$, the ^{16}N decay rates are higher in conceptual design utilizing central channel as the transient time becomes shorter.

For both proposed conceptual designs the maximal flow rate is estimated to be around $1\,l/s$ due to the pipe sizes. At this flow rate the ^{16}N decay rate in the conceptual design utilizing the central channel is expected to be around $3 \cdot 10^8\ s^{-1}$, which is only by a factor two higher compared to the conceptual design utilizing radial piercing port despite an order of magnitude higher activation in the reactor core.

The uncertainty of the water flow rate is the higher contributor to the uncertainty in the ^{16}N decay rate in the irradiation facility as the power of the reactor can be kept constant by the reactor automated systems. Due to this effect the conceptual design utilizing the radial piercing port has an advantage compared to the conceptual design utilizing the central channel. A small change in the water flow rate has a small effect in the ^{16}N decay rate in the irradiation facility due to this region of flow rate being in the saturated part of the ^{16}N decay rate curve, while for the conceptual design utilizing the central channel a small change in water flow rate can change the ^{16}N decay rate significantly, as the ^{16}N decay rate curve is still in the ascending part.

A. Irradiation facility

Different experiments utilizing activated water as a source of high energy gamma rays are going to be performed in an irradiation facility to minimize background radiation and shield the reactor hall during operation of water loop. Currently the irradiation facility is considered to be a closed room, $2\,m$ high and $3\,m$ long and wide. The shielding of the facility will be achieved by $30\,cm$ thick high density concrete blocks. A schematic model of the propose irradiation facility is presented in Fig. 8.

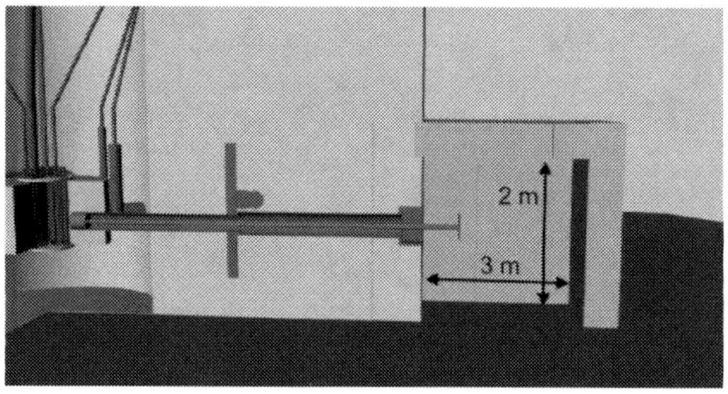

Fig. 8: Conceptual design of the irradiation facility for experiments utilizing high energy gamma rays from activated water decay. The depicted conceptual design utilizes the radial piercing port for water activation.

To asses the biological dose rates (H*(10) ambient dose equivalent) and air kerma in the proposed irradiation facility, a MCNP model was constructed. The source of gamma rays was modelled as an isotropic plane as it is currently considered as the best source for uniform activity with the use of activated water. The room in the model was filled with air and the biological dose rates and air kerma were calculated at different distances from the source. To calculate the biological dose rates from gamma rays flux-to-dose conversion factors from the standard ICRP-21 [12] were used while the air kerma was calculated using the energy deposited in the air due to gamma rays. The results for conceptual design utilizing radial piercing port for water activation at full reactor power and water flow rate of $1\,l/s$ are presented in Fig. 9. The biological dose rates close to the source are in the order of several mSv/h and air kerma is in the order of several mGy/h while these values fall significantly at greater distance from the source. At $1\,m$ distance from the source the biological dose rates are in the order of $0.1\,mSv/h$ and air kerma $0.1\,mGy/h$. For the conceptual design utilizing central channel for water activation the biological dose rates and air kerma are higher by around a factor of two at reactor full power and water flow rate of $1\,l/s$ due to higher ^{16}N decay rate.

These preliminary calculations on water activation rates, ^{16}N decay rates, biological dose rates and air kerma serve as a basis for design of the proposed water activation irradiation facility.

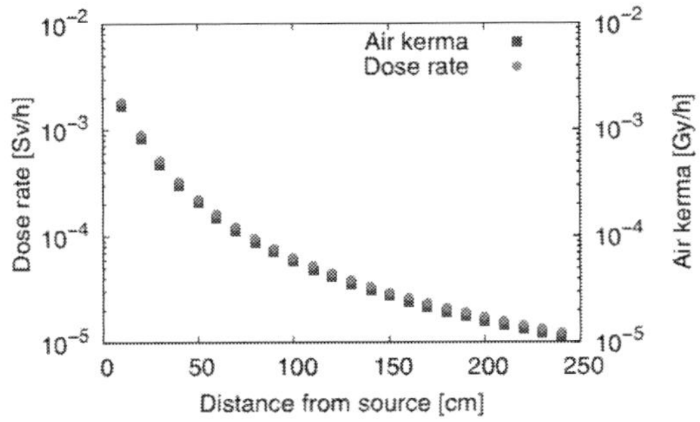

Fig. 9: Calculated biological dose rate and air kerma due to gamma rays in the proposed irradiation facility utilizing activated water as a source of high energy gamma rays. The values represent results for conceptual design utilizing radial piercing port at reactor full power and $1\,\mathrm{l/s}$ flow rate.

V. Conclusion

In the paper two concepts to utilize activated water for characterization of detector responses and shielding at high energy gamma rays at the JSI TRIGA reactor are presented. The first conceptual design utilizes the central channel for water activation while the second concept utilizes the radial piercing port for water activation. Using the Monte Carlo code MCNP the water activation in both conceptual designs was calculated. In the first conceptual design the water activation is an order of magnitude higher than in the second, however the transit time is longer by a factor of around 3. Also, the first conceptual design would require thick shielding of the water pipe in order to keep the reactor hall accessible.

The water activation results were used to obtain the ^{16}N decay rates in the irradiation facility for comparison. Despite an order of magnitude higher activation of water in the central channel the ^{16}N decay rate in the irradiation facility is just short of two times higher compared to conceptual design utilizing the radial piercing port at reactor full power and water flow rate of $1\,\mathrm{l/s}$, which is in the order of several 10^8 decays per second. However, for water flow rates below $0.5\,\mathrm{l/s}$ the ^{16}N decay rates from water activated in radial piercing port are higher compared to central channel due to longer transient times. In the irradiation facility water activity will produce biological dose rates in the order of $2\,\mathrm{mSv/h}$ to $4\,\mathrm{mSv/h}$ depending on the water flow rate, reactor power and which conceptual design is used for water activation.

The conceptual design utilizing the radial piercing port currently presents the best option for the irradiation facility due to the simpler design of the irradiation loop, already present shielding of the loop and comparable ^{16}N decay rates in the irradiation facility. Since the water activation is in saturation in the proposed design small changes in water flow rate will small effect on ^{16}N decay rate in the irradiation facility compared to the conceptual design utilizing central channel.

References

[1] A. Žohar and L. Snoj, "Gamma Dose Field due to Activated Cooling Water in a Typical PWR," in *Proceedings, 26th International Conference Nuclear Energy for New Europe - NENE 2017*, 2017.

[2] H. Iida *et al.*, "Three-Dimensional Analysis of Nuclear Heating in the Superconducting Magnet System Due to Gamma-rays from 16N in the ITER Water Cooling System of the Shielding Blanket," in *17th IEEE/NPSS Symposium Fusion Engineering*, vol. 2, 1997, pp. 837–840.

[3] D. Brown *et al.*, "ENDF/B-VIII.0: The 8th Major Release of the Nuclear Reaction Data Library with CIELO-project Cross Sections, New Standards and Thermal Scattering Data," *Nuclear Data Sheets*, vol. 148, pp. 1 – 142, 2018.

[4] A. Žohar and L. Snoj, "On the dose fields due to activated cooling water in nuclear facilities," *Progress in Nuclear Energy*, vol. 117, 2019.

[5] R. Jeraj and M. Ravnik, *TRIGA Mark II Reactor: U(20)-Zirconium Hydride Fuel Rods in Water with Graphite Reflector. IEU-COMP-THERM-003*, International Handbook of Evaluated Criticality Safety Benchmark Experiments, NEA/NSC/DOC(95)03, OECD-NEA, Paris, France, 2010.

[6] T. Goorley, *MCNP6.1.1-Beta Relese Notes*. LA-UR-14-24680, 2014.

[7] OECD/NEA Data Bank, "JEFF-3.3 Evaluated Data Library," 2017. [Online]. Available: http://www.oecd-nea.org/dbdata/jeff/jeff33/

[8] A. Koning and A. Trkov, "FENDL-3.1b: Fusion Evaluated Nuclear Data Library Ver.3.1b," 2016. [Online]. Available: https://www-nds.iaea.org/fendl/

[9] A. Koning and D. Rochman, "Modern nuclear data evaluation with the talys code system," *Nuclear Data Sheets*, vol. 113, no. 12, pp. 2841 – 2934, 2012.

[10] A. Žohar and L. Snoj, "Analysis of the Primary Water Activation in a Typical PWR," in *Proceedings, 25th International Conference Nuclear Energy for New Europe - NENE 2016*, 2016.

[11] Ž. Štancar and L. Snoj, "An improved thermal power calibration method at the TRIGA Mark II research reactor," *Nuclear Engineering and Design*, vol. 325, pp. 78 – 89, 2017.

[12] ICRP, *Data for Protection Against Ionizing Radiation from External Sources: Supplement to ICRP Publication 15*. ICRP-21, International Commission on Radiological Protection, Pergamon Press, 1973.

THERMOCOAX rhodium SPND sensitivity dispersion and validation of the sensitivity calculation model

L. Vermeeren[1] *, W. Leysen[1], L. Pichon[2], V. Salou[2] and G. Helleux[2]
[1] SCK•CEN
[2] THERMOCOAX
* lvermeer@sckcen.be

Abstract— **This paper describes the neutron irradiation tests of 7 THERMOCOAX Self-Powered Neutron detectors (SPNDs) in the BR1 reactor at SCK•CEN. The SPNDs were fabricated according to the same specifications, but from different fabrication batches. The SPND signals were recorded during stepwise power-up of the reactor, proving the linearity of the SPND response within a wide thermal neutron flux range: from as low as $0.8{\cdot}10^9$ n/(cm s) to $2.6{\cdot}10^{11}$ n/(cm s). Intercomparison of the SPND signals shows a very small spread, confirming the repeatability of the THERMOCOAX fabrication processes. The overall neutron sensitivities of the seven SPNDs agreed within a 1% margin.**
The experimental data were analyzed in terms of prompt and various delayed responses. Prompt contributions to the signal are due to external gamma induced processes and to processes involving gamma rays emitted instantaneously upon neutron capture. The main contribution in a rhodium SPND is due to activation of the rhodium emitter and beta emission during decay of the activated rhodium and leads to a delayed response with a characteristic time of a few minutes. Activation and subsequent beta decay in other materials present in the SPND lead to additional minor delayed signal contributions. The partial SPND sensitivities due to all these processes were calculated using an MCNPX based model and were compared with experimental sensitivities based on the recorded data. The results were in fair agreement; for the overall SPND neutron sensitivity an agreement within a 1% margin was achieved.

***Keywords*—Nuclear measurements, neutron detectors, reactor instrumentation, modelling**

I. Introduction

A crucial parameter in the monitoring of the behavior of research reactors and power reactors is the neutron flux distribution within the reactor core. Fission chambers can provide instantaneous data on in-core reactor neutron flux measurements. But thanks to improved simulation tools, there is a growing interest in self-powered neutron detectors (SPNDs) as a valuable and cheaper alternative for fission chambers for thermal neutron flux monitoring. They can be implemented as fixed in-core sensors for applications in which mobile in-core systems are not acceptable and in which ex-core sensors cannot ensure all required functions [1].

This paper deals with the characterization of SPNDs with rhodium emitter, fabricated by THERMOCOAX. A set of seven SPNDs was tested in the BR1 reactor at SCK•CEN. All SPNDs were fabricated according to the same specifications (see Fig. 1): a 250 mm long, 0.50 mm diameter rhodium emitter, surrounded by a 0.185 mm thick alumina insulator and enclosed in a 0.245 mm thick SS316L sheath (outer diameter 1.36 mm). The emitters parts of the SPNDs were integrated without external welds in 1.36 mm diameter twin core cables, with two 0.21 mm diameter SS316L core wires. The seven SPNDs were taken from different fabrication batches.

A first goal of the test was to investigate the linearity and the spread in neutron sensitivity of the SPND in order to realize a complete characterization in terms of linearity, sensitivity repeatability and deviation. Secondly, the experimental sensitivities were also compared with results from detailed Monte Carlo modelling of the various contributions to the SPND signal in order to validate the calculation approach.

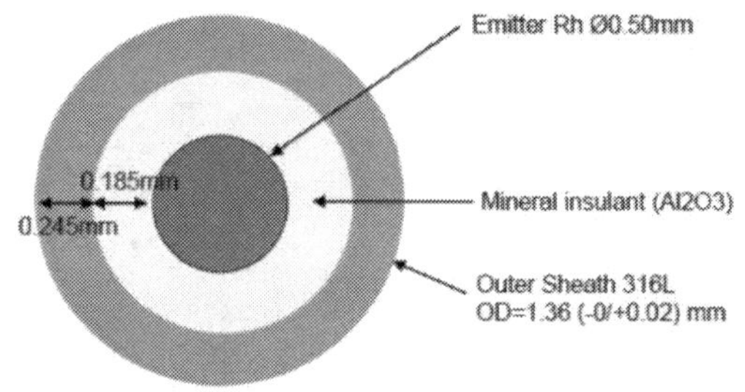

Fig. 1. Cross section of the sensitive part of the SPNDs with indication of the dimensions

II. Experimental set-up

The SPNDs were irradiated in the Y7 channel of the BR1 reactor (see Figure 2). BR1 is an air cooled graphite moderated thermal reactor with a nominal thermal power of 700 kW; for short periods the power can be raised to 1 MW. An activation dosimetry campaign performed just before the test irradiations discussed here, yielded an unperturbed conventional thermal neutron flux of $1.80{\cdot}10^{11}$ n/(cm /s) at a nominal BR1 power of 700 kW. The SPNDs were positioned with the emitters centered at the reactor mid-plane (so 12.5 cm emitter length at each side of the mid-plane). So the SPND emitters are fully exposed to the maximum flux (less than 1 % flux gradient over the emitter length).

L. Vermeeren and W Leysen are with the Nuclear Research Centre SCK•CEN, B-2400, Mol, Belgium (e-mail: lvermeer@sckcen.be); L. Pichon, V. Salou and G. Helleux are with Thermocoax, 8 Rue du Pré Neuf, 61100 St Georges des Groseillers, France

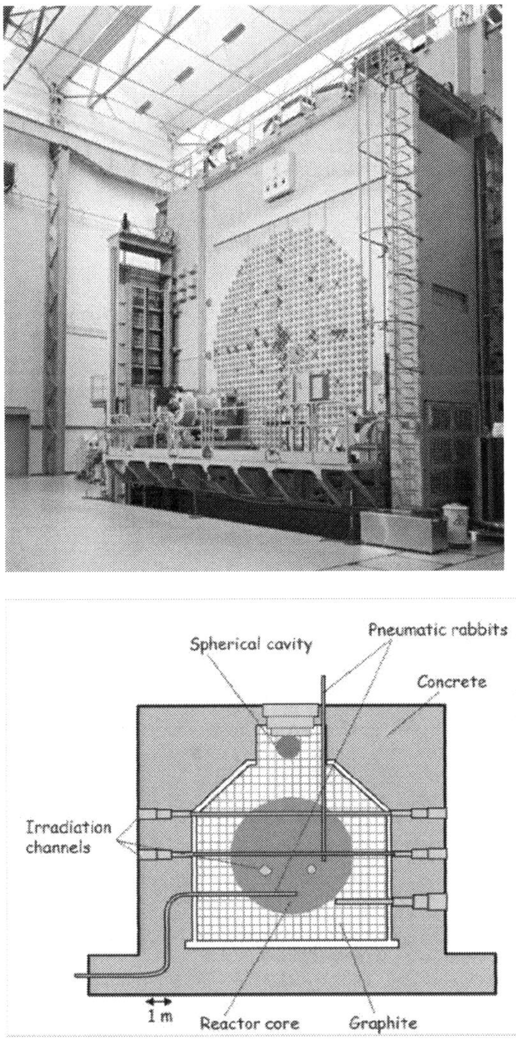

Fig. 2. Picture of the BR1 reactor (top) and schematic cross section of the BR1 (bottom); the SPNDs were loaded in the Y7 channel, which is a horizontal channel perpendicular to the plane of this sketch (indicated as a yellow square) and parallel with the reactor fuel.

The neutron and the gamma spectra in channel Y7 as calculated by means of MCNP [2] are shown in Figs. 3 and 4. At a nominal BR1 power of 700 kW, a total neutron flux of $4.06\cdot10^{11}$ n/(cm s) was obtained and a conventional neutron flux of $1.81\cdot10^{11}$ n/(cm s), in very good agreement with the dosimetry results; the total gamma flux is $5.43\cdot10^{10}$ γ(/cm s). These spectra were used for all partial SPND sensitivity calculations.

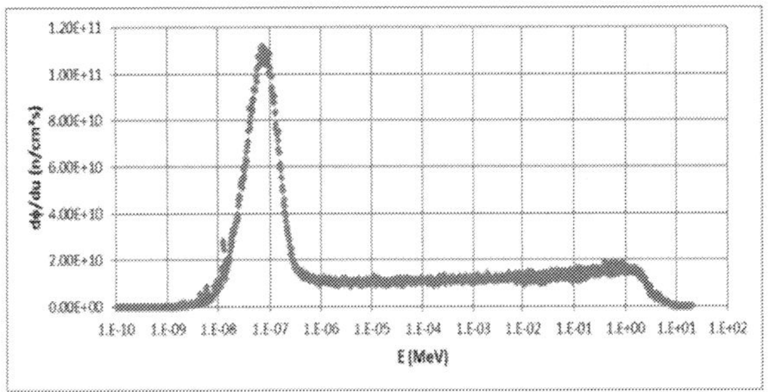

Fig. 3 – Calculated neutron spectrum in BR1/Y7. The data are represented as absolute flux values (at nominal BR1 power of 700 kW) per unit lethargy; this means that the total flux can be obtained directly via the surface under the curve.

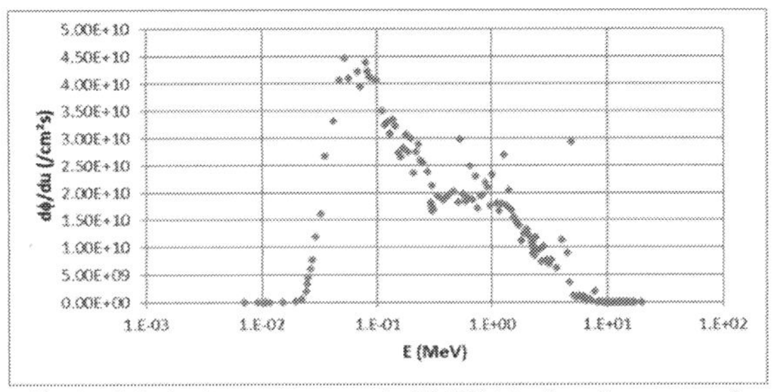

Fig. 4 – Calculated gamma spectrum in BR1/Y7 (also formally represented as absolute flux values per unit lethargy for 700 kW reactor power)

Figure 5 shows a few pictures of the SPND holder. The seven SPNDs are placed on fixed distance from the center of the aluminum box (where a thermocouple was positioned). They were guided through 1.5 mm holes in two thin aluminum plates (at both ends of the emitter part). At the exit of the box, the cables were strapped together to avoid any axial movement during manipulations. The cables were guided further through the long aluminum tube and through the shielding chicane at the border of the reactor. The acquisition system (connection box with 1 MΩ resistances, Keysight multimeter (34972A) with scanner cards (34902A) and pc) was positioned next to the reactor.

Fig. 5 – Pictures of the sample holder (cover removed) with the 7 SPNDs and a thermocouple

III. SPND SIGNAL COMPONENTS

The SPND current is composed of various components. For a rhodium SPND, the major component is due to neutron capture by the rhodium nuclei (^{103}Rh) in the emitter. Neutron capture by ^{103}Rh leads to the formation of ^{104}Rh in two different states: the excited state ^{104m}Rh, that decays with a half-life of 4.4 minutes to the ground state, or immediately the ground state ^{104}Rh, which decays with a half-life of 42 seconds towards (stable) ^{104}Pd.

The de-excitation from ^{104m}Rh to ^{104}Rh occurs mainly via internal transition, which will lead to a negligible signal contribution. Ultimately, all activated rhodium nuclei will decay towards ^{104}Pd, during which a beta ray is emitted, as well as some gamma rays. This beta (actually a fast electron) has a significant probability to escape from the emitter and reach the collector after crossing the insulator. This corresponds to a charge transfer from emitter to collector, which can be measured as a current when externally closing the circuit (with an ampere meter). This component has a delayed character related to the half-lifes of ^{104m}Rh and ^{104}Rh.

The neutron capture also leads to a signal component with a prompt character. This component results from the gammas emitted instantaneously during neutron capture. Interaction of these gammas with electrons (photoelectric effect, Compton scattering, pair formation,…) can lead to sufficiently energetic electrons so that they can cross the insulator and contribute to the current. Since this is a two-step process (neutron interaction + gamma-electron interaction), the resulting current is typically an order of magnitude lower than the beta induced current.

Neutron capture in other parts of the SPND (insulator and sheath) and in the materials surrounding the SPND can lead to additional (usually minor) contributions to the detector signal.

Similarly, external gammas can interact with electrons in the SPND (and its surroundings) and generate a current. It is known that this current contribution depends strongly on the materials and the geometry of the surroundings.

IV. MCSS MODELLING

All relevant signal contributions were calculated according to the MCSS model described in [3] and using the spectra from Figs. 3 and 4:

- the main rhodium beta related contribution
- aluminum beta contribution (from the insulator)
- manganese beta contribution (1.74% mass fraction in the sheath material)
- prompt neutron contribution
- gamma contribution

All related MCNPX calculations were performed for an infinitely long geometry (results per cm length), which included the seven SPNDs and the aluminum box (9.6 cm * 9.6 cm; wall thickness 0.2 cm). The neutron (or gamma) source was defined on a cylindrical surface at radius 6.9 cm in order to cover the complete geometry. The results for all seven SPNDs were identical within the statistical uncertainties, so average values were taken.

The calculations lead to the following partial SPND currents (at 700 kW reference reactor power):

- due to Rh betas : 5.18 nA (94.4%)
- due to Al betas : 0.0050 nA (0.1%)
- due to Mn betas : -0.0095 nA (-0.2%)
- prompt neutron contribution : 0.29 nA (5.3%)
- prompt gamma contribution : 0.023 nA (0.4%)
- total prompt contribution : 0.31 nA (5.6%)
- total SPND current: 5.49 nA (100%)

V. EXPERIMENTAL DATA ANALYSIS

A. Example of experimental data

Emitter and compensator current data were recorded sequentially for all seven SPNDs (1 complete data set every 20 seconds) during stepwise BR1 power increase (3-10-30-100-200-400-600-800-1000 kW) followed by a power reduction to 300 kW and scram (see Fig. 6). The power range [3-1000 kW] corresponds to a conventional thermal neutron flux range of [$8 \cdot 10^8$ - $2.6 \cdot 10^{11}$ n/(cm s)].

Additionally, in order to test the prompt response fraction and the repeatability, three tests at 300 kW reactor power were performed with continuous measurement of the SPND1 differential current (one point each 400 ms) (see Fig. 7).

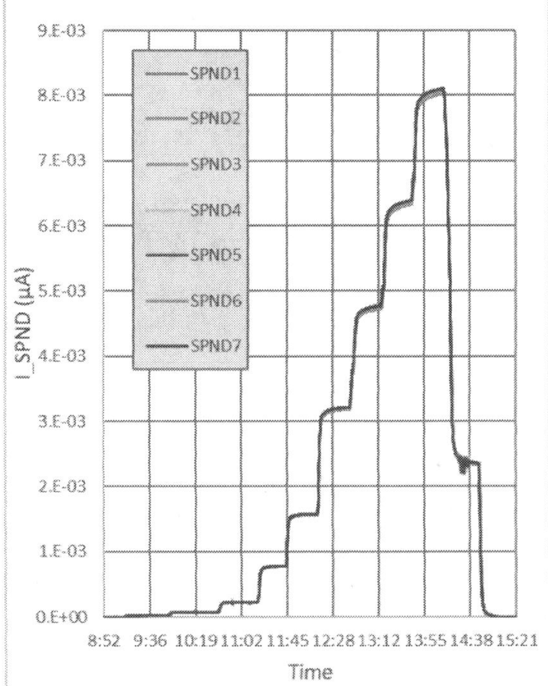

Fig. 6 – Raw differential signals (emitter minus compensator) of the seven SPNDs during stepwise BR1 power increase and scram. The signals are so close to each other that they can hardly be distinguished in this representation.

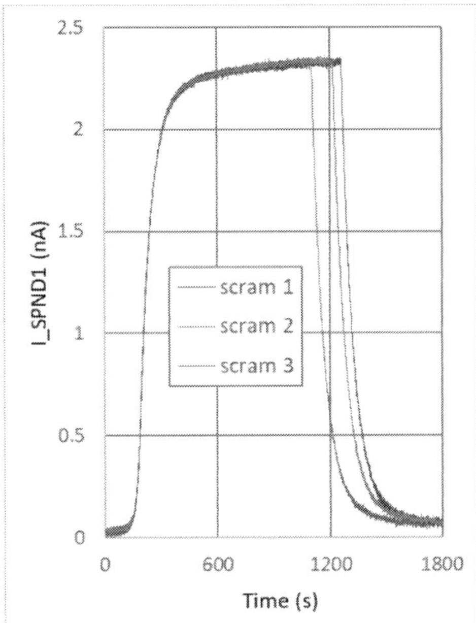

Fig. 7 – Three examples of differential signals (emitter minus compensator) of the SPND1 with BR1 operating at 300 kW followed by scram. Apart from the slightly different power-on time in the three cases, all three curves are nearly coinciding.

B. Comparison between signals of all SPNDs

The recorded signals of all seven SPNDs were compared with their average value. The percentage deviations were found to be:

- SPND1: +0.2%
- SPND2: -1.0%
- SPND3: -0.4%
- SPND4: +0.9%
- SPND5: +0.2%
- SPND6: +0.2%
- SPND7: -0.1%

So all signals are coherent within a margin of 1%. The flux gradients are expected to be smaller than 1%, but they are not known more accurately; so part of the differences between the SPND signals could even be due to flux gradients. The sensitivity dispersion between the seven SPNDs is within 1%.

C. Linearity

Figure 8 shows the average recorded net SPND currents (emitter currents minus compensator wire currents) as a function of the BR1 reactor power as deduced from the count rate of the reactor power monitoring fission chamber, from startup to the 1000 kW level. The deviations from a linear relation are within the combined uncertainty of the power determination (statistical uncertainty on the fission chamber count rate) and of the SPND current. This proves the proper functioning of the SPNDs in the complete range studied, from $0.8\cdot10^9$ n/(cm s) to $2.6\cdot10^{11}$ n/(cm s). It should be noted that the studied SPNDs gave already a readily measurable signal for a thermal neutron flux smaller than 10^9 n/(cm s).

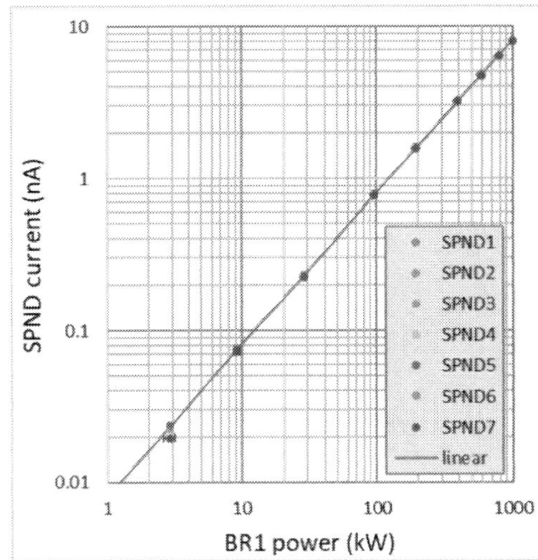

Fig. 8 – Signals of the 7 tested SPNDs vs. BR1 reactor power (in logarithmic scale). For one data set, all horizontal and vertical error bars are plotted; apart from the lowest flux point, all error bars fall within the dimensions of the data points, and the deviation from linearity is also less than these dimensions.

D. Reproducibility

Some test conditions were repeated several times in order to verify the reproducibility of the SPND response. As illustrated in Fig. 7, the obtained signals were very coherent, with mutual deviations below the 1% level.

E. Prompt signal component analysis

The distinction between the prompt and delayed components can be made in the clearest way by analyzing the scram data taken in fast acquisition mode, combining a high acquisition rate with a fast flux change. Fig. 9 shows part of the experimental data of Fig. 7, analyzed as a sum of a prompt component and a rhodium beta component. The relative prompt fractions deduced from the data of three subsequent tests are 4.0%, 4.1% and 4.0%, so on average 4.0%, which is somewhat lower than the 5.3% obtained by MCNP calculations (see Section 4). In the analysis of the slow acquisition data we will keep this ratio fixed at 4.0% for all seven SPNDs.

F. Analysis of the delayed signal components

In the slow acquisition rate data, the small contribution due to aluminum betas cannot be reliably disentangled from the rhodium beta contribution (similar characteristic time). Therefore, the calculated partial aluminum beta current of 5 pA (for 700 kW reactor power) was applied in the analyses, in which the rhodium beta sensitivity and the manganese beta sensitivity were left as free parameters. The average equilibrium rhodium beta related current (normalized to 700 kW reactor power) was found to equal 5.21 nA and the average equilibrium manganese beta related current -9.0 pA. The dispersion between the rhodium beta currents in the different SPNDs was limited to 1%, while the individual manganese beta current values obtained for the different SPNDs ranged from -8.8±0.7 to -9.6±0.7 pA. Fig. 10 illustrates the analysis for the data of Fig. 6, averaged over all seven SPNDs.

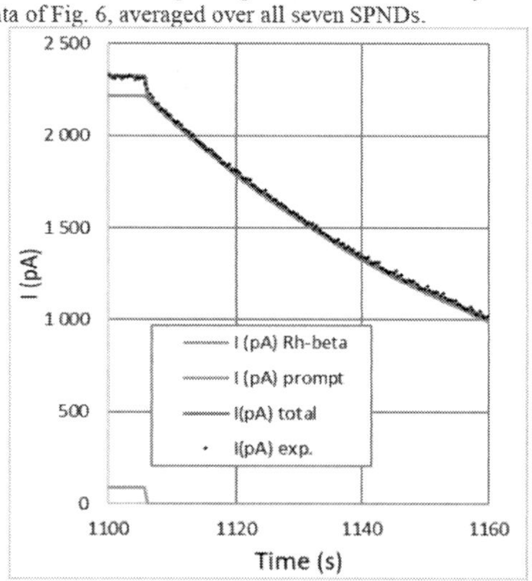

Fig. 9 –Analysis of the data of Fig. 7 in terms of prompt and rhodium beta contributions.

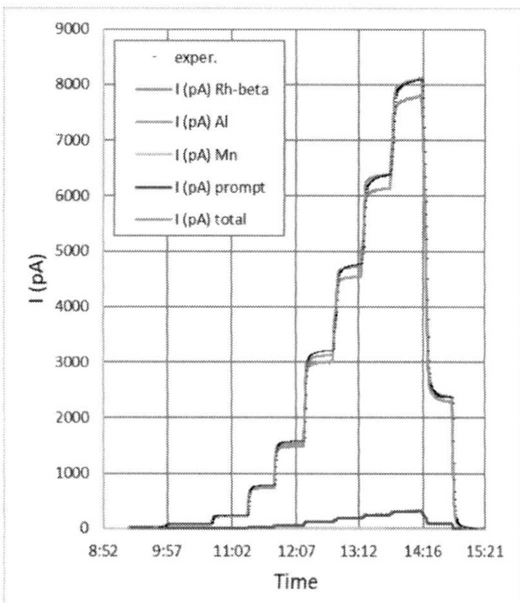

Fig. 10–Analysis of the data of Fig. 6 in terms of prompt and rhodium beta contributions.

VI. COMPARISON BETWEEN EXPERIMENTAL AND MCNP DATA

Table I summarizes the experimental and calculated partial current contributions, normalized to a reactor power of 700 kW. The total current and the rhodium beta related current values from experiment and calculations agree within 1%. The calculated total prompt current amounts to 0.31 nA, of which the majority (0.29 nA) would be due to neutrons and only 0.023 nA due to gammas. This value is significantly higher than the observed value of 0.22 nA (4.0% of the total current). The manganese beta related current is small and negative; the experimental and calculated values agree within the experimental uncertainty. The observed total current differs by only 1 % from the calculated value.

TABLE I
Experimental and calculated partial SPND currents and the percentage differences

	Exper.	Calc.	Rel. diff.
Rh beta	5.21 nA	5.18 nA	0.6 %
Mn beta	-0.009 nA	-0.0095 nA	5 %
Total prompt	0.22 nA	0.31 nA	29 %
Total	5.43 nA	5.49 nA	1 %

VII. CONCLUSIONS

Seven Thermocoax rhodium SPNDs from different fabrication batches were tested under irradiation in the BR1 reactor. The SPND signals were recorded during stepwise power-up of the reactor, proving the linearity of the SPND response down to a thermal neutron flux level below 10^9 n/(cm s). Intercomparison of the SPND signals showed a very small spread, confirming the repeatability of the fabrication process. The data were analyzed in terms of prompt and various delayed responses and the partial sensitivities were compared with theoretical sensitivities based on MCNP calculations. For the overall SPND neutron sensitivity an agreement within a 1% margin was achieved.

REFERENCES

1. M. Giot, L.Vermeeren, A. Lyoussi, C. Reynard-Carette, C. Lhuillier, P. Mégret, F. Deconinck and B. Soares Gonçalves, EPJ Nuclear Sci. Technol. 3, 33 (2017)

2. E. Malambu, "Characterization of neutron fields in support of dosimetry measurements in the BR1 reactor using MCNP/S code simulations", SCK•CEN Internal Report SCK•CEN-IR-182 (2010)

3. L. Vermeeren, IEEE Conf. Proc. 4th Intl. Conf. on Advancements in Nuclear Instrumentation Measurement Methods and their Applications (ANIMMA), IEEE Xplore (2016); doi: 10.1109/ANIMMA.2015.7465531

Design and Simulation of Gamma Spectrometry Experiments in the CROCUS Reactor

O. Pakari*, V. Lamirand*†, B. Vandereydt*, F. Vitullo*, M. Hursin*†, C. Kong‡, and A. Pautz*†
*Laboratory for Reactor Physics and Systems Behaviour
Ecole polytechnique fédérale de Lausanne, Lausanne, Switzerland
oskari.pakari@epfl.ch
†Laboratory for Reactor Physics and Thermal-Hydraulics
Nuclear Energy and Safety Division
Paul Scherrer Institut, Villingen, Switzerland
‡Ulsan National Institute of Science and Technology
Ulsan, South Korea

Abstract*—Gamma rays in nuclear reactors, arising either from fission or decay processes, significantly contribute to the heating and dose of the reactor components. Zero power research reactors offer the possibility to measure gamma rays in a purely neutronic environment, allowing for validation experiments of computed spectra, dose estimates, reactor noise and prompt to delayed gamma ratios. This data then contributes to models, code validation and photo atomic nuclear data evaluation. In order to contribute to aforementioned experimental data, gamma detection capabilities are being added to the CROCUS reactor facility. The CROCUS reactor is a two-zone, uranium-fueled light water moderated facility operated by the Laboratory for Reactor Physics and Systems Behaviour (LRS) at the Swiss Federal Institute of Technology Lausanne (EPFL). With a maximum power of 100 W, it is a zero power reactor used for teaching and research, most recently for intrinsic and induced neutron noise studies. For future gamma detection applications in the CROCUS reactor, an array of four detectors - two large 5"x10" Bismuth Germanate (BGO) and two smaller Cerium Bromide ($CeBr_3$) scintillators - was acquired. The BGO detectors are to be arbitrarily positioned in the core reflector and out of the vessel for measurements at arbitrary distances. The $CeBr_3$ detectors on the other hand are small enough to be set in the guide tubes of the control rods for in-core measurements. We present a study of the neutron and gamma flux in the core and reflector using the MCNP 6.2 and Serpent 2 Monte Carlo codes for coupled neutron and photon transport criticality calculations. More specifically, we investigate and compare predicted spectra as well as reactivity worth of different envisioned experimental setups. We further predict pulse height spectra as well as doses to the crystals with and without cadmium shielding to estimate allowable reactor powers with respect to detector radiation hardness. The results serve as basis for calibration and aid in the design and regulatory approval of the experiments.

I. Introduction

Gamma radiation as a by-product of fission and radioactive decay is a major component of a nuclear reactors radiation field. It contributes to the heating of structures, the degradation of materials, and the overall dose rate around a reactor. The characteristics of a reactors gamma field is thus important for operation and safety. Accurate predictions are desired in the design of experiments and in the inception of new reactors.

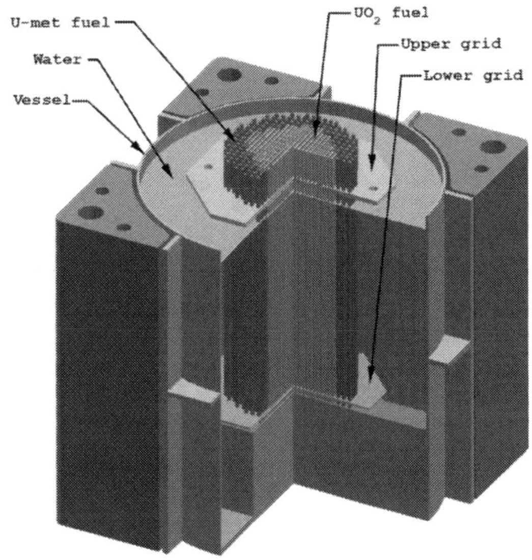

Fig. 1. The CROCUS reactor vessel with basic internal structures. Two fuel zones, an inner UO_2 and outer U_{met} zone, held in an approximately cylindrical array by two grids. Water is filled from the bottom to establish criticality.

Quantitative experiments on the spectra and fluxes of gamma radiation have become of recent interest with the introduction of coupled photon and neutron transport options in common Monte Carlo codes, such as Serpent and MCNP. On top of a relatively well characterized neutron field, the predicted photon fluxes and spectra lack experimental data for validation. The CROCUS reactor, presented in Section II, offers a zero power environment to provide such data.

To date, only TLD measurements of the gamma field of CROCUS have been conducted in the context of activation analysis. With a variety of experimental programs (e.g. [1], [2]) that so far only exploited neutron detection, the acquisition of gamma detectors intends to extend the capabilities of CROCUS.

II. The Research Reactor CROCUS

The description of CROCUS follows [2]. The CROCUS reactor is a two-zone, uranium-fueled, H_2O-moderated critical assembly operated by the Laboratory for Reactor Physics and Systems Behaviour (LRS) at the Swiss Federal Institute of Technology Lausanne (EPFL). It is a zero power reactor, with a maximum power of up to 100 W. The core is approximately cylindrical in shape with a diameter of about 58 cm and a height of 100 cm. Two different kinds of fuel rods make up the reactor core of CROCUS (see Fig. 1 and 3). The central zone is loaded with 336 UO_2 fuel rods (1.806 wt.%-enriched), set in a square lattice with a pitch of 1.837 cm. The peripheral zone is loaded with up to 176 thicker, U_{met} fuel rods (0.947 wt.%-enriched) with a pitch of 2.917 cm, also in a square lattice. All fuel rods have an aluminum cladding and are maintained in a vertical position by the upper grid and lower grid plates spaced 100 cm apart. Both grid plates incorporate a cadmium layer with a thickness of 1 mm to limit axial thermal leakage to surrounding structures. The active fuel length starts at the top surface of the lower cadmium layer and extends to 100 cm. The core is located in an aluminum water tank, its diameter is 130 cm and thickness is 1.2 cm. Demineralized light water is used as moderator and reflector. Reactivity is nominally controlled by a variation of the water level using a spillway with an accuracy of 0.1 mm (equivalent to 0.4 pcm) and optionally by means of two control rods containing naturally enriched boron carbide (B_4C) sintered pellets located in diagonal symmetry within the outer fuel zone.

III. New gamma detection experiment for CROCUS

A. Detector specification

The new gamma detection array to be used in the CROCUS core was chosen to be on scintillator basis. The most prominent measurement locations in CROCUS are, due to their accessibility, the two control rod tubes and the water reflector.

In the control rod position we expect a higher overall photon flux and due to the proximity to the fuel. Thus, a fast scintillator usable for spectroscopy - Cerium(III) Bromide ($CeBr_3$) - was chosen. The control rod tubes constrain the cylindrical crystal size to 13 mm in diameter and 15 mm in length. Two detectors to measure simultaneously in both control rod tubes were acquired from Scionix Holland [3]. Both house a Hamamatsu Type R12421 PMT.

For the periphery location, which is less constrained due to the large vessel radius holding the water reflector, a high Z scintillator for maximal efficiency - Bismuth Germanate (BGO) - was chosen. The size of the cylindrical crystal is 127 mm in diameter to increase the absorption efficiency to above 95% of 10 MeV photons. The height was constrained by weight and is 250 mm. Two detectors of this type were acquired. Both house a Photonis 5" Type XP4578 PMT. The detectors CAD models are shown in Figure 2.

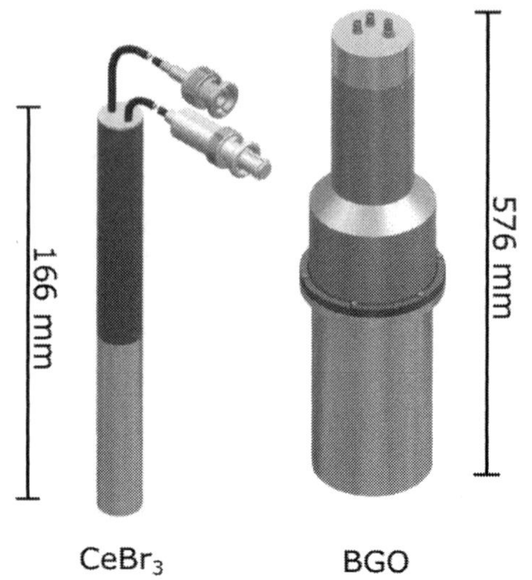

Fig. 2. CAD drawing of the detectors to be used in CROCUS. **Left:** $CeBr_3$ detector with 0.6"x2" cylindrical crystal (silver) and PMT with I/O cables (red). **Right:** BGO detector with 5"x5" cylindrical crystal (silver) and PMT with connectors (grey).

B. Envisioned detector locations in CROCUS

The $CeBr_3$ detectors can be moved axially in the control rod guide tubes, but also set in the reflector if deemed relevant. The BGOs are envisioned to be set in arbitrary locations in the reflector to measure effects at a distance. For the purpose of modeling the reactivity impact of the detector array and dose estimates, a reference setup at mid-core height and close to the fuel was used (see Figure 3).

IV. Monte Carlo simulation of the experiment

To model the future experiments using the acquired detectors, we used the to-date newest releases of two commonly used Monte Carlo transport codes: Serpent 2 (Version 2.1.30) [4] and MCNP 6.2 [5]. MCNP, with its long development history, is often used as reference for criticality and shielding calculations. Since around 2000 [6], coupled neutron photon transport is a standard feature of MCNP. Serpent features photon transport since Version 2.1.24 [7]. Both codes are frequently used at LRS for neutron transport in CROCUS - we hence chose to compare the codes for future reference.

The examined setups are listed in Table I. As the control rod tubes are already in place, no further measures need to be taken for the $CeBr_3$ usage. The BGO detectors are not watertight and require a housing. For the simulations we

https://doi.org/10.1051/epjconf/202022504016

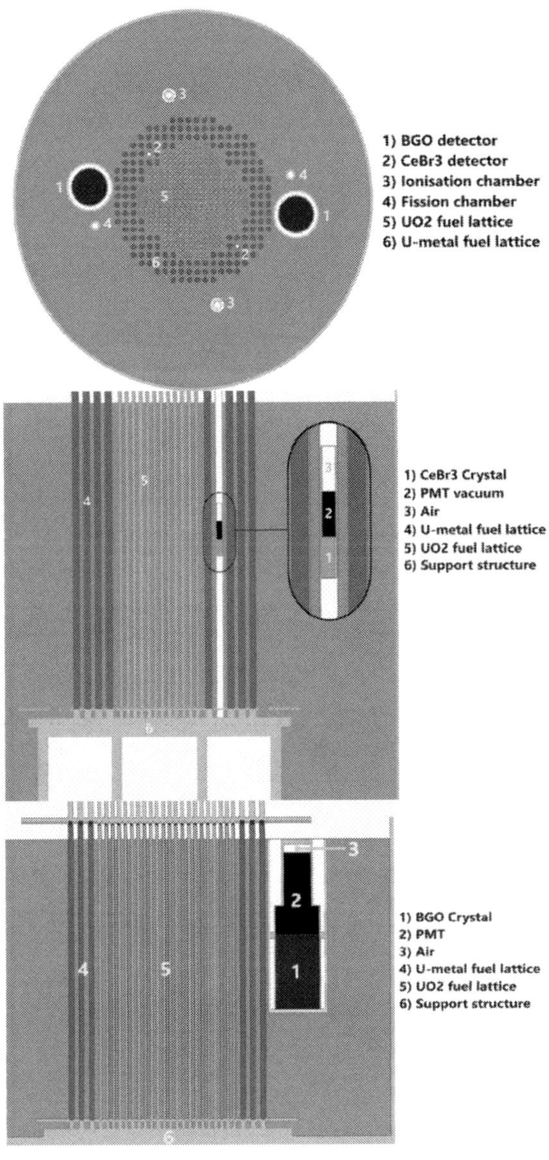

Fig. 3. **Top:** XY-view of the CROCUS Serpent 2 model at mid core height. The $CeBr_3$ are set in the control rod positions (2), while the BGOs are set into the reflector close to the fuel. **Middle:** ZX-view of the CROCUS core control rod tube with inserted $CeBr_3$. **Bottom:** ZX-view of the CROCUS core, showing a BGO detector.

will assume a Plexiglas cylinder to fill said role. The actual realization will depend on mechanical tests, as the crystal weighs more than 20 kg and thus might require custom made aluminium or steel solutions. In addition to the full array of four detectors, we examined single detector setups as well as setups with the detector crystals shielded by 1 mm cadmium. We present the comparison of reactivity worth, integral neutron/photon fluxes, and spectra. Pulse height or dose deposition tallies are not available in Serpent, and MCNP results will be presented without comparison. For all calculations we used a water level of 955 mm and the nuclear data library ENDF/B-7.1.

TABLE I
DETECTOR SETUP ABBREVIATIONS.

Detector set-up	**Name**
Reference (no detectors)	Ref
No Cd, 1 BGO	N1B
No Cd, 1 CeBr3	N1C
No Cd, 1 BGO, 1 CeBr3	N1B1C
No Cd, 2 BGO, 2 CeBr3	N2B2C
With Cd, 1 BGO	Y1B
With Cd, 1 CeBr3	Y1C
With Cd, 1 BGO, 1 CeBr3	Y1B1C
With Cd, 2 BGO, 2 CeBr3	Y2B2C

TABLE II
SERPENT VS. MCNP: RELATIVE REACTIVITY IMPACT FOR THE DIFFERENT SETUPS WITH RELATIVE DIFFERENCE BETWEEN THE TWO CODES.

	$\Delta\rho_{Serpent}$	$\Delta\rho_{MCNP}$	$\Delta_{\Delta\rho}$
Ref	0	0	
N1B	-27 $\pm$ 3	-22 $\pm$ 4	19 %
N1C	-3 $\pm$ 3	-11 $\pm$ 4	-267 %
N1B1C	-35 $\pm$ 3	-23 $\pm$ 4	34 %
N2B2C	-47 $\pm$ 3	-39 $\pm$ 4	17 %
Y1B	-67 $\pm$ 3	-45 $\pm$ 4	33 %
Y1C	-73 $\pm$ 3	-60 $\pm$ 4	18 %
Y1B1C	-143 $\pm$ 3	-108 $\pm$ 4	24 %
Y2B2C	-245 $\pm$ 3	-178 $\pm$ 4	27 %

A. Reactivity of different setups

One of the initial quantities of interest for the gamma experiment is the relative reactivity impact of their introduction into the core. CROCUS is constrained within regulatory boundaries to an excess of 200 pcm. Any proposed experiment requires thus, on top of calculations showing the reactivity impact, further considerations regarding mechanical stability and water incursion scenarios. As shown in Table II, a significant re-structuring and associated re-approval of CROCUS will not be necessary. Only in the case of fully Cd-shielded detectors does Serpent predict a reactivity < -200 pcm. As discussed in Section IV.3, the dose to the detectors was found to be small enough to not require shielding. As such, the proposed setups are within regulatory limits and ready to be installed once the detectors are available.

The comparison between MCNP and Serpent, however, is not as promising. The respective reactivity worths were found to deviate from each, with Serpent over-predicting MCNP by

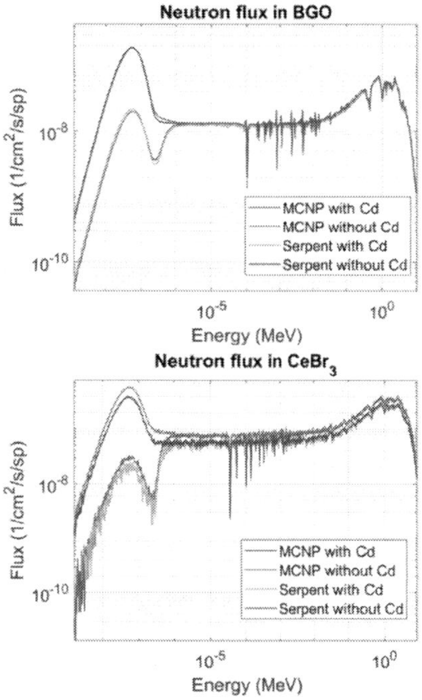

Fig. 4. Neutron fluxes predicted by MCNP and Serpent. **Top:** Flux tallied in the BGO detector with and without Cd shielding. **Bottom:** Flux tallied in the $CeBr_3$ detector with and without Cd shielding.

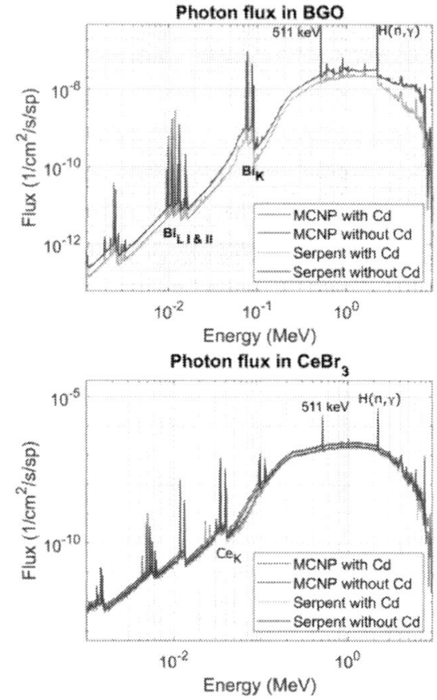

Fig. 5. Photon fluxes predicted by MCNP and Serpent. **Top:** Flux tallied in the BGO detector with and without Cd shielding. **Bottom:** Flux tallied in the $CeBr_3$ detector with and without Cd shielding.

around 30%, and by -267% in an extreme case. A deviation of the absolute values of k_{eff} was expected and indeed found to be consistently 150 pcm. This is nonetheless an atypical high difference. The relative worths, however, should agree within statistical uncertainties. This points to a deficiency in the models and requires a re-evaluation. This will also be reflected in the thermal neutron spectrum predicted, as shown in the next Section.

B. Neutron and photon flux

Part of the analysis was the calculation and comparison of neutron and photon flux spectra. Due to the symmetry of the setup we focused on one $CeBr_3$ and BGO. The results for neutron spectra with and without Cd shield, using Serpent and MCNP, are displayed in Figure 4. Overall, we find a good agreement between the codes, with differences arising in the thermal region for the $CeBr_3$. These differences could not be explained by statistical means and once more point to minor differences in the respective models.

The resulting photon spectra of both detectors are displayed and compared in Figure 5. Here we find similarly agreeing spectra between Serpent and MCNP. Note that any Cd(n,γ) peaks are not visible as the production cross-sections are missing in ENDF/B-7.1.

TABLE III
INTEGRAL NEUTRON AND PHOTON FLUXES IN FLUX PER SOURCE PARTICLE, Φ_n AND Φ_γ, WITH RELATIVE STATISTICAL UNCERTAINTIES AND RELATIVE DIFFERENCE BETWEEN THE CODES.

	Serpent	MCNP	Δ
Φ_n(BGO)	$3.103 \cdot 10^{-5} \pm 1\%$	$3.108 \cdot 10^{-5} \pm 0.05\%$	0.1%
Φ_n($CeBr_3$)	$1.398 \cdot 10^{-4} \pm 1\%$	$1.449 \cdot 10^{-4} \pm 0.05\%$	3.6%
Φ_γ(BGO)	$1.029 \cdot 10^{-5} \pm 1\%$	$1.005 \cdot 10^{-5} \pm 0.05\%$	2.3%
Φ_γ($CeBr_3$)	$7.652 \cdot 10^{-5} \pm 1\%$	$7.644 \cdot 10^{-5} \pm 0.05\%$	0.1%

The integral neutron and photon fluxes in comparison are shown in Table III. Overall we can state that the photon flux to neutron flux ratio is higher in-core than ex-core, and that Serpent and MCNP only differ by up to 3%. This largest difference is again for the $CeBr_3$, probably indicating a modeling error.

C. Pulse height spectrum

The predicted spectra in the detectors are naturally not an observable, and serve mostly as baseline for what can be expected to be seen in an optimal case. The true response of a detector will depend on the energy deposition of the individual photons that enter the detector. Pulse height tallies allow to capture this behavior. In MCNP this includes the explicit

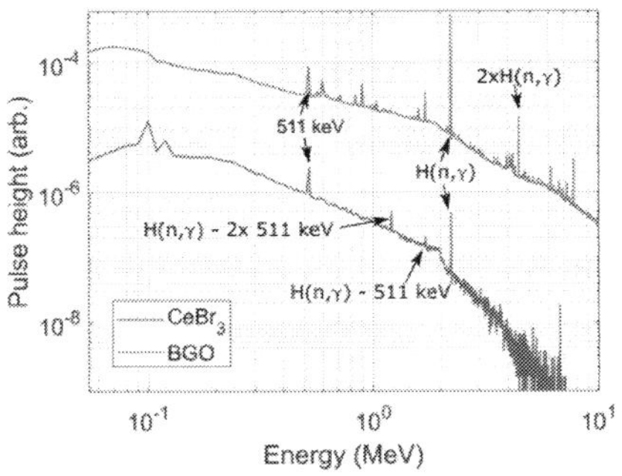

Fig. 6. Pulse height spectra estimated using the F8 tally of MCNP. Prominent peaks, such as annihilation, absorption in water and (double) escape, are highlighted.

modeling of secondary particles (electrons and photons) and their energy deposit. For both detector types we produced said tallies and display them in Figure 6. Note that the pulse height spectra are an approximation of the energy deposited in the crystal volume and do not take individual scintillation properties into account. We observe that from the initial photon spectrum we expect to observe mostly a lumped spectrum. The peaks due to the H(n,γ) and annihilation reactions are nonetheless visible. Single and double escapes as well as sum deposits of H(n,γ) are also visible. This spectrum provides a first prediction that will be compared to experiments.

D. Deposited dose

The crystals are damaged by the radiative environment and over time lose light yield. For this purpose, we investigated the dose in dependence of energy for neutrons and photons respectively as well as integral doses, with and without cadmium shielding.

The dose rate was estimated using the dose deposition tally in MCNP. More specifically, we used an ICRP-60 dose function that uses stopping power dependent tables to convert the deposited energy. We display the neutron dose over energy results in Figure 7, whilst Figure 8 displays the results for photons. We can state that the photon dose approximately follows the spectral shape, as is expected due to quality factors being defined at unity for photons. A general decrease in flux due to attenuation, as expected, can be seen. The neutron doses reveal that ex-core BGO measurements do not benefit from a Cd shield, as most of the dose is received from fast neutrons. In-core measurements using the $CeBr_3$ would see a larger positive impact of a shielding. In Table IV we summarize the individual integral dose components.

The integral results were weighted with the core average source per watt, as determined in [8], to receive values per Watt, as presented in Table IV.

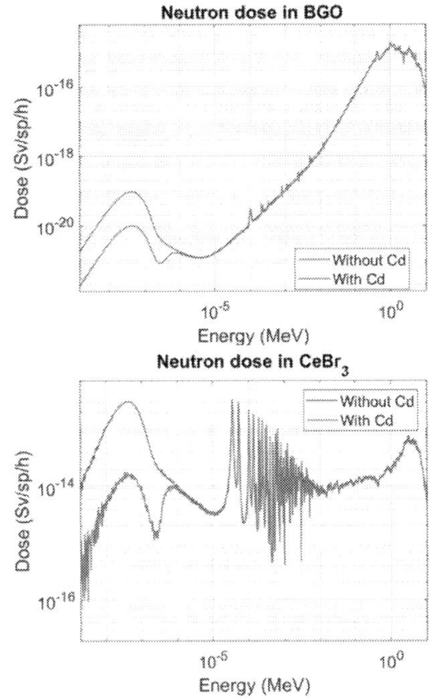

Fig. 7. Neutron dose over neutron energy in the detectors of the gamma detection array, with and without cadmium shield.

TABLE IV
INTEGRAL NEUTRON, PHOTON, AND TOTAL DOSES IN SIEVERT PER HOUR PER WATT, D_n, D_γ, AND D_{tot}, WITH AND WITHOUT CADMIUM SHIELDING. RESULTS ARE DISPLAYED WITH RELATIVE STATISTICAL UNCERTAINTIES.

	MCNP (Sv/h/W)
D_n(BGO)	$0.011 \pm 0.1\%$
D_n(BGO+Cd)	$0.010 \pm 0.1\%$
D_γ(BGO)	$0.024 \pm 0.1\%$
D_γ(BGO+Cd)	$0.014 \pm 0.1\%$
D_n($CeBr_3$)	$3.310 \pm 0.1\%$
D_n($CeBr_3$+Cd)	$1.136 \pm 0.1\%$
D_γ($CeBr_3$)	$0.114 \pm 0.1\%$
D_γ($CeBr_3$+Cd)	$0.102 \pm 0.1\%$
D_{tot}(BGO)	0.035
D_{tot}(BGO+Cd)	0.024
D_{tot}($CeBr_3$)	3.424
D_{tot}($CeBr_3$+Cd)	1.238

The Cd shield resulted in a total dose reduction by 30% in the BGO, and by 64% in the $CeBr_3$. Literature revealed that BGO shows a 15% yield decrease at 102 kGy [9], whilst $CeBr_3$ suffer a 7% decrease in yield at 111 kGy [10]. The unshielded BGO could thus withstand 2.91 MWh until it reaches 15% yield decrease, while the $CeBr_3$ 32 kWh until 7%. This allows us to conclude that given the maximum

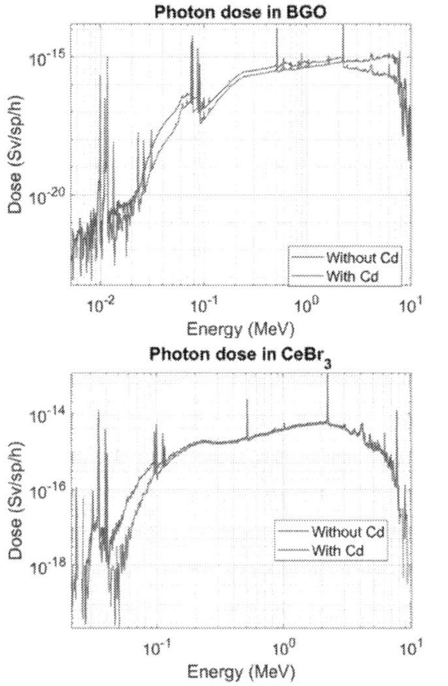

Fig. 8. Photon dose over photon energy in the detectors of the gamma detection array, with and without cadmium shield.

power of CROCUS, and that the detectors are to be used in <1 W or sub-critical configurations, that a Cd shield will not be necessary. The reactivity concerns mentioned before are hence also alleviated. The Cd shield could nonetheless be used to force Cd(n,γ) reactions and possibly monitor neutrons simultaneously to gammas.

V. Conclusion

In this study, we presented the simulation results of a new gamma detection experiment for the CROCUS reactor. A recently acquired array of four scintillation based gamma detectors are to extend the capabilities of the CROCUS facility: Two large BGO, to be set in the spacious water reflector, and two smaller $CeBr_3$ detectors, to be set in the control rod guide tubes. Various setups were modeled in common Monte Carlo transport codes. We compared the codes MCNP 6.2 and Serpent 2.1.30 with respect to their coupled neutron photon transport capabilities, using the ENDF/B-7.1 library.

Relative reactivities were found to deviate, with Serpent over-predicting MCNP by ~30%. The codes showed overall good agreement with regards to photon and neutron flux predictions, with exceptions found in the thermal neutron spectrum in-core. Both deviations point to minor differences in the models around the control rod tube. The reflector region fluxes showed < 3% relative difference between the codes, affirming that the codes produce consistent results.

Using MCNP only, we also predicted the dose received and pulse height spectrum of the detectors. The dose estimations allowed us to conclude that a Cd shielding will not be necessary, as the radiation hardness of both crystal types should suffice even in extreme conditions of CROCUS operation. The Cd shield was in fact shown to only contribute to a 30% dose reduction in the reflector, as most energy is deposited by fast neutrons. The pulse height spectra will be used as expectation for eventually observable spectra.

This study will be continued by examining the possible errors in either model. Next we will supply calibration and characterization data of the detectors. Finally, first in core measurements will be compared to the presented results.

References

[1] Lamirand, V. P., Hursin, M., Perret, G., Frajtag, P., Pakari, O., & Pautz, A. (2016). Future experimental programmes in the CROCUS reactor. In Conference proceedings of RRFM/IGORR 2016 (No. CONF, pp. 284-292).
[2] Pakari, O., Lamirand, V., Perret, G., Frajtag, P., & Pautz, A. (2018). Kinetic Parameter Measurements in the CROCUS Reactor Using Current Mode Instrumentation. IEEE Transactions on Nuclear Science, 65(9), 2456-2460.
[3] Scionix Holland. Mechanical, optical and scintillation properties. https://scionix. nl/scintillation-crystals/#tab-id-4. Accessed: 11/12/18.
[4] Leppänen, J., et al. (2015) "The Serpent Monte Carlo code: Status, development and applications in 2013." Ann. Nucl. Energy, 82 (2015) 142-150.
[5] C.J. Werner, et al., "MCNP6.2 Release Notes", Los Alamos National Laboratory, report LA-UR-18-20808 (2018).
[6] White, M. C. (2000). Development and implementation of photonuclear cross-section data for mutually coupled neutron-photon transport calculations in the Monte Carlo N-particle (MCNP) Radiation Transport Code.
[7] Kaltiaisenaho, T. (2016) "Implementing a photon physics model in Serpent 2." M.Sc. Thesis, Aalto University, 2016.
[8] Pakari, O., Lamirand, V., Perret, G., Braun, L., Frajtag, P., & Pautz, A. (2018). Current mode neutron noise measurements in the zero power reactor CROCUS. In EPJ Web of Conferences (Vol. 170, p. 04017). EDP Sciences.
[9] Sahu, S. K., Peng, K. C., Huang, H. C., Wang, C. H., Chang, Y. H., Hou, W. S., ... & Wei, Y. Y. (1997). Radiation hardness of undoped BGO crystals. Nuclear Instruments and Methods in Physics Research Section A: Accelerators, Spectrometers, Detectors and Associated Equipment, 388(1-2), 144-148.
[10] Drozdowski, W., Dorenbos, P., Bos, A. J. J., Bizarri, G., Owens, A., & Quarati, F. G. A. (2008). CeBr 3 Scintillator Development for Possible Use in Space Missions. IEEE Transactions on Nuclear Science, 55(3), 1391-1396.

EPJ Web of Conferences **225**, 04017 (2020) https://doi.org/10.1051/epjconf/202022504017
ANIMMA 2019

Comparison of neutron/gamma separation qualities of various organic scintillation materials

Zdeněk Matěj
Faculty of Informatics
Masaryk University
Brno, Czech Republic

Filip Mravec
VF, a.s.
Černá Hora, Czech Republic

Aleš Jančář
VF, a.s.
Černá Hora, Czech Republic

Michal Košťál
Research Center Řež Ltd
Husinec-Řež, Czech Republic

František Kučera
Central European Institute of Technology
University of Technology
Brno, Czech Republic

František Cvachovec
Faculty of Informatics
Masaryk University
Brno, Czech Republic

Václav Přenosil
Faculty of Informatics
Masaryk University
Brno, Czech Republic

Zdeněk Kopecký
VF, a.s.
Černá Hora, Czech Republic

Jiří Čulen
VF, a.s.
Černá Hora, Czech Republic

Abstract— **In this work we compare the pulse-shape discrimination (PSD) properties of EJ-299-33A, BC-501A, stilbene, p-terphenyl and Hidex Aqualight in neutron field generated by the LVR-15 reactor with silicon filter utilization. Pulses from the scintillators are processed by Neutron-Gamma spectrometer. This spectrometric system with fast digitizer card contains two analog-digital (A/D) converters with a resolution of 12 bits and sampling frequency 500 MHz. For photomultiplier (PMT) linearity improvement active divider has been used. Measured data from scintillators have been processed using the integration method and compared. Results are presented.**

Keywords— ***organic scintillators, neutron, gamma, impulse separation, FPGA, FoM, LVR-15***

I. Introduction

The scintillators that enable the separation of neutron-gamma radiation are widely used for measurements of mixed radiation fields. Several competing types of PSD capable scintillators were compared (EJ-299-33A, BC-501A, stilbene, p-terphenyl and Hidex Aqualight).The reactor LVR-15 with well-known radiation field has been used for measurements in the energy range 1 to 10 MeV [1, 2]. Measured data have been processed using digital Neutron-Gamma spectrometer which we developed. The digital spectrometric system minimizes the drawbacks of analog devices and provides substantial simplification for the operation and setting of the system. The spectrometer is connected with a computer via Ethernet.

II. Experimental Setup

Laboratory with experimental reactor LVR-15 has been used for the measurement, see Fig. 1.

We utilized well defined moderate neutron beam from the horizontal channel HK-1, see Fig. 2. Reactor thermal power output was 10 MW. Neutron beam has been moderated via 1 m wide silicon single crystal which provides spectrum with characteristic energy peaks, see unfolded spectrum in Fig. 3 [4].

The experimental arrangement is shown in Fig. 2. Influence of thermal neutrons and gammas was reduced via filter composed of ^{6}Li, Cd, Bi and Pb.

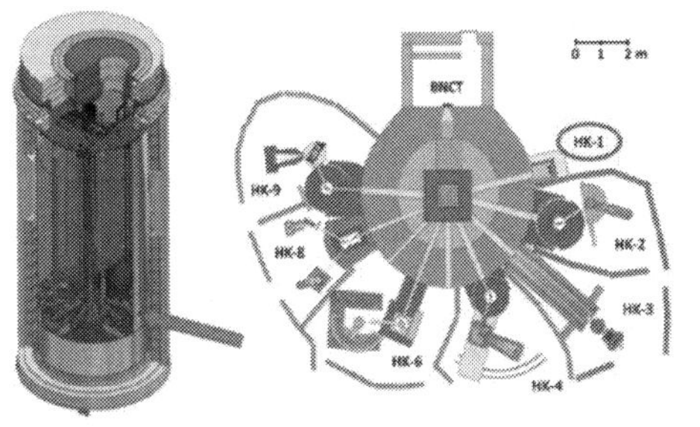

Fig. 1. - LVR-15 reactor (left) and its horizontal channels (right) [1]

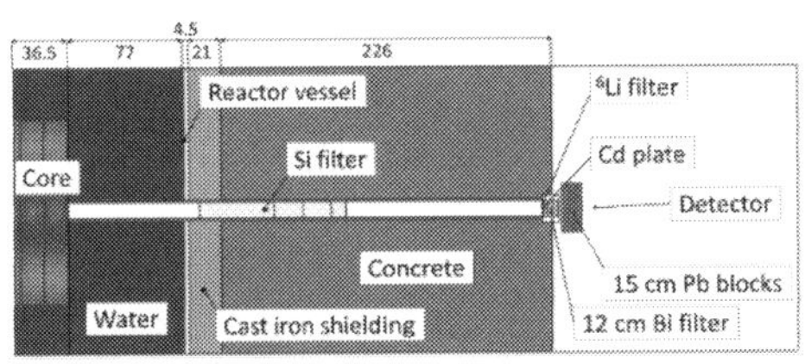

Fig. 2. The experimental arrangement in the reactor laboratory. The detectors were placed along the beam axis at 1.5 m distance from the silicon filter target [1].

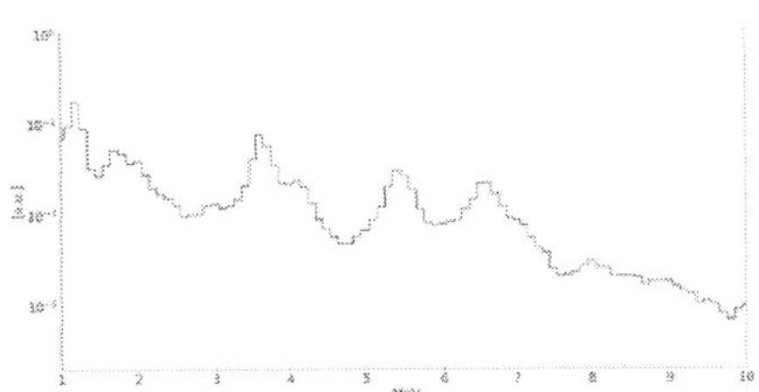

Fig. 3. Unfolded silicon moderated neutron spectrum measured with stilbene.

III. Results

Integration method based on the pulse charge comparison was implemented in FPGA. Separation parameters for various scintillators calculated with this method are shown below in PSD matrices.

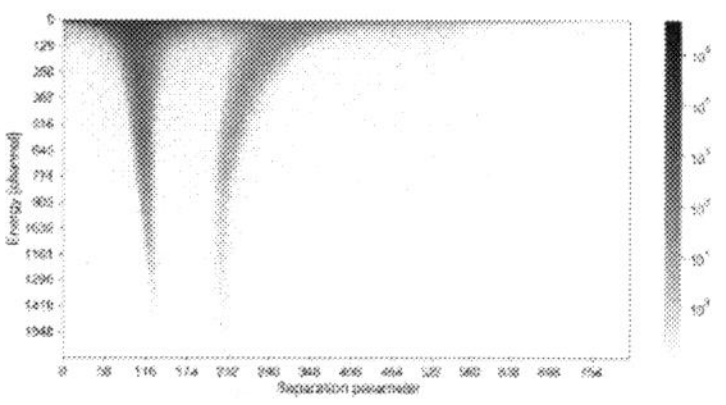

Fig. 4. Stilbene

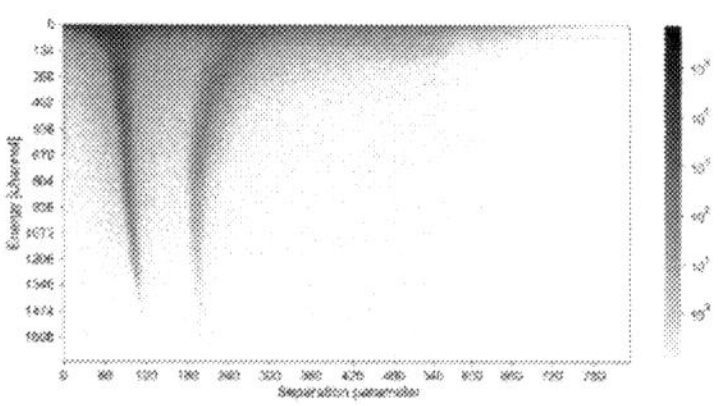

Fig. 5. BC-501A

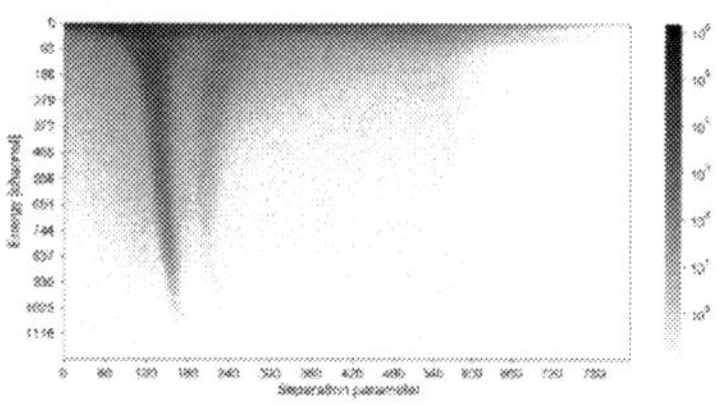

Fig. 6. EJ-299-33A

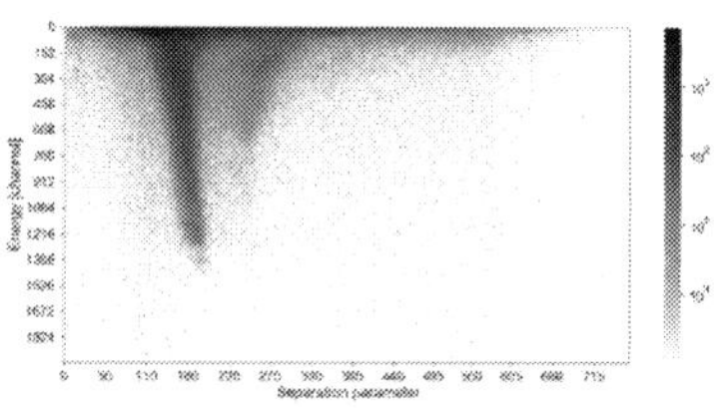

Fig. 7. P-terphenyl

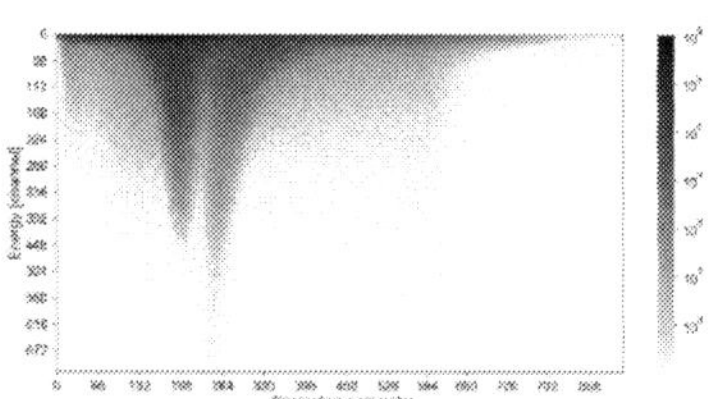

Fig. 8. Hidex Aqualight

Neutron-Gamma spectrometric system has been used for the measurements. The input analog signal from the detector was digitized by two fast A/D converters working with a sampling frequency of 500 MHz. Digital signal processing is implemented in FPGA Xilinx Virtex-6.

IV. FIGURE OF MERIT

The quality of neutron–gamma discrimination for a given scintillator is characterized by figure of merit (FOM), see Fig. 9.

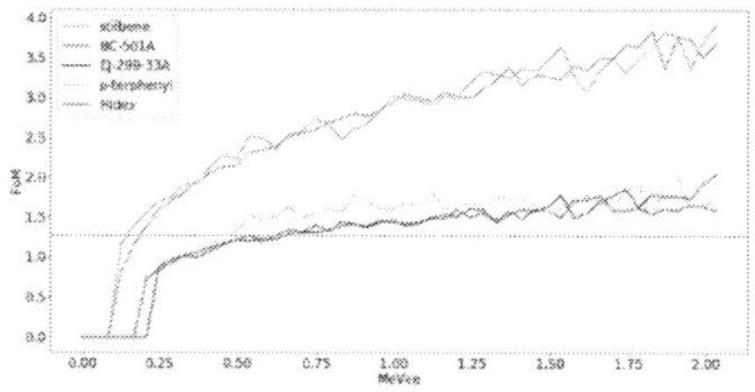

Fig. 9 FoMs for examined scintillators

Selected results of neutron-gamma separation quality are shown in Tab. 1. The critical value of FoM 1.27 is an indicator of good neutron-gamma separation [3]. Yellow fields indicate gamma energies where FoM values exceed the critical value. According to the combination of FoM and energy parameters we can evaluate the quality of neutron-gamma separation. Stilbene detector achieves the best quality of separation because the FoM exceeds the critical value at the smallest energy.

MeVee	0,12	0,17	0,2	0,29	0,5	0,62	0,66
stilbene	1,16	1,4	1,55	1,75	2,23	2,36	2,54
BC-501A	0,82	1,15	1,39	1,72	2,15	2,39	2,51
p-terphenyl	0	0	0	0,92	1,33	1,51	1,63
EJ-299-33	0	0	0,73	0,94	1,21	1,27	1,35
Hidex	0	0	0	0,97	1,21	1,22	1,3

Tab 1. Selected critical values of FoM from examined scintillators.

Among the examined scintillators the stilbene detector is considered to have the best energy resolution. Apparatus spectra from the two best scintillators from Tab. 1 were compared with stilbene detector in Fig 10. The x-axis contains channel number, y-axis contains ratio E_s/E_x, where E_s is apparatus spectrum of stilbene and E_x is a spectrum for BC-501A and p-terphenyl.

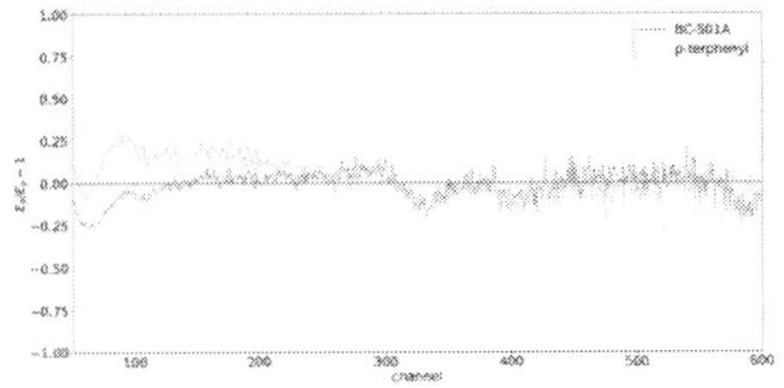

Fig. 10 E_s/E_x - 1 comparison of BC-501A and p-terphenyl with stilbene

ACKNOWLEDGMENT

The presented work has been supported by Ministry of Industry and Trade, within the project TRIO, No. FV20453. Presented results were obtained with the use of the infrastructure Reactors LVR-15 and LR-0, which is financially supported by the Ministry of Education, Youth and Sports - project LM2015074 and also financially supported by the Ministry of Education, Youth and Sports Czech Republic - project LQ1603 Research for SUSEN. This work has been realized within the SUSEN Project (established in the framework of the European Regional Development Fund (ERDF) in project CZ.1.05/2.1.00/03.0108 and of the European Structural Funds and Investment Funds (ESIF) in the project CZ.02.1.01/0.0/0.0/15_008/0000293), which is financially supported by the Ministry of Education, Youth and Sports - project LM2015093 Infrastructure SUSEN.

REFERENCES

[1] KOŠŤÁL, M., M. SCHULC, J. ŠOLTÉS, E. LOSA, L. VIERERBL, Zdeněk MATĚJ, F. CVACHOVEC a V. RYPAR. Measurements of neutron transport of well defined silicon filtered beam in lead. Applied Radiation and Isotopes, Elsevier, 2018, roč. 2018, č. 142, s. 160-166. ISSN 0969-8043.

[2] KOŠŤÁL, Michal, Jaroslav ŠOLTÉS, Ladislav VIERERBL, Zdeněk MATĚJ, František CVACHOVEC, Vojtěch RYPAR a Evžen LOSA. Measurement of Neutron Spectra in a silicon filtered neutron beam using stilbene detectors at the LVR-15 research reactor. Applied Radiation and Isotopes, 2017, roč. 128, s. 41-48.

[3] N. Zaitseva, B. L. Rupert, I. PaweŁczak, A. Glenn, H. P. Martinez, L. Carman, M. Faust, N. Cherepy, S. Payne, Plastic scintillators with efficient neutron/gamma pulse shape discrimination, Nuclear Instruments and Methods in Physics Research Section A: Accelerators, Spectrometers, Detectors and Associated Equipment, Volume 668, 2012, Pages 88-93

[4] CVACHOVEC, Jiří a František CVACHOVEC. Maximum likelihood estimation of a neutron spectrum and associated uncertainties. Advances in Military Technology, Brno: Univerzita obrany, 2008, Vol. 3, No. 2.

EPJ Web of Conferences **225**, 04018 (2020) https://doi.org/10.1051/epjconf/202022504018
ANIMMA 2019

Developing and testing a miniature fiber-coupled scintillator for in-core neutron counting in CROCUS

Fanny Vitullo*, Vincent Lamirand*, Jean-Baptiste Mosset†, Pavel Frajtag*,
Oskari Pakari*, Gregory Perret† and Andreas Pautz*, †
*École polytechnique fédérale de Lausanne, Lausanne, Switzerland
†Paul Scherrer Institut (PSI), Villigen, Switzerland
Email: fanny.vitullo@epfl.ch

Abstract—**An advanced neutron detection system for highly localized measurements in nuclear reactor cores was developed and tested in the Laboratory for Reactor Physics and System Behaviour (LRS) at the École polytechnique fédérale de Lausanne (EPFL), Switzerland, in close collaboration with the Detector group of the Laboratory for Particle Physics (LTP) at the Paul Scherrer Institute (PSI), Switzerland. The miniature-size detector is based on the coupling of a ZnS:^{6}LiF scintillator/converter screen of 1 mm^2 and 0.2 mm thickness with a 10-m optical fiber, the latter being connected to a silicon photomultiplier (SiPM). In this development version, the output signal is processed via analog read-out electronics. The present work documents the characterization of a detection system prototype in the mixed-radiation fields of the CARROUSEL facility and its testing in the CROCUS zero-power reactor operated at LRS. The fiber-coupled scintillator shows a linear response with the reactor power increase up to 6.5 W (i.e. around 10^8 cm^{-2}·s^{-1} total neutron flux), with a subsequent loss of linearity due to electronic dead time of the analog system. Nevertheless, the detector shows excellent neutron counting capabilities whether compared to other localized detection systems available at LRS, e.g. miniature fission chambers and an sCVD diamond detector.**

Keywords—**scintillator, in-core neutron detection, optical fibers, CROCUS reactor.**

I. Introduction

Neutron measurements with high spatial resolution are paramount for the investigation of complex in-core phenomena and for the validation of modern high-fidelity neutron transport codes. Highly-resolved experiments have been performed in the past in the zero-power research reactor PROTEUS, operated at Paul Scherrer Institut (PSI) from 1968 to 2011, by means of neutron activation techniques [1], [2], [3] or standard reactor instrumentation like miniature fission chambers [4]. The idea to perform extensive investigations of three-dimensional static and dynamic in-core phenomena in research reactors [5], [6] triggered the development of advanced neutron detection techniques for highly localized measurements. These applications require instruments with high sensitivity and flexibility, as well as the possibility to be parallelized and to perform online measurements. Scintillation materials have recently come to the fore in the field of neutron detection [7] due to the large availability of screens of reduced size with good efficiency and low price, together with the possibility to couple them with cutting-edge light collection technologies and fast electronics. A first detection system prototype for in-core applications using a scintillator/converter ZnS:^{6}LiF screen was developed at Nagoya University in 1994 by Mori et al. [8] and tested in the Studsvik-Nuclear 1 W reactor [9]. However, the scintillator was connected with an optical fiber to a photomultiplier tube (PMT) that limited the possibility of parallelization of the designed detector. Over the past decade photodetector technologies have seen a shift from the use of classical PMTs to solid-state devices such as silicon photomultipliers (SiPMs), which are robust, compact and electronically stable instruments [10]. The coupling of a ZnS:^{6}LiF scintillator with a SiPM has been already tested by the Detector group of the Laboratory for Particle Physics (LTP) at PSI, Switzerland, in the framework of the upgrade of a neutron diffractometer [11], [12]. Although this detection system was not developed for in-core reactor applications, an adapted version of the latter would represent an efficient solution to perform highly-resolved experiments in research reactors and to engineer advanced non-invasive reactor diagnostic techniques [13].

In the current work a cutting-edge neutron detection system has been designed for the investigation of highly localized in-core phenomena in the CROCUS zero-power research reactor [14], operated by the Laboratory for Reactor Physics and System Behaviour (LRS) at the École polytechnique fédérale de Lausanne (EPFL), Switzerland. The advanced system, built in collaboration with the LTP group at PSI, features a miniature ZnS:^{6}LiF scintillator coupled to a SiPM via a 10-m optical fiber. An analog read-out electronics is employed for signal processing. The present work documents the characterization and testing of the new neutron detection system. The first part of the paper is dedicated to a detailed description of the detection system and its components. The second part of the paper presents the characterization of a first prototype with a Pu-Be neutron source. In the last section, the experimental setup employed during the testing campaign in the CROCUS reactor is described. Finally, the performances of the fiber-coupled scintillator during in-core measurements are analyzed and compared with the ones of other localized neutron detection systems available at LRS, such as miniature fission chambers and an sCVD diamond detector.

II. Experimental setup

A. Scintillator screen

The commercially available ND2:1 scintillator screen, manufactured by Scintacor [15] and used in the present work, is

TABLE I: Specifications of the ND2:1 scintillator screen by Scintacor [15], [16].

Formula	ZnS(Ag):^{6}LiF
Components state	Particulate blend
Mass ratio ZnS:^{6}LiF	2:1
Sensitive area	1 mm^2
Thickness	0.2 mm
Density	2.2 g/cm^3
^{6}Li atom density	1.4 × 10^{22} atoms/cm^3
Peak wavelength	450 nm
Light yield	160000 photons/n
Decay to 10%	80 μs

composed of a mixture of lithium-6 fluoride (^{6}LiF) and zinc-sulphide (ZnS) polycrystalline powders bonded with Plexiglass. The interaction between an incoming neutron and a ^{6}Li atom is described by Eq. (1). The two products of the reaction, an α particle and a tritium nucleus interact with the ZnS inorganic scintillator grains mixed into the same matrix. The scintillation light is generated as a result of the de-excitation of the luminescence centers in the ZnS molecules [18].

$$ {}^{6}_{3}Li + {}^{1}_{0}n \rightarrow {}^{4}_{2}He + {}^{3}_{1}H \qquad Q = 4.78MeV \tag{1}$$

An overview of the characteristics of the ND2:1 screen is reported in Tab. I. The mass ratio between the two components ZnS:^{6}LiF of 2:1 and the high cross section in the thermal energy region for the reaction in Eq. (1) allows reaching a good thermal neutron detection efficiency. Although ^{6}Li has a smaller thermal cross section than ^{10}B [17], the first isotope has been preferred as a converter material for the large energy released in the reaction (4.78 MeV) that translates into a higher light production by the ZnS scintillator [18]. The emitted light has a typical emission spectrum lying in the blue zone of the visible light. The scintillator sample used in the current work has a total sensitive area of 1 mm^2 and a thickness of 0.2 mm. Because of its small thickness, it shows a low intrinsic sensitivity to gamma contributions [18], which is an essential feature for its use in mixed radiations fields such as a nuclear reactor core. At the same time, the slow decay time of the emitted light by fluorescence, in the order of tens of μs, requires signal processing techniques.

B. Optical fiber

The scintillator has been optically coupled to a polished SH8001 optical fiber, commercialized by ESKA [19], by means of the BC-630 grease of St-Gobain [20] to reduce the loss of scintillation photons. The schematic of the front end of the probe is shown in Fig. 1, where an aluminum cap protects the sensitive area of the scintillator and a black tape (visible in Fig. 2) completes the coupling with the fiber, minimizing light parasites and dispersion. In this prototype version, the optical fiber is composed of a poly(methyl methacrylate) (PMMA) core with an average diameter of 1.96 mm covered with a 0.04 mm thick fluorinated polymer cladding with a lower refractive index for light wave guiding by total internal

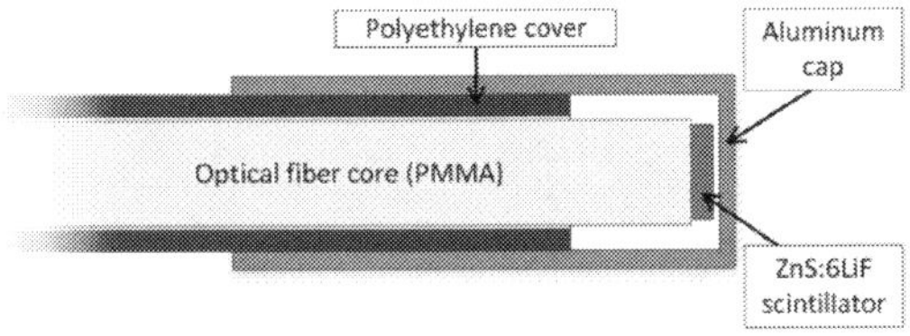

Fig. 1: Schematic of a single optical fiber front end.

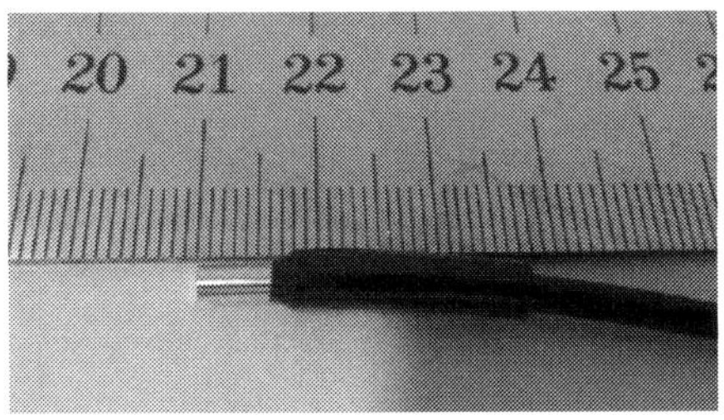

Fig. 2: Picture of a prototype of a single optical fiber coupled with ZnS:^{6}LiF scintillator, protected by an aluminum cap and wrapped with black tape.

refraction. The core refractive index is 1.49. A 1-mm thick jacket covers the whole length of the 10-m fiber to protect it from external damages and harsh environments.

C. Silicon photomultiplier (SiPM)

The rear end of the fiber was connected to a silicon photomultiplier, commonly called SiPM. A SiPM is single-photon sensitive solid-state devices made of an array of avalanches photodiodes (APDs) working in Geiger mode (called pixels and microcells) and connected in parallel on a monolithic silicon crystal [21]. In this work, the MPPC (Multi-Pixel Photon Counter) SiPM produced by Hamamatsu [21], whose specifications are described in Tab. II, was used. The SiPM was kept in the dark by being wrapped in several layers of a black cloth to avoid that external light contributions would influence the SiPM performances. A black plastic guide was designed to ensure a perfect connection between the optical fiber and the SiPM active area, minimizing parasitic light.

D. Analog read-out electronics

The associated analog read-out electronics is described in Fig. 3 with a detailed description of the output signals. The light photons emitted by the scintillator propagate through the optical fiber and reach the SiPM surface. The SiPM output, typically few tens of ns long, is integrated in a fast pre-amplifier designed by the LTP group at PSI. As shown in Fig. 3, the fast pre-amplifier generates a short negative output signal in the range of ns with an amplitude in the order of hundreds of mV. A connected fast discriminator produces a square signal with a fixed amplitude when a signal corresponding to the detection of one or more photons reaches its input. This first discriminator stage is thought to cut out

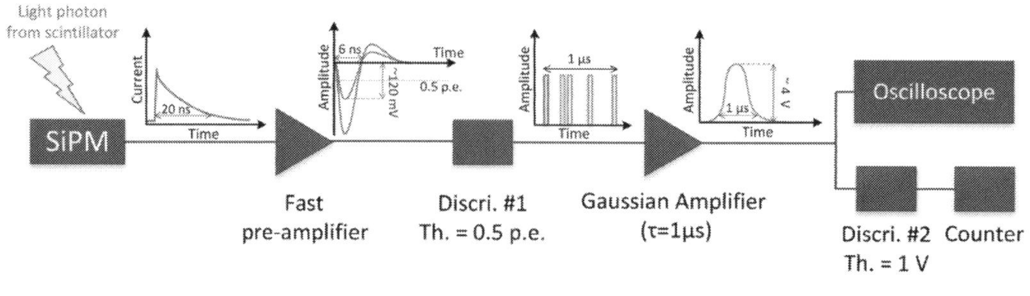

Fig. 3: Signal processing system with detailed description of the output signal after each stage.

TABLE II: Specifications of the MPPC S13360-3050PE silicon photomultiplier by Hamamatsu [21]. (*) Photon detection efficiency does not include crosstalk or after pulses.

Pixel pitch	50 μm
Effective photosensitive area	3 × 3 mm
Number of pixels	3,600
Fill factor	74%
Window material	Epoxy resin
Operating temperature	-20 to 60 °C
Spectral response range	320 to 900 nm
Peak sensitivity wavelength	450 nm
Photon detection efficiency (*)	40%
Gain	1.7×10^6
Breakdown voltage V_{BR}	53 ± 5
Recommended operating voltage	V_{BR} + 3 V
Crosstalk probability	3%
0.5 p.e. dark count	500 kcps (max. 1500 kcps)

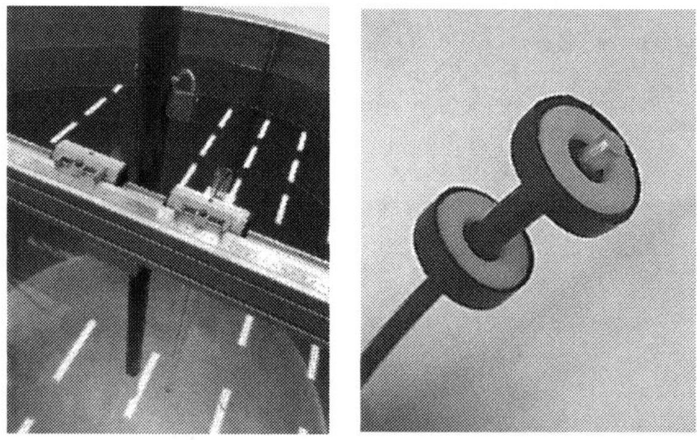

(a) (b)

Fig. 4: (a) CARROUSEL water tank with the Pu-Be source placed at the bottom of the black tube visible in the picture. The prototype detector is placed in the transparent plexiglass tube. (b) The fiber front end with two centering tools.

noise and cross-talk events in the SiPM. The result is a train of equal square pulses, each one corresponding to the detection of one or more photons at the same time. The density of pulses increases when a neutron interacts with the scintillator.

In the current work the train of pulses has been read by analog electronics. The analog system is composed by a Gaussian amplifier (CANBERRA Model 2022 [22]) generating a gaussian output signal of 1 μs width (corresponding to the amplifier shaping time), whose amplitude (in the range of few volts) depends upon the density of the train of pulses. The latter signal can be either directly read by an oscilloscope or sent to a second discriminator stage. The second discrimination, performed by a single channel analyzer (SCA) by Canberra (CANBERRA Model 2030 SCA [23]), produces fixed square signals only when the signal coming from the amplifier exceeds the defined threshold limit. The scope is to cut noise and stand-alone gamma contributions out and to count only the neutron events by means of a simple counter.

III. Prototype characterization in CARROUSEL

The CARROUSEL facility installed at LRS consists of a water tank containing a Pu-Be neutron source located in its center. The aim of the test in CARROUSEL was to prove that the system composed by the miniature-size scintillator coupled to the SiPM through the optical fiber was properly designed and that the read-out mechanism was able to detect and count neutrons.

In CARROUSEL, neutrons are generated by the ^{9}Be(α,n)^{12}C reaction happening in the Pu-Be neutron source, where the α particles are produced by the decay of plutonium. The generated neutron rate is 8.9×10^6 s^{-1}. Neutrons are produced with an average energy of 3.5 MeV and a maximum energy of 10.6 MeV. In addition, the ^{12}C originating from the (α,n) reaction is generated at an excited state and thus it emits de-excitation gamma rays at 4.44 MeV.

The water tank is equipped with a rail for the placement and movement of a dedicated measurement tube in plexiglass. The fiber has been positioned in a the detector tube, as shown in Fig. 4a, and moved at different distances from the Pu-Be source. A centering tool was used for the fiber positioning inside the tube, so that the uncertainty related to the position is ± 0.5 mm. The plastic centering tools are visible in Fig. 4b.

The output signals from the different electronic components have been studied in detail. At first, the output signal from

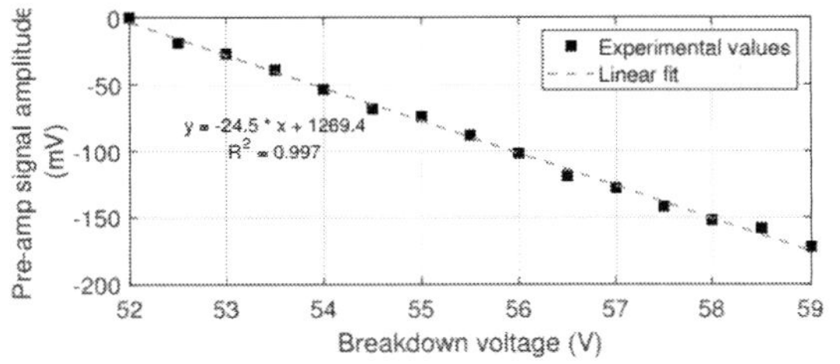

Fig. 5: Amplitude of the pre-amplifier output signal for the detection of a single photon (1 p.e.) as a function of the breakdown voltage applied to the SiPM.

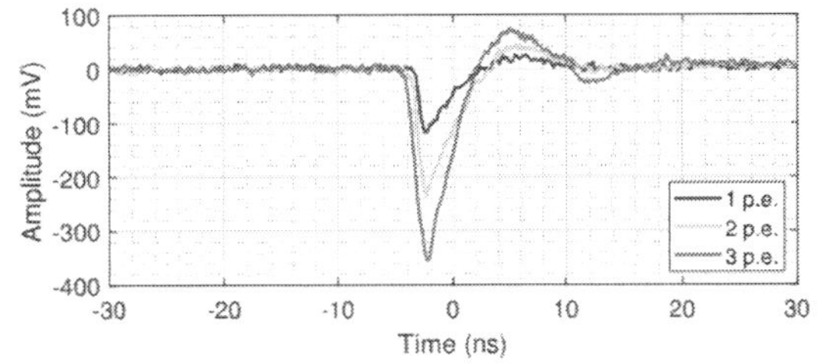

Fig. 6: Example of the output signals from the fast pre-amplifier recorded with the Teledyne LeCroy oscilloscope with a sampling rate of 10 GS/s.

the fast pre-amplifier was studied for the fiber placed at 10.00 ± 0.05 cm distance from the neutron source. Secondly, the full read-out system was installed as described in the previous section and, according to the first preliminary results and to the amplifier output signal, it was properly characterized. In particular, the amplifier signal was thoroughly investigated with both the oscilloscope and the MCB multichannel analyzer produced by ORTEC [24].

A. Electronic setup configuration and optimization

The amplitude variation as a function of the breakdown voltage applied to the SiPM was recorded in order to choose the appropriate high voltage to be supplied. According to the results presented in Fig. 5, the voltage 56.50 ± 0.05 V was chosen and employed in all the following measurements because it resulted in the first voltage value giving an amplitude of the output signal greater than 100 mV. The chosen value complies also with the range of recommended breakdown voltages declared by the SiPM manufacturer, as previously shown in Tab. II.

For the chosen over-voltage and the distance of 10.00 ± 0.05 cm, Fig. 6 shows the short (~20 ns) negative signals exiting the fast pre-amplifier. The amplitude assumes discrete values in multiples of approximately 120 mV per photon (p.e.). The discrete nature of the signal amplitude derives from the proportionality with the number of photodiodes triggered at the same time in the SiPM after a scintillation event, whilst each photodiode generates the same amount of charges whether being hit by single or multiple incoming photons. When a neutron interacts with the scintillator, a large quantity of light reaches the SiPM cells so that the pre-amplifier signal amplitude may reach values up to 2 V. Figure 6 served also to determine the threshold for the first discriminator stage following the pre-amplifier. Its value was set to half the amplitude of the single photon signal (0.5 p.e.), corresponding approximately to -60 mV. However, the electronic module employed to perform the first discriminator stage had a non-calibrated regulation of the threshold level. Therefore, the threshold was set testing the discriminator capabilities to digitize all the signals corresponding to one or more photon detection.

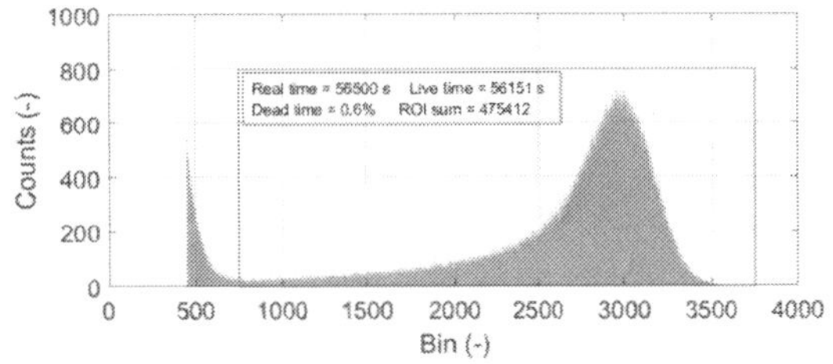

Fig. 7: Pulse height spectrum of the amplifier output signal, acquired at a distance of 10.00 ± 0.05 cm distance from the Pu-Be source. The horizontal axis is not calibrated.

Based on the knowledge acquired in the previous tests, the fast pre-amplifier output was sent to a first discriminator and an amplifier, as shown in Fig. 3. A pulse height spectrum of the amplifier output signal was recorded at the distance of 10.00 ± 0.05 cm for 56500 s (15 hours and 41 minutes) with the MCB module connected to a computer. The results are visible in Fig. 7. Aside from the noise contribution, the spectrum shows a clear peak and a tail towards lower amplitudes given by a lower density of digital pulses, probably due to light dispersion in the fiber or in the coupling between the fiber and the SiPM. Considering a region-of-interest (ROI) starting after the noise region, the sum of the counts is 475412 ± 689 in a time interval of 56500 s, the count rate being 8.414 ± 0.012 cps. It is important to remember that the amplitude analyzed with the spectrum in Fig. 7 is proportional to the density of digital pulses generated by the first discriminator in the time interval of 1 μs and thus it contributes to a small dead time. Correcting for the detector dead time, estimated to be approximately 0.6%, the real count rate of events is 8.467 ± 0.012 cps.

The calibration of the spectrum was performed through the acquisition of 3300 signals with the Le Croy Wavesurfer 10 oscilloscope [25] and their post-processing in Matlab. With this method it is possible to reconstruct the full spectrum, as shown in Fig. 8, but the amount of data to be stored and the computational power required for the data analysis makes this approach less practical compared to the use of a MCB module.

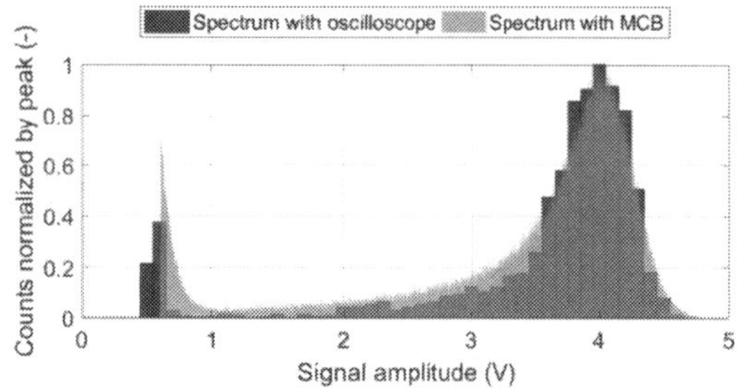

Fig. 8: Pulse height spectra obtained from oscilloscope acquisitions and from the MCB module.

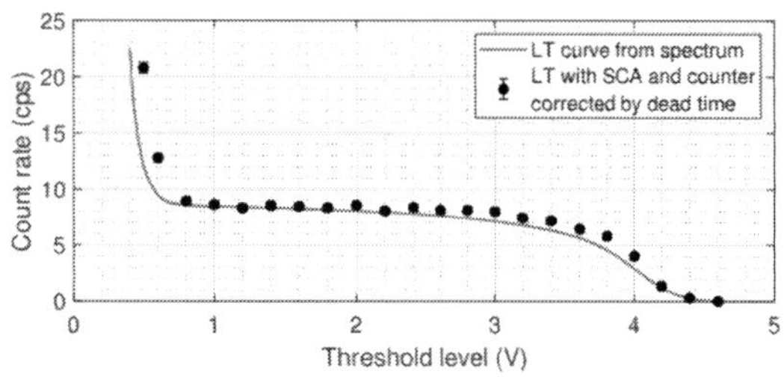

Fig. 9: Low-threshold curves reconstructed from the spectrum and acquired with a single channel analyzer (SCA) and a counter.

In addition, this method does not allow retrieving the time information and thus calculating the count rate. Nevertheless, it served for the calibration in amplitude of the spectrum that presents one peak with an average signal amplitude of 4 V.

The lower threshold curve was reconstructed by post-processing the spectrum acquired with the MCB and it was compared with the count rate recorded by the counter for every selected threshold of the SCA. Every signal coming from the amplifier with an amplitude above the selected threshold is transformed into a square signal that, if sent to the counter, increases the counts by one. The results, presented in Fig. 9, show a good agreement between the two sets of count rate values. Consequently, the threshold of the second discriminator stage was chosen to be 1 V in order to include all the events related to neutron interactions.

An example of the output signals from every component during a single neutron detection is reported in Fig. 10.

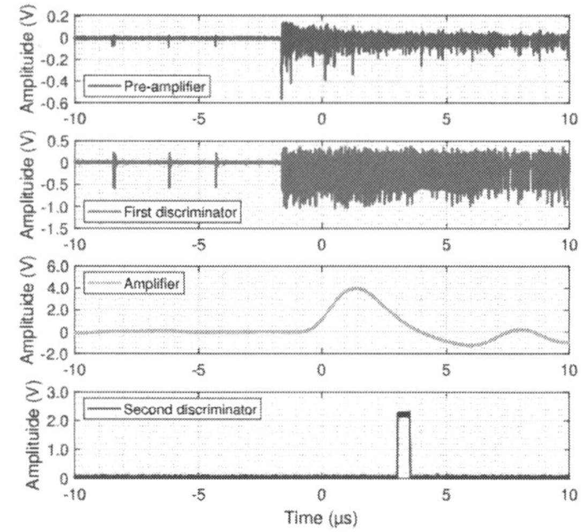

Fig. 10: Example of the output signal from each electronic component during the same neutron detection event.

IV. Experimental campaign in the CROCUS reactor

The CROCUS reactor [14] is an experimental zero-power reactor operated at EPFL and dedicated to education and research activities. The reactor is a two-zone, uranium-fueled and light-water-moderated core. The fuel is composed of an inner uranium dioxide zone with 336 rods enriched to 1.806% and an outer uranium metal zone with 172 to 176 rods enriched to 0.947%. The core height is 1 m and its diameter is around 60 cm. The allowed fission power is limited to 100 W, corresponding to a maximum achievable neutron flux of 2.5×10^9 $cm^{-2}{\cdot}s^{-1}$. Criticality can be achieved in two different ways in CROCUS: by adjusting the core water level and by moving the two absorbing control rods made of B_4C. In the following test, the first method will be used, while the control rod guide tubes will serve as experimental channels.

A. Installation of the experimental setup

The fiber front end was inserted into the guide tube of the North West control rod (see Fig. 11) of the CROCUS reactor. It was positioned so that the scintillator lies in the center of the control rod guide tube in correspondence of the active core geometrical mid-height equal to 50.0 ± 0.5 cm from the lower cadmium grid (corresponding to the beginning of the fuel rods height). At this position the neutron flux is approximately at its maximum. The 10-m fiber passes through the reactor experimental channel and reaches the electronics, from the SiPM to the oscilloscope and multi-channel analyzer, placed outside the biological shield of the reactor. The setup was installed and set with the parameters previously chosen during the characterization in the CARROUSEL facility (see Section III).

The neutron count rate was acquired several times with the counter: in the shutdown configuration (i.e. without water in the reactor pool); after the insertion of the external Pu-Be neutron source while the reactor was filled with water up to a level of 800 ± 0.1 mm; while the reactor was approaching its critical level by increasing the water level. Once the reactor reached its critical configuration, the neutron count rate was estimated at different power levels to test the linearity of the response of the detector with an increasing reactor power.

https://doi.org/10.1051/epjconf/202022504018

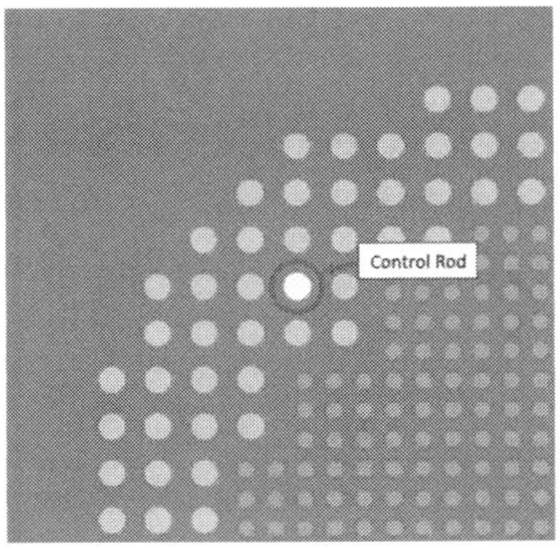

Fig. 11: Horizontal cut of the CROCUS reactor schematic (taken from the model in Serpent 2) with indication of the position for the fiber placement.

TABLE III: Neutron count rate in CROCUS after the neutron source insertion, while the water was filled up to 800 ± 0.1 mm.

Time after source insertion (min)	Count rate (cps)
0	0.38 ± 0.06
2	3.00 ± 0.17
5	5.5 ± 0.2
7	6.0 ± 0.2
9	6.0 ± 0.2

B. Results and discussion

Before conducting the measurements in the reactor at power, a background acquisition with the reactor in shutdown state has been performed. It has been measured that the background count rate in the CROCUS cavity (in the control rod position) is equal to 1.7×10^{-4} cps with a 20% statistical error. This in-core blank measurement lasted for 38.5 hours and it suggested that the detection system is not counting events related to background radiation or fuel activation.

Once the external neutron source was inserted, the neutron count rate (measured with the counter) started to increase due to the water filling-up procedure, until a stable configuration was reached at the water level of 800 ± 0.1 mm (see Tab. III). The water level was further increased in steps to approach and determine the reactors critical level. The measured count rate is reported in Tab. IV. The final critical water level was set by the operator at 957.8 ± 0.1 mm for a power of 40 mW. At this power, the count rate acquired with the counter is 168.4 ± 1.0 cps (0.6% error).

The linearity of the detector response has been tested increasing the reactor power from 40 mW up to 10.8 W, corresponding approximately to the range from 3.8×10^5 to 1.0×10^8 $cm^{-2} \cdot s^{-1}$ in the detector position. The count rate acquired with the counter is reported in Tab. V and

TABLE IV: Neutron count rate in CROCUS during the approach to critical. The water level is increased in steps until the critical level was found.

Water level (mm)	Count rate (cps)
910.0 ± 0.1	22.2 ± 0.3
920.0 ± 0.1	28.3 ± 0.2
933.0 ± 0.1	43.1 ± 0.4
941.0 ± 0.1	63.5 ± 0.6
Critical 40 mW	168.4 ± 1.0

TABLE V: Neutron count rate acquired at different power levels of the CROCUS reactor.

Power (W)	Count rate (kcps)	Error
0.125	0.441 ± 0.002	0.5%
0.2	0.677 ± 0.003	0.5%
0.5	1.717 ± 0.007	0.4%
0.8	2.778 ± 0.010	0.4%
1	3.368 ± 0.018	0.5%
1.25	4.37 ± 0.02	0.5%
1.57	5.42 ± 0.02	0.4%
2.035	6.99 ± 0.03	0.4%
3	10.33 ± 0.03	0.3%
4.9	16.57 ± 0.04	0.2%
6.5	21.35 ± 0.05	0.2%
8.85	25.48 ± 0.05	0.2%
10.8	26.70 ± 0.05	0.2%

plotted in Fig. 12 as a function of the reactor power. The reactor power is obtained from the out-core monitor fission chamber of CROCUS, which are calibrated using the gold foil activation technique [26]. The evolution of the count rate shows a good capability of the system to linearly respond to the increase of the reactor power up to 6.5 W, corresponding to approximately 6.2×10^7 $cm^{-2} \cdot s^{-1}$ in the North West control rod guide tube at core mid-height. The subsequent loss of linearity is caused by common detection dead time, given by the fact that analog signal processing is employed by the read-out electronics. Indeed, the Gaussian amplifier processes the pulses reaching its input in the time interval of 1 µs and thus trains of pulses deriving from different neutron interactions might be read together contributing to a single amplifier pulse. This effect becomes relevant at higher reactor power, where the neutron count rate increases significantly. This limitation might be partially overcome by reducing the shaping time of the gaussian amplifier to have a better discrimination between different neutron events at higher neutron flux rates.

C. Comparison with different in-core instrumentation

Other measurements in the same position and irradiation conditions, i.e. criticality and same neutron spectrum, have been performed in the past in CROCUS with different types of neutron detectors. All the neutron detectors have been placed in the guide tube of the North West control rod of the CROCUS reactor, at approximately 50 cm from the bottom of the core active zone. The neutron count rate shown by a BF_3 detector (type Transcommerce MN-1) in the aforementioned position is about 180000 cps/W [27], whereas

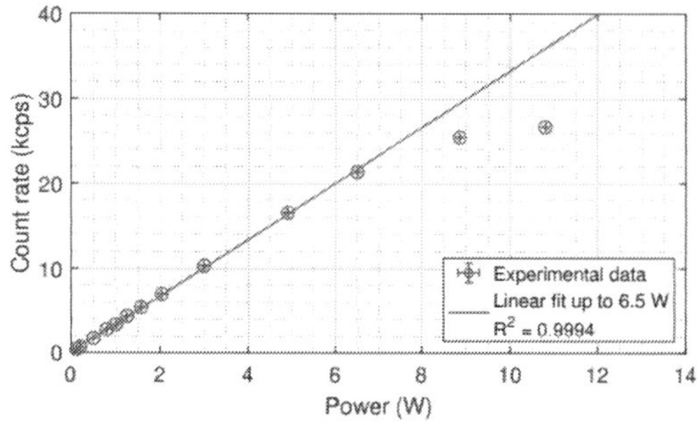

Fig. 12: Neutron count rate in CROCUS as a function of the reactor power. Data are linearly fitted up to 6.5 W.

TABLE VI: Neutron count rate per unit power in the North West control rod guide tube of CROCUS for different neutron detection technologies [27], [28].

Detector type	Neutron count rate per unit power (kcps/W)
Cividec sCVD	0.15
Photonis MFC	0.56
Scintillator + fiber	3.37
Transcommerce BF_3	180

the value registered by the Cividec sCVD diamond detector is 150 cps/W [28] and a miniature fission chamber (MFC, Photonis CFUR44/EP-M3 type with 130 μg of ^{235}U) displays 560 cps/W. In the same position, the detection setup composed by the scintillator screen coupled with the optical fiber and the described read-out electronics gives a neutron count rate equal to 3368 cps/W. The results are summarized in Tab. VI.

Despite the limitations due to the read-out electronics (as described before), the neutron count rate obtained with the system designed in the current work is very high for such a small scintillator volume (1 mm^2 area and 0.2 mm thickness), which makes this technology of interest for the investigation of localized and three-dimensional phenomena in research reactors. The detector shape and arrangement might be also designed in order to increase the detection efficiency and to optimize the light collection. In addition, the use of the advanced digital electronics (currently under testing) will allow parallelizing the designed detector and to perform multi-channel online measurements.

V. Conclusions

The current test proved the feasibility of neutron detection with a ZnS:^{6}LiF screens coupled to silicon photomultiplier (SiPMs) via optical fibers in mixed neutron and gamma fields. The miniature-size detector and its read-out electronics have been thoroughly characterized in the CARROUSEL facility. The detection system, tested in the CROCUS zero-power research reactor, showed a linear behaviour up to the power of 6.5 W (in order of $\sim 10^8$ cm^{-2}·s^{-1} total neutron flux). It also shows an excellent neutron sensitivity with respect to other instruments of reduced size such as miniature fission chambers and sCVD diamond detectors, both available at LRS. The lack of detected events in the reactor shutdown configuration suggests a good gamma-blindness of the system that will be further investigated. The first testing campaign proved that the fiber-coupled scintillator system developed and tested in this work is an efficient and practical solution for the investigation of complex in-core phenomena. The excellent performances combined with its simplicity will allow us to use it for multiple simultaneous localized measurements in the CROCUS reactor. Additional developments and tests are foreseen to be performed in the future in order to fully-characterize the miniature detector, improve its performances and broaden its applications in the CROCUS reactor.

References

[1] C. Pralong Fauchre, O. Joneja, R. Chawla, P. Grimm, F. Jatuff and M. Murphy, "Reactor analysis methods 5 Within-Pin Reaction Rate Distributions in a SVEA-96+ Fuel Assembly," Transactions of the American Nuclear Society, 211-212, 2001.

[2] C. Pralong Fauchre, M. Murphy, F. Jatuff and R. Chawla, "Within-pin Reaction Rate Distributions: CASMO-4 and HELIOS Compared Against Tomographic Measurements at the PROTEUS Reactor," Nuclear Science and Engineering, 150:1, 27-36, 2005.

[3] K. Macku, F. Jatuff, M. Murphy, M. Plaschy, P. Grimm, O. P. Joneja and R. Chawla, "Radial and Azimuthal ^{235}U Fission and ^{238}U Capture Distributions in BWR UO_2 Pins: CASMO-4 and MCNP4C versus Activation Foil Measurements," Nuclear Science and Engineering, 155:1, 96-101, 2007.

[4] M. F. Murphy, M. Plaschy, F. Jatuff, U. Bergmann and R. Chawla, "Fission and capture rate measurements in a SVEA-96 OPTIMA2 BWR assembly compared with MCNPX predictions," United States: American Nuclear Society - ANS, 2006.

[5] V. Lamirand, M. Hursin, P. Frajtag, G. Perret, O. Pakari and A. Pautz, "Future experimental programmes in the CROCUS reactor," European Research Reactor Conference (RRFM) 2016: Conference Proceedings, (p. 1154). Belgium: European Nuclear Society, 2016.

[6] F. Vitullo, V. Lamirand and A. Pautz, "A novel neutron detector for localized in-core measurements in the CROCUS reactor for high-fidelity code validation purposes." The poster will be presented at the Poster session at the FISA2019 conference, 4-7 June 2019, Pitesti, Romania.

[7] R. T. Kouzes, A. T. Lintereur, and E. R. Siciliano, "Progress in alternative neutron detection to address the helium-3 shortage," Nucl. Instruments Methods Phys. Res. Sect. A Accel. Spectrometers, Detect. Assoc. Equip., vol. 784, pp. 172-175, Jun. 2015.

[8] C. Mori, T. Osada, K. Yanagida, T. Aoyama, A. Uritani, H. Miyahara, Y. Yamane, K. Kobayashi, C. Ichihara and S. Shiroya, "Simple and Quick Measurement of Neutron Flux Distribution by Using an Optical Fiber with Scintillator," Journal of Nuclear Science and Technology, 31:3, 248-249, 1994.

[9] Y. Yamane, A. Uritani, and T. Misawa, "Measurement of the thermal and fast neutron flux in a research reactor with a Li and Th loaded optical fibre detector," Nucl. Instruments Methods Phys. Res. Sect. A Accel. Spectrometers, Detect. Assoc. Equip., vol. 432, pp. 403-409, 1999.

[10] M. Stagliano, A. Chierici, L. Abego, and F. dErrico, "Silicon photomultiplier current and prospective applications in biological and radiological photonics," se, vol. 4, no. 10, pp. 10-29, Oct. 2018.

[11] J.-B. Mosset, A. Stoykov, U. Greuter, A. Gromov, M. Hildebrandt, T. Panzner and N. Schlump, "A 16-ch module for thermal neutron detection using ZnS:6LiF scintillator with embedded WLS fibers coupled to SiPMs and its dedicated readout electronics," Nucl. Instruments Methods Phys. Res. Sect. A Accel. Spectrometers, Detect. Assoc. Equip., vol. 845, pp. 494-498, 2016.

[12] A. Stoykov, J. -B. Mosset, and M. Hildebrandt, "Trigger Efficiency of a ZnS:6LiF Scintillation Neutron Detector Readout With a SiPM," IEEE Trans. Nucl. Sci., vol. 63, no. 4, pp. 2271-2277, 2016.

[13] International Atomic Energy Agency, "Advanced Surveillance, Diagnostic and Prognostic Techniques in Monitoring Structures, Systems and Components in Nuclear Power Plants," IAEA Nuclear Energy Series No. NP-T-3.14, Vienna, 2013.
[14] V. Lamirand, "Installation nucléaire CROCUS - Rapport de sécurité version 1.2," 2017.
[15] Scintacor, "Neutron Screens,". [Online]. Available: https://scintacor.com/wp-content/uploads/2015/09/Datasheet-Neutron-Screens-High-Res.pdf
[16] C. W.E van Eijk, "Neutron PSDs for the next generation of spallation neutron sources," Nuclear Instruments and Methods in Physics Research Section A: Accelerators, Spectrometers, Detectors and Associated Equipment, Volume 477, Issues 13, 2002, Pages 383-390.
[17] N. Soppera, M. Bossant and E. Dupont, "JANIS 4: An Improved Version of the NEA Java-based Nuclear Data Information System," Nuclear Data Sheets, Volume 120, June 2014, Pages 294-296.
[18] G.F. Knoll, "Radiation Detection and Measurement," 4th edition, 2010.
[19] ESKA, "Specification sheet SH-8001," 2001. Available: http://www.i-fiberoptics.com/pdf/sh8001.pdf.
[20] Saint-gobain, "Safety data sheet BC-630 Silicone Grease," 2017. Available: https://www.crystals.saint-gobain.com/sites/imdf.crystals.com/files/documents/bc-630.pdf
[21] A. Ghassemi, K. Sato, and K. Kobayashi, "MPPC - Technical note," Hamamatsu Photonics K.K., Solid State Division, 2017. Available: https://www.hamamatsu.com/resources/pdf/ssd/mppc_kapd9005e.pdf
[22] CANBERRA, "Model 2022 Spectroscopy Amplifier," 2010. Available: https://www3.nd.edu/ wzech/Model-2022-SS-M3833.pdf
[23] CANBERRA, "Model 2030 SCA." Available: http://web.mit.edu/8.13/8.13d/manuals/canberra/canberra-2030-sca.pdf
[24] ORTEC, "926-M32-USB Multichannel Analyzer," 2015. Available: http://www-old.pd.infn.it/elettronica/Strumenti/926-m32-usb.pdf
[25] TELEDYNE LECROY, "WaveSurfer 10 oscilloscope," 2015. Available: http://cdn.teledynelecroy.com/files/pdf/wavesurfer10-datasheet.pdf
[26] V. Lamirand, O. Pakari and A. Pautz, "Power calibration methodology at the CROCUS reactor, " presented at ANIMMA2019 conference, 17-21 June 2019, Portoroz, Slovenia.
[27] G. Perret, G. Girardin, P. Frajtag, and M. Hursin, "Decay constant and delayed neutron fraction measurements in CROCUS," Paul Scherrer Institut, TM-41-14-02 Rev. 1, 2014.
[28] M. Hursin, C. Weiss, P. Frajtag, V. Lamirand, G. Perret, P. Kavrigin, A. Pautz, and E. Griesmayer, "Testing of a sCVD diamond detection system in the CROCUS reactor," Eur. Phys. J. A, 54: 82, 2018.

Building and Use of Infrastructure for Detector Development and Testing at VR-1 Reactor Facility

Tomas Bily
Dept. of Nuclear Reactors, FNSPE
Czech Technical University in Prague
Prague, Czech Republic
tomas.bily@fjfi.cvut.cz

Jan Rataj
Dept. of Nuclear Reactors, FNSPE
Czech Technical University in Prague
Prague, Czech Republic
jan.rataj@fjfi.cvut.cz

Ondrej Huml
Dept. of Nuclear Reactors, FNSPE
Czech Technical University in Prague
Prague, Czech Republic
ondrej.huml@fjfi.cvut.cz

***Abstract*— Activities related to detector development, testing, characterisation and applications belong to key research objectives of the VR-1 reactor facility. The contribution gives a review of related improvements, achievements, used approaches, methods, and trends.**

Keywords—VR-1 reactor, neutron detection, detector development

I. Introduction

Neutron and gamma detection and spectrometric systems have plenty of research and industrial applications. Development of such systems and adaptation of their use for specific application often requires extensive experimental effort. At VR-1 reactor facility operated by the Czech Technical University in Prague the infrastructure for detector development and testing is being developed to provide experimental support in all phases of detection system development. This includes conditions for initial detector testing, detection system characterisation and calibration, and application oriented testing.

II. VR-1 Reactor Facility

The VR-1 reactor facility is primarily oriented on education and training with significant international impact in this field [1,2]. The utilization for research activities covers typically about 20 – 30 % of operational time. The research utilisation has been widening since 2012 when the facility was included among the large research infrastructures supported by the Ministry of Education, Youth and Sports of the Czech Republic. Since then the open access to the facility for the purposes of research, development and innovation has been enabled which made the utilization of the facility by external users significantly easier and more effective.

The reactor itself is of open-pool light-water type with zero power and low enriched uranium fuel. It has a simple access to the core through a set of dry vertical channels with diameters ranging from 16 mm to 90 mm and one horizontal radial channel with diameter up of 250 mm which can be reduced to 90 mm. The reactor can operate in a wide range of neutron fluxes (up to some 10^9 n/cm^{2}s) and may provide various ratios of neutron and gamma component. The facility further possesses several dedicated adjacent labs: gamma spectrometry lab, radiation interaction lab, neutron lab and electronics lab. Gamma spectrometry lab is used mainly for neutron activation analysis of samples irradiated in the reactor. Radiation interaction lab is used mainly for education on basics of radiation detection. It can also be used for initial detection testing covering detector response to alpha, beta or gamma radiation. Weak calibration sources are mostly shared with gamma spectrometry lab. Medium activity Cs-137 source (0.8 GBq) is available to study gamma sensitivity of neutron detectors. Neutron lab is dedicated for utilisation of radionuclide neutron sources of Am-Be and Cf-252 type and future plans incorporates the set-up with DD generator as well.

III. Development of Infrastructure to Support Detector Development and Testing

The infrastructure is gradually being improved utilizing a strong synergy between own research, education & training activities, and activities of open access users. The own research is to a high degree aimed on development of research capabilities of the facility through broader offer of neutron source types, dedicated characterized radiation fields for specific tests, and development of methods related to radiation measurements and signal processing. With respect to neutron sources, the facility originally utilized beside the VR-1 reactor two radionuclide neutron sources of Am-Be type. The weaker one with neutron emission rate of ca. 2×10^5 n/s serves mainly to study response of detection systems to neutron radiation; the stronger one with emission rate of ca 1×10^7 n/s was used for advanced tests and experiments in neutron lab. In 2013, a portable DD generator with maximal neutron emission rate of 7×10^6 n/s has been acquired; it is aimed to be used coupled with the VR-1 as well as an independent facility. Beside different neutron energy spectra it enables to provide pulsing ranging from 250 Hz to 20 kHz with adjustable pulse width. Its neutron output can be modified based on high voltage and beam current set-up. Due to its portable and compact design it may be used in various set-up and geometries. In 2014, the 0.8 MBq Cs-137 source was bought to enable studies of gamma sensitivity of neutron detection systems and to enlarge the abilities in education training with gamma radiation detectors. In 2018, the new neutron lab was developed and a Cf-252 neutron source was acquired. The source enables experiments in fission neutron spectrum. Simultaneously, the storage container for neutron radionuclide sources was built in the lab. Later dedicated irradiation positions have been developed inside storage container to enable long term irradiation in close geometry with achievable dose rate of ca. 8 mGy/h. The irradiation can be performed in tailored capsules that can be attached to the front-end of the shielding plug of the source storage container. To further improve the flexibility of available neutron sources and neutron radiation fields, a portable DT generator is being acquired. Again, a portable, flexible device with a choice of continuous or pulsing operation is envisaged. Thereafter, the facility would be able to cover various needs of testing in all common neutron energy spectra at a single site. This can be advantageously used in development of both neutron detection and spectrometric systems.

Together with variety of sources the variety of neutron radiation fields and their characterisation belongs to factor improving the facility capability. For neutron field characterisation techniques based on foil activation [3] and Monte Carlo calculations are standardly being used. Recently, in frame of open-access to the facility, a testing of on-line neutron spectrometric system has been performed which helped to further characterize the radiation field in radial channel of the reactor [4, 5].

Further, next to reactor shielding the polyethylene cube with several vertical channels have been developed. The central channel can be connected with the reactor fast post systems. This enables for studies in radiation field coming from irradiated samples (typically delayed neutrons, or gamma radiation from uranium samples). However, the cube design enables to insert other neutron source inside the central channel and detectors or samples into the remaining to perform specific tests. Specific set-ups and experiments in various geometries and environments, incl. polyethylene, water, or graphite are prepared in the neutron lab. They can serve for optimization of neutron detector performance, and its basic characterisation (e.g. sensitivity or angular dependence). In the field of complex gamma radiation fields, the capability of the facility has been significantly improved during 2013-2015 when the MONTE project was realized together with National Radiation Protection Institute. In frame of the project special set-up has been developed to study detector response in complex gamma field of irradiated nuclear fuel [6,7,8]. This supports testing and experiments in gamma detection and spectrometric system research and applications.

The strengthening and development of own research of the facility is further supported to improve the capabilities and attractiveness of the facility for open access users. One of the three key activities is studies related to behaviour of detection systems in a nuclear reactor core and external neutron sources environments. Standardly, besides testing at various steady-state power levels, the reactor enables to perform various transients that can be advantageously used for detector testing. The reactor is further equipped with several experimental devices capable to change the core conditions: e.g. device for dynamic experiments (oscillator), device for bubble-boiling simulation, and instrumentation for temperature-reactivity effects studies. Using these devices research for neutron detector applications covering in-core diagnostics or noise techniques is enabled. Further development is being performed with shut-down (subcritical) reactor with multiple external neutron sources in the core. This can simulate conditions similar to those in power reactor during core reconfigurations where the self-fissioning of used fuel may challenge detector applications e.g. in reactivity-meters.

IV. USERS

The backbone of users utilizing the infrastructure for detector development, testing (incl. application oriented testing) has been formed within the university and covers activities related to neutron detector development, improvements of methods for neutron detector characterisation (e.g. non-linearity characterisation, pulse processing, etc.), neutron detector testing and characterisation (e.g. in the frame of competence centre for advanced detection system of ionizing radiation). The facility is used by other local research institutions such as Research Centre Rez [4, 5] or National Radiation Protection Institute, by local companies producing and developing neutron detectors and as well by detection systems users such as radiation units of Czech Fire & Rescue Service. From the international users that have recently utilized the infrastructure for detectors related research, Defense Academy of the UK, or CEA France [9] can be mentioned.

V. CONCLUSIONS

At VR-1 reactor facility, the infrastructure for detector development and testing is being gradually developed utilizing synergies between own research, external user research and education and training. The facility tends to provide support in all steps of development, from initial functional tests, through device characterisation up-to application oriented testing at one site.

ACKNOWLEDGMENT

The paper was supported by the project "Strengthening and development of research at Czech Technical University in Prague with the use of research infrastructure VR-1 Training Reactor for research activities" (reg. nr. CZ.02.1.010.00.016_0130001790) supported by Operational Programme Research, Development and Education co-financed from European Structural and Investment Funds and from state budget of the Czech Republic.

REFERENCES:

[1] Ľ. Sklenka, M. Kropík, "Training courses at VR-1 reactor". Proceedings of the International Conference Nuclear Energy for New Europe 2006

[2] Ľ. Sklenka, J. Rataj, "Education and Training in the Field of Reactor Physics and Technology through Research Reactor Networks & Coalitions". Transaction of the American Nuclear Society, Vol. 107, San Diego, California 2012. La Grande Park, Illinois: American Nuclear Society, 2012. pp. 1243-1246. ISSN 0003-018X.

[3] M. Štefánik, K. Katovský, M. Vinš, L Závorka, J. Šoltés, "Neutron field for activation experiments in horizontal channel of training reactor VR-1". Radiation Physics and Chemistry. 2014, 104 302-305. ISSN 0969-806X.

[4] M. Košťál, E. Losa, Z. Matěj, V. Juříček, D. Harutyunyan, O. Huml, M. Štefánik, F. Cvachovec et al., "Characterization of mixed N/G beam of the VR-1 reactor". Annals of Nuclear Energy. 2018, 122 69-78. ISSN 0306-4549.

[5] Z. Matěj, M. Košťál, E. Losa, O. Huml, M. Štefánik, F. Cvachovec, F. Mravec, V. Přenosil "Measurement of fast neutron spectra in radial channel of VR-1 research reactor" Radiation Physics and Chemistry. 2019, ISSN 0969-806X.

[6] O. Huml, P. Žlebčík, Ľ. Sklenka, J. Rataj, M. Štefánik, "Monte Carlo Simulation of Radionuclide Inventory of Irradiated Nuclear Fuel for Generation of Complex Fission Gamma Ray Fields". 2015 IEEE Nuclear Science Symposium & Medical Imaging Conference, San Diego, 31 October – 7 November 2015

https://doi.org/10.1051/epjconf/202022504019

[7] P. Žlebčík, H. Malá, P. Rulík, O. Huml, Ľ. Sklenka, "MONTE-1 experimental device for detection system testing in fission radionuclide fields" (Experimentální zařízení MONTE-1 pro testování detekčních systémů v polích štěpných produktů) BJE. 2016, 24(62)(1/2), 28-32. ISSN 1210-7085.

[8] P. Žlebčík, O. Huml, H. Malá, P. Rulík, M. Hýža, J. Škrkal, B. Bulánek, J. Hůlka, et al., "Comparison of various detection systems in real fission field". In: FOURTH EUROPEAN IRPA CONGRESS, Geneve, Book of Abstracts, 2014. pp. 249. ISBN 978-2-8399-1520-5

[9] M. Pomorski, Ch. Mer-Calfati, F.Foulon Ľ. Sklenka, J. Rataj, T. Bily, "Diamond as a solid state micro-fission chamber for thermal neutron detection at the VR-1 research reactor". Conference: ANIMMA 2015: 4. International Conference on Advancements in Nuclear Instrumentation Measurement Methods and their Applications, Lisboa (Portugal), 20-24 Apr 2015.

The COLIBRI experimental program in the CROCUS reactor: characterization of the fuel rods oscillator

Vincent Lamirand[1,2] *, Pavel Frajtag[1], Daniel Godat[1], Oskari Pakari[1], Axel Laureau[1], Adolfo Rais[1], Mathieu Hursin[1,2], Grégory Perret[2], Carlo Fiorina[1], Andreas Pautz[1,2]
1 Ecole Polytechnique Fédérale de Lausanne (EPFL)
2 Paul Scherrer Institut (PSI)
* vincent.lamirand@epfl.ch

***Abstract*—The present article presents the mechanical characterization of the fuel rods oscillator developed for the purposes of the COLIBRI experimental program in CROCUS. COLIBRI aims at investigating the radiation noise related to fuel vibrations. The main motivation is the increased amplitudes in the neutron noise distributions recorded in ex- and in-core detectors that have been observed in recent years in Siemens pre-Konvoi type of pressurized water reactors. Several potential explanations have been put forward, but no definitive conclusions could yet be drawn. Among others, changes in fuel assembly or pin vibration patterns, due to recent modifications of assembly structural designs, were pointed out as a possible cause. Computational dynamic tools are currently developed within the Horizon 2020 European project CORTEX, to help with understanding the additional noise amplitude. The COLIBRI program is used for their validation. An in-core device was designed, tested, and licensed between 2015 and 2019 for fuel rods oscillation in CROCUS, in successive steps from out-of-pile tests with dummy fuel rods to critical in-core tests. The characterization of its mechanical behavior is presented, in air and in water, and as a function of the load, for safety and experimental purposes. The device allows simultaneously oscillating up to 18 fuel rods. The maximum oscillation amplitude is 5 mm, while the maximum allowed frequency is 2 Hz, i.e. in the frequency range in which the induced neutron flux fluctuations are most pronounced in nuclear power plants.**

***Keywords*—Core monitoring and diagnostics, fuel vibration, noise analysis, research reactor experiment, zero-power reactor**

I. Introduction

THE aim of the COLIBRI (*CROCUS Oscillator for Lateral Increase Between U_{metal} Rods and Inner zone*) experimental program in CROCUS is to investigate the radiation noise related to fuel vibrations. It consists in carrying out experiments on rod displacement (static) and oscillation (dynamic) with a variation of both number and configuration of oscillated rods at relevant ranges of amplitudes and frequencies. The main motivation for this investigation is the increased amplitudes in the neutron noise distributions recorded in ex- and in-core detectors that have been observed in recent years in Siemens pre-Konvoi type of pressurized water reactors. Several potential explanations have been put forward, but no definitive conclusions could be drawn yet. Among others, changes in fuel assembly or pin vibration patterns, due to recent modifications of assembly structural designs, were pointed out as a possible cause. Such mechanical noise is suspected to arise from vibration of groups of fuel assemblies. Computational dynamic tools are currently developed at the Paul Scherrer Institute, as well as within the Horizon 2020 European project CORTEX [1], to help with understanding the additional noise amplitude. The COLIBRI program is used for their validation [2]–[4].

An in-core device was designed for fuel rods oscillation in CROCUS, and manufactured, tested, and licensed between 2015 and 2019 [5]. Limited modifications of core structures and fuel rods were required. Licensing and tests were conducted in successive steps from out-of-pile tests with dummy fuel rods to critical in-core tests. In this article, the tests of the fuel rods oscillator are presented, focusing on the mechanical characterization for safety and experimental purposes.

II. Experimental setup of COLIBRI

In this part we briefly present CROCUS, and the mechanical design of the fuel rods oscillator for the COLIBRI program.

A. The CROCUS reactor

CROCUS at the École Polytechnique Fédérale de Lausanne (EPFL) is an experimental zero-power reactor, uranium-fueled and water-moderated, dedicated to teaching radiation and reactor physics, and to research [6]–[24]. A complete description of the reference core can be found in the International Reactor Physics Experiments Handbook (IRPhE) [25], [26]. It has been licensed for operating at a maximum power of 100 W, i.e. a total neutron flux of ~$2.5\cdot10^9$ $cm^{-2}\cdot s^{-1}$ at the core center. Criticality is controlled either by water level using a spillway, or by two B_4C absorber control rods, with an accuracy of ±0.1 mm (equivalent to approximately ±0.4 pcm) and ±0.5 mm (up to ±0.2 pcm), respectively. CROCUS operates at room temperature using a controlled water loop with secondary and tertiary circuits, two heat exchangers and an electrical heater.

The core is located in an Al-6060 grade vessel of 130 cm in diameter, 160 cm in height, and 1.2 cm in thickness. The vessel is filled with demineralized light water used as both moderator and reflector. The core active part has the approximate shape of a cylinder of 100 cm in height and about 60 cm in diameter. It consists of two interlocked fuel zones with square lattices of different pitches:

- an inner zone of 336 UO_2 rods with an enrichment of 1.806 wt.% and a pitch of 1.837 cm;
- an outer zone of 176 U_{metal} rods for these experiments, 0.947 wt.% and 2.917 cm;
- a varying water gap between the two zones because of the two different pitches.

A picture of the facility and core configuration is shown on Figure 1. Both uranium fuels consist of a 1-m pile of cylindrical pellets cladded in aluminum. The rods are maintained vertically by two octagonal aluminum grid plates spaced 1 m apart. In the COLIBRI program, the grids have a 1 mm cadmium layer to limit axial neutron leakage to the environment, i.e. structures activation, with the active zone of the fuel starting in the middle of the lower cadmium layer.

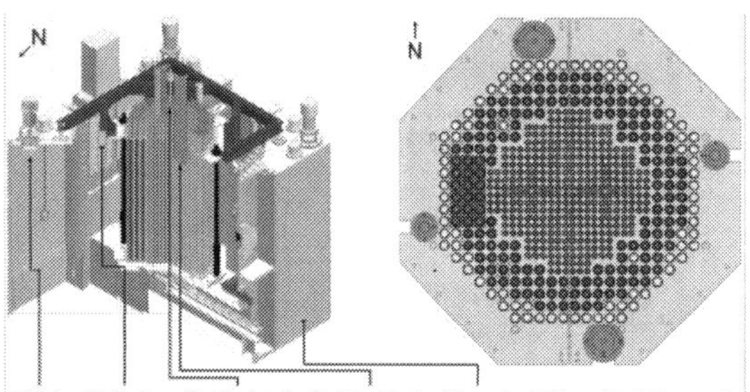

Fig. 1. View of the CROCUS reactor (left), and the core superior grid and configuration with the oscillation location (U_{met} rods in green).

B. *Mechanical design of the fuel rods oscillator*

The fuel rods oscillator is designed to simultaneously oscillate any of 18 metallic uranium fuel rods laterally in the west region of the core periphery zone. Its design, development and licensing is presented in [5]. It consists of two moving plates set above and below the core grids, and rigidly connected by an aluminum beam (see Figure 2). Each plate carries an extremity of the fuel rods, top and bottom respectively. The top moving plate is fixed on the superior grid *via* gliders. Its oscillation is produced by a motor: the motor rotation is converted to a linear translation using an eccentric sheave and a rod. The oscillation is transferred to the bottom moving plate *via* the aluminum beam. The bottom moving plate is not constrained by gliders, it is therefore displaced only because of its connection to the transmission beam.

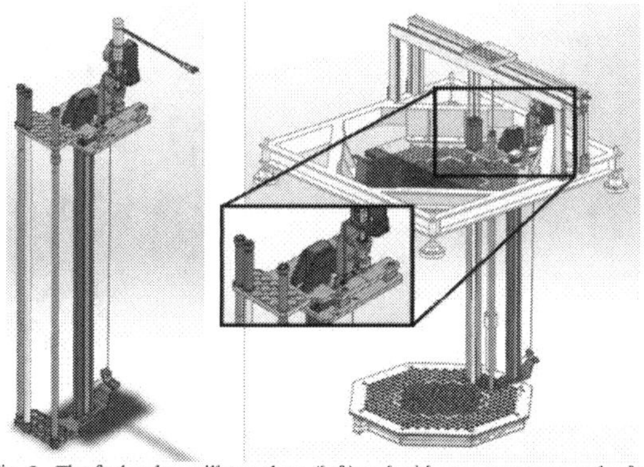

Fig. 2. The fuel rods oscillator alone (left) and with core structures and a few rods inserted in the device (right).

The selection of the moving fuel rods is performed by letting the rods lay on the reactor base plate (i.e. static), or suspend them up 10 mm above the base plate to insert them in the moving plates (i.e. moving). Top and bottom end caps are fixed to each rod to allow the insertion in the enlarged holes of either the static grids or the moving plates. Bumpers with soft rubber O-rings and hard metallic core are added to absorb any shock at the interface between grid and moving plate to prevent propagation to the fuel rods. The weight of the oscillating rods is supported by a platform. The amplitude of the oscillation is precisely tuned by changing the eccentricity of the sheave with calibration plates, 0.5 by 0.5 mm from 0 to ±2.5 mm (i.e. 5 mm total) around their nominal positions in the lattice. Its frequency is depending on the speed of the motor.

The oscillation is controlled and monitored *via* a LabVIEW-developed software with 10 ms time-steps. An inductive captor is set at the rotation axis (i.e. at the top), which detects the actual movement of the motor by detecting the passage of four metallic pins per rotation. A cable coder is used to measure the displacement of the moving plate, i.e. at the bottom, with a 0.1 mm precision. The software produces a csv file output with the recordings of the motor position and speed, the signal of the inductive captor, and the position measurement of the cable. The inductive captor signal is also extracted for live and synchronized recording with the detection instrumentation. The motor, inductive captor, and cable coder are presented in Figure 3, as well as a typical recording of an oscillation in Figure 4.

a) b) c)

Fig. 3. From left to right, a) setup of the motor on the rotation axis, b) inductive captor with a focus on the pin being read, and c) connection of the cable to the transmission beam.

https://doi.org/10.1051/epjconf/202022504020

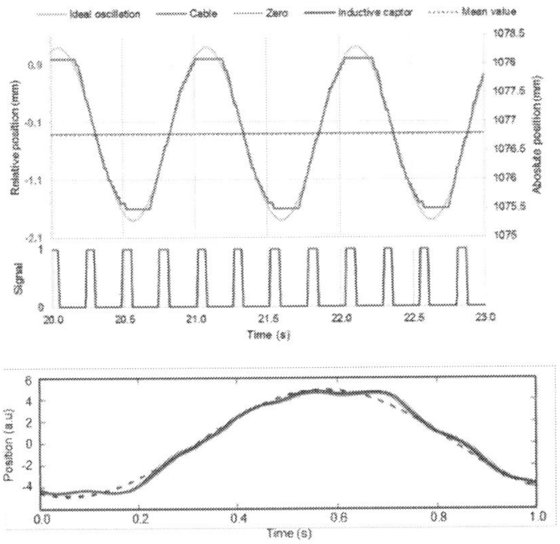

Fig. 4. Signals from one-rod oscillation in air, ±1.5 mm and 1 Hz: on the top, signals from the cable (blue) and inductive captor (bottom, red) as provided by the control system; at the bottom, signal extracted from a video analysis.

III. Mechanical characterization of the oscillator

A. Motivation and means

The aim of the mechanical characterization is to define operation limits for safety purposes, as well as to provide a knowledge of the introduced perturbation for the experiments. As the study was realized for commissioning purposes in the prospect of the oscillator's licensing, a strict step by step procedure was followed, defined in the authorization request and approved in the official authorization delivered by the Swiss regulator, IFSN/ENSI [27]. It is important to note that in the case of a safety shutdown in CROCUS, the water is dumped from the vessel to expansion tanks: the oscillator has to prove safe in both air and water conditions. In this prospect, the behavior of the oscillator has been characterized in air and in water, out-of-pile as a start and in-pile in the end, with empty, 1-rod and 18-rods loading.

At the top of the fuel rods, the transmission of the oscillation depends only on the rotation to translation conversion, and the limited plays associated. They were measured to be fixed with a dial gauge, and their impact is visible in Figure 4 on the flattening of the cable measurement (blue) as compared to an ideal sinus (yellow). Thus the present study focuses on the behavior at the bottom of the fuel rods, which is subject to transmission and inertia effects. The cable coder was used for measuring the oscillation, and determining its behavior in amplitude and frequency. The demonstration of the sufficiency of the cable coder proves important for the final in-pile operation. However, these data were complemented by video analysis for verification purposes when needed, mainly for frequency assessment, and identification of possible harmonics.

B. Characterization in air, out-of-pile and in-pile

The oscillator's behavior was characterized in air at amplitude and frequency ranges from 0.5 to ±2.5 mm, and 0.5 to 2.0 Hz, respectively. The results are represented for both frequency and amplitude in Figures 5 and 6. The frequency results prove sound for all cases, as the set frequency corresponds to the measured one, and therefore, they are not discussed further.

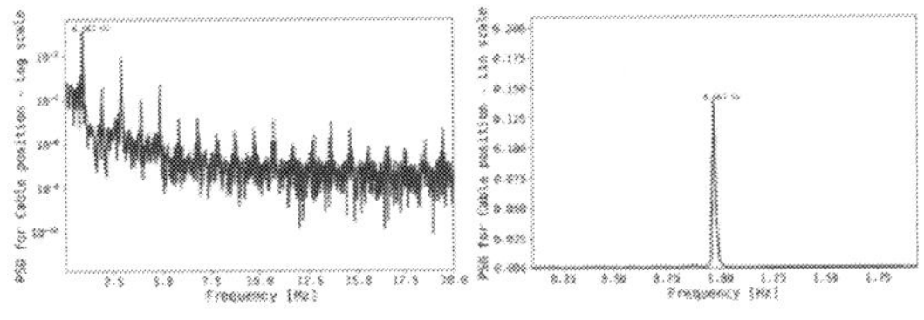

Fig. 5. Example of estimated power spectral density based on the cable data, for the same case than Figure 4 of 1 rod in air, ±1.5 mm and 1 Hz.

The tests empty-loaded proved to be representative of the top position, and thus were taken as a reference for comparison with loaded configurations. The tests with one rod demonstrated a behavior equivalent to the empty case within the ±0.1 mm uncertainties. However, the tests with 18 oscillating rods demonstrated an increase of the measured amplitude with requested frequency and amplitude, i.e. applied force. As the increase quickly reaches a plateau, it seems that it is due to the compression of parts, and most probably the gliders, although they were selected for withstanding compression.

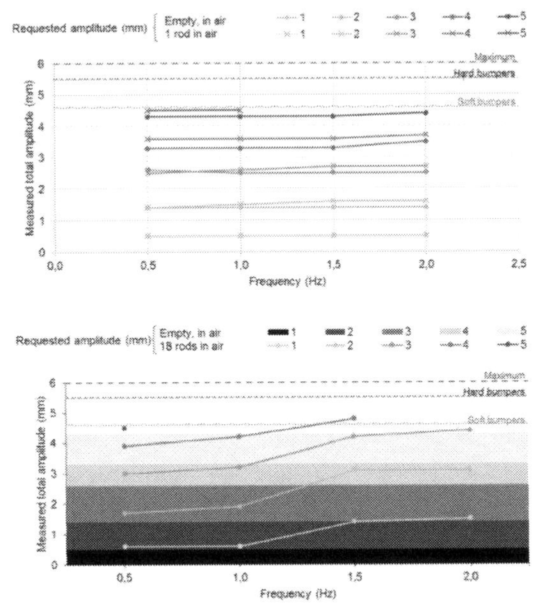

Fig. 6. Behavior of the oscillator in air, compared between empty- and one-rod loads (top), and between empty- and 18-rods loads (bottom). The brilliant red dot at 4 mm and 1.5 Hz indicates the observed cushioning by the soft bumpers. The points are connected for readability purposes.

C. Characterization in water in-pile

The tests were repeated in water in more details, and with a frequency range extended to the low frequency of 0.1 Hz for the 18-rods load. The results are presented in Figure 7 in comparison with the case in air. The oscillator behavior in air and in water is comparable within the uncertainties.

The addition of measured frequencies, and their extension to lower frequencies for 18 rods, shows that the amplitude increase stabilizes above a certain threshold. In addition, we observe that the increase is gradual when the requested amplitude is smaller. It supports the probable compression hypothesis. We also observe a general saturation for the ±2.5 mm case, which confirms that the oscillation is limited by the bumper at this amplitude. As a consequence, it was proposed that the operation would be limited by the conservation of a stable oscillation shape, i.e. without observation of any bouncing on the soft bumpers.

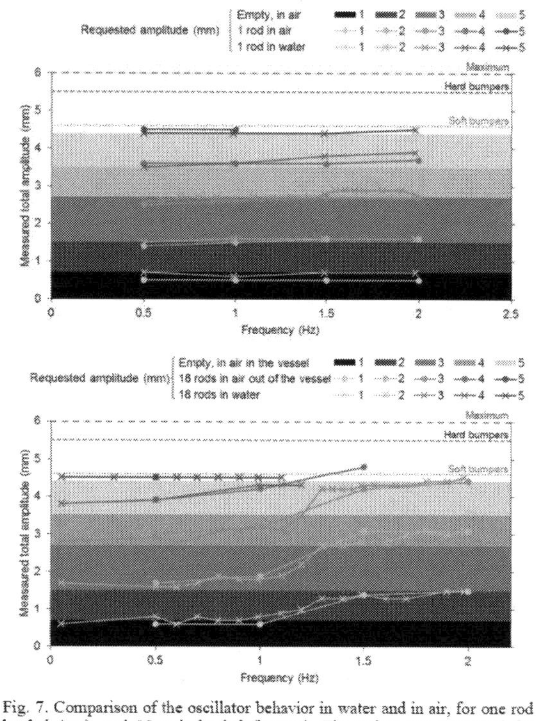

Fig. 7. Comparison of the oscillator behavior in water and in air, for one rod loaded (top), and 18 rods loaded (bottom). The points are connected for readability purposes.

D. Discussion

The results in amplitude for all tests in water are the reference for the COLIBRI experiments. They are summarized in Figure 8. Regarding the impact of the load, we observe a clear separation between the empty/1-rod loads and the 18-rods load: it is more progressive with frequency at lower amplitudes, but it always leads to a maximal addition of around 1.5 mm in the total course. As a consequence, the ±2.0 and ±2.5 mm case would be limited in operation for safety purposes at 1.0 and 1.1 Hz, respectively, for all loads to be conservative.

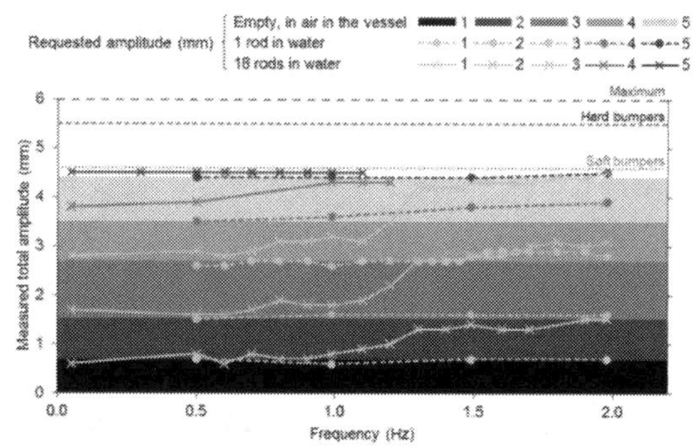

Fig. 8. Comparison of the behavior of the oscillator in water for one and 18 fuel rods loaded. The points are connected for readability purposes.

The defined operation range is presented in Table I. We assume that the behavior of the fuel rods oscillator is validated and characterized in amplitude and frequency within this whole range. Frequencies above 2 Hz were not considered, as the frequency of interest for nuclear power plants is around one Hz, and the transfer function of CROCUS does not vary significantly in the closest superior range.

TABLE I
VALIDATED AND CHARACTERIZED OPERATION RANGE

Amplitude (mm)	Frequency					
	Below	0.5	1.0	1.5	2.0	Above
±0.5	✓	✓	✓	✓	✓	?
±1.0	✓	✓	✓	✓	✓	?
±1.5	✓	✓	✓	✓	✓	?
±2.0	✓	✓	✓(1.1)	✗	✗	✗
±2.5	✓	✓	✓(1.0)	✗	✗	✗

✓: Validated; ✗: Invalidated; ?: Not tested; (value) : maximum frequency.

IV. CONCLUSION AND OUTLOOK

The mechanical behavior of the fuel rods oscillator developed for the purposes of the COLIBRI experimental program in CROCUS was characterized, out-of-pile and in-pile, both in air and in water, and as a function of the load, for safety and experimental purposes. The experimental device allows simultaneously oscillating up to 18 fuel rods. The behavior in frequency is sound, without any observations of abnormal harmonics. In amplitude, it demonstrates increased amplitudes at the bottom of the fuel for the full 18-rods load case and above 1 Hz. Based on the obtained results, an operation range was defined, awaiting final approval from the Swiss regulator, IFSN/ENSI. The maximum allowed amplitude is ±2.5 mm, i.e. 5 mm total, while the maximum allowed frequency is 2 Hz, i.e. in the frequency range in which the induced neutron flux fluctuations are most pronounced in nuclear power plants. The first COLIBRI neutron noise experimental campaign was successfully carried out in 2018 within the framework of the Horizon 2020 European project CORTEX. Following the on-going analysis of the collected data, the next CORTEX campaign is planned for the end of 2019.

ACKNOWLEDGMENT

The authors would like to thank the CROCUS operation team for the help and dedication to the experiments.

REFERENCES

[1] C. Demazière, P. Vinai, M. Hursin, S. Kollias, and J. Herb, "Overview of the CORTEX project," in *PHYSOR 2018: Reactor Physics Paving The Way Towards More Efficient Systems*, 2018, pp. 2971–2980.

[2] V. Lamirand, M. Hursin, A. Rais, S. Hübner, C. Lange, J. Pohlus, U. Paquee, C. Pohl, O. Pakari, and A. Laureau, "Experimental report of the 1st campaign at AKR-2 and CROCUS," 2018.

[3] V. Lamirand, A. Rais, S. Hübner, C. Lange, J. Pohlus, U. Paquee, C. Pohl, O. Pakari, P. Frajtag, D. Godat, M. Hursin, A. Laureau, G. Perret, C. Fiorina, and A. Pautz, "Neutron noise experiments in the AKR-2 and CROCUS reactors for the European project CORTEX," in *ANIMMA 2019*, 2019.

[4] A. Rais, O. Pakari, A. Laureau, J. Pohlus, C. Pohl, V. Lamirand, M. Hursin, and A. Pautz, "Towards the validation of neutron noise simulators: qualification of data acquisition systems."

[5] V. Lamirand, P. Frajtag, D. Godat, M. Hursin, G. Perret, O. V. Pakari, A. Laureau, A. Rais, C. Fiorina, M. Chadwick, and A. Pautz, "The COLIBRI programme in CROCUS: development and licensing of a fuel rods oscillator," in *RRFM/IGORR 2019*, 2019.

[6] A. Rais, D. Siefman, G. Girardin, M. Hursin, and A. Pautz, "PARCS Few-group Homogenized Parameters Generation using Serpent Monte Carlo code at the CROCUS Reactor," in *IGORR 2014*, 2014, pp. 1–11.

[7] M. Hursin, A. Rais, D. Siefman, G. Girardin, and A. Pautz, "Verification of a reactor physics calculation scheme for the CROCUS reactor," in *International Technical Meeting on Small Reactors*, 2014, pp. 3–8.

[8] A. Rais, D. Siefman, G. Girardin, M. Hursin, and A. Pautz, "Methods and Models for the Coupled Neutronics and Thermal-Hydraulics Analysis of the CROCUS Reactor at EFPL," *Sci. Technol. Nucl. Install.*, vol. 2015, 2015.

[9] D. J. Siefman, G. Girardin, A. Rais, A. Pautz, and M. Hursin, "Full Core modeling techniques for research reactors with irregular geometries using Serpent and PARCS applied to the CROCUS reactor," *Ann. Nucl. Energy*, vol. 85, pp. 434–443, 2015.

[10] V. Lamirand, M. Hursin, G. Perret, P. Frajtag, O. Pakari, and A. Pautz, "Future experimental programmes in the CROCUS reactor," in *RRFM/IGORR 2016*, 2016, no. 02–2016, pp. 284–292.

[11] A. Rais, M. Hursin, G. Perret, and A. Pautz, "Experimental validation of control rod reactivity worth and fission rate distributions for the CROCUS Reactor," *Phys. React. 2016, PHYSOR 2016 Unifying Theory Exp. 21st Century*, vol. 6, pp. 1–10, 2016.

[12] A. Rais, "Performance Assessment of a 3-D Steady-State and Spatial Kinetics Model for the Crocus Reactor," Ecole Polytechnique Fédérale de Lausanne (EPFL), 2017.

[13] O. Pakari, V. Lamirand, G. Perret, and A. Pautz, "Current Mode Neutron Noise Measurements in the Zero Power Reactor CROCUS," in *ANIMMA 2017*, 2017.

[14] O. Pakari, V. Lamirand, G. Perret, P. Frajtag, and A. Pautz, "Kinetic Parameter Measurements in the CROCUS Reactor Using Current Mode Instrumentation," *IEEE Trans. Nucl. Sci.*, vol. 65, no. 9, pp. 2456–2460, 2018.

[15] V. Lamirand, G. Perret, S. Radman, D. Siefman, M. Hursin, P. Frajtag, A. Gruel, P. Leconte, P. Blaise, and A. Pautz, "Design of separated element reflector experiments in CROCUS: PETALE," *React. Dosim. 16th Int. Symp. ASTM STP1608*, p. 7, 2018.

[16] O. Pakari, V. Lamirand, G. Perret, D. Godat, M. Hursin, P. Frajtag, and A. Pautz, "Investigation of spatial effects on neutron noise measurements in the zero power reactor CROCUS," in *Proceedings of PHYSOR 2018*, 2018, p. 12.

[17] V. Lamirand, G. de Izarra, A. Krasa, G. Perret, O. V. Pakari, M. Hursin, P. Blaise, J. Wagemans, and A. Pautz, "Intercomparison of neutron noise measurement systems in the CROCUS reactor," in *Proceedings of PHYSOR 2018*, 2018, p. 11.

[18] M. Hursin, C. Weiss, P. Frajtag, V. Lamirand, G. Perret, P. Kavrigin, A. Pautz, and E. Griesmayer, "Testing of a sCVD diamond detection system in the CROCUS reactor," *Eur. Phys. J. A*, vol. 54, no. 5, 2018.

[19] A. Laureau, V. Lamirand, D. Rochman, and A. Pautz, "Total Monte Carlo acceleration for the PETALE experimental programme in the CROCUS reactor," *EPJ Web Conf.*, vol. 211, p. 03002, Jun. 2019.

[20] V. Lamirand, A. Laureau, D. Rochman, G. Perret, A. Gruel, P. Leconte, P. Blaise, and A. Pautz, "An Experimental Programme optimized with Uncertainty Propagation: PETALE in the CROCUS Reactor," *EPJ Web Conf.*, vol. 211, p. 03003, Jun. 2019.

[21] Y. Jiang, A. Laureau, V. Lamirand, P. Frajtag, and A. Pautz, "In-core dosimetry for the validation of neutron spectra in the CROCUS reactor," in *ANIMMA 2019*, 2019.

[22] O. Pakari, V. Lamirand, B. Vandereyt, F. Vitullo, M. Hursin, C. Kong, and A. Pautz, "Design and simulation of gamma spectrometry experiments in the CROCUS reactor," in *ANIMMA 2019*, 2019.

[23] F. Vitullo, V. Lamirand, J.-B. Mosset, P. Frajtag, O. Pakari, G. Perret, and A. Pautz, "A miniature fiber-coupled scintillator for in-core neutron counting in CROCUS," in *ANIMMA 2019*, 2019.

[24] V. Lamirand, "Experiments in a ZPR in the 2010s : Harvesting data in CROCUS," in *ANIMMA 2019 - Workshop N°1 "The coupling of experiments and modelling to enhance experiments in research reactors,"* 2019.

[25] U. Kasemeyer, R. Früh, J. M. Paratte, and R. Chawla, "Benchmark on Kinetic Parameters in the CROCUS Reactor," in *International Reactor Physics Experiments Handbook (IRPhE)*, no. 4440, OECD, Ed. 2007, p. 94.

[26] J. M. Paratte, R. Früh, U. Kasemeyer, M. a. Kalugin, W. Timm, and R. Chawla, "A benchmark on the calculation of kinetic parameters based on reactivity effect experiments in the CROCUS reactor," *Ann. Nucl. Energy*, vol. 33, no. 8, pp. 739–748, May 2006.

[27] K. Macku, "Permis d'exécution pour l'installation d'un dispositif d'oscillation du combustible dans CROCUS, programme expérimental COLIBRI," IFSN MAK/GUJ-25/18/001, 2018.

EPJ Web of Conferences **225**, 04021 (2020) https://doi.org/10.1051/epjconf/202022504021
ANIMMA 2019

In-core dosimetry for the validation of neutron spectra in the CROCUS reactor

Y.Jiang, A.Laureau, V. Lamirand*, P. Frajtag, A.Pautz
Ecole Polytechnique Fédérale de Lausanne (EPFL)
* vincent.lamirand@epfl.ch

***Abstract*—The present article describes the preliminary validation study of simulated in-core and reflector n eutron spectra in preparation of oncoming experimental programs in the zero-power reactor CROCUS at EPFL. For this purpose, a set of activation foils were irradiated at three characteristic positions in the CROCUS reactor, and the subsequent activities were analyzed *via* γ spectrometry. The experimental setup was then modeled with the Monte Carlo neutron transport code Serpent2 and associated with an analysis tool to include the effect of the reactor power history during experiments.**

The comparison of calculated and measured reaction rates (C/E) indicates a general consistency (at 2σ) between calculated and measured spectra. However, offsets of C/E values were observed in (n,γ) reactions, up to 18% for ^{115}In and 8% for ^{63}Cu dosimeters. This could be caused by an unexpected isotopic composition, uncertainties in nuclear data, or the spectrometry analysis.

In addition, a 100-groups spectrum unfolding was performed using the experimentally determined reaction rates and the Serpent2 spectra as the prior knowledge. The unfolded spectra were mainly adjusted in the thermal and fast ranges, while few modifications w ere m ade i n t he e pithermal r egion d ue t o the low contribution of epithermal neutrons in activation processes. Moreover, within energy groups where the capture reactions show resonant behavior, flux depletion (up to 38% as compared to the prior spectra) is observed due to the absence of self-shielding effect in the unfolding process. For this purpose, an unfolding method based on energy groups weighting is developed and tested.

Keywords—**Neutron spectrum measurement, in-core dosimetry, gamma spectrometry, spectrum unfolding**

I. Introduction

The development of the knowledge on nuclear data have largely contributed to the advances of reactor technology, while continuous improvement is still required for further optimization, new designs and waste management for example.

The nuclear data of stainless steel components (Fe, Cr and Ni) are of great interest for the performance assessment of the Gen-III EPR reactor. Validation of the iron nuclear data was accomplished in the EOLE reactor for this purpose [1]. In continuation of this work, the PETALE program [2]–[4] will be performed in the zero-power reactor CROCUS. For such program, the interpretation of experimental data would require a precise knowledge of neutron energy spectra.

In this paper is described the experimental study conducted in CROCUS, for the validation of simulated in-core and reflector neutron spectra by means of activation dosimetry.

II. Method

A. Experimental setup

CROCUS is a an experimental reactor operating at a maximal power of 100 W. The reactor is light water-moderated with two interlocked fuel zones in a square lattice, separated by a water gap, as shown in fig. 1. The power of the reactor is monitored by two ex-core fission chambers and two γ compensated ionization chambers. Further information on the reference experimental configuration of the reactor can be found in [5].

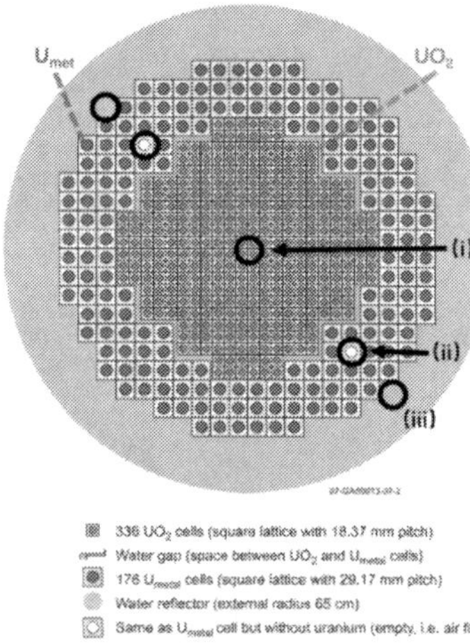

Figure 1: CROCUS core configuration with irradiation positions for the present study (after [5])

For the current study, disk-form dosimeters are placed at three different positions (indicated in fig. 1) within the reactor: (i) at the center of the core, (ii) in one of the empty control rod tubes, (iii) in two empty aluminium rods placed in the near periphery of the core (surrounded by the water reflector). Dosimeters are prepared and laminated in a thin plastic film to avoid possible deformation and contamination during experiments. These dosimeters are then attached to Plexiglas plates for positioning.

All authors are with the Laboratory for Reactor Physics and Systems behaviour, Ecole Polytechnique Fédérale de Lausanne, CH-1015 Lausanne, Switzerland. Y.Jiang is the corresponding author, currently with the Experimental Physics Service, French Atomic Energy Commission, 13108 Saint-Paul-lez-Durance, France (e-mail: yifeng.jiang@cea.fr).

V.Lamirand is with the Laboratory for Reactor Physics and Thermal Hydraulics, Paul Scherrer Institut, CH-5232 Villigen, Switzerland.

A.Pautz is with the Nuclear Energy and Safety Research Division, Paul Scherrer Institut, CH-5232 Villigen, Switzerland.

At the end of the experiments, safety systems are activated to shut down the reactor. The activation of dosimeters are then analyzed *via* γ spectrometry, using High Purity Germanium (HPGe) detectors [6], [7].

B. Experimental design

The experimental design consists of selecting dosimeters, reactor power and operation time. It was accomplished with simulated activation experiments using the Monte Carlo neutron transport code Serpent2 [8], under criticality configuration based on the modeling in [9].

For illustration purpose, some simulated flux profiles (in 100 energy groups) at different experimental positions are shown in fig. 2, while the shaded areas represent the 1σ uncertainty of the simulation. The calculation is made in a 0.005 cm^3 detector, which is representative of the size of dosimeters.

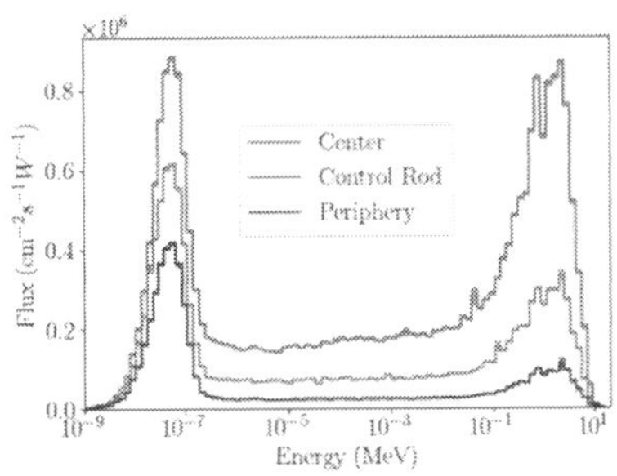

Figure 2: Simulated flux in different experimental positions

One can observe a symmetrical profile for the thermal and the fast contributions in the core center, while the fast component drops significantly outside the fuel zone. Based on the flux profiles, it was decided to use two types of dosimeter: (1) resonant dosimeters presenting capture reactions (n,γ) (2) threshold dosimeters with threshold reactions such as (n,p) or (n,n').

The dosimeters are selected according to the contribution of the calculated reaction rates (RR) in different energy groups between 1×10^{-5} eV and 20 MeV. Figure 3 illustrates the sensitivity of different dosimeters in terms of distributed and integral RR. Resonant dosimeters are sensitive to the thermal flux and certain groups of energy where there are resonant peaks in the cross section. On the other side, threshold dosimeters are meant to target the fast flux since the threshold energy of the reaction is generally above 0.5 MeV. Additionally, resonant dosimeters covered with neutron poison (cadmium boxing) are also used to target epithermal contribution. The set of dosimeters (reactions and associated decay data) used for spectrum characterization is listed in Table 1.

C. Analysis with simulations

An effort has been made to accurately reproduce the experimental configurations in simulation for comparison purpose. In the Serpent2 calculations, dosimeters are modeled with their exact geometry and position within the reactor during

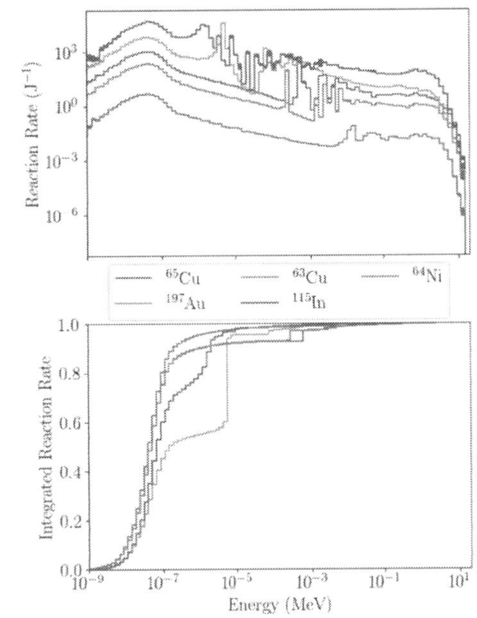

(a) Capture reactions

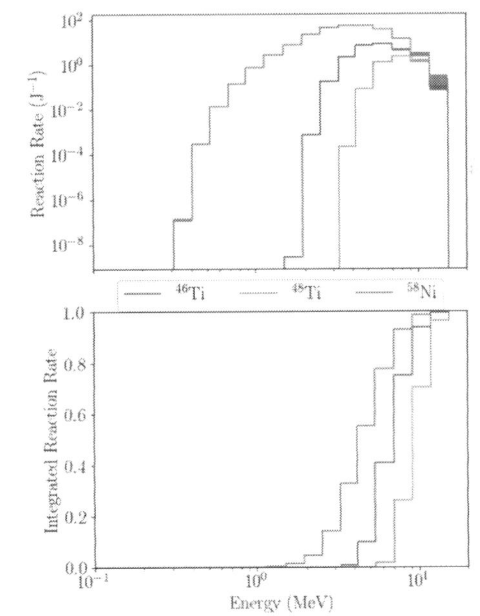

(b) Threshold reactions

Figure 3: Distribution of reaction rates for different types of dosimeters: reaction rate (top) and integrated reaction rate (bottom) for the considered capture and threshold activation reactions

experiments. The isotopic composition of the materials are assumed to be as the natural abundance. The activity $A_i(t)$ for a certain isotope i is estimated using the calculated RRs R_i and the reactor power history $P(t)$.

For precise activity estimation of short-lived isotopes, the power history is of importance. Consistently, the influence of power history is checked for all isotopes of interest. This is implemented in a numerical code based on a simplified Bateman equation, by neglecting reaction rates in isotopes once activated:

$$A_i(t_n) = \lambda_i \Delta t \sum_{k=0}^{n} [C_0 R_i P(t_k) - A_i(t_k)] \quad (1)$$

where λ_i is the decay constant, Δt the elementary time step and C_0 the normalization constant. This normalization is achieved with a reference ^{197}Au (n,γ) dosimeter in each experiment.

A direct comparison can therefore be made between the measured dosimeter activity A_E and the simulated one A_C. These quantities are proportional to the measured and calculated RRs R_E and R_C. The ratio C/E between RRs in calculation (C) and experiment (E) can be estimated as:

$$\frac{C}{E} = \frac{A_C}{A_E} = \frac{R_C}{R_E} \quad (2)$$

In addition to the validation based on C/E, the calculated neutron spectrum can be adjusted by the experimental feedback of measurements (R_E). This is accomplished *via* an unfolding procedure that compares the measured RR and that for an initial guess spectrum (the one calculated in Serpent2 in the present study). The best approximated spectrum is obtained by minimizing the residual on RRs in the χ^2 sense. The unfolding is done with a reference code MAXED [10] based on a maximum entropy criterion, and two non-linear optimization methods with the CVXPY package [11] in Python. The first method uses augmented response matrix (of cross sections) and the information of the spectrum (referred as the unweighted method hereafter). The other method uses the spectral distribution of RRs calculated in Serpent2 to weight the contribution of the response in each energy group. Thus, it will be referred as the weighted method. The unweighted method uses a response matrix of constant cross section values, therefore the self-shielding is not taken into account. The weighted method, by contrast, aims at introducing implicitly the self-shielding effect into the unfolding by assigning different importance to energy groups.

III. Results and discussion

In the present part, the comparison of RRs in simulations and experiments is summarized. The coherence between calculation and measurement indicates that the Serpent2 modeling using a common purpose JEFF 3.1.1 library [12] is in general consistent with outcomes of the activation dosimetry. The result of the comparison also allowed a study on spectrum unfolding. However, discrepancies between calculated and measured RRs are observed for some dosimeters. Possible sources of these discrepancies will be discussed.

A. *C/E in thermal and epithermal ranges*

The overview of C/E values for resonant dosimeters, therefore sensitive to thermal and epithermal ranges, is presented in fig. 4.

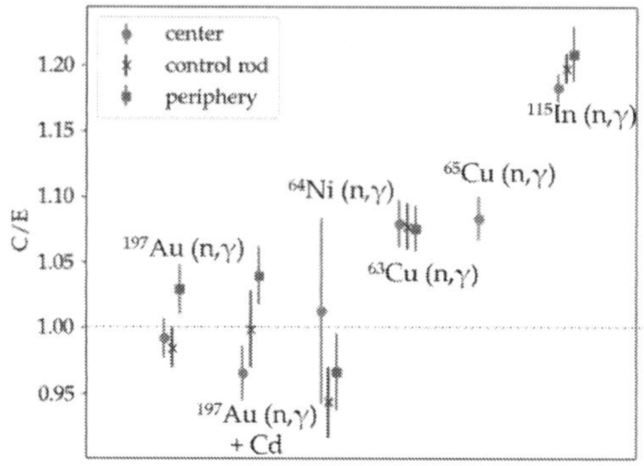

Figure 4: C/E values for resonant dosimeters at different positions

For each reaction at the three experimental positions, one representative and consistent value throughout a number of experiments is shown with its uncertainty. The C/E values indicate the agreement between simulations and measurements either with the absolute value or the relative one. It should be noted that uncertainties associated to resonant dosimeters are mostly small as their high activation allows a precise determination of the activity. The relatively large uncertainty for ^{64}Ni is partly due to its low isotopic proportion in nature ($<$ 1%), partly due to its short half-life (2.5 hours). Discrepancies for ^{63}Cu, ^{65}Cu and ^{115}In dosimeters were observed to be an offset, regardless of the position of the dosimeter across different experiments. For ^{63}Cu and ^{65}Cu, the uncertainty in cross section data and branching ratio of γ decay emission is of the same order of magnitude [12]–[15]. In the case of ^{115}In the source of the offsets could be a possible isotopic composition that is different from natural abundance, or biases and uncertainties in γ spectrometry analysis not taken into account in this study.

It is worth noticing that the activity measurement of ^{63}Cu is made on a particular γ ray: the 511 keV photon created through the annihilation of a positron over an electron. It is however difficult to confirm that the background annihilation in the acquisition system during measurements is fully subtracted from the target decay emissions. Therefore, the applicability of spectrometry analysis is questionable by definition. It was also observed that the acquired peak shape in γ spectrum, illustrated in fig. 5, resembles that of a Gaussian (Poisson distribution of a large number of events), but the width of the peak is generally two times larger than ordinary peaks measured in this work.

B. *C/E in the fast range*

Despite the general agreement between simulation and experiment for threshold dosimeters as shown in fig. 6, large

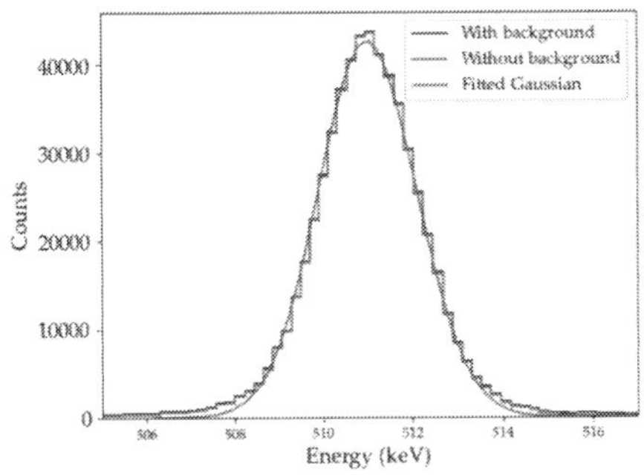

Figure 5: Illustration of measured annihilation peak of ^{64}Cu decay γ spectrum

uncertainties were observed due to difficulties in measuring the activity and high statistical uncertainties in the RR calculations.

In addition to the dosimeters presented in fig. 6, the inelastic scattering reaction (n,n') of ^{115}In was also measured in experiments, through the 336 keV photon emission during the transition from the first metastable level to its fundamental level. However, the results show important discrepancies between calculation and measurement: the calculated RRs is about 6% of the measured one. This issue was also reported by [16]. A possible explanation is that, in general purpose nuclear data libraries, inelastic scattering cross section is defined for different metastable levels while in γ spectrometry the activity is measured as the sum of all transitions that would contribute to the metastable level of interest. The complexity of the decay scheme makes inconceivable to compare calculations with measured inelastic scattering RR through common libraries.

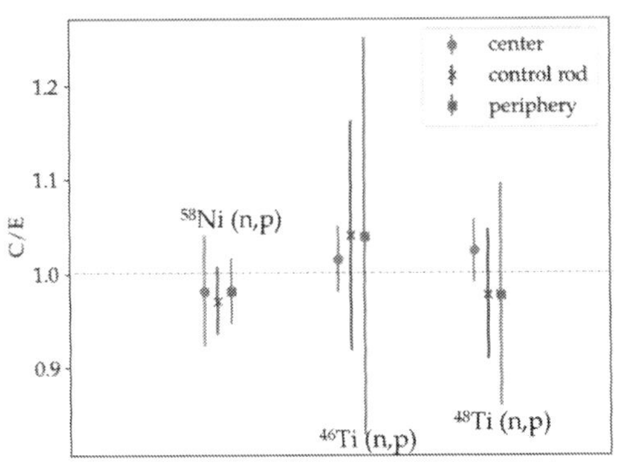

Figure 6: C/E values for threshold dosimeters at different positions

Specialized libraries for dosimetry, for instance IRDFF, provide dosimetry cross section files, which estimate in terms of inelastic scattering cross section the overall yield of the metastable level being considered. Thus, the use of specialized libraries for (n,n') reactions will be considered in future analysis.

C. Spectrum unfolding

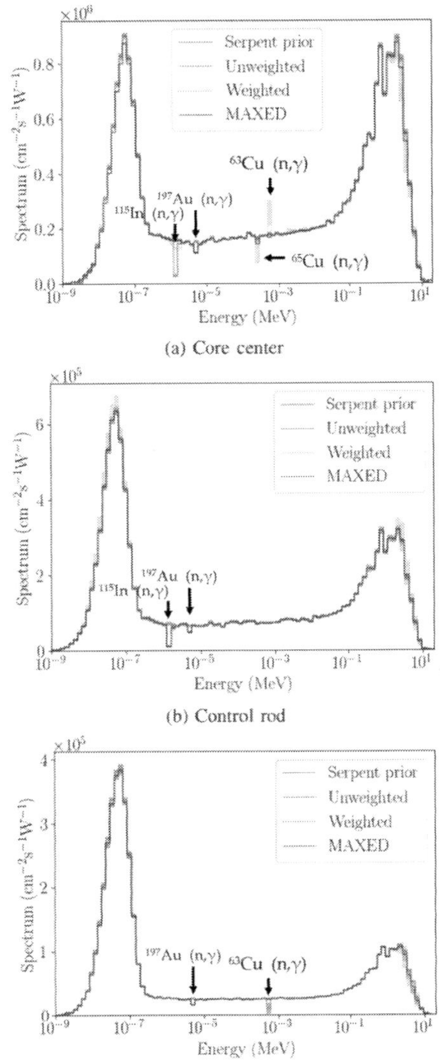

(a) Core center

(b) Control rod

(c) Periphery

Figure 7: Unfolded neutron spectra for experimental positions

As indicated in fig. 7, the unfolding results are generally in agreement with the initial Serpent2 spectra. Sharp depletion (up to 38% as compared to the prior spectrum) or augmentation of the flux is observed in the epithermal range, which is caused by the large value of resonant capture cross sections. Energy groups with strong flux variation are identified to be associated with resonance peaks of different reactions (fig. 7). In reality, the self-shielding limits the actual RR in the resonance region, while in the unfolding this mechanism has to be integrated numerically. This situation can be illustrated through the comparison of unfolded spectra by different methods: The unweighted method takes the cross section values as constant and independent of the spectrum, therefore the unfolded spectra show serious flux variation (fig. 7a). The weighted method shows positive results for the self-shielding correction. With a relatively high cost associated to flux adjustment in resonance regions, the self-shielding effect is represented to a certain extent. Nevertheless, in the case of the spectrum at the control rod position (fig. 7b) the flux depletion is important regardless of the method, possibly caused by numerical issues. Additionally, calculated and measured uncertainties in the RR of threshold dosimeters lead to large uncertainty in the unfolded fast spectra. Further improvement on the experimental procedure and more accurate reaction rate estimation in simulation are expected to optimize the unfolding results.

IV. Conclusion

This paper summarizes the used methodology to validate simulated neutron spectra in the zero-power reactor CROCUS for dosimetry applications. A set of activation dosimeters were chosen, so that the energy spectrum from thermal to up to 20 MeV is covered as much as possible. In-core experiments were designed by Monte Carlo modeling, and performed at different locations in CROCUS and the consequent activation is measured *via* γ spectrometry. The experimental setup were then accurately modeled to compare calculations with experimental results in terms of C/E. The general agreement on C/E indicates the consistency between calculated and measured spectra, and thus the validation of spectra and experimental methodology.

Spectrum unfolding was performed using experimental feedbacks. Two unfolding methods were implemented, which show larger uncertainty in the solution spectrum, but are consistent with the MAXED reference solution.

Inconsistency and offsets in C/E were observed in the analysis, which could be caused by different sources of biases in simulation, experiments and nuclear data. Further improvements are to be made on the design of activation experiments, along with a dedicated model for corrections in spectrometry analysis.

References

[1] C Vaglio-Gaudard, A Santamarina, P Blaise, O Litaize, A Lyoussi, G Noguère, J. M. Ruggieri, and J. F. Vidal, "Interpretation of PERLE Experiment for the Validation of Iron Nuclear Data Using Monte Carlo Calculations," *Nuclear Science and Engineering*, vol. 166, no. 2, pp. 89–106, 2010.

[2] V. Lamirand, G. Perret, S. Radman, D. Siefman, M. Hursin, P. Frajtag, A. Gruel, P. Leconte, P. Blaise, and A. Pautz, "Design of separated element reflector experiments in CROCUS: PETALE," in *International Symposium on Reactor Dosimetry*, 2017, p. 7.

[3] A. Laureau, V. Lamirand, D. Rochman, and A. Pautz, "Total Monte Carlo acceleration for the PETALE experimental programme in the CROCUS reactor," Accepted in WONDER 2018 - 5th edition of the International Workshop On Nuclear Data Evaluation for Reactor Applications, 2018.

[4] V. Lamirand, A. Laureau, D. Rochman, G. Perret, A. Gruel, P. Leconte, P. Blaise, and A. Pautz, "An Experimental Programme optimized with Uncertainty Propagation: PETALE in the CROCUS Reactor," Accepted in WONDER 2018 - 5th edition of the International Workshop On Nuclear Data Evaluation for Reactor Applications, 2018.

[5] OECD/NEA, *Benchmark on Kinetic Parameters in the CROCUS Reactor*, 4440. 2007, vol. 9, p. 94.

[6] Canberra Industries, "Genie ™ 2000 Spectroscopy Software Operations," pp. 61–97, 2006.

[7] R Venkataraman, F Bronson, V Atrashkevich, M Field, and B. M. Young, "Improved Detector Response Characterization Method in ISOCS and LabSOCS 1 . General Philosophy of ISOCS Detector Characterization 2 . Response Characterization Methodologies," vol. 264, no. Marc Vi, pp. 1–21, 2003.

[8] J. Leppänen, *SERPENT– a Continuous-energy Monte Carlo Reactor Physics Burnup Calculation Code*. 2015, p. 162.

[9] D. J. Siefman, G. Girardin, A. Rais, A. Pautz, and M. Hursin, "Full Core modeling techniques for research reactors with irregular geometries using Serpent and PARCS applied to the CROCUS reactor," *Annals of Nuclear Energy*, vol. 85, pp. 434–443, 2015.

[10] M. Reginatto, P. Goldhagen, and S. Neumann, "Spectrum unfolding, sensitivity analysis and propagation of uncertainties with the maximum entropy deconvolution code MAXED," *Nuclear Instruments and Methods in Physics Research Section A: Accelerators, Spectrometers, Detectors and Associated Equipment*, vol. 476, no. 1-2, pp. 242–246, 2002.

[11] S. Diamond and S. Boyd, "CVXPY: A Python-embedded modeling language for convex optimization," *Journal of Machine Learning Research*, vol. 17, no. 83, pp. 1–5, 2016.

[12] A Santamarina, D Bernard, P Blaise, M Coste, A Courcelle, T. D. Huynh, C Jouanne, P Leconte, O Litaize, S Mengelle, G Noguère, J. M. Ruggiéri, O Sérot, J Tommasi, C Vaglio, J. F. Vidal, and Y Rugama, *The JEFF-3.1.1 Nuclear Data Library*, 22. 2009, pp. 1–62.

[13] R. Capote, K. Zolotarev, V. Pronyaev, and A. Trkov, "Updating and extending the irdf-2002 dosimetry library," 2012.

[14] D. Rochman, A. Koning, J. Sublet, M. Fleming, E. Bauge, S. Hilaire, P. Romain, B. Morillon, H. Duarte, S. Goriely, S. Van Der Marck, H. Sjöstrand, S. Pomp, N. DzysiGBR, O. Cabellos, H. FerroGBRhi, and A. Vasiliev, "The TENDL library: Hope, reality and future," *EPJ Web of Conferences*, vol. 146, 2017.

[15] M. Berger, J. Hubbell, S. Seltzer, J. Chang, J. Coursey, R. Sukumar, D. Zucker, and K. Olsen. (). Nist stan-

 https://doi.org/10.1051/epjconf/202022504021

dard reference database 8, Com: Photon cross sections database.

[16] M. Tesinsky, C. Berglöf, T. Bäck, B. Martsynkevich, I. Serafimovich, V. Bournos, A. Khilmanovich, Y. Fokov, S. Korneev, H. Kiyavitskaya, and W. Gudowski, "Comparison of calculated and measured reaction rates obtained through foil activation in the subcritical dual spectrum facility YALINA-Booster," *Annals of Nuclear Energy*, vol. 38, pp. 1412–1417, 2011.

Table I: Activation characteristics for different foils

Reaction	Abundance (%)	Half-life (days)	Threshold Energy (MeV)	Main γ raies (keV)	Activity [1] $(Bq.g^{-1}.W^{-1})$
$^{46}Ti(n,p)^{46}Sc$	8.25	83.8	1.6	889 1020	0.237
$^{48}Ti(n,p)^{48}Sc$	73.72	1.82	3.3	983 1037	2.73
$^{56}Fe(n,p)^{56}Mn$	91.75	0.107	0.35	847	1.55×10^2
$^{58}Ni(n,p)^{58}Co$	68.08	70.8	0.4	811	20.0
$^{64}Ni(n,\gamma)^{65}Ni$	0.927	0.105	-	1482	2.09×10^3
$^{63}Cu(n,\gamma)^{64}Cu$	69.15	0.529	-	511	9.74×10^4
$^{65}Cu(n,\gamma)^{66}Cu$ [1]	30.85	3.54×10^{-3}	-	1345	3.89×10^5
$^{115}In(n,\gamma)^{116}In$	95.72	0.0376	-	1294 1097	5.23×10^6
$^{115}In(n,n')^{115m}In$	95.72	0.187	0.34	336	1.95×10^2
$^{197}Au(n,\gamma)^{198}Au$	100	2.67	-	412	3.12×10^5

[1] The activity corresponds to the activation after 1 hour of irradiation, when the foil is positioned in the central axis of CROCUS, at the middle height of the critical water level.

EPJ Web of Conferences **225**, 04022 (2020) https://doi.org/10.1051/epjconf/202022504022
ANIMMA 2019

Power calibration methodology at the CROCUS reactor

Vincent Lamirand[1,2]*, Axel Laureau[1], Oskari Pakari[1], Pavel Frajtag[1], Andreas Pautz[1,2]
[1] Ecole Polytechnique Fédérale de Lausanne (EPFL)
[2] Nuclear Energy and Safety Research Division (NES), Paul Scherrer Institut (PSI)
* vincent.lamirand@epfl.ch

***Abstract*—In the present article, we detail the method used to experimentally determine the power of the CROCUS zero-power reactor, and to subsequently calibrate its ex-core monitor fission chambers. Knowledge of the reactor power is a mandatory quantity for a safe operation. Furthermore, most experimental research programs rely on absolute fission rates in design and interpretation – for instance, tally normalization of reaction rate studies in dosimetry, or normalization of power spectral density in neutron noise measurements. The minimization of associated uncertainties is only achieved by an accurate power determination method. The main experiment consists in the irradiation, and therefore, the activation of several axially distributed Au-197 foils in the central axis of the core, which activities are measured with a High-Purity Germanium (HPGe) gamma spectrometer. The effective cross sections are determined by MCNP and Serpent Monte Carlo simulations. We quantify the reaction rate of each gold foil, and derive the corresponding fission rate in the reactor. The variance weighted average over the distributed foils then provides a calibration factor for the count rates measured in the fission chambers during the irradiation. We detail the calibration process with minimization of respective uncertainties arising from each sub-step, from power control after reactivity insertion, to the calibration of the HPGe gamma spectrometer. Biases arising from different nuclear data choices are also discussed.**

***Index Terms*—Neutron dosimetry, neutron activation analysis, in-core dosimetry, integral experiment, zero-power reactor**

I. Introduction

THE CROCUS reactor is a two-zone, uranium-fuelled light water moderated facility operated by the Laboratory for Reactor Physics and Systems Behaviour (LRS) at the Swiss Federal Institute of Technology Lausanne (EPFL). With a maximum power of 100 W, it is a zero-power reactor used for teaching and research purposes, most recently for studies on intrinsic and induced neutron noise, highly-localized measurements, and nuclear data [1]–[19]. Knowledge of the reactor power is a mandatory quantity for a safe operation. Furthermore, most experimental research programs rely on absolute fission rates for design and interpretation [20] – for instance, tally normalization of reaction rate studies in dosimetry, or normalization of power spectral density in neutron noise measurements. The minimization of associated uncertainties is only achieved by an accurate power determination method. We present hereafter the method used to determine the reactor power and to subsequently calibrate the ex-core monitor fission chambers [21]–[23]. The main experiment consists in the irradiation, and therefore, the activation of several axially distributed ^{197}Au foils in the central axis of the core, which activities are measured with a HPGe gamma spectrometer. The effective cross sections are determined using MCNP6.1 [24] and Serpent2 [25] Monte Carlo simulations. We quantify the reaction rate of each gold foil, and derive the corresponding fission rate in the reactor. The variance weighted average over all foils then provides a calibration factor for the count rates measured in the fission chambers during the foil irradiation. We detail the calibration process with minimization of respective uncertainties arising in each sub-step, from power control after reactivity insertion, to the calibration of the HPGe gamma spectrometer. Biases arising from different nuclear data choices are also discussed.

II. Experiments

In this part we briefly present CROCUS, the methodology and experimental setup for foil activations and measurements, and the dedicated experimental campaign.

A. The CROCUS reactor

A complete description of the reference core of CROCUS is available in the International Reactor Physics Experiments Handbook (IRPhE) [26], [27]. The reactor has been licensed for operating at a maximum power of 100 W, i.e. a total neutron flux of ~$2.5\cdot10^9$ $cm^{-2}\cdot s^{-1}$ at the core center. Criticality is controlled either by water level using a spillway, or by two B_4C absorber control rods, with an accuracy of ±0.1 mm (equivalent to approximately ±0.4 pcm) and ±0.5 mm (up to ±0.2 pcm), respectively. CROCUS operates at a core temperature from 17.5°C to 22.5°C using a controlled water loop with secondary and tertiary circuits, two heat exchangers and an electrical heater.

The core is located in an Al-6060 grade open vessel of 130 cm in diameter, 160 cm in height, and 1.2 cm in thickness. The vessel is filled with demineralized light water used as both moderator and reflector. The core active part has the approximate shape of a cylinder of 100 cm in height and about 60 cm in diameter. It consists of two interlocked fuel zones with square lattices of different pitches:

- an inner zone of 336 UO_2 rods with an enrichment of 1.806 wt.% and a pitch of 1.837 cm;
- an outer zone of 176 U_{metal} rods for these experiments, 0.947 wt.% and 2.917 cm;

https://doi.org/10.1051/epjconf/202022504022

- a varying water gap between the two zones because of the two different pitches.

A picture of the reactor and its critical configuration are shown on Figure 1. Both uranium fuel rods consist of a 1-m pile of cylindrical pellets cladded in aluminum. The rods are maintained vertically by two octagonal aluminum grid plates spaced 1 m apart. In the current experimental configuration for the COLIBRI program [28]–[30], the grids have a 1-mm cadmium layer to limit axial neutron leakage to the environment, i.e. structures activation, with the active zone of the fuel starting in the middle of the lower cadmium layer.

The neutron flux is monitored using two Merlin Gerin CC54 compensated ionization chambers (north and south), and two Photonis CFUM21 10-mg ^{235}U fission chambers (east and west), which are the calibrated power and safety monitors.

The reactor possesses six independent shutdown mechanisms to bring it to a subcritical state in less than one second: two cruciform cadmium (Cd) safety blades in the inner zone, and four expansion tanks to drain the moderator, set in the vessel corners and controlled each by a valve. The safety systems are also used for shutdown under normal operation conditions.

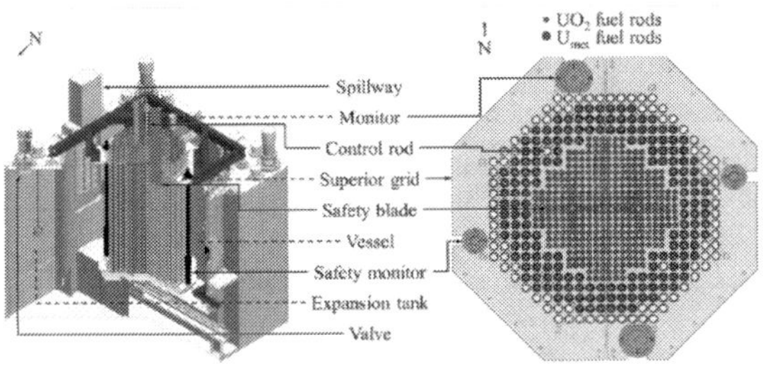

Fig. 1. Schematic isometric view of the CROCUS reactor (left), and top view of the core configuration (right).

B. *Methodology and experimental setup*

The used calibration methodology consists in three steps:

- in-core irradiation of dosimeters, monitored by the fission chambers to be calibrated,
- determination of the dosimeters' absolute activities with an High-Purity Germanium gamma spectrometer,
- simulation of the irradiation by Monte Carlo codes, to extract each dosimeter's reaction rate, and the corresponding fission rate, i.e. power of the reactor.

Gold dosimeters are employed as the $^{197}Au(n,\gamma)$ cross section is a standard for neutron flux measurements in the thermal energy range, with a high cross section and low uncertainties [31]. In this study, 14 disc-form foil dosimeters with a diameter of ø15 mm, and a thickness of 10 µm were used. For each irradiation, six dosimeters were axially distributed on a PMMA plate (see Figure 2). They are laminated in a plastic film to avoid deformation and contamination in the vessel's water.

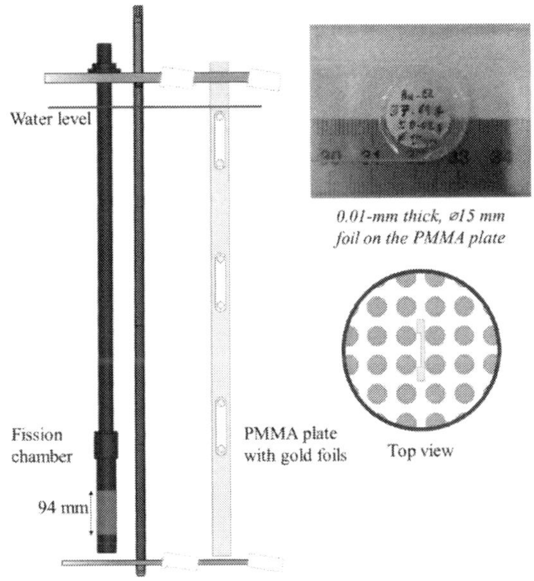

Fig. 2. On the left, schematic side view of one periphery fuel rod set in the superior and inferior grids, aside a monitor fission chamber with the active zone represented, and the PMMA plate inserted in the core. On the right, photo of a dosimeter set on the plate (top), and top view of the plate inserted in the lattice of the core inner zone.

The irradiation plate is positioned above the core center while the reactor is brought to criticality at the desired power by increase of the water level. Once the reactor is stabilized (~10 min for the delayed neutron contribution), the plate is dropped in the core central axis. The enf-of-irradiation is a fast process, less than a second due to the safety systems, which also drain around half of the vessel's water. The dosimeters with the irradiation plate are recovered by opening the lid of the reactor's cavity immediately after the shutdown. Then, the dosimeters activities are measured one by one using a Canberra HPGe gamma spectrometer (detector: GC4518/S; shielding: VG-BB-98/16D1-2). The HPGe detector is calibrated by a point-like calibration source of ^{152}Eu ($\sigma_{\text{efficiency at 411.8 keV (Au)}} = 0.5\%$).

As the monitors' count rates are recorded during the irradiation, based on the measured activities and calculated reaction and fission rates, we derive the monitors' count rates per fission, or per unit of power.

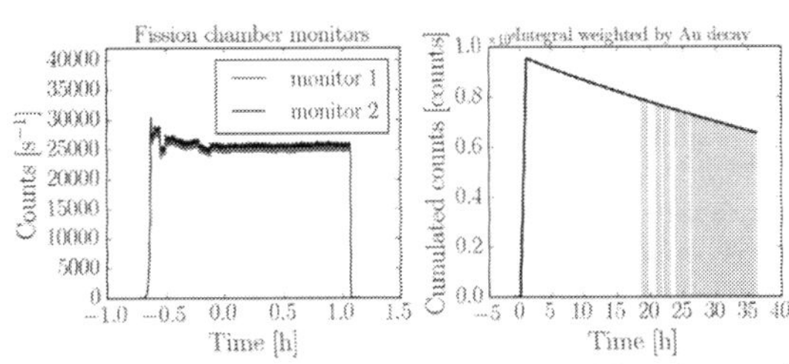

Fig. 3. Power history followed by both safety monitors (left), and computed activation with a representation of the foils' measurements on the HPGe station. t_0=0 h corresponds to the start of the irradiation (#6).

C. *Experimental campaign*

In order to investigate possible biases, as well as to quantify and reduce uncertainties, a set of irradiations was performed with selected repetitions and change of parameters (see Table I). All irradiations were carried out without control rods (guide tubes were left empty), so criticality was controlled only by changing the water level. At criticality, the PuBe start-up source is sent back to its storage shielding. This configuration allows the flux distribution to be as symmetrical as possible, without any local perturbations. The primary water-cooling circuit was engaged for at least one hour before the irradiation, in order to stabilize the water and temperature in the reactor vessel.

The irradiation in the reference configuration at (20±0.1)°C was carried out three times at different power levels (10, 20 and 40 W), and with different sets of dosimeters. One irradiation was performed with additional instrumentation in the reflector, in order to modify the critical water level. Two irradiations were accomplished in the reference configuration but at two different temperatures, at 18 and 22°C, in order to assess the temperature dependence. A supplementary irradiation was undertaken at a subcritical level with a presence of the start-up source, at a water level of 800 mm, to observe the deformation of the flux distribution.

TABLE I
EXPERIMENTAL CAMPAIGN

Irradiation #	Power [W]	Duration [s], ±1 s	Temperature [°C], ±0.1°C	Critical level [mm], ±0.1 mm
1	9.7	3776	20.0	954.4
2	20	4136	20.0	954.6
3	40	1040	20.0	954.4
4	10	3597	20.0	966.4
5	10	3595	18.0	952.6
6	10	3847	22.0	956.65
7	~3.10^{-3}	17580	20.0	800.0 (subcrit.)

III. MONTE CARLO CALCULATIONS

The irradiations were simulated with the MCNP6 and Serpent2 Monte Carlo codes, and using JEFF3.3 [32], ENDF/B-VII.1 [33], and IRDFF-v1.05 [34] nuclear data libraries for comparison purposes. The detailed modelling included each individual dosimeter, and the water level of each irradiation. Models were used to compute reaction rates for each dosimeter at their location during the irradiation, and the reactor's fission rate. The reactor power was derived by multiplying the latter with the effective energy released per fission, and it was directly exectuted by the codes (e.g., for Serpent2, $3.245507 \cdot 10^{-11} \pm 0.0025\%$ J/fission). In the case of the IRDFF library, the calculations were performed with Serpent2 only, by replacing only the dosimetry cross sections. The use of the IRDFF gold files for propagating the uncertainties, for all foils axially distributed during one experiment, namely irradiation #1, from a cross section to individual reactions rates as a function of the neutron energy, is presented in Figure 4.

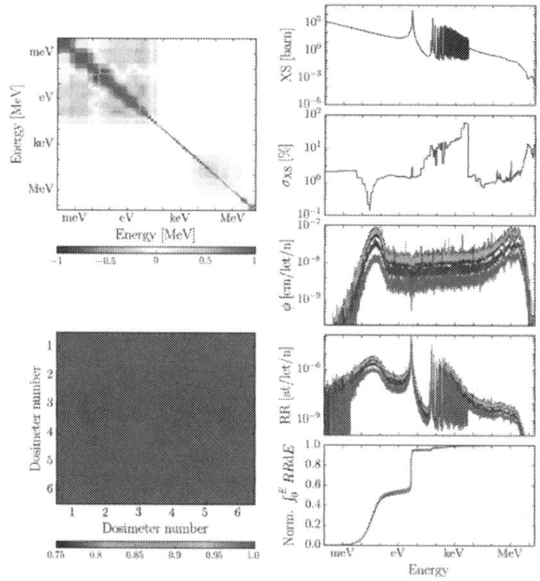

Fig. 4. Propagation of the IRDFF files for the $^{197}Au(n,\gamma)$ reaction for uncertainty quantification of the reference irradiation #1. On the top left, covariance matrix according to the IRDFF library. On the bottom left, correlation matrix of the six dosimeters of the experiment, 100% correlated as expected. On the right, top to bottom, cross section, cross section uncertainty, computed spectra in the dosimeter, corresponding reaction rates as a function of the energy, and reactions rates integrated over the energy.

IV. RESULTS

The experimentally determined reaction rates were averaged by weighting in order to calculate a single calibration factor for each monitor and irradiation. The dispersion has been used to estimate biases and uncertainties, but the results used by the facility are those from the first reference irradiation. As MCNP is the validated code for CROCUS by a regulatory body, the preliminary calibration factors based on MCNP6.2 with JEFF3.3 for the 2019 campaign are (2708 ± 54) cps/W for Monitor 1 (east) and (2784 ± 56) cps/W for Monitor 2 (west). The dosimeters at the top positions were discarded, as they present systematic errors, probably due to their position close to the water level. We observe a consequent shift in the distribution, which would require further investigation. For illustration purposes, the measured reaction rates are compared to the calculated ones using Serpent2 and JEFF3.3 with an arbitrary normalization in Figure 5.

Results obtained with the support of calculations with MCNP6.2 and Serpent2 are in a good agreement, in the order of their statistical uncertainty (σ_{stat}=0.5 %). We observe a discrepancy of the order of +1% between JEFF3.3 and ENDF/B-VII.1, which is low but still significant as compared to the statistical uncertainty (σ_{stat}=0.2%). There is also a systematic discrepancy of (+1.1±0.1)% between JEFF3.3 with or without IRDFF for gold, when compared using Serpent2.

TABLE II
RESULTS FOR MONITOR 1 DEPENDING ON CODE AND LIBRARY [CPS/W]

Monte Carlo code	JEFF3.3	ENDF/ B-VII.1	JEFF3.3 +IRDFF
MCNP6.1	2687.6	2659.1	-
Diff. /JEFF3.3	0	-1.1%	-
Serpent2	2685.3	-	2714.8
Diff. /JEFF3.3	0	-	+1.1%
Diff MCNP6.1/Serpent2	+0.1%	-	-

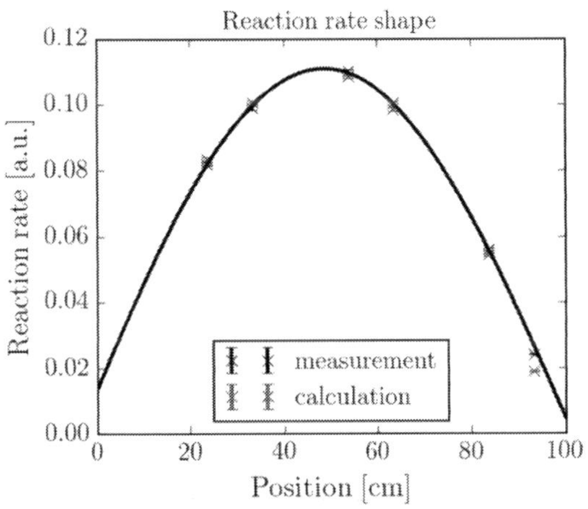

Fig. 5. Fitted axial distribution of the measured and calculated (Serpent2/JEFF3.3) reaction rates for the irradiation #6.

V. CONCLUSION AND OUTLOOK

In the present article, we introduce the methodology used in CROCUS for the power calibration of the safety monitors. It is based on the in-core neutron dosimetry technique using the $^{197}Au(n,\gamma)$ reaction for an activation analysis, complemented with detailed Monte Carlo calculations. For the 2019 calibration campaign, a set of irradiations was carried out to identify possible biases and quantify uncertainties. Preliminary calibration factors based on MCNP6.2 with JEFF3.3 are 2708 ± 54 cps/W for Monitor 1 (east) and 2784 ± 56 cps/W for Monitor 2 (west). Good agreements were found between codes (within the low statistical uncertainties), as well as limited discrepancies between nuclear data libraries (in the order of %). Further investigations shall include the understanding of discrepancies at the water-air interface, as well as the detailed study of temperature effects, and flux distributions at subcriticality.

ACKNOWLEDGMENT

The authors would like to thank the CROCUS operation team for the help and dedication to the experiments.

REFERENCES

[1] A. Rais, D. Siefman, G. Girardin, M. Hursin, and A. Pautz, "PARCS Few-group Homogenized Parameters Generation using Serpent Monte Carlo code at the CROCUS Reactor," in *IGORR 2014*, 2014, pp. 1–11.

[2] M. Hursin, A. Rais, D. Siefman, G. Girardin, and A. Pautz, "Verification of a reactor physics calculation scheme for the CROCUS reactor," in *International Technical Meeting on Small Reactors*, 2014, pp. 3–8.

[3] O. Pakari, V. Lamirand, G. Perret, D. Godat, M. Hursin, P. Frajtag, and A. Pautz, "Investigation of spatial effects on neutron noise measurements in the zero power reactor CROCUS," in *Proceedings of PHYSOR 2018*, 2018, p. 12.

[4] V. Lamirand, G. de Izarra, A. Krasa, G. Perret, O. V. Pakari, M. Hursin, P. Blaise, J. Wagemans, and A. Pautz, "Intercomparison of neutron noise measurement systems in the CROCUS reactor," in *Proceedings of PHYSOR 2018*, 2018, p. 11.

[5] M. Hursin, C. Weiss, P. Frajtag, V. Lamirand, G. Perret, P. Kavrigin, A. Pautz, and E. Griesmayer, "Testing of a sCVD diamond detection system in the CROCUS reactor," *Eur. Phys. J. A*, vol. 54, no. 5, 2018.

[6] A. Laureau, V. Lamirand, D. Rochman, and A. Pautz, "Total Monte Carlo acceleration for the PETALE experimental programme in the CROCUS reactor," *EPJ Web Conf.*, vol. 211, p. 03002, Jun. 2019.

[7] V. Lamirand, A. Laureau, D. Rochman, G. Perret, A. Gruel, P. Leconte, P. Blaise, and A. Pautz, "An Experimental Programme optimized with Uncertainty Propagation: PETALE in the CROCUS Reactor," *EPJ Web Conf.*, vol. 211, p. 03003, Jun. 2019.

[8] Y. Jiang, A. Laureau, V. Lamirand, P. Frajtag, and A. Pautz, "In-core dosimetry for the validation of neutron spectra in the CROCUS reactor," in *ANIMMA 2019*, 2019.

[9] O. Pakari, V. Lamirand, B. Vandereyt, F. Vitullo, M. Hursin, C. Kong, and A. Pautz, "Design and simulation of gamma spectrometry experiments in the CROCUS reactor," in *ANIMMA 2019*, 2019.

[10] F. Vitullo, V. Lamirand, J.-B. Mosset, P. Frajtag, O. Pakari, G. Perret, and A. Pautz, "A miniature fiber-coupled scintillator for in-core neutron counting in CROCUS," in *ANIMMA 2019*, 2019.

[11] V. Lamirand, "Experiments in a ZPR in the 2010s : Harvesting data in CROCUS," in *ANIMMA 2019 - Workshop N°1 "The coupling of experiments and modelling to enhance experiments in research reactors,"* 2019.

[12] A. Rais, D. Siefman, G. Girardin, M. Hursin, and A. Pautz, "Methods and Models for the Coupled Neutronics and Thermal-Hydraulics Analysis of the CROCUS Reactor at EFPL," *Sci. Technol. Nucl. Install.*, vol. 2015, 2015.

[13] D. J. Siefman, G. Girardin, A. Rais, A. Pautz, and M. Hursin, "Full Core modeling techniques for research reactors with irregular geometries using Serpent and PARCS applied to the CROCUS reactor," *Ann. Nucl. Energy*, vol. 85, pp. 434–443, 2015.

[14] V. Lamirand, M. Hursin, G. Perret, P. Frajtag, O. Pakari, and A. Pautz, "Future experimental programmes in the CROCUS reactor," in *RRFM/IGORR 2016*, 2016, no. 02–2016, pp. 284–292.

[15] A. Rais, M. Hursin, G. Perret, and A. Pautz, "Experimental validation of control rod reactivity worth and fission rate distributions for the CROCUS Reactor," *Phys. React. 2016, PHYSOR 2016 Unifying Theory Exp. 21st Century*, vol. 6, pp. 1–10, 2016.

[16] A. Rais, "Performance Assessment of a 3-D Steady-State and Spatial Kinetics Model for the Crocus Reactor," Ecole Polytechnique Fédérale de Lausanne (EPFL), 2017.

[17] O. Pakari, V. Lamirand, G. Perret, and A. Pautz, "Current Mode Neutron Noise Measurements in the Zero Power Reactor CROCUS," in *ANIMMA 2017*, 2017.

[18] O. Pakari, V. Lamirand, G. Perret, P. Frajtag, and A. Pautz, "Kinetic Parameter Measurements in the CROCUS Reactor Using Current Mode Instrumentation," *IEEE Trans. Nucl. Sci.*, vol. 65, no. 9, pp. 2456–2460, 2018.

[19] V. Lamirand, G. Perret, S. Radman, D. Siefman, M. Hursin, P. Frajtag, A. Gruel, P. Leconte, P. Blaise, and A. Pautz, "Design of separated element reflector experiments in CROCUS: PETALE," *React. Dosim. 16th Int. Symp. ASTM STP1608*, p. 7, 2018.

[20] Ž. Štancar, L. Snoj, V. Radulović, A. Trkov, L. Barbot, and C. Domergue, "Evaluation of the Axial Absolute Power Profile Measurements at the JSI Triga Mark II Reactor," *Nucl. Energy New Eur. 2012*, no. October 2011, pp. 1–8, 2012.

[21] D. V. Ba, "Mesures absolues du flux neutronique et de la puissance

du réacteur CROCUS par activation des feuilles d'or," 1971.

[22] P. Chartier, "Mesure de la puissance du réacteur CROCUS," 1974.

[23] R. Früh, "Réacteur CROCUS - Détermination de la puissance absolue," EPFL-IGA-LPR 179, 1990.

[24] J. F. Briesmeister and others, *MCNP--A general Monte Carlo code for neutron and photon transport*. Los Alamos National Laboratory Los Alamos, 1986.

[25] J. Leppänen, "Serpent--a continuous-energy Monte Carlo reactor physics burnup calculation code," *VTT Tech. Res. Cent. Finl.*, vol. 4, 2013.

[26] U. Kasemeyer, R. Früh, J. M. Paratte, and R. Chawla, "Benchmark on Kinetic Parameters in the CROCUS Reactor," in *International Reactor Physics Experiments Handbook (IRPhE)*, no. 4440, OECD, Ed. 2007, p. 94.

[27] J. M. Paratte, R. Früh, U. Kasemeyer, M. a. Kalugin, W. Timm, and R. Chawla, "A benchmark on the calculation of kinetic parameters based on reactivity effect experiments in the CROCUS reactor," *Ann. Nucl. Energy*, vol. 33, no. 8, pp. 739–748, May 2006.

[28] V. Lamirand, P. Frajtag, D. Godat, M. Hursin, G. Perret, O. V. Pakari, A. Laureau, A. Rais, C. Fiorina, M. Chadwick, and A. Pautz, "The COLIBRI programme in CROCUS: development and licensing of a fuel rods oscillator," in *RRFM/IGORR 2019*, 2019.

[29] V. Lamirand, P. Frajtag, D. Godat, M. Hursin, G. Perret, O. Pakari, A. Rais, C. Fiorina, and A. Pautz, "The COLIBRI programme in CROCUS : characterisation of the fuel rods oscillator," in *ANIMMA 2019*, 2019.

[30] V. Lamirand, A. Rais, S. Hübner, C. Lange, J. Pohlus, U. Paquee, C. Pohl, O. Pakari, P. Frajtag, D. Godat, M. Hursin, A. Laureau, G. Perret, C. Fiorina, and A. Pautz, "Neutron noise experiments in the AKR-2 and CROCUS reactors for the European project CORTEX," in *ANIMMA 2019*, 2019.

[31] B. Asselineau, F. Trompier, C. Texier, C. Itié, R. Médioni, D. Tikunov, H. Muller, and G. Pelcot, "Reference dosimetry measurements for the international intercomparison of criticality accident dosimetry SILENE 9-21 June 2002," *Radiat. Prot. Dosimetry*, vol. 110, no. 1–4, pp. 459–464, 2004.

[32] A. J. Koning, M. Avrigeanu, V. Avrigeanu, P. Batistoni, E. Bauge, M.-M. Bé, P. Bem, D. Bernard, O. Bersillon, A. Bidaud, O. Bouland, A. Courcelle, C. J. Dean, P. Dos-Santos-Uzarralde, B. Duchemin, I. Duhamel, M. C. Duijvestijn, E. Dupont, U. Fischer, R. A. Forrest, F. Gunsing, W. Haeck, H. Henriksson, A. Hogenbirk, T. D. Huynh, R. Jacqmin, C. Jouanne, J. Keinert, M. A. Kellett, I. Kodeli, J. Kopecky, H. Leeb, D. Leichtle, J. Leppanen, O. Litaize, M. J. Lopez Jimenez, M. Mattes, E. Menapace, R. W. Mills, B. Morillon, C. Mounier, A. L. Nichols, G. Noguere, C. Nordborg, A. Nouri, R. L. Perel, P. Pereslavtsev, R. J. Perry, M. Pescarini, M. Pillon, A. J. M. Plompen, D. Ridikas, P. Romain, Y. Rugama, P. Rullhusen, C. de Saint Jean, A. Santamarina, E. Sartori, K. Seidel, O. Serot, S. Simakov, J.-C. Sublet, S. Tagesen, A. Trkov, S. C. van der Marck, and H. Vonach, "The JEFF evaluated nuclear data project," pp. 3–8, 2008.

[33] M. B. Chadwick, M. Herman, P. Obložinský, M. E. Dunn, Y. Danon, A. C. Kahler, D. L. Smith, B. Pritychenko, G. Arbanas, R. Arcilla, R. Brewer, D. A. Brown, R. Capote, A. D. Carlson, Y. S. Cho, H. Derrien, K. Guber, G. M. Hale, S. Hoblit, S. Holloway, T. D. Johnson, T. Kawano, B. C. Kiedrowski, H. Kim, S. Kunieda, N. M. Larson, L. Leal, J. P. Lestone, R. C. Little, E. A. McCutchan, R. E. MacFarlane, M. MacInnes, C. M. Mattoon, R. D. McKnight, S. F. Mughabghab, G. P. A. Nobre, G. Palmiotti, A. Palumbo, M. T. Pigni, V. G. Pronyaev, R. O. Sayer, A. A. Sonzogni, N. C. Summers, P. Talou, I. J. Thompson, A. Trkov, R. L. Vogt, S. C. van der Marck, A. Wallner, M. C. White, D. Wiarda, and P. G. Young, "ENDF/B-VII.1 nuclear data for science and technology: Cross sections, covariances, fission product yields and decay data," *Nucl. Data Sheets*, vol. 112, no. 12, pp. 2887–2996, 2011.

[34] R. Capote, K. I. K. I. Zolotarev, V. G. V. G. Pronyaev, and A. Trkov, "Updating and Extending the IRDF-2002 Dosimetry Library," *J. ASTM Int.*, vol. 9, no. 4, p. 104119, 2012.

Neutron noise experiments in the AKR-2 and CROCUS reactors for the European project CORTEX

Vincent Lamirand[1,2], Adolfo Rais[1], Sebastian Hübner[3], Carsten Lange[3], Joachim Pohlus[4], Uwe Paquee[4], Christoph Pohl[5], Oskari Pakari[1], Pavel Frajtag[1], Daniel Godat[1], Mathieu Hursin[1,2], Axel Laureau[1], Grégory Perret[2], Carlo Fiorina[1], Andreas Pautz[1,2]

Abstract—**The present article gives an overview of the first experimental campaigns carried out in the AKR-2 and CROCUS reactors within the framework of the Horizon 2020 European project CORTEX. CORTEX aims at developing innovative core monitoring techniques that allow detecting anomalies in nuclear reactors, e.g. excessive vibrations of core internals. The technique will be mainly based on using the fluctuations in neutron flux, i.e. noise analysis. The project will result in a deepened understanding of the physical processes involved. This will allow utilities to detect operational problems at a very early stage, and to take proper actions before such problems have any adverse effect on plant safety and reliability. The purpose of the experimental campaigns in the AKR-2 and CROCUS reactors is to produce noise-specific experimental data for the validation of the neutron noise computational models developed within this framework. The first campaigns at both facilities consisted in measurements at reference static states, and with the addition of mechanical perturbations. In the AKR-2 reactor, perturbations were induced by two devices: a rotating absorber and a vibrating absorber, both sets in experimental channels close to the core. In CROCUS, the project benefited from the COLIBRI experimental program: 18 periphery fuel rods were oscillated at a maximum of ±2 mm around their central position in the Hz range. The present article documents the experimental setups and measurements for each facility and perturbation type.**

Keywords—**Core monitoring and diagnostics, noise analysis, research reactor experiment, zero-power reactor**

I. Introduction

THE Horizon 2020 European project CORTEX aims at developing innovative core monitoring techniques that allow detecting anomalies in nuclear reactors, such as excessive vibrations of core internals, flow blockage, or coolant inlet perturbations [1]. The technique will be mainly based on using the inherent fluctuations in neutron flux recorded by in-core and ex-core instrumentation, from which the anomalies will be differentiated depending on their type, location and characteristics. The method is non-intrusive and does not require any external perturbation of the system. The project will result in a deepened understanding of the physical processes involved. This will allow utilities to detect operational problems at a very early stage, and to take proper actions before such problems have any adverse effect on plant safety and reliability.

In this framework, computer methods and models are developed. The purpose of the experimental campaigns at the AKR-2 and CROCUS reactors is to produce high quality noise-specific experimental data for the validation of these neutron noise computational models. The AKR-2 reactor at Technische Universität Dresden (TUD) is a thermal, homogeneous, polyethylene moderated zero-power reactor with an allowed maximum continuous power of 2 W. The CROCUS reactor at the École Polytechnique Fédérale de Lausanne (EPFL) is a two-zone, uranium-fueled, and light water-moderated facility. With a maximum power of 100 W, it is a zero-power reactor used for teaching and research purposes, currently with a focus on studies on intrinsic and induced neutron noise [2]–[4].

The first campaigns at both facilities consisted in measurements at reference static states, and with the addition of mechanical perturbations. In the AKR-2 reactor, perturbations were induced by two devices: a rotating absorber and a vibrating absorber, both sets in experimental channels close to the core. In CROCUS, the project benefited from the COLIBRI experimental program: up to 18 periphery fuel rods are oscillated at a maximum of ±2.5 mm around their central position in the Hz range. The present article documents the experimental setups and measurement campaign for each facility and perturbation type.

The research leading to the results hereby documented has received funding from the Euratom research and training programme 2014-2018 under grant agreement No 754316 (i.e., CORTEX).

*V. Lamirand, A. Rais, O. Pakari, P. Frajtag, D. Godat, M. Hursin, A. Laureau, C. Fiorina, and A. Pautz are with the Laboratory for Reactor Physics and Systems behaviour (LRS), Ecole Polytechnique Fédérale de Lausanne (EPFL), CH-1015 Lausanne, Switzerland (*corresponding author: vincent.lamirand@epfl.ch).

S. Hübner and C. Lange are with the Institute of Power Engineering, Technische Universität Dresden, 01062 Dresden, Germany.

J. Pohlus and U. Paquee are with the Institut fur Sicherheitstechnologie GmbH (ISTec), 85748 Garching, Germany.

C. Pohl is with TÜV Rheinland Industrie Service GmbH (TUV), 51105 Cologne, Germany.

V. Lamirand, M. Hursin, G. Perret, and A. Pautz are with the Nuclear Energy and Safety Research Division (NES), Paul Scherrer Institut (PSI), CH-5232 Villigen, Switzerland.

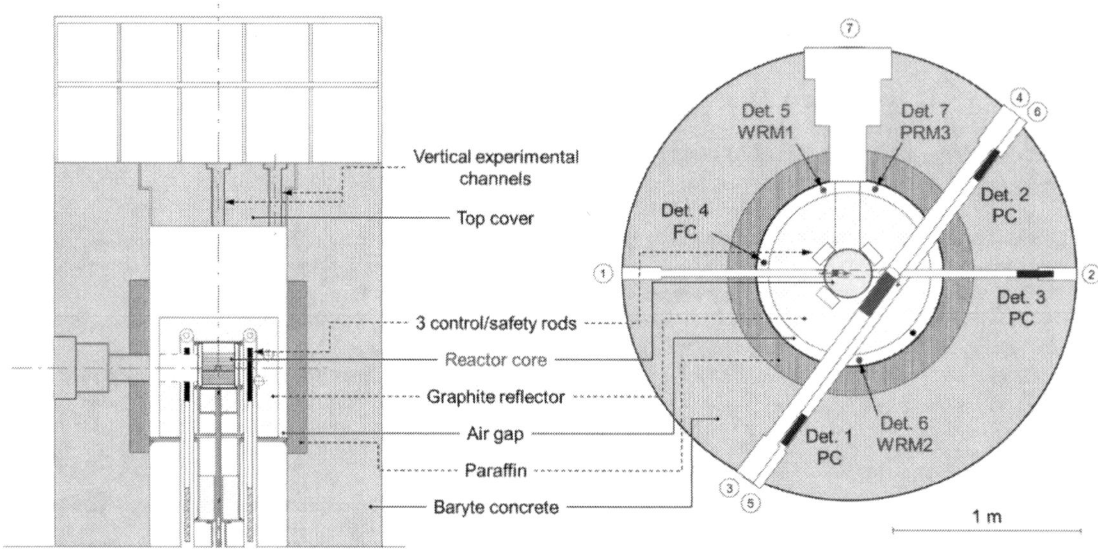

Fig. 1. The AKR-2 reactor: on the left, side cross section; on the right, cross section showing components, experimental channels (numbers), perturbation devices (green), and neutron detection instrumentation (blue and red).

II. First experimental campaign in AKR-2

The first experimental campaign in the AKR-2 reactor took place from 6 to 15 March 2018. Two types of experiments were carried out to induce neutron flux oscillations: rotation of a thermal neutron absorber (of variable absorption with respect to the rotation angle) and linear moving of an absorber oscillating back and forth in the reactor core (a.k.a. vibrating absorber).

A. The AKR-2 reactor

The AKR-2 reactor of TU Dresden (TUD) is a thermal, homogeneous, polyethylene moderated zero-power reactor with an allowed maximum continuous thermal power of 2 W [5]. The core has cylindrical shape with a diameter of 250 mm and a critical height of 275 mm. The disk-shaped fuel elements consist of a homogeneous dispersion of polyethylene and uranium oxide –19.8 % enriched in ^{235}U, with and O-to-U ratio of 2.27. The ^{235}U density in the fuel elements amounts to 0.060 g/cm^3. The core is surrounded by a graphite reflector with a density of 1.75 g/cm3 (see Figure 1). The axial and radial thicknesses are 20 and 32 cm, respectively. Therefore, the critical mass is relatively small (about 790 g of ^{235}U). Within certain constraints, it is a minimum-critical-mass reactor.

For safety reasons, the core consists of two separable sections. The fuel elements of each section are enclosed in a hermetically sealed aluminum container. In the shut-down position the distance between the lower and the upper core section is about 50 mm. The lower section is lifted by means of a core drive mechanism including an electromagnetic holder of the core section. Through a tube within this mechanism the start-up neutron source (Am-Be, neutron yield $2.2\cdot10^6$ n/s) is moved from the source container to the bottom side of the core. The distance between the bottom side of the core and the middle of the source container measures about 800 mm.

The reactor is controlled by three cadmium absorber plates. These plates are moved vertically in a gap within the reflector outside the reactor tank. They are designed as combined control and safety rods. An automatic reactor power control is available using the control rods. The lower core section and the control rods are held by electromagnets in their normal-operation positions. Any scram signal opens the direct-pull holding magnets allowing the rods and the lower core section to fall in the shut-down position by gravity.

The reactor core, reactor tank, graphite reflector and control rods are arranged on a horizontal steel plate (diameter 1000 mm, thickness 30 mm) inside the reactor vessel. This is a cylindrical double-walled vessel containing the biological shield and having an outer diameter of 2500 mm and a height of 2800 mm.

There are four horizontal and two vertical experimental channels with different diameters and shapes. They provide adequate in-pile irradiation volume with different neutron spectra. At 2 W power level, the maximum flux density of thermal neutrons in the central experimental channel 1-2 (see Figure 1) is about $2.5\cdot10^7$ cm$^{-2}\cdot$s^{-1}.

B. Neutron detection instrumentation

The nuclear instrumentation used for the first campaign of measurements at the AKR-2 reactor is comprised by seven detection channels from which the neutron flux level in the reactor is inferred, their locations are shown in Figure 1 (right):

- Three ^{3}He proportional counters (PC)
- One ^{235}U fission chamber (FC)
- Two logarithmic wide-range monitors (WRM1 and 2), ^{235}U fission chambers
- One linear power-range monitor (PRM3), gamma-compensated ionization chamber

It is worth noting that the three latter channels comprise the reactor instrumentation for safety and reactor monitoring purposes. These detectors are shown in red color in Figure 1.

Three independent data acquisition systems (DAQ) were provided by TUD, EPFL and ISTec and used during the measurements for signal readout and storage. The complete neutron detection sequence with the associated electronics is available in [6].

C. Experiments

The experiments at the AKR-2 reactor consisted in introducing two different sources of periodic reactivity perturbations that induce neutron flux oscillations. The first type of perturbation is generated by a rotating neutron absorber that has a varying absorption cross-section with respect to the rotation angle, and the second one by a linearly oscillating absorber that is moved back and forth inside the reactor core (vibrating absorber). In addition to the seven neutron detection signals, three other signals were measured and recorded:

- linearly moving absorber position as a function of time,
- inductive proximity sensor's signal to detect the rotation,
- external TTL signal for synchronization purposes.

1) Rotating absorber experiments

In order to maximize the perturbation induced to the neutron flux, the rotating absorber shaft was inserted in a location where the gradient of the neutron flux is maximum: i.e., the tangential line (channel 3-4). The reactivity worth of the absorber was measured and is presented in Appendix 7.6.1.

The geometry description and mechanical arrangement of the rotating absorber are summarized below:

- absorber material: natural cadmium (with impurities),
- dimensions of cadmium sheet (cm): 25 × 2 × 0.02,
- rotation axis location: tangential channel 3-4,
- rotation axis distance with respect to core center: 22 cm in the radial direction and 8 cm above in height,
- Cd sheet distance with respect to rotation axis: 2.98 cm.

The experimental arrangement is illustrated in Figure 2, presenting two section views of the rotating absorber with geometrical details. The static reactivity worth of the absorber was measured and is about 11 ct in total. The rotation speed was kept constant while measuring. The experiments were carried out at a frequency range from 0.2 to 2 Hz. The rotation was measured by an inductive proximity sensor, which yielded a pulsed signal per revolution.

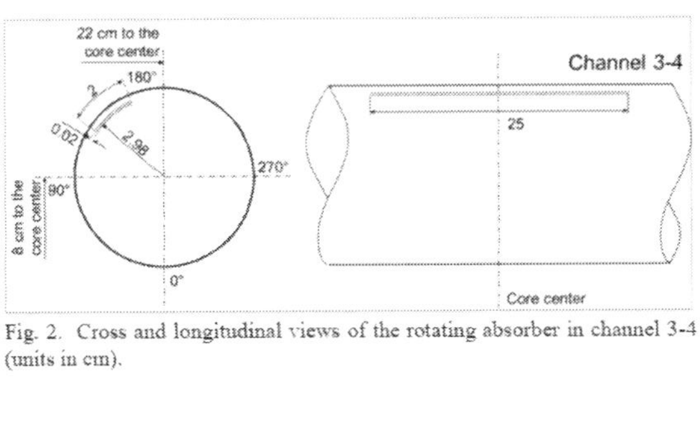

Fig. 2. Cross and longitudinal views of the rotating absorber in channel 3-4 (units in cm).

2) Vibrating absorber experiments

A pneumatic-driven vibrating absorber was employed to linearly oscillate a neutron absorber back and forth inside the reactor core. The absorber is moved along the experimental channel 1-2, between a position inside the core and a position outside the core as illustrated in Figures 1 and 3. The static reactivity worth of the absorber was measured at 13 ct in total. The material and geometry data for the absorber are summarized below:

- linear motion axis: radial channel 1-2 at core mid-height,
- absorber thickness: 1.016 mm,
- absorber diameter: 12.7 mm,
- absorber material: 99.97% natural Cd plus impurities.

The periodic motion of the vibrating absorber is defined by two main parameters: the waiting and moving time as shown in Figure 3. The pneumatic system allows to modify the absorbers motion by varying the moving time (t_m) from 0.4 to 2.0 seconds, and/or the waiting time (t_w) from 0.1 to 4.0 seconds. By tuning these parameters, the experiments were carried out at a frequency range from 0.08 to 0.71 Hz. The relative position of the vibrating absorber was recorded along.

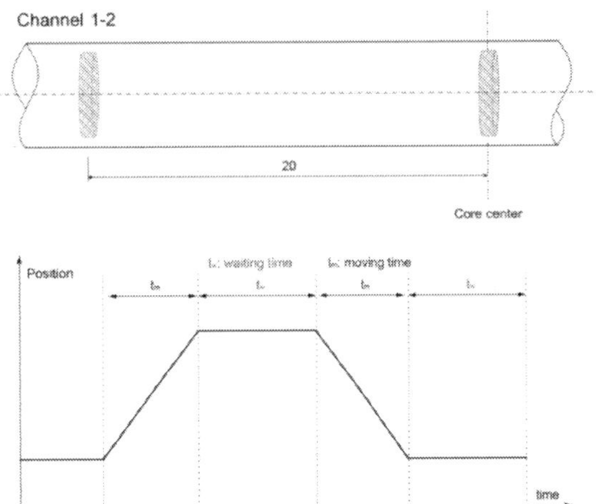

Fig. 3. On the top, side view of the absorber motion path in channel 1-2 (units in cm). On the bottom, trapezoidal motion followed by the absorber in the experiments.

III. First experimental campaign in CROCUS

The first experimental campaign in the CROCUS reactor took place from 17 to 21 September 2018. Vibrating fuel rods experiments were carried out using an in-core device that was specifically developed for that purpose [7], [8]. This section describes the CROCUS reactor, the experimental setup, and the vibrating fuel rods experiments.

A. The CROCUS reactor

CROCUS at the École Polytechnique Fédérale de Lausanne (EPFL) is an experimental zero-power reactor, uranium-fueled and light water-moderated, dedicated to teaching radiation and reactor physics, and to research. A complete description of the

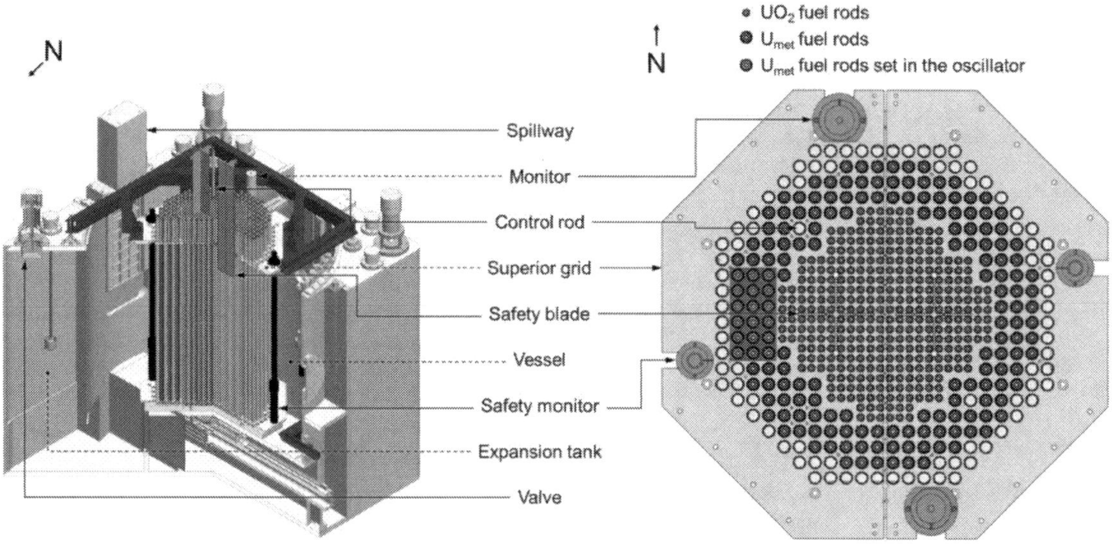

Fig. 4. The CROCUS reactor: isometric view of the vessel (left), and top view of the core superior grid and configuration, with the indication of the location of the fuel rods set in the oscillator (green).

core can be found in the International Reactor Physics Experiments Handbook (IRPhE) [9], [10]. It has been licensed for operating at a maximum power of 100 W, i.e. a total neutron flux of ~$2.5\cdot10^9$ $cm^{-2}\cdot s^{-1}$ at the core center. Criticality is controlled either by water level using a spillway, or by two B_4C absorber control rods, with an accuracy of ±0.1 mm (equivalent to approximately ±0.4 pcm) and ±0.5 mm (±0.2 pcm), respectively. CROCUS operates at room temperature using a controlled water loop with secondary and tertiary circuits, two heat exchangers and an electrical heater.

The core is located in an Al-6060 grade vessel of 130 cm in diameter, 160 cm in height, and 1.2 cm in thickness. The vessel is filled with demineralized light water used as both moderator and reflector. The core active part has the approximate shape of a cylinder of 100 cm m in height and about 60 cm in diameter. It consists of two interlocked fuel zones with square lattices of different pitches: an inner zone of 336 UO_2 rods with an enrichment of 1.806 wt.% and a pitch of 1.837 cm; an outer zone of 176 U_{metal} rods for these experiments, 0.947 wt.% and 2.917 cm; a varying water gap between the two zones because of the two different pitches. The picture of the reactor and critical assembly configuration is shown on Figure 4.

Both uranium fuels consist of a 1-m pile of cylindrical pellets cladded in aluminum. The rods are maintained vertically by two octagonal aluminum grid plates spaced 1 m apart. In the COLIBRI program, the grids have a 1 mm cadmium layer to limit axial neutron leakage to the environment, i.e. structures activation, with the active zone of the fuel starting in the middle of the lower cadmium layer.

B. The COLIBRI experimental setup

The COLIBRI fuel rods oscillator is designed to simultaneously oscillate any of 18 metallic uranium fuel rods laterally in the west region of the core periphery zone. It consists of two moving plates set above and below the core grids, and rigidly connected by an aluminum beam (see Figure 5). Each plate carries an extremity of the fuel rods, top and bottom respectively. The top moving plate is fixed on the superior grid *via* gliders. Its oscillation is produced by a motor: the motor rotation is converted to a linear translation using an eccentric sheave and a rod. The oscillation is transferred to the bottom moving plate *via* the aluminum beam. The bottom moving plate is not constrained by gliders, it is displaced only because of its connection to the transmission beam.

The selection of the moving fuel rods is performed by letting the rods lay on the reactor base plate (non-moving), or suspend them up 10 mm above the base plate to insert them in the moving plates. Top and bottom end caps are fixed to each rod to allow the insertion in the enlarged holes of either the static grids or the moving plates. The weight of the oscillating rods is supported by a platform. The amplitude of the oscillation is precisely tuned by changing the eccentricity of the sheave with calibration plates, 0.5 mm by 0.5 mm from 0 to ±2.5 mm. Its frequency is depending on the speed of the motor.

The oscillation is controlled and monitored *via* a LabVIEW-developed software with 10 ms time-steps. An inductive captor is set at the rotation axis (i.e. at the top), which detects the actual movement of the motor by detecting the passage of four metallic pins per rotation. A cable coder is used to measure the displacement of the moving plate, i.e. at the bottom, with a

0.1 mm precision. The software produces a csv file output with the recordings of the motor position and speed, the signal of the inductive captor, and the position measurement of the cable. A typical recording of an oscillation is presented in Figure 6. The inductive captor signal is also extracted for live and synchronized recording with the detection instrumentation.

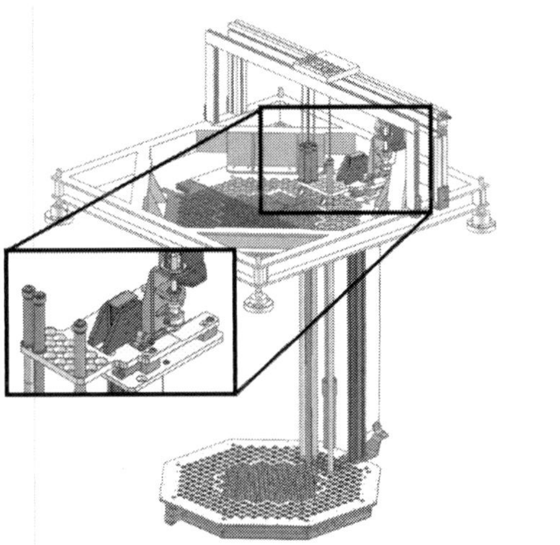

Fig. 5. The fuel rods oscillator with core structures and a few rods inserted in the device (right).

The behavior of the oscillator has been characterized in air and in water, out of the vessel and in-core, empty, 1-rod and 18-rods loaded [11]. The behavior in frequency is sound. In amplitude, the device is rigid at the top, depending only on the rotation to translation conversion, and limited plays associated. At the bottom, it demonstrates inertia effects inducing an increase of the amplitude for the full 18-rods load case and above 1 Hz. This effect cannot be corrected, and thus has to be taken into account as is. The static reactivity worth of the 18 rods was measured at about 10 ct in total for a maximal amplitude of ±2.5 mm.

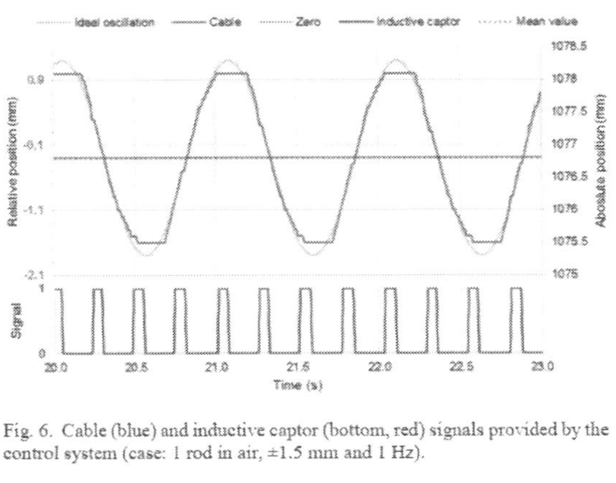

Fig. 6. Cable (blue) and inductive captor (bottom, red) signals provided by the control system (case: 1 rod in air, ±1.5 mm and 1 Hz).

C. Neutron detection instrumentation

The detection instrumentation comprised the facility monitors, and different types of additional ex-core and in-core detectors, for a total of eleven detectors of various sizes and sensitivities:

- two operation monitors ^{10}B-coated compensated ionization chambers (CIC): Merlin-Gerin CC54,
- two safety monitors ^{235}U-coated fission chambers (FC): Photonis CFUM21,
- two ex-core large ^{235}U-coated fission chambers: Photonis CFUL01,
- two in-core small BF_3 proportional counters (PC): Transcommerce International MN-1,
- two in-core smaller BF_3 proportional counters
- one in-core ^{235}U-coated miniature fission chamber (MFC): Photonis CFUF34.

All detectors are based on prompt detection processes. The location of each of them with regards to the core and COLIBRI are presented in Figure 7. All four monitor detectors are set at reference positions, as presented in the CROCUS benchmark. The other detectors were set vertically at core mid-height. On the horizontal plane, the large CFUL01 fission chambers were set in the reflector at west (close to COLIBRI) and east positions. The BF_3 counters were set at positions within the lattice as presented in Figure 7, in the control rod guide tube for the NW one (detector 3), and in aluminum channels for the others. The miniature fission chamber was set at core center. The general uncertainty on detector position is ±1 mm.

The operation ionization chambers and the large fission chambers were operated in current mode [2], whereas all the other detectors were operated in pulse mode using spectrometry charge preamplifiers. The detectors and their electronics were connected to the ISTec, TUD, and both EPFL pulse and current mode instrumentations, with a number of detectors depending on the number of possible inputs [6]. In addition, the safety monitors were used in MCS mode for power monitoring.

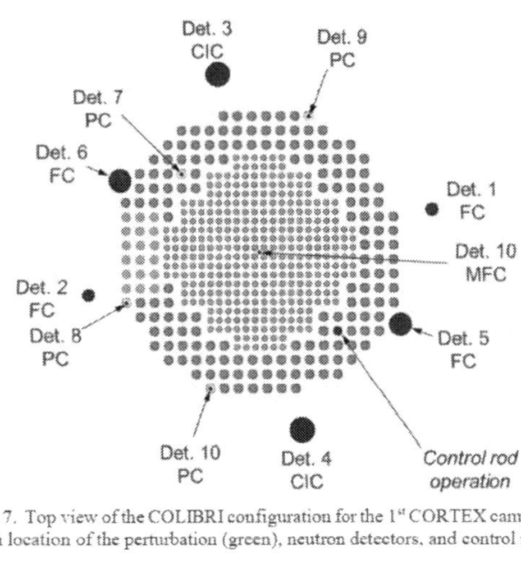

Fig. 7. Top view of the COLIBRI configuration for the 1st CORTEX campaign with location of the perturbation (green), neutron detectors, and control rod.

D. *Experiments*

Two types of measurements were carried out: static, and a range of measurements in oscillation. The reactor was operated with the south control rod (see Figure 7), at a fixed water level of 1000 ± 0.1 mm. Following linearity tests, the power was set around 100 mW, and the water temperature was controlled at 20.0 ± 0.1 °C. In all cases, the 18 rods were lifted in the oscillation position: in the static case, the motor was off and the position was set at zero, i.e. in their nominal position within the lattice. The global reactivity effect of the oscillations was compensated to follow a stable power. In practice, the oscillation was started first, then the reactor was stabilized in power with the control rod before starting the measurement. The control rod insertion was updated along the operation and measurement.

The experiments are listed in Table I. 20 measurements were carried out in total, with limited repetition. All measurements were acquired by all four systems. The duration of each measurement was chosen to cover enough cycles of the perturbation oscillation, around 1000 except for 0.1 Hz, with a minimum time of 30 min for resolving the zero power noise. Amplitudes of limited waves at the water surface induced by oscillations were measured using the reactor instrumentation (INUS ultrasonic sensor). The oscillation is listed here in requested displacement amplitude and frequency, with a difference between requested and measured ones due to the system limitations [11]. In addition to the neutron detection signals, three other outputs were recorded:

- the synchronized signal from the inductive captor,
- the oscillator's structured output, including motor, inductive captor, and cable,
- the two power monitors outputs.

TABLE I
LIST OF OSCILLATION EXPERIMENTS IN CROCUS

Amplitude (mm)	Measurement duration for each frequency (min)				
	0.1 Hz	0.5 Hz	1.0 Hz	1.5 Hz	2.0 Hz
±0.5	60	30	30	-	-
±1.0	60	30	30	30	30
±1.5	60*	30	120*	30	30
±2.0	60	30	30	-	-

* with repetition.

IV. CONCLUSION AND OUTLOOK

The first CORTEX experimental campaigns in the AKR-2 and CROCUS reactors were successfully carried out in 2018. The focus was on the acquisition of reference data set to qualify the acquisition systems, start the validation of the modelling tools developed within the Consortium, and orient the next campaigns. The analysis of the experiments is on-going for their comparison with the simulation results. The next campaigns are planned in 2019 and 2020 in CROCUS and AKR-2, respectively.

ACKNOWLEDGMENT

The authors would like to thank the operation teams at both reactor facilities for their help and dedication to the experiments.

REFERENCES

[1] C. Demazière, P. Vinai, M. Hursin, S. Kollias, and J. Herb, "Overview of the CORTEX project," in *PHYSOR 2018: Reactor Physics Paving The Way Towards More Efficient Systems*, 2018, pp. 2971–2980.

[2] O. Pakari, V. Lamirand, G. Perret, P. Frajtag, and A. Pautz, "Kinetic Parameter Measurements in the CROCUS Reactor Using Current Mode Instrumentation," *IEEE Trans. Nucl. Sci.*, vol. 65, no. 9, pp. 2456–2460, 2018.

[3] V. Lamirand, G. de Izarra, A. Krasa, G. Perret, O. V. Pakari, M. Hursin, P. Blaise, J. Wagemans, and A. Pautz, "Intercomparison of neutron noise measurement systems in the CROCUS reactor," in *Proceedings of PHYSOR 2018*, 2018, p. 11.

[4] M. Hursin, O. Pakari, G. Perret, P. Frajtag, V. Lamirand, I. Pazsit, V. Dykin, G. Por, H. Ferroukhi, and A. Pautz, "Measurement of the gas velocity in a water-air mixture in CROCUS by neutron noise technique," in *International Conference on Mathematics and Computational Methods applied to Nuclear Science and Engineering (M&C 2019)*, 2019.

[5] B. Boehmer, M. Grantz, W. Hansen, D. Hinke, J. Konheiser, H.-C. Mehner, K. Noack, R. Schneider, I. Stephan, and S. Unholzer, "Investigation of Gamma and Neutron Energy Fluences in Iron - Water Benchmark Configurations for the Verification of Monte Carlo Calculations and their Application in Reactor Material Dosimetry," *J. Nucl. Sci. Technol.*, vol. 39, no. sup2, pp. 947–950, 2002.

[6] A. Rais, V. Lamirand, O. Pakari, A. Laureau, J. Pohlus, C. Pohl, M. Hursin, C. Demazière, and A. Pautz, "Towards the validation of neutron noise simulators: qualification of data acquisition systems," in *International Conference on Mathematics and Computational Methods applied to Nuclear Science and Engineering (M&C 2019)*, 2019.

[7] V. Lamirand, M. Hursin, G. Perret, P. Frajtag, O. Pakari, and A. Pautz, "Future experimental programmes in the CROCUS reactor," in *RRFM/IGORR 2016*, 2016, no. 02–2016, pp. 284–292.

[8] V. Lamirand, P. Frajtag, D. Godat, M. Hursin, G. Perret, O. V. Pakari, A. Laureau, A. Rais, C. Fiorina, M. Chadwick, and A. Pautz, "The COLIBRI programme in CROCUS: development and licensing of a fuel rods oscillator," in *RRFM/IGORR 2019*, 2019.

[9] U. Kasemeyer, R. Früh, J. M. Paratte, and R. Chawla, "Benchmark on Kinetic Parameters in the CROCUS Reactor," in *International Reactor Physics Experiments Handbook (IRPhE)*, no. 4440, OECD, Ed. 2007, p. 94.

[10] J. M. Paratte, R. Früh, U. Kasemeyer, M. a. Kalugin, W. Timm, and R. Chawla, "A benchmark on the calculation of kinetic parameters based on reactivity effect experiments in the CROCUS reactor," *Ann. Nucl. Energy*, vol. 33, no. 8, pp. 739–748, May 2006.

[11] V. Lamirand, P. Frajtag, D. Godat, M. Hursin, G. Perret, O. Pakari, A. Rais, C. Fiorina, and A. Pautz, "The COLIBRI programme in CROCUS : characterisation of the fuel rods oscillator," in *ANIMMA 2019*, 2019.

 https://doi.org/10.1051/epjconf/202022504024

Usage of multiple fission cells for neutron flux measurements during rod-insertion method

Igor Lengar*, Sebastjan Rupnik, Andrej Žohar, Vid Merljak, Marjan Kromar, Luka Snoj
Jožef Stefan Institute, Ljubljana, Slovenia
* igor.lengar@ijs.si

Abstract— **The measurements of physical parameters of the TRIGA reactor and Nuclear power plant Krško (NEK) reactor cores have been in the past performed on hand of the neutron flux signal obtained from uncompensated ionization cells and by employment of the a digital meter of reactivity (DMR). At the TRIGA reactor only one ionization cell is currently used for flux measurements. During the insertion of one control rod the neutron flux distribution is significantly altered affecting the flux measurements of inserting different control rods. The problem is presently solved by assigning a correction factor to each control rod what introduces an additional uncertainty.**

In the present paper the implementation of four fission cells for reactivity measurements is presented. In this way determining the correct gamma background and its subtraction, performed by DMR algorithms, becomes less important as previously by using ionization chambers. The larger number of detectors also reduces the flux redistribution effects on the signal during individual control rod movements.

Keywords—**Fission cells, control rod worth measurements, research reactor, combining signals**

I. INTRODUCTION

THE control rod worth in research reactors and power plants can be determined by different methods [1 - 7]. In the present work the rod-insertion method, by employing measurement of the neutron flux with fission cells, is studied. This method is particularly convenient because it is very quick and simple to perform. The principle of the rod-insertion method is to start from a critical reactor operating at low power and to measure the time-dependent reactivity change while a control rod is inserted into the core with the drive mechanism at normal speed. By analyzing the flux trace, not only the total rod worth but also the differential and the integral control rod worth curves are obtained.

During the rod-insertion measurement the flux may drop by several orders of magnitude. The analysis is performed by transferring the data of the neutron flux from nuclear instrumentation to a digital reactivity meter (DMR) consisting of a high-quality electrometer and a computer using special software for analysis of the signal. The DMR software uses point kinetics equations for calculation of the reactivity from the flux trace [3, 7]. The procedure, which will not be repeated here in detail, involves several stages from smoothing of the signal to using a Fourier transform for fitting the reactivity trace during the rod-insertion. The final result are the integral and differential worth curves of the inserted control rod; an example is presented in Figure 3.

During the rod-insertion measurement the flux is, however, redistributed and a flux depression forms at the position of the inserted control rod. Due to this re-distribution the signal at the detector is not necessarily proportional to the average flux in the core. The effect can be compensated by the usage of a neutron flux depression factor, what has been done in the past. Another possibility is to use multiple detectors, symmetrically positioned around the core. Most of the available studies of the described method are performed for NPPs, usually having two sets of detectors, at the top and bottom of the core and with the absorbing control rod clusters distributed around the core [5-7]. The present study was performed on a small research reactor with one set of detectors at the equatorial plane and four localized control rods. The possibility to what extent the axial dependence of the flux depression factor can be omitted is performed, and comparison between the method of using a single detector and the average of four detectors is performed in the scope of this work.

II. REDISTRIBUTION OF NEUTRON FLUX

Control rods are large neutron absorbers and have a large impact on the neutron flux spatial distribution. In the TRIGA research reactor of the "Jožef Stefan" Institute four control rods are employed [3]. The particular core configuration, on which measurements in the scope of this work were performed, is presented in Figure 1.

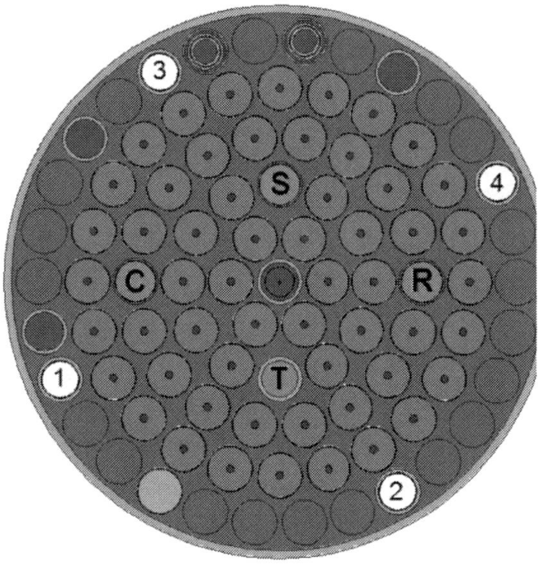

Figure 1: TRIGA core configuration. The control rods are labeled as (T) transient, (R) regulating, (C) shim, (S) safety. The ionization chamber is located behind the reflector (both are not visible in the figure) on the left side of the core. The Fission cells are numbered.

The DMR software is based on point-kinetic equations and for this reason a signal, proportional to the core average flux is needed as the input. Since the integral neutron flux in the core $\Phi(t)$ [4] cannot be measured directly, we generally measure the neutron flux $\Phi(\boldsymbol{r},t)$ at one or more points $\boldsymbol{r}$ outside the core. The first approximation is to assume that this signal is proportional to the integral of the flux in the core. This assumption is acceptable if the relative change between $\Phi(t)$ and $\Phi(\boldsymbol{r},t)$ is small during the measurement. Otherwise, a correction due to the flux redistribution is necessary.

The correction depends on the positions of the control rods and of the detector. An ionization chamber neutron detector measures essentially the flux of neutrons thermalized in the vicinity of the detector. The relative thermal flux for a core in which a control rod in the vicinity of the detector is inserted is much lower at the detector location and correspondingly higher at a location far from the inserted rod and the detector, compared to the unrodded core. Such a flux redistribution, with a visible flux depression corresponding to the insertion of a control rod (the compensating rod (C), is graphically presented in Figure 2. It can be observed from the figure that fission cell locations 1 and 3, in the vicinity of the control rod, read a lower signal than fission cells 2 and 4 on the opposite side. All of the readings deviate from the signal, which would be proportional to the core integral flux.

This flux redistribution during the insertion of a control rod can be compensated with a numerically determined flux depression factor F_Y; it represents the required correction of the neutron flux radial redistribution for rod Y [4]. There are several limitations for the usage of F_Y, among others it has to be calculated for each combination of control rods and ionization chamber positions.

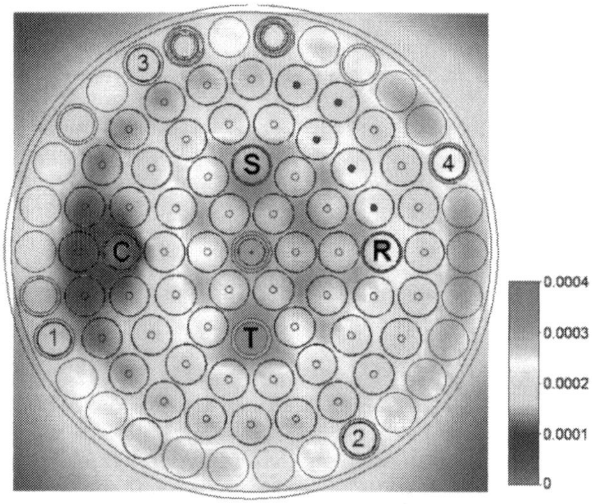

Figure 2. Graphical presentation of the thermal neutron flux (E_n < 1eV) distribution (normalized to n/cm^2/fission source neutron) for the case of the full insertion of the compensating control rod (C); calculation with MCNP6.1.

III. Using signal from multiple fission cells

As outlined in the previous section one of the drawbacks of using a signal from one ionization chamber is the necessary calculation of the flux depression factors, which can be dependent also on core configuration. For this reason, in the frame of this work, the rod insertion measurements were performed by using four identical fission cells instead, each with 165 mg of ^{235}U fission material. They were located symmetrically around the core (numbered in Figure 1). The fission cells were operated individually and the signal recording was synchronized and averaged by the computer.

The differential values of the control rods were measured and calculated with the rod-insertion method by using signals from both described sources – a) signal from one ionization chamber or b) signal from four fission cells, located symmetrically around the core. Both results were compared in order to evaluate the modified technique.

The differential control rod curves for the compensating control rod (C), measured and calculated by the rod insertion method for the four individual fission cells, are presented in Figure 3. On the figure the curve by using the combined signal from all fission cells is also presented. It should be noted that the curve, derived from the average over all signals is not equal to the average of the four curves (i.e. the curve *average* in Figure 3 is not equal to the average over the curves FC1 – FC4); their calculation is subject to the complex nature of the rod-insertion method.

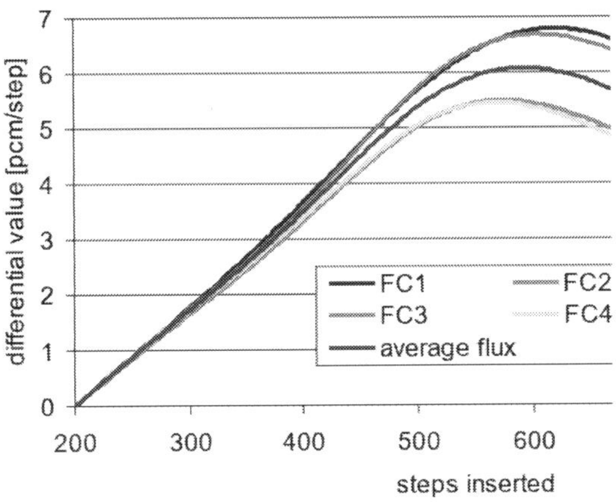

Figure 3: The differential control rod curves for the compensating control rod (C), measured and calculated by the rod-insertion method for the four individual fission cells, and the curve calculated from the combined (average) signal from all fission cells.

It is observed that usage of four symmetrically positioned fission cells reduces the need for the flux depression factor employment. The data presented in Figure 3 are compared in Table 1.

Table 1: Integral values for the transient rod as obtained by using individual fission cells and the combined signal from all fission cells. The standard measurement with one ionization chamber and employment of the flux depression factor is given for comparison.

	integral value of rod (C)
Fission cell 1	2872
Fission cell 2	2367
Fission cell 3	2869
Fission cell 4	2362
combined (averaged) signal	2627
ionization chamber including flux depression factor	2707

The standard method for determination of a control rod worth is by using the ionization chamber, located behind the reflector and with the associated flux depression factor - for the compensating control rod (C) such value amounts to 2707 pcm. The comparison with the control rod worth obtained from the average signal shows a deviation of 3% and is smaller than deviations due to measurements with any single fission cell. It is also on the same order as the uncertainty of the method itself. Introducing multiple fission cells thus enables quick measurements of the integral and differential value without having to calculate the flux form factor, which is configuration specific.

As anticipated and observed also from Figure 3 and Table 1, the location of the fission cell with respect to the measured control rod influences the integral value: without the usage of the flux depression factor the two fission cells which are on the same side of the reactor as the measured compensating rod, experience an excessive large change in neutron flux and hence a too large value of the calculated integral value, whereas for the other two cells, located on the far side of the regulating rod (R), the situation is reversed.

IV. Measurement Background / Noise

In case of rod-insertion measurements with one uncompensated ionization chamber, which has been the practice up to now, the value of the gamma background is substantial; since the reactor becomes highly subcritical during the measurement the background is often a few times larger than the signal due to the neutron flux – hence the background noise and exact determination of the background also become important. In case of measurements with fission cells the gamma background is much less important due to their smaller sensitivity to gammas. Figure 4 displays a typical reactivity trace, calculated with DMR during rod-insertion for the two cases – for obtaining the signal from an ionization chamber or from fission cells.

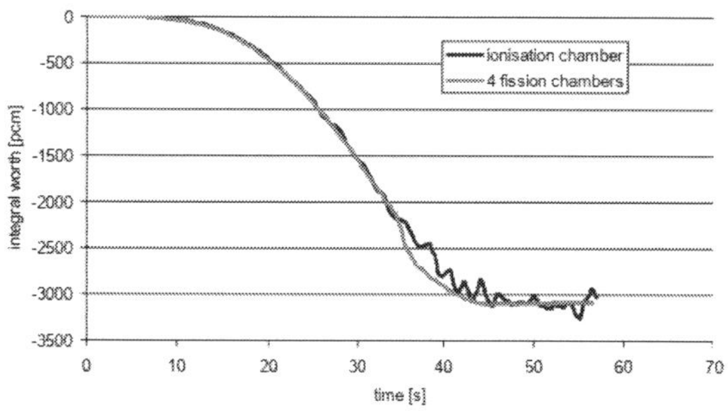

Figure 4: A typical calculation of the reactivity with DMR for the two cases of obtaining the signal from ionization chambers or fission cells. The larger noise in the former measurement is clearly visible.

As can be observed from Figure 4 the noise due to fluctuations in gamma background is greatly reduced by using the signal from fission cells and it is reflected in a smaller noise in the reactivity signal. A larger noise of the flux signal due to gammas is reflected in the lower accuracy of the rod-in method. Usage of fission instead of ionization chambers thus increases the overall accuracy of the method.

V. Conclusions

The rod insertion method for measuring integral and differential worth of control rods using a digital meter of reactivity was extended by collecting the signal from four fission cells rather than the usual signal source from a single uncompensated ionization chamber. The larger number of detectors was found to reduce the flux redistribution effects on the signal during individual control rod movements. Four fission cells, symmetrically positioned around the reactor core, were found to be a suitable alternative for control rod measurements using the rod insertion method even without calculation of flux redistribution factors. Additionally due to usage of fission rather than ionization chambers the correct gamma background determination and subtraction, performed by DMR algorithms, became less important.

References

[1] J. Shaw, *Reactor Operation.* Pergamon Press, New York (1969).
[2] V. Merljak, I. Lengar, A. Trkov, Proc. Int. Conf. *Comparison of Measured and CalculatedIntegral and Differential Reactivity Worth of Control Rods in a TRIGA Reactor*, Proceedings of the RRFM, 30 March - 3 April, Ljubljana, Slovenia (2014).
[3] A. Trkov, M. Ravnik, H. Wirnmer, B. Glumac, H. Böck, H, "Application of the rod-insertion method for control rod worth measurements in research reactors", *Kerntechnik* **60**, pp 255–261 (1995).
[4] I. Lengar, A. Trkov, M. Kromar and L. Snoj, "Digital meter of reactivity for use during zero-power physics tests at the Krško NPP", *Journal Of Energy Technology - JET*, **vol. 5**, pp. 13-26 (2012).
[5] E.K. Lee, H.C. Shin, S.M. Bae, Y.K. Lee, "New dynamic method to measure rod worths in zero power physics test at PWR startup", Annals of Nuclear Energy **32** pp. 1457–1475 (2005).
[6] Chao, Y.A. et al, "Dynamic rod worth measurement", Nucl. Tech. **132**, p. 403(2000).
[7] B.Glumac, G. Skraba, "Rod Insertion Method for Rod Worth Measurement", Proc. Technical Committee, IAEA/TC-650, p. 280, International Atomic Energy Agency (1989).
[8] J.J. Duderstadt, L.J. Hamilton, *Nuclear Reactor Analysis*, John Wiley & Sons, (1976).
[9] R. Jeraj, M. Ravnik, "TRIGA Mark II reactor: U(20) - Zirconium Hydride fuel rods in water with graphite reflector", *IEU-COMP-THERM-003, Nuclear Energy Agency, NEA/NSC/DOC(95)03, Paris* (1999).
[10] L. Snoj, M. Ravnik, *Calculation of Power Density with MCNP in TRIGA Reactor*, Proc. Int. Conf. Nuclear Energy for New Europe 2006, Portorož, Slovenia, September 18-2, p.p. 109.1-6 (2006).
[11] V. Merljak, A. Trkov, M. Kromar, L. Snoj, *Rod insertion method analysis – dynamic vs. static reactivity*, Proc. this conference
[12] P. Blaise, F. Mellier, P. Fougeras, "Modified Source Multiplication (MSM) technique", IEEE Transactions On Nuclear Science, **58**, No.3 (2011).

Utilization of a zero power reactor for examination of operator influence on reactor control

Tomas Bily
Dept. of Nuclear Reactors, FNSPE
Czech Technical University in Prague
Prague, Czech Republic
tomas.bily@fjfi.cvut.cz

Alice Darbyshire
Nuclear Faculty
Defence Academy of the United Kingdom
Gosport, United Kingdom
alice.m.darbyshire@gmail.com

Ondrej Huml
Dept. of Nuclear Reactors, FNSPE
Czech Technical University in Prague
Prague, Czech Republic
ondrej.huml@fjfi.cvut.cz

***Abstract*— The reactor operator is an inseparable part of reactor safety and his actions influence the course of the reactor transients. Understanding the operator's effect on reactor control thus contributes to the knowledge of the safe operation of nuclear reactors. Complementary to studies performed on simulators, or utilizing standard reactor or plant operational data, the dedicated experiments at research reactors can help to improve the insight into human aspects of reactor control. The potential of research reactors to study the human operator aspects of reactor control is broad; the related activities thus may also help to increase the utilization of research reactor facilities.**

The paper summarizes the considerations that had to be assessed during the implementation of such experiments at the VR-1 research reactor from the viewpoint of the reactor facility. Further, it recapitulates the activities that have been performed in this field at the VR-1 reactor in the collaboration of the Czech Technical University in Prague with the Defence Academy of the United Kingdom in the previous years.

***Keywords*— operator effect, reactor control, reactor dynamics, zero power reactor, control engineering**

I. Introduction

Analyses of nuclear reactor transients are usually performed solely by codes describing the time dependent behaviour of a nuclear reactor as a physical system. However, in reality the course of a reactor transient is ruled not only by reactor physics, but also by a response of a reactor operator. Understanding the influence of the reactor operator on reactor control may thus improve our knowledge of safety cases, and may serve for operators' evaluation during their training and retraining as well. The potential fields of applications thus include, among others, improvements in operator training programs, including, quantitative evaluation of an individual operator's progress, understanding the operator-reactor system dynamic response, defining the manual controllability boundaries, and the evaluation of the human-machine interface, including effects of its modifications on the system controllability [1].

Although, historically they have not been commonly used for this purpose, zero power reactors seem to be the proper tool for the related research. This paper summarizes the related considerations, tools and approaches, and the gained experience from a reactor facility perspective.

II. VR-1 Reactor

The VR-1 is a zero power pool type reactor operated by the Czech Technical University in Prague. The reactor core consists of 16 to 21 fuel assemblies of IRT-4M type enriched to 19.7 % of U-235. The reactor has seven neutron absorbing rods, out of which two are used for power regulations, the remaining being the safety and shim rods. The reactor can be operated in manual or in automatic mode. In the manual mode, an operator can directly change the rod positions using dedicated buttons. In automatic mode, an operator controls the desired power and an appropriate rod movement is achieved by the control computer.

The installed human machine interface [2] enables the operator to observe the course of time dependent reactor operation parameters, such as reactor power or rod positions.

Fig. 1. Human-machine interface of the VR-1 reactor (before HMI desk upgrade in mid-2018)

Fig. 2. Control room of the VR-1 reactor (before HMI desk upgrade in mid-2018)

https://doi.org/10.1051/epjconf/202022504025

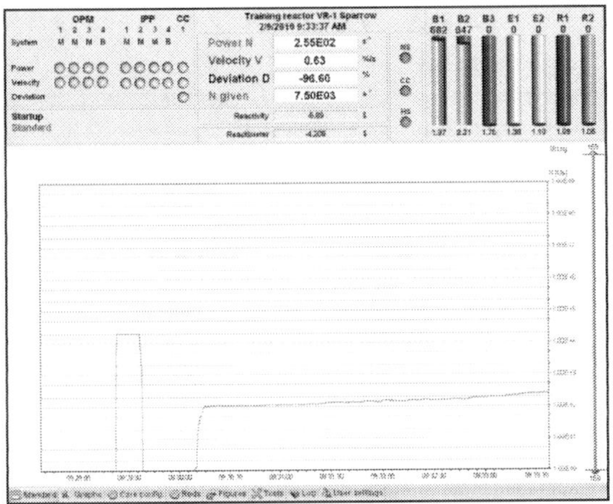

Fig. 3. Human machine interface of the VR-1 reactor

III. EXAMINATION OF OPERATOR INFLUENCE ON REACTOR CONTROL AT THE VR-1 REACTOR

Experiments aimed at operator aspects of reactor control have been discussed as a potential field of cooperation between the Defence Academy of the United Kingdom and the Czech Technical University in Prague since 2015, at which time the first experiments were performed. Since the beginning the experiments have been designed to utilize as much as possible of the available reactor infrastructure. The aspects which had to be considered in order to introduce the required type of operator examination included: core physics, human characteristics variations, input and output data acquisition, testing procedures, required equipment and the external conditions.

A. Core physics aspects

Studies of the operator influence on reactor control may be performed at any type of nuclear reactor. However, for initial experiments it was beneficial to run the experiments in the zero power operational region of the reactor (for the VR-1 reactor this covers practically the whole power range). At zero power the core heating and core temperature changes are negligible, thus internal reactivity feedbacks are negligible as well. Hence, the core physics can be described by a simple point kinetics model relating the reactivity and the power changes; and the operator effects can be well separated from the core physics effects.

Operator performance studies utilize various reactor transients. Thus their design needs to be optimized with respect to related reactor operation limitations. In the case of the VR-1 reactor the limitations were the allowable power rate and allowable value of reactivity insertion rate. The former parameter is limited by 4 % of actual power per second for warning signalization, and 6 % of actual power per second for scram signalization. For the latter, it is only allowed to introduce reactivity lower than 0.1 $ per second. On the other hand, the studied transients have to be high enough to enable efficient data analysis, and this can be achieved by manipulation of low reactivity worth control rods. In the case of the VR-1, the control rod with a reactivity worth around 1 $ was used for the experiments. The rod can be moved from 0 (bottom position) to 680 mm with a step size of 0.25 mm. For the tests in which external reactivity is used, the size and rate of its introduction to the core has to be assessed with respect to the above mentioned limitations.

B. Methods for operator effect examination, and experimental equipment

In principal, a zero power reactor enables two kinds of operator tests to be performed:

1) Reacting operator: the operator had to react in a prescribed way to a known, or unknown reactivity input and his response is being evaluated, or

2) Acting operator: the operator is asked to follow a prescribed input and the discrepancies between the expected and the real output are evaluated.

The first approach examines how an operator reacts to external reactivity inputs (i.e. external to those achievable by the reactor control system). To establish such kind of experiments a device that enables the production of reactivity (power) perturbations that are independent to reactor operator is needed. In the case of the VR-1 reactor experiments, the existing 'device for dynamics studies' [3] has been used. The device is based on a pneumatic system which enables fast movements of an inserted specimen (typically a piece of cadmium). The device can produce three types of movements: a) pulses, i.e. movements from the starting position to another defined position and then returning to initial position, b) steps, i.e. movements from the starting position to another defined position, and cycling, i.e. periodic movements between two defined positions, with an adjustable starting position. In all cases the speed of movement can be adjusted between 1 mm/s up to few hundreds mm/s. For operator influence examinations the range of introducible reactivity has been optimized by the proper choice of the sample reactivity worth and the initial sample position. The optimization was aimed at producing fast and well recognizable power changes by the sample movement, which remains reasonably below the power rate warning signalization of the reactor control system. The optimal sample reactivity worth was found about 0.06 $. The sample speed was set to 100 mm/s. The examined operator had to react to a set of external reactivity changes produced by the device for dynamics studies using standard operational means, i.e. using control rod movement.

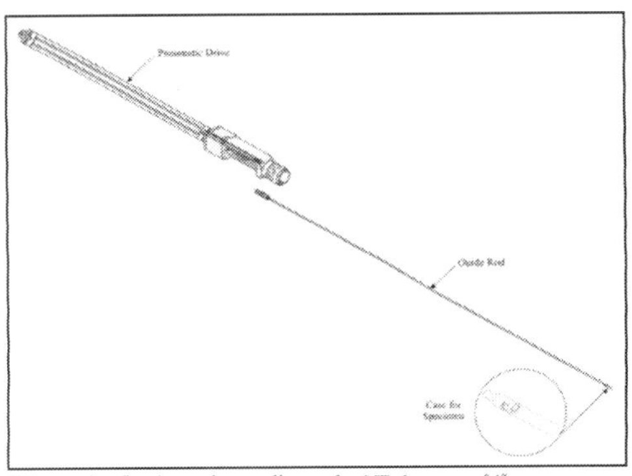
Fig. 4. Device for dynamics studies at the VR-1 reactor [4]

In the second approach, an operator has to use standard operational means, i.e. a control rod, to follow a prescribed

 https://doi.org/10.1051/epjconf/202022504025

input. This approach may require specific software which provides information about the desired operator actions, or the required operator actions may be given verbally, by way of an instruction or command to the operator (by a shift leader).

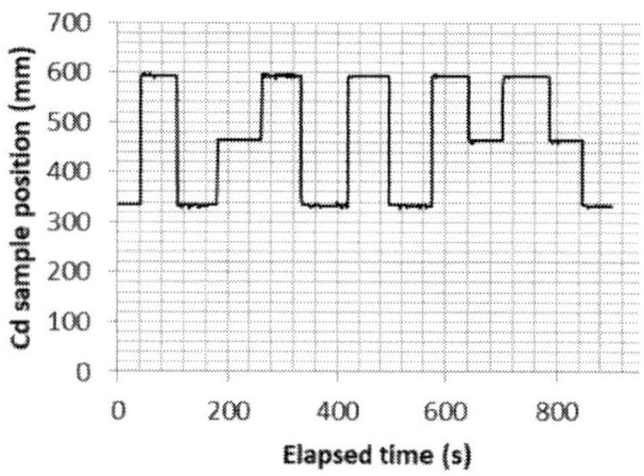

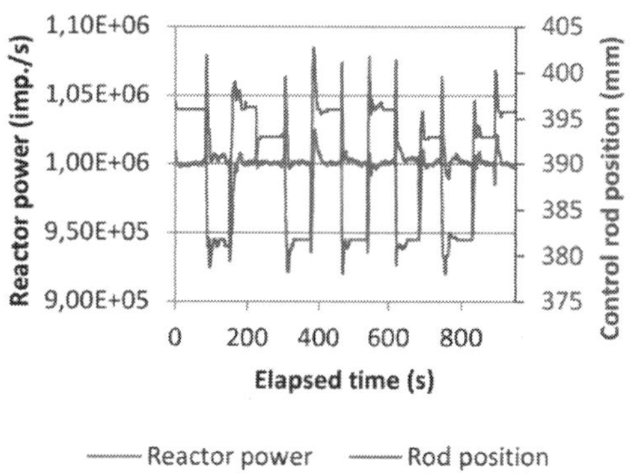

Fig. 5. Example of a disturbance rejection test. The operator was asked to keep the power at a constant level. The device for dynamics studies provided external reactivity changes which were to be compensated by an operator using control rod movements. Up: Movement of Cadmium (Cd) sample in device for dynamics studies; Down: operator response by control rod movement and reactor power evolution. The rods can be moved from 0 to 680 mm, and the active part of fuel assemblies is 600 mm.

C. Human characteristics variations

Contrary to the core physics, the operator effects exhibit operator-to-operator variations. Thus the experiments can be aimed at either the performance evaluation of a single operator, or aimed at a set of operators, to estimate the effect of human factors dispersion on the studied operator related parameters that could influence the reactor control. Such factors may include age, experience, workload, or the professional background of the operator.

The experiments at the VR-1 reactor benefits from the fact that the facility is the university's training reactor with a large amount of licensed operators (more than 10). The operators significantly differ in their reactor control workload, professional background (either technician based or scientific based), age and experience (from a few years to more than twenty years). The facility enables students and trainees operators within their education or training sessions. Although evaluating their performance may provide valuable information for early career, low-experienced operators, they have not been tested or evaluated, thus far.

D. Data acquisition system

To study an operator's influence on reactor control using control engineering techniques and models one needs to have digitalized time-dependent information of the input and output parameters of operator examinations. This includes the reactor control system data, with as a minimum, the value of the reactor power and the control rod position. If external devices are used (e.g. for external reactivity changes, or for information on the desired prescribed power profile that should be followed by the operator), they need to allow for the acquisition of digital data that can be synchronised with the reactor operator data. In the case of the VR-1 reactor, the Instrumentation and Control (I&C) system has been fully digital since the reactor was first operated in 1990. The I&C system has since been completely upgraded in 2009 [5]. In 2011, dedicated software [6] was developed to enable the online and offline transport of selected reactor I&C data outside the control system. Later, the system was upgraded to allow for the acquisition of data from various experimental devices, together with reactor I&C data.

The sampling rate of the observed parameters was defined based on the needs of the data processing algorithms that were used in the r control engineering models that describe the human operator behaviour. The time step of 0.1 s was found to be adequate for the experiments performed.

E. External conditions

The performance of the reactor operator may depend on the external conditions under which the tasks are to be performed. Thus, including variations in external conditions may help to understand the range of operator response. The external conditions should include, in particular, varying the amount of stress and the workload conditions. Obviously, studies on stress dependent operator performance have to stay within the safe practices of nuclear reactor operation. In the case of the VR-1 reactor experiments it was decided that they should correspond to situations which may typically occur during daily reactor operation. Still, the space to study stress dependent operator performance remains broad and may also include:

a) Single isolated task vs. complex task examination,

b) Response to a new, or to already known, transient,

c) Presence/absence of visual/audible information about the external reactivity change,

d) Perturbations/commands that are known in advance or are new at the time of request,

e) Full operator concentration on the task-at-hand during the examination period, or the addition of one or more distracting requirements on the operator during the examination. This may include periodic or on-demand observation and recording of selected reactor parameters.

F. Other aspects

The presence of an operator response evaluator during the experiments has been found to be important in understanding all of the behaviours and characteristics of an operator's response during individual tests, and in discovering the correlations with other factors that can influence operator response or concentration. This may somehow limit the capability of the facility for performing these types of tests. At the VR-1 reactor, this aspect was improved during the reactor operator desk upgrade, which was performed in mid-2018, when a videoconference system was installed in the control room to support remote education and training capabilities. The videoconference system is connected to multiple cameras inside and outside the reactor control room. Thus, video and audio signal from the reactor control room may be transmitted to remote sites, an enabling an evaluator to make on-line observations and discussions regarding the course of the operator performance experiments and, if required, a record of operation actions and experimental processes can be obtained.

IV. History of operator influence on reactor control at the VR-1 reactor

The development of the experiments to evaluate the operator influence on reactor control has been progressed through the close cooperation between the control engineers with their initial requests for tests and the reactor operating team who provide feedback on the feasibility of the proposed tests, and who also contribute specific knowledge of the conditions in which the operators have to work. The quality of the experimental outputs is continuously improving along with the mutual understanding of both points of view. The first discussions regarding the feasibility of performing operator experiments at the VR-1 reactor facility started in 2015. Disturbance rejection experiments, i.e. experiments in which an operator has to react to an externally induced reactivity/power change, were found to be quickly and easily implemented using the existing reactor equipment. At the end of 2015, the first experiments were performed with the aim of understanding the most important factors in the experimental design and implementation , and also to gain experience in applying control engineering methods to the real experimental data. Based on these tests, a four-stage procedure of operator tests was defined, with a gradual increase in task complexity and operator workload. Three operators of the VR-1 reactor were selected for these tests, each with a significantly different background (age, knowledge, experience, or reactor operator workload) in 2016. Duration of a single test typically varied between two and fifteen minutes. Some results have been already published in [1]. In 2017, tracking experiments i.e. experiments in which an operator is asked to follow a prescribed input were discussed and a test procedure based on a shift-leader's audible commands was developed. The procedure was tested using three reactor operators. These initial tests revealed an uncertainty in the specific timing of the audible commands, during off-line evaluation of the operators' performance. In 2018, an in-house software was developed, with the aim of decreasing this uncertainty in audible command timing. The bespoke software provides the operator with a real-time graphical display of the input signal (e.g. desired reactor power). The software will also use the reactor operation data, which is shared through the Experimental studio software [3] and it will enable a complex time-dependent desired power input to be set for the operator to follow; also it will have the capacity for on-line visualization the desired reactor power and the actual reactor power achieved by an operator.

V. Conclusion

The VR-1 reactor has been utilized for the unique examination of operator influence on reactor control since 2015. During the last few years, the capability of the facility for such studies has been proven, and most aspects of experimental setups and procedures have been understood. A significant potential to undertake related future studies has been identified.

The reported research was mainly achieved with the existing instrumentation and equipment at the facility. The identification of operator dynamic characteristics have significantly benefited from the education and training orientation of the facility. The early success of the experiments have been due to, in particular, the existing bespoke experimental instrumentation for reactor dynamics studies, the reactor operation data acquisition system, and the broad range of experienced operators at the VR-1 reactor facility.

Acknowledgment

The paper was supported by the project "Strengthening and development of research at Czech Technical University in Prague with the use of research infrastructure VR-1 Training Reactor for research activities" (reg. nr. CZ.02.1.010.00.016_0130001790) supported by Operational Programme Research, Development and Education co-financed from European Structural and Investment Funds and from state budget of the Czech Republic. References

References

[1] A.M. Darbyshire, T. Bily, J.S. Hatherall, K.D. Atkinson, "Manual control during reactivity transients at the VR-1 reactor - I. Disturbance rejection," Nuclear Engineering and Design, vol. 325, pp. 178-183, 2017, DOI:10.1016/j.nucengdes.2017.09.025.

[2] M. Kropík, J. Rataj, M. Juříčková "New human-machine interface at VR-1 reactor" In: 18th International Conference on Nuclear Engineering. New York: American Society of Mechanical Engineers - ASME, 2010. ISBN 978-0-7918-4929-3

[3] M. Kropík, J. Rataj, "New Control and Data Acquisition of Experiments" In: Nuclear Energy for New Europe 2011. Ljubljana: Nuclear Society of Slovenia, 2011. pp. 1-12. ISBN 978-961-6207-32-4 2011

[4] J. Rataj et al., "Reactor Physics Course at VR-1 Reactor". 2nd edition, Prague, Czech Republic, CTU in Prague, 2014.

[5] M. Kropík, M. Juříčková "Calibration of new I&C at VR-1 training reactor, Nuclear Engineering and Design, Volume 241, Issue 4, 2011, Pages 1090-1096, ISSN 0029-5493, https://doi.org/10.1016/j.nucengdes.2010.05.049. (http://www.sciencedirect.com/science/article/pii/S0029549310003754)

[6] J. Rataj, Ľ. Sklenka, "New Experimental Instrumentation for Study of Reactor Dynamics at Zero Power Reactor VR-1" In: RRFM 2011 Transactions. Brusel: European Nuclear Society, 2011. pp. 177-181. ISBN 978-92-95064-11-9.

Validation of Neutron Evaluated Data Based on The Experimental Reactivity Worth of Tungsten Target in CiADS

Wei Jiang[1,2], Long Gu[1,2*], Lu Zhang[1], Qi Zhou[3], Liang Chen[1], Jin-Yang Li[1], Rui Yu[1], Qing-Fu Zhu[3], Yong Dai[2]

[1]Institute of Modern Physics, Chinese Academy of Sciences, Lanzhou, 730000, China
[2]Paul Scherrer Institute, Villigen, 5232, Switzerland
[3]China Institute of Atomic Energy, China National Nuclear Corporation, Beijing, 102413, China
* gulong@impcas.ac.cn

Abstract—**Measurement of a cylindrical tungsten target reactivity worth has been performed on the light water zero-power reactor of VENUS-II at China Institute of Atomic Energy (CIAE) in order to verify the neutron evaluated data related to the engineering design of Chinese initiative Accelerator Driven Systems (CiADS). The reactivity worth of the tungsten target was measured and processed as -1.234±0.114mk by a period method. The experimental result was compared with the simulation ones calculated by MCNP with five different libraries, i.e., ENDF/B-VII.0, ENDF/B-VII.1, JENDL-4.0, CENDL-3.1 and JEFF-3.2. By comparing the results of experiment and simulation, the simulated results from ENDF/B-VII.0, JENDL-4.0 and JEFF-3.2 are higher than the experimental result, however that from CENDL-3.1 is lower. The result from ENDF/B-VII.1 library shows better agreement with the experiment one and the relative deviation is less than 2%. Through the analysis of the differences of the results, non-tungsten elements cross sections in the ENDF/B-VII.1 mainly affect the tungsten radiation capture and elastic scattering reaction rates in the energy range of 10^{-9}-10^{-7} MeV, which results in a better simulated tungsten target reactivity worth value. Therefore, it is recommended that the tungsten target reactivity worth should be calculated with the ENDF/B-VII.1.**

Keywords—**Tungsten target reactivity worth, nuclear data library, MCNP, VENUS-II, period method**

I. Introduction

Recently, Chinese initiative Accelerator Driven Systems (CiADS) project has been started in China [1]. For CiADS, the primary goal is to achieve the integration of three systems, i.e. a subcritical lead-bismuth reactor, a heavy metal spallation target and an intense-beam proton linac [2]. During the early stages of ADS projects, some experiments [3-7], which coupled the proton accelerator with the spallation target, have been performed by the researchers. However, the experimental studies about the coupling behaviors between the nuclear reactor and the spallation target have been rarely investigated. In order to ultimately build the CiADS experimental setup, it is necessary that the nuclear reactor and spallation target coupling experiments should be performed. The tungsten granular spallation target has been proposed by Institute of Modern Physics, Chinese Academy of Sciences (IMPCAS) as an innovative spallation target in the world. The tungsten granular target, which is placed in the center of the reactor, is impinged by the high-energy protons to produce spallation neutrons to maintain the stable operation of the system [8]. Meanwhile, the spallation target, as the structural material, has an effect on the reactor physical parameters, such as neutron flux, neutron spectra, system reactivity and so on. The mass fraction of tungsten is 93% in the spallation target, therefore tungsten cross sections in evaluated data libraries can impact directly on the accuracy of reactor physical calculations in CiADS. In the past experiment on tungsten data evaluation, tungsten materials are generally used as reflecting layer or fuel diluent materials and almost placed in fast-spectrum benchmark experiment setups [9-15]. However, the relevant studies on the tungsten spallation materials in thermal-spectrum reactor are still scarce. In this paper, the measurement of the cylindrical tungsten target reactivity worth was performed on VENUS-II light water reactor by the period method. And then the experimental tungsten target reactivity worth was compared with the results calculated by MCNP with the different libraries, i.e. ENDF/B-VII.0 [16], ENDF/B-VII.1 [17], JENDL-4.0 [18], JEFF-3.2 [19] and CENDL-3.1 [20]. The difference between the experimental and simulated results is discussed, which provides some reference for the selection of nuclear data libraries on CiADS neutronics simulations.

II. Experiment

II.1 Experimental facility

The VENUS-II zero-power facility, which consists of a light water reactor and a lead reactor, firstly reached the criticality on December 2016 through the cooperation of China Institute of Atomic Energy (CIAE) and Institute of Modern Physics, Chinese Academy of Sciences [21]. The structures and materials in the VENUS-II light water reactor are relatively simple, which is helpful in validating the neutron nuclear data libraries. Therefore, the present experiments were carried out on the light water reactor. The physical photos of VENUS-II light water reactor are shown in Fig. 1. Light water was used as the moderator and reflector, in which eight γ compensation ionization chambers were arranged to monitor nuclear reactor power. Fuel rods of U_3O_8 with the enrichment 20% were arranged outside the spallation zone of ϕ170mm×H804 mm in the concentric manner. The experimental tungsten target, with the dimensions of ϕ100mm×H100mm, was placed on the bottle of the spallation target zone in the center of the core. The tungsten target purity and density were respectively 99.95% and 18.2 g/cm^3. The 304 stainless steel target clad with the thickness of 2mm was helpful to place the cylindrical tungsten target in and out the spallation target zone.

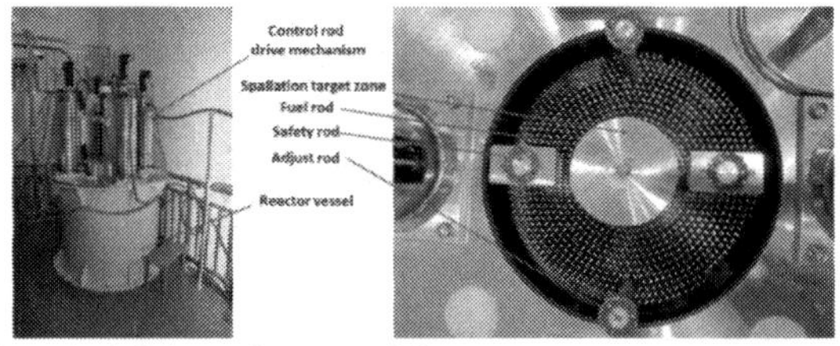

Fig. 1. Physical photos of VENUS-II light water reactor

II.2 Experimental principle and data processing

The reactivity ρ values of the VENUS-II light water reactor are measured by the period method [22]. Base on the point reactor dynamics equation, in-hour equation is derived and as followed:

$$\rho = \frac{\Lambda}{T/\ln 2} + \sum_{i=1}^{m} \frac{\beta_{ieff}}{1 + (\lambda_i T / \ln 2)} \quad (1)$$

Where, ρ is the reactivity; Λ is the neutron generation time; T is the nuclear reactor doubling time; m is the number of effective delayed neutron groups; $\beta_{i,eff}$ and λ_i are respectively the ith group effective delayed neutron fraction and the decay constant of the ith group delayed neutron precursor.

The double period T values were measured by stopwatch and reactor power monitor [23] in different supercritical states of the reactor, and the supercritical states were maintained by adjusting the fuel rods number in the reactor. During the experiment, the four supercritical states to be measured are as follows: (1) 961 fuel rods without anything in the spallation zone, (2) 960 fuel rods without anything in the spallation zone, (3) 963 fuel rods with the target clad in the spallation zone, and (4) 968 fuel rods with tungsten target (include cylindrical tungsten target and target clad) in the spallation zone. More details on the experiments were reported in the Ref. [24]. Meanwhile, the reactor kinetic parameters (Λ, $\beta_{i,eff}$ and λ_i) were respectively calculated by MCNP6 with different libraries, i.e., ENDF/B-VII.0, ENDF/B-VII.1, JENDL-4.0, CENDL-3.1 and JEFF-3.2. And then the reactivity ρ values were calculated by substituting the measured T and the calculated kinetic parameters into the Eq. (1). At last, according to the formula $\Delta\rho = \rho_{968} - [\rho_{963} + 5\times(\rho_{961} - \rho_{960})]$, the tungsten target reactivity worth Δρ was calculated as −1.234±0.114mk, where ρ_{961}, ρ_{960}, ρ_{963} and ρ_{968} represent the measured reactivity values under the above conditions (1)-(4), respectively. The experimental error comes from the deviations of T measurement and the uncertainties of the simulated kinetic parameters [25].

III. Monte Carlo simulation

III.1 Simulation method

In this paper, the reactivity worth of the cylindrical tungsten target is simulated by MCNP software, which is produced and developed by Los Alamos National Laboratory. In MCNP simulations, VENUS-II light water reactor model was created accurately at real geometry and material. Meanwhile, the loading number of fuel rods in the light water reactor was set as the number (totally 961 rods) of initial fuel rods in the experiment and kept the constant. And then the simulation process had two steps: (1) The target clad was placed in the spallation target zone; (2) The tungsten target replaced the target clad in the spallation target zone. Therefore, the reactivity worth of the tungsten target was calculated as the difference between the reactor reactivity values with the tungsten target and the target clad in the spallation target. The simulation method can reduce the simulated error of the tungsten target reactivity worth results and satisfy the accuracy demands.

III.2 Details description

The effective multiplication factor k_{eff} values of the two steps in Part 3.1 were calculated by KCODE card in MCNP with 3000 cycles and $5*10^5$ histories per cycle, which caused the estimated standard deviation of about 2pcm (1pcm=10^{-5}) for the k_{eff} results. The temperature values of the material cross sections were set as the room temperature 293k and remained unchanged. Considering the upward scattering of the thermal neutrons, the thermal neutron scattering library S (α,β) was used to achieve the accurate neutron transportation calculations in the light water reactor. Then, the k_{eff} was converted to the reactivity ρ through the equation $\rho = (k_{eff} - 1)/k_{eff}$. The simulated tungsten target reactivity worth was calculated as the difference between the reactivity values with the tungsten target and the target clad in the spallation zone.

To examine the different neutron nuclear data libraries, the cross sections of all materials in the reactor, i.e. target, fuels, moderator and among others, are respectively from ENDF/B-VII.0, ENDF/B-VII.1, CENDL-3.1, JENDL-4.0 and JEFF-3.2. And then five corresponding tungsten reactivity worth values were obtained and the errors originated from the statistical errors of MCNP simulation.

IV. Results and discussion

The experimental and calculated results of tungsten target reactivity worth values are shown in Fig. 2, where the shaded area indicates the experimental results within the error range, and the five dots respectively represent the MCNP simulated results with five cross section libraries. In Table I, the calculated results in the second column are respectively simulated by MCNP with the different cross section libraries. Meanwhile, the ratio values of the calculated and experimental results (C/E) are presented in the fourth column. From the obtained results, it can be observed that:

(1). Tungsten target reactivity worth values, obtained by the experiment and simulation, are all negative, which causes the reduction of the reactor reactivity.

(2). Comparing with the experimental result, the simulated results of ENDF/B-VII.0, JENDL-4.0 and JEFF-3.2 are higher, whereas the one of CENDL-3.1 is lower.

(3). The simulated result with ENDF/B-VII.1 has a good agreement with the experiment one and the relative deviation is less than 2%.

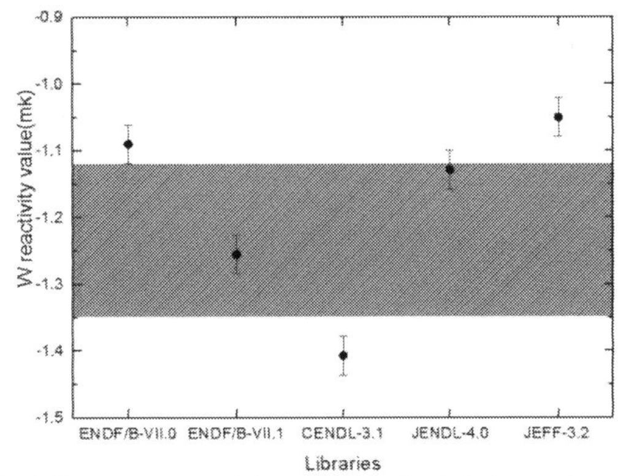

Fig. 2. Tungsten target reactivity worth values from both the experiment and the simulation.

TABLE I
TUNGSTEN TARGET REACTIVITY WORTH VALUES BETWEEN THE MCNP CALCULATIONS AND PERIOD METHOD

Libraries	Tungsten target reactivity worth (mk)		C/E
	MCNP	Period method	
ENDF/B-Ⅶ.0	-1.090±0.029		0.883
ENDF/B-Ⅶ.1	-1.255±0.029		1.017
CENDL-3.1	-1.407±0.029	-1.234±0.114	1.14
JENDL-4.0	-1.129±0.029		0.915
JEFF-3.2	-1.050±0.029		0.851

In order to find out the reason of the differences among the tungsten reactivity worth values from five libraries, the affecting factors of tungsten reactivity worth were divided into two aspects: the tungsten element and non-tungsten elements. Meanwhile, the simulation results from ENDF/B-Ⅶ.0, JENDL-4.0 and JEFF-3.2 are basically in agreement within the error ranges (see Fig. 2), so that the results are grouped into the same group and one calculated by ENDF/B-Ⅶ.0 was chosen to represent the group. Therefore, the simulated result from the ENDF/B-Ⅶ.1 is only compared with those from the ENDF/B-Ⅶ.0 and CENDL-3.1. First, the reason for the difference between the tungsten reactivity worth results with the ENDF/B-Ⅶ.1 and CENDL-3.1 is mainly the discrepancy of the tungsten radiative capture cross sections at the epithermal neutron region [24].

Next, the study focused on analyzing the difference between the tungsten reactivity worth values from ENDF/B-Ⅶ.1 and ENDF/B-Ⅶ.0. Therefore, the tungsten reactivity worth values in the three cases are calculated and shown in Table II. The table shows the results in the case 1 and case 2 are almost consistent with each other within errors. However, the result in case 1 is remarkably different from that in case 3. It can be concluded that the non-tungsten element cross sections are the main reason for the discrepancy of the results between ENDF/B-Ⅶ.0 and ENDF/B-Ⅶ.1, whereas the impact of the tungsten element cross sections on the difference is almost negligible.

TABLE II
THE SIMULATED TUNGSTEN TARGET REACTIVITY WORTH WITH NUCLEAR DATA IN THE ENDF/B-Ⅶ.0 AND ENDF/B-Ⅶ.1

Case No.	Tungsten cross sections	Non-tungsten cross sections	Target worth (mk)
1	ENDF/B-Ⅶ.1	ENDF/B-Ⅶ.0	-1.110±0.029
2	ENDF/B-Ⅶ.0	ENDF/B-Ⅶ.0	-1.090±0.029
3	ENDF/B-Ⅶ.1	ENDF/B-Ⅶ.1	-1.255±0.029

In the neutron field of the reactor, the reaction channels in the tungsten target mainly include the radiative capture, elastic scattering, inelastic scattering, (N, 2N) and (N, 3N). To analyze the effect of the each reaction channel on the tungsten reactivity worth, the reaction rates of each reaction channel were respectively increased by 1.3 times the original values for the sensitivity analysis. More specifically, the input card in case 3 of Table II was calculated by the perturbation function of MCNP and the variations of the reactivity ρ values are shown in Table III. For the specific reaction type, it can be seen that the radiative capture causes a negative reactivity in the system, whereas the elastic scattering and inelastic scattering respectively cause a positive reactivity. (N, 2N) and (N, 3N) have little impact on the reactivity value. Therefore, it can be concluded that the reaction rates discrepancies of the radiative capture, elastic scattering and inelastic scattering in the tungsten target cause the difference of the tungsten reactivity worth values. Moreover, the reactivity variation caused by the radiative capture is the largest, that is, the radiative capture reaction rates have the greatest impact on the tungsten reactivity worth.

TABLE III
CALCULATION RESULTS OF SENSITIVITY ANALYSIS FOR THE MAIN REACTION CHANNELS OF TUNGSTEN ELEMENT

Reaction channel	Reactivity ρ variation (mk)
Radiative capture	-0.28
Elastic scattering	0.13
Inelastic scattering	0.06
N,2N	0
N,3N	0

According to the above sensitivity analysis results, the reaction rates for the radiative capture, elastic scattering and inelastic scattering in the tungsten target were calculated with FM card in MCNP for case 1 and case 3 in Table II. Integral values of the whole neutron energy interval for the three reaction channel rates are given in the second and third columns of Table IV. From the relative difference RED [RED= $(^{ENDF/B\text{-}VII.1}value - ^{ENDF/B\text{-}VII.0}value)/^{ENDF/B\text{-}VII.0}value$] in the fourth column of Table IV, it can be seen that comparing with case 1, integral value of radiation capture reaction rates in case 3 is larger, whereas that of elastic scattering and inelastic scattering rates in case 3 is smaller. And the relative difference RED of radiation capture reaction rates is maximal among the three reaction channels. Together with the results in Table III, the changes of all three reaction channel rates result in the lower tungsten reaction worth value in the case 3 and the most significant impact comes from radiation capture reaction channel. Therefore, the cross sections of non-tungsten elements mainly affect the radiative capture, elastic scattering and inelastic scattering rates, which results in the difference of tungsten reaction worth values from case 1 and case 3. Meanwhile, the difference of radiation capture reaction rates is the main reason for the lower result in case 3.

TABLE IV
INTEGRAL VALUES OF THE REACTION RATES FOR THE MAIN REACTION CHANNELS IN THE TUNGSTEN TARGET

Reaction channel	Case 1	Case 3	RED (%)
Radiative capture	7.73E-06	7.91E-06	2.23
Elastic scattering	5.33E-05	5.31E-05	-0.40
Inelastic scattering	7.54E-06	7.51E-06	-0.43

*Unit: number/source neutron/ cm^3

In order to further analyze the influence of non-tungsten elements on main reaction channels, the whole neutrons energy region (10^{-9}-10 MeV) was divided into 10 energy groups in geometric progression manner of common ratio 10. The relative differences RED of the integral values for the radiation capture, elastic scattering and inelastic scattering reaction rates were respectively calculated in each energy group for case 1 and case 3. As the results in Fig. 3(a-c) shows the maximum RED values of the radiation capture and elastic scattering between the case1 and case 3 are both presented in the energy interval of 10^{-9}-10^{-7} MeV, whereas the RED of inelastic scattering reaction rates is very low and only found in the energy interval of 10^{-2}-1 MeV. Therefore, it can be concluded that the difference of non-tungsten elements cross sections between ENDF/B-Ⅶ.1 and ENDF/B-Ⅶ.0 most mainly results in the discrepancy of the radiation capture and elastic scattering reaction rates in the energy range of 10^{-9}-10^{-7} MeV.

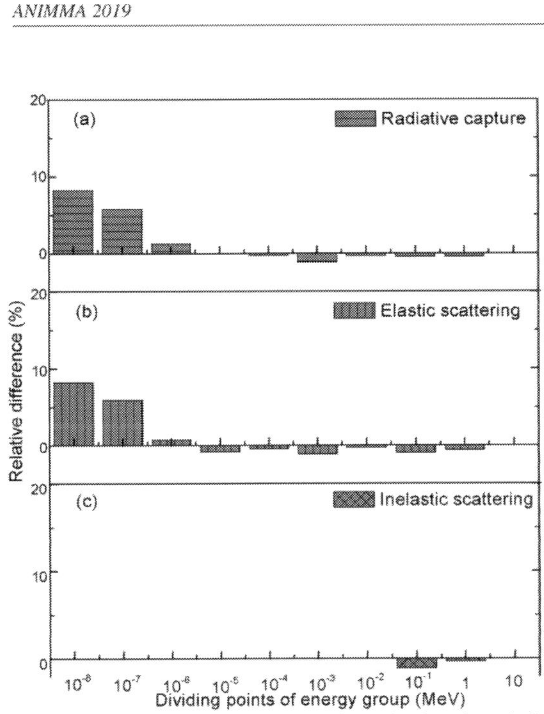

Fig. 3. Relative difference values of tungsten main reaction channel rates in the different energy groups.

V. CONCLUSIONS AND FUTURE WORK

The reactivity worth value of tungsten cylindrical target has been measured as -1.234±0.114mk by period method on VENUS-II light water reactor at China Institute of Atomic Energy. Meanwhile, the MCNP simulations were performed with ENDF/B-VII.0, ENDF/B-VII.1, JENDL-4.0, JEFF-3.2 and CENDL-3.1, respectively. By comparing the calculated results with the experimental one, it is found that the results of ENDF/B-VII.0, JENDL-4.0 and JEFF-3.2 are higher, whereas the one of CENDL-3.1 is lower. The calculated result with ENDF/B-VII.1 has a good agreement with the experiment one and the relative deviation is less than 2%. An investigation was made to mainly analyze the difference reason of the tungsten target reactivity worth values from ENDF/B-Ⅶ.1 and other libraries (ENDF/B-VII.0, JENDL-4.0 and JEFF-3.2). Through the analysis, it can be concluded that the main reason for the difference is due to the non-tungsten elements cross sections in ENDF/B-Ⅶ.1, which mainly results in the difference of the tungsten radiation capture and elastic scattering reaction rates in the energy range of 10^{-9}-10^{-7} MeV. Therefore, it is recommended to adopt ENDF/B-VII.1 to calculate the tungsten target reactivity worth. In the future, more integral experiments related to the key materials in CiADS will be done, i.e., lead, lead-bismuth and among others to satisfy the needs of the CiADS engineering design.

ACKNOWLEDGEMENTS

The work was completed under the support of Forerunner Special Foundation of Chinese academy of sciences (NO. XDA03010200). The authors express their gratitude to all members in the experimental groups of the Institute of Modern Physics, Chinese Academy of Sciences and the China Institute of Atomic Energy.

REFERENCES

[1] *L. Yang, W. Zhan. A new concept for ADS spallation target: gravity-driven dense granular flow targets. Proceedings of the ThEC13 Conference, CERN, Globe of Science and Innovation, Geneva, Switzerland, 27-31 October 2013. doi: 10.1007/978-3-319-26542-1_47*

[2] *G. Xiao, H. Xu, S. Wang, HIAF and CiADS National Research Facilities: Progress and Prospect, Nucl Phys Rev, 34(3), 275-283(2017).*

[3] *X. Zhang, Y. Zhang, F. Ma and et al., Neutron production for 250 MeV protons bombarding on thick grain-made tungsten target, EUR PHYS J A, 51(106) (2015). doi: 10.1140/epja/i2015-15106-0*

[4] *Y. Ju, X. Zhang, F. Ma et al., The production of residual radionuclides by a 250MeV proton beam, J. Phys. G: Nucl. Part. Phys., 42(12) (2015). doi: 10.1088/0954-3899/42/12/125102*

[5] *Y. Li, X. Zhang, Y. Ju et al., Study of neutron spectra in a water bath from a Pb target irradiated by 250 MeV protons, CHINESE PHYS C, 39(4) (2015). doi: 10.1088/1674-1137/39/4/044001*

[6] *H. Zhang, X. Zhang, F. Ma et al., Residual Nuclides Induced in Cu Target by a 250 MeV Proton Beam, CHINESE PHYS LETT, 32(4) (2015). doi: 10.1088/0256-307X/32/4/042501*

[7] *L. Chen, F. Ma, X. Zhang et al., Spallation yield of neutrons produced in thick lead target bombarded with 250 MeV protons, Nucl. Instrum. Methods Phys. Res., Sect. B, 342, 87~90 (2015). doi: 10.1016/j.nimb.2014.09.020*

[8] *W. Zhan, H. Xu, Advanced Fission Energy Program - ADS Transmutation System, CHINESE SCI BULL, 27(3): 375-381 (2012). doi: 10.3969/j.issn.1000-3045.2012.03.017*

[9] *A.C. Kahler, R.E. MacFarlane, R.D. Mosteller et al. ENDF/B-VII.1 neutron cross section data testing with critical assembly benchmarks and reactor experiments. NUCL DATA SHEETS, 112(12), 2997-3036 (2011). doi:10.1016/j.nds.2011.11.003*

[10] *S. Zhang, Z. Chen, Y. Nie et al., Measurement of leakage neutron spectra for Tungsten with D-T neutrons and validation of evaluated nuclear data, FUSION ENG DES, 92, 41–45 (2015). doi: 10.1016/j.fusengdes.2015.01.010*

[11] *S. Zhang, Y. Nie, J. Ren et al., Benchmarking of JEFF-3.2, FENDL-3.0 and TENDL-2014 evaluated data for tungsten with 14.8 MeV neutrons, NUCL SCI TECH, 28:27 (2017). doi: 10.1007/s41365-017-0192-0*

[12] *K. Seidel, R. Eichin, U. Fischer et al. Benchmark experiments to validate the nuclear data of tungsten, FUSION ENG DES, 81, 1211-1217 (2006). doi: 10.1016/j.fusengdes.2005.09.050*

[13] *P. Batistoni, M. Angelone, et al. Neutron-nuclear data benchmark for copper and tungsten by slab assembly transmission experiments with D-T, FUSION ENG DES, 58-59, 617-621 (2001). doi: 10.1016/S0920-3796(01)00518-X*

[14] *I. Murata, T. Nishio, L. Petrizzi et al. Neutronics benchmark experiment on tungsten, J NUCL MATER, 329-333, 683-686 (2004). doi: 10.1016/j.jnucmat.2004.04.191*

[15] *O. Cabellos, Processing of the JEFF-3.2T3 Cross Section Library with the NJOY Code System into Various Formats for Testing Purposes (2013). JEFF Meeting, Spain, November 27-28, 2013. https://www.researchgate.net/publication/321731952*

[16] *M.B. Chadwick, P. Oblozinsky, M. Herman at al., ENDF/B-VII.0: next generation evaluated nuclear data library for nuclear science and technology. NUCL DATA SHEETS 107, 2931–3060 (2006). doi: 10.1016/j.nds.2006.11.001*

[17] *M.B. Chadwick, M. Herman, P. Oblozinsky at al., ENDF/B-VII.1 nuclear data for science and technology: cross sections, covariances, fission product yields and decay data. NUCL DATA SHEETS 112: 2887–2996 (2011). doi: 10.1016/j.nds.2011.11.002.*

[18] *Z. Ge, Z. Zhao, H. Xia et al., The updated version of Chinese evaluated nuclear data library (CENDL-3.1). J. Kor. Phys. Soc. 59, 1052 (2011). doi: 10.1051/ndata:07570*

[19] *K. Shibata, O. Iwamoto, T. Nakagawa et al., JENDL-4.0: a new library for innovative nuclear energy systems. J. Kor. Phys. Soc., 59(23), 1046 (2011). doi: 10.3938/jkps.59.1046*

[20] *A. Koning, R. Forrest, M. Kellett et al., JEFF Report 21: the JEFF-3.1 Nuclear Data Library (OECD, 2006). https://www.researchgate.net/publication/267362554*

[21] *Q. Zhu, Q. Zhou, S. Liang et al., VENUS-II ADS Zero Power Reactor in CIAE, Proceedings of the Reactor Physics Asia Conference, Chengdu, China, 24-25 August 2017.*

[22] *S. Hu, R. Wang, X. Liang et al., Method and Data Processing for HTR-10 Physics Experiment, Nucl Power Eng, 23(1), 8-11 (2002).*

[23] *J. Khorsandi, Z. Nasr, R. Salimi et al., Heavy Water Zero Power Reactor (HWZPR) mixed core first criticality, calculation and experiments. PROG NUCL ENERG, 85, 613-616 (2015). doi: 10.1016/j.pnucene.2015.08.007*

[24] *J. Wei, G. Long, Z. Qi el al. Measurement of tungsten reactivity worth on VENUS-II light water reactor and validation of evaluated nuclear data. PROG NUCL ENERG, 108, 81-88 (2018). doi: 10.1016/j.pnucene.2018.05.002*

M.K. Jaradat, V. Radulovic, C.J. Park et al., Verification of MCNP6 model of the Jordan Research and Training Reactor (JRTR) for calculations of neutronic parameters. ANN NUCL ENERGY, 96, 96-103 (2016). doi: 10.1016/j.anucene.2016.06.00

EPJ Web of Conferences **225**, 04027 (2020) https://doi.org/10.1051/epjconf/202022504027
ANIMMA 2019

Utilisation of JSI TRIGA Pulse Experiments for Testing of Nuclear Instrumentation and Validation of Transient Models

Ingrid Vavtar
Reactor Physic Division F-8
Jožef Stefan Institute
Ljubljana, Slovenia
ingrid.vavtar.93@gmail.com

Anže Pungerčič
Reactor Physic Division F-8
Jožef Stefan Institute
Ljubljana, Slovenia
anze.pungercic@ijs.si

Luka Snoj
Reactor Physic Division F-8
Jožef Stefan Institute
Ljubljana, Slovenia
luka.snoj@ijs.si

Abstract—A pulse experiment model was validated in order to support future pulse experimental campaigns. All pulse experiments data was collected and are publicly available at http://trigapulse.ijs.si/. A comparison of the measured pulse physical parameters (maximal power, total released energy and full width at half maximum) and theoretical predictions (Fuchs-Hansen and the Nordheim-Fuchs models) was made.

Keywords—pulse experiments, Fuchs-Hansen model, Nordheim-Fuchs model, maximum power, total energy released, full width at half maximum

I. Introduction

A vital phase in the development of nuclear instrumentation detectors and associated electronic data acquisition systems is experimental testing and qualification in a well-characterized and representative radiation field in a reference irradiation facility. The 250 kW Jožef Stefan Institute (JSI) TRIGA Mark II research reactor is a very well characterized reactor in terms of the knowledge of the neutron and gamma fields, a product of the work performed at the JSI over the last decade, mostly in collaboration with the Instrumentation, Sensors and Dosimetry Laboratory at CEA, Cadarache [1]-[5]. The neutron flux level in the JSI TRIGA reactor in steady state mode is on the order of 1E13 n/cm^{2}s, which is too low for do MTR relevant testing. However in pulse mode operation the JSI TRIGA reactor can achieve 1 GW for approximately 5-10 ms thus achieving MTR relevant testing conditions for a short time. Moreover pulse operation can be used for experimental validation of reactor transient modelling.

In order to support pulse experimental campaigns data from all pulse experiments was collected and is publicly available at *http:// trigapulse.ijs.si/*. In addition a comparison of measured pulse physical parameters (maximal power, total released energy and full width at half maximum) was made with theoretical predictions from the Fuchs-Hansen and the Nordheim-Fuchs models. The FH model was used for evaluation of experimental uncertainties. It was shown that experimental data follows the theoretical models, but in some cases there is a large deviation, which can be attributed to large experimental uncertainties and uncertainties of the theoretical models.

The validated models will be used to support future experimental campaigns, where nuclear instrumentation and data acquisition systems for fast transient measurements will be tested.

II. TRIGA Mark II

The TRIGA research reactor at JSI is a 250 kW TRIGA Mark II reactor. It is a light water pool-type reactor cooled by natural convection. The reactor core is placed at the bottom of a 6.25 m high open aluminum tank, measuring 2 m in diameter (see Figure 1) and is filled with demineralized water. The core shape is cylindrical and there are 91 locations in the core, which can be occupied by fuel rods, neutron source, irradiation channels and control rods (see Figure 2).

The fuel elements are cylindrical shape with a homogeneous mixture of uranium and zirconium hydride coated in stainless steel. In the TRIGA reactor are four control rods, three control rods with fuel extensions and transient control rod with void instead fuel extension. In the control rods the absorber is boron carbide (B_4C)). Elements in the core are arranged in six concentric rings with 1, 6, 12, 18, 24 and 30 available locations. Each location corresponds to a hole in the aluminum upper and lower grid plates of the reactor, which confine the reactor core. The core is surrounded by a graphite reflector enclosed in aluminum casing. The pulsed experiment is performed using a pulse rod system where the inserted transient control rod is ejected from the reactor core by pneumatic system. [6]-[7]

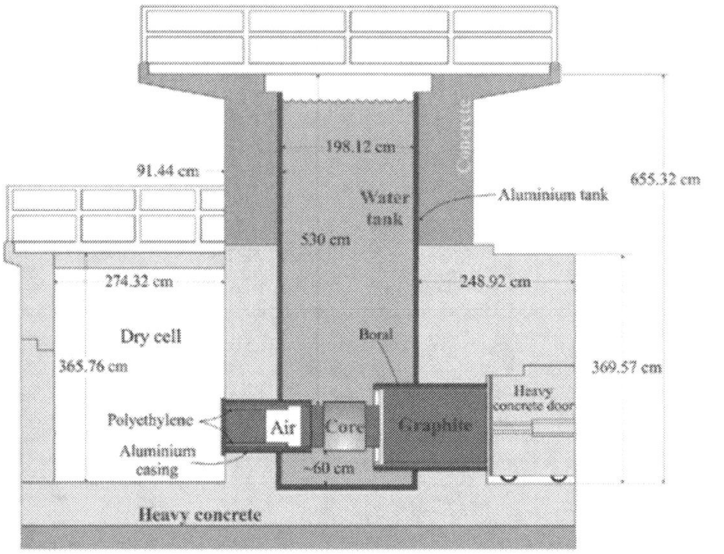

Figure 1: Side view of the TRIGA Mark II reactor. [7]

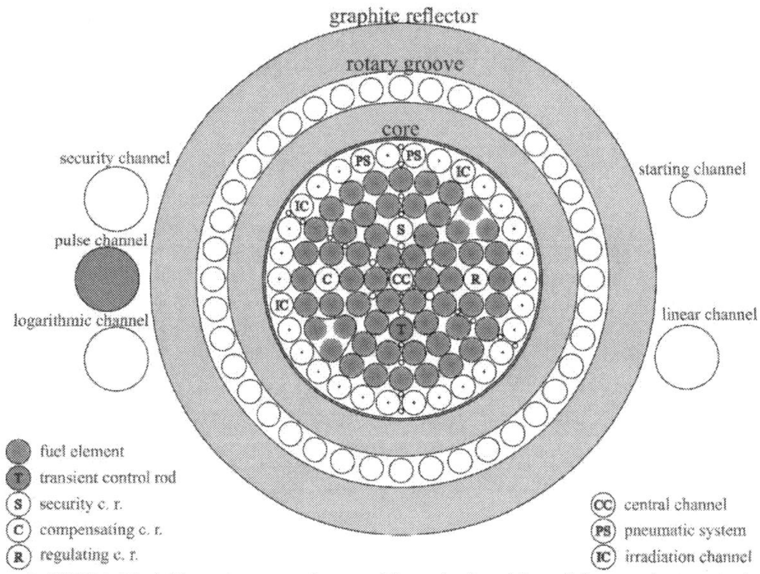

Figure 2: TRIGA Mark II reactor core scheme with marked position of the transient control rod. [7]

III. Pulse Experiment

After a quick ejection of the transient control rod from the reactor at pulse experiment, the reactor becomes prompt supercritical in a short time, a few 10 ms, and the power begins to increase exponentially. The change in power and consequently, the fuel temperature affects the decrease of the reactivity due to the prompt negative temperature reactivity coefficient of the fuel, which makes the reactor establish a new equilibrium state quickly and efficiently.

Due to the decrease of the reactivity the chain reaction is slowed down or interrupted, resulting in a decrease in power. The peak power of the pulse reaches some MW and the total released pulse energy is relatively small (typically a few MJ) due to the short pulse time.

In the TRIGA reactor, the temperature reactivity coefficient of fuel is the strongest and most important feedback effect of the reactor state on reactivity. The temperature reactivity coefficient of fuel α_g is defined as:

$$\alpha_g = \frac{\Delta\rho}{\Delta T_g}, \qquad (1)$$

where $\Delta\rho$ is the change in the reactivity of the reactor and ΔT_g is the change in the average fuel temperature in the entire reactor core.

The reason for an effective and prompt temperature feedback mechanism is in the special fuel composition, which is a homogeneous mixture of 20 % enriched uranium and zirconium hydride (ZrH ratio 1.6). Since hydrogen in the zirconium hydride serves as moderator, most moderation takes place in the fuel element itself and only a small part in the water surrounding the fuel elements. Consequently, any change in power and therefore also the change in fuel temperature immediately reflects on the moderator in the fuel element. Therefore, both the fuel and the moderator immediately affect back the reactivity of the core. Since the change in fuel temperature affects reactivity in different ways, the negative temperature reactivity coefficient of fuel is the sum of several contributions. The Doppler effect in the fuel and the shift of the thermal part of the neutron spectrum to higher energies contribute the most to the prompt negative temperature reactivity coefficient of the fuel in the TRIGA reactor.

The basis of the Doppler's phenomenon is broadening of resonances (see Figure 3) for the capture of the neutron in the uranium ^{238}U, which increases the absorption of neutrons and thus reduces reactivity.

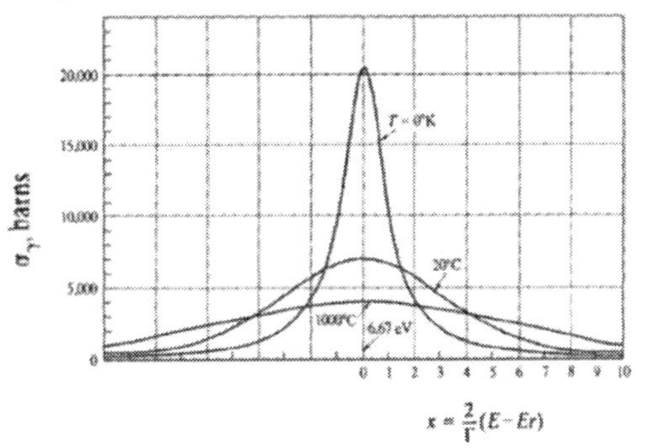

Figure 3: Shape of the resonance with mean energy 6.67eV in uranium ^{238}U depending on energy at three different temperatures. [8]

Another important contribution is the shift of the thermal part of the neutron spectrum to higher energies or spectrum hardening. Since the peak of the neutron spectrum (see Figure 4) is moved to higher energies, where the microscopic cross section for fission on the uranium ^{235}U is smaller, the neutron flux is reduced, followed by a decrease in reactivity.

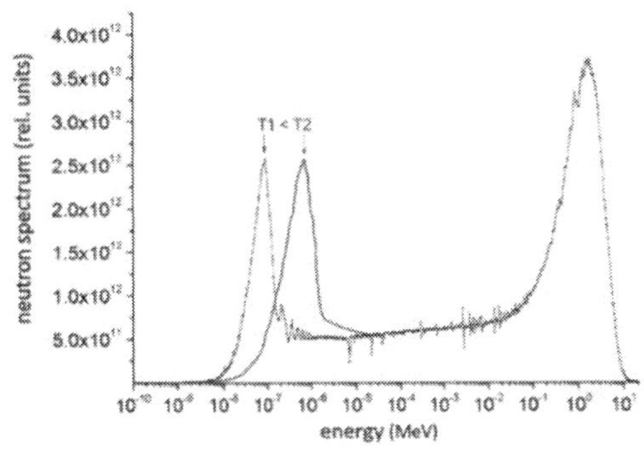

Figure 4: Lethargy neutron spectrum in standard (12 w/o) fuel in TRIGA Mark II reactor at JSI for two temperatures. Lethargy is defined as $u = ln\left(\frac{E_0}{E}\right)$ and lethargy neutron spectrum is neutron spectrum multiplied with energy.

IV. Theoretical Model

Theoretical model for the description of pulses is the Fuchs-Hansen adiabatic model [10], derived from the equations of the point reactor kinetics on the basis of four assumptions. The first assumption is that the system is adiabatic or that the fuel does not exchange heat with the environment and consequently all the energy released is used to heat the fuel. Another assumption is that all delayed neutrons can be ignored. Both of these assumptions are good, because the pulse time is short and there is no significant heat transfer during this time and also the time is short compared to the time when the delayed neutrons are released. The third assumption is that the power of the reactor before the pulse was low or equal zero. This assumption is valid, as the reactor is initially subcritical or at low power and the source of the neutrons in the reactor is not strong. The last assumption is instant reactivity change, which is valid due to rapid withdrawal of the transient control rod from the reactor. If we take into account the first two assumptions in the equations of point kinetics, one equation of point kinetics is obtained:

$$\frac{dP(t)}{dt} = \frac{\rho(t) - \beta}{\Lambda} P(t), \qquad (2)$$

where P is power, ρ is inserted reactivity, β effective delayed neutron fraction (TRIGA: β = 1\$ = 0.007 [6]), Λ is average generation time or time between the birth of a neutron and death with fission. Average generation time can be calculated as: $\Lambda = l/k$, where l is average lifetime of prompt neutrons or time between the birth of a neutron and death with absorption or leakage and k is the multiplication factor. In derivation, it is also important to define the prompt reactivity ρ':

$$\rho' = \rho - \beta. \qquad (3)$$

From the first assumption, it follows that the reactivity during the pulse decreases proportionally to the released energy or the resulting fuel temperature:

$$\rho(t) - \beta = \rho' - \gamma E(t), \qquad (4)$$

where γ is effective temperature reactivity coefficient of fuel and $E(t)$ total energy released during the pulse:

$$E(t) = \int_0^t P(t')dt'. \quad (5)$$

By solving the Fuchs-Hansen adiabatic model described above, the time dependence of power $P(t)$ and released energy $E(t)$ can be determined:

$$P(t) = \frac{2\left(\left(\frac{\rho'}{\Lambda}\right)^2 + 2\frac{\gamma}{\Lambda}P_0\right)\left(\frac{\sqrt{\left(\frac{\rho'}{\Lambda}\right)^2 + 2\frac{\gamma}{\Lambda}P_0} + \frac{\rho'}{\Lambda}}{\sqrt{\left(\frac{\rho'}{\Lambda}\right)^2 + 2\frac{\gamma}{\Lambda}P_0} - \frac{\rho'}{\Lambda}}\right)e^{-\left(\sqrt{\left(\frac{\rho'}{\Lambda}\right)^2 + 2\frac{\gamma}{\Lambda}P_0}\right)t}}{\frac{\gamma}{\Lambda}\left(1 + \left(\frac{\sqrt{\left(\frac{\rho'}{\Lambda}\right)^2 + 2\frac{\gamma}{\Lambda}P_0} + \frac{\rho'}{\Lambda}}{\sqrt{\left(\frac{\rho'}{\Lambda}\right)^2 + 2\frac{\gamma}{\Lambda}P_0} - \frac{\rho'}{\Lambda}}\right)e^{-\left(\sqrt{\left(\frac{\rho'}{\Lambda}\right)^2 + 2\frac{\gamma}{\Lambda}P_0}\right)t}\right)^2}, \quad (6)$$

$$E(t) = \frac{\frac{\rho'}{\Lambda} + \sqrt{\left(\frac{\rho'}{\Lambda}\right)^2 + 2\frac{\gamma}{\Lambda}P_0}}{\frac{\gamma}{\Lambda}} \cdot \frac{\left(1 - e^{-\left(\sqrt{\left(\frac{\rho'}{\Lambda}\right)^2 + 2\frac{\gamma}{\Lambda}P_0}\right)t}\right)}{1 + \left(\frac{\sqrt{\left(\frac{\rho'}{\Lambda}\right)^2 + 2\frac{\gamma}{\Lambda}P_0} + \frac{\rho'}{\Lambda}}{\sqrt{\left(\frac{\rho'}{\Lambda}\right)^2 + 2\frac{\gamma}{\Lambda}P_0} - \frac{\rho'}{\Lambda}}\right)e^{-\left(\sqrt{\left(\frac{\rho'}{\Lambda}\right)^2 + 2\frac{\gamma}{\Lambda}P_0}\right)t}}. \quad (7)$$

Considering the third assumption in equations (6) and (7) the maximum power P_{max} can be predicted:

$$P_{max} \approx \frac{\rho'^2}{2\gamma\Lambda} \quad (8)$$

and total energy released during pulse E:

$$\lim_{t\to\infty} E(t) \cong \frac{2\rho'}{\gamma}. \quad (9)$$

The total energy released does not depend on the average generation time and is therefore the same for fast and thermal reactors. Unlike the total energy released, the maximum power is inversely proportional to the average generation time and is therefore higher in fast reactors than in thermal reactors, although the total energy released is comparable. The average generation time in the thermal reactor is $\Lambda_t = 10^{-4} - 10^{-3}$s and in the fast reactor is $\Lambda_f = 10^{-8} - 10^{-7}$s. In the case of an effective temperature reactivity coefficient of fuel the inverse proportionality of both the maximum power and the total energy released can be seen, which is desirable from a safety point of view, because this part minimizes the effect of an increase in reactivity.

In addition to the maximum power and the total energy released, there is another limit value of the pulse experiment, full width at half maximum (FWHM). Full width at half maximum is defined as:

$$FWHM = t_2 - t_1, \quad (10)$$

where t_1 and t_2 represents times, when pulse reaches half of the maximum power:

$$\frac{P_{max}}{2} = P(t_1, t_2). \quad (11)$$

Full width at half maximum is determined as:

$$FWHM = \frac{3.525\Lambda}{\varphi'}. \quad (12)$$

In addition to the Fuchs-Hansen model, there is also the Nordheim-Fuchs model [11], which on the basis of the same assumptions with a slightly different execution procedure, leads to the same limit values.

V. Comparison of Theoretical Models and Experimental Values

The data from pulse experiments that have been carried out on the TRIGA Mark II reactor at the Jožef Stefan Institute has been collected and are publicly available at *http://trigapulse.ijs.si/*. The purpose was to analyse all the pulses that have been performed so far and to make a comparison between the theoretical models (Fuchs-Hansen model and Nordheim-Fuchs model) and performed pulse experiments. Limit values, maximum power, total energy released and full width at half maximum, are compared for theoretical models and experiments.

In Figure 5, the maximum power is represented depending on the square prompt reactivity. The black line shows the theoretical predictions and the blue dots show performed pulse experiments. The grey area shows uncertainty of the theoretical prediction. It was estimated that the uncertainties of the parameters of the theoretical predictions are uncorrelated. Relative uncertainty of effective delayed neutron fraction (β) has been estimated at 9 % [12], of average lifetime of prompt neutrons (l) was estimated at 8 % [12], relative uncertainty of inserted reactivity (ρ) has been estimated at 5 % [13] and relative uncertainty of effective temperature reactivity coefficient of fuel has been estimated at 50 % [6]. Where is five or more experimental data at a certain reactivity a statistical analysis is made. Statistical analysis indicates maximal (max) and minimal (min) values, average, median and deviation 1σ. It can be seen that experimental data are mostly within estimated uncertainty of theoretical prediction. In the case of experimental data, there is a considerable uncertainty in determining the inserted reactivity and between the various pulses the core composition changes (number and position of fuel elements), which affects the maximum power between the pulse and consequently total released energy and full width at half maximum.

experimental measurements, is determined by an integral of the power curve. It is important to set the proper boundaries for integration to capture only the response of prompt neutrons and omit the delayed neutrons as foreseen in the theoretical models. The boundaries were set in three different ways (see Figure 6). The first method is 1000 pts, where the power curve is integrated 1000 measurements before and after the maximum power value. The second method is the 1 %, where the integration of the power curve is performed from 1 % of maximum power to 1 % of maximum power. The third method is the tangent, where the result is an integral of the power curve between the points determined by the tangents of the power graph on 1/10 of the maximum power value.

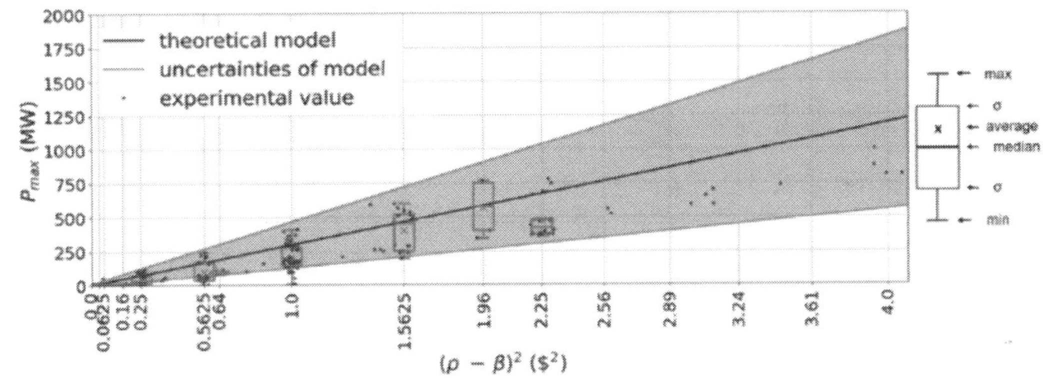

Figure 5: Comparison of the theoretical models with experimental data for maximum power.

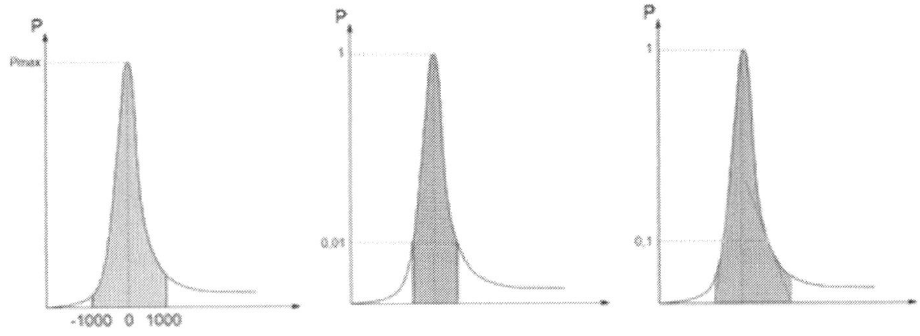

Figure 6: The determination of the total energy released using three methods: 1000pts (left), 1% (centre) and tangent (right).

The experimental values of maximum power and total energy released can be divided into two parts because of different number of fuel elements in the core.

A comparison between the theoretical models and the measurements was also made for the total energy released. The total energy released, in the case of

In Figure 7, the results of the experimental data of the total energy released (1000pts green, 1% blue and tangent red) and the theoretical models (black line) of the total released energy flow in dependence of the inserted reactivity are shown. In case of five or more experimental data at a certain reactivity a statistical analysis is made and with grey area uncertainty of theoretic model is shown. As in the case of maximum power, it can be observed that the experimental values follow the theoretical predictions, but at low inserted reactivities the experimental data are outside of determined uncertainties of theoretical model. It can also be observed that all three modes of determination of total energy released are in good agreement among themselves, especially in the case of large inserted reactivity.

statistical analysis is made. It can be noted that the match between theoretical predictions and experimental data is rather poor. Many experimental data are outside the estimated uncertainty of the theoretical models. The worst are data at low inserted reactivities, which was expected. The reason for bad matching is the high uncertainty in determining the full width of half maximum from the experimental data, which is very high especially at low inserted reactivities.

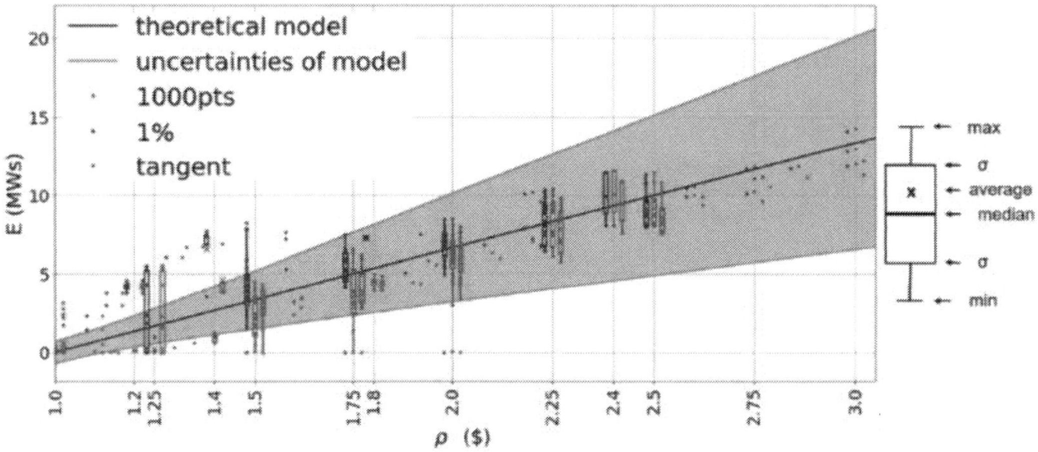

Figure 7: The comparison of the theoretical models with the experimental data for total energy released with three different integration models.

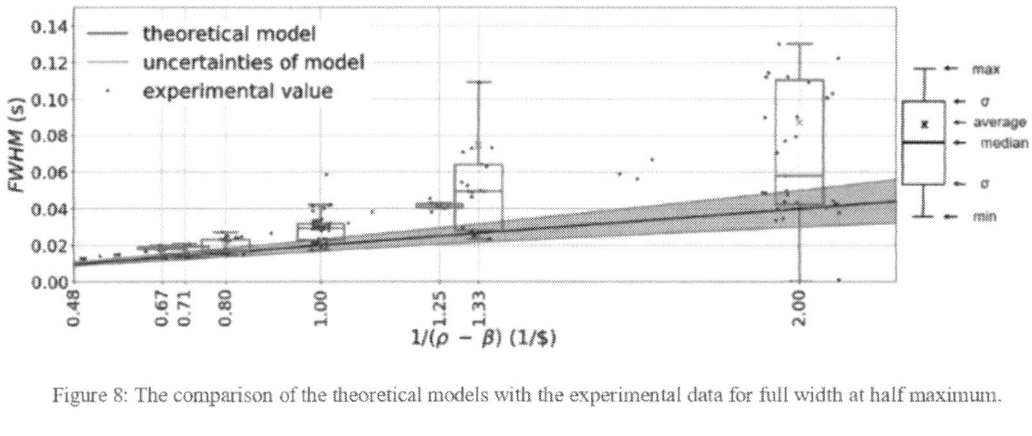

Figure 8: The comparison of the theoretical models with the experimental data for full width at half maximum.

The comparison of theoretical models and experiments for full width at half maximum is presented in Figure 8. As in case of maximum power the black line shows the theoretical predictions and the blue dots show performed pulse experiments. The grey area shows uncertainty of the theoretical prediction. Where is five or more experimental data at a certain reactivity a

CONCLUSION

Purpose of utilisation was to collect all data from pulse experiments in order to support pulse experimental campaigns where pulse operation can be used for experimental validation of reactor transient modelling. Also, to achieve MTR relevant testing conditions at the

JSI TRIGA Mark II reactor a pulse mode operation is used where can be achieved 1GW for approximately 5 ms - 10 ms. The pulse mode is a well-characterized and representative radiation field which will be used to support future experimental campaigns, where nuclear instrumentation and data acquisition systems for fast transient measurements will be tested. All experimental data of pulse experiments are publicly available at *http://trigapulse.ijs.si/*. A comparison of measured pulse physical parameters (maximal power, total released energy and full width at half maximum) was made with theoretical predictions from the Fuchs-Hansen and the Nordheim-Fuchs models.

It was shown that the experimental data of maximum power and total energy released are mostly within the estimated uncertainty of the theoretical model. They differ most at low inserted reactivities, where background noise is disturbing at determination of experimental data, and theoretical models can only be applied at higher reactivities. At full width at half maximum the deviations are greater due to the greater uncertainty in determining values from the experimental data.

REFERENCES

[1] Goričanec, T., et al., Evaluation of neutron flux and fission rate distributions inside the JSI TRIGA Mark II reactor using multiple in-core fission chambers, Ann. Nucl. Energy, 111 (2018) 407-440.

[2] Štancar, Ž., et al., Computational validation of the fission rate distribution experimental benchmark at the JSI TRIGA Mark II research reactor using the Monte Carlo method, Ann. Nucl. Energy, 112 (2018) 94-108.

[3] Radulović, V., et al., Measurements of miniature ionization chamber currents in the JSI TRIGA Mark II reactor demonstrate the importance of the delayed contribution to the photon field in nuclear reactors. Nucl. Instrum. Meth. A, 804(2015) 149-154.

[4] Ambrožič, K., et. al., Delayed gamma determination at the JSI TRIGA reactor by synchronous measurements with fission and ionization chambers. Nucl. Instrum. Meth. A, 911(2018) 94-103.

[5] Žerovnik, G., et al., Validation of the neutron and gamma fields in the JSI TRIGA reactor using in-core fission and ionization chambers, Appl. Radiat. Isot. 96(2015) 27-35.

[6] Pregl, Varnostno poročilo za reaktor TRIGA Mark II v Podgorici, Revision 7, IJS-DP-10675, Ljubljana, March 2017.

[7] Ž. Štancar, L. Snoj, L. Barbot, Reaction Rate Distribution Experiments at the Slovenian JSI TRIGA Mark II Research Reactor, TRIGA-FUND-RESR-002, International Handbook of Evaluated Reactor Physics Benchmark Experiments. Paris: NEA/NSC/DOC(2006)1, OECD NEA, 2017.

[8] L. Snoj, A. Trkov, October 2012. Temperaturni koeficient reaktivnosti. Ljubljana.

[9] A. Trkov, L. Snoj, M. Ravnik, 2.10.2013. Reaktorska in radiacijska fizika, študijsko gradivo. Ljubljana.

[10] G. I. Bell, S. Glasstone, 1970. Nuclear Reactor Theory. New York, p. 517.

[11] D. L. Hetrick, Dynamics of Nuclear Reactors, American Nuclear Society, Illinois, 1993.

[12] Filliatre, Experimental assessment of the kinetic parameters of the JSI TRIGA reactor, Annals of Nuclear Energy, pages 236-245, September 2015.

[13] A. Trkov, M. Ravnik, H. Wimmer, B. Glumac, H. Boeck, Application of the rod-insertion method for control rod worth measurements in research reactors, Kerntechnik (1987), pages 255-261, 1995.

EPJ Web of Conferences **225**, 04028 (2020) https://doi.org/10.1051/epjconf/202022504028
ANIMMA 2019

Comparison of neutron/gamma separation qualities of various organic scintillation materials

Zdeněk Matěj
Faculty of Informatics
Masaryk University
Brno, Czech Republic

Filip Mravec
VF, a.s.
Černá Hora, Czech Republic

Aleš Jančář
VF, a.s.
Černá Hora, Czech Republic

Michal Košťál
Research Center Řež Ltd
Husinec-Řež, Czech Republic

František Kučera
Central European Institute of Technology
University of Technology
Brno, Czech Republic

František Cvachovec
Faculty of Informatics
Masaryk University
Brno, Czech Republic

Václav Přenosil
Faculty of Informatics
Masaryk University
Brno, Czech Republic

Zdeněk Kopecký
VF, a.s.
Černá Hora, Czech Republic

Jiří Čulen
VF, a.s.
Černá Hora, Czech Republic

Abstract— **In this work we compare the pulse-shape discrimination (PSD) properties of EJ-299-33A, BC-501A, stilbene, p-terphenyl and Hidex Aqualight in neutron field generated by the LVR-15 reactor with silicon filter utilization. Pulses from the scintillators are processed by Neutron-Gamma spectrometer. This spectrometric system with fast digitizer card contains two analog-digital (A/D) converters with a resolution of 12 bits and sampling frequency 500 MHz. For photomultiplier (PMT) linearity improvement active divider has been used. Measured data from scintillators have been processed using the integration method and compared. Results are presented.**

Keywords— organic scintillators, neutron, gamma, impulse separation, FPGA, FoM, LVR-15

I. Introduction

The scintillators that enable the separation of neutron-gamma radiation are widely used for measurements of mixed radiation fields. Several competing types of PSD capable scintillators were compared (EJ-299-33A, BC-501A, stilbene, p-terphenyl and Hidex Aqualight).The reactor LVR-15 with well-known radiation field has been used for measurements in the energy range 1 to 10 MeV [1, 2]. Measured data have been processed using digital Neutron-Gamma spectrometer which we developed. The digital spectrometric system minimizes the drawbacks of analog devices and provides substantial simplification for the operation and setting of the system. The spectrometer is connected with a computer via Ethernet.

II. Experimental Setup

Laboratory with experimental reactor LVR-15 has been used for the measurement, see Fig. 1.

We utilized well defined moderate neutron beam from the horizontal channel HK-1, see Fig. 2. Reactor thermal power output was 10 MW. Neutron beam has been moderated via 1 m wide silicon single crystal which provides spectrum with characteristic energy peaks, see unfolded spectrum in Fig. 3 [4].

The experimental arrangement is shown in Fig. 2. Influence of thermal neutrons and gammas was reduced via filter composed of ^{6}Li, Cd, Bi and Pb.

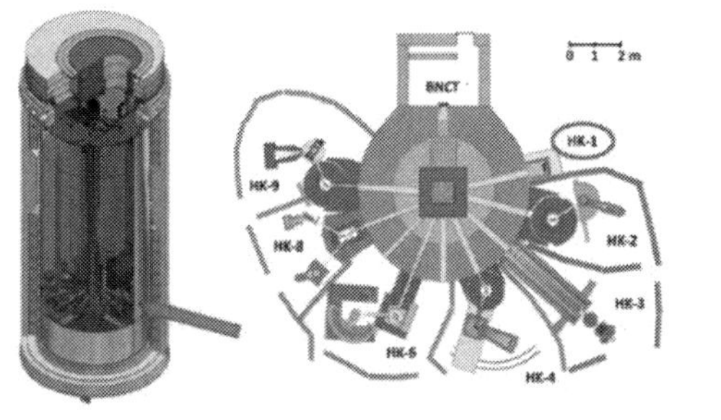

Fig. 1. - LVR-15 reactor (left) and its horizontal channels (right) [1]

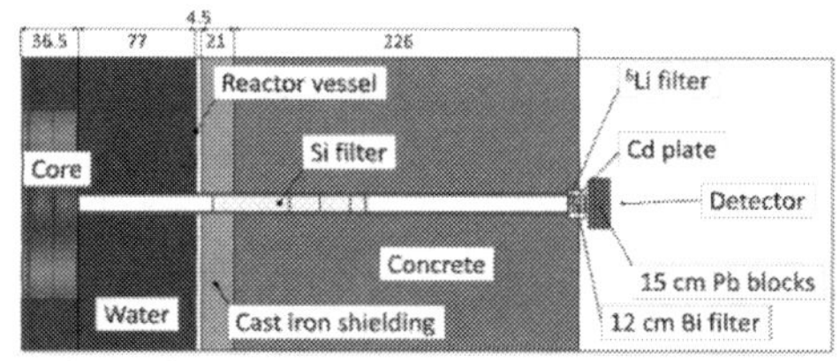

Fig. 2. The experimental arrangement in the reactor laboratory. The detectors were placed along the beam axis at 1.5 m distance from the silicon filter target [1].

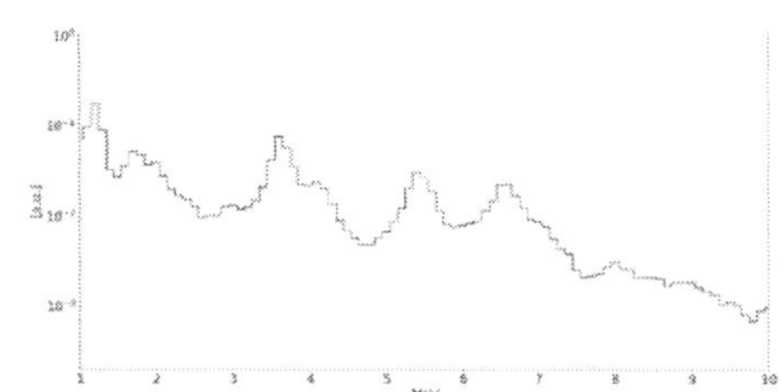

Fig. 3. Unfolded silicon moderated neutron spectrum measured with stilbene.

III. Results

Integration method based on the pulse charge comparison was implemented in FPGA. Separation parameters for various scintillators calculated with this method are shown below in PSD matrices.

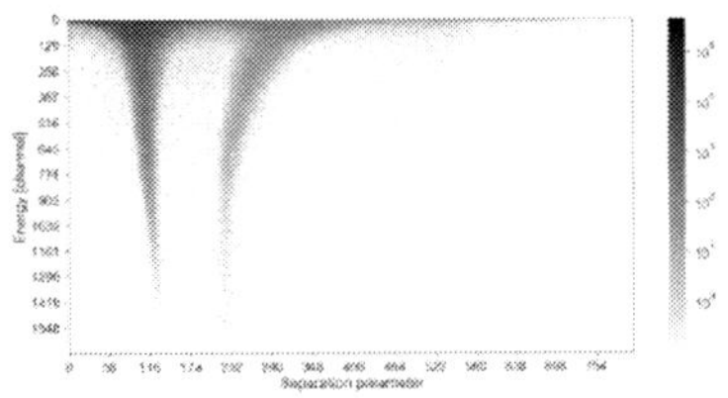

Fig. 4. Stilbene

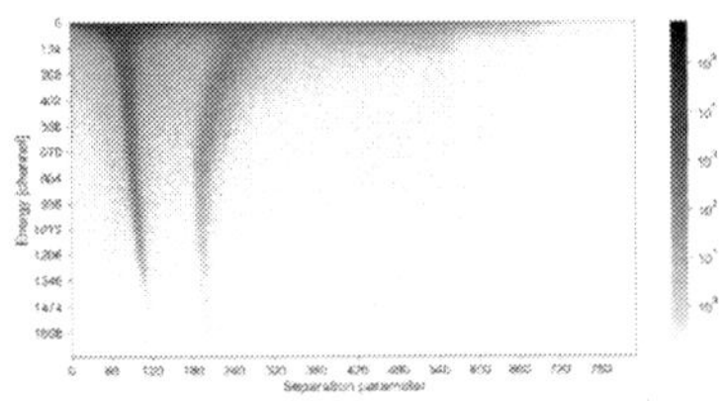

Fig. 5. BC-501A

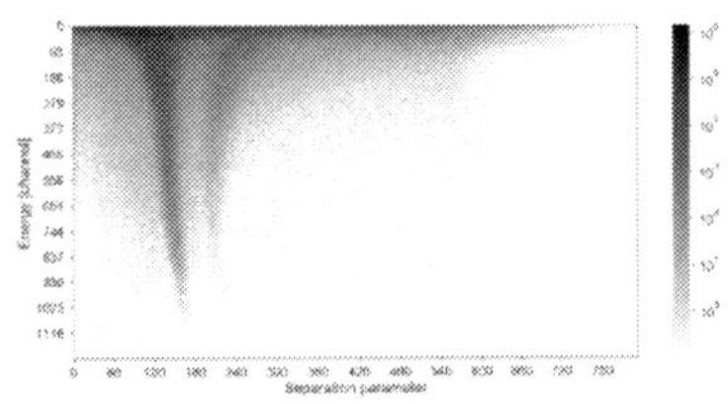

Fig. 6. EJ-299-33A

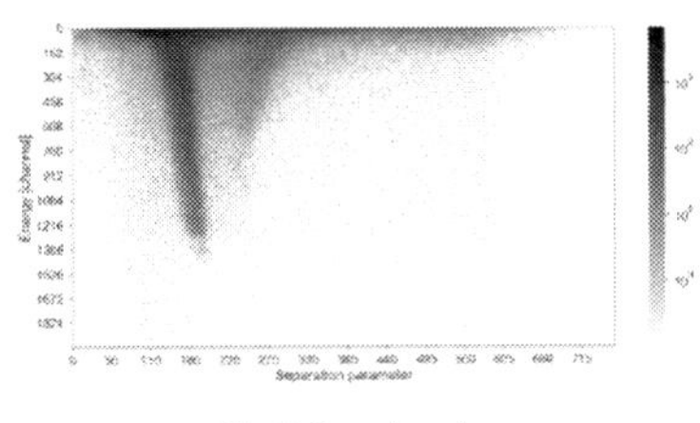

Fig. 7. P-terphenyl

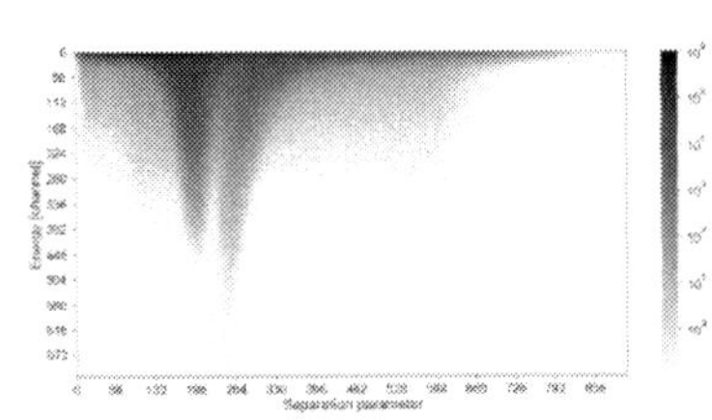

Fig. 8. Hidex Aqualight

Neutron-Gamma spectrometric system has been used for the measurements. The input analog signal from the detector was digitized by two fast A/D converters working with a sampling frequency of 500 MHz. Digital signal processing is implemented in FPGA Xilinx Virtex-6.

IV. FIGURE OF MERIT

The quality of neutron–gamma discrimination for a given scintillator is characterized by figure of merit (FOM), see Fig. 9.

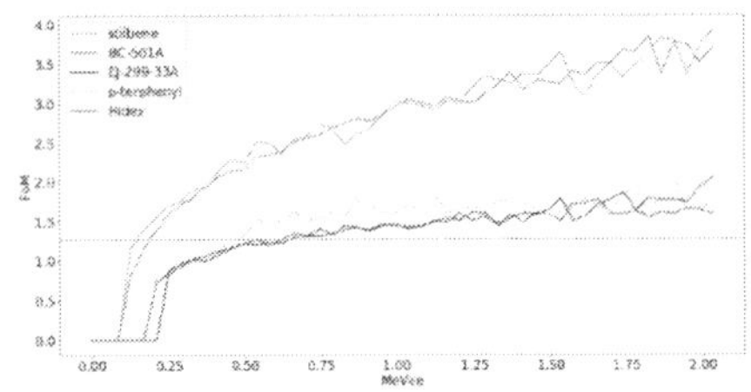

Fig. 9 FoMs for examined scintillators

Selected results of neutron-gamma separation quality are shown in Tab. 1. The critical value of FoM 1.27 is an indicator of good neutron-gamma separation [3]. Yellow fields indicate gamma energies where FoM values exceed the critical value. According to the combination of FoM and energy parameters we can evaluate the quality of neutron-gamma separation. Stilbene detector achieves the best quality of separation because the FoM exceeds the critical value at the smallest energy.

MeVee	0,12	0,17	0,2	0,29	0,5	0,62	0,66
stilbene	1,16	1,4	1,55	1,75	2,23	2,36	2,54
BC-501A	0,82	1,15	1,39	1,72	2,15	2,39	2,51
p-terphenyl	0	0	0	0,92	1,33	1,51	1,63
EJ-299-33	0	0	0,73	0,94	1,21	1,27	1,35
Hidex	0	0	0	0,97	1,21	1,22	1,3

Tab 1. Selected critical values of FoM from examined scintillators.

Among the examined scintillators the stilbene detector is considered to have the best energy resolution. Apparatus spectra from the two best scintillators from Tab. 1 were compared with stilbene detector in Fig 10. The x-axis contains channel number, y-axis contains ratio E_s/E_x, where E_s is apparatus spectrum of stilbene and E_x is a spectrum for BC-501A and p-terphenyl.

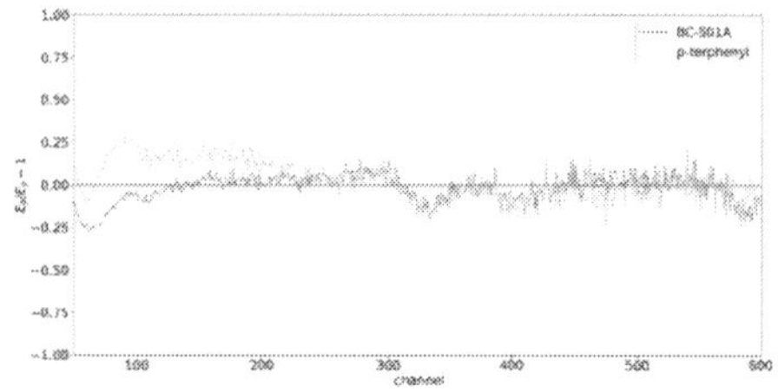

Fig. 10 E_s/E_x - 1 comparison of BC-501A and p-terphenyl with stilbene

ACKNOWLEDGMENT

The presented work has been supported by Ministry of Industry and Trade, within the project TRIO, No. FV20453. Presented results were obtained with the use of the infrastructure Reactors LVR-15 and LR-0, which is financially supported by the Ministry of Education, Youth and Sports - project LM2015074 and also financially supported by the Ministry of Education, Youth and Sports Czech Republic - project LQ1603 Research for SUSEN. This work has been realized within the SUSEN Project (established in the framework of the European Regional Development Fund (ERDF) in project CZ.1.05/2.1.00/03.0108 and of the European Structural Funds and Investment Funds (ESIF) in the project CZ.02.1.01/0.0/0.0/15_008/0000293), which is financially supported by the Ministry of Education, Youth and Sports - project LM2015093 Infrastructure SUSEN.

REFERENCES

[1] KOŠŤÁL, M., M. SCHULC, J. ŠOLTÉS, E. LOSA, L. VIERERBL, Zdeněk MATĚJ, F. CVACHOVEC a V. RYPAR. Measurements of neutron transport of well defined silicon filtered beam in lead. Applied Radiation and Isotopes, Elsevier, 2018, roč. 2018, č. 142, s. 160-166. ISSN 0969-8043.

[2] KOŠŤÁL, Michal, Jaroslav ŠOLTÉS, Ladislav VIERERBL, Zdeněk MATĚJ, František CVACHOVEC, Vojtěch RYPAR a Evžen LOSA. Measurement of Neutron Spectra in a silicon filtered neutron beam using stilbene detectors at the LVR-15 research reactor. Applied Radiation and Isotopes, 2017, roč. 128, s. 41-48.

[3] N. Zaitseva, B. L. Rupert, I. PaweŁczak, A. Glenn, H. P. Martinez, L. Carman, M. Faust, N. Cherepy, S. Payne, Plastic scintillators with efficient neutron/gamma pulse shape discrimination, Nuclear Instruments and Methods in Physics Research Section A: Accelerators, Spectrometers, Detectors and Associated Equipment, Volume 668, 2012, Pages 88-93

[4] CVACHOVEC, Jiří a František CVACHOVEC. Maximum likelihood estimation of a neutron spectrum and associated uncertainties. Advances in Military Technology, Brno: Univerzita obrany, 2008, Vol. 3, No. 2.

EPJ Web of Conferences 225, 05001 (2020) https://doi.org/10.1051/epjconf/202022505001
ANIMMA 2019

Performance assessment of amplification and discrimination electronic devices for passive neutron measurements

Ben Mosbah M.[1*], Eleon C.[1], Passard C.[1], Loridon J.[1], Perot B.[1], Barbot L.[2] and Grassi G.[3]

***Abstract*– The knowledge of the fissile material mass is a key challenge to enhance radioactive waste management and to ensure a high level of safety in nuclear industry. Data is analyzed according to the principles of the neutron measurement techniques. As proportional counters filled with ^{3}He gas display high neutron detection efficiency and a good gamma-ray discrimination, they are the reference detector for passive neutron coincidence counting. A charge preamplifier or a current amplifier, depending on applications, collects the electric pulse produced by neutron interaction in the ^{3}He gas and a threshold discriminator produces a logic pulse used for neutron counting. This paper describes the performance assessment of different commercially available electronics from Mirion Technologies, Precision Data Technology (PDT), Mesytec, as well as MONACO electronics originally developed by CEA LIST for fission chamber measurements in experimental reactors. Comparative passive neutron measurements are carried out with these electronics at CEA/DEN Nuclear Measurement Laboratory in Cadarache. Overall, PDT and Mesytec electronics show similar detection efficiency as the ACH-NA98 charge amplifier, which is commonly used in our laboratory for such applications. However, MONACO electronics have a lower detection efficiency, similar to Mirion 7820 current amplifier used in specific high-count rate applications. An optimisation of MONACO settings would probably be necessary to adapt to ^{3}He counters instead of fission chambers.**

I. INTRODUCTION

The knowledge of the fissile material mass is a key challenge to enhance radioactive waste management and to ensure a high level of safety in nuclear industry.

Proportional counters filled with ^{3}He gas display high neutron detection efficiency and good gamma-ray discrimination. Therefore, they are an obvious choice for neutron detection, especially for radioactive waste characterization [1]. An amplifier and a threshold discriminator transform the analog pulse of the counter into a logic signal for neutron counting purpose. The amplification and discrimination electronics is essential to ensure linear counting with respect to the quantity of plutonium, especially to avoid count losses due to dead time and pile-up [2].

This paper describes an experimental benchmark of commercially available electronics. The 7820 current amplifier and the ACH-NA98 charge amplifier from Miron Technologies, which have been widely qualified at the Nuclear Measurement Laboratory (LMN) of CEA/DEN in Cadarache , are used as reference electronics. The ACH-NA98 module allows the best performances in terms of detection efficiency, while 7820 amplifier is more robust in case of high-count rate and high gamma radiation level. The PDT10A and PDT10M+PDT12S modules from Precision Data Technology, as well as MRS2000-1 and MRS2000-2 amplifiers from Mesytec, have been compar]ed to our reference electronics. CEA's prototype amplifier MONACO is also studied: this device was originally designed for fission chamber measurements in experimental reactors [3]. All tested electronics are plugged to the same ^{3}He detectors to compare their performances.

The experimental programme has been carried out at the LMN. The first section describes the experimental setup, the detection system, and the electronics. Experiments are presented in the second section. The detector efficiency and die-away time are determined for each configuration.

II. EXPERIMENTAL SETUP

Two polyethylene detection blocks wrapped with a 2 mm cadmium envelop and covered with a 5 mm B_4C plug are used (see Fig. 1 and 2). They can host up to seven 65NH45 ^{3}He proportional counters from Canberra (currently Mirion Technologies). However, for these tests, each detection block contains only two ^{3}He detectors, the other cavities being filled with polyethylene cylinders. The sensitive length of the counters is 45 cm, their diameter is 2.5 cm (1 inch) and their internal ^{3}He pressure is 4×10^5 Pa (4 bar).

A 32 channels MEDAS acquisition board from Cesigma dates logic TTL signals from the amplification/discrimination electronics with minimum time bins of 100 ns [4]. Its clock frequency is 80 MHz.

*Corresponding author: mehdi.benmosbah@cea.fr
[1] CEA, DEN, Cadarache, DTN, SMTA, Laboratoire de Mesures Nucléaires, F-13108 Saint-Paul-Lez-Durance, France.
[2] CEA, DEN, Cadarache, DER, SPESI, LDCI, F-13108 Saint-Paul-Lez-Durance, France.
[3] Orano Group, Recycling Business Unit, F-92084 Paris-La Défense, France.

The distance between the two detection blocks is 40 cm. A ^{252}Cf source placed between the two blocks is used to determine their detection efficiency and die-away time.

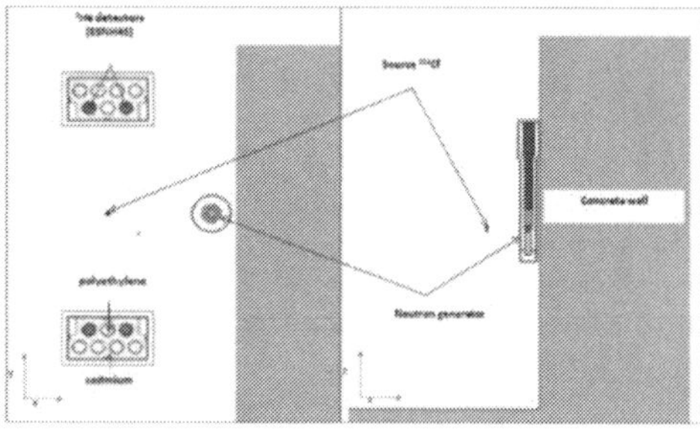

Figure 1: Setup of passive detection tests with a ^{252}Cf source placed in the middle and at mid height of the two detection blocks.

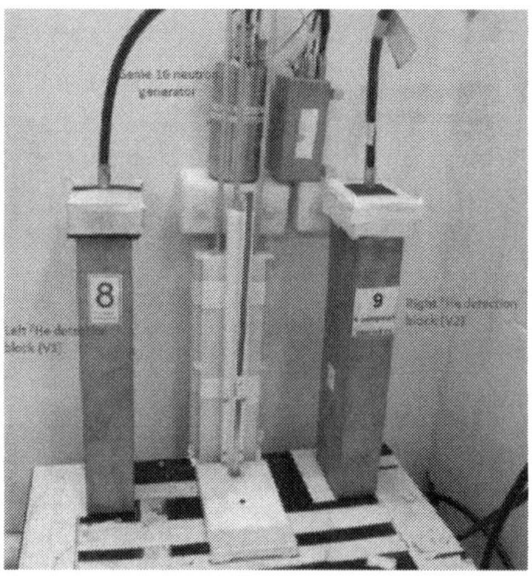

Figure 2: Picture of the experimental setup.

III. Electronics settings

A. The 7820 current amplifier

The 7820 current amplifier is combined with 7821 power supply from Mirion Technologies. This amplifier is designed for high-count rate (up to 5.10^5 counts per second) and high gamma dose rates (up to 200 mGy/h with ^3He detectors). In addition, ^3He detectors can be connected to such current amplifier with very long cables (up to 100 m) while with charge amplifiers short cables are mandatory (maximum 1 m).
The connection between the 7820/7821 electronics and the detector is done with a unique coaxial cable with a high immunity to electromagnetic perturbation. The cable is adapted at an impedance of 50 Ω at the input of the amplifier.

The operating High Voltage is 1250 V and the discrimination threshold is 400 mV.

B. PDT electronics

PDT10A and PDT12S+PDT10M (Figure 3) are compact neutron modules [5] designed for high radiation environments. The PDT10A is directly connected to ^3He detectors. The PDT12S module is directly connected to detectors and linked with a cable (30 m maximum length) to PDT10M module. This last allows remote settings of the gain and the discrimination threshold. Each instrument contains a precision wide band charge sensitive pre-amplifier followed by an amplifier, pulse shaping and discriminator circuits. An U8001A Agilent device provides the low voltage + 12V power supply of PDT modules. A Cyberstar multichannel power supply provides the High Voltage.

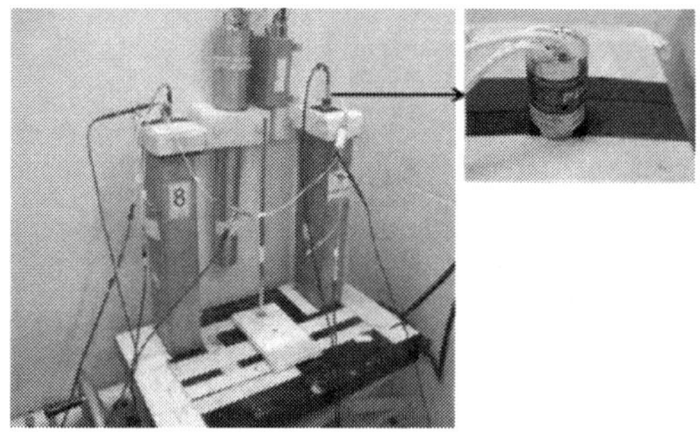

Figure 3 : PDT10A module connected directly to ^3He detection block #8 (on the left), and PDT12S module connected to detection block #9 (on the right) linked to the PDT10M module. The right picture is a zoom on the PDT12S module.

The PDT modules provide an analog monitoring test point allowing output signal visualization with an oscilloscope, and a TTL pulse used for neutron counting and dating with MEDAS acquisition card.
The operating High Voltage for PDT10A and PDT12S+PDT10M are respectively 1225 V and 1175 V and the discrimination threshold is not adjustable however it seems that it is appropriate for the measurements (good discrimination between neutron pulses and noise/gamma pulses, see next sections).

C. Mesytec electronics

Both the Mesytec MRS-2000-1 and MRS2000-2 devices (see Figure 4) include a preamplifier, a pulse shaper, a window discriminator and the corresponding TTL output [6].

 https://doi.org/10.1051/epjconf/202022505001

The Mesytec MRS-2000 model is a square shape amplifier with baseline restorer. The experimental tests were carried out with MRS2000-1 designed for very fast counting (pulse FWHM = 1 μs for a maximum count rate of 150,000 s^{-1} indicated by the manufacturer) and with MRS2000-2 designed for standard use (pulse FWHM = 3 μs FWHM for a maximum count rate of 50,000 s^{-1}).

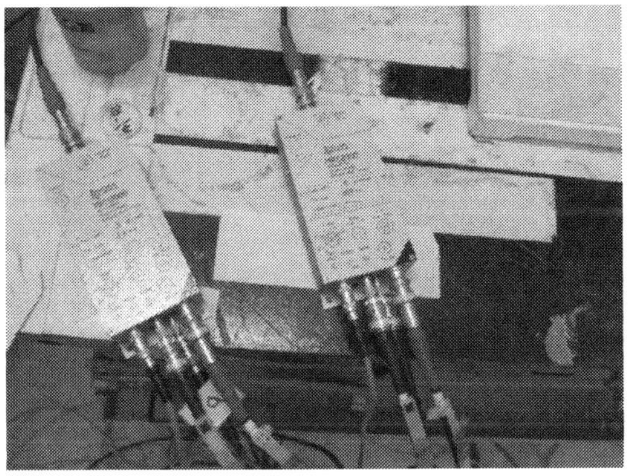

Figure 4 : Mesytec modules (MRS2000-1 on the left and MRS2000-2 on the right).

We set a 1075 V operating high voltage for both MRS2000-1 and MRS2000-2 modules. The discrimination threshold was set to 100 mV.

D. MONACO electronics

We tested the performance of a MONACO ("Multichannel Online Neutron Acquisition in Campbell mOde") prototype, which was originally developed for fission chamber measurements in experimental reactors. This electronics works in different modes (pulse, current, Campbell) and we use here the pulse-counting mode.
The MONACO device (see Figure 5) is packed in a 19-inch standard NIM mechanics. Four HN connectors – High Voltage of the Neill connector – allows the connection of up to four detectors. MONACO also provides the High Voltage for the ^{3}He detectors. An analog output enables to display the signal with an oscilloscope. Additionally, a control interface allows adjusting the HV and threshold discrimination values. The reference [3] outlines further details about the MONACO prototype.

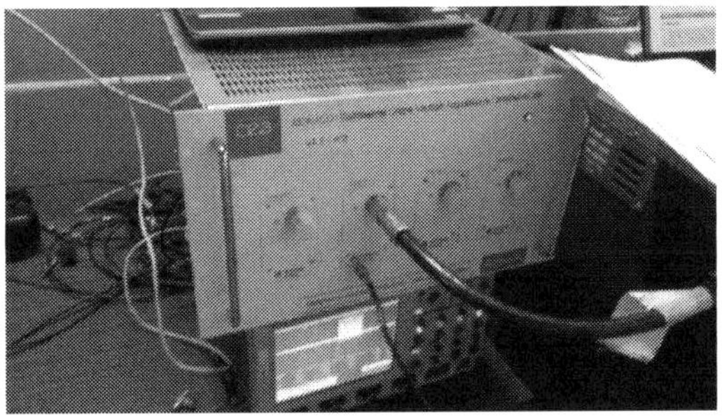

Figure 5 : MONACO prototype electronics.

As MONACO does not provide any TTL output for neutron counting, a Single Channel Analyzer (SCA) was used to convert the analog signal into a TTL pulse that is counted by MEDAS acquisition card. An inverter module also converts the negative analog signal into positive pulses that can be processed by the SCA.
The High Voltage is 1200 V.
Regarding discrimination threshold settings, we use a 10 mV value to eliminate electronics noise. Note that MONACO electronics delivers a very small analog pulse height of less than 50 mV because its gain was not optimized for ^{3}He detectors but for fission chambers. Therefore, discrimination threshold is very small (10 mV) compared to other tested electronics (larger than 100 mV with 7820, Mesytec, and PDT modules).

IV. COMPARATIVE TESTS

A. Setup and measured parameters

Measurements are carried out with a ^{252}Cf source placed between the two ^{3}He detection blocks. Its neutron emission rate at the date of measurements is 44893 ± 823 n/s. We first report here below the main features of the analog output pulse which are: average pulse height, time width (FWHM), overshoot or undershoot, pulse height distribution and discrimination threshold position with respect to electronic noise and gamma background, on the one hand, and neutron pulses on the other hand. Then we present the main parameters for total and coincidence counting [7], i.e. neutron detection efficiency and die-away time in the detection blocks.

B. ACH-NA98 charge amplifier

The ACH-NA98 charge amplifier, which is the reference for neutron detection efficiency in this work, was not calibrated and tested in the same conditions because all its characteristics (pulse amplitude and width, Rossi-α curve, etc.) were already available. They were obtained from previous measurements [8] and are reported below (see Table I). Only its detection efficiency was determined in this work, for further comparisons.

TABLE I : CHARACTERISTICS OF ACH-NA98 CHARGE AMPLIFIER

Parameter	Value
Operating High Voltage	1400 V
Discrimination threshold	200 mV
Pulse FWHM	1 µs
Mean pulse amplitude	1 V
Undershoot	200 mV
Die-Away Half Time	32 µs
Detection efficiency	1 (reference device)

In the following section, we present detailed results for the PDT devices only. For other devices, results are summarized in section V.

C. PDT electronics

The oscilloscope screenshot of Figure 6 shows the output signals of the PDT10A module. The average pulse height of the analog signal is 161 mV and its width (FWHM) is 1.41 µs, to be compared to the FWHM of 350 ns obtained with the 7820/7821 electronics. Note that these average values for PDT10A, have been determined for 1226 pulses.

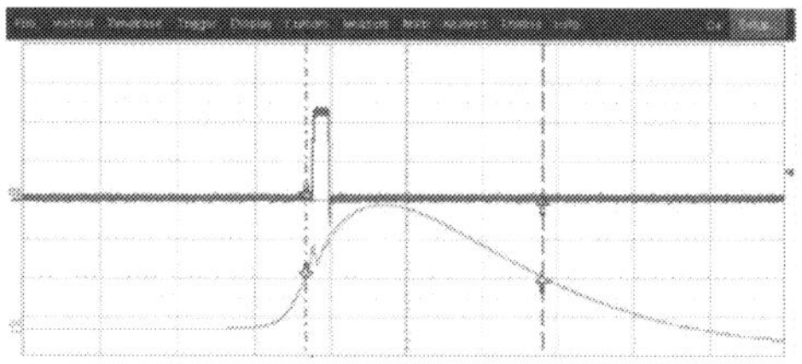

Figure 6 : Average analog signal (green; 50 mV/div) and logical TTL signal (pink; 2 V/div) for the PDT10A module (time base 500 ns/div).

As observed in Figure 7, the analog pulse has a noticeable undershoot of about 20 mV and returns to the baseline only after 45 µs. This undershoot can induce count losses because another pulse arriving in this time interval might mistakenly be rejected by the low discrimination threshold due to its smaller apparent height.

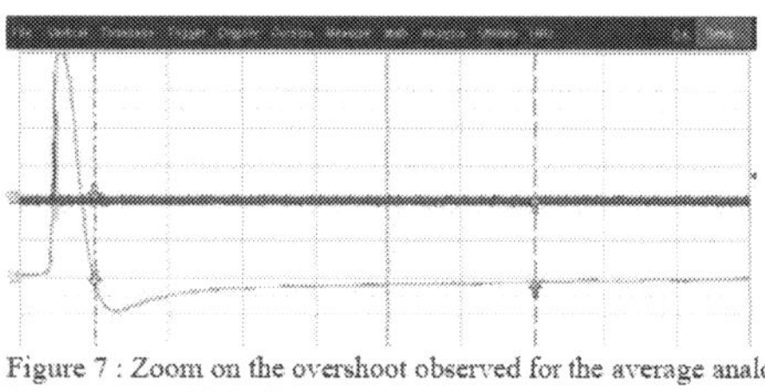

Figure 7 : Zoom on the overshoot observed for the average analog pulse (green; 20 mV/div) of the PDT10A module (time base 5 µs/div).

Figure 8 presents the pulse height distribution of the analog signal of the PDT10A module, which varies from 65 to 375 mV. This Figure illustrates that the discrimination threshold (corresponding to the factory value for ^{3}He detectors) allows removing the electronic noise.

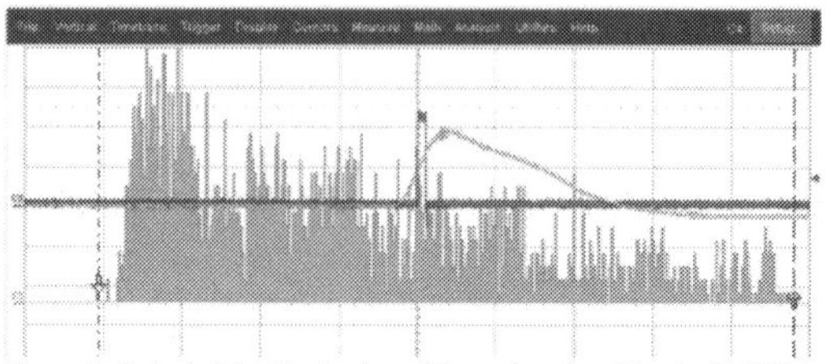

Figure 8 : Pulse height distribution of the analog signal for the PDT10A module starting at 65 mV (3 events/div; 35 mV/div for the horizontal scale).

Figure 9 displays the average analog pulse of the PDT12S+PDT10M module for 1197 events. The average pulse height is 190 mV and its average FWHM is 1.6 µs.

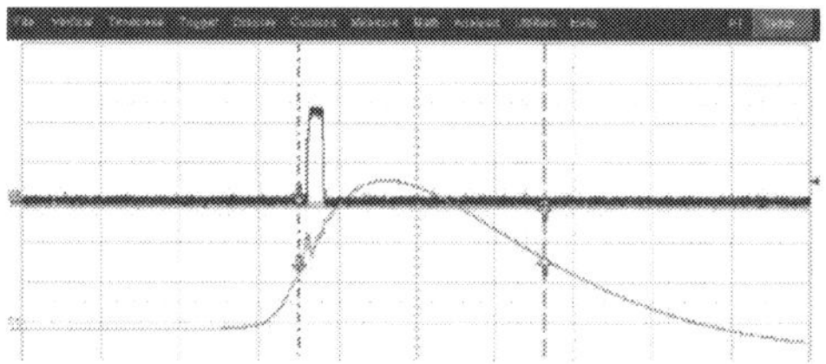

Figure 9 : Average analog signal (yellow; 50 mV/channel) and logical TTL signal (blue; 2V/channel) for the PDT12S+PDT10M module (time base 500 ns/channel).

Figure 10 shows that the PDT12S+PDT10M pulse also exhibits an undershoot of 50 mV and returns to the baseline only after 60 µs.

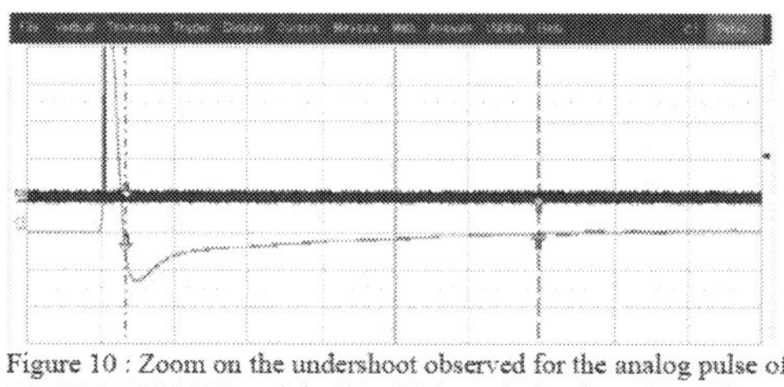

Figure 10 : Zoom on the undershoot observed for the analog pulse of the PDT12S+PDT10M module (20 mV/channel; time base 10 µs/channel).

The pulse height distribution, shown in Figure 11 varies from 67 mV to 549 mV. As observed for the PDT10A module, the pulse height distribution seems to indicate that the discrimination threshold (corresponding to the factory value, which is not specified by the manufacturer) to cut the electronic noise.

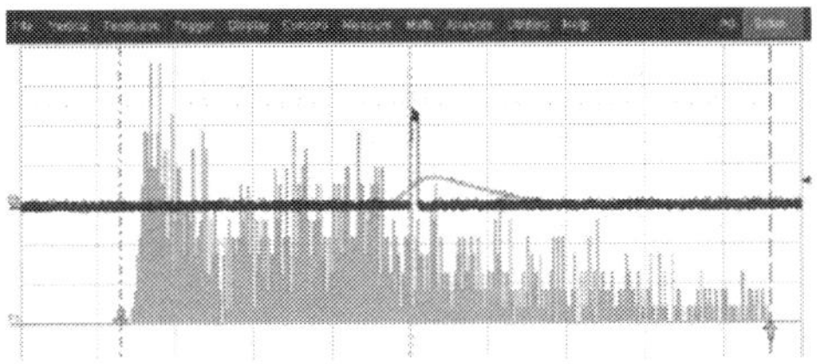

Figure 11 : Pulse height distribution of the analog signal for the PDT12S+PDT10M module starting at 67 mV (2.25 events/channel ; 44 mV/div).

In conclusion, as PDT charge amplifiers deliver much longer pulses (~ 1.4 to 1.6 µs FWHM) than 7820 current amplifier (80 ns FWHM), they are more sensitive to pile-up effects at high neutron or gamma count rates.

The experimental Rossi-α curves obtained with a ^{252}Cf source are presented in Figure 12. The die-away half times are obtained with the fit of a decreasing exponential function: 35.98 ±1.18 µs [1σ] for PDT10A module and 35.10 ±1.15 µs [1σ] for PDT12S+PTD10M module. These values of the same order of magnitude as the Die Away Half Time measured for the 7820 amplifier: 45.1 ± 5.5 µs [1σ].

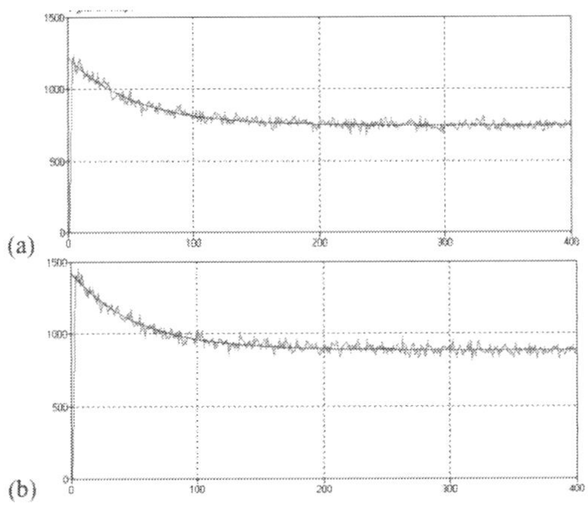

Figure 12 : Rossi-α curves (histogram of pulse arrival times after a first detection, in µs): (experimental in red; adjustment in blue) for PDT10A (a) and PDT12S+PTD10M (b).

We finally determine the experimental detection efficiency reported in with the ^{252}Cf neutron source. For PDT12S+PDT10M module, it is similar to the reference efficiency obtained with ACH-NA98 charge amplifier, which provides a very good signal-to-noise ratio and therefore small neutron count losses due to electronics threshold. In contrast, it is not the case for the PDT10A device, which displays a detection efficiency that is 30% lower (ratio vs. ACH-NA98 of 0.7).

TABLE II : PDT vs. ACH-NA98 detection efficiencies. the measured data of ACH-NA98 are available in table II.

Parameter	PDT10A	PDT12S+PDT10M
Measured count rate (c/s)	171.59 ± 0.18	258.4 ± 0.21
Detection efficiency	0.38 ± 0.01	0.57 ± 0.01
Ratio vs. ACH-NA98	0.7	1

V. Summary

Table III summaries the main characteristics of the amplification/discrimination electronics tested in this paper.

TABLE III: COMPARISON OF MAIN CHARACTERISTICS OF TESTED ELECTRONICS.

Device	Pulse FWHM (µs)	Pulse Amplitude (mV)	Undershoot (mV)	Die Away Half Time (µs)	Detection Efficiency Ratio (ACH-NA98 = 1)
7820 amplifier	0.35	1177	100	33	0.7
ACH-NA98	1	1000	200	32	1
PDT10A	1.41	161	20	35.98	≈1
PDT12S+ PDT10M	1.57	186	50	35.1	≈1
MRS2000-1	1.55	1531	5	37	≈1
MRS2000-2	3.44	658	10	40	≈1
MONACO	1.5	69	none	35.4	0.7

The time width (FWHM) of the output pulse of the PDT10A, PDT12S+PDT10M, MRS2000-1 and MONACO devices is similar, around 1.5 µs. The FWHM of the 7820 current amplifier pulse is much shorter (350 ns) and therefore, it can be operated at high count rate (up to 5 × 10^5 n/s [7]). Additionally, thanks to limited pulse pile-up, it can be operated at high gamma count rate (up to a 200 mGy.h^{-1} gamma dose rate [7]; note that neither high count rate, nor high gamma level tests have been performed in this work). Finally, the MRS2000-2 analog pulse FWHM is the longest (3.4 µs) of all tested devices and would leads to the largest signal processing dead time.

Detection efficiencies of all tested electronics have been compared to that of the reference ACH-NA98 charge amplifier. The PDT10A, PDT12S+PDT10M, MRS-2000-1 and MRS2000-2 modules give similar efficiencies as ACH-NA98, whereas MONACO is comparable to 7820 current amplifier with about 30 % count losses.

Finally, the neutron die-away half time in the detection blocks varies between 35 and 45 µs for all tested electronic devices, which shows that all these electronics can be used for passive neutron coincidence counting.

VI. Conclusion

In the field of a technology survey, five electronics have been tested for passive neutron counting with ^{3}He gas proportional counters and compared to 7820 current amplifier and ACH-NA98 charge amplifier, both from Mirion technologies, which are commonly used at Nuclear Measurement Laboratory of CEA, DEN, Cadarache. The new tested electronics are the

PDT10A and PDT12S+PDT10M modules from Proportional Technologies, the MRS2000-1 and MRS2000-2 modules from Mesytech, and finally the MONACO prototype electronics from CEA LIST.

The PDT and Mesytech devices lead to an equivalent detection efficiency to that of ACH-NA98 charge amplifier, which allows the best signal-to-noise ratio. However, detection efficiency obtained with MONACO is smaller, equivalent to that of Mirion 7820 current amplifier designed for high-count rate measurements with very short pulses to limit pile-up. Note that MONACO is designed for fission chamber measurements and its settings need to be adapted to ^{3}He counters.

Analog signals of all tested electronics display pulse widths of a few µs (FWHM), which is much longer than the 7820 current amplifier (350 ns), but without noticeable influence on neutron die-away time in the detections blocks made of polyethylene and cadmium surrounding ^{3}He counters. Consequently, all these electronics can be used for passive neutron coincidence counting with ^{3}He detectors at low and intermediate count rates. However, Mirion 7820 is recommended for high count rate applications, to limit pulse pile-up and improve gamma rejection.

This series of comparative tests will be continued in active mode with a DT pulsed neutron generator, in view of active neutron interrogation applications with the Differential Die-Away Technique [9] [10].

REFERENCES

[1] G.F. Knoll, Radiation Detection and Measurement. New York, John Wiley and Sons, 1999.

[2] A. Boffa, A. Fazzi, C. Pirovani and V. Varoli. "A single-supply, high rate, small size and cheap electronic chain for 3He neutron counters". Nuclear Science Symposium, 1996, IEEE Conference Record.

[3] M. Thevenin, L. Barbot, G. Corre, R. Woo, C. Destouches and S. Normand. "Digital Real-Time Multiple Channel Multiple Mode Neutron Flux Estimation on FPGA-based Device". Conference Paper in The European Physical Journal Conferences. May 2014.

[4] Cesigma Signal and Systems. [Online] Available : http://www.cesigma.net/ang/carte-medas-A.html

[5] Precision Data technology. "PDT10A-HN General Specifications". http://www.pdt-inc.com/products/10series/10series_datasheet.htm

[6] Mesytec. "MRS2000 data sheet Version 1.2_01". [Online] Available: http://www.mesytec.de/products/datasheets/MRS-2000.pdf.

[7] B. Pérot, F. Jallu, C. Passard, O. Gueton, PG. Allinei, L. Loubet, N. Estre, E. Simon, C. Carasco, C. Roure, L. Boucher, H. Lamotte, J. Comte, M. Bertaux, P. Fichet, F. Carrel, A. Lyoussi, "The characterization of radioactive waste: a critical review of techniques implemented or under development at CEA, France", EPJ Nuclear Sci. Technol. 4, 3 (2018) 1-24

[8] C. Passard. CEA Technical Report, unpublished. 2015.

[9] R. Antoni, C. Passard, B. Perot, F. Guillaumin, C. Mazy, M. Batifol, G. Grassi, Reduction of the uncertainty due to fissile clusters in radioactive waste characterization with the Differential Die-away Technique, Nuclear Inst. and Methods in Physics Research, A 895 (2018) 144–149

[10] R. Antoni, C. Passard, B. Perot, M. Batifol, J-C. Vandamme, G. Grassi, First Industrial Tests of a Matrix Monitor Correction for the Differential Die-away Technique of Historic Waste Drums, ANIMMA 2015, Advancements in Nuclear Instrumentation Measurement Methods and their Applications, 20-24 April 2015, Lisbon, Portugal.

EPJ Web of Conferences **225**, 05002 (2020) https://doi.org/10.1051/epjconf/202022505002
ANIMMA 2019

Improved electronics for ^{3}He based neutron counters

Eric Boogers, Alessandro Borella*, Riccardo Rossa
SCK-CEN
* aborella@sckcen.be

***Abstract*— Several non-destructive assays techniques have been developed for the measurement of fissile materials in the fields of dismantling, decommissioning, nuclear security, and nuclear safeguards. Among these techniques, neutron coincidence counting is based on the detection of time-correlated neutrons from induced and spontaneous fissions. ^{3}He Tubes have been the primary choice for neutron coincidence counting due to their high detection efficiency, rather low sensitivity to gamma-rays and proven field reliability. This paper covers the implementation of a new electronic setup to a Canberra WM3400 neutron coincidence counter.**

First we describe the properties of the used detectors, with focus on the characteristics of the default electronics and highlight its limitations such as the high input capacitance, short shaping time and the necessity for selected tubes. We then propose the new electronic setup to overcome these limitations. This setup includes a dedicated preamp for every tube , the possibility to adjust for gain differences between the tubes and a better optimised shaping time for ^{3}He detectors.

We carried out measurements with the two electronic systems to compare their performances in terms of gamma-ray sensitivity, efficiency and die-away time. The gamma ray sensitivity was measured with calibrated ^{137}Cs and a ^{60}Co sources at the Laboratory for Nuclear Calibration of the Belgian Nuclear Research Centre with dose rates between 10 µSv/h and 50 mSv/h. Measurements with a ^{252}Cf source were used to determine the die-away time of the system and the total measurement efficiency for the considered geometry.

The measurements showed that, with the default electronics, neutron count-rates are already affected by gamma radiation at a dose rate of 10÷30 µSv/h. On the other hand the neutron coincidence counter equipped with the new electronics proved to be insensitive to gamma-radiation up to a dose rate of at least 20 mSv/h. The high-voltage set with the new electronics is lower than in the case of the default electronics and is within the range recommended by the tubes manufacturer. The die-away time was not affected by the used electronics. A reduction of about 20% in the neutron detection efficiency due to the used discriminator threshold was observed.

***Index Terms*— ^{3}He counters, Electronics, Neutron coincidence counting, Non destructive assay**

I. Introduction

Several non-destructive assays (NDA) have been developed for the measurement of fissile materials in the fields of dismantling, decommissioning, nuclear security, and nuclear safeguards. Among the NDA techniques, neutron coincidence counting (NCC) is based on the detection of time-correlated neutrons from induced and spontaneous fissions [1]. Tubes filled with ^{3}He gas have been the primary choice for NCC due to their high detection efficiency, typically low sensitivity to γ-rays, and proven in-field reliability [2].

This paper covers the implementation of a new electronic system to the Canberra WM3400 neutron coincidence counter [3]. We describe the characteristics of the detectors used, with focus on the default electronics and the new system. We carried out measurements with the two electronic systems to compare the performance in terms of gamma-ray sensitivity, efficiency and die-away time. The conclusions from the comparison and the plans for future work are included at the end of paper.

II. WM3400 Neutron Coincidence Counters

Two used Canberra WM3400 neutron coincidence counters as well as two Canberra JSR-12 shift registers [4] for the data acquisition and analysis were acquired recently by SCK•CEN in the context of disposal of irradiated material containing small amounts of fissile material. Each detector has six ^{3}He tubes with 2.54 cm diameter and active length of 91.44 cm; the ^{3}He tubes are aligned in a single row and the ^{3}He tubes are embedded in polyethylene to ensure neutron moderation. The moderator is enclosed in an aluminium cover, and a 1-mm layer of Cd is applied to five sides of the moderator below the aluminium cover. Only the side facing the radioactive source is not covered by Cd.

A. Default Electronics

In the default electronics, all six ^{3}He tubes are connected in parallel in order to facilitate the distribution of the high voltage. The output signal of the tubes is capacitive coupled to a JAB-01 Amptek board. This board is equipped with an Amptek A111 hybrid preamplifier-discriminator with its threshold adjusted to minimize the gamma sensitivity [5] and some driver-logic. The output-signal is a 50 ns wide TTL signal which can be used for data acquisition e.g. with a neutron coincidence analyser such as the JSR-12 shift register [6] or the MCA-527 with time stamp processing firmware [7]. Both systems can provide the high voltage (HV) for the detectors and the +5V for the JAB-01.

The first problem with this configuration is the strong increase in input capacitance due to the fact that all the tubes

E. Boogers (eboogers@sckcen.be), A. Borella (aborella@sckcen.be) and R. Rossa (rrossa@sckcen.be) are with the SCK•CEN, Belgian Nuclear Research Centre, Boeretang 200 - BE-2400 Mol (Belgium).

are mounted together. This setup increases the input capacitance, reduces the signal/noise ratio and makes the preamplifier more prone to oscillations.

A second problem is related to the shaping time. The shaping time of the Amptek A111 is only 0.15 μs which is adequate for high count rates but results in a non-optimal separation of neutron and gamma signals. To overcome the consequent lack of gain, the manufacturer decided to operate the tubes well beyond their specification. While the normal high voltage operating range according to the manufacturer of the tubes is 1000 ÷ 1250 V, the recommended value in [5] is 1700 V. This fact further worsens the discrimination between neutron and gamma signals.

Finally, because there is no option to adjust the gain of the individual tubes, they were selected to match their responses at the same high voltage.

To address all these issues and to operate the tubes at a lower high voltage, we decided to develop new dedicated electronics.

B. New Electronics

In the new electronics the signal processing is split into a preamp unit and a processing unit as shown in Fig. 1. In the preamp unit each tube is equipped with its own preamp in a charge amplifier configuration. This setup reduces the input capacitance for each preamp giving a better signal to noise ratio and allowing to calculate the proper feedback capacitance/resistor to ensure enough phase-margin for stability. It also makes it easier to see if one of the tubes fails.

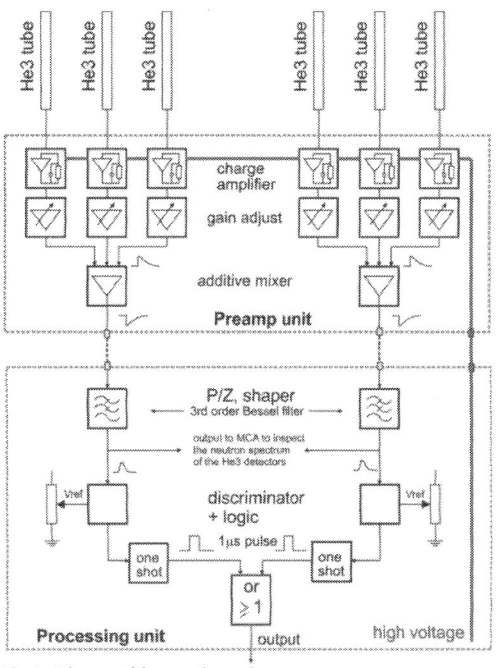

Fig. 1. Diagram of the new electronics.

The next step is an adjustable non-inverting amplifier. The resulting analogue output signals of three tubes are added together in the additive mixer. Adding the signals of three tubes allows to reduce the number of cables to the processing unit and the number of shapers. Adding more tubes together would increase the risk of pile-up. The resulting signals are then sent to a 3rd order Bessel filter (150 kHz) indicated as P/Z shaper in the processing unit.

The gains are adjusted so that all tubes have the same output-level. This setting is verified by measuring the signal with a Multi-Channel Analyser (MCA) just after the shaper. All tubes are adjusted so that the thermal peak of the ^{3}He spectrum ends up in the same channel. By doing so, we have just one discrimination level for all tubes and there is no need to use specially selected tubes anymore.

A measured energy deposition spectrum with a ^{252}Cf source is shown in Fig. 2. The peak at 765 keV corresponds to the energy deposition of tritons and protons following (n,p) reaction on ^{3}He. Noise and gamma signals are present in the low energy part of the spectrum.

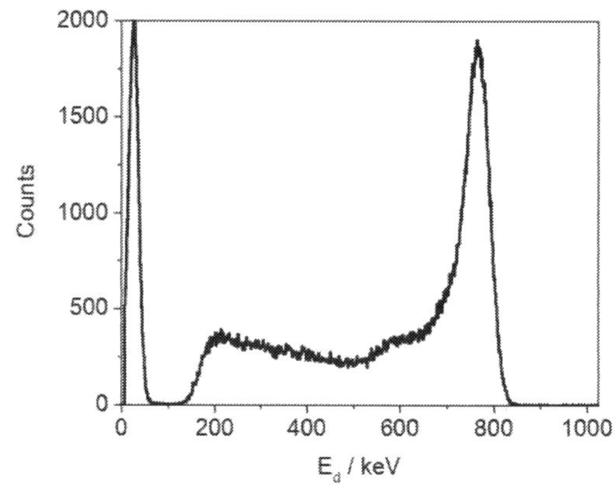

Fig. 2. Measured energy deposition spectrum. E_d stands for deposited energy.

The discrimination, as shown in Fig. 3, is done in the processing unit which is physically far away from the preamps in order to reduce crosstalk between them. The discriminator value should be set such that all events below a deposited energy of approximatively 191 keV are rejected. This value correspond to the minimum energy typically observed in the energy deposition spectrum and corresponds to the energy deposition of triton following (n,p) reaction on ^{3}He. By setting the discrimination threshold at 191 keV all neutrons interacting by (n,p) reaction on ^{3}He are detected. Adjusting the discriminator level is done by using an MCA. By delaying the signal and gating it with the output pulse of the discriminator we can visualise the effect of the adjustment. The output of the discriminator is fed to a non retriggerable one shot. As shown in Fig. 1, both channels are OR-ed and the output can go to the shift register. The high voltage is reduced to +1350 Volt, which is closer to the nominal operating range recommended by the

https://doi.org/10.1051/epjconf/202022505002

tubes manufacturer. At this voltage we observed a change in gain of 10% when the high voltage is changed by 10 Volt; therefore, it is important to use a high voltage supply unit with sufficient stability.

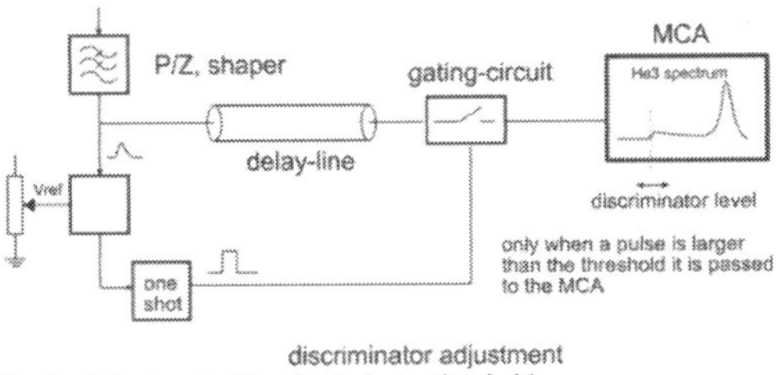

Fig. 3. Adjustment of the discriminator threshold.

III. MEASUREMENTS

We measured the gamma sensitivity with calibrated ^{137}Cs and ^{60}Co sources at the Laboratory for Nuclear Calibration of SCK•CEN [8] with dose rates between 10 μSv/h and 50 mSv/h. We recorded the total neutron counts as well as the number of counts in the R+A and A gate as a function of the dose rate. All measurements at different dose rates were carried out with and without a ^{252}Cf source (37 kBq on 15 August 2015) positioned in front of the detector in a fixed geometry. Background measurements were also carried out.-Most of the measurements lasted for 600 s.

In addition, measurements with the ^{252}Cf were used to determine the die-away time of the system and the total measurement efficiency for the considered geometry. The considered geometry consisted in positioning the source and geometrical centre of the front face of the WM3400 counter.

Both the new and the default electronics were tested with the JSR-12 and the MCA527.

IV. RESULTS

The neutron count rate and reals rate in absence of an external gamma source were taken as a reference. The change in the total neutron count rate without ^{252}Cf source as a function of the dose rate for the system with the default and new electronics are shown in Fig. 4. The variation in reals rate with the ^{252}Cf source, defined as the difference between the counts in R+A gate and in the A gate divided by the measurement time, as function of the dose rate is shown in Fig. 5. In the absence of an external gamma source, the total count rate without ^{252}Cf source was 1.5±0.1 cps, while the reals rate with the ^{252}Cf source was 9.6±0.5 cps.

For the default electronics, the gamma sensitivity data indicate that the neutron counts are affected already at a gamma dose rate of 10÷30 μSv/h. The reals rate is almost not affected up to a dose rate of about 1 mSv/h although a bias seems to be present.

With the new electronics the neutron count rate is unaffected by gamma radiation up to 20 mSv/h and the maximum available dose rate of 50 mSv/h was not sufficient to change the reals rate.

The results of these measurements provide confidence in our interpretation of the limitations of the default electronics.

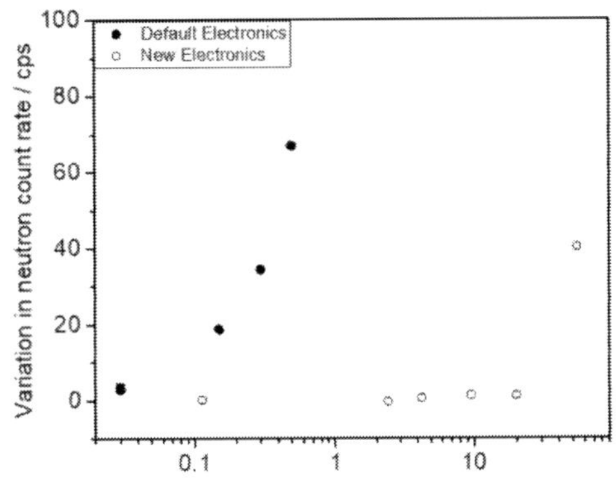

Fig. 4. Change in the total neutron count rate without ^{252}Cf source as a function of the dose for the system with the default and new electronics.

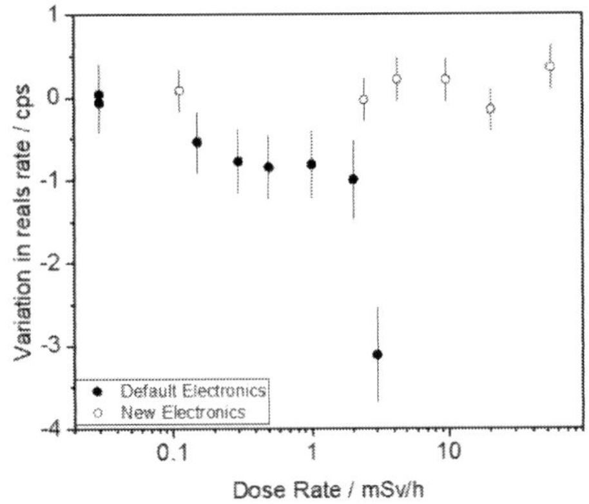

Fig. 5. Change in the reals count rate without ^{252}Cf source as a function of the dose for the system with the default and new electronics.

The total count rate and reals rate with a ^{252}Cf source can be used to determine the efficiency and the absolute source strength at the time of the measurement by applying the point model equation given in [10]. For the default electronics, we obtained an efficiency of 6.6%±0.1%. We used the nuclear data given in [11]. The die-away was obtained by applying Eq. 16-16 of Ref. [1] and the obtained value was 57.1±0.1 μs. The source strength was about 2150 n/s, consistent with the declared activity after accounting for its decay.

For the new electronics, the obtained source strength was in agreement with the results of the default electronics when taking into account the decay of the source. The obtained values for the efficiency and die away are respectively 5.1%±0.1% and 57.6±0.1 μs. While the die-away time is consistent with the one obtained with the default electronics, the efficiently is about 20% lower. Further investigations revealed that the discrimination threshold was set higher than the desired value of 191 keV.

V. Conclusions

This work focussed on an improved electronics for a ^{3}He neutron coincidence counter. The performance in terms of gamma-ray sensitivity, die-away time and efficiency were determined for a WM3400 neutron coincidence counter with a default and with a new signal processing electronics. The obtained results indicate that the neutron coincidence counter is insensitive to gamma-radiation up to a dose rate of at least 20 mSv/h. Due to the used value of the discrimination threshold a reduction of about 20% in the neutron detection efficiency was observed..

References

[1] N. Ensslin, "Principles of Neutron Coincidence Counting" in Passive Nondestructive Assay of Nuclear Materials, NuREG/CR-5550, LA-UR-90-732, 1991
[2] H.O. Menlove, "Neutron coincidence instruments and applications". in Passive Nondestructive Assay of Nuclear Materials, NuREG/CR-5550, LA-UR-90-732, 1991
[3] http://www.canberra.com/products/waste_safeguard_systems/pdf/WM3400-SS-0258.pdf. Last accessed 05/03/2018.
[4] http://www.canberra.com/products/waste_safeguard_systems/pdf/JSR12-SS-C28919.pdf. Last accessed 05/03/2018.
[5] Model 3400 Slab Neutron Counter – Hardware Reference Material, Canberra Industries Inc. 2009
[6] Model JSR-12 Neutron Coincidence Analyzer User's Manual, Canberra Industries Inc. 2006
[7] https://www.gbs-elektronik.de/media/download_gallery/mca527_neutron.pdf . Last accessed 19/04/2018
[8] http://www.amptek.com/pdf/a111.pdf
[9] F. Vanhavere, M. Coeck and F. Vermeersch, Radiat. Prot. Dosim. 93(1), 5 (2001).
[10] D.M. Cifarelli, W. Hage, "Models for a three-parameter analysis of neutron signal correlation measurements for fissile material assay", Nucl. Instr. and Meth. A, Vol. 251, Issue 3, Pages 550-563 (1986)
[11] W. Hage, K. Caruso, "A Preliminary Procedure for the Treatment of Time Correlation Analyser Measurement Data for Non-Destructive Assay on Pu", Technical Note No. 1.07.C3.86.115 (Nov. 1986)

EPJ Web of Conferences **225**, 05003 (2020) https://doi.org/10.1051/epjconf/202022505003
ANIMMA 2019

The use of self-induced X-ray fluorescence in gamma-ray spectroscopy of uranium ore samples

T. Marchais[1]*, B. Pérot[1], C. Carasco[1], J-L. Ma[1], P-G. Allinei[1], H. Toubon[2], R. Goupillou[3], J. Collot[4]
[1] DEN/CAD/DTN/SMTA/LMN
[2] ORANO Mining
[3] SEPA/SET ORANO Mining
[4] Laboratory of Subatomic Physic and Cosmology, Université Grenoble Alpes, CNRS/IN2P3
* thomas.marchais@cea.fr

***Abstract* — Gamma logging for uranium exploration are currently based on total counting with Geiger Müller gas detectors or NaI (Tl) scintillators. However, the total count rate interpretation in terms of uranium concentration may be impaired in case of roll fronts, when the radioactive equilibrium of the natural ^{238}U radioactive chain is modified by differential leaching of uranium and its daughter radioisotopes of thorium, radium, radon, etc. Indeed, in case of secular equilibrium, more than 95 % of gamma rays emitted by uranium ores come from ^{214}Pb and ^{214}Bi isotopes, which are in the back-end of ^{238}U chain. Consequently, these last might produce an intense gamma signal even when uranium is not present, or with a much smaller activity, in the ore. Therefore, gamma spectroscopy measurements of core samples are performed in surface with high-resolution hyper-pure germanium HPGe detectors to directly characterize uranium activity from the 1001 keV gamma ray of ^{234m}Pa, which is in the beginning of ^{238}U chain. However, due to the low intensity of this gamma ray, i.e. 0.84 %, acquisitions of several hours are needed. In view to characterize uranium concentration within a few minutes, we propose here a method using both the 92 keV gamma ray of ^{234}Th and the 98.4 keV uranium X-ray. This last is due to uranium self-induced fluorescence caused by gamma radiations of ^{214}Pb and ^{214}Bi, which create a significant Compton scattering continuum acting as a fluorescence source and resulting in the emission of uranium fluorescence X-rays. The comparison of the uranium activity obtained with the 92 keV and 98.4 keV lines allows detecting a uranium heterogeneity in the ore. Indeed, in case of uranium nugget, the 92 keV line leads to underestimated uranium concentration due to gamma self-absorption, but on the contrary the 98.4 keV line leads to an overestimation because of increased fluorescence. In order to test this new approach, several tens of uranium ore samples have been measured with a handheld HPGe FALCON 5000 detector.**

***Index Terms*— Self-induced X-ray fluorescence, gamma-ray spectroscopy, germanium detector, uranium mining, MCNP6**

I. Introduction

Current gamma measurements used to characterize the uranium content of ore samples are mainly based on two estimators. The first one is the total gamma count rate, which is approximately 90 % due to gamma radiations of ^{214}Pb and ^{214}Bi daughters (see further Fig. 1), in the back-end of ^{238}U chain. It allows short acquisition times but it is subject to potential imbalances in this decay chain. The second one is the 1001 keV gamma ray of ^{234m}Pa daughter in the beginning of ^{238}U chain, which is therefore not subject to a potential disequilibrium of the decay chain. Moreover, the sample mineralogy has little impact on attenuation of this radiation with a relatively high energy. However, its emission intensity is very small (0.847%) and its detection requires acquisition times of several hours, see Fig. 1. The objective of this study is to measure uranium content using more intense lines at lower energy, i.e. the 92 keV gamma ray of ^{234}Th and ^{235}U, and the 98 keV line due to uranium self-induced fluorescence (and to a lesser extend to a ^{234m}Pa gamma ray). This study is performed with samples of crushed ore filling a PVC holder. The gamma spectra are measured with a Falcon 5000 [1] handheld spectrometer from Mirion-Canberra, equipped with a HPGe planar detector (BEGE 2830 model). The method is studied by MCNP6 Monte Carlo calculation by the Nuclear Measurement Laboratory of CEA, DEN, Cadarache, and validated with experimental data obtained by ORANO Mining during a measurement campaign intended to characterize uranium content and imbalance of the samples [2].
The two rays at 92 keV and 98 keV are systematically present with a high intensity on uranium ore spectra (see example of a small concentration sample in Fig. 1), which makes it possible to envisage a significant measurement time reduction with respect to 1001 keV line.

II. Analysis of the Gamma Spectrum

A. Typical gamma spectrum of a uranium ore sample

Fig. 1 shows a typical HPGe high-resolution gamma spectrum of a real ore sample with a 419 ppm uranium concentration. The experimental setup is described further in Section II. In this section, we develop the theoretical formulae used to determine the uranium concentration from the net area of 1001 keV, 92 keV and 98 keV peaks, respectively. The concentration obtained with the 1001 keV gamma ray is considered in this work as the reference because it is less sensitive to attenuation effects compared to the two other low-energy radiations.

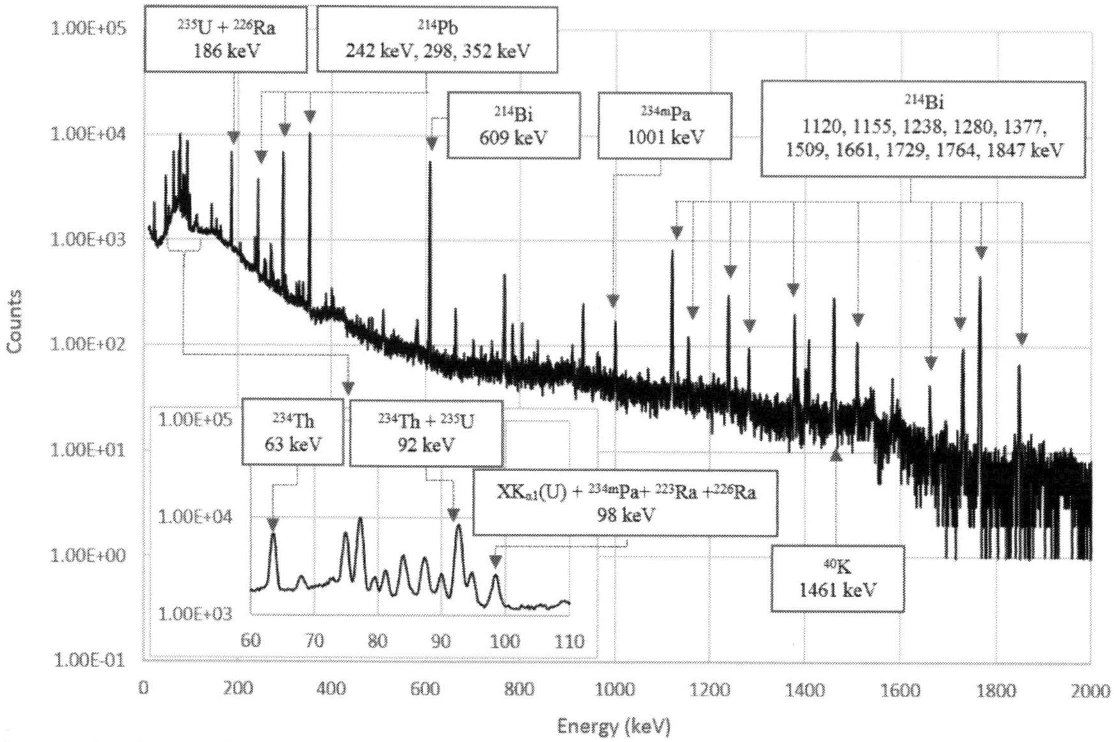

Fig. 1. Experimental spectrum of a 479 g ore sample with an uranium concentration of 419 ppm$_U$ and a density of 1.46 g.cm^{-3} with for 7503 seconds of acquisition

B. Radioactive gamma emission at 1001 keV

Equation (1) allows converting the net area of the 1001 keV peak into uranium mass concentration in ppm (i.e. in mg of uranium per kg of ore, dimensionless unit). This equation has been already used and validated during a measurement campaign in Bessines (France) calibration facility [3].

$$Cm_u(1001\ \text{keV}) = \frac{S_n(1001\ keV)}{Tc \times M_s \times CE_{1001}} \quad (1)$$

Where :

$$CE_{1001} = \frac{Eff_{1001keV} \times I_{1001keV}(^{234m}_{91}Pa) \times Na \times \ln(2)}{(1+\eta) \times 10^6 \times M(^{238}_{92}U) \times T_{\frac{1}{2}}(^{238}_{92}U)}$$

With:

- $Cm_u(1001\ \text{keV})$: uranium mass concentration (in ppm) of the sample calculated thanks to the gamma ray at 1001 keV,
- $S_n(1001\ keV)$: net area of the peak at 1001 keV measured on the gamma spectrum after subtraction of the Compton continuum under the peak (number of counts, dimensionless number),
- Tc: acquisition "live time" (i.e. real time corrected for dead time) of the gamma spectrum (s),
- M_s : mass of the sample (g),
- $Eff_{1001keV}$: detection efficiency calculated with MCNP [4] (number of count in the full-energy peak at 1001 keV per photon of 1001 keV emitted in the sample),
- $I_{1001keV}(^{234m}_{91}Pa)$: intensity of the 1001 keV gamma ray of ^{234m}Pa (number of gamma radiation emitted per disintegration: 0.847 % [5]) in the ^{238}U decay chain,
- Na: Avogadro constant ($6{,}023 \cdot 10^{23}$ atoms per mole, dimensionless number),
- η: mass fraction of ^{235}U in natural uranium (0.72 %),
- $M(^{238}_{92}U)$: molar mass of ^{238}U (g),
- $T_{\frac{1}{2}}(^{238}_{92}U)$: ^{238}U radioactive period ($4.5\ 10^9$ a ≡ $141.0\ 10^{15}$ s).

C. Radioactive gamma and X-ray emissions at 92 keV

The "92 keV" peak is already used in ORANO Mining laboratories for fast uranium characterization. It is composed of close-in-energy lines mainly due to ^{235}U and ^{234}Th:

❶ the α-decay of ^{235}U towards ^{231}Th is followed by the emission of a thorium X-ray at 93.35 keV, after internal conversion of the excited ^{231}Th daughter. This X-ray contributes to the signal at 92 keV due to the finite resolution of the detector, with a tabulated emission intensity $I_{93keV}(^{235}_{92}U) = 5.75\ \%$ [5],
❷ ^{234}Th (in the beginning of the ^{238}U decay chain) emits two gamma rays at 92.38 keV and 92.80 keV following its β^- radioactive decay into ^{234}Pa. We use here the total intensity $I_{92keV}(^{234}_{90}Th) = 4.33\%$ [5],
As ^{235}U and ^{234}Th are located at the top of the uranium decay chains, the net area of the 92 keV peak is not subject to potential imbalances and correctly reflects the uranium concentration.

Equation (2) allows converting the net area of the 92 keV peak into uranium mass concentration.

$$Cm_U(92\,keV) = \frac{S_n(92\text{ keV})}{Tc \times M_{ech} \times CE_{92}} \quad (2)$$

Where:

- $CE_{92} = CE_{^{234}_{90}Th} + CE_{^{235}_{92}u}$
- $CE_{^{234}_{90}Th} = \frac{Eff_{92keV} \times I_{92keV}(^{234}_{90}Th) \times \ln(2) \times Na}{1.0072 \times 10^6 \times M(^{238}_{92}U) \times T_{\frac{1}{2}}(^{238}_{92}U)}$
- $CE_{^{235}_{92}u} = \frac{\frac{A(^{235}_{92}U)}{A(^{238}_{92}U)} \times Eff_{93keV} \times I_{93keV}(^{235}_{92}U) \times \ln(2) \times Na}{(1+\eta) \times 10^6 \times M(^{238}_{92}U) \times T_{\frac{1}{2}}(^{238}_{92}U)}$

With:

- $Cm_U(92\,keV)$: uranium mass concentration in ppm of sample calculated thanks to the 92 keV ray,
- Tc : acquisition "live time" (i.e. real time corrected for dead time) of the gamma spectrum (s),
- M_s : mass of the sample (g),
- $S_n(92\text{ keV})$: net area of the peak at 92 keV (number of counts, dimensionless number),
- $I_{92keV}(^{234}_{90}Th)$: total intensity of the two 92 keV close peaks of ^{234}Th (4.33 % [5]), in the ^{238}U decay chain,
- $Eff_{92\text{ keV}}, Eff_{93\text{ keV}}$: detection efficiency calculated with MCNP (number of count in the full-energy peak at 92 keV or 93 keV, respectively per photon of 92 keV or 93 keV emitted in the sample),
- $\frac{A(^{235}_{92}U)}{A(^{238}_{92}U)}$: activity ratio of ^{235}U and ^{238}U (0.046 in case of natural enrichment, i.e. 0.72 % of ^{235}U),
- $I_{93keV}(^{235}_{92}U)$: emission intensity at 93 keV of ^{235}U (5.75% [5]).

D. *Radioactive emission at 98 keV*

The 98 keV peak is dominated by four contributions:

❶ the X-$K_{\alpha 1}$ fluorescence line of uranium at 98.44 keV. The net area of this peak does not follow linearly uranium concentration, contrary to the above radioactive-decay emissions, but it rises in a quadratic way because of the combined increase of fluorescence source intensity (gamma emitters of the uranium chain) and uranium quantity itself (which undergoes fluorescence).

❷ the same 98.44 keV X-ray but emitted after the β^- disintegration of ^{234m}Pa (at the beginning of the ^{238}U decay chain) into ^{234}U, followed by internal conversion. The tabulated emission intensity is $I_{98keV}(^{234m}_{91}Pa) = 0.316\ \%$,

❸ two close 97.53 keV and 97.85 keV X-rays linked to ^{223}Ra (^{235}U decay chain) α-decay into ^{219}Rn, followed by internal conversion of the excited daughter ^{219}Rn. The tabulated total intensity for these two X-rays is $I_{92keV}(^{223}_{88}Ra) = 2.72\ \%$ [5],

❹ the same 97.53 keV and 97.85 keV X-rays but linked to ^{226}Ra (^{238}U decay chain) α-decay towards ^{222}Rn, again followed by internal conversion. The tabulated total intensity is here $I_{92keV}(^{226}_{88}Ra) = 0.036\%$ [5].

Concerning the first contribution, each gamma ray emitted by the uranium decay chain can cause the fluorescence of uranium after undergoing Compton scattering and photoelectric absorption in the sample. In order to quantify the ability of a photon of energy E to induce uranium fluorescence, the $\eta_{fluo}(E)$ yield defined in equation (3) is calculated by MCNP (in g^{-1} units). This fluorescence yield is specific to the geometry of the sample: uranium content, filling height, density, mineralogy.

$$\eta_{fluo}(E) = \frac{S_n(XK_{\alpha 1})}{S_n(E) \times Cm_U \times M_s} \quad (3)$$

With:

- $S_n(XK_{\alpha 1})$: net area of the $XK_{\alpha 1}$ fluorescence line of uranium at 98.4 keV (number of counts, dimensionless number),
- $S_n(E)$: net area, on the MCNP spectrum, at the simulated energy E (number of counts, dimensionless number),
- Cm_U: uranium mass concentration (in ppm) in the MCNP model of the sample,
- M_s : mass of the sample (g).

A typical fluorescence yield calculation with a mono-energy 200 keV photon is shown in Fig. 2.

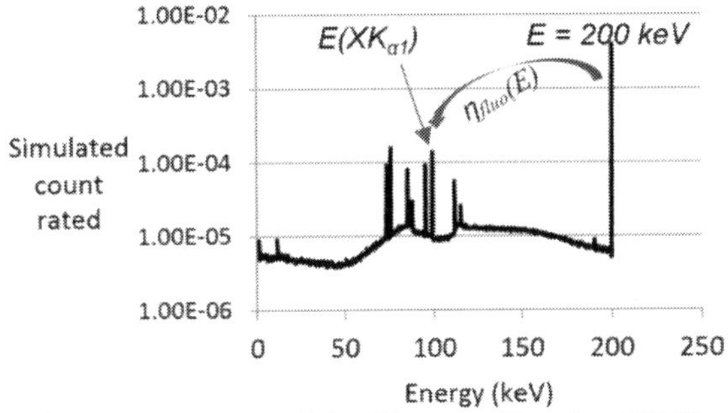

Fig. 2. Gamma spectrum and induced fluorescence X-rays for a 200 keV mono-energy photon source.

The fluorescence yield and the three radioactive decay emissions listed above allow us to convert the net area of the 98 keV peak into a uranium mass concentration using equation (4). It is important to notice that, in order to limit the number of simulations, only the six more intense gamma rays were taken into account to calculate the fluorescence yield: 185 keV (^{214}Pb), 242 keV (^{214}Pb), 298 keV (^{214}Pb), 352 keV (^{214}Pb), 609 keV (^{214}Bi) and 1120 keV (^{214}Bi). As they represent 78 % of the total fluorescence signal, a correction factor F equal to 0.78 is used in equation (4).

$$Cm_U(98\,keV) = \frac{S_n(98\text{ keV})}{Tc \times M_S \times CE_{98}} \quad (4)$$

Where:

- $CE_{98} = CE_{^{226}_{88}Ra} + CE_{^{223}_{88}Ra} + CE_{^{234m}_{91}Pa} + \frac{\sum_{E_\gamma}\left[\eta_{fluo}(E_\gamma, d, h) \times S_n(E_\gamma)\right]}{Tc \times F}$
- $CE_{^{226}_{88}Ra} = \frac{\frac{A(^{226}_{88}Ra)}{A(^{238}_{92}U)} \times Eff_{98keV} \times I_{98keV}(^{226}_{88}Ra) \times \ln(2) \times Na}{(1+\eta) \times 10^6 \times M(^{238}_{92}U) \times T_{\frac{1}{2}}(^{238}_{92}U)}$

- $CE_{{}^{223}_{88}Ra} = \frac{\frac{A({}^{223}_{88}Ra)}{A({}^{238}_{92}U)} \times Eff_{98keV} \times \ln(2) \times \mathcal{N}a \times I_{98keV}({}^{223}_{88}Ra)}{(1+\eta) \times 10^6 \times M({}^{238}_{92}U) \times T_{\frac{1}{2}}({}^{238}_{92}U)}$
- $CE_{{}^{234m}_{91}Pa} = \frac{Eff_{98keV} \times I_{98keV}({}^{234m}_{91}Pa) \times \ln(2) \times \mathcal{N}a}{(1+\eta) \times 10^6 \times M({}^{238}_{92}U) \times T_{\frac{1}{2}}({}^{238}_{88}U)}$

With:

- $Cm_U(98\ keV)$: uranium mass concentration (in ppm) of the sample calculated thanks to the 98 keV ray,
- Tc: acquisition "live time" (i.e. real time corrected for dead time) of the gamma spectrum (s),
- M_s : mass of the sample (g),
- $S_n(98\ \text{keV})$: net area of the peak at 98 keV (number of counts, dimensionless number),
- Eff_{98keV} : detection efficiency calculated with MCNP (number of count in the full-energy peak at 98 keV per photon of 98 keV emitted in the sample),
- $I_{98keV}({}^{223}_{88}Ra)$: emission intensity at 98 keV of ^{223}Ra (2.72 % [5]),
- $I_{92keV}({}^{226}_{88}Ra)$: emission intensity at 98 keV of ^{226}Ra (0.0351% [5]),
- $I_{92keV}({}^{234m}_{91}Pa)$: emission intensity at 98 keV of ^{234m}Pa (0.316% [5]),
- E_γ = [185, 242, 295, 352, 609, 1120 keV], energy of the 6 gamma rays selected to calculate the fluorescence yield,
- $\eta_{fluo}(E, d, h)$: fluorescence yield (in g^{-1}) as a function of gamma energy, sample density and filling height,
- $\frac{A({}^{223}_{88}Ra)}{A({}^{238}_{92}U)}$: activity ratio between ^{223}Ra and ^{238}U (0.046 in the case of natural enrichment in ^{235}U),
- F: correction factor to take into account the limited number the lack of gamma rays modeled in the fluorescence source (here F = 0.78).

III. Experimental qualification

In order to validate the new approach described above, we use 76 gamma spectra of uranium ore samples recorded by ORANO Mining (such as the spectrum of Fig. 1) to study radioactive disequilibrium in Dulann Uul and Zoovch Ovoo deposits, Mongolia [2]. The samples have different densities and filling heights. The uranium ore was coarsely crushed with a hammer and placed in a flask, as shown in Fig. 3. The crushed samples have an average density of 1.4 g.cm^{-3}. The average filling height is 6 cm and the sample holder diameter is 11 cm. The sample is placed on PVC shims of variable height to ensure the contact between the detector cover and the ore sample. In addition, a copper shielding was used to absorb fluorescence X- rays of the lead shielding (Fig. 3). The BEGE 2830 planar HPGe crystal of the FALCON 5000 handheld detector (from MIRION-CANBERRA) has a 60 mm diameter and a 30 mm length. In order to calculate the fluorescence yield, the measurement geometry was modeled with MCNP6 Monte-Carlo transport code [4], see Fig. 3.

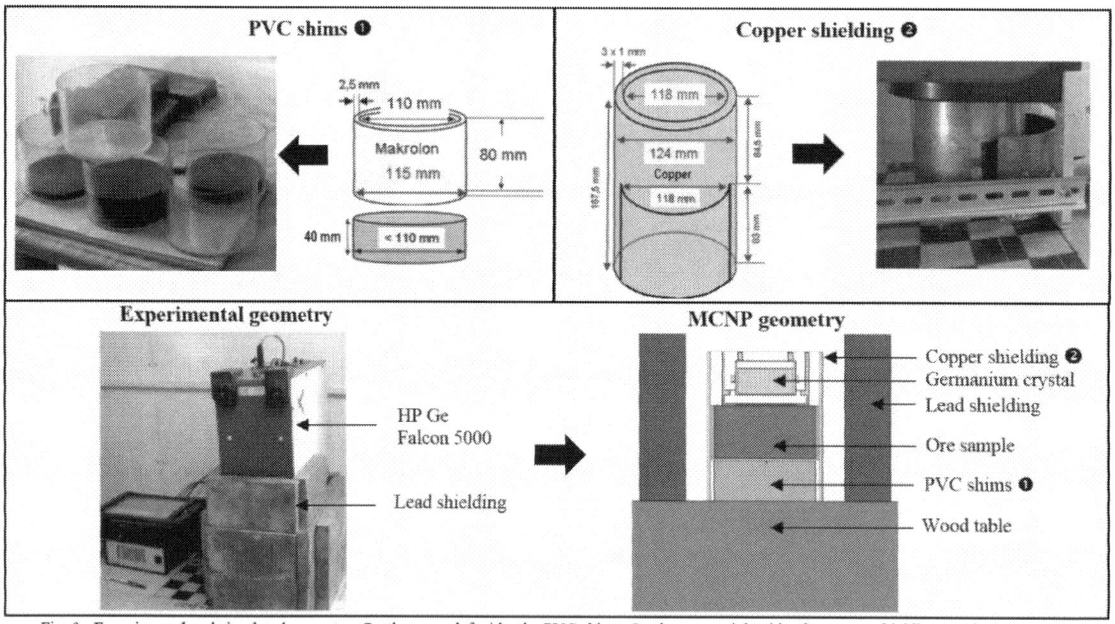

Fig. 3. Experimental and simulated geometry. On the upper left side, the PVC shims. On the upper right side, the copper shielding. At the bottom, the experimental geometry and the MCNP geometry.

https://doi.org/10.1051/epjconf/202022505003

Efficiencies at 92 keV, 98 keV and 1001 keV are those calculated with ISOCS software [6] for a typical FALCON 5000 geometry. The uncertainty associated to calculated efficiencies is 10 % for 92 keV and 98 keV peaks and 5 % for 1001 keV peak.
The following graph represents the $Cm_U(92\ keV)$ and $Cm_U(98\ keV)$ concentrations calculated with the net area of the 92 keV and 98 keV peaks, respectively, as a function of the concentration calculated with the 1001 keV peak, which is considered as the reference.

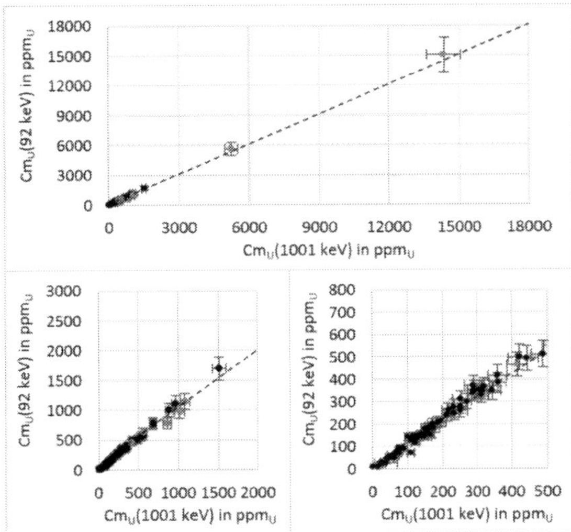

Fig. 4. Uranium content $Cm_U(92\ keV)$ as a function of $Cm_U(1001\ keV)$.

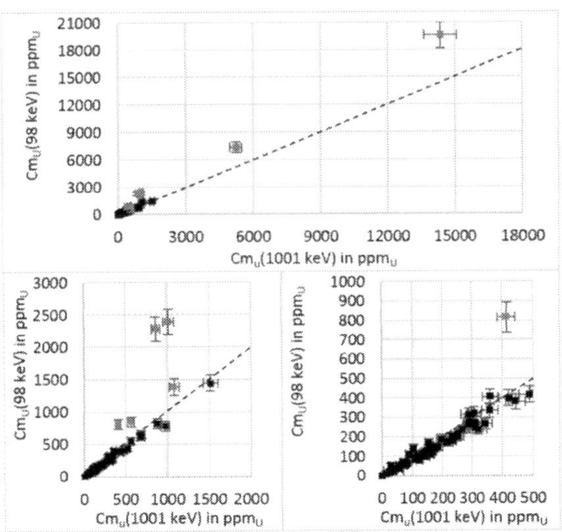

Fig. 5. Uranium content $Cm_U(98\ keV)$ as a function of $Cm_U(1001\ keV)$.

The uncertainty calculated for each sample (considered as homogeneous) includes:

- detector calibration and detection efficiency calculation by ISOCS [6],
- the statistical uncertainty on the net area of the 1001 keV, 92 keV and 98 keV peaks,
- the relative uncertainties on the intensities of the gamma or X-rays of interest are given in the nuclear databases [5]: 0.94 % for the 1001 keV gamma ray; for the 92 keV peak it is 6.21% for the ^{234}Th contribution and 1.91 % for ^{235}U part; for the 98 keV line, it is 1.26 % for ^{234m}Pa, 2.85 % for ^{226}Ra, and 2.57% for ^{223}Ra,
- the relative uncertainty on the fluorescence yield simulated for the calculation of $Cm_U(98\ keV)$ is estimated to be around 20 % for the six modeled gamma emissions,
- the relative uncertainty on the F = 0.78 correction factor (for the non-modeled gamma rays inducing about 22 % of the fluorescence) is estimated to be 5% based on several simulations with different samples.

An average relative uncertainty of 10.34 % is thus calculated for $Cm_U(1001\ keV)$, 11.69 % for $Cm_U(92\ keV)$ and 12.85% for $Cm_U(98\ keV)$. The uncertainty on $Cm_U(1001\ keV)$ is mainly due to the poor counting statistics, leading to large uncertainties in both the Gaussian adjustment of the gamma ray (broadened by the detector resolution) and in the estimation of the Compton continuum under the peak. Concerning $Cm_U(92\ keV)$ and $Cm_U(98\ keV)$ the relative uncertainty on the efficiency calculated with ISOCS software dominates (10 %). It could be reduced by using a MCNP refined detector model based on a fine geometrical characterization with collimated calibration sources, which will be reported in a future article. For most of the 76 samples, $Cm_U(92\ keV)$ and $Cm_U(98\ keV)$ are consistent with the reference $Cm_U(1001\ keV)$, but for seven samples (in red on Fig. 4 and Fig. 5), discrepancies are observed that might be due to heterogeneities. Indeed, for all these samples, we observe the following inequality:

$$Cm_U(98\ keV) > Cm_U(1001\ keV) \geq Cm_U(92\ keV)$$

As explained above, in case of uranium nuggets in the sample, X and gamma emissions following radioactive decays are more self-absorbed than with a uniform distribution, while on the contrary, X-ray fluorescence is enhanced. This phenomenon was demonstrated by simulating with MCNP a uranium nugget of 0.343 g_U in a homogeneous matrix of uranium of 0.455 g_U. This simulation, equivalent of a homogeneous sample of 1000 ppm_U, gives $Cm_U(92\ keV)$ = 690 ppm_U and $Cm_U(98\ keV)$ = 3645 ppm_U.
Beyond detection of possible heterogeneities, the advantage of the method is to reduce measurement time. The gain in measurement time was calculated for each sample by comparing the statistical uncertainty of the 1001 keV with that of 92 keV and 98 keV peaks, taking in account the estimation and subtraction of the Compton continuum under the peaks. Than the gain in measurement time is estimated with the aim to have the same relative uncertainty in the 92 and 98 keV lines, during a short acquisition, as in the 1001 keV line during the long acquisition. In other words, based on Poisson law, the ratio of the 1001 keV relative uncertainty to that of 92 or 98 keV lines is squared to be converted in time gain. The average gain

is 65 for $Cm_U(92\ keV)$ and 2 for $Cm_U(98\ keV)$, but they vary a lot for each sample depending of many factors such as geometry, uranium content and possible radioactive disequilibrium in the uranium chain. For instance, no gain is observed for $Cm_U(98\ keV)$ for very small uranium contents (namely less than 200 ppm_U) because self-fluorescence if too weak. On the other hand, for large uranium contents, the gain for $Cm_U(98\ keV)$ increases quickly because, as explained above, the intensity of fluorescence X-ray rises with the square of uranium concentration. For instance, the gain in measurement time for the highest uranium content sample (14,000 ppm) is about 8 for Cm_U(98 keV).

The gain for Cm_U(92 keV) fluctuates from one sample to another, but it is little dependent on uranium concentration in this range. In addition, if sample homogeneity is guaranteed (e.g. thanks to fine crushing), the time gain is a factor 70 by using the 92 keV line only, as heterogeneity detection with the 98 keV peak is no more necessary.

IV. CONCLUSION AND PROSPECTS

In this work, we have shown the possibility to measure the uranium concentration of uranium ore samples by gamma-ray spectroscopy using the 92 keV and 98 keV lines, which lead to a large measurement time gain (minutes vs. hours) compared to the reference 1001 keV peak. The total relative uncertainty on $Cm_U(92\ keV)$ and $Cm_U(98\ keV)$ mass concentrations measured in a short acquisition remains lower than 13 %, while it was about 10 % for the long measurement of $Cm_U(1001\ keV)$.

In addition, a difference in the uranium concentrations assessed with the 92 and 98 keV peaks, respectively, alerts the operator of a possible heterogeneity of the sample.

In next studies, the geometric MCNP model of the Falcon 5000 detector will be optimized to reduce the uncertainty on $Cm_U(92\ keV)$ and $Cm_U(98\ keV)$, which is so far dominated by modeling uncertainty. This detector is equipped with a planar HPGe crystal, which will be finely characterized with a multi-energy highly collimated ^{152}Eu gamma source. This narrow photon beam will allow precisely estimating the dead layers vs. active area of the germanium crystal [3].

We will also measure 38 additional ore samples provided by ORANO Mining, on a wider range of uranium concentrations, to fully qualify the method in the Nuclear Measurement Laboratory of CEA, DEN, Cadarache.

In parallel, we also study a low-resolution spectroscopy approach with easy-to-use room temperature NaI(Tl) or $LaBr_3$(Ce) detectors, reducing again measurement time, and system cost as well.

REFERENCES

[1] "Falcon 5000 Portable HPGe-Based Radionuclide Identifier." [Online]. Available: https://www.mirion.com/products/falcon-portable-hpge-based-radionuclide-identifier. [Accessed: 22-Jan-2019].

[2] T. Boulesteix *et al.*, "Ilmenites and their alteration products, sinkholes for uranium and radium in roll-front deposits after the example of South Tortkuduk (Kazakhstan)," vol. To be published.

[3] T. Marchais *et al.*, "Detailed MCNP Simulations of Gamma-Ray Spectroscopy Measurements With Calibration Blocks for Uranium Mining Applications," *IEEE Trans. Nucl. Sci.*, vol. 65, no. 9, pp. 2533–2538, Sep. 2018.

[4] "MCNP6TM, User's manual – Version 1.0 - LA-CP-13-00634, Rev. 0 – May 2013 – Denise B. Pelowitz, editor." Los Alamos National Laboratory.

[5] "NUCLÉIDE-LARA on the web (2018)." [Online]. Available: http://www.nucleide.org/Laraweb/index.php. [Accessed: 22-Jan-2019].

[6] MIRION Technologies, "ISOCS– Model S573 ISOCS, Calibration Software." Apr-2013.

 https://doi.org/10.1051/epjconf/202022505004

Study of gamma-ray background noise for radioactive waste drum characterization with plastic scintillators

V. Bottau[1]*, L.Tondut[2], P.G. Allinei[1], B. Perot[1], C.Eleon[1], C.Carasco[1], R. De Stefano[1], G. Faussier[2]
[1] CEA, DEN, Cadarache, DTN, SMTA, Nuclear Measurement Laboratory
[2] Orano Business Unit Recycling
* vincent.bottau@cea.fr

Abstract– **In the framework of the radioactive waste drum characterization using neutron coincidence counting, the Nuclear Measurement Laboratory of CEA Cadarache is studying plastic scintillators as an alternative to ideal but costly ^{3}He gas proportional counters. Plastic scintillators are at least 5 times cheaper for the same detection efficiency, and in addition, they detect fast neutrons about three orders of magnitude faster than ^{3}He detectors. However, they are sensitive to gamma rays, which implies the necessity to identify precisely gamma background sources that may affect the useful signal. This paper presents a detailed analysis of the gamma-ray spectrum of a radioactive waste drum containing glove box filters contaminated by plutonium dioxide. Gamma emissions accompanying inelastic scattering (n,n') and (α,n) reactions that can lead to neutron-gamma coincidences parasitizing useful coincidences from plutonium spontaneous fissions are identified. Some of these parasitic gamma rays having energies up to several MeV, we plan to reject high-energy scintillator pulses with an electronics rejection threshold above 1 MeV, which should preserve the major part of useful fission neutron pulses.**

I. Introduction

In the context of passive neutron measurements for nuclear material characterization, conventional coincidence collars based on helium 3 gas counters permits a temporal discrimination of spontaneous fissions neutrons and parasitic neutrons from (α, n) reactions. ^{3}He proportional counters are the golden standard because they show a very high capture cross section for thermal neutrons, leading to high detection efficiency when surrounding detectors with polyethylene moderator, while being practically insensitive to gamma rays. However, the global shortage on ^{3}He gas [1] has greatly increased their cost in the past 15 years, leading to worldwide initiatives to look for cheaper alternatives with equivalent performances [2]. The Nuclear Measurements Laboratory of CEA, DEN, Cadarache, is currently studying the use of low-cost Polyvinyl Toluene (PVT) plastic scintillators as a potential alternative [3] [4] that have the advantage of being sensitive to fast neutrons, thus avoiding thermalizing materials and offering a nanosecond response time. Consequently, very short coincidence windows of a few tens of nanoseconds allow greatly reducing accidental coincidences compared to three order of magnitude slower ^{3}He detection systems.

However, plastic scintillators are very sensitive to gamma rays. Some of them are emitted in coincidence with (α,n) neutrons often encountered with plutonium oxides and alpha bearing waste, leading to neutron-gamma coincidences that penalize the detection of plutonium spontaneous fission coincidences. Other correlated gamma emissions may also occur in neutron-induced reactions, such as (n,n') inelastic scattering. Pulse Shape Discrimination (PSD) liquid organic scintillators are commonly used to parry gamma sensitivity, but they would not constitute a cost-effective alternative to ^{3}He counters for radioactive waste package characterization. Indeed, large detection volumes are needed to preserve coincidence counting statistics and therefore, a large number of PSD scintillators should be implemented as their size is limited. Actually, the difference between neutron and gamma ray pulse tails is not preserved in large PSD scintillators due to light multiple scattering [5] [6]. Finally, a large array of small (maximum 5 inches) PSD scintillators would increase the cost to a similar level as ^{3}He detectors, not to mention flammability and toxicity of organic liquid scintillators. In addition, PSD information lies in pulse tails and would be damaged by pulse pile-up in case of high count rate, as it is the case with some radioactive waste.

In view to find mitigations to gamma sensitivity of plastic scintillators, this paper reports the gamma spectrum analysis of a 120 L waste drum containing glove box filters contaminated by plutonium dioxide powder with a focus on gamma rays emitted by (α,x), (n,n') and (n,γ) reactions.

II. Gamma Spectrum

The High Purity Germanium (HPGe) detector used for the measurement is a planar Broad Energy Germanium (BEGe) from Mirion Technologies (Canberra), equipped with a CP5 Cryo-Pulse. It was in quasi-contact with a 120 L radioactive waste drum containing glove box filters contaminated with plutonium dioxide (PuO_2).

A 1 mm tin plate was placed in front of the detector to clean very low energy gamma rays from the spectrum, especially the very intense 59.5 keV line of ^{241}Am. The acquisition electronics is a DSALX from Mirion-Canberra, including high-voltage, amplification, pulse digitization and processing, and gamma spectrum storage. Data acquisition and processing is performed with Genie2000 gamma spectrometry software from Mirion-Canberra. The measurement lasted over 53 h, from which 38,9 h of real acquisition time with a dead time of 27 %. Registered pulses are coded on 16384 channels and the resulting spectrum recorded on the [0-6400] keV range is given in Figure 1.

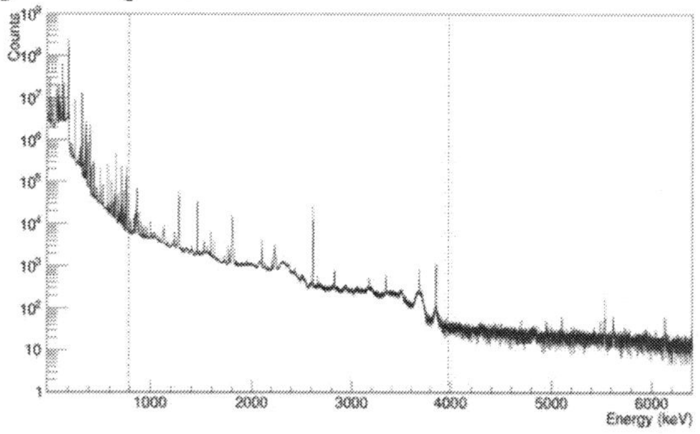

Figure 1 : Gamma spectrum of the 120 L plutonium waste drum resulting from a 38.9 hours real time measurement with a BEGe planar HPGe. Three main regions separated by dotted lines (from 0 to 800 keV, 800 to 4000 keV and above 4000 keV) are analysed

Three main regions of interest are identified.

- The region below 800 keV includes classical gamma rays following americium and plutonium decays. Note that the low-energy threshold setting will be discussed in section III, in particular to cut the residual 59.5 keV peak.
- The region between 800 keV and 4000 keV includes gamma rays emitted by natural radionuclides (^{40}K, thorium chain), (α,x) reactions on light nuclei, and (n,n') inelastic scattering reactions. This region represents the very heart of this work, because reaction gamma rays emitted in correlation with a neutron can lead to real but parasitic neutron-gamma coincidence, which are indistinguishable from useful spontaneous fission coincidences. It corresponds also to the energy range in which the high-energy threshold will be adjusted as discussed in section III. Expansions of the 800-1800 keV and of the 1800 – 4000 keV regions are shown in Figure 2 and Figure 3, respectively, showing a large number of peaks, some of which broadened by Doppler effect due to the short life time of the reaction product excited state. This excited nucleus is indeed still slowing down when the de-excitation gamma ray is emitted.
- a region between 4000 keV and 6400 keV with few peaks due to neutron inelastic scattering, which can also generate parasitic neutron-gamma coincidence. The high-energy region of the spectrum reported in Figure 4 shows neutron-induced radiative capture and inelastic scattering gamma rays on carbon and oxygen nuclei present inside the waste drum. The artefact peak near 5500 keV is due to DSA-LX electronics. This peak is also present during background noise acquisitions. The 4946.31 keV radiative capture gamma ray on ^{12}C nuclei is weak but present, as well as the 6128.63 keV inelastic scattering gamma ray on ^{16}O accompanied by its single escape (SE) and double escape (DE) peaks. The $^{16}O(n,n')$ reaction may also lead to 6917 keV and 7117 keV gamma rays [7] but they are beyond the 6400 keV end-energy of the spectrum.

III. Thresholds settings for plastic scintillators

In view to limit parasitic non-fission coincidences in plutonium measurements with plastic scintillators, we plan to use low- and high-energy thresholds to cut as many gamma rays as possible. Uncorrelated gamma rays from alpha and beta disintegrations, and from (α,x) reactions do not create coincidences, except from cross talk between close detectors but such events can be discarded by time rejection as explained in [3]. However, as mentioned above, prompt gamma rays accompanying (α,n) and neutron inelastic scattering reactions, as well as cascade gamma rays following disintegrations or reactions, can create non-fission coincidences. In order to find a trade-off between neutron preservation and gamma-ray elimination using low- and high-energy cutoffs, is it important to take into account the difference in scintillation light output between neutron and gamma detection in plastic scintillators. Indeed, the quenching effect reduces the light output generated by heavy particles, such as recoil protons after neutron elastic scattering on hydrogen nuclei, compared to the same energy deposition by a gamma ray [8]. The energy deposition of a proton is therefore converted in MeV equivalent electron (MeVee), which is the energy that an electron would release by slowing down in the plastic scintillator to produce the same light pulse. A previous study concerning BC-420 plastic scintillators [9] gives the following conversion in case for recoil protons:

$$E\ (MeVee) = 0.0364E_n^2 + 0.125E_n$$

and in case of recoil carbon nuclei [5] [10] the relationship is:

$$E\ (MeVee) = 0.02E_n$$

where E_n is the energy deposited by the neutron, in MeV.
The only consider here fast neutron reactions inducing recoil protons in plastic scintillators.

Figure 5 shows a Maxwell spectrum representing the energy distribution of neutrons emitted by spontaneous fission of ^{240}Pu, which is the major spontaneous fission emitter [11]:

$$N(E) = \sqrt{E}\exp(-E/1.32\ MeV)$$

Table 1 reports fission neutron losses when applying different low- and high-energy thresholds. It is important to remind that neutron elastic scattering leads to uniform energy deposition between 0 and E_n (incident neutron energy) on hydrogen nuclei in plastic scintillators. Consequently, neutron detection is still possible when E_n is larger than the high-energy threshold, in case of energy deposition smaller than this high-energy threshold. On the other hand, a neutron with E_n larger than the low-energy threshold can be discarded if its energy deposition is below this low-energy threshold. The same is true for partial photon energy deposit in plastic scintillators, in which Compton scattering is the major interaction process. Therefore, Table 1 just provides a qualitative idea of neutron losses. The real thresholds will be determined experimentally based on signal-to-noise ratio (i.e. spontaneous fission vs. parasite coincidences using ^{252}Cf and AmBe sources, respectively) and detection limit considerations (using Pu samples).

Following the gamma spectrum analysis, three low-energy thresholds are considered

- a 60 keVee threshold allows cutting the contribution of the intense 59.54 keV gamma ray of ^{241}Am in plastic scintillators. In fact, this gamma ray mainly deposits a much lower energy than 59 keV by Compton scattering. The 60 keVee threshold would cut energy depositions below 426.9 keV for neutrons, which leads to about 4 % neutron losses of the Maxwell spectrum;

- a 130 keVee threshold would cut the intense gamma and X-ray signal from americium and plutonium disintegrations but also neutron energy depositions up to 836.3 keV, i.e. 11 % of spontaneous fission neutrons;

- a 210 keVee threshold would cut the intense ^{237}U and plutonium gamma rays, especially the 208 keV peak of $^{241}Am/^{237}U$, which is the most intense line of the whole gamma spectrum, and its associated Compton continuum. In fact, the 208 keV peak Compton edge is close to 93 keV and even if PVT scintillators have a poor energy resolution, 208 keV gamma rays can easily be eliminated even with the previous 130 keV threshold. Just for information, the 208 keV threshold would cut neutron energy depositions up to 1236 keV, i.e. 20 % of spontaneous fission neutrons

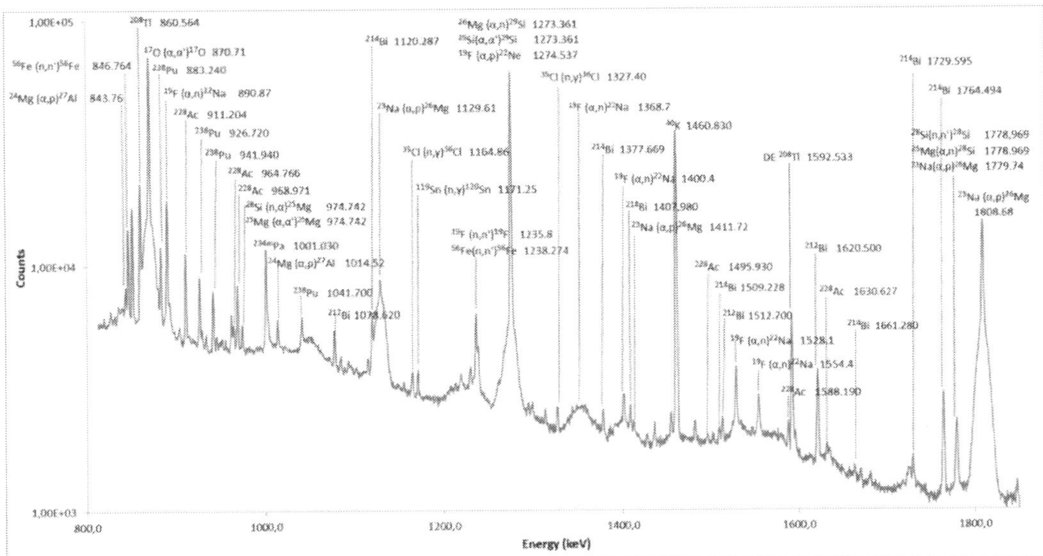

Figure 2: Gamma rays from natural radionuclides (^{40}K, uranium and ^{232}Th decay chains), (α,x) and inelastic scattering reactions in the 800 - 1800 keV energy range

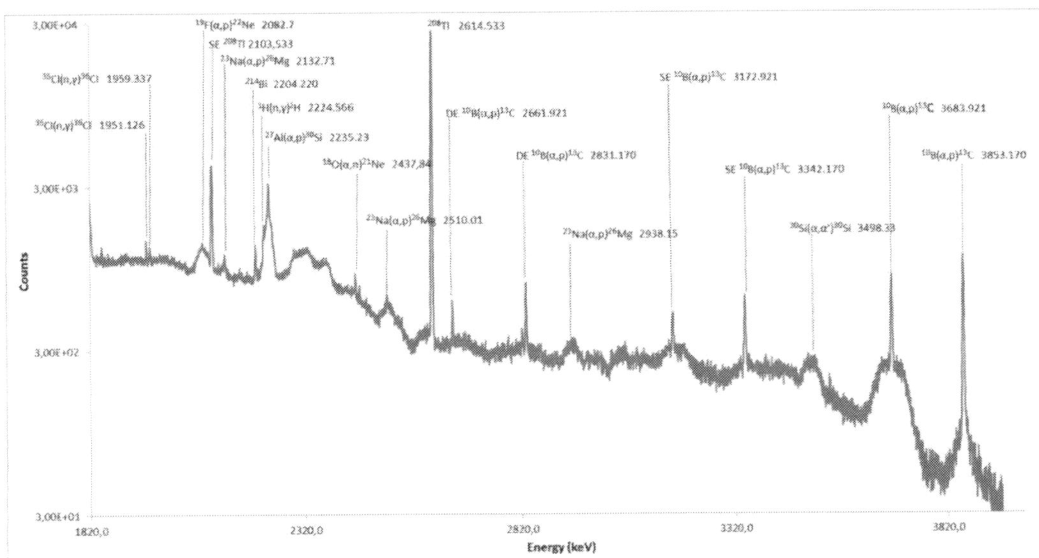

Figure 3: Gamma rays from natural radionuclides (uranium and ^{232}Th decay chains), (α,x) and inelastic scattering reactions in the 1800 - 4000 keV energy range

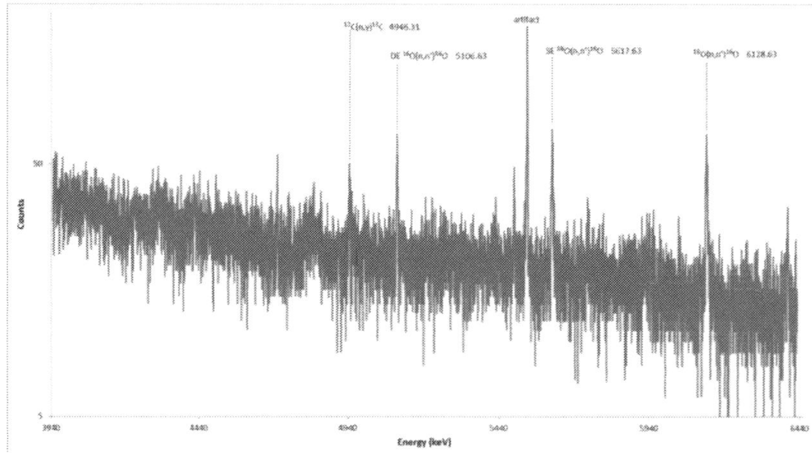

Figure 4: Gamma rays from neutron radiative capture and inelastic scattering reactions in the high-energy range (4000-6400 keV)

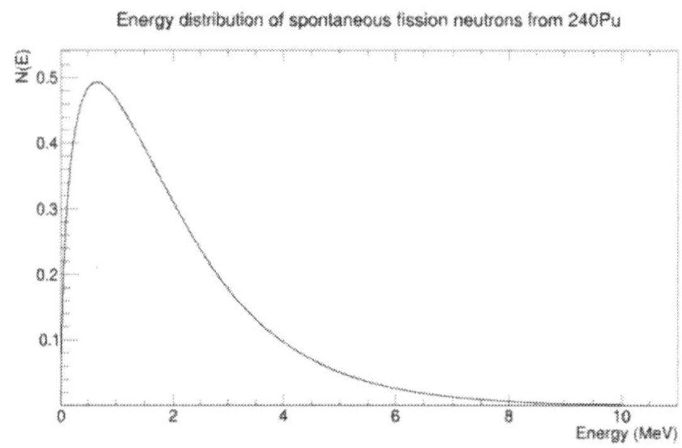

Figure 5: Prompt neutron spectrum from 240Pu spontaneous fissions, approximated as a Maxwell distribution N(E)

Table 1: Different high- and low-energy threshold combinations, and corresponding percentage of ^{240}Pu spontaneous fission neutrons between the two thresholds in the above Maxwell spectrum. In the two last columns, percentage below and above the low- and high-energy thresholds, respectively).

Low-E threshold		High-E threshold		% of neutrons between the low- and high-E thresholds	% of neutrons below the low-E threshold	% of neutrons above the high-E threshold
keVee	keV n	keVee	keV n			
60	426.9	800	3276	60.92	3.96	35.39
60	**426.9**	**1000**	**3798**	**68.95**	**3.96**	**27.36**
130	836.3	800	3276	53.32	11.29	35.39
130	836.3	1000	3798	61.35	11.29	27.36
210	1236	800	3276	44.42	20.19	35.39
210	1236	1000	3798	52.45	20.19	27.36

Considering the high-energy threshold, the following settings are considered:
- a 800 keVee threshold eliminates some gamma rays coming from (α,x) reactions and neutron inelastic scattering (as 1129 keV from ^{26}Mg, 1238 from ^{56}Fe, 1274 keV from ^{29}Si and ^{22}Ne, 6128 keV from ^{16}O) but also neutron energy deposits above 3.276 MeV, i.e. 35 % of spontaneous fission neutrons;
- a 1000 keVee threshold would no more cut 1129 keV gamma rays from ^{26}Mg (Compton Edge at 922 keV) , but as it corresponds to a higher 3.798 MeV neutron energy deposit, only 27 % of spontaneous fission neutrons would be lost.

Different combinations of these low and high-energy thresholds are reported in
Table 1, along with the percentage of preserved or lost ^{240}Pu fission neutron losses. The 60 and 1000 keVee low and high-

energy thresholds, respectively, seems to be a good trade-off because they cut the intense 59.54 keV ray of ^{241}Am and a large part of high-energy gamma rays, while keeping more than 69 % of spontaneous fission neutrons.

IV. Conclusion

In order to propose plastic scintillators as a viable alternative to replace helium 3 gas counters in plutonium characterization coincidence collars, it is important to find mitigation means of their high sensitivity to gamma rays. These last, when emitted in correlation with other particles like in (α,n) and (n,n') reactions, or in gamma cascades, lead to parasitic coincidence that can be misidentified as coming from plutonium spontaneous fissions. The gamma spectrum of a 120 L waste drum filled with plutonium bearing glove box filters allowed us identifying, besides natural background rays (^{40}K, U and Th chains) and well- known rays due to plutonium and americium radioactive decays, a number of less common gamma rays. These last are due to nuclear (α,x) reactions induced by the high alpha activity of Pu and Am, as well as by neutrons from spontaneous fissions and (α,n) reactions. The analysis of the residual nuclei excited levels produced in alpha- and neutron-induced reactions allowed us identifying many such gamma rays, most of them with an energy larger than 1 MeV, which could lead to spurious coincidences in a measurement system with plastic scintillators. This analysis will lead us to use a high-energy rejection threshold to reject such gamma rays, while preserving the largest possible fraction of useful fission neutron pulses.

References

[1] R.Kouzes, J.Ely, L. Erikson, W. Kernan, A. Lintereur, E. Siciliano, D. Stephens, D. Stromwold, R. Van Ginhoven, M. Woodring, *"Neutron detection alternatives to ³He for national security applications"*, Nuclear Instruments and Methods in Physics Research A 623 (2010) 1035-1045 , January 2010.

[2] D.Henzlova, R. Kouzes, R.McElroy , P. Peerani, M. Aspinall, K. Baird, A. Bakel, M. Borella, M. Bourne, L. Bourva, F. Cave, R. Chandra, D. Chernikova, S. Croft, G. Dermody, A. Dougan, J. Ely, E. Fanchini, P. Finocchiaro, V. Gavron, M. Kureta et al. *"Current Status of 3He Alternatives Technologies for Nuclear Safeguards"* LA-UR-15-21201 Ver. 3, PNNL-24307, July 2015.

[3] B. Simony, C. Deyglun, B. Pérot, C. Carasco, N. Saurel, S. Colas, J. Collot, Cross-talk characterization in passive neutron coincidence counting of radioactive waste drums with plastic scintillators, IEEE Transactions on Nuclear Science, Vol. 63, No. 3, June 2016, pp. 1513-1519.

[4] B. Simony, B. Pérot, C. Carasco, F. Jallu, N. Saurel, S. Colas, P. Girones, J. Collot, Passive neutron coincidence counting with plastic scintillators for the characterization of radioactive waste drums, IEEE Transactions on Nuclear Science, Vol. 64, No. 10 (2017) 2719-2724

[5] C. Carasco, B. Pérot, S. Normand, G. Sannié, *"POLITRANI, A New Toolkit to Simulate Organic Scintillator Pulses"*, IEEE Transactions on Nuclear Science, Vol.61, No.4 , August 2014.

[6] G. Corre, K. Boudergui, G. Sannié, V. Kondrasovs, *"Neutron Detection with Large Plastic Scintillators for RPM Applications"*, 2015 4th International Conference on Advancements in Nuclear Instrumentation Measurements Methods and their Applications (ANIMMA), April 2015, Lisbon, Portugal. IEEE Engineers Inc, pp.7465625, 2015.

[7] S. P. Simakov, A. Pavlik, H. Vonach, S. Hlavac, *"Status of experimental and evaluated discrete gamma-ray production at E_n = 14.5 MeV"*, INDC(CCP)-413, September 1998.

[8] G. F. Knoll "Radiation Detection and Measurement" 4th ed, september 2010.

[9] S. A. Pozzi, J. A. Mullens, J. T. Mihalczo, *"Analysis of neutron and photon detection position for the calibration of plastic (BC-420) and liquid (BC-501) scintillators"*, Nuclear Instruments and Methods in Physics Research A254 (2004) 92 101.

[10] S. A. Pozzi, M. Flaska, A. Enqvist, I. Pazsit, *"Monte Carlo and analatycal models of neutron detection with organic scintillation detectors"*, Nuclear Instruments and Methods in Physics Research A582 (2007) 629-637

[11] N. Ensslin,*"Passive Nondestructive Assay of Nuclear material"*, NUREG/CR-5550, LA-UR-90-732, march 1991

Neutron spectrometry at low fluence rates for radiation protection at the AMANDE facility

V. Lacoste, M. Petit, and V. Gressier
IRSN, F-13115 Saint-Paul les Durance, BP3, Cedex.

***Abstract*— A neutron spectrometry campaign was carried out in the AMANDE accelerator control room. These measurements had several objectives, one of which was to verify the possibility of determining, with the HERMEIS Bonner sphere system, very low fluence and ambient dose equivalent rates. These measurements were also expected to provide comparison values with calculations, performed with MCNPX 2.6.0, used for modeling the whole facility and to verify the radiological zoning implemented. Neutrons of 3.3 MeV then 15 MeV were produced in the experimental hall of AMANDE and measurements were made in the control room, behind a 40 cm thick concrete wall. The ambient dose equivalent rates derived from the measured spectra are respectively of the order of 0.2 and 10 µSv/h, which are in agreement with a LB6411 surveymeter data also involved.**

I. Introduction

The Laboratory for Micro-irradiation, Neutrons Metrology and Neutrons Dosimetry develops, operates and maintains various facilities producing neutron fields, in accordance with ISO 8529-1 [1]. These fields are dedicated mainly to the calibration of radiation protection devices according to the ISO 8529-2 standard [2]. Among these facilities, AMANDE [3] allows to study the response of measuring instruments to neutron energies between a few keV and 20 MeV. These mono-energetic fields are produced by means of nuclear reactions, involving incident particles of selected energy (ion beams) and target particles, in the form of a thin layer deposited on a metallic backing (the assembly being called "target "). Depending on the nuclear reactions and the current delivered on the target, neutron emission rates may be higher or lower and the ambient dose equivalent rates around the facility may vary by much, depending on the neutron energies produced.

Given the large variety of possible irradiation configurations, radiation protection measurements can not be conducted exhaustively; therefore, a Monte Carlo simulation model, using the MCNPX code version 2.6.0 [4], was developed to evaluate the values of ambient dose equivalent rates outside and inside the building [5]. In order to validate the geometrical model and the Monte Carlo calculations, an experimental measurement campaign was conducted using the HERMEIS (High Energy Range Multisphere Extended IRSN System) Bonner sphere system [6] [7] and the Berthold LB6411 neutron surveymeter [8]. These measurements were made in the immediate vicinity of the AMANDE accelerator control room. Neutron fields of 3.3 MeV then 15 MeV (energies at 0° with respect to the ion beam direction) were generated in the AMANDE experimental hall, and the energy distributions of the neutron fluence corresponding to each irradiation configuration were determined behind the concrete wall separating the experimental area from the control room.

II. Measurement with the Hermeis Bonner sphere system

A. Measurement position

The HERMEIS system was installed behind the separation wall between the irradiation hall and the control room. The thickness of this concrete wall is 40 cm, to ensure the biological protection required for operators in the control room against the neutrons produced in the irradiation hall. The HERMEIS system was at an angle of about 150° with respect to the ion beam direction. At this angular position, the mean energies of the expected direct neutrons, according to the kinematics of the neutron production reactions, were 2.05 MeV and 13.33 MeV, respectively for the energies of 3.3 MeV and 15 MeV at 0°. The neutron fields will be called in the following, 3.3 MeV field and 15 MeV field.

At the measurement position, behind the wall, a relatively large contribution of thermal neutrons was expected due to scattering reactions on the light atoms composing concrete.

The center of the Bonner spheres was positioned at a height of 150 cm from the ground and at a radial distance of 50 cm from the concrete wall.

B. Beam conditions

A first measurement was made with a 3.3 MeV neutron field. The nuclear reaction used was $^{2}H(d,n)^{3}He$, with an incident deuteron energy of 432 keV. The deuterated titanium (TiD) target was 792.9 ± 10.2 µg.cm^{-2} thick, deposited on a 0.5 mm silver backing. The current on the target was about 14 µA.

The second series of measurements was performed with the 15 MeV field. The reaction used is $^{3}H(d,n)^{4}He$, with an incident deuteron energy of 432 keV on a tritiated titanium (TiT) target. The thickness of the target was 776 ± 10 µg.cm^{-2} on a 0.5 mm thick silver backing. The current on the target remained constant, around 3 µA.

The different measurement systems of the AMANDE

facility routinely used to monitor the neutron production in the experimental hall consist of three precision long counters, equipped with ^{3}He cylindrical counters, and a current integrator to check the stability of the incident beam intensity on the target.

C. Experimental data

The HERMEIS system consists of 13 spheres, ten of which being made of polyethylene, with an external diameter of 3", 3.5", 4", 4.5", 5 ", 6", 7", 8", 10" and 12", and three equipped with metal shells for high energies (> 20 MeV), made of tungsten or lead. For these measurement campaigns, the metal spheres were not used given the energy range of the expected neutron field. The other spheres were used in turn. The acquisition time for each sphere was chosen so as to integrate about 10,000 counts into the pulse height spectrum of the charged particles resulting from the reaction ^{3}He(n,p)T.
The quality and the correlation of the data can be checked with an adjustment of the distribution of the sphere counting rates as a function of the sphere radius, by a five-degree polynomial function. When no data deviates from the polynomial, they can be all considered for the unfolding step, which was the case for the two series of data.

III. Monte-Carlo simulations

A. Nominal materials and geometry

The AMANDE facility was modeled with the MCNPX Monte-Carlo code. The material of the 40 cm thick wall, separating the experimental area and the control room, was considered as a mixture of concrete and steel, with a nominal mean density of 2.4 g/cm^3. A dedicated analysis of a sample of this concrete had been performed at the CENBG [9] to determine the detailed composition. In the nominal version, Monte-Carlo calculations were undergone considering an hydrogen rate of 0.62% in the concrete composition. In the model, the ground and the air were included with realistic compositions whereas, an homogeneous material was considered for the control room equipment. The geometry of the MCNPX model is shown in figure 1.

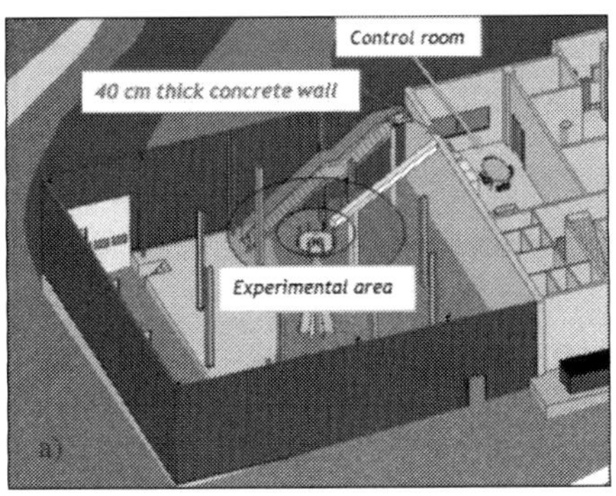

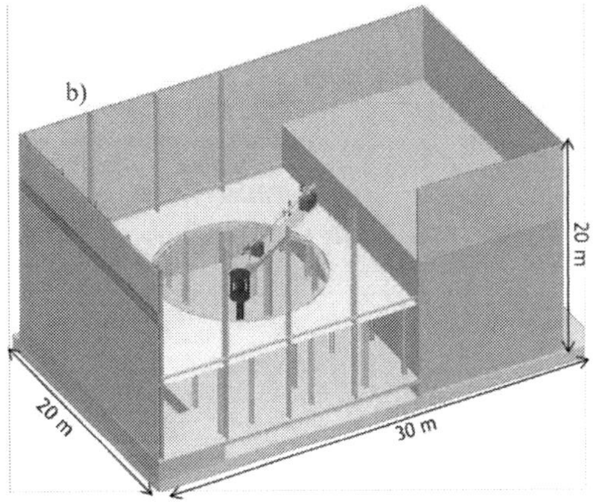

Fig. 1. Geometry of the AMANDE facility; a) from technical drawing b) from MCNPX modeling.

B. Neutron source and cross-sections libraries

The neutron sources induced by a nuclear reactions with an accelerator have a double differential distribution, an angular distribution (with respect to the beam propagation axis), and for each solid angle, a specific neutron fluence energy distribution, which has to be normalized to the emission rate within the corresponding solid angle. In the present cases, the double differential energy and angle fluence distributions were calculated with two different softwares, NeuDesc [10] (JRC-Geel) and then Target [11] (PTB).
The neutron source was positioned at the coordinates (0, 0, 720) and the center of the scoring volume was at (955, -62.5, 300). The neutron cross-sections library ENDF (.60c) and the evaluated data library (.42c) were considered.

IV. Data unfolding

A. Unfolding codes

Two codes were used for data unfolding, WinBUGS and GRAVEL [12] [13]. In the first case, the energy distribution of the neutron fluence is considered as the sum of three components, a thermal part, an intermediate region and a high energy domain. Each component is determined using parameters, such as the amplitude, the high energy peak position or the slope for the intermediate component. The WinBUGS code determines a density of probabilities for each of the parameters, by using the Bayesian method [14]. The probability densities for the integral values of the energy distribution, such as the fluence or the ambient dose equivalent are also calculated.

The energy distribution of the neutron fluence is described using three components; a thermal component (Maxwell-Boltzmann type), an intermediate component (represented by a function in 1/E), and a fast component (represented by a fission spectrum of Maxwell type).

In the case of GRAVEL, the deconvolution is carried out by an iterative process to generate a solution spectrum from an initial guess spectrum (called a priori spectrum or Default Spectrum DS). Unlike WinBUGS, GRAVEL can generate a

 https://doi.org/10.1051/epjconf/202022505005

solution spectrum whose shape and number of components are not predefined, but which depends to a certain extent, of the initial spectrum. The solution spectra are given in the next section; they were obtained with Gravel using MCNPX spectra then WinBUGS spectra as prior information.

B. Solution spectra

The solution spectra, shown in Figures 2 and 3 for the 3.3 MeV and 15 MeV fields respectively, from Gravel using either MCNPX or WinBUGS as default spectra, are in agreement, which suggests that the solution spectra are independent of the prior information to some extent.

The default spectra, represented in dotted lines are also shown; the shape of the MCNPX spectrum is in rather good agreement with Gravel, but some discrepancies are observed, depending on the initial neutron energy: an overestimate for the 3.3 MeV field and a slight underestimate for the 15 MeV one. The integral quantities of neutron fluence and ambient dose equivalent rates, are given for each neutron spectrum, in table 1. These MCNPX discrepancies are confirmed in terms of integral fluence rates, which is about 44% higher and about 17% less respectively for the 3.3 MeV and 15 MeV fields.

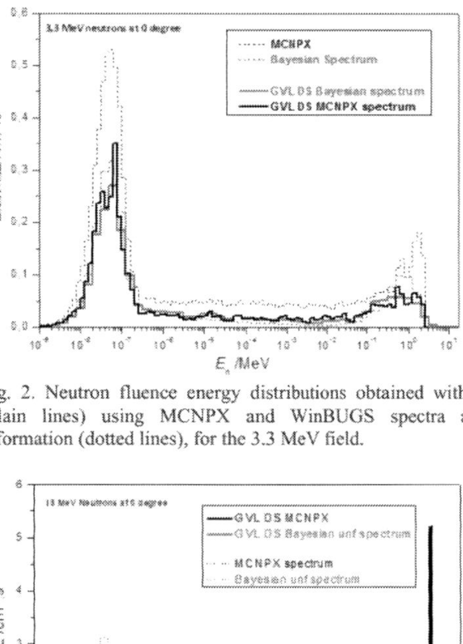

Fig. 2. Neutron fluence energy distributions obtained with Gravel (plain lines) using MCNPX and WinBUGS spectra as prior information (dotted lines), for the 3.3 MeV field.

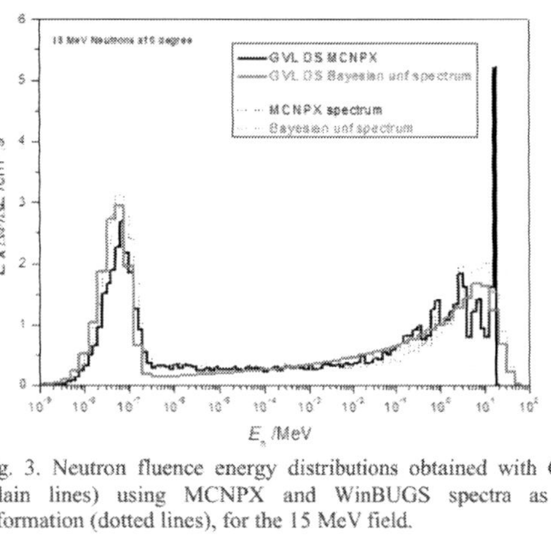

Fig. 3. Neutron fluence energy distributions obtained with Gravel (plain lines) using MCNPX and WinBUGS spectra as prior information (dotted lines), for the 15 MeV field.

Tab. 1. Integral quantities for each neutron field, experimentally determined and calculated with MCNPX. The statistic uncertainties associated to the WinBUGS results correspond to the standard deviation of the integral quantities probability density distributions.

3.3 MeV field				
Neutron spectrum	**WBGs**	**GVL + WBGs**	**MCNPX**	**GVL+ MCNPX**
dΦ/dt /cm^{-2}.s^{-1}	1.00 ± 0.04	0.93	1.40	0.97
dH*(10)/dt /μSv.h^{-1}	0.22 ± 0.01	0.28	0.35	0.21
15 MeV field				
dΦ/dt /cm^{-2}.s^{-1}	15.76 ± 0.37	16.1	12.8	15.4
dH*(10)/dt /μSv.h^{-1}	8.70 ± 0.34	9.6	7.4	10.2

V. MCNPX SIMULATION SENSITIVITY STUDIES

A sensitivity study using MCNPX was undergone to determine the parameters playing main roles in the discrepancies observed between the experimental and the nominal simulated spectra. In turns, some parameters were varied and the consequences studied. The sensitivity study is fully detailed in reference [5]. The following parameters were considered:

- the density of the concrete was varied from 2.4 to 2.3 g/cm^3
- the hydrogen rate in the concrete was varied from 0.62%, to 0.8% and to 1%)
- air was considered as material in the measurement room instead of an homogenous mixture
- the perturbations of the neutron fields due to the presence of the polyethylene spheres were studied
- the MCNPX geometry model was extended to 50 m beyond the facility
- the concrete was considered as inhomogeneous within the 40 cm thick wall (2 configurations)

The consequences of each parameter variation, in terms of integral fluence and ambient dose equivalent rates, are shown in table 2 for the 3.3 MeV case and in table 3 for the other.

Tab. 2. Fluence and ambient dose equivalent rates from calculated spectra at 3.3 MeV where different parameters are varied, and comparison with the experimental quantities. Uncertainties associated to the MCNPX data are less than 1% in statistics and are not reported here.

Configurations **Reaction D(d,n)^{3}He**	Calculated fluence cm^{-2}.s^{-1}	Measured fluence cm^{-2}.s^{-1}	Calculated ADE μSv/h	Measured ADE μSv/h
Configuration 1 (reference)	1,4	0,97±0,06	0,35	HERMEIS 0,24±0,01 / Berthold : 0,24±0,05
Configuration 2 (TARGET)	1,3		0,35	
Configuration 3 (d=2,3 g/cm^3)	1,5		0,41	
Configuration 4 (H=0,8%)	1,0		0,24	
Configuration 5 (H=1,0%)	0,7		0,17	
Configuration 6 (SC vacuum)	1,3		0,35	
Configuration 7 (effect of BSS)	1,4		0,34	
Configuration 8-9 (+outside)	1,4		0,35	
Configuration 10 (inhomogeneity 1)	1,6		0,46	
Configuration 11 (inhomogeneity 2)	1,6		0,45	

https://doi.org/10.1051/epjconf/202022505005

Tab. 3. Fluence and ambient dose equivalent rates from calculated spectra at 15 MeV where different parameters are varied, and comparison with the experimental quantities. Uncertainties associated to the MCNPX data are less than 1% in statistics and are not reported here.

Configurations **Reaction T(d,n)⁴He**	Calculated fluence $cm^{-2}.s^{-1}$	Measured fluence $cm^{-2}.s^{-1}$	Calculated ADE µSv/h	Measured ADE µSv/h
Configuration 1 (reference)	12,8	15,7±0,9	7,4	HERMEIS 9,3±0,6 Berthold : 8,2±0,1
Configuration 2 (TARGET)	12,6		7,3	
Configuration 3 (d=2,3 g/cm³)	14,1		8,4	
Configuration 4 (H=0,8%)	10,8		6,5	
Configuration 5 (H=1,0%)	9,3		5,8	
Configuration 6 (SC vacuum)	12,1		7,4	
Configuration 7 (effect of BSS)	12,7		7,4	
Configuration 8-9 (+outside)	13,0		7,7	
Configuration 10 (inhomogeneity 1)	14,2		8,6	
Configuration 11 (inhomogeneity 2)	14,4		8,9	

VI. Discussion and Conclusion

A. MCNPX sensitivity studies

A better agreement between the calculated and the measured integral values is obtained for an hydrogen rate of 0.80% instead of 0.62% for the 3.3 MeV field, whereas for the 15 MeV field, the agreement is obtained when the concrete is considered as inhomogeneous. The neutron fluence energy distributions corresponding to these cases are shown in Figures 4 and 5. In the 3.3 MeV case, the decrease of the hydrogen rate content in the concrete composition seems to have an impact on the amplitude of the whole spectrum, but does not modify the shape of the structures in the MeV energy region. Further studies are thus needed to determine what parameter could mainly play a role on the shape and amplitude of these two high energy peaks.

In the 15 MeV case, the concrete, composed of sand, gravel, cement and water, was modeled as an inhomogeneous compound, made of clusters of sand and gravel of two different sizes and shapes, placed into a mixture of cement and water. The calculated neutron fluence energy distributions are shown in Figure 5. Even if the agreement between the integral quantities is better, none of these configurations explain the discrepancies observed in the 100 keV-1 MeV region.

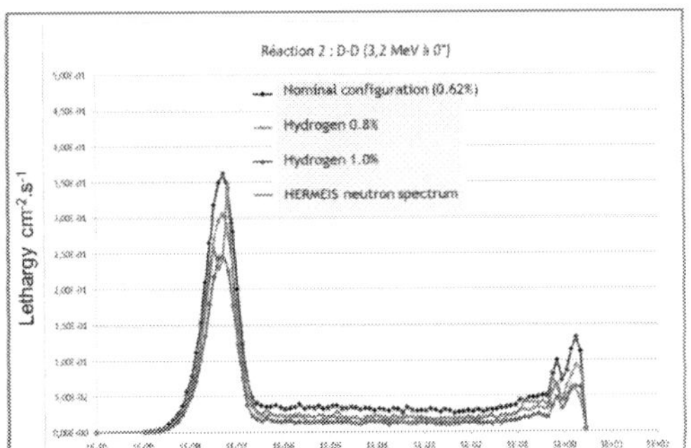

Fig. 4: Neutron fluence energy distributions calculated with MCNPX with different values of the hydrogen content in the concrete, for the 3.3 MeV case, compared to the HERMEIS Bonner spheres spectrum.

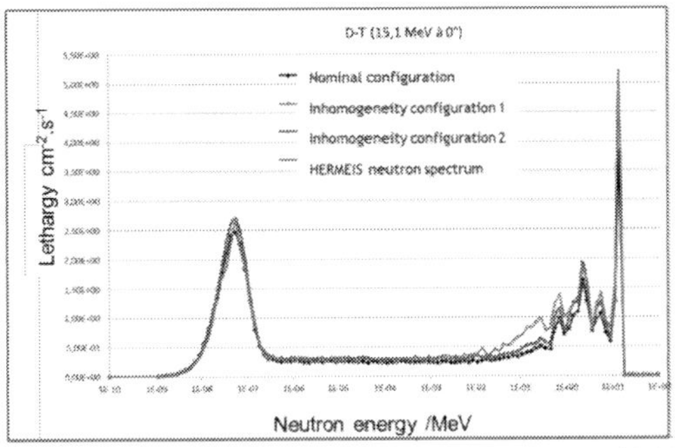

Fig. 5: Neutron fluence energy distributions calculated with MCNPX with different values of the hydrogen content in the concrete, for the 15 MeV case, compared to the HERMEIS Bonner spheres spectrum.

Further investigations are needed; in both cases, the discrepancies between the experimental spectra and the experimental ones concern the structures and shapes in the high energy region. In this energy range, the cross-sections energy distributions of most of materials have many resonances and structures. The comparison and the use of different cross-section libraries in the MCNPX simulations might help to understand the discrepancies observed with the experimental data.

B. Final experimental results

The HERMEIS system, equipped with a ^3He proportional counter of 10 atm, was used for a neutron background measurement campaign in the AMANDE control room. Two configurations were studied, with a neutron source respectively 3.3 MeV and 15 MeV, produced in the experimental hall at 0° relative to the axis of propagation of the incident particles on the target. The final integral values, detailed in [15], are respectively (0.97 ± 0.07) $cm^{-2}.s^{-1}$ and (0.24 ± 0.02).$\mu Sv.h^{-1}$, and (15.7 ± 0.9) $cm^{-2}.s^{-1}$ and (9.5 ± 0.7).$\mu Sv.h^{-1}$. The ambient dose equivalent rates from the LB6411 surveymeter are respectively (0.24 ± 0.03) $\mu Sv.h^{-1}$ and (8.2 ± 0.1) $\mu Sv.h^{-1}$, which is in very good agreement with both HERMEIS results.

C. Conclusion

This measurement campaign allowed validating the capability of the HERMEIS system to perform neutron spectrometry expertise within a few hours (between two and eight hours in total) at workplaces where very low dose rates are expected.

In parallel, Monte-Carlo simulation of the AMANDE facility was performed including realistic geometry and detailed material compositions. Despite the good knowledge of these parameters, some discrepancies were observed and further studies should be undertaken to determine their origin.

VII. Bibliography

[1] International Organization for Standardization, Reference neutron radiations — *Part 1: Characteristics and methods of production*, ISO 8529-1 (2001).

[2] International Organization for Standardization, Reference neutron radiations — *Part 2: Calibration fundamentals of radiation protection devices related to the basic quantities characterizing the radiation field*, ISO 8529-2 (2000).

[3] V. Gressier et al., *International Key Comparison of neutron fluence measurements in mono-energetic neutron fields —* CCRI(III)-K11 Metrologia 51 (Tech. Suppl.) 06009 (2014).

[4] MCNPX-TM User's manual « version 2.6.0 April 2008 »

[5] M. Petit, *Simulation MCNPX d'une spectrométrie des neutrons dédiée à la radioprotection. Etudes de sensibilité du modèle*, Rapport PSE-SANTE/SDOS/2018-00022 (2018).

[6] Serre, S., Castellani, K., Paul, D. and Lacoste, V., *Optimization using Monte Carlo calculations of a Bonner sphere spectrometer extended to high energies for the neutron environments characterization*, IEEE Transactions on Nuclear Science, Vol. 56 (6), pp 3582-3590 (2009).

[7] A. Cheminet, V. Lacoste, V. Gressier, G. Hubert, A. Martin, M. Pépino, *Characterization of the IRSN neutron multisphere spectrometer (HERMEIS) at European standard calibration fields*, Journal of Instrumentation (2012) http://iopscience.iop.org/1748-0221/7/04/C04007/

[8] A. Klett and B. Burgkhardt, *The new remcounter LB6411: measurement of neutron ambient dose Equivalent H*(10) according to ICRP60 with high sensitivity*. IEEE Transactions on Nuclear Science, vol.44, N°.3, June 1997.

[9] H. Guégan, *Analyse par techniques IGA et PIXE de bétons issus de l'installation AMANDE/MIRCOM*, Rapport d'analyse N°05MP12-14, Atelier Régional de Caractérisation par Analyse Nucléaire Elémentaire ARCANE/CENBG 2015.

[10] NeuSDesc – Neutron Source Description Software Manual by Evert Birgersson and Göran Lövestam 2007.

[11] Dietrich Schlegel, *TARGET USER's MANUAL*, PTB-6.42-05-2, Braunschweig, April 2005.

[12] M. Matzke, *Unfolding of Pulse Height Spectra: The HEPRO Program System*. Report PTB-N-19 (1994).

[13] M. Reginatto, *The "few-channel" unfolding programs in the UMG package, MXD_FC33, GRV_FC33 and IQU_FC33, UMG package*, version 3.3. (2004).

[14] D.S. Sivia, *Data Analysis – A Bayesian Tuturial*, Calendon Press, Oxford, 1996.

[15] V. Lacoste, *Spectrométrie des neutrons pour la radioprotection en salle de commande d'AMANDE – Etude de postes de travail sur AMANDE*, Rapport PSE-SANTE/SDOS/2018-00021 (2018).

EPJ Web of Conferences **225**, 05006 (2020)
ANIMMA 2019
https://doi.org/10.1051/epjconf/202022505006

Perspective liquid scintillators for spectrometry of neutron and gamma radiation

Kawa Haji Mahmoud[1]*, Vaclav Prenosil[1], Zdenek Matej[1], Frantisek Kucera[2], Filip Mravec[3], Michal Kostal[4], Frantisek Cvachovec[5], David Valis[5]

[1] Masaryk University, Brno, Czech Republic
[2] Faculty of Chemistry, Brno University of Technology, Czech Republic
[3] VF, Černá Hora, Czech Republic
[4] Research Center Rez Ltd, Czech Republic
[5] University of Defence, Brno, Czech Republic
* 395772@mail.muni.cz

***Abstract*—This paper presents some results of the development of two new scintillation detectors usable for the detection and spectrometry of neutron and photon components of the mixed field. It is a series of liquid organic scintillators containing carbon, hydrogen, nitrogen and oxygen. Elements of individual scintillator designs are arranged in various chemical bonds defining physical and chemical properties of the scintillators. These scintillators have advantageous useful properties. They are non-hygroscopic, not poisonous and are not sensitive to contact with the outside atmosphere and can be stored in conventional laboratory containers.**

***Index Terms*—scintillators, scintillation detectors, pulse shape discrimination, Figure of merit**

I. Introduction

The authors list basic properties of two types of new liquid scintillators LSB205 and LSB206 that are compared with the characteristics of the BC 501 A (NE-213) liquid scintillator, which can be considered the standard used for many years for detection and spectrometry in the mixed field of neutron and photon radiation. Characteristic data of the scintillators developed are focused on the energy range useful, e.g. reactor dosimetry, i.e., neutron energy $\leq$ 15 MeV. This paper mainly contains qualitative and quantitative data on the resolution of neutron and photon particles detected by the scintillator. The resolution of the charged particles is done by pulse shape discrimination, which is based on a variant of the integration method [1], [2]. Quantitatively, the quality of the resolution of charged particles is characterized by the generalized function of Figure of Merit - FOM [2], [3]. If FOM > 1, neutron and gamma radiation responses are perfectly separated. If FOM < 1 then the responses of both particles overlap. Besides, the article contains data on energy resolution and light output of scintillator. The interesting data were observed when investigating the relations of characteristics in a temperature range of 0 – 50 °C. In order to use these scintillators for gamma/neutron mixed field spectrometry, it is necessary to obtain light output function at least for protons. The obtained scintillator temperature dependencies described in this article will enable to refine the measurement of the amplitude resolution of the neutron spectrum measured at the collimated reactor output beam moderated by silicon. This spectrum contains several neutron lines with well-defined energies which could substitute hard to obtain mono-energetic neutron sources [4], [5]. The next step towards the usability for neutron and photon spectrometry in the mixed field is the calculation of the response functions needed for the unfolding of energy spectra.

The properties of newly developed liquid scintillators give them the potential to be used in various areas of nuclear physics, including nuclear safety.

II. Description of the measuring object

There are many types of scintillators for mixed neutron and gamma radiation fields. The main requirement for such a scintillator is the ability to distinguish between neutron and gamma radiation interactions. In this paper, we want to focus on comparing the temperature dependencies of the standard BC-501A scintillator and two versions of the newly developed LSB 205 to LSB 206 scintillators containing 88.3 to 91.1% C; 7.6 to 8.7% H; 0.1 to 1.9% N and 0.1 to 2.2% O in the CF^{252} source pool.

III. Methodology

We compare the thermal properties of the experimental scintillators and the standard liquid scintillator BC-501A. This scintillator has very good separation properties. However, it loses its properties when exposed to air oxygen. The composition of the liquid scintillators tested is not public - it is still protected.

IV. Experimental Setup

The NGA-01 spectrometric system was used for the measurement. This two-parameter system uses

2” PMT Hamamatsu R329-02 equipped with an active tapered voltage divider as a detector. This configuration allows the device to be used even with large amplitude ranges of input signals while maintaining excellent output linearity. This configuration ensures excellent results even at high count rates per second. The device can measure up to 700,000 pulses per second.

The NGA-01 measures detector output pulses using two differently amplified branches for two 12-bit ADC inputs with a sample rate of 500 MS/s - see Fig. 1. Two channels are used for better signal-to-noise ratio, especially for smaller amplitudes. The data are digitally filtered, summed and evaluated by Field Programmable Gate Array (FPGA).

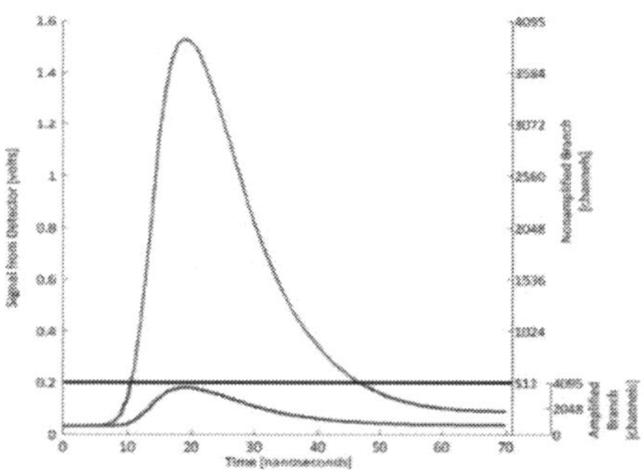

Fig. 1. Amplified and non-amplified impulse selection [3], [6]

The Cs^{137} and Co^{60} emitters were used for calibration of the measurement. The Compton edges in the photon instrument spectrum are for the Cs^{137} a 0.477 MeV energy edge and for the Co^{60} emitter a 1.17 MeV energy edge.

For all scintillators, we used the same PSD (pulse shape discrimination) method based on Charge Comparison [1], [2] implemented directly in the FPGA of the spectrometer NGA-01.

The individual pulse parameters used in the relationship to quantify the PSD are ‘energy delta’ and ‘discrimination delta’ values are determined empirically for optimal delimitation of areas with maximum neutron and photon responses in the scintillator being measured.

V. Description of the temperature chamber

The Vötsch Industrietechnik GmbH VT4004 type chamber was used for temperature dependence testing. Temperatures of 0, 5, 10, 20, 30, 40 and 50 degrees Celsius were used for the experiments. A photomultiplier with an active resistor divider was placed in the chamber. The measuring apparatus was outside the air-conditioned space. The connection was made by coaxial cables. The interior of the temperature chamber was shielded from ambient light.

All tested scintillators were air-conditioned at the same time. Calibration sources and test source were mounted on scintillators from their side.

VI. Temperature test results

The results of the experiments were CSV matrices expressing temperature dependence of individual scintillators. The FOM temperature dependencies of each scintillator for the selected test temperature repertoire and concrete energy are shown in the following graphs for each scintillator separately.

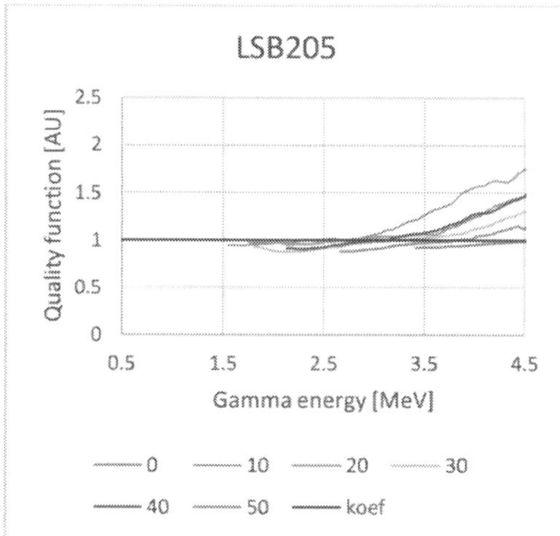

Fig. 2. FOM temperature dependencies of the LSB 205 scintillator

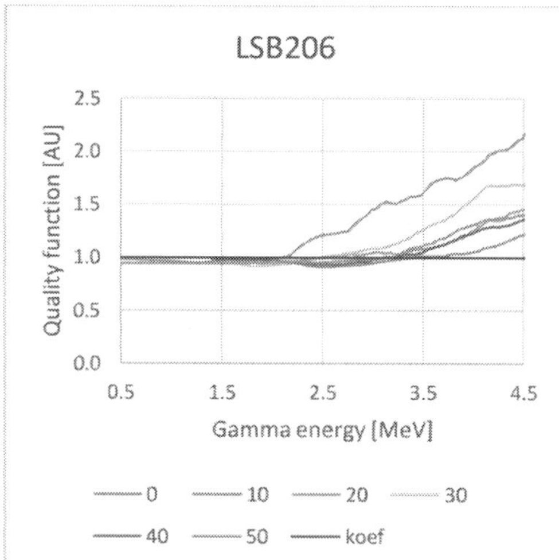

Fig. 3. FOM temperature dependencies of the LSB 206 scintillator

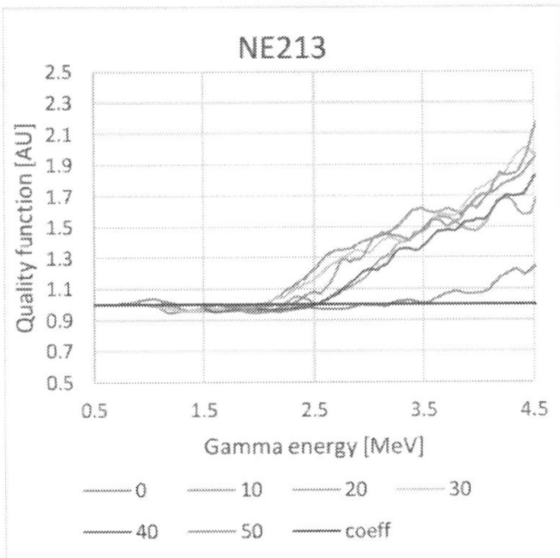

Fig. 4. FOM temperature dependencies of the NE213 scintillator

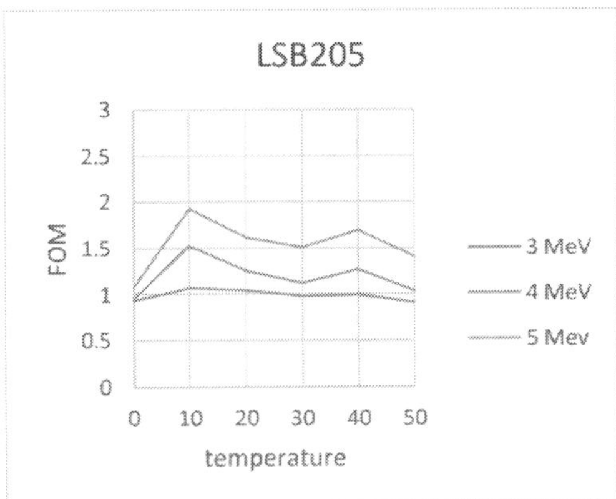

Fig. 5. FOM temperature dependence of the LSB 205 for some energy

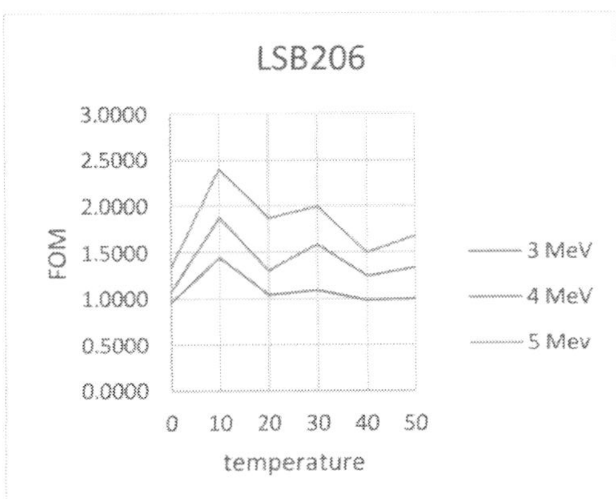

Fig. 6. FOM temperature dependence of the LSB 206 for some energy

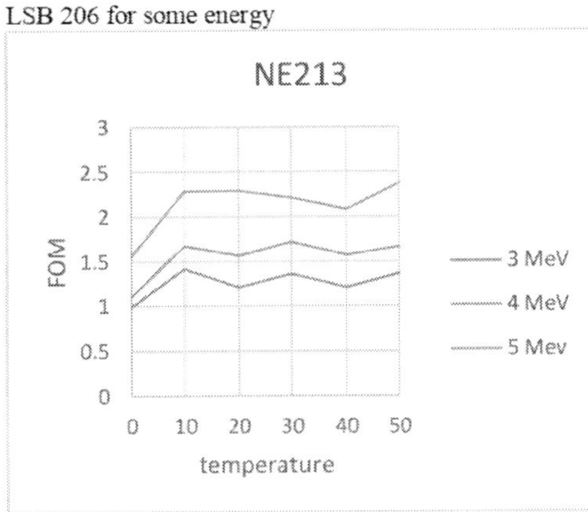

Fig. 7. FOM temperature dependence of the NE213 for some energy

VII. Conclusion

It can be seen from the graphs that the FOM temperature dependence is neither linear nor monotonic. The reason for the non-linearity will be the subject of further research. Knowledge of the FOM temperature dependence of scintillators allows eliminating this dependence in calculations of energy spectra of measured neutron-gamma mixed radiation.

References

[1] M. Amiri, V. Prenosil, F. Cvachovec, Z. Matej, and F. Mravec. "Quick algorithms for real-time discrimination of neutrons and gamma rays," *Journal of Radioanalytical and Nuclear Chemistry*, 2015; 303(1): 583–599. Published online 2014 Aug 24. Doi: 10.1007/s10967-014-3406-5

[2] Z. Matej, M. Amiri, F. Cvachovec, V. Prenosil, F. Mravec, Z. Kopecky, "Digital two-parametric processing of the output data from radiation detectors," *Progress in Nuclear Science and Technology*. Volume 4 (2014) pp. 670-674. DOI:10.15669/pnst.4.670

[3] M. Amiri, F. Cvachovec, Z. Matej, V. Prenosil, F. Mravec, "Digital data processing of stilbene," *3rd International Conference on Advancements in Nuclear Instrumentation, Measurement Methods and their Applications (ANIMMA)*. June 2013. DOI: 10.1109/ANIMMA.2013.6728010

[4] M. Kostal, M. Schulc, J. Soltes, E. Losa, L. Viererbl, Z. Matej, F. Cvachovec, V. Rypar, "Measurements of neutron transport of well-defined silicon filtered beam in lead. "*Applied Radiation and Isotopes*, Elsevier, 2018, roč. 2018, č. 142, s. 160-166.

[5] M. Kostal, Z. Matej, E. Losa, O. Huml, M. Stefanik, F. Cvachovec, M. Sulc, B. Jansky, E. Novak, D. Harutyunyan, V. Rypar, "On similarity of various reactor spectra and 235U prompt fission neutron spectrum," *Applied Radiation and Isotopes*, Elsevier, 2018, roč. 135, May, s. 83-91. ISSN 0969-8043 doi: 10.1016/j.apradiso.2018.01.028.

[6] M. Pavelek, Z. Matej, O. Herman, F. Mravec, M. Veskrna, F. Cvachovec, M. Kostal, V. Prenosil, "Fast Digital Spectrometer for Mixed Radiation Fields,". *2017 IEEE SENSORS Proceedings*. IEEE, 2017. s. 436-438, 3 pp. ISBN 978-1-5386-4056-2. doi:10.1109/ICSENS.2017.8234012

Estimation of fissile material content in irradiated In-Pile Sections using neutron coincidence counters

Riccardo Rossa*, Alessandro Borella, Sven Boden, Wouter Broeckx
SCK-CEN
* rrossa@sckcen.be

***Abstract*— A set of In-Pile Sections (IPS) has been irradiated in the BR2 reactor at SCK•CEN in Belgium during the 1970's and 1980's. The primary goal of the IPS was to replicate the thermo-hydraulic loop of a sodium-cooled fast reactor in order to study severe accident scenarios. The top part of the IPS contained the sodium-cooled loop whereas the lower part contained the fuel element. Due to the experimental conditions, the rupture of the fuel pins contained in the IPS occurred and fuel fragments may have been deposited in the rest of the IPS loop. The part of the IPS containing the fuel pins has been cut from the rest of the IPS and underwent post-irradiation examinations at specialized EU laboratories, while the top parts remained stored at SCK•CEN. To prepare for future transport, dismantling and conditioning, a reliable estimation of the total fissile content in the stored parts of the IPS is indispensable.**

In this framework, two IPS were measured with a Canberra WM3400 neutron coincidence counter with customized electronics. The measurements of the IPS were challenging due their length (roughly 6 m) and intense gamma-ray radiation background. For each IPS an axial scan was carried out with a series of short measurements (600-700 s each) recording the Totals rate and Reals rate. Based on the results of the axial scans, measurements with longer measurement time were conducted for the axial positions with the larger values of Reals rates.

A system of equations was then established to quantify the ^{240}Pu content in the different sections of the IPS from the Reals rates in each measurement position and account for cross-talk between the neutron emission associated to the different sections. A set of Monte Carlo simulations was carried out to estimate the probability to record a Real count in the detector due to spontaneous fission events occurring in a given section of the IPS. The ^{240}Pu content in each section of the IPS was calculated by combining the measured Reals rates and the detection probabilities calculated with the simulations. The total fissile content in the IPS was then determined with scaling factors based on burnup calculations for the irradiated fuel assemblies in the IPS. The results indicate that both IPS measured with the neutron coincidence counters have a fissile content lower than the limit for transport. It is expected that the envisaged segmentation of the IPS in shorter sections required to fit into 200L drums will provide an additional safety margin on this limit.

***Keywords*— Neutron coincidence counters, decommissioning, In-Pile Sections, fissile material, non-destructive assay**

I. Introduction

IN the frame of research on the safety analysis of Liquid Metal Fast Breeder Reactors, several research projects have been performed at the SCK•CEN. A set of so-called In-Pile Sections (IPS) has been irradiated in the BR2 reactor during the 1970's and 1980's, testing the behavior of fuel pins during severe accidents [1]. Due to the experimental conditions, the rupture of the fuel pins contained in the IPS occurred and fuel fragments may have been deposited in the rest of the IPS loop. The part of the IPS containing the fuel pins has been cut from the rest of the IPS and underwent post-irradiation examinations at specialized EU laboratories, while the top parts remained stored at SCK•CEN.

To prepare for future transport, dismantling and conditioning, a reliable estimation of the total fissile content in the stored parts of the IPS is indispensable. Two IPS were measured with a Canberra WM3400 [2] neutron coincidence counter with customized electronics [3].

Irradiated material such as the IPS emits neutrons by (α,n) reactions, spontaneous fission, or induced fission. In the first case only one neutron is emitted per reaction, whereas usually more than one neutron is emitted in the fission process. Coincidence counters exploit this difference to estimate the presence of fissile material in the measured sample [4].

In this paper the Canberra WM3400 detector is described in Section II with the associated Monte Carlo model, whereas the geometry of the IPS is described in Section III. The details of the measurement campaign and the approach used for the estimation of the total fissile content in the IPS are discussed in Section IV. The results of the measurements are presented in Section V and the conclusion is drawn in Section VI.

II. Canberra WM3400 detector

A. Detector design

Two used Canberra WM3400 coincidence counters were purchased for the measurement of the IPS. A picture of the WM3400 detector with a ^{252}Cf source used for testing is shown in Figure 1.

The dimensions of the detector are 104 cm (height) x 41 cm (width) x 16 cm (depth), and the total weight is approximately

R. Rossa (rrossa@sckcen.be), A. Borella (aborella@sckcen.be), S. Boden (sboden@sckcen.be), and W. Broeckx (wbroeckx@sckcen.be) are with the SCK•CEN, Belgian Nuclear Research Centre, Boeretang 200 - BE-2400 Mol (Belgium).

50 kg. An aluminum cover protects the internal parts of the detector.

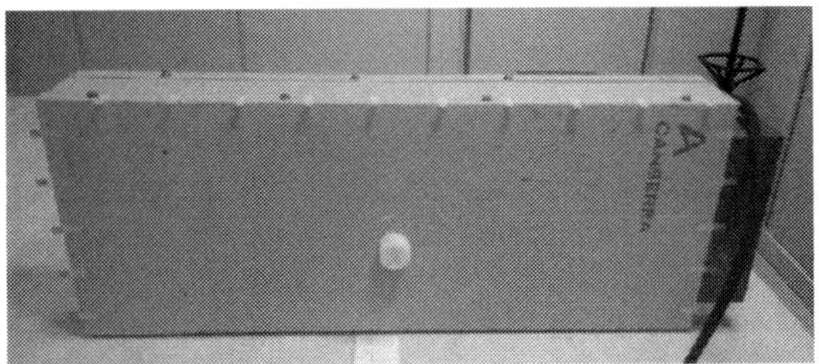

Figure 1: Picture of the WM3400 detector with a ^{252}Cf source used for testing.

The WM3400 detector contains six ^{3}He tubes with active length of 91.44 cm and diameter of 2.54 cm. The ^{3}He tubes are longer than the active length, and an air gap of 7.4 cm and 1.27 cm precedes and follows the section filled by gas, respectively. A Cd layer of 1 mm covers five sides of the detector according to the arrangement observed by opening the outer aluminum layer of the WM3400 detector. Only the side of the detector facing the radioactive source is not covered by Cd.

The original electronics of the detector was replaced to reduce the sensitivity to gamma-rays and time fluctuations observed during a previous measurement campaign. The new electronic system is described in detail in [3].

The WM3400 detector was connected to a MCA-527 multi-channel analyzer with upgraded firmware to record the time stamps of the neutron detections [5]. The time stamps are processed off-line to determine Totals (*T*), Reals+Accidentals (*R+A*), and Accidentals (*A*) counts and the multiplicity distribution [4] as well as to remove system instabilities due to high voltage fluctuations or interferences.

B. Monte Carlo model

A Monte Carlo model with the MCNPX code [6] was developed from the technical characteristics of the WM3400 coincidence counter. Figure 2 shows the inner section of the model to highlight the position of the ^{3}He tubes and the Cd layers within the polyethylene.

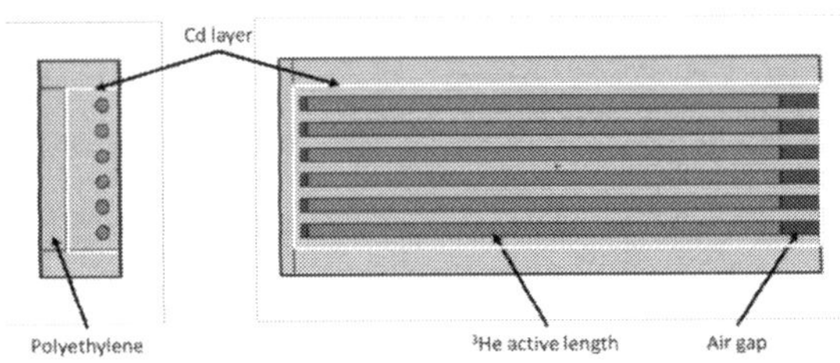

Figure 2: 2D-views of the Canberra WM3400 counter developed with the MCNPX Monte Carlo code.

The MCNPX-PoliMi V.2.0 code [7] and the MPPost post-processor [8] were used for the data analysis to obtain the Rossi-Alpha and multiplicity distributions. The MCNPX-PoliMi code requires to perform the Monte Carlo simulation in the so-called analog mode, in which each particle has a weight equal to 1. This is set by the following lines:

$$CUT\!:\!N \quad 2j \; 0 \; 0 \qquad (1.1)$$

$$PHYS\!:\!N \quad j \; 100 \; 3j \; -1 \qquad (1.2)$$

The source term in the MCNPX-PoliMi code was defined with the *IPOL* card.

$$IPOL \;\; +2 \; 2 \; 0 \; 0 \; 2j \; 6 \; 201 \; 202 \; 203 \; 204 \; 205 \; 206 \qquad (2)$$

The first term (*+2*) in the *IPOL* card sets ^{240}Pu spontaneous fission as the modelled source term, while the second term (*2*) defines the angular distribution in case of induced fission. The third term (*0*) is used to neglect the correlation between neutron and photon production, and the fourth term (*0*) applies no delay in the photon emission following a fission event. Two other entries are set to default values (*2j*), and finally the six cells with the ^{3}He tubes are listed (*201, ..., 206*).

In addition, the threshold for neutron energy release to be collected was set to 1 keV using the first term of the *RPOL* card. The second term of the *RPOL* card is the threshold for gamma-rays and was kept to the default value (*j*).

$$RPOL \;\; 0.001 \; j \qquad (3)$$

The neutron flux tally (so-called F4 type) was calculated for each ^{3}He tube with the FM treatment [6] to simulate the detector response taking into account the *(n,p)* reaction cross-section for ^{3}He and a ^{3}He pressure of 4 atm [2].

The *(R+A)* and *(A)* scalers from the Rossi-Alpha distribution were calculated with the MPPost post-processor from the output file of the Monte Carlo simulation specifying a pre-delay time of 4.5 µs, a gate width of 64 µs, and a long delay of 1000 µs.

III. In-Pile Sections

Two IPS (named MFBS6 and MOL7C7) were measured in May-June 2018 with one WM3400 neutron coincidence counter. The MFBS series of IPS were irradiated in the BR2 reactor during the 1970's, whereas the MOL7C IPS were used in the period 1977-1989 [1]. Some information about the IPS designs are included in Table 1.

Table 1: Design information of the IPS taken from [1]

	MFBS6	MOL7C7
Irradiation year	1971	1989
Fuel type	6 double MOX pins; 1 double UO_2 pin	30 MOX pins
Length	500 cm (approx.)	608 cm
Diameter	36 cm (maximum)	74 cm (body) 165 cm (maximum)

During the measurement campaign we noticed that the actual length of the MFBS6 IPS appeared to be longer than the 500 cm reported in literature. Both IPS therefore were modelled in MCNPX as cylinders with total length of 600 cm. The diameter of each IPS was varied according to the literature data.

A thickness of 5 mm of stainless steel was chosen as outer case of the IPS. The inner section of the IPS was a cylinder filled with a homogeneous material with same composition of stainless steel, but density reduced to 1.955 g/cm^3 (25% of normal density). The reduction in density took into account that the IPS is not completely filled with material. Both IPS presented a lead shielding on the top part and this was included in the MCNPX models as well. The thickness of the lead shielding was 2 cm for the MFBS6 IPS and 10 cm for the MOL7C7 IPS.

IV. Measurement campaign

A. Measurement setup

Both IPS are stored under water in the BR2 channel at SCK•CEN so the WM3400 detector was placed horizontally on the side of the channel. A picture of the measurement setup is shown in Figure 3. The position of the detector was fixed for the whole measurement campaign and for each measurement the IPS was moved to the desired height and distance from the detector.

Figure 3: Picture of the measurement setup.

A set of short measurements (600-700 s each) was carried out to obtain an axial profile of the IPS both in terms of Totals rate and Reals rate.

The starting position (*POS01*) was approximately set with the IPS completely out of the water and the bottom of the IPS at the same height of the neutron detector. The IPS was then lowered in steps of 40 cm until it was entirely immersed in water. A set of longer measurements was then carried out for the positions with the largest Real rates recorded during the axial scan.

B. Estimation of the fissile material content

A system of equations was established to first quantify the ^{240}Pu content in the different sections of the IPS from the Reals counts in each position.

$$\begin{cases} R_1 = S_1P_{1,1} + S_2P_{1,2} + S_3P_{1,3} + \cdots + S_{15}P_{1,15} \\ \quad\quad\quad\quad \cdots \\ \quad\quad\quad\quad \cdots \\ R_{15} = S_1P_{15,1} + S_2P_{15,2} + S_3P_{15,3} + \cdots S_{15}P_{15,15} \end{cases} \quad (4)$$

For each measurement position i, the Reals count R_i is related to the ^{240}Pu mass in each section S_i, through the importance function $P_{i,j}$ which is the probability to have a Real count in section j from a ^{240}Pu spontaneous fission event in section i. The importance function was calculated with a set of Monte Carlo simulations with the WM3400 detector facing the IPS in section i and the neutron source term only in section j. Different simulations were carried out to take into account the distance between the detector and the IPS.

Only the three sections above and below the detector position were considered in the analysis, since it was assumed that the importance function was negligible for the positions further away from the detector. This assumption is supported by the consideration that the section facing the detector contributes for about 50% of the Totals rate and 75% of the Reals rate considering all source positions. Considering also the positions above and below the detectors the combined values account for about 88% and 98% of the Totals rate and Reals rate, respectively.

The Reals rates are then converted in ^{240}Pu mass using the ^{240}Pu specific neutron emission [9]. This conservative approach assumes that the measured Reals rates are only due to ^{240}Pu. The total fissile material content in the IPS was finally determined with scaling factors based on burnup calculations for the irradiated fuel pins in the IPS [1].

V. Results

A. MFBS6

The Totals rate and Reals rate measured during the axial scan and long measurements are shown in Figures 4 and 5, respectively. The plots include also the uncertainty due to counting statistics.

The trends for the Totals rate and Reals rate are similar, with the maximum rates that were obtained in central sections of the IPS (*POS04*, *POS07*, and *POS08*). For these positions the values for the long measurement were larger than the values obtained in the quick axial scan. The values of the Reals rate in *POS13* are also larger than in the neighboring positions, but this is probably due to the presence of lead shielding in that axial region of the IPS. Cosmic radiation induces spallation reactions in lead which cannot be easily distinguished from spontaneous fission events.

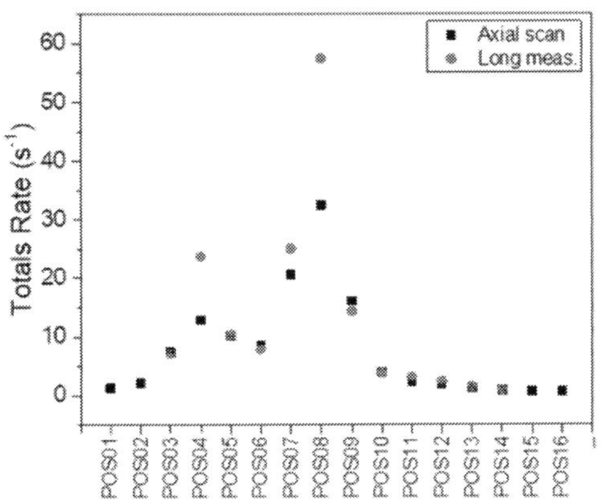

Figure 4: Measured Totals rate from the axial scan and the long measurements of the MFBS6 IPS.

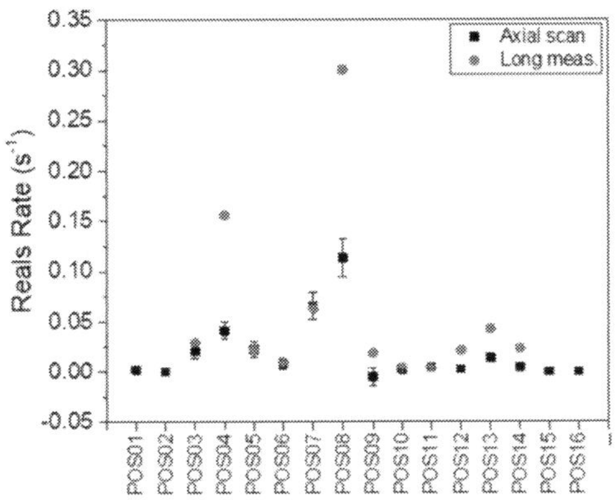

Figure 5:Measured Reals rate from the axial scan and the long measurements of the MFBS6 IPS.

The results from the long measurements are reported in Table 2 in terms of Totals, Reals, and Accidentals rates. The estimate of the ^{240}Pu content in each section is also included. According to these measurements the ^{240}Pu content is concentrated in the central section of the IPS, with a total ^{240}Pu mass for the whole IPS of 0.8 g. Using the scaling factors based on the burnup calculations the estimated total fissile material content in the IPS is 14 g. This estimate is comparable to a conservative upper limit based on the ^{137}Cs content [1].

The total fissile material estimated in the IPS is lower than the limit of 45 g set by current regulation for transport [10], and the further segmentation of the IPS in several 220L drums will introduce an additional margin.

B. MOL7C7

The Totals rate and Reals rate measured during the axial scan and long measurements are shown in Figures 6 and 7, respectively. The plots include also the uncertainty due to counting statistics.

Also for this IPS the Totals rate and Reals rate are strongly correlated, but in this case the maximum Totals and Reals rates were obtained at the extremities of the IPS (*POS02*, *POS11*, and *POS12*). The values for the long measurements are comparable to the values obtained in the quick axial scan, with the Totals rates that are outside the statistical uncertainty only for a couple of positions. One should stress that the positioning of the IPS was not necessarily the same for the axial scan and the long measurements. As for the previous IPS, the increase of the Reals rate in *POS11* and *POS12* are probably due to the presence of lead shielding in that axial region. The Reals rates obtained for the MOL7C7 IPS are in general lower than those obtained during the MFBS6 measurements.

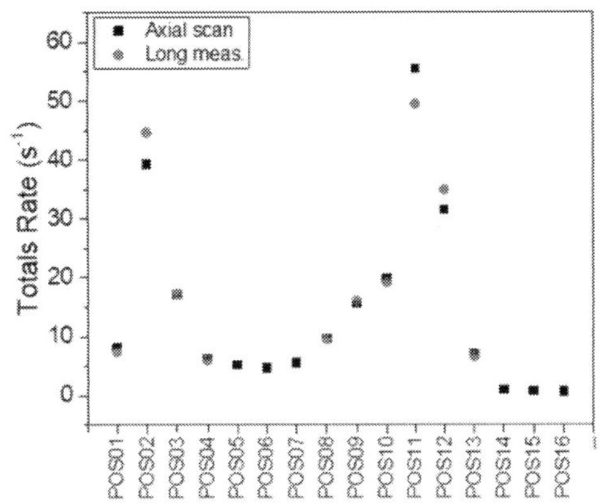

Figure 6: Measured Totals rate from the axial scan and the long measurements of the MOL7C7 IPS.

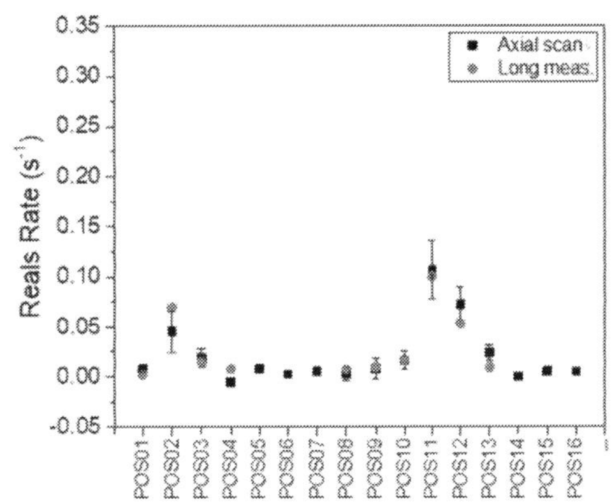

Figure 7:Measured Reals rate from the axial scan and the long measurements of the MOL7C7 IPS.

The results from the long measurements are reported in Table 3 in terms of Totals, Reals, and Accidentals rates. The estimate of the ^{240}Pu content in each section is also included. According to these measurements the ^{240}Pu content is concentrated at the extremities of the IPS, with a total ^{240}Pu mass for the whole IPS of 1.8 g. Using the scaling factors based on the burnup calculations the estimated total fissile material content in the IPS is 22 g. This estimate is comparable to the detection limit based on an independent total neutron counting using a Berthold LB6411 portable neutron dose rate probe that is used for ambient dose equivalent for neutrons [1].

VI. Conclusion

A Canberra WM3400 neutron coincidence counter was used for the measurement of two irradiated IPS in view of their future transport, dismantling and conditioning.
In addition, a Monte Carlo model of the detector and of the IPS was developed and used for the estimation of the total fissile material in the IPS.
The IPS were measured with a short axial scanning and long measurements for the positions with the largest values of the Reals rate. The results obtained during the two sets of measurements were comparable for both IPS.
The total fissile material estimated for the two IPS is 14 g for the MFBS6 and 22 g for the MOL7C7. Both estimates are in line with results obtained with other measurement techniques. For both IPS the total fissile material estimated is lower than the limit of 45 g set by current regulation for transport, and the further segmentation of the IPS in several 220L drums will introduce an additional margin.

References

[1] S. Boden, et al., "Remaining In-Pile Sections of the Fast Breeder Reactor Programme - Estimation of the fissile material present" SCK•CEN internal report I-0642, 2018

[2] http://www.canberra.com/products/waste_safeguard_systems/pdf/WM3400-SS-0258.pdf. Last accessed 05/03/2018.

[3] A. Borella et al, "Improved electronics for 3He based neutron counters" to be presented at the 2019 ANIMMA conference, 2019.

[4] N. Ensslin, "Principles of Neutron Coincidence Counting" in Passive Nondestructive Assay of Nuclear Materials, NuREG/CR-5550, LA-UR-90-732, 1991

[5] https://www.gbs-elektronik.de/media/download_gallery/mca527_neutron.pdf. Last accessed 19/04/2018

[6] D. Pelowitz, "MCNPX user's manual version 2.7.0". Los Alamos National Laboratory LACP-11-00438, 2011.

[7] E. Padovani, et al., "MCNPX-PoliMi User's Manual". Radiation Safety Information Computational Center, Oak Ridge National Laboratory, 2012.

[8] E. Miller, et al., "MCNPX-PoliMi Post-Processor (MPPost) Manual". Radiation Safety Information Computational Center, Oak Ridge National Laboratory, 2012.

[9] A. J. Koning, et al., "Status of the JEFF Nuclear Data Library". Proceedings of the International Conference on Nuclear Data for Science and Technology, 2010, pp.1057-1062.

[10] "European Agreement concerning the International Carriage of Dangerous Goods by Road". United Nations Economic Commission for Europe (UNECE), 2007.

Table 2: Results from the long measurements and estimate of the ^{240}Pu content for each measurement position of the MFBS6 IPS.

Position ID	Meas. time (s)	γ-ray dose rate (μSv/h)	Totals rate (s^{-1})	Reals rate (s^{-1})	Accidentals rate (s^{-1})	^{240}Pu content (g)
POS03	11100	7000	7.211 ± 0.025	0.029 ± 0.002	0.004 ± 0.001	<0.01
POS04	58380	8000	23.771 ± 0.020	0.156 ± 0.002	0.036 ± 0.001	0.28
POS05	13500	21000	10.343 ± 0.028	0.022 ± 0.002	0.007 ± 0.001	<0.01
POS06	14220	1000	7.976 ± 0.024	0.010 ± 0.001	0.005 ± 0.001	0.01
POS07	59340	15000	25.117 ± 0.021	0.063 ± 0.002	0.042 ± 0.001	0.03
POS08	211260	370	58.491 ± 0.017	0.307 ± 0.002	0.218 ± 0.001	0.38
POS09	10620	8800	14.344 ± 0.037	0.019 ± 0.002	0.013 ± 0.001	<0.01
POS10	12347	320	3.861 ± 0.018	0.004 ± 0.001	0.001 ± < 0.001	0.01
POS11	14880	170	3.081 ± 0.014	0.004 ± 0.001	0.001 ± < 0.001	<0.01
POS12	14557	8	2.418 ± 0.013	0.021 ± 0.001	0.000 ± < 0.001	0.01
POS13	58560	3	1.523 ± 0.005	0.045 ± 0.001	0.000 ± < 0.001	0.07
POS14	10020	8	0.960 ± 0.010	0.023 ± 0.001	0.000 ± < 0.001	0.03

Table 3: Results from the long measurements and estimate of the ^{240}Pu content for each measurement position of the MOL7C7 IPS.

Position ID	Meas. time (s)	γ-ray dose rate (μSv/h)	Totals rate (s^{-1})	Reals rate (s^{-1})	Accidentals rate (s^{-1})	^{240}Pu content (g)
POS01	12480	8	7.471 ± 0.024	0.002 ± 0.001	0.004 ± 0.001	< 0.01
POS02	60420	145	44.623 ± 0.027	0.070 ± 0.002	0.128 ± 0.001	0.12
POS03	60060	40	17.234 ± 0.017	0.015 ± 0.001	0.019 ± 0.001	< 0.01
POS04	211260	20	6.056 ± 0.005	0.008 ± < 0.001	0.002 ± < 0.001	0.01
POS08	25140	180	9.563 ± 0.020	0.007 ± 0.001	0.006 ± < 0.001	0.01
POS09	25917	170	16.005 ± 0.025	0.009 ± 0.001	0.017 ± 0.001	0.02
POS10	26460	90	19.198 ± 0.027	0.016 ± 0.002	0.024 ± 0.001	< 0.01
POS11	59040	3.5	49.446 ± 0.029	0.100 ± 0.003	0.158 ± 0.002	0.56
POS12	61020	8	34.923 ± 0.024	0.053 ± 0.002	0.079 ± 0.001	1.08

EPJ Web of Conferences **225**, 06002 (2020) https://doi.org/10.1051/epjconf/202022506002
ANIMMA 2019

The remote-controlled spectrometric system for searching and characterization of high-level radioactive waste

Alexey Danilovich
Dept. Rehabilitation
NRC "Kurchatov Institute"
Moscow, Russia
Danilovich_AS@nrcki.ru

Victor Potapov
Dept. Rehabilitation
NRC "Kurchatov Institute"
Moscow, Russia
Potapov_VN@nrcki.ru

Sergey Smirnov
Dept. Rehabilitation
NRC "Kurchatov Institute"
Moscow, Russia
Smirnov_SV@nrcki.ru

***Abstract* - In 2003, the work on environmental rehabilitation of the former coastal technical base of the Russian Navy in Gremikha town, with technical and financial support of the State Atomic Energy Corporation "Rosatom" and the French Alternative Energies and Atomic Energy Commission had begun. In 2017, the task of searching, characterization and packaging in transport containers for export to the long-term storage of high-level waste containing ^{152}Eu was set as part of this program. These wastes were presented as fragments of reactor control and safety system rods of first generation nuclear submarine reactors and were placed in temporary storage containers mixed with other radioactive waste and construction garbage on the territory of the Gremikha branch of the North-West Center "SevRAO".**

To accomplish the task, the spectrometric system "Gamma-Pioneer" was developed and made at National Research Center "Kurchatov Institute". The system consist of a collimated spectrometric gamma-ray CdZnTe detector 500 mm^3 volume, a dosimeter for continuous measurement of ambient equivalent dose rate at the instrument location and a color video camera with zoom lens. The device is placed in a protective frame and installed as an attachment implements on robot Brokk-90. Communication between the measuring system and the control computer is carried out using USB extender and 100 meters cable. The video camera is connected to the control computer via Ethernet.

To search for the most active sources of gamma radiation, an equivalent dose rate calculation program using the results of measured instrumental spectra was developed. The characterization of the extracted fragments was carried out by activity evaluation of the main radionuclides polluting the samples under investigation. The paper describes design and operating principles of the system, as well as methods for the equivalent dose rate calculation and the activity evaluation of the fragments of reactor control and safety system rods based on spectral measurements obtained using the developed radiometric system

Keywords—radiation detection, gamma spectrometry, radioactive waste, detectors.

I. Introduction

Environmental rehabilitation of the former coastal technical base of Russian Navy in Gremikha town (today, this is Gremikha branch of the North-West Center "SevRAO" branch of FSUE "RosRAO", the closed city Ostrovnoy) is currently one of the largest projects addressed to solve the "nuclear legacy" problems in the North-West region of Russia. The works began in 2003 under the Global Partnership G-8 program with the financial and technical support of Atomic Energy Commission of France and the Ministry of Atomic Energy (Rosatom) [1].

At the first stage, the arrangements to improve the radiation situation, modernize the infrastructure and ensure the personal safety were implement. Specialists of NRC "Kurchatov institute", NIKIET, VNIPIET and other conducted a comprehensive engineering and radiation survey of buildings, structures, territory and coastal waters to obtain initial data. In conditions of an increased radiation situation, in building 19 and in the open site for storage of spent nuclear fuel (SNF) and solid radioactive waste (SRW), remote identification of γ-radiation sources was carried out. Fragments of SNF were located in the open site. In concrete containers and in two packages with radioactive waste highly active sources with ^{152}Eu lines in spectra were placed. This suggest that the founded sources are fragments of reactor control and safety system rods of first generation nuclear submarine reactor [2].

In the period from 2008 to 2013, more than 800 SNF assemblies were removed from the temporary storage site to Production Association "Mayak" for reprocessing [3]. From 2013, Gremikha branch became the first facility among the former coastal technical bases, freed from nuclear fuel of submarine's water-cooled reactors.

In 2017, with the financial support of the State Corporation Rosatom, works aimed at solving the task of handling high-level radioactive waste (HLW) accumulated during the operation of nuclear submarines, including the preparation and removal of SRW and HLW began. The main objects of search were reactor control and safety system rods stored in abnormal conditions in temporary packages and concrete containers in building 19 and in the open area of temporary storage of SRW. For the searching and sorting of radioactive waste by category and activity, NRC "Kurchatov institute" developed the spectrometric system "Gamma-Pioneer".

II. Spectrometric System "Gamma-Pioneer"

A. System designated purpose

Spectrometric system "Gamma-Pioneer" is designed to search for and identify highly active objects in conditions of intense gamma radiation and conducting surveys and scouting of contaminated premises. Using the CdZnTe detector allows not only determine the radionuclide composition of contamination, but also to calculate the dose rate generated by objects in the field of view of the system.

B. System composition

Spectrometric system "Gamma-Pioneer" consist of (Fig.1):

A. The measuring part, which is mounted on the boom of the Brokk-90 robot using a standard mount;

B. Control computer for managing the system and processing results;

https://doi.org/10.1051/epjconf/202022506002

C. Protective case with installed inside power supply unit and USB extender.

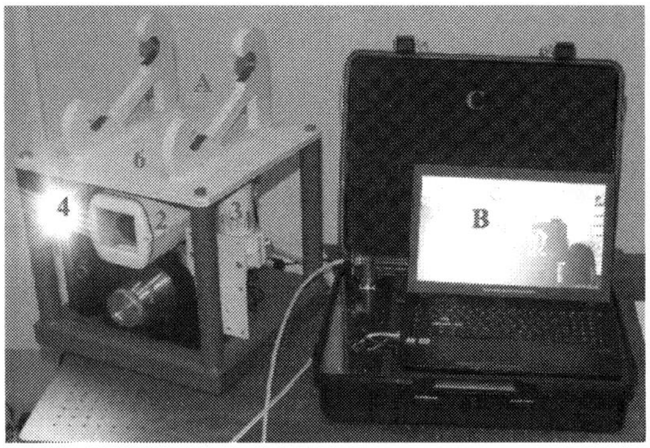

Fig.1. Photo of the "Gamma-Pioneer" system with a control computer and a protective case.

The connection between the control computer and the measuring system is carried out using a USB extender and a twisted pair cable 100 meters length.

The measuring part of system consist of (Fig.1):

1. Collimator with internally mounted spectrometer µSPEC500 [4];
2. Color video camera with a trans focal lens;
3. Gamma radiation detector DBG-S11D for measuring the dose rate at the point of the system location [5];
4. LED floodlight;
5. Waterproof box with USB port extender and power supply, installed inside;
6. Protective frame that serves to protect the measuring system's components from damage due to accidental impact.

The collimator of the measuring system is made of lead and placed in a stainless steel body. To eliminate the influence of the characteristic radiation of lead on the measurement result, the micro spectrometer is placed in a brass case with a wall thickness of 2 mm, and only then is installed in a collimator. Also, to reduce the influence of the characteristic radiation of lead, and in addition, to change the collimation angle and to reduce the flux of gamma quanta on the detector surface, two interchangeable brass inserts into collimator with an internal diameter of 10 mm and 5 mm were made.

When examining contaminated objects under conditions of strong radiation fields, it may be necessary to take into account background radiation. For this, a lead plug covering the collimator hole at the time of the background spectrum collection, was made.

To reduce the number of gamma quanta, determined by scattered radiation, and falling on to the detector, a filter consisting of two glued together round plates of lead and brass, is used. If necessary, the filter is installed in front of the collimator hole with a brass plate inwardly.

In operation, the measuring part of system, installed on robot Brokk-90, scans the inspected object in order to find the most active fragments (in our case, control and safety system rods have been). The video image of the object under examination, as well as gamma spectrum, the dose rate values, measured by collimated and "opened" detectors are displayed on the screen of the control computer. Next, another robot extracts the fragment of interest to clarify its radionuclide composition, measure activity and package it into a transport container.

III. The method of dose rate calculation

The result of the registration of radiation by spectrometric detector is the instrumental spectrum *N(E)*. This information is sufficient to estimate the dose rate generated by the registered radiation, and the formula for calculating the equivalent dose rate can be represented as follow [6]:

$$\dot{H} = A_{calibr} * \sum_{i=0}^{N} \frac{N_i * K_{ei}}{t}, \tag{1}$$

where A_{calibr}– calibration coefficient; N_i – number of pulses in channel i; K_{ei} – weighting coefficient for channel i; N – number of channels in spectra; t – lifetime in seconds.

The weighting coefficient K_{ei} allows taking into account the energy dependence of the registration efficiency of a spectrometric detector. This dependence is calculated based on the data obtained by the Monte-Carlo method, and can be written in the form of a six-order polynomial:

$$K_{ei} = \sum_{j=0}^{6} a_j \, E_i^j, \tag{2}$$

where a_j – polynomial coefficients; E_i – energy of channel i, calculated by formula:

$$E_i = (i - a_c)/b_c,$$
$$a_c = N_1 - b_c * E_1,$$
$$b_c = (N_2 - N_1)/\Delta E,$$
$$\Delta E = E_2 - E_1,$$

N_1 and N_2 – the channel numbers of the instrumental spectrum corresponding to the positions of the total absorption peaks of gamma quanta with energy E_1 and E_2. In this case, photo peaks ^{137}Cs (661.657keV) и ^{60}Co (1332.501keV) were used for the energy calibration of the spectrometer.

To determine the calibration coefficient A_{clibr}, a ^{137}Cs source with a known activity of $1.5*10^{10}$ Bq was used. The source was installed at a distance of 10 to 100 centimeters from spectrometer's detector, placed in collimator of system, in steps of five centimeters. Measurements of the dose rate generated by the source at the detector's location were carried out for different distances, as well as for different collimator variants. Collimators with an internal diameter of 5 mm and 10 mm were used. In addition, measurements were carried out with a lead-brass filter installed in front of the collimator hole, and without it. The theoretical dependence of the dose rate generated by the source at the detector's location on the distance between them was calculated for all four options. According to the measurement results and theoretical data, calibration coefficients were calculated. Further, they were taken into account in the program for calculating the dose rate of radiation registered by the spectrometric detector.

IV. The method for calculating the activity of radionuclides

The radionuclides of ^{152}Eu and ^{154}Eu contained in reactor control and safety system rods have complex emission spectra consisting of a large number of lines with different quantum yield for decay.

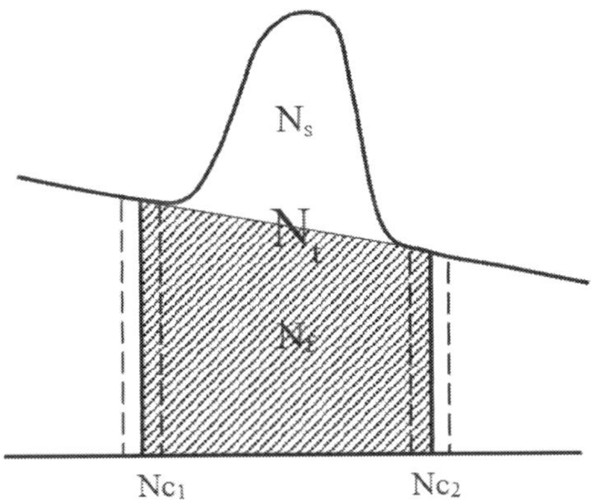

Fig.2. Graphic representation of the peak of the radionuclide in spectrum.

It has known that for a specified measurement geometry, the activity of a radionuclide is proportional to counting rate in the total absorption peak, i.e. equal to the ratio of the useful signal to the live time of measurement multiplied by the calibration coefficient:

$$A = C_{calibr} * \frac{N_S}{T}, \tag{3}$$

where N_S – useful signal, C_{calibr} – calibrating coefficient in Bq/(impulse/s), T – life time of measurement in seconds.

The useful signal is equal to the difference of the total signal area in selected range (the number of registered impulses in the total absorption peak) and the background signal:

$$N_s = N_t - N_f, \tag{4}$$

where N_t – number of registered impulses in the total absorption peak in selected range, N_f – background signal.

To estimate the calibration coefficients, a mathematical model of the radiometric system, in which it was assumed that the source of gamma radiation corresponded to the size, material and radionuclide composition of the object of study (reactor control and safety system rods), was built.

Since it was supposed to measure the objects at different distances, the calibration coefficient was presented in form of a polynomial:

$$C_{calibr} = \sum_{i=0}^{2} b_i\, x^i, \tag{5}$$

where x – distance between the object and the detector in meters; b_i - polynomial coefficients.

The determination of the number of pulses in the total absorption peak minus the background substrate was carried out in standard way (according to formula (4)). The total number of pulses in the range where the peak is located, was determined by the sum:

$$N_t = \sum_{i=N_{c1}}^{N_{c2}} N_i, \tag{6}$$

and the background signal was calculated using the trapezoid area formula (Fig.2):

$$N_f = (a + b) * c\,, \tag{7}$$

$$c \equiv (N_{c2} - N_{c1} + 1)/2\,,$$

$$a = \left(N_{j-1} + N_j + N_{j+1}\right) / 3, \text{ where } j=Nc_1\,,$$

$$b = \left(N_{k-1} + N_k + N_{k+1}\right) / 3, \text{ where } k=Nc_2\,,$$

Nc_1 and Nc_2 – numbers of the initial and final channels of the selected range.

In radiometric measurements, when the activity of radionuclides is estimated by the counting rate at the total absorption peak, it is also important to know the statistical error of this quantity:

$$\Delta N_s = 2 * \sqrt{N_t + c^2 * (N_a + N_b)} \tag{8}$$

$$N_a = \frac{1}{9}\sum_{i=N_{c1-1}}^{N_{c1+1}} N_i\,,$$

$$N_b = \frac{1}{9}\sum_{i=N_{c2-1}}^{N_{c2+1}} N_i\,.$$

In the case when the statistical error of measurement is greater than the useful signal ($N_s < \Delta N_s$) or the useful signal is less than zero, the activity is considered equal to zero.

V. The results of spectrometric system using

In the summer of 2018, the spectrometric system "Gamma-Pioneer" installed on the Brokk-90 robot was used to search and characterize high-level waste containing ^{152}Eu, housed in two storage facilities, placed on the territory of branch Gremikha.

Fig.3. The location of the equipment during the measurements of the rods activity in the process of sorting radioactive waste. 1 – measuring part of system "Gamma-Pioneer", 2 – Brokk-90 robot, 3 – Brokk-160 robot, 4 – rod fragment, 5 – inspected container, 6 – construction waste conteiner, 7 – low level waste container, 8 – transport container for high level waste.

In the first storage, the waste was located in two metal containers. It was necessary to find highly active fragments of control and safety system rods among the medium and low active waste and construction debris contained in containers. The Brokk-160 robot was used to extract the contents from the inspected container.

Figure 3 shows the disposition of the equipment during the measurement of activity of one of the rods fragments found in the container. An example of the instrumental spectrum is presented in Figure 4.

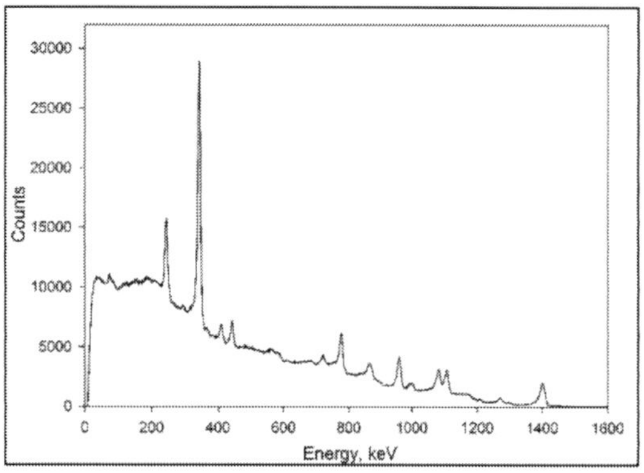

Fig.4. An example of a instrumental spectrum of radiation of control and safety rod.

As a result, twenty-three fragments of control and safety system rods with a total activity of 3.55*10^{12} Bq were found and placed in a transport container. The results of the activity evaluation of radionuclides ^{152}Eu, ^{154}Eu, ^{60}Co и ^{137}Cs, contained in the rods loaded in the transport container are presented in Table I.

TABLE I.

A(^{152}Eu), Bq	*A(^{154}Eu), Bq*	*A(^{60}Co), Bq*	*A(^{137}Cs), Bq*	*AΣ, Bq*
3.17*10^{12}	3.57*10^{11}	2.25*10^{10}	0.00	3.55*10^{12}

The purpose of the work in the second storage was the extraction of high-level waste in the form of fragments of control and safety system rods from three containers, which, according to the results of previous surveys, contained the most active radioactive waste and made a significant contribution to the dose rate at the site. Preliminary spectrometric measurements showed that these containers contain ^{152}Eu.

As result, twenty-two rods fragments with total activity of 9.96*10^{12} Bq were loaded into the second transport container. The results of the activity evaluation of radionuclides ^{152}Eu, ^{154}Eu, ^{60}Co и ^{137}Cs, contained in the rods loaded in the transport container are presented in Table II.

TABLE II.

A(^{152}Eu), Bq	*A(^{154}Eu), Bq*	*A(^{60}Co), Bq*	*A(^{137}Cs), Bq*	*AΣ, Bq*
8.83*10^{12}	9.14*10^{11}	2.16*10^{10}	0.00	9.96*10^{12}

VI. Conclusions

Spectrometric system "Gamma-Pioneer" can be effectively used for the exploration and survey of objects in conditions of high-levels of radioactive contamination. It is especially necessary during emergency abnormal operations at HLRW and SNF storage facilities, decommissioning of nuclear facilities, searching for lost sources (for example RITEGs) and emergency response.

References

[1] N. N. Ponomarev-Stepnoi, B. S. Stepennov, and V. N. Kovalenko, "Ecological rehabilitation of the temporary storage site for spent nuclear fuel and radwastes in Guba Chervyanaya," Bezop. Okruzh. Sredy, No. 1, 92–97 (2008).

[2] K. E. Ivanov, N. N. Ponomarev-Stepnoi, B. S. Stepennov, Yu. A. Teterin, "Remote measurement of the radiation dose rate on solid radwaste temporary storage site", Atomic Energy, 2008,Vol. 105, No. 2, pp.133-137.

[3] https://cegdb.iaea.org/pages/ProjectDetails.aspx?ProjID=45

[4] http://ritec.lv/uspec.html

[5] http://www.doza.ru/eng/catalog/detection_bloks/1135/

[6] A.V. Chesnokov, V.I. Fedin, A.A. Gulyaev, V.N. Potapov, S.V. Smirnov, S.B. Shcherbak, L.I. Urutskoev, A.G. Volkovich. Surface Activity and Dose Rate Distribution inside 4-th Reactor Hall of Chernobyl NPP. IEEE Trans. Nucl. Sci., 1998 Vol. 45, No. 3, part I, pp. 986-991.

EPJ Web of Conferences **225**, 06003 (2020) https://doi.org/10.1051/epjconf/202022506003
ANIMMA 2019

Calorimetric Non-Destructive Assay of Large Volume and Heterogeneous Radioactive Waste Drums

W. R. Kubinski[1], C. Carasco , D. Kikola[1], C. Mathonat , D. Ricard[4], D. Tefelski[1], H.Tietze-Jaensch[5]

[1] Warsaw University of Technology, Plac Politechniki 1, Warszawa 00-661, Poland
[2] CEA/DEN/DTN Cadarache, F-13108 Saint-Paul-Lez-Durance, France
[3] KEP Technologies, Rue de l'Oratoire 7, 69300 Caluire et Cuire, France
[4] Andra, 1/7 rue Jean Monnet, Parc de la Croix-Blanche, F-92298 Châtenay-Malabry, France
[5] Forschungszentrum Jülich, Wilhelm Johnen Strasse, Jülich 52428, Germany

Abstract—**The EU-CHANCE project aims at the issue of the characterization of conditioned radioactive waste (CRW) and one objective of CHANCE is to focus on: Calorimetry as a comprising non-destructive technique to reduce uncertainties on the inventory of radwaste containing shielded and hidden material difficult to be measured by other means.**

A MCNP6-based numerical study comprising the particle flux out of a 200L mock-up drum in a Large Volume Calorimeter (LVC) currently manufactured by KEP Nuclear (France) will be presented and discussed. For the analyses, the particle flux and energy deposition in each layer of the calorimeter were determined. The results yield that a significant fraction of the radiation would leave the system and not contribute to the measurable heat deposition. The expected energy deposition is obtained and cumulated for each layer over the whole energy range revealing the fraction of particles actually escaping the LVC calorimeter. While this escape fraction needs and can be determined, the LVC is a very suitable apparatus for the anticipated experiments on large and heterogeneous waste drums that possibly contain deeply buried beta-emitters (e.g. Sr/Y-90) or shielded alpha-sources hidden inside the drum with a significant level of gamma and neutron radiation background radiation. The high-energy part of this gamma and neutron flux may even reach the reference chamber of the calorimeter and deposit some energy there, compromising the calibration and may cause a double-bias.

I. Introduction

In order to manage, securely store and efficiently process radioactive materials, their precise characterization is essential and mandatory. However, conditioned waste, unlike raw materials, is sealed and embedded in a solid material matrix making access difficult. In addition, radioactive waste may contain materials of various origin, and the spectrum of their radiation can be quite complex. Often waste drums, in particular after years of interim storage, are no more in their state of initial state, thus unfit for final disposal. The comprehensive characterisation of radioactive waste is a requisite for its safe final disposal, and it is often quite complicated. It requires a set of specific assessment techniques, preferably non-destructive methods.

Amongst others calorimetry is one of the methods for the determination of nuclear material masses (or activities) by the quantitative measurement of heat flux, emerging from a sample. Calorimetry is used for non-destructive assay (NDA) which means that no sample needs be tempered with or destroyed by the analysis. Heat-flow calorimeters have been used successively for measuring a wide range of thermal powers i.e. from 0.5 mW (0.2g of low-burnup plutonium) up to 1kW for samples varying in size from about 2.5 cm up to 60 cm wide and 100 cm long [3].

As a non-destructive method for the quantitative measuring of the heat emission, calorimetry is used to determine the masses/activities of nuclear material hidden inside nuclear waste compounds. In combination with gamma and neutron emission measurement, calorimetry is well suited for the multi-nuclide assay of conditioned waste drums of large-volumes and possibly heterogeneous content,

All radioactive decay energy converts into heat, eventually. On the one hand, calorimetry encompasses all the radionuclides contained in a sample and emitting radiation that converts into heat by the interaction of the emitted particles with surrounding matter. On the other hand, calorimetry is by no means specific for one or the other radionuclide. Moreover, any radiation that escapes from the calorimetric apparatus remains undetected and unaccounted for. To employ this method as a comprehensive and useful tool for the characterization of nuclear waste, it is therefore essential to support calorimetric assay not only with complementary gamma and neutron emission measurements but also with detailed numerical simulation of the experimental set-up. Then, this method can be turned into a viable one, and potential bias can be eliminated by proper instrument calibration. None-the-less, a possible reason of diminished accuracy of calorimetry is the energy leakage caused by fugitive gamma and neutron radiation. This radiation is produced by gamma emitters (e.g. ^{60}Co, ^{137}Cs), as an effect of radioactive decay and various nuclear interaction inside the drum, e.g. bremsstrahlung radiation resulting from intense beta-emitters (e.g. $^{90}Sr/^{90}Y$). As the accuracy of the assay is one of the key elements of reliable measurements, the radiation leakage must be considered as a major source of uncertainty, which is definitely worthwhile to be studied and analysed in detail [5].

At times of comfortable computing power, numerical calculations can mimic the experimental set-up and radiation vs. matter interactions and so support and supplement the uncertainty analysis associated with the radiation leaking from the calorimeter. One of those numerical methods is the Monte Carlo method based on a probabilistic approach and random numbers. It is becoming more and more frequently used in the analysis of nuclear systems because of the possibility of an easy-play studying the parameters, without unjustified simplifications. Nowadays computing power allows us to simulate a full, three-dimensional nuclear system, made of detectors, fuel assemblies and even whole reactor cores as well as material configurations in radioactive waste drums, and providing reliable results at satisfactorily short calculation times.

The studies presented here were undertaken in the frame of the European research and development project "CHANCE" (Characterization of conditioned nuclear waste for its safe disposal in Europe), which focuses on specific issues of conditioned radioactive waste characterisation. One of the aims of the project is to design and construct a new, improved Large Volume Calorimeter with reduced detection limits and improved accuracy. The goals of this work were to support the instrument's evaluation and validation by numerical calculations and simulation. This study's aim is to set-up an appropriate computational environment and to perform a series of simulations in order to understand and estimate the influence of the radiation leakage on the final results and their precision.

II. THE MODEL

Matching the design of the calorimeter, a simplified but still viable numerical model was implemented using MCNP6 code. The numerical model is presented in Figure 1. It consists of different layers, just alike the real KEP-LVC calorimeter. There is a cylindrically shaped sample drum at the centre of the measurement chamber which is filled with air. The next layer is the octagon-shaped structure with the heat flux detectors mounted on each wall. Then, there is the package of homogenization (red layers in Figure 1) and alternating insulation layers (blue layers), up to sixth layer each. Underneath the measurement chamber there is the reference chamber (or ghost chamber) with a phantom aluminium block.

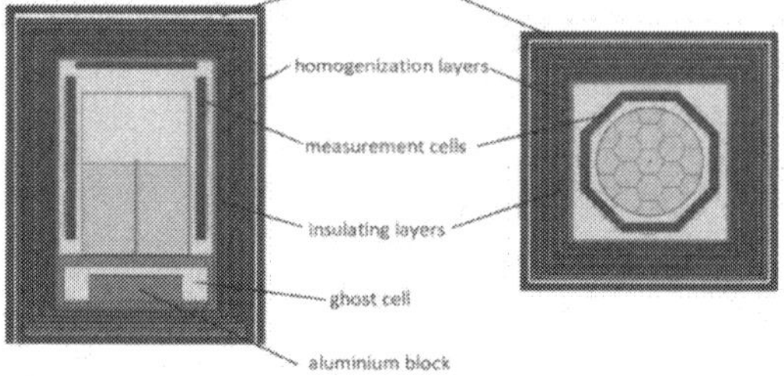

Figure 1. Simplified model of the calorimeter.

III. CONSIDERED SCENARIOS

To check how the distribution of radioactive material inside the drum influences the radiation leakage, three scenarios were realised in the simulations. For all the scenarios the drum was filled with a virtual radioactive composition up to 50 cm (active part of the drum). The remaining part was filled with air.

1st scenario: most conservative

The first source configuration, hereafter called the "most conservative" set-up was a drum filled with sand and one pin placed in its centre. In the middle of the pin there was a container with the virtual sampled source. In this scenario it was assumed that all the radioactivity was buried deep inside the drum, thus on average, particles needed the longest possible path to escape the system.

2nd scenario: homogenous source

In this scenario, hereafter called the "homogenous" set-up, it was assumed that the virtual radioactive material is uniformly distributed within the whole volume of the active part of the drum. The active part was filled only with sand and particles were sampled inside this whole volume. The amount of escaping radiation was, of course, higher in this case, as the particles could appear nearby the wall of the drum as well.

3rd scenario: least conservative

A third kind of source configuration, hereafter called "least conservative" set-up was a drum, similar to case I, but with only one pin filled and this one placed closest to the edge of the drum. In addition, the container with the virtual radioactive material was located at the bottom of the pin. In this scenario the particles, on average, had the shortest pathway to leave the system.

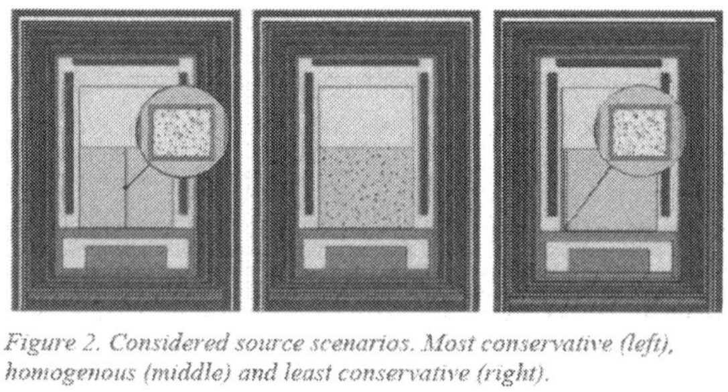

Figure 2. Considered source scenarios. Most conservative (left), homogenous (middle) and least conservative (right).

https://doi.org/10.1051/epjconf/202022506003

Figure 2 shows the differences between sampling particles scenarios in all three kinds of the source configurations. The radioactive composition was modelled by a virtual gamma source and a virtual neutron source at energies from 10 keV up to 10 MeV. In addition, some calculations of waste composition for beta and alpha emission were carried out, too, but as expected, no escape radiation has been found.

IV. RESULTS

A. Particles flux

In order to understand how the particles behave inside the calorimeter, the particle flux was checked using tally 4 with the mesh option and normalized to 10^6 particles source strength. The flux was calculated on a XY mesh made of 10000 rectangular cells, each 1.4cm wide and 188cm high. The flux was also determined along Z axis, on a mesh made of 300 rectangular cells, 72cm wide and 0.6 cm high. The obtained 3D fluxes are illustrated in Figure 3. The fluxes were analysed for the three scenarios and three different source energies (0.1MeV, 1MeV and 5 MeV). For the clarity of the results presented, the charts were zoomed up to the range of $(0,10^{-4})$ #/particle/cm^2.

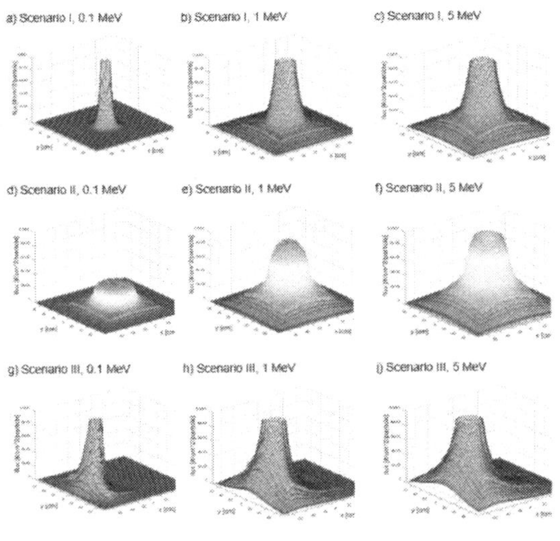

Figure 3. Flux distributions. Gamma radiation.

The results confirm that for the 0.1 MeV source almost all the radiation is deposited inside the system. Despite all particles being absorbed and no radiation escaping the system, even for low energy gammas not all the heat is deposited inside the Peltier elements where it is detected. For the homogenous and least conservative scenario, still some layers of the calorimeter are visible, which means that some part of the radiation penetrates past the detection zone and the homogenization and insulation parts. Only the heat deposited inside the drum and measurement parts are detectable, thus not all the total deposited energy can be measured. And for higher energies, certainly even more particles leave the system, a results that has been expected.

The charts (Figure 3) are meant to be just illustrative. To quantify and compare the fluxes, especially the tails which describe the radiation leakage, two-dimensional charts are plotted along the X and Z axes. Figure 4 and Table 1 show a juxtaposition of fluxes at mid-energy (i.e. 1MeV) for all three scenarios.

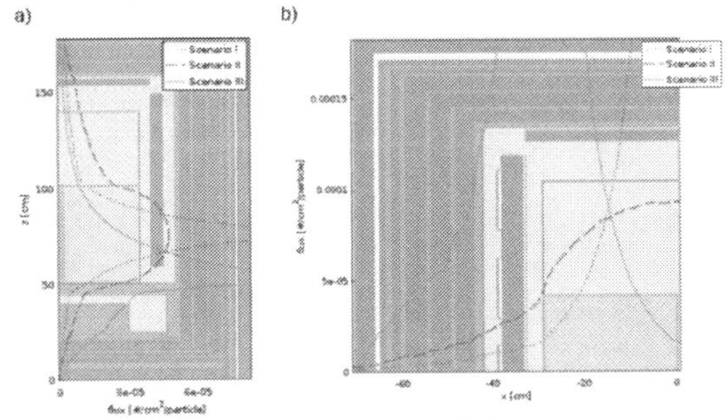

Figure 4. Gamma particles flux.

One can see that the energy of the source has significant influence on the flux distribution. The flux plotted along Z axis (Figure 4a) shows that some part of it reaches into the ghost chamber and beyond. In general, for higher energies, the flux penetrates more layers of the calorimeter, as well as the reference parts which cause inevitably some double bias.

TABLE I
GAMMA FLUX ESCAPE [#/PARTICLE/CM^2]

	Escape along Z-axis			
scenario:	bottom:	rel. error:	top:	rel. error:
I	1.93·10-7	2.7%	4.63·10-7	1.4%
II	5.08·10-7	1.9%	1.60·10-6	1.0%
III	1.76·10-6	0.7%	2.13·10-7	2.1%
	Flux in the aluminum block:	rel. error:	escape along X-axis	rel. error:
I	2.48·10-6	0.1%	9.47·10-7	6.7%
II	5.82·10-6	0.5%	2.75·10-6	4.3%
III	2.03·10-5	0.4%	1.07·10-5	2.0%

The results show that source distribution pattern has a strong influence on the flux distribution. In case of the flux along X-axis (Figure 4b) and the source close to the wall of the drum, the flux leakage is around four times higher than compared to the homogeneous case and around 11 times higher compared to the most conservative scenario I.

B. Energy deposition

As it was mentioned, not all the heat is likely to be detected by the calorimeter's heat detectors (Peltier elements). In order to fully simulate the actual situation the heat measured by the device, the energy/heat deposition in all the layers was determined.

TABLE II
ENERGY DEPOSITION IN THE CALORIMETER'S LAYERS
HOMOGENOUS 1 MEV GAMMA SOURCE

source energy [MeV/particle]:	1	
Layer:	energy deposition:	rel. error:
thermal block	6.8%	0.1%
insulation layer 1	0.0%	0.2%
homogenization layer 1	4.0%	0.2%
insulation layer 2	0.0%	0.2%
homogenization layer 2	3.2%	0.2%
insulation layer 3	0.0%	0.2%
homogenization layer 3	2.5%	0.2%
insulation layer 4	0.0%	0.3%
homogenization layer 4	2.0%	0.2%
insulation layer 5	0.0%	0.3%
cold plate	1.7%	0.3%
insulation layer 6	0.0%	0.3%
drum	**(+) 64.5%**	**0.1%**
measurement chamber	**(+) 0.0%**	**0.2%**
ghost chamber walls	5.5%	0.3%
ghost cell and phantom	**(-) 0.5%**	**1.1%**
ghost cell reference part	**(+) 0.0%**	**0.8%**
measurement plates	**(+) 1.9%**	**0.2%**
octagon structure	0.0%	0.2%
reference plates	**(-) 0.4%**	**0.4%**
total:	**93.2%**	**0.1%**
detected:	**65.5%**	**0.1%**

As expected, for the uniformly distributed radioactive material, the radiation leakage increases and not all the heat is detected. Additional simulations showed that even in the case of low energy gamma radiation (i.e. 100keV), and despite almost 100% of the energy being deposited inside the system, more radiation penetrates the layers and only around 83% of the energy would be detected. For higher energies half of the energy may escape from being detected.

V. SUMMARY

The results showed that some part of gamma radiation of energies >100keV is not deposited inside the LVC calorimeter, this especially, for the scenarios of the radiation source being placed nearby the edge of the drum (i.e. homogenous and least conservative sources). Despite a good part of the energy is deposited inside calorimeter measurement chamber, some gamma radiation penetrates the outer layers and escaped the deposited energy detection. Some other part of radiation may also be deposited inside the reference parts of the calorimeter and cause a double-bias.. In order to summarise and create an indicator for the future measurements, the simulated total errors of the measurement for each of the scenarios at different energies are presented in Figure 5. Bolded energies will contribute to the calorimeter's response. These marked with a plus (+) increase the final result, those with a minus (-) reduce it. Remining deposition do not significantly affect the final response.

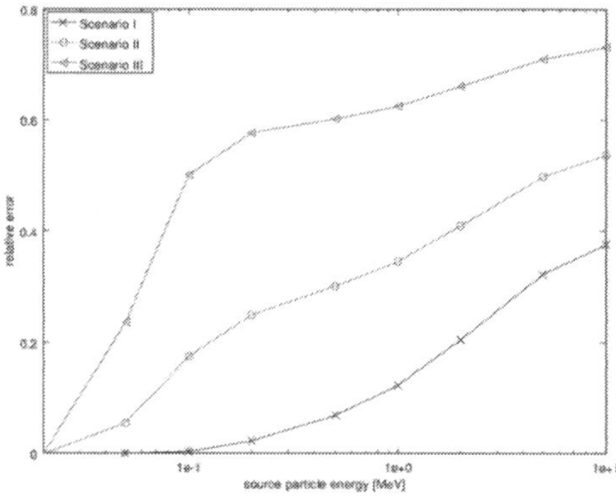

Figure 5. Maximal share of missed heat, gamma radiation.

One can see that the total error of the measurement, connected with the penetrative character of the gamma radiation, is more or less proportional to logarithm of the source energy. As for the energy deposition and detection probability, the results suggest that even for a 50 keV gamma source placed close to the edge of the drum around 24% of the emitted energy cannot be detected. For the least conservative scenario at the highest simulated energy, even around 73% escape detection.

VI. CONCLUSION

The result showed that the radiation leakage has a significant influence on the heat flux measurement, namely for the high energy gamma and neutron particles. Even if the energy of the radiation is not high enough to leave the system, the penetrative character of the radiation impacts the accuracy of the measurement by heat deposition inside the reference parts and outside the detector sphere. The accuracy is being compromised up to ~50% because of the unwanted and undesirable outside deposition of both, neutrons and photons. In general, it is fair to say that the radiation leakage is a complicated problem, and for some isotopes and configurations it can either impair the precision of the heat measurement or have no influence at all. Therefore, and especially for radioactive waste characterization, it is an interesting issue worthwhile further analyses, measurements and simulations. It also provides the chance to improve or at least understand the processes which cause the uncertainties of calorimetric assay, which is one of the main goals of the CHANCE project.

The model is implemented and works properly. It can be used as a tool for the uncertainties assessment support or can to indicate the outcome of any measurements anticipated. Therefore, the model will certainly be used again and may be modified as to adjust it to the real experiments, by changing the drum composition, the embedding material matrix, the energy spectrum and source location. In addition, the model can also be used to improve the existing LVC construction by adding some extra radiation shielding structures or filling the measurement chamber with a proper material to decrease the influence of the radiation leakage.

REFERENCES

[1] Status and Trends in Spent Fuel and Radioactive Waste Management, IAEA Nuclear Energy Series, No. NW-T-1.14, 2018 https://www-pub.iaea.org/MTCD/Publications/PDF/P1799_web.pdf

[2] The CHANCE project "Characterization of conditioned nuclear waste for its safe disposal in Europe". DEM 2018 - Dismantling Challenges: Industrial Reality, Prospects and Feedback Experience, D. Ricard, D., Plumeri, S., Boucher, L., Rizzo, A., Tietze-Jaensch, H., Mathonat, C., Bruggeman, C., Velthuis,J., Thompson, L., Genoud, G., Bucur, C., Kikola, D., Zakrzewska-Koltuniewicz, G France, Avignon – 2018 and CHANCE project official website, http://chance-h2020.eu, accessed on 05.08.2018

[3] Principles and applications of calorimetric assay, D.S. Bracken, C.R. Rudy, Los Alamos Report LA-UR-07-5226, 2007, https://www.lanl.gov/orgs/n/n1/panda/10.%20Calorimetry.pdf

[4] Deliverable report D3.1 Applicability of Calorimetry to Real Waste Characterisation, CHANCE project deliverable, C. Carasco, B. Perot et. al., 2018, publ. on CHANCE 2020: http://chance-h2020.eu

[5] Application Guide to Safeguards Calorimetry, D.S. Bracken, R.S. Biddle et. al., LA-13867-M, January 2002

[6] Performance values for non-destructive assay (NDA) techniques applied to safeguards: The 2002 evaluation by Esarda, P. Schwalbach, M.T.Swinhoe, ResearchGate, 2676830081, 2002

[7] NDA quantification of Nuclear Waste Containers, G. Jossens, P. Girones, C. Mathonat, A. Godot, ResearchGate, 278025167, July 2015

[8] https://www.nucleonica.com, accessed on 30.08.2018

[9] Initial MCNP6 Release Overview - MCNP6 version 1.0, J. T. Goorley, M. R James, et. al., LA-UR-13-22934, April 2013

[10] Development of a New Monte Carlo Reactor Physics Code, Jaakko Leppänen, VTT PUBLICATIONS 640, June 2007

[11] https://www.researchgate.net/figure/Flow-diagram-to-show-the-simplified-transport-process-for-a-single-neutron-history-in_fig7_277027129, accessed on 30.08.2018

[12] An MCNP primer, J. K. Shultis, R. E. Faw, Kansas State University, 2011, https://www.mne.k-state.edu/~jks/MCNPprmr.pdf

[13] Expressing Precision and Bias in Calorimetry, D.K. Hauck, S. Croft, D.S. Bracken, NM 87545, 2010

[14] https://www-nds.iaea.org/exfor/endf.htm accessed on 05.09.2018

EPJ Web of Conferences **225**, 06004 (2020)
ANIMMA 2019 https://doi.org/10.1051/epjconf/202022506004

Design by numerical simulation of an *in situ* alpha spectrometer operating at ambient air pressure

D. Degrelle*, J. Venara, M. Ben Mosbah, M. Cuozzo, C. Mahé, R. Serrano
* CEA, DEN, SEAD/LSTD
deborah.degrelle@safetechnologies.fr

***Abstract* — Alpha emitters are usually identified and quantified by alpha spectrometry measurements in a vacuum chamber performed in laboratory environments. This study shows that transuranic elements can be distinguished under ambient conditions using a grid collimator. The aim of this work was to use numerical simulations with the MCNP6 code to design a grid with a resolution high enough to differentiate the same radionuclide combinations as alpha spectrometry in a vacuum chamber, namely $^{239}Pu + ^{240}Pu$, $^{241}Am + ^{238}Pu$ and ^{244}Cm. Results show that a compromise is required to obtain the best performances in terms of energy resolution and detection efficiency, leading to the choice of two hexagonal grid collimators. The first has a collimation height of 0.5 cm and an apothem of 1 mm. Laboratory tests on electrodeposited sources show that the target radionuclides can be identified without prior deconvolution, with an energy resolution of about 70 keV and a detection efficiency of 0.74% at incident energies of 5–6 MeV. The second grid has the same collimation height but a coarser mesh with an apothem of 2 mm. In this case, the alpha peaks are still distinguishable, but with a lower resolution of 125 keV. The detection efficiency is three times higher however.**

***Index Terms* — Ambient conditions, *In situ* alpha spectrometry, Monte-Carlo simulations, Nuclear waste, Radiological characterization.**

I. Introduction

THE reconversion of the French Alternative Energies and Atomic Energy Commission (CEA) site in Fontenay-aux-Roses began in 1995. Two nuclear installations are still awaiting decommissioning. The first has been used for research and development on nuclear fuel reprocessing and the production of transuranic elements. The second is dedicated to nuclear waste management, notably legacy waste from the first installation and recent operational and decontamination and dismantling waste.

The challenges are to restore and process this legacy waste before evacuation to appropriate storage sites, which requires the establishment of a precise radiological inventory.

Nuclear waste containing beta and gamma emitters are readily characterized by gamma spectrometry. The waste from Fontenay-aux-Roses, in contrast, has a high proportion of alpha emitters ($^{238,239,240}Pu$, ^{237}Np, $^{241,243}Am$, ^{244}Cm). These transuranic elements need to be precisely quantified because of criticality safety concerns. In general, fissile material is quantified using the PNA (passive neutron assay) technique. However, the high concentration of ^{244}Cm, a strong emitter of neutrons by spontaneous fission, skews PNA results and makes them difficult to use.

The technique considered here to overcome these difficulties is *in situ* alpha spectrometry at ambient pressure. This is a non-destructive method with measurements performed on smear samples of nuclear waste obtained beforehand. The main benefit of this approach compared with alpha spectrometry performed in laboratory (under vacuum) is that it does not require lengthy and complex sample preparation, which increases the throughput for the analysis and thus the processing of nuclear waste.

The spectrometric selectivity of this technique should be sufficient to identify the following three groups of radionuclides:

- $^{239}Pu + ^{240}Pu$
- $^{241}Am + ^{238}Pu$
- ^{244}Cm

For ^{244}Cm, there is no interference from other radionuclides. The ability to discriminate is valuable notably to complete PNA measurements.

Since the measurements are performed on smear samples, alpha spectrometry under ambient conditions is only sensitive to the labile part of the contamination. The purpose of this approach is not to quantify the alpha emitters present in nuclear waste but to determine the relative proportions of alpha emitting radionuclides or groups of radionuclides.

II. Principle

The nature of alpha particles makes them difficult to detect by spectrometry. The successive ionizations they induce as they interact with their environment lead to such large energy losses that 5 MeV incident particles travel no further than 4 cm in air. The difficulty of measuring alpha particles in air lies in their isotropic emission by the source or sample. Depending on their incidence angles indeed, the emitted alpha particle will each travel a different distance and thus loose more or less energy, degrading the resolution of the spectrum. This is why alpha spectrometry is usually carried out in a vacuum to avoid interactions with the surrounding air.

Pöllänen *et al.* [1][2] have shown that certain alpha emitters can be distinguished at ambient pressure by placing a wire grid

Submitted July 12, 2019.
D. Degrelle, J. Venara, M. Cuozzo and C. Mahé are with the CEA, DEN, DE2D, SEAD, Dismantling Technics and Simulation Laboratory F-30207 Bagnols-sur-Cèze, France (e-mail: julien.venara@cea.fr).

M. Ben Mosbah is with the CEA, DEN, DTN, SMTA, Nuclear Measurements Laboratory, F-13108 Saint-Paul-lez-Durance, France
R. Serrano is with the CEA, DEN, UADF, F-91191 Gif-sur-Yvette, France.

collimator between the sample or source and the window of the detector to select alpha particles with a small angle of incidence. Once the particle beam is collimated, the alpha particles with too large an angle of incidence, i.e. those with the longest path lengths in air, cannot reach the detector (Fig. 1). The particles that are detected will all have traveled a similar distance, improving the energy resolution of the spectra.

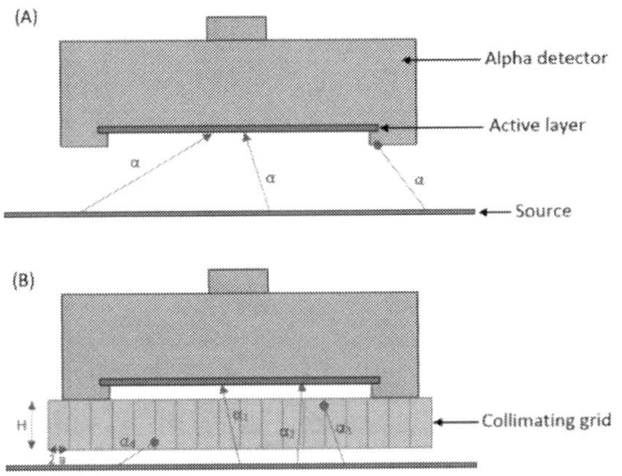

Fig. 1. Schematic representation of alpha particle selection using a collimator grid [2]. In the absence of the grid (A), all particles are detected, regardless of their angle of incidence. With the grid, only particles with a small angle of incidence are detected (α_1 et α_2). *H* is the collimation height and 2a is the cell width.

It has been shown [1] that ^{239}Pu, ^{241}Am and ^{244}Cm can be distinguished. This alpha ray detection technique was developed in the 1980s in the context of assessing atmospheric contamination [3]. It is still used to identify the proportion of alpha radiation from radon in certain air-monitoring beacons with radial fin grids [4]. The resolution can reach 120–130 keV for the alpha peak of ^{241}Po at 7.686 MeV [2].

The aim of the present study was to optimize the collimator grid design using numerical simulations with the Monte-Carlo MCP6 (Monte-Carlo N-Particle) code [5] so that the resolution is high enough to discriminate between certain alpha emitters. Laboratory tests were performed to validate the design of the chosen grids.

III. Materials and Simulation

A. *Materials*

Laboratory alpha spectrometers in vacuum chambers commonly employ PIPS (Passivated Implanted Silicon) detectors [6]. These silicon semiconductor detectors are generally chosen for their low cost and the energy resolution they offer, on the order of 30–80 keV for routine measurements [7]. The thinner the active layer is, the better the resolution is. This high performance is due to a thin dead layer (50 nm of Si).

The detectors used in this study were CAM (Continuous Air Monitoring) PIPS detectors [8]. They are specifically designed for continuous use in air, notably in air-monitoring beacons. Because of the conditions in which they are used and because of the light-sensitivity of silicon, CAM PIPS detectors have more protective layers than standard ones [8]. These extra layers increase the thickness of the dead layer on the detection window by 1.5–2 µm Si-equivalent [6][8].

CAM PIPS detectors are available in a wide range of sizes. Those used in air-monitoring beacons are usually small. The detector used in this work was the CAM2000 produced by MIRION Technologies. It has an active surface of 2000 mm^2 and a thickness of 300 µm Si. It active area is one of the largest available. It was chosen to optimize the studied surface. The recommended operating voltages for this type of detector are +24 V or +70 V, giving theoretical resolutions of 34.4 keV and 28 keV, respectively (performance tests performed under vacuum [8]). At +70 V, the layer is fully depleted.

The detection chain was provided by MIRION Technologies. The detector is associated with a 2018EB preamplifier and a LYNX multichannel analyzer. This combination offers a maximum theoretical count rate of 2 x 10^6 MeV/s according to the manufacturer. The alpha spectra are analyzed using the software Genie 2000 with the alpha analysis module [9].

Electrodeposited reference sources were used for the laboratory measurements. They consisted of an active layer 15 mm in diameter electrodeposited on a 0.5 mm thick stainless-steel disk. The energy broadening due to deposition process is less than 20 keV for these sources [10]. Self-absorption in the sources is assumed negligible. This assumption was verified by simulating ^{241}Am alpha rays broadened by 20 keV. The resulting peak's resolution differed by 1.4% with respect to a simulation performed with discrete lines. The radionuclides available for this study were ^{239}Pu, ^{241}Am and ^{244}Cm, with respective activities of 2979 Bq, 3295 Bq and 2717 Bq on 4 March 2019. The relative uncertainty on the source activities is 1%. Table I lists the energies of the main alpha particles emitted by these radionuclides [11].

TABLE I
Main Alpha Particle Energies of Electrodeposited Reference Sources

^{239}Pu		^{241}Am		^{244}Cm	
Energy (keV)	Intensity (%)	Energy (keV)	Intensity (%)	Energy (keV)	Intensity (%)
5 156.59 5 143.82 5 105.81	70.75 17.14 11.87	5 486.46 5 442.86 5 388.25	84.45 13.23 1.66	5 804.77 5 762.65	76.7 23.3

B. *Simulation*

The detector/grid/source setup was modeled using the MCNP6 calculation code [5]. The detector dimensions were set to the values specified by the manufacturer (section III.A).

The grid was modeled using a lattice structure (repeated hexagonal elements with LAT=1 in a mesh universe created with the U card). This approach reduces modeling and calculation times.

The most appropriate tally to model the response of the CAM2000 detector is the F8 tally, which gives the distribution of energies deposited in the detector characteristic of an alpha spectrum. Each calculation was launched with 5 x 10^7 histories and the statistical error was no more than 5%.

Simulations were also performed to energy calibrate the

experimental spectra. Care was taken to ensure they were representative. Comparisons between simulations and experimental measurements were used in this study to determine the systematic error associated with the simulated spectra, as discussed in the following.

IV. Design by Monte-Carlo simulation

Table II lists the different parameters used in this section to describe the geometry of the collimator grid. A "cell" is defined as one complete element in the lattice.

TABLE II
LIST OF PARAMETERS CHARACTERIZING A MESH CELL

Symbol	Definition
e	Wall thickness of a cell
H	Collimation height or height of a cell
r	Radius of a circular cell
c	Side length of a square cell
d	Diagonal length of a square cell
a	Apothem of a hexagonal cell; radius of the inscribed circle
r_c	Radius of the circumcircle of a hexagonal cell; side length of a regular hexagon
P	Perimeter of a cell
A	Area of a cell

The geometric shapes correspond to the face of the cell perpendicular to the alpha particle beam and parallel to the detector. The hexagonal mesh is composed of regular hexagons.

A. Choice of mesh geometry

The first stage involved choosing the mesh geometry best adapted for alpha particle detection. It should not restrict the energy resolution of the detector or its energy efficiency.

Hexagonal, square and circular meshes with identical surface areas were compared. Contrary to the hexagonal and square meshes, a circular mesh does not give complete surface coverage (Fig. 2). This is because the walls of round cells require extra material to fill the intercellular space. With a circular mesh, the alpha particles are thus more likely to encounter the cell walls, reducing the detection efficiency. The circular mesh was therefore not chosen for the rest of the study.

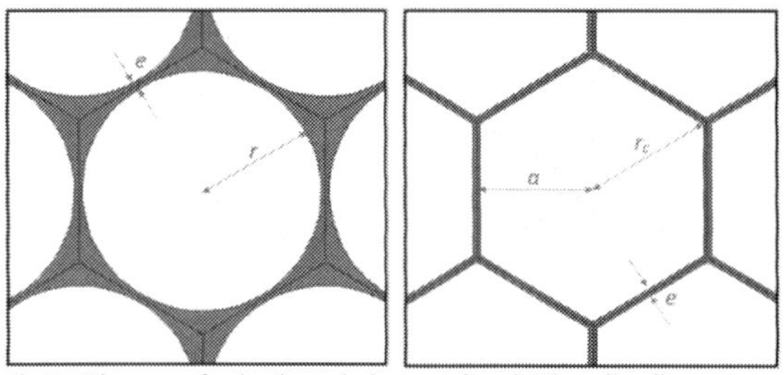

Fig. 2. Diagrams of a circular and a hexagonal mesh. The cell walls are shown in grey. The walls of the circular mesh are thicker because the intercellular space is larger.

The choice between a hexagonal and a square mesh can be decided mathematically. The perimeter and area of these two shapes are respectively:

$$P(\text{hexagon}) = 6 \cdot r_c \quad (1)$$

$$P(\text{square}) = 4 \cdot c \quad (2)$$

$$A(\text{hexagon}) = \frac{3\sqrt{3}}{2} \cdot r_c^2 \quad (3)$$

$$A(\text{square}) = c^2 \quad (4)$$

Setting the areas of the two shapes equal so that each one accommodates the same number of particles gives, using (3) and (4) :

$$c^2 = \frac{3\sqrt{3}}{2} \cdot r_c^2 \quad (5)$$

Inserting this value into (2) gives:

$$P(\text{square}) = 4 \cdot \sqrt{\frac{3\sqrt{3}}{2}} \cdot r_c \quad (6)$$

Which in comparison with (1) yields,

$$P(\text{hexagon}) < P(\text{square}) \quad (7)$$

A hexagon has a shorter perimeter than a square with the same area. This area/perimeter ratio is preferable to optimize the detection efficiency. A hexagonal mesh was therefore chosen for the collimator grid.

This choice is also justified by manufacturing constraints. For equivalent areas indeed, the side length of a square (c) is shorter than that of a hexagonal cell ($2 \cdot r_c$). For the fabrication of grids with very narrow cells, machining size limits will be reached sooner with square cells. A hexagonal mesh appears to be the best alternative. Pöllänen *et al.* also chose this geometry [1].

B. Cell dimensions

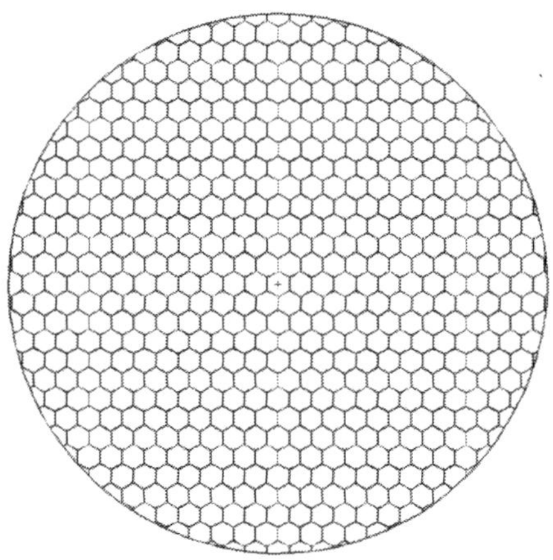

Fig. 3. Cross-sectional representation of the grid modeled with MCNP6, as visualized in VisEd. The diameter of the grid is 5 cm. The walls are 64 µm thick.

The following characteristics of the hexagonal cells were optimized:

- The collimation height, H,
- The apothem of the hexagonal cells, a,

The simulated setup is shown in Fig. 1 (B). A number of parameters were kept fixed:

- The grid material, stainless steel.

- The thickness of the cell walls, 64 µm. This is the value used by Pöllänen *et al.* in their study [1]. This thickness ensures that alpha particles of up to 9 MeV are completely blocked (about 20 µm are required to stop 9 MeV alpha particles).
- A surface source was modeled. Its active diameter was 4.7 cm to ensure that a large proportion of the detector's active surface was illuminated. This is also the standard diameter of the filters used for swabbing.
- The grid diameter was 5 cm.
- There was a 1 mm gap between the window of the detector and the collimator grid, as required by the CAMP PIPS setup.
- The source/grid distance was set to 4 mm.
- The thickness of the dead layer was set to the most unfavorable value (2 µm).

Fig. 3 shows a cross-sectional top view of the collimator prepared using VisEd (Visual Editor for MCNP) [12].

The influence of H and of a on energy loss and the energy resolution and detection efficiency of the device was studied by simulation. The energy of the incident alpha particles was varied from 3 to 9 MeV (the emission energy range of alpha particles) in 2 MeV steps. This article only reports the data obtained with 7 MeV incident particles.

Fig. 4 shows how the relative resolution varies as a function of the apothem of hexagonal cells for different collimation heights. The relative resolution is defined as

$$\mathrm{FWHM}(\%) = \frac{\Delta c}{c} \cdot 100 \quad (8)$$

Where Δc is the full-width at half maximum of the alpha peak in number of channels and c is the channel of the center of the peak.

The highest resolutions, less than 1.2%, are achieved with configurations in which the apothem is small (a = 1 mm). The energy resolution can also be improved by increasing the collimation height. The larger H is indeed, the more the incidence angle of the particles tends toward 0° and the closer the distances traveled by each of the particles through air are.

Fig. 4. Evolution of the relative resolution as a function of the apothem of the cells and the collimation height, *H*, for incident particles with an energy of 7 MeV. The curves are 2nd degree polynomial fits.

Fig. 5 shows the evolution of the absolute detection efficiency as a function of the apothem of the cells for different collimation heights. The energy of the incident particles was once again 7 MeV. The absolute efficiency is given by:

$$\varepsilon_{\mathrm{abs}} = \frac{N_{\mathrm{net}}}{A \cdot I \cdot t} \quad (9)$$

With N_{net}, the number of counts in the interval of interest around the alpha peak, A, the activity of the source in Bq, I the emission intensity of the alpha particles (here, 100%), and t the acquisition time in s.

The larger the cells are (a = 4 mm), the higher the detection efficiency is. Small collimation heights also improve the efficiency.

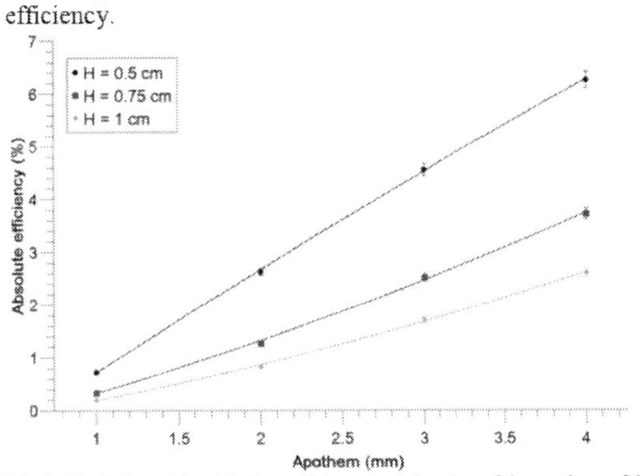

Fig. 5. Evolution of the detection efficiency as a function of the apothem of the cells and the collimation height, *H*, for incident particles with an energy of 7 MeV. The curves are 2nd degree polynomial fits.

These results show that a compromise is required. Given the activity of the nuclear waste in Fontenay-aux-Roses (several thousand Bq per radionuclide), detection efficiency does not seem to be the most important criterion. A small collimation height (H = 0.5 cm) can nevertheless be chosen to limit the loss of efficiency and limit the energy loss of the alpha particles, particularly for those with low energies (Table III).

TABLE III
ENERGY LOSS OF INCIDENT ALPHA PARTICLES AS A FUNCTION OF THE GRID'S COLLIMATION HEIGHT

Incident energry	3 MeV	5 MeV	7 MeV	9 MeV
H = 0.5 cm	1.806	1.242	0.963	0.795
H = 1 cm	2.650	1.757	1.333	1.094

Energy losses are given in MeV. The uncertainty on the energy loss values is ± 0.005 MeV. The amount of energy lost does not depend on the size of the cells.

With a collimation height of 0.5 cm, the energy resolution of the detector is optimal with an apothem of 1 mm (Fig. 4 and Fig. 6). Fig. 6 shows simulated alpha spectra for incident particles of 7 MeV. They highlight the effect of the detection geometry on the shape of the alpha peak. With a large apothem (a = 4 mm), the characteristic alpha peak shape is observed, namely a double exponential convoluted with a Gaussian [13]. Decreasing the apothem reduces the left tail of the peak, until an almost pure Gaussian is obtained (the gray spectrum for a = 1 mm). Obtaining this type of spectrum is promising for the discrimination of transuranic elements that emit alpha particles at similar energies. Furthermore, the removal of the left tail should allow the deconvolution of the alpha spectra to be optimized. This was confirmed by simulating the alpha spectrum of a legacy waste package from Fontenay-aux-Roses (Fig. 7), assuming a surface source.

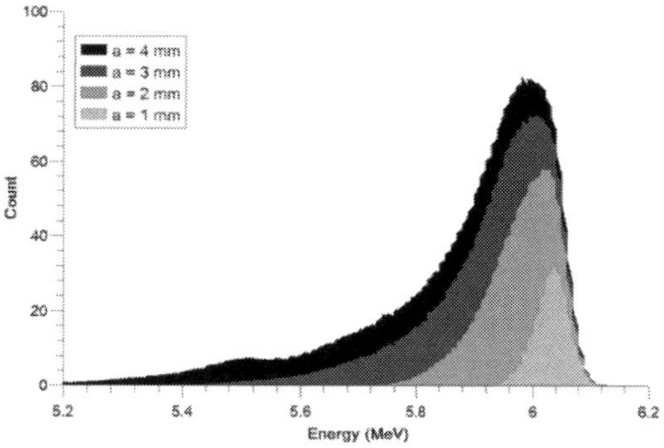

Fig. 6. Ambient pressure alpha spectra simulated using MCNP6 for different apothems and a collimation height of 0.5 cm. The energy of the incident alpha particles was 7 MeV.

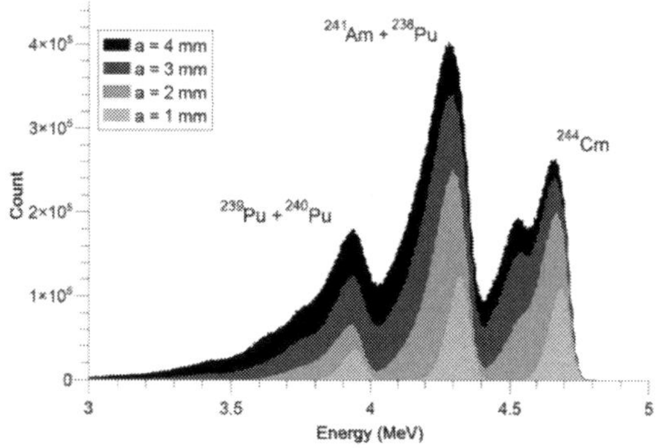

Fig. 7. Simulated ambient pressure alpha spectra of a waste package from Fontenay-aux-Roses. The grid had a collimation height of 0.5 cm. Simulations were performed with different apothem values.

The major alpha emitters in this characteristic spectrum are ^{238,239,240}Pu, ^{241}Am and ^{244}Cm. Their daughter nuclei make negligible contributions (less than 10^{-9}%) and were not included in the simulation. Fig. 7 shows that certain grid geometries allow the following groups of radionuclides to be distinguished from each other:

- ^{239}Pu + ^{240}Pu
- ^{241}Am + ^{238}Pu
- ^{244}Cm

When the apothem is very small (a = 1 mm), the three alpha peaks can be analyzed without deconvolution. To allow for potential operational imperfections (self-absorption in the source, additional material between the sample and the detector, etc.), it is preferable to choose the setup with the highest energy resolution.

Two grids were therefore chosen: one to facilitate the discrimination between alpha peaks, with H = 0.5 cm and a = 1 mm, and the other, with a higher detection efficiency, with H = 0.5 cm and a = 2 mm. The detection efficiency is three times higher with the second grid and the resolution remains good enough to distinguish the three peaks, with deconvolution in some cases. The theoretical energy resolutions of the two grids are presented in Table IV. Note that since the source/grid distance is 4 mm, the source is 10 mm away from the detector. The thickness of the dead layer was set to 2 µm (the most unfavorable value).

TABLE IV
THEORETICAL ENERGY RESOLUTIONS WITH THE TWO GRIDS

Grids	Radionuclide groups	FWHM (%)	FWHM (keV)
H = 0.5 cm, a = 1 mm	^{239}Pu + ^{240}Pu	2.41 ± 0.25	95.0 ± 9.9
	^{238}Pu + ^{241}Am	2.19 ± 0.23	94.8 ± 10.0
	^{244}Cm	1.92 ± 0.21	90.1 ± 9.9
H = 0.5 cm, a = 2 mm	^{239}Pu + ^{240}Pu	3.56 ± 0.25	140.0 ± 9.8
	^{238}Pu + ^{241}Am	3.49 ± 0.23	150.0 ± 9.9
	^{244}Cm	2.78 ± 0.21	129.6 ± 9.8

The resolution values in keV are calculated from the detected energies (after attenuation).

V. EXPERIMENTAL DEVICE

A. Making of the collimator grids

Fig. 8. Stainless steel collimator grids prepared by electrical discharge machining. In the grid on the left, the cells have an apothem of 2 mm and in the grid on the right, the apothem is 1 mm.

The radioactive operating environment of the collimator grid prevented the use of certain materials, notably aluminum, which releases gaseous H_2 when corroded [14]. The properties required for the grid material are that it should be simple to machine, sufficiently hard to resist possible impacts, easily decontaminated and compatible with the waste.

The collimator grids were therefore produced in stainless steel using electrical discharge machining. The uncertainty on the cell walls with this technique was ± 0.01 mm. The dimensions chosen for the grid meant that the thickness of the cell walls had to be increased to 200 µm to ensure the grid remained rigid and avoid mesh breakage. This increased thickness does not have a significant effect on the detection efficiency compared with the 64 µm considered initially. The two electrical-discharge machined grids are shown in Fig. 8

B. Acquisition parameters

A 2000 mm² CAM PIPS detector was used. The data were acquired with a bias voltage of 70 V to maximize the energy resolution of the device [6][8]. The parameters chosen for the acquisition of the spectrum were:

- The Pole Zero (P/Z) was set to automatic. Its value

was less than 400. Adjusting the P/Z is critical to avoid pile-up and alpha peak distortion.

- The Rise Time was set to 4 μs, the value that optimizes the energy resolution in our setup, irrespective of the applied voltage (70 or 24 V). The corresponding dead time is 0.04 %.
- The Flat Top was set to 0.2 μs.
- The alpha spectra were acquired over 1024 channels to improve the counting statistics.
- The source/grid distance was set to 2 mm (not 4 mm as previously) to improve the resolution, giving a source/detector distance of 8 mm.

Data were acquired with electrodeposited ^{239}Pu, ^{241}Am and ^{244}Cm sources, separately for each radionuclide. The final spectrum was assembled by summing the three spectra obtained. The presence of impurities in the source is apparent, notably ^{239}Pu impurities in the ^{244}Cm source. This leads to a systematic error of -1.965 % in the combined spectrum for number of counts in the ^{239}Pu alpha peak.

The acquisition time was 15 min per radionuclide. Under these conditions the detection efficiency variation was 0.662 % and the variation of the full-width at half maximum, 5.194 %. The latter can be reduced by acquiring the alpha spectra over 2048 channels.

C. *Energy calibration*

Energy calibration cannot be performed directly on experimental spectra because of the energy lost by the alpha particles between the source and the window of the detector.

The energies actually detected can be determined either by Monte-Carlo simulation, with the MCNP6 calculation code for instance, or by using the SRIM (Stopping and Range of Ions in Matter) [15] or the NIST ASTAR (Stopping power and range tables for helium ions) [16] database.

In both cases, with MCNP6 and with SRIM/ASTAR, precise knowledge of the CAMP PIPS detector geometry is required, in particular the thickness of the dead layer. It is listed as being between 1.5 and 2 μm Si-equivalent. In the setup considered here, the alpha particles also travel 8 mm before reaching the detector.

An initial rapid energy calibration was carried out using the SRIM database, and Table V shows the theoretical detected energies of the alpha particles emitted by ^{239}Pu, ^{241}Am and ^{244}Cm obtained in this way. This calibration was also used to characterize the energy resolution of the device (in keV).

TABLE V
THEORETICAL ENERGIES CALCULATED USING THE SRIM DATABASE

	Maximum incident energy	Detected energy
^{239}Pu	5.156	4.162 ± 0.040
^{241}Am	5.486	4.536 ± 0.038
^{244}Cm	5.804	4.893 ± 0.036

The energies are given in MeV and represent the energies detected by the CAM PIPS detector for alpha particles that have passed through 8 mm of air and (1.75 ± 0.25) μm Si-equivalent of dead layer. The maximum incident energies are the main emission lines of the radionuclides in Table I.

Subsequently, all the spectra were energy calibrated using MCNP6, the difference between the SRIM and MCNP6 calibrations being just 0.5 %.

D. *Experimental results and comparison with simulations*

Experimental and simulated spectra were compared to verify the reliability of the model and thus validate its use in future investigations, such as understanding the effects of the operational parameters.

The reliability of the model was first tested in a setup without the grid. Individual measurement was made for the three sources, ^{239}Pu, ^{241}Am and ^{244}Cm, so that the integrated area of the alpha peaks could be calculated rapidly. The final spectrum was obtained by co-adding the three individual spectra. In this case, the sources were placed 1.34 cm from the active surface of the CAMP PIPS detector. The acquisition time was 15 min. Fig. 9 shows a comparison of the simulated and the experimental spectrum.

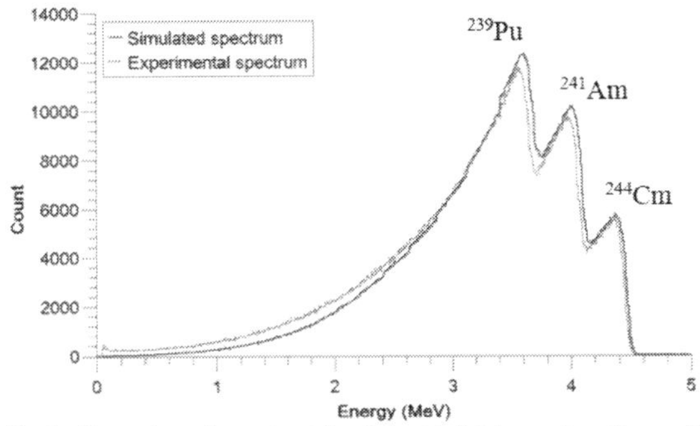

Fig. 9. Comparison of experimental and simulated alpha spectra with no grid in place between the source and the detector. The measurements were performed under ambient conditions.

The experimental energy resolutions of the ^{239}Pu, ^{241}Am and ^{244}Cm peaks are (722 ± 42) keV, (740 ± 43) keV and (695 ± 40) keV, respectively. The uncertainties are large here because the data were acquired over 1024 channels to increase the counting statistics. The signals from the three radionuclides cannot be quantified separately without using spectral deconvolution software [17].

The systematic error on the detection efficiency for the simulated spectrum is less than 2%. The systematic error on the energy resolution is less than 5% for all three radionuclides.

The analysis was repeated using the two grids presented in Fig. 8. The sources were placed 8 mm from the detector. Fig. 10 presents the spectra obtained and Table VI gives the energy resolution of the three alpha peaks as a function of the grid used.

TABLE VI
EXPERIMENTAL ENERGY RESOLUTION AS A FUNCTION OF THE GRID USED

	H = 0.5 cm, a = 1 mm	H = 0.5 cm, a = 2 mm
^{239}Pu	69.61 ± 15.90	124.18 ± 16.86
^{241}Am	69.73 ± 15.93	124.57 ± 16.90
^{244}Cm	77.92 ± 16.11	125.00 ± 16.95

The energy resolutions are expressed in keV. The corresponding uncertainties were calculated using (8) and the actually detected energies given in Table V.

The experimental results show that adding a grid allows the peaks from the three radionuclides ^{239}Pu, ^{241}Am and ^{244}Cm to be distinguished, particularly the finer grid (a = 1 mm) which improves the resolution by 70–77 keV for 5–6 MeV incident particles. In this case, no deconvolution of the spectrum is required. Furthermore, the absolute experimental efficiency is as predicted by simulation (Fig. 5) with a mean efficiency of about (0.740 ± 0.012) % for a = 1 mm and (3.030 ± 0.039) % for a = 2 mm.

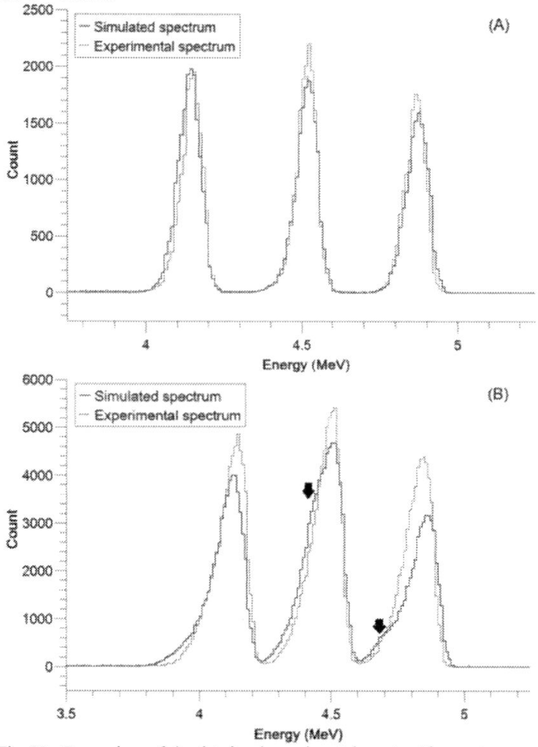

Fig. 10. Comparison of simulated and experimental spectra. The spectra were acquired with grids whose apothem was (A) 1 mm and (B) 2 mm. The black arrows indicate distortions in the simulated alpha spectra.

The energy resolutions in the alpha spectra simulated using MCNP6 are representative of those in the experimental spectra, except for the ^{241}Am peak with the coarser grid, for which the relative error is 33%.

The models are less reliable in reproducing the detection efficiency. Although the spectra in Fig. 10 (A) appear close on visual inspection, the systematic error for the simulated detection efficiency is 14.2% for the ^{239}Pu peak. This error is less 3.3% for the other peaks. There are clear differences in Fig. 10 (B) between the simulated and experimental data, with a systematic error of 20 % for the ^{244}Cm peak. However, distortions in the simulated spectrum are visible for the ^{241}Am and ^{244}Cm peaks. They are similar to those observed in Fig. 7, particularly for ^{244}Cm, which are more prominent for the coarser grid. They are not present on all the simulated peaks (^{239}Pu does not seem to be affected in Fig. 10 (B)) and no distortions are observed in the experimental spectrum. These distortions are most probably the cause of the 33% relative error on the resolution of the simulated ^{241}Am peak with the coarser grid.

The origin of these distortions is presumably not the model of the CAM PIPS detector because the experiment/simulation comparison without the grid showed that it was reliable. Moreover, adjusting the thickness of the dead layer to 1.5 and 2 µm does not reduce the systematic error in the presence of the grid. The origin of the error must therefore be the grid model.

The grid was modeled using a lattice structure. Modeling an isolated cell (i.e. without a lattice) showed that the resulting alpha spectrum was not distorted. A simulation of the device was performed with each cell of the larger grid (only those illuminated by the source) modeled individually. Fig. 11, Fig. 12 and Table VII show the results obtained in comparison with the experimental spectrum.

TABLE VII
SYSTEMATIC ERROR OF THE SIMULATED DETECTION EFFICIENCY WITH RESPECT TO EXPERIMENT

	Simulated efficiency	Experimental efficiency	Systematic error
^{239}Pu	3.160 ± 0.036 %	2.990 ± 0.039 %	5.69 %
^{241}Am	3.145 ± 0.035 %	3.033 ± 0.039 %	3.70 %
^{244}Cm	3.187 ± 0.036 %	3.074 ± 0.040 %	3.69 %

Results shown for the coarser grid only. The cells in the grid were modeled individually.

These results confirm that the use of repeated structures in MCNP6 can create problems in the simulations, such as distortions of the spectrum, which potentially affect the energy resolution and the detection efficiency. These problems can nonetheless be overcome by modeling each cell in the grid individually. In this case, the simulation reproduces:

- The detection efficiency with a systematic error of less than 6 % over the energy range of interest.
- The energy resolution, with differences between simulation and experiment within the uncertainty range.

MCNP6 is sufficiently reliable to be used for the rest of the project, in particular to study the potential negative effects of operational parameters on the energy resolution of the *in situ* alpha spectrometer, in particular source self-absorption.

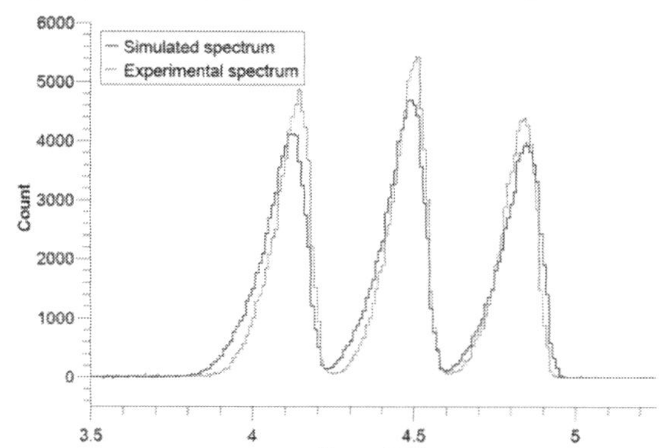

Fig. 11. Comparison of simulated and experimental spectra. The apothem of the grid was 2 mm. Each cell in the grid was modeled individually. There are no longer any distortions in the simulated spectrum.

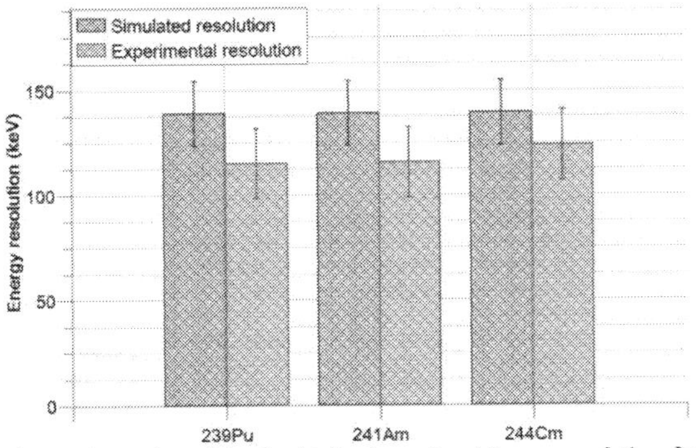

Fig. 12. Comparison of the simulated and experimental energy resolutions for the alpha peaks of ^{239}Pu, ^{241}Am and ^{244}Cm. Each cell in the grid was modeled individually. The apothem of the grid was 2 mm.

VI. Conclusion

These investigations have shown that alpha spectrometry under ambient conditions is experimentally feasible. The collimation method employed here allows the isotropic emissions to be controlled by only selecting the particles that reach the detector at close to perpendicular incidence. This reduces differences between the path lengths of each alpha particle and improves the energy resolution.

The collimator grids with hexagonal meshes were designed using the MCNP6 calculation code, revealing an inevitable compromise in their dimensions to optimize the energy resolution and the detection efficiency. A small collimation height, *H*, was thus chosen to preserve efficiency and limit the source/detector distance over which the alpha particles lose energy before reaching the active zone. The simulations identified two values for the apothem, $a = 1$ mm and $a = 2$ mm, to respectively favor the energy resolution and the detection efficiency.

Laboratory tests of the grids using electrodeposited sources validated their theoretical performances. The alpha peaks of ^{239}Pu, ^{241}Am and ^{244}Cm can thus be distinguished without deconvolution and with and an energy resolution of about 70 keV for the alpha peaks of ^{239}Pu and ^{241}Am, and 77 keV for the peak from ^{244}Cm with the finest collimator grid. The energy resolution of the coarser grid is poorer, no better than 125 keV for any of the alpha peaks, but the detection efficiency is three times higher.

Alpha spectrometry under ambient conditions with a collimator grid can distinguish the same groups of radionuclides as vacuum alpha spectrometry in a laboratory. The objective of this project is to use this method for the *in situ* analysis of nuclear waste. The technique requires no sample preparation, which will help increase throughput for the analyses and thus for the evacuation of the nuclear waste. Nonetheless, since the measurements are performed on smear samples, self-absorption of alpha particles during the swab process but also in the swab will have to be taken into account. Operational conditions will inevitably degrade the resolution obtained in this study (70 keV in optimal conditions for electrodeposited sources); to allow for these degradations, it is preferable to choose the grid that provides the best resolution.

References

[1] R. Pöllänen et al., "High-resolution alpha spectrometry at ambient air pressure – Towards new applications," *Nucl. Instr. Meth. A*, vol. 694, pp. 173—178, 2012.

[2] R. Pöllänen et al., "In-situ alpha spectrometry from air filters at ambient air pressure," *Radiations Measurements*, vol. 53—54, pp. 65—70, 2013.

[3] J.-C. Carossi et al., "Dispositif de contrôle de contamination atmosphérique due à des aérosols émetteurs de particules alpha," France. Patent 2 563 633, April, 27, 1984.

[4] MIRION TECHNOLOGIES: characteristics sheet of ABPM 203M, Online at < https://mirion.app.box.com/s/eihrxkex0v45s2933ztf > (accessed on May 2019)

[5] D. B. Pelowitz (Ed.), Los Alamos National Laboratory Report LA-CP-11-01708, Los Alamos, 2011.

[6] MIRION TECHNOLOGIES: characteristics sheet PIPS Detectors, Online at < https://mirion.s3.amazonaws.com/cms4_mirion/files/pdf/spec-sheets/c39313_passivated_pips_super_spec_1.pdf?1557861393 > (accessed on May 2019).

[7] *Handbook of Radioactivity Analysis*, 3rd ed., Elsevier Inc., Oxford, UK, 2012, pp. 380—390.

[8] MIRION TECHNOLOGIES: characteristics sheet CAM PIPS detectors, Online at < http://www.canberra.com/literature/detectors/application_notes/CAM-PIPS-Detector-Properties-App-Note-C39167.pdf > (accessed on May 2019).

[9] Genie 2000 3.3, Canberra Industries Inc., March 2013.

[10] Eckert & Ziegler: Alpha Spectroscopy sources, Online at < https://www.ezag.com/home/products/isotope_products/isotrak_calibration_sources/reference_sources/alpha_spectroscopy_sources/ > (accessed on June 2019).

[11] LNHB, Nucleide-Lara: Library for gamma and alpha emissions, Online at < www.nucleide.org/Laraweb/index.php > (accessed on April 2019).

[12] Vised version 2.7.E, generated using MCNPX, April 2011.

[13] G. Bortels et al., "Analytical function for fitting peaks in alpha-particle spectra from Si detectors", *Appl. Radiat. Isot.*, vol. 38, pp. 831 – 837, 1987.

[14] K. Fujiwara et al., "Corrosion behavior of aluminium under simulated environmental conditions of low-level waste," *Corrosion Engineering, Science and Technology*, vol. 52, pp. 162—165, 2017.

[15] J. F. Ziegler et al., "SRIM – The stopping and range of ions in matter (2010)", *Nucl. Instr. Meth. B*, vol. 268, pp. 1818—1823, 2010.

[16] M. J. Berger et al., ASTAR: Stopping-power and range tables for helium ions, NIST, Online at < https://physics.nist.gov/PhysRefData/Star/Text/ASTAR.html > (accessed on April 2019).

[17] MIRION TECHNOLOGIES: Interactive Peak Fit option in Genie 2000 software, Online at < https://www.mirion.com/products/genie-2000-interactive-peak-fit >

Radioactive Hot-spot Detection Using Unmanned Aerial Vehicle Surveillance

Yoeri Brouwer, Alberto Vale*, Duarte Macedo, Bruno Gonçalves, Horácio Fernandes
* Instituto Superior Técnico
avale@ipfn.tecnico.ulisboa.pt

Abstract—**This work proposes a solution to identify the number of sources of radiation, as well as their respective intensities and locations based on data acquired by Global Positioning System (GPS) receivers and affordable radiological sensors, such as Geiger-Müller counters (GMC). An optimization algorithm is required to minimize the estimation error in terms of location, intensity and number of sources of radiation given all the intensity measurements acquired in different locations, taking into account the sensors' models, background radiation intensity values and noise. Experimental results were achieved in a laboratory with controlled sources of radiation. The solution was also tested with real data gathered by a GMC connected to a mobile phone with a software application developed by the authors to synchronize the sensor readings with GPS data. The sensor and the mobile phone are attached to a quadcopter flying over the scenario with sources of radiation.**

Keywords—**UAV, Radioactive Source, Radioactivity, Hot-spot Detection, Collimator.**

I. Introduction

THE defence against Chemical, Biological, Radiological and Nuclear (CBRN) threats is an increasing demand due to wars, terrorist attacks, disasters, or simply due to negligence. Recent projects have achieved improvements, but we are still far from safeguarding humankind against CBRN threats. NATO already has a Task Force working towards this objective and the National Academy of Engineering presented in 2008 "The 14 Grand Challenges for Engineering", where the prevention of nuclear terrorism was identified as one of the great challenges for the 21st century [1]–[3].

Radiological and nuclear threats are associated with radioactive elements that release large amounts of energy through alpha particles, beta particles, X-rays, gamma rays, and neutrons. This paper is particularly focused on these threats, which arise due to the emerging applications of nuclear technologies in industry, power production, medicine, and the military. It is also important to address radiological threats given the catastrophic consequences of nuclear bombs of the Second World War and the nuclear incidents which occurred in Chernobyl in 1986 and in Fukushima in 2011. In future work, the proposed approaches will be extended to chemical and biological threats.

Radiological and nuclear threats are often solely associated with nuclear bombs and nuclear fission power plants. However, apart from terrorist threats, nuclear materials can be commonly found in nuclear medicine waste products (e.g., radiation therapy for cancer treatment), abandoned mine sites, or even decommissioned radioactive material storage facilities.

As mentioned above, there are a number of projects which aim to address the issues of radiological and nuclear threats. MANUELA [4] is a project of the AREVA company, currently ORANO, a French multinational group in nuclear and renewable energy, where the mobile robot is equipped with LIDAR systems and it performs a 3D reconstruction of internal scenarios with the indication of the radiation levels of hotspots. Project REWARD [5] extends the radiological survey to large outdoor environments using radiological sensors installed on cars equipped with GPS receivers. The used sensors are multiple miniaturized solid state radiological sensors: Cadmium, Zinc and Tellurium (CZT) for gamma radiation detection and a high efficiency neutron detector based on new semiconductor technologies. SEAGULL [6] is a project based on Unmanned Aerial Vehicles (UAV) for surveillance and monitoring in the marine environment using video and thermal cameras. RIANA [7], also from AREVA, is a project with wheeled mobile robots with the capacity to operate in nuclear scenarios to provide radiological readings and take physical samples from hazardous scenarios. GAMMAex [8] is a project which uses UAVs to perform monitoring and recognition of chemical and radiological patterns in scenarios where biological, chemical, and radiological threats are identified.

The application of UAVs has been extended to scenarios where the danger of explosive atmospheres exists, through the construction of an ATEX (ABS Explosive) system. This project was mainly focused on the sensory part, being open to the application of new AI techniques. The ENCIRCLE project [9], funded by the European Union (EU), aims to strengthen the European CBRNe (CBRN and Explosive Materials) industry and its capabilities.

The use of Unmanned Aerial Vehicles (UAVs) enables further expansion of the total area coverage for a given scenario. Systems comprised of UAVs mounted with radiological sensors have already been tested as a solution to cover broader areas [10], [11], and commercial solutions already exist, such as [12] and [13]. Preliminary tests were performed by the authors of this paper applying Simultaneous Localization and Mapping (SLAM) techniques, commonly used in robotics, resulting in a 3D map of a scenario with radiological information [14]. This paper addresses the problem of estimating the quantity, intensity, and location of different sources of radiation based on observations and sensor models.

The rest of this paper is organized as follows. Section II

Y. Brouwer is with the Department of Electrical and Computer Engineering, Instituto Superior Técnico, Lisbon, Portugal, e-mail: yoeri.brouwer@ipfn.tecnico.ulisboa.pt.

A. Vale, B. Gonçalves and H. Fernandes are with the Instituto de Plasmas e Fusão Nuclear, Instituto Superior Técnico, Lisbon, Portugal.

D. Macedo is with the Department of Physics, Instituto Superior Técnico, Lisbon, Portugal.

Manuscript received May 13, 2019; revised [month day], 2019.

introduces the variables used and describes the behaviour of the radiological sources. Section III describes the approaches which were studied in order to address the given problem, as well as the approaches which were selected for testing. Section IV illustrates and analyses the results obtained via simulation and real world experiments. Finally, section V summarizes the main conclusions of the work done.

II. Problem Statement

The problem is illustrated in Figure 1. The scenario may contain a number of radiological sources $i = 0, 1, \ldots, n$, represented by the symbol s_i. Each source has a position $\mathbf{p}_{s_i} = (x_{s_i}, y_{s_i}, z_{s_i})$ and an activity a_{s_i} representing the average number of particles emitted per second. The following four assumptions are made: first, the radiological sources are isotropic point sources; second, radiation intensity measurements are independent random variables each following a Poisson distribution; third, all the measurements have an equal exposure time; and fourth, the activity of each source remains constant over time, i.e., the duration of any measurement session is negligible compared to the half-life of the sources.

A set of observations are made in this scenario, represented by the symbol o_j, $j = 1, 2, \ldots, m$. Similarly, each observation has a position $\mathbf{q}_{o_j} = (u_{o_j}, v_{o_j}, w_{o_j})$ and a measured intensity c_{o_j} corresponding to the number of particles detected within the exposure time interval. Since the sources are isotropic point sources, the measured intensity only depends on the distances $d_{\mathbf{p}_{s_i}\mathbf{q}_{o_j}}$, between the position of observation $\mathbf{q}_{o_j}$ and the position of each source $\mathbf{p}_{s_i}$, and the background radiation intensity β_{o_j}, as expressed by (1) and (2). In particular, (2) describes the radiation field created by source s_i for all space by combining the inverse square law model with an exponential absorption model (to describe the absorption of radiation particles in the air).

$$c_{o_j} = \sum_{i=1}^{n} f_{s_i o_j} + \beta_{o_j} \tag{1}$$

$$f_{s_i o_j} = s_i \frac{e^{-\alpha d_{s_i o_j}}}{d_{s_i o_j}^2} \tag{2}$$

The problem is to estimate the number of sources N and the parameters (position and activity) of each source $\hat{s}_l = (\mathbf{p}_{\hat{s}_l}, a_{\hat{s}_l})$, $l = 1, 2, \ldots, N$, given a set of observations.

III. Proposed Approach

Several approaches have been proposed to solve the problem described in Section II. In this section we will describe a number of them, and justify the selection of the approaches which will be implemented and tested in Section IV.

A. Studied approaches

1) Maximum Likelihood approach: The Maximum Likelihood approach (MLA) is based on the estimation of the source parameters which have the maximum likelihood for explaining the observations. The likelihood is calculated through the joint probability distribution of N Poisson distributions (one for each potential source), where the samples correspond to the observed radiological intensity, and the distribution parameters are the positions and activities of the sources. Collecting the source parameters into a stacked vector $\mathbf{s} = [\mathbf{s}_1^\intercal \ldots \mathbf{s}_N^\intercal]^\intercal$, as well as the observations $\mathbf{o} = [o_1 \ldots o_m]^\intercal$, the maximum likelihood estimator is obtained from (3), where $p(\mathbf{o}|\mathbf{s})$ is the likelihood function.

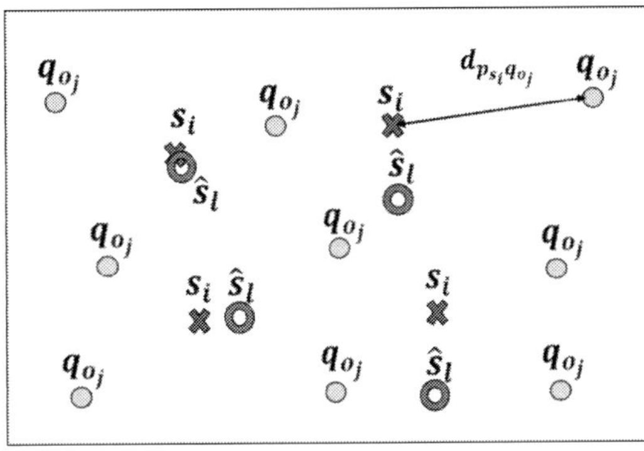

Fig. 1. Schematic representation of sources, observations, and estimated sources, illustrating the nomenclature used.

$$\hat{\mathbf{s}} = \arg\max_{\mathbf{s}} \log p(\mathbf{o}|\mathbf{s}) \tag{3}$$

In the work described in [15], this approach is applied to problems containing a single source. The authors use a small number of sensor measurements to localize and quantify the unknown source in a computational simulation. However, real world scenarios may have more than one source, so it is also necessary to determine the number of sources, in addition to estimating the parameters of each source.

In the work described in [16] the maximum likelihood approach is used to estimate the position and activity of an unknown number of sources. This work assumes that the radioactive sources are independent from each other and uses the sum of the contributions of each source in a computational simulation. However, this does not solve the problem that the number of sources remain unknown a priori. The proposed solution is to estimate the source parameters using the maximum likelihood for a different number of sources and then choose the solution with the highest score. The score of each solution is computed using (4), where $\hat{\mathbf{s}}_{\mathrm{ML},r}$ is the maximum likelihood estimated source parameter vector, $\mathbf{J}$ is the Fisher information matrix, and r is the number of sources being tested.

$$\beta_r = \log p(\mathbf{o}|\hat{\mathbf{s}}_{\mathrm{ML},r}) - \frac{1}{2} \log |\mathbf{J}(\hat{\mathbf{s}}_{\mathrm{ML},r})| \tag{4}$$

2) Bayesian approach: This approach uses an initial guess for the source parameters which, through an iterative process, is updated until a convergence threshold is met or the limit of the number of iterations is reached.

In the work described in [16], the authors apply importance sampling to the Bayesian approach, drawing samples (particles) of the parameter vector and calculating their importance weight to approximate the integral in (5). In each iteration, particles are re-sampled according to their weight, which is proportional to the likelihood of that sample, given the observations which were made. Subsequently, each particle is updated according to a kernel density function.

$$\hat{\mathbf{s}}_{\mathrm{B}} = \mathbb{E}\left[\mathbf{s}|\mathbf{o}\right] = \int \mathbf{s}\,\pi(\mathbf{s})\,d\mathbf{s} \tag{5}$$

After repeating this process a fixed number of times, the estimated parameter vector is taken to be the weighted average of all the particles. Similarly to what was described in the maximum likelihood approach, the number of sources is determined by repeating these steps for different numbers of sources, and choosing the solution which has the highest score. In this case, the score is given by the Bayesian information criterion, also known as the Schwartz criterion.

3) Robust Localization with Clustering approach: The Robust Localization with Clustering approach (RLCA) is based on a Monte Carlo technique which uses a set of particles and is centered around the idea of updating the set of particles biased towards the best-classified particles.

In the works described in [17] and [18], a multi-layer sequential Monte Carlo method coupled with suitable clustering algorithms is used to estimate the number, activities, and positions of sources. Each particle is a three-value vector representing a hypothetical source, containing the source's position (in $\mathbb{R}^2$) and activity. The initial set of particles is sampled from a uniform distribution. At each iteration of the recursive Bayesian estimation, the particles are classified by their weight, which is proportional to their respective likelihood given the observations. The particles with the lowest weight are replaced by particles with higher weights, and some are replaced with new randomly generated particles. In addition to the resampling process, a clustering process is applied to the particle set. The clustering process attempts to resolve the sources by estimating their parameters with the cluster centroid. The authors use three different clustering methods: mean-shift clustering, hierarchical clustering, and ID-based clustering. These two steps (resampling and clustering) are repeated until the confidence score is larger than a predefined threshold value.

4) Least Square Error approach: The Least Square Error approach (LSEA) attempts to fit a model to the data while minimizing the sum of square errors, as expressed by (6). In the work described in [19], the LSEA is used to estimate the position of one source using four sensors. The proposed solution is also able to track the position of the source over time. Once again, to determine the number of sources, (6) is solved for an increasing (but limited) number of sources, and the solution which produces the smallest sum of square errors is selected as the best estimator.

$$\hat{\mathbf{s}} = \arg\min_{\mathbf{s}} \sum_{j=1}^{m} \left(c_{o_j} - \sum_{l=1}^{N} f(\mathbf{s}_l, \mathbf{q}_{o_j}) \right)^2 \tag{6}$$

B. Selected approaches

Due to differences in implementation complexity, not all of the approaches described in Subsection III-A were applied to test scenarios. The selected approaches are:

- MLA: it is simple to apply, can be used to identify the number of sources and their parameters, it is simple to adapt to the collimator model and it works with smaller numbers of observations, compared with other approaches.
- RLCA: it was tested in 3D simulations, it does not require repeating the estimation process for different numbers of sources, since it estimates the number of sources directly, and it is simple to adapt to the collimator model.
- LSEA: it produces good results with few observations in scenarios with one source, it requires fewer observations than other approaches in general, and it is simple to adapt to the collimator model.

IV. Results

In this section the experimental results are presented and discussed. Presented first are the results obtained in simulation, which, due to their quality, reduced the viable approaches to MLA and LSEA. Then, experimental results with real sources and sensors are presented, including the analysis of the application of a collimator.

The scenarios of experimentation are depicted in Figure 2 and explained as follows. The radiation sources are assumed to be static and located on the ground. The 2D experiments are associated to a human operator walking in the scenario, equipped with a mobile phone connected to the GMC, and with an approximately constant distance between the sensor and the ground. The 3D experiments are associated to the ones performed by UAV flying over the scenario at different altitudes, but assuming that each mission is performed with an approximately constant altitude, i.e., also a constant distance between the sensor and the ground. The 2D and 3D experiments are performed in simulated and real scenarios, as presented in the following subsections.

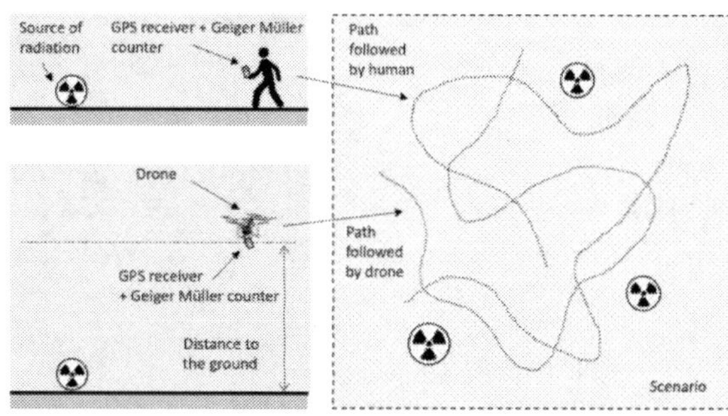

Fig. 2. 2D and 3D experiments associated to a human walking and a drone flying over the scenario, respectively (left images). The scenario, the sources of radiation and the described paths (right image).

https://doi.org/10.1051/epjconf/202022506005

A. *Simulated Results*

A simulation environment was developed for the initial testing of each approach. In this environment, each source has a position in $\mathbb{R}^3$. The results compare the number of estimated sources with the number of simulated sources, and each test scenario was simulated several times to calculate a mean and a variance. Figure 3 shows an example graph to explain the axes and error bars. Under ideal conditions we would expect to see a linear relationship with the number of estimated sources equal to the number of simulated sources (i.e., the slope of the curve would be equal to 1).

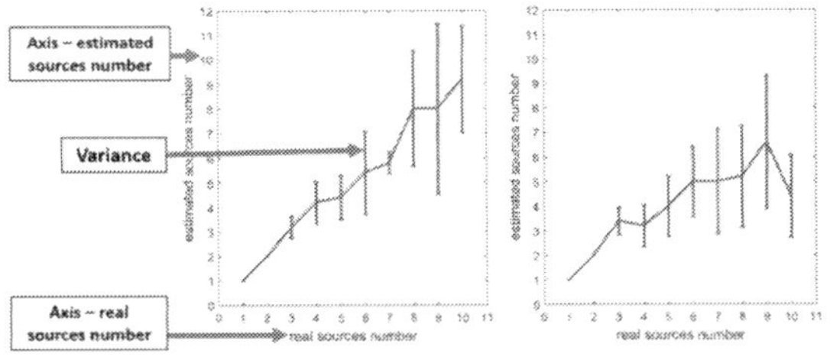

Fig. 3. Explanation of the elements of the result graphs. The horizontal axis represents the number of simulated sources and the vertical axis represents the number of estimated sources. Each simulation run has a fixed number of sources between 1 and 10 and is repeated several times to calculate the mean and the variance of the number of estimated sources.

We will start by comparing the results obtained when making observations in a grid pattern and the results obtained when making observations in random positions. The top image of Figure 4 shows observations made in a grid pattern in a scenario containing 3 sources, and the bottom image shows observations made in random positions in the same scenario. Illustrated in Figure 5 are the results of the estimation of the number of sources for MLA, RLCA and LSEA. It can be seen immediately that the RLCA produces poor results both in the case of grid positions and in the case of random positions. The MLA and LSEA produce similar results, with the MLA estimates having a smaller variance in the case of random positions. Due to the poor performance of the RLCA, it was abandoned in future tests.

Simulation results were also obtained for measurements made in 3 planes at $z = 50$ cm, $z = 60$ cm, and $z = 70$ cm. The results for the estimated number of sources in this scenario are presented in Figure 6. Here we can see that the MLA performs better than the LSEA, especially for a greater number of sources.

Following the results of Figures 5 and 6, all subsequent tests were performed using only the MLA, since this approach produced the best results.

Up to this point no uncertainty was introduced in the observations so they were taken as perfect observations, both in regards to the position $\mathbf{q}_{o_j}$ and in regards to the measured intensity c_{o_j}. Thus, in order to test the robustness of the Maximum Likelihood Approach, tests were made with noise in the measured intensity, with noise in the observation position, and with noise in both. The results are presented in Figure 7. It can be seen that the results deteriorate considerably, especially when there is noise in the observation position. In general, except for the case where there is only one source being simulated, the MLA greatly underestimates the number of sources in the scenario. The most striking example is the case of observations made in a grid pattern with noise in position and measured intensity, where the MLA estimates 1.5 sources on average when there are between 1 and 4 sources being simulated, and approximately 2.5 sources on average when there are 5 sources being simulated. We conclude then that the MLA is quite sensitive to noise in the observations, and that, in general, it underestimates the number of sources.

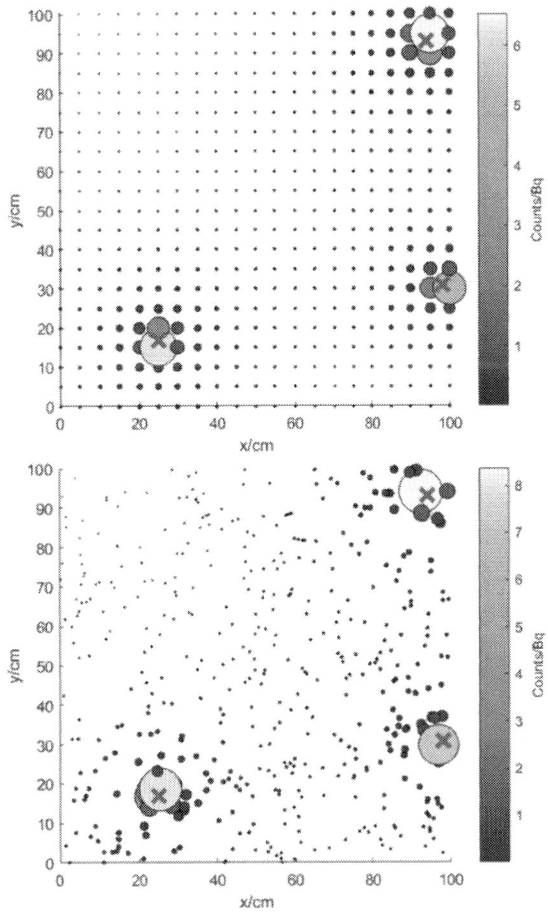

Fig. 4. Simulation of observations in a grid pattern and in random positions (top and bottom images, respectively). The blue crosses represent the source positions and the color and radius of the circles represent the measured intensity.

1) Application of a collimator: One way to improve the results observed in Figure 7 is with the application of a collimator. A collimator gives an effective orientation to the simulated measurements by only taking into consideration the sources which are inside a cone defined by the collimator angle ω. Figure 8 illustrates the collimator's effect.

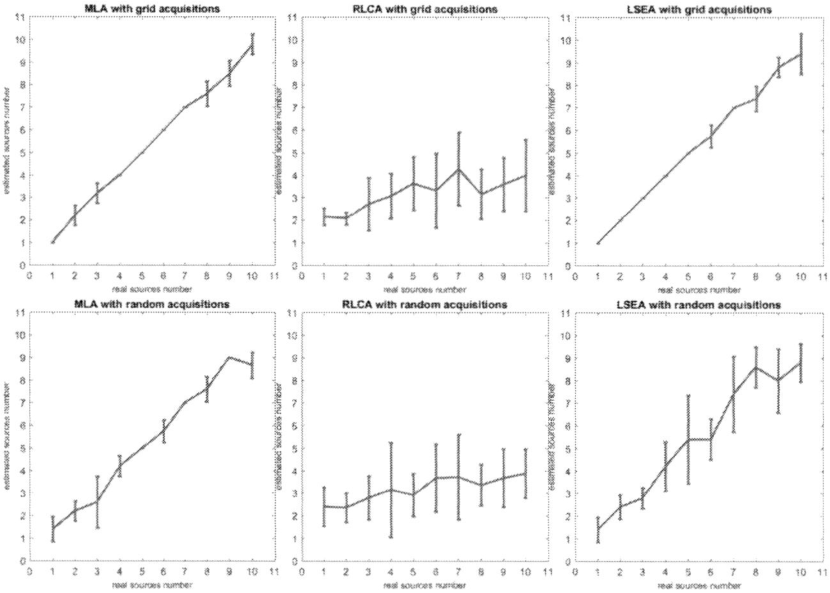

Fig. 5. Simulation results with measurements made at a height of 0 cm. The top row refers to measurements made in a square grid pattern. The bottom row refers to measurements made in random positions. In all figures the horizontal axis is the real number of sources, and the vertical axis is the estimated number of sources.

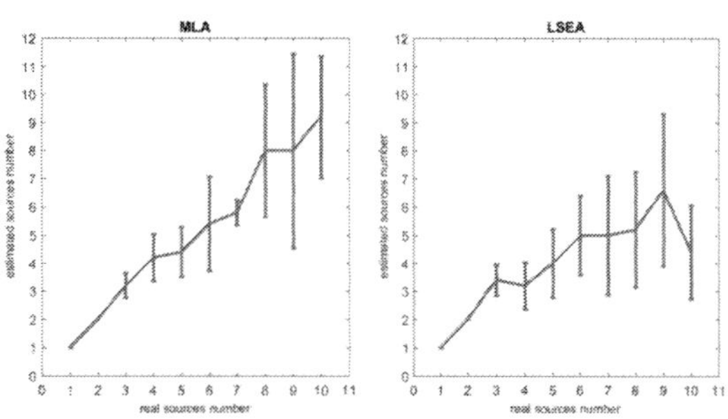

Fig. 6. Simulation results with measurements made in a square grid pattern at heights of 50, 60, and 70 cm. Left: results for the Maximum Likelihood Approach. Right: results for the Least Square Errors Approach.

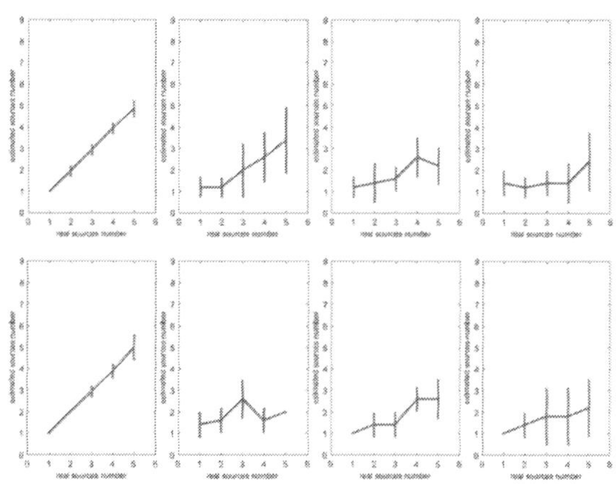

Fig. 7. Effect of intensity and position noise on the estimation of the number of sources. All noise levels are 20%. Top row: grid pattern. Bottom row: random positions. Columns from left to right: no noise, intensity noise, position noise, intensity and position noise.

Thus, by simulating the effect of a collimator with $\omega = 15°$ the MLA results are improved, as shown in Figure 9. It can be seen that the estimated number of sources is greatly improved when compared to the results in Figure 7, especially in the case of observations with position noise (third column from the left). However, in the case of noise both in position and in measured intensity, results are still poor and generally have a large variance.

B. Experimental Results

In order to further validate the application of the MLA to radiological source detection and characterization, tests were also performed in laboratory conditions and in a real-world scenario containing radioactive sources. Additionally, a lead collimator was constructed to assess the potential improvement in source characterization, since the GMC is approximately cylindrical and, without a collimator, the measured intensity depends almost exclusively on the distance to the sources.

The collimator which was built for testing is shown in Figure 10. The collimator is made out of lead (which has a large absorption coefficient) and has the shape of a satellite

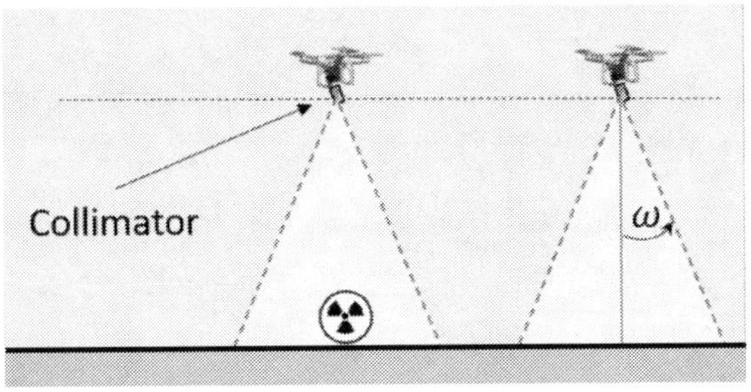

Fig. 8. Schematic representation of the collimator's effect. Only the sources inside the cone defined by ω contribute to the intensity measured.

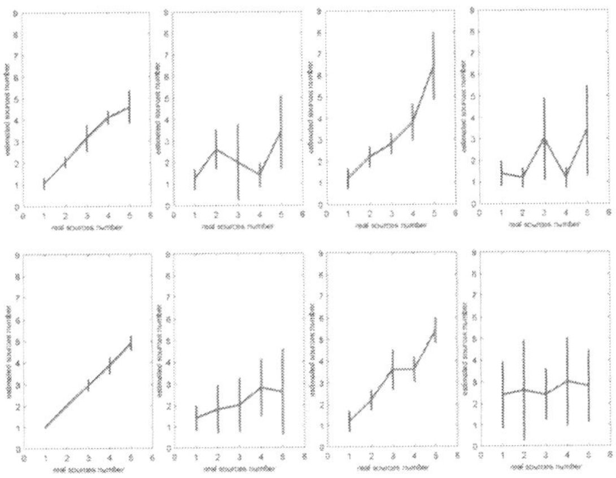

Fig. 9. Application of a collimator: effect of intensity and position noise on the estimation of the number of sources. Observations are simulated in the plane $z = 5$ cm and the collimator has an angle $\omega = 15°$. All noise levels are 20%. Top row: grid pattern. Bottom row: random positions. Columns from left to right: no noise, intensity noise, position noise, intensity and position noise.

dish. The hole in the center allows radiation particles to reach the detector. However, since the collimator only covers the "front" of the Geiger-Müller tube, particles can also be detected from the sides. The measured intensity as a function of the GM tube's orientation towards the source is shown in Figure 11. This figure compares the intensity vs angle curves with and without the collimator. It can be seen that the collimator has the desired effect for angles smaller than 60°, allowing radiation to be detected for very small angles and then absorbing radiation for larger angles. But, for angles larger than 60° the collimator's effect starts to become negligible.

Experiments were then made to test whether the collimator improved the results of the MLA using a real radioactive source in a laboratory environment. Measurements were made with a GMC in a grid pattern in two scenarios: with and without the collimator. The results are illustrated in Figure 12. These results confirm that the collimator improves the localization of the source, although this improvement comes at a price: the measured intensity is an order of magnitude smaller, which decreases the signal-to-noise ratio, i.e., the

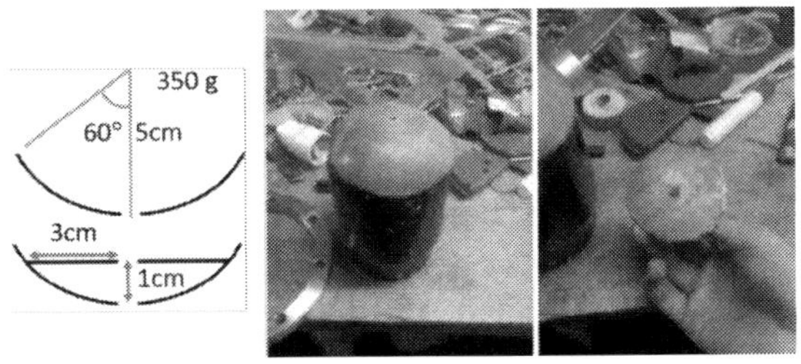

Fig. 10. Left: collimator design. Right: photographs of the collimator which was built for testing.

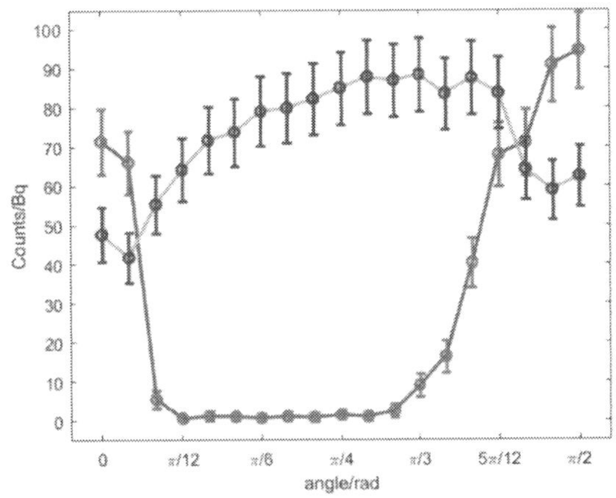

Fig. 11. Comparison between sensor readings with (blue/orange) and without (yellow/purple) a collimator as a function of the orientation towards the source. Due to the "satellite dish" shape of the collimator, intensity readings begin to increase above 60° since the sensor is not shielded on its sides.

measurements become more sensitive to noise. Nonetheless, we can conclude that the collimator does increase the accuracy of the source's estimated location.

Finally, experiments were made in a real-world setting containing two radioactive sources. The GMC was connected to a smartphone for recording and georeferencing the measurements, and two situations were tested: first, a human operator carried the sensor and smartphone and walked around the scenario (Figure 13, left); second the sensor and smartphone were mounted on a UAV (Figure 13, right). Due to the limited payload of the UAV, it was not possible to include the collimator. The radioactive sources were placed on the ground in a courtyard at IST, as shown in Figure 14. One test was performed with a human operator and two tests were performed with the UAV, represented in Figure 15.

V. Conclusions

Three approaches for quantification of radiological sources were tested: robust localization with clustering (RLCA), least square error (LSEA), and maximum likelihood (MLA). The MLA provided the best results when considering 2D and 3D scenarios, and was therefore selected for further testing. It was

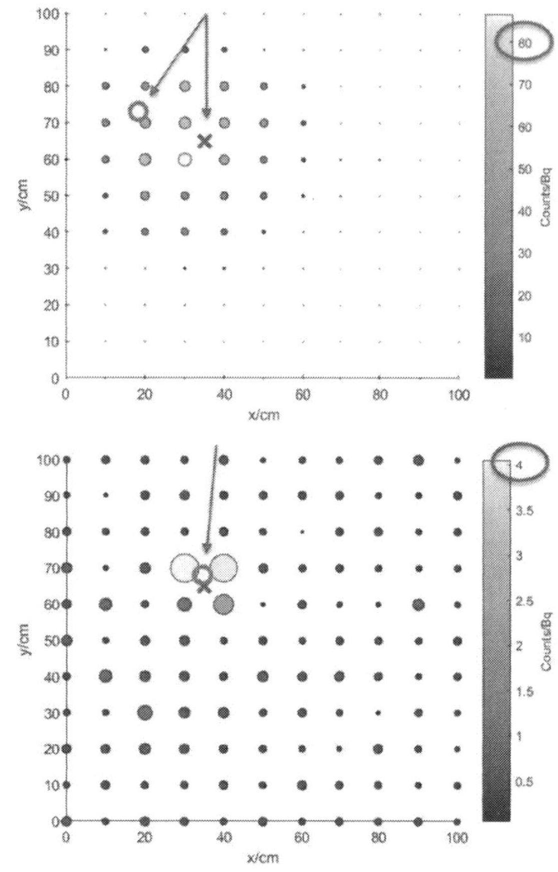

Fig. 12. Comparison of the estimated map without and with a collimator (top and bottom images, respectively).

shown that a collimator provides considerable improvement in the localization of sources, but significantly reduces the payload of a UAV and also decreases the signal-to-noise ratio. For future work the authors will perform new experiments with a more efficient Geiger-Müller counter (namely, the Mazur PRM-9000), and will develop new research into path planning and area coverage to maximize information gain and reduce mission length.

Acknowledgment

This paper is a result of the project the project FRIENDS - Fleet of dRones for radIological inspEction, commuNication anD reScue, [20] supported by the "Fundação para a Ciência e Tecnologia" (FCT), Compete 2020 and Lisboa 2020 under the PORTUGAL 2020 Partnership Agreement, through the European Regional Development Fund (ERDF). IPFN activities also received financial support from FCT though project UID/FIS/50010/2019.

Fig. 13. Left: a human operator carrying the radiological sensor connected to a smartphone. Right: a UAV mounted with the radiological sensor connected to a smartphone.

Fig. 14. Map view illustrating the source positions and the origin of the reference frame (yellow circle). Map provided by Google Maps^{TM}.

References

[1] J. Chamot and R. Atkins. (2008, February) Leading engineers and scientists identify advances that could improve quality of life around the world. National Science Foundation. [Online]. Available: https://www.nsf.gov/news/news_summ.jsp?cntn_id=111158
[2] National Academy of Engineering, "NAE Grand Challenges for Engineering," 2008. [Online]. Available: https://www.nae.edu/187212/NAE-Grand-Challenges-for-Engineering
[3] ——. Grand Challenges - Prevent Nuclear Terror. National Academy of Engineering. [Online]. Available: http://www.engineeringchallenges.org/challenges/nuclear.aspx
[4] Orano DS, "Press Release - Dismantling Robots 2018," 2018. [Online]. Available: https://www.orano.group/en/press/dossiers-de-presse/robots-dismantling-and-services
[5] M. Lozano and A. Fernandez. Reward Project. [Online]. Available: http://www.reward-project.eu/
[6] Academia da Força Aérea. (2015) Project seagull. [Online]. Available: http://www.academiafa.edu.pt/subPagina-10D00-019.005.003.011-seagull
[7] Framatome. (2016) WNE: AREVA Awarded in the Fields of Safety and Innovation. [Online]. Available: http://www.framatome.com/EN/businessnews-211/wne-areva-awarded-in-the-fields-of-safety-and-innovation.html
[8] ISQ. (2015) Gamma-Ex. [Online]. Available: https://www.isq.pt/projeto/gamma-ex/
[9] ENCIRCLE Consortium. ENCIRCLE: European CBRN Innovation for the Market Cluster. European Union. [Online]. Available: http://encircle-cbrn.eu/

[10] C. Cai, B. Carter, M. Srivastava, J. Tsung, J. Vahedi-Faridi, and C. Wiley, "Designing a radiation sensing uav system," in *2016 IEEE Systems and Information Engineering Design Symposium (SIEDS)*. IEEE, 2016. [Online]. Available: http://ieeexplore.ieee.org/document/7489292/
[11] K. Boudergui, F. Carrel, T. Domenech, N. Guénard, J.-P. Poli, A. Ravet, V. Schoepff, and R. Woo, "Development of a drone equipped with optimized sensors for nuclear and radiological risk characterization," in *2011 2nd International Conference on Advancements in Nuclear Instrumentation, Measurement Methods and their Applications*. IEEE, 2011. [Online]. Available: http://ieeexplore.ieee.org/document/6172936/
[12] Mirion Technologies. SPIR-Explorer Sensor. [Online]. Available: https://www.mirion.com/products/spir-explorer-sensor-light-wide-range-radiological-detection-and-identification-sensor
[13] FlyCam UAV. Drone UAV Aerial Radiation Detection. [Online]. Available: https://www.flycamuav.com/aerial-radiation-detection/
[14] H. Carvalho, A. Vale, R. Marques, R. Ventura, Y. Brouwer, and B. Gonçalves, "Remote inspection with multi-copters, radiological sensors and slam techniques," *EPJ Web of Conferences*, vol. 170, 2018. [Online]. Available: https://www.epj-conferences.org/10.1051/epjconf/201817007014
[15] H. E. Baidoo-Williams, "Maximum likelihood localization of radiation sources with unknown source intensity," *ArXiv e-prints*, 2016. [Online]. Available: http://arxiv.org/abs/1608.00427
[16] M. Morelande, B. Ristic, and A. Gunatilaka, "Detection and parameter estimation of multiple radioactive sources," in *2007 10th International Conference on Information Fusion*. IEEE, 2007. [Online]. Available: http://ieeexplore.ieee.org/document/4408094/
[17] J. Chin, D. K. Y. Yau, and N. S. V. Rao, "Efficient and robust localization of multiple radiation sources in complex environments," in *2011 31st International Conference on Distributed Computing Systems*. IEEE, 2011. [Online]. Available: http://ieeexplore.ieee.org/document/5961730/
[18] D. Shah and S. Scherer, "Robust localization of an arbitrary distribution of radioactive sources for aerial inspection," *Computing Research Repository*, 2017. [Online]. Available: http://arxiv.org/abs/1710.01701
[19] J. W. Howse, L. O. Ticknor, and K. R. Muske, "Least squares estimation techniques for position tracking of radioactive sources," *Automatica*, vol. 37, no. 11, 2001. [Online]. Available: https://www.sciencedirect.com/science/article/pii/S0005109801001340
[20] IPFN. (2018) FRIENDS: Fleet of drones for radiological inspection, communication and rescue. Instituto Superior Técnico. [Online]. Available: https://www.ipfn.tecnico.ulisboa.pt/FRIENDS/

Fig. 15. Experimental tests in a controlled scenario of IST campus with two sources of radiation. From top to down: walking in the scenario with the sensor on hand and two tests with a UAV flying over the scenario.

Compton imaging reconstruction methods: a comparative performance study of direct back-projection, SOE, a new Bayesian algorithm and a new Compton inversion method applied to real data with Caliste

G. Daniel[1,2], *O. Limousin*[1], *D. Maier*[1], *A. Meuris*[1], *F. Carrel*[3]

[1] AIM, CEA, CNRS, Université Paris-Saclay, Université Paris Diderot, Sorbonne Paris Cité, Gif-sur-Yvette, France
[2] École des Ponts ParisTech, Champs-sur-Marne, France
[3] CEA, LIST, Sensors and Electronic Architectures Laboratory, Gif-sur-Yvette, France

***Abstract*— Compton imaging is one of the main methods to localize radioactive hotspots, which emit high-energy gamma-ray photons, above 200 keV. Most of the Compton imaging systems are composed by at least two detection layers or one 3D position sensitive detector. In this study, we demonstrate the application of a new miniature pixelated single plane detector to Compton imaging. In this configuration, we do not have the information on interaction depth but we successfully test its ability to perform Compton localization by means of comparing different Compton reconstruction algorithms applied to real data measured with our single plane detection system.**

Index Terms **– Compton imaging, SOE, Bayes, Caliste, Compton inversion.**

I. Introduction

COMPTON imaging systems are used to localize radioactive sources which emit high-energy gamma-ray photons. This information is necessary in different domains such as nuclear safety, in accidental situations or decommissioning contexts [1], in medicine for hadron therapy dose monitoring [2] or in space observations [3] for instance.

These systems rely on the Compton scattering effect, which is the main interaction process at work for photons with an energy above 200 keV, in most of materials.

The principle of the method for gamma-ray imaging is to apply Compton kinematics [4] to find the possible origin of an impinging photon. If we note E_0 the initial energy of a photon, E_1 the energy of the scattered photon, E_2 the energy deposited by Compton interaction and θ the scattering angle, the energy and total relativistic momentum conservations lead to the formula:

$$\cos(\theta) = 1 - m_e c^2 \frac{E_1}{E_2 E_0} \qquad \text{(eq. 1)}$$

Where $m_e c^2 = 511$ keV is the energy mass of an electron at rest.

Regarding the reconstruction method, Compton kinematics give only the angle θ between the direction of the original photon and the scattered photon, which gives a cone with an angle θ of possible origins in 3D.

Usual Compton imaging systems measure the Compton scattering interaction, which deposits E_2, followed by the photoelectric interaction of the scattered photon.
Most of the systems use two detection layers, one scatter layer and one absorption layer, which fully absorbs the scattered photon [5]. Other systems perform Compton imaging with three or more layers, allowing reconstruction even if the final photon is not fully absorbed, assuming that the photon is scattered at least two times [6]. Finally, 3D-position sensitive systems, which are able to localize the interactions in the detector in 3D, are also used as Compton camera [7].

In this paper, we demonstrate the feasibility of Compton imaging with a single plane pixelated miniature detector, namely Caliste [8], by exploiting its advantages, especially its spectral and spatial resolution through Compton event selection. We study different Compton reconstruction algorithms to achieve the best localization performance with our detector: Direct Back-Projection (DBP) of the cones obtained by Compton kinematics, Stochastic Origin Ensemble with Resolution Recovery [9] and we introduce a new Bayesian approach and a new inversion of the Compton DBP.

https://doi.org/10.1051/epjconf/202022506006

Other algorithms for Compton reconstruction exist, such as Filtered Back-Projection or List-Mode Maximum Likelihood Expectation Maximization (LM-MLEM) [10]. LM-MLEM has been unsuccessfully tested for our particular detection configuration.

The paper is structured as follows:
Section II focuses on the detection system properties. Section III is dedicated to the Compton events selection. Section IV describes the different algorithms we have developed and implemented for reconstruction and Section V presents the results we get from real data acquisitions.

II. Caliste: a miniature CdTe spectro-imager

Caliste is a hybrid CdTe detector module, initially developed for high-energy astronomy in space, declined in different versions depending on its applications. The version used for this study is Caliste-HD [11], [12].

Caliste-HD is a pixelated imaging spectrometer made of a 16×16 pixels array with 625-μm pitch patterned on a 1 cm² monolithic CdTe crystal, 1 mm thick. The CdTe crystal is placed on top of a 3D electronics package where eight ultra-low noise read-out ASICs IDeF-X HD [13] are embedded. In such an architecture, each pixel is a miniature and independent spectrometer channel. The package size is 10×10×16.5 mm³. Caliste-HD performance has an excellent spectral response with 670 eV FWHM at 60 keV and 4.1 keV at 662 keV for single events in the summed spectrum of all pixels, with an energy range from 2 keV to 1 MeV for single events. An operating temperature of -10°C and an electric field of 300 V/mm are appropriate to reach these performances.

Figure 1 shows a picture of Caliste-HD.

Figure 1: Caliste-HD

We use these characteristics to perform Compton imaging with this detector integrated in a very compact system, namely WIX-HD, with a total weight of 1 kg. WIX-HD includes one Caliste-HD module installed into a small vacuum tight enclosure connected to a data acquisition electronics system. The cooling is realized by miniature thermos-electric Peltier coolers.

TABLE I
CHARACTERISTICS OF CALISTE-HD

PARAMETER	CALISTE-HD
Number of pixels	256 (16×16)
Pixel Pitch	625 μm
Crystal thickness	1 mm
Dimension without CdTe	10×10×16.5 mm³
Power Consumption	200 mW
Energy Range	2 - 1000 keV
FWHM at 60 keV	670 eV
FWHM at 662 keV	4.1 keV

This system is slightly different to the usual Compton imaging gamma-ray cameras. It is indeed composed of one planar layer of detection without interaction depth measurement, which causes disadvantages in terms of angular resolution because of the uncertainty on the interaction depth. If the two deposited energies are compatible with Compton scattering, we are not able to distinguish the Compton scattering interaction and the photoelectric absorption at first glance. Moreover, the planar configuration of our detection system implies that the detector plane is a symmetric plane in the reconstructed image, so that the field of view is not 360°×180° but 180°×180° instead.

Like every Compton imaging system, the localization resolution is essentially limited by the Doppler broadening effect [14].

However, we use the other advantages of the system, its spectroscopic performances and its small size, associated to specific algorithms, in order to perform Compton reconstruction.

III. Events selection

The first step of Compton reconstruction is the selection of the relevant events to perform the reconstruction. In a measurement with Caliste-HD, we first select multiple events, which are events, registered in time coincidence into the detector. The time between the Compton scattering interaction and the second interaction is indeed small compared to the time resolution of the detector of 20 μs.

The second rule is that the two events do not occur in neighbouring pixels, so that there is no possible confusion with charge sharing caused by a single interaction between the two pixels [15]. The events are also accepted if they are recorded in two separated clusters of pixels. Figure 2 shows some different and typical multiple events detection patterns.

The position in 2D associated with the events is considered to be the center of the pixel, when only single pixel hits are detected such as illustrated as type a on Figure 2. Conversely,

https://doi.org/10.1051/epjconf/202022506006

the interaction position is considered to be the centroid of the cluster weighted by the energy depositions in the cluster's pixels, see Figure 2, types b and c. For instance, in the case of events type c, the centroids of each clusters are calculated separately to determine the most probable interaction coordinates in the detector array.

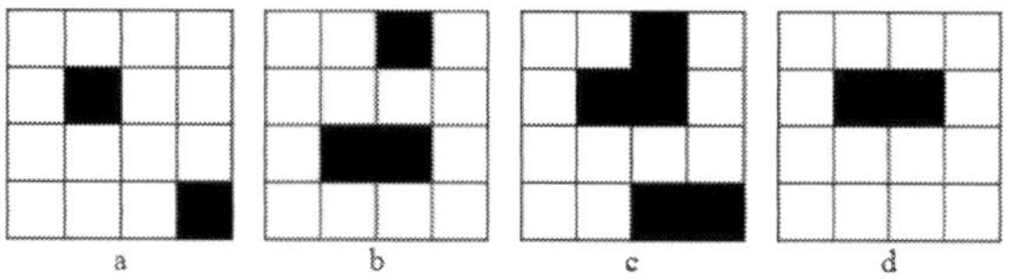

Figure 2: Multiple event configurations, a: Two pixels separated, accepted pattern; b: One single pixel and one cluster, accepted pattern; c: Two clusters separated, accepted pattern; d: One cluster, rejected pattern.

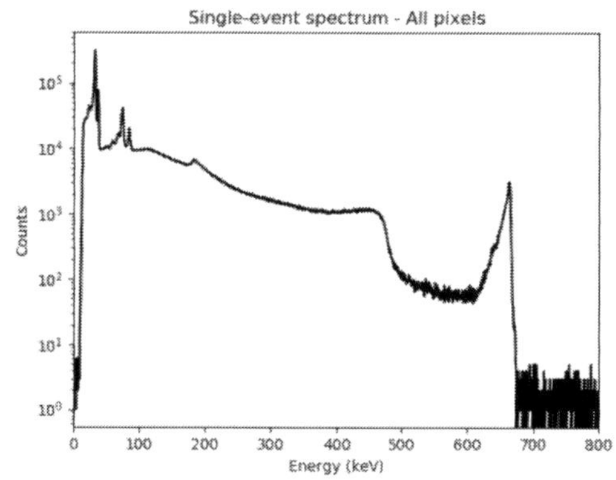

Figure 3: Single-event spectrum of ^{137}Cs source

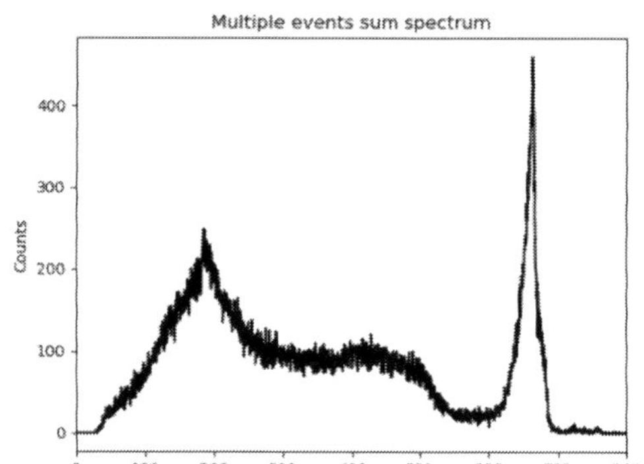

Figure 4: Multiple events sum spectrum of a ^{137}Cs source measurement with Caliste-HD

Once these events are discriminated, we use energy conservation to select events with one Compton scattering interaction with an energy deposition E_2 and a photoelectric absorption of the scattered photon E_1.

$$E_0 = E_1 + E_2 \qquad \text{(eq. 2)}$$

In our case, we assume having a prior knowledge of E_0 as a result of a spectral source identification performed on a single-event spectrum such as shown in Figure 3. This information can be obtained with a Bayesian Convolutional Neural Network [16], especially in case of small counting statistics.

Figure 4 shows a spectrum of summed multiple events, separated by at least one pixel, for a ^{137}Cs source measurement. The 662 keV line corresponds to reconstructed double events with a Compton scattering in the detector followed by a photoelectric absorption of the scattered photon, so that the total energy deposited corresponds to the full energy of the original photon.

The cases of single 662 keV Compton events whose scattered photons escape from the detector form the Compton edge at 478 keV as illustrated in Figure 3. However, as expected in our simulations, 662 keV double Compton events whose second scattered photon escapes from the detector tend to shift the Compton edge towards 500 keV as shown in Figure 5. In that case, the maximum reachable energy deposition is $E_{\max} = 555$ keV as expressed in eq. 3.

$$E_{\max} = E_0 - \frac{E_0}{1 + \frac{4E_0}{m_e c^2}} \qquad \text{(eq. 3)}$$

Whatever the Compton interaction is, single or double, the backscatter peak remains at 184 keV for 662 keV photons emitted by a ^{137}Cs source (eq. 4).

$$E_c = \frac{E_0}{1 + \frac{2E_0}{m_e c^2}} \qquad \text{(eq. 4)}$$

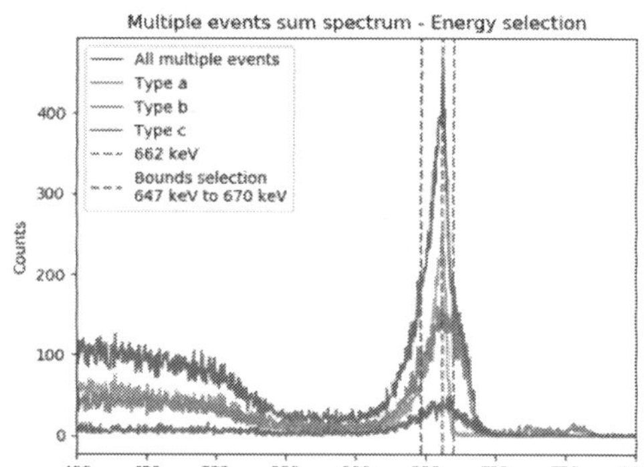

Figure 5: Energy selection with a ^{137}Cs source measurement on double events with Caliste-HD. The shape of the line includes a narrow peak populated with double events type a (45% of the events), a larger bump populated with double events type b (43%) and an even broader bump populated with double events type c (12%)

As shown in Figure 5, the line shape at 662 keV is composed of three event populations: the narrow peak structure is populated with events of type a (45%) summed up with two other bumpy populations from events of type b and c. The energy of the main line emission of the source is 662 keV. Because of the slightly non-linear detector response [17] and charge loss due to split events, we use a selection window, between 647 keV and 670 keV, which is considerably larger than the detector resolution for single events. The selection is made in the main peak of the spectrum so that we keep only the best events in energy to reduce the influence of the energy uncertainty on the computation of the scattering angle θ.

When these events are selected, we have to determine the order of the two measured interactions, i.e. which of the interactions is most likely corresponding to the Compton interaction. When one of the energies recorded is below the minimal energy of a Compton scattered photon E_c, this interaction is associated to the Compton scattering and the other one to the photoelectric absorption.

In this case, we distinguish the two interactions. However, the Doppler broadening effect may result in an energy of the scattered photon lower than E_c, which induces an error in the identification of the interaction order.

When the two deposited energies are between E_c and $E_0 - E_c$, we are not able to distinguish the two interactions with this rule. This situation happens when the energy of the original photon is above $\frac{m_e c^2}{2} = 255.5$ keV. One solution is to compute the two possibilities accepting that one is necessarily wrong, which causes noise in the reconstructed image, making the treatment more challenging. Alternatively, considering that the probability of photoelectric interaction is lower when the energy of the photon is higher, we attribute the photoelectric absorption to the lower deposited energy.

Writing E_A and E_B as the two measured energy deposits, the selection rule is:

- if $E_A > E_0 - E_c$ or $E_c \leq E_A < \frac{E_0}{2}$, E_A is attributed to the photoelectric absorption;
- otherwise, E_B is attributed to the photoelectric absorption.

Eventually, the formula for Compton kinematics (eq. 1) can be used with different parameters, considering the energy conservation (eq. 2). We can use $E_1 + E_2$ instead of E_0, $E_0 - E_1$ instead of E_2 and $E_0 - E_2$ instead of E_1. Analyzing the sensitivity of θ in (eq. 1) on the energy measurements E_1 and E_2 according to the different forms of the expression, we conclude that using $E_0 - E_1$ instead of E_2 leads to the minimization of the maximal error on θ. Thus, we use the formula:

$$\cos(\theta) = 1 - m_e c^2 \frac{E_1}{(E_0 - E_1)E_0} \qquad \text{(eq. 5)}$$

IV. Reconstruction algorithms

After Compton event selection, we perform Compton imaging using different algorithms. In this study, we have implemented four algorithms that we explain in this section and we evaluate their performance with our detector system on real data in Section V.

A. Direct Back-Projection (DBP)

Also called Simple Back-Projection, this algorithm is the direct application of the Compton kinematics. We note A, the position of the Compton interaction, B the position of the photoelectric absorption and θ the angle computed according to equation 5. We determine the equation of the cone of axis $[BA)$, of apex A and angle θ.

We compute the intersection of the cone with a sphere of radius r, where r is the expected distance from the detector to the source. In most cases the distance is unknown and r is set to a sufficiently high value so that we consider an infinite radius length comparatively to the detector size. This intersection is called the projection of the cone.

This algorithm is applied to each individual event and the cones are added iteratively, so that we get a back-projection map in the end. For a point source, this algorithm localizes the source at the intersections of the cones, with an uncertainty caused by the spatial and spectral resolutions of the detector.

However, when two point-sources or more with the same emission energy E_0 are present in the field of view, the cones created by one source turns to be a noise source for the other sources, wherever they are on the reconstructed image. The problem is naturally more important when the sources are close to each other, as we will see in the results in Section V. Consequently, more advanced algorithms are required to be able to localize and separate multiple point sources emitting at the same energy.

B. Stochastic Origin Ensemble with Resolution Recovery (SOE-RR)

Stochastic Origin Ensemble [18] is a Monte-Carlo Markov Chain algorithm applied to Compton imaging. It associates each event to a pixel of the reconstructed image. For each iteration, each event might move to another pixel according to a transition probability.

For initialization of this iterative method, one particular azimuthal direction on the back-projected cone is randomly selected. The selection can be uniform along the projection of the cone or it is also possible to use a prior distribution, based on a DBP calculation for instance.

A density map $d(x)$ is then computed on the reconstructed image, counting the number of events attributed to each pixel x.

At each iteration, for each event i, another position is randomly selected on the back-projected cone. n_i being the new selected position, o_i the old position, the transition probability is:

$$p_i = \min\left(1; \frac{(d(n_i)+1)^{d(n_i)+1}(d(o_i)-1)^{d(o_i)-1}}{d(n_i)^{d(n_i)}d(o_i)^{d(o_i)}}\right) \quad \text{(eq. 6)}$$

In eq. 6, if the detection efficiency of each pixel is known, we can multiply this term by the ratio of the efficiency of the new and the old pixel.

After a sufficient number of iterations, depending on the counting statistics, the sources positions, and the number of sources in the field of view, the algorithm converges to a stationary state, which gives the most probable positions of the sources.

The SOE version with Resolution Recovery [9] constructs the back-projection cone randomly according to the spatial and spectral resolutions of the detector. In this work, we consider only the spatial resolution: for each interaction, we pick up a position randomly around the estimated position of the event, uniformly in a size of one detector pixel, and we randomly select the interaction depth, uniformly in the detector volume. The measured energy is kept as measured, without any modification.

Finally, all the events are processed at once, by parallelizing the computation in order to accelerate the calculation as many iteration are needed to converge, typically 20 000).

C. *Bayesian algorithm*

We propose a new application of a Bayesian algorithm [19] to Compton reconstruction. The idea of this algorithm is to draw each cone by accumulating the information thanks to the previous events and by applying Bayes' theorem.

On a pixelated reconstructed image, we attribute a counter $k(j)$ to each pixel j, firstly initialized to 0. We also attribute a prior probability $p(j)$ of the origin of the photons, which could be either uniformly initialized or initialized with a prior knowledge depending on the geometry of the detector. For instance, in the case of Caliste and its direct environment, we use a uniform prior in the field of view of the detector.

For each event i, we compute the following posterior probability for each pixel j to be the origin of this measurement i thanks to Bayes' theorem:

$$p(j|i) = \frac{p(j)p(i|j)}{p(i)} \quad \text{(eq. 7)}$$

Where $p(j)$ is the current prior probability of observing an event originated from the pixel j.

$p(i|j)$ is the probability to observe the event i among all of the observable Compton events originated from the pixel j. The rigorous computation of this probability requires a heavy simulation of the detector response in terms of Compton events for each position of the reconstructed image. With this approach, a histogram of all recorded events is made, considering the positions of the first and the second interaction and the energy deposited in the second interaction. The bin is defined according to the detector configuration. For instance, a bin of one pixel for the two interaction positions and 1 keV in energy can be used.

However, this simulation is very long to compute to get a sufficient number of examples. For $E_0 = 662$ keV, $E_0 - E_c = 478\,keV$ and considering that neighbouring interactions are excluded, the number of bins required is:

$$256 \times (256 - 9) \times 478 = 30\,224\,896$$

This is the order of magnitude of the number of events to simulate in order to get a fair approximation of $p(i|j)$. Moreover, this simulation must be performed for every pixel j, i.e., 2025 times for a pixelization of 4°×4° in a field of view of 180°×180°.

Consequently, we chose to approximate $p(i|j)$ by considering the function $f(i|j)$, which is chosen equal to 1 in the pixels j belonging to the cone defined by i and equal to 0 for the other pixels. This approach is similar to the approximation made in some implementations of LM-MLEM [10].

In eq. 7, $p(i)$ is the probability to observe the event i. Its computation is not necessary by using the rule that $\sum_j p(j|i) = 1$.

A normalization constant C is computed as:

$$C = \sum_j p(j)f(i|j) \quad \text{(eq. 8)}$$

Consequently, $p(j|i)$ is computed as:

$$p(j|i) = \frac{p(j)f(i|j)}{C} \quad \text{(eq. 9)}$$

Then, the counter k is updated as: $k(j) \coloneqq k(j) + p(j|i)$.

We update the prior probability with:

$$p(j) = \frac{k(j)}{\sum_{j'} k(j')} \quad \text{(eq. 10)}$$

This update is made after accumulating some events to avoid that the first events put a probability of 0 to some pixels. We empirically chose to update $p(j)$ after processing 10 % of the total number of events.

After processing all events, $p(j)$ gives a proportion map of the presence of radioactive sources.

Reference [19] gives more details on possible variations of this Bayesian based algorithm.

D. Compton inversion

The last algorithm we propose is a Compton inversion algorithm of the image initially computed by a Direct Back-Projection method.

A passing matrix M is obtained by simulating the presence of a source in each pixel of the source space image, and applying the DBP to these simulated data, so that we derive, at each possible position in space, the expected DBP pixelated with the same resolution as the source space image.
The line j of M contains the simulated DBP, in flattened data, of a source placed in the pixel j of the source space image.

If we note Y, the DBP of a real acquisition and X, the expected intensities of the source in each pixel of the source space image, the equation to be solved is written as:

$$Y = MX \qquad \text{(eq. 11)}$$

M is a matrix with high dimensions, of a typical size of 8100^2 for a field of view of 180°×180° with a discretization of 2°×2°. Its inversion is not feasible; therefore, we may approach an optimal solution of this inverse problem minimizing instead:

$$L(X) = \|Y - MX\|_2^2 \qquad \text{(eq. 12)}$$

The minimization process is constrained to positive elements of X. A gradient descent algorithm computes this minimization, since this is a convex problem.

Once the optimal solution X is found, we apply a TV-regularization [20], which is used in Compton imaging with other algorithms [21], in order to denoise the source space image by minimizing $\|\nabla X\|_1$.

V. TESTS AND PERFORMANCES

A. Test configuration

We take real data with Caliste-HD to apply and test the algorithms. A 1.7 MBq ^{137}Cs source is placed in a distance of 30 cm from the detector at different longitudes and latitudes. The corresponding dose rate is ~1.4 µSv/h at the detector level.

Two separate measurements are acquired: one measurement with a source directly placed on axis, at longitude $\varphi = 0°$ and latitude $\theta = 0°$ for a total exposure time of 35 h. The corresponding count rate at Caliste level is 50 counts/s/cm². A second measurement with the source moved to the right at longitude $\varphi = 10°$ and $\theta = 0°$ for a total exposure time of 57 h. Count rate is the same. The coordinates system is explained on Figure 6.

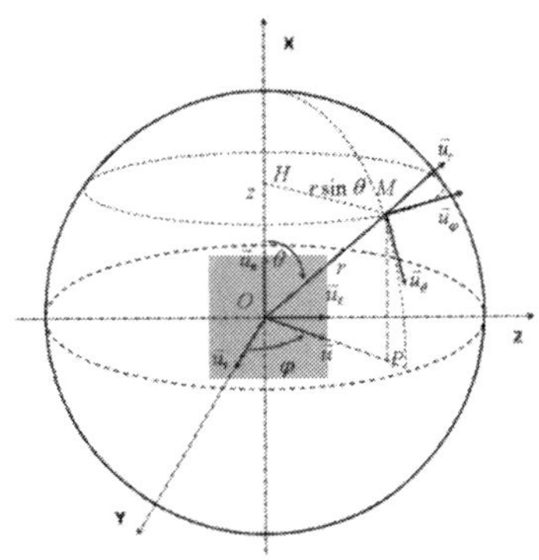

Figure 6: Spherical coordinates system - The square represents the detector

From these data, we can build a third augmented data set combining randomly all events from the two independent measurements into a single list.
The count rate being low, (50 counts/s at 30 cm), the chance of spurious double events from the two sources - that we cannot record combining the two event lists - is considered to be negligible (<< 1 count/s). This way, we obtain a representative data set that we would have recorded with two ^{137}Cs sources simultaneously installed in the field of view the detector. Choosing the number of events in the two lists, allows to virtually modulate the sources relative intensities. It also allows to test our methods with an arbitrary number of sources in the field of view even having only one source available in the lab.

In the results section, we study the Compton reconstruction processed on the measurement with the source at coordinates (10°,0°) and the "augmented data set" with two sources at coordinates (0°,0°) and (10°,0°). Other configurations have been successfully tested with similar performance of the algorithms.

The algorithms are applied with a field of view of 180°×180°. To help the reader to see the results, the images of Figure 7A-D are zoomed in the region of interest, between -30° and 30° for the longitude and between -15° and 15° for the latitude. Based on experience, we apply a discretization of 2°×2° for DBP and Compton inversion, 4°×4° for the Bayesian algorithm and 5°×5° for SOE-RR. We apply a spline interpolation for the representation on the reconstructed image. SOE-RR is performed with 20 000 iterations.

B. Results

The total number of detected events with the acquisition at (10°,0°) is 5 423 829. Applying the energy selection according to the method described in section III, 8119 Compton events are extracted, corresponding to 0.15 % of the total number of recorded events.
With the two sources mix augmented data set, the total number of events is 8 614 247 and 13 076 Compton events are extracted, also corresponding to 0.15 % of the events.

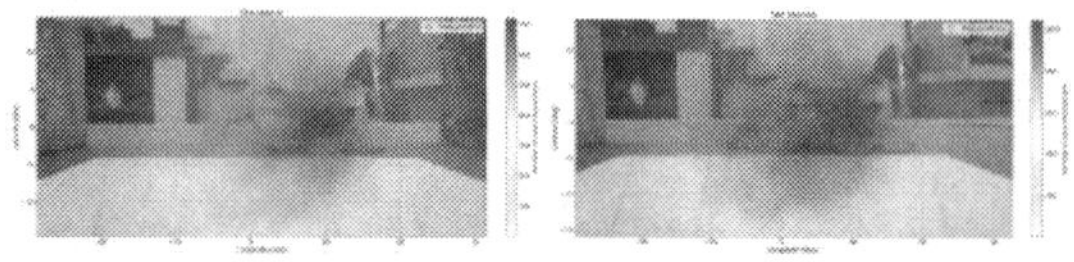

Figure 7A: Results with DBP

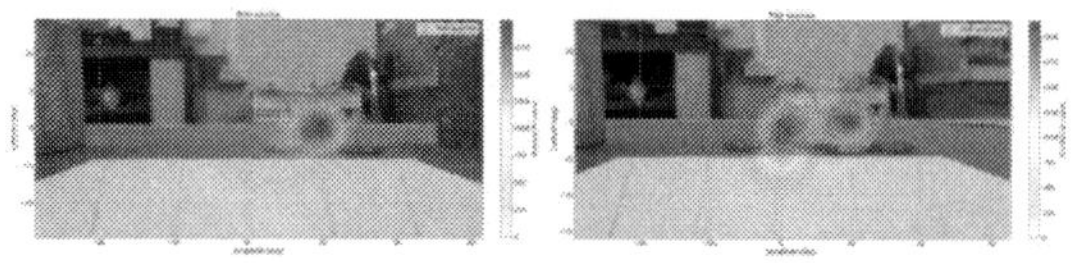

Figure 7B: Results with SOE-RR

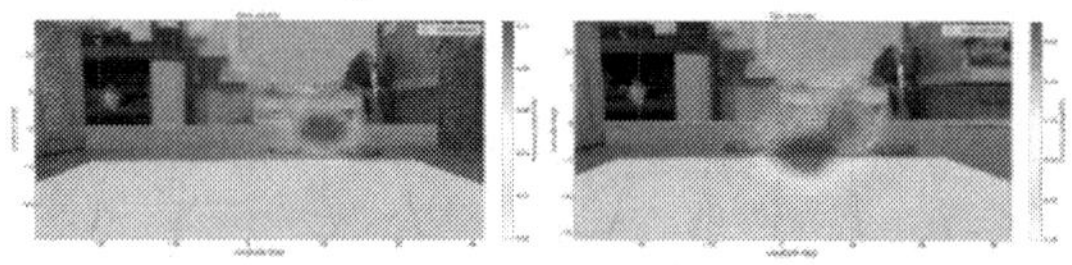

Figure 7C: Results with Bayesian algorithm

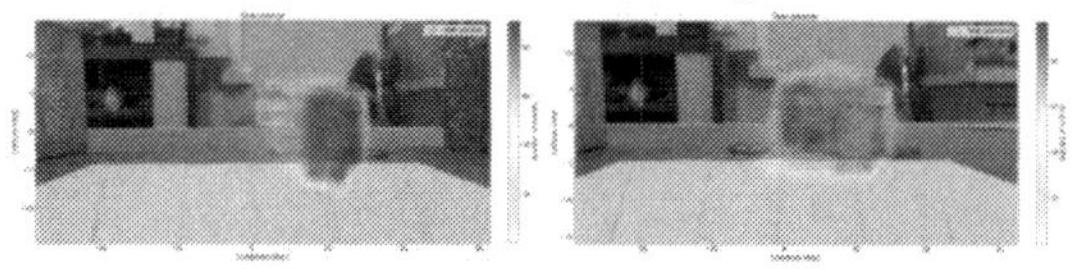

Figure 7D: Results with Compton inversion

Table II: Algorithms performance: computation time, angular resolution FWHM for one source at (10°,0°) and separating power for two sources separated by 10°.

TABLE II
PERFORMANCES OF THE ALGORITHMS

PARAMETER	DBP	SOE-RR	BAYESIAN ALGORITHM	COMPTON INVERSION
FWHM one source (10°,0°)	37°	7,9°	8,6°	14,5°
Separation of two sources (10°)	No	Yes	Yes	No
Computation time (4500 events)	12 s ± 1 s	110 s ± 2 s	12 s ± 1 s	65 s ± 2 s (12 s DBP + 53 s gradient descent)
Complexity according the number of events	Linear	Linear	Linear	Linear (DBP) + Constant (gradient descent)

Computation time is evaluated with the following hardware setup: Intel® Xeon® Silver 4114, CPU @ 2.20 GHz. The algorithms are implemented in Python using NumPy library.

C. Comments

The Direct Back-Projection algorithm shows the worst performances in terms of angular resolution. However, it shows the fastest computation time, 12 s to process 4500 events.

SOE-RR shows a way better localization performance, especially with the power to separate two sources 10° apart. However, the main limitation is related to the computation time, nine to ten times slower than DBP. Moreover, as the number of iterations to compute is not pre-determined, the computation time depends on the number of iterations. Nevertheless, this algorithm considers the uncertainties in the measurement thanks to the Resolution Recovery version, which is important to get a more robust result.

The Bayesian algorithm shows intermediate performances in terms of localization. On Figure 7C right, the left source is not localized at the exact position: this is a discretization effect in the 4°×4° pixelization reconstructed image. In fact, the probability of presence is higher in the neighbouring pixel of the real position of the source. The two sources are however well separated, with a quick computation time similar to the Direct Back-Projection algorithm. Actually, the number of operations is the same as in DBP, since the list of events is scanned only once.

We conclude here that SOE-RR is the most promising algorithm to solve our Compton reconstruction problem with a single layer Compton imager as long as computation time is not an issue. Our Bayesian algorithm gives a fast and relatively accurate result, possibly used in other methods as a better prior than the Direct Back-Projection reconstruction.

Compton inversion algorithm is less convincing than SOE-RR and the Bayesian algorithm for the detection of point sources. However, a preliminary study based on simulations shows that this algorithm will reach better performances in the case of extended sources.

VI. CONCLUSION AND OUTLOOKS

Compton imaging can be performed with a single plane miniature pixelated detector with fine pitch and high energy resolution. It requires a particular attention on the selection of the Compton events, given the configuration and the characteristics of the Caliste-HD detector, followed by a reconstruction algorithm. We tested four algorithms on real data and demonstrated that SOE-RR and a new Bayesian algorithm, are able to separate two sources of ^{137}Cs separated by only10° which a classical Direct Back Projection cannot do.

In forthcoming studies, we envision to further develop the advanced SOE-RR algorithms, noticeably including the energy resolution and iterative event selection. As computation time might be a severe limitation for use in real time applications, we will cascade several algorithms using one as a prior for the next. An evaluation of the sensitivity of the system must be carried out, since we use here the whole statistics of acquired photons leading to more than 8000 Compton events, but it is important to evaluate the necessary number of measured Compton events to perform the reconstruction algorithms. In addition, the case of extended sources will be studied, with real data acquisition, in order to evaluate the capability of our system and algorithms to image extended sources.

VII. References

[1] Y.-s. Kim, J. H. Kim, H. S. Lee, H. R. Lee, J. H. Park, J. H. Park, H. Seo, C. Lee, S. H. Park and C. H. Kim, Development of Compton imaging system for nuclear material monitoring at pyroprocessing test-bed facility, 2016: Journal of Nuclear Science and Technology, 53:12, 2040-2048, DOI:.

[2] A. Koide, J. Kataoka, T. Masuda, S. Mochizuki, T. Taya, K. Sueoka, L. Tagawa, K. Fujieda, T. Maruhashi, T. Kurihara and T. Inaniwa, Precision imaging of 4.4 MeV gamma rays using a 3-D position sensitive Compton camera, Scientific Reportsvolume 8, Article number: 8116, 2018.

[3] P. Bloser, R. Andritschke, G. Kanbach, V. Schonfelder, F. Schopper and A. Zoglauer, The MEGA advanced Compton telescope project, New Astronomy Reviews 46 611–616, 2002.

[4] W. Leo, Techniques for Nuclear and Particle Physics Experiments: A How-to Approach, Springer-Verlag, 1993.

[5] S. Takeda, Y. Ichinohe, K. Hagino, H. Odaka, T. Yuasa, S.-n. Ishikawa, T. Fukuyama, S. Saito, T. Sato, G. Sato, S. Watanabe, M. Kokubun and T. Takahashi, Applications and imaging techniques of a Si/CdTe Compton gamma-ray camera, TIPP 2011 - Technology and Instrumentation in Particle Physics 2011. Physics Procedia 37 (2012) 859-866, 2012.

[6] G. Llosa, M. Trovato, J. Barrio, A. Etxebeste, E. Muñoz, C. Lacasta, J. F. Oliver, M. Rafecas, C. Solaz and P. Solevi, First Images of a Three-Layer Compton Telescope Prototype for Treatment Monitoring in Hadron Therapy, Frontiers in Oncology 6:14, 2016.

[7] C. Lehner, Z. He and F. Zhang, 4Pi Compton Imaging Using a 3-D Position-Sensitive CdZnTe Detector Via Weighted List-Mode Maximum Likelihood, IEEE: Transactions on Nuclear Science, Vol. 51, NO. 4, 2004.

[8] O. Limousin, The story of Caliste: CdTe based pixel detectors for Hard X-Ray astronomy in space, Habilitation à Diriger des Recherches, Université Paris-Diderot, CEA, 2016.

[9] A. Andreyev, A. Celler, I. Ozsahin and A. Sitek, Resolution recovery for Compton camera using origin ensemble algorithm, Medical Physics 43, 2016.

[10] D. Xu, Z. He, C. E. Lehner and F. Zhang, 4 Pi Compton imaging with single 3D position sensitive CdZnTe detector, Hard X-Ray and Gamma-Ray Detector Physics VI, Proc. of SPIE Vol. 5540 (SPIE, Bellingham, WA, 2004) · 0277-786X/04/$15 · doi: 10.1117/12.563905, 2004.

[11] A. Meuris, O. Limousin, O. Gevin, F. Lugiez, I. L. Mer, F. Pinsard, M. Donati, C. Blondel, A. Michalowska, E. Delagnes, M.-C. Vassal and F. Soufflet, Caliste HD: a New Fine Pitch Cd(Zn)Te Imaging Spectrometer from 2 ke V up to 1 Me V, 2011: IEEE Nuclear Science Symposium Conference Record.

[12] S. Dubos, O. Limousin, C. Blondel, R. Chipaux, Y. Dolgorouky, O. Gevin, Y. Ménesguen, A. Meuris, T. Orduna, T. Tourette and A. Sauvageon, Low Energy Characterization of Caliste HD, a Fine Pitch CdTe-Based Imaging Spectrometer, IEEE: Transactions on Nuclear Science, Vol. 60, No. 5, 2013.

[13] A. Michalowska, O. Gevin, O. Lemaire, F. Lugiez, P. Baron, H. Grabas, F. Pinsard, O. Limousin and E. Delages, IDeF-X HD: A low power Multi-Gain CMOS ASIC for the readout of Cd(Zn)Te Detectors, IEEE Nuclear Science Symposuim & Medical Imaging Conference, 2010.

[14] C. E. Ordonez, A. Bolozdynya and W. Chang, Doppler Broadening of Energy Spectra in Compton Cameras, IEEE Nuclear Science Symposium Conference Record, 1998.

[15] A. Meuris, O. Limousin and C.Blondel, Charge sharing in CdTe pixilated detectors, Nuclear Instruments and Methods in Physics Research Section A: Accelerators, Spectrometers, Detectors and Associated Equipment, Volume 610, Issue 1, Pages 294-297, 2009.

[16] G. Daniel, O. Limousin, D. Maier, F. Ceraudo and A. Meuris, Application of Bayesian Convolutional Neural Network to spectral identification of radionuclides for nuclear monitoring, To be submitted TNS 2019, 2019.

[17] D. Maier and O. Limousin, Energy calibration via correlation, Nuclear Instruments and Methods in Physics Research Section A, DOI: 10.1016/j.nima.2015.11.149, 2016.

[18] A. Andreyev, A. Sitek and A. Celler, Fast image reconstruction for Compton camera using stochastic origin ensemble approach, Medical Physics 38, 2010.

[19] C. Bobin ; O. Bichler ; V. Lourenço ; C. Thiam ; M. Thévenin, Applied Radiation and Isotopes, Vol. 109, Pages 405-409, 2015.

[20] A. Chambolle, V. Caselles, M. Novaga, D. Cremers and T. Pock, An introduction to Total Variation for Image Analysis, hal-00437581, 2009.

[21] V. Maxim, Enhancement of Compton camera images reconstructed by inversion of a conical Radon transform, Inverse Problems 35 014001, 2018.

Simulation of delayed gamma rays from neutron-induced fissions using MCNP 6.1

R. De Stefano*, B. Pérot, C. Carasco, E. Simon
* Nuclear Measurements Laboratory
roberto.destefano@cea.fr

***Abstract*– As part of its R&D activities in the fields of radioactive waste drum storage and homeland security, the Nuclear Measurement Laboratory of CEA Cadarache has started studies related to the detection of induced delayed fission gamma rays as a signature of U/Pu presence either in radioactive wastes or in cargo containers and luggage. The study described in the present paper explores the feasibility of detecting fission delayed gamma rays of nuclear materials interrogated by a pulsed neutron generator. For this purpose, Monte Carlo simulations have been performed with ACT, the MNCP6 Activation Control Card. Simulated results have been compared with experimental data to validate the numerical model. Samples of uranium and plutonium have been irradiated for 2 hours with a pulsed D-T neutron generator delivering 14 MeV neutrons with an average emission of 8.10^7 n/s, which are thermalised in a graphite cell called REGAIN. At the end of irradiation, activated nuclear materials were placed in a low-background, high-resolution gamma spectroscopy station in order to detect delayed gamma rays emitted by fission products. Anomalies have been observed in the calculated time decay curve of fission delayed gamma rays with MCNP6 ACT card, but the time behavior is correct for non-fission activated materials like aluminum or copper. On the other hand, the number of counts recorded in the main simulated gamma ray lines from activated nuclear material fission products is consistent with the experimental results, thus validating the simulation scheme in view of further studies on the characterization of radioactive waste drums or special nuclear material detection in cargo containers.**

I. Introduction

Delayed gamma rays detection emitted after neutron-induced fission or photo-fission is being studied in the fields of radioactive waste drum characterization but also for homeland security [1] [2] [3] [4]. Indeed, the detection of well-known delayed gamma rays emitted by uranium and plutonium fission fragments can be a signature of nuclear materials presence either in legacy drums or in cargo containers. In addition, the relative intensities of these delayed gamma rays provide additional information on the nature of the interrogated isotopes, such as ^{235}U and ^{239}Pu, which have different fission yields [5]. In this context, it is important to test and qualify simulation tools in view to further design non-destructive measurement systems based on the detection of induced-fission delayed gamma rays.

Manuscript submitted on June 4, 2019. This work takes place in the framework of Grenoble Institute of Technology, Physics, Electronics and Materials engineering cursus partnership with CEA Cadarache.
Roberto De Stefano, Bertrand Pérot, Cédric Carasco, Eric Simon are with CEA, DEN, Cadarache, DTN, SMTA, Nuclear Measurement Laboratory, F-13108 Saint-Paul-Lez-Durance, France.

With the implementation of the new Activation Control Card (ACT Card) [6], evolution calculations are now possible with MCNP 6.1 and higher versions allowing the creation and transport of delayed particles from non-nuclear materials and fissionable isotopes. The study presented in this paper mainly focuses on the detection of delayed gamma rays emitted by fission fragments of nuclear materials (^{235}U and ^{239}Pu) or by non-nuclear materials activated during neutron irradiation. The creation of delayed gamma rays from fissionable isotopes is done with the option FISSION=P, and for non-nuclear materials with the option NONFISSION=P. In order not to calculate delayed gamma rays, the options to be used are FISSION=NONE for nuclear materials, and NONFISSION=NONE for non-nuclear materials. The main delayed gamma rays of interest for nuclear materials in this study based on LARA nuclear database [7] are reported in Tab. 1.

γ ray energy	885 keV	1011 keV	1384 keV	1436 keV	1768 keV	2218 keV
Fission fragment	^{104}Tc	^{101}Mo	^{92}Sr	^{138}Cs	^{138}Xe	^{138}Cs
Half-life $T_{1/2}$	18.3 min	14.61 min	2.65 h	33.4 min	14.14 min	33.4 min

Table 1: *Main uranium and plutonium delayed gamma rays of interest*

The 1778 keV delayed gamma ray emitted by ^{28}Al ($T_{1/2}$= 2.24 min) and the 1345 keV line from ^{64}Cu ($T_{1/2}$=12.7 h) are also studied, which are produced by neutron radiative capture on ^{27}Al and ^{63}Cu, respectively.

The simulation results obtained for these delayed gamma rays were confronted to experimental data in order to validate a calculation scheme adapted to further feasibility studies related to delayed gamma-ray measurements after neutron irradiation.

II. Experiment

The experimental campaign described in [5] is used to validate the calculation scheme based on MCNP 6.1 ACT card. Highly Enriched Uranium (HEU) and plutonium samples have been irradiated for 2 hours in a graphite cell named REGAIN (see Fig. 1 and Fig. 3). REGAIN was equipped with a pulsed D-T neutron generator (GENIE 16 model from SODERN) delivering 14 MeV neutrons with an average emission of 8.10^7 s^{-1}, and with a coaxial N-type HPGe detector (from CANBERRA) of 33 % relative efficiency inserted in one of the graphite walls. REGAIN was dedicated to PGNAA (Prompt Gamma Neutron Activation Analysis) studies, for instance to characterize toxic chemicals in radioactive waste drums [2] by detecting prompt gamma rays during (inelastic scattering) or between (radiative capture) the pulses of the neutron generator, as well as post-irradiation delayed gamma rays. For the experiments reported in [5], the pulse width was 10 ms and the pulse frequency 20 Hz, i.e. 50 ms between pulses. The delayed gamma rays emitted by ^{235}U and ^{239}Pu fission fragments with half-lifes shorter than 10 min were measured between the neutron pulses. However, in this paper, we only focused on post-irradiation measurements of delayed gamma rays with half-lifes longer than 10 min, because of the non-physical ACT results presented in section III (see further Fig. 7). The samples were moved just after the 2 h irradiation into a low-background spectrometer (see Fig. 2) to limit the background due to non-fissionable material activation. One can note for instance the 1778 keV gamma ray from aluminum frame bars of REGAIN cell (see Fig. 1 and Fig. 3). The sample transfer time from REGAIN to the low-background station was around 1 min, then delayed gamma-ray spectra of the ^{235}U and ^{239}Pu samples were recorded during 24 hours. In order to verify the half-life associated to the different fission products, acquisition was segmented in one spectrum every 15 min. The low-background spectrometer is made of thick lead walls covered by thin copper layers, and it holds a type-P HPGe coaxial detector with a 25 % relative efficiency. A thin 3 mm lead shield was placed between the metallic nuclear material samples and the HPGe detector in order to limit the detection of low-energy gamma rays from ^{241}Am. Both lead shield and fissile material samples were placed 12.5 cm from the surface of the detector.

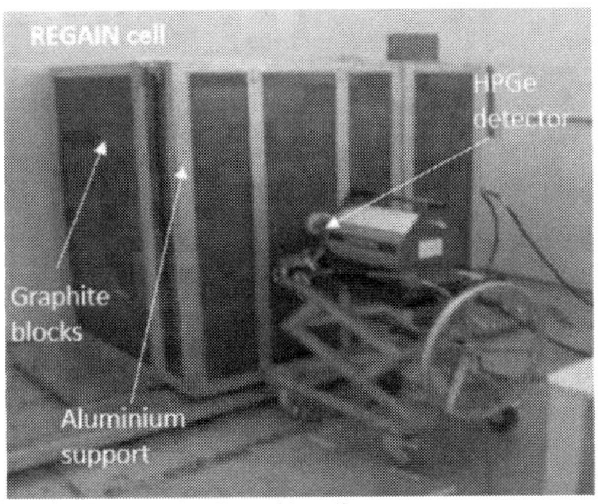

Figure 1: *PGNAA REGAIN graphite cell*

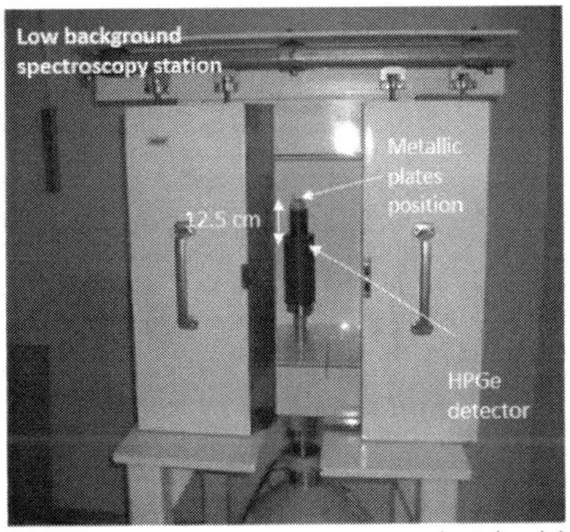

Figure 2: *Low background spectroscopy station for induced-fission delayed gamma rays measurement*

III. SIMULATION VS. EXPERIMENT

In order to validate the use of MCNP 6 ACT card, experimental and simulated results were confronted. Simulations were done based on REGAIN MCNP model presented in Fig. 3 and Fig. 4.

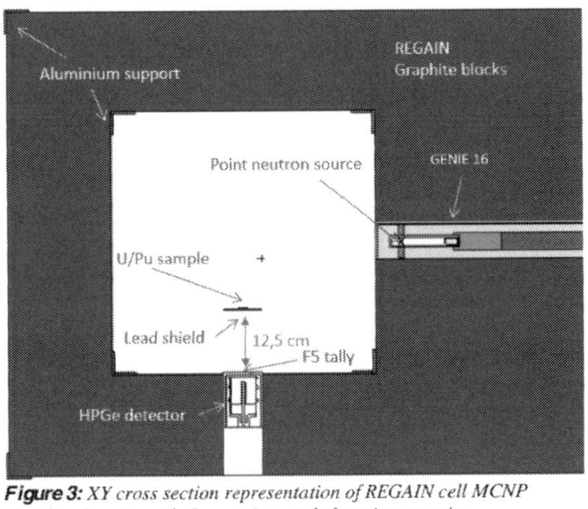

Figure 3: *XY cross section representation of REGAIN cell MCNP model and cavity with the uranium and plutonium samples*

https://doi.org/10.1051/epjconf/202022506007

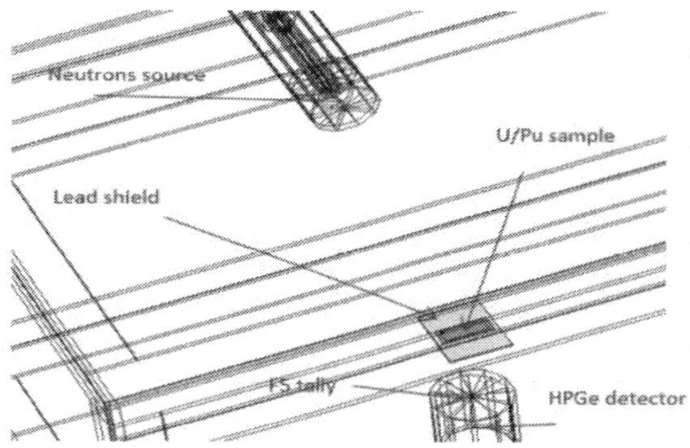

Figure 4: *3D representation of REGAIN cell MCNP model and cavity with the uranium and plutonium samples.*

In the MCNP model, the 14 MeV neutrons are emitted in a point source located 10 cm away from the neutron generator endpoint. The graphite blocks are hold by aluminum frame bars with a total mass of 121 kg (note that depending on the information source, REGAIN being a legacy measurement system dismantled today, the mass of the aluminum bars ranges between 121 and 151 kg). In order to simulate the sample transfer from REGAIN cell to the low-background spectrometer, the detection of delayed gamma rays emitted by the samples was calculated one minute after the end of the 2 h neutron irradiation. In order to simulate the low-background cell detection conditions, the fissile material samples and the 3 mm lead shield were placed 12.5 cm from the surface of the HPGe 33 % relative efficiency detector, and activation gamma rays from not fissionable materials were not calculated by using the NONFISSION=NONE option of ACT card. The time binning used for delayed gamma ray calculation corresponds to a whole 24 h measurement with one spectrum recorded every 15 min, as in the experiment.

For MCNP calculations, a F5:P point tally was placed on the surface of the detector, which allowed obtaining the full delayed gamma-ray spectra with the 33 % relative efficiency HPGe detector response function implemented in MODAR software [9]. The response function of the 25 % HPGe detector used in the low-background spectrometer has not been parametrized with MODAR. Therefore, to calculate the signal expected for a 25 % HPGe, the number of counts obtained with the MODAR 33% relative efficiency HPGe is corrected by the ratio of full peak efficiency of the 25 % HPGe over the full peak efficiency of the 33 % HPGe for all delayed gamma rays of interest. This ratio, reported in Tab. 2, has been obtained by simulating mono-energetic photon point sources placed at a 12.5 cm distance from the top of both detectors for each delayed gamma ray of interest.

γ ray energy	885 keV	1011 keV	1384 keV	1436 keV	1768 keV	2218 keV
Efficiency for a 25% HPGe	1.471×10^{-3}	1.334×10^{-3}	1.014×10^{-3}	0.982×10^{-3}	0.820×10^{-3}	6.582×10^{-4}
Efficiency for a 33% HPGe	1.875×10^{-3}	1.686×10^{-3}	1.317×10^{-3}	1.277×10^{-3}	1.054×10^{-3}	0.880×10^{-3}
Efficiency ratio (25% to 33%)	0.784	0.791	0.770	0.769	0.778	0.748

Table 2: *Ratios of the HPGe 25 % and HPGe 33 % detectors relative efficiencies calculated with MCNP*

Simulated and experimental spectra are shown in Fig. 5 and Fig. 6 for uranium and plutonium samples, respectively.

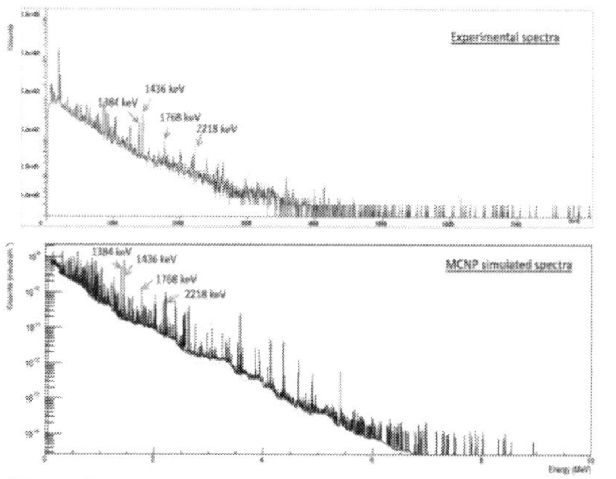

Figure 5: *Experimental (above) and simulated (below) delayed gamma spectra for the first 15 min following the end of the uranium sample irradiation in REGAIN cell*

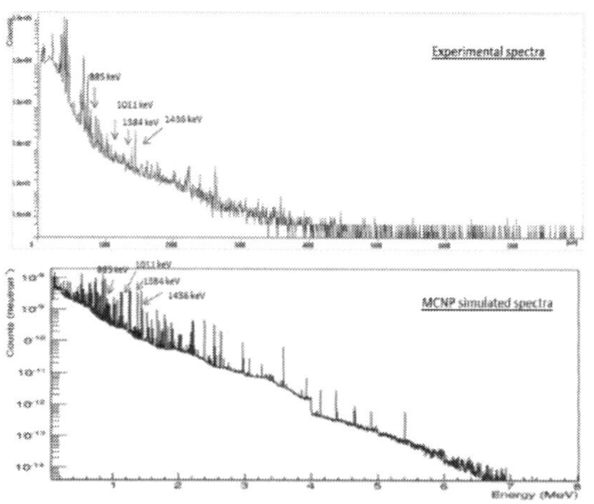

Figure 6: *Experimental (above) and simulated (below) delayed gamma spectra for the first 15 min following the end of the plutonium sample irradiation in REGAIN cell*

Although a non-physical continuum is observed in the decay curves of the main delayed gamma rays of interest 12 h after the end of neutron irradiation (see Fig. 7), the first hours seem to be properly calculated and the first 15 min have been used in order to estimate the simulated net counts. The non-physical continuum was subtracted, as illustrated in Fig. 7 for the delayed gamma ray at 1384 keV emitted by the uranium sample. Then the net area in the first 15 min after the end of irradiation was multiplied by the neutron generator emission (8.10^7 n.s^{-1}), by the 2 h irradiation time and multiplied by the corrections factors reported in Tab. 2 for each delayed gamma ray of interest.

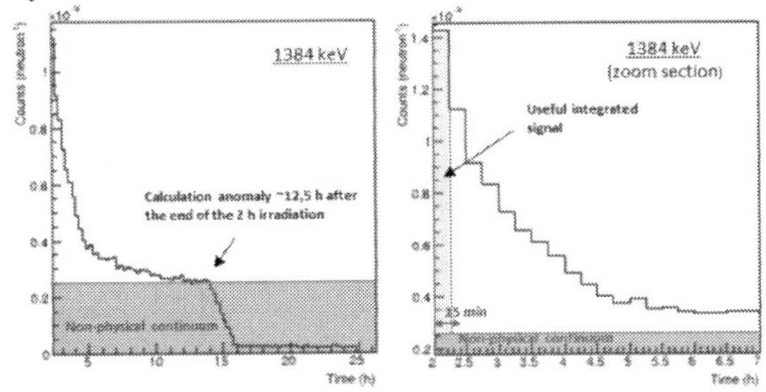

Figure 7: *Simulated number of counts (per source neutron) as a function of time after the end of the 2h irradiation, for the 1384 keV delayed gamma ray of the ^{235}U sample, showing also the principle of the non-physical continuum (in red) subtraction from the signal of the first 15 min (in green)*

Simulated and experimental net counts obtained for the main delayed gamma rays of ^{235}U and ^{239}Pu are shown in Tab.3.

γ ray energy	885 keV	1011 keV	1384 keV	1436 keV	1768 keV	2218 keV
Experimental net counts with U	790 ± 67	918 ± 98	612 ± 40	1211 ± 54	172 ± 25	102 ± 24
MCNP 6.1 net counts with U	648 ± 8	744 ± 23	597 ± 10	877 ± 12	183 ± 4	130 ± 3
Experimental net counts with Pu	1464 ± 106	760 ± 66	347 ± 36	1057 ± 48	102 ± 29	198 ± 37
MCNP 6.1 net counts with Pu	873 ± 6	552 ± 14	238 ± 3	567 ± 4	118 ± 2	85 ± 1

Table 3: *Simulated vs. experimental net counts of the main delayed gamma rays of interest from uranium and plutonium samples after a 15 min measurement following a 2 h irradiation in REGAIN neutron interrogation cell.*

Uncertainties for experimental counts were calculated as $\sqrt{S+2B}$, where B corresponds to the Compton background under the delayed gamma peak of interest and S to the net peak area. In the case of simulated results, for which about 5.10^9 source neutrons were computed, only the corresponding statistical uncertainties of around 1% are reported. However, as detailed in [5], the uranium and plutonium samples used for this study have important self-shielding factors of respectively 3.8 and 5.7, with a large uncertainty due to imprecise knowledge of the geometry of their fissile core. In addition, uncertainties related to delayed gamma ray nuclear data might also be significant. Taking into account these important sources of uncertainties, simulated and experimental net counts are quite consistent, thus validating the use of MCNP 6.1 Activation Control Card for predicting the order of magnitude of induced-fission delayed gamma ray signal.

The experiment vs. simulation comparison is also performed for the net counts of the 1778 keV activation gamma ray due to the aluminum frame bars of REGAIN cell. The experimental net counts measured during a 30 ms gate between the pulses of the neutron generator for a whole 2 h neutron irradiation as presented in Fig. 8 are evaluated to 153937 ± 443 counts.

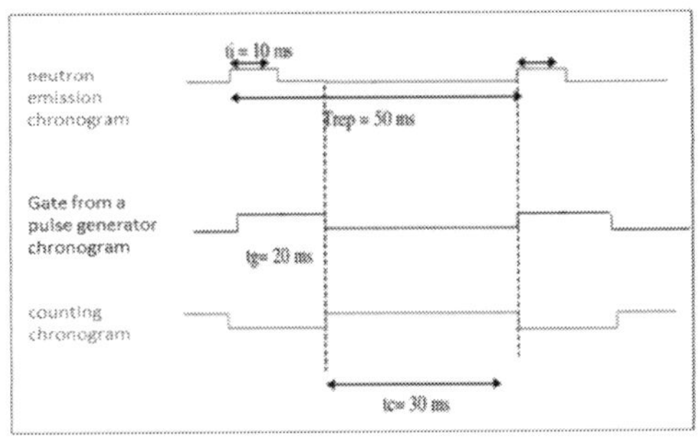

Figure 8: *counting chronogram for the PGNAA experiment for the 1778 keV total net counts evaluation*

The aluminum support MCNP REGAIN model is presented in Fig. 9.

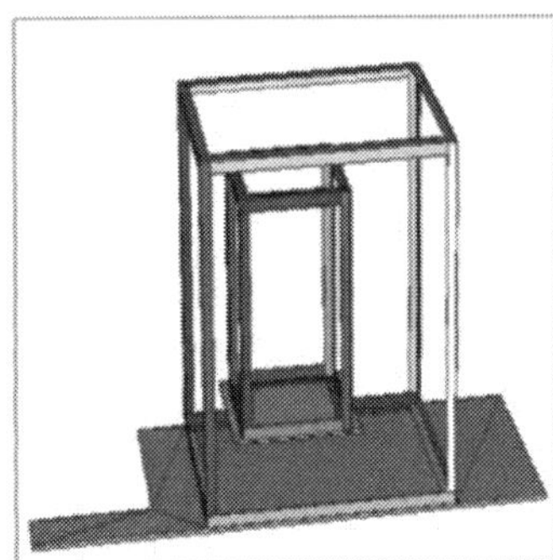

Figure 9: *REGAIN aluminum support MCNP model (121 kg mass)*

Contrary to delayed gamma rays from uranium and plutonium fission fragments, no anomaly is observed after the end of irradiation in the decay curve of the ^{28}Al ($T_{1/2}$ = 2.24 min) 1778 keV activation gamma ray (see Fig. 12 in section V). Furthermore, the decay curve seems also properly calculated for longer half-life activation products such as the 1345 keV activation gamma ray of ^{64}Cu ($T_{1/2}$ = 12.7 h), see Fig. 12.

The simulated net counts were calculated as the net area during the 2 h total neutron irradiation, multiplied by the 2 h irradiation time and the 8.10^7 $n.s^{-1}$ average neutron generator emission, and finally normalized by a 3/5 factor corresponding to the ratio of the 30 ms gate out of the 50 ms pulse repetition period (Fig. 8). Net simulated counts with this model are evaluated to 253038 ± 227 counts, which is twice the measured counts. Note that calculation uncertainty only corresponds to MCNP simulation statistical uncertainty, i.e. 0.09 % here.

The over-estimation by MCNP calculation has probably several origins. A possible cause could be the simple MNCP model of the irradiation casemate where REGAIN was located (for instance a paraffin door that was modeled as concrete), as well as non-modelled objects and laboratory equipment located near REGAIN cell, which may have an important impact on the thermal neutron flux around REGAIN. Indeed, the largest aluminum mass comes from the external frame bars as shown in Fig. 1 and 9, in which the $^{27}Al(n,\gamma)$ activation rate highly depends on the external thermal flux. It is also important to recall that the total aluminum mass is not known with a good precision (between 121 and 151 kg). On the other hand, the badly known bore concentration of REGAIN graphite blocks, which varies at least from 4 ppm to 16 ppm according analyses of a few samples taken in REGAIN, might also be a significant cause of calculation over-estimation. Indeed, modelling the whole REGAIN graphite blocks with 16 ppm bore concentration rather than 4 ppm decreases the simulated 1778 keV signal from 253038 ± 227 counts to 208457 ± 833 counts. As previously, only MCNP statistical fluctuations are reported in these results, which are of the order of 0.09 % and 0.4 % for each computation, respectively.

Taking into account these possible causes of biases in the calculation, the MCNP activation control card provides an order of magnitude of the activation signal of aluminum.

IV. Complementary ACT calculation tests

In order to better understand the anomaly observed 12 h after the end of irradiation in the delayed gamma rays decay curves (Fig. 7), we performed complementary calculations with a simpler geometry of nuclear and non-nuclear materials.

As calculations with ACT card demand an important computation time (several days with a cluster of few hundred cores for the study presented in former section), a simpler geometry corresponding to a spherical sample with a point source emitting 0.025 eV thermal neutrons in its center has been modeled, as presented in Fig. 10. All source neutrons are emitted at time zero.

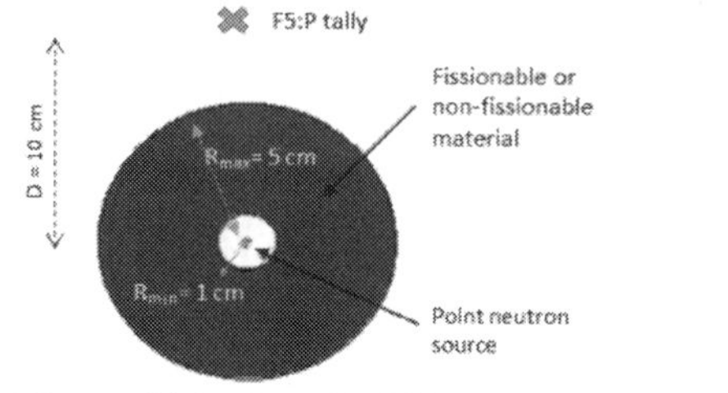

Figure 10: *MCNP model for the study of delayed gamma rays and activation gamma rays created with ACT card*

The first step of this study concerns the main delayed gamma rays of interest induced in a ^{235}U or a ^{239}Pu sphere. As for the previous study of section III with a 2 h neutron irradiation in REGAIN cell, the calculated decay curve for a theoretical neutron irradiation peak at time zero presents several anomalies, see Fig. 11. The most visible anomaly corresponds to a spectra discontinuity about 10.5 h after the irradiation peak, which might correspond to the 2 h long non-physical decay observed in Fig. 7 between 12 h and 14 h approximately, after the 2 h irradiation. Fig. 11 shows the decay of the 1384 keV and 1436 keV delayed gamma rays of ^{235}U and ^{239}Pu, which are also present in the curves of the other delayed gamma rays of interest emitted by uranium and plutonium fission fragments.

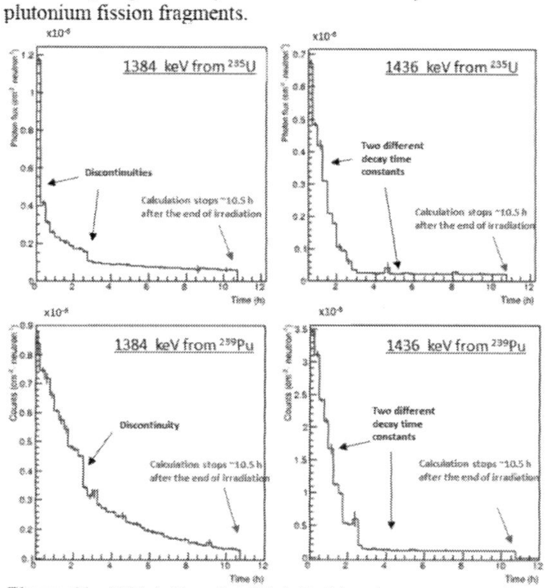

Figure 11: *1384 keV and 1436 keV delayed gamma rays from uranium (above) and plutonium (below) time decay curves*

The same study with the spherical geometry of Fig. 10 has been done with aluminum and copper. As for the 2 h neutron irradiation in REGAIN, no calculation discontinuities are observed here in the 1778 keV decay curve with a neutron emission peak at time zero, see Fig. 12 that also shows a normal decay for the 1345 keV activation gamma ray of copper. The decay curves of the 1778 keV activation gamma

https://doi.org/10.1051/epjconf/202022506007

ray from ^{28}Al and 1345 keV activation gamma ray from ^{64}Cu are presented in Fig. 12.

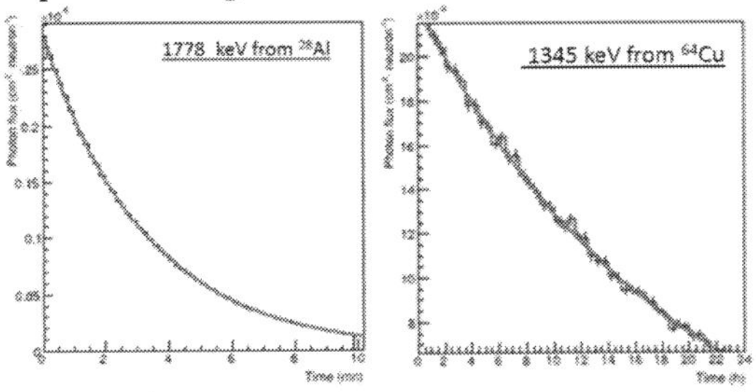

Figure 12: *1778 keV (left) and 1345 keV (right) activation gamma rays from aluminum and copper, respectively.*

The exponential fits of the 1778 keV and 1345 keV decay curves gives ^{28}Al and ^{64}Cu half-lifes consistent with theoretical data from LARA nuclear database : $T_{1/2}^{sim} = 2.24 \pm 0.006$ min vs. $T_{1/2}^{th} = 2.2414(12)$ min for aluminum, and $T_{1/2}^{sim} = 12.74 \pm 0.03$ h vs. $T_{1/2}^{th} = 12.700(2)$ h for copper.

V. Conclusion

The ACT card allowing evolution transport calculation in MCNP 6.1 was used to model neutron-induced fission delayed gamma rays, as well as activation gamma rays from non-fissionable materials. Although the decay curves of delayed gamma rays from nuclear materials show non-physical discontinuities, the simulated net counts are consistent with experiment for uranium and plutonium samples irradiated in REGAIN graphite cell. On the other hand, the decay curves of activation gamma rays from non-fissionable materials such as aluminum and copper show a more physical exponential shape with half-lifes consistent with ^{28}Al and ^{64}Cu ones.

Quantitatively, the number of counts predicted with MCNP 6 ACT card is consistent with experimental data, taking into account significant modeling uncertainties. Therefore, the calculation scheme presented in section III can be used to perform feasibility studies related to the measurement of delayed gamma rays from neutron induced fission in the fields of radioactive waste drum characterization as presented in [10], or for homeland security with the detection of special nuclear materials. The use of MCNP ACT card is also considered to model systems based on the detection of delayed gamma rays induced by photo-fission for radioactive waste drum characterization as presented in [11] [12].

References

[1] M. Agelou *et al.*, « Detecting special nuclear materials inside cargo containers using photofission », in *Nuclear Science Symposium Conference Record (NSS/MIC), 2009 IEEE*, 2009, p. 936–939.

[2] B. W. Blackburn *et al.*, « Utilization of actively-induced, prompt radiation emission for nonproliferation applications », *Nucl. Instrum. Methods Phys. Res. Sect. B Beam Interact. Mater. At.*, vol. 261, n° 1-2, p. 341-346, août 2007.

[3] J. Stevenson, T. Gozani, M. Elsalim, C. Condron, et C. Brown, « Linac based photofission inspection system employing novel detection concepts », *Nucl. Instrum. Methods Phys. Res. Sect. Accel. Spectrometers Detect. Assoc. Equip.*, vol. 652, n° 1, p. 124-128, oct. 2011.

[4] M. Gmar *et al.*, « Detection of nuclear material by photon activation inside cargo containers », présenté à Defense and Security Symposium, Orlando (Kissimmee), FL, 2006, p. 62130F.

[5] T. Nicol *et al.*, « Feasibility study of 235 U and 239 Pu characterization in radioactive waste drums using neutron-induced fission delayed gamma rays », *Nucl. Instrum. Methods Phys. Res. Sect. Accel. Spectrometers Detect. Assoc. Equip.*, vol. 832, p. 85-94, oct. 2016.

[6] J. W. Durkee, « MCNP6 DELAYED-PARTICLE VERIFICATION AND VALIDATION Rev 5: January 26, 2012 », p. 54.

[7] « NUCLEIDE-LARA on the web (2018) ». [online]. available on : http://www.nucleide.org/Laraweb/index.php.

[8] J.L Ma, C. Carasco, B. Perot, E. Mauerhofer, J. Kettler, et A. Havenith, « Prompt gamma neutron activation analysis of toxic elements in radioactive waste packages », *Appl Radiat Isot 70 2012 1216-1263*.

[9] C. Carasco, « MCNP output data analysis with ROOT (MODAR) », *Comput. Phys. Commun.*, vol. 181, n° 12, p. 2210-2211, déc. 2010.

[10] R. De Stefano, B. Perot, C. Carasco, E. Simon, T. Nicol, et E. Mauerhofer, « Characterization of 870 L radioactive waste drums using delayed gamma rays from neutron-induced fissions », vol. presented in the Seminar of Activation Analysis and Gamma Spectrometry SAAGAS27, submitted to the Journal of Radioanalytical and Nuclear Chemistry.

[11] E. Simon, F. Jallu, B. Pérot, et S. Plumeri, « Feasibility study of fissile mass quantification by photofission delayed gamma rays in radioactive waste packages using MCNPX », *Nucl. Instrum. Methods Phys. Res. Sect. Accel. Spectrometers Detect. Assoc. Equip.*, vol. 840, p. 28-35, déc. 2016.

[12] E. Simon, F. Jallu, B. Perot, et S. Plumeri, « Fissile mass quantification in radioactive waste packages using photofission delayed gamma rays », in *2016 IEEE Nuclear Science Symposium, Medical Imaging Conference and Room-Temperature Semiconductor Detector Workshop (NSS/MIC/RTSD)*, Strasbourg, France, 2016, p. 1-4.

EPJ Web of Conferences **225**, 06008 (2020) https://doi.org/10.1051/epjconf/202022506008
ANIMMA 2019

Non-destructive assay of nuclear waste containers using muon scattering tomography in the Horizon2020 CHANCE project

Anna Kopp[1*], Ahmad Alrheli[2], Daniel Kikoła[3], Mohammed Mhaidra[13], Patrick Stowell[2], Holger Tietze-Jaensch[4], Lee Thompson[2], Jaap Velthuis[1], Michael Weekes[2]

[1] School of Physics, University of Bristol, UK
[2] Department of Physics and Astronomy, University of Sheffield, UK
[3] Warsaw University of Technology, Poland
[4] Forschungszentrum Jülich, Germany
* anna.kopp@bristol.ac.uk

***Abstract*—Methods for the non-destructive assay of nuclear waste drums are of great importance to the nuclear waste management community, especially where loss in continuity of knowledge about the content of drums happened or chemical processes altering the contents of the drums may occur. Muon scattering tomography has been shown to be a promising technique for the non-destructive assay of nuclear waste drums in a safe way. By measuring tracks of muons entering and leaving the probed sample and extracting scattering angles from the tracks, it is possible to draw conclusions about the contents of the sample and its spatial arrangement. Within the CHANCE project, a newly built large-scale mobile detector system for scanning and imaging the contents of nuclear waste drums using atmospheric muons is currently undergoing commissioning.**

I. Introduction

Non-destructive methods to assay nuclear waste drums are of great interest to the nuclear waste management community. It has been observed that chemical processes like oxidation of metals may occur within drums, possibly leading to the formation of gas bubbles or cracks. Furthermore, knowledge about the contents of legacy waste drums is not always preserved. Muon scattering tomography (MST) is a promising technique to address these problems. It allows to scan and image nuclear waste drums in a safe, non-destructive way using natural background radiation. Compared to other methods like X-ray or gamma-ray scanning it does not introduce additional artificial radiation or any additional hazards to personnel or equipment. Fitting the tracks of muons entering and leaving the probed sample allows to reconstruct approximated scattering vertices and to e.g. differentiate between various materials.

II. Muon Scattering Tomography

Muon scattering tomography uses secondary cosmic radiation to probe volumes from a safe distance. Compared to other scanning methods using e.g. X-rays or gamma rays it does not rely on a radiation source but uses atmospheric muons, particles resulting from primary cosmic radiation. These are ubiquitous and abundant at a rate of about $10\,000/(\text{m}^2\text{minute})$ at sea level, spread over a wide range of momenta and incidence angles. Muons are highly penetrating particles; it is almost impossible to stop them and they are thus ideally suited for scanning nuclear waste drums, where the nuclear waste is often embedded in concrete.

As charged particles, they undergo multiple Coulomb scattering processes when traversing matter. The projected scattering angle distribution depends on the atomic number Z of the traversed material and can be approximated as a Gaussian distribution [1] with mean zero and a standard deviation σ_μ of

$$\sigma_\mu \approx \frac{13.6\,\text{MeV}}{pc\beta}\sqrt{X/X_0}(1 + 0.038\ln(X/X_0)), \quad (1)$$

where p is the muon's momentum, βc its velocity, X the thickness of the scattering material and X_0 the material-specific radiation [2] length given by

$$X_0 \approx \frac{A \cdot 716.4\,\text{g/cm}^2}{Z(Z+1)\ln(287/\sqrt{Z})}. \quad (2)$$

Here, A is the atomic weight given in g/mol. As can be seen from equations 1 and 2, the width of the projected scattering angle distribution varies approimately with Z, making the technique particularly sensitive to materials with high atomic numbers.

Thus scattering angles from muons scattering off materials with large atomic numbers Z are more likely to be large than those from scatters in low-Z materials.

By placing multiple detector planes above and below the probed sample and measuring where the muons hit these, the trajectories of the incoming and outgoing muons are reconstructed and fitted. A scattering vertex is then reconstructed where these two trajectories meet. The assumption here that for each muon all scattering processes happen in the same location, the vertex, is not strictly correct but has been shown to be a good approximation. Scattering angles for all muons are derived from their fitted tracks.

Then, the volume under investigation is divided into voxels as described in ref. [4]. In each voxel with at least N

 https://doi.org/10.1051/epjconf/202022506008

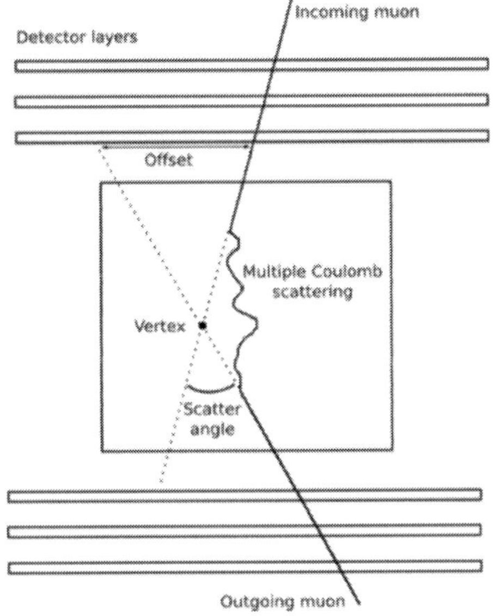

Fig. 1. Illustration of the muon trajectory and the reconstructed scattering vertex [3].

reconstructed vertices, a metric discriminator m_{ij} is calculated for each pair of tracks i and j of the N most scattered tracks,

$$m_{ij} = \frac{||\vec{v}_i - \vec{v}_j||}{(\theta_i p_i)(\theta_j p_j)}. \quad (3)$$

Here, $\vec{v}_i$ is the location of the vertex pertaining to muon track i with momentum p_i and θ_i is the associated scattering angle. The median of this weighted metric distribution per voxel is an indicator of the predominant material in that voxel: large values are more likely to result from materials with high Z than from materials with low Z.

In the past, this has been exploited for various discrimination and identification studies in simulation: to differentiate between different high-Z materials [3], [5] or to locate gas bubbles in bitumen-filled drums [6].

III. The CHANCE Project

The CHANCE project ("Characterization of conditioned nuclear waste for its safe disposal in Europe", http://www.chance-h2020.eu), is funded by the EU's Horizon2020 research programme. It started on June 1st 2017, runs over 4 years and has 11 partners in 7 countries. Collaborators include universities and research institutes as well as government agencies and industrial partners. Within the project, three different techniques to assay nuclear waste drums in non-destructive ways are investigated: calorimetry, muon scattering tomography and cavity ring-down spectroscopy.

IV. Muon Scattering Tomography within the CHANCE Project

Three universities mainly contribute to the MST program within CHANCE: University of Sheffield, University of Bristol, both in the UK and Warsaw University of Technology, Poland. A large, mobile muon scattering tomography detector shown in figure 2 has been recently built in a non-laboratory environment outside of Bristol, UK. It is placed in a building, so it is rain-protected but the building is neither temperature- nor humidity-controlled. Thus the environmental conditions are the same as they are likely to be in an actual field deployment. The active area measures approximately $1.8 \times 1.8\,\mathrm{m}^2$. As the detector system is modular, it is easy to dis- and reassemble and even to change the configuration to adapt it to the requirements in a potential deployment.

The detector system combines two different kinds of gas detectors, namely drift chambers and resistive plate chambers (RPCs). A coincidence between two layers of plastic scintillators is used as trigger to start the read-out process.

The RPCs were designed and built at Univeristy of Bristol. A smaller prototype of these chambers has been operated on a mixture of freon (R134-a) and isobutane in the past. Due to recent legislation restricting the usage of freon, the chambers are currently run on CO_2. Other, environmentally friendly gas mixtures will be tested in the future. The drift chambers are operated on a mix of methane, argon and CO_2. They were built at University of Sheffield.

During the construction of the detector system the data analysis effort had been focussed on simulation studies. The detector layout was optimized and a figure of merit was developed for comparing the smallest observable separations between features or the smallest observable features in concrete-filled waste drums using different reconstruction algorithms [7]. The latter is especially relevant considering the heterogeneity of real nuclear waste drums. Commissioning of the full detector system has started in early June and is ongoing, with first test data having been taken. Full data taking is expected to start soon while further simulation studies are continuing in parallel.

The Detector System

Both RPCs and drift chambers provide information in 3D space about where a muon hit the detector. As the muon traverses a detector chamber, the gas inside it gets ionized. With a high voltage applied between the top and bottom side for the RPCs or the anode wire and cathode for the drift chambers, respectively, this creates a signal, which is read out and digitized. The spatial resolution of the RPCs is in the sub-millimeter range [8], while that of the drift chambers is in the order of mm. The time resolution for RPCs with a 2 mm gap as the ones used here is on the order of nanoseconds.

In both the detector stack above and below the sample space, two space points from the RPCs and three from the drift chambers are measured and read out. Hence it is possible to fit tracks and extract scattering angles with high precision. The detectors are fast, with data acquisition per event taking

Fig. 2. The muon scattering tomography detector system of the CHANCE project. Shown are the detector systems above and below the empty sample space consisting of three layers of drift chambers and two RPC layers. The top stack additionally contains two layers of plastic scintillators used as triggers. The detectors cover an area of approximately $1.8 \times 1.8\,\mathrm{m}^2$.

on the order of a few 10μs and have a low cost per unit area, making them ideal for large-scale detectors. For both the RPCs and drift chambers, three individual detectors with widths of 58 and 60 cm, respectively, are placed next to each other to cover the whole active area. The support structure holding the individual chambers can be seen as empty space in the data as shown for one layer of RPCs in figure 3. Since the footprint of the active area is larger than the drums that will be scanned, it is nonetheless possible to scan whole drums and avoid dead areas by moving the drum to different locations within the sample space.

V. Conclusions and Outlook

Muon scattering tomography is a non-destructive technology well suited to investigate the contents of nuclear waste drums in a safe way, without introducing any additional radiation to the samples or personnel. A large, mobile MST detector has been built in a non-laboratory environment close to Bristol, UK, within the CHANCE project and first test data have been taken. Once the detector system is fully commissioned, full data taking will commence.

As the scattering behavior of atmospheric muons in many materials is well known from simulation studies, data with blocks of high-Z materials like lead or tungsten will be taken in a first step. Subsequently, 'blind tests' with drums mimicking nuclear waste drums will be carried out. Our collaboration partners at SCK·CEN, Belgium, filled drums

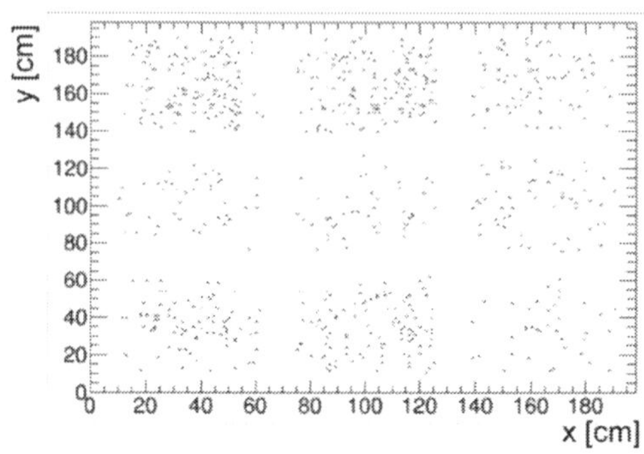

Fig. 3. Hits in one layer of RPCs. The pattern is due to the support structure holding the individual detector chambers. Each entry corresponds to one hit, showing the x and y coordinates.

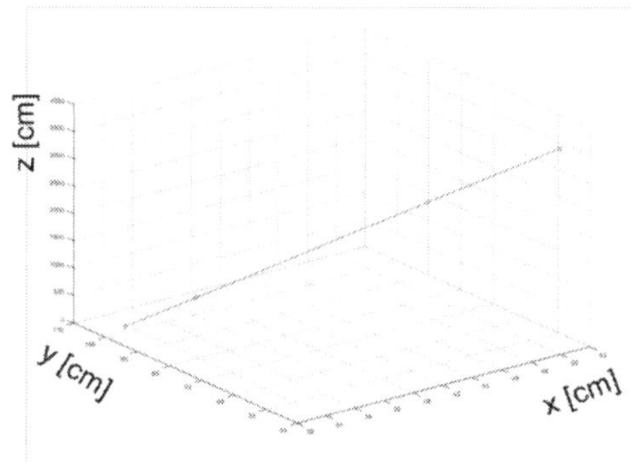

Fig. 4. Muon track fitted through all four layers of RPCs. No sample is placed between the upper and lower detectors so no scattering is expected or seen.

with non-radioactive materials otherwise similar to what could be found in nuclear waste drums but did not reveal the contents. Scanning the drums and applying the reconstruction algorithm described above will then show their contents. The access and application to real waste drums is currently being investigated.

Acknowledgment

This project has received funding from the European Union's Horizon2020 research and innovation programme under grant agreement number 755371.

References

[1] M. Tanabashi et al., "Review of particle physics," *Phys. Rev. D*, vol. 98, p. 030001, Aug 2018. [Online]. Available: https://link.aps.org/doi/10.1103/PhysRevD.98.030001

[2] T. A. Lasinsky et al., "Review of particle properties particle data group," *Rev. Mod. Phys.*, vol. 45, pp. S1–S175, Apr 1973. [Online]. Available: https://link.aps.org/doi/10.1103/RevModPhys.45.S1

https://doi.org/10.1051/epjconf/202022506008

[3] L. Frazão, J. Velthuis, C. Thomay, and C. Steer, "Discrimination of high-z materials in concrete-filled containers using muon scattering tomography," *Journal of Instrumentation*, vol. 11, no. 07, pp. P07020–P07020, Jul 2016. [Online]. Available: https://doi.org/10.1088/1748-0221/11/07/p07020

[4] C. Thomay, J. J. Velthuis, P. Baesso, D. Cussans, P. A. W. Morris, C. Steer, J. Burns, S. Quillin, and M. Stapleton, "A binned clustering algorithm to detect high-z material using cosmic muons," *Journal of Instrumentation*, vol. 8, no. 10, pp. P10013–P10013, Oct 2013. [Online]. Available: https://doi.org/10.1088/1748-0221/8/10/P10013

[5] C. Thomay, J. Velthuis, T. Poffley, P. Baesso, D. Cussans, and L. Frazão, "Passive 3d imaging of nuclear waste containers with muon scattering tomography," *Journal of Instrumentation*, vol. 11, no. 03, pp. P03008–P03008, Mar 2016. [Online]. Available: https://doi.org/10.1088/1748-0221/11/03/p03008

[6] M. Dobrowolska, J. Velthuis, L. Frazão, and D. Kikoła, "A novel technique for finding gas bubbles in the nuclear waste containers using muon scattering tomography," *Journal of Instrumentation*, vol. 13, no. 05, pp. P05015–P05015, May 2018. [Online]. Available: https://doi.org/10.1088/1748-0221/13/05/p05015

[7] J. Stowell, J. Velthuis, H. Tietze-Jaensch, A. Kopp, A. Alrheli, M. Mhaidra, L. Thompson, and D. Kikoła, "Figures of merit for the application of muon tomography to the characterization of nuclear waste drums," *Waste Management Symposia 2019*, no. 19253, March 2019.

[8] P. Baesso, D. Cussans, C. Thomay, and J. Velthuis, "Toward a RPC-based muon tomography system for cargo containers," *Journal of Instrumentation*, vol. 9, no. 10, pp. C10041–C10041, Oct 2014. [Online]. Available: https://doi.org/10.1088/1748-0221/9/10/c10041

Fast neutron and γ-ray backscatter radiography for the characterization of corrosion-born defects in oil pipelines

Mauro Licata*, Helen M. O. Parker, Michael D. Aspinall, Manuel Bandala, Frank Cave, Sebastian Conway, Domas Gerta and Malcolm J. Joyce
* Lancaster University
m.licata@lancaster.ac.uk

***Abstract.* — In this research, a combined fast-neutron/γ-ray backscatter imaging technique is described. The aim of this work is to understand corrosion defects in pipelines by measuring differences in the scattered radiation flux, generated when different steel thicknesses are irradiated by a neutron and γ-ray focused beam. A californium-252 radiation source is used to produce fast neutron and γ rays, exploiting its spontaneous fission. This mixed radiation field is collimated and directed towards the steel samples. Backscattered neutrons and γ rays are measured as a function of the steel thickness using 4 liquid organic scintillation detectors linked to a real-time, pulse-shape discrimination system, which separates and retains the neutron and γ-ray event data. In this paper, we describe how, using a single radiation source and detection system, it is possible to perform and combine two complementary imaging modalities. This research is validated by an MCNP6 computer simulation study. The backscatter imaging system developed for this research and the experimental results of the measurements carried out using the National Physical Laboratory neutron low-scatter facility are also presented in this paper.**

***Keywords*— Fast Neutron tomography, combined neutron-γ imaging, backscatter, scintillation detectors, corrosion, pipelines.**

I. Introduction

BACKSCATTER radiography and backscatter computer tomography (BCT) are non-destructive testing (NDT) techniques that exploit the scattered radiation in contrast to that which is transmitted. The latter is used in the more common and well-known computerized axial tomography (CT). X-ray backscatter imaging has been studied for the last three decades and utilizes the scattered radiation arising by virtue of the Compton Effect, the extent of which depends on the electron density in the scatterer rendering this technique particularly effective for heavy metals. The common procedure for BCT is to probe the object under scrutiny with a collimated pencil-beam and to detect the back-scattered radiation. The main advantage of this particular NDT technique is the investigation of large objects that cannot be scanned with the more usual axial CT approach and furthermore, this technique helps to interpret misleading transmission images that are more difficult to comprehend. X-ray BCT has applications in non-destructive testing particularly for quality control and for security inspection, such as the detection of dangerous materials, border inspections, biomedical science, several engineering fields, above all, oil and gas sector and aerospace. Conversely, there are few if any papers in the literature reporting the use of fast neutrons in place of X-ray as the probing radiation for backscatter radiography. Fast neutrons interact differently compared to X- and γ rays and, in particular, elastic nuclear scattering of neutrons and the corresponding cross section can be exploited, as neutrons have high penetration capabilities, which allow high-density materials to be investigated.

Elastic neutron scattering is a phenomenon exploited for measurements of hydrogen (usually as part of water) content in rocks as well as density and porosity measurements for oil exploration, and for the measurements of the level/presence of water or gas in pipelines. In this research, a simultaneous, fast neutron and γ-ray backscattered radiography system will be described. Both an extensive Monte Carlo simulation study, computed with the tool MCNP6 [**1**], and experimental results will be presented. In this particular MCNP6 study, the entire device has been simulated. Such a system will be applied to investigate and characterize defects and possibly corrosion in the walls of oil pipelines, both for the case of no insulation and under insulation (e.g., concrete or polyethylene etc.). Pipelines are subject to several types of corrosion, for example as a consequence of the action of chemical agents, high pressure gases and vapors, crude oil, hydrogen sulfide, and especially water and carbon dioxide. The quantification of corrosion-born defects is, therefore, a key issue in the industry, particularly for all the companies that rely on thousands of kilometres of pipes for transporting their feedstock. Nowadays, ultrasound is the most widespread technology used in pipeline inspections, via inspection gauges deployed inside pipes.

This work was supported by Innovate UK and coordinated by Hybrid Instruments Ltd.. M. Licata, H. Parker, M. Aspinall, M. Bandala and Malcolm J. Joyce are with the Engineering Department of Lancaster University. F. Cave, S. Conway and D. Gerta are with Hybrid Instruments Ltd. Corresponding author contact: m.joyce@lancaster.ac.uk

However, despite the advantage of having high resolution and the possibility of performing measurements quickly, it has the disadvantages that it requires a coupling medium and that it only allows measurements of internal corrosion and not corrosion under insulation. The technique we present is, to some extent, complementary to ultrasound since it does not need a coupling medium; rather it probes the pipes from the outside *and* has the potential to measure corrosion under insulation.

II. Experimental Methods

Different steel thicknesses have been reproduced and covered with 1-cm layer of two different types of insulation: concrete and polyethylene. The neutron and γ-radiation field produced by a californium-252 (^{252}Cf, emission rate: 8.727×10^{6} neutrons/second) source has been collimated by means of different layers of lead and high-density polyethylene drilled with 1-cm diameter pinhole in such a way as to produce a pencil-like beam to be focused on the steel surface, as shown in fig. 1a and b.

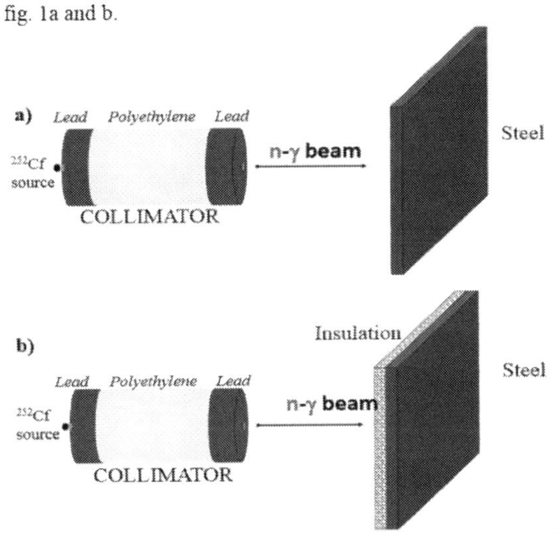

Fig. 1. Illustration of the collimator and the steel slab with (a) and without (b) the presence of insulation.

The presence of defects in the steel, or the manifestation of corrosion and rust, produces a variation of the mean density and a variation of the steel thickness itself, therefore changing the atomic number density, on which elastic scattering by neutrons is heavily dependent. These changes produce a different transmitted-reflected neutrons flux ratio. Scattered fast neutron and γ-ray fluxes have been measured with an array of EJ-301 (Scionix, Netherlands) scintillation detectors, as a function of the different steel thicknesses. The whole system developed, shown schematically in Fig. 2, comprises a mechanical rig with the aforementioned collimator and detectors. The rig is an aluminum frame of dimensions 600 × 400 × 340 mm, mounted with two stepper motors configured to move the sub-system comprising the source-collimator-detectors in the X-Y plane to allow the scan of different thicknesses of material. These motors are controlled and driven by an Arduino® board that receives coordinates from the user.

The optimum collimator and detector position were studied extensively via the Monte Carlo tool MCNP6. The geometry and scintillator location were selected in such a way as to have similar collimation factor of fast neutrons and γ rays, to maximize the flux on the steel surface and to isolate the detector radiation-sensitive volume from the beam. With this particular configuration, the scintillators are capable of measuring only the backscattered radiation flux generated by the beam focus point. The detectors are placed in a cylindrical assembly using adjustable arms. Such a configuration can accommodate up to 8 detectors, however, in the measurements described in this conference, only 4 detectors were used, since this arrangement maximizes the total geometrical efficiency. The scintillators were connected to a 4-channel mixed radiation field analyzer [**2**] (MFA, Hybrid Instruments Ltd., UK) capable of providing real-time, pulse-shape discrimination via pulse gradient analysis [**3**] (PGA), thus retaining separate datasets of both neutrons and γ rays. The MFA can generate two transistor-transistor-logic (TTL) signals per channel, depending on whether a neutron or a photon has been detected. TTL signals are redirected to an Intel Cyclone V FPGA / ARM Processor system, in which a 16-channel counter has been configured. A program written in C# for a PC coupled with a Graphical User Interface (GUI) is capable of monitoring the counters, coordinate and synchronize the collimator-detector position with the respective measured counting rate.

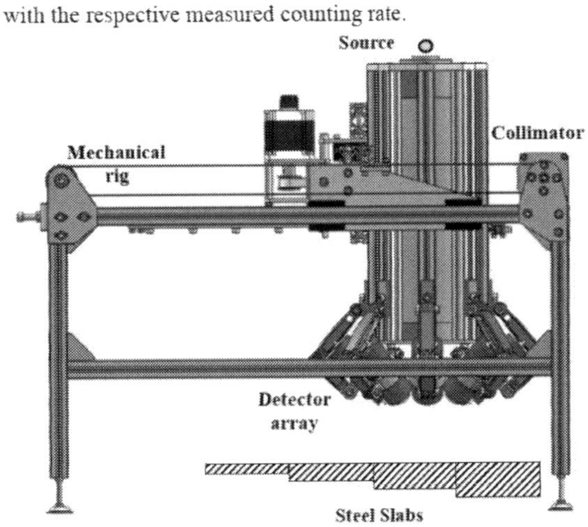

Fig. 2. Schematic, elevation (X-Z plane) of the mechanical rig and the system source-collimator-detectors.

Experiments were carried out at the National Physical Laboratory low-scatter neutron metrology facility (London, UK), in order to minimize the background level. Moreover, a wide range of bare neutron sources of different activities is available at this facility.

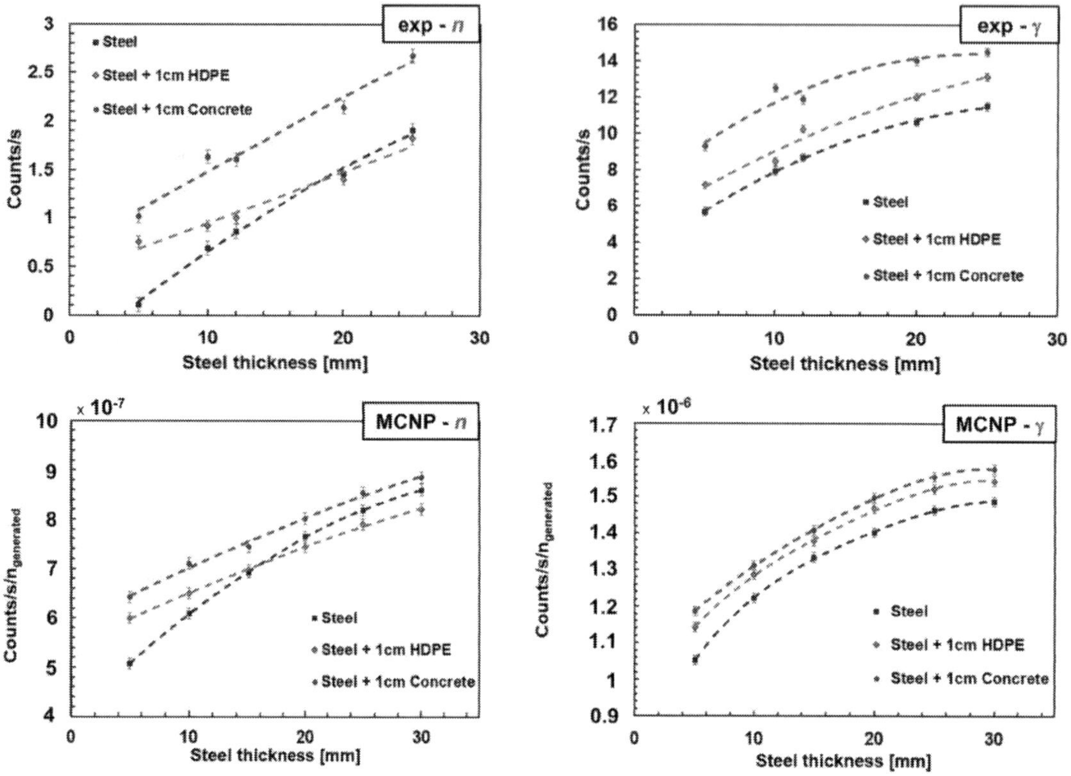

Fig. 3. Top: experimental results of the neutron (left) and γ-ray (right) backscattered flux for bare steel, steel and polyethylene insulation, and steel and concrete insulation. Bottom: MCNP6 Monte Carlo simulation results of backscatter flux, as a function of the steel thickness, for bare steel, steel and polyethylene insulation, and steel and concrete insulation [**5**].

III. Results

Reflected neutrons and γ rays were detected over a period of 20 minutes for every single slab. The measurement was carried out three times respectively with: bare steel, steel and polyethylene, steel and concrete. The experimental results and the MCNP6 Monte Carlo simulations are shown in Fig. 3. Results are presented with an uncertain of $\pm 1\sigma$ standard deviation and a second-order polynomial function was chosen to fit the data. Qualitative agreement between experimental results and computer simulations is observed and different steel thicknesses can be clearly discriminated with each other in the range 5 - 25 mm. As expected, the backscattered flux increases as the steel thickness increases, for both neutrons and γ rays, reaching a saturation plateau for the latter, after 20-25 mm. The results also show a difference in the backscatter counting rate as well as a difference in slope for the case of 1-cm layer of polyethylene and concrete. These measurements can be taken as a basis to develop an imaging technique, for instance, radiography or tomography, since the measured backscattered counting rate can be converted, as per analogy with the common transmission imaging, in an image contrast index. With this technique, both neutrons and γ rays can be exploited simultaneously and merged with each other, as done in previous research [**4,5**] for transmission tomography.

IV. Conclusion

In this research, neutron and γ-ray backscatter are exploited to discern different steel thicknesses. The potential of combining different imaging modalities is also described, providing a valid alternative to the well-known technique of X-ray transmission imaging as well as to the ultrasound technology techniques. Furthermore, an advantage of the system described in this research, is the possibility of performing two different imaging modalities using the same source and detection system. The methods presented in this work have the potential to be applied in several fields. In particular, the use of a radiation source (neutron generators can also be used) opens the system for potential *in-situ* NDTs and applications, for instance, in safety assessments, quality controls and security inspections.

ACKNOWLEDGMENTS

The authors thank the National Physical Laboratory for the use of the low scatter facility, in particular N. Roberts and G. Taylor for the help during the experimental campaign and with the use of the radiation sources. The authors wish to acknowledge: Lancaster University, the Engineering and Physical Sciences Research Council (EPRSC), the Next Generation Nuclear Centre for Doctoral Training (NGN-CDT), the Lloyd's Register Foundation and Innovate UK, for supporting this project. M. J. J. acknowledges the Royal Society for the award of a Wolfson Research Merit Award.

REFERENCES

[1] Los Alamos National Laboratory. Los Alamos, NM, USA. https://mcnp.lanl.gov/
[2] Joyce, M. J. et al. *The design build and test of a digital analyzer for mixed radiation fields*. IEEE. Trans. Nucl. Sci. 57, 2625–2630 (2010).
[3] D'Mellow, B. et al. *Digital discrimination of neutrons and γ-rays in liquid scintillators using pulse gradient analysis*. Nuclear Instruments and Methods in Physics Research. Section A. Vol. 578, no. 1, pp. 191-197.
[4] Licata, M. and Joyce M. *Concealed nuclear material identification via combined fast-neutron/γ-ray computed tomography (FNGCT): a Monte Carlo study*. Journal of Instrumentation, Vol. 13, Feb. 2018.
[5] Mauro Licata, Helen M. O. Parker, Michael D. Aspinall, Manuel Bandala, Frank Cave, Sebastian Conway, Domas Gerta and Malcolm J. Joyce. *Depicting corrosion-born defects in pipelines with combined neutron / γray backscatter: a biomimetic approach*. Under review, Scientific Reports, March 2019.

Improving Activity Estimation in Passive Gamma Scanning for Radioactive Waste Drums

E. Vax[1]*, E. Marcus[1], T. Mazor[1], Y. Kadmon[1], A. Osovizky[1,2]
[1] NRCN, Israel
[2] Rotem Ind., Israel
* vax.eran@gmail.com

***Abstract*— A method to improve radioactive waste drum activity estimation in Segmented Gamma Scanning (SGS) systems was developed for homogenous content. We describe a method to quantify the activity of spatially distributed gamma-emitting isotopes ('hot spots') in homogenous content waste drums without the use of a collimator. Instead of averaging all the detector's readings we treat it as many different spatial samples as if we have multiple detectors surrounding the waste drum ("virtual detectors"). From these readings, we form a general linear model. Next, we derive the Maximum Likelihood Estimator (MLE) for the multiple sources position and activity. We solve this hyper-dimensional search problem using an Alternating Projections (AP) technique which transforms the problem into a simpler one-dimensional maximization problem. We tested this method using a mathematical simulation with a various number of sources, at random activities and positions for several energy bands. The preliminary results are consistent and show large improvement of the accuracy with comparison to industrial SGS systems and the same accuracy as new methods which exploits the spatial samples. Furthermore, since this method eliminates the need for heavy led collimator, none of the sources is blocked for the whole measurement period, which provides increased count rates and decreases the total measurement time.**

***Keywords*—Radioactive waste drum, Alternating Projections, Projection algorithms, Maximum likelihood estimation, Passive scanning.**

I. Introduction

GHIS Gamma-ray spectrometry uses the fact that each isotope emission gamma-rays in proportion to the mass of that isotope in the sample. However, not all emitted radiation will be detected by the detector. This is caused due to absorption and dispersion of the photons in the materials in its path to the detector, and in the source itself (self-absorption). Another effect is the energy efficiency of the detector. In Classical SGS some of these effects may be corrected with efficiency and transmission calibrations. In order to estimate the matrix density, many systems use a preliminary step of using an RTR system (real-time radiography) with X-ray, known external gamma or neutron source.

A standard SGS procedure for more accurate materials estimation is Isotope Specific (Mass) Calibration, which is the original SGS procedure, uses standards of known radionuclide masses to determine detector response in a mass versus corrected count rate calibration and [2] Efficiency Curve Calibration, typically uses non-SNM radionuclide sources to determine system detection efficiency vs. gamma energy. In this paper, the activity estimated is relative to the radiation flux reaching the detector and do not take into account the detector efficiency nor translate the counts to the mass of the radiating material.

Assaying waste drums that have a heterogeneous matrix and that contain a non-uniform radionuclide distribution is usually done with Tomographic Gamma Scanning (TGS) rather than SGS systems. However, when the matrix is homogenous with known density SGS can be adequate, especially after calibration.

In the classic Segmented Gamma Scanning (SGS) method the 208-liter drum is rotated in order to smooth spatial distributions of the activity and the matrix. The SGS method assumes a uniform activity distribution for each segment visible to the detector through a slit collimator. However, the waste activity distribution in the drum is concentrated at 'hot-spots'. The different 'hot spots' distances from the center of the drum are the main reason for wide reading variations when the drum is rotated. The different positions of the 'hot spots' are the main cause of activity estimation errors [1]. The first standard correction procedure handles gamma attenuation using density estimation. Next, the counts are accumulated while the drum is turning, in order to average the source distance from the detector. To minimize the reading variations the distance of the detector from the drum is increased. For many cases, this procedure gives adequate results, but in some cases the accuracy is poor, depending on the spatial distribution of the hot-spots in the drum. Industrial SGS systems for estimating the activity of 208-liter standard waste drums, declare on accuracy of 10-30% but with the worst case of up to 500% mistake [2]. A Recent method developed by [3] exploits the angular dependent count rate distribution and improved by [4], but requires long measurement time to get good counting statistics for the analysis.

The goal of this work is to improve the accuracy results while reducing the total measurement time. To increase the count rate, we remove the narrow physical collimator and apply a digital filter on the measurements from the different spatial samples from the virtual detectors array surrounding the waste drum.

II. METHOD

A. General Description

In this paper, we describe a method to quantify the activity of spatially distributed gamma-emitting isotopes ('hot spots') in homogenous content waste drums without the use of a detector collimator. We use two steps to estimate the total activity in the drum. First, instead of minimizing the reading variations, the variations are exploited to localize the 'hot spots' with Maximum Likelihood Estimator (MLE) for multiple sources. Next, we calculate the activity based on the estimated hot spots positions using Least Squares (LS).

The ML estimator for the location of the sources, shown on part B, is a type of a Beam Former (BF), taken from antennas array spatial signal processing [5]. However, solving this multidimensional maximization requires to search for M sources in K discrete optional positions volume, which accumulate to $\frac{K!}{M!\cdot(K-M)!}\xrightarrow{K>>M}K^M$ search options. A non-optimal but fast way is to develop the ML for a single source and then do a search in the drum volume and picking the M largest local maximums of the Likelihood. As a good compromise between the two options, we use an iterative technique referred to as "Alternating Projection" (AP) [6]. The AP technique, described in part D, is not optimal but performs better than the M single sources search.

B. Problem formulation

Under the assumption of known homogenous absorbing material, we define the attenuation factor $a_{nm}(\gamma)$, for the total attenuation between the m-th source in the drum, with activity I_m, to the counts in the n-*th* detector position surrounding the drum envelope for energy γ (referred to as the n-th detector). The counts at detector n originated from source $\boldsymbol{m}$ is marked as $y_{nm}(\gamma)$:

$y_{nm}(\gamma) = a_{nm}(\gamma)\cdot I_m(\gamma)$ where

$$a_{nm}(\gamma) = \frac{1}{L_{nm}^{\ 2}}\cdot e^{-\mu_i(\gamma)\cdot l_{i_{nm}}}\cdot e^{-\mu_w(\gamma)\cdot l_{w_{nm}}}$$

L_{nm} is the distance between source n to detector m. the distance is comprised of $l_{i_{nm}}$ and $l_{w_{nm}}$, the partial distances the photons travel inside the drum, and crossing the drum wall respectively, the rest of the photon path through the air is ignored due to the small attenuation over the air. The attenuation coefficient inside the drum, and of the drum wall are marked $\mu_i(\gamma)$ and $\mu_w(\gamma)$. In vector form the total counts in the detectors are:

$$\vec{y}_{t_i}(\gamma) = A(\gamma, D, S)\cdot\vec{I}(\gamma) + \vec{w}_{t_i}(\gamma), \qquad t_i = 1, \dots T$$

Where $\vec{y}_j(\gamma) = [\boldsymbol{y_1}(\gamma), \dots, \boldsymbol{y_N}(\gamma)]_j^T$ is the vector of the N detectors readings, with contribution from all M sources in each detector at energy γ, at time t_i and $\vec{I}(\gamma) = [\boldsymbol{I_1}(\gamma), \dots, \boldsymbol{I_M}(\gamma)]^T$ is the M sources activity vector. The noise $\vec{w}_{t_i}(\gamma)$ is independent identical additive white Gaussian noise vector and $A(\gamma, D, S)$ is the attenuation coefficients matrix as defined in [2]. Matrixes D and S are the detectors and sources 3-dimensional coordinates, size 3xN and 3xM respectively.

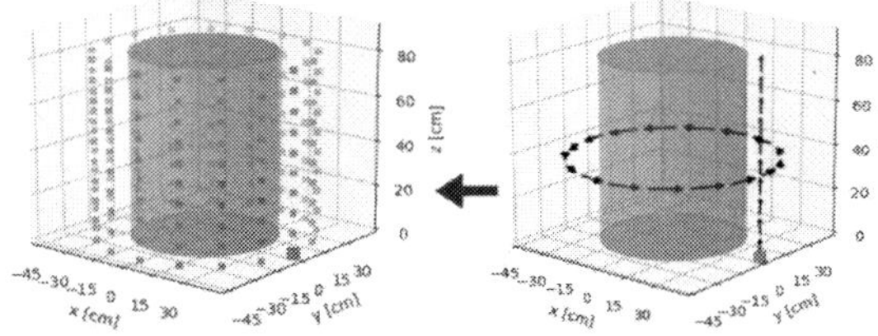

Fig. 1. Translating the rotation and upward motion into N spatial sample positions ('virtual detectors')

C. Maximum Likelihood Estimator

From the linear model we derive the Maximum Likelihood Estimator (MLE) of the multiple sources position (marked as S_M) and activities ($\vec{I}_M$).

$$LL = \log f_{\vec{y}/I,\vec{s}}(\vec{y}/\vec{I}_M, S_M)\xrightarrow[\vec{I}_M, S_M]{} max$$

Under the assumption of knows sources positions ($\hat{S}$) we get the estimator for the sources activities:

$$\hat{\vec{I}}(\gamma) = (A^T\cdot A)^{-1}\cdot A^T\cdot\vec{y}(\gamma)$$

We reinstate the estimated activity into the log-likelihood and get:

$$LL(\gamma, D, \hat{S}) = \frac{J}{\sigma^2}\cdot tr(P_A(\gamma, D, \hat{S})\cdot\hat{R}_y)$$

where P_A is the projection matrix to the space spanned by the M attenuation vectors and $\hat{R}_y$ is the sample covariance matrix calculated as:

$$P_A(\gamma, D, \hat{S}) = A\cdot(A^T\cdot A)^{-1}\cdot A^T$$

$$\hat{R}_y = \frac{1}{J}\sum_{j=1}^{J}\left(\vec{y}-\frac{1}{J}\sum_{j=1}^{J}\vec{y}\right)\cdot\left(\vec{y}-\frac{1}{J}\sum_{j=1}^{J}\vec{y}\right)^T$$

For a case with only a single source, the LL is reduced to:

$$\hat{\vec{s}}_{ml} = \arg\max_{\vec{s}}\left\{\frac{\vec{a}^T(\vec{s})\hat{R}_y\vec{a}(\vec{s})}{\|\vec{a}(\vec{s})\|^2}\right\}$$

Pay attention that there is no constraint on the sources activities to be positive in [5] since our final goal is not estimating each source activity but the total activity in the drum. This will be demonstrated with the example in the results section.

D. Alternating Projections

The ML solution is to search in the drum space for the source coordinate that maximizes the LL expression. In the single source, it is easily done by calculating K uniformly spread positions and finding the maximum. However, in the general M sources problem, we need to do a 3-dimensional search of the M optimal positions that maximize the LL expression which is a resource and time-consuming task.

We solve this hyper-dimensional search problem for Maximum Likelihood localization of multiple sources using an Alternating Projections (AP) technique introduced by [6]. The AP has two steps. In the first step, we add source every search and keep the previous sources in fixed positions until we reach M sources.

On the second step we eliminate a random source, then we redo the search for the deleted source position while keeping the other M-1 sources fixed.

We start the first step by searching for a single source using [8] over the different source positions by changing the attenuation vector $\vec{a}(\vec{s})$. After finding the first source position $\vec{s}_1$, we search for the second source using [7], where the attenuation matrix A is built from two attenuation vectors – the first vector is from the position we found in the previous search, $\vec{a}(\hat{\vec{s}}_1)$, and we search over the possible positions of the second source attenuation vector. In the same manner, we repeat the search for the m-th source, by searching only for one source every time and keep the first m-1 sources positions fixed, e.g. $A = [\vec{a}(\hat{\vec{s}}_1), \vec{a}(\hat{\vec{s}}_2), \ldots, \vec{a}(\hat{\vec{s}}_3), \vec{a}(\vec{s}_m)]$. After every step, we calculate the sources activities using [5]. Stopping condition for adding more sources is when the change in the sources activities is small.

The second step purpose is to refine the M sources positions estimation. We randomly pick one source and delete it. Then we search for a new source position with all the rest of the positions fixed. Again, the stopping condition is as before when the change of the positions and activities in each iteration is very small.

III. Preliminary results

Mathematical simulation enabled to easily create the detectors reading vectors for many different configurations. These configurations had variable detectors spatial sample grid and different attenuations. Multiple simulations were executed for each configuration with variable radioactive sources quantity, activity and distribution.

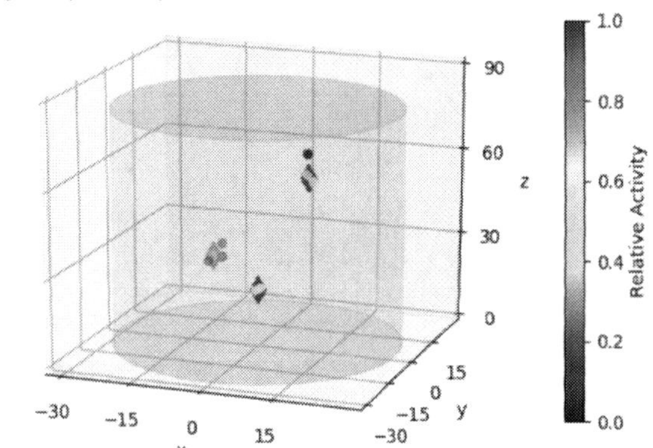

Fig. 2. Example of location and activity estimation for 3 sources with unit relative activity (red diamonds), using 10 sources (colored dots).

Figure 2 shows an example of the estimation result for simulation of 3 random positioned sources (blue diamonds), each with equal activity normalized to 1. The estimation was made using a preset of 10 sources. The plot shows that each source was estimated using two, very close, sources with approximately half activity each (orange dots), while the other further four sources activity is close to zero (green dots). Notice that although there is no constraint on the sources to be positive, there are no large negative sources only small sources that are close to zero, which are compensation for the measurement noise, since the true sources were estimated very closely in position and activity.

IV. Discussion and conclusions

Traditionally, when designing a facility for LLW there are two main options - SGS or TGS. This work has shown a new type of system which has the advantage of simple mechanics, no need for heavy shielding, thus having higher efficiency and a shorter total time of measurement. From preliminary simulations results, it shows potential to be located between SGS and TGS in terms of activity estimation accuracy.

The first challenge was importing a technique which uses the phase difference in the arriving of an electrical field to antennas, to the radiation detectors which only measures the radiation field intensity. Next, we needed to find a way to dramatically reduce the K^M searches to find M sources in *K* discrete optional positions inside the drum volume. This was done using the AP technique. Last we needed to find a way to estimate roughly the number of hot-spots inside the drum. For this, there are a few estimation methods that were tested such as

Based on our preliminary results, we predict that a large improvement in the accuracy of comparing to the standard SGS systems can be expected. The biggest advantage of the suggested system is the lack of necessity for heavy led collimator. Since no source is blocked by the collimator, the count rates are increased and the total measurement time decrease.

References

[1] Dung, Tran Quoc. Calculation of the systematic error and correction factors in gamma waste assay system. *Ann. Nucl. Energy, Vol. 24, No1.* 1997, Vol. 24, 1, pp. 33-47.

[2] ORTEC. Comparison of Gamma-Ray Nondestructive Assay Measurement Techniques. s.l. : ORTEC, ANTECH Corp.

[3] "An improved method for the non-destructive characterization of radioactive waste by gamma scanning", Baia, Y.F., et al. 2009, Applied Radiation and Isotopes 67, pp. 1897–1903.

[4] "Reconstruction of the activity of point sources for the accurate characterization of nuclear waste drums by segmented gamma scanning", Krings, Thomas and Mauerhofer, Eric. 2011, Applied Radiation and Isotopes 69, pp. 880–889.

[5] Trees, H. L. van., "Arrays and Spatial Filters" in *Optimum array processing – Part IV of detection, estimation, and modulation theory*, John Wiley & Sons, 2004.

[6] "Maximum Likelihood Localiztion of Multiple Sources by Alternating Projection", Ziskind, I. and Wax, M. ASSP-36, 1988, IEEE Trans.on ASSP, pp. 1553-1560.

[7] Group N-1. *Segmented Gamma-Ray Scanner.* Group N-1, Los Alamos National Laboratory. s.l. : Los Alamos National Laboratory, 1991. Application Note. LALP-91-011.

[8] Standard Test Method for Nondestructive Assay of Special Nuclear Material in Low-Density Scrap and Waste by Segmented Passive Gamma-Ray Scanning. s.l. : ASTM International, 2018. C1133.

Active Neutron Monitoring of Nuclear Fuel Cycle Fissile Materials

V.F. Batyaev, M.D. Karetnikov, S.V. Sklyarov

Dukhov Research Institute of Automatics (VNIIA), 127055 Moscow, Russia

***Abstract*— A decommissioning of nuclear fuel cycle facilities is inseparable from the problems of radioactive waste disposal. One of these problems is the categorization of a waste according to the content of beta- and alpha-emitters. Beta-emitters can be identified by existing technologies; however, the trouble arises when detecting alpha-emitting elements, primarily the long-lived members of the actinium chain with the specific activity of kBq/kg when they are spread inside a structural material. The report considers an application of an active neutron method-a differential die-away technology for reliable control of small quantities of FM.**

The essence of this method consists in sounding the interrogated item by pulsed thermal neutrons and recording the induced fission neutrons. The ratio of the number of fission neutrons to the number of source neutrons gives the normalized number of fission neutrons that is linked to the FM mass in the interrogated object. The work presents the scheme and principle of operation of an experimental device, as well as the results of measurement of concrete structures that contain internal traces of fissile materials. Analysis of the results shows that the proposed method allows the detection of ~ 6 mg of fissile material per kg of concrete with possible localization (cartogram) of the contaminated area.

***Keywords*— Fissile materials, Neutron generator, radioactive wastes, neutron detector.**

I. Introduction

Detection of small amounts of fissile materials (FM) is a key mission for determining the degree of contamination of decommissioned buildings and structures where FM were handled [1]. Since the cost of burial of low-level (LLW) and high-level (HLW) radioactive wastes differs by more than an order of magnitude, measuring the FM content in the structures of decommissioned buildings is a very urgent task. To classify such structures as LLW, the minimum content of alpha emitters (not transuranium) shall be as low as 100 kBq/kg, which corresponds to ~ 30 mg of ^{235}U per kg of concrete taking into account the full range of alpha emitters. It should be noted that the ^{235}U enrichment has insignificant influence on the total activity of alpha emitters in concrete.

In the existing practice, the activity measurement usually involves a spot sample from the LLW package, passing preliminary multi-stage preparation: grinding, dissolving, evaporation, averaging, etc. Sampling of a representative quantity is difficult for LLW in form of metal structures, construction debris, etc. The correctness of the results of measurement and their duration leaves much to be desired. There is a problem of temporary storage of waste batches, which is particularly urgent for enterprises with large volumes of LLW circulating at various stages of processing.

Passive methods of FM control in LLW based on detection of spontaneous fission products are much more rapid and do not require sampling. However, the gamma radiation of ^{235}U or ^{239}Pu has a weak penetrability and often a relatively low intensity for practical measurements. In addition, the spectral peaks of ^{235}U and ^{239}Pu interfere with the characteristic spectral lines of many radioactive isotopes in LLW (e.g., ^{133}Ba, ^{137}Cs, ^{131}I), which can completely mask gamma-radiation from FM even when HPGe high-resolution spectrometers are used. ^{235}U and ^{239}Pu feature small neutron yield of spontaneous fission, therefore, unlike ^{240}Pu and other even isotopes, they cannot practically be detected by neutron radiation. Thus, when small amounts of FM are distributed in the volume of a concrete structure, it is very difficult or practically impossible to detect low content of ^{235}U or ^{239}Pu using passive control methods.

Active methods based on detecting the products of nuclear reactions in FM sounded by penetrating radiation can potentially provide reliable control of small quantities of fissile materials. More than dozens nuclear-physical methods have been proposed that are suitable to different extent for solving the missions [3]. The analysis shows that the best accuracy, rate and reliability of the results are provided by the photonuclear and the differential die-away technologies. Choice of the method depends, first of all, on the type of matrix (material) where FM are concealed. It is associated with the features of penetration of the sounding and response (useful) radiation. For a matrix with a small atomic number and a high content of hydrogen atoms, it is preferable to use the photonuclear technology [4]; for a matrix with a medium and large atomic number (iron, concrete, lead), the differential die-away technology provides the best sensitivity. Moreover, differential die-away technology is insensitive to the fertile (^{238}U, ^{232}Th) materials that is essential for our task. The essence of this method consists in sounding the interrogated item by pulsed thermal neutrons and recording the induced fission neutrons. The neutron detector registration threshold is 0.4 eV due to the cadmium screening, thus the detectors responde to fission neutrons but insensitive to sounding thermal neutrons. The neutron moderator, measurement chamber and detector assembly are configured in such a way that the sounding neutrons and fission neutrons are time-correlated and they can be separated by measuring their time behavior [2]. The counting rate of neutron detector can be represented as

$$\Phi(t) = A\cdot exp(-\alpha t) + B\cdot exp(-\beta t) + \Phi_0, \quad (1)$$

where A, B are pre-exponential factors; α is the neutron decay constant determined by the moderator thickness behind the cadmium shield; β is the decay constant of the thermolized neutrons of FM determined by the neutron properties of the set-up; Φ_0 is the sum of the responses of delayed neutrons, spontaneous fission neutrons, neutrons from (α,n) reactions, as well as the background of the premises. The ratio of the number of fission neutrons to the number of source neutrons gives the normalized number of fission neutrons (normalized amplitude coefficient *B/A*), that does not depend on the neutron generator yield. The specified value of *B/A* is linked to the FM mass placed in the interrogated object [2].

II. Experimental setup of FM detector in concrete

To study the possibility of detecting small amounts of FM in a concrete matrix (a model of buried LLW) the experimental FM detector based on the differential die-away technology (Fig. 1) was designed using the ING-07T D-T neutron generator (produced by VNIIA, Russia) as a source of sounding neutrons. The generator provided a pulsed flux of 14-MeV neutrons with a 800 Hz pulse frequency and a 14 μs neutron pulse width. An average intensity was as high as $5\cdot10^8$ n/s. Fission neutrons were detected by five boron gas-filled counters of thermal neutrons of the SNM-KCh type (manufactured by NIITFA, Russia) surrounded by a moderator and a cadmium filter. One boron counter without cadmium filter was used to monitor the neutron flux from the neutron generator. The choice of boron counters is due to their low sensitivity to gamma radiation, since measurements in the realistic environment can take place under a strong gamma background. The contaminated and pollution-free parts of concrete constructions were imitated by: a concrete 850x530x280 mm^3 cube; 100x100x50 mm^3 concrete samples (12 pieces without uranium; 12 pieces with a ^{235}U content uniformly distributed with an average concentration of 6 mg/kg); 42 pieces of 3.5 mg point samples of ^{235}U (147 mg of ^{235}U in total)). The signals were detected using a MAVS-16 time analyzer (produced by VNIIA, Russia) which allowed detecting the number of counts in 256 consecutive time windows of arbitrary width (the window number is counted from the sync pulse from the neutron generator). In the experiments the width of each time window was chosen to be 4 μs for a complete analysis of the counting rate of counters as function of time.

Neutron counters were installed in a chamber that simultaneously served as a moderator for 14 MeV source neutrons and a moderator for fission neutron counters. The configuration of the camera has been optimized to ensure maximum sensitivity of fission neutron detectors and a minimum background level.

Fig.1 Experimental detector of FM in concrete structures.

An example of the experimental data processed by smoothing function is given in Fig. 2. The figure of merit *FOM*=—— was used as a criterion for the effect F_1 above the background F_2. To maximize the effect-to-background ratio the optimal time interval $\Delta T = t_2 - t_1 = 500$ μs for integrating the effect (number of counts in the presence of FM) and the background was applied.

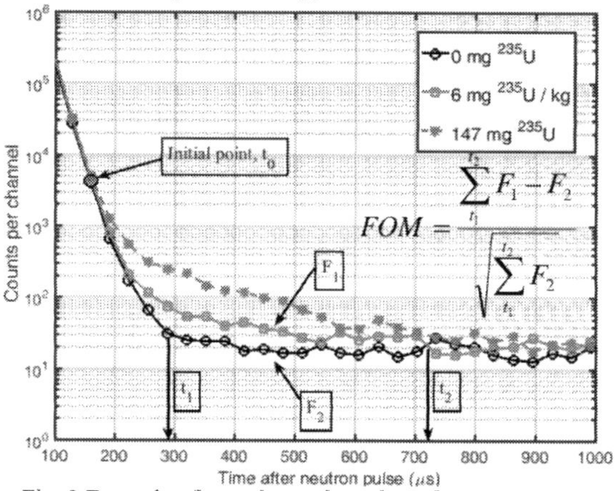

Fig. 2 Example of experimental number of counts per 16-μs channel as function of measurement time.

The sensitivity of FM detector in case of a point distribution of ^{235}U on the concrete surface was close to 6 mg for FOM=3.

The FM detector allowed localization the FM, which is very important for the measurement of concreted LLW with non-homogeneous density of FM. This is physically possible due to the geometric factor ($1/R^2$) and the attenuation of the diffuse neutron flux with a distance. Measurements with the stepwise shift of the FM detector from FM were carried out to demonstrate the possibility of determining the FM position coordinates in the concrete structure. FM was in a form of a long tape with target indicators applied on it (a total weight of ^{235}U of 147 mg), and was placed in a 6-cm-deep hole in concrete. The dependence of *FOM* on the shift of the FM detector is demonstrated in Fig. 3. It can be seen that *FOM*=27 when the FM detector is placed at 7 cm from FM. In this case, the minimum detectable mass of FM is 15 mg. The sensitivity drops by 3.2 times at a distance of about 30 cm.

III. Numerical simulation of FM detection

To further substantiate the possibility of detecting the volumetric distribution of FM inside a concrete structure, a numerical simulation of FM detection with a geometry similar to the experiments performed was carried out. 147 mg of ^{235}U was placed inside the 850x530x280 mm^3 concrete block in a tunnel 400-mm long at a 60-mm distance from the surface. The neutron flux density was calculated in the cells of a computational grid with cell sizes of 20x20x20 mm^3 to study the spatial distribution of neutrons from FM with energies above 1 eV. The choice of this neutron energy range is due to a relatively high probability of their detection by the boron counters located behind the cadmium shield. The neutron source with the Watt spectrum was set as an extended one, geometrically coincident in position and dimensions with FM. This made it possible to determine the neutron flux density of FM in the entire volume of the concrete (including the surface corresponding to the location of the FM detector) of the structure and compare the results of epithermal neutron flux attenuation with the experimental results of measuring the minimum detectable FM mass. Fig. 4 illustrates the results of simulation for one surface of the concrete block.

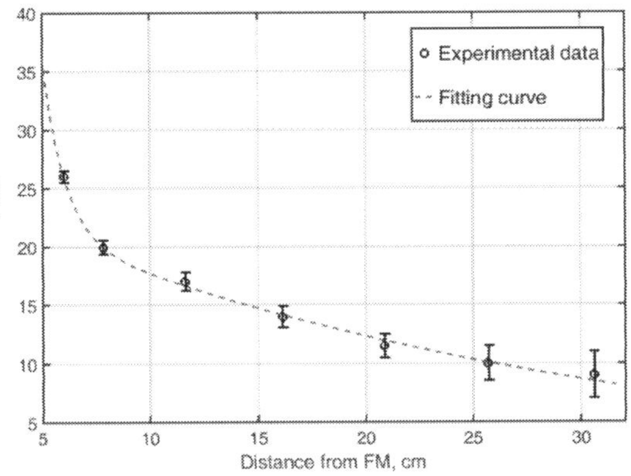

Fig. 3 The dependence of *FOM* on the shift of the FM detector.

The data obtained show that the attenuation of the epithermal neutrons flux along the *X* axis along the surface of the concrete structure is ~ 3.2 times over a length of 30 cm from the center of the FM location. This result correlates well with the measured function of the FM detector sensitivity (Fig. 3) vs the distance to FM. Moreover, to improve the convergence of the experimental and calculated results, it is necessary to obtain the distribution of the neutron flux density not on the surface of the concrete structure, but after passing through the cadmium and polyethylene shield, thereby simulating the neutrons hitting the detector.

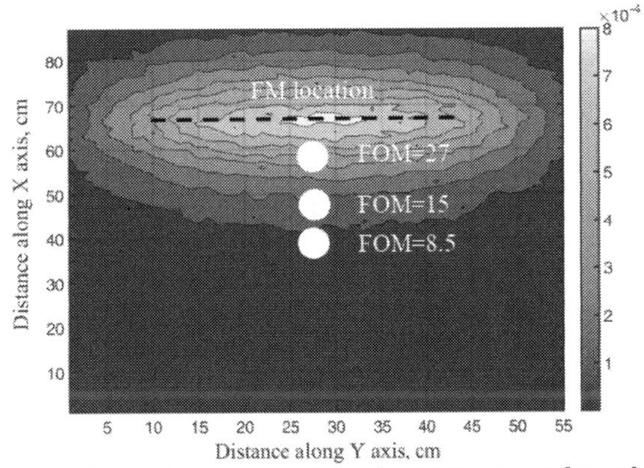

Fig. 4 Flux density of epithermal neutrons on surface of concrete structure per one source neutron. The value of FOM corresponds to the position of FM detector.

The algorithm for spatial detection of the FM position in concrete can be based on the representation of the experimentally obtained count rate of detector, E_i, as a product of the counters response matrix by a unit amount of uranium in the previously known cells and the vector of weight factors, M_{Uj}, corresponding to the amount of FM in the cells.

Count rate of counter i:

$$E_i = w_{i_j} \cdot r_j = w_{i_j} \bullet [\sigma_U \cdot \Phi_j \cdot M_{Uj} \cdot \frac{N_a}{A}] = R_{ij} \bullet M_{Uj}$$

Where j is number of cell in object; w_{ij} is counter sensitivity matrix as a measure of counter i response to neutrons from one uranium fission in cell j; r_j is fission rate of uranium in cell j; σ_U is uranium fission cross-section; Φ_j is neutron flux density in cell j during response measurement; M_{Uj} is uranium mass in cell j; N_A is Avogadro number; A is uranium atomic mass; R_{ij} is counter response (count rate) matrix to a unit of uranium in cells; $\bullet$ is matrix multiplication operation

This expression in the matrix form can be represented as follows:

$$\begin{bmatrix} R(1,1) & \cdots & R(1, j_{max}) \\ \cdots & \cdots & \cdots \\ \cdots & \cdots & \cdots \\ \cdots & \cdots & \cdots \\ R(i_{max},1) & \cdots & R(i_{max}, j_{max}) \end{bmatrix} \times \begin{bmatrix} M(1) \\ \cdots \\ M(j_{max}) \end{bmatrix} = \begin{bmatrix} E(1) \\ \cdots \\ \cdots \\ \cdots \\ E(i_{max}) \end{bmatrix}$$

IV. Conclusion

Experiments have shown that the minimum detectable mass of ^{235}U by five counters is 6 mg/kg with a uniform distribution over the concrete samples. In the future, our intension is to increase the number of counters up to 64 to improve the sensitivity of the setup and to study its applicability for measuring the amount and position of FM in various matrices. In addition, the paper presents an analytical method for determining the FM distribution in a concrete structure, which shall be further verified using a cellular concrete structure and a set of samples with different FM content.

The authors acknowledge participation of S.N. Florya, D.I. Vlasov (FSUE "RosRAO"), V.F.Danilov and G.V.Poletov (NIITFA industrial JSC) regarding their help in preparing and arranging demonstration experiments under process conditions and valuable discussions.

References

[1] Radioactive waste.-[electronic resource.] URL: www.norao.ru/waste/ (accessed 20.02.2019).

[2] Viacheslav F.Batyaev, Sergey S. Sklyarov., Comparative Analysis of Non-destructive Methods to Control Fissile Materials in Large-size Containers. International Conference Nuclear Data 2016, Bruges, Belgium, 2016

[3] Gozani T. Non Destructive Assay of Spent Fuel for Determination of Residual Fissile Content // Journal of Instruments for Nuclear Materials Management, V. 5, 1976, P. 514-524.

[4] Gmar M., Jeanneau F., Laine F. et al. Assessment of Actinide Mass Embedded in Large Concrete Waste Packages by Photon Interrogation and Photofission // Applied Radiation and Isotopes, V. 63, 2005, P. 613–619.

EPJ Web of Conferences **225**, 06012 (2020)
ANIMMA 2019

https://doi.org/10.1051/epjconf/202022506012

Assessment of the impact of selected changes in the deep geological repository model on its long-term safety

Dorota Flamíková
Slovak University of Technology in Bratislava,
Institute of Nuclear and Physical Engineering
Bratislava, Slovakia
dorota.flamikova@stuba.sk

Vladimír Nečas
Slovak University of Technology in Bratislava,
Institute of Nuclear and Physical Engineering
Bratislava, Slovakia
vladimir.necas@stuba.sk

***Abstract*— The deep geological repository system provides long-term protection against the undesirable effects of ionizing radiation on the population and the environment. An important part of the long-term safety strategy is development of a monitoring program that collects information about the behaviour of the deep geological repository throughout its whole lifetime. A simplified model of the disposal system, geosphere, and biosphere was developed using the GoldSim simulation tool to demonstrate the behaviour of the hypothetical deep geological repository located in crystalline rocks. Also an initial model of the reference biosphere was created based on the scenario of an agricultural habitation (normal evolution scenario) and it was developed based on the recommendations provided in the BIOMASS methodology. After a significant period of time, disposal containers will be degraded and evolution changes in the repository system will occur. Several important parameters appear in the annual effective dose calculation for an individual from critical exposure group within the reference biosphere model. One of them are, for example, distribution coefficients and so-called translocation factors that define the transported rate of released radionuclides into the environment. This paper provides a view into the selected part of the deep geological repository through the data obtained by monitoring during the selected period of time. Simulations describing changes in the repository system. The aim of this contribution is to evaluate the impact of selected changes on the annual effective dose for an adult individual from a critical exposure group while it is assumed, that the respondent consumes contaminated crops and animal products. This model includes various biosphere components and multiple exposure pathways such as inhalation, ingestion and external exposure.**

Keywords — deep geological repository, biosphere model, annual effective dose, GoldSim, long-term safety

I. Introduction

The disposal of spent nuclear fuel and other highly radioactive waste originating mainly in the nuclear energy sector is a topic with huge social importance. The basic requirement of repositories for radioactive waste is the prevention of the leakage of radionuclides into the biosphere. The safe disposal of spent nuclear fuel in deep geological repositories (DGR) is based on the so-called multi-barrier concept. In order to evaluate the long-term safety of the deep repository, a comprehensive model of the reference biosphere, geosphere and disposal system should be created to obtain information on radionuclide migration in a given environment, radiation load, etc. The model serves as a tool for calculation of the annual effective dose for the individual representative from the critical exposure group caused by radionuclides released into the environment.

This contribution has been partially supported by the Slovak Grant Agency for Science through grant VEGA 1/0863/17 and by the Slovak Research and Development Agency through grant APVV-15-0558.

II. Monitoring of the Deep Geological Repository

Geological repositories for the disposal of high level radioactive waste and spent nuclear fuel are designed to provide isolation of these type of waste from the population and environment for many thousands of years.

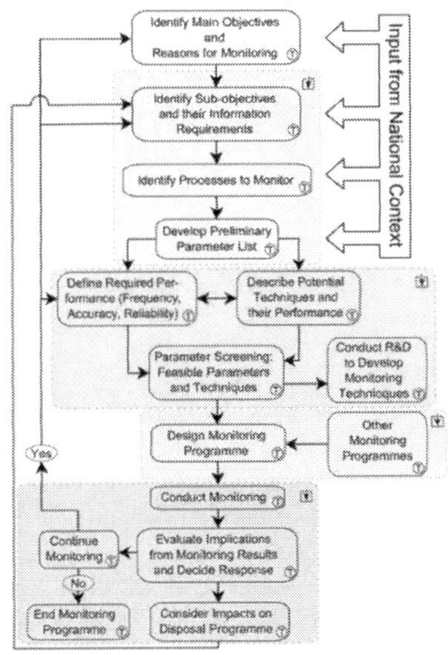

Fig. 1. Structural approach of the DGR monitoring process [1].

Monitoring the environment of such repositories is not therefore expected to reveal any increase of radioactivity due to the repository, it is planned and intended for other purposes. These include providing reassurance that the repository is fulfilling its intended purpose, but there can be other reasons of a technical nature.

Monitoring is expected to play an important role in both development and execution of geological disposal programmes. In particular, monitoring will contribute essential information for the satisfactory completion of the various phases of the geological repository programme and, in doing so, will strengthen confidence in long-term safety, which is the key objective of radioactive waste disposal. The example of an structural approach of the monitoring programme of deep geological repository shows Fig. 1. For the purpose of the present paper the following definition of monitoring has been employed: continuous or periodic observations and measurements of engineering, environmental or radiological parameters, to help evaluate the behaviour of components of the repository system, or the impacts of the repository and its operation on the environment. The process leading to the commissioning of geological disposal facilities for long lived radioactive waste is quite lengthy and involves a number of decision points. Once authorized to operate the facilities are expected to remain in activity for long periods of time before their eventual decommissioning and closure. Monitoring of various aspects of the disposal system is likely to be an important support to decision-making at all stages of the repository development programme. After closure, the long term safety of geological disposal facilities, due to the duration of the hazard associated with the waste, cannot rely on institutional controls, including monitoring. However, continuing monitoring is likely to be a societal demand for some time after repository closure. Such monitoring, besides showing that the process of decommissioning surface facilities has been successfully completed, would also strengthen the confidence, at least in some sectors of society, that the evolution of the waste isolation system is in accordance with expectations. The need to safeguard nuclear materials contained in the repository may also cause some form of monitoring or surveillance to continue in the post closure phase.

Using of monitoring information

Certain monitoring activities should begin at the earliest possible time within a repository development programme, before the perturbations caused by repository construction and operation begin to accumulate. This early information is important because it allows an understanding to be developed of the nature and properties of the natural, 'undisturbed' environment of the disposal system. Parameters obtained from monitoring process will be used for example to evaluate changes that occur in the rock and groundwater system during the construction and operational stages and, in the post-closure period, to evaluate any impacts that the presence of the repository may have on natural processes and the environment. In practice, the monitoring programme will begin during the site investigation stage. The characteristics of primary interest in the context of establishing baseline information are [2]:

- the groundwater flow field in the host rock and in the surrounding geological environment (groundwater pressure distributions, hydraulic gradients, regions of recharge and discharge, etc.);
- geochemical characteristics of groundwater (redox, salinity, major and trace element concentrations, natural radionuclide content, etc.);
- levels of radioactivity in groundwater, surface waters, air, soils and sediments, animal and plant life;
- meteorological and climatic conditions;
- hydrology of surface water systems, including drainage patterns and infiltration rates;
- ecology of natural habitats and ecosystems.

Baseline data should be established as part of the site characterisation activity, e.g. measurements from local and regional boreholes and surface investigations. Where important parameter values are found to follow an increasing or decreasing trend, baseline monitoring will need to be continued until that trend is established with confidence and the reasons for the trend are sufficiently well understood. The establishment of baseline values for surface environmental indicators is relatively straightforward, because the process of measurement will, in general, not affect the parameters being measured. However, it is to be appreciated that invasive investigations will themselves perturb the natural groundwater system to a degree based on site specific conditions. In order to establish baseline conditions with which to judge later impacts, e.g. changes to groundwater pressures and hydro-chemical conditions in response to repository construction, sufficient information needs to be collected in the surface exploration phase to have confidence that the undisturbed conditions have been adequately characterised both spatially and temporally. This data will further serve as background information for monitoring during other lifecycle phases of the deep geological repository.

III. CREATION OF THE MODEL

For evaluation of long-term safety of the deep geological repository the conceptual model of such a facility was developed. Main part of the model is reference biosphere, which is developed based on the BIOMASS methodology (IAEA, 1996). The role of the BIOMASS document is to provide a methodology in the process of creating practical examples demonstrating ways in which models of the biosphere for long-term safety assessment can be developed and appropriately assessed for the specific purposes. A relatively simple interface between the geosphere and the biosphere considers the natural flow of contaminated groundwater into the aquifer. This water passes through other transport pathways and then it is discharged into the surface. The ultimate goal is to calculate the effective dose for the individual without considering the evolution of the biosphere system [3].

A. Disposal system and geosphere

The modelled disposal system is based on a multi-barrier concept, which consists of a combination of engineering and natural barriers providing sufficient protection for the environment and the population over a very long period of time. All individual elements of the multi-barrier protection system are interacting with each other in such a way to isolate radionuclides from the biosphere. One of the barriers is the

https://doi.org/10.1051/epjconf/202022506012

waste itself – very form of spent nuclear fuel. Another important barrier is disposal container inspired by Czech proposal, where spent fuel assemblies (from VVER-440 reactors) are considered to be disposed in. The disposal capacity of one container is 7 fuel assemblies. This container is constructed of stainless steel (inner part) and carbon steel (outer part) with an outer diameter of 650 mm and a length of 3 670 mm. Fuel assemblies are inserted into the inner profiled pipes made of aluminium alloy [4].

Within this assessment the calculations were performed for the spent fuel with an initial average enrichment of 4.25 % of U-235 and burnup 60 MWd/kgU. The calculations of spent nuclear fuel inventory was performed for the radial profiled VVER-440 fuel assembly of the 2nd generation by using the depletion module TRITON included in SCALE 6.1.3 code system. The model is 2D assembly model developed at our institute [5]. The time of storage before the final disposal was set at 50 years. Disposal container is surrounded by a bentonite buffer with a wall thickness of 300 mm.

It is considered, that the disposal container fails due to normal evolution processes after 1 000 years and water comes into contact with the source term (fuel and structural material), released radionuclides start to migrate through the repository system. After the radionuclides penetrate all the barriers in the near field represented by the spent nuclear fuel matrix itself, the disposal container, bentonite buffer and excavation disturbed zone (EDZ) they reach the geosphere. In our model, the geosphere is represented by a crystalline host rock environment where radionuclides reach the geospheres through the fractures and faults. The interconnection of the geosphere with the biosphere is represented by an aquifer, from which the contaminated water flows through the surface flow into the water reservoir.

B. Reference biosphere

The biosphere in this model is represented by a set of interconnected biosphere objects. A biosphere object is defined as an area of the landscape that can receive radionuclides released, either through discharge of deep groundwater or in contaminated surface water, at any time.

The primary medium through which radionuclides come into the biosphere is water. Contaminated water from the considered water reservoir is used as drinking water for the population, water for livestock and also for irrigation of the agricultural land. In this water reservoir also, fish are being farmed. On such land are cultivated cereals, vegetables, grassland and trees. Radionuclides are transported through the root or leaf system into the plants, hence either directly into the human organism, or into the animal organism from which the animal products enter the human body. Receiving of the radionuclides is also partially caused by the consumption of contaminated fish meat.

For livestock in our model, we consider bovine animals whose animal products - meat and milk are consumed by residents from the critical exposure group. In the process of creating a reference biosphere, some input elements were selected, representing the constituent elements of the environment within the biosphere model. In the case of crops, cereals, grassland and trees are considered in the model.

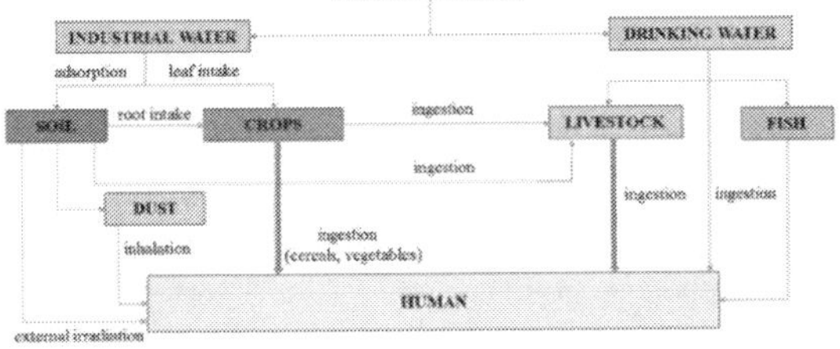

Fig. 2. Scheme of the conceptual model of the reference biosphere.

In the case of crops, cereals, grassland and trees are considered in the model. The schema of the conceptual model of biosphere and exposure pathways is shown in Fig. 2. Conservatively, it is assumed that all consumed environmental components are contaminated with radionuclides, with the primary exposure pathways such as ingestion, inhalation and external exposure.

We consider, that the source of an external exposure is contaminated soil. On the surface of the soil farmers are moved carrying out, for example, agricultural work. The value of the effective dose from external exposure is therefore dependent on the time spent on the contaminated soil. The exposure pathway of external exposure also includes external irradiation of the person's skin from the dust that is deposited on it.

The effective dose caused by inhalation depends, inter alia, on the amount of inhaled dust arising from the contaminated soil and the respiration rate of the exposed object, assuming a constant content of soil particles in the air. The exposure is caused by inhalation of particles of contaminated dust. The effective dose caused by ingestion depends on the amount of ingested contaminated water and contaminated food. The model for long-term safety assessment of the deep geological repository was developed using the GoldSim simulation tool [6].

IV. TRANSPORT CHARACTERISTICS IN BIOSPHERE

Radionuclide transport processes included to the model are diffusion and sorption in the repository near-field and advection, matrix diffusion and sorption in the geosphere. A simplified approach to handle solubility limited release of the nuclide from the waste canister is also included into the model.

To assess radiological risks, the annual individual effective dose for an individual from a critical exposure group has been calculated. This value can be related to the risk of damage to health caused by ionizing radiation. For this purpose, a generally accepted method of calculating the radiation dose based on differential equation is used, the results of which determine the concentrations of individual radionuclides in the environmental components (crops, animal products, etc.) based on known, respectively calculated values of individual radionuclide concentrations in water and agricultural soil. For these calculations so-called concentration and translocation factors are used. These factors determine the transport of contaminants from soil and water into plants, respectively from plants consumed by livestock to animal products.

A. Definition of selected parameters

Within the model, changes were made to selected parameters within the geosphere and biosphere. The selected parameters are diffusion coefficients for different soil types and translocation factors for selected environmental components.

1) Distribution coefficients

Sorption is the process, where dissolved contaminants partition from the ground-water and adhere to the particles of the bentonite and host rock matrix. Sorption of the dissolved radionuclides onto the matrix results in slowing (retardation) of contaminants relative to the average advective ground-water flow velocity and a reduction in dissolved concentrations in ground water.

The rate of sorption is described by distribution coefficient K_d, which is defined as the ratio of the sorbed contaminant concentration to the dissolved contaminant concentration. For systems described by a linear isotherm, K_d is a constant [7].

$$K_d = \frac{m_S/M_S}{m_f/V_f}, \tag{1}$$

where: K_d - distribution coefficient (m^3/kg),
ms - weight of the nuclide sorbed on the solid material (kg),
Ms - weight of the solid material (kg),
mf - weight of the nuclide dissolved in the liquid medium (kg),
Vf - volume of the liquid medium (m^3).

Dissolved radionuclide ions can bind to solid surfaces by a number of processes that are often classified under the broad term of sorption. The behaviour and ultimate radiological impacts of radionuclides in soils are largely controlled by their chemical form and speciation, which strongly affect their mobility, the residence time within the soil rooting zone and uptake by biota.

The degree of radionuclide sorption on the solid phase is often quantified using the solid-liquid distribution coefficient, K_d, which can be used when making assessments of the overall mobility and likely residence times of radionuclides in soils. K_d is the ratio of the concentration of radionuclide sorbed on a specified solid phase to the radionuclide concentration in a specified liquid phase, as describes following equation:

$$K_d = \frac{activity\ concentration\ in\ solid\ phase\ [Bq/kg]}{activity\ concentration\ in\ liquid\ phase\ [Bq/l]} \tag{2}$$

The K_d approach takes no explicit account of sorption mechanisms but assumes that the radionuclide on the solid phase is in equilibrium with the radionuclide in solution and that exchange between these phases is reversible.

2) Translocation factors

For the purpose to assess the radiological risks, the annual individual effective dose for an individual from a critical exposure group needs to be calculated. For doing so, a generally accepted method of calculating the radiation dose based on differential equation is used, the results of which determine the concentrations of individual radionuclides in the environmental components (crops, animal products, etc.) based on known, respectively calculated values of individual radionuclide concentrations in water and agricultural soil. For these calculations so-called concentration and translocation factors are used. These factors determine the transport of contaminants from soil and water into plants, respectively from plants consumed by livestock to animal products.

Radionuclide concentration in vegetables and cereals is calculated as follows [8]:

$$C_{food} = (CF_{food} + (1 - f_{prep}) \cdot S_{food}) \cdot C_{soil} + \mu_{food} \cdot q_{irr} \cdot C_{water} \cdot \frac{1 - f_{prep} + f_{trans}}{Y_{food} \cdot \lambda_{weather}}$$

where:

CF_{food} [-] is a concentration factor for vegetables and cereals,
f_{prep} [-] is an external contamination loss due to processing,
S_{food} [kg/kg] is crop contamination by soil,
C_{soil} [Bq/m^3] or [Bq/kg] is soil concentration of radionuclide in soil,
μ_{food} [-] is fraction of contamination caused by irrigation,
q_{irr} [mm/year] is amount of irrigation water,
C_{water} [Bq/kg] concentration of radionuclides in kilogram, resp. litters of water,
f_{trans} [-] is fraction of activity passing from the surface to the interior of the plant,
Y_{food} [kg/m^2] is crop yield,
$\lambda_{weather}$[d/kg] is contamination loss coefficient due to weathering.

B. Relationship between Kd and other parameters characterizing radionuclide mobility

In a porous medium such as soil, the radionuclide diffusion process differs from diffusion in free water. Also, the ability of radionuclides to be sorbed onto porous media such as soil is different than for host rock itself. This feature is defined by the so-called retardation factor, which is defined as follows.

$$f_{ret} = 1 + \left(\frac{\rho}{\varepsilon}\right) . K_d \tag{3}$$

where: f_{ret} - retardation factor (-),
ρ - dry bulk density of the soil (kg/m^3),
ε - total porosity of the medium,
Kd - distribution coefficient (m^3/kg).

It determines what amount of radionuclide is bound to the medium during the migration process and how much of it will be transported further. Radionuclides are transported from the soil to the roots of plants and other parts of the environmental components from where they either directly or indirectly enter the human body.

V. EXPOSURE ASSESSMENT FOR CONSIDERED SCENARIOS

In order to evaluate the long-term safety of the deep geological disposal system of spent nuclear fuel, our aim was to determine the values of the annual effective doses for individuals from the critical exposure group for considered deep geological repository. The representative individual from a critical exposure group is considered to be an adult person (over the age of 17 years). The geosphere and the disposal system, where disposal containers for spent nuclear fuel are considered to be disposed in and sealing material (bentonite), are modelled conservatively. The scenario is based on the National Program for Spent Nuclear Fuel and Radioactive Waste Management in Slovak Republic [9], which requires the allocation of 32 658 spent nuclear fuel assemblies produced from the 60-year operation of Slovak nuclear power plants in the repository. Since it is assumed that the capacity of one disposal container is 7 fuel assemblies, we assume that approximately 4 666 containers will be disposed in the deep geological repository. The conservative scenario further assumes that each disposal container will fail after 1 000 years due to degradation of the materials used in repository.

It is assumed that the main transport pathway of radionuclide migration is groundwater, which passes through fractures and faults in the geosphere and into the aquifer and then to the water reservoir. Radionuclide transport processes included into the model are diffusion and sorption in the repository near-field and advection, matrix diffusion and sorption in the geosphere. Contamination of the environmental media is caused by the transport of released radionuclides from the source term and is significantly dependent on the defined boundary conditions of the initial reachable parts of the biosphere.

For each disposal container, same boundary conditions are assumed (same container distance from host rock fractures, the same transport pathway of released radionuclides etc.). The chosen scenario for assessing the long-term safety of the deep geological repository considers a normal evolution scenario. It is based on the assumption that the conditions for which the repository was built are unchanged throughout its whole existence and that the properties of engineering barriers correspond to project prerequisites. It is not considered the introspection of the person into the disposal area, the climatic conditions remains unchanged, the use of the environmental components associated with the agricultural activity and other activities is identical throughout the evaluated period to the current habits.

For purposes of our analysis, investigation of the impact of selected parameters changes on the long-term safety of the deep geological repository was performed. Selected parameters are diffusion coefficients for various soil types and transfer factors for selected environmental media (root and leafy plants, cereals and crops). Diffusion coefficient K_d [m^3/kg] represents ability of radionuclide to be sorbed (retented) within the media. Transfer factors t_f [-] are defined as the ratio of the contaminant concentration in the plants and the concentration in the specific soil layer. Within the model, ingestion of selected crops is considered to be the dominant intake pathway. Distribution coefficients for various soil types and ingestion rate of selected environmental components can be seen from Table I. and Table II.

TABLE I. DISTRIBUTION COEFFICIENTS FOR VARIOUS SOIL TYPES [10]

DISTRIBUTION COEFFICIENTS [m^3/kg]			
ELEMENT	***Loamy soil***	***Clay soil***	***Sand***
Ac	47	24	2.77
Am	17	84	4.4
C	0.0007	0.09	0.0086
Cl	7.3	20	0.014
Cm	12.6	54	2.4
Cs	0.46	1.9	0.029
I	0.004	0.01	0.0003
Nb	15.6	92	5.1
Np	12.3	98	3.24
Pa	4.7	27	10
Pb	0.01	0.94	0.0074
Pd	1.5	6.7	2.7
Pu	1.4	3.6	0.86
Ra	0.049	0.53	0.0038
Se	0.55	0.89	0.34
Sm	2.9	11.1	6.1
Sn	13	33	3
Tc	12	18.2	4.8
Th	0.0018	0.028	0.00062
U	0.08	0.0441	0.0026
Zr	6	1.2	8.2

TABLE II. INGESTION RATE OF SELECTED ENVIRONMENTAL COMPONENTS [11]

	ADULTS (> 17 years)
Milk and milk products [l/year]	15.3
Meat [kg/year]	6
Cereals [kg/year]	157
Vegetables [kg/year]	75
Crops [kg/year]	43

These different types of soil were chosen because of the fact, that all these soil types are present in prospective localities for the siting of the deep geological repository in Slovakia and therefore they may be the subject of further research in this area in the future. The higher the value of the respective distribution coefficient is, the more the given nuclide can be retarded within the engineering and natural barriers, the later it enters the environment and thus the more optimistic scenario is assumed. The rate of selected ingestion rate of selected environmental components was determined on the basis of data from the Statistical Office of the Slovak Republic.

Result of the simulations

The calculated maximal annual effective dose for representative individual from the critical exposure group shows Fig. 3.

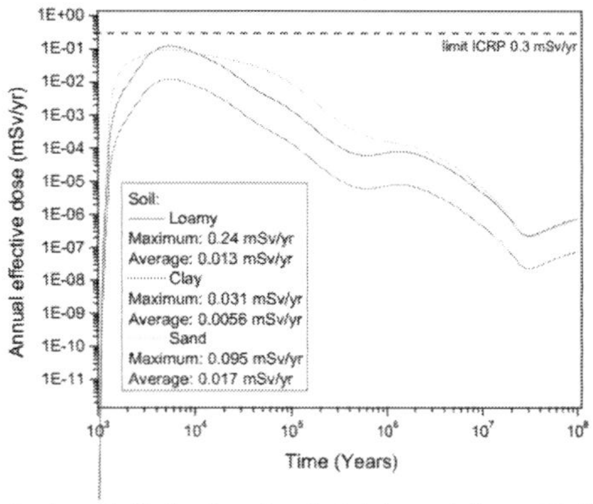

Fig. 3. Annual effective doses for selected changes in the model of deep geological repository.

From this figure, we can see that the maximal value of the annual effective dose was calculated for the loamy soil and the minimum value for the sand. Within these results it can be said, that the retardation factor depends significantly on the soil porosity and retention properties of the individual radionuclides. We can conclude, that from the dose point of view, the most important is the first achieved part of the biosphere, which is in our case represented by the soil. Changing translocation coefficients for individual soil types and selected environmental components has only a negligible effect on the calculated dose value.

Maximal values of achieved doses had been also compared to the limit established by the International Commission on Radiological Protection (ICRP) for waste disposal facilities. Radiological limits for the long-term assessment of the safety of a waste disposal facility recommended by ICRP is set to an annual dose constraint for the population of 0.3 mSv/year [12]. The values of the maximum annual effective dose are in all considered cases lower than this limit.

In all the considered cases, the same radionuclides affect total annual effective dose. Contribution of selected radionuclides for an adult representative shows Fig. 4. The release rate of many nuclides is reduced due to the transport of nuclides through the host rock and the major water-conducting fault. Actinides are relatively strongly sorbed on the bentonite as well as on the host rock matrix and therefore their release rates are very low. The largest contribution to total annual effective dose represent activation and fission products C-14, Cl-36, I-129, Se-79 and Cs-135.

I-129 and Cl-36 are assumed to have very poor retentive properties and that's why these nuclides dominate the total release rate for a long period of time. In the early years, the most significant nuclide dominating the total release rate is C-14. It is caused by its relatively high specific activity in comparison with long-lived radionuclides.

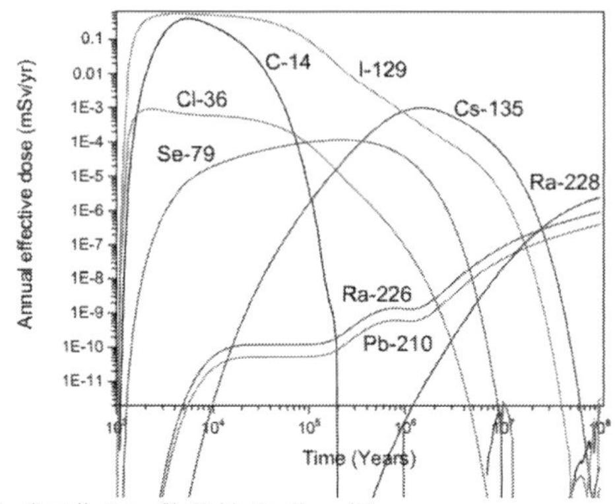

Fig. 4. Contribution of individual radionuclides.

In the last years of the analysis the increase in dose is due to the contribution of specific radionuclides. Pb-210 accumulates in the soil at a later time of the analysis and from there passes into the plant leaves. The maximum far field dose from Ra-226 increases because of the greater amount of uranium in spent fuel. Ra-226 migrates appreciably from the repository system only as a result of the migration of Th-230, which is not at a high enough concentration to be solubility limited.

CONCLUSION

To evaluate the long-term safety of the deep geological repository, a model of the disposal system, geosphere and the initial model of the reference biosphere has been developed. Using the GoldSim simulation tool, the annual effective dose for individuals of different ages from critical exposure group was calculated. Within the analysis, the impact of changes in distribution coefficients for different soil types and relevant translocation factors expressing the rate of contaminant transition into environmental compartments was investigated, when ingestion is considered to the dominant exposure pathway. From the results we can see, that the properties of the first achieved part of the biosphere have a significant influence on the dose value. Subsequent changes in other contaminant spreading parameters have a negligible effect on the overall dose. The maximum dose value is in all cases lower than the recommended limit set by the ICRP. In all considered cases, same radionuclides affect annual effective dose. In the last years of the analysis the increase in dose is due to the contribution of specific radionuclides. Pb-210 accumulates in the soil at a later time of the analysis and from there passes into the plant leaves. The maximum far field dose from Ra-226 increases because of the greater amount of uranium in spent fuel. Ra-226 migrates appreciably from the repository system only as a result of the migration of Th-230, which is not at a high enough concentration to be solubility limited.

ACKNOWLEDGMENT

This study has been partially supported by the Slovak Grant Agency for Sciences through grant VEGA 1/0863/17 and by the Slovak Research and Development Agency through grant APVV-15-0558.

REFERENCES

[1] M. Jobmann, „MODERN 2020," Second Technical Meeting of the Working Group on the Use of Monitoring Programme in the Safe Development of Geological Disposal Facilities for Radioactive Waste, Vienna, Austria, December 2018.

[2] OECD/NEA, „Preservation of Records, Knowledge and Memory across Generations, Monitoring of Geological Disposal Facilities - Technical and Societal Aspects,", February 2014, pp. 15-18.

[3] International Atomic Energy Agency (IAEA). „Reference Biospheres" for Solid Radioactive Waste Disposal. Report of BIOMASS Theme 1 of the BIOsphere Modelling and ASSessment (BIOMASS) Programme," ISBN 92-0-106303-2, Vienna, Austria, 2003.

[4] Radioactive Waste Repository Authority, "Reference Project of Geological Repository: Technological Part,". Czech Republic. 1999, pp. 50-55

[5] B. Vrban, J. Lüley, D. Barátová, J. Haščík, V. Nečas, "Parametric Study of Burnup Modelling Aspects of VVER-440/V213 Fuel," 26th International Conference Nuclear Energy for New Europe, Nuclear Society of Slovenia, Bled, Slovenia, September 2017.

[6] GoldSim Technology Group, LLC. GoldSim Contaminant Transport Module, User's Guide. version 7.0., Washington, USA, 2017.

[7] V. M. Pulkkanen, H. Nordman, "Modelling of Near-Field Radionuclide Transport Phenomena in a KBS-3V Type of Repository for Nuclear Waste with GoldSim Code – and Verification Against Previous Methods.". Finland: Posiva Oy, March 2010, pp. 37.

[8] Nordén S. et. al., "Element-specific and constant parameters used for dose calculation in SR-Site", Svensk Kärnbränslehantering AB, Sweden, 2010.

[9] National Nuclear Fund for Decommissioning of Nuclear Facilities and for the Management of Spent Nuclear Fuel and Radioactive Waste, "National Program for Spent Nuclear Fuel and Radioactive Waste Management in Slovak Republic," , Slovak Republic, August 2014.

[10] International Atomic Energy Agency (IAEA). Technical Report Series no, 472., "Handbook of Parameter Values for the Prediction of Radionuclide Transfer in Terrestrial and Freshwater Environments," ISBN 978-92-0-113009-9, Vienna, Austria, 2010, pp. 42-62.

[11] Statistical Office of Slovak Republic, 2017.

[12] Annals of the ICRP, ICRP Publication 122, "Radiological Protection in Geological Disposal of Long-lived Solid Radioactive Waste,". 2013, pp. 14.

EPJ Web of Conferences **225**, 07001 (2020)
ANIMMA 2019

https://doi.org/10.1051/epjconf/202022507001

Preliminary Study on Improving Resolution of D-T Neutron Radiography based on Associated Alpha and Coded Source Imaging Methods

Sheng Wang*, Yang Wu, Heyong Huo, Hang Li, Chunlei Wu, Li An, Bin Tang, Zhenghong Li
* Institute of Nuclear Physics and Chemistry, CAEP
mritn5851@gmail.com

***Abstract*—Limitations of fast neutron radiography include low detection efficiency and poor spatial resolution. D-T neutron radiography is one compact fast neutron radiography method. Based on D-T associated alpha particle method and coded source imaging method, we indicate one new method to improve resolution of D-T neutron radiography. This method could get distribution of D-T neutrons by detecting alpha particles. Without real coded mask, the D-T radiography structure is considered as coded source imaging of fast neutrons. With reconstruction method, the real object could be reconstructed from projections. One prospect setup of D-T associated alpha neutron source has been carried out with Monte-Carlo simulation. The projection images of two different situations are collected and reconstruction results show that it's possible to improve image quality of D-T neutron radiography.**

***Keywords*—D-T Neutron Radiography, Associated Alpha Method, Coded Source Imaging Method, Reconstruction**

I. Introduction

NEUTRON radiography (NR) is one of the most important nondestructive testing methods, which is sensitive to low density materials. Especially, D-T Neutron radiography could penetrate heavy shielding materials of large thickness and distinguish flaws of low density materials [1, 2]. Traditional D-T neutron radiography could supply L/D ratio with several hundred limited by the size of neutron source and distance between source and detector, which restricts the resolution of D-T neutron radiography.

Based on the associated alpha particles of D-T source, one could acquire distribution of D-T neutron source with a small pinhole, and then subdivide the whole neutron source into smaller neutron sources, which could improve the L/D ratio with several multiples. With method of coded source imaging (CSI) [3-6], then one could reconstruct images from raw projections.

This method is similar as associated particle imaging (API) [7, 8], but it's not interrogation method based on time of flight of D-T neutrons and scattering neutrons. In this method, one needs an alpha particle detector to collect alpha particle distribution without time detection ability. The neutron detector only need position detection ability too. It means that requirement of detectors could be simpler than API method.

We indicate principle of this method, and carry out Monte-Carlo simulation with a glancing incidence deuterium beams, getting the distribution of D-T neutron sources and reconstruction results from projection images. The results show that it's possible to improve image quality of D-T neutron radiography.

Manuscript received July 4, 2019. This work was supported by National Natural Science Foundation of China (Grant no. 11605168).

Sheng Wang is with Institute of Nuclear Physics and Chemistry, CAEP, Mianyang of Sichuan Prov., 621900, China. (telephone: 0086 0816 249 4238, e-mail: mritn5851@gmail.com).

II. Principle

The idea of this method is shown in Fig. 1. With a pinhole between D-T neutron source and Alpha particle detector, one could collect the distribution of alpha particles, which means that the distribution of D-T neutron could be recorded. Then neutron source is subdivided as multi smaller neutron sources. The projection of D-T neutron radiography could be collected by one neutron imaging detector. The projection is convolution of neutrons distribution and object distribution. With reconstruction method, we could get the object's reconstruction images from projection images.

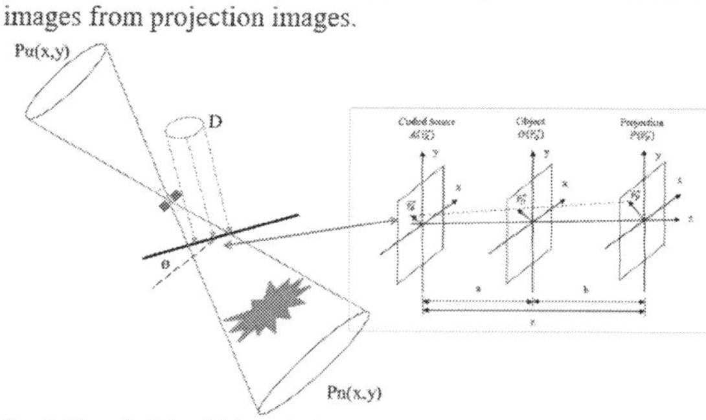

Fig. 1. The principle of this method

III. D-T Neutron Distribution

Based on theory analysis we could get the relation as follow:

$$\frac{d_\alpha}{d_h} = \frac{x}{y} \tag{1}$$

$$l_\alpha = x + y \tag{2}$$

$$\frac{d_h}{d_s} = \frac{y}{y + l_s} \tag{3}$$

Where D beam incidence angle is θ_s, pinhole diameter is d_h, alpha detector resolution is d_α. The distance between alpha detector and pinhole as is l_a, and distance between pinhole and neutron source is l_s.

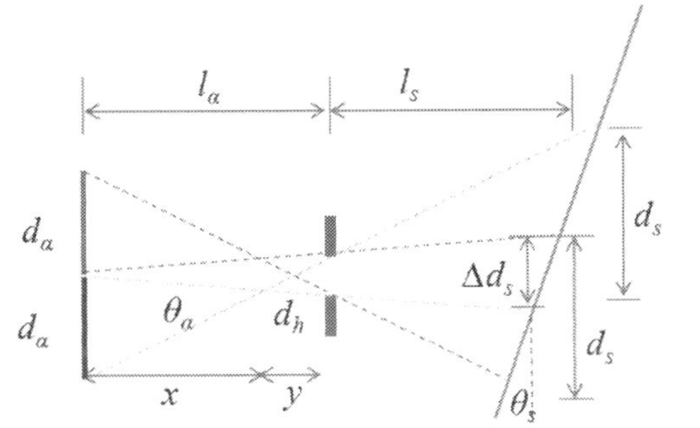

Fig. 2. The sketch map of neutron source resolution calculation

The resolution d_s of neutron source as Eq. (4), and difference Δd_s as Eq. (5).

$$d_s = d_h + \frac{l_s}{l_a}(d_a + d_h) \tag{4}$$

$$\Delta d_s = d_h \frac{l_a + l_s}{l_a} \tag{5}$$

With Monte-Carlo simulation, we have calculated D-T neutron source distribution, with D beam as diameter 10mm and incidence angle θ_s as 20 degrees, pinhole diameter d_h as 0.1mm, alpha detector resolution d_α as 1 mm. The distance l_α between alpha detector and pinhole is 10 cm, and distance l_s between pinhole and neutron source is 5 cm. Under conditions above, the largest distance differ between T target and α detector is 10mm×sin20°≈3.4 mm, compared with total 150 mm could been ignored. The slope of T target will affect much smaller between T target and neutron detector.

The resolution d_s of neutron source is 0.65 mm as Eq. (4), and difference Δd_s is 0.15 mm as Eq. (5). The image of neutron source is shown in Fig. 3. It has improved the L/D ratio almost 15 times. Especially, in this setup, one single pixel of neutron is 0.65 mm×0.65 mm=0.4225 mm^2, which include parts belonging to itself with 0.35 mm×0.35 mm=0.1225 mm^2 and parts sharing with adjacent pixels with 0.15 mm×0.5 mm×4=0.3 mm^2. In this paper we assume neutron distribution is even, and the affection is small. But for other neutron distribution, one needs to decrease diameter of mask to get a smaller Δd_s.

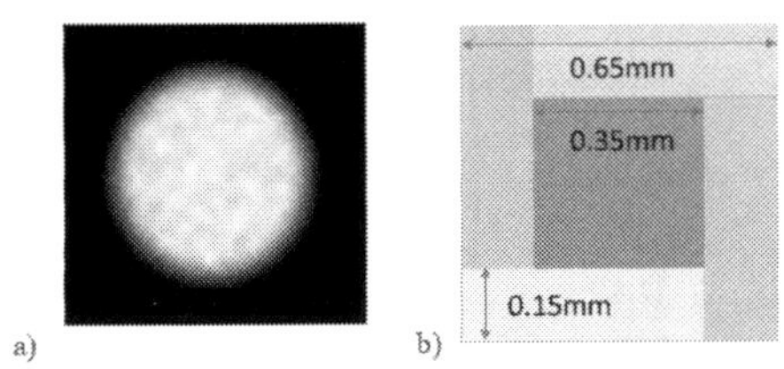

Fig. 3. The distribution of D-T neutron source, a) Neutron distribution, b) Pixel distribution

IV. Coded Source Imaging

The traditional NR images will choose a suitable distance between neutron source and neutron detector when the size of neutron source has been determined, which could supply enough neutron flux at imaging plane. CSI could increase L/D ratio by subdivide one big neutron source into smaller neutron sources [9]. It is shown in Fig. 4. The object is compressed into 2D distribution as an idea matrix.

The projection of CSI is a matrix with neutron source distribution and object distribution as follow:

$$P = A \otimes O \tag{6}$$

Where P is the matrix of projection, A is the neutron distribution, O is the object matrix, $\otimes$ is the symbol of convolution.

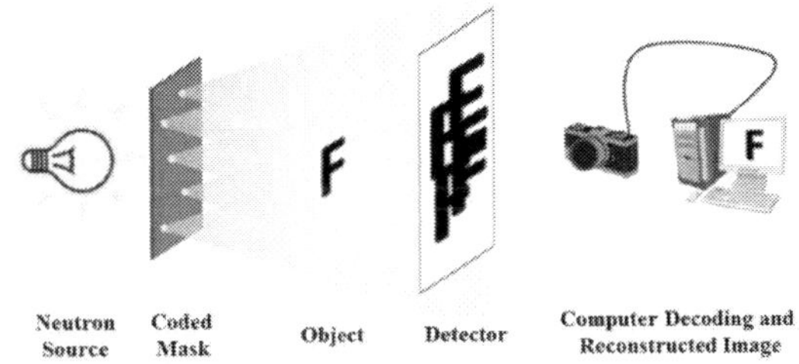

Fig. 4. The sketch map of CSI

If the object is too large, the fro and back of object couldn't been considered in on pixel at neutron. According the analysis result [10], the pixel zoom as ω, a, b and z as Fig. 1, the largest object thickness differs from center is da, as follow:

$$\frac{d\omega}{\omega} = \pm \frac{z}{ab} da \tag{7}$$

Three usually used reconstruction methods of correlation reconstruction method, Wiener filter deconvolution method and maximum likelihood iteration method [5, 6], could be used for coded source imaging. Considering that neutron distribution matrix is much smaller than object matrix in this work, and correlation reconstruction need the mask satisfy correlation requirement [11], we only use Wiener filter deconvolution method and maximum likelihood iteration method.

Wiener filter deconvolution method is shown in Eq. (8). C_{wiener} is a constant value relating with the noise level, FT and FT^{-1} is Fourier transform and inverse Fourier transform, and the equation is point-wise [5].

$$O_{deco} = FT^{-1}\left\{\frac{FT(P)}{FT(A)} \cdot \frac{|FT(A)|^2}{|FT(A)|^2 + C_{wiener}}\right\} \tag{8}$$

Maximum likelihood iteration method of Richardson-Lucy is shown in Eq. (9) [5].

$$O_{guess}^{new}(x,y) = O_{guess}^{old}(x,y) \cdot \frac{P(x,y) \otimes A(-x,-y)}{\left[A(x,y) \otimes O_{guess}^{old}(x,y)\right] \otimes A(-x,-y)} \tag{9}$$

V. Monte-Carlo Simulation

The projection of D-T neutron radiography also has been

https://doi.org/10.1051/epjconf/202022507001

calculated with neutron field as $5.0\times10^{10}s^{-1}$, acquiring time as 10 seconds when detector efficiency is 10%, neutron detector resolution as 1 mm, distance between neutron source and detector as 1 m.

The ideal object is composed of polyethylene and iron as Fig. 5, with different size of dots (diameter of 1 mm, 2 mm, 4 mm, 6 mm, 8 mm and 10 mm) and slits (width of 1 mm, 2 mm, 3 mm, 4 mm, 5 mm and 10 mm) in the polyethylene part, thickness as 1 cm and length as 8 cm. The thickness of iron part is 3 cm, and the length is 10 cm. The total object is enwrapped by one square diaphragm with thickness as 10 cm, inside length as 10 cm and outside length as 20 cm.

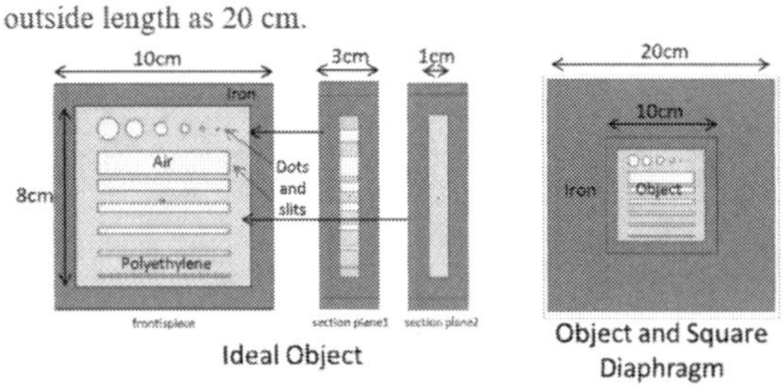

Fig. 5. The sketch map of object

Based on the resolution of neutron source and detector, distance between object and detector is determined as [6]. The distance b between object center and detector should be 60.6cm. The projection image could be calculated with Monte-Carlo simulation, it size is 130×130. With Eq. (7), supposed that $d\omega/\omega$ is $\sqrt{0.001}\approx3.16\%$, with total distance as 1000 mm, pixel of neutron as 0.65 mm, pixel of detector as 1 mm, then da is about 7.5 mm. The total thickness of object could be 15mm, which is larger than light materials of object.

Based on deconvolution and maximum likelihood iteration method of Richardson-Lucy, we could get the reconstruction image from projection image. The reconstruct images show that distribution of dots could be distinguished by this method, and image quality could be improved from projection image as Fig. 6.

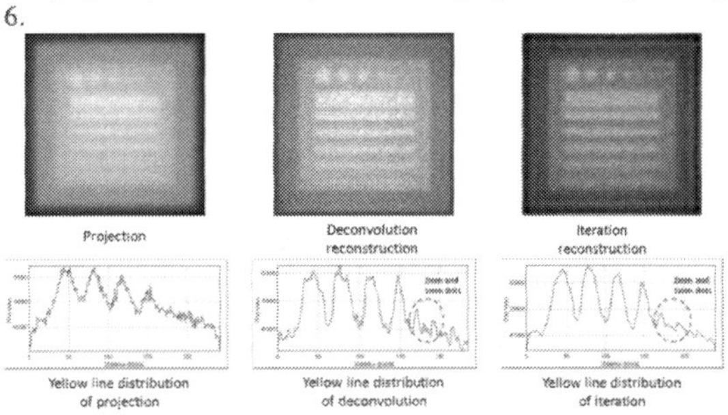

Fig. 6. The results of Monte-Carlo simulation when distance between object and detector as 60.6cm

Compared with traditional D-T neutron radiography, the distance between object and neutron detector is much farther. It will extend projection images and impair L/D ratio improvement. We could subdivide the projection pixel into smaller pixel to fulfill requirement of coded source imaging. On condition that distance between object center and detector as 3 cm, pixel of neutron detector should be 0.02 mm. With calculation resolution of neutron detector as 1mm, projection image matrix will be enlarged for 50 times. The other conditions are same as above. Two setups is shown in Fig. 7.

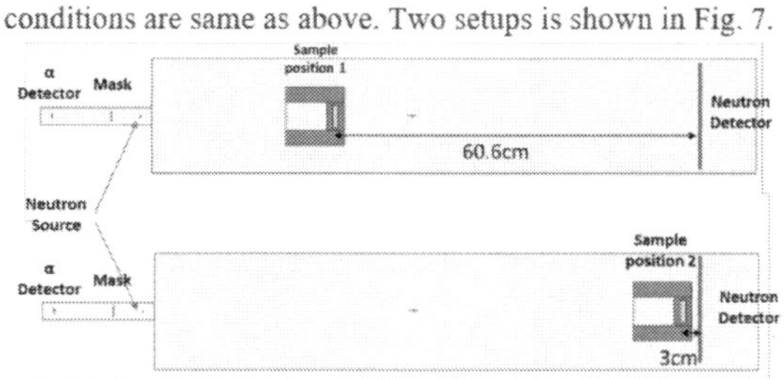

Fig. 7 Two setups of geometer

The projection and reconstruct images are shown in Fig. 8. With newly method of reconstruction, the image's quality also improves. Limited by calculation speed, SNR of projection images is not so high enough, and reconstruction images could be better when we increase account of neutrons.

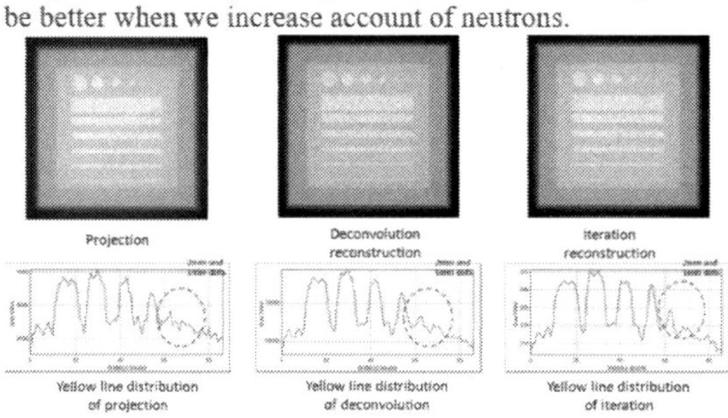

Fig 8. The results of Monte-Carlo simulation when distance between object and detector as 3cm

VI. Conclusion

With theory analysis and Monte-Carlo simulation, we indicate one new method to improve D-T neutron radiography quality. With associated alpha particle distribution detection and D-T neutron projection images detection, we could get the reconstruction results of object distribution by coded source imaging method. The key component of this method is an associated alpha particle D-T neutron source. The neutron detection could use common neutron imaging detectors. It supply one possibility to improve L/D ratio of D-T neutron radiography with more than 10 times. On the other side, keeping L/D ratio as traditional D-T neutron radiography, we could enlarge size of D-T neutron source by this method. It will decrease power intensity in T target, which is beneficial to improving neutron source lifetime and neutron flux.

This paper only analysis the ideal principle and carry out Monte-Carlo simulation, it need much more work to push it into application, such as α detector and neutron target structure improvement. On the other hand, this method still has some limitations. The CSI structure limited the testing size of object, and α detection for neutron distribution increase the noise, it need more analysis and simulation to estimate its feasibility.

VII. Acknowledge

This work was financially supported by National Natural Science Foundation of China (Grant no. 11605168).

References

[1] B. Tang, "The plot experimental study of 14MeV fast neutron digital radiography," *Chinese Science*, vol. 52, no. 9, pp. 1330-1336, 2009.
[2] B. Tang, "The physics analysis and experiment study of zinc sulphide scintillator for fast neutron radiography," *Nucl. Ins. and Meth. A*, vol. 729, no. 327-333, 2013.
[3] A. L. Damato, "Coded Source Imaging for Neutrons and X-Rays," *IEEE NSS*, vol. 11, no. 6, pp. 199-203, 2006.
[4] A. L. Damato, "Progress Report on a Neutron Coded Source Phase Contrast Imaging System at the MIT Reactor," *IEEE NSS*, vol. 22, no. 440, pp. 1725-1728, 2007.
[5] F. Grünauer, "Design, Optimization, and Implementation of the New Neutron Radiography Facility at FRM-II," Techische Universitat Munchen, Munich, Germany, 2005.
[6] S. Wang, "Coded source imaging simulation with visible light", *Nucl. Ins. and Meth. A*, vol. 65, pp. 187-191, 2011.
[7] J.T. Mihalczo, "Physical description of nuclear materials identication system (NMIS) signatures," *Nucl. Ins. and Meth. A*, vol. 450, pp. 531-555, 2000.
[8] A. J. Antolak, "Negative ion-driven associated particle neutron generator," *Nucl. Ins. and Meth. A*, vol. 806, pp. 30-35, 2016.
[9] S. Wang, "Preliminary Study of Nuclear Fuel Element Testing Based on Coded Source Neutron Imaging," ANIMMA 2015 Conference.
[10] S. WANG, "Study on Coded Source Neutron Imaging Technology based on a Compact Accelerator," Peking University, Beijing, China, 2013.
[11] A. L Damato, "Capabilities and limitation of phase contrast imaging techniques with X-rays and neutrons," Massachusetts Institute of Technology, Cambridge, USA, 2001.

EPJ Web of Conferences **225**, 07002 (2020) https://doi.org/10.1051/epjconf/202022507002
ANIMMA 2019

Design and Optimisation of a Three Layers Thermal Neutron, Fast Neutron and Gamma-Ray Imaging System

H. Al Hamrashdi, S. D. Monk, and D. Cheneler
(Lancaster University, Engineering Department)

***Abstract*—The design and configuration of a multi-layered imaging system with the ability to detect thermal neutrons, fast neutrons and gamma rays has been developed and its efficacy demonstrated. The work presented here numerically determines the systems efficiency and spatial resolution, using ^{252}Cf and ^{137}Cs as a case study. The novelty of this detection system lies in the use of small form factor detectors in a three-layer design, which utilises neutron elastic scattering and Compton scattering simultaneously. The current configuration consists of 10 mm thick natural lithium glass (GS10) scintillator integrated with a 20 mm thick plastic scintillator (EJ-204) in the first layer, a 15 mm thick lithium glass (GS10) scintillator in the second and a 30 mm thick CsI(Tl) scintillator forming the final layer. Each of these layers is backed with an 8 x 8 silicon photomultiplier diode (SiPM) array. The overall size of the imaging system is 27 mm x 27 mm x 135 mm. MCNPv6.1 and Geant4-10.04 were alternatively used to optimise the overall configuration and to investigate detection modalities. Results show promising performance with high precision source localisation and characterization abilities. Measurements were virtually obtained of two gamma-ray sources within steel enclosures at angles of 15°, 30° and 50° separation in order to test spatial resolution ability of the system. With the current active size of the system and the 8x8 SiPM configuration, the results estimate the spatial resolution to be close to 30°. The ability of the system to characterise and identify sources based on the type and energy of the radiation emitted, has been investigated and results show that for all radiation types the system can identify the source energy within the energy range of typical reported sources in literature.**

***Keywords*—Neutrons, Gamma-rays, Lithium glass, EJ-204, CsI(Tl), Nuclear Materials Security, MCNPv6.1, Geant4.**

I. INTRODUCTION

The detection and localization of special nuclear materials (SNM) and radioactive materials in general, is a significant problem for national security agencies across the globe and international organisations such as the IAEA [1, 2]. Over the last two decades, much research has involved the investigation of the design and the build of passive neutron/gamma mobile detection systems for safeguard and security applications [3]. The size of these systems range from vehicle scale down to wearable technology scale [4-10]. Many vehicle size systems, such as [4-6], are designed to detect strong-orphaned sources at long distances with a very wide field of view, although these systems have limited portability. Handheld devices offer an efficient solution for personal and luggage inspection at security checking points and within potentially crowded areas such as airports, but current systems have limited functionality. Commercially available handheld devices, such as Thermo Scientific™, PackEye Radiation Detection Backpack [11] and The Bruker Radiation Backpack Sentry [12] offer localisation abilities for gamma sources only, and have a relatively narrow field of view. To enhance systems abilities to identify and characterise radioactive materials and SNM, the focus on simultaneous detection of both neutrons and gamma-rays has grown noticeably in the last decade. An example of a handheld device that employs the characteristics of modern scintillation materials with dual particle detection abilities is a Cs_2LiYCl_6:Ce (CLYC) based monitoring device [7] which claims radioisotope identification and localisation of thermal neutron and gamma–ray sources. Some coded aperture based imaging systems that are related to homeland security and safeguard applications with dual-mode capability offer enhanced sensitivity and source localization abilities. Examples include the three-dimensional localisation of radioactive sources using liquid scintillators [8] and the CLYC based RadCam imaging system which are based on the RadCam gamma imaging system [9]. Multi-layered dual-particles imaging systems use neutron scattering and Compton scattering techniques to offer a better field of view and higher detection efficiency [10, 13]. For example, the dual-particle imager by Polack et al. [10] allows for accurate detection and source localisation even in the presence of shielding. This offers a larger range of detection materials, shorter acquisition time and a wider range of targets.

The aim of the work here is to design and optimise an innovative handheld, real-time dual-particle imaging system that is capable of simultaneously detecting thermal neutrons, fast neutrons and gamma rays using Monte Carlo simulations. The design utilises multi-layered system techniques to enhance the spectral and temporal abilities of the system, while keeping the weight and size flexible. The proposed design will allow the system to image, localise and

https://doi.org/10.1051/epjconf/202022507002

characterise a wide range of radioactive sources and SNM in a real time frame. The system is capable of detecting three different types of radiation, allowing a wider range of radioactive sources and SNMs monitoring capabilities.

II. IMAGING SYSTEM CONCEPT

The multi-layered imaging system described here is based on the combination of three different modes of radiation detection in one handheld device: thermal neutron capture reactions in ^{6}Li nuclei, neutron scattering physics for fast neutrons detection and Compton scattering physics for gamma-ray detection. Scintillation crystal candidates along with their interaction probabilities and proposed thicknesses for a multi-particle detection system were extensively studied in previous work [14]. The work presented here goes beyond this earlier study in order to better understand the interaction of these detection subsystems and to verify the efficacy of their combination to form a more comprehensive radiation detection system.

The first layer consists of a 10 mm thick Lithium glass (GS10) scintillator (Scintacor Ltd) combined with a 20 mm thick EJ-204 plastic scintillator. The second layer features a 15 mm thick Lithium glass (GS10) scintillator (Scintacor Ltd), with a 30 mm thick CsI(Tl) scintillator forming the final absorption layer. The rational behind the use of GS10 (natural Lithium content) rather than GS20 (95% ^{6}Li content) in the first and second layers is mainly due to technical and financial reasons. GS10 will serve the purpose of proving the ability of the system to capture thermal neutrons with 10 mm thick crystal [15] and sufficiently scatter fast neutrons and gamma rays [14]. Secondly, GS10 was found to be more practical in the current prototype due to financial restrictions. It is assumed here that SensL's ArrayJ-30035-64P-PCB 8 x 8 SiPM is to be used as the photodetector throughout the design [16]. SiPM's are compact in size and require a relatively low operating voltage (~30V) offering a solution for small size imaging systems. This 64-pixel avalanche photodiode offers a segmented detection of pulses generated on the area covered by the detector. Hence, the geometry used to define the pixels in the simulation were based on these SiPMs. A schematic of the system arrangement is shown in figure 2.1.

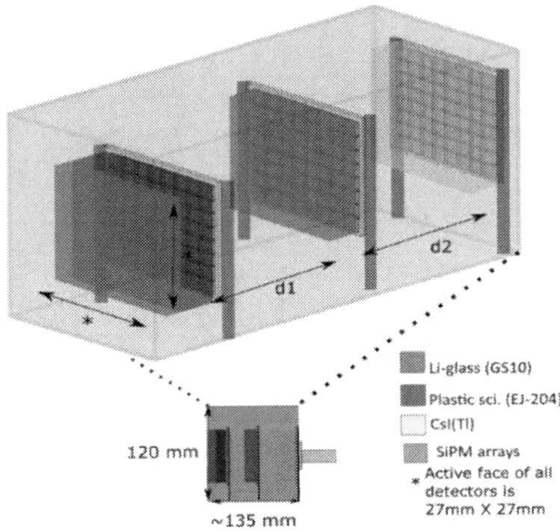

Fig. 2.1 Schematic of the multi-layered system configuration.

The thermal neutron detection mechanism in this imaging system is based on the ^{6}Li(n,α)^{3}H capture reaction in the first layer lithium glass detector, releasing alpha and triton particles. Lithium glass will emit detectable light as a result of the energy deposited by the resultant ionising particles [17]. The thermal capture cross section for this particular reaction is 940 barns with a Q value of +4.78 MeV. The aim of adding thermal neutron detection in the first layer is to enhance the energy range of detected neutron source. Although most SNM is characterised with neutrons in the fast energy range (>1 keV), shielding around smuggled or hidden sources can moderate the energy of these fast neutrons to be in the thermal neutron energy range.

Fast neutron imaging in the system is based on the physics of neutron elastic scattering in hydrogen rich EJ-204, followed by a second elastic scattering event in the Li-glass detector. The principle of fast neutron imaging in scattering cameras is thoroughly discussed within the literature [10, 13, 18-20]. The elastic scattering physics proportionally relates the neutron scattered angle in the first detector, θ, with the energy of the recoil proton, Ep and the energy of the scattered neutron, En' (equation 1)

$$\tan^2\theta = Ep/En' \quad (1)$$

The energy of the recoiled proton is measured in the first scattering plane. The energy of the scattered neutron is calculated using the time of flight (*ToF*) between the two neutron scattering planes as illustrated in equation 2.

$$En' = \frac{1}{2}\, m \left[\frac{d_1}{ToF}\right]^2 \quad (2)$$

Where m is the mass of the neutron and d_1 is the separation between the two scattering planes. The probability cone of the location of the source can be reconstructed using the scattering angle of each incident neutron. The intersection of the cones generated by all incident neutrons will form an image of the neutron source. Figure 2.2, shows a schematic of a neutron scattering event in the system.

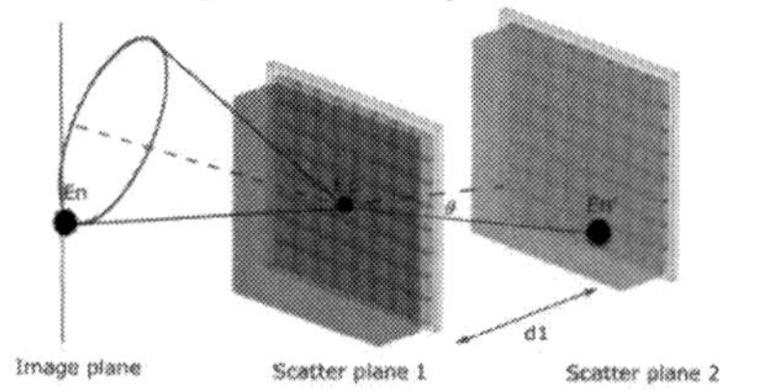

Fig. 2.2 Schematic of a neutron scattering system, with monolithic detectors and segmented photodetectors.

Gamma-ray detection in this system is based on Compton scattering kinematics; examples of the analysis of Compton camera in literature can be found abundantly [21, 22]. This method of gamma-ray detection has been studied extensively for the last four decades and has proved to offer a wider field of view and improved sensitivity over conventional mechanically collimated systems [23-25]. In principle, and as illustrated in figure 2.3, gamma rays scatter in the first plane depositing part of their energy before undergoing photoelectric absorption in a second high density scintillator. The energy and position of the two interactions are then utilised to find the angle between the incident and scattered gamma-rays, φ. Compton kinematics relates these parameters and allows the calculation of φ as in equation 3:

$$\cos\varphi = 1 + 0.511\text{MeV}\left[\frac{1}{E_1 + E_2} - \frac{1}{E_2}\right] \tag{3}$$

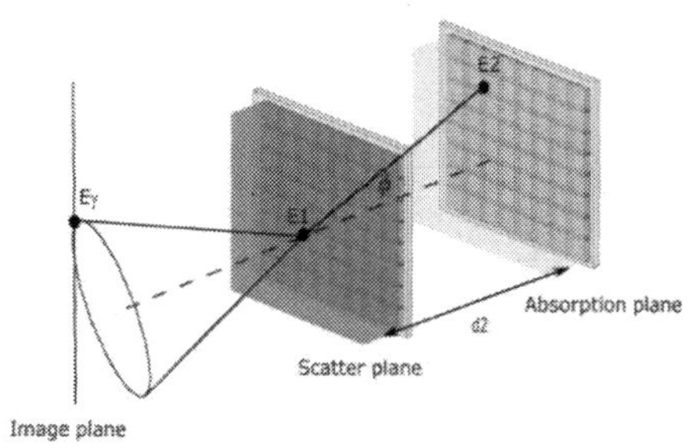

Fig. 2.3 Schematic of Compton scattering system with monolithic detectors and segmented photodetectors.

Where d2 is the separation between layers two and three. This will allow the generation of a probability cone of the incident gamma-ray location. The intersection of the cones created by all detected Compton scattering events is used to obtain the source location.

III. PROTOTYPE SETUP

MCNPv6.1 [26] and Geant4-10.04 [27] Monte Carlo codes have been used to investigate the optimum configuration of detectors and layers and then to test the feasibility and efficacy of this design. In the MCNP environment, neutron simulations utilised data from the ENDF70a file which is based on the ENDF/B-VII cross-section data [26]. The Evaluated Photon Data Library (EPDL) and the Activation Library (ACTL) compilations from Livermore data libraries were mainly used in the simulation of gamma photon transport [26]. All default options in physics cards are applied except for the neutrons physics card where analog capture is invoked. In Geant4, additional gamma simulations used the Electromagnetic (EM) physics package. The model is based on the emstandard physics list with the Livermore evaluated library [27]. For all simulations a point source configuration was used in testing the response of the system. The exact position and energy of the point sources used in the simulations is investigation specific and details are discussed during each investigation in the following sections.

In the current design and for all simulations, all layers have the same active front face size of 27 mm x 27 mm. Simulations were first used to investigate the effect of increasing the separation distance between detection layers, d1 and d2, and detection efficiency of the system. For the neutron scattering layers, a bare ^{252}Cf source 1 cm away from the system was used in MCNP6.1.0 simulations with no intervening materials and no scattering objects. Although simulations started at separation distance d1=10 mm, 30 mm is the minimum d1 in the current porotype configuration because of the physical size of the SiPM array and the associated breakout board. The relationship between the numbers of successive scattering events relative to the total number of fast neutrons entering the system and the separation distance is plotted in figure 3.2. The efficiency decreases with increasing plane separation, with the maximum efficiency observed at a 30 mm plane separation.

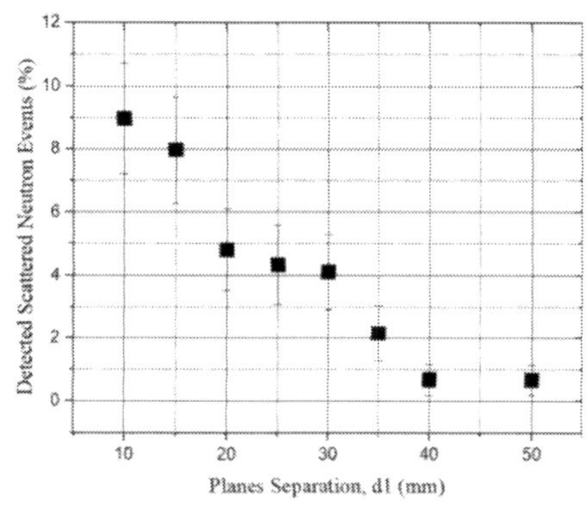

Fig. 3.2 Detected fast neutron scattering events as function of planes separation for ^{252}Cf source.

Optimising the value of d2 in the Compton camera subsystem was based on Geant4-10.04 simulations of four different gamma-ray energies in point source configuration with no intervening materials and no scattering objects. The relation between the fractions of Compton scattering events relevant to the total number of gamma rays entering the system was plotted in figure 3.3 as a function of the plane separation. As above, in reality the minimum plane separation starts at 30 mm because of the physical size of the SiPM and the associated breakout board in the current configuration.

https://doi.org/10.1051/epjconf/202022507002

The efficiency decreases with increasing planes separation with maximum efficiency at 30 mm.

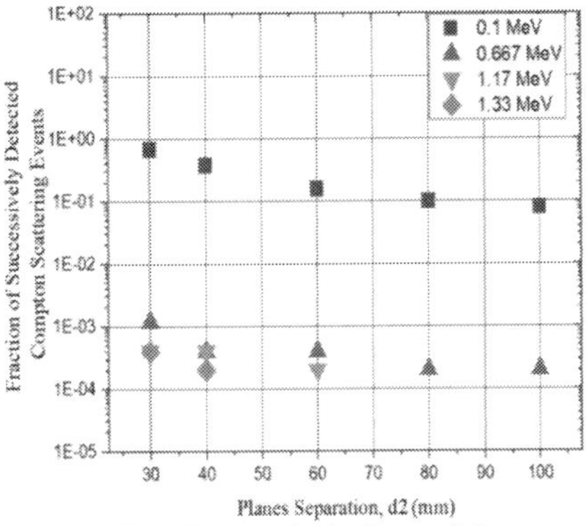

Fig. 3.3 Fraction of successively detected Compton events as function of planes separation.

The overall trend of the results shows that increasing the intra-plane distance between detectors for this multi-layered system decreases the probability of successively detected scattering events especially for fast neutrons. However, this flexible prototype configuration allows changing the size of the front face of layers which can consequently increases detection probabilities and improves geometric efficiency.

IV. Systems Response and Detection Characteristics

4.1 Systems response to a thermal neutron source.

Here, the ability of the system to locate and identify thermal neutron point sources in a 10 cm x 10 cm carbon steel enclosure was investigated. The source was located 1 cm away from the system at the centre of the detector with no intervening materials and no scattering objects. An example of the simulation made in the MCNPv6.1 environment is shown in figure 4.1. The 2D figure shows a top down view of thermal neutrons tracks inside the box and their trajectory in the detector. The 3D right hand figure shows the distribution of particles inside the carbon steel box.

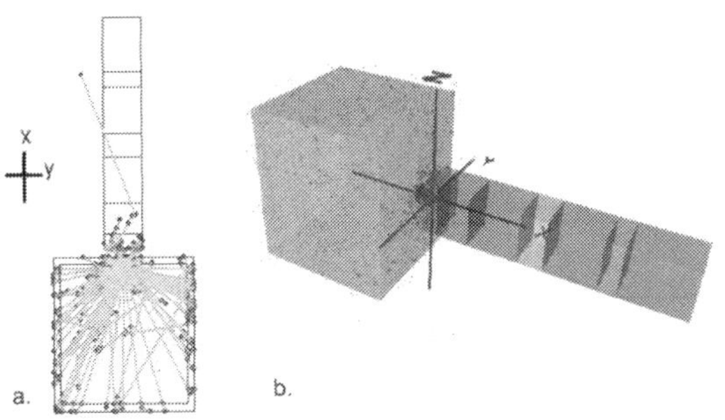

Fig. 4.1 a. 2D and b. 3D representation of the MCNPv6.1 simulation of a bare thermal neutrons source.

The results in this section are mainly presented as reconstruction images of the system response in an 8 x 8 array format (the size of the SiPM photodetector utilised in the design), which epitomises the configure of light output of each layer.

Figure 4.2 show image reconstruction of the signal generated in the lithium glass detector in the first layer (MCNPv6.1.0). The calculated intrinsic efficiency was found to be (79 ± 3)%.

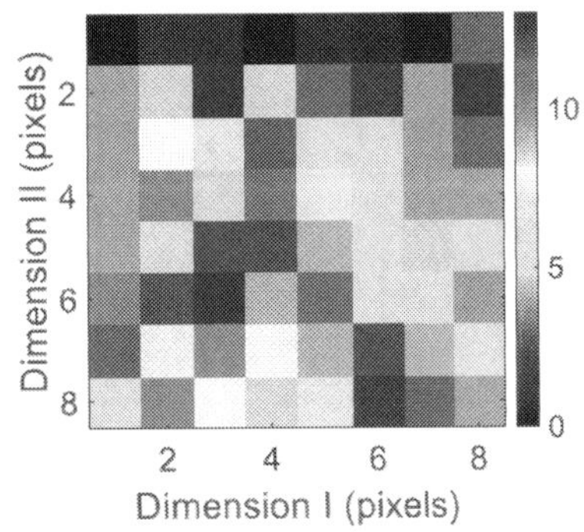

Fig. 4.2 System response to thermal neutron source located at pixels a. (4 ,4) cm, colour bar shows pixel intensity.

4.2 Systems response to fast neutron sources.

The response of the system to fast neutrons is based on neutron scattering kinematics in EJ-204 within the first layer and lithium glass (15mm) in the second layer. Using the PTRAC file generated in MCNPv6.1 simulations, energies of scattered fast neutrons and *ToF* in these two layers was found. Based on equations (1) and (2), estimation of the source location was obtained. In the first testing scenario, the response of the system to a bare ^{252}Cf source placed at different positions was investigated in order to evaluate the detection ability of the system as well as the spatial resolution of the system. The source to detector distance was 1 cm with no intervening materials and no scattering objects. The results in figure 4.5 illustrate the ability of the system to identify the direction of incident radiation using neutron scattering kinematics.

https://doi.org/10.1051/epjconf/202022507002

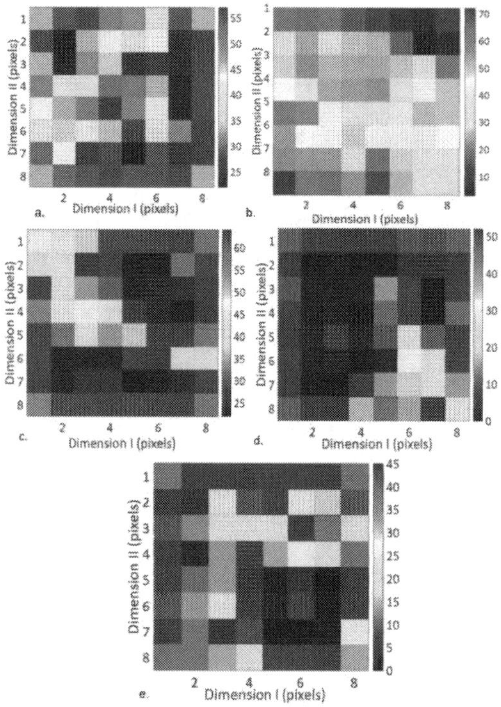

Fig. 4.5 The response of the system to a ^{252}Cf source located at five different positions at pixels a. (4, 4), b. (7,2) cm, c. (2,2) d.(7,7) e.(2,7) (Colour bars indicate image intensity).

The second test involved simulations utilising ^{252}Cf and $^{241}AmBe$ sources located 1cm from the detection plane, in order to test the ability of the system to identify two fast neutron sources based on their energies. Table 4.1 list reported average energies and calculated average energies of these two sources. These results illustrate the system's ability to identify different neutron sources based on their energies.

Table 4.1 Systems average energy calculations for different fast neutrons sources [17, 28].

Source	*Average Energy in Literature (MeV)*	*Average energy calculated from simulations (MeV)*
$^{241}AmBe$	*4.2*	*4.2 ± 1.3*
^{252}Cf	*2.35*	*2.0 ± 0.8*

4.3 Systems response to a gamma source.

The ability of the system to identify the location of gamma sources was investigated using a bare ^{137}Cs point source. In the simulations (Geant4), the source was positioned 30 cm away from the system, with no intervening materials and no scattering objects. Using equation 3 where E1 and E2 are the energies deposited in Li-glass and CsI(Tl) respectively, probability cones were constructed and the image intensities map of SiPM was constructed as illustrated in figure 4.6. The results in this figure show a high ability to precisely locate the position of a gamma-ray source.

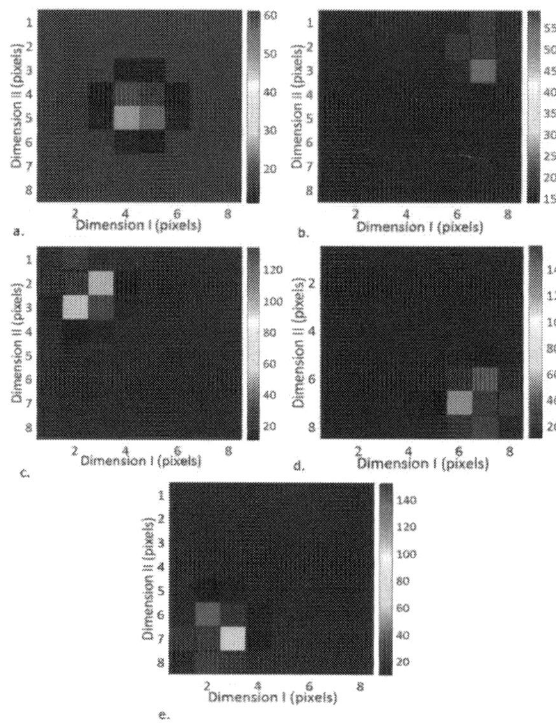

Fig. 4.6 The response of the system to a ^{137}Cs source located at five different pixel positions a. (5,5), b. (7, 2), c. (2,2) d.(7,7) e.(2,7) (Colour bars indicate image intensity).

In addition, the ability of the system to differentiate between closely spaced sources was investigated using two ^{137}Cs sources. The sources were placed 1 mm away from the imaging system. The results of this investigation are shown in figure 4.7 for 15°, 30° and 50° sources separation, with spatial resolution close to 30°.

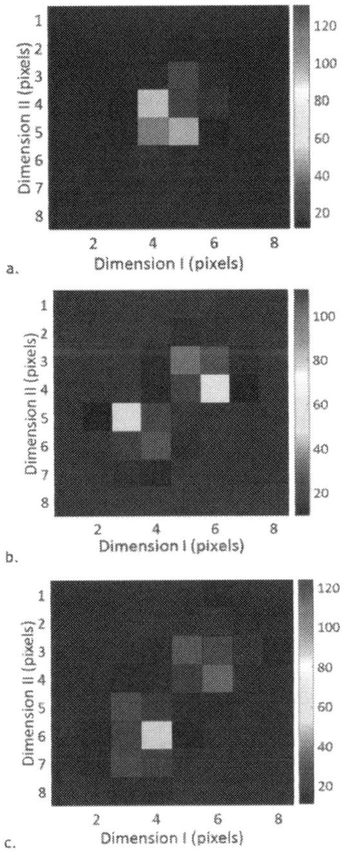

Fig. 4.7 System's response to two ^{137}Cs sources at a. 15° and b. 30° and c. 50° separation distance (Colour bars indicate image intensity).

The system's capability to identify and characterise gamma-ray sources according to their energy was investigated using the simulations of ^{137}Cs and ^{60}Co point sources. The sources were placed 1 mm away from the system. Using equation 3 with E1 and E2 being the eneries deposited in Li-glass and CsI(Tl) respectively, the average gamma-ray energy peaks were calculated. Table 4.2 shows calculated average energies from simulations and reported average energies in literature.

Source	*Average Energy in Literature (MeV)*	*Average energy calculated from simulations (MeV)*
^{137}Cs	*0.667*	*0.669 ± 0.006*
^{60}Co	*(1) 1.17*	*1.17± 0.01*
	(2) 1.33	*1.33±0.05*

V. Conclusion

This work discusses briefly the designing parameters and computational testing of the feasibility of a multi particle detection system that is based on three detection layers. Results show the prototype configuration is capable of detecting and localising thermal neutrons, fast neutrons and gamma-ray sources. The results from this study have been directly implemented to build the porotype imaging system which is at the moment under experimental investigation.

References

1. IAEA, *IAEA Inicdents and Trafficking Database (ITDB) Incidents of nuclear and other radioactive material out of regulatory control*, in *2018 Fact Sheet*. 2018, International Atomic Energy Agency Vienna, Austria.
2. D. Reilly, N.E., H. Smith, and S. Kreiner, *Passive Nondestructive Assay of Nuclear Materials,* , N.E. D. Reilly, H. Smith, and S. Kreiner Editor. 1991: National Technical Information Service, U.S. Department of Commerce, Washington, DC,.
3. Al Hamrashdi, H., S.D. Monk, and D. Cheneler, *Passive Gamma-Ray and Neutron Imaging Systems for National Security and Nuclear Non-Proliferation in Controlled and Uncontrolled Detection Areas: Review of Past and Current Status*. Sensors, 2019. **19**(11): p. 2638.
4. Zelakiewicz, S., et al., *SORIS—A standoff radiation imaging system*. Nuclear Instruments and Methods in Physics Research Section A: Accelerators, Spectrometers, Detectors and Associated Equipment, 2011. **652**(1): p. 5-9.
5. Ziock, K.P., et al. *Source-search sensitivity of a large-area, coded-aperture, gamma-ray imager*. in *Nuclear Science Symposium/Medical Imaging Conference*. 2004. Rome, ITALY.
6. Christian, J.F., et al. *Nuclear material detection techniques*. in *SPIE Defense and Security Symposium*. 2008. SPIE.
7. Budden, B.S., et al., *A Cs2LiYCl6:Ce-based advanced radiation monitoring device*. Nuclear Instruments and Methods in Physics Research Section A: Accelerators, Spectrometers, Detectors and Associated Equipment, 2015. **784**: p. 97-104.
8. Gamage, K.A.A., M.J. Joyce, and G.C. Taylor, *Investigation of three-dimensional localisation of radioactive sources using a fast organic liquid scintillator detector*. Nuclear Instruments and Methods in Physics Research Section A: Accelerators, Spectrometers, Detectors and Associated Equipment, 2013. **707**: p. 123-126.
9. Soundara-Pandian, L., et al., *Characterization of Large Volume CLYC Scintillators for Nuclear Security Applications*. IEEE Transactions on Nuclear Science, 2017. **64**(7): p. 1744-1748.
10. Polack, J.K., et al. *Dual-particle imager for standoff detection of special nuclear material*. in *2011 IEEE Nuclear Science Symposium Conference Record*. 2011. Valencia, Spain.
11. *PackEye Radiation Detection Backpack*. [cited 2018 26/10/2018]; Available from: https://www.thermofisher.com/order/catalog/product/FHT1377.
12. *Radiation Backpack Sentry*. 2012 [cited 2018 26/10/2018]; Available from:

https://www.bruker.com/fileadmin/user_upload/8-PDF-Docs/CBRNE_Detection/Literature/Backpack-Brochure-en.pdf.

13. Mascarenhas, N., et al., *Results With the Neutron Scatter Camera.* IEEE Transactions on Nuclear Science, 2009. **56**(3): p. 1269-1273.
14. Al Hamrashdi, H., D. Cheneler, and S.D. Monk, *Material optimization in dual particle detectors by comparing advanced scintillating materials using two Monte Carlo codes.* Nuclear Instruments and Methods in Physics Research Section A: Accelerators, Spectrometers, Detectors and Associated Equipment, 2017. **869**(Supplement C): p. 163-171.
15. Wraight, L.A., D.H.C. Harris, and P.A. Egelstaff, *Improvements in thermal neutron scintillation detectors for time-of-flight studies.* Nuclear Instruments and Methods, 1965. **33**(2): p. 181-193.
16. *SensL J-series.* [cited 2018 01/11/2018]; Available from: https://sensl.com/products/j-series/
17. Knoll, G.F., *Radiation detection and measurment.* Fourth ed. 2010, New York, NY, USA: John Wiley and Sons.
18. Vanier, P.E. and L. Forman. *Demonstration of a directional fast neutron detector.* in *IEEE Nuclear Science Symposium Conference Record, 2005.* 2005.
19. Vanier, P.E., et al. *Calibration and testing of a large-area fast-neutron directional detector.* in *2007 IEEE Nuclear Science Symposium Conference Record.* 2007. Honolulu, HI, USA.
20. Macri, J.R., et al. *The Fast Neutron Imaging Telescope (FNIT) - hardware development and prototype testing.* in *2007 IEEE Nuclear Science Symposium Conference Record.* 2007.
21. Todd, R.W., Nighting.Jm, and D.B. Everett, *PROPOSED GAMMA CAMERA.* Nature, 1974. **251**(5471): p. 132-134.
22. Herzo, D., et al., *A Large Double Scatter Telescope for Gamma Rays and Neutrons.* Nuclear Instruments and Methods, 1975. **123**(3): p. 583-597.
23. Kamae, T., N. Hanada, and R. Enomoto, *PROTOTYPE DESIGN OF MULTIPLE COMPTON GAMMA-RAY CAMERA.* Ieee Transactions on Nuclear Science, 1988. **35**(1): p. 352-355.
24. Everett, D.B., et al. *Gamma-radiation imaging-system based on compton-effect.* in *Proceedings of the Institution of Electrical Engineers-London.* 1977. IET Digital Library: London, UK.
25. Singh, M. and R.R. Brechner, *Experimental Test-Object Study Of Electronically Collimated SPECT.* Journal of Nuclear Medicine, 1990. **31**(2): p. 178-186.
26. Goorley, J.T.J., Michael R.; Booth, Thomas E.; Brown, Forrest, et al., *Initial MCNP6 Release Overview - MCNP6 version 1.0.* 2012.
27. Allison, J., et al., *Recent developments in GEANT4.* Nuclear Instruments & Methods in Physics Research Section a-Accelerators Spectrometers Detectors and Associated Equipment, 2016. **835**: p. 186-225.
28. Gozani, T., *Active Nondestructive Assay of Nuclear Materials, Principles and Applications* 1981, Washington DC: US Nuclear Regulatory Commission.

EPJ Web of Conferences **225**, 07003 (2020) https://doi.org/10.1051/epjconf/202022507003
ANIMMA 2019

Image texture analysis and colorimetry for the classification of uranium ore concentrate powders

Mara Marchetti*, Klaus Mayer, Maria Wallenius, Antonio Bulgheroni, Thierry Wiss, Klaus Lützenkirchen and Lorenzo Fongaro
* European Commission, Joint Research Centre (JRC)
mara.marchetti@ec.europa.eu

***Abstract*—In the context of nuclear security, uranium ore concentrates (UOCs) play an important role: they are traded in large quantities and this makes their use "out of regulatory control" a possible scenario.**
Once an incident of illicit trafficking o f n uclear m aterial is detected, an understanding of its origin and production process is required; this implies the necessity to use analytical techniques able to measure characteristic parameters (e.g. physical, chemical, isotopic characteristics of the nuclear materials) which are referred to, in the field of t he n uclear f orensics, a s signatures.
The present study investigates the potential of image texture analysis (i.e. the angle measure technique), combined with the spectrophotometric determination of colours for the evaluation of the origin of several UOCs. The use of different multivariate statistical techniques allows the categorization of about 80 different samples into a few groups of UOCs powders, which makes this approach a promising method complementing the already established methods in nuclear forensics.

Keywords—**Nuclear forensic science, Image texture analysis, spectrophotometry, PCA, SVM, AMT**

I. INTRODUCTION

THE illicit trafficking of nuclear and other radioactive materials represents a potential threat to the citizens, hence efforts are required in order to deter, detect, investigate and prosecute their illicit movement. In this context, nuclear forensics plays an important role. Since the early seizures of nuclear materials back in the 1990's an investigative approach has been gradually established for the identification of the origin and history of the material in a manner to support legal proceedings to the actors involved in the illicit trafficking [1]. This approach is the core of nuclear forensic science. The key in the tracing process is the identification of characteristic parameters or "signatures" which can relate a particular nuclear or other radioactive material to a production process, date and possibly to a production site. Examples of characteristic parameters are: chemical composition, isotopic abundances of U and Pu, metallic impurities or physical morphologies.
Incidents involving nuclear materials out of "regulatory control" in the last decades have raised the necessity to include the early stages of the nuclear fuel cycle in the nuclear forensics investigations: in this scenario increasing interest has been attributed to the uranium ore concentrates (UOCs). They are produced after milling, leaching and precipitating uranium from the uranium ores. UOCs contain about 60-80% uranium in different chemical composition (e.g. ammonium diuranate, uranyl hydroxide, uranyl peroxide, or uranium oxide [2]) and are available in large quantities; this may lead to an increased risk of thefts for diversion purposes.
Different signatures have been identified in the field of nuclear forensics for UOCs which are related to their origin or production process, namely the isotope ratio of U, in particular $^{238}U/^{234}U$ and $^{235}U/^{234}U$ [3]–[5], or that of Pb, Sr, Nd [6]–[9]), the measurement of trace level of ^{232}Th and ^{228}Th [10] or the impurities content [9], [11]. Recent works have highlighted the potential of discriminating different uranium ore concentrates on the basis of morphological information extracted by using image analysis. Keegan and co-workers have supported their conclusion on an unknown UOC sample thanks to a visual comparison performed via scanning electron microscopy (SEM) with the powder of the Mary Kathleen uranium mine [9]; Ho Mer Lin [12] and Fongaro et al. [13] have tested the potentiality of image texture analysis for the classification of 14 uranium ore concentrates characterised by different chemical compositions and production processes.
The present work aims to further investigate the approach used in [13]. Here, a batch of 79 commercial UOCs, produced in different facilities around the world (see section II), was used as training-samples; they were firstly classified on the basis of their colours by using diffuse-reflected spectrophotometry, cluster analysis and support vector machine (SVM) algorithm for classification. This allowed obtaining six different colour-groups of UOCs, hence performing a first differentiation among the powders. A further discrimination was then obtained by applying the Angle Measure Technique (AMT) algorithm within two colour-groups of samples; this last approach represents the preliminary phase of the image texture classification within each colour-group for the entire batch of 79 samples.

II. MATERIALS AND METHODS

A. *Uranium ore concentrate powders*

The 79 uranium commercial uranium ore concentrates investigated in the present experimental campaign are listed in Table I.

Mara Marchetti is with the European Commission, Joint Research Centre (JRC), Ispra, Italy e-mail: mara.marchetti@ec.europa.eu
Klaus Mayer, Maria Wallenius, Antonio Bulgheroni, Thierry Wiss, Klaus Lützenkirchen and Lorenzo Fongaro are with the European Commission, Joint Research Centre (JRC), Karlsruhe, Germany
Manuscript received May 13, 2019

TABLE I
UOC POWDERS MEASURED IN THE PRESENT EXPERIMENTAL CAMPAIGN.

1	England-Wheal Edward	41	Canada-North Span
2	USA-Kerr MgGee	42	USA-Utah
3	Spain-Jen	43	Australia-Ranger
4	Australia-Yeelirrie	44	Germany-Brunhilde
5	Namibia-Rossing	45	Germany-Ellweiler
6	Zambia-Mindola	46	Niger-Somair
7	USA-Cotter	47	Portugal
8	Gabon-EFI(Mouand)	48	USA-Lucky McGill
9	USA-Pathfinder	49	USA-Everest Black
10	Canada-Stanrock	50	Canada-Rio Algom
11	Holland-Delft	51	Canada-Rabbit Lake
12	Mozambique-Maruzi	52	USA-Everest Yellow
13	S.Africa-Palabora	53	Sweden-Ranstadt
14	Belgium-Belgian Congo	54	USA-El Mesquite
15	Brazil-Nuclebras	55	USA-Union Carbide
16	Spain-Enusa	56	Canada-Denison
17	Australia-Queensland	57	USA-Atlas
18	Germany-Wismut	58	Australia-Mary Kathleen
19	USA-Yankee Yellow	59	USA-United Uranium
20	Canada-Dyno	60	USA-South Dakota
21	Canada-Key Lake	61	Argentina
22	China-Hengyang	62	USA-Federal American Partners
23	USA-Petromic	63	USA-Dawn
24	Canada-Blind River	64	Canada-Milliken Lake
25	Canada-Gunnair	65	Australia-Rum Jungle
26	Yugoslavia-Spisak Black	66	Canada-El Dorado
27	Canada-Faraday	67	Canada-Ray Rock
28	Australia-Olympic Dam	68	USA-Chevron Hill
29	Yugoslavia-Spisak Yellow	69	S.Africa-Nufcor
30	USA-Mulberry	70	Russia-Techsnab
31	USA-Falls City	71	USA-United Nuclear
32	Canada-Madawaska	72	Yugoslavia-Rudnik
33	USA-Irigaray	73	Australia-South Alligator
34	USA-ESI	74	USA-Yankee Black
35	Canada-Macassa	75	USA-Sesquehanna
36	USA-Anaconda	76	Romania
37	USA-Shirley Basin	77	USA- White Mesa
38	USA-Mobil	78	USA-Homestake
39	Canada-ESI	79	Canada-Stanleigh
40	Australia-Radium Hill		

The set of UOCs consists of $\sim$ 30 ammonium diuranate, six sodium diuranate, $\sim$ 10 uranyl hydroxide, $\sim$ five uranyl peroxide, seven are triuranium octoxide and then two are of non-conventional chemical composition. Mixtures of ammonium diuranate-oxide or oxide-hydroxide complete the list.
For spectrophotometric investigations the samples were introduced in vials made of quartz (75 x 10 mm) or in vials of borosilicate glass (45 x 14.7 mm).
In the case of the SEM acquisitions, the powder was not coated due to the necessity of reusing it for future analyses. It was inserted inside a disc made of graphite and gently pressed with a weight of 1 g. In general, three different repetitions were used for each UOC powder (to consider the variability introduced with the preparation); this number was reduced to two when the available amount of sample was insufficient.

B. Spectrophotometry

The colour of the samples was measured by means of a portable Konika Minolta CM-700d sphere-type spectrophotometer, which is characterized by a silicon photodiode array detector and an integrating sphere having a diameter of 40 mm. This spectrophotometer uses a xenon lamp equipped with a UV cut filter and operates in the visible range of the electromagnetic spectrum ($\sim$ 400-700 nm).
The light reflected by the sample is collected in the integrating sphere and then is normalized to the zero reflection condition and to a pure white standard (100 % reflection). The reflectance values as a function of the wavelength were collected, together with the "tristimulus values" in terms of L^*, a^*, b^* in the L*a*b colour space, where [14]:

$$L^* = 116f(\frac{Y}{Y_n}) - 16 \tag{1}$$

$$a^* = 500(f(\frac{X}{X_n}) - f(\frac{Y}{Y_n})) \tag{2}$$

$$b^* = 200(f(\frac{Y}{Y_n}) - f(\frac{Z}{Z_n})) \tag{3}$$

X, Y, Z are the tristimulus values of the considered object; X_n, Y_n and Z_n are those of the perfect reflecting standard.

C. Scanning electron microscopy image acquisition and angle measure technique (AMT)

SEM images of the UOC powders: 10, 11, 33, 38, 51, 52, 54, 63, were acquired using a FIB/SEM FEI Versa 3D in low-vacuum mode (pressure = 10 Pa) equipped with a concentric backscattered detector for backscattered electrons.
In order to reduce the electrostatic charging phenomenon, due to the interaction between the electrons and the UOC powders, a voltage of 5 kV was applied to the electron beam and the images were acquired using a maximum magnification of 500 x, with a dimension of 3072 x 2048 pixels.
The acquisition was performed in integration mode (16 frames to obtain an image), with a dwell time of 300 ns. For each UOC powder 15 independent images per magnification were acquired and stacked, then mean-centred with a plugin for ImageJ 1.51j8 [15], in a manner to obtain a more uniform contrast prior to the analysis [13].
The texture of the images, which is strictly linked to the physical surface texture, is usually defined as the relationship of the intensity values, frequencies and spatial distribution among all the pixels in the image [16]. Different measures of the image texture are possible, one of these is the AMT.
The angle measure technique was introduced by Andrle [17] to characterize the complexity of geomeorphic lines. In image texture analysis, AMT operates in the following manner:

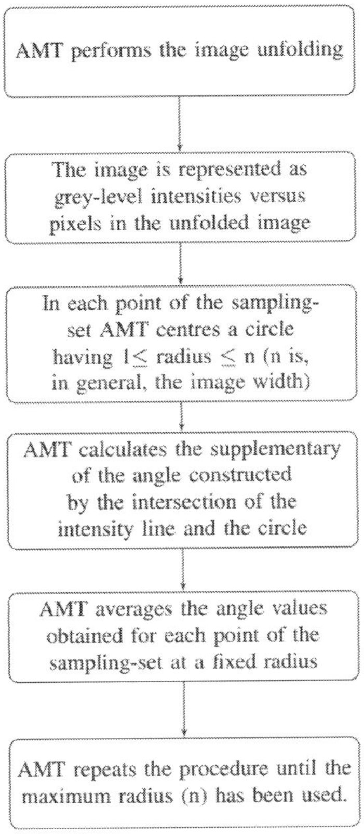

The reader can find detailed explanation of the AMT algorithm in [13], [17]–[19].

D. Multivariate statystical analysis: cluster analysis, principal components analysis (PCA) and support vector machine (SVM)

In the present work reflectance spectra and L^*, a^*, b^* values were analysed by means of cluster analysis and support vector machine algorithms. Principal components analysis was applied to the AMT spectra of two groups of samples. For multivariate statistical analysis we used a PLS Toolbox version 8.6.1 (Eigenvectors Research, Inc., USA) for Matlab 2017a (The Matworks Inc., Natick, MA, USA).

1) Cluster analysis: The aim of the cluster analysis is to group a large number of objects on the basis of an appropriate criterion applied to a matrix of multivariate measurements of these objects. The clustering must be performed in a manner to maximize the differences among the groups while minimizing those within them [20]. Different algorithms are available; here, the hierarchical Ward's method was used. This method joins two clusters that minimise the sum of the square errors [21], hence the increase of variation inside the groups.

2) Principal components analysis: PCA is a technique that allows extracting information from the covariance or correlation matrix in a manner that a group of variables can be reduced in dimensionality. PCA performs linear combinations of the observed variables that represent the so-called "principal components" and are able to explain most of the variance in the original variables. The coefficients of the linear combination are the loadings corresponding to each extracted principal component. [22]. PCA is a data reduction technique that is widely used in the explorative analysis.

3) Support vector machine: Support vector machine is a kernel-based algorithm for classification and regression analysis. In classification, SVM uses a function generated by learning from the original data which finds the hyperplane maximizing the distance among the available classes of data.

III. Results

A. Colour-groups

The 39 variables, corresponding to the reflectance values in the range 360 - 740 nm, were coupled with the L^*, a^*, b^* values for each powder in Table I and statistically treated. Fig. 1 shows the clustering representation obtained by applying Ward's method on the data pre-processed with the autoscale function. It is possible to choose different combinations, however, that with 6 colour-groups was chosen.
Once performed the colour-classes selection, these were renamed in: *Whyello* (white-yellow), *Yellow*, *Dyello* (Dark-yellow), *Orange*, *Brown* and *Black*.
It is worthwhile to notice that peroxides belong mostly to the first g roup, a mmonium d iuranate, s odium d iuranate and hydroxides to *Yellow*, *Dyello* and *Orange* (differences in the yellow nuances are most likely linked to the production processes). Mixed compounds are in the *Brown* class whilst the *Black* group contains oxides. This result highlights the capability of the statistical analysis in identifying links between the powders production process and their colours.

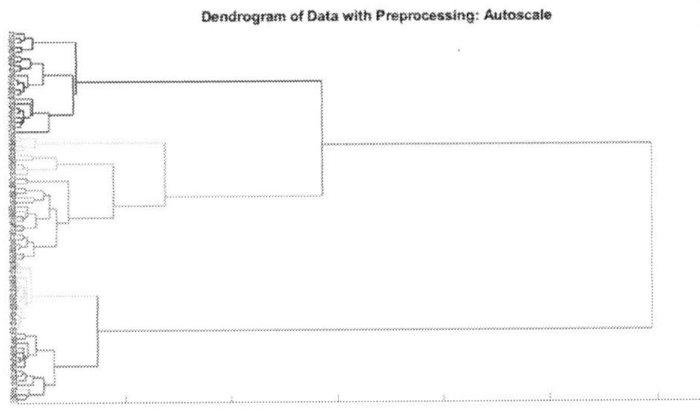

Fig. 1. Hierarchical cluster analysis of the entire samples collection. Different clusters are possible by selecting the position on the dendogram. In this work the 6 different colour-groups were chosen.

The above-mentioned classes were used as category variables for the SVM colour-based classification, hence each sample was renamed according to the corresponding class. To achieve the classification, SVM implemented with LIBSVM library and the Radial Basic Function Kernel were employed; data were then pre-processed by means of the autoscale function whilst a maximum cost of 50 was selected (50 was

chosen among the lower costs having the best performance). Table II lists the sensitivities and the specificities obtained with the present model which has a Root Mean Square Error Cross Validation (RMSECV) of 0.420969. The model is able to attribute each sample in Table I in the correct class.
Figures 2 and 3 provide an example of the results obtained: all the samples belonging to the *Whyello* and *Yellow* categories are correctly classified. These samples were successively used for the preliminary image texture analysis and sub-classification. The model was validated by using a set of additional measures of 25 UOC powders, which represents the validation dataset. The model attributes 24/25 samples to the correct class, while misclassifies one *Orange* sample as *Yellow*. This result appears satisfactory, especially considering the low RMSECV obtained.

TABLE II
SENSITIVITIES AND SPECIFICITIES IN CALIBRATION AND CROSS-VALIDATION FOR THE SVM CLASSIFICATION PERFORMED ON THE 6 COLOUR-CLASSES OF SAMPLES. THE ROOT MEAN SQUARED ERROR IN CROSS-VALIDATION (RMSECV): 0.420969

Sensitivity (Cal)	1.000	1.000	1.000	1.000	1.000	1.000
Specificity (Cal)	1.000	1.000	1.000	1.000	1.000	1.000
Sensitivity (CV):	0.933	1.000	0.955	1.000	0.800	1.000
Specificity (CV):	1.000	0.967	0.982	1.000	1.000	1.000
Class. Err (Cal):	0	0	0	0	0	0
Class. Err (CV):	0.0333	0.0164	0.0315	0	0.1	0

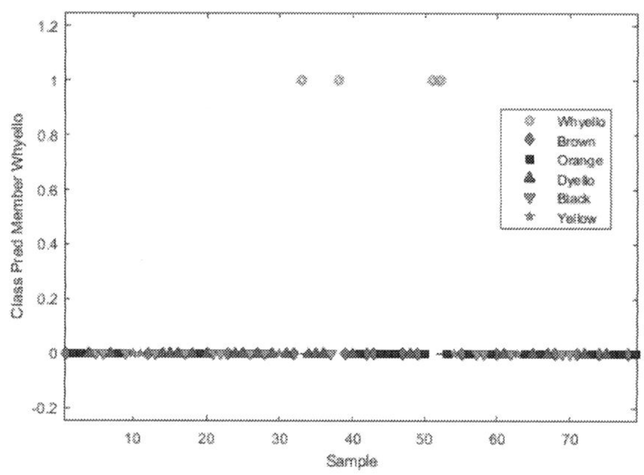

Fig. 2. Classification results for the samples pre-classified as *Whyello*. The classification prediction membership is measured either by a 0 (non-membership) or by a 1 (membership). These values are assigned on the basis of the highest probability of belonging to a particular group. Specifically, each sample in the *Whyello* class is correctly assigned.

B. Subgroups exploration

The determination of the colour-groups enabled us to perform first discrimination among the UOC powders available in the present study. To enhance the differentiation capability, the samples within a specific colour-group must be further separated by applying a technique capable to tackle possible

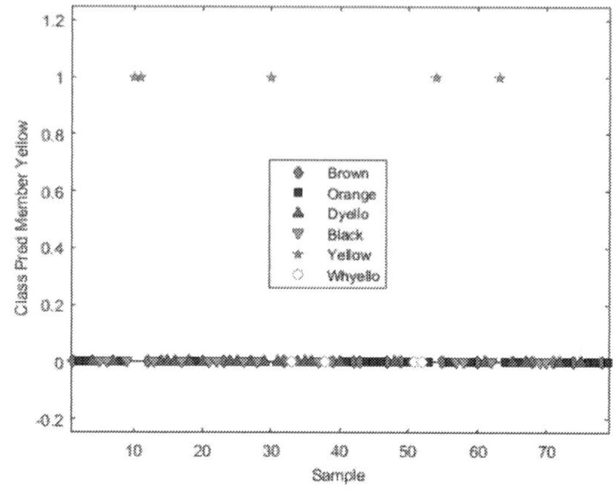

Fig. 3. Classification results for the samples pre-classified as *Yellow*. The classification prediction membership is measured either by a 0 (non-membership) or by a 1 (membership). These values are assigned on the basis of the highest probability of belonging to a particular group. Specifically, each sample belonging to the *Yellow* class is correctly assigned.

additional differences.
Most of the samples in one colour-group have the same chemical composition, however, the production route causes differences in the morphology of the powders, in particular in the powder's texture (i.e. particle size, size distribution, homogeneity and shape), which is reflected on the texture of the acquired images.
The images at 500 x of the samples in subsection II-C and belonging to the *Whyello* and *Yellow* groups were selected. In general acquisition of the images of peroxides was more difficult. The images of 51, 38, 52 and 33 (*Whyello*) were stacked together and pre-processed as already explained in subsection II-C; the same operation was performed on 10, 11, 54 and 63 (*Yellow*). The images were then reduced in dimension (as suggested by Kucheryavski and co-workers) [23] to 600 x 400 and a sampling of 2% was chosen for shorter calculation time. The mean angle values of the five replicates obtained for each sample repetition were averaged which provided three mean angle distributions for each facility-class. Fig. 4 shows the averaged mean angle spectral data for the *Yellow* category samples, together with their textural images. AMT spectra are in general complex and interpretation is not straightforward, however, high mean angle (MA) values can be interpreted as a high variation in the image texture complexity, while low values are more related to a lower variation [18]: in the *Yellow* category the highest complexity is displayed by the specimen number 10 (Canada Stanrock), the lowest by number 63 (USA Dawn).
These spectra were successively analysed using PCA (autoscale preprocessing and leave-one-out cross-validation): Figures 5 and 6 show respectively the score-plots of the *Whyello* and *Yellow* samples. In both cases two components were chosen: for *Yellow* UOCs this choice explains 99.35% of the variance (PC1= 94.6%; PC2= 4.7%); for *Whyello* UOCs the

explained variance is 86.19% (PC1= 48.6%; PC2= 37.59%). In both cases outliers in terms of Q-residuals and Hotelling number T^2 [24] were absent, only one average value of the sample 51 was borderline in terms of Q-residuals, but this value was included in the exploratory analysis. In both Figures 5 and 6 it is possible to identify samples clusters within the same colour-group: in the *Whyello*, overlapping between Canada Rabbit Lake (51) - USA Irigaray (33) and USA Mobil (38) - USA Everest Yellow (52) are clearly visible; in the *Yellow* group the facilities-classes are separated mostly along the PC1, where it is possible to recognize the "trend" observed in Fig. 4: USA Dawn (63) has the largest negative score along PC1 (lowest mean angle values), whilst Canada Stanrock the larger positive score along the same PC (higher mean angle values). The overlapping among clusters observed in the *Whyello* analysis is most probably caused by the difficult image acquisitions of the UOC powders containing peroxides, this would suggest that actions must be taken to limit the charging phenomena such as coating the powder with a conducting material.

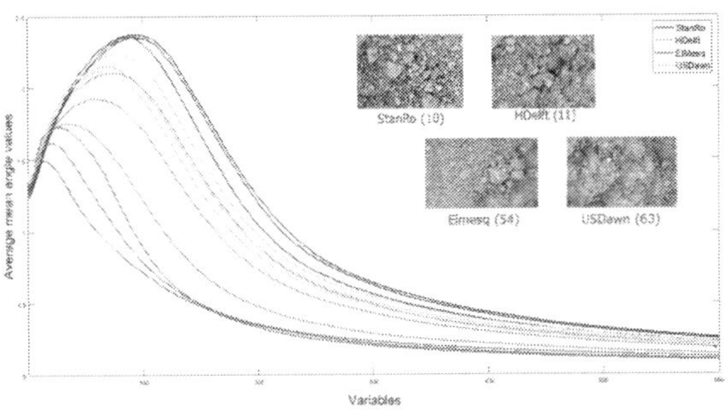

Fig. 4. Average mean angle spectra for the UOCs belonging to the *Yellow* group together with SEM images of the specimens displaying their texture

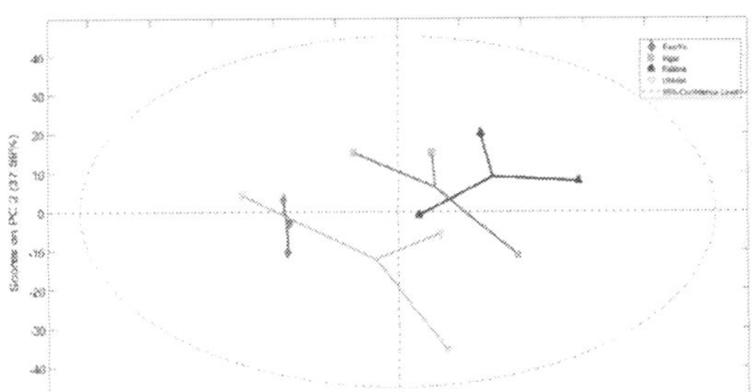

Fig. 5. PCA of the average mean angle spectra obtained with the UOC powders in the *Whyello* group

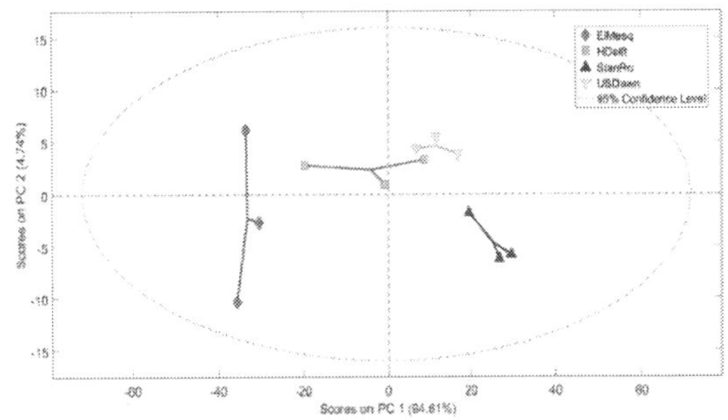

Fig. 6. PCA of the average mean angle spectra obtained with the UOC powders in the *Yellow* group

IV. Conclusion

The present study aims to investigate a new methodology in nuclear forensic science, capable to provide hints on the origin of uranium ore concentrates in a fast and simple manner. The route explores the potentiality of multivariate statistical analysis applied to techniques that could highlight differences among the powders which are based on the chemical composition and which were obtained through the same generic production process. The first part of the study provides a model based on the combination of spectrophotometry and support vector machine algorithm which is capable to restrict the field of investigation among a few facilities only on the basis of the colour of the sample. The model was validated and, then, finalised. The second part of the study is a preliminary attempt to distinguish among samples belonging to the same colour-group: to achieve this goal AMT image texture analysis was used and its potentiality, as signature in nuclear forensics, was investigated by using the Principal Components Analysis. In this last case results are promising, a first discrimination among the determined facilities-classes, thus the different production processes, appears possible. However, different algorithms or combination of other image texture analyses will be further investigated in order to obtain a model which can provide reliable nuclear forensics findings.

Acknowledgment

The authors convey their gratefulness to Alessandro Benedetti and Markus Ernstberger for the assistance received when using the FIB/SEM, Adrian Nicholl and Martin Vargas Zuniga for their openness toward helping out when needed. The present work has been financially supported by the European Commission in the frame of the Exploratory Research programmes.
The authors thank the Scientific Development Unit of the Joint Research Centre Ispra, for supporting the activities of the IDENTICLASS project.

References

[1] IAEA, "Nuclear forensics support," International Atomic Energy Agency, Tech. Rep., 2006.
[2] M. J. Kristo, A. M. Gaffney, N. Marks, K. Knight, W. S. Cassata, and I. D. Hutcheon, "Nuclear forensic science: Analysis of nuclear material out of regulatory control," *Annual Review of Earth and Planetary Sciences*, vol. 44, no. 1, pp. 555–579, 2016.

[3] E. Keegan, S. Richter, I. Kelly, H. Wong, P. Gadd, H. Kuehn, and A. Alonso-Munoz, "The provenance of Australian uranium ore concentrates by elemental and isotopic analysis," *Applied Geochemistry*, vol. 23, no. 4, pp. 765 – 777, 2008.
[4] G. A. Brennecka, L. E. Borg, I. D. Hutcheon, M. A. Sharp, and A. D. Anbar, "Natural variations in uranium isotope ratios of uranium ore concentrates: Understanding the 238U/235U fractionation mechanism," *Earth and Planetary Science Letters*, vol. 291, no. 1, pp. 228 – 233, 2010.
[5] T. L. Spano, A. Simonetti, E. Balboni, C. Dorais, and P. C. Burns, "Trace element and U isotope analysis of uraninite and ore concentrate: Applications for nuclear forensic investigations," *Applied Geochemistry*, vol. 84, pp. 277 – 285, 2017.
[6] Z. Varga, M. Wallenius, K. Mayer, E. Keegan, and S. Millet, "Application of lead and strontium isotope ratio measurements for the origin assessment of uranium ore concentrates," *Analytical Chemistry*, vol. 81, no. 20, pp. 8327–8334, 2009.
[7] J. Švedkauskaite LeGore, K. Mayer, S. Millet, A. Nicholl, G. Rasmussen, and D. Baltrunas, "Investigation of the isotopic composition of lead and of trace elements concentrations in natural uranium materials as a signature in nuclear forensics," *Radiochimica Acta*, vol. 95, no. 10, pp. 601–605, 2007.
[8] J. Krajkó, Z. Varga, E. Yalcintas, M. Wallenius, and K. Mayer, "Application of neodymium isotope ratio measurements for the origin assessment of uranium ore concentrates," *Talanta*, vol. 129, pp. 499 – 504, 2014.
[9] "Nuclear forensic analysis of an unknown uranium ore concentrate sample seized in a criminal investigation in Australia," *Forensic Science International*, vol. 240, pp. 111 – 121, 2014.
[10] Z. Varga, M. Wallenius, K. Mayer, and E. Hrnecek, "Alternative method for the production date determination of impure uranium ore concentrate samples," *Journal of Radioanalytical and Nuclear Chemistry*, vol. 290, no. 2, pp. 485–492, Nov 2011.
[11] A. K. Kennedy, D. A. Bostick, C. R. Hexel, R. R. Smith, and J. M. Giaquinto, "Non-volatile organic analysis of uranium ore concentrates," *Journal of Radioanalytical and Nuclear Chemistry*, vol. 296, no. 2, pp. 817–821, May 2013.
[12] D. M. L. Ho, "Study on the applicability of structural and morphological parameters on selected uranium compounds for nuclear forensic purposes," Ph.D. dissertation, 2015.
[13] L. Fongaro, D. M. L. Ho, K. Kvaal, K. Mayer, and V. V. Rondinella, "Application of the angle measure technique as image texture analysis method for the identification of uranium ore concentrate samples: New perspective in nuclear forensics," *Talanta*, vol. 152, pp. 463 – 474, 2016.
[14] J. Schanda, *Colorimetry: understanding the CIE system*. John Wiley & Sons, 2007.
[15] C. A. Schneider, W. S. Rasband, and K. W. Eliceiri, "Nih image to imagej: 25 years of image analysis," *Nature methods*, vol. 9, no. 7, p. 671, 2012.
[16] J. C. Russ, *The image processing handbook*, 3rd ed. CRC press, 1999, ch. 4.
[17] R. Andrle, "The angle measure technique: A new method for characterizing the complexity of geomorphic lines," *Mathematical Geology*, vol. 26, no. 1, pp. 83–97, Jan 1994.
[18] J. Huang and K. H. Esbensen, "Applications of Angle Measure Technique (AMT) in image analysis: Part I. a new methodology for in situ powder characterization," *Chemometrics and Intelligent Laboratory Systems*, vol. 54, no. 1, pp. 1 – 19, 2000. [Online]. Available: http://www.sciencedirect.com/science/article/pii/S0169743900001003
[19] K. H. Esbensen, K. H. Hjelmen, and K. Kvaal, "The AMT approach in chemometrics-first forays," *Journal of Chemometrics*, vol. 10, no. 5-6, pp. 569–590, 1996.
[20] W. Härdle and L. Simar, *Applied multivariate statistical analysis*. Springer, 2007, vol. 22007.
[21] T. Strauss and M. J. von Maltitz, "Generalising wards method for use with manhattan distances," *PloS one*, vol. 12, no. 1, p. e0168288, 2017.
[22] D. J. Denis, *Applied univariate, bivariate, and multivariate statistics*. John Wiley & Sons, 2015.
[23] S. V. Kucheryavski, K. Kvaal, M. Halstensen, P. P. Mortensen, C. K. Dahl, P. Minkkinen, and K. H. Esbensen, "Optimal corrections for digitization and quantification effects in angle measure technique AMT texture analysis," *Journal of Chemometrics: A Journal of the Chemometrics Society*, vol. 22, no. 11-12, pp. 722–737, 2008.
[24] K. A. Bakeev, *Process analytical technology: spectroscopic tools and implementation strategies for the chemical and pharmaceutical industries*. John Wiley & Sons, 2010.

EPJ Web of Conferences **225**, 07004 (2020)
ANIMMA 2019

https://doi.org/10.1051/epjconf/202022507004

APPLICATION OF CdZnTe QUASI-HEMISPHERICAL DETECTORS IN STRONG GAMMA-RADIATION FIELDS

Anatoli Loutchanski
ZRF RITEC SIA
Riga, Latvia
a.loutchanski@ritec.lv

Viktors Fjodorovs
ZRF RITEC SIA
Riga, Latvia
v.fjodorof@ritec.lv

Victor Ivanov
ZRF RITEC SIA
Riga, Latvia
v.ivanov@ritec.lv

Vadims Ogorodniks
ZRF RITEC SIA
Riga, Latvia
v.ogorodnik@ritec.lv

***Abstract*—The results of a study of some ways to improve spectroscopy characteristics of the CdZnTe quasi-hemispherical detectors when working in high gamma radiation fluxes are presented. It was shown that the use of IR illumination with a wavelength of 1050 nm or 1200 nm or at slight warm-up of the detector to +30°C ... +40°C can significantly improve spectroscopy performance of the CdZnTe detectors of size 3.5 mm × 3.5 mm × 1.75 mm when operating in a tested gamma-radiation field with a dose rate up to 590 mGy/h.**

Keywords*—CdZnTe (CZT), gamma-radiation detectors, semiconductor radiation detectors, room temperature semiconductor detectors, infra-red (IR) illumination, IR LED, high flux gamma-radiation, leakage current, voltage-current characteristic

I. INTRODUCTION

Presently room temperature CdZnTe (CZT) nuclear radiation detectors of various designs and sizes are widely used for spectrometric measurements of X- and gamma-radiations. In most cases, the spectrometric detectors with highest efficiency are required. However, there are tasks where good spectroscopic CZT detectors with a low efficiency should be used. These tasks are connected with spectrometric measurements in high gamma-radiation fluxes. Small size quasi-hemispherical CZT detectors [1–6] are already successfully used to control nuclear spent fuel or/and high-level radiation waste. Recently a new system for verification of nuclear spent fuel assemblies using Passive Gamma-Emission Tomography (PGET) [7] was developed and tested. The system contains about 200 detector modules with CZT quasi-hemispherical detectors of size 3.5 mm × 3.5 mm × 1.75 mm. More than 1000 such detectors were tested. Approximately, only half of the measured detectors successfully passed tests under high gamma-radiation fluxes and were found suitable for subsequent use in the PGET system. Therefore, study of causes of the detectors' characteristics deterioration in high gamma-radiation fluxes and search for ways to eliminate these causes is an urgent task. It was previously shown [8] that IR illumination with infrared light with wavelengths close to the absorption edge of the CdZnTe can significantly improve the performance of the quasi-hemispherical detectors operating at room and low temperatures. In this work, we studied the effects of IR illumination and heating of the quasi-hemispherical detectors operating in high gamma-radiation fluxes from the ^{137}Cs.

II. RESULTS AND DISCUSSION

For measurements, experimental setup based on application of a calibrated measurement bench with a high activity gamma-radiation source ^{137}Cs was used. The stand allows changing distance between the source and the detector. The maximum achievable dose rate at the measurement point with the minimal distance between the source and the detector (R=28 cm) was about 590 mGy/h. Detector fixture allowing installation and quick change of the detectors and IR LED. CZT quasi-hemispherical detectors of size 3.5 mm × 3.5 mm × 1.75 mm fabricated of different raw CdZnTe crystals were tested. Changeable IR LED with wavelengths 940 nm, 1050 nm and 1200 nm were used at measurements. Used heater allows heating the detector up to the +50°C. During the measurements LED forward current, heater consumption currents and operation temperature were controlled. Detector's leakage currents and spectra of ^{137}Cs were recorded. Schematic experimental setup is shown on Fig. 1.

Intense gamma irradiation significantly affects leakage currents and spectrometric characteristics of the CZT detectors. Fig. 2 show typical voltage-current characteristics of CZT quasi-hemispherical detectors measured at room

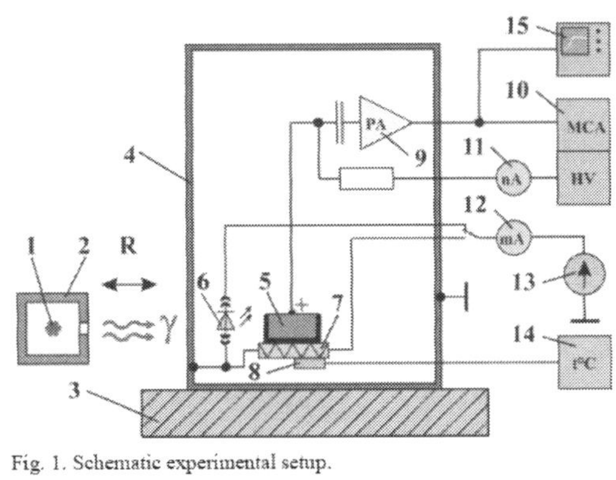

Fig. 1. Schematic experimental setup.

1-highly active radiation source ^{137}Cs; 2-lead container/collimator; 3-movable stand; 4-detector fixture; 5-CZT detector; 6-IR LED; 7-heater; 8-temperature sensor; 9-charge sensitive preamplifier; 10-digital multichannel analyzer type MCA527 with high voltage power supply; 11-detector leakage current meter (resolution 1 nA); 12-LED forward current or heater consumption current meter; 13-stabilized current source for IR LED or heater; 14-temperature monitor; 15-digital oscilloscope.

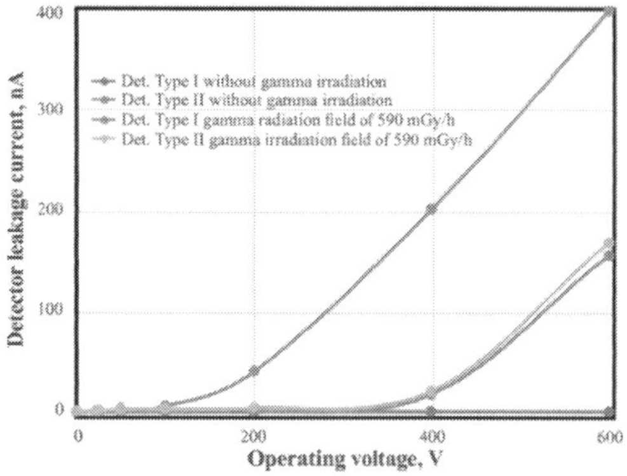

Fig. 2. Voltage-current characteristics of CZT quasi-hemispherical detectors measured at room temperature without and under influence of high gamma-radiation fluxes.

temperature without and in high gamma-radiation flux. Processing of measurement results allows distinguishing two types of detectors differing in a behavior of leakage currents under gamma irradiation. Type I – are detectors that have a significant (20 … 100 times) increase in leakage current under influence of strong gamma irradiation. Large increasing of the leakage currents were accompanied by a significant deterioration or even complete disappearance of the spectrometric capabilities of such detectors. These detectors, as a rule, had insignificant leakage currents < 5 nA at room temperature at 400 V without exposure to intense gamma irradiation. Type II – are detectors that did not have or had a slight increase in leakage current under to influence of gamma irradiation. These detectors, as a rule, had significant leakage currents 20 … 300 nA at 400 V without exposure to intense irradiation and did not allow increasing the operating high voltage to acceptable values due to a significant increase in the noise level with increasing voltages.

For detectors of the first type, operating in high gamma-radiation fluxes, complex dependencies of the leakage currents on IR illumination intensity with IR LED wavelength of 1050 nm were obtained, Fig. 3 in the range of the LED operating currents (⩽ 100 mA), the LED relative radiant intensity is approximately directly proportional to the LED forward current. With increasing of the LED forward current,

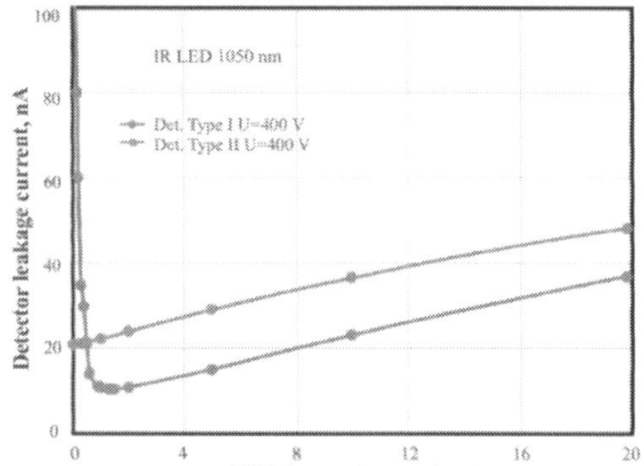

Fig. 3. Detectors leakage currents versus IR LED (1050 nm) forward current measured in gamma-radiation field of 590 mGy/h.

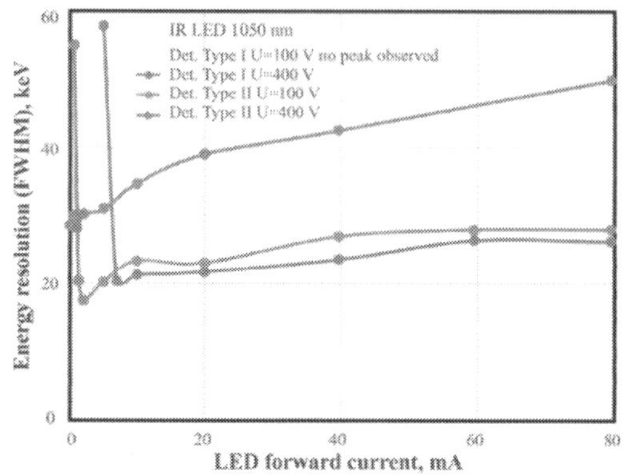

Fig. 4. Energy resolution (FWHM) at 662 keV versus IR LED forward current at different operating voltages measured in gamma-radiation field of 590 mGv/h.

the leakage currents of such detectors first decreases, then reaches the minimum, and then begins to rise slowly. The leakage current of the second type detectors increases with rising of the LED forward current.

The behavior of the energy resolution of the detectors of both types under IR illumination also had own peculiarities, Fig. 4. First, it improves noticeably with increasing of IR LED forward current and then starts to deteriorate. Moreover, such a dependence of energy resolution on values of the LED direct current for detectors of the second-type was observed only at relatively low operating voltages < 200 V. At higher operating voltages, the energy resolution of the second-type detectors deteriorated with increasing of IR illumination intensity.

Illumination by IR light with wavelengths of 1050 nm or 1200 nm as well was found to greatly improve spectrometric characteristics of the quasi-hemispherical CZT detectors operating in high gamma-radiation fluxes. This allows expanding the range of count rates in which the normal operation of the detectors is observed. Fig. 5 demonstrates dependence of energy resolution versus count rate obtained with different levels of IR illumination. Count rate was varied by changing the distance between radioactive source and tested detector. Illumination by IR light with wavelengths of 940 nm did not produce positive effect.

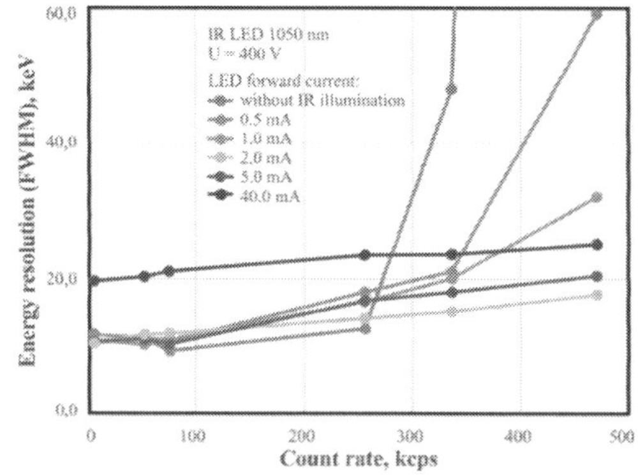

Fig. 5. Energy resolution (FWHM) at 662 keV versus recorded count rate measured without and with IR LED illumination with different LED forward currents.

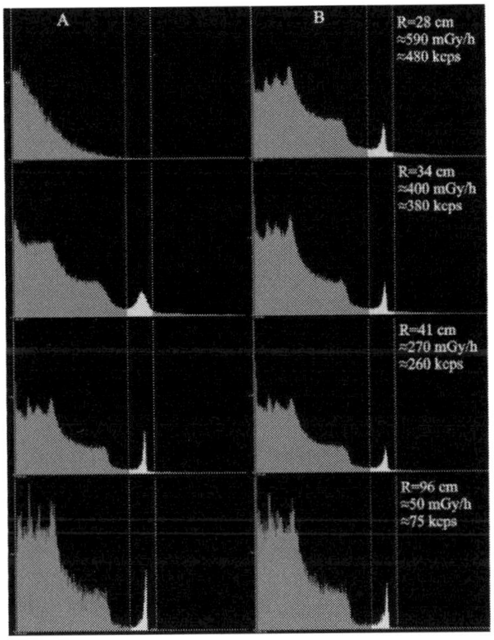

Fig. 6. Spectra of ^{137}Cs measured without (A) and with (B) IR illumination at different distances between radiation source and detector.

Fig. 6 illustrates changes in the shape of ^{137}Cs spectra measured at different distances between the radiation source and the detector without and with IR illumination (1050 nm) The LED forward current 2 mA and detector operating voltage 400 V were applied.

A study of temperature dependence of the detector's leakage currents and energy resolutions of the detectors operating in a high gamma radiation fluxes was carried out. A decrease (quenching) of the detector's leakage currents and energy resolution improvement with a slight increase in the operating temperature by + 5°C … + 10°C were found, Fig. 7 dependences were measured in gamma-radiation field of 590 mGy/h without IR illumination.

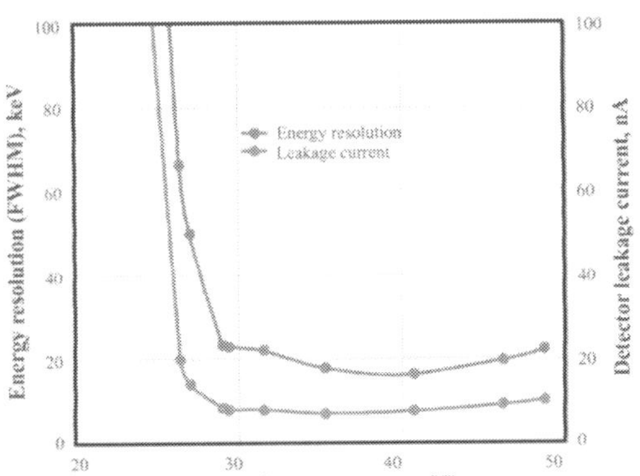

Fig. 7. Temperature dependencies of energy resolution (FWHM) at 662 keV and leakage current at 400 V.

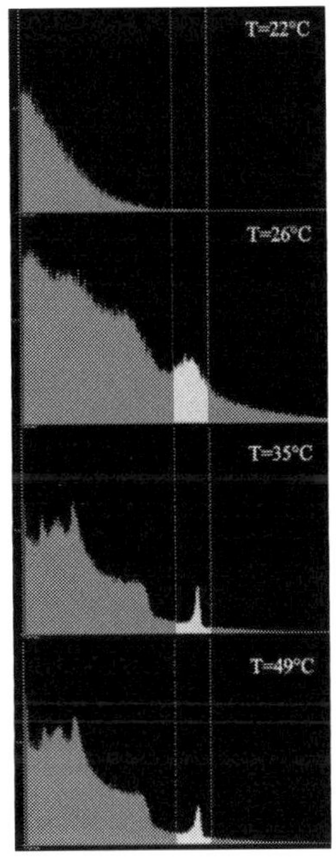

Fig. 8. Spectra of ^{137}Cs measured without IR illumination at different operating temperatures in radiation field of 590 mGy/h. Operating voltage 400 V.

The Table I shows the measurement results of energy resolutions of two types CZT quasi-hemispherical detectors of 3.5 mm × 3.5 mm × 1.75 mm without and with IR illumination measured in a weak and in a high gamma-radiation flux.

TABLE I.

Detector type	Operating voltage, V	Energy resolution (FWHM) at 662 keV, keV			
		Count rate <5 kcps	*Count rate ≈ 500 kcps*		
		room temperature, without IR illumination	*room temperature, without IR illumination*	*room temperature, with IR illumination*	*increased temperature (+30°C … +50°C), without IR illumination*
I	50	19	no peak	no peak	58
	100	15	no peak	no peak	15
	200	13	no peak	14	13
	400	12	no peak	18	16
	600	10	no peak	24	17
II	50	16	no peak	26	24
	100	15	no peak	24	28
	200	13	28	25	23
	400	16	44	37	52
	600	29	62	57	61

III. Main Results

Operating in high gamma-radiation fluxes can significantly worsen spectroscopy performance of the tested quasi-hemispherical detectors.

Tested detectors can be divided into two groups. The first are detectors with rather low leakage currents < 5 nA at 400 V without gamma-irradiation but with strongly increasing leakage currents (to 20 … 100 times) under influence of the high intensity gamma-radiation flux. In second group are detectors with leakage currents of about 20 … 300 nA at 400 V without gamma-irradiation, which only slightly increasing in high gamma-radiation flux.

Illumination by IR light with wavelengths of 1050 nm or 1200 nm as well as heating up to + 30°C … + 40°C was found to greatly improve spectrometric characteristics of the quasi-hemispherical CZT detectors operating in high gamma-radiation fluxes.

For detectors of the first type, complex dependencies of the leakage currents and the energy resolutions on forward currents of the IR LED and operating temperatures were obtained. With increasing IR illumination intensity or with increasing operating temperature, the leakage current of such detectors first decreases, reaches the minimum, and then begins to rise slowly. The behavior of the energy resolution is similar – first it improves noticeably and then starts to deteriorate.

Such complex behavior of leakage currents and energy resolution under the influence of high gamma-radiation fluxes and IR illumination can be associated with a change in an electric field distribution in the detector caused by formation of a space charge due to trapping of photo generated holes at deep levels and their detrapping under the influence of IR illumination [9]. A detailed explanation of the effects found requires additional research.

References

[1] K. Zanio, "Use of Various Device Geometries to Improve the Performance of CdTe Detectors," Rev. de Phys. Appl., vol. 12, pp. 343-347, 1977.

[2] L. A. Alekseeva, P. G. Dorogov, V. I. Ivanov, A. K. K husainov, "Cadmium telluride gamma radiation detectors with a high energy resolution," Pribory i Teknika Eksperimentya, no. 1, pp. 54-58, 1985.

[3] V. Ivanov, P. Popov, A. Loutchansky, L. Aleksejeva, E. Mozchaev, "Further Development of emispherical CdZnTe Detectors for Safeguards Applications," Proc. 21th Annual ESARDA Symp. on Safeguards and Nuclear Material Management, Seville, Spain, pp. 479 -484, 1999.

[4] C. Szeles, D. Bale, J, Grosholz, G. L. Smith, M. Blostein, J. Eger, "Fabrication of High Performance CdZnTe Quasi-Hemispherical Gamma-ray CAPuretm Plus Detectors," Hard X -Ray and Gamma-Ray Detector Physics VIII, Proceedings of SPIE, Vol. 6319 , (SPIE, Bellingham, W A, 2006

[5] V. Ivanov, P. Dorogov, A. Loutchansky, and L. Aleksejeva, "CZT quasi-hemispherical detectors with improved spectrometric characteristics," in Proc. IEEE NSS-MIC Conf. Rec., Orlando, FL, Oct. 25-31, 2009.

[6] Spectrometric detection probes Types SDP310 and SDP500, ZRF RITEC SIA Data Sheets, http://www.ritec.lv.

[7] T. White, M. Mayorov, A. Lebrun, P. Peura, T. Honkamaa, J. Dahlberg, J. Keubler, V. Ivanov, A. Turenen, "Application of Passive Gamma Emission Tomography (PGET) for the Verification of Spent Nuclear Fuel", INMM 59th Annual Meeting, Baltimore, Maryland, USA, July 22-26, 2018.

[8] V. Ivanov, P. Dorogov, A. Loutchansky, L. Grigorjeva, D. Millers, "Improving the Performance of Quasi-Hemispherical CdZnTe Detectors Using Infrared Stimulation", IEEE Trans. Nucl. Sci., vol. 59, no. 5, pp. 2375-2382, 2012.

[9] Jakub Pekárek, Václav Dědič, Jan Franc, Eduard Balas, Martin Rejhon, Pavel Moravec, Jan Touš, Josef Voltr, "Infrared LED Enhanced Spectroscopic CdZnTe Detector Working under High Fluxes of X-rays", Sensors 2016, 16(10), 1591.

Modular and Integrated Sensor Network of Intelligent Radiation Monitor Systems for Radiological and Nuclear Threat Response

M. Corbo*, M. Morichi, E. Fanchini, G. Mini, A. Pepperosa, G. Mangiagalli
* CAEN SyS s.r.l., Viareggio, Italy
m.corbo@caensys.com

Abstract— **Increased sensitivity to nuclear safety and security issues has prompted public entities and private institutions to maximize their capability to rapidly assess risks and intervene in the case of accident or threat. Quick intervention and response are achieved through nuclear measurements via airborne, land, and underwater systems. A network of cohesive, well-integrated and easy deployable radiation monitoring systems combined with real-time analysis of data is essential to facilitate and enhance the decision-making process during these most critical moments enhancing the quality of the management plan.**

The presented radiation monitoring systems can be integrated in several form factors which depend mainly on operational needs and internal battery for autonomous operation. Compact ARM based computers is embedded, which can store large amount of data in their non-volatile memory, run automatic data analysis and trigger alarms in case of exceeding radiation levels. All the systems can communicate with redundant interfaces in failover configuration and upload the acquired environmental information in a central database. The same monitoring systems can alert the emergency response personnel on the field as well, through wireless connection to common tablets or cellphone or SMS, guaranteeing a prompt response in case an illicit transportation of radiological or nuclear material is detected.

Keywors—**Radiological Monitoring, Environmental Monitoring, Spectrometer, Nuclear Security, Autonomous Stations.**

I.INTRODUCTION

AUTORITIES have the necessity to monitor the radiological situation and implement actions to minimize radiological effects on the population. One common issue, in case of an alarm, is that the available measurement systems do not communicate directly to the operator causing important delays in the decision-making process necessary to define the assessment and mitigation plans. Furthermore, some operations require the use of several measurement systems and in case these systems were not designed to be integrated into a network, results cannot be analysed together in real time, requiring a time consuming offline post-processing. From here the necessity to have a real time network of integrable systems for a precise and immediate evaluation of the situation.

Current embedded commercial electronics, as compact ARM based computers, allow the integration of intelligent radiation monitoring systems meeting the needs of integrated solutions for radiological and nuclear security. The building blocks of the radiation monitoring systems can be easily programmed for different operations remotely or locally. The working condition can be monitored real time and in case of malfunctioning the optimization of the data acquisition can be performed remotely as well. Access credentials and secured data transmission avoid that unauthorized people could read or delete the acquired data and avoid that operators on field could accidentally modify hardware or software configuration.

Web based software instruments ease, at the same time, the design of user-friendly graphical interfaces and guarantees a good level of data security as mentioned. A common software platform can manage the information from different radiation detectors giving the possibility to arrange the network of deployed sensors in a modular way.

II. DESCRIPTION OF HARDWARE

The intelligent radiation monitoring systems are composed by some elements that are reported here below.

- The low power consumption and compact microprocessor
- The spectrometer composed by inorganic scintillators for gamma radiation identification; thanks to innovative scintillation detectors it is possible to identify neutron and gamma radiation at the same time as well
- The gamma and neutron counters as Geiger-Muller or 3He tubes.

The intelligent radiation monitoring systems are configured in different form factors to match the needs of all possible operation scenarios: compact and discrete to be carried in a backpack; lightweight and provided with wireless data transmission to be transported on unmanned aerial drones; rugged designed with high efficiency detectors to be transported on vehicles; tolerant to a wide temperature range and harsh weather conditions to be installed as autonomous stations in open field; waterproof to monitor the contamination in water or in the sediments of rivers, lakes, seas and oceans.

A. Microprocessor

The microprocessor is based on an ARM which stores the data in an internal non-volatile memory of more than 8 GB. Like other commercially available microprocessors it is provided with multiple communication interfaces, like Ethernet, USB and serial ports, and multiple inputs and outputs, which can be programmed for connecting peripherals and connect radiation counters. The operator can select the preferred communication interface and backup interfaces, which guarantee the communication thanks with failover capability.

Commonly available rain sensors, meteorological stations, GPS receivers, antitampering sensors can be connected to the microprocessor to provide additional functionalities and improving the ability to respond to a wide range of operational activities.

The microprocessor runs a web interface which gives the user the possibility to easily configure the data acquisition and the isotopic analysis.

B. Spectrometer

The gamma spectrometer is composed by an inorganic scintillation crystal which can be either NaI(Tl), $CeBr_3$ or $LaBr_3(Ce)$. NaI(Tl) is preferable its high light yield and moderate cost, $LaBr_3(Ce)$ for its excellent energy resolution and $CeBr_3$ for its good energy resolution and low internal radioactivity.

The innovative NaIL™ [1] detector can be integrated in the radiation monitoring systems to obtain high efficiency thermal neutron detection combined with gamma ray spectrometry. This allows the system to detect the presence of both gamma ray emitters and identify special nuclear material. E.g. Fig. 1 shows the data acquired in presence of 238Pu source with a neutron yield of 6.5 n/s with 5 cm of Polyethylene.

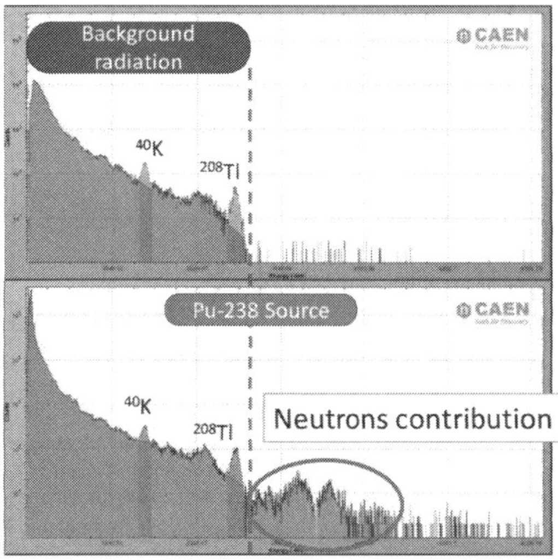

Fig. 1: Spectra acquired with a NAIL detector 2"x2" at 0.5% Lithium doping. The spectrum above represents the background, while the spectrum below shows the data acquired in presence of 238Pu source with a neutron yield of 6.5 n/s with 5 cm of Polyethylene

Signals from scintillation detector is conditioned by a preamplifier, the pulse is digitized by a 12-bit 62.5 MHz flash Analog to Digital Converter and the corresponding data flow is analysed by a Field Programmable Gate Array performing a Digital Pulse Processing (DPP) [2]. The DPP can be optimized for different detector types thanks to the programmable parameters of the digital discriminator and the digital shaping for pulse height analysis.

The spectroscopy detector is configured to collect gamma interactions in the energy range from 50 keV to 3 MeV, even though alternative energy ranges can be easily configured on request. The spectrometer runs isotope identification and isotope related dose rate evaluation automatically while acquiring the data. The real-time data processing and the defined isotope-based alarm allow system to autonomous detect the presence of artificial nuclides in a short time window and more effectively compared to traditional dose rate meters. The user can select the isotopes to be identified from the library and adjust the thresholds of the isotope related alarms. It provides statistically accurate dose rate measurements in the wide range from 1 nSv/h to 100 μSv/h.

The spectrum stabilization is obtained with the identification of natural occurring radionuclides as the ^{40}K and rescaling the calibration of the spectrum to maintain the centroid of ^{40}K peak at the right position in the spectrum.

C. Gamma or Neutron Counters

Geiger-Muller or neutron counters can be connected to the microprocessor inputs and outputs. Compensated Geiger-Muller could effectively measure the Ambient Dose Equivalent, while ^{3}He or boron lined tubes could effectively detect neutrons from special nuclear material.

III. SOFTWARE

One of the main advantages of the commonly used web application is the multiplatform compatibility, which means that the intelligent radiation monitoring systems can be monitored from different types of devices and their information can be received from tablets or smartphones. This gives the security or emergency response personnel the possibility to promptly detect threats on field.

The web-based application is a webserver which may run either in the deployed monitoring systems or in a central monitoring facility with gathers together the information from a network of systems. A database is present both in the monitoring system and in the central monitoring facility as well, enabling the possibility to select and download data reports depending on user defined criteria. This capability is fundamental for operators to promptly and effectively diagnose an emergency scenario or to easily select the data of monitoring systems in a comprehensive nationwide network.

A. Web Application of the Intelligent Radiation Monitoring Systems

The intelligent radiation monitoring systems are provided with a web server application running in the internal central unit processor. The radiological information is stored in an internal database and in the wide non-volatile memory, allowing autonomous data taking even in case of sudden connection lost. Spectra, acquisition reports, data history can be displayed and downloaded through a common web browser. Users can monitor the data or modify the acquisition settings depending on the level of access credentials: user, technician or administrator. E.g. Fig. 2 represents the administrator page for setting the operating parameters of the spectrometer.

Spectral analysis and statistical data treatment are performed, and extended isotope library is used for identification routines to address the robust spectrum stabilization algorithms are implemented. The user can perform long term quality checks and even recalibrate the system in case of need.

The station can signal radiological alarms or malfunctions automatically to the central monitoring facility or directly to operators through phone messages or emails. Different levels of alarm can be configured on the dosimeter measurement, on the neutron counting, on the spectrometer count rate and on the dose rate associated to identified peaks in the spectrum.

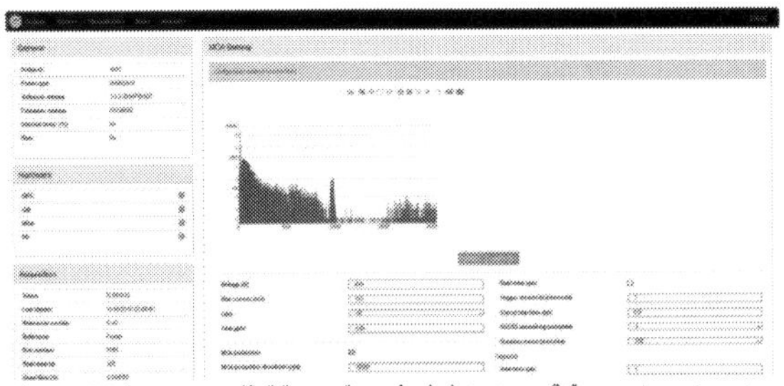

Fig. 2: Web page available to the administrator of the system, to set the operational parameter of the spectrometer

B. Web application for the Network of Systems

The web application for the network is an intuitive and easy handling tool for the operator to supervise all intelligent radiation monitoring systems deployed in the field by a user-friendly web interface. The intuitive graphical interface allows the user to fully control the situation guaranteeing fast and reliable decision making for prompt emergency response.

The software visualizes all the network of systems on a map showing the status of all deployed systems and the radiological situation, alerting the operator in case of alarm. The dosimetric heat map is obtained using data fusion algorithms using real time measurements for all georeferenced systems.

The network software is the perfect tool for wide area coverage applications or the monitoring of sensitive locations or events to have a safer and clearer monitoring overview.

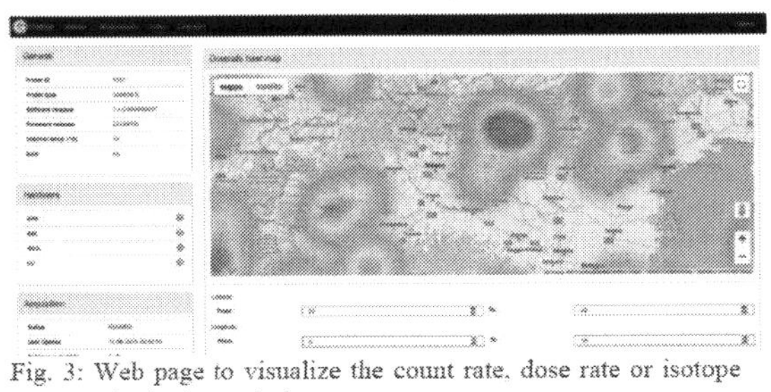

Fig. 3: Web page to visualize the count rate, dose rate or isotope presence in the form of a heat map

IV. GAMON-S CHARACTERIZATION

The GAMON-S System has been characterized under fast and extreme temperature changes and the radiological measurements have been verified in laboratory and Monte Carlo method. In this section we report the most relevant features and performance that have been measured.

A. Environmental Chamber Test

The GAMON-S System has been tested in environmental chamber to assess the temperature stabilization capability of the software algorithms. The environmental chamber has been configured to obtain a stress test with a wide range temperature cycle, with extreme temperatures -40 and +60 °C, to be performed in 24 hours cycle.

A radioactive source of Cs-137 has been used to create a reference peak for the spectrum stabilization technique. Internal temperature of the crystal has been monitored and the position of the stabilization peak has been correlated to the measured temperature. Fig. 4 represents the trend of the reference peak position over time, together with the monitored temperature. On the overall the peak stabilization algorithm guarantees the stability over a total range of ±0.5% respect to the central reference peak position. This performance allows the algorithm of radioisotope identification to not fail in case of quick changes of external temperature.

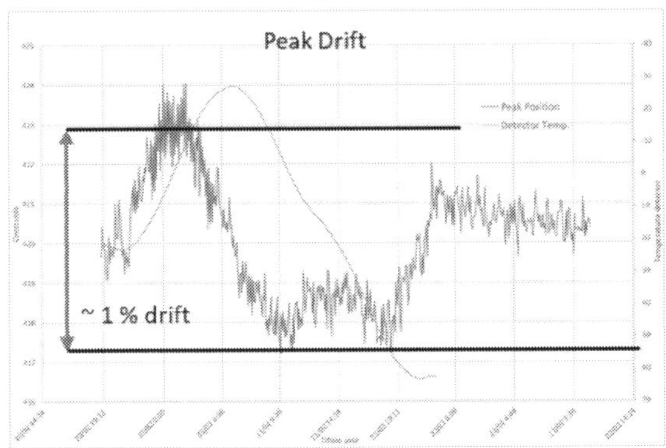

Fig. 4: In blue the reference peak position over the 24 h temperature cycle. In orange the measured temperature of the CeBr3 scintillator crystal (ranging from -30 to +65 °C)

B. Spectrometer Response to Ambient Equivalent Dose H(10)*

The GAMON-S has been characterized for its response to the Ambient Equivalent Dose H*(10) through laboratory measurements and a comparative study with a simulation performed with MCNP 6.1.1.

The simulation at MCNP 6.1.1 has been done using the geometry of a 1.5"x1.5" CeBr3 crystal together with the aluminum case, reflective material and mu-metal cover in the proximity of the crystal itself. The scintillator crystal has been inserted inside sections of pipes having the thickness and material type of the GAMON-S cases. The flux of radiation has been generated perpendicular to the CeBr3 crystal axis with uniform distribution of parallel gamma rays ranging from 30 keV to 3 MeV. This simulation has been used in conjunction with the data reported in the ICRU Report N.57 to compute the conversion factors between the number of counts in the gamma ray photopeak and the associated Ambient Equivalent Dose H*(10).

The result of the simulation has been compared with experimental measurements performed with ^{241}Am and ^{137}Cs point sources at 20 and 30 cm distance from the center of the detector axe perpendicular to its axis. The result of the comparison is reported in the table below, which demonstrate a good agreement at the energy of the 661.7 keV gamma line, better than 5 %, and sufficient agreement at the energy of 59.5 keV, with a disagreement of the order of 20%.

Nuclide (Act.)	Energy [keV]	D. [cm]	Measured [nSv/h / cps]	Simulated [nSv/h / cps]	Discrepancy
^{241}Am (3,43 kBq)	59,5	20	0.2387 ± 2.0%	0.2061	13.7 %
		30	0.2670 ± 3.1%	0.2061	22.8 %
^{137}Cs (23.0 kBq)	661,7	20	3.9090 ± 1.1%	3.7927	3.0 %
		30	3.9881 ± 1.7%	3.7927	4.9 %

C. Spectrometer Linearity

The GAMON-S has been tested at the secondary metrology laboratory at Polytechnical University of Milan to monitor the system linearity as function of the incident instantaneous radiation equivalent dose H*(10). The test has been performed at the hot chamber at Milan with the use of a ^{137}Cs source. In the table below the results of the measurement.

Nominal H*(10)	H*(10) measured	Discrep.	Peak position [keV]	Peak Net Area [cps]
13.90 μSv/h ± 5.10%	13.90 μSv/h ±0.20%	+0.0%	656.9	2604.9
24.70 μSv/h ± 5.42%	24.82 μSv/h ± 0.17%	+0.5%	659.1	4652.2
103.00 μSv/h ± 6.01%	104.72 μSv/h ±0.11%	+1.7%	664.5	19624.9
148.00 μSv/h ± 6.42%	147.88 μSv/h ±0.11%	-0.1%	667.9	27712.3

V. SYSTEMS OF THE PLATFORM

The radiation monitoring systems can be integrated in several form factors which depend mainly on type of deployment. From this perspective, the radiation monitoring systems can be compact to be carried in a backpack or handheld, lightweight to be transported on unmanned aerial drones, rugged designed to be transported on vehicles, tolerant to a wide temperature range and water immersion to be installed in open field.

Here below the description the various version of the intelligent radiation monitoring systems is reported.

A. Gamma Radiation Spectroscopy System for Real-Time Radiation Monitor

The gamma radiation spectroscopy system represented in Fig. 5 is designed for early environmental warning and emergency response. It runs automatic real-time analysis of surrounding gamma emitting isotopes for detecting radiological threats.

On-field installation allows the system to easily monitor wide surrounding area, detect small deviation from natural background thanks to the high detector efficiency, to provide detailed reports on the radioisotopes found. It has been designed and assembled with a robust case designed to guarantee IP68 and for operating outdoor in extreme weather conditions from -40 to +60 °C.

To be operated as a station it mounts batteries, optional solar panels and is provided with wired and wireless communication interfaces: USB 2.0, Ethernet, WiFi and 3G/4G LTE.

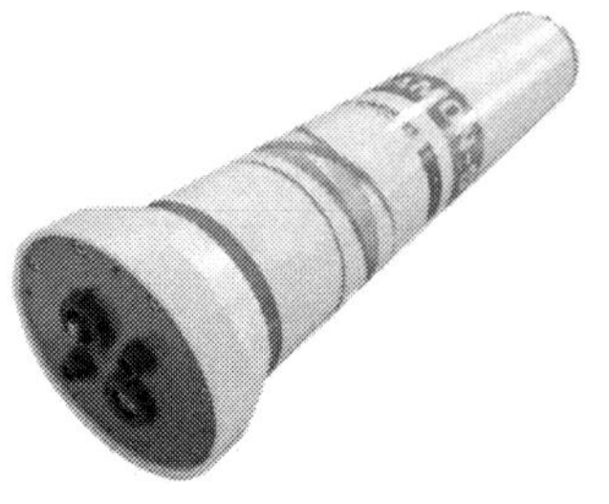

Fig. 5: Picture of the gamma radiation spectroscopy system for real-time radiation monitoring

B. Gamma Dose Rate Monitoring System for Real-Time Radiation Monitoring

The system is a versatile gamma dose rate monitoring system designed for early environmental warning and emergency response with wide measurement range from 10 nSv/h to 10 Sv/h with subtraction of intrinsic background. The gamma dose rate monitoring system integrates a low power consumption CPU, to guarantee autonomous operations, including the activation of backup communication interfaces for data transfer in any environmental condition.

C. Spectroscopic Mobile and Networkable Station

Spectroscopic mobile station is a fast and easy deployable gamma and neutron detection system with the capability to be networked wirelessly with other systems.

The station allows response personnel to quickly and safely address radiation threats. It can be positioned in sensitive locations to monitor the situation in real time and in parallel, by both infield personnel and operators at remote command centre.

It provides remote monitoring and turnkey data reach back telemetry capabilities, thereby greatly reducing exposure risk to personnel in compromised environments and allows a discrete monitoring of sensitive locations and events.

D. Portable and Discrete Radionuclide Identifier

The portable and discrete radionuclide identifier represented in Fig. 6 is a portable detection unit which allows the detection and identification of radiation sources in crowded or sensitive areas or locations forbidden to vehicles.

It is designed to perform undercover measurements in strategic areas like airports or railway stations where the probability for a terroristic attack is higher. It can detect and identify orphan source or radiological dispersal device classifying them in NORM, medical or industrial.

The data are displayed on a visualization device (smartphone or tablet) that allows to remain anonymous in the crowd. In case

 https://doi.org/10.1051/epjconf/202022507005

of alarm an acoustic signal, a short vibration or a message can be configured as warning notification.

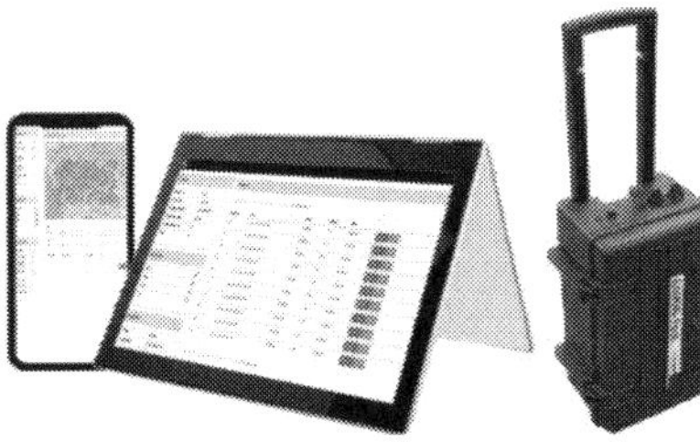

Fig. 6: Picture of the portable and discrete radionuclide identifier

E. Vehicle Mountable Gamma Spectrometric System

The vehicle mountable gamma spectrometric system is embedded in a rugged and IP66 container designed to resist to external stresses (e.g. shocks and vibrations) and easy to mount and fix on the vehicle chosen for the inspection (boat, helicopter or car), with rechargeable batteries for extended operation.

It is designed to make radiometric measurement and identification of gamma emitters on wide areas providing a map of the radioactivity detected. The real time identification algorithm can recognize artificial gamma emitters beyond the natural background and mark the hot spot over the trajectory travelled by the moving vehicle. This feature allows the user to make a map of the radioactivity of the scanned area.

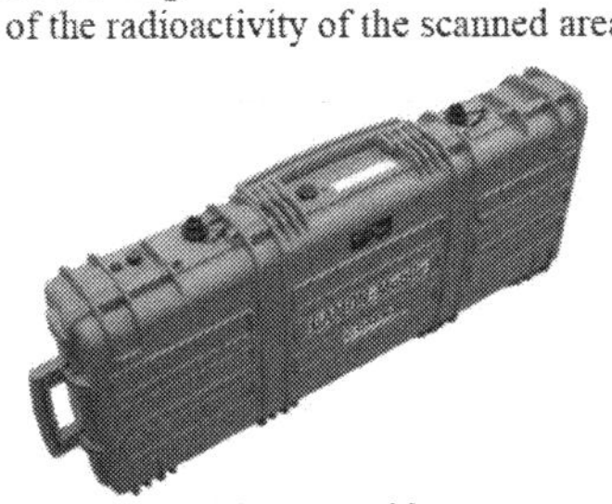

Fig. 7: Picture of the vehicle mountable gamma spectrometric system in its rugged case

F. Compact Radionuclides Identification Unit

The compact radionuclides identification unit represented in Fig. 8 is an autonomous system designed for radiological threat search. Weight and dimensions allow an easy installation on an UAV or UGV to conduct measurements over large areas to improve the operator safety or for access difficulties.

The system is composed by spectrometric and dosimetry probes for gamma detection integrated in a rugged housing for outdoor environments. It communicates wireless directly to the ground control station where a GIS map visualizes trajectory and measurements without affecting the unmanned vehicle connections and performances.

The system is designed to maximize portability, with a weight lower than 3 kg that may be further optimized reducing the volume of the installed detector.

Fig. 8: Picture of the compact radionuclide identification unit in its rugged transportable case

G.Portal Monitor Family

The radiation monitor systems can be setup to operated portal monitor family for infield real time radiation control of elements passing through the portal for security controls in sensitive access points.

The family is made by two main types of portals: the Pedestrian Portal Monitors (PPM) and the Radiation Portal Monitors (RPM) for the monitoring of vehicles in transit. The systems detect both gamma and neutron emitters thanks to the innovative NaIL™ detector which gives the possibility to detect and identify Special Nuclear Material with unprecedented sensitivity, while performing high efficiency gamma spectrometry.

The systems take advantage of a real-time data processing combined to automated background subtraction algorithms to alarm in presence of both artificial nuclides and SNM in a short time window.

H. Underwater System for Radionuclides Identification

The underwater system for radionuclide identification, represented in Fig. 9, is specifically designed for submerged radiometric measurement and radiological alerts. The system can be used as a quick response measurement device, or it can be installed as a permanent monitoring device for sensitive underwater locations or for oil&gas applications.

The submersible hermetic case and its weight allow an easy gamma spectroscopic measurement in real time from small boat or watercraft.

The system was designed to offer the best combination of portability, low power consumption and performance.

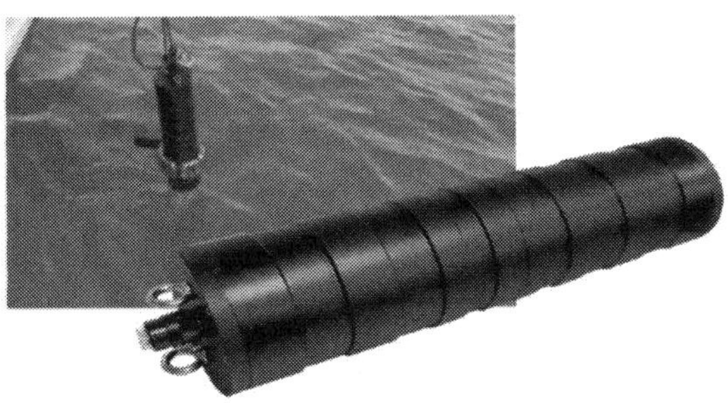

Fig. 9: Picture of the underwater system for radionuclides identification and an on-field use of the system for monitoring the see contamination

VI. Conclusion

A platform of intelligent radiation monitoring systems with a common web based software gives the possibility to arrange the network of radiation sensors in a modular way. This allows the emergency response teams to choose which type of system shall be deployed and to relocate the monitoring systems depending on the environment, the type of infrastructure or public building to be monitored, the logistic or the geometry of transported material. The real time processing and the connectivity of the systems enhance prompt intervention in case of alarms and allow the decision makers to quickly organize the actions on a contaminated area limiting the operator exposure.

References

[1] YANG K., IEEE Transactions on Nuclear Science PP(99): 1-1 · June 2017

[2] Gamma stream, Digital MCA Tube Base for Gamma-Ray Spectroscopy at http://www.caen.it/csite

EPJ Web of Conferences **225**, 07006 (2020) https://doi.org/10.1051/epjconf/202022507006
ANIMMA 2019

A Novel Portable Device For Gamma And Neutron Spectroscopy With Special Nuclear Material Identification

Massimo Morichi
CAEN SpA
Viareggio, Italy
m.morichi@caen.it

Erica Fanchini
CAEN SpA
Viareggio, Italy
e.fanchini@caen.it

Giacomo Mangiagalli
CAEN SpA
Viareggio, Italy
g.mangiagalli@caen.it

***Abstract*— The development of new types of detectors and the increased performance of the electronics have paved the way for developing advanced systems for the measurement and identification of radioactive material that can be involved, for example, in illicit trafficking. Radioactive isotope identifiers are today commercially available. Nowadays those systems make use of inorganic scintillators as Sodium Iodide (NaI(Tl)) or, for enhanced resolution, Lanthanum Bromide (LaBr3) to identify the gamma emitters through their characteristic gamma lines. The most complete systems usually include an additional ^{3}He proportional counter for neutron detection and counting. The performances of such devices are compliant with standards as the IEC 62327 Hand Held Instruments for the Detection and Identification of Radionuclides. This paper presents a new type of portable radioactive isotope identifier. This device, based on an organic liquid scintillator with excellent Pulse Shape Discrimination (PSD) proprieties for the simultaneous detection of gamma rays and neutrons, detects radioactive source as Special Nuclear Material (SNM), medical, industrial and Naturally Occurring Radioactive Material. The exclusive feature of this instrument is the identification of neutron sources with discrimination between fission sources (like Californium ^{252}Cf) and alpha-n type sources (like Americium Beryllium Am-Be) from Plutonium and Uranium through an innovative dedicated algorithm. Individual thresholds for neutron and gamma counts are calculated to allow detection with 95% detection probability for a dose rate on the front face of the scintillator of at least 50 nSv/h. Alarms are triggered separately when the respective rate exceed these thresholds. The neutron source detection has also been proved in a gamma ray field up to 100 μSv/h. The electronics is equipped with two analog inputs and two high voltage power supplies in a small form factor thus becoming an enabling technology for higher performance yet portable radioactive isotope identifier devices, which can include more detectors and perform data fusion analysis. The addition of a second detector allows to detect a masked neutron source through the PSD algorithm performed by the liquid scintillator detector while the added inorganic scintillator identifies the masking gamma emitters. The inorganic scintillator allows also the calculation of Pu and U enrichment grade through characteristic gamma emission line.**

Keywords—Special Nuclear Material (SNM), Uranium, Plutonium, neutron spectroscopy, homeland security

I. INTRODUCTION

The development of new types of detectors and the increased performance of the electronics have paved the way for developing advanced systems for the measurement and identification of radioactive material that can be involved, for example, in illicit trafficking. Radioactive isotope identifiers are today commercially available. Nowadays those systems make use of inorganic scintillators as Sodium Iodide NaI(Tl) or, for enhanced resolution, Lanthanum Bromide LaBr3 to identify the gamma emitters through their characteristic gamma lines. The most complete systems usually include an additional ^{3}He proportional counter for neutron detection and counting. The performances of such devices are compliant with standards as the IEC 62327 Hand Held Instruments for the Detection and Identification of Radionuclides.

II. PSD ALGORITHM FOR SNM IDENTIFICATION

This paper presents a new type of portable radioactive isotope identifier. This device, based on an organic liquid scintillator with excellent Pulse Shape Discrimination (PSD) proprieties for the simultaneous detection of gamma rays and neutrons, is able to detect radioactive source as Special Nuclear Material (SNM), medical, industrial and Naturally Occurring Radioactive Material (NORM). The main ability of this instrument is not limited to the detection, but also includes the identification of neutron sources with discrimination between fission sources (like 252Cf) and alpha-n type sources (like Am/Be) from Plutonium and Uranium through an innovative dedicated algorithm.

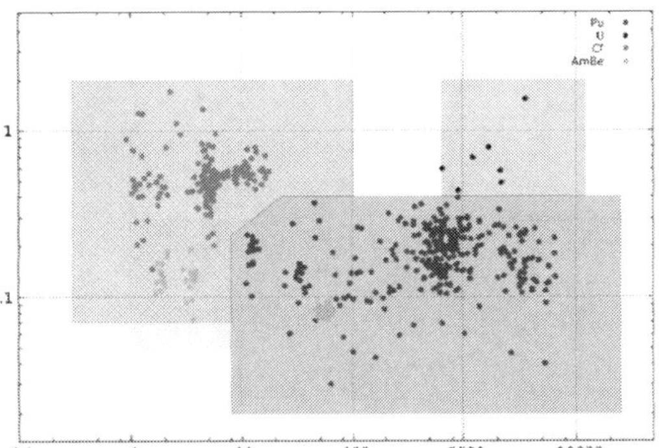

Fig. 1. *Neutron source identification algorithm results. Experimental points for Cf and Pu correspond to: naked source, Poly shielded source and Lead shield source. The parameters used in the patented algorithm are plotted on x and y axis. Each point corresponds to the results obtained by the algorithm run on a spectrum collected in 1 min of identification. Each point can fall in the colored boxes which correspond to neutron sources present in the library. For Am-Be sources a third parameter is used in the algorithm*

A moving average is used to integrate the last three minutes of natural background in order to set individual thresholds for neutron and gamma counts. Alarms are triggered separately when the respective rate exceed the alarm thresholds.

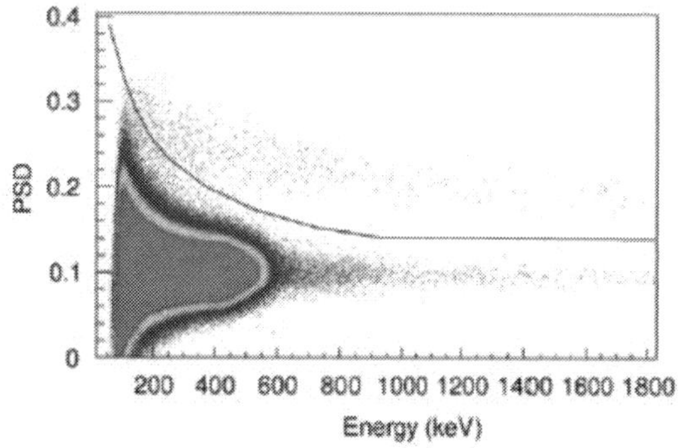

Fig. 2. . *Neutron-Gamma discrimination with gamma dose rate of 0.1 mSv/h on the front face of the detector. The black line is a polynomial function used to separate the gamma region from the neutron region. Pulse Shape Discrimination Plot obtained by a CAEN desktop digitizer with a EJ309 liquid scintillator detector coupled to a flat panel Hamamatsu.*

These thresholds are calculated to allow detection with 95% detection probability at 95% confidence level for a dose rate on the front face of the scintillator of at least 50 nSv/h. The neutron source detection has been tested also in high gamma ray fields up to 0.1 mSv/h. The liquid scintillator detector is able to discriminate neutrons from gamma rays by using the PSD technique. Such signal processing is currently performed by using complex and expensive read-out systems. Today, thanks to the development of compact and fully digital pulse analyzers, it is possible to design compact systems with enhanced performance. The system here presented makes use of advanced digital electronics including a Multi Channel Analyzer (MCA) and a FPGA running Digital Pulse Processing (DPP) firmware for PSD. This firmware allows for online fast discrimination of gamma rays from neutrons thus enabling individual alarms for each kind of particle. The PSD firmware is based on an advanced online digital dual gate charge integration technique able to sustain high counting rates.

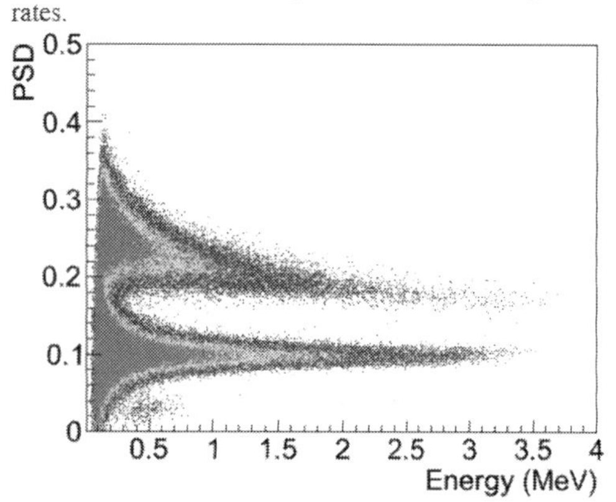

Fig. 3. *Pulse Shape Discrimination Plot obtained by a CAEN desktop digitizer with a EJ309 liquid scintillator detector coupled to a flat panel Hamamatsu. The gamma region is between 0.0 < PSD < 0.2 and the neutron region is between 0.2 < PSD < 0.4. PSD = (long gate integration – short gate integration)/ long gate integration..*

It performs input signal baseline calculation, self-triggering, double integration of both prompt and total charge for PSD and pedestal subtraction for energy calculation and pile-up rejection.

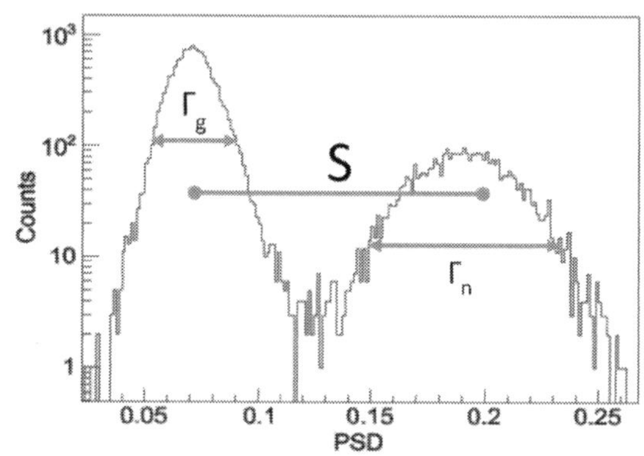

Fig.4. *Typical example of FoM plot obtained by the projection of the PSD plot on the x axis. FWHMs and separation between centroids are shown in the plot. Centroids and FWHM are obtained by Gaussian fit of the plot. The fit is not shown*

Moreover, with a novel discrimination algorithm, a neutron source classification is also possible. An index of the performance of the Pulse Shape Discrimination is the Figure of Merit (*FoM*). The graph for the FoM is obtained by projecting the y axis of the PSD plot on the x axis in a PSD vs count plot thus obtaining two Gaussian distribution, one for the gamma and one for the neutron (see Fig. 4). The FoM is defined as:

$$\mathrm{FOM} = S / (\Gamma_n + \Gamma_g) \quad (1)$$

where S is the distance between the two centroids of neutron and gamma gaussian distribution and $(\Gamma_n + \Gamma_g)$ is the sum of gamma and neutron FWHMs [1]. The FoM obtained with the configuration described above in the caption of Fig. 3 and with an energy window of 480±75 keV is 1.73.

III. NEUTRON SOURCE IDENTIFICATION ALGORITHM

The results described in the above section were obtained with the digitizer and the detectors arranged in a laboratory set-up. The algorithm was run with ad-hoc program run on a desktop pc. To target the engineering of this set up, a more complicated scenario was set up.

Fig. 5. *Aluminum box set-up containing the two detectors, the DT5790, the $LiFePO_4$ Battery and the Single Board Computer (SBC). The data are transmitted through Wi-Fi.*

The system is packaged in an aluminum box containing the digitizer and the detectors. It is powered with a $LiFePO_4$ battery with capacity of 25Ah at 12V. The computer is replaced with Single Board Computer (SBC) equipped with ARM processor which run the optimized algorithms. In this configuration the system becomes portable. Some homeland security standards were identified to better understand where this kind of set up can be included. There isn't a specific standard for this kind of device but the most suitable are:

- IEC 62327 Hand Held instruments for the detection and identification of radionuclides
- ANSI 42.34 Hand Held instruments for the detection and identification of radionuclides

Considering the European standard the requirements for gamma and neutron alarm and identification are:

- Gamma alarm raised in 1 second for a source moving at a speed of 0.5 m/s producing a radiation field of 500 nSv/h above the bakground at the reference point of the RID at a closest distance of 1 m.
- Neutron alarm raised in 2 seconds for a Cf-252 source having an emission rate of 20.000 neutrons/s surrounded by 4cm of High-density poliethilene at 25 cm of distance from the reference point of the RID
- 1 minute or less for source identification

Tests were repeated with the new and more complicated set-up. Most of the measurement were focused on the neutron alarm and the neutron source identification performed by the patented algorithm (see Fig. 1). The identification measurement is performed on a 1 minute spectra as set as maximum in the standard. Bigger distance respect to the standard were also tested. The Legnaro laboratories use a stronger Cf-252 respect to the standard, so the distance of 25 cm was scaled proportionally and the new reference standard distance is set to 45.5 cm.

Many test were performed with different thickness of lead, poliethilene and mixed shields.

Table I. *Cf-252 source identification test performed with EJ309 and DT5790 packaged in a aluminum box and powered by a $LiFePO_4$ battery. Identification patented algorithm is run by a SBC computer with ARM processor*

distance (cm)	Shielding	Success/ trials
standard	Pb 5 cm + Poly 6 cm	10/10
2 x standard	Poly 10 cm	10/10
4.5 x standard	-	15/15

Other tests were performed with different neutron source by scaling the neutron emission rate proportionally to standard definition. The best result is obtained with a special mixed source enriched in Pu-239 at 61%. With this source a 100% success of identification (30 success on 30 trials) is obtained at 23 times the standard distance.

IV. DOUBLE CHANNEL ELECTRONIC FOR GAMMA RADIATION DETECTOR

The electronics is equipped with two analog inputs and two high voltage power supplies in a very small form factor thus becoming an enabling technology for higher performance yet portable radioactive isotope identifier devices, which can include more detectors and perform data fusion analysis. The double channel capabilities of the electronics are exploited by adding another detector to the system for improving the identification performance of the gamma emitters. A high-resolution inorganic scintillator, such as $LaBr_3$ or $CeBr_3$, has been easily integrated into the system to perform an enhanced gamma radioisotope identification by characteristic gamma lines recognition. This solution allows for a simultaneous identification of multiple radionuclides, exceeding therefore the limits of gamma identification performed with organic liquid scintillator and Compton spectra libraries only. The added benefit of this solution is the possibility to detect a masked neutron source through the PSD algorithm performed by the liquid scintillator detector while the inorganic scintillator identifies the masking gamma emitters. The inorganic scintillator enhances also the neutron source identification. In fact, thanks to the high resolution of the crystal, it is possible to calculate the Pu and U enrichment grade through their characteristic gamma emission rate ratio as well.

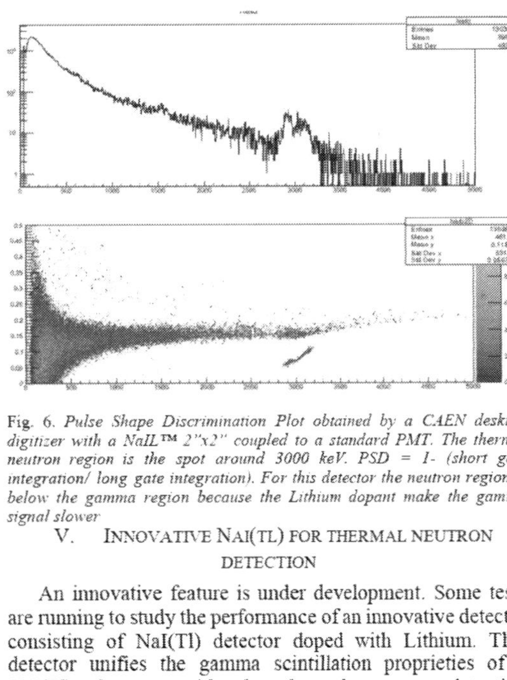

Fig. 6. *Pulse Shape Discrimination Plot obtained by a CAEN desktop digitizer with a NaIL™ 2"x2" coupled to a standard PMT. The thermal neutron region is the spot around 3000 keV. PSD = 1- (short gate integration/ long gate integration). For this detector the neutron region is below the gamma region because the Lithium dopant make the gamma signal slower*

V. INNOVATIVE NAI(TL) FOR THERMAL NEUTRON DETECTION

An innovative feature is under development. Some tests are running to study the performance of an innovative detector consisting of NaI(Tl) detector doped with Lithium. This detector unifies the gamma scintillation proprieties of a NaI(Tl) detector with the thermal neutron detection capabilities. NaIL™ detector is provided by Saint-Gobain and with this detector neutrons and gamma rays are easily distinguished through pulse shape discrimination (PSD). With the addition of Li into the NaI matrix, gamma ray scintillation pulses become longer than neutron reaction pulses. The effect is dramatic, and simple PSD techniques create complete separation [2]. Some test has been performed with a 2"x2" NaIL™ detector coupled with a standard PMT tube and analysed with a CAEN DT5725 digitizer (14-bit @ 250 MS/s) with the DPP-PSD firmware. With a dynamic range of 0.5 Vpp and a long gate integration of 4000 ns the best resolution obtained at 662 keV with a Cs-137 source is 6.6% which is comparable with standard NaI(Tl) detector of the same dimension. The detector was also tested with a Cf-252, a source of fast neutrons and gamma rays, placed at 30 cm and shielded with different thickness of polyethylene (0, 2, 4 and 6 cm) adjacent to the source. Using the Pulse Shape Discrimination, the signals produced in the detector by the interaction of polyethylene slowed down neutron were identified. The 2-D plot of PSD vs energy is shown is Fig. 6. Around the 3 MeV energy the contribution of thermal neutron can be noted. This contribution is produced by thermal neutrons that release energy inside the detector by interacting with Lithium according to the reaction:

$$n + {}^{6}Li \rightarrow {}^{3}H + {}^{4}He \qquad (2)$$

The maximum thermal neutron efficiency is obtained in the configuration with 6 cm of polyethylene and the neutron efficiency obtained for NaIL™ is the 89% with respect to the EJ-420 detector, with a diameter of 2" and placed in the same geometrical configuration and with the same activity of the source. The EJ-420 can be considered as a standard for the measurement of thermal neutrons. Replacing the $CeBr_3$ with this innovative detector can add a lot of benefits to the device. First of all, the system with two detectors can detect and identify gamma, thermal neutron and fast neutron sources. Further information on the polyethylene shielding thickness can be retrieved by the thermal to fast neutron ratio. The only deterioration is the worse resolution of a NaI(Tl) compared to the $CeBr_3$. This feature don't affect the gamma identification algorithm dramatically; in fact most of the Radiation Isotope Identification Device is equipped with a NaI(Tl) detector and can easily identify gamma radiation source.

REFERENCES

[1] J.R.M Annand, Nuclear Instruments and Methods A 262 (1987) 371.
[2] K. Yang, P. R. Merge at al, a large volume neutron-gamma scintillator with exceptional pulse shape discrimination, IEEE NSS-MIC.

https://doi.org/10.1051/epjconf/202022507007

E-SiCure Collaboration Project: Silicon Carbide Material Studies and Detector Prototype Testing at the JSI TRIGA Reactor

Vladimir Radulović
Jožef Stefan Institute
Ljubljana, Slovenia
vladimir.radulovic@ijs.si

Klemen Ambrožič
Jožef Stefan Institute
Ljubljana, Slovenia
klemen.ambrozic@ijs.si

Luka Snoj
Jožef Stefan Institute
Ljubljana, Slovenia
luka.snoj@ijs.si

Ivana Capan
Rudjer Bošković Institute
Zagreb, Croatia
ivana.capan@irb.hr

Tomislav Brodar
Rudjer Bošković Institute
Zagreb, Croatia
tomislav.brodar@irb.hr

Zoran Ereš
Rudjer Bošković Institute
Zagreb, Croatia
zoran.eres@irb.hr

Željko Pastuović
ANSTO
Lucas Heights, Australia
zkp@ansto.gov.au

Adam Sarbutt
ANSTO
Lucas Heights, Australia
ars@ansto.gov.au

Takeshi Ohshima
QST
Takasaki Gunma, Japan
ohshima.takeshi@qst.go.jp

Yuichi Yamazaki
QST
Takasaki Gunma, Japan
yamazaki.yuichi@qst.go.jp

José Coutinho
University of Aveiro
Aveiro, Portugal
jose.coutinho@ua.pt

Abstract—**In 2016, the "E-SiCure" project (standing for Engineering Silicon Carbide for Border and Port Security), funded by the NATO Science for Peace and Security Programme, was launched. The main objective is to combine theoretical, experimental and applied research towards the development of radiation-hard SiC-based detectors of special nuclear materials (SNM), and by that way, to enhance border and port security barriers. Along the plan, material modification processes are employed firstly to study, and secondly to manipulate the most severe electrically active defects (which trap or annihilate free charge carriers), by specific ion implantation and defect engineering. This paper gives an overview of the experimental activities performed at the JSI TRIGA reactor in the framework of the E-SiCure project. Initial activities were aimed at obtaining information on the radiation hardness of SiC and at the study of the energy levels of the defects induced by neutron irradiation. Several Schottky barrier diodes were fabricated out of nitrogen-doped epitaxial grown 4H-SiC, and irradiated under Cd filters in the PT irradiation channel in the JSI TRIGA reactor with varying neutron fluence levels. Neutron-induced defects in the material were studied using temperature dependent current-voltage (I-V), capacitance-voltage (C-V) and Deep-Level Transient Spectroscopy (DLTS) measurements. Our prototype neutron detectors are configured as 4H-SiC-based Schottky barrier diodes for detection of secondary charged particles (tritons, alphas and lithium atoms) which are result of thermal neutron conversion process in ^{10}B and ^{6}LiF layers above the surface of the 4H-SiC diodes. For field testing of neutron detectors using a broad beam of reactor neutrons we designed a standalone prototype detection system consisting of a preamplifier, shaping amplifier and a multichannel analyser operated by a laptop computer. The reverse bias for the detector diode and the power to electronic system are provided by a standalone battery-powered voltage source. The detector functionality was established through measurements using an ^{241}Am alpha particle source. Two dedicated experimental campaigns were performed at the JSI TRIGA reactor. The registered pulse height spectra from the detectors, using both ^{10}B and ^{6}LiF neutron converting layers, clearly demonstrated the neutron detection abilities of the SiC detector prototypes.**

***Keywords*—Silicon carbide, Neutron detection, Neutron converter, JSI TRIGA reactor**

NATO Science for Peace and Security Programme

I. Introduction

Increasingly complex risks like geopolitical instability or decentralized terrorism threats, have led to the urge for deploying nuclear screening systems for detection of illicit trafficking of nuclear materials, and from that, to a growing interest in the field of research and development of new radiation detection technologies suitable for homeland security applications. Recent progress in the manufacturing of high-quality bulk and epitaxial silicon carbide (SiC) and processing technologies for fabrication of SiC-based electronic devices, could enable unprecedented opportunities for future SiC-based detection of neutron and alpha particle emissions. Unlike existing and commonly used gas-based neutron detectors, SiC-based devices have the potential to be simultaneously portable, operable at room temperature and radiation hard. The main objective of the E-SiCure project is to combine theoretical, experimental and applied research towards the development of radiation-hard SiC-based detectors of special nuclear materials (SNM), and therefore to enhance border and port security barriers, which is achieved through studies of material radiation hardness and modification processes in order to manipulate the most severe electrically active defects which trap or annihilate free charge carriers, by specific ion implantation and defect engineering. This paper presents the experimental activities performed at the JSI TRIGA reactor in the framework of the E-SiCure project. Section 2 presents the study of radiation-induced defects in neutron irradiated SiC diodes with Deep-Level Transient Spectroscopy. Section 3 presents the realization of prototype SiC detectors and initial functional testing with a ^{241}Am radiation source. Section 4 presents a dedicated experimental campaign performed at the JSI TRIGA reactor aimed at demonstrating the neutron detection capability of SiC-based detector prototypes.

II. Neutron-Induced Defects in SiC

To study neutron induced defects in SiC, n-type silicon carbide Schottky barrier diodes (SBDs) were fabricated onto nitrogen-doped epitaxial grown 4H-SiC single crystal layers approximately 25 μm thick, their lateral dimensions being 1 mm by 1 mm [1]. The Schottky diodes were irradiated either upon bare exposure or inside Cd thermal neutron filters with a wall thickness of 1 mm in the 250 kW JSI TRIGA reactor in Ljubljana, Slovenia. The irradiations were performed in the Pneumatic Tube (PT) irradiation facility, located in the F24 position in the outermost F ring of the reactor; the neutron fluences ranged from 10^8 n cm^{-2} to 10^{15} n cm^{-2}. The neutron spectrum in the PT irradiation channel was obtained from Monte Carlo neutron transport calculations with the MCNP6 code [2] in conjunction with the ENDF/B-VII.1 nuclear data library [3]. A verified and validated computational model of the JSI TRIGA reactor was employed, based on the JSI TRIGA criticality benchmark model [4], featured in the International Handbook of Evaluated Critical Safety Benchmark Experiments [5]. The computational model is in permanent use at the Reactor Physics Department of the JSI for research purposes (computer code and nuclear data validation) and in support of experimental campaigns performed at the reactor [6][7][8]. Over time, the model has been considerably updated and refined. It has been validated against experimental measurements for calculations of the multiplication factor k_{eff} [4], reactor kinetic parameters [9], neutron flux and reaction rate distributions [10][11][12]. The neutron spectrum obtained from Monte Carlo calculations underwent a subsequent process of neutron spectrum adjustment to measured reaction rate ratios. The adjustment process was performed using the JSI-developed GRUPINT code package [13]. Fig. 1 displays the adjusted bare and Cd-filtered neutron spectrum in the PT facility.

A total of 8 samples were irradiated at different reactor power levels (2.5 W, 25 W, and 250 kW), specifying different irradiation times, in order to span a range of 7 orders of magnitude in the neutron fluence delivered to the samples (10^8 n cm^{-2} to 10^{15} n cm^{-2} nominal fluence range). Activation measurements for the ^{197}Au(n,γ) reaction were performed for each power level used for the irradiations in order to deduce the total neutron flux. The subcadmium flux was obtained from the characterized neutron spectrum and the Cd cutoff energy for the thickness of 1 mm, i.e. 0.55 eV. Table 1 reports

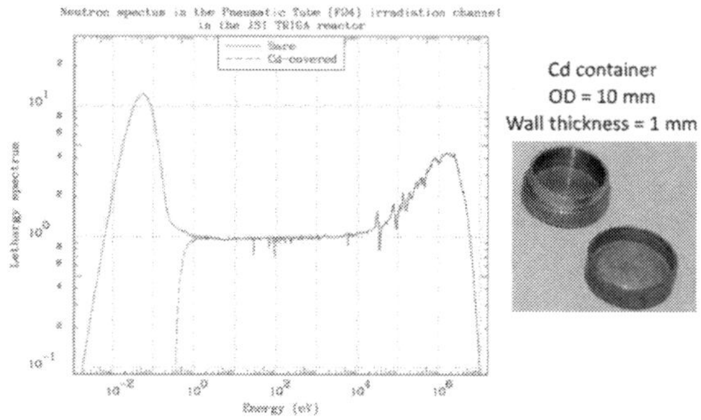

Fig. 1. Neutron spectra in the Pneumatic Tube (PT) irradiation channel, located in the F24 position in the core of the JSI TRIGA reactor. Red: bare spectrum. Blue: spectrum under Cd cover.

details of the neutron irradiation conditions for the selected samples. The calculated uncertainties in the neutron fluences were obtained from uncertainties in the neutron flux and the irradiation time.

TABLE I
IRRADIATION DETAILS, SUBCADMIUM NEUTRON FLUX LEVELS AND 1-σ EXPERIMENTAL UNCERTAINTIES, SAMPLE NEUTRON FLUENCE VALUES AND 1-σ EXPERIMENTAL UNCERTAINTIES.

ID	**Irr. time** [s]	**P** [W]	**Sub-Cd flux** [n cm^{-2}s^{-1}]	**Fluence** [n cm^{-2}]
R6	3.0	2.5	$3.66 \times 10^7 \pm 2.7\%$	$1.10 \times 10^8 \pm 9.8\%$
R10	30	2.5	$3.66 \times 10^7 \pm 2.7\%$	$1.10 \times 10^9 \pm 2.9\%$
R11	300	2.5	$3.66 \times 10^7 \pm 2.7\%$	$1.10 \times 10^{10} \pm 2.7\%$
R9	273	25	$3.44 \times 10^8 \pm 2.9\%$	$9.39 \times 10^{10} \pm 2.9\%$
R12	2730	25	$3.44 \times 10^8 \pm 2.9\%$	$9.39 \times 10^{11} \pm 2.9\%$
1E13	2.73	250k	$3.66 \times 10^{12} \pm 2.7\%$	$1.00 \times 10^{13} \pm 10.7\%$
IE14	27.3	250k	$3.66 \times 10^{12} \pm 2.7\%$	$1.00 \times 10^{14} \pm 2.9\%$
1E15	273	250k	$3.66 \times 10^{12} \pm 2.7\%$	$1.00 \times 10^{15} \pm 2.7\%$

Neutron induced defects in the samples were studied using temperature dependent current-voltage (I-V), capacitance-voltage (C-V) and Deep-Level Transient Spectroscopy (DLTS) measurements. I-V and C-V measurements were carried out using a Keithley 6487 Picoammeter/Voltage Source and a Keithley 4200 Semiconductor characterization system. DLTS and Laplace-DLTS measurements were carried out using a Boonton 7200 capacitance meter, a NI PCI-6251 DAQ and Laplace DLTS software. DLTS and Laplace-DLTS measurements were performed in the temperature range from 100 K up to 380 K. Temperature dependent I-V measurements revealed that the fast neutron irradiation did not affect the transport properties of SBDs for neutron fluences lower than 10^{12} n cm^{-2}. An increase in the series resistance and decrease of the capacitance of neutron irradiated samples was observed first after the neutron fluence reached the value of 10^{12} n cm^{-2}, followed by even more pronounced changes in the sample irradiated with the fluence of 10^{13} n cm^{-2}. The quality of SBDs was not satisfactory for DLTS measurements above

 https://doi.org/10.1051/epjconf/202022507007

the neutron fluence of 10^{14} n cm^{-2}. Temperature dependent C-V measurements revealed uniform changes in the free carrier concentration, showing a small decrease for the highest fluences.

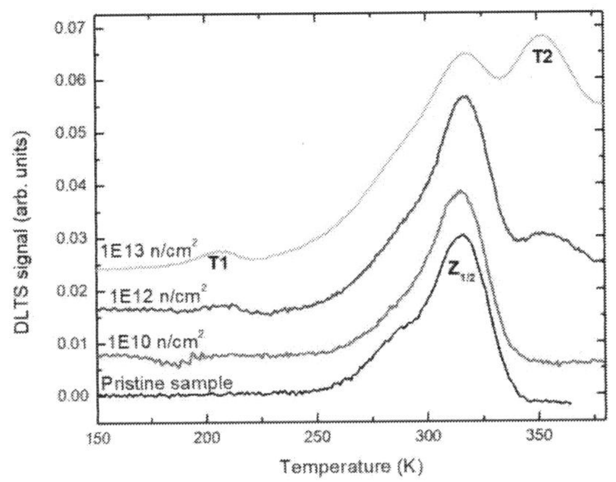

Fig. 2. DLTS spectra of pristine and selected neutron irradiated SBDs. Measurement setting are: emission rate 50 s^{1}, reverse bias V_R = 10 V, pulse bias V_P = 0.1 V, and pulse width τ_p = 10 ms. The spectra are shifted vertically for clarity.

Fig. 2 shows DLTS spectra for the pristine SBD and selected neutron irradiated SBDs. In the pristine sample, only one broad and asymmetric DLTS peak is observed. The estimated activation energy for electron emission was 0.64 eV. This defect, known as $Z_{1/2}$ has already been reported and was ascribed to the carbon vacancy V_C(=/0) transition [14]. V_C is known as one of the most stable defects in SiC, acting as a strong minority carrier recombination center in n-type material. The intensity of $Z_{1/2}$ slightly increases with the neutron fluence. Additional DLTS peaks, which we have labelled as T1 and T2, have been detected for neutron fluences of 10^{12} n cm^{-2} and 10^{13} n cm^{-2}, respectively. The activation energies for electron emission are estimated at 0.36 eV and 0.70 eV. The T1 and T2 traps have been reported in the literature in numerous instances, however, information regarding these defects, which regularly appear in implanted or irradiated 4H-SiC samples are very limited and controversial. We suggest that T1 and T2 are intrinsic, simple radiation induced point-like defects.

III. Prototype Detection System and Initial Testing

Two main components of any detector system are 1) the detector of particle events and 2) the electronic system for processing and recording of events. The prototype detectors were based on hexagonal 4H-SiC-based Schottky diodes with several active layer thicknesses, ranging from 25 μm to 170 μm, their lateral dimensions being 1 mm by 1 mm. The detectors consisted of SBDs surface-mounted onto chip carriers with electrical contacts. The chip carriers were enclosed in 3D-printed plastic holders in order to completely enclose the electrical contacts under high voltage (Fig. 3a). The holders had an opening above the SBD to enable charged particles to reach the SBD surface. The electronic system for particle event processing and recoding consisted of a preamplifier and a shaping amplifier PCB modules (CR-150-R5 and CR-160-R7) manufactured by CREMAT, and a digital signal processing multichannel analyser (DSP-MCA) manufactured by AMPTEK (model no. 8000D), connected to a laptop computer (Fig. 3b). Power to the electronic system (preamplifier and shaping amplifier) was provided by a standalone battery-powered voltage source, in order to avoid the use of mains power. The latter often carries electronic noise which can negatively affect the measurement performance. The voltage source also provided power to a separate high voltage bias module. Reverse negative bias was connected to the front Schottky contact, and the back Ohmic contact of the prepared prototype detectors was grounded. All initial tests have been performed in low-AC-noise environment of the ANSTO detector lab.

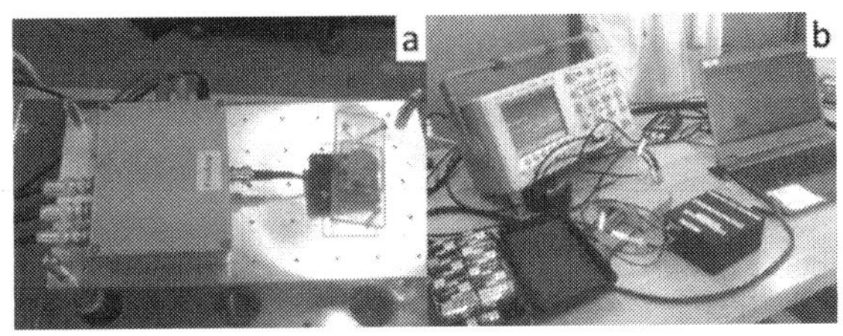

Fig. 3. Detector system under development: a) A SBD detector being tested using a ^{241}Am source. b) Electronic system for operation of a detector and data acquisition during bench testing at ANSTO.

The detection set-up was tested using a precision pulse generator and a ^{241}Am source of energetic alpha particles. For a low capacitance input ($\sim$10 pF) to the preamplifier stage we used a CREMAT CR-110-R2 preamplifier chip and tested the detection energy resolution with the shaping PCB module equipped with the CR-200-R2 chip with shaping times of 0.5 and 1.0 μs. The total gain for event signal amplification was kept constant. Pole/Zero (P/Z) was adjusted for different shaping constants. Recorded events were sorted by the MCA in 2k-channel energy spectra. The pulser peak and the ^{241}Am peak were fitted with Gaussians, given by the following expression:

$$y = y_0 + \frac{A}{w\sqrt{\frac{\pi}{4\ln 2}}} \exp\left(-\frac{4\ln 2(x-x_c)^2}{w^2}\right) \quad (1)$$

where y is the number of detected counts in the channel x, x_c is the centroid, A determines the height of the peak and w the peak width. Fig. 4 shows recorded alpha particle spectra. Although the energy resolution for line-pulse detection is very similar for reverse bias voltages of 50 V and 100 V, the calculated width, FWHM and corresponding energy resolution for reverse bias of 50 V have lower values compared to reverse bias of 100 V. Therefore we performed neutron detection tests using a reverse bias of 50 V and a shaping constant of 1 μs. The initial testing results demonstrated well the functionality of the detector prototypes and the electronic measurement system.

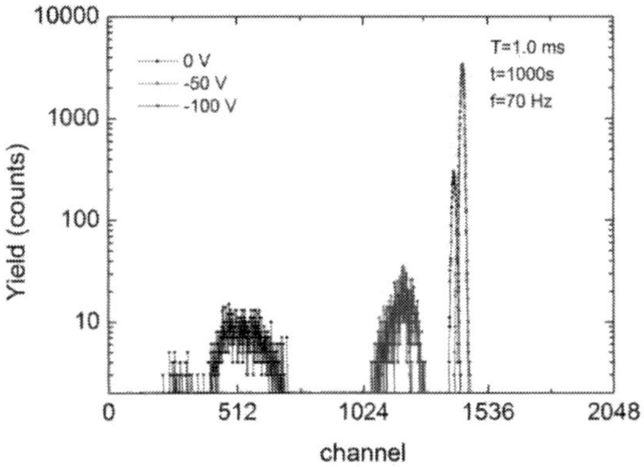

Fig. 4. Recorded ^{241}Am alpha particle spectra during initial testing with the prototype detector system and the following settings: precision pulse generator frequency: 70 Hz, reverse bias voltages: 0 V (black), -50 V (red) and -100 V (blue), total acquisition time: 1000s.

IV. Test and the JSI TRIGA Reactor

Following the results obtained from the initial testing, an experimental campaign was performed at the JSI TRIGA reactor in May 2018 aiming at demonstrating the neutron detection capabilities of the prototype detectors. For the test, four new detectors were realized, enclosed in 3D-printed plastic holders, and housed in turn in custom-made aluminium enclosures. The detector prototypes were equipped with thermal neutron converting materials for neutron detection. Due to the very large thermal neutron cross sections of ^{10}B and ^{6}Li isotopes (around 3843 b and 938 b respectively, at an incident neutron energy of 0.0253 eV [15]), the chosen converting materials were enriched ^{10}B, ^{6}LiF and $^{10}B_4C$ powders. The ^{10}B and ^{6}LiF powders were applied onto plastic film and mounted above the openings of the 3D-printed SBD holders. One of the SBDs was realized with a capping layer of $^{10}B_4C$, in physical contact with the SBD. For the preparation of the $^{10}B_4C$ capping layer, firstly, the whole surface of a SiC SBD mounted onto a chip carrier was covered with an adhesive material in order to prevent leakage current. After drying, $^{10}B_4C$ powder was dropped onto the adhesive surface on the SiC SBD and encased in a second adhesive layer. Fig. 5 shows two realized prototype detectors used in the experimental tests (fully enclosed and open).

Firstly, the detectors were tested with no neutron converters (except for the detector with the $^{10}B_4C$ capping layer) and no radiation sources. All the obtained spectra showed a large peak at low channel numbers, which was attributed to electronic noise, and no counts appeared at higher channels. It was also

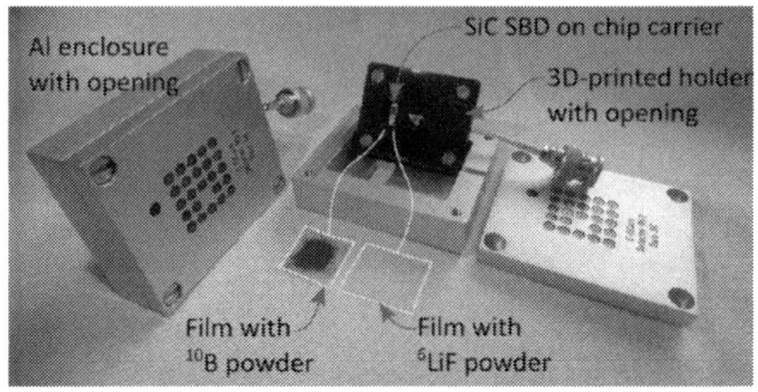

Fig. 5. E-SiCure prototype detectors. Left: assembled detector prototype in aluminium enclosure. Right: prototype detector components: SiC SBD mounted onto chip carrier with contacts, installed in 3D printed holder with opening, converter films (with ^{10}B and ^{6}LiF powder), open aluminium enclosure with opening.

observed that counts started to appear at higher channels if the laptop computer was connected to mains power or if the mousepad of the laptop computer was touched. Therefore, all the subsequent measurements were performed with the laptop computer running only on battery power and using a wireless mouse for control. The detectors were then tested using a ^{241}Am source, its nominal activity being 416 kBq (reference date 29.5.2018). For the three detectors without the $^{10}B_4C$ capping layer a clear peak due to the alpha particles emitted from ^{241}Am at high channels was observed, confirming the functioning of the SBDs as detectors of charged particles. The peak was not observed for the detector with the $^{10}B_4C$ capping layer, indicating that the detector was not functioning as expected. Following the functioning tests, the neutron converter films were applied to the detectors and neutron irradiations were performed in the Dry Chamber of the JSI TRIGA reactor. The Dry Chamber is a large irradiation room in the concrete body of the reactor, connected with the reactor core by a graphite thermalizing column [16]. It is mostly used for radiation hardness testing of detectors, electronic components and systems [17][18]. Measurements were taken firstly with a 25 μm thick prototype detector equipped with ^{10}B converter film at the following power levels: 10 kW, 50 kW, 100 kW, 180 kW and 250 kW. In all the recorded spectra we observed a significant number of counts at higher energy channels, attributed to alpha particles and recoil ^{7}Li particles originating from $^{10}B(n, \alpha)$ reactions. We also observed a distinctive structure in the spectra, attributed to the different energies of the secondary particles. Fig. 6 shows the recorded spectra using the ^{10}B converter film.

Measurements were continued with a 69 μm thick prototype detector equipped with ^{6}LiF converter film, at the following reactor power levels: 10 kW, 50 kW, 100 kW, 170 kW and 250 kW. As for the ^{10}B converter film, with the ^{6}LiF converter film we also observed a significant number of counts at higher energy channels, attributed to tritons and alpha particles originating from $^{6}Li(n, t)$ reactions. We also observed a distinctive structure in the spectra, which was qualitatively different from

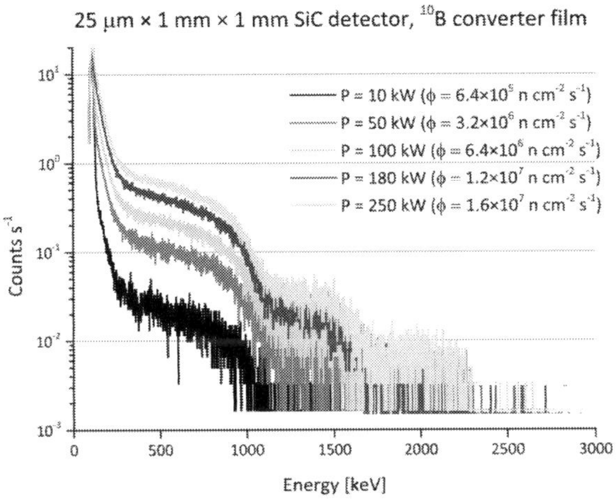

Fig. 6. Measured spectra with a 25 μm SiC detector equipped with ^{10}B converter film in the JSI TRIGA reactor Dry Chamber at different reactor power levels.

the one observed using ^{10}B converter film. Fig. 7 shows the recorded spectra using the ^{6}LiF converter film.

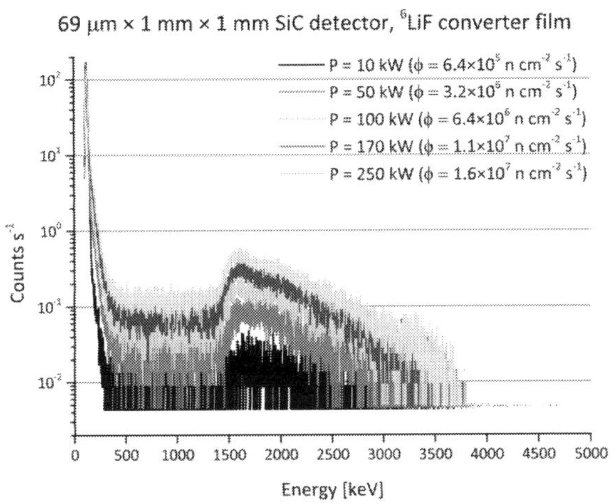

Fig. 7. Measured spectra with a 25 μm SiC detector equipped with ^{6}LiF converter film in the JSI TRIGA reactor Dry Chamber at different reactor power levels.

V. Conclusions and Future Work

This paper gives an overview of the activities performed at the JSI TRIGA reactor in the framework of the E-SiCure project. The project aims at studying SiC material properties for the detection of SNM for border and port security. SiC-based SBDs displayed very good radiation hardness properties. Irradiation tests of detector prototypes, equipped with ^{10}B and ^{6}LiF converter films were performed in the Dry Chamber of the reactor, clearly demonstrating the ability of small SiC SBDs (side dimensions of 1 mm × 1 mm, thickness below 200 μm) to detect alpha and recoil charged particles originating from neutron interactions. Further studies and irradiation testing will be carried within the E-SiCure project with the objective to optimize the detector and converter properties for maximal neutron detection efficiency.

Acknowledgment

The present work was financially supported by the NATO Science for Peace and Security Programme, under project no. SPS 985215 - Engineering Silicon Carbide for Border and Port Security E-SiCure. The JSI project team would like to acknowledge the support from the Slovenian Ministry of Education, Science and Sport (project no. P2-0073 - Reactor Physics). The RBI project team would like to acknowledge financial support from the European Regional Development Fund for the Center of Excellence for Advanced Materials and Sensing Devices (Grant No. KK.01.1.1.01.0001). JC acknowledges the Fundação para a Ciência e a Tecnologia (FCT) through project UID/CTM/50025/2013.

References

[1] M. Ito, L. Storasta, H. Tsuchida, "Development of 4HSiC Epitaxial Growth Technique Achieving High Growth Rate and Large-Area Uniformity" Appl. Phys. Expr. 1 (2008) 015001.

[2] J.T. Goorley et al., "Initial MCNP6 Release Overview - MCNP6 version 1.0", LA-UR-13-22934 (2013).

[3] M.B. Chadwick, et al., ENDF/B-VII.1 nuclear data for science and technology: cross sections, covariances, fission product yields and decay data, Nucl. Data Sheets, 112, 2011, pp. 2887-2996.

[4] R. Jeraj, M. Ravnik, TRIGA mark II reactor: U(20)-zirconium hydride fuel rods in water with graphite reflector, IEU-COMP-THERM-003, International Handbook of Evaluated Criticality Safety Benchmark Experiments, NEA/NSC/DOC(95)03, OECD-NEA, Paris, France, 2010.

[5] International Handbook of Evaluated Critical Safety Benchmark Experiments, Organization for Economic Cooperation and Development, Nuclear Energy Agency, NEA/NSC/DOC(95)03, Paris (published on DVD).

[6] A. Kolšek et al., Using TRIGA Mark II research reactor for irradiation with thermal neutrons, Nucl. Eng. Des, 283, 2015, pp. 155-161.

[7] V. Radulović et al, Qualification of heavy water based irradiation device in the JSI TRIGA reactor for irradiations of FT-TIMS samples for nuclear safeguards, Nucl. Inst. Meth. A, 885, 2018, pp. 139-144.

[8] V. Radulović et al, Report on experimental tests of a combined SPND and thermocouple sensor for the DISCOMS project at the JSI TRIGA reactor, JSI report No. 12547, Jožef Stefan Institute, 2018.

[9] L. Snoj et al., Calculation of kinetic parameters for mixed TRIGA cores with Monte Carlo, Ann. Nucl. Energy, 37 (2), 2010, pp. 223-229.

[10] L. Snoj et al., Analysis of neutron flux distribution for the validation of computational methods for the optimization of research reactor utilization, Appl. Radiat. Isot., 69 (1), 2011, pp. 136-141.

[11] V. Radulović et al., Validation of absolute axial neutron flux distribution calcula- tions with MCNP with ^{197}Au$(n,\gamma)^{198}$Au reaction rate distribution measurements at the JSI TRIGA Mark II reactor, Appl. Rad. Isot., 84, 2014, pp. 57-65.

[12] L. Snoj, G. Žerovnik, A. Trkov, Computational analysis of irradiation facilities at the JSI TRIGA reactor, Appl. Rad. Isot., 70 (3), 2012, pp. 483-488.

[13] A. Trkov, V. Radulović, L. Snoj, The GRUPINT neutron spectrum adjustment code - general features and characterization of the spectra in three irradiation channels of the JSI TRIGA reactor, Program and Book of abstracts, 16th International Symposium on Reactor Dosimetry, May 7 - 12, 2017, Santa Fe, USA.

[14] I. Capan et al., Double negatively charged carbon vacancy at the h- and k-sites in 4H-SiC: combined Laplace-DLTS and DFT study. Journal of applied physics, 123 (16), 2018, pp. 161597-1-161597-6.

[15] R. Capote et al., "Updating and Extending the IRDF-2002 Dosimetry Library," Journal of ASTM International, 9 (4), 2012, pp. 1-9.
[16] V. Radulović et al., Characterization of ex-core irradiation facilities of the JSI TRIGA Mark II reactor, Proceedings, 21st International Conference Nuclear Energy for New Europe, September 5-7, 2012, Ljubljana, Slovenia.
[17] I. Mandić et al., "Bulk damage in DMILL npn bipolar transistors caused by thermal neutrons versus protons and fast neutrons", IEEE Transactions on Nuclear Science, 51 (4), 2004, pp. 1752-1758.
[18] I. Mandić, V. Cindro, A. Gorišek, G. Kramberger and M. Mikuž, "Online Integrating Radiation Monitoring System for the ATLAS Detector at the Large Hadron Collider", IEEE Transactions on Nuclear Science, 54 (4), 2007, pp. 1143-1150.

https://doi.org/10.1051/epjconf/202022507008

Testing and performances of Spectroscopic Radiation Portal Monitor for homeland security

C. Deyglun, IRSN

—Each year States report to IAEA loss, theft or out of regulatory control radioactive materials. Most incidents are minor, but material is potentially available for criminal acts. Measures to reduce the radiological and nuclear threat are many-faceted. An important component is the ability to detect illicit transport of radioactive material. Spectroscopic Radiation Portal Monitors are deployed around the world to detect illegal radioactive material traffic. A combination of experimental data collected during testing campaigns and simulations is a good way to study the performance of Spectroscopic Radiation Portal Monitors in realistic conditions. The paper presents a process to evaluate the performances of a portal, based on a combination of experimental data and MCNP simulations to calculate the detection probability and the false alarm rate. IRSN developed platforms for testing Spectroscopic Radiation Portal Monitors for pedestrian control. Experimental data were collected from an available commercial Spectroscopic Radiation Portal Monitor, tested in the framework of the Illicit Trafficking Radiation Assessment Program phase II Round Robin Test. Many scenarios were tested with different sources using realistic setups and many experimental data were collected. The tested equipment was then simulated with MCNP only based on the data provided in the user manual and the standards found in the industry. To get a realistic idea of the uncertainty, all the variables inherent in the measurement were considered, their relative contributions were identified and quantified, then propagated to predict an overall uncertainty. The combination of experimental data, numerical simulations and uncertainty evaluation showed good agreement with experimental assays. The results were used to test the sensitivity of a Spectroscopic Radiation Portal Monitor to special nuclear materials for different alarm thresholds. This process applied to different scenarios according to defined targets should help in the selection of operating characteristics of the portal.

***Keyword* — Safeguards, Homeland security, Nuclear security, Gamma-ray detection, Radioactive source detection**

I. Introduction

In recent years, the threat of nuclear and radiological terrorism have raised at the international level [1]. A significant number of incidents of trafficking of nuclear materials have been recorded in the IAEA Incident and Trafficking Database [2]. For example, in 2013 incident reported by the IAEA [3] included details of a group of traffickers convicted in Moldova for attempting to sell quantities of uranium and plutonium that had been transported in shielded lead canisters in an attempt to lure detection systems. Measures to reduce the radiological and nuclear threat are many-faceted. An important component is the ability to detect illicit transport of radioactive material. Spectroscopic Radiation Portal Monitors (SRPM) are deployed at front-line sites around the world to detect illegal radioactive material traffic. The Illicit Trafficking Radiation Assessment Program (ITRAP+10) is a program initiated by the European Union and the United States [4, 5]. From 2011 to 2016, the first phase aimed at evaluating the performance of available commercial radiation detection equipment against the American National Standards Institute/Institute of Electrical and Electronics Engineers (ANSI/IEEE) and the International Electrotechnical Commission (IEC) published consensus standards. During the second phase (ITRAP+10 phase II Round Robin Test), started in 2017, an available commercial Spectroscopic Radiation Portal Monitor was tested by five European laboratories including IRSN. For this purpose, IRSN developed testing platforms and collected experimental data. In parallel, a MCNP model of the portal was developed based on the data provided in the user manual and the standards found in the industry in order to study the sensitivity of such portal to special nuclear materials. The paper presents an evaluation of the performances of a SRPM based on few experimental data and simulations. At first the tested SRPM is simulated with MCNP6 [6] and it's response to point sources is compared to experimental results to adjust the model. Then based on such model, the response to HEU source is simulated and uncertainties are calculated. Finally the Detection probability is calculated through simulations.

C. Deyglun works for the Institute for Radiological Protection and Nuclear Safety, Fontenay-aux-Roses, BP 17 - 92 262, France (e-mail: clement.deyglun@irsn.fr).

II. Performance Evaluation By Simulation

A. Modelling the SRPM response

The SRPM is based on inorganic scintillator NaI(Tl) to detect gamma-rays. The energy resolution of the full energy peak of a scintillator coupled to a photomultiplier tube depends on the intrinsic resolution of the crystal and the photomultiplier tube contribution. The Full Width at Half Maximum (FWHM) curve according to gamma-rays energy is expressed by the following function:

$$FWHM = a + b\sqrt{E + cE^2} \tag{1}$$

Where E is incident gamma rays energy (MeV) and a, b, c are

constants from the fitting function. A set of measurements with Am-241, Cs-137, Co-60 and Eu-152 sources were performed (see Fig. 1) in order to determine the energy resolution, and the curve's coefficients were introduced in the function provided by the GEB command of MCNP6 code that fits a Gaussian to the spectrum to make the proper corrections.

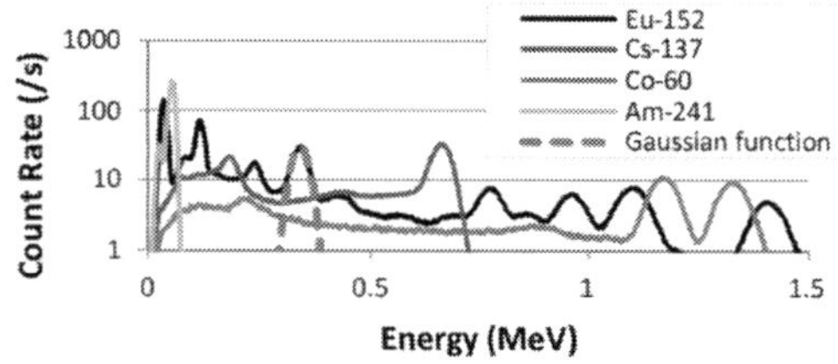

Fig 1. Gamma spectra fitting with a Gaussian function.

The SRPM model was developed based on the data of the user manual (for example: mechanical parts, shields, detectors dimensions), the standards found in the industry (e.g. NaI(Tl) crystals are generally machined and mounted in a housing of aluminum with standard thickness of 0.5 mm) and some simple assumptions (for example: the presence of a photomultiplier and electronic component inside the SRPM). Different model of the SRPM were tested, the first one is only composed of the NaI(Tl) bloc, the second is composed of the NaI(Tl) bloc and structure of the portal, and the last model is composed of the complete SRPM and its surroundings. These models allow estimating the best compromise between the complexity of the model and a good agreement with experimental measurement. They are presented in Fig. 2

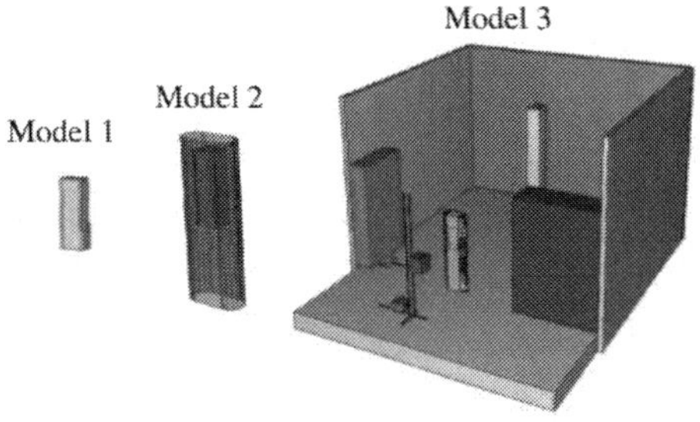

Fig.2. Tested models of the SRPM.

A pulse height distribution estimate (F8 tally), was used to obtain the deposited energy distribution per incident photon in the detector volume. A comparison between empirical spectrum and MCNP6 simulated spectra of the NaI(Tl) detector in the case of a Eu-152 source is shown in Fig. 3.

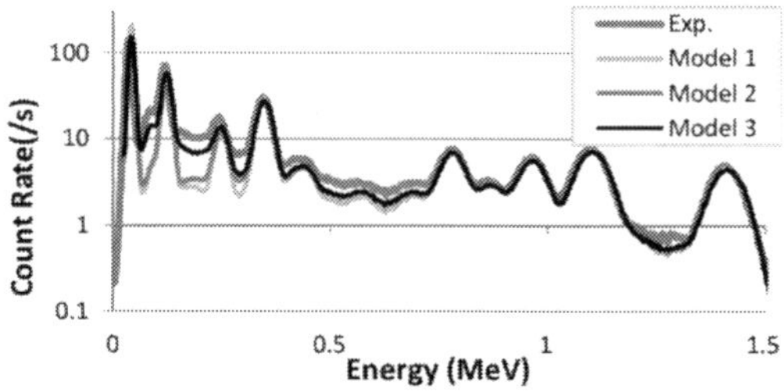

Fig 3. Comparison between experimental and simulated pulse height distributions of a Eu-152 source.

A good agreement between simulated and measured energy spectra can be seen. While for the high-energy gamma rays the measured and calculated spectra match well, for the low-energy gamma-rays however a systematic difference and an under-estimation of the models are observed. The first one is due to non-proportional scintillation efficiency of the NaI(Tl) crystals at low energy region [7]. The scintillation efficiency is defined as the light output per unit deposited energy, is not a constant value and is not considered in MCNP6 simulation. The second is due to the bad simulation of peaks arising from Compton scattering of gamma rays in the room's walls and in the shield surrounding the detector. The results presented in Fig. 3 indicate that:

- A complete setup model is important at low energy;
- The data of the user manual and basic knowledge about the composition of the SRPM are enough to obtain good agreement between simulated and experimental spectra;
- There is a good agreement for medium and high energy peaks;
- Gamma rays from Compton scattering of gamma rays in the surrounding environment are difficult to model.

B. Alarm Threshold Calculation

False-alarms are caused by statistical variations in background counting. In a defined region of spectra (Region Of Interest: ROI), count varies according to a Poisson distribution, with a standard deviation, which is the root square of the number of counts. The alarm threshold (AL) is a compromise between sensitivity and the false alarms. AL can be expressed as a distance above the mean background (BKG) value in terms of standard deviation:

$$AL = BKG + M\sqrt{BKG} \tag{2}$$

M is a multiple that provides a constant false-alarm rate at any background intensity. The probability for one acquisition without radioactive source to exceed AL (i.e. the false alarm probability) is presented in TABLE I.

TABLE I
PROBABILITY OF EXCEEDING THE ALARM THRESHOLD AL [8]

Multiplier M	False alarm probability per acquisition
0	0.5
1	0.1586
2	2.275×10^{-2}
3	1.35×10^{-3}
4	3.15×10^{-5}
5	2.85×10^{-7}

The Alarm is triggered if the dose rate excesses the dose rate threshold or if the number of counts inside one Region of Interet of the spectrum (ROI 1 30 keV – 300 keV, ROI 2 300 keV-800 keV, and ROI 3 800 keV – 3000 keV) raises the threshold AL. The paper only focuses on the detection through the counting inside ROIs.

C. Model a static source

Based on model 3, many sources were tested with a static platform presented in Fig. 4, with 2 s per acquisition (see spectrum in Fig. 5, 6 and 7). In order to take into account the ambient background radiations, the experimental spectrum of background is added in the simulated spectrum.

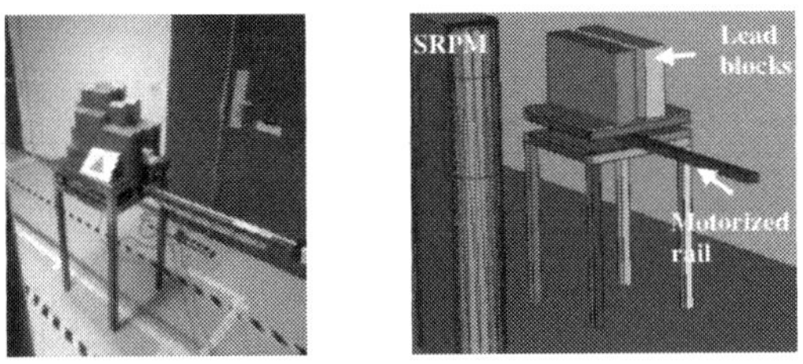

Fig. 4 Static platform (on the left : experimental setup, on the right model).

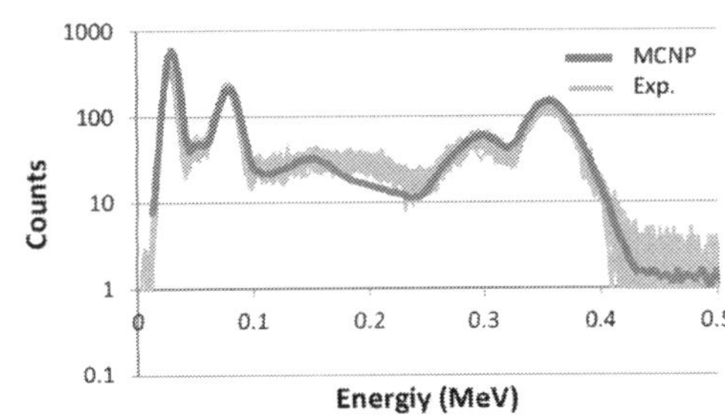

Fig.5. *Comparison between 30 experimental and one simulated pulse height distributions from a Ba-133 source for 2s acquisition time.*

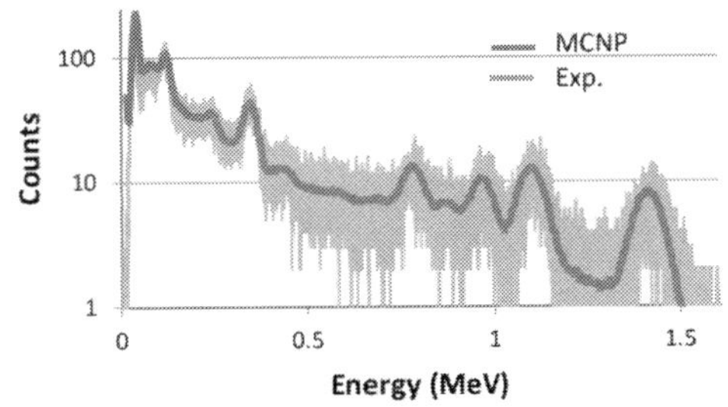

Fig.6. *Comparison between 30 experimental and one simulated pulse height distributions from a Eu-152 source inside PEHD container (thick 8 cm) for 2s acquisition time.*

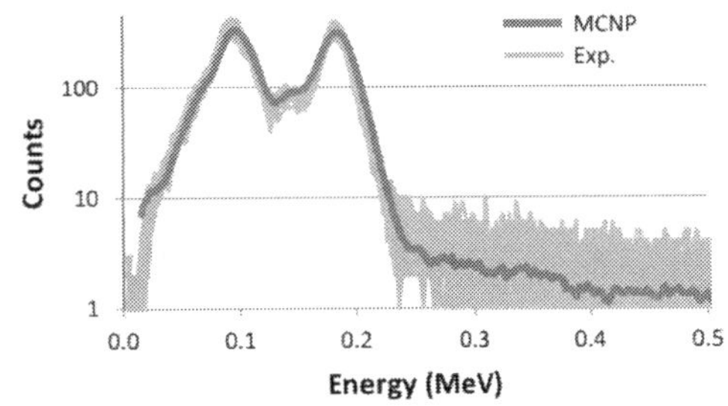

Fig.7. *Comparison between 30 experimental and one simulated pulse height distributions from HEU source for 2s acquisition time.*

TABLE II summarizes the comparison between experimental and simulated measurements for the three tested sources.

TABLE II
COMPARAISON BETWEEN EXPERIMENTZL AND SIMULATED RESULTS $\frac{MS_{im\,ul.}-E_{xp.}}{Exp}$

ROI	HEU	Eu-152	Ba-133
ROI 1 [30 keV - 300 keV[	4.12%	20.66%	17.47%
ROI 2 [300 keV - 800 keV[	14,75%	3.41%	13.29%
ROI 3 [800 keV - 3000 keV]	6.82%	7.43%	4.61%

HEU does not have high energy gamma-rays, so the difficult-to-simulate continuum Compton is low and there are no very low energy gamma-rays which are very sensitive to attenuation simulation. Therefore there is a good agreement between experimental and simulated spectra in in ROI1.

1) Uncertainties

To get a realistic idea of the measurement uncertainty, all the variables inherent to the measurement were considered,

 https://doi.org/10.1051/epjconf/202022507008

their uncertainties were identified and quantified, and they were propagated to predict an overall uncertainty. The paragraph focuses on the detection of HEU source.

a) Location uncertainty

The position of the source has a significant impact on the results. An area of "possible locations" is defined in Fig. 8. It corresponds to the middle of an "average" human body, the bottom of this legs and the top of the torso.

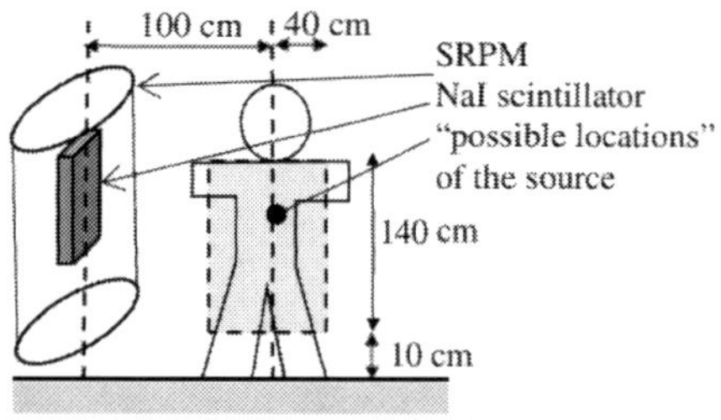

Fig. 8. Source locations.

It was assumed that the HEU source could be uniformly moved within the area described in Fig. 8. These locations are simulated in Fig. 9 to calculate its impact and counts in ROI 1 distribution is presented in Fig. 10.

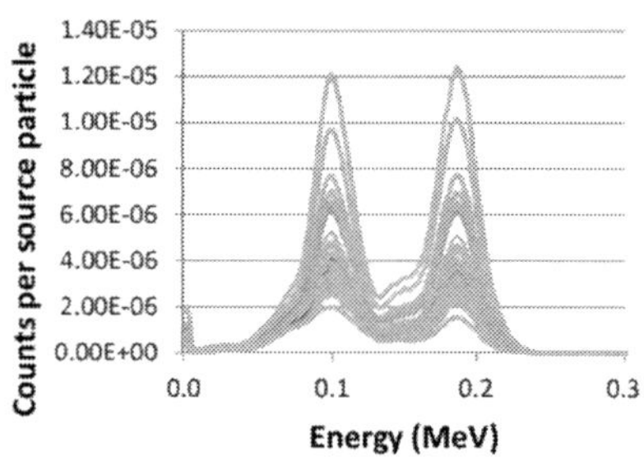

Fig.9. Simulation of the impact of the source's location on spectra (31 spectra).

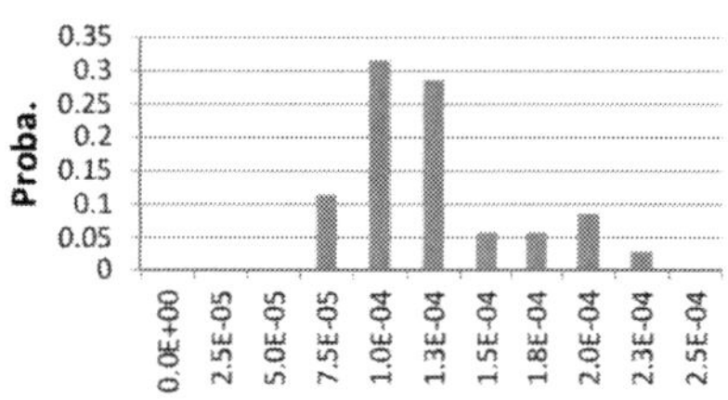

Fig. 10. Distribution of the counts in ROI 1 for HEU source simulation.

In the case of the HEU source, most of the gamma-rays are emitted below 300 keV. Hence, only ROI 1 is relevant. The variance distribution of the counts inside ROI 1 gives the uncertainty $\sigma_{Loca.}(ROI_1)$ and the relative uncertainty $\sigma_{Loca.\ Relat.}(ROI_1) = 46.4\%$.

The uncertainty related to source location is the main contributor to the useful signal of the radioactive source. The useful signal U can be expressed as follows:

$$U = U_0 + CF_{Location} \quad (3)$$

Where U_0 is the signal induced by the source at reference location (in the study 100 cm away from the center of the detector and 100 cm height) and $CF_{Location}$ is a correction factor which depends on the source location. The uncertainty related to the source location can be expressed as follows:

$$\sigma_{U\ Relat.} = \sqrt{\sigma_{stat}{}^2 + \sigma_{Loca.Relat.}{}^2} \quad (4)$$

b) Background uncertainty

As presented in Equation 2, detection efficiency of the SRPM is influenced by the background radiation and variation of its intensity with time. Variations in background intensity can be caused by people walk-trough, natural background radiation process or movement of radioactive materials. The last one is ignored; in the study it is supposed a perfect control of the radioactive materials in the vicinity of the SRPM area.

During the crossing through the SRPM detection area, the pedestrian partly absorbs the ambient background radiations and decreases the measured background. For example a truck in the detection area of a vehicle monitor reduces the measured background from 10% to 25% [9]. Because the monitor's alarm threshold is defined before the presence of an occupant, background decrease is not taken into account and a much larger signal is required to alarm an occupied monitor than an unoccupied one.

To estimate the impact of people walk-through, several acquisitions were performed, with and without a polyethylene phantom in the detection area of the SRPM (see Fig. 11). The variation in intensity has been calculated in the three ROIs. Assuming a uniform distribution of the variation between the minimum (BKG_{min}) and maximum (BKG_{max}) measured background, the background uncertainty is given by the following formula:

$$\sigma = (BKG_{max} - BKG_{min})/2\sqrt{3} \quad (5)$$

The background uncertainty is estimated within the three ROIs:

$\sigma_{BKG\ relat.}^{Occupant}(ROI_1) = 0.0030\%$
$\sigma_{BKG\ relat.}^{Occupant}(ROI_2) = 0.0021\%$
$\sigma_{BKG\ relat.}^{Occupant}(ROI_1) = 0.0015\%$

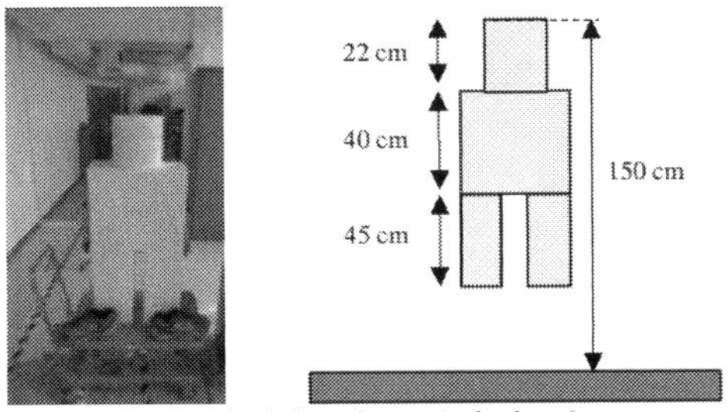

Fig. 11. Polyethylene phantom in the detection area.

The natural background radiation originates from a variety of sources, cosmic radiation and environmental radioactivity from naturally occurring radioactive materials (such as radon and radium). It varies with location and time. Many short background acquisitions (122 s) were done with "stable" environmental conditions (temperature, atmospheric pressure and humidity) in Fig. 12.

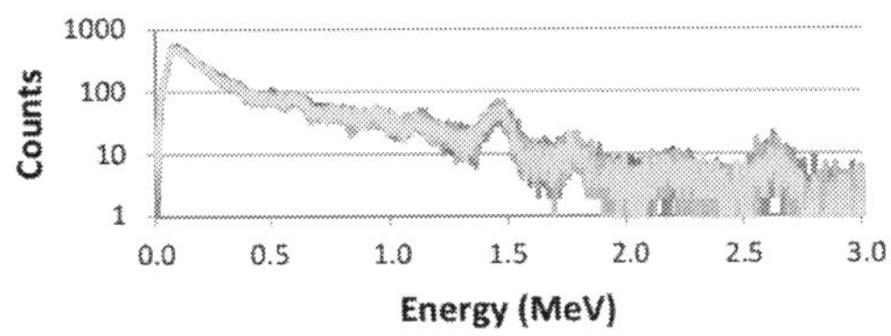

Fig. 12. Background spectra for 122s acquisitions.

The variance distribution of the counts inside ROI 1, 2 and 3 is estimated:

$\sigma_{BKG\ relat.}^{Natural}(ROI_1) = 1.89\%$

$\sigma_{BKG\ relat.}^{Natural}(ROI_2) = 1.92\%$

$\sigma_{BKG\ relat.}^{Natural}(ROI_3) = 1.42\%.$

The total background uncertainty is calculated based on these two contributions, people walk-trough and natural variations. Within each ROI i (i=1, 2, 3), the background contribution can be expressed as the sum of a "stable" background $BKG_0(ROI_i)$, variation due to people walk-trough $\Delta_{Occupant}(ROI_i)$ and natural variation $\Delta_{Natural}(ROI_i)$:

$$BKG(ROI_i) = BKG_0(ROI_i) + \Delta_{Occupant}(ROI_i) + \Delta_{Natural}(ROI_i) \quad (6)$$

Hence background uncertainty is expressed as follows:

$$\sigma_{BKG}(ROI_i) = \sqrt{\sigma_{BKG}^{statistic}(ROI_i)^2 + \sigma_{BKG}^{Occupant}(ROI_i)^2 + \sigma_{BKG}^{sNatural}(ROI_i)^2} \quad (7)$$

Where $\sigma_{BKG}^{statistic}(ROI_i)$ is the statistical uncertainty.

c) Other sources of uncertainties

There are many other sources of uncertainties. For example, the internal absorption of gamma-rays depends on the material composition, density and shape. Thin uranium sample powders emit most of this radiation, whereas more compact shapes such as sphere are impacted by auto-absorption and absorb most of it [10]. The assumption "only one kind of nuclear material is available" is done in the study to neglect this contributor.

The presence of metallic shield, as modeled in Fig.13, is also an important uncertainty. It is assumed that iron shield and thick lead shield shall be detected by metallic detector portal. The impact of thin lead shield is simulated in Fig 14.

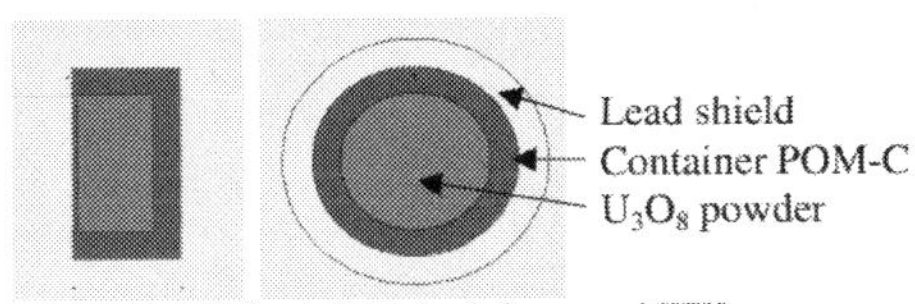

Fig. 13. Lead shield simulation around HEU source.

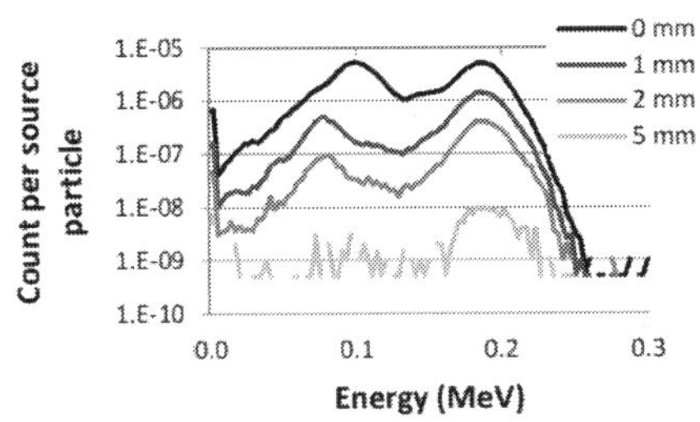

Fig. 14. Simulated spectra of shielded HEU sources.

Detection sensitivity is calculated for bare source and a correction factor is applied according to the scenario. For example if the signal from 1 bare source of 100g HEU is equivalent to 6 shielded 100g HEU sources wrapped by 1 mm lead. The shortcut "100g of HEU produces the same signal than 600g of 1mm lead shielded HEU" is not correct due to auto-absorption.

d) Conclusion about uncertainties

The measured signal S is the sum of the source signal (the useful signal U) and the background contribution BKG:

$$S = U + BKG \quad (8)$$

Therefore, the total uncertainty can be expressed as follows:

$$\sigma_S = \sqrt{\sigma_U{}^2 + \sigma_{BKG}{}^2} \quad (9)$$

https://doi.org/10.1051/epjconf/202022507008

Where σ_U and σ_{BKG} have been previously studied.

2) Detection probability

The coefficient M of the alarm threshold (see Equation 2) is fixed to limit false alarm rate and not disturb the flow of people through the SRPM. Fig. 15 illustrates the compromise between false-alarm rate and detection efficiency.

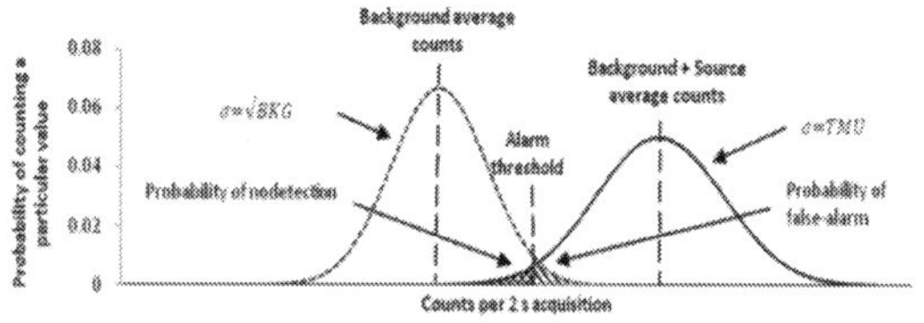

Fig. 15. Illustration of the background distribution and the distribution after addition of a source.

We assumed that the counting distribution from a SRPM is a Gaussian distribution:

$$P_{Measure}(c) = \frac{1}{\sigma\sqrt{2\pi}} exp\left(-\frac{(\bar{c}-c)^2}{2\sigma^2}\right) \quad (10)$$

Where $\bar{c}(m)$ is the average counts induced by m grams of HEU and σ the counting uncertainty. The detection probability is the area from $P_{Measure}$ above the alarm threshold AL, expressed as follows:

$$P_{Detection}(m) = \frac{1}{2} \times \left(1 - erf\left(\frac{AL-\bar{c}}{\sigma\sqrt{2}}\right)\right) \quad (11)$$

Where *erf* is the Gauss error function. To simplify calculations, in first approximation we neglect the variation of auto-absorption with the mass, a linear relation is assumed between the mass m of HEU and the average counts $\bar{c}(m)$ induced in the SRPM:

$$\bar{c}(m) = a \times m \quad (12)$$

Where the coefficient a is estimated with the MCNP model of the SRPM. Finally, the detection probability is calculated. Fig. 16 shows the detection probability calculated in ROI 1 for different M values (i.e. different alarm thresholds).

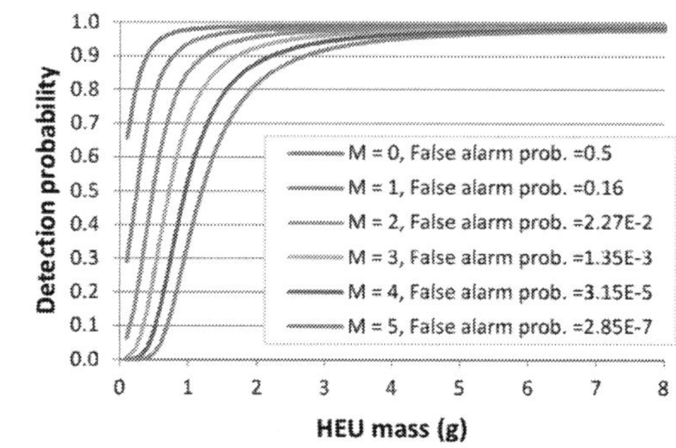

Fig. 16. Detection probability of the tested setup with a bare source of U_3O_8 (HEU).

The detection probability described in Fig. 16 was calculated based on simulation of U_3O_8 (HEU). The variation of auto-absorption with the mass, density or shape was not taken into account to simplify calculations. Fig. 16 is just an approximation of the detection efficiency of the SRPM, small source emit most of their radiations and the real detection efficiency should be better than the calculated one. Additional calculations should be done to take into account the variation of auto-absorption. The detection probabilities calculated in Fig. 16 are nevertheless useful. The detection probability is good for mass above 5g, but for lower mass the threshold alarm has an important impact and the operator must choose between false alarm rate and the SRPM detection efficiency.

III. CONCLUSIONS

Evaluate the performance of a radiation portal monitor is not a simple task. Many scenarios and setups are possible, the environment has an important impact on the results and there are many sources of uncertainties. A combination of experimental data, collected during ITRAP+10 Phase II RRT testing campaign, and simulations is a good way to study the performance of a SRPM in realistic condition. In the paper, the code MCNP6 was used to calculate the detection efficiency of a SRPM. This model showed a good agreement with ITRAP+10 experimental assays. The detection probability was calculated for one source of nuclear material, one shape, and one density but new scenarios must be tested. At the end, the results shall help in the selection of the operating characteristics of the SRPM to choose between false alarm rate and the detection efficiency.

REFERENCES

[1] Ministerial Declaration at the International Conference on Nuclear Security: Enhancing Global Efforts (2013). http://www-pub.iaea.org/MTCD/Meetings/PDFplus/2013/cn203/cn203MinisterialDeclaration.pdf

[2] Incident and Trafficking Database managed by the IAEA. http://www-ns.iaea.org/security/itdb.asp

[3] G. TUDOR, "HEU seizure highlights Moldova's strong work in nuclear security", IAEA Division of Public Information (2013).

[4] https://ec.europa.eu/jrc/en/publication/illicit-trafficking-radiation-assessment-program-itrap10-test-campaign-summary-report

[5] C. CARRAPICO, "Illicit Trafficking Radiation Detection Assessment Program ITRAP+10 Phase II Round Robin, Spectroscopic Radiation Portal Monitor, Test Methods", JRC Technical Reports (2017).

[6] MCNP6, version 1.0., https://mcnp.lanl.gov/

[7] G. F. KNOLL, Radiation Detection and Measurement, 3rd ed. (2000).

[8] Statistique appliquée à l'exploitation des mesures, CETAMA, 2nd edition (1978).

[9] P. E. FEHLAU, "An applications guide to pedestrian special nuclear material monitors",LA-10633-MS (1986).

[10] D. REILLY, N. ENSSLIN, and H. SMITH, "Passive Nondestructive Assay of Nuclear Materials", NUREG/CR-5550, Washington, D.C. (1991).

EPJ Web of Conferences **225**, 07009 (2020) https://doi.org/10.1051/epjconf/202022507009
ANIMMA 2019

Development of a Coated-Micro-Particle Neutron Detector Based on LiF/ZnS Scintillator

Faruk Logoglu
Ken and Mary Alice Lindquist
Department of Nuclear Engineering
Pennsylvania State University
State College, PA, USA
ful51@psu.edu

Marek Flaska
Ken and Mary Alice Lindquist
Department of Nuclear Engineering
Pennsylvania State University
State College, PA, USA
mxf5309@psu.edu

Douglas E. Wolfe
Department of Material Science and Engineering
Pennsylvania State University State College, PA, USA
Dew125@arl.psu.edu

***Abstract*—^{6}LiF:ZnS(Ag) micro-particle neutron detectors are a promising technology to further improve neutron detection capabilities for a variety of applications. Specifically, we have been investigating ^{6}LiF micro-particles coated with ZnS(Ag) to increase the neutron detection efficiency, light production, and light collection efficiency when compared to the existing powder-based technology (EJ-426 from Eljen Technology). Extensive radiation and light transport simulations with single micro-particles have been performed to find the optimal ^{6}LiF diameter and ZnS(Ag) coating thickness. Full-scale multi-particle simulations also have been performed to determine the optimal pitch (particle-to-particle distance) and detector thickness. Randomizations of ^{6}LiF radius, ZnS(Ag) coating thickness, position of particles, as well as shape of particles and partial coating have been performed to account for possible manufacturing imperfections. EJ-426 sheets have been modeled for reference purposes by defining spherical grains of ^{6}LiF and ZnS(Ag) and compared against experiments. The simulation results show that the coated micro-particles should dramatically increase the neutron detection efficiency, light production, and light collection efficiency when compared to the existing EJ-426 technology.**

Keywords*—*Micro-particles, ^{6}LiF:ZnS(Ag), Neutron-detection optimization, Geant4

I. INTRODUCTION

Neutron detectors based on ^{6}LiF:ZnS(Ag) are used for various applications including neutron imaging, high efficiency cold neutron detection [1,2] and neutron multiplicity counting [3]. Current existing technology employs mixing of ^{6}LiF and ZnS(Ag) powders to create the active volume for neutron detection. Although this technique works well, it has some serious shortcomings such as clustering of the same type of grains and non-uniform grain size, which reduce both charged particle energy deposition in ZnS(Ag) and light propagation through the detection medium. In order to overcome these problems and increase the neutron detection efficiency, it was proposed to coat ^{6}LiF particles with ZnS(Ag) and distribute these coated micro-particles throughout the volume to create the active detection medium.

Optimization of ^{6}LiF radius, ZnS(Ag) coating thickness, pitch (particle-to-particle distance), and detector thickness is required to find the ideal geometry for neutron capture, charged particle energy deposition (light production) and light propagation. Geant4 and MCNP6 have been used to optimize the ^{6}LiF radius and Geant4 has been used to determine the optimal ZnS(Ag) coating thickness, micro-particle pitch, and detector thickness since it has a capability of simulating light photons, as well as neutrons and charged particles.

After determining the optimal dimensions, randomizations of particle position, particle size, particle shape, and partial coating have been performed within a specified tolerance to account for the possible future manufacturing imperfections.

Four different EJ-426 sheets have been modeled to compare the performance of coated micro-particle detectors to existing technology.

II. OPTIMIZATION

Extensive simulations were performed to determine the various optimal parameters of the geometry and individual optimization steps are described in the following sections.

An illustration of the proposed geometry is shown in Fig. 1.

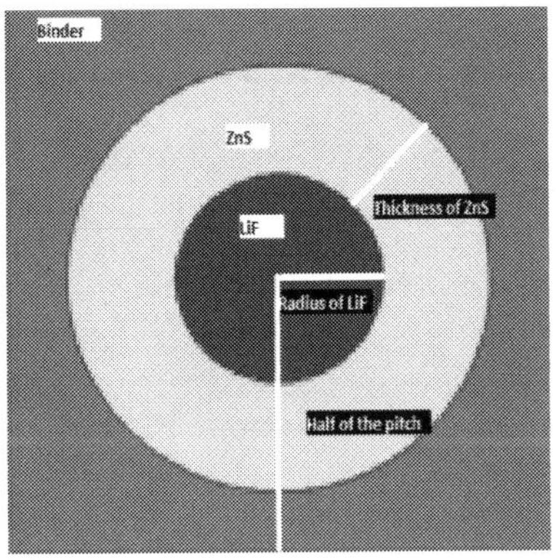

Fig. 1. Illustration of a single coated micro-particle (created with MCNP6).

A. Optimization of LiF radius

First, the radius of ^{6}LiF particles was optimized. ^{6}LiF radius should be chosen large enough to capture as many neutrons as possible, but small enough so that the alpha and triton particles upon neutron capture can escape the neutron converter volume without substantial energy loss.

Coating thickness was set constant at a relatively large value (90 μm) and ^{6}LiF radius was varied. Pencil beam of thermal neutrons were incident on the single particle and the energy deposition in scintillator medium per primary neutron was tallied in both Geant4 and MCNP6. The results are shown in Fig. 2.

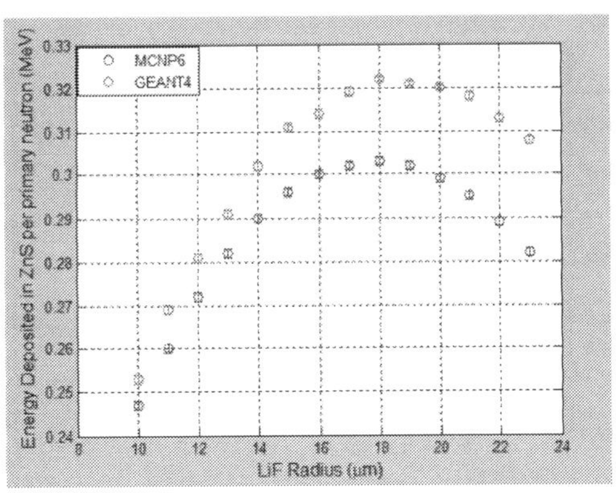

Fig. 2. Energy deposited in ZnS per primary thermal neutron as a function of LiF radius simulated in MCNP6 and Geant4.

Although there is a relatively small disagreement between the codes regarding how much energy is deposited in ZnS(Ag), both codes have agreed upon the optimal radius for ^{6}LiF, which was found to be 18μm.

B. Optimization of ZnS coating thickness

After determining the optimal ^{6}LiF radius, the ZnS(Ag) coating thickness was optimized. Although ZnS(Ag) is a bright scintillator (50000 photons/MeV) [4], its relative opacity to its own luminescence limits its practical size when it is used as the scintillating medium for radiation detection. Therefore, the optimization is required to find the ideal thickness. The coating thickness should be large enough to capture most of the energy from the charged particles, but small enough so that the light produced in ZnS(Ag) can escape without significant internal attenuation.

^{6}LiF radius was set constant at its optimal value of 18μm and ZnS(Ag) thickness was varied. The number of light photons escaping a single micro-particle per primary thermal neutron was tallied in Geant4. The results are shown in Fig. 3.

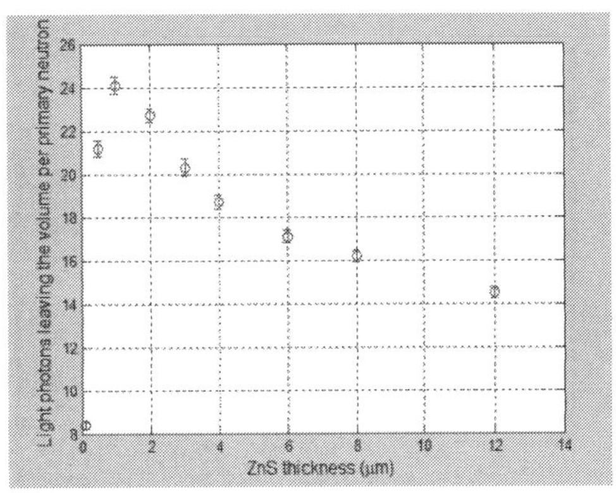

Fig. 3. Number of light photons leaving the micro-particle as a function of coating thickness in Geant4.

Optimal ZnS(Ag) thickness was found to be 1μm. Although approximately 85% of the charged particle energy is not deposited in ZnS(Ag) at this thickness, the produced light photons escape more efficiently. Besides the small absorption length of the ZnS(Ag), another problem is the higher refractive index of

ZnS(Ag) (2.49) compared to the refractive index of ^{6}LiF (1.39). In addition, the refractive index of the organic binder simulated is 1.51. This refractive index mismatch causes substantial internal reflection, which results in less efficient light collection.

C. Optimization of pitch and detector thickness

Optimization of micro-particle pitch and detector thickness requires the simulation of a large number of micro-particles. Pitch is the particle-to-particle distance and it is measured from the centers of the particles. The pitch needs to be small enough to maximize the concentration of ^{6}Li, but large enough to allow for efficient light propagation without significant attenuation. Therefore, the binder material needs to have favorable optical properties; e.g., it has to be transparent to light.

In simulations, micro-particles were placed at a constant pitch inside the detector medium and circular thermal neutron beam was made incident on the detector surface. Keeping the detector cross-sectional area constant, pitch and detector thickness were varied to find the optimal geometry. The active detector volume was placed between two polyester sheets and the light photons were collected on the photocathode surface with the same cross-sectional area as that of the scintillator.

An illustration of the simulation geometry is shown in Fig. 4.

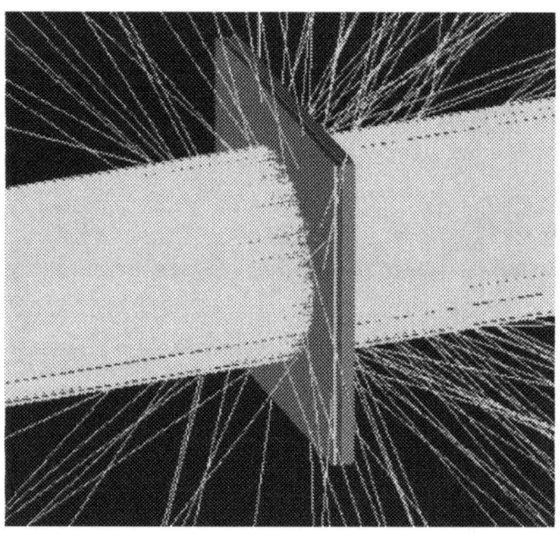

Fig. 4. Illustration of full-scale simulation geometry. Green, blue, and red surfaces represent polyester, active detector volume, and photocathode surface, respectively.

The ideal detector thickness and the micro-particle pitch are correlated since the specification of a certain pitch size dictates the optimal detector thickness. If the pitch is large, the detector thickness can also be relatively large since the areal density of ZnS(Ag) in the detector would be less than in the case of a smaller pitch. The results are shown in Fig.s 5 and 6.

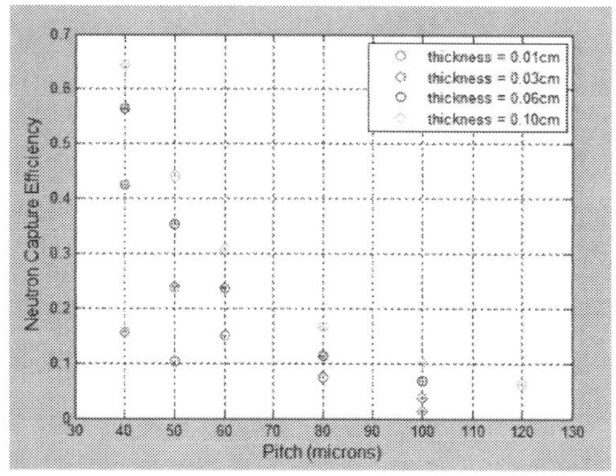

Fig. 5. Neutron capture efficiency as a function of pitch for different detector thicknesses.

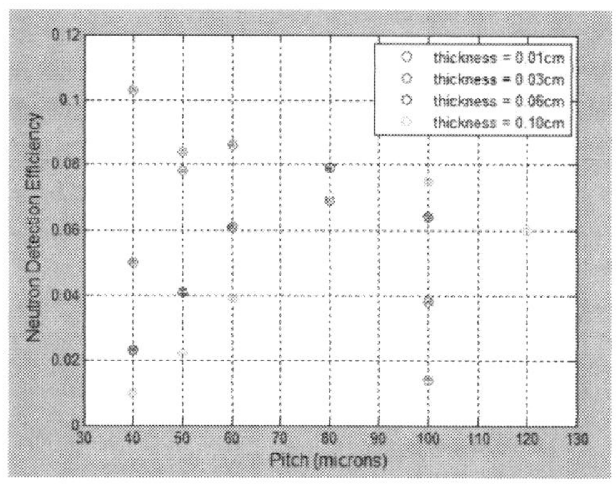

Fig. 6. Neutron detection efficiency as a function of pitch for different detector thicknesses with a light photon threshold of 20.

Neutron capture efficiency decreases with increased pitch and with smaller detector thickness, because of smaller amount of ^{6}Li in the active detection volume. Neutrons are attenuated exponentially in the micro-particle geometry according to

$$I = I_0 \exp(-\Sigma t) \qquad [Eq. 1]$$

where I, Σ, and t are the neutron intensity, the total macroscopic cross section, and the effective detector thickness, respectively. In Eq.1, t is not the overall detector thickness but rather the thickness that is occupied by ^{6}LiF, the neutron capturing material. The attenuation in ZnS(Ag) and binder can be neglected for small thicknesses.

Neutron detection efficiency is defined as the number of neutrons that are captured and subsequently result in the detection of 20 light photons on the photocathode surface. Detection of 20 photons include quantum efficiency of the photocathode surface so it is the number of photoelectrons generated upon a neutron capture event. This number of detected photons is somewhat arbitrary; however, it is well above the single photoelectron noise signal level. It was observed that decreasing this threshold increases the neutron detection but doesn't affect the overall trends.

It is shown in Fig. 6 that there is no single optimal pitch size that applies to each detector thickness. Instead, it is a function of detector thickness and it decreases as the detector thickness increases since the detector thickness compensates for the sparsely placed micro-particles. However, when the thickness is made smaller and smaller, the overall neutron detection efficiency increases and peaks at 0.1mm detector thickness, suggesting that photon propagation properties dominates over the neutron capture probability.

III. RANDOMIZATION

The aforementioned optimization of the micro-particle parameters was performed assuming an ideal geometry with perfectly spherical particles, symmetric placement of these particles throughout the detector volume and perfect coverage of ^{6}LiF with ZnS(Ag) without any bare surfaces.

It is possible to have coated micro-particles with non-ideal dimensions, different shapes and partial coverage in the manufacturing process. Moreover, these microparticles may not be aligned perfectly to have constant pitch between them. It is expected that the deviation from optimal dimensions and shape would result in worsened charged particle energy deposition and light propagation through the detection medium.

Therefore, randomization of these parameters was performed in Geant4 within certain tolerance levels. An illustration of randomization steps is shown in Fig. 7.

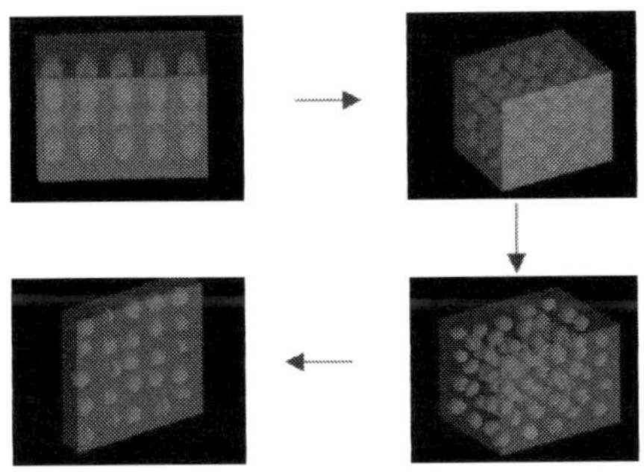

Fig. 7. An illustration of several randomization steps: a) Ideal geometry with symmetrically placed particles and fixed LiF and ZnS size, b) randomization of position of micro-particles, c) randomization of shape, rotation, and position of micro-particles, and d) randomization of coating coverage.

Since there are thousands of micro-particles in a given detector volume, due to the complexity of simulation models it is not feasible to randomize all of them individually. Instead, a conventional unit cell was defined with multiple micro-particles in it and only those particles were randomized. After that randomization step, the conventional unit cell was repeatedly placed throughout the detector medium. The effect of the number of randomized micro-particles is shown in Fig. 8.

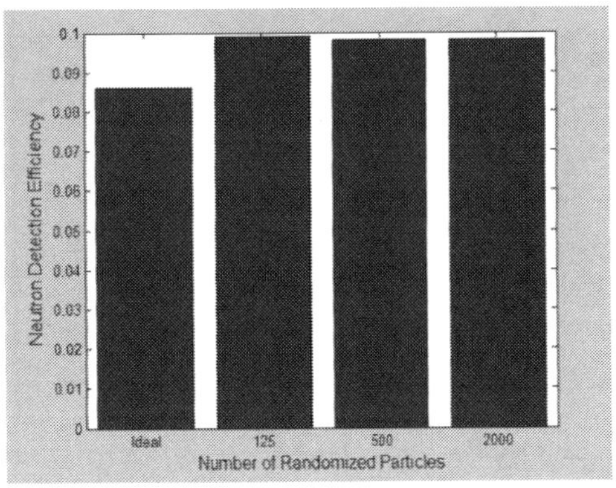

https://doi.org/10.1051/epjconf/202022507009

Fig. 8. Neutron detection efficiency as a function of number of particles in a conventional unit cell.

As shown in Fig. 8, the neutron detection efficiency quickly saturates at 125 randomized micro-particles. It is also observed that the neutron detection efficiency is somewhat increased when the geometry deviates from ideal. The main reason is that there are empty spaces in the ideal geometry where a neutron has no chance of undergoing a capture reaction with ^{6}Li. Since circular thermal neutron beam is directly incident on the detector some of the source neutrons travel through these empty spaces. However, when the particles are randomly placed, some of these empty spaces are filled with ^{6}LiF, increasing the neutron capture probability.

After the randomization of the position of particles, the ^{6}LiF radius, and the ZnS(Ag) coating thickness was carried out. The overall particle size was kept constant and only ^{6}LiF and ZnS(Ag) dimensions were randomized. Again, the particles were randomly placed inside the conventional unit cell. The results are shown in Fig. 9.

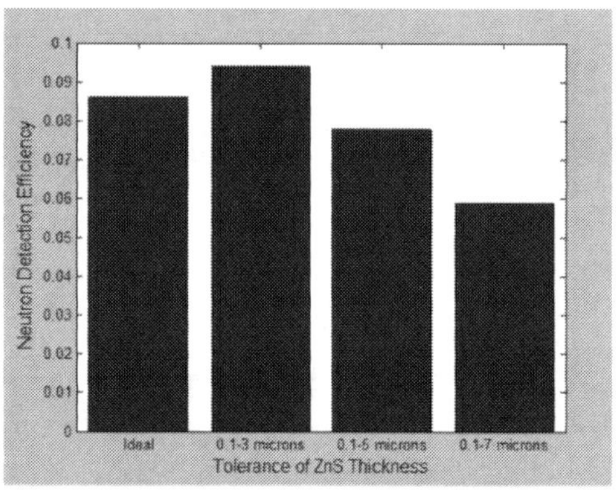

Fig. 9. Neutron detection efficiency as a function of different tolerance intervals of ZnS thickness.

Although the deviation from the optimal ZnS(Ag) thickness result in deterioration of the neutron detection efficiency, a random placement of micro-particles surpasses the effect of non-ideal ZnS(Ag) thickness when the tolerance is between 0.1µm-3µm. However, when the tolerance is increased, neutron detection efficiency clearly suffers.

In addition, the randomization of shape and coating coverage was performed, and their effects are shown in Fig. 10 and 11.

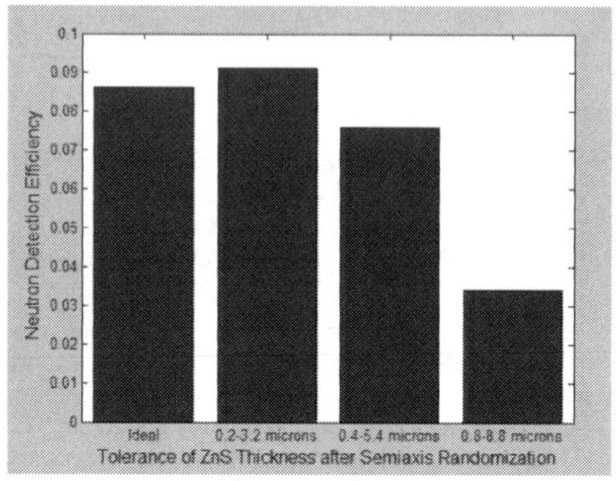

Fig. 10. Neutron detection efficiency as a function of different tolerance intervals of ZnS thickness. The difference is that the shapes are not spherical but ellipsoid.

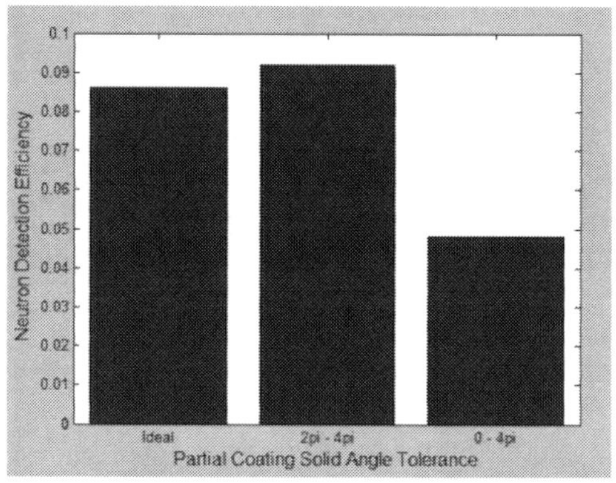

Fig. 11. Neutron detection efficiency as a function of different tolerance intervals of solid angle for partial coating.

Again, there is a small increase in the neutron detection efficiency compared to the ideal case when the tolerance interval is narrow. When the tolerance intervals become wider, the particles become more distorted and the detection efficiency suffers significantly.

When the partial coating tolerance is between 2π-4π steradians, the neutron detection efficiency increases compared to the ideal case. However, the position randomization is always implicit so the partial coating itself actually decreases the neutron detection efficiency. The effect is much pronounced when the tolerance interval is narrow. When the tolerance interval is made 0-4π, the detection efficiency decreases rapidly.

IV. EJ-426 SIMULATIONS

In addition to the micro-particle simulations, EJ-426 sheets were simulated in Geant4. ^{6}LiF and ZnS(Ag) grains were modeled as spheres and placed in a unit cell respecting the LiF:ZnS mass ratio and LiF concentrations of these sheets. A unit cell represents the building block of the sheets which comprise multiple unit cells.

An illustration of a unit cell is given in Fig. 11.

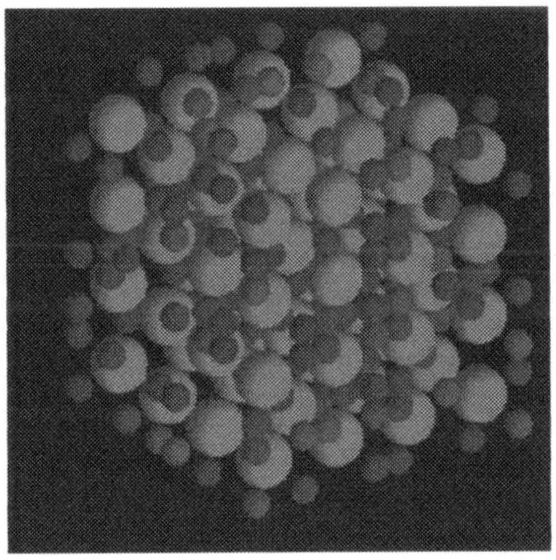

Fig. 12. An illustration of a unit cell in an EJ-426 sheet.

Again, a circular thermal neutron beam was incident on the detector sheet as shown in Fig. 4. Four different EJ-426 sheets with different LiF:ZnS mass ratios and thicknesses were tested and the results are given in TABLE I.

Detector Thickness (mm)	LiF:ZnS Mass Ratio	Neutron Detection Efficiency (%)
0.32	1:3	1.9
0.32	1:2	2.6
0.60	1:3	1.5
0.60	1:2	1.6

TABLE I. Neutron detection efficiencies for different EJ-426 sheets.

With the same cross sectional detector area and light photon threshold, it is observed that the best performing EJ-426 sheet is the 0.32mm thick one with the LiF:ZnS mass ratio of 1:2. As seen from Fig. 6, most of the coated micro-particle detectors have higher neutron detection efficiencies than 2.6%.

While 0.1mm thick micro-particle detector performs almost 4 times better, the 0.32mm thick micro-particle detector performs 3 times better than EJ-426.

V. CONCLUSION

Coated micro-particles were optimized and the ideal ^{6}LiF radius, ZnS(Ag) coating thickness, pitch, and detector thickness were found by performing extensive simulations of neutrons, charged particles, and light photons in Geant4. EJ-426 sheets were also modeled and simulated in Geant4. The neutron detection efficiencies of coated-micro-particle detectors for thermal neutrons were found higher than the best performing EJ-426 sheet, by a factor of 2-4. Non-idealities were also tested in coated micro-particle geometries and it was observed that the proposed technology under imperfect conditions still performs better than the existing technology.

ACKNOWLEDGEMENT

The authors would like to thank the Penn State's College of Engineering and the Applied Research Laboratory for supporting this research.

REFERENCES

[1] A. Osovizky et al. "^{6}LiF:ZnS(Ag) Mixture Optimization for a highly Efficient Ultrathin Neutron Detector" IEEE Trans. Nucl. Science, Volume:65, Issue: 4 (2018)

[2] A. Osovizky et al. "Design of an ultrathin cold neutron detector" Nucl. Inst. Methods A, 893 (2018), pp. 1-9

[3] Sean Stave et al. "LiF/ZnS neutron multiplicity counter" Nucl Inst. Methods A, 784 (2015), pp 208-212

[4] Y. Yehuda-Zada et al. "Optimization of ^{6}LiF:ZnS(Ag) scintillator light yield using GEANT4" Nucl, Inst. Methods A, 892 (2018), pp. 59-69

EPJ Web of Conferences **225**, 07010 (2020) https://doi.org/10.1051/epjconf/202022507010
ANIMMA 2019

Characterization of Neutron Detection Systems with Confidence via Statistical Parameter Correction

Phillip Boone
Nucsafe, Inc.
Oak Ridge, USA
pboone@nucsafe.com

***Abstract*—Neutron detection systems utilize statistical alarm techniques where a measured false alarm rate (FAR) can vary drastically from the FAR predicted by a theoretical model. The ability to set an alarm threshold that results in a practically controlled FAR is crucial to characterize detector sensitivity with both accuracy and precision. A generalized and automated method is presented to statistically evaluate FAR performance by assuming that the FAR itself is not deterministic, but a normal stochastic process over a specific parameter to be corrected that will hereafter be referred to as the *correction*. In this manner, a specific correction results in not only a point estimate of FAR, but also a confidence interval. The central objective is focused exclusively on characterization assuming that experiments are executed in a tightly controlled environment so that an accurate comparison is enabled across detectors. Once a correction is calculated, the estimated FAR is only assumed accurate in a similar environment for sensitivity evaluation. Initially, the calculated correction factor was used to compare FARs across various distributions including normal, corrected normal, Poisson, and a simplified normal distribution. Later verification data sets were used to empirically demonstrate the rate of containment of measured confidence coefficients using two detectors of different technology. A second application uses the correction method to improve the signal-to-noise ratio metric to agree more with dynamic sensitivity results. Finally, a third application studies the effect of altering the duration of background acquisition on FAR performance.**

I. Introduction

By partitioning a notably large set of background data into potentially thousands of mutually exclusive periods, a distinct alarm threshold can be calculated using a specific correction and statistics on an individual background period. By applying this distinct threshold, a single alarm rate sample can be calculated over the remaining background data. Each subsequent background period is used likewise to create additional alarm rate samples for the entire data set. A maximum likelihood problem is formulated in section II-D1 assuming that a detector's false alarm rate (FAR) is a normal random process of correction. Using numerical optimization in section II-E, a first-order search is utilized to determine a correction to meet a target FAR.

False alarm performance of the corrected model is compared to three statistical models in section III-C. In section IV, additional background data is gathered to verify that the estimated correction results in the desired FAR with statistical confidence.

In section V three neutron detection systems demonstrate that the signal-to-noise ratio (SNR) metric may result in a theoretical sensitivity that contradicts sensitivity measured with a more rigorous dynamic test [1] according to American National Standards Institute (ANSI). Section V-A2 introduces a correction to SNR and section V-C demonstrates that the correction results in a sensitivity that agrees more with the rigorous ANSI test.

Section VI studies the effect of reducing the duration of background acquisition from 5 minutes down to a minute. As a result, system sensitivity decreased via the increase in critical value required to meet a target FAR. In addition, the overall precision of FAR drastically decreased via the estimated increase in standard deviation of FAR by a factor of 28.

In the context of the experiments performed, FAR is simply a scaled version of significance level, critical value indicates a specific alarm threshold, and standard critical value represents the number of standard deviations above the mean for an alarm threshold. FAR is used in place of significance level for convenience when presenting measured false alarm rates.

II. Statistical Model

A. Data Collection Periods

1) Large Experimental Set of Data: Assume that count rate samples for a neutron radiation detector under natural ambient background conditions follow a normal distribution with random variable $D \sim \mathcal{N}(\mu_D, \sigma_D^2)$ and a realization of N_D random samples is the vector $\vec{D} \triangleq [D_1, \ldots, D_{N_D}]^T$.

2) Background Partitions of Data: These samples are partitioned into N_{PB} non-overlapping background periods of length N_B with the i-th period defined as $\vec{B}_i \triangleq [D_{N_B(i-1)+1}, \ldots, D_{N_B(i-1)+N_B}]^T \triangleq [B_{i,1}, \ldots, B_{i,N_B}]^T$ with $i \in \{1, 2, 3, \ldots, N_{PB}\}$.

For each background period i the statistic that is a minimum variance unbiased estimator (MVUE) of mean μ_D is

$$\hat{\mu}_{B,i} \triangleq \frac{1}{N_B} \sum_{j=1}^{N_B} B_{i,j}$$

For each background period i the statistic that is a MVUE of variance σ_D^2 is

https://doi.org/10.1051/epjconf/202022507010

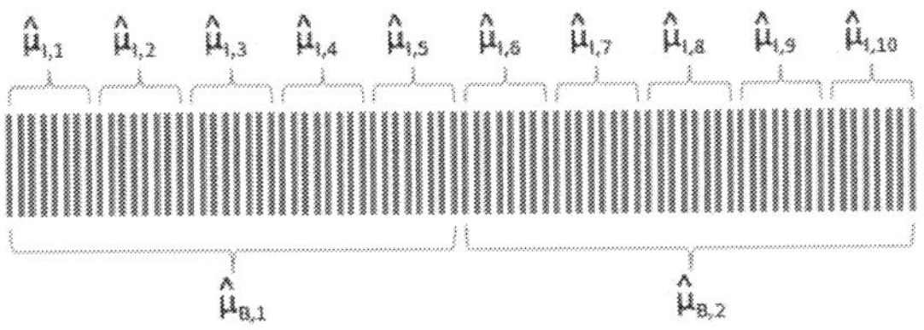

Fig. 1. Example Data Partitions : $N_D = 80$ Samples Partitioned into $N_{PB} = 2$ Background Periods of $N_B = 40$ Samples and $N_{PI} = 10$ Integration Periods of $N_I = 8$ Samples

$$\hat{\sigma}_i^2 \triangleq \frac{1}{N_B - 1} \sum_{j=1}^{N_B} (B_{i,j} - \mu_D)^2$$

The number of samples N_B should be chosen as large as possible within system initialization in a known environment without artificial sources.

3) Integration Partitions of Data: The samples in vector $\vec{D}$ are partitioned again into $N_{PI} << N_{PB}$ non-overlapping integration periods of length N_I with the i-th period defined as $\vec{I}_i \triangleq \left[D_{N_I(i-1)+1}, \ldots, D_{N_I(i-1)+N_I}\right]^T \triangleq \left[I_{i,1}, \ldots, I_{i,N_I}\right]^T$ with $i \in \{1, 2, \ldots, N_{PI}\}$.

For each period i the statistic that's a MVUE of mean μ_D is

$$\hat{\mu}_{I,i} \triangleq \frac{1}{N_I} \sum_{j=1}^{N_I} I_{i,j}$$

The number of samples N_I is typically chosen to optimize system detection sensitivity for a given source strength and speed.

B. Determine the Theoretical Parameters of a Hypothesis Test

Since it can be shown that the test statistic, a linear combination of normal samples, also follows a normal distribution according to $\hat{\mu}_{I,i} \sim \mathcal{N}(\mu_D, \frac{\sigma_D^2}{N_I})$, if a one-sided hypothesis test is used then either one of the following choices are possible:

1) **Null Hypothesis H_0 (radiation is background)**
   a) $\hat{\mu}_{I,i} = \mu_D$
2) **Alternate Hypothesis H_1 (radiation is significantly higher than background)**
   a) $\hat{\mu}_{I,i} > \mu_D$

The critical region of statistic $\hat{\mu}_{I,i}$ is the region where background radiation is determined to erroneously contain an artificial source (H_0 is rejected) and is defined for some constant $\bar{Z}_\alpha > 0$ such that

$$C_i \triangleq \left\{\hat{\mu}_{I,i} : \hat{\mu}_{I,i} > \mu_D + \left(\bar{Z}_\alpha\right) \frac{\sigma_D}{\sqrt{N_I}}\right\} \quad (1)$$

A desired significance level $\alpha \in [0, 1]$ together with the probability P_{H_0} of the error region of H_0 constrain a specific solution for constant $\bar{Z}_\alpha$ according to the equation

$$P_{H_0}(C_i) = \alpha \quad (2)$$

1) Determining a Value for the Standard Critical Value Constant: The standardized test statistic for the i-th period of an integration partition is

$$Z_i \triangleq \frac{\hat{\mu}_{I,i} - \mu_D}{\left(\frac{\sigma_D}{\sqrt{N_I}}\right)}$$

and the critical region C_i can be rewritten in terms of the probability of a standard normal distribution (or the cumulative distribution function $\Phi(Z)$) as

$$C_i \triangleq \left\{Z_i : P\left(Z_i > \bar{Z}_\alpha\right) = \alpha\right\} = \left\{Z_i : 1 - \Phi\left(Z_i\right) = \alpha\right\}$$

where the threshold $\bar{Z}_\alpha$, referred to as the standard critical value of the hypothesis test, and corresponding significance level α can be looked up in a standard normal Z table.

2) Critical Value of the Hypothesis Test: Once the standard critical value $\bar{Z}_\alpha$ of the standard normal distribution is found, the corresponding critical value $\bar{\mu}_{I,i}$ for the statistic $\hat{\mu}_{I,i}$ is

$$\bar{\mu}_{I,i} = \mu_D + \left(\bar{Z}_\alpha\right) \frac{\sigma_D}{\sqrt{N_I}} \quad (3)$$

3) Verification of Significance Level: In order to facilitate calculating the value of the error probability in (2), an indicator function $1_\mu(\mu) : \mathbb{R} \to \{0, 1\}$ is defined as

$$1_\mu(\mu) = \begin{cases} 1 & \text{when } \mu > \mu_D + \left(\bar{Z}_\alpha\right) \frac{\sigma_D}{\sqrt{N_I}} \\ 0 & \text{when } \mu \leq \mu_D + \left(\bar{Z}_\alpha\right) \frac{\sigma_D}{\sqrt{N_I}} \end{cases}$$

where the expected value

$$E\left[1_\mu\left(\hat{\mu}_{I,i}\right)\right] = P_{H_0}(C_i) = \alpha$$

can be estimated using a sample mean of $1_\mu\left(\hat{\mu}_{I,i}\right)$ where

$$\hat{\alpha} = \frac{1}{N_{PI}} \sum_{i=1}^{N_{PI}} 1_\mu\left(\hat{\mu}_{I,i}\right) \quad (4)$$

C. Determine the Practical Significance Level of a Hypothesis Test

Note that in section II-B the population mean μ_D and variance σ_D^2 of a detector are not known exactly. Therefore, assume that a background period of sufficient length N_B is used so that $\hat{\mu}_{B,i} \approx \mu_D$ and $\hat{\sigma}_i^2 \approx \sigma_D^2$ are substituted in the formulas of the previous section.

1) Execution of a Very Large Number of Experiments via Background Partitions: Typically, when a user performs a false alarm test using a radiation detector the user programs a standard critical value, a single background period is acquired, statistics are performed resulting in an alarm threshold, and this threshold is applied using a live system for a long duration in order to verify a specified FAR.

Instead, if a large number N_D of data samples are recorded and if these samples are partitioned as in II-A, then thresholds can be calculated and a false alarm experiment can be performed in post-processing for every possible background period in the entire data set. Also, if the theoretical standard

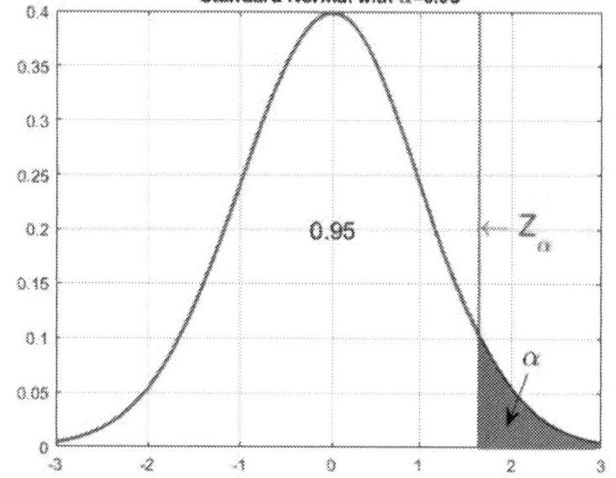

Fig. 2. Standard Critical Value

critical value $\bar{Z}_\alpha$ results in a measured significance level that doesn't agree with the theoretical significance α within acceptable tolerance, then adjusted standard critical values can be used to iterate through the whole process again.

Furthermore, significance level estimates $\hat{\alpha}$ can be computed for each background and statistics can be calculated on the resulting set to obtain realistic false alarm performance metrics across backgrounds.

2) Verification of Significance Level using Empirical Data: The indicator function defined in II-B3 is redefined using the j-th actual background parameter estimates as $1_{\mu,j}(\mu) : \mathbb{R} \to \{0,1\}$ according to

$$1_{\mu,j}(\mu) = \begin{cases} 1 & \text{when } \mu > \hat{\mu}_{B,j} + (\bar{Z}_\alpha)\frac{\hat{\sigma}_j}{\sqrt{N_I}} \\ 0 & \text{when } \mu \leq \hat{\mu}_{B,j} + (\bar{Z}_\alpha)\frac{\hat{\sigma}_j}{\sqrt{N_I}} \end{cases}$$

where the corresponding estimated significance level for the j-th background period is

$$\hat{\alpha}_j = \frac{1}{N_{PI}} \sum_{i=1}^{N_{PI}} 1_{\mu,j}(\hat{\mu}_{I,i}) \quad (5)$$

Theorem 1 (Unique Significance): For a one sided upper hypothesis test, if significance level α is written as a function of standard critical value Z_α in the form $\alpha(Z_\alpha) : \mathbb{R} \to [0,1]$ then the function is strictly decreasing over the domain of the function as illustrated in Fig.2. Therefore, given a unique significance level $\alpha_0 \in [0,1]$ there exists a corresponding unique standard critical value Z_{α_0}.

3) Experimental Error: Assume that the theoretical standard critical value $\bar{Z}_\alpha$ does not result exactly in the desired significance level α due to experimental data not fitting the model exactly, but instead $\bar{Z}_\alpha$ results in the significance value $\tilde{\alpha}$. The significance level experimental error is then defined as

$$\epsilon_\alpha \triangleq \tilde{\alpha} - \alpha$$

Assume that if experimental data approximately retain the property of Theorem 1, then there exists a unique experimental standard critical value $\hat{Z}_\alpha$ that results in the theoretical significance level α within tolerable precision. The standard critical value experimental error is then defined as

$$\epsilon_Z \triangleq \bar{Z}_\alpha - \hat{Z}_\alpha$$

4) Random Variable of an Experimental Standard Critical Value: Assume that the significance level estimates $\{\hat{\alpha}_j : j \in \{1,\ldots,N_{PB}\}\}$ are a realization of N_{PB} random samples from a random variable with a standard critical value $\hat{Z}_\alpha$ that results in the ideal significance level α. Samples are assumed to approximately follow a normal distribution [2][pp.216] for sufficiently large N_{PB} according to

$$\hat{\alpha}_j \dot{\sim} \mathcal{N}(\alpha, \sigma_\alpha^2) \quad (6)$$

5) Practical Significance Level Estimate: Therefore, the sample mean of $\hat{\alpha}_j$ is the MVUE of significance level

$$\hat{\mu}_\alpha \triangleq \frac{1}{N_{PB}} \sum_{j=1}^{N_{PB}} \hat{\alpha}_j \quad (7)$$

The sample variance of $\hat{\alpha}_j$ is the MVUE of significance level variance

$$\hat{\sigma}_\alpha^2 \triangleq \frac{1}{N_{PB}-1} \sum_{j=1}^{N_{PB}} (\hat{\alpha}_j - \hat{\mu}_\alpha)^2 \quad (8)$$

D. Determine the Practical Standard Critical Value of a Hypothesis Test

Since samples of $\hat{\alpha}_j$ are assumed to approximately follow a normal distribution with mean α and variance σ_α^2, the probability density function (PDF) is

$$f(\hat{\alpha}_j) = \frac{1}{\sigma_\alpha\sqrt{2\pi}} \exp\left\{\frac{-(\hat{\alpha}_j - \alpha)^2}{2\sigma_\alpha^2}\right\} \quad (9)$$

1) Maximum Likelihood Estimate of Standard Critical Value: The only quantity in (9) that is a function of standard critical value Z_α is $\hat{\alpha}_j$, which can be thought of as either a continuous-time stochastic process [3] over parameter Z_α or a random variable over the family of PDFs [2] indexed on the parameter Z_α written as $\hat{\alpha}_j(Z_\alpha)$ when used in the likelihood functions.

The joint probability of N_{PB} random samples taken from the PDF in (9) results in the likelihood function

$$\begin{aligned} L(Z_\alpha) &= \prod_{j=1}^{N_{PB}} f(\hat{\alpha}_j(Z_\alpha)) \\ &= \frac{1}{\sigma_\alpha^{N_{PB}}(2\pi)^{\frac{N_{PB}}{2}}} \exp\left\{\frac{-\sum_{j=1}^{N_{PB}}(\hat{\alpha}_j(Z_\alpha) - \alpha)^2}{2\sigma_\alpha^2}\right\} \end{aligned}$$

and the corresponding log likelihood function is

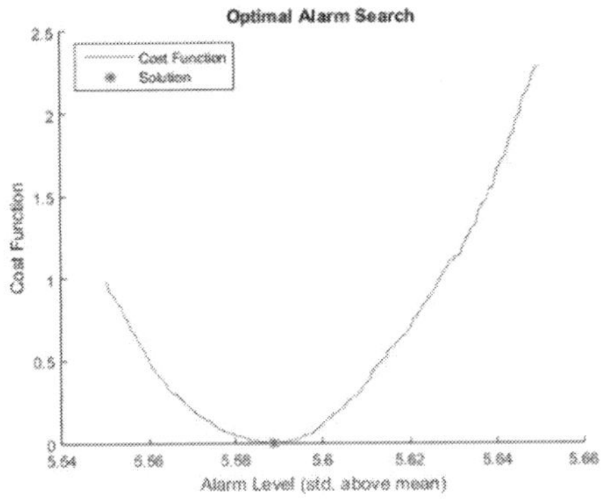

Fig. 3. Solution to the Cost Function

$$l(Z_\alpha) = -\frac{N_{PB}}{2}\Bigg[\log(\sigma_\alpha^2) + \log(2\pi) + \frac{1}{N_{PB}\sigma_\alpha^2}\sum_{j=1}^{N_{PB}}(\hat{\alpha}_j(Z_\alpha) - \alpha)^2\Bigg]$$

Therefore the maximum likelihood estimate of standard critical value is

$$\hat{Z}_\alpha = \arg\max_{Z_\alpha}\{l(Z_\alpha)\} \tag{10}$$

E. Optimization Problem Formulation

1) Cost Function: In order to facilitate computation of the likelihood function in (10), all constants independent of Z_α are eliminated and the negative sign is removed resulting in the following minimization cost function

$$C(Z_\alpha) = \sum_{j=1}^{N_{PB}}(\hat{\alpha}_j(Z_\alpha) - \alpha)^2 \tag{11}$$

Therefore, the maximum likelihood estimate of (10) is equivalent to minimizing the mean squared error of significance level α over standard critical value.

2) Optimization Problem: Using the cost function from (11) in a minimization problem results in the following optimization problem formulation:

$$\begin{aligned}\hat{Z}_\alpha &= \arg\min_{Z_\alpha}\{C(Z_\alpha)\} \\ &= \arg\min_{Z_\alpha}\left\{\sum_{j=1}^{N_{PB}}(\hat{\alpha}_j(Z_\alpha) - \alpha)^2\right\}\end{aligned} \tag{12}$$

3) Optimization Problem Solver: The optimization problem formulated by (12) can be solved using various first order numerical methods, but care must be taken to use a method that is tolerant of noise as illustrated by the solution in Fig.3. In this case, the problem was solved via simulated annealing [4].

F. Generalization of Parameter Correction

The upper hypothesis test utilized normal samples $\{\hat{\mu}_{I,i} : i \in \{1, \ldots, N_{PI}\}\}$ while adjusting the parameter Z_α to yield a target theoretical significance α. However, the model can be made even more flexible by slightly altering definitions including the indicator function in (5). Instead let $\mu \in \mathbb{R}$ be a measurable random variable that follows all axioms of probability with the indicator function $1_{\mu,j}(\mu, \boldsymbol{\beta}_j, \theta_\alpha) : \mathbb{R} \rightarrow \{0, 1\}$ where $\boldsymbol{\beta}_j$ is a set of statistics on background period j and the parameter $\theta_\alpha \in \mathbb{R}$ is the target parameter to correct. The indicator function must also be a simple inequality for an upper hypothesis test, $\hat{\alpha}_j$ must be approximately normal according to (6), and experimental samples must approximately adhere to Theorem 1. In this case, a much more flexible framework results allowing modification to other system factors such as distribution, correction parameter θ_α, sampling techniques as in section III-B, or another system parameter as in section VI. As factors are adjusted the impact on measured FAR performance values $\hat{\mu}_\alpha$ and $\hat{\sigma}_\alpha^2$ could be evaluated.

III. Error Analysis using a Neutron Detector

A. Theoretical Probability Distribution vs. Detector Probability Histogram

In order to qualitatively illustrate the disparity between model and detector data, in Fig.4 a histogram was created for neutron data and scaled theoretical PDFs were overlaid for the normal models. Curie's criteria was used where variance is assumed equal to the sample mean of data [5][pp.75]. Matlab's histogram function was used with the "Normalization" parameter set to "probability" in order to create a probability mass function (PMF). Both PDFs were scaled to have the same maximum value as the PMF. Not only is the histogram data highly discretized but the overall shape does not match either normal model at lower count rates.

B. Correlated Sampling

During problem formulation in section II-A the term "random samples" was used to describe the N_D number of data samples acquired. This is a fundamental assumption often referred to as a set of independent, identically-distributed samples of data. The integration samples as defined in section II-A3 follow this convention of sampling. However, the neutron detector data acquired are not sampled independently with non-overlapping segments, but in a temporally correlated and overlapping manner with a slightly different definition of $\vec{I}_i \triangleq \left[D_{(i-N_I)+1}, \ldots, D_{(i-N_I)+N_I}\right]^T \triangleq [I_{i,1}, \ldots, I_{i,N_I}]^T$ with $i \in \{N_I, N_I + 1, \ldots, N_D\}$. In other words, every detector sample in $\vec{D}$ starting with number N_I is processed as an average over a rolling buffer where at each iteration the oldest

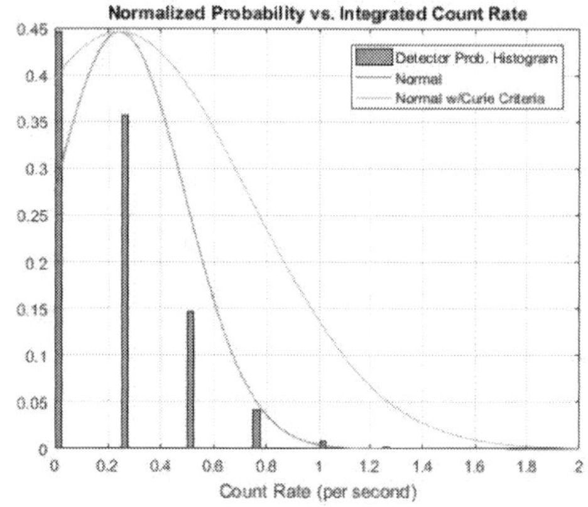

Fig. 4. Qualitative Probability Comparison

sample is discarded from the beginning and a new sample is inserted to the end. Using this method of sampling is likely a significant contributing factor to the discrepancy between the theoretical standard critical value $\bar{Z}_\alpha$ and the practical value $\hat{Z}_\alpha$.

If independent sampling were performed as originally defined the sample rate would be decimated by a factor of N_I. This would result in under sampling where a neutron detector could miss a moving source at sufficient speed that would have been otherwise detected using correlated sampling. Correlated sampling allows the increased sensitivity of averaging periods while retaining the sample rate.

1) Alternate Alarm Definition to Compensate for Correlation: A single outlier D_O in $\vec{D}$ will persist in the $N_I - 1$ correlated samples before and $N_I - 1$ correlated samples after D_O. Correlated samples $\hat{\mu}_{I,O-N_I+1}$ through $\hat{\mu}_{I,O+N_I+1}$ all contain the outlier D_O. When processing false alarms for a set of data this is evident when false alarms themselves are observed as temporally correlated.

Consequently, when counting alarms all alarms that occur within a distance of $N_I - 1$ samples of another alarm are counted as a single alarm event. The simple heuristic is employed where all correlated samples are first compared as normal to the threshold $\bar{\mu}_{I,j}$ for the j-th background period and each alarm index is appended to a list. Alarms are processed in sequential order. Assuming that the i-th sample is an alarm then if any alarm exists after the i-th sample in the interval $[\hat{\mu}_{I,i+1}, \ldots, \hat{\mu}_{I,i+N_I-1}]$ then the i-th alarm is removed from the alarm set. The same heuristic continues at the subsequent alarm. The whole process is repeated for all background periods.

The alarm heuristic described is incorporated into a class that is used for all presented false alarm measurements of correlated samples.

TABLE I
FALSE ALARM RATE (per hour)

Distribution	Critical Value	Target FAR	Measured FAR
Normal	1.2459	0.5	5.5195
Poisson[a]	10	0.5	0.4952
Normal w/Curie Criteria	2.2258	0.5	0
Corrected Normal	1.6673	0.5	0.1311

C. Theoretical Model Significance vs. Measured False Alarm Rates

In order to demonstrate the discrepancy between theoretical and measured significance levels, several theoretical models were evaluated. For convenience estimated significance levels were scaled to false alarm rates measured per hour. An 8 segment neutron detector was used to collect 68 hours of data with a sample period of 0.2 seconds, an integration period of 4 seconds, and a background period using the first 5 minutes of data. The requisite statistics for each model were computed on a single background period. The resulting critical values were employed on integrated samples using a similar hypothesis test of section II-B and compensated according to section III-B1. Each distribution was configured with the theoretical critical value to meet a target FAR of 0.5 alarms per hour.

Refer to the results compiled in Table I according to distribution. Since a Poisson model is a discrete distribution of natural numbers, measurements of count rates included in the correlated sampling of III-B are not possible. In this case the unprocessed integer counts were used at the 0.2 second sampling interval. Again, results include a normal model assuming the Curie criteria [5][pp.75]. The last entry labeled "Corrected Normal" uses a corrected standard critical value that was estimated using the theory developed in this paper. The value $\bar{Z}_\alpha = 4.0309$ that results in a false alarm every 2 hours was corrected to $\hat{Z}_\alpha = 5.7235$.

The results in Table I indicate that the statistical model with the closest FAR to the desired rate is actually a Poisson distribution. However, justification for correlated sampling becomes apparent when the resulting thresholds are all placed on the same scale in the critical value column. The Poisson threshold is approximately six times that of the corrected normal rendering the Poisson model relatively insensitive.

IV. STATISTICAL VERIFICATION OF FALSE ALARM RATE USING EMPIRICAL DATA

In this section the statistical model of section II is applied to hundreds of background periods using a single data set referred to as the training data. A statistical inference is made during training concerning FAR performance and the training data is set aside. Verification of FAR utilizes a large number of individual false alarm tests [1][pp.17] each of duration 10 hours where the verification library consists of over 1000 hours

[a]The critical value was actually 2 counts but was scaled up to count rate per second for comparison purposes.

of background data per scintillating detector for both a ^{6}Li optical fiber detector and a Li^6 coated waveguide detector. Using the statistics estimated during training, confidence intervals were constructed at various coefficients and applied across all verification false alarm tests.

In this section several common experimental procedures are followed. Data is acquired with a sample period of 0.2 seconds, an integration period of 4 seconds, and a background period of 5 minutes. Again, the standard critical value is tuned to meet a target FAR of 0.5 alarms per hour.

A. *Confidence Interval using Training Data*

Each training data set is acquired such that the number of samples N_D results in about 70 hours of data to estimate a corrected $\hat{Z}_\alpha$ in tandem with the FAR performance statistics $\hat{\mu}_\alpha$ and $\hat{\sigma}_\alpha$ of section II-C5.

Instead of using the training statistic $\hat{\mu}_\alpha$ over N_{PB} background periods, the target statistic $\hat{\alpha}_j$ over a single background is used in verification to estimate a confidence interval that models an individual ANSI false alarm experiment [1][pp.17]. The standardized statistic for $\hat{\alpha}_j$ is $Z_j \triangleq \frac{\hat{\alpha}_j - \hat{\mu}_\alpha}{\hat{\sigma}_\alpha}$ where the corresponding standard critical value is $\bar{Z}_{\alpha/2}$, confidence coefficient is $(1-\alpha)$ for $\alpha \in [0,1]$, and probability of containment is

$$1 - \alpha = P\left(-\bar{Z}_{\alpha/2} < \frac{\hat{\alpha}_j - \hat{\mu}_\alpha}{\hat{\sigma}_\alpha} < \bar{Z}_{\alpha/2}\right) \tag{13}$$

Note that the critical value $\bar{Z}_{\alpha/2}$ can be looked up in a standard normal Z table accounting for a two-sided test while $\bar{Z}_\alpha$ in (2) uses a single-sided hypothesis test. A realization of a confidence interval [2][pp.216] of FAR for a specific training set is

$$\left(\hat{\alpha}^l, \hat{\alpha}^u\right) \triangleq \left(\hat{\mu}_\alpha - \left(\bar{Z}_{\alpha/2}\right)\hat{\sigma}_\alpha, \hat{\mu}_\alpha + \left(\bar{Z}_{\alpha/2}\right)\hat{\sigma}_\alpha\right) \tag{14}$$

and is tested for accuracy via the verification data set described next.

B. *Containment of Verification Data*

Verification data is segmented into background periods and processed like training data as in section II except that $\hat{Z}_\alpha$, $\hat{\mu}_\alpha$, and $\hat{\sigma}_\alpha$ were already calculated via training data. Each verification data set is acquired such that the number of samples N_D results in exactly 10 hours of data. For a given verification data set and j-th background period, $\hat{\alpha}_j$ is calculated according to (5) using $\hat{Z}_\alpha$ from training while the selected verification data set is used to calculate both background statistics $\hat{\mu}_{B,j}$ and $\hat{\sigma}_j$. The value $\hat{\alpha}_j$ is tested to determine if contained within the interval in (14). All verification data sets are processed similarly for the j-th background period. In this manner a measured confidence coefficient is calculated as a simple decimal fraction of successful containment. In order to account for background variation, a confidence coefficient is calculated for every possible background period in addition to the j-th and the result is averaged. $\ddot{\mu}_\alpha$ is the sample mean and $\ddot{\sigma}_\alpha^2$ is the sample variance of $\hat{\alpha}_j$ over every background period and every verification data set.

TABLE II
COATED DETECTOR TRAINING DATA SETS

Hours	68.7	71.5	71.8	71.9	71.4	72.4	72.1	95.1
$\hat{\mu}_D$	0.24	0.22	0.23	0.22	0.23	0.22	0.22	0.24
$\hat{\sigma}_D$	0.25	0.24	0.24	0.24	0.24	0.24	0.24	0.25
$\hat{Z}_\alpha$	5.70	5.74	5.83	5.77	5.75	5.68	5.83	5.67
$\hat{\mu}_\alpha$	0.5	0.49	0.49	0.51	0.51	0.5	0.5	0.5
$\hat{\sigma}_\alpha$	0.58	0.55	0.57	0.62	0.62	0.58	0.54	0.61
$\hat{\alpha}^l$	-0.63	-0.58	-0.62	-0.71	-0.70	-0.64	-0.56	-0.69
$\hat{\alpha}^u$	1.63	1.56	1.61	1.73	1.72	1.63	1.56	1.69

TABLE III
COATED DETECTOR VERIFICATION RESULTS FOR EACH TRAINING DATA SET

Ver. Sets	$\ddot{\mu}_\alpha$	$\ddot{\sigma}_\alpha$	85%	90%	92.5%	95%	97.5%	99%
128	0.52	0.63	87.6	89.8	91.8	94.0	95.1	96.3
127	0.48	0.60	86.5	89.2	90.2	92.3	94.8	95.8
127	0.40	0.54	92.6	93.6	94.9	95.6	96.4	97.9
127	0.46	0.58	91.8	93.5	94.6	95.5	97.5	98.6
127	0.48	0.61	88.8	92.2	93.0	94.6	95.9	98
127	0.53	0.66	86.7	89.4	91.2	92.5	94.1	96.3
127	0.42	0.56	89.5	91.2	92.8	94.2	96.2	96.8
125	0.54	0.66	86.6	88.4	91.2	92.3	95.8	97.6

TABLE IV
COATED DETECTOR VERIFICATION RESULTS OVER ALL TRAINING DATA SETS

Statistic	$\hat{\alpha}_j$	85%	90%	92.5%	95%	97.5%	99%
Sample Mean	0.48	88.8	90.9	92.5	93.9	95.7	97.2
Sample Std.	0.61	3.34	2.90	2.53	2.24	1.90	1.64

TABLE V
FIBER DETECTOR TRAINING DATA SETS

Hours	67.8	69.7	71.5	71.8	71.9	71.4	72.4	72.1
$\hat{\mu}_D$	2.28	2.30	2.26	2.28	2.27	2.26	2.25	2.24
$\hat{\sigma}_D$	0.81	0.82	0.81	0.81	0.81	0.81	0.80	0.80
$\hat{Z}_\alpha$	4.66	4.80	4.71	4.59	4.66	4.59	4.63	4.58
$\hat{\mu}_\alpha$	0.5	0.5	0.5	0.5	0.5	0.5	0.5	0.5
$\hat{\sigma}_\alpha$	0.35	0.30	0.31	0.41	0.36	0.44	0.41	0.39
$\hat{\alpha}^l$	-0.18	-0.09	-0.10	-0.30	-0.20	-0.36	-0.30	-0.27
$\hat{\alpha}^u$	1.18	1.10	1.10	1.29	1.21	1.36	1.30	1.26

TABLE VI
FIBER DETECTOR VERIFICATION RESULTS FOR EACH TRAINING DATA SET

Ver. Sets	$\ddot{\mu}_\alpha$	$\ddot{\sigma}_\alpha$	85%	90%	92.5%	95%	97.5%	99%
125	0.49	0.41	87.6	90.8	92.8	92.8	94.7	96.1
125	0.35	0.30	85.7	85.7	97.0	97.0	97.3	98.1
125	0.44	0.37	84.3	93.3	93.3	95.1	95.1	96.3
125	0.58	0.46	86.5	88.3	91.2	91.2	95.4	96.2
125	0.49	0.40	90.9	90.9	92.7	94.4	96.3	97.2
125	0.59	0.46	88.2	91.6	91.6	93.8	95.5	96.9
125	0.53	0.42	89.0	91.4	93.8	95.6	96.8	97.4
125	0.60	0.47	84.9	87.8	87.8	90.8	93.4	95.9

C. *Confidence Testing Results*

A total of 8 training files were used per detector with all results listed in each column of Tables II and V. $\hat{\mu}_D$ is the

TABLE VII
FIBER DETECTOR VERIFICATION RESULTS OVER ALL TRAINING DATA SETS

Statistic	$\hat{\alpha}_j$	85%	90%	92.5%	95%	97.5%	99%
Sample Mean	0.51	87.1	90.0	92.5	93.9	95.6	96.7
Sample Std.	0.42	3.43	3.33	3.23	2.81	2.04	1.63

sample mean of all samples measured in counts per second. $\hat{\sigma}_D$ is the sample standard deviation of all samples measured in counts per second. $\hat{Z}_\alpha$ is an estimate of the standard critical value that will achieve the desired false alarm rate measured in the number of standard deviations above the mean. $\hat{\mu}_\alpha$ is the estimated significance level of threshold $\hat{Z}_\alpha$ scaled to FAR measured in alarms per hour. $\hat{\sigma}_\alpha$ is the estimated standard deviation of $\hat{\mu}_\alpha$. $\hat{\alpha}^l$ and $\hat{\alpha}^u$ are the lower and upper bounds of FAR, respectively, for a 95% confidence interval.

The results of testing each training file with verification data is summarized in Tables III, IV, VI, and VII. Each training data set column number of Table II corresponds to the verification row number result listed in Table III. Each training data set column number of Table V corresponds to the verification row number result listed in Table VI. Measured confidence is in percent and is a function of not only $\hat{\alpha}^l$ and $\hat{\alpha}^u$, but also $\hat{Z}_\alpha$.

V. CORRECTED SIGNAL-TO-NOISE RATIO

A. Corrected Signal-to-Noise Ratio Definition

1) Signal to Noise Ratio Definition: With a neutron detector setup with ANSI moderation [1], a sample mean of count rate measurements denoted μ is acquired using a background period of 5 minutes. Using a ^{252}Cf source with an emission rate of 20,000 n/s placed at 1.5m from the detector [1], a sample mean of count rate measurements notated ρ is acquired using a period of 5 minutes. It is assumed in [5][pp.94] that data follow a normal distribution where $\sigma^2 \approx \mu$.

SNR is then defined [5][pp.94] as

$$\mathrm{SNR} \triangleq \frac{S^2}{B} = \frac{(\rho-\mu)^2}{\mu} \tag{15}$$

2) Sigma Correction Factor: When utilizing a neutron detector to alarm using a specific threshold, the threshold is set using a critical value that assumes that the detector will result in a specific FAR. Section III-C demonstrates how a theoretical standard critical value may result in a measured error rate that is drastically different than the rate that the model predicts.

When comparing two detectors a simple metric often used is SNR. The noise value encompassed by the denominator parameter σ represents a certain spread of data from the mean that occurs for a specific percent of samples. In other terms, this spread in data can be viewed as occurring at a specific rate for a real-time system. Let's define a correction factor γ^2 to adjust the measured variance σ^2 to account for the disparity between theoretical and measured rates. Using the alarm threshold developed in (3) with the additional correction factor and notation of this section results in a new alarm threshold

$$T_C = \mu + \left(\bar{Z}_\alpha\right)\frac{\gamma\sigma}{\sqrt{N_I}} \tag{16}$$

Using the notation of this section and the practical standard critical value of section II-D, the practical threshold results in

$$T_P = \mu + \left(\hat{Z}_\alpha\right)\frac{\sigma}{\sqrt{N_I}} \tag{17}$$

To correct the theoretical critical value according to (16) this threshold should be equal to the practical or measured threshold in (17) resulting in

$$\mu + \left(\bar{Z}_\alpha\right)\frac{\gamma\sigma}{\sqrt{N_I}} = \mu + \left(\hat{Z}_\alpha\right)\frac{\sigma}{\sqrt{N_I}}$$

Therefore, if using the theoretical threshold $\bar{Z}_\alpha$ then σ could be corrected using the factor

$$\gamma = \frac{\hat{Z}_\alpha}{\bar{Z}_\alpha} \tag{18}$$

3) Corrected Signal to Noise Ratio: Using the sigma correction factor in (18) and the approximation $\sigma^2 \approx \mu$ [5][pp.94], SNR of section V-A1 is adjusted to

$$\mathrm{SNR}_\gamma = \frac{(\rho-\mu)^2}{\gamma^2\mu} \tag{19}$$

B. Corrected Signal-to-Noise Ratio Application

1) Backpack Test Overview: With respect to false alarms, the system should tolerate no more than five alarms due to natural background in a 10 hour period [1]. In addition, once the system is configured with an alarm threshold to meet the mentioned error rate, a source that is moving at a specific rate is required to be detected successfully in 96 out of 100 trials [1]. Note that an alarm threshold equal to or greater than the threshold to meet the specified FAR must be used in the moving source test to meet all system requirements.

In the subsequent sections several detector technologies are evaluated using the same parameters. The sampling periods described in section III were used so that the value $\bar{Z}_\alpha = 4.0309$ results in a false alarm every 2 hours.

2) Fiber Detector Version 1 Test Results: The first fiber detector tested labeled D1 required modification due to an inherent flaw. A standard critical value of $\hat{Z}_\alpha = 5.87$ resulted in a measured FAR of 0.5 alarms per hour. Using the corresponding threshold in a moving source test yielded a detection rate of 90%. Detector D1 measured a background mean of $\mu = 4.94$ according to section V-A1. When measuring the ^{252}Cf source a flaw with D1 occurred where the measurement slowly increased in time. Consequently, the initial value of count rate $\rho = 20$ was used instead of a 5 minute average. A sigma correction factor of $\gamma = 1.46$ resulted.

TABLE VIII
SIGNAL-TO-NOISE RATIO RESULTS

Detection System	Detection %	SNR	SNR_γ
Fiber D1	90	45.9	21.6
Fiber D2	95	22.8	17
Coated D3	97	43.8	22.6

3) Fiber Detector Version 2 Test Results: An improved detector tested was labeled D2. A standard critical value of $\hat{Z}_\alpha = 4.67$ resulted in a measured FAR of 0.5 alarms per hour. Using the corresponding threshold in a moving source test yielded a detection rate of 95%. Detector D2 measured a background mean of $\mu = 1.75$ according to section V-A1. When measuring the ^{252}Cf source the detector resulted in a count rate of $\rho = 8.07$ after a 5 minute average. A sigma correction factor of $\gamma = 1.16$ resulted.

4) ^{6}Li Coated Detector Test Results: A third detector consisting of a ^{6}Li coated waveguide was labeled D3. A standard critical value of $\hat{Z}_\alpha = 5.61$ resulted in a measured FAR of 0.5 alarms per hour. Using the corresponding threshold in a moving source test yielded a detection rate of 97%. The detector measured a background mean of $\mu = 0.26$ according to section V-A1. When measuring the ^{252}Cf source the detector resulted in a count rate of $\rho = 3.63$ after a 5 minute average. A sigma correction factor of $\gamma = 1.39$ resulted.

C. Signal-to-Noise Ratio Test Results

The conventional method resulted in a SNR value for D1 about *twice* that of the value for D2 despite sensitivity testing [1] indicating D2 is marginally more sensitive than D1. Likewise, the conventional method resulted in a SNR value for D1 that was higher than the value for D3 despite sensitivity testing [1] indicating the contrary. The conventional method also indicated a SNR value for D3 that was substantially higher than D2 despite similar detection sensitivity.

Notice that in Table VIII in both cases the relative spread of SNR_γ values more accurately reflected the corresponding spread in measured detection percent compared to the conventional SNR metric. However, the value of SNR_γ for detector D1 remained higher than the value for detector D2 despite D1 proving less sensitive in a dynamic test.

VI. IMPLICATIONS OF SYSTEM PARAMETER MODIFICATION

Utilizing the method developed to process many background periods has additional advantages with respect to the analysis of practical system parameters. Parameters such as sample period, background duration, or even alarm methods discussed in section II-F could be adjusted in order to measure the effects on system response such as $\hat{Z}_\alpha$ in (10) or the observed FAR in (6).

For instance, the background period was altered from 1 minute to 5 minutes in 1 minute increments using the same detector, configuration, and data set as in section III-C. The resulting maximum likelihood estimate of $\hat{Z}_\alpha$, mean estimate $\hat{\mu}_\alpha$, and variance estimate $\hat{\sigma}^2_\alpha$ are listed in Table IX where significance levels are scaled to FAR per hour for convenience.

TABLE IX
BACKGROUND PERIOD (minutes) VS. FAR (per hour)

Period	$\hat{Z}_\alpha$	Estimated FAR	Estimated FAR Std. Dev.
5	5.72	0.51	0.55
4	5.8	0.50	0.69
3	5.91	0.50	0.83
2	6.41	0.46	4.14
1	8.55	0.44	15.31

With FAR approximately constant, notice as background period decreases system sensitivity decreases via the estimated standard critical value $\hat{Z}_\alpha$ increase. Each estimated FAR data point remains approximately equal to the target rate but slightly lower when the background period falls below 3 minutes. However, note the drastic increase in the standard deviation of FAR when the background falls below 3 minutes.

VII. CONCLUSION

When utilizing all practical aspects of a radiation detection system to recognize neutron sources including the background period, integration period, correlated sampling, and the sparse nature of acquired data, the measured noise performs at a FAR significantly different than the FAR that the theoretical model predicts. This practical error performance of a neutron detection system must be normalized to some known reference in order to properly characterize practical detection sensitivity.

Modifications to detection algorithms, manufacturing process, or physical composition of neutron detectors could result in a deviation in the necessary $\hat{Z}_\alpha$ to meet a required FAR. A method was introduced to adjust theoretical alarm parameters to more accurately quantify and meet a desired FAR using a static measurement and an automated algorithm.

Performing an additional static measurement with a source while applying the correction factor γ results in a SNR metric that better agrees with more involved dynamic detection tests.

In conclusion, the estimate of standard critical value in (10) and the sigma correction factor in (18) should be considered as methods to characterize how the measured error rate and detection sensitivity are affected by differences in neutron detection systems.

REFERENCES

[1] "American national standard performance criteria for backpack-based radiation-detection systems used for homeland security," *ANSI N42.53-2013*, pp. 1–62, Aug 2013.
[2] R. Hogg, J. McKean, and A. Craig, *Introduction to Mathematical Statistics*, ser. Always learning. Pearson, 2013. [Online]. Available: https://books.google.com/books?id=xFlVYAAACAAJ
[3] A. Papoulis, *Probability, Random Variables, and Stochastic Processes*, ser. Communications and signal processing. McGraw-Hill, 1991. [Online]. Available: https://books.google.com/books?id=4lwQAQAAIAAJ
[4] S. Kirkpatrick, C. D. Gelatt, and M. P. Vecchi, "Optimization by simulated annealing," *SCIENCE*, vol. 220, no. 4598, pp. 671–680, 1983.
[5] G. F. Knoll, *Radiation Detection and Measurement; 3rd ed.* New York, NY: Wiley, 2000. [Online]. Available: https://cds.cern.ch/record/441925

Measurements and simulations to investigate the feasibility of neutron multiplicity counting in the current mode of fission chambers

Lajos Nagy
Institute of Nuclear Techniques
Budapest University of Technology and Economics
H-1111 Budapest, Hungary
nagy.lajos@reak.bme.hu

Imre Pázsit
Division of Subatomic and Plasma Physics
Chalmers University of Technology
SE-412 96 Göteborg, Sweden
imre@chalmers.se

Lénárd Pál
Centre for Energy Research
Hungarian Academy of Sciences
H-1525 Budapest 114, POB 49, Hungary
pal.lenard@wigner.mta.hu

Gergely Klujber
Institute of Nuclear Techniques
Budapest University of Technology and Economics
H-1111 Budapest, Hungary
klujby@reak.bme.hu

Máté Szieberth
Institute of Nuclear Techniques
Budapest University of Technology and Economics
H-1111 Budapest, Hungary
szieberth@reak.bme.hu

***Abstract*—In two earlier papers [1], [2] we investigated the possibility of extracting traditional multiplicity count rates from the cumulants of fission chamber signals in current mode. It was shown that if all neutrons emitted from the sample simultaneously are also detected simultaneously, the multiplicity rates can be retrieved from the first three cumulants of the currents of up to three detectors, but the method breaks down if the detections of neutrons of common origin take place with a time delay spread wider than the pulse shape. To remedy these shortcomings, in this work we extended the theory to two- and three-point distributions (correlations). It was found that the integrals of suitably chosen two- and three-point moments with respect to the time differences become independent of the probability density of the time delays of detections. With this procedure, the multiplicity rates can be retrieved from the detector currents for arbitrary time delay distributions. To demonstrate the practical applicability of the proposed method, a measurement setup was designed and built. The statistics (shape and amplitude distribution) of the detector pulse were investigated as important parameters of the theoretical model. Simulations were performed to estimate the expected value of the multiplicity rates in the built setup. Measurements were performed and two types of moments (the mean and the covariance function) of the recorded detector signals were calculated. Values of singles rates were successfully recovered.**

The research presented in the paper has received funding from the following sources: National Research, Development and Innovation Fund, Hungary, proj. No. VKSZ 14-1-2015-0021; Foundation for Nuclear Engineers, MVM Paks Nuclear Power Plant Ltd., Hungary; Ringhals Nuclear Power Plant, Vattenfall AB, Sweden, contr. No. 677353-003; Royal Society of Arts and Sciences in Gothenburg, Sweden; Nemzeti Tehetség Program, Ministry of Human Capacities, Hungary. The pre-amplifiers used in the measurements were designed and built by István Barth from the Institute of Nuclear Techniques at the Budapest University of Technology and Economics.

Keywords—**nuclear safeguards, multiplicity counting, fission chambers**

I. Introduction

The primary objective of multiplicity counting is to determine the mass of small fissile samples which is extracted from the detection rates of single, double and triple coincidences (S, D and T count rates) with pulse counting techniques [3]. The development of a method that extracts these multiplicity rates from the first three cumulants of the current of fission chambers is the subject of this paper and its two predecessors [1], [2]. The basic idea was introduced in Ref. [1], in which the neutrons emitted simultaneously from the sample were assumed to be detected simultaneously. It was shown that the S, D and T count rates could be retrieved from the first three cumulants of the detector current, provided that certain parameters of the detector (such as its pulse shape and amplitude distribution is known. In reality, though, detection of neutrons of common origin does not take place simultaneously. Incorporation of this phenomenon into the theory was made in the sequel paper [2], by assuming a random arrival time to the detector. The results showed that when the width of the density function of the time delay is much wider than the pulse width (which is a typical case is thermal detection systems), the coefficients of the D and T rates (analogues of the doubles and tripes gate factors in pulse counting methods) will become vanishingly small, hence only the S rates can be unfolded from the cumulants. To remedy this problem, the theory has recently been extended to the use of the double and triple covariance functions of the detector signals derived from their two- and three-point distributions (in time). In this paper, the main results of this extension are reported. In order to demonstrate the use of the proposed method and to investigate its practical applicability, an experimental configuration has recently been prepared in the Training Reactor Facility at the Budapest University of Technology and Economics in Hungary. Extensive measurements are currently in progress in various configurations. At the end of the paper, some preliminary experimental results will be presented as well.

II. Traditional Multiplicity Counting

In this section the traditional method of multiplicity counting is summarized briefly using the terminology of [4]. In a multiplicity counting measurement, the detection rates of the first three k-tuplets (k detected neutrons originating from the same sample emission) are determined. These rates are called the *singles* (S), *doubles* (D) and *triples* (T) rates, respectively and they can be written as:

$$\begin{aligned} S &= F\,\varepsilon\,\widetilde{\nu}_1 \\ D &= \frac{F\,\varepsilon^2\,f_d\,\widetilde{\nu}_2}{2}, \\ T &= \frac{F\,\varepsilon^3\,f_t\,\widetilde{\nu}_3}{6}, \end{aligned} \tag{1}$$

Here F is the intensity of spontaneous fission in the sample, ε is the detection efficiency, $\widetilde{\nu}_i$ is a modified form of the so-called Böhnel moments, the factorial moments of the number of emitted neutrons per one source event. f_d and f_t the so-called doubles and triples "gate factors" which are introduced empirically in order to account for the non-coincident detection of neutrons of common origin, and the loss of detection outside the measurement windows.

Using these equations, the sought sample quantities (including the fissile mass of the sample) can be obtained from the measured values of the S, D and T rates using algebrain inversion [4].

III. Multiplicity counting using fission chamber signals

The theory of the newly proposed method of multiplicity counting is based on a formalism describing the fluctuating signals of neutrons detectors [5]. The key element of the formalism is a stochastic model of the detector response, which describes the pulse by a deterministic (constant) pulse shape $f(t)$ with a random amplitude **a**, whose nth moment is denoted by $\langle a^n \rangle$. It is assumed that the detection of neutrons occurs with a random time delay τ, which is independent and identically distributed for each neutron, and is characterized by a density function $u(\tau)$.

Using these quantities as building blocks, various low order cumulants of the detector signals can be calculated with a master equation formalism. In [1] and [2] expression were derived for the one-point cumulants. In this paper calculations for the two and three point cumulants will be presented. We shall see that as in the one-point case, the two- and three point cumulants can also be expressed with the Böhnel moments hence they are also related to the traditional multiplicity rates. As it was already discussed in the previous papers, when interpreting the cumulants in terms of the multiplicity rates, with the purpose of unfolding the Böhnel moments from the cumulants, one has to substitute $f_d = f_t = 1$ in (1). This is because the effect of non-coincident detection of neutrons of common origin, which is described by the empirical gate factors in the traditional method, is explicitly included in the theory of the new method. Nevertheless, as it will be seen, the analogies of the gate factors appear in the cumulants as well.

A. One-point Distributions

To serve as a reference of comparison with the expressions of the two- and three point cumulants presented in the following subsection, a brief overview of the theory concerning the one-point distribution is given here based on [2].

By omitting the details of the derivations, the first three cumulants (the mean, variance and skewness) of the detector signal are given by

$$\kappa_1 = S\,\langle a \rangle I_1, \tag{2}$$

$$\kappa_2 = \left(S\,\langle a^2 \rangle + \xi_{1,1}\,2D\,\langle a \rangle^2\right)\,I_2, \tag{3}$$

$$\kappa_3 = \left(S\,\langle a^3 \rangle + \xi_{1,2}\,6D\,\langle a \rangle \langle a^2 \rangle + \xi_{1,1,1}\,6T\langle a \rangle^3\right)\,I_3. \tag{4}$$

Following similar considerations, the expressions for the double and triple cross-covariances between two and three detectors read as

$$\kappa_{1,1} = 2D\,\langle a \rangle^2\,\xi_{1,1}\,I_2, \tag{5}$$

$$\kappa_{1,1,1} = 6T\,\langle a \rangle^3\,\xi_{1,1,1}\,I_3. \tag{6}$$

Here, I_n as well as $\xi_{1,1}$, $\xi_{1,2}$ and $\xi_{1,1,1}$ are integrals of the pulse $f(t)$ shape and the delay density function $u(\tau)$ and their definitions can be found in [2]. Because it will appear in the two-and three point cumulants, we include here the definition of I_1:

$$I_1 = \int_0^\infty f(t)\,\mathrm{d}t. \tag{7}$$

We will refer to the ξ's as "gate factors" because they play a similar role in the above formulas as the traditional gate factors in (1). From Equations (2)–(6) it is seen that if I_n and the ξ's are known from calibration, the detection rates can be obtained from the measured cumulants by simple algebraic inversion. However, as it was also shown in [2], if the spread of the density of the time delay is much larger than the pulse width (which is typical in thermal detection systems), the gate factors become vanishingly small. In such a case, only the S rates can be extracted.

B. Two- and Three-point Distributions

To explore the temporal correlations in the detector signals, their distribution must be described at two or even three points in time: besides time t, a second time $t - \theta$ and a third time

$t-\theta-\rho$ is considered as well. Regarding a single detector, our goal is to determine the integrals

$$\mathrm{Cov}_2 = \int_0^\infty \mathrm{Cov}_2(\theta)\,\mathrm{d}\theta \tag{8}$$

and

$$\mathrm{Cov}_3 \equiv \int_0^\infty \int_0^\infty \mathrm{Cov}_3(\theta,\rho)\,\mathrm{d}\theta\mathrm{d}\rho \tag{9}$$

of the second and the third order cumulants of the signal, the covariance function $\mathrm{Cov}_2(\theta)$ and the bi-covariance function $\mathrm{Cov}_3(\theta,\rho)$ defined as

$$\begin{aligned}\mathrm{Cov}_2(\theta) = \lim_{t\to\infty} \mathbb{E}\left[(y(t)-\mathbb{E}[y(t)])\right. \\ \left.\times\,(y(t-\theta)-\mathbb{E}[y(t-\theta)])\right],\end{aligned} \tag{10}$$

and

$$\begin{aligned}\mathrm{Cov}_3(\theta,\rho) = \lim_{t\to\infty} \mathbb{E}\left[(y(t)-\mathbb{E}[y(t)])\right. \\ \times\,(y(t-\theta)-\mathbb{E}[y(t-\theta)]) \\ \left.\times\,(y(t-\theta-\rho)-\mathbb{E}[y(t-\theta-\rho)])\right].\end{aligned} \tag{11}$$

Detailed calculation of the moments can be found in [6]; here only the final results will be provided. One finds that the integral of the covariance function $\mathrm{Cov}_2(\theta)$ has the form

$$\mathrm{Cov}_2 = \frac{1}{2}\left(S\,\langle a^2\rangle + 2\,D\langle a\rangle^2\right)\,I_1^2, \tag{12}$$

whereas the integral of the bicovariance function $\mathrm{Cov}_3(\theta,\rho)$ has the form

$$\begin{aligned}\mathrm{Cov}_3 = \frac{1}{6}\left(S\,\langle a^3\rangle + 2D\,\langle a\rangle\,\langle a^2\rangle\,(\xi_A+\xi_B+\xi_C)\right. \\ \left. +\,6T\,\langle a\rangle^3\right)\,I_1^3.\end{aligned} \tag{13}$$

Here I_1 is the same as (7) in the previous Section, whereas the three point doubles gate factors ξ_A, ξ_B and ξ_C are also integrals of the pulse shape and the delay density function; their definitions can be found in [6]. The corresponding integrals of the covariance function $\mathrm{Cov}_{1,1}(\theta)$ of two detectors and the bi-covariance function $\mathrm{Cov}_3(\theta,\rho)$ of three detectors read as

$$\mathrm{Cov}_{1,1} = D\langle a\rangle^2\,I_1^2 \tag{14}$$

$$\mathrm{Cov}_{1,1,1} = T\,\langle a\rangle^3\,I_1^3. \tag{15}$$

One can see that the expressions (12)–(15) have, with minor differences, the same form as the one-point cumulants (3)–(6). The difference is that, except in the doubles term of Cov_3, all the gate factors (hence the density of the unknown time delay distribution) have disappeared.

With the results presented above, it possible to design an experimental procedure such that all selected moments (cumulants and covariances) are independent of the time delay distribution. This makes the method applicable for multiplicity counting with thermalised neutrons, for which the detector efficiency is much higher than for fast neutrons. In particular, one can use the first cumulant κ_1 to determine the singles rates, the covariance $\mathrm{Cov}_{1,1}$ or Cov_2 to determine the doubles rate, and the bi-covariance $\mathrm{Cov}_{1,1,1}$ to determine the triples rate.

IV. Experimental Investigation

An experimental configuration has been prepared in the Training Reactor Facility at the Budapest University of Technology and Economics in Hungary in order to demonstrate the use of the newly proposed method. Strictly speaking, the applied measurement setup corresponds to the active version [7] of multiplicity counting as opposed to the passive version described in the first part of this paper. The reason for this is that no spontaneous fissionable materials are available at our facility, therefore the only option is to use low enriched uranium samples irradiated by an isotopic neutron source to induce fission in the sample. Although the two versions of multiplicity counting provide different formulas for the detection rates [7], the relationship between the moments of the detector signals and the detection rates is expected to be the same in the two cases. Therefore as our primary goal is not to unfold the fissile mass of an unknown sample, but only to show the possibility of extracting detection count rates from the time-resolved signals of fission chambers, we shall disregard the differences of the two versions and use the formulas for the signal moments as presented in the preceding sections.

A. Description of the measurement set-up

Fig. 1 shows the MCNP 6.1 [8] model of the measurement geometry whereas Fig. 2 shows a picture of the built setup.

The central element of the setup is an EK-10 type fuel assembly. Different variants of the same assembly type are available in the facility which differ in their shape, in the number of fuel rods they contain, and the arrangement of the rods within the assembly. The one shown in the figures is a square shaped assembly with 68 mm side length containing 16 fuel rods in a square lattice. Each rod is filled with fresh fuel pins of 10% enriched uranium oxide in metal magnesium matrix yielding 7.94 grams of ^{235}U per rod. The rods have a 50 cm active length and a 7 mm active diameter. The cladding of the fuel pin is 99.5% purity aluminum alloy, with an outer diameter and thickness of 10 mm and 1.5 mm, respectively.

The fuel assembly is surrounded by three KNT-31-1 type fission chambers labeled with A–C on the Figure. The fission chambers have a 17.6 cm length, a 32 mm outer diameter and 500 cm^2 area of sensitive layer covered with 90% enriched uranium. The thermal neutron sensitivity of the detectors is 0.25 pulse per one neutron/cm^2.

A ^{241}Am–Be source, labeled with S on the Figure, is located close to the fuel assembly to provide neutrons that cause induced fission in the fuel. The isotopic source has an emission intensity of $2\cdot 10^6$ neutrons/s. Both the source and the detectors are covered (at least partially) by a paraffin-wax coating which serves as a moderator for the fast neutrons originating from the isotopic source and the fuel. The thickness of the coating was chosen to maximize the fission intensity in the detectors; based on MCNP simulations with different thicknesses, 3 cm was found to be optimal.

The signal of each detector is sent to an in-house-built high-frequency pre-amplifier which produces a voltage signal ranging between −1 and 1 V. The pre-amplifier circuit has a small time constant (compared to the charge collection time of the detector), hence the shapes of the amplified voltage pulses reflect the shapes of the current pulses in the detector. The voltage signals of three detectors are then registered by a pair of Red Pitaya STEMLab 125-14 type FPGA-based A/D converters. Each converter has two analogue inputs and provides a 14 bit vertical resolution as well as 125 MHz maximal sampling frequency (corresponding to a 8 ns maximal resolution in time).

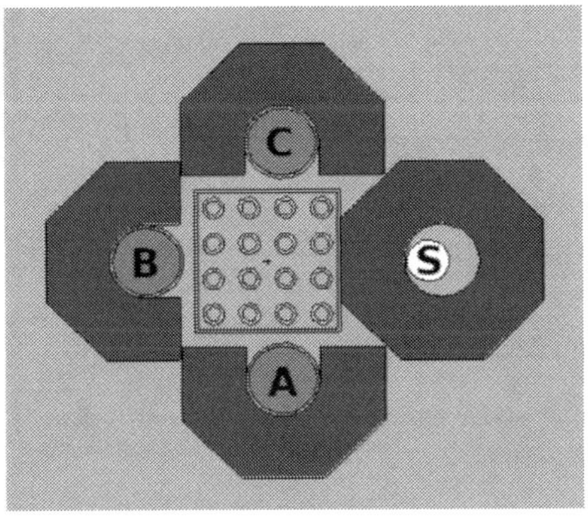

(a) Horizontal view.

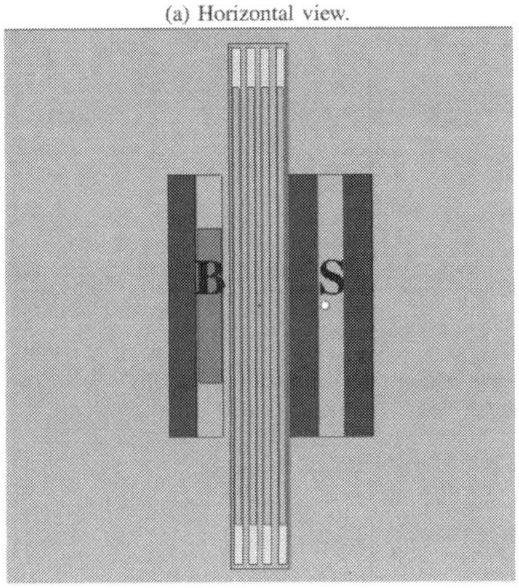

(b) Vertical view.

Fig. 1: The MCNP model of the measurement setup.

The complete experimental investigation is divided into three steps:

1) First, by recording relatively short segments of the detector signals, the relevant properties of the pulses were determined: the mean amplitude $\langle a \rangle$, as well as the integral I_1 of the pulse shapes.

Fig. 2: A picture of the measurement setup.

2) Next, by running MCNP simulations, the detection efficiency ε of the measurement configuration is determined, furthermore, the expected number of singles, doubles and triples detection events is estimated.
3) Finally, by running long measurements and analyzing signals, the mean value and the covariance function of the detector signal is determined. From the mean value, the singles rate is also recovered.

B. The properties of the detector pulse

The properties of the impulse of the detectors (their shape and amplitude distribution) were determined by analyzing a large number of recorded pulses. Signals were recorded for 5 minutes with a time-resolution of 48 ns; Fig. 3 shows a short segment of the recorded signal of two detectors. From all the registered pulses, the non-overlapping ones were selected and analyzed. An other set of signals were recorded without the isotopic source being present in order to characterize the background noise. The results are summarized on Figures 4 and 5 for one of the three detectors. Fig. 4 shows the probability density function of the amplitudes of the selected pulses, i.e. the pulse amplitude spectrum. There is a consider-able background below 150 mV which originates partly from the α-decay in the fissile deposit of the fission chamber and partly from electronic noise. The mean amplitude of the pulses generated by neutrons is around 272 mV. On Fig. 5 the shapes of three individual registered pulses is shown together

 https://doi.org/10.1051/epjconf/202022507011

with the average pulse shape calculated from the complete set of selected pulses. One can see that the average pulse has a characteristic length of around 10 μs. The integral of the pulse shape is 3.09 μs.

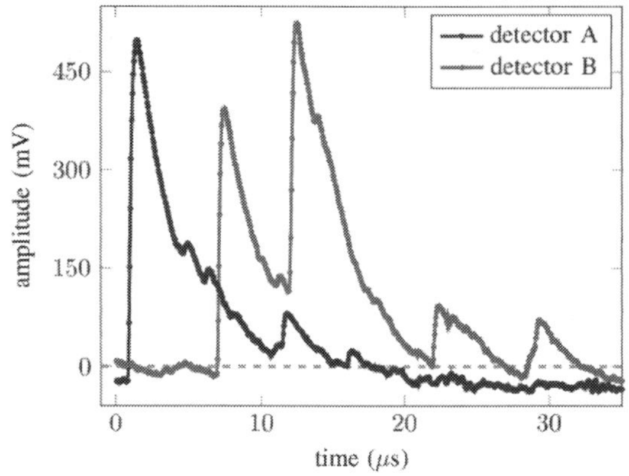

Fig. 3: Short segments of the recorded signals of two detectors.

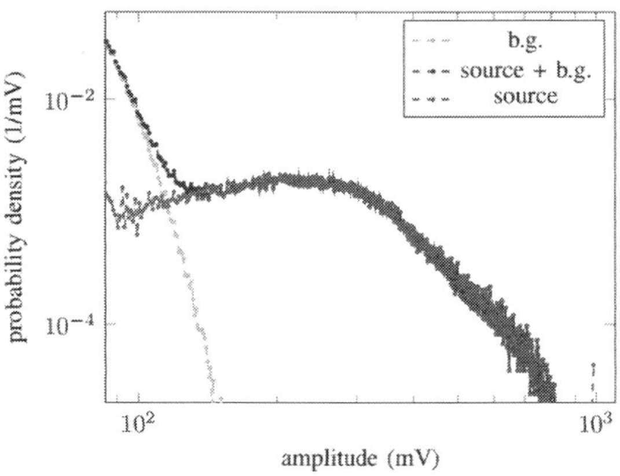

Fig. 4: Probability density function of the amplitude of the detector impulse.

C. The efficiency of detection

The efficiency of detecting single neutrons, as well as doubles and triples was investigated with MCNP simulations using the model shown on Fig. 1. Each simulation was performed with the PTRAC option which, provided listed information on (among many others) the location and time of fission events. This data was then processed and used to determine the quantities discussed in this subsection.

In order to estimate the efficiency of the detectors, that is, the probability of detecting a neutron coming from the fuel assembly after an induced fission, source neutrons were initiated from the fuel region. Their position and direction

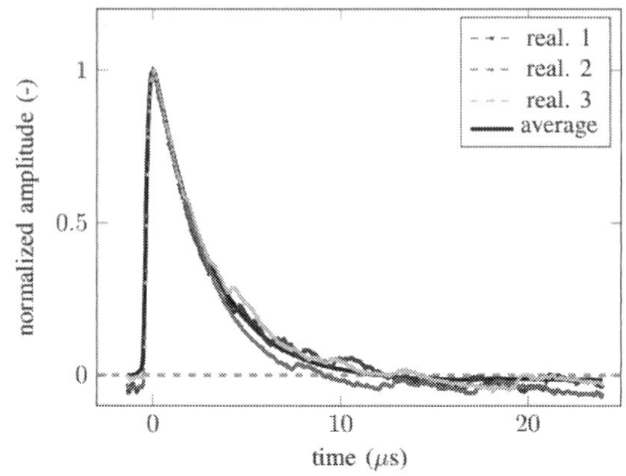

Fig. 5: Some realizations of the shapes of the detector impulse and the calculated average shape.

followed uniform and isotropic distributions, respectively, whereas their energy followed the Watt spectrum specific to ^{235}U. To eliminate the effect of internal multiplication, which is of no interest in determining the detection efficiency, the fuel pins were replaced by void space. After initiating $3 \cdot 10^9$ source particles in total, the number of fission events in the detectors were counted, from which the detection efficiency could be easily estimated. The results are summarized in Table I. One can see that the efficiency of all three detectors is around 0.07% yielding an overall detection efficiency of around 0.21%. This is a very low value (especially when compared with the usual 40–60% efficiency of traditional multiplicity counters [3]) primarily caused by the unfavorable geometrical conditions in the system. As a consequence, by recalling expressions (1) of the detection rates, we expect that the observed doubles and triples rates will be several orders of magnitudes smaller than the singles rates, which can make their determination difficult or even impossible.

TABLE I: The probability of detecting neutrons arriving from the fuel.

detector	detection probability (%)
A	0.0710
B	0.0703
C	0.0707
total	0.212

There is an additional effect related to the detection of neutrons, which is not included in the theoretical model presented in the first part of the paper, nevertheless it affects the statistics of the measured signals. Namely, in a fission chamber the detection of a neutron generates further neutrons, which might then get detected in this particular detector or in a neighboring one, thus increasing observed rates of singles, doubles and triples detections. To estimate the probability of such events, simulations were performed in which source neutrons were initiated from one of the detectors (distributed uniformly in space, isotropically in direction and with Watt energy spectrum), and the number of fission events in all detectors were counted. This same procedure was repeated with all three detectors, and the probabilities of detecting neutrons originating from any of the three detectors were estimated. The results are shown in Table II, where the rows represent the detectors where the neutrons starts from, and the columns represent detectors where the neutrons arrives to. Considering a neutron generated during the detection process in one of the detectors, the probability that it will be detected by the same detector is around 0.033–0.035%, whereas the probability that it will be detected by an other detector is around 0.016%. Since these probabilities are quiet low, the number of such "false counts" and hence their contribution to the singles, doubles and triples is expected to be negligible. Nevertheless, in order to gain a better understanding of this process and its consequences, we are planning to incorporate it into the theoretical model in the near future.

TABLE II: The probabilities of detecting neutrons born in the detection process in one of the detectors.

	detection probability (%)		
	A	B	C
A	0.035	0.016	0.016
B	0.016	0.033	0.016
C	0.016	0.016	0.035

The detection probability presented above is a simple way to characterize the detection efficiency in the measurement setup. An other, more direct approach is to estimate the expected number of single, double and triple detection events in a given period of time. To get this information, neutrons were initiated from the ^{241}Am–Be isotopic source region, and the number of events when one, two and three neutrons were detected (excluding the detection of neutron born in the fission chambers) were counted. The simulations were performed with $3 \cdot 10^9$ source neutrons in total which, taking into account the strength of the isotopic source, corresponds to a measurement time of 25 minutes. This simulation was repeated with four different number of fuel rods (8, 10, 13 and 16) in the assembly, in order to investigate the sensitivity of the detection rates on the amount of fissile material. The results are summarized in Table III. Two types of singles are shown: source singles are neutrons detected directly from the isotopic source, whereas sample singles originate from induced fission in the fuel assembly. No values are presented for the triples, because their number was zero in each case due to the very low detection efficiency of the system. As expected, with increasing amounts of fuel the number of sample singles and sample doubles increases, because the probability of inducing fission in the fuel also increases. The expected number of source singles shows the opposite tendency which can be explained by the shielding property of the fuel: the more fuel is present, the lower the probability that source neutrons reach the detectors without causing induced fission in the fuel. In general, as it is expected from the detection probabilities shown earlier, the expected number of singles is several order of magnitudes higher than the expected number of doubles. Additionally, the expected number of source singles is much higher than that of the sample singles. As a result, one expects that the source single counts will have a dominating contribution to the measured detector current and to its moments, as we will see in the following subsection.

TABLE III: Number of detection events with different amounts of uranium in the fuel assembly.

# rods (-)	number of detection events per 25 min		
	source singles	sample singles	sample doubles
8	3152978 ± 1775.7	261434 ± 511.3	227 ± 15.1
10	2989616 ± 1729.1	301605 ± 549.2	272 ± 16.5
13	2804619 ± 1674.7	346319 ± 588.5	279 ± 16.7
16	2623345 ± 1619.7	379215 ± 615.8	331 ± 18.2

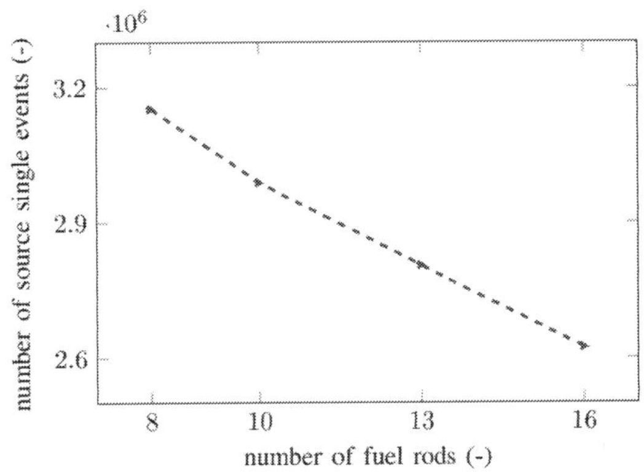

Fig. 6: Number of source single events per 25 min as function of the number of fuel rods in the assembly.

D. The mean value and the covariance function of the signals

As a simple preliminary demonstration of the proposed method, measurements were performed using three different types of fuel assemblies, containing 0, 5 and 16 rods. The mean values and the covariance functions of the registered signals were then determined which, according to (2) and (14), are related to the (sample) singles and (sample) doubles rates, respectively. The bicovariance functions were not calculated due to the low expected value of the triple events.

Since, as we saw earlier, detection events (especially doubles) are rare, long measurements lasting several hours are necessary to gain usable statistics. Considering the 48 ns time resolution of the AD converter, the recorded signals would require a large amount of disk space. In order to save space, a compression technique was developed and applied. The technique utilizes the fact that – due to the low detection intensity – neighboring pulses in the signal are often fare from each other and separated by background noise which, however, contains

https://doi.org/10.1051/epjconf/202022507011

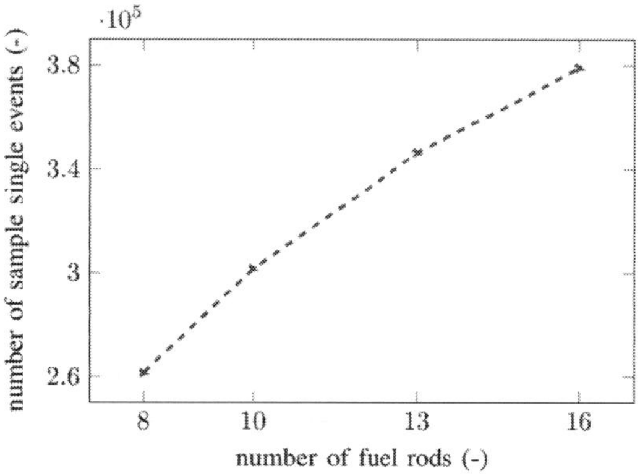

Fig. 7: Number of sample single events per 25 min as function of the number of fuel rods in the assembly.

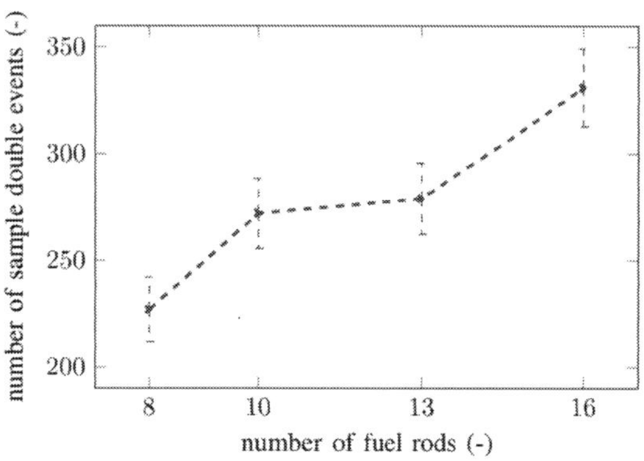

Fig. 8: Number of sample double events per 25 min as function of the number of fuel rods in the assembly.

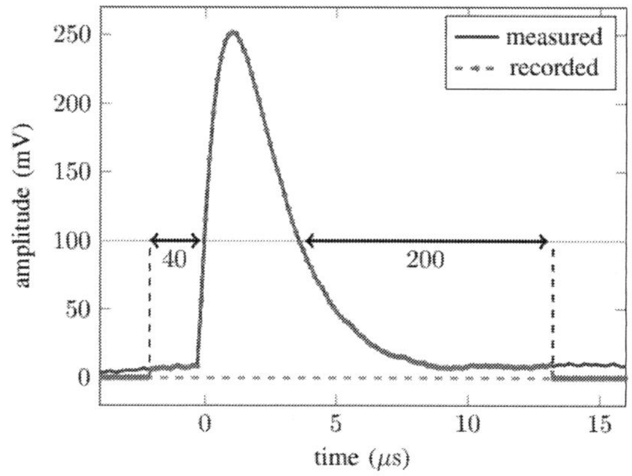

Fig. 9: Illustration of the triggering technique applied in the recording of the detector signals.

no useful information on the system. Therefore, a triggering procedure was implemented on the FPGA module of the AD converter to disregard the unnecessary parts of the signal. The procedure is illustrated on Fig. 9 and is explained in the following. An arbitrary threshold level is defined for the input analogue signal coming from the detector to the converter. By default, the digitized signal values are not written to files, instead they are collected in a memory buffer within the FPGA. When the signal goes above threshold, the values start being written to files from 40 samples earlier. When the signal goes below threshold and stays there for at least 200 samples, they stop being written to files. During analysis of the signals, these unrecorded segments were considered to be zero.

Using the described compression technique, a single 8 hour long measurement was performed with each of the three assemblies. The threshold values for the triggering were determined individually for each detector based on their amplitude spectrum, just like the one seen on Fig. 4.

The mean values of the registered signals of all three detectors are shown on Fig. 10. One immediately sees, that they show a decreasing tendency with increasing amounts of fuel, just as the source singles in Table III. This is not surprising if we recall that the mean value is proportional to the singles detection rate and – as shown in the previous subsection – singles events are dominated by the so called source singles. Table IV lists the singles rates (number of singles per 1 second) calculated with formula (2) from the mean values of detector A. To serve as a comparison with the simulated values shown in Table III, the number of singles per 25 minutes were also calculated. We see that the measured singles rates are in the same order as the simulated (and dominating) source singles.

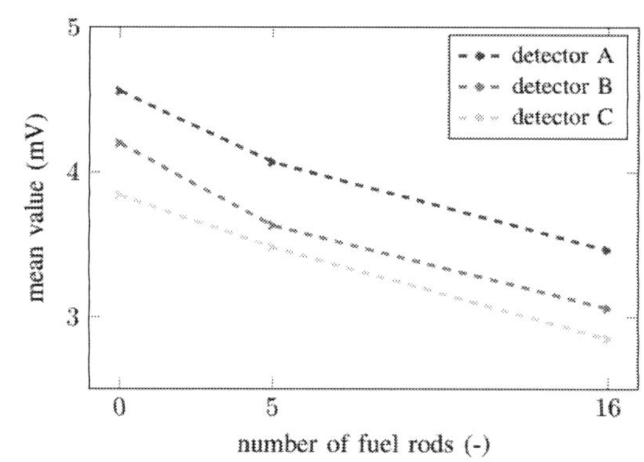

Fig. 10: The mean value of the recorded detector signals at different number of fuel rods in the assemblies.

The covariance function of the signals of detectors A and B are shown on Fig. 11. Regardless of the number of fuel rods in the assembly, the covariance function appears to be a

TABLE IV: Singles rates estimated from the mean value of detector A with different amounts of uranium in the fuel assembly.

# rods (-)	number of singles (-)	
	per 1 second	per 25 minutes
0	5425.5	8138206.7
5	4842.5	7263706.5
16	4116.7	6175042.8

constant zero function buried in noise, whose integral will also practically be zero, making it impossible to recover the doubles detection rate using formula (14). This is most likely a result of the low rate of doubles events, which are suppressed by the much more frequent singles events and by the background noise.

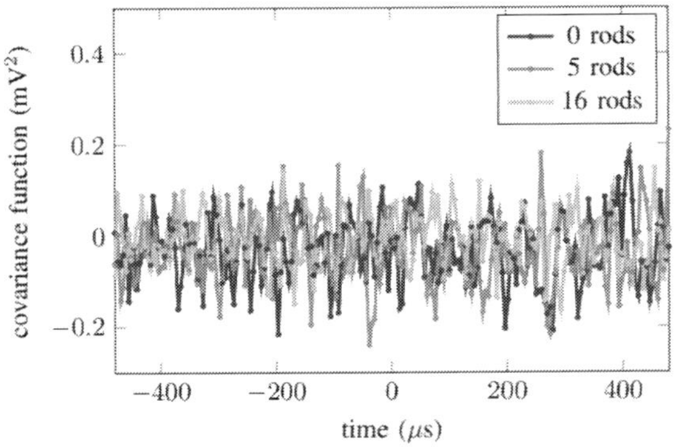

Fig. 11: The covariance function of the signals of detectors A and B at different number of fuel rods in the assemblies.

V. Conclusions and Future Plans

A new form of neutron multiplicity counting has been developed with the possibility of extracting traditional multiplicity count rates (namely the singles, doubles and triples rates) from the cumulants of fission chamber signals in current mode. It was shown that, at least in theory, by using two- and three-point statistics of the currents of one to three fission chambers, the detection rates can be recovered even in the case when the detection of neutrons of common origin does not take place simultaneously. The proposed method has the advantage that it does not suffer from the dead time problem, on the other hand it requires the knowledge of the detector pulse statistics (shape and amplitude distribution), which are though only the properties of the detector and can be determined from calibration. An experimental setup was designed and built to demonstrate the practical usability of the method. The properties of the detector pulse required by the theoretical model have been successfully determined. Monte Carlo simulations were performed to estimate the detection efficiency in the built setup. It was found that the probability of detecting a single neutron is much smaller than in a traditional multiplicity counter and as a consequence, the expected number of doubles and triples events is almost negligible compared to that of the singles events. Measurements were performed and the mean values as well as the covariance functions of the signals were estimated. Values for the singles rates were successfully extracted from the mean current, but the measured covariance functions were found to be zero, making it impossible to recover any doubles rates.

These latter results indicate that in order to effectively demonstrate the use of the proposed method of multiplicity counting, the detection efficiency should be significantly increased. There is two obvious choice to achieve this: one possibility is to use three groups of detectors where each group contains several small-sized detectors whose signals are unified; the other possibility is to use three large fission chambers. Coordination is in progress with partner institutes to realize such measurements.

References

[1] I. Pázsit, L. Pál, and L. Nagy, "Multiplicity counting from fission chamber signals in the current mode," *Nuclear Instruments and Methods in Physics Research Section A: Accelerators, Spectrometers, Detectors and Associated Equipment*, vol. 839, pp. 92–101, 2016.

[2] L. Nagy, I. Pázsit, and L. Pál, "Multiplicity counting from fission detector signals with time delay effects," *Nuclear Instruments and Methods in Physics Research Section A: Accelerators, Spectrometers, Detectors and Associated Equipment*, vol. 884, pp. 119–127, 2018.

[3] N. Ensslin, W. C. Harker, M. S. Krick, D. G. Langner, M. M. Pickrell, and J. E. Stewart, "Application guide to neutron multiplicity counting," *Los Alamos Report LA-13422-M*, 1998.

[4] I. Pázsit, A. Enqvist, and L. Pál, "A note on the multiplicity expressions in nuclear safeguards," *Nuclear Instruments and Methods in Physics Research Section A: Accelerators, Spectrometers, Detectors and Associated Equipment*, vol. **603**, pp. 541 – 544, 2009.

[5] L. Pál, I. Pázsit, and Z. Elter, "Comments on the stochastic characteristics of fission chamber signals," *Nuclear Instruments and Methods in Physics Research Section A: Accelerators, Spectrometers, Detectors and Associated Equipment*, vol. 763, no. 0, pp. 44–52, 2014.

[6] L. Nagy, I. Pázsit, and L. Pál, "Two- and three-point (in time) statistics of fi
ssion chamber signals for multiplicity counting with thermal neutrons," *Nuclear Instruments and Methods in Physics Research Section A: Accelerators, Spectrometers, Detectors and Associated Equipment*, vol. 929, pp. 148–155, 2019.

[7] M. Krick, N. Ensslin, D. Langner, M. Miller, R. Siebelist, J. Stewart, R. Ceo, P. May, and L. Collins Jr, "Active neutron multiplicity analysis and monte carlo calculations," Los Alamos National Lab., NM (United States), Tech. Rep., 1994.

[8] J. T. Goorley, M. R. James, T. E. Booth, F. B. Brown, J. S. Bull, L. J. Cox, J. W. Durkee Jr, J. S. Elson, M. L. Fensin, R. A. Forster III *et al.*, "Initial mcnp6 release overview," *Nuclear Technology*, vol. 180, pp. 298–315, 2012.

EPJ Web of Conferences **225**, 07012 (2020) https://doi.org/10.1051/epjconf/202022507012
ANIMMA 2019

Optimization of Neutron Detection Module based on Li-Glass scintillator and an array of SIPMs

Udi Wengrowicz*, Alon Osovizky, Aaron Ocherashvili, Dimitry Ginzburg, Yair Ifergan, Eli Volasky, Yagil Kadmon, Avi Raveh, Itzhak Orion
* NRCN and Ben-Gurion University of the Negev
udiwe@post.bgu.ac.il

I. Introduction

NEUTRONS do not interact directly with electrons in matter; therefore, the mechanisms for detecting neutrons are based on indirect methods through nuclear reactions in converter materials. Lithium-6 (^{6}Li) is considered among one of the most effective converters for thermal neutrons. In this work we investigated the performances of ^{6}Li-Glass scintillators coupled to state-of-the-art light sensors, arrays of Silicon Photomultipliers (SiPM) are presented. According to set-up models described here, PMT light sensors, used for decades, can be effectively replaced with SiPM sensors while maintaining and even improving the detection efficiency and reducing dimensions, weight and energy consumption.

Neutron detection is a key component of applications in national and homeland security, industry, and science. In the last two decades, the need for highly sensitivity neutron detectors for Homeland Security (HLS) applications has increased significantly. Therefore, novel technologies and devices are constantly being investigated. Specifically, neutron detectors are important because of their ability to detect small amounts of Special Nuclear Materials (SNMs) that could be used in hostile actions. SNMs are radioactive, and the radiation types that might be detectable a few meters from a hidden source are gamma rays and neutrons. Gamma-ray emissions are plentiful and span a wide range of energies, but they are difficult to distinguish from background gamma radiation and can be easily confused with medical, industrial or Naturally Occurring Radioactive Material sources. In contrast, neutron detection possesses two intrinsic benefits, a unique signature and long attenuation lengths, also in the gamma-absorbing shield. Therefore, the mere presence of neutrons above background levels is an excellent indicator of SNMs [1].

The essential requirements for all types of neutron detectors are high intrinsic efficiency, large sensitive area at a reasonable cost and effective rejection of gamma rays. For HLS applications, slow neutron detectors are generally used. In these instruments, the detection of neutrons is based on moderating (slowing down) their kinetic energy and then absorbing the neutrons with a target material called a converter. Desirable converter characteristics include high neutron absorption cross-section, minimal gamma-ray interactions and high Q-value. In reactions where all or most of the Q-value energy is delivered to heavy charged particles, there are several pulse processing methods that allow the discrimination of low-amplitude pulses related to gamma interactions and reliable measurement of pulses related to neutron absorption reactions.

Gas proportional detectors based on ^{3}He isotope possess most of the required qualities for slow neutron detection. Therefore, this technology has been for decades the traditional "gold standard" for neutron detection. By far, the most common source of ^{3}He is the decay of tritium. Starting in about 2001, consumption has risen rapidly, greatly surpassing production. Due to the extreme use of this isotope, the global reserves have been practically exhausted. The shortage of ^{3}He has triggered the search for an effective neutron detection alternative[2]. ^{6}Li is considered among one of the most effective converters for thermal neutrons, having cross-section of 940 barn via the reaction ^{6}Li (n, α) ^{3}H + 4.78 MeV, a capture reaction in which all the Q-value is released to charged particles. Hence, in this work, we investigated the use of ^{6}Li-Glass scintillators coupled to arrays of SiPMs for personal and mobile neutron detectors. Scintillation counters based on Li-Glass scintillators coupled to PMT are suitable for harsh environments. The newest developments based on SiPMs enable high gain and the low-light detection capabilities of the PMT while offering the benefits of a solid-state sensor: low-voltage operation, insensitivity to magnetic fields, reduced dimensions, mechanical robustness and excellent uniformity of response[3]. These traits make the SiPM of great interest for applications in portable radiation

***Index Terms* – Gamma rays, MCNP, Neutron detection, Numerical Simulations, Radiation Detectors, Scintillation counters.**

U. Wengrowicz is with the NRCN (1), P.O. Box 9001, Beer-Sheva, 84190, Israel. Now he is also with the Nuclear Eng. unit, Ben-Gurion University of the Negev(3), Beer-Sheva, 84105, Israel (e-mail: udiwe@post.bgu.ac.il).
A. Ocherashvili, Y. Ifergan and Y. Kadmon are with the NRCN (1), P.O. Box 9001, 84190, Beer-Sheva,
A. Osovizky, E. Volasky, D. Ginzburg and A. Raveh are with Rotem Industries Ltd.(2), Israel.
I. Orion is with the Nuclear Eng. unit, Ben-Gurion University of the Negev(3), Beer-Sheva, 84105, Israel (e-mail: iorion@bgu.ac.il).

https://doi.org/10.1051/epjconf/202022507012

detection instrumentation based on scintillation material. According to existing commercial technologies, the effective detection area of a single SiPM is less than 0.4 cm^2. Therefore, when working with large scintillators, the collection efficiency of the crystal output light decreases, which affects the sensitivity of the measured signal. To overcome these, in this study, we investigated the use of arrays of SiPMs with a combined sensing area that fits the scintillation dimensions. Models of neutron sensors consisting of Li-Glass Scintillators coupled to arrays of SiPMs are presented in this work. These models enable effective gamma-ray rejection and a neutron efficiency that comply and even exceed the requirements of the IEC 62534 standards[4] for portable neutron search detectors for homeland security applications.

II. METHODS

The detector developed in this work consists of a 50 mm × 50 mm Li-Glass scintillator of different thicknesses attached to an array of SiPMs. Neutron detection performances and gamma rejection evaluation of 1 mm and 2 mm thick scintillators, type GS20® made by Scintacor Ltd.[5], are presented. These scintillators consist of a Ce-activated lithium aluminosilicate glass with a lithium content of 6.6%, where the lithium is isotopically enriched to 95% ^{6}Li. Since only about 6000 photons are emitted per absorbed neutron, a photo-sensor with an effective light detection is required. To improve the scintillator light collection, we used arrays of the newest series of SiPM, type Sensl® ARRAYJ-60035-64P®. This device, that consists of an array of 8 × 8 SiPMs, 6 mm × 6 mm each, has an overall sensitivity equal to or even higher than the sensitivity of most of commercial PMTs. The array is assembled in a 51 mm × 51 mm board. That way, one face of this board totally covers one face of the scintillator; the other side of this board includes two 80-way connectors. Via a mating board-to-board connector, bias voltage is applied to all SiPMs, but the output of each SiPM can be measured separately. Since we need to measure the response of the scintillator as a whole unit all the 64 SiPM output signals were connected together to an electronic card that includes amplifier, voltage comparator and neutron threshold set-up. The module also includes a polyethylene moderator in front of the scintillator and a reflector in the back. The developed neutron sensor probe, presented in Figure 1, is more rigid than similar sensors coupled with PMTs, reducing the dimensions and power consumption and preserving and even improving the physical properties of the detector.

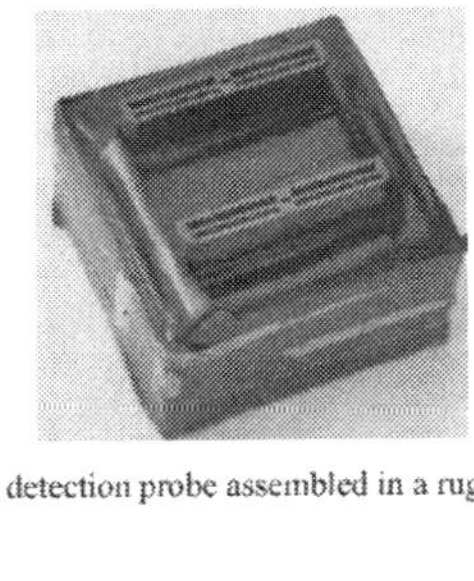

Figure 1: Neutron detection probe assembled in a rugged configuration.

III. SIMULATIONS AND MEASUREMENTS

For the detection and measurement of neutron sources, the scintillator composition, its thickness, the moderator materials and dimensions play an important role in the sensitivity of the instrument. By MCNP numerical simulations [6], we evaluated the sensitivity of the scintillators used in this work, Li-Glass type GS20®. The simulated detector consists of 5 × 5 cm^2 Li-Glass scintillator and 3 mm thick electronic cards including SiPM array and connection card. The simulations define also a reflector, 2 cm thick on each side of the scintillator and 4 cm thick behind the scintillator and the electronic cards. For a ^{252}Cf neutron source, the sensitivity of 1 mm and 2 mm thick scintillators was evaluated for different moderator thicknesses. The highest efficiencies were obtained with a 4 cm moderator. For thicker moderators, efficiency is reduced due to self-absorption.

The performances required for the hand-held neutron search detectors were defined by the IEC 62534 standard in the previous decade. Test methods defined by the standard require the detection within a period of 2 seconds of a 20000 n/s ^{252}Cf source placed 125 cm from the instrument. The specified maximum dimensions of the instrument are 300 mm × 200 mm × 150 mm, with a maximum allowed weight of 6 kg. To meet IEC 62534 required sensitivity, the instrument efficiency must be at least 25 cps/nv, and the average background rate must be less than 0.05 cps. Until the previous decade, it was possible to meet these requirements with a detector based on ^{3}He probes. Due to the scarcity of ^{3}He, in this work, we look for an alternative based on Li-Glass scintillators.

Using SiPM light sensors instead of PMTs, in the same detector dimensions, the volume of the moderator and the reflector can be increased, thus increasing the sensitivity. To achieve an optimal configuration, we performed MCNP simulations. Among the best results, it is worth mentioning a detector that includes a 24 cm × 19 cm × 9 cm polypropylene moderator. Behind the 4 cm moderator layer, there is a 200 mm × 150 mm, 1 mm thick Li-Glass scintillator. Arrays of SiPMs attached to one face of the scintillator, collect the output light producing a measurable voltage pulse. The described dimensions include also a 2 cm reflector layer on each side of the scintillator and a layer of 4 cm polypropylene behind the SiPMs. MCNP simulations show a sensitivity of approximately 46 cps/nv. Although in practice, in approximately 80% of the neutron interactions, the output pulse has sufficient amplitude to register as a neutron pulse, the described configuration and simulation results show a good alternative to meet IEC 62534 sensing performances instead of previous solutions based on ^{3}He tubes.

Measurement Results: The performances of the developed probe were evaluated with a 13000 n/s ^{252}Cf source. The source, located 30 cm in front of the detector, produced a flow of 1.15 n/s; maintaining the same measurement conditions we examined scintillators of 1 mm and 2 mm thickness. Measurement results show very low differences in their efficiencies: 2.4 cps/nv compared to 2.5 cps/nv for the 1 mm and 2 mm thick scintillators, respectively.

One of the most important parameters in the performance of a neutron search detector is its immunity to the gamma radiation. In Li-Glass scintillators, the gamma equivalent energy of the neutron photo-peak is about 1.6 MeV. Since the measured full width of this photo-peak is greater than 30%, there is the possibility that high-energy gamma rays are incorrectly measured as neutrons. it should be noted that in high-intensity gamma fields, there is pulse pile-up. Hence, some of these pulses are erroneously registered as neutron pulses. Since the effect increases with the scintillator thicknesses, in many applications, it may be preferable to use a relatively thin, 1 mm thick scintillator to improve the gamma radiation immunity and increase the volume of the moderator to improve the neutron sensitivity.

IV. CONCLUSIONS

Models of neutron detectors consisting of Li-Glass scintillators coupled to arrays of SiPMs and appropriate moderator configuration are presented. Measurements performed with these models show that Li-Glass scintillators coupled to arrays of SiPMs are an effective alternative to the PMT light sensors while maintaining and even improving the detection efficiency and reducing dimensions, weight and energy consumption. Neutron sensitivity measurements with a 1 mm thick scintillator were 10% lower than with a 2 mm thick scintillator, but the gamma rejection is better than twice that with the thin scintillator, which greatly improves the compatibility of the presented technology with the IEC 62534 standard requirements. According to measurements made with the presented neutron sensors and the results of MCNP simulations for a larger detector, using the described technology, it is possible to develop detectors that comply with the dimension and weight allowed by the IEC 62534 standard for these types of instruments, and as well, exceeding the efficiency requirements for neutron detection.

ACKNOWLEDGMENT

We would like to thank Rotem Industries Ltd., for their help in assembling the detectors and conducting experiments with their equipment and installations.

REFERENCES

[1] R. C. Runkle, A. Bernstein, and P. E. Vanier, "Securing special nuclear material: Recent advances in neutron detection and their role in nonproliferation," J. Appl. Phys., vol. 108, no. 11, 2010, Art. no. 111101, doi: 10.1063/1.3503495.

[2] R.T. Kouzes et al., "Neutron detection alternatives to ^{3}He for national security applications", Nucl. Instrum. Methods Phys. Res. A vol. 623, pp. 1035–1045, 2010, doi: 10.1016/j.nima.2010.08.021.

[3] "Introduction to SiPM Technical Note, Rev 6.0," SensL, Feb. 2017. Accessed: Mar. 2019. [Online]. Available: tps://www.sensl.com/downloads/ds/TN%20-%20Intro%20to%20SPM%20Tech.pdf

[4] Radiation Protection Instrumentation — Highly Sensitive Hand-held Instruments for Neutron Detection of Radioactive Material, IEC-62534, International Electrotechnical Commission, Geneva, June 2010.

[5] "6-Lithium glass bespoke to your application," Scintacor, Accessed: Mar. 14, 2019. [Online]. Available: https://scintacor.com/products/6-lithium-glass/

[6] J.F. Briesmeister, "MCNP — A Monte Carlo N-Particle Transport Code," Los Alamos Nat. Lab., NM, USA, , LA-13709-M, version 4C2, 2001.

https://doi.org/10.1051/epjconf/202022508001

On Line Measurement of Organic Iodine During a Severe Nuclear Accident

M. Chebbi[1], D. Doizi[1]*, L. Manceron[2], A. Perrin[3], J. Vander Auwera[4], F. Kwabia Tchana[5]

[1] DEN-Service d'Etude du Comportement des Radionucléides (SECR), CEA, Université Paris-Saclay, France
[2] Ligne AILES, Synchrotron SOLEIL, Gif sur-Yvette, France
[3] Laboratoire de Météorologie Dynamique/IPSL, UMR CNRS 8539, Ecole Polytechnique, Université Paris-Saclay, France
[4] Service de Chimie Quantique et Photophysique, Brussels, Belgium
[5] Laboratoire Interuniversitaire des Systèmes Atmosphériques (LISA), UMR CNRS 7583, Université Paris Est Créteil et Université Paris Diderot, Institut Pierre Simon Laplace, France
* denis.doizi@cea.fr

Abstract– **A severe nuclear accident may induce a dramatic dissemination of radioactive species into the environment. In that respect, improving the nuclear safety remains an important challenge to improve the society acceptability towards this energy. A solution may consist on implementing robust and reliable measurement systems operating near the Containment Venting Systems (CVS). These devices should be able to provide real time monitoring of the emitted fission products (FPs) in the course of a hypothetical accidental sequence. In the present study, a peculiar attention was devoted to iodine species (namely CH_3I) measurement by complementary techniques (photoacoustic spectroscopy and gas chromatography). The most important results will be described here.**

Index Terms—**severe accident, fission products, monitoring, methyl iodide, photoacoustic spectroscopy, gas chromatography.**

I. Introduction

After the Fukushima disaster, one of the main objective of the nuclear industry has been to evaluate mitigation strategies to limit the radiological consequences of a severe accident [1]. A significant lack of instrumentation was particularly highlighted during this accident. Indeed, dose rates were only measured twelve days after the event beginning. Hence, an improvement may consist on implementing reliable measurement systems, operating on line at the earliest and closest to the emission source of Fission Products (FPs). The real time monitoring of these FPs as well as their speciation may give valuable information on the core reactor state, resulting in more appropriate severe accident management procedures. Accordingly, the French DECA-PF ANR program was launched from 2013 with the aim to solve these issues.

For instance, the monitoring of radioiodine releases is of specific interest owing to the high radiological impact of its isotopes (namely ^{131}I), and its ability to exist in volatile forms (I_2 and CH_3I) [2]. Here, a peculiar attention will be devoted to CH_3I measurement by complementary techniques. The main objective is to develop specific sensors for the detection of methyl iodide traces. Our methodology is based on a combined approach. On the one hand, some promising sensors (namely micro-GC and miniaturized photoacoustic cells) will be tested. On the other hand, some innovated materials will be investigated for the selective and reversible capture of CH_3I. Those materials can be used in a pre-concentration stage before the sensor. Finally, some emphasis will also be attempted in order to extrapolate the behavior of these sensors in more representative conditions (temperature, presence of water and other contaminants…).

II. Gas Chromatography

II.1. Principle and optimization of parameters

A commercial compact device *(Agilent 490 μGC)* has been tested for the detection of methyl iodide. This instrument was already validated within the DECA-PF program, for the detection and quantification of carbon oxides (signature of MCCI occurrence [3]), as well as fission gases namely Xe and Kr, in representative conditions of severe nuclear accident. This apparatus is classically composed of three main parts (Fig. 1): (i) micro-injector (ii) set of four micro-columns and (iii) a micro detector of TCD type. The set of multiple columns affords the possibility to separate and analyze a large panel of gaseous compounds. Giving the polar nature of the investigated molecule, a polar column (*PoraPlot U* with a length of 10 m) was then selected in our case.

At first, preliminary tests were performed in order to optimize the detector response towards CH_3I detection. For instance, the column temperature was varied from 90 to 150°C. As stated from the resulted CH_3I chromatograms (C = 1000 ppm, Fig. 1), higher temperature leads to a better analysis efficiency in terms of both detection limits and time response.

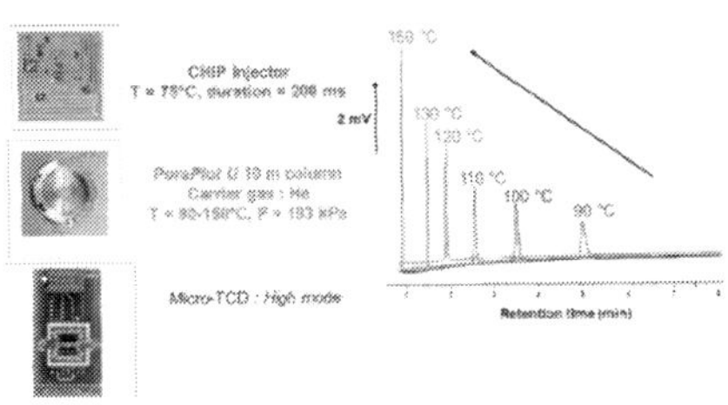

Figure 1: Evolution of CH_3I peaks as a function of column temperature: C_0 = 1000 ppm/N_2.

Thus, the column temperature was fixed at 150°C, in order to have good detection performances (retention time of about 1 min, high signal/noise ratio).

II.2. Detection Performances

The ability of the optimized µGC was then explored towards the detection of CH_3I traces. Reproducible concentrations were generated using a certified CH_3I/N_2 bottle (Air Products, 1000 ppm) and Bronkhorst mass-flow controllers (N_2 balance). A calibration curve was then validated in the range of 5 and 600 ppm (Fig. 2). Hence, a detection limit of 600 ppb was deduced.

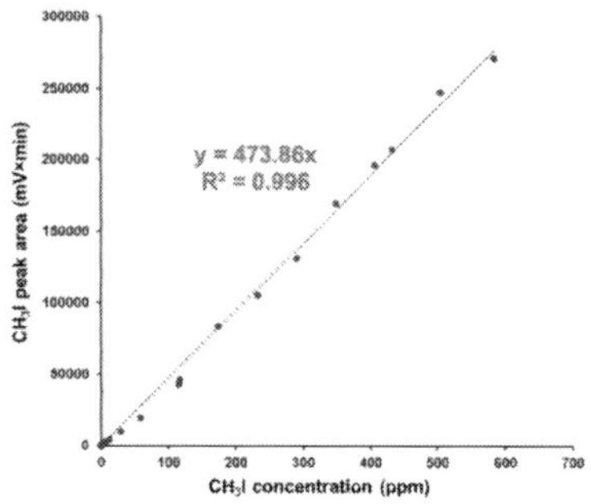

Figure 2: Calibration of CH_3I signals by Gas Chromatography (Agilent 490 µGC).

On the other hand, the ability of this instrument to detect CH_3I in the presence of some gaseous species potentially present in accidental conditions (CO_x, hydrocarbons, H_2 and Kr) has also been validated. It was found that CH_3I chromatograms were not affected by the presence of these contaminants.

It is worth recalling that CH_3I concentrations range from 1 to 100 ppb [4]. Thus, a pre-concentration filter stage is required to be implemented before this instrument in order to reach that level. Preliminary works were carried out in order to investigate the retention behavior of some innovated materials towards CH_3I capture. These works will be more detailed in a forthcoming study.

In the following section, the behavior of an example of absorption spectroscopic technique will be proposed for the first time to detect CH_3I.

III. PHOTOACOUSTIC SPECTROSCOPY

III.1. Principle

Photoacoustic spectroscopy (PA) was found to be one of the most powerful techniques to detect gas traces [5]. The principle of this spectroscopy consists on the excitation of the studied molecule with a light source (QCL: *Quantum Cascade Laser* as an example) emitting at the wavelength of an absorption line of the molecule. Furthermore, this light source should be modulated at the acoustic frequency of a resonant cavity containing the targeted gas. During the molecules relaxation, kinetic energy exchanges create local periodic variations of the temperature. Thus, an acoustic wave is generated and detected by microphones [6,7].

From the description of these techniques, it can inferred that analysis sensitivity towards a given molecule can be function of several parameters namely: power of the emitted light source, absorption coefficient, microphone sensitivity… [5].

In the case of the studied context, a peculiar attention should be devoted to the selection of the studied absorption line. This latter should be specific for CH_3I in order to avoid interference phenomena with species present in severe accident conditions (namely water vapor [4]). In addition, the finally selected absorption line should be sufficiently intense in order to reach the detection of methyl iodide traces (1-100 ppb) [4].

In that respect, preliminary FTIR spectroscopic investigations (experimental and theoretical) were performed within the DECA-PF project [8]. It was found that absorption line located at 885 cm^{-1} is of particular interest for CH_3I detection thanks to its specificity and its high intensity (cf. Fig. 3).

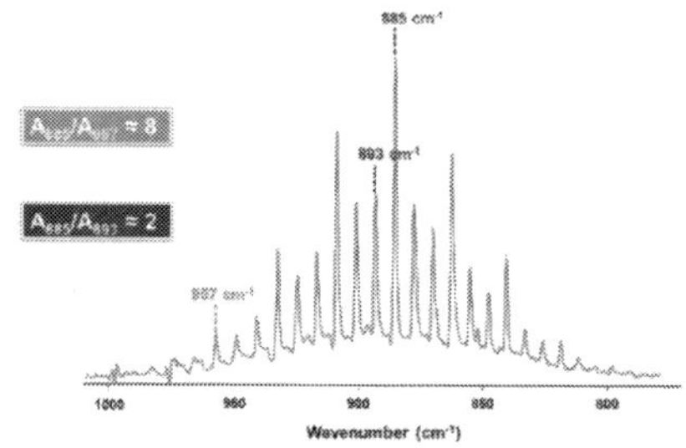

Figure 3: CH_3I FTIR spectrum in the region of interest (C=1000 ppm, Resolution of 0.5 cm^{-1}, 64 scans).

However, it was found that QCL technology associated to this specific wavelength is not available on the market. Hence, we will propose in the next section to discuss the obtained performances of this sensor using other absorption lines.

III.2. Results

Two complementary setups were tested in order to assess the PA spectroscopy ability towards the detection of CH_3I traces. On the one hand, a lab-scale configuration (Fig. 4 (A)) was used. It consists on a large setup composed by separate items (continuous CO_2 laser, photoacoustic cell, high sensitivity microphone, signal generator…). Despite the complexity and the bulkiness of this apparatus, a very promising detection limit of about 150 ppb was determined experimentally for CH_3I analysis. On the other hand, a commercial miniaturized detector (Fig. 4 (B)) was tested also for CH_3I monitoring. This second configuration seems to be more convenient for remote sensing fission products in the studied context. However, this centimeter sized instrument displayed very poor detection performances with a detection limit of 1% vol.

Several factors can explain these completely different results between the two tested configurations. The most important factor is the laser wavelength used in each apparatus. In the large setup, a continuous CO_2 laser with a high power was used (P about 200 mW). Furthermore, the selected absorption line was located at 893 cm^{-1}, and presented a rather high intensity ($A_{885}/A_{893} \approx 2$, Fig. 3). However, a pulsed QCL laser with a very low power (about 2 mW) was used in the second setup. In addition, the targeted absorption line displayed very low intensity (wavelength of 957 cm^{-1}, $A_{885}/A_{957} \approx 8$) as stated from Fig. 3. Other factors related to the performances of the photoacoustic cell as well as the light source focusing may also probably explain these differences.

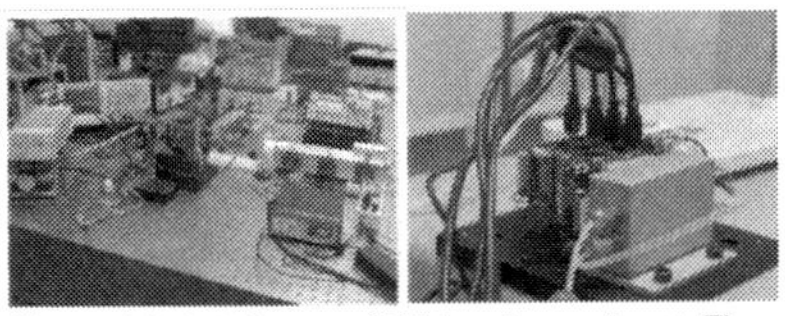

Figure 4: Photos of the tested setups: (A) lab-scale experiment, (B) commercial miniaturized sensor.

IV. Conclusions and perspectives

For the first time, the detection performances of different sensors were investigated towards the detection of methyl iodide traces. On the one hand, the tested μGC apparatus was found to display interesting results in terms of detection limit and also of its ability to detect CH_3I in presence of other compounds. Nevertheless, additional efforts have to be made in order to improve the detection limit using some specific pre-concentration filter stages for CH_3I. On the other hand, different results were obtained for PA spectroscopy, between the lab scale experiment and the commercial miniaturized instrument. Those differences were mainly explained by the nature of the light source. These preliminary studies have highlighted the necessity to perform additional tests at an intermediate scale before the ultimate miniaturization of the instrument. These findings are expected in the future to be important towards the enhancement of FPs monitoring in the context of severe nuclear accident.

Acknowledgment

The DECA-PF project, launched end of 2013 for 3 years, is sponsored by the French government "Investments for the future" program through the grant ANR-11-RSNR-0003 supervised by the French National Research Agency (ANR) under the "Research in Nuclear Safety and Radioprotection" (RSNR) research initiative.

References

[1] International Atomic Energy Agency, Nuclear Safety Review, IAEA/NSR, 2017.

[2] B. Clément, L. Cantrel, G. Ducros, F. Funke, L. Herranz, A. Rydl, G. Weber, C. Wren, *OCDE Report.*, NEA/CSNI/R(2007)/1, 2007.

[3] J.J. Foit, Nucl. Eng. Des. 170 (1997) 73-79.

[4] L.E. Herranz, T. Lind, K. Dieschbourg, E. Riera, S. Morandi, P. Rantanen, M. Chebbi, N. Losch, "State of the Art Report : Technical Bases for Experimentation on Source Term Mitigation Systems." Passam-Theor-T04 [D2.1], 2013.

[5] A. Miklós, P. Hess, and Z. Bozóki, Rev. Sci. Instrum. 72 (2001) 1937-1955.

[6] J. Rouxel et *al.* Procedia Engineering 120 (2015) 396 – 399.

[7] J. Rouxel et *al.* Sensors and Actuators B 236 (2016) 1104–1110.

[8] A. Perrin, I. Haykal, F. KwabiaTchana, L. Manceron, D. Doizi, G. Ducros, J. Mol. Spectrosc. 324 (2016) 28-35.

Using laser remote heating to simulate extreme thermal heat loads on nuclear fuels

T. Vidal*, L. Gallais, J. Faucheux, H. Capdevila, Y. Pontillon
* Aix Marseille University, CNRS, Centrale Marseille, Institute Fresnel, Marseille, France
vidal@fresnel.fr

***Abstract*— Up to now, predicting accurately the Fission Gas Release (FGR) from high burn up UO_2 and/or MOX (Mixed Oxide) fuels at off-normal conditions, such as power transient, reactivity-initiated accident (RIA) and loss-of-coolant accident (LOCA), is still a significant and very challenging task. For this purpose, different R&D programs have been carried out in France, as well as in other countries. This has been done with a specific emphasis on mechanisms which promote the FGR under accidental conditions. These studies can be performed thanks to dedicated integral experiments conducted in-pile (i.e. in Materials Testing Reactor) with the corresponding cost and constraints, or at the laboratory scale with annealing tests which allow to be representative of specific parameters (thermal history for instance). During these annealing tests under well-known conditions (temperature, atmosphere), both the absolute level and the time dependence of the released gases should be monitored, together with the corresponding fuel micro-structural changes, since experimental knowledge of fission gas release alone is not efficient enough. This approach requires more and more accurate on-line measurements. This corresponds to the driving force of the present work.**

In this contribution, we will present our progress in developing an experimental platform that can submit nuclear fuel and cladding samples to annealing tests involving very high temperatures (up to 2500°C) and very fast temperature ramp (up to thousands of °C/s) with controlled thermal gradients and temporal dynamics. This new platform implements innovative instrumentation, such as optical diagnostics to measure fuel fragmentation kinetics and infrared pyrometry for temperature monitoring. This experiment is based on a high-power laser (1.5kW) coupled to an experimental chamber with controlled atmosphere (Ar, N_2, or vacuum) and specific optical components.

Based on the spatial beam profile and temporal power function of the laser, it is possible which such a system to produce complex spatio-temporal temperature gradients, relevant for addressing different research needs. It provides access to extreme conditions that are very difficult to reach with other means. Particularly, one of main objectives of this work is to investigate conditions of Reactivity Initiated Accident (RIA).

The first experiments performed on inactive materials, non-irradiated uranium dioxide, is presented in order to highlight the capabilities of this technique.

Keywords*— *Nuclear fuel, Laser Heating, Simulation, Experimentation, High temperature, Instrumentation

I. Introduction

The understanding of the behaviour of nuclear materials in relation to the different thermal loads to which they can be subjected, i.e. from nominal operation through extreme conditions, is of paramount importance in the nuclear field. By improving this knowledge, it is possible to improve nuclear reactor safety and to precisely define specific on-site emergency plans in the case of an extreme event. Up to now, predicting correctly the fission gas (FG), fission product (FP) releases (i.e. the source term) and the corresponding fuel fragmentation from nuclear fuels at off-normal conditions, such as power transient, reactivity initiated accident (RIA) and loss-of-coolant accident (LOCA), remains a significant and important challenge. In the specific case of the so-called RIA [1], a fast rise in fuel power and temperature occurs. The power excursion may lead to failure of the nuclear fuel rods and release of radioactive material into the primary reactor coolant. In the more severe cases, the fuel rods may be shattered and large amount of the fuel pellet inventory could be dispersed into the primary coolant. To prevent those potential consequences, safety criteria are usually set up to limit the energy injection into the fuel. In such a situation the temperature gradient evolves from nominal situation to an almost flat top radial profile with a maximum localized in an outer ring closed to the pellet periphery [2]. Typical increase above 2000 °C occurs in few tens of millisecond on the edges of the fuel as described on the simulation reported in [2].

Historically, the worst reactivity-initiated accident took place on April 26 1986 in reactor 4 of the Chernobyl nuclear power plant in Ukraine. The Chernobyl accident reminded the nuclear community of the destructive potential of RIAs, and it prompted much research into the subject. In order to study the impact of this particular accident, in the early 1990s, experimental programs, in the form of pulse irradiation tests performed in Material Testing Reactor (MTR), were launched in France, Japan and Russia in order to check high burn-up fuel behaviour under RIA. These test programs were primarily intended to check the adequacy of regulatory acceptance criteria for RIA, which were mainly based on experiment on low or un-irradiated fuel. Therefore, up to now, integral experiments (i.e. MTR type) are most often the only solution for this category of R&D. This means that the potential number of experiments that can be performed is very limited both in terms of their cost and their complexity. To overcome these difficulties, the present reported work aims at exploring a concept to reproduce the so-called RIA (Reactivity Initiated Accident) thermal conditions to study the behaviour of nuclear fuel in such situations. The goal is to provide complementary results to those obtained under integral experiment such as the CIP program (Cabri

This work has been partly funded by EdF and Framatome under the CEA PJIPG's program. The authors would like to thank S. Gossé and A. Quaini from CEA Saclay for their help with temperature profile measurements.

T. Vidal and L. Gallais are with the Institut Fresnel, Aix Marseille Univ, CNRS, Centrale Marseille, France (e-mail: laurent.gallais@fresnel.fr).
J. Faucheux, H. Capdevila, Y. Pontillon are with the CEA DEN, Saint-Paul-Lez-Durance, France (e-mail: yves.pontillon@cea.fr).

International Program) which studies the capacity of Pressurized Water Reactor (PWR) fuel based on uranium dioxide (UO_2) and mixed oxide fuel (MOX) to withstand sudden power transients that may occur during an accident scenario [3].

II. Concept of Laser controlled RIA thermal transient experiment

A. Laser heating approach

In order to produce the described RIA thermal transients on nuclear fuel samples at the laboratory scale, we have explored the possibility to use a laser heating approach. Laser remote heating has many very interesting characteristics for such application: (1) it is a contactless technique which can easily drive materials to extreme temperatures with very high spatial precision and temporal control; (2) with focused beam application, the sample being tested essentially acts as its own containment vessel, avoiding any wall contaminants, and virtually without affecting the surroundings because only a small part of the material is being treated; (3) in combination with infrared imaging the measurement is less affected by thermal radiation that should occur from the containment vessel; (4) because of the possible optical access to the sample, laser heating can be combined to several non-contact diagnostics such as infrared imaging to derive meaningful thermo-physical properties avoiding interferences that would normally occur with direct contact measurements of temperature. In the field of nuclear fission research, there are several works dealing with the use of high power laser / material interactions for the investigation of the properties of fuel materials at high temperature, starting from the 80's [4,5]. Particularly, laser heating devices together with analysis diagnostics and temperature monitoring systems have been applied for the measurement of thermal properties (heat capacity, diffusivity, conductivity) [6] and thermodynamic properties (melting point, evaporation point, vapor pressure, equation of state) [7-15]. Combined laser heating with sample levitation and synchrotron x-rays has also allowed studying molten uranium dioxide structure at very high temperature (up to 3270K) [16]. There is clearly a high potential of using such techniques for fuel material research and we report in the following on our first steps in that direction.

B. The proposed concept

Nuclear fuels being highly absorbing materials to visible/IR wavelengths, the laser flux can be efficiently absorbed by the material. Based on the optical properties of UO_2 [17] laser absorption is localized only at the surface of the material and the laser heat source can only be used as an incoming heat flux on the sample surface. Experiments have then to be conceived accordingly. On the thermal point of view, considering a thermal diffusion length of $2\sqrt{Dt}$ with D the diffusivity of the material and t the time of applied laser heating, it occurs that with a diffusivity of $2x10^{-6}$ m^2s^{-1} [18], homogenous heating in the depth of the material at the time scale of few tens of millisecond, corresponding to RIA temporal dynamics, is only obtained if the width of the sample is sub-millimetric. Therefore, in order to conduct relevant laser heating experiments on fuel materials, the samples should be thin disks of less than 1 mm thickness cut from a fuel pellet if one want to reach homogenous heating in the depth. In such conditions, the proposed scenario to reproduce the RIA spatio-temporal thermal transient is described schematically in Fig. 1: Based on a section of a fuel cladding assembly, the sample is maintained in nominal temperature conditions which can be done through laser heating with flat-top or Gaussian intensity distribution and adapted insulation/cooling. Then annular laser beams, coming from both sides, can be applied to reproduce the temperature gradients encountered in RIA.

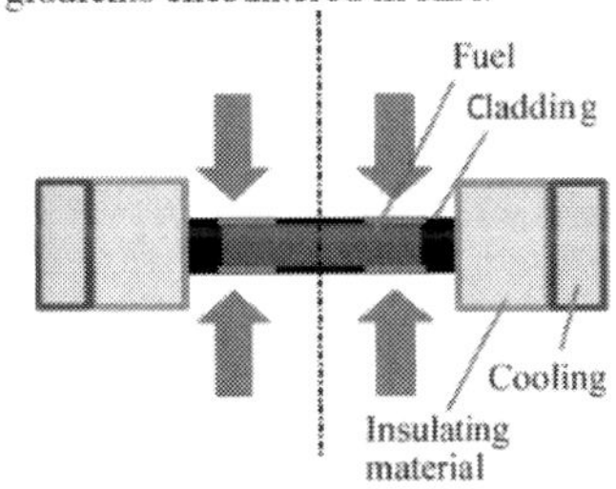

Fig. 1. Concept of laser simulated RIA conditions (The axis shows the axial symetry)

C. Required laser power

The dimensioning of such an experiment has been achieved by numerical simulations with COMSOL Multiphysics Software (Finite Elements Method). Depending on the required spatial gradients and temporal dynamics, the laser beam and temporal profile can be adjusted to reach the required conditions. With a typical thickness of the disk sample in the range of 0.5-1mm, few hundreds Watts of laser power for the transient annular laser beam heating are required to reach temperature close to 2000°C in few tens of milliseconds (Example Fig. 2).

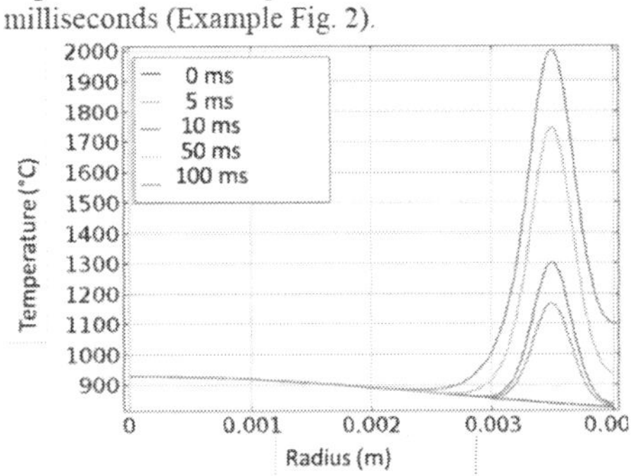

Fig. 2. Simulation of radial temperature distribution at the surface of a UO_2 disk submitted to 300W annular laser beam heating during 100ms (in the configuration shown on Fig. 1). The initial tempeature in this example is 900°C.

III. Experimental platform

A. Experimental configuration

Following previous description of the targeted experiment a high power laser delivering hundreds of Watts is required, with the ability to generate millisecond pulses. We have chosen to work with high power Ytterbium fiber lasers (wavelength of 1080nm) that can deliver monomode laser beams (high optical quality for annular beam generation) with kW output power and typical rising time of few microseconds.

We have designed a specific sample holder for the experiments, described on Fig. 3. The thin disk sample is placed at the center of an experimental vacuum chamber in a mechanical assembly that has been designed to hold the sample vertically, with insulating materials, zirconia and alumina, that can withstand high temperatures (up to 2000 °C). Sapphire optical windows are used to maintain the sample and potentially retain the fragments in place in case of loss of integrity, while simultaneously allowing optical access from both sides for laser beams and optical instrumentation.

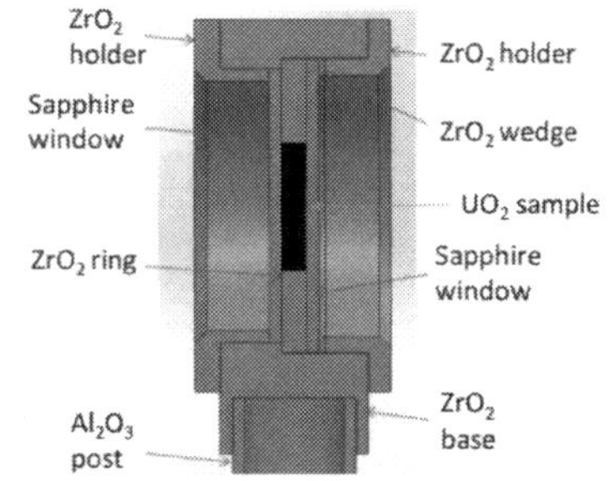

Fig. 3. Schematic of the sample holder assembly at the center of the vacuum chamber. A UO_2 disk, 8mm diameter and 1mm thick is visible in the sample holder.

B. Laser beams

On each side of the sample two different laser beams are used for (1) heating the sample at a nominal temperature corresponding to initial conditions before the transient and (2) generate a transient heat load with annular distribution on the edges of the sample. In order to generate the annular beam for controlled energy deposition on the edges of the sample, a classical lens and positive axicon combination has been used [19]. A 8mm beam diameter with a 150μm width is obtained with a 13mm incoming Gaussian beam (diameter at $1/e^2$), a 500mm focal length and a 1 degree axicon. The image of the obtained ring in such conditions is shown on Fig 4 as measured with a laser beam profiler placed in the sample plane. By defocusing the lens it is possible to adjust the width of the annular beam and accordingly modify the temperature gradients.

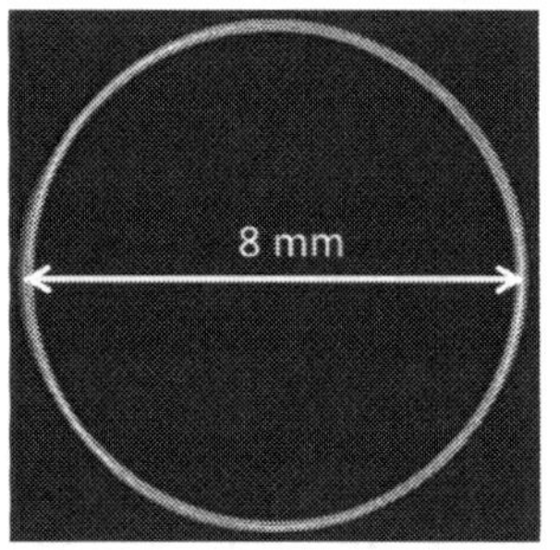

Fig. 4. Image of the beam profile at 150W in the sample plane obtained with a high power laser beam profiler.

C. Instrumentation

The optical instrumentation around the sample is a key point to monitor and control the temperature, therefore optical diagnostics such as thermal imaging with high speed cameras and optical pyrometry have be implemented through different viewports of the experimental chamber. A picture of the experimental configuration is given on Fig. 5.

Fig. 5. Picture of the experimental configuration. (a) Sample holder in the center of the opened experimental chamber ; (b) High speed MWIR thermal camera ; (c) Pyrometer ; (d) High power Laser collimator for transient heating ; (e) Optical systems for annular beam generation ; (f) long distance microscope ; (g) Medium power laser collimator for continuous heating ; (h) Observation system for beam alignement.

The instrumentation is based on:

a) *Optical access to the sample* with transparent parts of the sample holder (sapphire windows) and specific viewports in the experimental chamber (Anti-reflection coated silica, BaF2 or ZnSe windows).

b) *A High speed thermal camera* which operates in the 3-5μm spectral range, corresponding to the transparency range of the Sapphire windows. This system allows measurements up to 3000°C at 1kHz.

c) *A pyrometer* with fast response time (<1ms) operating in the 2-2.6μm spectral range. This system is used to monitor the surface temperature at the center of the sample, which is used in a PID regulation loop to control the laser power and ensure stable temperature.

d) *Laser control* with a dedicated software that allows generation of arbitrary temporal functions for the transient pulse heating.

e) *Beam shaping systems* to produce the annular beams. The beam profiles are measured at the sample location before the experiments. Knowledge of the beam profile is required for numerical simulations of the experiments.

f) *A long distance microscope* coupled with a visible range high speed camera (>1000 frames per second) is used to observe the sample surface in real time and provide additional information to thermal measurements (for instance micro structure evolution and fragmentation). This system has been specifically developed for UO_2 observation and is described in [20].

g) *PID regulation* of a second medium power laser, based on the temperature monitoring of the sample with the pyrometer, in order to ensure stable temperature conditions.

h) *An optical system* (CMOS camera + objective) is used to observed the positioning of the laser beam on the sample with respect to the sample. This alignment is indeed critical for the experiments

IV. Results

As a preliminary study, we have firstly worked on graphite samples to validate the experimental choices and compare experimental results to numerical ones. This part is not reported in this paper, we refer to [21]. In the following we describe the performances of the system to generate control thermal transients on UO_2 samples. The samples used in this work were synthesized in the UO_2 laboratory at the CEA Cadarache Center using standard general process: they were synthesized with depleted UO_2 from batch produced through wet process, followed by conventional pre-compaction (50 MPa), pressing (450 MPa) and sintering at 1650 °C for 2h under flowing H_2. The reported experiments have been done in vacuum environment with a base pressure of $2x10^{-2}$ mbar.

A. Initial conditions

The RIA experiments require to reach the initial temperature conditions before the transient heating, with an accurate monitoring and control of the sample temperature. This is achieved with the heating laser and PID feedback loop described before. The laser beam used for this purpose has a Gaussian spatial profile with 8 mm diameter (at $1/e^2$) at the sample location in order to cover the whole sample surface and ensure homogeneous heating. The sample is heated only from one side since due the low thickness (<1mm) and long irradiation time it is sufficient to obtain a homogeneous temperature in the sample. A stable temperature with variations below 1% was measured (see example Fig. 6), with heating ramps up to 100°C/s to reach the temperature plateau.

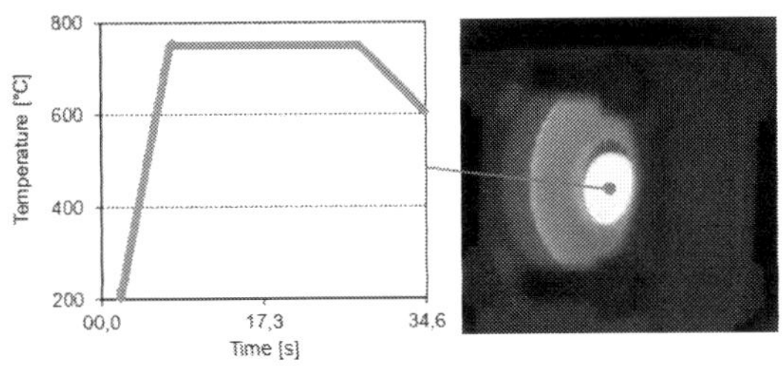

Fig. 6. Example of a regulated laser heating on a UO_2 sample. Left: measured temperature with the pyrometer at the sample surface in case of 100°C/s heating ramp followed by 20s plateau at 750°C. Right: Observation of the sample with the thermal camera during the temperature plateau.

B. Transient heating

Once the initial temperature is reached, the two laser beams with annular spatial profiles can be launched. In the reported experiments the two beams are coming from the same laser source that is divided in two beam paths with a 50% beam splitter. This ensures synchronization of the 2 beams and power balance. The temperature distribution however was only measured on one side with the thermal camera. We report on Fig. 7 on typical experimental results for illustration of the concept. This experiment has been done with a 20 ms pulse transient and 600W incident power (300W on each face).

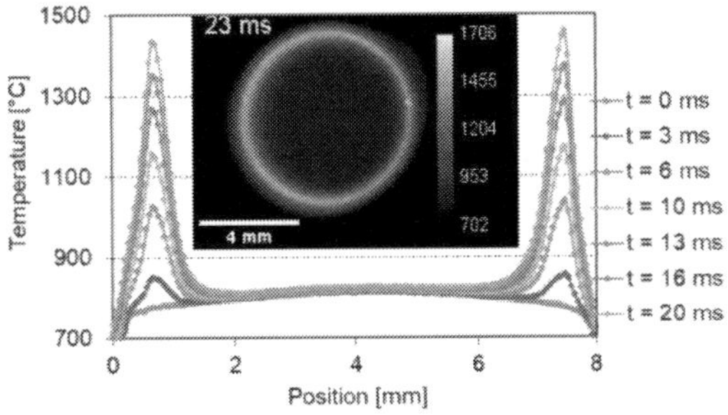

Fig. 7. Measured temperature at the surface of a UO_2 sample submitted to 300W annular beam irradiation during 20ms on both sides. Image: thermal camera snapshot at the end of the laser pulse ; Data plot: extraction of the temperature along vertical profile of the thermal images at different times.

The temperature distribution during the transient has the expected radial symmetry, as illustrated by the good correspondence of the extracted temperature profiles, with peak temperatures corresponding to the required ones to reach RIA conditions. It illustrates the fact that in the proposed experimental conditions, the temperature gradients can be well controlled by the proposed laser technique.

In the different set of experiments that were conducted, fracturation of the samples was however observed in some cases, as shown on the Scanning Electron Microscope images reported in Fig. 8. Fragmentation of edges of the sample was also observed in some extreme cases (transient temperature > 2200 °C). The experiments have then also some interest to study thermo-mechanical behaviour of the materials. Thermo-mechanical simulations conducted with the ALCYONE code indicate that under such laser exposure conditions, sample rupture should occur in agreement with observations.

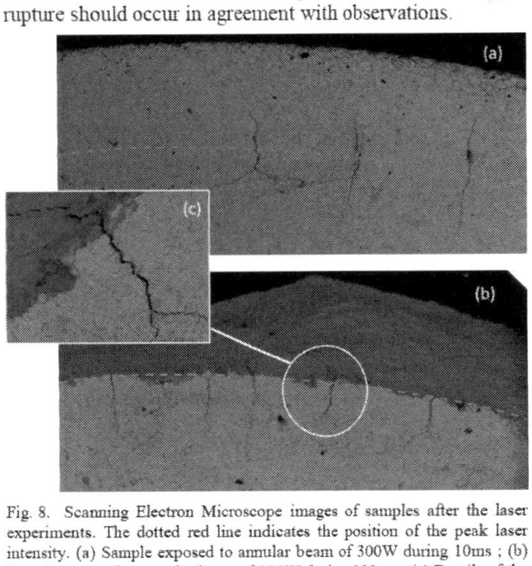

Fig. 8. Scanning Electron Microscope images of samples after the laser experiments. The dotted red line indicates the position of the peak laser intensity. (a) Sample exposed to annular beam of 300W during 10ms ; (b) Sample exposed to annular beam of 300W during 100ms ; (c) Details of the observed fracture.

V. Conclusion and perspectives

A concept based on the use of high power lasers has been proposed for the generation of temperature transients on nuclear fuels corresponding to Reactivity Initiated Accident conditions. A pilot platform has been developed for the validation of this concept, then as the first phase of hot-cell integration. This pilot platform will be used for the proposal of devices and experimental programs that can be integrated into the hot laboratories at the CEA Cadarache to perform experiments that can meet emerging research needs. Its successful completion will permit, with some further developments, a full real time analysis of the irradiated nuclear fuel behaviour in case of a major mishap, including the release of radioactive gases and particles at high temperature. This will provide crucial experimental data for physical model validation implemented in fuel performance codes.

(1) Acknowledgment

This work has been partly funded by EdF and Framatome under the CEA PJIPG's program.

The authors would like to thank S. Gossé and A. Quaini from CEA Saclay for their help with temperature profile measurements.

(2) References

[1] F. Schmitz and J. Papin, "High burnup effects on fuel behavior under accident conditions: the tests CABRI REP-Na," Journal of Nuclear Materials, vol. 270, pp. 55, 1999.

[2] J I. Guenot-Delahaie, J. Sercombe, T. Helfer, P. Goldbronn, E. Federici, T. Le Jolu, A. Parrot, C. Delafoy and C. Bernaudat, "Simulation of reactivity-initiated accident transients on UO_2-M5 fuelrods with ALCYONE V1.4 fuel performance code," Nuclear Engineering and Technology, vol. 50, pp 268, 2018.

[3] J. Papin, B. Cazalis, J.M. Frizonnet, J. Desquines, F. Lemoine, V. Georgenthum, F. Lamare and M. Petit, "Summary and interpretation of the CABRI REP-Na program," Nuclear Technology, vol. 157, pp 230, 2007.

[4] R.W. Ohse, J.F. Babelot, G.D. Brumme, and P.R. Kinsman, 'Extension of vapor-pressure measurements of nuclear oxide fuels on UO_2 and $(U,Pu)O_2$ for fast-reactor safety analysis by laser techniques up to 5000K', Revue internationale des hautes temperatures et des réfractaires, vol. 15, pp 319, 1978.

[5] S.R. Yagnik, D.R. Olander, 'Surface temperature transients from pulsed laser heating of UO_2', Journal of Nuclear Materials, vol. 154, pp 253, 1988.

[6] C. Ronchi, M. Sheindlin, M. Musella, G. J. Hyland, 'Thermal onductivity of uranium dioxide up to 2900 K from simultaneous measurement of the heat capacity and thermal diffusivity', Journal of Applied Physics, vol. 85, pp 776, 1999.

[7] C. Ronchi, M. Sheindlin, "Laser-Pulse Melting of Nuclear Refractory Ceramics," Int. J. of Thermophysics, vol; 23, pp 293, 2002.

[8] R. Pflieger, J.-Y. Colle, I. Iosilevskiy, M. Sheindlin, "Urania vapor composition at very high temperatures," Journal of Applied Physics vol. 109, pp 033501, 2011

[9] F. De Bruycker, K. Boboridis, P. Pöml, R. Eloirdi, R.J.M. Konings, D. Manara, "The melting behaviour of plutonium dioxide: A laser-heating study," Journal of Nuclear Materials, vol. 416, pp 166, 2011.

[10] D. Manara, C. Ronchi, M. Sheindlin, "Pressure Dependence of UO_2 Melting Measured by Double-Pulse Laser Heating," International Journal of Thermophysics, vol. 23, pp 47, 2002.

[11] M. Joseph, N. Sivakumar, P. Manoravi, "Studies on equation of state of high temperature nuclear materials," Annals of Nuclear Energy, vol. 31, pp 1163, 2004.

[12] P. D. W. Bottomley, Th. Wiss, A. Janssen, B. Cremer, H. Thiele, D. Manara, M. Scheindlin, M. Murray-Farthing, P. Lajarge, M. Menna, D. Bouexière, V. V. Rondinella, "Characterisation of high temperature refractory ceramics for nuclear applications," Materials Science and Engineering, vol. 32, pp 012003, 2012.

[13] D. Manara, R. Böhler, K. Boboridis, L. Capriotti, A. Quaini, L. Luzzi, F. De Bruycker, C. Guéneaud, N. Dupine, R. Konings, "The melting behaviour of oxide nuclear fuels: effects of the oxygen potential studied by laser heating," Procedia Chemistry, vol.7, pp 505, 2012.

[14] F. Cappia, R. Jovani-Abril, J. Spino, L. Luzzi, A. Janßen, D. Manara, "Laser melting of nano-crystalline uranium dioxide," Progress in Nuclear Energy, vol. 72, pp 11-16, 2013.

[15] D. Prieur, F. Lebreton, M. Caisso, P.M. Martin, A.C. Scheinost, T. Delahaye, D. Manara, "Melting behaviour of americium-doped uranium dioxide," J. Chem. Thermodynamics, vol. 97, pp 244, 2016.

[16] L. B. Skinne, C. J. Benmore, J. K. R. Weber, M. A. Williamson, A. Tamalonis, A. Hebden, T. Wiencek, O. L. G. Alderman, M. Guthrie, L. Leibowitz, J. B. Parise, "Molten uranium dioxide structure and dynamics," Science, vol. 345, pp 984, 2014.

[17] M.Bober, J.Singer, K.Wagner, "Determination of the optical constants of molten nuclear fuels," Journal of Nuclear Materials, vol. 124, pp 120, 1984.

[18] L. Vlahovic, D. Staicu, A. Küst, R.J.M. Konings, "Thermal diffusivity of UO2up to the melting point," Journal of Nuclear materials, vol. 499, pp 504, 2018.

[19] M. Rioux, R. Tremblay,P. A. Belanger, "Linear, annular, and radial focusing with axicons and applications to laser machining," Applied Optics, vol. 17, pp 1532, 1978.

[20] T. Vidal, L. Gallais, R. Burla, F. Martin, H. Capdevila, S. Clément, Y. Pontillon, "Optical system for real-time monitoring of nuclear fuel pellets at high temperature," Nuclear Engineering and Design, in press.

[21] T. Vidal, L. Gallais, J. Faucheux, H. Capdevila, J. Sercombe, Y. Pontillon, "Simulation of Reactivity Initiated Accident thermal transients on nuclear fueld with laser remote heating," submitted.

https://doi.org/10.1051/epjconf/202022508003

Study of online measurements techniques of metallic phase spatial distribution into a corium pool

D. Tisseur*, M. Cavaro, F. Rey, K. Paumel, N. Chikhi, J. Delacroix, P. Fouquart, R. Le Tellier, V. Bouyer
* CEA Cadarache
david.tisseur@cea.fr

Abstract— **In the context of in-vessel retention (IVR) strategy in order to better assess the risk of reactor vessel failure, the knowledge related to the kinetics of immiscible liquid phases stratification phenomenon needs to be further improved. So far, only one medium-scale experiment (MASCA-RCW, in the frame of the OECD MASCA program) gives direct information regarding the transient relocation of metal below the oxide phase through post-mortem measurements. No experimental characterization of the stratification inversion kinetics when heavy metal becomes lighter and relocates at the top exists. Further investigation of these hydrodynamic and thermochemical processes could be made possible thanks to on line instrumentation enabling to follow displacement of oxidic and metallic phases into the corium pool. At CEA Cadarache, studies are under progress to set up innovative technologies for corium stratification monitoring which would be integrated to a cold crucible induction melting furnace. Based on space and time resolution specifications, three on-line measurements techniques were selected and studied. The first one is an ultrasonic technique using a refractory material waveguide and based on a time-of-flight measurement. We present the feasibility approach with the preliminary results obtained during experiments at high temperature on VITI facility. The second method consists in electromagnetic characterization of the corium pool thanks to an excitation by a magnetic field induced by surroundings coils and measurement of magnetic response by sensors placed around the crucible. A modelling study has enabled to define an appropriate experimental configuration. An experimental set up has also been tested to verify the calculation results. The third technique is 2D X-rays imaging. A feasibility study for a real-time X-ray imaging with a framerate of 1 image/s has been performed using home-made simulation software MODHERATO, accounting for scattering, based on corium behavior previsions. Results on the detection of interfaces between different type of corium phases (oxide, light metal, heavy metal) are shown.**

Index Terms— **PLINIUS-2, Severe Accidents, Corium, Ultrasonic testing, Electromagnetic testing, X-ray imaging**

I. Introduction

In case of a nuclear severe accident, when core meltdown occurs, the main objective is to limit the consequences on environment and populations, in terms of contamination and radiation. One of the strategies of severe accident management is the in vessel retention (IVR) strategy which aims at minimizing the risk of vessel failure by reflooding the reactor pit and whose likelihood of success is determined by the assessment of the heat flux from the corium pool to the vessel. This is why there is an interest in the knowledge related to the kinetics of immiscible liquid phases stratification phenomenon.

Today, the understanding and modelling of this kinetic only lies on the conclusions and timescales deduced from post-test examination of MASCA-RCW test [Asmolov, 2003], a medium-scale experiment giving limited information regarding the transient relocation of metal below the oxide phase. There is no experiment regarding the stratification "inversion" kinetics, when addition of metal leads to a light metallic layer on top of the oxide pool.

In the framework of PLINIUS 2 CEA project dealing with the building of a large corium facility [Journeau, 2017], the development of an experimental device in order to measure the kinetic of stratification of the different liquids layers is under consideration. The set-up is based on the induction melting technique in a cold crucible in order to perform medium-scale experiments with a ~50 kg corium pool as in MASCA-RCW. The key feature of the experiments is the online measurement of the metallic phase spatial distribution with the following criteria: (1) measuring continuous phases axial thicknesses with spatial resolution of about 1 mm and time resolution of about 1 min; (2) quantifying a dispersed phase (i.e. droplets) distribution of about 1cm and time resolution of about 1s. Three techniques have been selected for this measurement:

- an ultrasonic technique using a refractory material waveguide and based on a time-of-flight measurement;
- an electromagnetic characterization of the corium pool thanks to an excitation by a magnetic field induced by surroundings coils and measurement of magnetic response by sensors placed around the crucible;
- 2D X-rays imaging of the corium pool.

All authors work at CEA, DEN, Cadarache, F-13108 Saint-Paul-lez-Durance, France.

 https://doi.org/10.1051/epjconf/202022508003

We present the feasibility study of each method, their specifications and their limitation regarding the objective of the experiments in

section II, III and IV. Section V will finally present the conclusion and perspectives of this work.

II. Ultrasonic testing

A. Principle

In order to monitor online the evolution of a stratified corium, measurements of ultrasonic pulse time-of-flight are investigated for the localization of the upper light metal / oxide interface. An ultrasonic pulse is emitted by a transducer. By crossing an interface, part of the energy of this pulse is reflected (see **Fig. 1**). It is then possible to deduce the position D of the interface thanks to the time-of-flight (*tof*) of this reflected pulse and the acoustic velocity v in the propagation medium thanks to the relation : $tof = \frac{2D}{v}$.

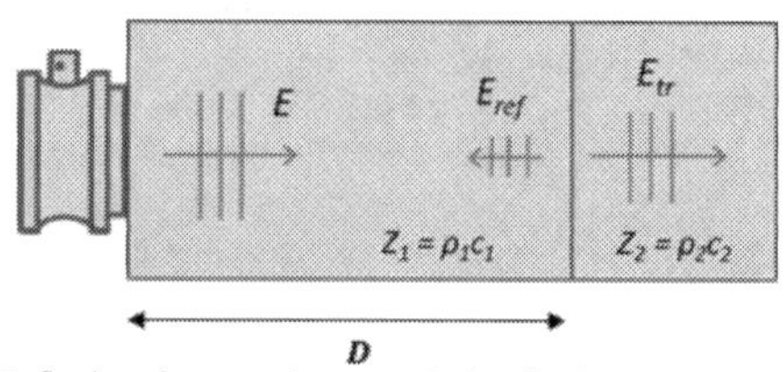

Fig. 1. Reflection of an acoustic wave at the interface between two materials.

In our case, the implementation of this technique requires:

- the knowledge – or an *in situ* measurement – of the speed of sound in the upper layer,
- an appropriate waveguide able to ensure the transmission of ultrasound in the molten metal (T ~ 2000°C),
- an appropriate ultrasonic transducer. Even if not directly immerged in molten metal, this should be a high temperature one (see **Fig. 2**).

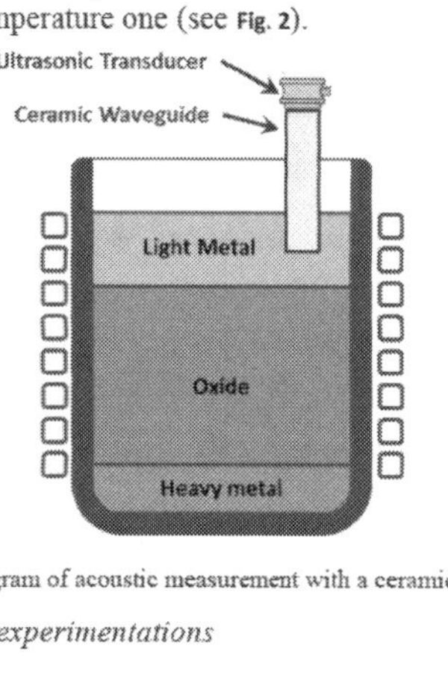

Fig. 2. Principle diagram of acoustic measurement with a ceramic waveguide.

B. Preliminary experimentations

In order to rule on the feasibility of the method and to confirm the simulation results, experiments were conducted on the VITI facility (see **Fig. 3**). Based on induction heating, VITI allows the melt of samples up to 3000 ° C. We performed tests in liquid steel with zirconia and hafnia waveguides. We used a Framatome® TUCSS high temperature transducer (able to operate at 200 °C).

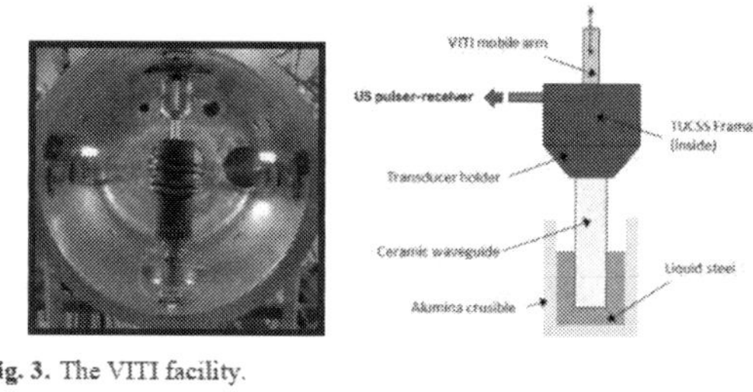

Fig. 3. The VITI facility.

These experiments allowed the demonstration of zirconia and hafnia acoustical ability in liquid steel. We also managed to perform speed of sound measurements in liquid steel between 1500 and 1850°C. The sensor-waveguide assembly is mounted on a motorized micrometer plate. We have shown that the tested acoustic technique allows a very precise positioning by tracking the echo returned by the liquid steel interface - alumina crucible (see **Fig. 4**).

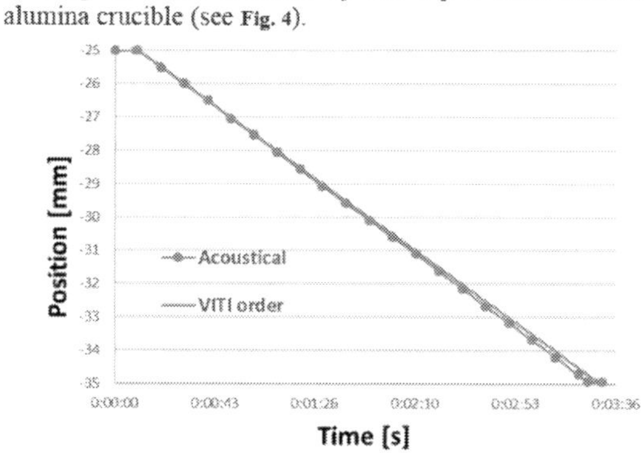

Fig. 4. Positioning by acoustic monitoring versus VITI order.

C. Prospects

The preliminary experiments carried out allowed the demonstration of the feasibility of the ultrasonic pulse time of flight measurement for the localization of an interface through a molten metal. Further tests will be necessary to understand phenomena that we have observed on the surface of ceramic waveguides. An application of the technique on real stratified corium is now under study. Solutions will be studied to be able to work in metal mixtures of unknown acoustic celerity.

III. Electromagnetic testing

Even if materials in corium are at a temperature above their Curie point, they can be discriminated by the way of an electromagnetically mean. They differ by their electrical conductivity that can be exploited for monitoring their nature and position in the bath. By the way of time varying electromagnetic induction, currents can be induced in the melted corium. Induced currents modify surrounding magnetic field that can be used to determine nature of materials and their geometrical distribution in space.

Finite element analysis is used to numerically test various configurations of excitation coils arrangements around the furnace before doing some experiments on relevant mock-up. The melted corium was sliced in relevant areas (light and heavy metal, oxide) (see **Fig. 5**).

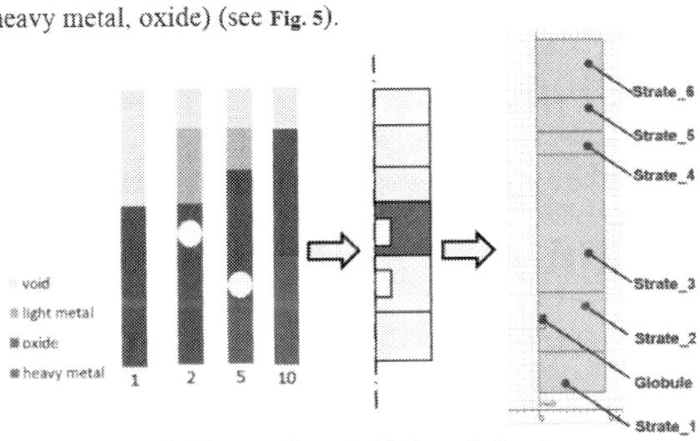

Fig. 5. Relevant volumes inside the melted corium

One of the challenges of this measurement is that the magnetic field must go into the corium through the surrounding cold crucible induction-melting furnace. Therefore, modelling has to take into account the furnace structure (see **Fig. 6**).

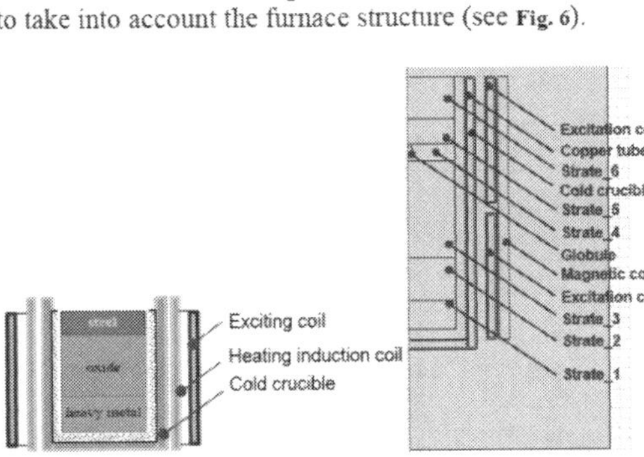

Fig. 6. Modeling of cold crucible induction furnace

Some of the several configurations that were tested, are promising (see **Fig. 7**). Major goal is to obtain a sufficient magnetic induction flux penetrating everywhere inside the melted corium.

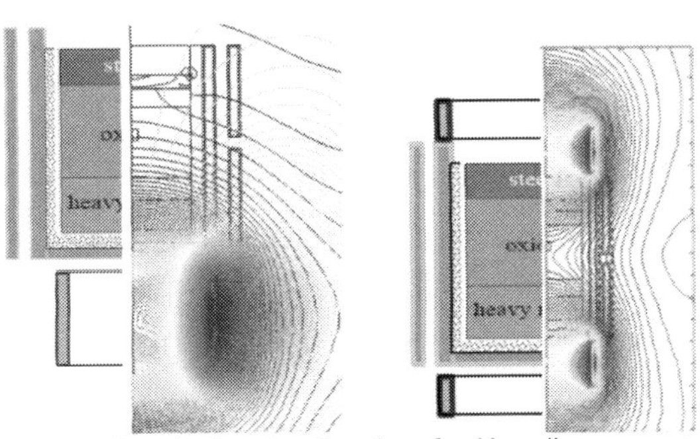

Fig. 7. Example of tested configurations of exciting coils arrangement (located only at the bottom and both at bottom and top)

Results of simulations let appear some perspectives in the possibility to locate materials of significantly to slightly different electromagnetic properties (electrical conductivity and magnetic permeability). For example, when stratum 5 is chosen with a magnetic permeability different from that of the others, results show local inflexion of magnetic induction measurement around the furnace (see **Fig. 8**).

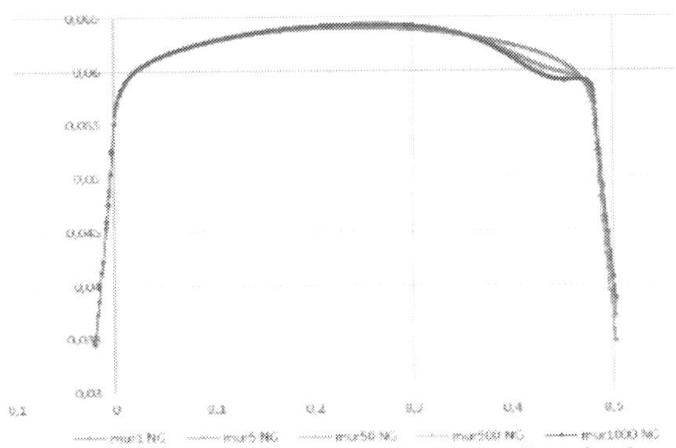

Fig. 8. Numerical measurments on a ficitve line of sensors disposed on the side of the furnace (along a vertical axis)

The simulations indicate the possibility to do the measurements but also confirm that the furnace structure acts as a very powerful shield against propagation of magnetic induction. The challenge is to obtain a sufficient induction everywhere in the melted corium in order to receive an exploitable distortion of inducting field. Next step will be to go on simulation and to experiment the measurement system on a relevant mock-up.

IV. High energy X-ray imaging

High energy X-ray imaging is investigated for the study of corium stratification. The experiment consists in observing the evolution of different corium phases inside a cold crucible. The feasibility study for a real-time X-ray imaging with a framerate of 1 image/s consists in determining the maximum thickness of the self crucible of the cold crucible, allowing the detection of the oxide/heavy metal interface. MODHERATO simulation code, developed by CEA/LMN for high-energy imaging applications in radiography and tomography, has been used with the last developments including scattering contribution on the radiography image simulation [Berge, 2018]. Test section is composed of a vessel, an inductor, a cold crucible, a self crucible and the corium bed. Fig. 9 and **Fig. 10** give a geometric description of the test section.

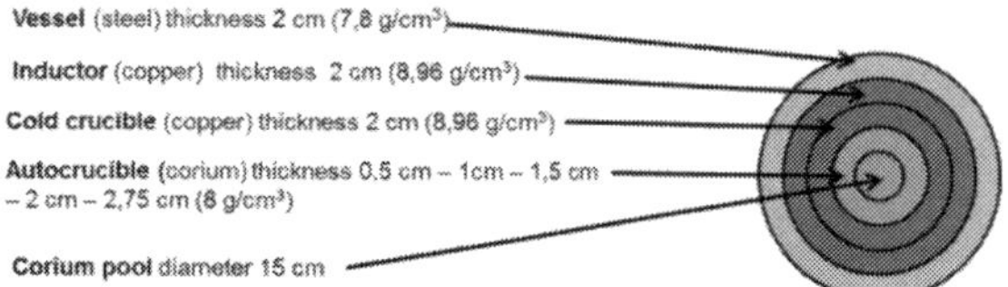

Fig. 9. Diagram of the test section: top view

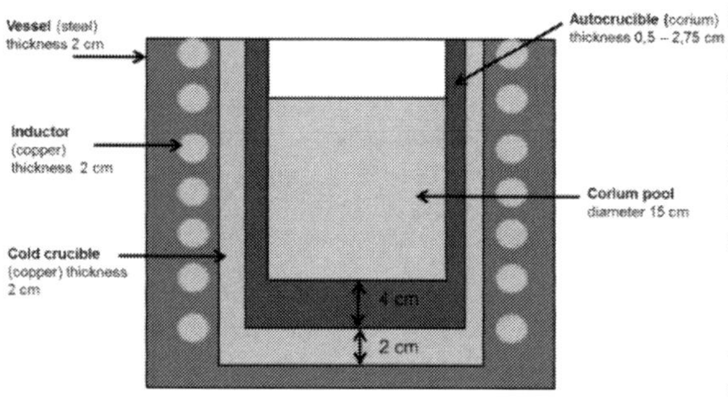

Fig. 10. Diagram of the test section: side view

The corium bed cylinder is decomposed into 33 slices of different chemical composition and density. The time evolution of these 33 slices is discretized into 17 configurations describing a temporal stratification process. Each slice has its own volumic mass and chemical composition. Fig. 11 shows image simulated with MODHERATO of using a 15 MeV LINAC (with a 4 µs pulse duration@250 Hz with an intensity of 100 mA), and the same detector as on KROTOS [Estre, 2016] (Gd_2O_2S scintillator coupled to a camera, pixel size 0.86 mm). Distance from high energy X-ray source to the detector is 1.3 m, distance from source to the center of the test section is 0.86 m.
The interfaces between different corium phases (oxide, light metal, heavy metal) are detectable, as shown on Fig. 11.

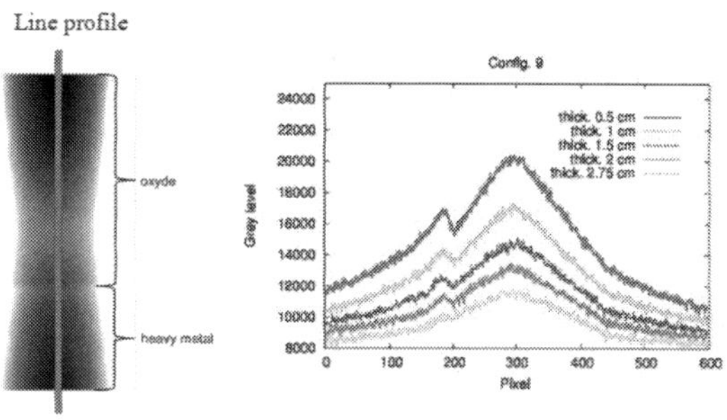

Fig. 11. On the left, simulated radiography of the corium bed with an oxide/heavy metal interface (configuration 9). On the right, comparison of vertical profiles of corium cylinder for different tickness of the autocrucible. The small peak corresponds to the interface visible on simulated radiography.

Simulations show that the contrast between the different layers (oxide, heavy metal and light metal) is sufficient to identify them if the thickness of the autocrucible does not exceed 1 cm (considering a cylinder of liquid corium diameter 15 cm thick, a 2 cm thick copper crucible, a 2 cm thick copper inductor and a 2 cm thick steel vessel).

V. CONCLUSION

The preliminary experiments carried out with ultrasonic testing allowed the demonstration of the feasibility of the ultrasonic pulse time of flight measurement for the localization of an interface through a molten metal. Simulations with electromagnetism testing indicate the possibility to do the stratification measurements. Work is under progress to go further on a relevant mockup. High energy X-ray imaging simulations show that stratification layers can be identify with a 15 MeV linear accelerator. Radiography imaging on a real mockup will allow confirming simulations results.

ACKNOWLEDGEMENTS

This work has been done in a tripartite agreement between CEA, EDF and FRAMATOME.

REFERENCES

1. C. Journeau, "*Severe accident research at the PLINIUS prototypic platform*" in ICAPP'05., Seoul., Korea, 2005, pp. 1074
2. M. Bonnet *et al.*, "*KROTOS FCI Programme at CEA Cadarache*",. ICAPP-05, Seoul, KOREA, May 15-19, 2005, Paper 5381
3. N. Estre *et al.*, "*Fast Megavoltage X-Rays Radioscopy*", proc. of IEEE-NSS/MIC, Strasbourg, 2016
4. S.-W. Hong et al., "*Status of the OECD-SERENA Project for the Resolution of Ex-vessel Steam Explosion Risks*", Journal of Energy and Power Engineering 7 (2013), pp 423-431
5. Ch. Journeau *et al.*, "*Corium-sodium and corium-water fuel-coolant-interaction experimental programs for the PLINIUS-2 prototypic corium platform*", NURETH-17, Xi'an, China, Sept. 3-8, 2017 (to be published)
6. J. Goorley *et al.*, "*Initial MCNP6 Release Overview - MCNP6 version 1.0*", LA-UR-13-22934 NEA report, 2013
7. M.J. Berger and J.H. Hubbell, XCOM: Photon cross-section database: https://www.nist.gov/pml/xcom-photon-cross-sections-database
8. V.G. Asmolov, S.S. Abalin, Yu.A. Veselkin, V.Yu. Vishnevsky, V.V. Vlasov, B.L.Gershman, Yu.G. Degaltsev, Ye.K. Dyakov, J.F. Isaev, A.N. Kiselev, N.P. Kiselev, A.M.Kovalev, A.G. Ol'khovsky, K.V. Pechalin, LM. Semenov, V.F. Strizhov, T.V.Trushkina, V.S. Uglov, Yu.M. Utkin, V.V.Chudanov, A.E.Aksenova, V.A.Pervichko & L.M. Khazanivich, "RCW Post-Test Analysis Results", Technical Report MP-TR-11, Russian Research Centre Kurchatov Institute, 2003.
9. L. Berge *et al.*, "*Fast High-Energy X-Ray Imaging for Severe Accidents Experiments on the Future PLINIUS-2 Platform,*" in IEEE Transactions on Nuclear Science, vol. 65, no. 9, pp. 2573-2581, Sept. 2018, doi: 0.1109/TNS.2018.2847460
10. N. Estre *et al.*, "*Fast Megavoltage X-Rays Radioscopy*", proc. of IEEE-NSS/MIC, Strasbourg, France, 2016

Remote monitoring of Molten Core-Concrete Interaction experiment with Optical Fibre Sensors & perspectives to improve nuclear safety – DISCOMS project

L. Maurin*, P. Ferdinand, V. Bouyer, A. Denoix, G. Jouvin, S. Rougeault, C. Journeau, D. Molina, P. Tena and Y. Ouerdane
* CEA
Laurent.MAURIN@cea.fr

Abstract—**The DISCOMS project (Distributed Sensing for Corium Monitoring and Safety) aimed at providing innovative solutions not requiring local electrical power supplies, for remote monitoring of a severe nuclear accident. The solutions are based on both long length SPNDs (Self Powered Neutron Detectors) and on distributed OFSs (Optical Fibre Sensors) capable to detect the onset of a severe accident, the corium pouring on the containment building concrete basemat, and its interaction with the concrete floor under the reactor vessel, until it spreads in the core catcher (EPR case). This paper mainly focuses on these last three detection targets achievable with distributed OFSs. It is based on the results of a Molten Core & Concrete Interaction (MCCI) experiment, namely VULCANO, held in June 2018 with a concrete crucible equipped with overall ~ 180 m long optical fibre sensing cables. This small scale experiment (50 kg of prototypical corium) has demonstrated the ability of distributed OFSs to remotely provide useful data during the MCCI run: *i)* temperature profiles images up to about 580°C (single wavelength Raman DTS reflectometer) until cooling down to room temperature, *ii)* high spatial-resolution frequency shifts profiles, due to combined (non-selective) strain and temperature influences (Rayleigh OFDR and Brillouin reflectometers), and *iii)* cables lengths ablated by the corium on sections weakened by the temperature (Raman DTS, Rayleigh OFDR, telecom and photon counting reflectometers).**

Index Terms—**NPP (Nuclear Power Plant), nuclear safety, severe nuclear accident, corium, concrete basemat, OFS (Optical Fibre Sensor), reflectometry, distributed measurement, remote monitoring, power-supply free sensors, optical sensing cables, MCCI (Molten Core-Concrete Interaction) experiment, advices.**

I. Introduction

In March 2011, an 9.1 magnitude earthquake struck off the north-east coast of Japan, and caused a powerful tsunami, with waves greater than 30 m height in some places. A wave with 15 m height reached the Fukushima-Daiichi NPP (Nuclear Power Plant) built between 6.5 m and 10 m above the sea level.

The first consequence of this seawater submersion for the NPP was the loss of external electrical power supplies and internal means of cooling the core of the 4 nuclear reactors on the sea front. This led to a severe nuclear accident with core meltdown, ranked at level 7 on the INES scale [1].

Left with almost no information coming from the reactor pits, TEPCO was unable to monitor in real-time the accident status and its progress, whose accident management operations were also complicated by the massive destructions on the site and its surroundings.

These major drawbacks led the French national authorities to organize a few months later a parliamentary session devoted to nuclear safety in order to evaluate the mitigation strategies currently in place for Gen II and Gen III (EPR) nuclear reactors, and planned for future Gen IV [2]. This work further led to a new and dedicated research program, the RSNR [3], to stimulate and fund new R&D topics in nuclear safety, with four main objectives: *i)* analyse the conditions that previously led to nuclear accidents, including the Fukushima-Daiichi event, taking into account the first conclusions of the Complementary Safety Assessments that followed [4], *ii)* analyse the severe accident management procedures to be considered by licensees and public authorities, for all major nuclear accidents to date, *iii)* analyse their impact, with regard to radioactive material releases, on human health and on the environment, and *iv)* enhance nuclear facilities robustness to extreme situations.

In this framework, the DISCOMS (Distributed Sensing for Corium Monitoring and Safety) project [5] aimed at providing additional *ex-core* remote monitoring solutions (free of local electrical power supply) able: *i)* to monitor the nuclear reactor during at least 60 years of normal operation, and to detect the

Acknowledgments—This work (reference number: ANR-11-RSNR-0007), was carried out within the framework of the RSNR (research on nuclear safety and radiation protection) research program launched after the Fukushima-Daiichi accident, and co-funded by the French Programme d'Investissements d'Avenir (PIA) and managed by the ANR (French National Research Agency).

L. Maurin, G. Jouvin and S. Rougeault are with Institut LIST, CEA, Université Paris-Saclay, F-91120 Palaiseau, France (e-mail: laurent.maurin@cea.fr).

V. Bouyer, A. Denoix, C. Journeau, D. Molina and P. Tena are with the CEA, DEN, Cadarache, DTN, SMTA, LEAG, F-13108 Saint-Paul-lez-Durance, France (e-mail: viviane.bouyer@cea.fr).

Y. Ouerdane is with Univ Lyon, UJM, CNRS, IOGS, Laboratoire Hubert Curien, UMR 5516, 18 rue Prof. B. Lauras, F-42000 Saint-Étienne, France (e-mail: ouerdane@univ-st-etienne.fr).

P. Ferdinand, previously with Institut LIST, is now with LMP S.A., 49-51 rue du Moulin des Prés, F-75013 Paris, France (e-mail: p-ferdinand-lmpsa@sfr.fr).

onset of any nuclear accident involving the reactor core and its vessel, and *ii)* to monitor such accident in order to facilitate the management of the situation by the licensee and the safety authorities, by providing information in real-time about the status of the vessel breakthrough, as well as the corium pouring on the concrete basemat, its progression through the concrete floor, and the potentially failure of the third containment barrier (leading to corium spreading in the core catcher for the EPR).

These innovative means rely on passive sensors, namely long lengths (up to 6 m long) ex-core SPNDs (Self Powered Neutron Detectors) [6] and distributed OFSs (Optical Fibre Sensors): SPNDs would be installed in the reactor pit whereas OFSs would be located at several depths in the concrete floor, and below the core catcher (EPR). These sensors could then be monitored in real-time from a remote location (from hundreds of metres to several km) where operators could work safely.

Such additional instrumentations would facilitate the situation management, thanks to decisions based also on real measurements coming from the reactor pit rather than on predictive models results taking for inputs the available measurements. This is particularly true for the concrete floor (last containment barrier), which does not benefit until now (at least for the French fleet of reactors) from any embedded sensor, but only a few thermocouples at its surface, which will be destroyed once the corium would have poured on them.

On the contrary, a few hundreds of metres of optical sensing cables embedded in the concrete floor would benefit from its radiation and thermal protections, and would withstand several additional hours during which more appropriate mitigation strategies could be implemented.

Equally important, public authorities' decision to evacuate or not the population, with regard to contamination risks, will be based on the real information coming from the NPP, otherwise provided by conservative simulation results computed during the accident. As a reminder, the global cost for the society of a severe nuclear accident which could occur in France has been estimated to 120 G€, with 11 G€ related to contaminated territories, depending on the scenario [7].

II. Monitoring the concrete floor with OFSs in the reactor pit does make sense

A. Modelling of radiation conditions in the reactor pit for Gen II and Gen III nuclear reactors

Numerical simulations permitted to assess the irradiation conditions in the reactor pit *versus* the concrete floor depth, for both Gen II and Gen III nuclear reactors, in terms of kerma (kinetic energy released per unit mass) gamma & fast neutrons ($E > 1$ MeV) fluence [8] [9].

For both reactors generations, results demonstrate that at any depth in the concrete floor (> 10 cm), the kerma gamma after 60 years of normal operation is greater than effects of the first 24 h of a severe accident. In addition, the kerma is drastically reduced with the floor depth, due to the concrete absorption.

Thus, *e.g.* at 50 cm below the surface, the fast neutrons fluence drops from 10^{16} n/cm^2 to 2×10^{14} n/cm^2 for Gen II, and from 10^{14} n/cm^2 to 10^{12} n/cm^2 for Gen III, while the kerma gamma drops from 5.5 MGy to 700 kGy for Gen II and from 2 MGy to 22 kGy for Gen III after 24 h of a severe accident occurring after 60 years of normal operation, these values being conservative. Concrete floor shielding can therefore be advantageously used to protect optical fibres from excessive damages caused by ionizing radiations.

B. Selection and qualification of radiation hardened optical fibres to be used in optical sensing cables

In parallel, several commercially available radiations resistant singlemode optical fibres were screen-tested, first as standalone fibres, then in sensing cables in order to determine the RIA (Radiation Induced Attenuation) generated by the cable compounds (polymers) in the optical fibre core. The irradiation tests performed within the industrial gamma-irradiator POSÉÏDON, and the research nuclear reactor ISIS, both located at CEA Saclay, have permitted to select rad-hard optical fibres that exhibit a maximum RIA < 50 dB/km at both 1310 nm and 1550 nm operating wavelengths, for cumulated gamma-dose close to 2 MGy [10].

It has also been proved that under irradiations, the cable polymer compounds release molecular hydrogen H_2, which migrates towards the optical fibre core, and is finally responsible for significant additional RIA, without highlighting any tendency to saturation. One solution, qualified during this project, is to use an optical fibre with a protective carbon layer at its surface, thus acting as a barrier against hydrogen [10].

C. OFSs in the reactor pit to remotely monitor the corium and its progression throughout the concrete floor

We allocate a maximum optical budget of 5 dB to the monitored parts, based on performances of today's optical fibre sensing systems. Combined with experimental RIA results *versus* cumulated dose, the modelling assesses that it makes sense to embed sections of 100 m of optical fibre(s) sensing cables at several depths in the concrete floor reactor pit. Minimum embedment depth of 30 cm and 5 cm below the top surface, for Gen II and Gen III respectively, will ensure correct measurements of cable lengths, strains & temperature profiles, as well as temperature changes from a reference trace [10].

The distributed sensing measurement principles rely on the analysis of the backscattered light by dedicated reflectometers as described in Table I. The useful information provided are profiles with typically 50 cm to 1 m raw spatial sampling over several kilometres for both Raman DTS and Brillouin instruments, with a measurement frequency ranging from one to several minutes (depending on the setting parameters). The Rayleigh OFDR technique provides better spatial sampling measurements and rates, typically from one to several centimetres, over much shorter sensor lengths (Table I).

Cable lengths are provided by reflectometers (telecom or photon counting OTDRs) through the detection of the Fresnel reflexion peak at the fibre tip, even if some distributed sensing reflectometers can also provide this information.

The main advantage of photon counting over traditional OTDR is the shorter raw spatial sampling: typically, 1 cm *versus* 1 m. For both techniques, measurement ranges exceed

20 km, with ~ 1 minute measurement time (depending on both setting parameters and the expected signal to noise ratio).

These five C-band singlemode measurement techniques can therefore be operated far away from the reactor pit (where operators can work safely), on the condition that the total RIA does not exceed the optical budgets. Instruments can be swapped from one cable to another one in case of failure.

III. Optical Fibre Sensing Cable Tests and Selection for the VULCANO Experiment

Three different kinds of optical fibre sensing cables, devoted to strain sensing, temperature sensing and security, were supplied on the market (Table II). Their compliance with EDF PMUC standard [11] was required in the call for tenders.

They were first qualified in temperature between 4°C and 85°C to reflect normal operation, then at higher temperatures (up to 750°C) to characterize their sustainability to corium vicinity during a severe nuclear accident. The strain cables were tested in elongation to characterize their strain transduction.

A. Tests between 4°C and 85°C – Normal operation scenario

For each cable, 10 m-sections were spliced together, then tested in a climatic chamber (between 4°C and 85°C) while their Rayleigh frequency shifts profiles were recorded for each temperature step. It could be noticed that both temperature and safety cables exhibited irregular profiles, and also more pronounced side effects at their ends, in comparison with strain sensing cables (Fig. 1). This is due to a higher temperature sensitivity in the latter case, the optical fibre being glued along its length in a small diameter metallic tubing (Fig. 6).

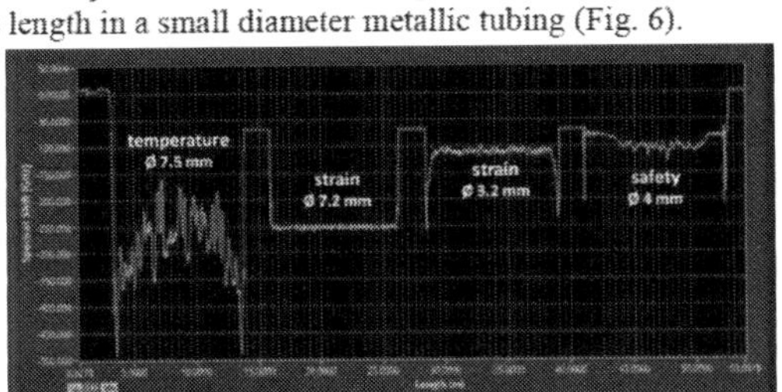

Fig. 1. Typical Rayleigh OFDR frequency shifts profiles between 4°C and 58°C for the 4 optical sensing cables parts spliced together.

Ø 7.2 mm-diameter strain and temperature sensing cables provide the highest temperature sensitivities due, for the latter, to a significant friction between the fibre in its plastic tubing with the gel (Table II). But the temperature cable is not the optimal choice, even if the optical fibre thermal expansion is expected to be driven by the concrete if cables are embedded.

Therefore, with more homogenous profiles at constant temperature, the Ø 7.2 mm strain sensing cable is the better choice for distributed temperature sensing.

B. Tests up to 750°C – Severe accident scenario

The cable sustainability in corium vicinity is of major importance for the severe accident monitoring in order to get measurements coming from the reactor pit as long as possible. It can also be noticed that the security cable, devoted to the fire detection market, has been especially designed to sustain 750°C

TABLE I
Optical Fibre Distributed Sensing During VULCANO Experiment

Scattering effect and Technique	Type of analysis and typical performances	Measurement principle
Raman DTS incoherent OFDR z scaling: set of beat frequencies + FFT	ratio of anti-Stokes *vs.* Stokes intensities I_{AS}/I_S • 1 m sampling • > 5 km range • minute to a few minutes rate • 1°C resolution	anti-Stokes band intensity I_{AS} is temperature-dependent, while Stokes band intensity I_S is practically free of temperature influence – local temperature T is derived from I_{AS}/I_S, the ratio of the anti-Stokes and Stokes band intensities.
$\frac{I_{AS}}{I_S}=\frac{C_{AS}}{C_S}\left(\frac{\lambda_S}{\lambda_{AS}}\right)^4.\exp\left(-\frac{h\Delta\nu_R}{k_BT}\right).\exp\left(-\int_0^z(\alpha_{AS}(u)-\alpha_S(u))du\right)$		
Brillouin B-OFDA z scaling: frequency domain multiplexing	Brillouin frequency shift $\Delta\nu_{Brillouin}$ • 1 m sampling • > 10 km range • minute to a few minutes rate • a few 10 µm/m and °C resolutions	Brillouin frequency $\nu_{Brillouin}$ in silica [10 GHz 12 GHz] depends on local strain ε and on local temperature T – local Brillouin frequency shift $\Delta\nu_{Brillouin}$ is related to both local temperature ΔT and longitudinal strain $\Delta\varepsilon$ changes.
$\nu_{Brillouin}=2.n_c.\nu_0.\frac{V_A}{c}$ and $\Delta\nu_{Brillouin}=C_T^B\,\Delta T+C_\varepsilon^B\,\Delta\varepsilon$		
Rayleigh OFDR z scaling: frequency domain multiplexing	Rayleigh frequency shift $\Delta\nu_{Rayleigh}$ • 1 cm sampling in sensing mode • 70 m max range in standard, up to 2 km over 70 m in extended mode • half-minute rate • 1 µm/m and 0.1°C resolutions	the state-of-polarisation of the Rayleigh backscattered light is analysed, and cross-correlated with a reference profile – local Rayleigh frequency shift $\Delta\nu_{Rayleigh}$ is related to both local temperature ΔT and longitudinal strain $\Delta\varepsilon$ changes.
$\Delta\nu_{Rayleigh}=C_T^R\,\Delta T+C_\varepsilon^R\,\Delta\varepsilon$		

Raman DTS: k_B is the Boltzmann's constant, h the Planck's constant, $C_{S(AS)}$ are defined by the Raman cross-section for the Stokes (anti-Stokes) lines and by the band-pass filters used to select them, $\Delta\nu_R$ is the Raman frequency shift (13.2 THz for silica), T is the temperature in Kelvin, $\alpha_{S(AS)}$ are the Stokes (anti-Stokes) attenuation coefficients including all fibre losses. $\lambda_S-\lambda_0=\lambda_0-\lambda_{AS}\cong 100$ nm with probe wavelength $\lambda_0=1550$ nm.

Brillouin: V_A is the speed of sound in silica (5.96 km/s), c the speed of light in vacuum, $n_c\cong 1.47$ the core refractive index, $C_T^B\cong 950$ kHz/°C and $C_\varepsilon^B\cong 50.55$ kHz/(µm/m) with probe wavelength $\lambda_0=1550$ nm.

Rayleigh OFDR: $C_T^R\cong -1.25$ GHz/°C and $C_\varepsilon^R\cong -0.150$ GHz/(µm/m) with probe wavelength $\lambda_0=1550$ nm.

during at least 90 minutes according to IEC 60331-25 standard in terms of optical waveguide continuity [12]. However, unlike the standard test which must be performed with a flame, the cables were placed in a metallic pipe, and the pipe itself in a tubular oven. The temperature profile along the tube was controlled thanks to 4 type-K thermocouples (Fig. 2), and it could be assumed that at least 40 cm of cable length were

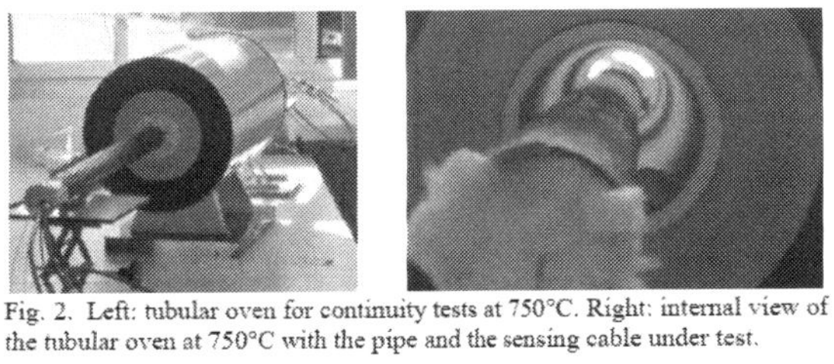

Fig. 2. Left: tubular oven for continuity tests at 750°C. Right: internal view of the tubular oven at 750°C with the pipe and the sensing cable under test.

systematically heated over the 750°C target during the tests.

Results are summarized in Table II. The optical continuity was controlled with a telecom OTDR.

As expected, the security cable passed the tests, even up to 2h30min at 750°C, but its sheath had burnt, and only the tight buffered parts were partially observable (Fig. 3). Indeed, the cable, nor the fibre, did not exhibit any significant residual mechanical strength after this test.

Fig. 3. Typical aspect of the security cable after more than two hours at 750°C.

The most surprising result concerned the temperature cable which also passed the test, although it was not specified for such a purpose by its manufacturer. Its polymer sheath has disappeared in fumes during the test, and only the very fragile bare optical fibre (without its polymer coating) remained. Nevertheless, this result should be considered with caution: in case of an embedded cable, the optical fibre would have probably been broken due to the mechanical stresses driven by the concrete thermal expansion.

The last interesting result is the early break of the optical fibre in the two strain sensing cables, which was systematically observed between 400°C and 450°C for both cables. This can be explained by the conversion of their thermal expansion $\alpha_{fibre}\Delta T$ into mechanical strain $\Delta\varepsilon_{fibre}$ according to:

$$\Delta\varepsilon_{fibre} \cong (\alpha_{cable} - \alpha_{fibre}) \times \Delta T$$

With $\alpha_{cable} - \alpha_{fibre} \cong 50 \times 10^{-6}\,°C^{-1}$, it leads to an equivalent mechanical elongation of the optical fibre greater than 2% (*i.e.*: twice the 1% standard telecom screen-test level). As expected, the polymer cable sheath has burned, and only the central metallic tubing remained (Fig. 4).

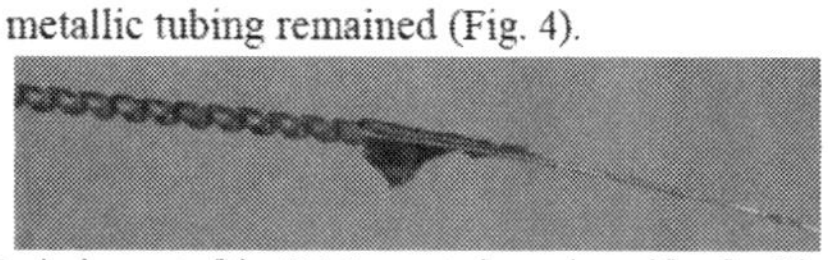

Fig. 4. Typical aspect of the Ø 3.2 mm strain sensing cable after 2 h at 750°C.

C. *Mechanical tests on strain sensing cables*

Only the two strain cables have been tested to characterize their ability to transduce the mechanical strains applied on their surface to the embedded optical fibre. Such cables could be used to detect corium pouring on the concrete floor below the vessel, through a progressive strain increase as far as the corresponding stresses rise, taking into account for the EPR an overall mass close to 300 t applied on ~ 30 m² of surface, *i.e.*: an equivalent stress close to 10^5 Pa (1 bar hydrostatic pressure equivalent), with up to 20 bar additional hydrostatic pressure in case of hydrogen release (before the reactor building breaks).

The experimental set-up consisted of one cable section attached between two translation stages. Step by step translations were applied, which resulted in homogeneous elongations applied to the cable (Fig. 5, left).

[1] For the sake of the explanation, the proportionality coefficient (close to 1), depending on the tubing and the optical fibre diameters, has been omitted.

TABLE II

OPTICAL FIBRE SENSING CABLES SELECTED FOR CORIUM MONITORING

Strain sensing	Temperature sensing	Security
• 1 optical fibre glued into hermetic metallic tubing • specific external structure for better anchorage in concrete • metallic reinforcement for the Ø 7.2 mm cable	• up to 12 optical fibres in a plastic loose tube filled with gel • no metallic reinforcement	• 2 tight buffered optical fibres • no metallic reinforcement • compliant with IEC 60331-25 for circuit integrity at 750°C
• provided in Ø 3.2 mm and Ø 7.2 mm	• provided in Ø 7.5 mm	• provided in Ø 4 mm
TEMPERATURE SENSITIVITY BETWEEN 4°C AND 85°C RAYLEIGH OFDR FREQUENCY SHIFTS MEASUREMENTS		
-1.90 GHz/°C Ø 3.2 mm -4.54 GHz/°C Ø 7.2 mm	-4.25 GHz/°C	-1.82 GHz/°C
OPTICAL FIBRE CONTINUITY DURING 90 min AT 750°C TESTS PERFORMED IN A TUBULAR OVEN, CABLE IN A PIPE		
optical fibre systematically broken between 400°C & 450°C	pass (but not a product specification)	pass

Strain profiles were measured with Rayleigh OFDR technique and were quite homogenous, even if discrepancies appeared, which also might be the result of the winding process on the delivery strands (Fig. 5, right).

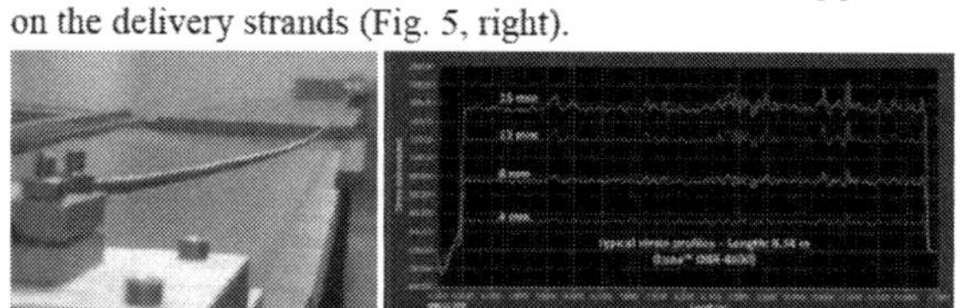

Fig. 5. Experimental set-up for strain cable characterization. Left: cable horizontally pulled between two translation stages. Right: strain profiles recorded for several translation steps (*e.g.*: 4 mm, 8 mm, 12 mm and 15 mm).

Several strain cycles were performed, and the average strain for the tested sections recorded *versus* the translation steps. The analysis led to typical hysteresis values smaller than 15 µm/m for 1600 µm/m strain range, *i.e.*: less than 1% full scale (Fig. 6, left), which is an excellent result for such a complex structure made of different materials layers.

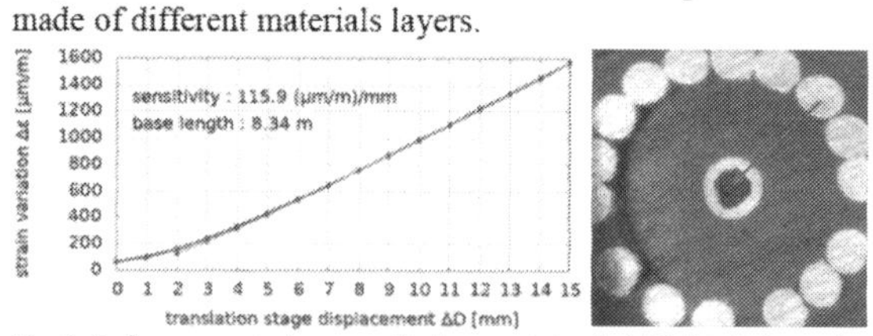

Fig. 6. Left: average strain *versus* displacements (cycles). Right: strain sensing cable section with the optical fibre glued along its length in the central tubing.

An explanation is the strain transduction mechanism, based on the shear $\Delta\gamma$ of a thin glue layer in the cable central tubing, this transduction efficiency being: *i)* inversely proportional to its thickness e (estimated smaller than 400 µm, Fig. 6, right), and *ii)* proportional to its stiffness characterized by its Young's modulus E. Assuming that the optical fibre diameter is small[1] in comparison with the central tubing inner diameter:

$$\Delta\tau \cong \frac{E}{2(1+\nu)}\Delta\gamma \quad \text{with:} \quad \Delta\gamma \cong \frac{\delta d}{e} \implies \delta d \cong \frac{2(1+\nu)e}{E}\Delta\tau$$

with $\Delta\tau$ the shear stress (proportional to the applied pulling force), ν the glue Poisson ratio and δd the *relative* optical fibre displacement in the tubing. If ΔD is the cable translation stage displacement and D the length of the section under test, the strain hysteresis can then be estimated of the same order of magnitude than $\delta d/\Delta D \cong \frac{2(1+\nu)e}{E\,\Delta D}\Delta\tau$.

This explanation for strain is compliant with the early fibre break stated for temperatures starting from 400°C, since the differential thermal expansion between the optical fibre and the cable is quasi-exclusively converted into mechanical strain with an amplification factor roughly equal to $\alpha_{cable}/\alpha_{fibre} \gg 1$.

D. *Cables candidates for corium monitoring*

Therefore, the most interesting sensing cables[2] for corium monitoring are: *1)* the strain sensing cables in a safety role, for their earlier corium vicinity detection capability (from 400°C) likewise their ability to detect strain profiles resulting both from corium pouring on the floor and hydrostatic pressure increase in the reactor pit, and *2)* the safety cable, for its ability to sustain 750°C during at least 90 minutes (Table II).

With a typical temperature gradient equal to 250°C/cm in the concrete, and a temperature front progression speed typically about 5 cm/h after 12 hours of a severe accident, it would take ~ 17 minutes to heat from 400°C up to 750°C. This would be an additional detection delay during which safety cables could still provide measurements in comparison with strain cables.

IV. The VULCANO small scale final MCCI experiment

The VULCANO experiment has consisted in obtaining a depleted UO_2 prototypical corium pool into a concrete crucible, and monitoring the resulting concrete ablation. The test section (Fig. 7) is 2D cylindrical (Ø 250 mm inner and Ø 500 mm outer diameters), with 300 mm inner and 650 mm outer heights.

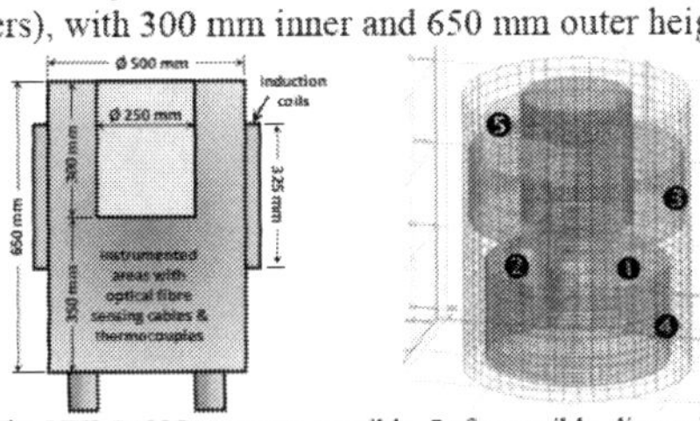

Fig. 7. The VULCANO concrete crucible. Left: crucible dimensions. Right: position of the 5 sensing cables wounded around and below the melted pool.

It is equipped with a hundred of type-K, plus 6 type-C thermocouples, located on 6 azimuths. Type-C thermocouples enable temperature measurements at the bottom of the melted pool, whereas the main role of type-K thermocouples (identified as "failure-thermocouples") is devoted to track the concrete ablation front; their measurements also provide information about temperature distribution in the concrete.

Five twin optical fibres sensing cables are incorporated in the

TABLE III
CABLES POSITION & DEDICATED MEASUREMENT TECHNIQUES

cable	cable position & sensing length	optical measurement technique & sensing parameters
❶	below melted pool 1 turn ~ 55 cm 19 m sensing length	photon counting OTDR / cable length telecom OTDR / cable length
❷	below melted pool 1 turn ~ 75 cm 25 m sensing length	Rayleigh OFDR / {strain & temperature} & cable length Raman DTS / {temperature} & cable length
❹	below melted pool 1 turn ~ 95 cm 22 m sensing length	Brillouin B-OFDA / {strain & temperature}
❺	around melted pool 1 turn ~ 125 cm 25 m sensing length	Raman DTS / {temperature} & cable length

Optical fibres in cable ❸ were both not exploitable for optical sensing due to elevated optical losses after concrete casting.

Sensing lengths do not take into account additional lengths incorporated in the concrete crucible to connect to the remote instruments.

concrete test section, for an overall length close to 180 m. Both thermocouples and optical fibre sensing cables were attached on nylon strings to lock their position before concrete casting.

A. *Optical fibre sensing cables arrangement in the crucible*

To overcome spatial sampling issues with a concrete crucible external circumference close to 1.5 m, each cable, around and below the pool, was arranged to make 2 turns at the same height to improve the vertical depth sampling (Table III), and the distance between two pairs of turns was set to 1 cm. At least 2 measurements at each depth could have been achieved for both Brillouin and Raman DTS instruments.

The security cable was selected for 3 main reasons: *i)* its small Ø 4 mm diameter to reduce risks of cracks propagation which could have led to a crucible break, *ii)* its compatibility with Eddy currents generated by the induction coils, and *iii)* its ability to sustain high temperatures to provide measurements as long as possible. The optical fibres were traditional (*i.e.*: non rad-hard) bend insensitive ITU-T G.657-type[3].

B. *Remote monitoring principle during experiment*

The five optical measurement systems depicted in Tables I & III were located 20 m away from the instrumented crucible, and the control station with all the staff in a separate building at ~ 100 m distance. These instruments were connected to a Local Area Network and synchronized with the same network clock. All the measurement files were stored on a network share, with a periodical replication on a remote mirror for security reasons: they could be retrieved in real-time for specific interfaces, without interfering with the optical measurement loops (Fig. 8).

[2] For embedded cables, these results are not fully representative, because they do not take into account the mechanical stresses (consequence of differential thermal expansions) applied by the concrete, nor the confinement of the fumes released by the cables polymer compounds (which might also induce some additional attenuation in the optical fibres).

[3] It has already been assessed during this project that the best commercial radiation resistant fibres, embedded into their sensing cables, can sustain the expected radiation levels, without exceeding the maximum allowed RIA [10].

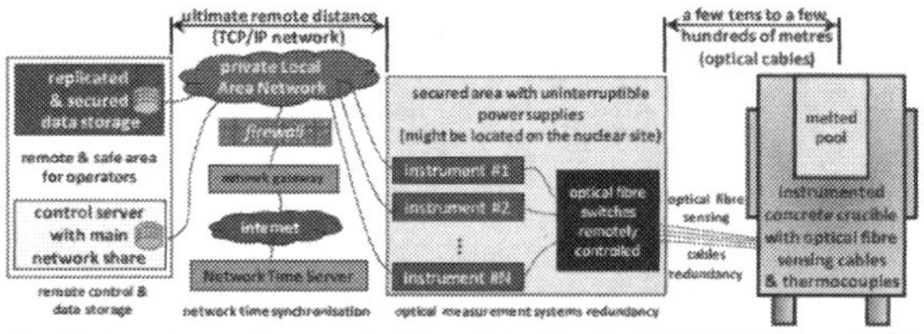

Fig. 8. Remote monitoring principle retained for the VULCANO experiment in DISCOMS project. Redundancy is achieved by switching instruments and/or cables thanks to several optical switches remotely controlled.

C. The VULCANO experiment held on the 26th of June, 2018

The 50 kg corium pool was obtained by thermite reaction (a self-propagating strongly exothermic high temperature synthesis method) directly into the concrete crucible (Fig. 9).

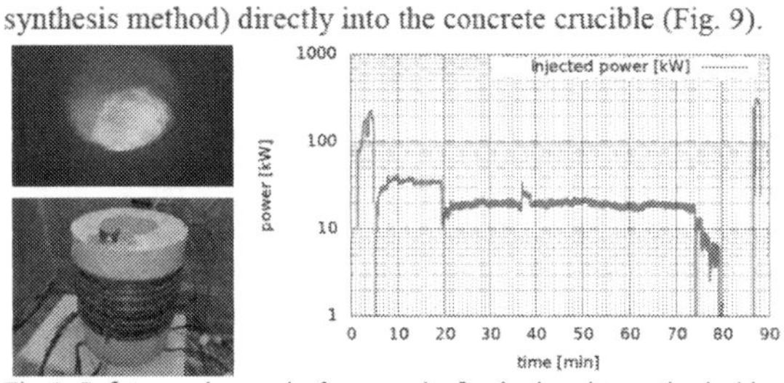

Fig. 9. Left top: corium pool a few seconds after the thermite reaction ignition. Left bottom: concrete crucible with its induction coils. Right: injected power during the experiment.

Induction coils surrounding the test section provided sustained heating power in the corium by Eddy currents into the melted pool. During this experiment, MCCI was carried out during ~ 90 minutes with an injected power varying from ~ 200 kW down to ~ 25 kW. Optical measurements were performed during the entire duration of this experiment, up to complete cooling at room temperature ~ 23 hours later.

V. CORIUM MONITORING RESULTS AND INTERPRETATION

A. Backscattered signal, cable length & damage process

The first analysis was performed on cable lengths given by the Fresnel's reflexion at the fibre tip. It highlighted an ablation process in 2 steps: *i)* first of all, strong damages due to high temperatures, visible on backscattered traces by increased optical losses, and then *ii)* the optical fibre break, a few minutes later, somewhere along the weakened section, leading to typically several metres of optical fibre loss (Fig. 10).

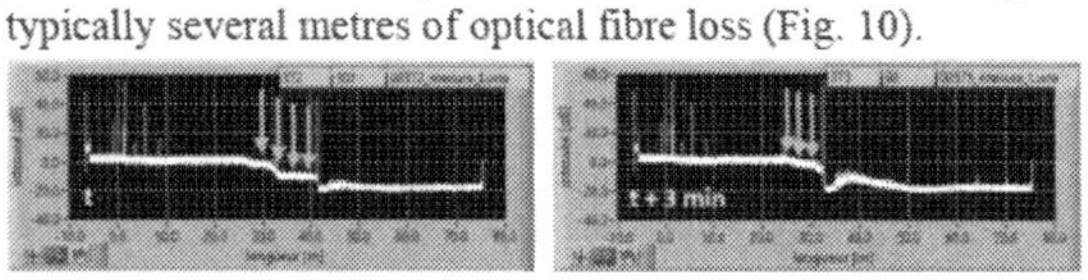

Fig. 10. Optical sensing cable damage in corium vicinity indicated by high optical losses, highlighted by arrows (Rayleigh OFDR).

This process may also lead to damages at several locations, and to a competition between Fresnel's reflexion peaks, which drastically complicates cable length detection (Fig. 11).

The ablation progress, based on the Fresnel's reflexion peak location as reference, continued during several hours even after the induction had stopped, thus probably indicating the effect of differential thermal strains during concrete cooling (Fig. 12).

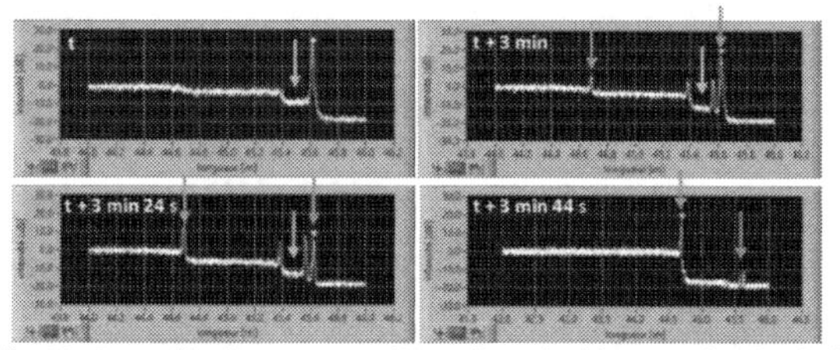

Fig. 11. Competition between several Fresnel's peaks (blue arrows) and optical fibre and cable damage process in corium vicinity (Rayleigh OFDR).

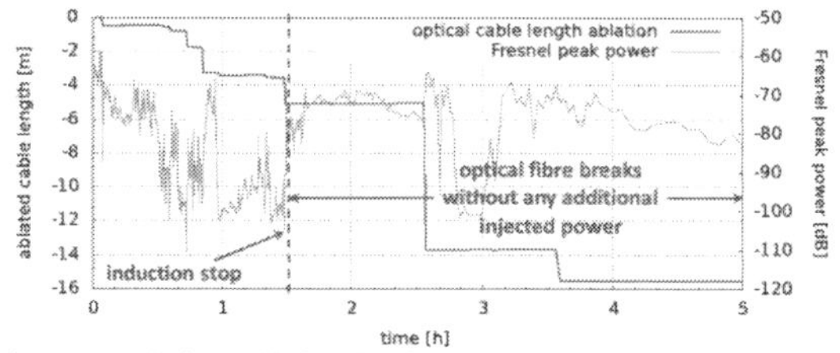

Fig. 12. Equivalent cable length ablation *versus* time. Optical fibre breaks continued several hours after injected power had stopped (Rayleigh OFDR).

Considering the optical fibre length as an image of the MCCI, ablated depths were compliant with thermocouples indications (Fig. 13).

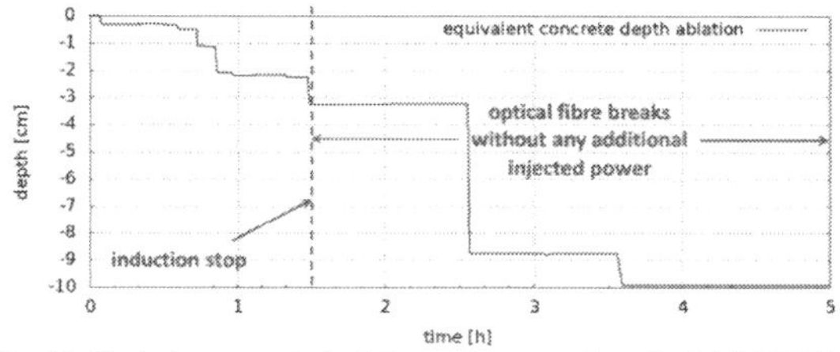

Fig. 13. Equivalent concrete depth interaction *versus* time (Rayleigh ODFR).

Telecom OTDR measurements analysis led to similar results, except for the Fresnel's reflexion peaks which were sometimes difficult to track, requiring manual detections. More reliable detections would require specific algorithms implementation, resilient enough to sustain high optical losses, and taking into account events history. It can also be noticed that most of the time, the Fresnel's peak power was greater at 1310 nm in comparison with measurements at 1550 nm (Fig. 14).

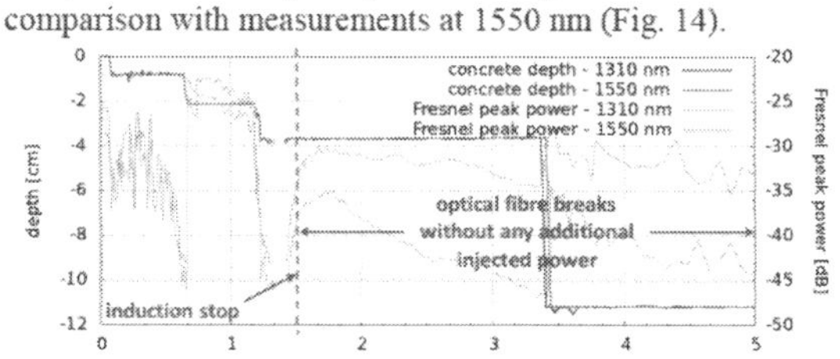

Fig. 14. Equivalent concrete depth interaction *versus* time (telecom OTDR, measurements at 1310 nm and 1550 nm).

But the worst case was encountered with photon counting OTDR, whose signal disappeared during several tens of minutes, making this technique not suitable for reliable corium detection during MCCI, but rather during cooling (Fig. 15).

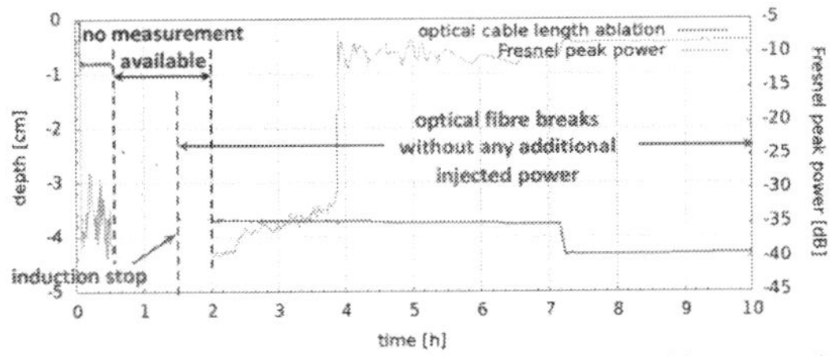

Fig. 15. Equivalent concrete depth interaction *versus* time (photon counting OTDR). Fresnel's peak disappeared during 90 min in corium vicinity.

Nevertheless, for the same sensing cable, comparison between telecom and photon counting OTDRs measurements showed the same results when measurements became again available (2 hours after induction stop). During cooling, some slight noticeable differences might be explained by differential thermal expansions affecting differently the 2 fibres (Fig. 16).

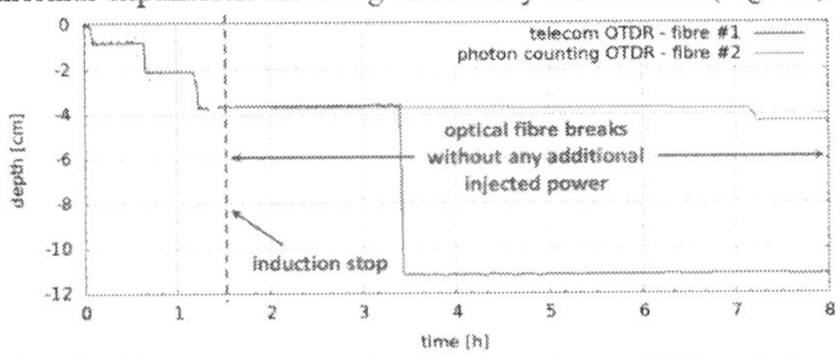

Fig. 16. Measurements comparison between telecom OTDR and photon counting OTDR (two different optical fibres in the same sensing cable).

B. Sensing – Rayleigh OFDR, Brillouin and Raman DTS

Rayleigh frequency shifts were computed by summation of frequency shifts steps $\Delta\nu_{R_{i \to i+1}}$ to avoid post-processing artefacts, but measurement traces still appeared noisy (Fig. 17).

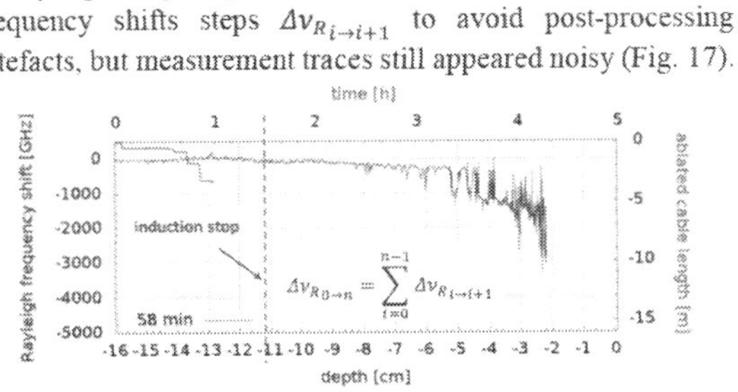

Fig. 17. Blue: typical Rayleigh frequency shift profile *versus* concrete depth after 58 min of interaction with corium. Red: ablated cable length.

Estimated temperatures, according to sensitivity given in Table I, have reached values greater than 1500°C, in the hypothesis of negligible mechanical effects.

Brillouin measurements gave realistic combined (non-selective) temperature and strain gradients indications in the concrete. However, in order to avoid to open the measurement loop (Brillouin B-OFDA sensing is stimulated), the sensing cable was located at 6 cm from the pool: the temperature gradient could be estimated about 25°C/cm (assuming negligible mechanical effects), which is one magnitude order less than expected while in close contact with corium (Fig. 18).

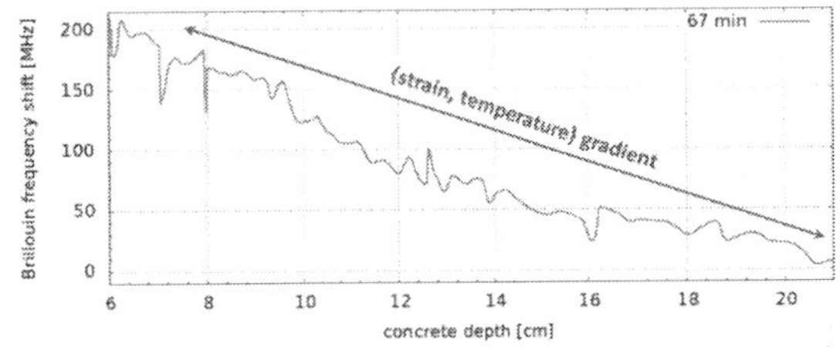

Fig. 18. Combined {strain & temperature} profile expressed in Brillouin frequency shift *versus* concrete depth, after 67 min of MCCI.

Raman DTS also provided realistic *short term* temperature profiles up to ~ 580°C in concrete[4]. Temperature gradients $|\nabla T|$ in corium vicinity were estimated close to 200°C/cm, and close to 25°C/cm, 6 cm deeper in concrete, compliant with Brillouin results (Fig. 19).

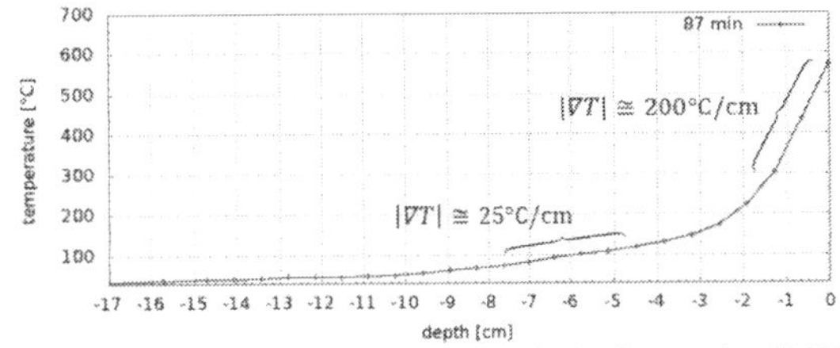

Fig. 19. Temperature profile *versus* concrete depth after 87 min of MCCI (Raman DTS).

Raman DTS measurements during cooling exhibited profiles also compliant with expected results, with ambient temperature reached ~ 23 hours later (Fig. 20).

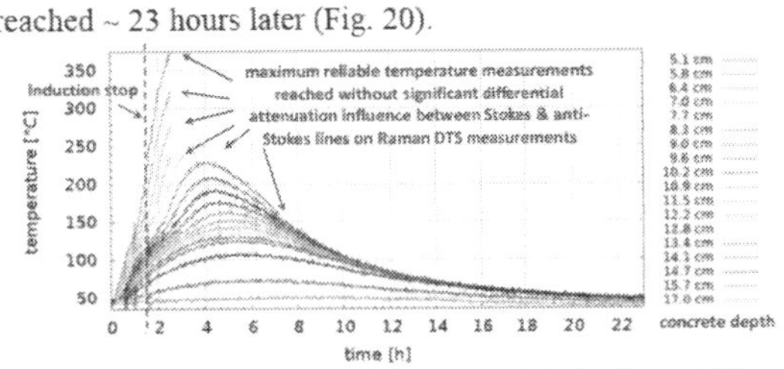

Fig. 20. Temperature *versus* elapsed time at several depths (Raman DTS).

Optical and thermocouples measurements exhibit the same tendencies. More accurate comparisons were not possible since optical measurements are integrated over ~ 1 m (Fig. 21).

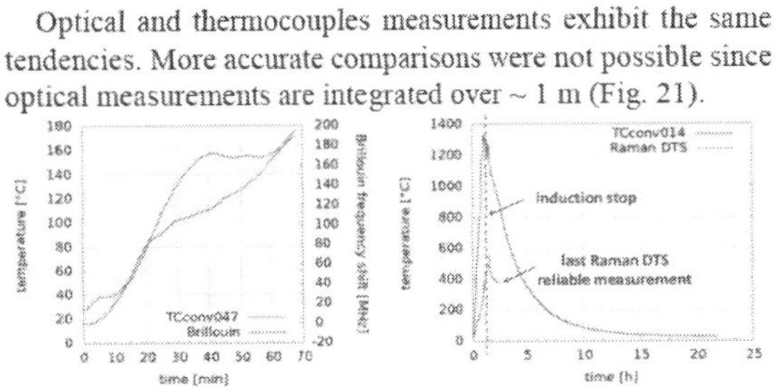

Fig. 21. Left: Brillouin frequency shift compared with thermocouple (depth: 70 mm). Right: Raman DTS temperature compared with thermocouple (depth: 25 mm).

However, with the hypothesis of the sole effect of temperature on the Brillouin frequency shift, measurements

[4] Higher temperatures were not achievable with this single-wavelength Raman DTS instrument, unable to compensate efficiently significant differential attenuations between Stokes and anti-Stokes lines: a dual-wavelengths Raman DTS instrument is mandatory for temperatures greater than 500°C [13].

appear to be compatible with thermocouple data (Fig. 21, left). Strong discrepancies with Raman DTS are probably due to its measurement principle, not efficient enough for differential Stokes & anti-Stokes attenuations compensation (Fig 21, right).

Post-mortem concrete analysis also showed the remaining optical sensing cables, apparently not damaged, only a few centimetres away from the melted pool (Fig. 22).

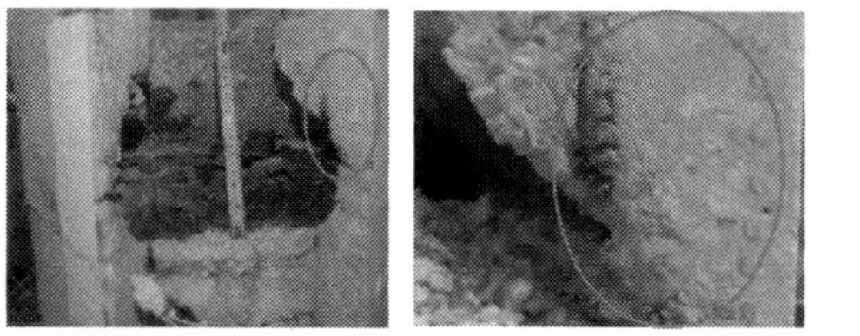

Fig. 22. *Post-mortem* analysis with remaining sensing cables apparently intact.

VI. Conclusion and perspectives

The DISCOMS project is clearly a credible answer to the lack of information coming from the reactor pit in case of a severe accident. Its final demonstration, carried out during a MCCI test on the VULCANO facility of the CEA Cadarache PLINIUS Platform, the 26th of June, 2018, has demonstrated the potential of singlemode Distributed Optical Fibre Sensing to provide valuable information (cable lengths, temperature and strain profiles) which could be advantageously taken into account by operators to monitor severe nuclear accidents from a remote and safe location, and improve nuclear safety.

DISCOMS consortium therefore recommends:

1) Radiation hardened Raman DTS technique to provide reliable temperature measurements up to 700°C [13],
2) Spontaneous Brillouin technique to provide strain and temperature shifts profiles up to 1000°C, even in case of optical fibre break, but also for its resilience to ionizing radiations [14]. Instrumentations capable to dissociate strains from temperature could also be preferred if this is of crucial importance [15].

Additional telecom OTDR can be used for cable (optical fibre) length detection, along with reliable Fresnel's reflexion peak detection algorithms (resilient to optical losses). Strain sensing and security cables, with embedded radiation resistant optical fibres with a protective carbon coating layer [10], should be used for respectively early detection of corium vicinity (from 400°C in a safety role) and monitoring up to at least 750°C.

Such corium remote monitoring system would not replace the existing ones, but could provide real-time information to improve the mitigation strategies during severe nuclear accidents, and also for long-term monitoring.

References

[1] INES, The International Nuclear and Radiological Event Scale, IAEA, 2008. [Online]. Available: https://www-pub.iaea.org/MTCD/Publications/PDF/INES2013web.pdf

[2] C. Bataille and B. Sido, "Rapport de la mission parlementaire sur la sécurité nucléaire, la place de la filière et son avenir – Rapport d'étape : la sécurité nucléaire", Office Parlementaire d'Évaluation des Choix Scientifiques et Technologiques (OPECST), Report n°701, June 2011. [Online]. Available: http://www.senat.fr/rap/r10-701/r10-7011.pdf

[3] ANR, "Investissements d'Avenir, Recherche en Matière de Sûreté Nucléaire et Radioprotection (RSNR)," 2012. [Online]. Available: https://anr.fr/fileadmin/aap/2012/ANR-AAP-RSNR-2012.pdf

[4] ASN, "Complementary Safety Assessments of the French Nuclear Power Plants (European "Stress Tests")", Dec. 2011. [Online]. Available: http://www.french-nuclear-safety.fr/content/download/53850/367799/version/1/file/12 0106-ASN-Report-CSA.pdf

[5] P. Ferdinand, L. Maurin, S. Rougeault, H. Makil, G. Cheymol, H. Maskrot, P. Le Tutour, D. Duval, Ch. Journeau, J.-F. Villard, J.-F. Haquet, Q. Souquet, L. Barbot, J. Miss, I. Duhamel, B. Dechenaux, M. Brovchenko, G. Musoyan, S. Balleydier, F. Pouillé, G. Helleux, S. Fourrez and L. Pichon, "DISCOMS: DIstributed Sensing for COrium Monitoring and Safety", *in CANSMART 2015 Conf.*, Vancouver, Canada, 2015. [Online]. Available: https://hal-cea.archives-ouvertes.fr/cea-02151741

[6] L. Barbot, J.-F. Villard, S. Fourrez, L. Pichon and H. Makil, "The Self-Powered Detector Simulation 'MATiSSe' Toolbox applied to SPNDs for severe accident monitoring in PWRs", *in ANIMMA 2017 Conf.*, Liège, Belgium, 2017. [Online]. Available: https://doi.org/10.1051/epjconf/201817008001

[7] L. Pascucci-Cahen and P. Momal, "Massive radiological releases profoundly differ from controlled releases", Eurosafe forum, 2012. [Online]. Available: https://www.irsn.fr/FR/Actualites_presse/Actualites/Documents/EN_Eurosafe-2012_Massive-releases-vs-controlled-releases_Cost_IRSN-Momal.pdf

[8] M. Brovchenko, B. Dechenaux, K.W. Burn, P. Console Camprini, I. Duhamel and A. Peron, "Neutron-gamma flux and dose calculations in a Pressurized Water Reactor (PWR)", *in ICRS-13 & RPSD-2016 Conf.*, Paris, France, 2016. [Online]. Available: https://doi.org/10.1051/epjconf/201715305008

[9] M. Brovchenko, I. Duhamel and B. Dechenaux, "Neutron-gamma flux and dose calculations for feasibility study of DISCOMS instrumentation in case of severe accident in a GEN 3 reactor", *in ICRS-13 & RPSD-2016 Conf.*, Paris, France, 2016. [Online]. Available: https://doi.org/10.1051/epjconf/201715307030

[10] G. Cheymol, L. Maurin, L. Remy, V. Arounassalame, H. Maskrot, S. Rougeault, V. Dauvois, P. Le Tutour, N. Huot, Y. Ouerdane and P. Ferdinand, "Tests under irradiation of optical fibers and cables devoted to remote corium monitoring in case of severe accident in a Nuclear Power Plant", *in ANIMMA 2019 Conf.*, Portorož, Slovenia, 2019.

[11] Spécifications chimiques des produits et matériaux utilisables en centrale – PMUC, EDF standard, 2008.

[12] Tests for electric cables under fire conditions – Circuit integrity – Part 25: Procedures and requirements – Optical fibre cables, IEC standard 60331-25, 1994.

[13] D. Di Francesca, S. Girard, I. Planes, A. Cebollada, G. Li Vecchi, A. Alessi, I. Reghioua, C. Cangialosi, A. Ladaci, S. Rizzolo, V. Lecœuche, A. Boukenter, A. Champavère and Y. Ouerdane, "Radiation Hardened Architecture of a Single-Ended Raman-Based Distributed Temperature Sensor", IEEE Transactions on Nuclear Science, Vol. 64, No. 1, January 2017. [Online]. Available: https://dx.doi.org/10.1109/TNS.2016.2631539

[14] D. Alasia, A. Fernandez Fernandez, L. Abrardi, B. Brichard and L. Thévenaz, "The effects of gamma-radiation on the properties of Brillouin scattering in standard Ge-doped optical fibres", Measurement Science and Technology, Volume 17, Number 5, 7 April 2006. [Online]. Available: https://dx.doi.org/10.1088/0957-0233/17/5/S25

[15] T.R. Parker, M. Farhadiroushan, R. Feced, V.A. Handerek, and A.J. Rogers, "Simultaneous Distributed Measurement of Strain and Temperature from Noise-Initiated Brillouin Scattering in Optical Fibers", IEEE Journal of Quantum Electronics, Volume: 34, Issue: 4, Apr 1998. [Online]. Available: https://doi.org/10.1109/3.663443

https://doi.org/10.1051/epjconf/202022508005

Measurement of reactor core temperature using multispectral infrared pyrometry in accidental conditions

A.Ladaci*, G. Cheymol, H. Maskrot, C. Destouches, L. Ferry, and C. Gonnier
* Den – Service d'Etudes Analytiques et de Réactivité des Surfaces (SEARS), CEA
Ayoub.ladaci@cea.fr

The Loss Of Coolant Accident (LOCA) is one of the most considered accidental scenario in nuclear power plant design [1]. It occurs after a break of the primary circuit, which leads to a strong pressure decrease inducing an overheating of the full cladding. The oxidation due to water steam and high temperature undermines the claddings and can lead to their burst with the release of fission products [2]. To simulate such accident, the Light-water One Rod Equipment for LOCA Experimental Investigations (LORELEI) test device will be implemented in the Jules Horowitz research reactor at the CEA Cadarache centre [3]. It will allow the study of full claddings behaviours under such conditions [4]. The cladding surface temperature monitoring is essential in this experiment; it allows linking the burst conditions with the temperature. However, this measurement needs to be non-invasive in order to minimize the perturbation and avoiding any change of the burst conditions, which exclude the use of thermocouples. In this case, the Pyrometry based temperature measurement techniques presents a suitable solution [5].

The aim of this work is to develop a non-contact temperature measurement technique allowing the monitoring of the full cladding temperature during a LOCA. In order to achieve that, we developed an infrared optical multispectral pyrometry-based system. This technique presents a good solution allowing to reach the required temperature range, accuracy and acquisition speed (700-1200 °C, >10 °C and >10 Hz, respectively). The multispectral analysis allows us to measure both emissivity and temperature by using a calibrated black body source as a reference. The device transfer function is obtained by a direct comparison between the theoretical black body spectrum and one measured on the black body source. The emissivity is considered quite flat on our working spectral range (1-1.6 µm) as it was demonstrated in previous works [6].

However, a main issue of this technique is the optical system weakness to the reactor core extremes conditions during the accident (high temperature, radiation flux and hydrogen emission) [7]. Indeed, the multispectral pyrometry measurement is very sensitive to the system spectral losses variations, which can be produced by the Radiation Induced Attenuation (RIA) due to the high gamma and neutron fluxes [8]. In order to avoid such issue, we used radiation-hardened materials such as diamond and pure silica glass [9]. The choice of these materials is based on the CEA expertise in developing optical devices for harsh environment. However, the device prototype has to be tested under those new extreme conditions in order to be qualified before its integration in the LORELEI device.

In this paper, we study the water steam influence on the temperature measurement using a prototype of the IR multispectral pyrometer on the EDGAR facility of the CEA (Figure 1.a and b.). The EDGAR facility allows the cladding ohmic heating up to 1200°C, with a temperature monitoring using an internal thermocouple in addition to a commercial optical monochromatic pyrometer [10]. The atmosphere can be switched from air to water vapor using a steam generator. This allows to validate the reliability of the system under the reactor accidental atmosphere where high steam flows are expected.

A. LADACI, G. Cheymol and H. Maskrot are with the Den – Service d'Etudes Analytiques et de Réactivité des Surfaces (SEARS), CEA, Université Paris-Saclay, F-91191, Gif sur Yvette, France (corresponding author: guy.cheymol@cea.fr).
C. Destouches is with the Den – Service de Physique Expérimentale (SPEX/LDCI), CEA, F-13108, St Paul lez Durance, France.
L. Ferry is with the Den – Service Réacteur Jules Horowitz (SRJH/LEDI), F-13108, St Paul lez Durance, France.
C. Gonnier is with the Den – Service Réacteur Jules Horowitz (SRJH), F-13108, St Paul lez Durance, France.

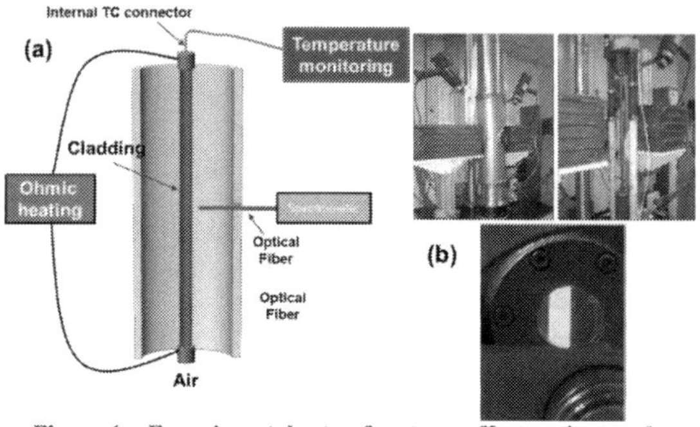

Figure 1.a Experimental setup for steam effect testing on the temperature measurement by IR pyrometry, b. Pictures of EDGAR test facility

The steam flows affect the measurement accuracy by introducing additional spectral losses to the signal acquisition chain; this attenuation will induce an important disturbance in a spectral band at about 1.38 µm, which will decrease the temperature measurement accuracy. In order to avoid such issue, this absorption band was removed from the tested spectral range; in this case, the measurements are performed on a [1-1.3 µm] spectral range.

The measured temperatures in air and steam environments using IR pyrometry are reported in Figure 2.

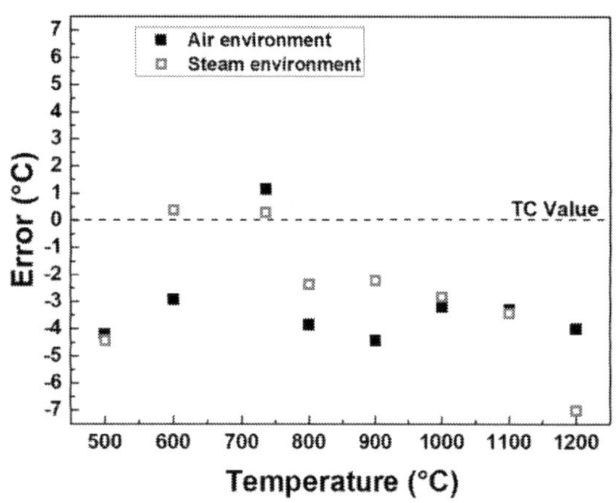

Figure 2. Temperature values measured in air (black marks) and steam (red marks) environments compared to internal thermo-couple values (red dash line) on Zircaloy-4 cladding

In both cases, the temperature is estimated with a good accuracy compare to the TC values, using the multispectral pyrometry in an adapted spectral range.

The uncertainty of the system remains less than 1% for both air and steam environments.

Acknowledgments

The authors are thankful to LC2M laboratory for providing the cladding samples; and in particular, to Q. Houmaire and M. Dumerval for their help during the EDGAR experiments.

References

[1] K.A. Terrani, D. Wang, L.J. Ott, R.O. Montgomery, 'The effect of fuel thermal conductivity on the behavior of LWR cores during loss-of-coolant accidents', Journal of Nuclear Materials, 448, 1–3,2014, 512-519.

[2] F. Erbacher, H. Neitzel, H. Rosinger, H. Schmidt, and K. Wiehr, "Burst Criterion of Zircaloy Fuel Claddings in a Loss-of-Coolant Accident," in Zirconium in the Nuclear Industry, ed. D. Franklin (West Conshohocken, PA: ASTM International, 1982, 271-283.

[3] L. Ferry, D. Parrat, C. Gonnier, C. Blandin, Y. Weiss, and A. Sasson, "The LORELEI test device for LOCA experiments in the Jules Horowitz reactor," in WRFPM 2014 conference, Japan, 2014.

[4] F. Erbacher and S. Leistikow, "Zircaloy Fuel Cladding Behavior in a Loss-of-Coolant Accident: A Review," in Zirconium in the Nuclear Industry, ed. R. Adamson and L. S. Van (West Conshohocken, PA: ASTM International, 1987, 451-488.

[5] B. Bouvry, G. Cheymol, L. Ramiandrisoa, B. Javaudin, C. Gallou, H. Maskrot, et al., "Multispectral pyrometry for surface temperature measurement of oxidized Zircaloy claddings," Infrared Physics and Technology, 83, 2017, 78-87.

[6] B. Bouvry , G. Cheymol , C. Gallou, H. Maskrot, C. Destouches , L. Ferry, and C. Gonnier.,'Theoretical and Experimental Analyses of the Impact of High-Temperature Surroundings on the Temperature Estimated by an Optical Pyrometry Technique,' in IEEE Transactions on Nuclear Science, 65, 9, 2018, 2593-2600.

[7] Y. Lee, T. J. McKrell, C. Yue and M. S. Kazimi, 'Safety Assessment of SiC Cladding Oxidation under Loss-of-Coolant Accident Conditions in Light Water Reactors' Nuclear Technology, 183:2, 210-227.

[8] G. Cheymol, H. Long, J.F. Villard, B. Brichard, "High Level Gamma and Neutron Irradiation of Silica Optical Fibers in CEA OSIRIS Nuclear Reactor" IEEE Trans. Nucl. Sci., 55, 2008, 252-2258.

[9] S. Girard, A. Morana, A. Ladaci, T. Robin, L. Mescia, J.J. Bonnefois, M. Boutillier, J. Mekki, A. Paveau, B. Cadier, E. Marin, Y. Ouerdane, and A. Boukenter, "Recent advances in radiation-hardened fiber-based technologies for space applications", Journal of Optics, 20, 9, 2018, 093001.

[10] T. Forgeron, J. Brachet, F. Barcelo, A. Castaing, J. Hivroz, J. Mardon, and C. Bernaudat, "Experiment and modeling of advanced fuel rod cladding behavior under LOCA conditions: alpha-beta phase transformation kinetics and EDGAR methodology," presented at the Zirconium in the Nuclear Industry: 12th international symposium, 2000.

https://doi.org/10.1051/epjconf/202022508006

Tests under irradiation of optical fibers and cables devoted to corium monitoring in case of severe accident in a Nuclear Power Plant

G. Cheymol*, L. Maurin, L. Remy, V. Arounassalame, H. Maskrot, S. Rougeault, V. Dauvois, P. Le Tutour, N. Huot, Y. Ouerdane and P. Ferdinand
* Den – Service d'Etudes Analytiques et de Réactivité des Surfaces (SEARS), CEA, Université Paris-Saclay, Gif-sur-Yvette, France
guy.cheymol@cea.fr

Abstract — **The DISCOMS project, which stands for "DIstributed Sensing for COrium Monitoring and Safety", considers the potential of distributed sensing technologies, based on remote instrumentations and Optical Fiber Sensing cables embedded into the concrete floor under the reactor vessel, to monitor the status of this third barrier of confinement. This paper focuses on the selection and testing of singlemode (SM) optical fibers with limited RIA (Radiation Induced Attenuation) to be compliant with remote distributed instruments optical budgets, the ionizing radiation doses to sustain, and their reduction provided by the concrete basemat shielding. The tests aimed at exposing these fibers and the corresponding sensitive optical cables, to the irradiation doses expected during the normal operation of the reactor (up to 60 years for the European Pressurized Reactor), followed by a severe accident. Several gamma and mixed (neutron-gamma) irradiations were performed at CEA Saclay facilities: POSÉÏDON irradiator and ISIS reactor, up to a gamma cumulated dose of about 2 MGy and fast neutron fluence (E > 1 MeV) of 6 x 10^{15} n/cm . The first gamma test permitted to assess the RIA at various optical wavelengths, and to select three radiation tolerant singlemode fibers (RIA < 5 dB/100 m, at 1550 nm operating wavelength). The second one was performed on voluminous strands of sensitive cables encapsulating the selected optical fibers, up to approximately the same accumulated dose, at two temperatures: 30°C and 80°C. A significant increase of the RIA, without any saturation tendency, appeared for fibers inserted into cables, correlated with the increase of the hydroxyl attenuation peak at 1380 nm. Molecular hydrogen generated by the radiolysis of compounds of the cable is at the origin of this phenomenon. A third gamma irradiation run permitted to measure the radiolytic hydrogen production yield of some compounds of a dedicated temperature cable sample. The efficiency of a carbon coating layer over the silica cladding, acting as a barrier against hydrogen diffusion, was also successfully confirmed. Finally, the efficiency of this carbon coating layer has also been tested under neutron irradiation, then qualified as a protection barrier against hydrogen diffusion in the optical fiber cores.**

Index Terms — **carbon coating layer, distributed measurement, hydrogen diffusion, gamma, neutron, optical fibers, *rad-hard* optical fiber, radiation effects, radiolysis.**

I. Introduction

In case of severe accident in a Nuclear Power Plant leading to both reactor vessel failure and corium spreading, the loss of power supply can lead most of the instrumentations to be out of order with no way to monitor the status and the evolution of the accident.

After the Fukushima-Daiichi nuclear accident in March 2011, the so-called RSNR program –dedicated to nuclear security R&D– has been launched in France to improve the mitigation solutions in case of severe accident for Gen II, Gen III (*i.e.* the EPR), and for future Gen IV nuclear reactors.

In this framework, the DISCOMS project aims at providing innovative solutions not requiring local electrical power supplies, for remote monitoring of such a severe nuclear accident [1].

One of the solutions investigated in the project [2] is based on distributed OFSs (Optical Fiber Sensors) to detect both the beginning of the severe accident, the corium pouring on the containment building concrete basemat, and its interaction with the concrete floor, until its spreads in the core catcher (in the

This project, with reference number ANR-11-RSNR-0007, was carried out within the framework of the RSNR (research on nuclear safety and radiation protection) research program launched after the Fukushima accident, co-funded by the French *Programme d'Investissements d'Avenir* and managed by the ANR (National Research Agency).

G. Cheymol, L. Remy, V. Arounassalame and H. Maskrot are with Den – Service d'Études Analytiques et de Réactivité des Surfaces (SEARS), CEA, Université Paris-Saclay, F-91191, Gif-sur-Yvette, France (+33(0)1.69.08.62.71; e-mail: guy.cheymol@cea.fr)

L. Maurin and S. Rougeault are with Institut LIST, CEA, Université Paris-Saclay, F-91120 Palaiseau, France.

V. Dauvois and P. Le Tutour are with Den – Service du Comportement des Radionucléides (SECR), CEA, Université Paris-Saclay, F-91191, Gif-sur-Yvette, France.

N. Huot was with Den – Service d'Exploitation du Réacteur OSiris (SEROS), CEA, Université Paris-Saclay, F-91191, Gif-sur-Yvette, France.

Y. Ouerdane is with Univ Lyon, UJM, CNRS, IOGS, laboratoire Hubert Curien, UMR 5516, 18 rue Prof. B. Lauras, 42000 Saint-Étienne, France

P. Ferdinand, previously with Institut LIST, is now with LMP S.A. 49-51 rue du Moulin des Prés, 75013 Paris, France.

case of EPR). The OFSs cables are expected to be embedded into the concrete floor.

Since ionizing radiations affect the optical transmission properties of singlemode optical fibers, it appeared essential for this nuclear application to evaluate the level of attenuation induced on long term by these radiations.

This paper focuses on the tests conducted in order to expose selected fibers and the corresponding sensitive optical cables to irradiation doses expected during the normal operation of the reactor (up to 60 years for the European Pressurized Reactor), followed by a severe accident occurring at the end of this period which corresponds to the worst case of radiation dose to be sustained. It is important to notice that one task of the project was precisely to better assess these expected irradiation doses, the reactor being in normal operation [3], and in case of severe accident [4].

Different gamma and mixed neutron/gamma irradiations have been conducted in lab facilities in the CEA research center based in Saclay: POSÉÏDON irradiator and ISIS reactor, up to a gamma dose of about 2 MGy and fast neutron fluence (E > 1 MeV) of 6 x 10^{15} n/cm .

In the paper, we describe the selected fibers and cables, as well as the irradiation conditions for four irradiation tests. Then we present the RIA measurements and the corresponding analysis, including the measurements of the radiolytic hydrogen production yield of some compounds of the cable.

II. Optical fibers and cables

In our safety related project, even if optical fibers sensing cables are intended to be embedded into the reactor pit concrete basemat, and even if quite short lengths (100 m) are considered for corium monitoring, it is necessary to assess the RIA (Radiation Induced Attenuation) levels and to select the less sensitive fibers to be compliant with the radiation doses to be sustained and the measurement systems optical budgets, in the frame of 60 years of normal operation of the NPP, followed by a severe accident.

The behavior of fibers under ionizing radiations, especially γ rays, has been widely studied during the last decades, highlighting the various parameters which influence the RIA levels and kinetics, see [5] for a recent review. The measurement wavelength, which is a critical parameter, is somehow imposed by the available optical technologies [6]: 1550 nm being the dominating telecom wavelength, for both singlemode fibers and distributed sensing instruments.

The tested SingleMode Fibers (SMFs) are typically 9 μm core diameter, and 125 μm optical cladding diameter, surrounded by a protective coating made of polymer or even metal, leading to an outside fiber diameter typically ranging from 155 μm to 250 μm. In general, optical fiber behavior under ionizing radiations mainly depends on the fiber core composition, where the light waves propagate mainly. From previous research, pure-silica and fluorine-doped cores (both with F-doped cladding) are the most radiation-hardened optical fibers in the domain of dose aimed, up to 100 kGy / 1MGy for steady state irradiation [5, 7, 8].

Selection of OFSs is an output of the project. So, three different rad-hard optical fibers, named "a", "b" and "c", have been tested under irradiation, as well as a standard Corning SMF28 fiber (non *rad-hard*).

Due to the relatively short sensing lengths intended to be used inside the reactor building in such an application (typ. < 100 m per cable), we stated that the maximum acceptable RIA shall be less than 5 dB at 1550 nm for 100 m, *i.e.* 50 dB/km, with regards to the typical optical budgets of todays distributed measurements systems.

After a first γ irradiation test with a batch of standalone optical fibers, the best 3 *rad-hard* fibers fulfilling these requirements have all been selected, in order to be embedded in 3 different types of sensing cables (distributed temperature and strain measurements cables, plus safety cables) [6] for complementary irradiation tests to evaluate any additional cable influence on the RIA. The irradiation tests were mainly focused on long length temperature sensing cables (a few mm diameters), requiring high volume irradiators.

Last, a fourth rad-hard fiber, named "d", incorporating a thin layer of carbon between the cladding and the coating, has also been irradiated.

Table 1 sums up the general characteristics of the five fibers tested, and mentions which ones have been inserted in a sensing temperature cable. Cable with the fiber "a" inside will be named "Ca", cable with the fiber "b" inside will be named "Cb", ... As far as possible, for clarity, all graphs related to fiber "a" or cable "Ca" are colored in green, the corresponding color is blue for fiber "b" or cable "Cb", the corresponding color is red for fiber "c" or cable "Cc", and black is associated with SMF28 fiber and fiber "d" or cable "Cd".

Table 1.
General characteristics of tested fibers and cables

Fiber	Type of fiber	Associated Cable
a	*rad-hard*	Ca
b	*rad-hard*	Cb
c	*rad-hard*	Cc
d	*rad-hard* with carbon coating	Cd
SMF28	Standard	

III. Irradiations – Experimental test procedure

The irradiations performed in the project aims at testing fibers and cables in a configuration which is similar to the aimed environment, or rather in a configuration able to produce the same effects on fibers. We first consider a radiation level at the basemat surface in a very conservative configuration, since the reactor concrete basemat will act as a radiation protection and will drastically decrease the cumulated doses to be sustained *vs.* optical cable embedment depth.

A. Irradiation level to withstand

Two different phases have to be simulated: an operational phase (nominal operation lasting up to 60 years) and an accident phase with corium spreading out of the reactor vessel. Modeling has shown that the doses are of the same order of magnitude for the operational and for the accidental phases, although the dose rates be quite different: in the range of Gy/h for the first, and 20 kGy/h for the second. The RIA depends on the level of cumulative irradiation dose, but also generally on

https://doi.org/10.1051/epjconf/202022508006

the dose rate and the complete history in terms of succession of the different dose rates undergone by the optical fibers, the latter being quite complex. In other terms, that is the expected cumulative dose which is first considered below.

On the basis of the information available at the beginning of the project, the nominal configuration –representative of the irradiation level at the top of the concrete basemat, below the reactor vessel– aims to simulate 250 kGy for the operational phase, and at least 450 kGy during the first 24 h of the accident.

Nevertheless, our fine modeling of the γ dose at the surface of the concrete and assessments made, considering both the operational phase [3] and the first 24 h of the accident, lead us to consider slightly higher values: about 1 to a few MGy. It has also been assessed that the operational phase may produce a fast neutrons fluence (E > 1 MeV) up to 10^{15} n/cm or more on the basemat top surface, leading us to consider an irradiation in the vicinity of the nuclear reactor core.

The dose reduction *vs.* the embedment depth has also been considered, it depends on several parameters (concrete type, corium composition…). The evaluation resulted in a reduction of one order of magnitude after the first 40 cm in the concrete floor for both Gen II and Gen III, based on the fact that in the reactor pit, the main damage is the result of 60 years of irradiation during normal operation [3].

B. Irradiation facilities

γ irradiations have been carried out in the POSÉÏDON industrial pool type irradiator of the Applied Radiation Laboratory (LABRA) [9] in Saclay. In the associated casemate, 2 source planes produce about 10,000 TBq (300,000 Ci), for a maximum dose rate of 6 kGy/h, homogeneous (± 10%) between the source planes in a volume of approximate dimensions: 1 m x 0.6 m x 0.4 m (Fig. 1). The irradiation runs were performed at room temperature (RT), 20°C to 30°C or at 80°C by using an adapted oven in place in the irradiator casemate.

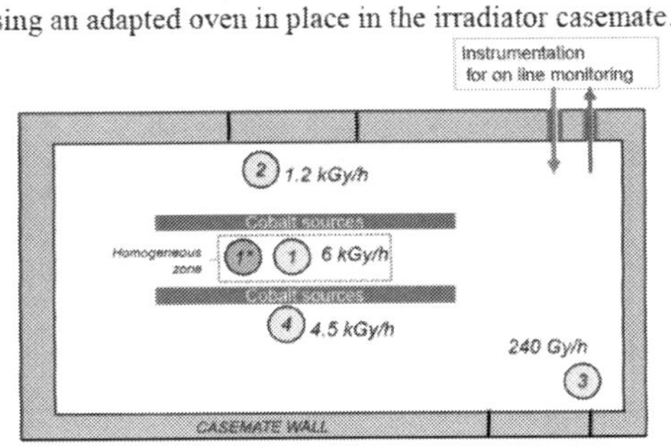

Fig. 1. Diagram of the POSÉÏDON casemate with positions used, and associated dose rate. 1 simply differs from 1 by a higher temperature (about 80°C instead of room temperature).*

Neutron irradiation were carried out in the ISIS research reactor [10] (installation which is now shut down), mainly dedicated to education and training, with a nominal power of 640 kW. Samples of optical fibers were placed in the vicinity of the reactor core in small capsules. Fast neutrons (E > 1 MeV) flux was 2.8×10^{11} n/cm /s. Thermal neutrons flux was 2.4×10^{12} n/cm /s, superposed with an associated 50 kGy/h γ dose rate. The temperature during the irradiation was about 40°C.

C. Four successive irradiation tests

1) Irradiation 1: γ irradiation of 100 m long fibers

The 100 m long fibers were wound on a 15 cm diameter strand and connected thanks to a feedthrough to an instrumentation placed outside the casemate for online monitoring. The optical fibers were tested in 4 configurations, with different dose rates or temperatures. In the nominal configuration, fibers were first in position ② (see Fig. 1) during 215 h, then moved in position ① for the next 113 h (this configuration being named "② to ①"), for a total fluence 5 times lower that tested with the configuration "③ to ②".

The maximum fluence achieved during the irradiation test was performed by leaving fibers all the time in the position ①, or (1*), which also permitted to assess the influence of increased temperature.

Table 2 gives the temperature, dose rate, cumulated dose for each phase of the irradiation 1, for the 4 configurations.

TABLE 2
IRRADIATION 1: γ IRRADIATION OF FIBERS, DOSE AND DOSE RATE IN 4 CONFIGURATIONS

Config.	Temp.	Operational Phase (215 h): dose *dose rate*	Accident (113 h): dose *dose rate*	Total (328 h): cumulated dose
② to ①	20°C	250 kGy *1.2 kGy/h*	650 kGy *5.7 kGy/h*	900 kGy
③ to ②	20°C	49 kGy *230 kGy/h*	131 kGy *1.2 kGy/h*	180 kGy
①	20°C	1.2 MGy *5.7 kGy/h*	700 kGy *5.7 kGy/h*	1.9 MGy
(1*)	80°C	1.2 MGy 5.7 kGy/h	700 kGy 5.7 kGy/h	1.9 MGy

2) Irradiation 2: γ irradiation of 100 m long optical cables

Optical sensing cables "Ca", "Cb" and "Cc" equipped with *rad-hard* selected optical fibers have been irradiated. This time, the 100 m long cables give a much more voluminous strand, typically 40 cm diameter with a section of 5 cm or greater. Hopefully, the POSEÏDON irradiator is well adapted to this run since it presents a relatively large irradiation zone with a homogeneous dose rate.

Here again, the cables were connected to remote distributed instrumentation outside the casemate for online monitoring. We kept the coiled cables in the same position during the 303 h irradiation time, with a dose rate of 5 kGy/h, resulting in a cumulated dose of 1.5 MGy.

It is mainly the positions ① and (1*) that have been tested. One cable was in the position ②, with a dose rate of 1.05 kGy/h resulting in a 320 kGy cumulated dose.

3) Irradiation 3: γ irradiation of short length of optical fibers and cable

Short lengths of optical fibers and cables, and bulbs with various components of the temperature cable, have been irradiated in the position ④, but in this case without any online measurement. The dose rate: 4.5 kGy/h during 400 h, lead to about 1.8 MGy cumulated dose.

4) Irradiation 4: Neutrons + γ irradiation of short optical fiber lengths

Shorts lengths of optical fibers have been irradiated in the vicinity of the ISIS reactor core, without online measurement. The fibers were wound in several strands of diameter ranging from 3 cm to 4 cm.

The power of the reactor was set to 640 kW, the irradiation lasted 6 hours. Dose rate and cumulated dose are displayed on Table 3.

TABLE 3
TEST 4: N + γ IRRADIATION OF OPTICAL FIBERS

	DOSE RATE	DOSE
Fast neutrons	2.8 x 10^{11} n/cm /s	6 x 10^{15} n/cm
Thermal neutrons	2.4 x 10^{12} n/cm /s	5.2 x 10^{16} n/cm
γ	50 kGy/h	300 kGy

Even if the 1550 nm nominal wavelength was especially considered, it was also interesting to assess the RIA on a larger spectral range. On that purpose, the beams of 2 SLED centred at 1550 nm and 1310 nm and a large spectrum halogen source were combined with optical couplers. In order to avoid any photobleaching, as far as possible, the total power injected in the fibers was in the range of 1 µW.

IV. RESULTS AND ANALYSIS OF TEST 1: γ IRRADIATION OF THE RAD-HARD SINGLEMODE OPTICAL FIBERS

The various configurations have been described in section III with a summary in Table 2.

For all optical fibers considered, the RIA increases during the irradiation, faster when the dose rate is higher, as shown on graph related to configuration ② to ① (Fig. 2).

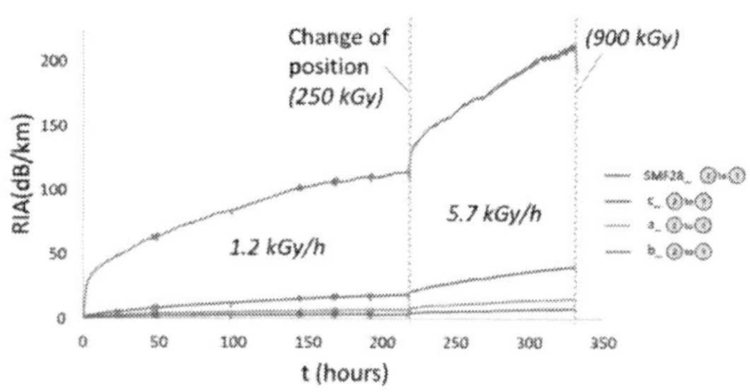

Fig. 2. RIA@1550 nm vs. time for fibers a, b, c and SMF 28 during irradiation 1/configuration "②to①". The dose rate is 1.2 kGy/h before the change of position and 5.7 kGy/h after. At the change of position, the cumulated dose is 250 kGy, at the end of irradiation, it reaches 900 kGy.

Fig. 3 shows the difference in RIA for one *rad-hard* fiber when it is irradiated at either 22°C or 80°C. The RIA decreases when the temperature increases, as already reported, *e.g.* in reference [11].

This first test confirmed that some *rad-hard* optical fibers have much less RIA than standard commercial Ge-doped fiber, in the range of 50 dB/km or less (down to 10 dB/km), at 1550 nm with a cumulated dose of 1 MGy or greater.

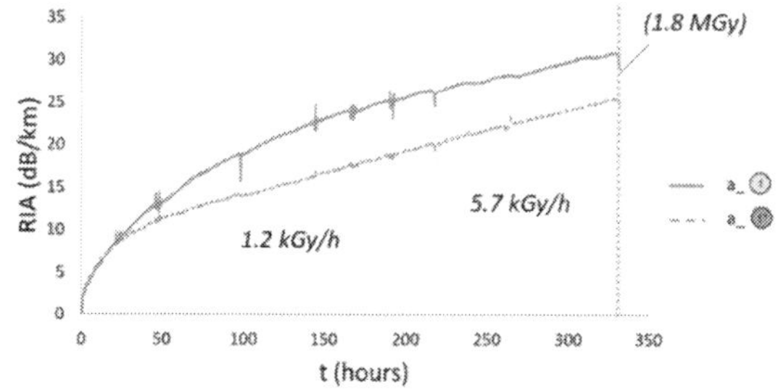

Fig. 3. RIA@1550 nm vs. time for fiber a during irradiation 1 in configuration ① (22°C) and ●(80°C). The dose rate is 5.7 kGy and the cumulated dose is 1.8 MGy at the end of irradiation.

V. RESULTS AND ANALYSIS OF TEST 2: γ IRRADIATION OF OPTICAL SENSING CABLES

The selected *rad-hard* optical fibers have been inserted in sensing cables and irradiated in the POSÉÏDON irradiator. We focus below on the tests of the temperature sensing cables in configurations ① and ● (Fig. 1). For comparison, we also irradiated the same fibers, conditioned in the same way (not into cable, but standalone), like the previous test based on configuration ①.

The RIA of the optical fibers inserted into cables happened to be higher than for the same standalone fibers, as shown on Fig. 4 for fiber a, without any tendency to saturation.

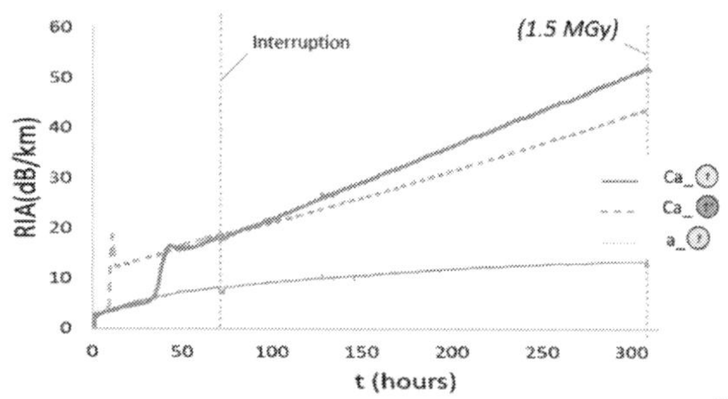

Fig. 4. Irradiation 2: RIA@1550 nm vs. time for fiber a in position ① (see Fig. 1), at room temperature (RT), and cables Ca, at RT (position ①) and in oven at 80°C (position ●). The dose rate is about 5 kGy/h. At the end of irradiation, the cumulated dose is about 1.5 MGy.

The graphs of RIA spectrum on a large spectrum range permitted to better understand the reason for the difference between RIA in standalone fibers with those inserted in cables. The increase of the RIA at 1550 nm is due to the large attenuation peak centred at 1380 nm (Fig. 5), which reveals the presence of hydroxyl groups in the fiber core.

Such peak is likely to be generated by the production of hydrogen by radiolysis of polymer components of cables; the hydrogen then diffuses towards the fiber core and induces the creation of hydroxyl groups.

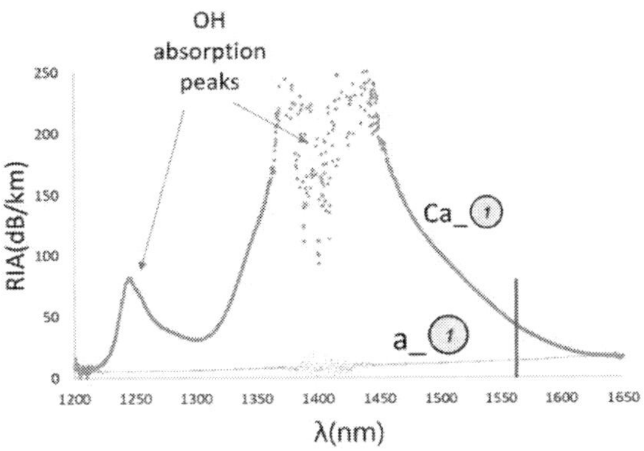

Fig. 5. RIA spectrum for fiber a and cable Ca at the end of irradiation 2 (cumulated dose: 1.5 MGy), in position ①(30°C/ 5 kGy/h).

VI. Results and analysis of test 3: γ irradiation of cables and carbon coated optical fiber parts

Even if the RIA observed with fiber into cables remains acceptable for this project, mainly because the embedment in the concrete will drastically decrease the dose received by the fibers, avoiding or reducing the amount of hydrogen diffusing towards the fiber core, it remains of scientific interest. In that objective, we have tested a cable equipped with a carbon coated fiber. Such a thin carbon layer is known to protect the silica-based fibers from external hydrogen which, without it, could diffuse towards the core and induce additional attenuation [12].

The tested cable incorporating this fiber was not made according to standard manufacturing procedure: the original optical fiber was replaced off-line in the temperature cable by a rad-hard carbon coated fiber "d", thus giving cable "Cd" (see Table 1). Then, 3.4 m of this cable Cd was irradiated, as well as the same lengths of cables Ca, Cb, and Cc placed in the position ④ within the irradiator POSÉÏDON (see Fig. 1), at the dose rate 4.5 kGy/h during 401 hours, leading to a cumulated dose of 1.8 MGy. Again, there was no on-line monitoring. The focus on the OH peak, with potentially high RIA, permitted to accommodate with the short fiber lengths tested.

Following the irradiation phase, the attenuation[1] of fibers inserted into cables was measured thanks to a strongly attenuated Supercontinuum source, and the same ANRITSU™ OSA as used in previous experiments. Corresponding graphs are depicted on Fig. 6, showing the absence of any OH peak for the carbon coated fiber irradiated into the cable, which demonstrates the efficiency of such carbon layer for the intended purpose.

To support the study of induced attenuations, some measurements were also carried out on specific components of the temperature sensing cable, in order to study the formation of volatile compounds, and in particular H2 dihydrogen, during the irradiation 3.

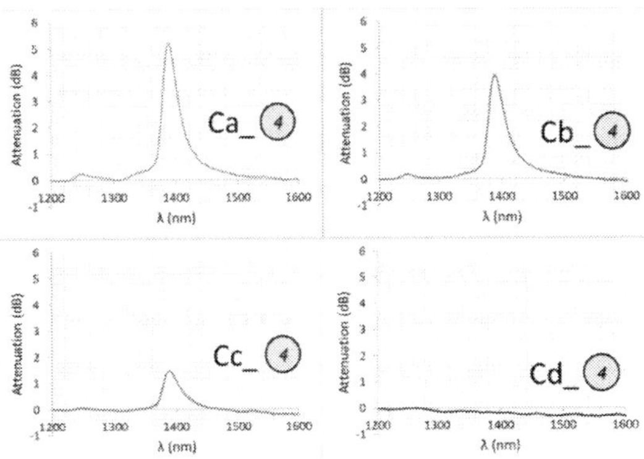

Fig. 6. Attenuation vs. λ for fibers into cables Ca, Cb, Cc, Cd after irradiation 3; cables in position ④ (22°C; 4.5 kGy/h; 1.8 MGy). To erase the attenuation due to connectors, all graphs are normalised to cancel attenuation at 1200 nm, where RIA is expected to be very low on such short lengths.

These materials have been irradiated into sealed glass bulbs. Extra components were also tested, including two kinds of gel surrounding the optical fiber used in temperature sensing cables. Such gels are used as a conductive thermal layer to improve thermal efficiency, but have also a tendency to induce some significant friction, which is a matter of concern when the measuring technique is sensitive to both strain and temperature, especially when only the latter information is wanted. Also, a piece of the overall cable, made mainly of one surrounding polymer material, was tested. For each material, 2 bulbs of about 20 cm³ were used with reconstituted air (with krypton as tracer). Two bulbs with only reconstituted air have been added for reference.

At the end of the irradiation, the gaseous mixture in each ampoule was analysed by direct injection mass spectrometry and magnetic sector, to obtain both characterization and quantification of all species. The formation of hydrogen, expressed in number of moles per unit length of fiber, or cable, is recorded in Table 4.

TABLE 4
PRODUCTION OF HYDROGEN FOR 1.8 MGy

Component	Bulb #	Hydrogen produced (mol_H_2/m)
Coated fiber	1	8.6×10^{-7}
	2	9.8×10^{-7}
Gel 1	3	6.6×10^{-4}
	4	7.2×10^{-4}
Gel 2	5	5.5×10^{-4}
	6	5.6×10^{-4}
Overall cable	7	4.2×10^{-3}
	8	5.4×10^{-3}

The overall optical cable and its surrounding polymer material produce the major part of hydrogen while irradiated.

[1] We may notice this is not a classical RIA measurement, since there is no online monitoring. The lack of repeatability in the loss of optical fiber connections prevents to make reliable measurements of the attenuation. We have normalised all graphs to get equivalent zero attenuation at 1200 nm, where RIA is expected to be very low for such short fiber lengths.

Dividing this production by the mass of each component permit to calculate the radiolytic hydrogen production yield. This time, the higher value is coming from the gels, with a yield of about 2 10^{-7} mol/J for about 1.8 MGy.

It does not come from this study that a specific element of the cable is mainly responsible for the production of hydrogen, and can be removed or substituted by another more favourable element with respect to this application.

VII. N + γ IRRADIATION OF SHORT FIBER LENGTHS AND CONCLUSION

Finally, the efficiency of the carbon coating layer has also been tested under neutron irradiation, showing its good efficiency as a protection barrier against hydrogen diffusion in the core of silica optical fibers.

Thanks to different tests conducted, we have been able to recommend the better *rad-hard* optical fiber for the aimed application, compliant with 60 years of normal operation of the NPP followed by a severe accident, and compliant with an installation respectively at minimum 30 cm and 5 cm below the surface of the concrete basemat for Gen II and Gen III reactors.

An optical fiber with carbon coating layer should therefore be systematically preferred for such application, or/and either any other means to protect the core of the fiber from the diffusion of hydrogen and its effects on attenuation. However, the long-term efficiency of the carbon coating under ionizing radiations should be further investigated.

REFERENCES

[1] P. Ferdinand *et al.*, "Optical Fiber Sensors to improve the safety of Nuclear Power Plants," Asia Pacific Optical Sensors, APOS2013, Wuhan, China, 17th-21st Oct. 2013.

[2] P. Ferdinand et al, "DISCOMS: DIstributed Sensing for COrium Monitoring and Safety," CANSMART, Vancouver, BC, Canada, 15-17 July 2015. https://hal-cea.archives-ouvertes.fr/cea-02151741

[3] M. Brovchenko *et al.*, "Neutron-gamma flux and dose calculations for feasibility study of DISCOMS instrumentation in case of severe accident in a GEN 3 reactor." EPJ Web of Conferences 153, 07030 (2017) - ICRS-13 & RPSD-2016. 25 Sept. 2017. https://doi.org/10.1051/epjconf/201715307030.

[4] TRIPOLI-4 Project Team, "TRIPOLI-4 v8 User *Guide*", CEA-R-6316 (Feb. 2013). www.oecdnea.org/tools/abstract/detail/nea-1716/.

[5] S. Girard, A. Morana, A. Ladaci, T. Robin, L. Mescia, J.J. Bonnefois, M. Boutillier, J. Mekki, A. Paveau, B. Cadier, E. Marin, Y. Ouerdane, and A. Boukenter, "Recent advances in radiation-hardened fiber-based technologies for space applications", Journal of Optics, vol 20, no. 9, Aug. 2018, Art. no. 093001.

[6] L. Maurin *et al.*, "Remote monitoring of Molten Core-Concrete Interaction experiment with Optical Fibre Sensors & perspectives for severe nuclear accident monitoring – DISCOMS project," 6th international conference on Advancements in Nuclear Instrumentation Measurement Methods and their Applications, ANIMMA 2019, 17-21 June 2019, Portorož, Slovenia.

[7] T. Wijnands, L. K. De Jonge, J. Kuhnhenn, S. K. Hoeffgen, and U. Weinand, "Optical absorption in commercial single mode optical fibers in a high energy physics radiation field," IEEE Trans. Nucl. Sci., vol. 55, no. 4, pp. 2216–2222, Aug. 2008.

[8] X. Phéron, S. Girard, A. Boukenter, B. Brichard, S. Delepine-Lesoille, J. Bertrand, and Y. Ouerdane, "High -ray dose radiation effects on the performances of Brillouin scattering based optical fiber sensors," Opt. Exp., vol. 20, pp. 26978–26985, 2012.

[9] http://www-centre-saclay.cea.fr/fr/Le-Labra-une-plateforme-d-irradiation-au-service-des-industriels-et-des-chercheurs

[10] F. Foulon, B. Lescop, and X. Lohleber, "Development of education and training programs using ISIS research reactor". Joint IGORR 2013 & IAEA Technology Meeting, Korea, 13-18 Oct 2013.

[11] J. Kuhnhenn, S. K. Hoffgen, and U. Weinand, "Quality Assurance for Irradiation Tests of Optical Fibers: Uncertainty and Reproducibility", IEEE Trans. Nucl. Science, vol. 56, no. 4, pp.2160-2166, August 2009.

[12] A. Mendez and T. F. Morse, "Specialty Optical Fibers Handbook"; https://www.elsevier.com/books/specialty-optical-fibers-handbook/mendez/978-0-12-369406-5

Comparative Evaluation of Three Wireless Sensor Network Transceivers in a High Radiation Environment

Q. Huang, *Member, IEEE*, J. Jiang*, *Fellow, IEEE* and Y. Q. Deng, *Student Member, IEEE*
University of Western Ontario
* jjiang@uwo.ca

Abstract **— This paper presents on the results of radiation studies for three commonly used wireless sensor nodes based on the following protocols: ZigBee, WirelessHART, ISA 100.11a, and network devices built with commercial off-the-shelf (COTS) components. The level of radiation considered is at par with that experienced at Fukushima Daiichi Nuclear Power Plant after the accident. An experimental setup is developed to monitor behaviors of each wireless device and network real-time under the ^{60}Co gamma radiator at The Ohio State University Nuclear Reactor Lab (OSU-NRL). The experimental results have indicated that the performance of the communication channels and wireless signal parameters do not degrade significant under such radiation. However, all the tested devices and networks can only survive for several hours under the high dose rate condition (20 K Rad/h). The results of these experimental studies have provided useful references to those who design and manufacture COTS-based wireless monitoring systems for use in high level radiation environments.**

Index Terms **— Total dose test, industrial wireless sensor networks, rad-hardened, severe accident, nuclear power plant**

I. Introduction

One of the lessons from Fukushima Daiichi nuclear disaster is the difficulty to gain up-to-date status about the key plant status, such as water levels in spent fuel bays, hydrogen concentration in reactor buildings, and temperatures in heat transport systems, after the accident due to loss of on-site power for monitoring instruments and strong radiation [1]. Considering various potential scenarios during a severe accident, a potential approach is to use wireless technologies to implement post-accident monitoring system (PAMS) which can provide much needed information about the plant conditions without relying on likely damaged communication infrastructure [1-3].

In past decades, a number of wireless transceivers and sensor networks have been developed to acquire data in a wide variety of applications [4-7]. A designer of industrial wireless systems has several wireless transceivers and network standards to choose from. However, in the event of a severe nuclear accident, a significant amount of radiation can be released due to failure of protection layers [8], to minimize the impact of radiation on these wireless systems, more considerations have to be given during component and protocol selection process for these monitoring instruments.

Taking Fukushima accident as an example, in March 2012, the level of radiation is estimated to up to 73 Sv/h inside the containment of No.2 reactor [9], and Feb. 2017, the level has shot up to 530 Sv/h [10]. Such high level of radiation can, no doubt, cause severe damage to wireless transceivers and networks if no precautions are taken. Before any protection measures can be taken, it is necessary to investigate how different wireless devices and network components react to radiation under similar radiation conditions. Unfortunately, there are only limited resources available on radiation responses of industrial wireless devices and networks in the literature. To fill this gap, the characteristics of commonly used industrial wireless devices and networks are studied. The results of this study provides a scientific basis in the selection of suitable wireless technologies and network devices in such applications.

Based on those studies and analysis, six industrial wireless nodes built with commercial off-the-shelf (COTS) components, CC2530 (2.4 GHz ZigBee standard), CC2520 (2.4 GHz WirelessHART standard), CC2530 (2.4 GHz ISA100.11a standard), are chosen to undergo irradiation tests in a gamma radiation environment. Since these are destructive tests, a group of diversified wireless devices are therefore built using different industrial wireless networks for the radiation tests. A total dose test is performed to measure the behaviors of the developed wireless devices and networks by directly exposing them to 60Co gamma irradiator at The Ohio State University Nuclear Reactor Laboratory (OSU-NRL) under a high dose rate (20 K Rad/h). The results of this test will provide guidelines to design wireless monitoring instruments for using in high level radiation environments. They serve as important references to assist other researchers and engineers to design and manufacture radiation-hardened (rad-hardened) monitoring system by using COTS components.

The reminder of the paper is organized as follows. The basics of the three industrial wireless devices and network standards are described in Section II. Subsequently, the chosen wireless sensor nodes and network devices are explained in Section III. Section IV presents experimental setups and summarizes the results with analysis. Finally, a number of observations are made and conclusions are drawn in Section V.

II. Industrial Wireless Devices and Networks

Wireless sensor network technologies have been used in many industrial applications, such as safety systems, control systems, and monitoring systems [4-6]. In a typical wireless sensor network system, a measured process variable from a sensor is transmitted from one node to another or to a base station using radio frequency (RF) waveforms. Several distinct frequency bands have been dedicated to wireless sensor networks for industrial applications based on industry and governmental regulations. They are, for example, at 433MHz, 868MHz, 915MHz, and 2.4GHz.

In the past few decades, a number of wireless transceivers have been developed with various communication functions. On the other hand, various functions and network protocols for sensor nodes have been defined in wireless sensor network standards for integration with the networks. Recently, several wireless sensor network standards based on the IEEE 802.15.4 have also been developed, such as ZigBee, WirelessHART, and ISA100.11a [4-6]. Brief description of these protocols are in order.

In general, radiation particles lose their energy through either non-ionization processes (displacement damage) and/or ionization processes when they pass through semiconductor devices. Radiation effects on electronics can generally be classified into three categories:

ZigBee: ZigBee network supports Star, Tree, and Mesh topologies, as shown in Fig. 1. Star topology is often choice for industrial applications. Within the network, the roles that the devices play can be categorized into the following: (1) Coordinator, which is used to initialize, maintain, and control the network; (2) Router, which is responsible for routing messages; and (3) End Device, which executes the intended functions associated with the application [5].

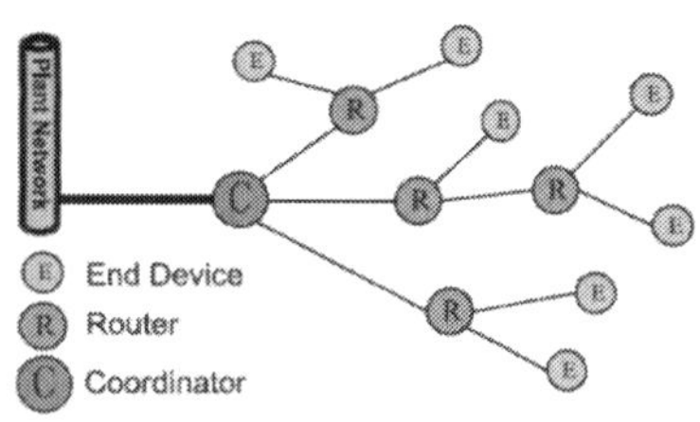

Fig. 1 ZigBee network [5]

WirelessHART: WirelessHART supports both Star and Mesh network topology as shown in Fig. 2. Basic elements in a WirelessHART network include: Field Devices, Gateway, Access Point, Network Manager, and Handheld Devices [5]. A sensor for measuring process variables is fed to the Field Device for transmission. An Access Point is used to connect the Field Devices to the Gateway. The network manager can be used to configure the network, schedule and overseeing communications among different devices[5]. WirelessHART systems have found many industry applications [6].

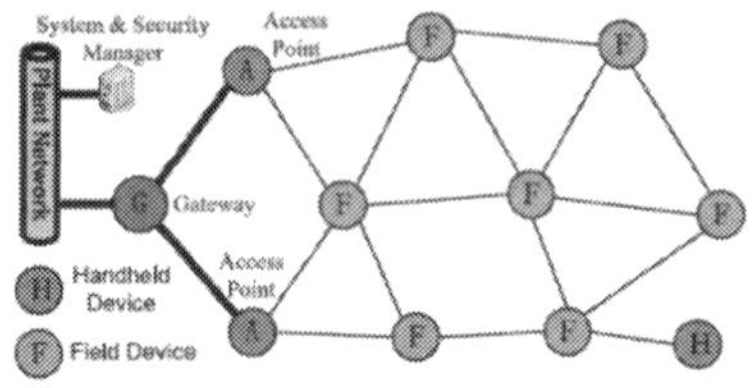

Fig. 2 WirelessHART network [5]

(3) **ISA100.11a**: ISA100.11a supports both Mesh and Star network topology as shown in Fig. 3. A typical ISA100.11a network includes two type devices: Field Devices and Infrastructure Devices. The former includes Routing Devices, I/O Devices, and Handheld Devices. The latter includes Backbone Routers, Gateways, and System and Security Managers [5]. ISA100.11a network has been shown to be a robust and secure communication system for industrial applications [6].

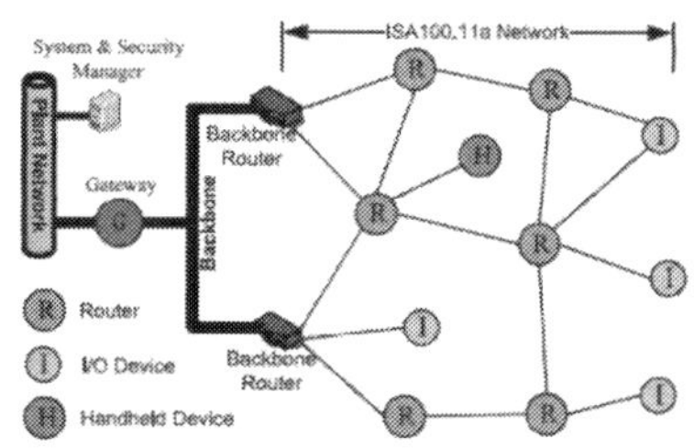

Fig. 3 ISA100.11a network [5]

III. Development of Devices for Radiation Tests

For radiation experiments for total ionizing dose conditions, three group of wireless devices, labelled as Sample#1 to Sample#3, have been designed and produced using two different transceivers and controller units. All three systems are operating at 2.4 GHz. The details are summarized in TABLE I:

TABLE I: WIRELESS DEVICES USED IN THE IRRADIATION TESTS

Sample ID	Transceivers	Protocols	Controller Units
Sample #1	CC2530	ZigBee	AVR ATMEGA 328
Sample #2	CC2520	WirelessHART	Proprietary
Sample #3	CC2530	ISA100.11a	STM32L486

These sample circuit boards are shown in Fig.4. The radiation tests are carried out when the circuit boards are on power, and the wireless networks have also been setup to relay the response of the tested wireless device out of the test rig and send to a Control Workstation through the RS232 protocol. The performance of each device is being monitored continually throughout the tests until it ceases operation. The logged data can then be used for determination of potential causes.

https://doi.org/10.1051/epjconf/202022508007

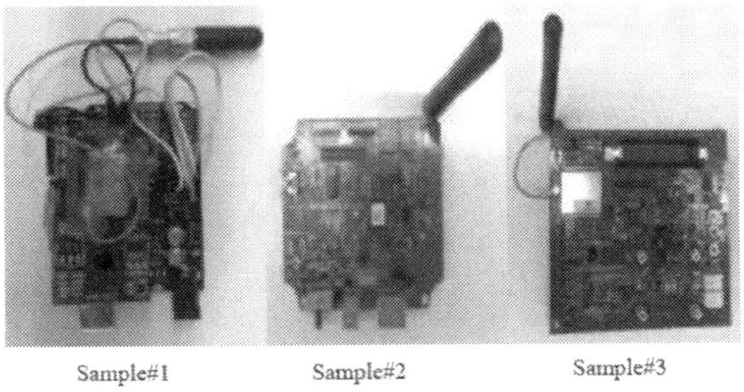

Fig. 4. Samples used in the total dose tests

For Sample#1, a star ZigBee network has been setup using E800-DTU (Z2530-485-20) as the Router and Coordinator and E18-MS1PA1-IPX as the transceiver for the End Device. Both are from Ebyte Electronic Company. Furthermore, AVR ATMEGA 328 is used as the controller for the End Device. End Device sends one message to the Router and Coordinator every two seconds.

For WirelessHART system of Sample#2, the Field Device used in the experiments is Awia Warrior 220 manufactured by AwiaTech Company. It communicates to the WirelessHART Gateway. The HART Server installed on the Control Workstation is used to configure the WirelessHART network.

In case of ISA100.11a system, i.e. Sample#3, the transceiver used as the Field Router and IO Device is VN210 from Nivis. The network is formed using VersaRouter 900 as the Backbone Router and Gateway which also perform the functions of System Manager, Security Manager, and System Time Source. A web application is developed to remotely check and configure the ISA100.11a network.

IV. Facilities and Environment for Radiation Tests

The radiation source is a 60Co irradiator tube at the OSU-NRL as illustrated in Fig. 5. The samples are placed in the 6-inch tube one at a time. A shielding elevator delivers the test sample to the irradiation location. The peak dose rate is estimated to be at 20 K Rad/h. The samples are powered through cables from a +5v power supply and a DC-DC Converter, which are located outside the irradiator. Several equipment, such as oscilloscope and spectrum analyzer, is used to measure electronic signals and communication channel performance.

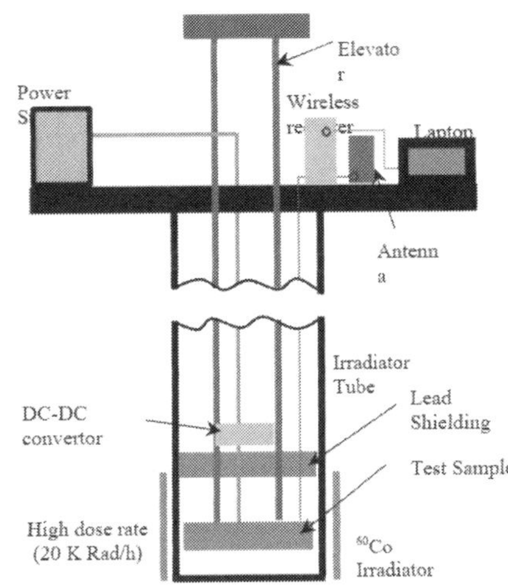

Fig. 5. Schematic of the experimental irradiator tube

The actual test environment is shown in Fig. 6 with all the facilities. It is important to point out that each test only involves one sample. Furthermore, the signal for the antenna of the wireless devices are routed outside before connecting to an antenna to avoid EM shielding of the test tube.

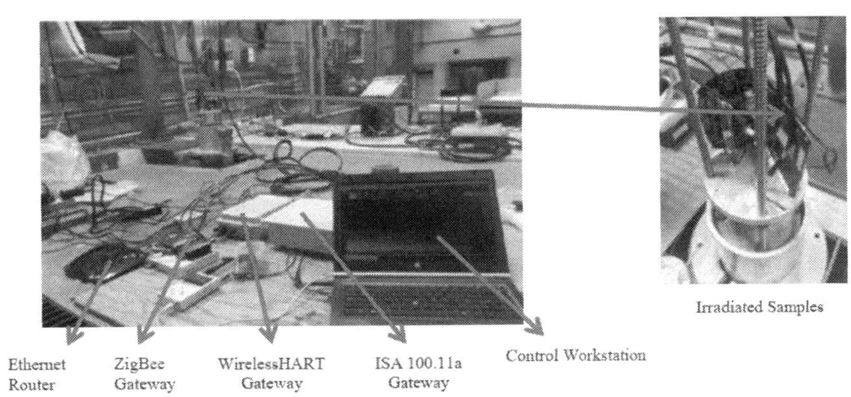

Fig. 6 Actual test environment for ZigBee, WirelessHART, and ISA100.11a devices

During each test, the performance of the communication channels and wireless signals are continuously monitored to capture the following performance indicators: loss packet rate, error packet rate, shifts in frequency, and Received Signal Strength Indicator (RSSI). The irradiation test continues as long as the device is functioning and transmitting wireless signals. When the device ceases operation and exact time of breakdown is recorded. Hence, the survival time of the device under a dose rate of 20 K Rad/h can be obtained. The total dosage (TD) of ionizing radiation deposited on to the wireless device can be calculated as:

$$= 20\ \text{K Rad/h} \times \text{survival time}$$

This variable represents how much total radiation a device can take before giving in and lead to a failure.

V. Experimental Results

After three destructive tests, the survival times for the three samples are recorded. The results are summarized in TABLE II.

TABLE II: RADIATION TEST RESULTS

	Sample#1	Sample#2	Sample#3
Survival time (h)	192m	50m	62m
First failed component	CPU and Transceiver	Transceiver	Transceiver
Total Dosage (TD) before the first failure (K Rad)	64	16.7	20.0

It can be observed from TABLE II that the ZigBee (Sample#1) device has the best performance in terms of radiation tolerance, as it survives the radiation environment much longer than the other two devices. This is followed by the ISA100.11a. However, when ZigBee device fails, both CPU and Transceiver have failed, but for the other two devices, the weakest component is the Transceiver.

To investigate how radiation affects the quality of the communication systems, two type of tests have been carried out: one is in an ideal laboratory environment free from ionizing radiation. These results provide a reference for comparison and the other is on-line and real-time measurements while the devices are at the bottom of the test tube. The results for Sample#1 to Sample#3 are summarized in TABLE III to V, respectively.

TABLE III
COMMUNICATION PERFORMANCE OF ZIGBEE DEVICE AND NETWORK UNDER NORMAL AND HIGH DOSE RATE CONDITIONS

Radiation condition	Duration (minutes)	Total packets	Loss packets	Loss rate (%)	Error packet	Error rate (%)	Frequency (GHz)	RSSI (dBm)
Normal condition	30	150	0	0.00	0	0.00	2.405	-43.00
	60	150	0	0.00	0	0.00	2.406	-32.00
	90	150	0	0.00	0	0.00	2.407	-53.50
	150	150	0	0.00	0	0.00	2.404	-35.00
	180	150	0	0.00	0	0.00	2.406	-48.00
20 K Rad/h	30	150	0	0.00	0	0.00	2.411	-22.50
	60	150	0	0.00	0	0.00	2.411	-53.50
	90	149	1	0.67	0	0.00	2.410	-45.50
	150	150	0	0	0	0.00	2.409	-35.50
	180	150	0	0	0	0.00	2.409	-39.50
	192				Fail			

TABLE IV
COMMUNICATION PERFORMANCE OF WIRELESSHART DEVICE AND NETWORK UNDER NORMAL AND HIGH DOSE RATE CONDITIONS

Radiation condition	Duration (minutes)	Device ID	Analog value (mA)	Percent range	Frequency (GHz)	RSSI (dBm)
Normal condition	30	131874	11.90	49.50%	2.402	-49.00
	60	131874	11.90	49.50%	2.407	-48.50
20 K Rad/h	30	131874	11.90	49.50%	2.402	-48.00
	50	131874	11.90	49.50%	-	-
	51			Fail		

TABLE V
COMMUNICATION PERFORMANCE OF ISA100.11A DEVICE AND NETWORK UNDER BOTH A NORMAL CONDITION AND A HIGH DOSE RATE CONDITION

Radiation condition	Duration (minutes)	Total packets	Loss packets	Loss rate (%)	Error packet	Error rate (%)	Frequency (GHz)	RSSI (dBm)
Normal condition	30	119	0	0.00	0	0.00	2.406	-33.50
	60	1375	0	0.00	0	0.00	2.403	-50.50
20 K Rad/h	30	117	2	1.68	0	0.00	2.403	-45.00
	60	1368	9	0.65	0	0.00	2.403	-51.00
	62				Fail			

As can be seen that radiation has hardly any effects to the performance of communication quality for both ZigBee and WirelessHART devices. For the ISA100.11a device, there are some minor loss of packets under high dose conditions, which can be easily dealt with through error correction codes.

As far as the quality of wireless signals is concerned, no effects have been observed on the central operating frequency of 2.4 GHz. However, fluctuations of RSSI are observed. Unfortunately, it is not conclusive to link such variations to the radiation, as there is no obvious trend observed from the data.

From a hardware and software complexity point of view, WirelessHART and ISA100.11a contain more hardware components and software functions, some may be more susceptible to radiation damage. On the other hand, ZigBee is relatively simpler in both hardware and software design. It turns out to be more tolerant to radiation.

Another important point to note is that none of these devices should be put in use for a severe accident monitoring system without proper protection and radiation-tolerant design, since they can be easily damaged in a high level radiation environment.

VI. Conclusion

To examine the suitability of using industrial wireless sensor network devices in a high level radiation environment, three most commonly used wireless sensor network devices, ZigBee, WirelessHART, and ISA100.11a, have been tested in a radiation environment during on-power operations. A 60Co gamma radiation source is used to provide a consistent radiation environment. The operating conditions of key components on the devices are monitored continuously so that the first failed component can be identified in the post-analysis. It is interesting to observe that the wireless device based on ZigBee protocol has the strongest radiation tolerance, while WirelessHART and ISA100.11a devices share similar outcomes. A very important conclusion is that these devices can only survive for a limited amount of time, so much shorter than needed for use in a monitoring system for severe accident. Hence, adequate protections against total radiation exposure have to be taken either through shielding or redundant design. This study provides a useful reference for the design of wireless devices and networks to be deployed in high level radiation environments.

Acknowledgment

This work is financially supported by the University Network of Excellence in Nuclear Engineering (UNENE) and the Natural Sciences and Engineering Research Council of Canada (NSERC). Authors would like to acknowledge technical support from the staff at The Ohio State University Nuclear Reactor Laboratory (NRL) for their assistance during the experiments.

References

[1] K. Nagatani, S. Kiribayashi, Y. Okada, K. Otake, K. Yoshida, S. Tadokoro, T. Nishimura, T. Yoshima, and E. Koyanagi. Emergency response to the nuclear accident at the Fukushima Daiichi nuclear power plants using mobile rescue robots. Journal of Field Robotics, vol. 30, pp. 44-63, 2013.

[2] M. Andreeva, M.P. Pavlova, and P.P Groudev, "Overview of plant specific severe accident management strategies for Kozloduy nuclear power plant, WWER-1000/320," Annals of Nuclear Energy, vol. 35, pp. 555-564, 2008.

[3] H. Kinoshita, R. Tayama, Y. Kometani, and Y. Kani, "Development of new technology for Fukushima Daiichi nuclear power station reconstruction," Hitachi Review, vol. 63, pp. 41-46, 2014.

[4] J. Yick, B. Mukherjee, and D. Ghosal. Wireless sensor network survey. Computer Networks, vol. 52, pp. 2292-2330, 2008.

[5] Q. Wang, and J. Jiang. Comparative examination on architecture and protocol of industrial wireless sensor network standards. IEEE Communication Survey & Tutorials, vol. 18, pp. 2197-2219, 2016.

[6] A. Ajith Kumar S., K. Øvsthus and L.M. Kristensen, An industrial perspective on wireless sensor networks – a survey of requirements, protocols, and challenges. IEEE Communications Surveys & Tutorials, vol. 16, pp. 1391-1412, 2014.

[7] O. Georgiou, and U. Raza. Low power wide area network analysis, Can LoRa Scale? IEEE Wireless Communications Letters, vol. 6, pp. 162-165, 2017.

[8] A.A. Adalja, E.S. Toner, A. Cicero, J. Fitzgerald, and T.V. Inglesby, Radiation at Fukushima: basic issues and concepts. Clinicians' Biosecurity News, Analysis of Advances and Challenges in Clinical Biosecurity.

[9] R. Eisler, The Fukushima 2011 disaster. U.S. Geological Survey, 2012.

[10] [Online] Available: The Fukushima nuclear reactor radiation at highest level since 2011 meltdown, https://www.theguardian.com/environment/2017/feb/03/fukushima-daiichi-radiation-levels-highest-since-2011-meltdown. Access date: Feb. 3, 2017.

Unmanned Radiation Monitoring System

Jakub Lüley
Institute of Nuclear and Physical Engineering
Slovak University of Technology in Bratislava
Bratislava, Slovakia
jakub.luley@stuba.sk

Branislav Vrban
B&J NUCLEAR ltd. Miloslavov, Slovakia
branislav.vrban@bjnuclear.eu

Štefan Čerba
B&J NUCLEAR ltd.
Miloslavov, Slovakia
stefan.cerba@bjnuclear.eu

Filip Osuský
Institute of Nuclear and Physical Engineering
Slovak University of Technology in Bratislava
Bratislava, Slovakia
filip.osusky@stuba.sk

Vladimír Nečas
Institute of Nuclear and Physical Engineering
Slovak University of Technology in Bratislava
Bratislava, Slovakia
vladimir.necas@stuba.sk

The absence of online radiation monitoring systems has been observed in the case of Fukushima nuclear accident. As the tsunami destroyed 23 of the 24 status monitoring points, almost no relevant radiation dose measurements data were available. A rapid deployment of a mobile radiological unit that can quickly determine the activity and direction of the radioactive cloud spread on the ground or in the air can prevent unnecessary deaths and related financial losses. Although the design of the current generation of NPPs incorporates features that minimize the risk of large radioactive releases outside the reactor, it is still important to focus on the development of systems that can mitigate the consequences of such events. In situations when the level of radiation does not permit the personal to perform the required measurements, online unmanned radiation monitoring systems may come to the play. For such a purpose the RMS-00x radiation monitoring system could be used, which is a modular system covering the functionality of dose rate measurement, air sampling and radiation map creation without requiring the human personnel to be present at the measurement site. The main purpose of the RMS-00x radiation monitoring system is the rapid deployment of unmanned monitoring devices to reduce the radiation burden on workers and on public. The system can be applied in the vicinity of a nuclear power plant (NPP) or at any location, where source of ionizing radiation could be present. Before this system is used in real conditions, its components must be thoroughly calibrated, based on certified measurement equipment and state-of-art simulation tools. This paper deals with the description of the RMS-00X sensor modules and demonstrates their functionality in combination with UAV. In addition, demonstration of the use of the developed technology was carried out as part of the regular emergency planning and preparedness of EBO NPP on 26th October 2017.

Keywords—RMS, UAV, radiation monitoring, accident, demonstration

I. Introduction

The absence of online radiation monitoring systems has been observed in the case of Fukushima nuclear accident. As the tsunami destroyed 23 of the 24 status monitoring points, almost no relevant radiation dose measurements data were available. The serious situation resulting from the accident led to the decision to evacuate the populations. This decision had been made on the basis of the parameters computed from the simulation models. The associated stress resulted from the invisible threat and the evacuation process caused significant socio-economic consequences and a death toll. The real radiological measurement proofed that the effective radiation dose had not exceed the daily limits in some areas and showed needless of evacuations. More information with the monitoring results can be find in [1]. A rapid deployment of a mobile radiological unit that can quickly determine the activity and direction of the radioactive cloud spread on the ground or in the air can prevent unnecessary deaths and related financial losses. Unmanned Aerial Systems (UASs) used for the monitoring of the radiation situation and the escape routes, help to assess the seriousness of the situation and provide the basis for the most appropriate decision making. UAS sensors operating in semiautomatic mode are capable of effectively measuring larger areas, while the ground personnel can be at a safe distance, and far, hard-to-reach locations, can be explored. UAS can also serve as a complement to periodic preventive dosimetry measurements around nuclear power plants. The currently used dosimetry monitoring stations are static, so they cannot dynamically respond to fast measurement requests in different locations. Illustrative is the example of the Sellafield Nuclear Power Plant in the UK, where the static monitoring station reported radiation leakage, which ultimately led to NPP shutdown. Subsequently, investigation revealed that the increased levels of the measured quantities had been caused by a natural radioactive source, located outside the nuclear power plant area, raised by the wind. In such a situation, a mobile detection unit located on an unmanned device would have effectively identify the source of the radiation [2].

II. Radiation Monitoring System

The main purpose of the Radiation Monitoring System (RMS) in combination with an Unmanned Aerial Vehicle (UAV) is the creation of UAS. Rapid deployment of the UAS in case of radiation event, which is its key feature, shall reduce the radiation burden on workers and on public. The system can be applied in the vicinity of a nuclear power plant (NPP) or at any location, where source of ionizing radiation could be present. Individual modules of RMS have been designed to operate with UAV without influencing functionality of each component, to provide sufficient information for the dosimetry staff and to achieve high standard of reliability. The key features of the RMS are the modularity, the custom independent in-house control software, the possibility to hibernate and subsequently activate the system upon specific incentives, the easy-to-wipe anti-contamination layer of sensors and the compatibility with various UAV carriers.

The originality of the RMS lies in optimized compatibility with UAVs and modularity enabling implementation of various radiation sensors. The proposed solution involves the parallel operation of several UAVs, their efficient cooperation with the ground station and the deployment of modular sensor nodes, which can be linked to a chain. The proposed management of multiple nodes ensures greater flexibility for fast deployment and more efficient mapping of large territories, compared to a single node system. A system of measuring nodes, optimized for minimal power consumption, distributed by the UAVs in the area of interest, can provide information on the local radiation situation for even several weeks. The deployment of this system is foreseen in situations and at locations where the risk of radiation exposure on personal responsible for radiation measurements is expected. The modularity of this system also makes possible to be used in the R&D field. The biggest advantage of the system is the low weight of the modules and extreme operation time, which in case of energy saving mode can achieve one month.

III. Detailed Description of the RMS Modules

The RMS is a modular system and is still in development. The current version consists of the RMS-000 communication and control module, the RMS-WASP communication software and three sensor modules (RMS-001, RMS-002, RMS-003). RMS-001 is the radiation measurement module, RMS-002 the air sampler module and RMS-003 the GPS tracker. The schematic view of the system is shown in Fig. 1.

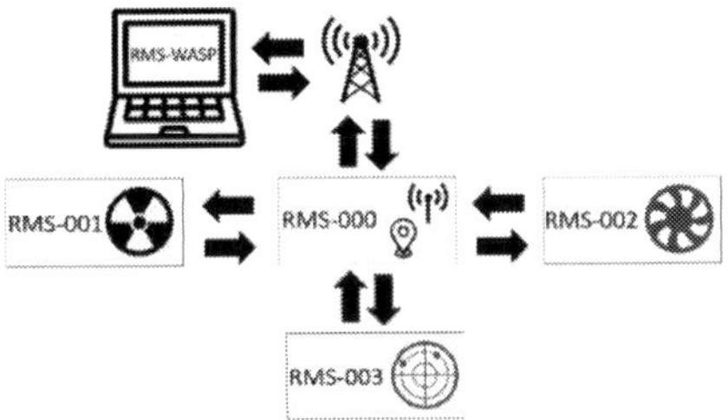

Fig. 1. 1. The schematic view of RMS modules.

A. RMS-000 Communication and control module

One of the essential parts of the RMS is the RMS-000 communication and control module. It serves to support the operation of the measurement modules and acts as a gateway between the ground station and the sensors. It makes efficient use of the programmable microcontroller and uses it as its brain. Considering that in a case of nuclear accident or other radiological threat the mobile networks might not be available, the emphasis in the design of the RMS-000 module was placed on securing independent communication and electricity supply channels. To achieve the maximum communication range and good penetration, 866 MHz ISM band and sequential communication mode were selected. The communication antenna, connected via the USB port of the computer, is used to receive the signal from the RMS-000 node. Both modules provide 2-way encrypted communication. The signal transmission is routed in regular time intervals, defined by the user. The RMS-000 module also provides device positioning with GPS co-ordinates and is transmitted to the ground station within a single message that includes also the measured parameters from the RMS-001 or RMS-002 modules. The main parameters of the RMS-000 module are listed in Table I.

TABLE I. Main parameters of the RMS-000 module [3]

Application	Modularization
Communication	Primary – Radio in 866 MHz EU ISM band
Range	Up to 20 km for 1 node
Positioning	External GPS, precision < 1m
Connectivity	RMS-001, RMS-002, RMS-003
Supply	Independent Li-Po batteries

B. RMS-001 Radiation sensor module

The RMS-001 radiation sensor module serves for the online measurement of effective dose rate of gamma and beta radiation. The major parts of the module are the radiation sensor board and the J305β GM detector. The sensor board, the RMS-000 and the battery are placed in a protective casing made of carbon fiber material. The casing is equipped by the connectors necessary for the communication of the RMS-000 module with ground station. The GM tube and the carbon fiber casing are placed in the outer casing, which provides IP56 protection against dust and humidity. The current configuration allows measurement of an effective dose rate of gamma and partly beta radiation, ranging from the natural background level to 277 μSv/h (up to 100 μSv/h without significant impact of dead-time). The operative replacement of the GM tube is possible for another unit meeting the dimensional and supply conditions of the design. The main parameters of the RMS-001 module are presented in Table II.

TABLE II. Main parameters of the RMS-001 module [3]

Application	Measurement of effective dose rate
Detector type	GM tubes
Type of radiation	gamma, beta
Active measurement	> 48 h (continuous with GPS)
Static measurement	> 1 week (continuous without GPS)
Control measurement	> 4 weeks („stand-by" and regular)
Measurement range	50 nSv/h – 277 μSv/h
Measurement range	200 nSv/h – 10 Sv/h
Weight of the node	0.5 kg

C. RMS-002 Air sampler module

The RMS-002 air sampler module serves to demonstrate the air sampling capability through a removable aerosol filter which intended use is to capture radioactive particles carried by aerosols in case of a nuclear accident. The filters can be subsequently used for spectral analysis to identify key radioisotopes. The module consists of a sensor board, removable filter, special LiPo battery, blower and flowmeter. The electronic parts (sensors and RMS-000) and the suction nozzle are located in a box made of carbon material. Currently the device is able to perform air sampling at speed that corresponds and surpasses to human respiration rate. The main parameters of the RMS-002 module are listed in Table III.

TABLE III. MAIN PARAMETERS OF THE RMS-002 MODULE [3]

Application	Aerosol collection for spectral analysis
Replaceable filter	Standard diam. 5 cm
Operation time	90 – 120 min.
Air flow range	0 – 30 l/min., upgradable to 55 l/min.
Measurement	Actual air flow, total air volume.
Operation	Continuous or sequential (remote control) Upgradable to GPS position.
Casing	Carbon fiber casing
Possible upgrade	External GM module, spectral analysis module
Weight of the node	< 1,5 kg

D. RSM-WASP Control software

The RMS-WASP control software has been developed to ensure the communication between nodes and to provide the user with a simple user-friendly environment, which however summarizes all the required parameters right on the screen, see Fig. 2. The software was developed in the ECLIPSE environment in C++ language using wxWidgets multi-platform libraries to create the GUI (graphical user interface) of the program. The communication between the measuring nodes is secured by configuring the receiving module, which requires a serial port configuration using the RMS-WASP software. The control message is encrypted using the "CRC MODBUS-16" cipher. The measurement data is treated in real time, and the results can be visualized in a form of radiation maps. The results are processed to the specific "kml" file that can be directly visualized using QGIS – Open Street Map, Google Earth or other mapping software. This functionality involves the creation of a contoured radiation map directly plotted into real geographic coordinates, indicating also the trajectory of the unmanned device flight. In the current version of the software, the radiation data is transformed to effective doses rate based on the count rare and the calibration constant of the used GM tube. The program interface also enables switching the signal processing on/off, which is an important feature in case of fast transport of the UAV between the measuring points.

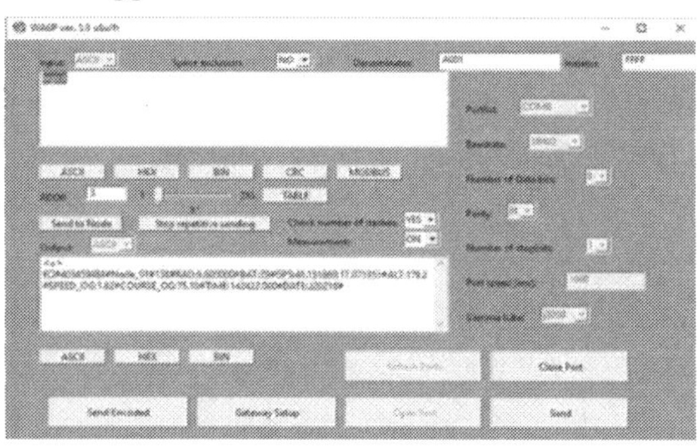

Fig. 2. RMS-WASP control software

E. Currently used UAV system

Although the RMS is compatible with almost every UAV that meets the required payload capacity, our reference UAV type is the DJI MATRICE – 600, owned and operated by UAVONIC ltd. This UAV was used in case of all testing and demonstration activities performed so far. The main parameters can be found in [4] and the UAV can be seen in Fig. 3.

Fig. 3. DJI matrice 600 with the RMS modules.

IV. LABORATORY TESTING AND CALIBRATION

In the development phase, the sensor modules and the RMS-WASP control software were intensively tested. First, to verify the functionality of components, the nodes were tested individually. To verify the gamma conversion factor of the J305β GM tube the Thermo-Fisher FH 40G-L10 [5] handheld multi-purpose digital survey meter was used. It is a wide range digital Geiger counter suitable for nearly all measurement tasks arising in radiation protection. It is a gas filled detector that measures H*(10) ambient dose equivalent. In order to provide stabilized results, the detector uses time integration constants that are functions of the measured dose rate.

For the experiments three sources of gamma radiation were used, Co-57, Cs-137 and Co-60. The last two sources are commonly used calibration sources. Co-57 was used, in order to evaluate the effectiveness of the sensor module to measure complicated gamma decay. The experimental setup (see Fig. 5) consists of a sliding bench, an adjustable holder, a lead shielding and of the detection position. The gamma sources were placed at exactly the same height in every case and 6 distances from the detector were investigated, 10 – 60 cm. The results from the FH 40G-L10 certified detector were read directly from the display and evaluated at regular time intervals. The results from the RMS-001 module were directly stored in text file on the measurement laptop. The results from the text file were averaged and normalized so as the results from both detectors were in the same time range. For the purpose of the results verification obtained by measurement and from the future development, the experimental setup was modeled in SCALE6 system and all measurements were simulated.

The obtained results showed that the precision of the RMS-001 radiation measurement module is comparable with the certified detector, however the death time of the detection tube plays important role. For real application, when the level of radiation exceeds 100μ Sv/h a GM tube designed for higher dose rate will have to be used. It was found out, that the first configuration was also sensitive to beta particles and X-ray. If the goal is to measure pure gamma radiation the detector must be shielded against beta particles and X-rays by material that has minimal cross section for the interaction with the abovementioned types of radiation.

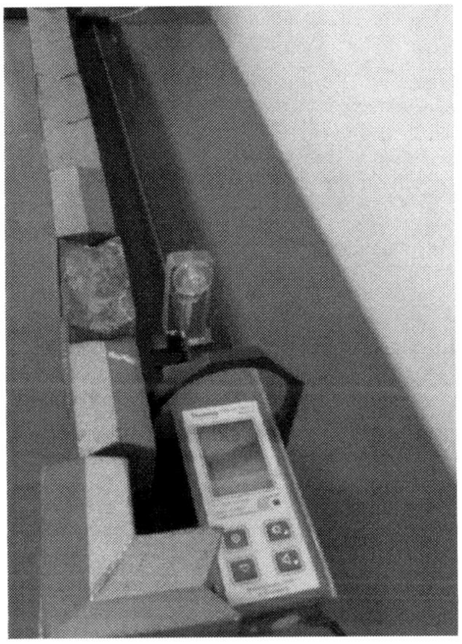

Fig. 4. *The experimental workspace*

V. Demonstration of RMS and UAV System

Demonstration of the use of the developed technology was carried out as part of the regular emergency planning and preparedness of EBO NPP on 26th October 2017. The demo flight was carried out by the staff and technology of UAVONIC ltd as a proposed scenario. The exercises were attended by the staff of the radiation protection, physical protection and dosimetry of SE a.s. and also the Director of EBO V2 NPP. The radiation map created from the measurements during the exercise is shown in Fig. 4. The specification of the locality, where the demonstration was carried out, is that in near area the dosimetry stations of Slovak Hydrometeorological Institute (SHMI) and NPP Jaslovské Bohunice are placed. Mean value of the dose rate measured during demonstration is compared to the values of SHMI and NPP JB from periodical monitoring of the locality in Table IV.

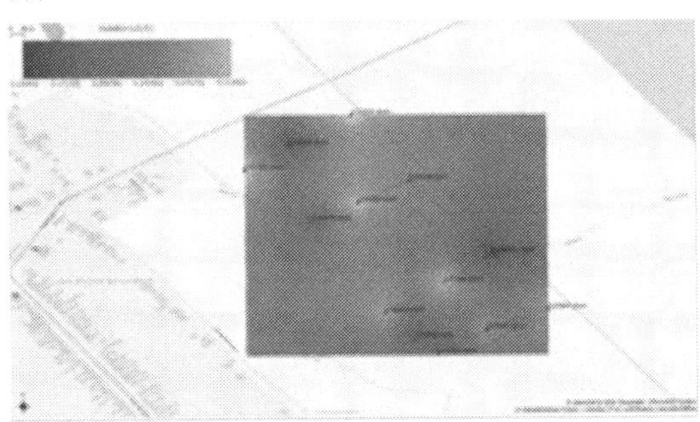

Fig. 5. Radiation map from the demonstration flight.

TABLE IV. Dose rate measurements (nSv/h)

RMS-001	SHMI [6]	NPP JB [6]
247.36 ± 61.34	124.19 ±7.41	82.25 ± 2.94

Differences between individual data are caused by different measurement technique and measurement conditions. Especially in our case the main factor is the short integration constant (25 seconds), the low number of measured intervals (13) and the calibration constant of the GM tube (123,15 cpm/μSv/h) where the statistical character of the background radiation plays an important role. Though the operational range of the J305β GM tube starts from 50 nSv/h, this GM tube was not designed for precise background measurement. In case of the average 247.36 nSv/h the GM tube accounted for only 0.45 counts per second, thus increase of measured counts by 1 cpm may cause significant deviation in the results. However, if precise background measurements are to be carried out, the J305β GM tube can be simply replaced, without requiring the RMS electronic and software parts to be modified. Data of SHMI and NPP JB are the mean value of monthly measurement with integration constant 10 min and 24 hours respectively.

VI. Conclusion

Although the design of the current generation of NPPs incorporates features that minimize the risk of large radioactive releases outside the reactor, it is still important to focus on the development of systems that can mitigate the consequences of such events. In situations when the level of radiation does not permit the personal to perform the required measurements, online unmanned radiation monitoring systems may come to the play. Such a system has been developed in the cooperation of the B&J NUCLEAR ltd. and UAVONIC ltd. companies. It is the RMS-00x technology and its main features are the modularity, long operation time, high level of reliability and the operation with an UAV system. Currently the system consists of the RMS-000 communication module, 3 measuring modules RMS-001, RMS-002 and RMS-003 and the control software RMS-WASP. This configuration makes possible to measure the effective dose of gamma and beta radiation up to the level of 277 μSv/h, air sampling with the speed twice to the human respiration rate, collection of the parameters of the measurement environment and the creation of online radiation maps. The RMS-001 module was tested and calibrated using various radioactive sources and certified measurement instruments at INPE FEI STU, later the testing was performed in environmental conditions. The obtained results showed that the precision of the RMS-001 radiation measurement module is comparable with the certified detector, however the death time of the detection tube plays important role. Similarly, the RMS-002 module was first calibrated in laboratory conditions. Later the data transfer capabilities were evaluated in real environmental conditions. Currently the radiation measuring module is being upgraded to perform the measurement of effective doses rate up to the level of 2 Sv/h. Also modules for neutron dose rate measurements and online gamma spectrometry are being developed. We believe that our system could contribute to enhance the level radiation safety of nuclear power plants in Slovakia and also abroad.

Acknowledgment

This work was supported by the B&J NUCLEAR ltd., the UAVNOC ltd. and; the Slovak Research and Development Agency [APVV-16-0288].

https://doi.org/10.1051/epjconf/202022508008

References

[1] A.J. Gonzáles, et al. "Radiological protection issues arising during and after the Fukushima nuclear reactor accident", J. Radiol. Prot., 33, 497-571 (2015).

[2] C. Beylone, Leveraging Drones to Improve Nuclear Facility Security and Safety, online: https://www.chathamhouse.org/expert/comment/16722, (2017)

[3] Š. Čerba, Callibration of the radiation monitoring module Martan (in Slovak), In B&J NUCLEAR ltd. internal report, 2017, undisclosed

[4] DJI, MATRICE 600 user manual, online: https://dl.djicdn.com/downloads/m600/20171017/Matrice_600_User_Manual_v1.0_EN.pdf, (2017)

[5] ThermoFisher Scientific, „FH 40 G Multi-Purpose Digital Survey Meter – product catalogue", 2017, online: https://www.thermofisher.com/order/catalog/product/4254002

[6] T. Melicherová, et al. Záverečná ročná správa Čiastkového monitorovacieho systému "Rádioaktivita životného prostredia" 2017, in SHMI and MŽP report, 2017, online: http://www.shmu.sk/File/radioaktivita/Zaverecna_sprava_CMS_Radioaktivita_2017_final.pdf

EPJ Web of Conferences **225**, 09001 (2020) https://doi.org/10.1051/epjconf/202022509001
ANIMMA 2019

The Potential of Photon Activation and Neutron Activation Techniques for Fast Soil Characterization

Adrien Sari
CEA List
Gif-sur-Yvette, France
adrien.sari@cea.fr

Sara Garti
CEA List
Gif-sur-Yvette, France
sara.garti@cea.fr

Frédéric Lainé
CEA List
Gif-sur-Yvette, France
frederic.laine@cea.fr

Frédérick Carrel
CEA List
Gif-sur-Yvette, France
frederick.carrel@cea.fr

Jonathan Dumazert
CEA List
Gif-sur-Yvette, France
jonathan.dumazert@cea.fr

Hamid Makil
CEA List
Gif-sur-Yvette, France
hamid.makil@cea.fr

Nicolas Dufour
CEA List
Gif-sur-Yvette, France
nicolas.dufour3@cea.fr

Caroline Mougel
CEA DEN
Gif-sur-Yvette, France
caroline.mougel@cea.fr

Adeline Masset
CEA DEN
Gif-sur-Yvette, France
adeline.masset@cea.fr

Lucile Mondon
VINCI Construction Terrassement
Nanterre, France
lucile.mondon@vinci-construction.com

Laurent Boutillon
VINCI Construction Grands Projets
Rueil-Malmaison, France
laurent.boutillon@vinci-construction.com

Jean-Pierre Lejeune
VINCI Construction Grands Projets
Rueil-Malmaison, France
jean-pierre.lejeune@vinci-construction.com

***Abstract*—In the frame of a partnership between CEA and VINCI, various measurement techniques are applied to soil analysis and tested in different laboratories located at CEA Saclay (France). This paper deals with two nuclear measurement techniques assessed in this project. More specifically, this paper presents the feasibility study carried out for two non-destructive active methods: photon activation and neutron activation. First, some atomic nuclides are activated either by photons or neutrons. Secondly, gamma-rays of specific energies are emitted by activated nuclides and gamma-ray spectrometry enables to identify these activated nuclides. Calibration of the full measurement system with reference samples would enable to quantify the mass of activated nuclides. Irradiations performed for photon activation measurements were conducted using a linear electron accelerator (linac) as the latter enables to generate high-energy photons by *Bremsstrahlung* thanks to its conversion target. Furthermore, irradiations performed for neutron activation measurements were also conducted with a linac. Indeed, photons may be converted to neutrons by photonuclear reactions using a secondary target. In the frame of this project, experiments were carried out at the SAPHIR platform (CEA Saclay) with a Linatron-M9 VARIAN linac. The electron energy was either 6 or 9 MeV. For neutron activation measurements, a secondary target made of heavy water has been used as neutron source and a polyethylene cell enabled to thermalize neutrons and increase the number of reactions of interest. In this paper, we present the different experimental setups and the measurement protocols established for this feasibility study. We show experimental results obtained with raw material samples coming from three construction sites.**

Keywords*—photon activation, neutron activation, linear electron accelerator (linac), gamma spectrometry, prompt gamma-rays, delayed gamma-rays, soil characterization.

I. Introduction

In the frame of a partnership between CEA and VINCI, various measurement techniques are applied to soil analysis and tested in different laboratories located at CEA Saclay (France). This paper deals with two nuclear measurement techniques assessed in this project. More specifically, this paper presents the feasibility study carried out for two nuclear non-destructive active methods: photon activation and neutron activation [1–3]. First, some atomic nuclides are activated either by photons or neutrons. Secondly, gamma-rays of specific energies are emitted by activated nuclides and gamma-ray spectrometry enables to identify these activated nuclides. In summary, the aims of this study are the following:

- Assessment of the potential of photon activation and neutron activation for fast soil analysis;
- Implementation using an HPGe detector and a linear electron accelerator (linac) operated at either 6 or 9 MeV;
- Tests on samples of raw material coming from three different construction sites.

The Linatron® M9 manufactured by VARIAN (now Varex Imaging) and located at the SAPHIR platform (CEA Saclay) was used for these experiments (shown in Fig. 1). This linac generates either 6 or 9 MeV electrons. High-energy photons are produced by *Bremsstrahlung* in its conversion target.

Fig. 1. The Linatron-M9 linac housed at the SAPHIR platform (CEA Saclay, France).

II. Photon interrogation

Photon activation measurements are illustrated by Figs. 2 and 3 and were carried out with the following measurement protocol:

- Electron energy: 9 MeV;
- Pulse frequency: 200 Hz;
- Irradiation time: 10 min;
- Cooling time: 2 min;
- Acquisition time: 10 min.

The gamma-ray spectrum of the background is shown in Fig. 4. The latter is due to the natural radioactivity and to the history of the SAPHIR platform. Indeed, presence of ^{137}Cs is explained by the fact that the building used to house a research reactor a few decades ago. Fig. 5 presents the delayed gamma-ray spectrum obtained for the sample No. 1. Two radioactive nuclei seem to be produced by photon activation at 9 MeV: $^{56}Fe^*$ and $^{28}Si^*$. Using a 30 MeV linac, much more elements would be identified in samples by photon activation [4]. However, deployment of such a linac would not be permitted on a construction site.

Fig. 2. Photon activation measurement: irradiation step.

Fig. 3. Photon activation measurement: delayed gamma-ray detection step.

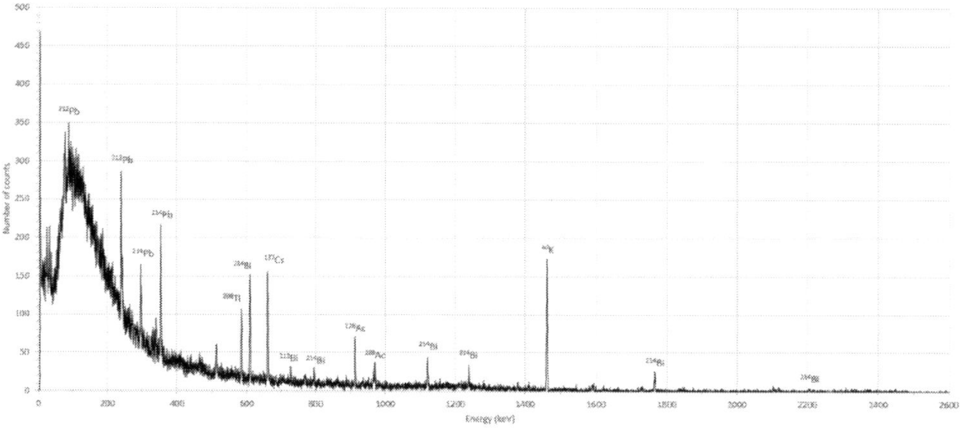

Fig. 4. Delayed gamma-ray spectrum of the background.

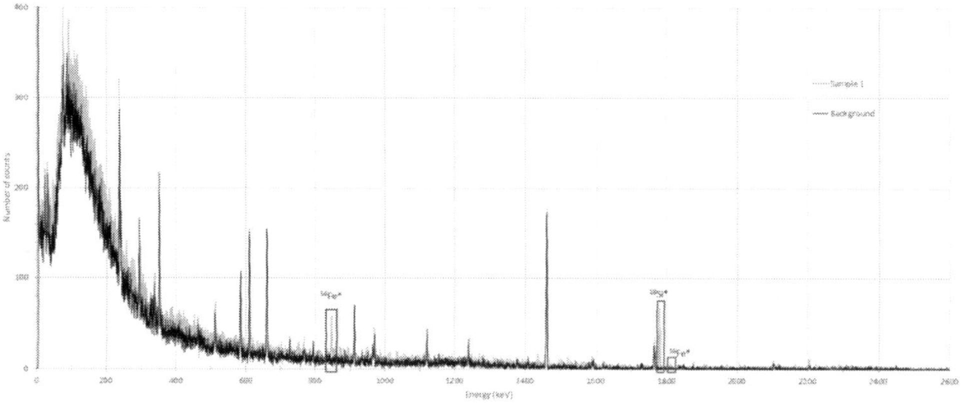

Fig. 5. Delayed gamma-ray spectrum of the sample No. 1.

III. Neutron interrogation

A. Production of photoneutrons

Photoneutrons are produced by (γ, n) reactions on deuterium using a secondary target made of heavy water (D_2O), as shown in Fig. 6. The electron energy was set at 9 MeV and the mass of heavy water was 16 kg. At 200 Hz, the average emission intensity is 1.16×10^{10} neutrons per second in 4π steradians [5]. Fast photoneutrons (energy spectrum shown in Fig. 7 [5]) are thermalized thanks to a polyethylene cell [6].

B. Detection of delayed gamma-rays: setup No. 1

The first experimental protocol dedicated to the detection of delayed gamma-rays was the following:

- Linac operated at a frequency of 200 Hz;
- 20 min of irradiation time;
- 2 min of cooling time;
- 20 min of acquisition time.

Fig. 8 shows both a sample of soil placed in a corner of the neutron cell (ouside the photon beam) and a sample of soil placed in front of the HPGe detector. Fig. 9 presents experimental results obtained. Titanium, manganese, magnesium, sodium, vanadium, and aluminum are detected.

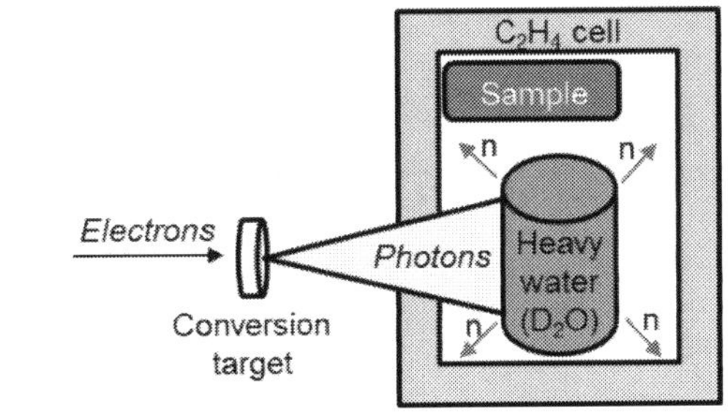

Fig. 6. Neutron activation measurement: production of photoneutrons.

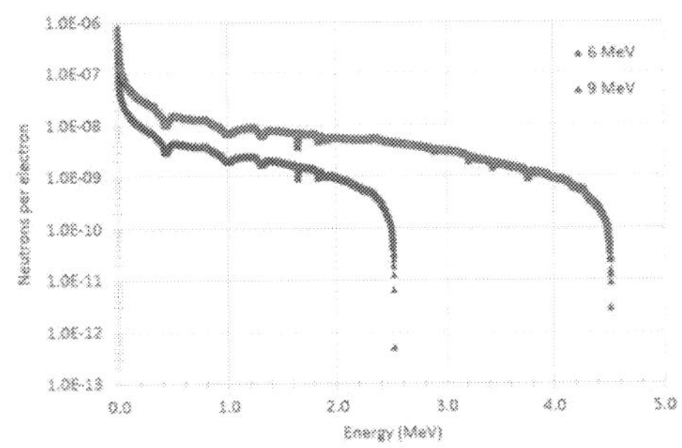

Fig. 7. Neutron energy spectra for two electron energies (MCNP6 simulation results) [5].

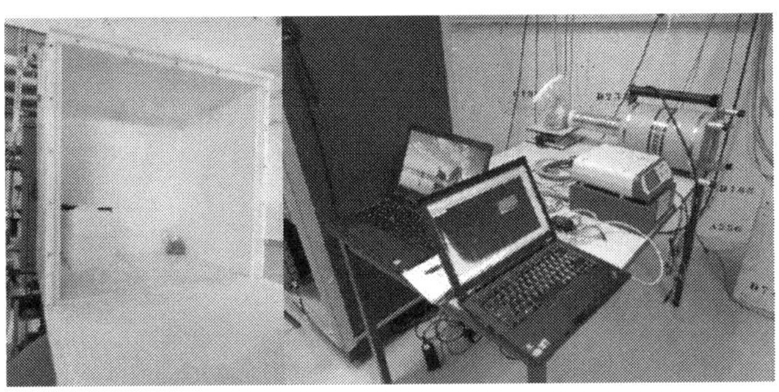

Fig. 8. On the left: irradiation step (neutron cell in open position). On the right: gamma-ray spectrometry detection step with an HPGe detector.

Fig. 9. Delayed gamma-ray spectra obtained for the three samples of soil with setup No. 1.

Fig. 10. Neutron activation measurement: view of the setup No. 2. HPGe detector placed on the side of the neutron cell.

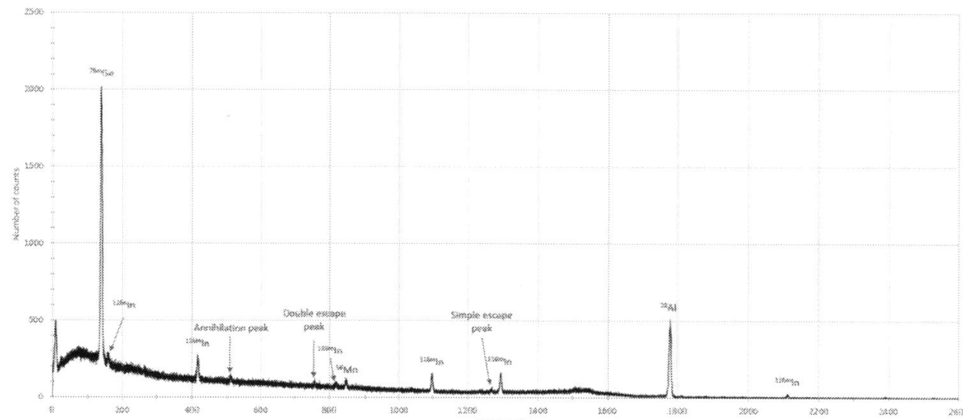

Fig. 11. Delayed gamma-ray spectrum of the background obtained with the setup No. 2.

C. Detection of delayed gamma-rays: setup No. 2

A second setup (shown in Fig. 10), more adapted to the detection of delayed gamma-rays emitted by radionuclides characterized by short half-lives, has been tested. The experimental protocol was the following:

- Linac operated at 6 MeV and 40 Hz;
- 16 kg of heavy water (1.62×10^9 neutrons per second in 4π sr [5]);
- 5 min of irradiation time;
- Gamma spectrum acquisition launched immediatly once linac switched off.

The delayed gamma-ray spectrum of the background is shown in Fig. 11. This spectrum shows the feasibility of such measurement (HPGe detector not degraded) but a shielding around the HPGe detector would be required to reduce activation of the detector itself [7] and thus the background.

D. Detection of prompt gamma-rays

Measurements aiming at detecting prompt gamma-rays from neutron activation between linac pulses have been explored. Indeed, as half-life of neutrons in the cell is on the order of 1 ms, neutron activation reactions occur after a few hundreds of microseconds up to a few milliseconds after the gamma flash from the linac. Prompt gamma-rays from neutron activation reactions are then emitted after the gamma flash and can be potentially detected.

Detection of prompt gamma-rays, which is complementary to the detection of delayed gamma-rays, is a technical challenge. Measurement of gamma-rays is carried out between linac pulses. The linac and the HPGe detector have to be synchronized. The HPGe detector has to be inhibited during the gamma flash (risk of saturation of the detector), and the detector must recover quite fast.

https://doi.org/10.1051/epjconf/202022509001

Fig. 12. View of the setup dedicated to the detection of prompt gamma-rays from neutron activation reactions.

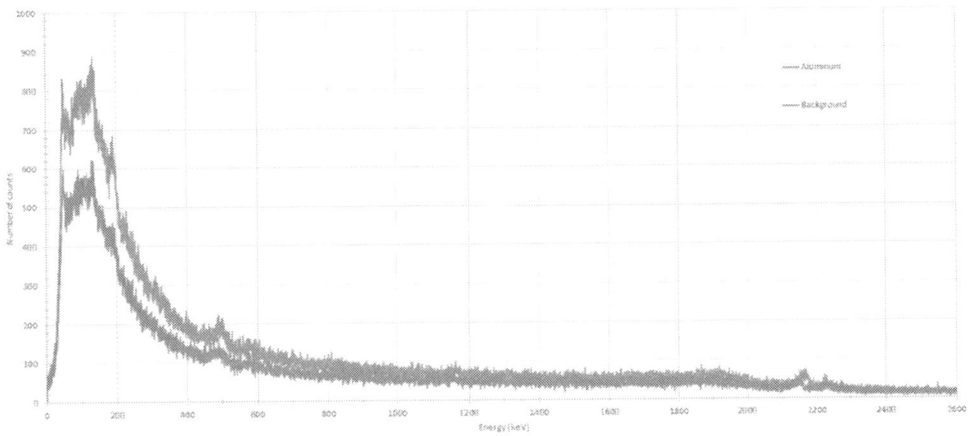

Fig. 13. Prompt gamma-ray spectrum obtained with aluminum at 6 MeV.

Fig. 12 shows a view of the experimental setup dedicated to the detection of prompt gamma-rays from neutron activation reactions. The experimental protocol was the following:

- Linac operated at either 6 or 9 MeV and 40 Hz;
- Acquisition between linac pulses during a few minutes;
- HV supply in a NIM rack located in the irradiation hall;
- Hexagon digitizer manufactured by CAEN located in the control room;
- First measurements on pure aluminum or vanadium foils of a few tens of grams.

In summary, experimental results obtained at 9 MeV lead to failure of the HPGe detector. Nevertheless, measurements performed at 6 MeV were successful. The inhibition signal was set at 1 ms and the dead time of the HPGe detector was stable and close to 70%. An example of spectrum measured at 6 MeV with a sample of aluminum is presented in Fig. 13. An excess of signal around 41% is obtained despite a low geometric detection efficiency.

Then, measurements were performed with various numbers of pure aluminum or vanadium foils. Figs. 14 and 15 present results obtained. A linear behavior can be seen between prompt gamma-ray signals and aluminum or vanadium masses. These results demonstrate the feasibility of the detection of prompt gamma-rays from neutron activation reactions by interpulse gamma spectrometry with a 6 MeV linac and an HPGe detector.

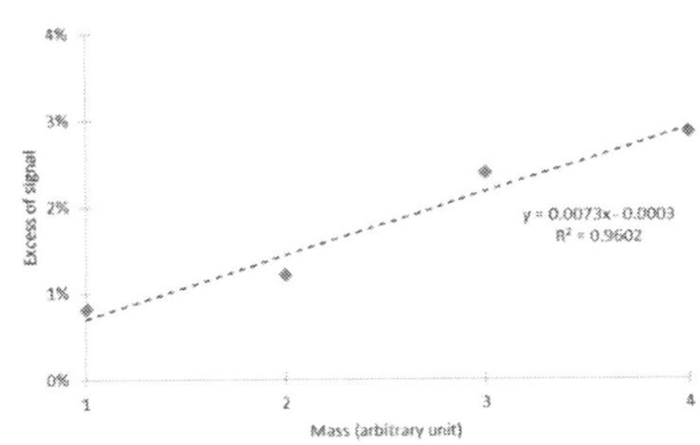

Fig. 14. Experimental results obtained for various masses of aluminum.

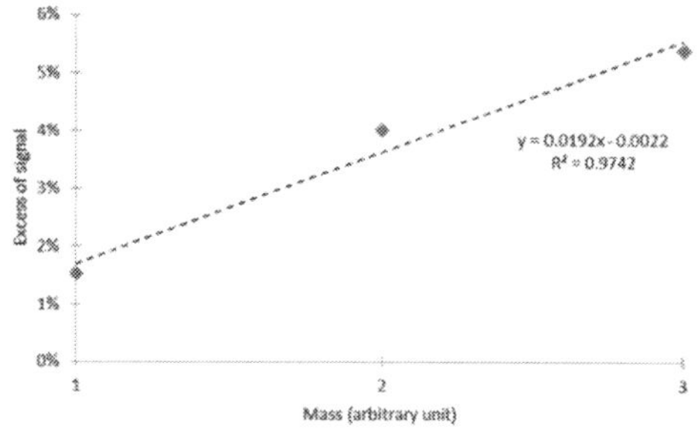

Fig. 15. Experimental results obtained for various masses of vanadium.

IV. Conclusion and Outlook

Finally, photon activation with a 9 MeV linac enabled to detect iron and silica in the samples of soil by delayed gamma-ray spectrometry using an HPGe detector. Using a 30 MeV linac, much more elements would be identified in samples by photon activation. However, deployment of such a linac would not be permitted on a construction site. Neutron activation using a 6 or 9 MeV linac and heavy water enabled to detect various elements in the samples of soil (titanium, manganese, magnesium, sodium, vanadium, and aluminum) also by delayed gamma-ray spectrometry with an HPGe detector. The feasibility of the detection of prompt gamma-rays from neutron activation measurements between linac pulses has been demonstrated using pure aluminum or vanadium samples.

Further, development of the interpulse gamma spectrometry technique with a linac will be pursued. A shielding of the HPGe detector will be designed, the blanking signal will be optimized, and new measurements will be performed with various materials.

References

[1] G. Revel, « Analyse par activation », Techniques de l'Ingénieur, Réf. P2565 V3, 1999.

[2] R. M. Lindstrom, "Prompt gamma neutron activation analysis (PGAA): recent developments and applications", Journal of Radioanalytical and Nuclear Chemistry (2017) 314:843–858.

[3] C. Engelmann, « Principe, possibilités et applications de l'analyse par activation aux photons gamma », rapport CEA, DRA/SAECNI/79-187/382/CE/MC, 1979.

[4] T. Kato, I. Morita, N. Sato, "Non-destructive photon-activation analysis of standard rocks with 30 MeV Bremsstrahlung", Journal of Radioanalytical Chemistry, Vol. 18 (1973) 97-110.

[5] A. Sari, F. Lainé, F. Carrel, "Characterization and optimization of the photoneutron flux emitted by a 6- or 9-MeV electron accelerator for neutron interrogation measurements", IEEE Trans. Nucl. Sci., Vol. 65, No. 9, 2018.

[6] A. Sari *et al.*, "Design of a neutron interrogation cell based on an electron accelerator and performance assessment on 220 liter nuclear waste mock-up drums," IEEE Trans. Nucl. Sci., vol. 61, no. 4, pp. 2144-2148, 2014.

[7] R. Pourimani *et al.*, "Resonance capture γ-ray spectrometry at Lead Slowing-down neutron spectrometer", Nuclear Instruments and Methods in Physics Research A 488 (2002) 226–239.

MAPSSIC, a communicating MAPS-based intracerebral positrons probe for deep brain imaging in awake and freely-moving rats

F. Gensolen*, L .Ammour, M. Bautista, J. Heymes, S. Fieux, M. Kachel, F. Lefebvre, P. Pangaud, L. Pinot, P. Lanièce, J. Baudot, P. Gisquet, L. Zimmer, M-A. Verdier, C. Morel
* Aix Marseille Univ, CNRS/IN2P3, CPPM, Marseille, France
gensolen@cppm.in2p3.fr

Abstract— Radioisotope imaging is a powerful tool to understand the biological mechanisms in-vivo, especially in the brain of small animals, providing a significant model to study the human brain.

In this context, we have developed and built a pixelated intracerebral positron probe to be embedded on awake and freely moving small animals, typically rats. This pixelated probe will represent a key instrument for neuroscientists to study neural mechanisms and correlate them to behavioral experiments.

We describe in this paper the simulations carried out to design the intracerebral sensor, its architecture, and the detection of positrons in a volume with a couple of sensors assembled back-to-back. We also depict the architecture of the wireless acquisition system. Finally, we present the first measurements performed in real-time by this miniaturized probe with sealed radioactive sources and a ^{18}F solution.

Index Terms— **Active pixel sensors, Animal behavior, CMOS Image sensors, Implants, Molecular imaging, Monte Carlo methods, Nuclear Imaging, Positron emission tomography**

I. INTRODUCTION

In the context of understanding the brain function, which is altered by diseases such as Parkinson, Alzeimer or by addictions, the correlation of neural activity measurements with behavioral experiments is of major interest for biologists. Small animals constitute preclinical models for this study. Therefore, we have started the development of a miniaturized and real-time intracerebral positrons probe for awake and freely moving small animals, typically rats. Taking advantage of the availability of new technologies and of the expertise we have acquired from former developments [1], [2], we propose a novel, robust, wireless, autonomous and lightweight active pixel positron probe.

We present in the following section the main results of the simulations carried out to validate the intracerebral sensor design with an emphasis on the modeling of the measurements with radioactive solutions (in particular ^{18}F).

Then we describe the architecture of the sensor and detection of positrons in a volume with the whole probe made of a couple of sensors assembled back-to-back. The wireless acquisition system is also described.

Finally, in the last section, we present the first measurements performed in real-time by this miniaturized probe and compare them to Monte Carlo simulations.

II. MONTE CARLO SIMULATIONS OF THE SENSOR DESIGN

In order to estimate the sensor performance regarding positron detection (referred to as *direct detection* in Fig 1) and rejection of annihilation photons and Compton electrons (referred to as *indirect detection* in Fig 1), we have performed Monte Carlo simulations using Gate [3] and studied the sensor performances including efficiency and sensitivity, and the deposited energy.

The pixel sensor circuit is developed in 180 nm CMOS technology, with an 18 μm thick high resistivity epitaxial layer sensitive to charged particles.

The sensor dimensions for the probe must take into account the biological, the electronics and the mechanical constraints. It results in a $12000 \times 610 \times 200$ μm^3 sensor with a sensitive volume of $6400 \times 480 \times 190$ μm^3. As shown in Fig 1, this model was placed at the center of a cylindrical and homogeneous aqueous solution of ^{18}F, ^{15}O, or ^{11}C. The simulations were carried out, without accounting for signal processing, nor charge diffusion. For more details about the Monte Carlo model of the sensor and the associated analysis, please refer to [4].

A. Sensor efficiency

The efficiency of the probe given by the count rate of the probe immersed in a ^{18}F solution is illustrated in Fig 2, where isocontours at 10% and 1% are drawn. It appears that the efficiency of the probe to the detection of positrons is far greater than for the detection of 511 keV annihilation photons

---

[1]This work was supported in part by the Project Instrumentation aux limites from CNRS in 2015 and 2017, in part by the France Life Imaging of the French Investissements d'Avenir program run by the Agence Nationale pour la Recherche under Grant FLI/ANR-11-INBS- 0006, and in part by IN2P3.

M. Bautista, F. Gensolen, P. Pangaud, and C. Morel are with the Aix Marseille Univ, CNRS/IN2P3, CPPM, Marseille, France.

L. Ammour, F. Lefebvre, L. Pinot, P. Laniece and M-A Verdier are with the Imagerie et Modélisation en Neurobiologie et Cancérologie (IMNC), CNRS, Univ Paris Sud, Université Paris-Saclay, F-91406 Orsay, France.

L. Ammour, F. Lefebvre, L. Pinot, P. Laniece and M-A Verdier are with Université de Paris, IMNC, F-91400 Orsay, France

J. Heymes, M. Kachel, and J. Baudot are with the Université de Strasbourg, CNRS, IPHC UMR 7178, Strasbourg, France.

S. Fieux and L. Zimmer are with the CERMEP-Imagerie du vivant, Université Claude Bernard Lyon 1, CNRS, INSERM, Hospices Civils de Lyon, Lyon, France.

P. Gisquet-Verrier is with the NeuroPSI CNRS/INSB, Univ Paris Sud, Orsay, France.

https://doi.org/10.1051/epjconf/202022509002

and Compton electrons. This is what we expect from the sensor, which is designed to determine the local distribution of positron emitters in the brain. The volume bounded by the 1% efficiency isosurface is $(17.34 \pm 0.05)\ \mathrm{mm}^3$. This is in accordance to the volumes of interest for the rat brain structures, hippocampus, or caudate putamen, which are in the order of 40 and 30 mm³, respectively.

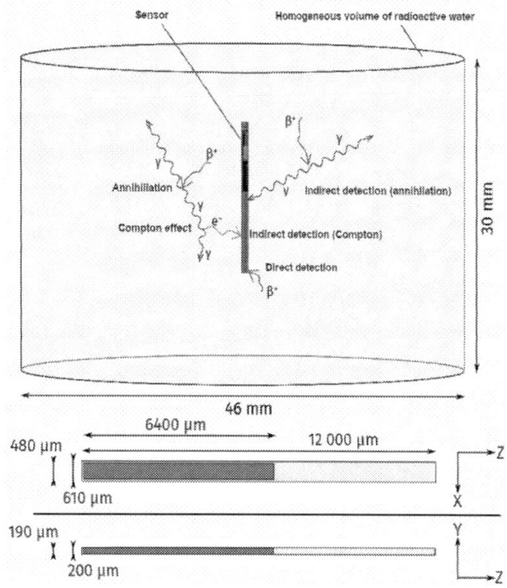

Fig 1. Sketch of the cylindrical radioactive solution (top) with the silicon sensor immersed in the center of the solution (bottom).

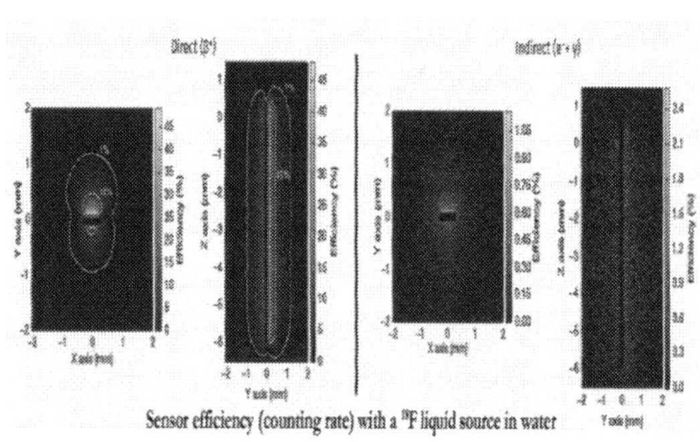

Fig 2. Spatial distribution of the efficiency given by the count rate of the sensor immersed in a ^{18}F water solution.

B. Probe sensitivity

The probe sensitivities to direct detection S_{direct} with ^{11}C, ^{15}O and ^{18}F are summarized in Table 1. We observed a greater sensitivity to direct detection for ^{11}C and ^{15}O than for ^{18}F solutions. This difference arises from the fact that positrons have a larger range in water for ^{11}C and ^{15}O than for ^{18}F. Nevertheless ^{18}F is the most used isotope and has the best definition in terms of brain structure due to the lower range of its positrons. In addition sensitivity with ^{18}F has been proven sufficient with the previous probe version [2]. Indirect detection sensitivity $S_{indirect}$ continuously increases because of the high penetration of 511 keV annihilation γ-rays in water.

Source	^{18}F	^{11}C	^{15}O
S^{max}_{direct} $(\mathrm{cps} \cdot \mathrm{Bq}^{-1} \cdot \mathrm{mm}^3)$	$(8.83 \pm 0.01) \times 10^{-1}$	1.79 ± 0.01	4.34 ± 0.01
$50\% S^{max}_{direct}$ radius (mm)	0.50	0.77	1.6
$99\% S^{max}_{direct}$ radius (mm)	1.73	2.85	5.96
$99.9\% S^{max}_{direct}$ radius (mm)	2.12	3.53	7.36
$\frac{S_{direct}}{S_{indirect}} = 20$ radius (mm)	1.53	2.96	5.70
$\frac{S_{direct}}{S_{indirect}} = 10$ radius (mm)	3.55	7.48	20.3
$\frac{S_{direct}}{S_{indirect}} = 5$ radius (mm)	8.28	20.8	> 23

Table 1. Maximum direct detection sensitivity values, radii for a given sensitivity value, and radii for different ratio of direct detection sensitivity over indirect detection sensitivity.

With a ^{18}F source of 5 mm radius, we get a S_{direct} over $S_{indirect}$ ratio of $8{,}28 \pm 0{,}01$, which also make this isotope the most interesting for our application.

C. Deposited energy spectra

The CMOS monolithic active pixel sensor (MAPS) features 16×128 pixels of $30 \times 50\ \mu\mathrm{m}^2$. The spectra of deposited energies by positrons, electrons and photons are presented in Fig 3. Deposited energies are maximum for 6.9 keV and 7.9 keV for positrons and electrons respectively, while 511 keV γ-rays involve relatively lower energy depositions.

Table 2 summarizes the deposited energy distribution for direct detection. This confirms that the MAPS with a detection threshold in the order of few keV can be suitable for direct detection of positrons.

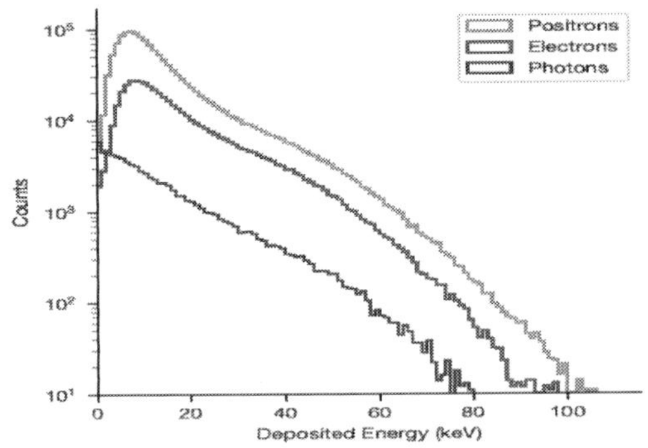

Fig 3. Deposited energy spectra for 16 × 128 pixels 30 × 50 µm² immersed in a homogeneous ^{18}F solution. In case multiple pixels are hit in the same event, the highest deposited energy in one pixel is considered.

Isotopes	^{18}F	^{11}C	^{15}O
E_{peak} (keV)	6.9	7.2	6.1
E_{median} (keV)	11.0	9.51	8.72
E_{mean} (keV)	15.1	12.9	11.4
fraction < 1 keV	0.37 %	0.44 %	0.59 %
fraction < 10 keV	44.3 %	52.9 %	59.0 %

Table 2. Peak, median, and mean deposited energy in the pixels for direct detection only. Sensor immersed in a homogeneous ^{18}F solution.

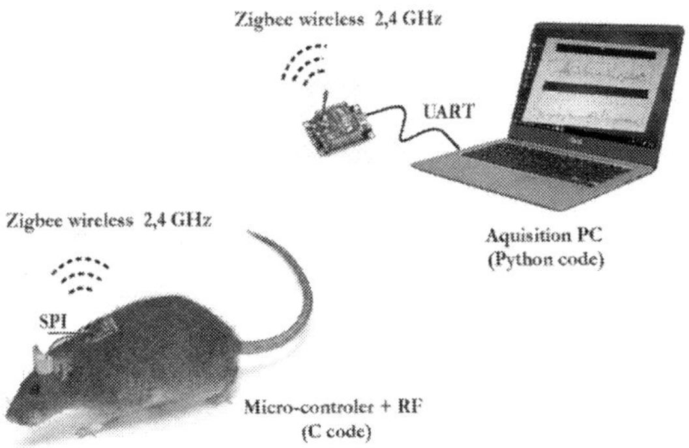

Fig 4: Artistic view of the whole acquisition system.

III. MAPSSIC PROBE DESIGN AND SETUP

At this point the positron sensor has been specified and modeled by simulations. We had then to build a wireless, lightweight, robust, autonomous, biocompatible and reliable acquisition system. For this, we realized the MAPS CMOS sensor circuit, then assembled a couple of these sensors back-to-back to constitute the probe, and designed a hybrid flex-rigid printed circuit board (PCB) to interface the probe with the embedded control and communication electronics. Finally, we coated the probe to make it biocompatible. The following sections detail all the design and manufacturing steps of the setup. Fig 4 shows the whole acquisition system.

A. MAPS CMOS circuit

Our positron sensor has been designed in a 0.18 µm CMOS technology on 18 µm thick high resistivity (> 1 kΩ·cm) epitaxial layer. It is composed of 30 × 50 µm² pixels arranged in a 16 × 128 matrix and has a power consumption lower than 55 nW/pixel [5]. The pixel readout is a rolling shutter scheme with an asynchronous memorization of the hit in the pixel (1 bit). Each pixel is reset right after it has been readout. The control of the pixel analog biasing and configuration (DACs values) is performed via SPI (Serial Peripheral Interface) protocol.

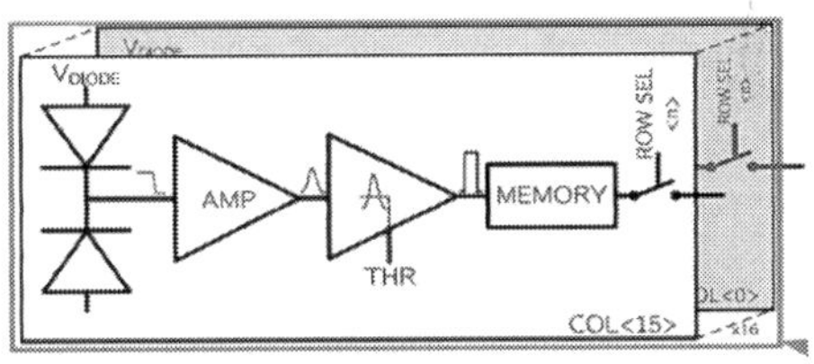

Fig 5: Architecture of a pixel.

B. 3D positron sensing and biocompatible coating

The positrons are detected mainly on the epitaxial layer side of the sensor. In order to build a probe able to detect the positrons on both sides while increasing the sensitivity of the probe as well as its robustness, we decided to assemble a couple of sensors back-to-back, as shown on Fig 6.

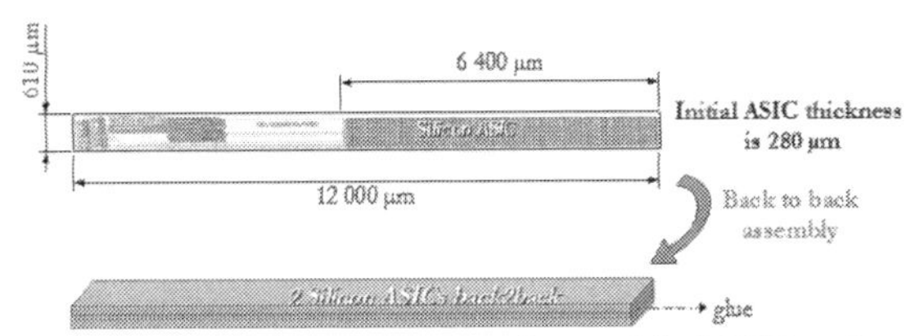

Fig 6: Dimensions and geometrical view of the MAPS probe.

Back-to-back assembly of a couple of MAPS probes wire-bonded to a thin hybrid flex-rigid PCB resulted in 3D devices featuring various thicknesses of 400, 500 and 560 µm.

An additional 6 µm biocompatible coating of parylene was then deposited over the whole probe surface to allow for its implantation in the brains (Fig 7).

https://doi.org/10.1051/epjconf/202022509002

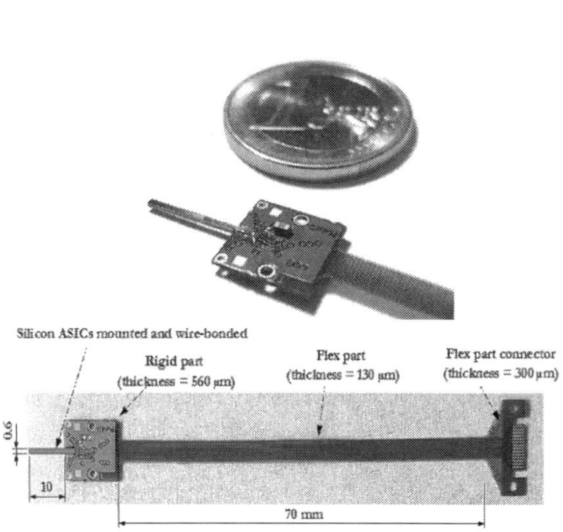

Fig 7: Back-to-back sensors assembly mounted and wire-bonded on an interface PCB in hybrid flex-rigid PCB process.

Fig 8: Encapsulated probe PCB interface.

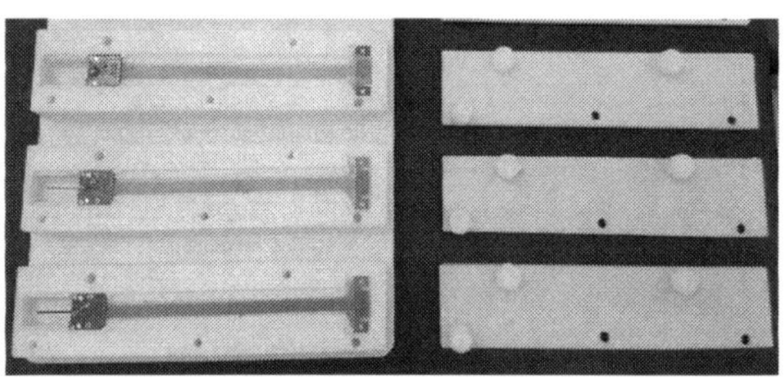

Fig 9: 3D-printing tool designed for the protection and safe transportation of the probes.

C. *Wireless embedded electronics*

As described in Fig 4, the control of the probe is carried out by a computer, which communicates with an embedded electronics through the Zigbee 2.4 GHz wireless protocol. This radio communication is performed via2two radio-frequency Xbee S2C modules from Digi-International. These modules are very well suited for our application since each module embeds two micro-controllers. One micro-controller is programmable and therefore set up to control the probe and readout the pixels, the other one is dedicated to the RF communication.

A battery and a power management is also embedded on the animal to provide energy. All the embedded electronics weights less than 20 g.

The micro-controller program is written in C for the implementation of low level instructions. The computer software has been developed in Python for the Human Machine Interface in order to control the probe and manage data handling.

IV. MEASUREMENTS RESULTS

To validate experimentally the probe, we have measured the counting rate of the probe with different sealed radioactive sources and its sensitivity in a radioactive solution.

A. *Radioactive setup*

Fig 10 bellow describes the setup for measurements with the probe immersed in a ^{18}F water solution. The probe flex connector is plugged in the corresponding male part with an interface board connected to the micro-controller PCB.

Fig 10: Experimental setup with the probe immersed in a ^{18}F water solution.

In the case of measurements with sealed radioactive sources, we used a trusquin to place the sources at a given and stable distance from the probe in its transportation tool (Fig 9).

 https://doi.org/10.1051/epjconf/202022509002

B. Results

Fig 11 reports some examples of frames acquisitions with different types of radioactive sources irradiating a single sensor.

With a pixel integration time of 224 ms, we measured a dark noise of 5.5×10^{-3} cps/mm² with a few noisy pixels removed. The mean pixel cluster size (typical for beta particles interactions) was 3,87 pixels.

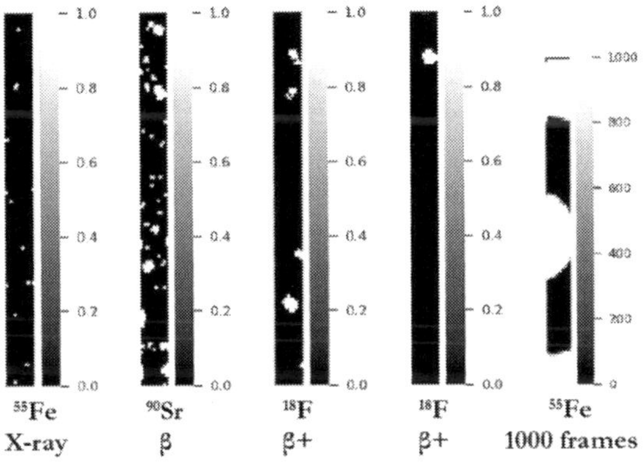

Fig 11: Examples of acquisitions with different types of radioactive sources.

We have also immersed the probe in a ^{18}F water solution and measured a source decay half-life of (110.7 ± 0.8) min., thus assessing the probe linearity (Fig 12).

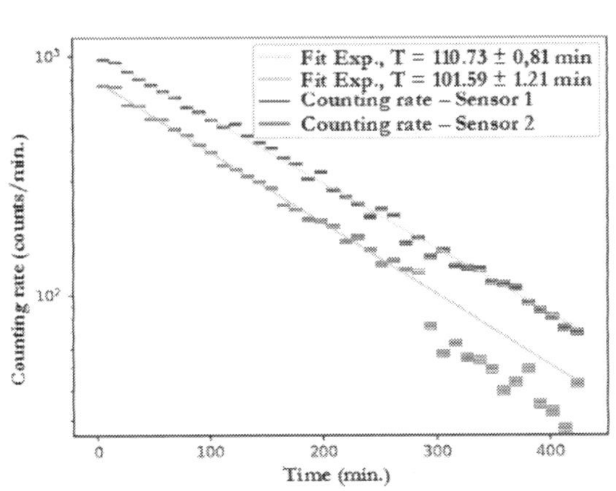

Fig 12: Probe count rate vs. activity.

However we experienced important instabilities leading to strong disparities in the count rate from one probe to the other.

Also the experimental sensitivity measured is 8.9×10^{-2} cps.Bq^{-1}.mm^{3}, to be compared to the simulated one which was (1.10 ± 0.02) cps.Bq^{-1}.mm^{3}.

These two surprising results are possibly explained by a problem on the sensor digital inputs/outputs pads, which suffer from a very important leakage current (tens of mA). This resulted in increasing the temperature of the sensor up to 60 °C, which affected the performances of the chip.

A new version of the MAPS CMOS circuit is currently under design to correct this problem.

V. CONCLUSION

Behavioral neuro-imaging is required to address the limits of anesthesia model and to correlate neuro-imaging with behavioral studies. In that context we developed MAPSSIC, the first CMOS active pixel sensor device for real-time positron imaging of freely moving small animals.

Monte Carlo simulation studies were performed to design the probe, which was then fabricated together with the associated wireless control and acquisition system.

Measurements carried out with sealed sources and also with the probe immersed in a radioactive solution have demonstrated that MAPSSIC can provide a robust tool for in-vivo imaging.

The short terms actions are the design of a new CMOS circuit to correct for high current on digital pads problem encountered with the first version of the probe that resulted in a significant probe temperature elevation.

REFERENCES

[1] F. Pain *et al.*, "SIC: an intracerebral radiosensitive probe for in vivo neuropharmacology investigations in small laboratory animals: prototype design, characterization, and in vivo evaluation." *IEEE Transactions on Nuclear Science*, vol. 49, no. 3, pp.822-826, 2002.

[2] L. Balasse *et al.*, "PIXSIC: A Wireless Intracerebral Radiosensitive Probe in Freely Moving Rats." *Molecular Imaging*, vol. 14, no. 43, pp. 484-489, 2015.

[3] S. Jan et al., "GATE V6: a major enhancement of the GATE simulation platform enabling modelling of CT and radiotherapy". Phys. Med. Biol., vol. 54, no. 4, pp. 881, 2011.

[4] L. Ammour *et al.*, "MAPSSIC, a Novel CMOS Intracerebral Positrons Probe for Deep Brain Imaging in Awake and Freely Moving Rats: A Monte Carlo Study." *IEEE Transactions on Radiation and Plasma Medical Sciences*, vol. 3, no. 3, pp. 302-314, 2019.

[5] J. Heymes et *al.*, "Implantable CMOS pixel sensor for positron imaging in rat brain." *Nucl. Instrum. Meth. A*, vol., 911, pp. 19-24, 2018.

EPJ Web of Conferences **225**, 09003 (2020)
ANIMMA 2019
https://doi.org/10.1051/epjconf/202022509003

Producing Useful Chemicals Using a Nuclear Reactor

Arran George Plant, Dr Vesna Najdanovic-Visak*, Professor Malcolm J. Joyce, Dr. Luka Snoj and Anže Jazbec
Engineering Department, Lancaster University
* v.najdanovic@lancaster.ac.uk

In this paper, the irradiation of glycerol and ethylene glycol by either mixed (neutron-γ) or γ-only (γ) fields at the TRIGA reactor of the Jožef Stefan Institute is described. This is highly relevant to future applications of fission reactor systems to produce useful feedstock derivatives from organic waste, beyond the production of heat and power.

Samples of glycerol and ethyl glycol have been exposed to neutron-gamma radiation with fast neutron fluxes ranging from 7.7×10^{10} to 3×10^{12} $cm^{-2}s^{-1}$ and gamma-only irradiation at maximum dose-rates of 492 and 10 kGy hr^{-1}, respectively. A study of the dependence of product yield versus absorbed dose has been conducted, for total dose ranges of 1 to 100 kGy. The products of these exposures have been identified through Gas Chromatography – Mass Spectrometry (GC-MS) techniques.

Analysis comparing neutron-gamma irradiated samples of ethylene glycol and glycerol with gamma-irradiated samples shows no detectable qualitative difference between either irradiation type. Although, additional radiolysis products were detected when compared with available literature; ethyl acetate from ethylene glycol and solketal from glycerol. Quantitatively, neutron-gamma irradiation seems to be less effective at producing acetaldehyde from ethylene glycol, compared with gamma-only which can be explained through neutron moderation and consequent heating effects due to the borosilicate vials.

***Keywords*— Chemical analysis, Chemical engineering, Neutron radiation effects, Organic chemicals, Radiolysis**

I. Introduction

RADIATION chemistry concerns the study of chemical processes within materials that are initiated by ionizing radiation, these processes can also be referred to as radiolysis. Many applications using highly ionizing radiation have been realised throughout history and implemented in the current commercial world, such examples include polymer synthesis, sterilization and cleaning of flue gases. [1]

In literature, there have a number of examples in that types of radiation have been used to synthesise organic chemicals through catalytic processes. [2] Such an example is the production of ethyl bromide which proved to be successful for a number of years. [3] This process which was implemented by the Dow Institute in 1963 used a cobalt-60 source which emitted prompt-gamma rays with an activity of $\approx10^{15}$ Bq. This large scale process managed to output approximately 500 kt·year^{-1} with a radiochemical yield, G of $\approx10^{5}$ per 100eV and fulfilled Dow's entire production of ethyl bromide to the common market.

Most radiolysis literature has focused on the use of electron, heavy ion beams and gamma rays for their interactions with materials. There has been very little published for the use of neutrons or 'pile' irradiation for the radiochemical synthesis of commodity chemicals. [4] A neutron-emitting device such as a commercial nuclear fission reactor would serve as a source which would be suitably powerful enough to initiate chemical transformations in a timely manner.

A recent economic study concluded that such an integrated chemical-nuclear process could be economically feasible within a commercial nuclear reactor's design, with relatively little additional capital costs associated with the inclusion of radiochemical process. [5,6]

Currently, there is still a heavy reliance on the petrochemical industry for the supply of commodity chemicals; therefore research into alternative, chemical processes that utilize waste natural feedstocks is vital to create a more sustainable future.

For this experimental radiolysis research, Glycerol can be identified as a potential, abundant renewable feedstock for radiation processing due to its large opportunity for valorisation.[7] Due to this low price, it has mainly been used for low-energy applications such as animal feed or incineration fuels. Due to research within the last decade, a number of thermochemical routes to upgrade the versatility of glycerol to value-added products have been explored.

However, there have only been a few attempts to explore the use of highly ionizing radiation to transform and valorise glycerol into a useful commodity. Radiation research has been limited to explore the effects of gamma and ion beams. [8-10] There has been no available literature for the interactions of neutrons on glycerol or many other organic compounds, therefore the use of neutron or 'pile' irradiation has potentially been overlooked as a source of intense energy to catalyse reactions.

In this paper, the use of neutron and gamma radiation for the

This Work was supported by the EPSRC.

Arran George Plant (email:a.plant@lancaster.ac.uk), Dr Vesna Najdanovic-Visak (email:v.najdanovic@lancaster.ac.uk) and Professor Malcolm J. Joyce (email:m.joyce@lancaster.ac.uk) are with Lancaster University, LA1 0PF UK

Dr Luka Snoj (luka.snoj@ijs.si) and Anže Jazbec (anze.jazbec@ijs.si) are with the Jozef Stefan Institute, Jamova 39, 1000 Ljubljana, Slovenia

processing of glycerol and ethylene glycol to more useful, valuable products is explored. Ethylene glycol was chosen as another feedstock due to its similar chemical structure to glycerol; it also offered more favourable physical properties such as lower viscosity and boiling point which is preferable for sample preparation for analysis. A comparative study between existing literature of gamma studies and this new experimental work is conducted, together with the description of the methods used to irradiate and chemically analyse samples of ethylene glycol and glycerol.

II. Experimental Methods

A. Irradiation Methods

Samples of ethylene glycol and glycerol were irradiated in a MKII TRIGA reactor located at the Jožef Stefan Institute (JSI). This light water, pool-type reactor can operate at a maximum power of 250 kW, giving neutron fluxes as high as 10^{13} $cm^{-2}s^{-1}$. [11] The inner reactor core itself, where the largest neutron fluences can be utilized has five different radiation channels; most notably is Triangular Irradiation channel (TriC) in which samples as large 4 cm diameter can be lowered down into the reactor's core. Compared with the other channels of this comparable size, this channel offers a higher percentage of fast neutrons flux (≈ 33%) and is utilized for these irradiations.

For a comparative study between irradiation types, organic samples were irradiated in two different reactor modes, current and shutdown mode. The 'current mode' allows samples to be irradiated with a combination of neutrons, delayed γ and prompt γ rays. In this mode, the majority of absorbed dose deposited in the organic material would be contributed by neutrons. When the reactor is in its 'shut down' mode, the samples are irradiated with only delayed gamma waves with the dose rates dependent on the exponential decay of the activated fuel after reactor operations.

For a radiochemical product comparison between the two types of irradiations of n+γ (neutrons and gamma) and γ-only (gamma-only), samples of ethylene glycol and glycerol were irradiated with 50 kGy of absorbed dose in either field.

For a quantitative radiochemical comparison, samples were irradiated with doses of 1, 10, 20, 50 and 100 kGy for each reactor mode/irradiation type respectively.

The samples of ethylene glycol and glycerol were crimp-sealed within 20 ml round-bottomed GC-MS (Gas Chromatography-Mass Spectrometry) headspace vials with a mostly inert nitrogen atmosphere. In existing radiolysis literature, [8-10] organic volatile losses were problematic when preparing samples for analysis.

Once irradiated, the sample vials were stored in a -18°C freezer until transport back to Lancaster University for chemical analysis. The samples were then stored in a 4°C refrigerator until time for analysis.

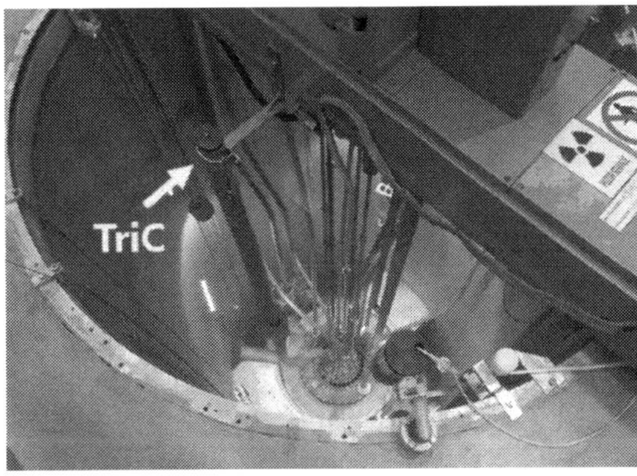

Fig. 1. Top View of the JSI TRIGA Reactor indicating the access port for the Triangular Irrational Channel.

B. Chemical Analysis Methods

To determine the difference in radiochemical products between n+γ and γ-only irradiation qualitatively, samples were analysed using gas chromatography-mass spectrometry (GC-MS) to identify the unknown products. The GC system that was used was a Shimadzu TQ8040 GC-MS, equipped with an AOC-6000 autosampler for repeatable, reproducible injections. Each separated peak response being referenced against the NIST 11 Standard Reference database for MS fragment patterns.

The GC separations of the organic samples were performed using a fused silica Zebron 624-Plus capillary column; with a length of 30 m x 0.25mm i.d. and a film thickness of 1.4 µm. To protect the stationary phase of the column, a 10m fused-silica column guard was installed at the beginning a connected to the analytical column. The oven temperature was programmed as follows; 40 °C (10mins); Ramp of 25 °Cmin^{-1} to hold at 300 °C (2.6 min). Split injections of a volume of 1 µl, with a split ratio of 20:1 with a constant column flow of 1.71 ml·min^{-1} during a run. The carrier gas type used was Helium with a purity of 99.999%. The injection port temperature was set at 300 °C, the MS ion source temperature at 250 °C and the GC to MS interface at 300 °C. A fragment mass range of m/z 30-300 was measured in scan mode with a scan rate of 1000 scans per second.

III. Results and Discussion

A. Qualitative Comparison & Radiolysis Products

Comparing the radiolysis products between the two different irradiation types, there did not seem to be any unique products formed from the n+γ irradiation type compared with the γ-only.

https://doi.org/10.1051/epjconf/202022509003

The following tables list the radiolysis products for both modes of irradiation;

TABLE I
N+Γ & Γ-ONLY RADIOLYSIS PRODUCTS OF ETHYLENE GLYCOL

Peak Number	NIST 11 Library Given Compound/ Radiolysis Product	% Similarity Match	Prior Literature? [10]
1	Carbon Dioxide	97	Y
2	Formaldehyde	97	Y
3	Methanol	97	Y
4	Acetaldehyde (Confirmed)	98	Y
5	Ethyl Acetate (Confirmed)	98	N.R. (Not Reported)
6 to 12	Miscellaneous Polymeric Structures	<89	N.R.

TABLE 2
N+Γ & Γ-ONLY RADIOLYSIS PRODUCTS OF GLYCEROL

Peak Number	NIST 11 Library Given Compound/ Radiolysis Product	% Similarity Match	Prior Literature? [8,9]
1	Carbon Dioxide	94	Y
2	Methanol	97	Y
3	Glycoaldehyde Dimer	96	Y
4	Acetol (Confirmed)	97	Y
5	Propylene Glycol	91	Y
6	1,2,3-Butanetriol	84	Y
7	2-(2-Methoxyethoxy) ethyl acetate	81	N.R.
8	Solketal (Confirmed)	97	N.R
9 to 11	Further Miscellaneous Polymeric Structures	<89	N.R.

Y indicates a mention in literature, N.R. indicates not reported radiolysis compounds.

Radiolysis Products that have been confirmed using internal standards have been indicated.

As shown by the analysis, there are a number of products from both the radiolysis of ethylene glycol and glycerol which agree with the available reported literature.

Some of the unreported products which are listed have been matched to chemicals with a low similarity match (<90%). It is thought that over time, the formation of larger, more complex polymeric chemicals have formed which cannot be referenced accurately against available fragmentation patterns. Such primary radiolysis products such as acetic acid and acetaldehyde can be highly reactive with other components within the mixture, allowing these higher-order products to form.

From ethylene glycol radiolysis, one previously unreported product that was detected was ethyl acetate, it is possible that it has been formed from the reaction between radiolysis-produced acetic acid and ethanol over time. Ethyl acetate is most commonly used as a solvent in a wide array of applications such as for cleaning electronics, coating formulations and pharmaceuticals.

From glycerol radiolysis, one notable product formed which is not listed in literature is solketal (isopropylidene glycerol). Solketal is a chemically protected form of glycerol which is commonly used as an intermediate in organic synthesis when attempting to selectively interact with a single hydroxyl functional group of glycerol. In this radiolysis research, Solketal is thought to have been formed through secondary processes over time with reactive primary products, similarly for the polymeric products. As a useful product, Solketal can be utilized as a green solvent with interesting properties [12] When compared to Glycerol, Solketal has a greater potential to be used as a renewable, green solvent with a boiling point of 188 °C, a viscosity of 11 cP and is miscible in most organic solvents. Glycerol derivatives such as Solketal also have great potential to be used as fuel additives.

B. Quantitative Comparison of Radiation Types

For an initial comparison between the two irradiation types, the concentration of a primary radiolysis product, acetaldehyde from ethylene glycol has been plotted against the absorbed dose.

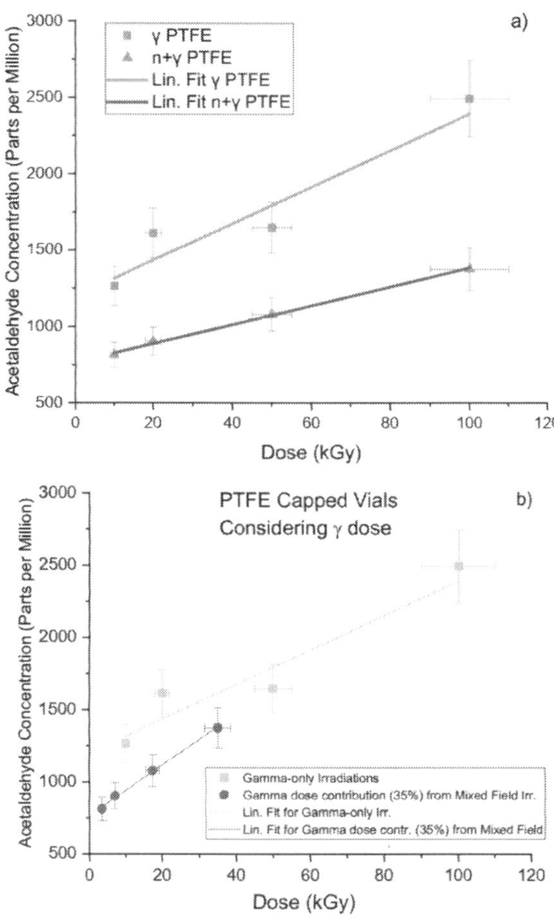

Fig. 2. Acetaldehyde concentration comparison between n+γ (red) and γ-only (green) radiation modes for the radiolysis of ethylene glycol in PTFE vials; a) Direct Comparison b) Considering γ-dose only for n+γ

When comparing the irradiation types, there seems to be a greater concentration detected from the γ-only irradiated samples for both septa types. This goes against the unsupported comments in the literature that the two radiation types are similar in terms of their radiolysis effectiveness. [1]

However, when considering just the gamma dose contribution (≈35%) from the mixed n+γ field irradiations, Fig. 2b can be generated.

From Fig. 2b, the two sets of data can be considered statistically similar in terms of their gamma-dose contribution. One explanation for the lower concentration observed is the boron content within the borosilicate vials themselves. The boron would absorb some of the absorbed dose intended for the organic samples. Boron-11 has a high cross-sectional area for thermal neutron (<0.025 eV) capture, acting as a neutron moderator. Also, Boron-10 has a high neutron cross-sectional area across the entire neutron energy spectra range which is unlike most isotopes. [13]

The second factor that could lead to reduced yields is the slight temperature increase accompanied by the boron acting as a neutron moderator, heating the vials during irradiation. A slight temperature increase could have a couple of unintended effects. One could be an increase in the rate of volatile loss from the vial. The second could be the increase in the rate of diffusion of the initial radical species, increasing the chance of recombination and decreasing molecular yields. [14]

IV. References

[1] Woods, Robert J., and Alexei K. Pikaev. "Applied radiation chemistry: radiation processing." John Wiley & Sons, 1993, pp. 37

[2] (IAEA), I.A.E.A. "Industrial uses of large radiation sources, vol.2. in Application of Large Radiation Sources in Industry" Vol II. . 1963. Austria.

[3] Harmer, D. E., et al. "The Dow ethyl bromide process: an industrial application of radiation chemistry." Industrial Uses of Large Radiation Sources. Proceedings of a Conference on the Application of Large Radiation Sources in Industry. Vol. II. 1963., pp. 205-230

[4] Collinson, E., and Albert J. Swallow. "The radiation chemistry of organic substances." Chemical Reviews 56.3, 1956 pp. 498, 506.

[5] McConnaughy, Thomas B., Matthew R. Shaner, and Eric W. McFarland. "A Techno-Economic Analysis of Chemical Processing with Ionizing Radiation." Chemical Engineering & Technology 40.6 (2017): 1196-1202.

[6] Schmeda-Lopez, Diego, Thomas B. McConnaughy, and Eric W. McFarland. "Radiation enhanced chemical production: Improving the value proposition of nuclear power." Energy 162 (2018): 491-504.

[7] Ciriminna, Rosaria, et al. "Understanding the glycerol market." European Journal of Lipid Science and Technology 116.10 (2014): 1432-1439.

[8] Baugh, Peter J., et al. "γ-Radiolysis of N2O-saturated aqueous glycerol solutions: Product yields and free radical mechanism." Radiation Physics and Chemistry (1977) 20.3 (1982): 215-222.

[9] Keough, T., et al. "Comparison of the stable products formed by fast atom bombardment and γ-irradiation of glycerol." Organic mass spectrometry 22.5 (1987): 241-247.

[10] Barker, S. A., J. S. Brimacombe, and E. D. M. Eades. "Action of gamma radiation on liquid ethylene glycol." Radiation research 22.2 (1964): 357-367.

[11] Snoj, Luka, Gašper Žerovnik, and Andrej Trkov. "Computational analysis of irradiation facilities at the JSI TRIGA reactor." Applied Radiation and Isotopes 70.3 (2012): 483-488.

[12] Garcia, José I., Héctor García-Marín, and Elísabet Pires. "Glycerol based solvents: synthesis, properties and applications." Green Chemistry 16.3 (2014): 1007-1033.

[13] N. Soppera, M. Bossant, E. Dupont, "JANIS 4: An Improved Version of the NEA Java-based Nuclear Data Information System", Nuclear Data Sheets, Volume 120, June 2014, Pages 294-296.

[14] Donoclift, Thomas. "The Radiolytic Steady-State and Factors Controlling H2 Production." Diss. University of Manchester, 2017.

https://doi.org/10.1051/epjconf/202022510001

Practical Demonstrations – Key to Efficient Explanations of Radioactivity to Pupils and Students

Matjaž Koželj
Milan Čopič Nuclear Training Centre
Jožef Stefan Institute
Ljubljana, Slovenia
matjaz.kozelj@ijs.si

Vesna Slapar Borišek
Milan Čopič Nuclear Training Centre
Jožef Stefan Institute
Ljubljana, Slovenia
vesna.slapar-borisek@ijs.si

Radko Istenič
Milan Čopič Nuclear Training Centre
Jožef Stefan Institute
Ljubljana, Slovenia
radko.istenic@ijs.si

***Abstract*— The radioactivity workshop and hands-on demonstrations in Milan Čopič Nuclear Training Centre enable us to effectively transfer some basic information about radioactivity, radiation and radiation effects to our young visitors. This activity has been well accepted and praised by teachers, who are aware of the subject importance for education.**

Keywords*— *radioactivity, demonstrations, workshop, pupils, students.

I. Introduction

Milan Čopič Nuclear Training Centre (ICJT), which is part of Jožef Stefan Institute (JSI), Ljubljana, was founded in 1989 to support training of Krško NPP workers. Since then, number of courses for control room staff and other technical personnel were prepared and implemented. Courses are also intended for members of technical support organisations, authorities and experts employed by Krško NPP subcontractors. ICJT also organise different courses and events in cooperation with international organisations or agencies like IAEA, ENS or EC.

Soon after successful conclusion of the initial courses, the decision has been made to expand our activities. At that time, the public opinion in Slovenia was still heavily influenced by Chernobyl accident and there were debates in media and among politicians about danger of nuclear energy and about the necessity to close Krško NPP. Explanations and clarifications related to safety of our plant provided by nuclear professionals were originally targeted to decision makers, and less to opinion makers. The information was presented and distributed within limited circles, also due to the limited interest of majority of media for, what was called at that time, "biased" opinion of nuclear experts.

Our aim was not to join those discussions, but to approach general public and to contribute to general opinion on long term basis. Since we were aware that the discussion on nuclear energy would follow into the following years, we have decided to establish nuclear technology information centre with permanent exhibition of nuclear technology. The vision was to become reliable and respected source of knowledge about nuclear technologies for general public. Since we had free basement at our premises, we were able to commission big lecture room and exhibition with posters and few mock-ups (Fig. 1) without huge investments and lasting constructions.

At the beginning (in the mid-nineties), emphasise was given to the Krško NPP technology and operation, but later a part related to radioactive waste management was added to the exhibition. In the recent decade, exhibition was complemented with overview of nuclear fusion technology research.

Fig. 1. Exhibition in nuclear technology information centre.

From the very beginning of the information centre operation, our most numerous and regular visitors are pupils and students from primary and secondary schools in Slovenia. In addition, other groups visit our centre – groups of university students, teachers, members of different professional associations, firefighters, groups of retirees, etc. Annually, our information centre visit more than 150 groups and more than 6500 visitors. Altogether, more than 3,500 school groups and more than 180,000 pupils, students, teachers and other persons visited our information centre since 1993 [1].

The exhibition was usually short addition to the lecture for our visitors at the beginning of information centre operation. Posters with information were prepared to support lectures with some additional data or visual material, and to provide explanation of some concepts from physics or engineering which are important for understanding of NPP operation. At that time, we have discovered that is that explanations of basic concepts of radioactivity and ionising radiation have de facto disappeared from school programmes. They were either pushed in schedule somewhere at the end of school year, in parallel with final exams like a filler, or were considered optional, leaving decision on presenting these contents to individual teacher.

It was also obvious that majority of teachers are not competent to speak about these subjects and that they are prone to avoided it. Radioactivity used to be one of the subjects discussed in Physics classes, but was later added to Chemistry classes. It would work in "old" times, but after Chernobyl accident radioactivity and ionising radiation were considered result of reactor operation and considering the consequences of accident, also extremely dangerous. The other problem was that just few schools had any equipment that can be used for classroom demonstration, and even if they had the equipment, teachers did not know how to use it properly.

What we have learned is that if we want to effectively transfer messages to our visitors, especially pupils and students, and if we want them to become active subjects in debates and decision process related to nuclear energy in Slovenia, we have to provide them with basic information about radioactivity, radiation and radiation effects to human beings. This knowledge should serve as a basis for evaluation and judgment of problems and questions that must be resolved if we want to continue living with nuclear energy in near future.

We felt that adding or expanding existing lectures would not be productive, and we decided to add some hands-on experiments and to prepare small radioactivity workshop with practical demonstrations of ionising radiation properties, demonstration of natural background radiation and radon. We were also considering idea to prepare hands-on experiments for all our demonstrations, but it would be costlier and we also had to comply with limited time that participants spent at our site. Therefore, we came to conclusion that the most effective approach would be to combine hands-on experiments at the exhibition with practical demonstrations in radioactivity workshop and to complement demonstrations with physical background explanation.

II. Hands-On Experiments And Demonstrations

A. Main Goals

There are two main goals behind hands-on experiments and demonstrations: first, we want to inform visitors that radioactivity is something natural and present everywhere in our environment, and second, we want to convince them that we know how to protect from excessive radiation.

The first message is the most important, since awareness of natural sources of radiation and the resulting exposure is essential for understanding of and discussion about the effects of ionising radiation in general. This is important not only when operation of existing (or planned) nuclear and radiation facilities are discussed and evaluated, but even more important when accidents occur, like the last one in Fukushima.

The importance of the second goal is related to acceptance of all radioactive sources, including nuclear or radiation facilities.

B. Approach and Design [2]

When we were designing exhibits and demonstrations our wish was to make them simple and attractive. Therefore, we decided to limit to basic information, which could be given in a short time (a few seconds!), without extensive explanation. Hands-on experiments should be self-explanatory (minimal necessary information should be written on the exhibit!) and visitors should be involved in experiment (they should "discover" information).

All instruments used should be very sensitive and must be equipped with loud acoustic indication (instrument read-outs are of minor importance), and only rate (counts per second or counts per minute) is allowed if some measurement result should be given. Abstract quantities, like dose, should be avoided. Old-fashioned detectors with End Window GM tubes and analogue displays satisfy these requirements and are used for demonstrations. For larger (and "noisier") groups of visitors, we use small web camera to project the instrument read-outs on the big TV screen.

For the hands-on experiments we use modern handheld instruments with sensitive pancake GM tubes and big digital displays as rate meters, but also with acoustic and optical (flashing light) indication.

C. Hands-on experiments

Two exhibits were originally prepared, one directly related to the demonstration of radioactivity, and the other more like a "teaser". The first exhibit is the radioactivity carousel, where different samples from the environment (fertiliser, potassium chloride, a watch with luminous dials, uranium glass, welding rods, gas mantle and an empty place – for background) are fastened on a round plate (Fig. 2). Visitors turns the plate and observe instrument response. Finding the most radioactive sample is a usual game that visitors play with the carousel, but nevertheless they also remember other samples. The carousel has also been copied (with our permission) as an applet on the Nucleonica.com site [3].

The second exhibit is just an instrument positioned above a table. Visitors can "measure" their own items, to check whether they are radioactive. As could be expected, nowadays the instrument is mostly used for checking mobile phones (Fig. 3).

Recently, we have also added two new historical exhibits: one exhibit is therapeutic ^{226}Ra source from the beginning of the twentieth century (Fig. 4), the oldest artificial radioactive source in Slovenia, which was donated by Faculty of Medicine, Ljubljana. The second exhibit are samples of uranium compounds (Fig. 5) prepared during pilot production of yellow cake in the 1980s at Jožef Stefan Institute, Ljubljana.

Fig. 2. Radioactivity carousel.

Fig. 3. Instrument for visitors.

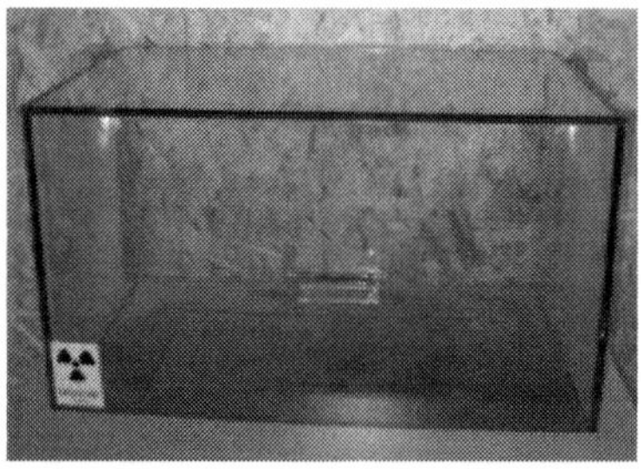

Fig. 4. Historical source: Therapeutic ^{226}Ra source.

Fig. 5. Samples of uranium compounds.

D. Radioactivity Workshop [2]

Demonstrations in the radioactivity workshop are intended to familiarise visitors with basic properties of radiation and principles of protection. The workshop takes place in a separate classroom with benches and a demonstration table with the measurement equipment. The walls of the classroom are covered with posters with some basic facts about radiation in support of the lecturer's explanations.

Fig. 6. Radioactivity workshop in Nuclear Technology Information Centre.

Three separate measurement stations were installed, each with a ratemeter and a sensitive End Window GM tube on a carrier which assures better visibility. Stations are dedicated to experiments with α, β and γ radiation.

Absorbers were prepared for each station: paper, cardboard (1 mm), and aluminium foil (10 μm) for α radiation, cardboard (1 mm), aluminium plates (0.5 mm), and acrylic glass plates (5 mm) for β radiation, and half-value layers of lead (10 mm), steel (16 mm), aluminium (45 mm) and concrete (70 mm) for γ radiation. Special rulers with visible marks were also prepared for demonstration of the range of α radiation in air and dose-to-distance dependency for γ radiation.

For demonstrations we use pure sources: ^{210}Po as α source, ^{90}Sr as β source, and ^{60}Co (with β shield) as γ source. It is important not to use mixed sources to avoid additional unconvincing explanations and justifications. All sources are school sources below exemption level – low activity is compensated with sensitive detectors. We also use an additional check sources – an old watch with luminous dials, similar to the watch on the radioactivity carousel, and Th-enriched Tungsten welding rods.

With this equipment we can demonstrate:

1) Natural background,
2) Absorption and range of α radiation in air, paper, and aluminium,
3) Absorption and range of β radiation in in cardboard, aluminium and acrylic glass,
4) Attenuation and half-value layers for γ radiation,
5) Dependency of count (dose) rate over distance,
6) Demonstration of radon progeny,

The intention of these demonstrations is to illustrate basic properties and differences between α, β, and γ radiation, and also to convince visitors that we know how to measure radiation and how to protect from it.

There is also one less sophisticated item in the classroom: an ordinary vacuum cleaner. We use it to demonstrate the presence of radon progeny in the air. Due to the elevated concentration of radon in the exhibition hall (it is in the basement of ICJT), it is possible to collect significant activity of radon progeny on a plain kitchen filter within a few minutes by pumping with the vacuum cleaner. The collected activity is usually much higher than the activity of the α-source, which is always a huge surprise for visitors.

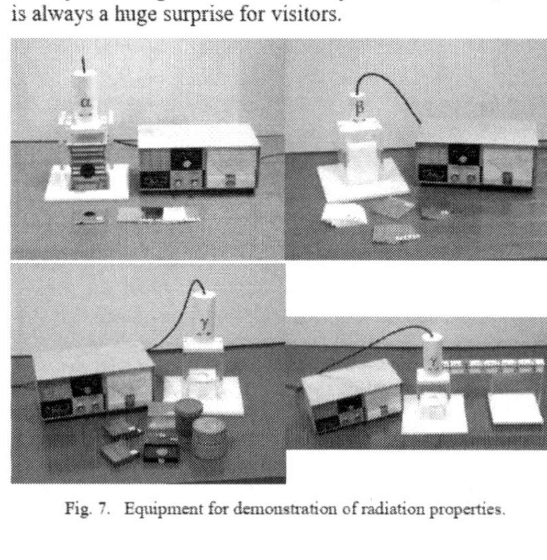

Fig. 7. Equipment for demonstration of radiation properties.

https://doi.org/10.1051/epjconf/202022510001

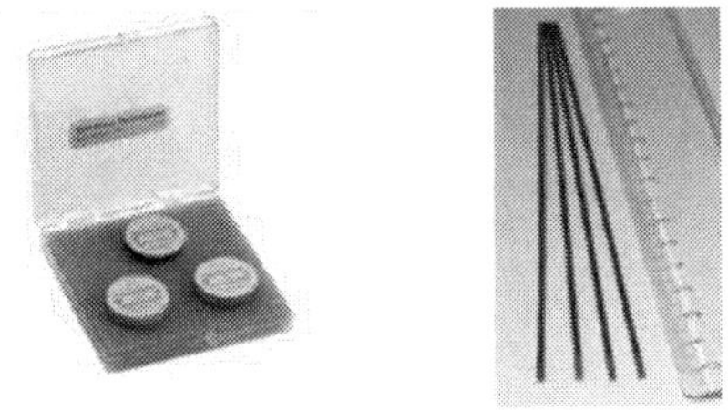

Fig. 8. Exempted sources are used for demonstrations.

There is even more "unusual" experimental gadget in the workshop: we use simple toy balloon for collection of radon progeny. After inflation, the surface of balloon becomes charged (some help with woollen rag is beneficial) and successfully collects radon progeny. Again, high radon concentration is helpful and collection of radon progeny in worst case requires no more than five to ten minutes.

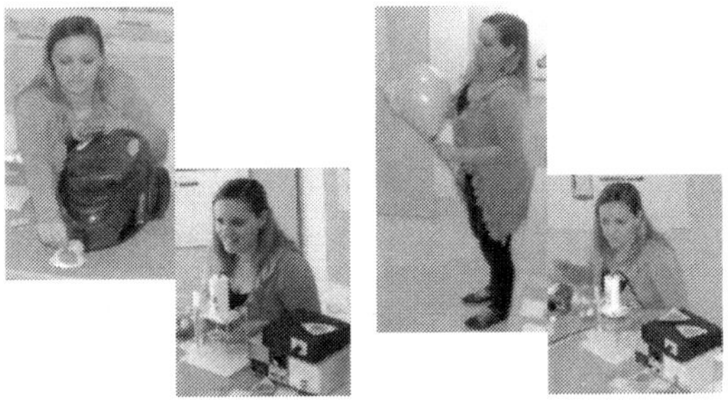

Fig. 9. Demonstration of radon progeny with filter and vacuum cleaner (left) and toy balloon (right).

Recent addition to our workshop is a small Cloud chamber, which we use in combination with web camera and TV screen. Visible traces of charged particles in the chamber on the big TV screen are illustrative support of our explanations about interaction of radiation in matter.

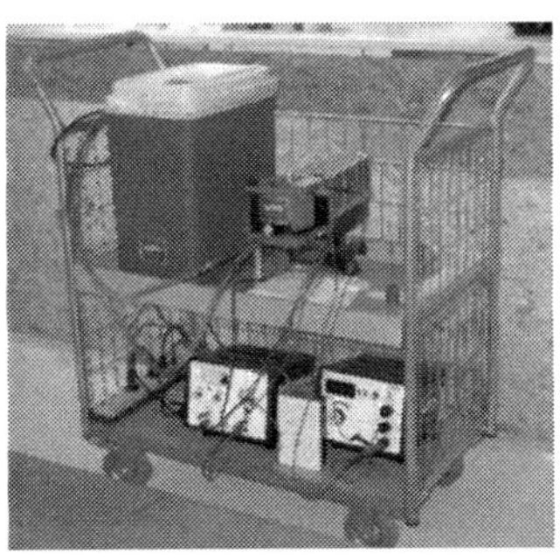

Fig. 10. Cloud chamber with cooler and power supplies.

III. Response From the Visitors

Almost all visitors of our Milan Čopič Nuclear Training Centre visit exhibition and radioactivity workshop. Most of them are teenagers, but we also have younger and also many older visitors, even retirees. Response from almost all visitors is positive, many of them have also express satisfaction for seeing "how things really work".

Our most faithful visitors are teachers who return every year with new groups of pupils or students. Teachers of Physics and Chemistry especially appreciate the radioactivity workshop as an addition to the regular classes. Most of them learned about radioactivity at University, but have no time and possibility to lecture and demonstrate radioactivity to their students. For other teachers (Biology, Math, Geography, History…) and students this is usually the first opportunity to observe radiation "at work".

In the radioactivity workshop demonstration of radon progeny is usually the most attractive and surprising. The visitors' reaction is almost always emotional, which also assures that it will stay in their memory. Other demonstrations in workshop are generally intriguing for the students with a more analytical mind and those, who already have some previous knowledge about radioactivity.

From time to time and for special interested groups, usually students from secondary schools, we prepare more advanced workshops where we perform measurements, and not only demonstrations. Although this is more demanding and time-consuming, it is also more rewarding.

IV. Conclusions

A number of teachers in elementary and secondary schools are returning visitors, who consider our workshop and demonstrations valuable addition to their lectures.

We can confirm that "A Picture is Worth a Thousand Words and an Experiment is Worth Fifty Slides" [4]. We have been very successful and efficient in the transfer of knowledge about radioactivity radiation and radiation effects to our visitors.

Probably one of the most important messages our visitors receive is related to the existence of natural background radiation and radon.

References

[1] M. Koželj, V. Slapar Borišek, "Spreading knowledge on radiation protection in Nuclear technology information centre," in ETRAP 2017 Conference Proceedings, Valencia, Spain, 30 May – 2nd June 2017, ENS, Brussels, 2017.

[2] M. Koželj, R. Istenič, "Radioactivity experiments for schools," in Proceedings, 22nd International Conference Nuclear Energy for New Europe - NENE 2013, Bled, September 9-12., 2013, L. Cizelj, M. Leskovar, and M. Uršič, Eds., Ljubljana, Nuclear Society of Slovenia, 2013, 1313.1-1313.9,

[3] https://www.nucleonica.com/wiki/index.php?title=Help%3ANatural_Radioactivity, July 2019,

[4] S. Landsberger, T. Tipping, W S Charlton, "A Picture is Worth a Thousand Words and an Experiment is Worth Fifty Slides: Development of Experiments in Health Physics to Reinforce Basic Radiation Protection Concepts", Contribution at 8th EUTERP Workshop, April 10-12, 2019, Qawra, St. Paul's Bay, Malta.

EPJ Web of Conferences **225**, 10002 (2020) https://doi.org/10.1051/epjconf/202022510002
ANIMMA 2019

A New Project dedicated to developing International MOBILity of APPrentices in the Master's Degree in Instrumentation, Measurement and Metrology (MOBIL-APP project)

C. Reynard-Carette
[1]Aix Marseille Univ, Université de Toulon, CNRS, IM2NP, Marseille, France
christelle.carette@univ-amu.fr

***Abstract*— The project concerns the outgoing international mobility of block-release apprentices preparing the Master's degree in Instrumentation, Measurement, Metrology of the Filière Instrumentation (Physics Department of the Science Faculty). It targets universities and research centers involved in nuclear energy and is based on ongoing partnerships with CFA Epure–Méditerranée (apprenticeship center), EDF, and CEA thanks to on a solid research background in the LIMMEX laboratory run jointly by CEA and AMU (IM2NP) and dedicated to Instrumentation and Measurements under EXtreme conditions, on the ANIMMA international conference and the EFMMIN Franco-Moroccan school. This has led to the development of a useful network of operators (partners involved in the project) in priority geographical areas for AMU's international strategy: Nuclear Reactor Laboratory at MIT, Moroccan Nuclear Center and Mohammed V University, Polish Nuclear Center, Slovenian Jožef Stefan Institute, Belgian Nuclear Center. The project will consist in sending a group of block-release apprentices for one week per year to a partner site who will propose visits to reactors, laboratories and companies, and participation in seminars, short courses, experimental projects and workshops together with local students. They will visit cultural places and prepare an activity report in English. When the project is completed its possible extension to other block-release courses at AMU will be examined because international mobility is a real advantage in terms of employability for young-people.**

Index Terms **Mobility, Education, Nuclear Energy, Research Reactor, Instrumentation.**

I. Introduction and Aims

First of all, the MOBIL-APP project concerns the outgoing international mobility of block-release training program apprentices preparing the Master's degree belonging to the "Filière Instrumentation" unit of the Physics Department of the Faculty of Sciences. The "Filière Instrumentation" was created in 1985 by AMU with the collaboration of the national nuclear agency CEA and industrial firms of South-East France.

This unit offers training courses in industrial R&D instrumentation, automation, testing means, process control and metrology. It has two main objectives: (1) to provide high-level scientific and professional programs and (2) to meet industrial and economic needs. It proposes two accredited degrees (Vocational Bachelor's and Master's degree) and includes over 150 students a year. The Filière Instrumentation unit has been ISO 9001 certificated since 2003 for the design and provision of initial, continuing and block-release training programs and the accreditation of prior experiential learning for all its training programs. Consequently, the courses may be taken in various ways: initial, continuing and block-release training programs, and the accreditation of prior experiential learning.

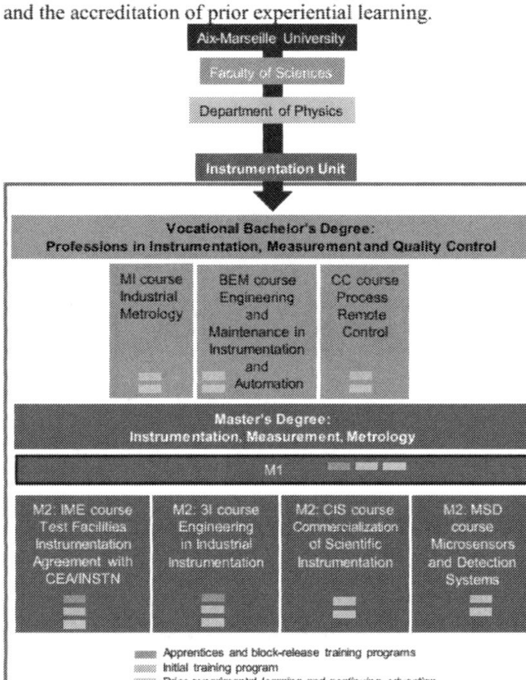

Fig. 1. Diagram of the degrees of the Instrumentation Unit ("Filière Instrumentation")

The MOBIL-APP project concerns the block-release training programs mainly with apprentices, but also with some students in the continuing program.

For students enrolled in this kind of block-release program (apprentice or continuing status), it is very difficult to take advantage of usual international mobility schemes for long periods. This is due to the specificities of the block-release schedule. Students enrolled in such programs must share periods between university and workplace as opposed to those in an initial program with long periods available for uninterrupted internship. In the case of our block-release students in M1 and M2, they begin with one month at university with the more fundamental courses. Then they pursue a period alternating every 2 weeks between their company and the university. Finally, they spend 1.5 months in the company during summer before defending their master's degree report on their activities there (cf. Fig.1). Our initial-program students in M1 and M2, however, have a 7-month period at university followed by a long continuous period devoted to their internship. The duration of the internship is from 3 to 6 months depending on the program year. The internship or the company alternation periods constitute a large education unit accounting for the most ECTS credits (12) in a year. In France, 43000 students accessed international mobility in 2017 as against 6 800 apprentices, for whom the duration was only from 2 to 4 weeks. Since 1987, 615000 French students have been involved the Erasmus program as against 25600 apprentices [1-2]. Students who choose the apprentice status often do so to receive a salary in order to finance their studies, and yet do not have an equal opportunity for international mobility with regard to students enrolled in initial training programs.

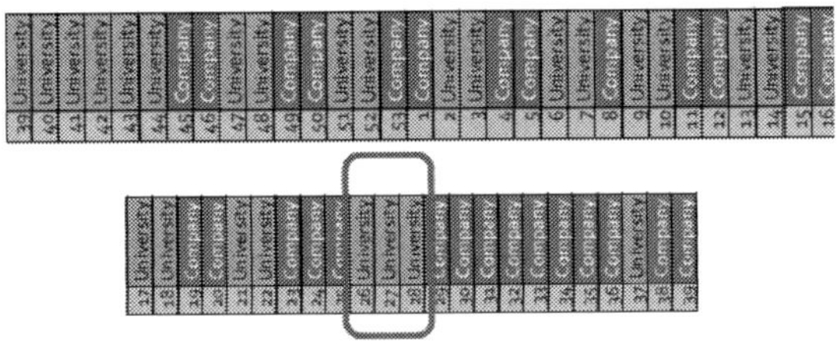

Fig. 2. Diagram of apprenticeship period schedule

Consequently, specific actions are required for apprentice students in order to increase their opportunities for outgoing international mobility as this is a crucial factor with various advantages. This is the aim of the MOBIL-APP project.
For instance, international mobility is a strong point on the curriculum vitae of alumni beginning their young professional career either in French companies with international activities or in international companies. It also provides the opportunity to practice English in real conditions. Students can gain in personal autonomy and acquire cultural experience. They can discover other educational systems. They can develop their capacity to adapt. All of which demonstrates the dynamism of the student in his CV. Moreover, outgoing international mobility performed in an academic environment with a high level of research activities can stimulate new vocations in the field of research for young students and thus a project for a PhD thesis. As they mainly work in industrial companies, few apprentices reach such an academic level, even if they have a capacity for hard work, possess organizational, scientific and technical skills, and obtain good academic results.
Furthermore, the project concerns the outgoing international mobility of apprentices with partners involved in a key field: the nuclear energy field. Nuclear energy field corresponds to an important job source [3]. It constitutes a major component of the energy mix and is in particular a recognized asset in terms of the reduction of greenhouse gas emissions. Nuclear energy could be a sustainable source of energy thanks in particular to continuous research and development on material and fuel behavior. The development and qualification of materials/fuels with relevant conditions for nuclear power plants contribute to the optimization and the demonstration of the safe operating of existing and upcoming nuclear power reactors - Generations II and III - (plant life extension and management, reinforcement of safety, waste and resource management, economic improvement), as well as the support for future generations of nuclear reactor design such as Generation IV reactors (improvement of the use of uranium resources and reduction of long-lived radioactive wastes).
Moreover, at present, major international nuclear fission and fusion facilities are under construction in the south of France (Cadarache area) such as the Jules Horowitz Reactor (JHR) and ITER.

Fig. 3. Photographs of major nuclear facilities under construction in the south of France

For instance, we can focus on the case of nuclear fission and of Material Testing Reactors (MTR) which offer the widest experimental capacity to study materials (accelerated ageing) and fuels under irradiation conditions (but also medical radioisotope and doped silicon carbide production). As all European Material Testing Reactors are over 40 years old and will be shut down in the next few years (for instance, the shutdown in 2016 of the only French MTR called OSIRIS near PARIS), the new JHR MTR is currently under construction at the CEA Cadarache. This new flexible experimental infrastructure will offer modern experimental capabilities with innovative on-line instrumentation, possess huge capacities with high performance and versatility, and provide harsher nuclear conditions. It involves an international consortium with various members (CIEMAT (Spain), SCK (Belgium), CVR (Czech Republic), VTT (Finland), IAEC (Israel), DAE (India), NNL (United Kingdom), STUDSVIK (Sweden), CEA, EDF and AREVA (France), JAEA (Japan)). It will be started up in 2022-2023.
Consequently, the MOBIL-APP project will target high-level international partners possessing specific nuclear facilities (in the USA, Europe and Mediterranean) in order to prepare students for present and future needs in the nuclear energy field while at the same time major research infrastructures now under construction require experts and young graduates with specific scientific and technological skills and international capacities.
As other European projects, such as the ENEN + project [4], the MOBIL-APP project aims to interest students for the nuclear field, to attract talents to careers in this field and to develop specific skills.

II. Project Characteristics

A. Partners and Funding

This project obtains a funding for two university years (2018-2020) from AMIDEX (Aix-Marseille University's foundation) in the framework of its Academy of Excellence program and from the CFA Epure Méditerranée.
The MOBIL-APP project involves:

✓ Two laboratories:

- IM2NP laboratory (UMR 7334), Detection department, Microsensors and Instrumentation research team,
- LIMMEX laboratory (laboratory run jointly by the CEA and AMU) dedicated to Instrumentation and Measurements under EXtreme conditions.

✓ And various partners:

- French partners such as the CFA Epure–Méditerranée (apprentice center), EDF and the CEA,
- International partners: Nuclear Reactor Laboratory at MIT (NRL, USA), Moroccan Nuclear Center (CNESTEN), Polish Nuclear Center (NCBJ), Jožef Stefan Institute (Slovenia), Belgian Nuclear Center (SCK-CEN)).

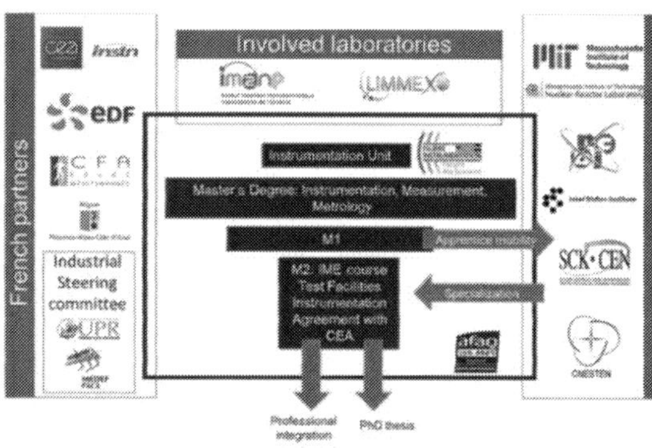

Fig. 4. Diagram of the MOBIL-APP project consortium

These partners were chosen and contacted because (except for the CFA) they are major research or industrial players in the nuclear energy sector or emergent player. They are involved in high level research works or research and development activities. They possess different and interesting kinds of nuclear facilities requiring dedicated instrumentation and measurement methods:

- ✓ CEA Cadarache: for instance the CABRI Reactor (pool type research reactor dedicated to fast power transients and therefore to simulating a reactivity insertion accident in a PWR electricity-generating reactor), JHR under construction (Material Testing Reactor, 100MWth), WEST Tokamak (Fusion Tokamak with actively cooled tungsten divertor), RES (research testing reactor for nuclear naval propulsion), and MASURCA (critical facility, zero power research reactor 5kWth)
- ✓ MIT: MIT-R, a pool type research reactor (6MWth) containing in-core experiment facilities and an instrumented water loop
- ✓ NCBJ: MARIA, a Material Testing Reactor (24MWth)
- ✓ SCK-CEN: BR2, a High Flux Material Testing Reactor (100MWth) and BR1 for research in reactor and neutron physics (700kWth)
- ✓ JSI and CNESTEN: TRIGA reactor (250kWth)
- ✓ EDF: Nuclear Power Plants from Gen II to Gen III

The CFA is involved in this project because it is an interuniversity (4 universities) center for the Region of Provence-Alpes-Côte d'Azur (PACA) dedicated to:
- Promoting and developing block-release training programs at university
- Improving the employability of young diploma holders
- Harmonizing the regional offer of university level apprenticeship training

The CFA is already involved in international mobility for apprentices belonging to engineering schools (in which case mobility is required in order to obtain the engineering diploma). Consequently, the CFA helps us especially at the administrative and regulatory level with the feedback from its experience with such diplomas.

B. Master's degree in Instrumentation, Measurement and Metrology

The Master's degree involved in the MOBIL-APP project is entitled Instrumentation, Measurement and Metrology and is composed of four specialities dedicated to Engineering, Test Facilities, Microsystems and Commercialization respectively (cf. Fig. 1). The project concerns the IME course on test facilities instrumentation. It was developed in collaboration with the CEA in 2004. It is now a diploma co-delivered by AMU and the national nuclear science and technology institute of CEA (INSTN). This training course comprises 545 hrs in the first year (or level M1) and 710 hrs in the second (M2). Each level is divided into several education units and comprises 60 ECTS credits. M1 is a common course year whatever the course chosen in M2. The IME course offers a specialisation in Nuclear Detection and Instrumentation. Currently this specialisation is obtained on the basis of preliminary courses in M1, a whole education unit in M2, partial education units in M2 and activities in a laboratory or company in M1 and M2. Teachers involved in this specialisation are mainly CEA researchers and research engineers, and teacher-researchers of the university belonging to the LIMMEX Laboratory and IM2NP Laboratory. This course (M1, M2) may be taken in various ways: initial, continuing and block-release training programs, and the accreditation of prior experiential learning.

C. Promotion 2018-2019

The project for the first mobility period realized this university year (2018-2019) consists in three phases corresponding to:

- ✓ The selection of the apprentices students and their preparation
- ✓ The mobility period for a group of selected students for 1 week in a partner
- ✓ The assessment

The two partners hosting the two groups of students were the Nuclear Reactor Laboratory of the MIT (USA) and the NCBJ (Poland).

 https://doi.org/10.1051/epjconf/202022510002

1) First step: selection and preparation in France

The selection is realized thanks to

- ✓ Different documents provided by students by respecting imperatively deadlines (a covering letter, a questionnaire in French and in English)
- ✓ A personal interview
- ✓ Academic results and English level
- ✓ Career project

For 2018-2019, the selection concerned 16 students out of 30 (8 for NRL mobility, 8 for NCBJ mobility). These selected students belong to several companies in different fields: ARCELORMITTAL, ARTELIA, CEA, CSTB, INEO NUCLEAIRE, IPSEN PHARMA BIOTECH, JTEKT AUTOMOTIVE LYON, M3E, PRECITEC, SECAUTO, SEMATEC METROLOGIE, TECHNICATOME.

The preparation for these 16 students consisted in:

- ✓ An Education Unit dedicated to an introduction to Research with international bibliographical works on Nuclear Detection and Instrumentation, then to the realization of a poster on selected topics (sensors, reactor specificities) and then a presentation associated to this poster (10.5 hours),
- ✓ Courses in scientific and technological communication, each student in particular having to prepare and give a talk presenting his/her educational background, company activities and professional responsibilities in English (14 hours)

Fig. 5. Photographs associated to the preparation step in France

- ✓ Different visits: a visit to Cadarache CEA Center, a visit to LIMMEX Laboratory (in particular research activities and associated experimental devices/benches dedicated to nuclear absorbed dose rate measurements by means of calorimeter in MTRs [5-8]) and a visit to EDF Company (Tricastin Power Plant)
- ✓ Talks given by representatives of EDF, the CEA (in particular talk dedicated to the Jules Horowitz Reactor) and the University
- ✓ Seminar with a scientist working on nuclear detection by means of semiconductors

2) Second step: mobility period

The mobility period consists in intense short stay (7 days). Each student-group is accompanied by two teachers-researchers belonging to the supervising and teaching team.

For 2018-2019, it included different activities:

- ✓ Partner talks for presenting their research activities and their reactor (MITR-II in USA and MARIA in Poland)
- ✓ Visits to reactors and associated laboratories
- ✓ Short courses and experimental projects (nuclear detection, communication, reactor control, gamma spectrometry)
- ✓ Seminars performed by students for presenting their academic cursus, their companies and their professional activities
- ✓ Visits to organisms and companies (Mirion Technologies (Canberra) Inc and RMD Inc in USA, POLATOM and The Maria Skłodowska Curie Memorial Cancer Centre and Institute of Oncology in Poland)
- ✓ Visits to university campus and museums
- ✓ Visits to other cultural places

Fig. 6. Photographs associated to the mobility period in USA

Fig. 7. Photographs associated to the mobility period in Poland

3) Third step: assessment

The supervising team will assess the mobility of each student by different means.

Each student will prepare an activity report in English including technical, scientific and cultural aspects.

Then student feedback will be done in various ways. A first assessment will be carried out just after their stay. A specific questionnaire will be delivered to each student with a 100% return targeted. Each student's degree of satisfaction concerning different points will thus be assessed: the duration and location of the mobility, organization, preliminary steps in Marseille, hosting partner, supervising team, training activities (seminar, short courses, facility visits), new skills acquired, enhancement of English level, other activities (cultural visits), impact on the professional/personal project.

Other assessments will be made during our surveys to establish the number of alumni who have become economically active 6 and 18 months after graduation (individual interview by phone). Questions will be added on their professional activity (advantage of mobility in finding a job or a thesis, international company or activities, job in or for the nuclear sector, opinion after several months about the international mobility experience), impact of the mobility period after a few months on the student's professional project and on the professional integration rate of diploma holders).

III. Conclusions

The MOBIL-APP project corresponds to a real opportunity for selected students (apprentices) enrolled in the block-release training program associated to the Master's degree in Instrumentation, Measurement, Metrology of Aix-Marseille University. An international mobility period owns various advantages inducing new soft and hard skills for each students that can be useful for their career. Moreover, as the international mobility is organized in the nuclear energy field, this mobility could confirm or generate a professional vocation in this field. The preliminary assessment for the first class will be realized before the end of September. The obtained output data will contribute to the enhancement of the second mobility period which will be done in June 2020 with two new groups of students.

Acknowledgements

The MOBIL-APP project leading to this publication has received funding from Excellence Initiative of Aix-Marseille University - A*MIDEX, a French "Investissements d'Avenir" programme.

References

[1] http://travail-emploi.gouv.fr/actualites/l-actualite-du-ministere/article/decryptage-5-cles-possibles-pour-davantage-de-mobilite-des-apprentis-en-erasmus

[2] S. Brunet, Report « le développement de l'apprentissage », January 2018.

[3] L. Parizot, Report « plan de programmation des emplois et des compétences », February 2019.

[4] https://plus.enen.eu/

[5] A. Volte, C. Reynard-Carette, A. Lyoussi, J. Brun, and M. Carette. Study of the Response of a New Compact Calorimetric Cell for Nuclear Heating Rate Measurements, IEEE TNS, Volume: 65, Issue: 9, Sept. 2018, Page(s): 2461 – 2470.

[6] Michel Giot, Ludo Vermeeren, Abdallah Lyoussi, Christelle Reynard Carette, Christian Lhuillier, Patrice Megret, Frank Deconinck, Bruno Soares Goncalves, Nuclear instrumentation and measurement: a review based on the ANIMMA conferences, EPJ Nuclear Sci. Technol. 3, 33 (2017) 49 pages.

[7] J. Brun, M. Tarchalski, C. Reynard-Carette, K. Pytel, A. Lyoussi, J. Jagielski, D. Fourmentel, J-F. Villard, M. Carette, Responses of Single-Cell and Differential Calorimeters: from Out-of-Pile Calibration to Irradiation Campaigns, IEEE Transactions on Nuclear Science, Vol. 63, NO. 3, June 2016, 1630-1639

[8] C. Reynard-Carette, G. Kohse, J. Brun, C. Carette, A. Volte, A. Lyoussi, Review Of Nuclear Heating Measurement By Calorimetry In France And USA, EPJ Web of Conferences 170, 04019 (2018)

EPJ Web of Conferences **225**, 10003 (2020) https://doi.org/10.1051/epjconf/202022510003
ANIMMA 2019

COSICAF, a fission chamber simulation tool for academic purposes

G. de Izarra, M. Lamotte, S. Bréaud, A. Pépino, P. Filliatre, C. Jammes
CEA, DEN, DER, Instrumentation, Sensors and Dosimetry Laboratory,
Cadarache, F-13108 St-Paul-Lez-Durance, France

Abstract—**Since a few years, simulation codes were built at CEA Cadarache to predict the signal of ionisation chambers and taylor detectors for specific applications. It is proposed here to present COSICAF, a tool developed for mainly academic purpose and rapid fission chamber prototyping. This numerical simulation, mostly based on semi-empirical models and Monte-Carlo method will help students to understand how ionisation chambers work. Through the paper, models and their numerical implementation will be discussed. A focus is made on recently implemented features like charge multiplication and correlated source which make the simulation of proportional counter possible. To demonstrate the interest of the code, simulations of a planar fission chamber is proposed.**

I. Introduction

Ionisation chambers sensitive to neutron flux and more specifically fission chambers play a key role in nuclear industry where they are used mainly for core neutron monitoring. Those detectors are constituted of a gastight body which contains at least two polarised electrodes, a volume of noble gas and an active material such as a fissile deposit. When a neutron travels trough the detector, it has a chance to react with the active material, producing energetic heavy ions. Heavy ions interact with the gas and electron/ion pairs are created. By gathering charged particles thanks to the polarised electrodes, current pulses are generated at the output of the detector.

Depending on the neutron flux and the ionisation chamber design, it is possible to operate such detector in three different modes [1]: pulse mode where pulses doesn't overlap; Campbell mode where pulses pile up but it is possible to link the count rate to signal variance; current mode, where the signal is high enough to produce a current proportional to the count rate[2].

Ionisation chamber simulation is complex since it involves neutron/matter interaction, heavy ion interactions with gases, and electromagnetics. Simulations are of great interest to prototype a detector, estimate their typical response in a given neutron flux, or even correct measurement performed during experiments.

To make simulation accessible, COSICAF [3], an open source software written in Octave/Matlab[4] was developed at the instrumentation, sensor and dosimetry laboratory of CEA Cadarache. This tool is part of the decade effort of our team on fission chamber simulation. Unlike previous softwares like CHESTER [5] and Pyfc [6] which were designed for research purpose only, COSICAF is dedicated mainly to rapid prototyping of detectors and for education purpose. Hence, included models are highly simplified and the core of the software is accessible to make practical work with minimum effort. Originally, the software was only able to simulate current pulses of fission chambers operating in saturation regime with Monte-Carlo method. It was recently enhanced to allow simulation of boron, lithium and helium 3 detectors. charge multiplication was also taken into account to handle proportional counters. At last, correlated sources were added to simulate detectors where more than one heavy ion is able to interact with the gas.

In this paper, the general structure of COSICAF will be presented; the models used during the simulation are summarized and implementation related problems like geometry description, particle tracking, electric potential computation are discussed in detail. In a second part, the new features of the code are described: It is shown how the signal generation procedure is modified to take into account the charge multiplication and how the COSICAF database can be expanded to simulate additional detectors. In the last part of the paper one example of simulation, a planar fission chamber built by the CEA fission chamber workshop is discussed.

II. General description of COSICAF

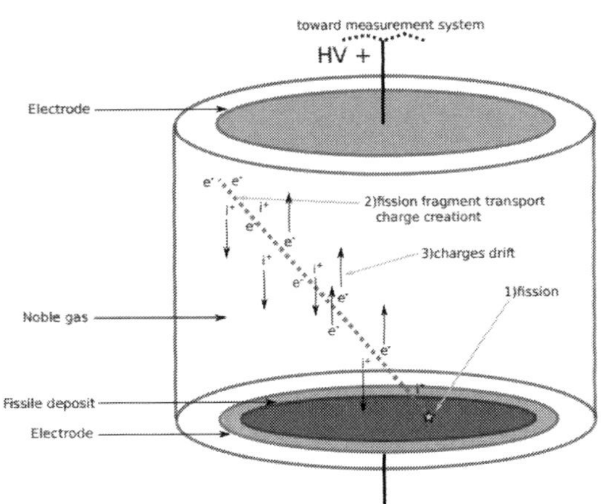

Fig. 1. Working principle of a fission chamber.

A scheme of ionisation chamber is shown in Fig. 1, it depicts the physical processes involved in a detector simulation which are:

1) Reaction between neutron and active material.
2) Emission of a heavy ion.
3) Transport of a heavy ion in the matter and generation of ion/electron pairs.
4) Transport of charge carriers.
5) Estimation of the signal induced by the charged particles.

For the sake of simplicity, since the simulation is performed in pulse mode, the interaction of neutrons with active material might be simplified into a uniform random generation of heavy ion inside the active material. Because of that hypothesis, the kinetic energy of the neutron is considered as negligible and no effects like flux depression or self shielding can be taken into account.

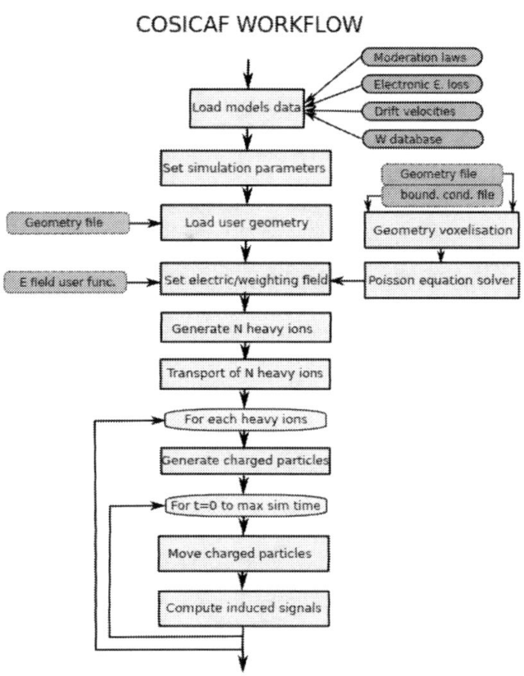

Fig. 2. COSICAF workflow. The core of the software uses a Monte-Carlo method to simulate N detector pulses.

The general structure of COSICAF is depicted in Fig. 2, each part of the software is detailed below.

A. Geometry definition

Source definition and particle tracking are highly dependent of the geometrical description of the problem. Before implementing any physics, it is mandatory to set a proper geometry representation. COSICAF was designed to handle geometry of arbitrary complexity with an undefined amount of heavy ion sources and electrodes. Because of its complexity, Constructive Solid Geometry used in software like MCNP is not applicable. Most of the fission chamber geometries are planar or cylindrical, we chose to reduce the geometry definition to cylinders and cylinder shells parallel to the z axis. All the volume are defined by a height, an outer and an inner radius plus the position of the cylinder base. Only the inclusion relation is implemented, it is handled by associating an inclusion number to each volume: Volumes with the higher inclusion numbers are contained in the volumes of lower inclusion numbers. Each piece of geometry is linked to a material, a volume mass, and an attribute that allows defining boundaries or how heavy ions are interacting with the volume. An example of a geometry is given in Fig.3 for a flat fission chamber.

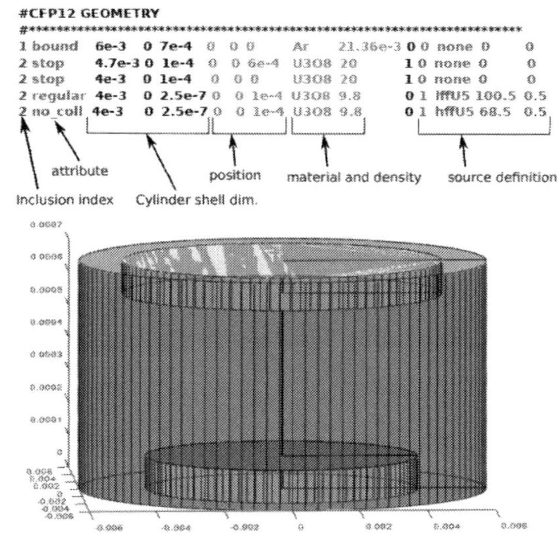

Fig. 3. Geometry description file and the resulting output generated by COSICAF.

B. Heavy ion generation and transport

Heavy ions are generated uniformly in source volume. For fission fragments (f.f.) emission, the use of the fission yield would lead to a large amount of data for heavy ions transport. Chung [7] and Mat'ev [8] showed it is possible to reduce the fission yield to only two mean fission fragments while keeping a rather good quantitative estimation of energy deposited by f.f. in material. From the JEFF 3.1.1 fission yield of U235, two mean fission fragments were estimated, their characteristic is available in Table I. Other fissile deposits might be considered but the original COSICAF code take only into account U235 fission fragments and alpha particles. At each randomly selected emission point, only one heavy ion is generated since it is assumed that one ion is able to exit the fissile deposit. Hence, cases with thin deposits or gaseous active materials could not be simulated until the inclusion of

TABLE I
DEFINITION OF THE TWO MOST PROBABLE FISSION FRAGMENT GENERATED DURING THE FISSION OF URANIUM 235.

kin. E(MeV)	Light f.f.			heavy f.f.		
	Z	M (uma)	E (MeV)	Z	M (uma)	E (MeV)
169.5 ± 3	38	95	100.5	54	138.5	68.5

correlated sources in the software. Then, a random propagation direction is associated to each generated heavy ion.

Many models are available to simulate the transport of heavy ion [9], [10]. Monte Carlo techniques and computation of the collision integral gives exact heavy ion trajectories and energy losses. It is also possible to compute energy losses on the ion projected trajectory toward the direction of propagation with PRAL algorithm [11]:

$$\frac{d\bar{x}}{dE} = \frac{1}{S(E)} - \frac{\mu}{2E}\frac{S_n}{S}\bar{x}, \quad (1)$$

with $S(E)$ and S_n (MeV/(mg.cm^2)) respectively the total and the nuclear stopping power, $\bar{x}$ the range and $\mu = M_t/M_p$ the ratio of the target and projectile particle mass. Other quantities might also be obtained with PRAL algorithm like straggling which compute the dispersion of the trajectory from the straight line.

In COSICAF, it is considered that heavy ion trajectories are straight lines. The error induced by this approximation is rather small for f.f. transport in noble gases: for light U5 fission fragment (U5 l.f.f.) in argon, the straggling correspond to 4.5% of the range. This hypothesis greatly simplify particle tracking: from the propagation direction vector it is possible to know all the volumes crossed by the heavy ion by computing only plane/line and cylinder/line intersections. To make the transport problem more tractable, Nguyen and Grossman [12] among others derived an analytical moderation law by solving equation 1 with a Bohr electronic stopping power, it gives:

$$E(x) = E_0\left(1 - \frac{x}{R_{E_0}}\right)^n, \quad (2)$$

where E(x) is the energy of the particle after a travel x, R_{E_0} is the range of the particle at the initial energy E_0 and n a semi empirical parameter related to the shape of the stopping power. A similar approach was chosen for COSICAF, but instead of the power law of eq.2 and the use of Bohr electronic stopping power, we decided to fit a five order polynomial on $f(1 - x/E_0) = E/E_0$ data computed with PRAL algorithm and total stopping powers provided by SRIM [13].

Considering the total stopping power is accurate to describe the transport of heavy ions, but it brings some errors while considering gas ionisation since only the electronic stopping power is related to charge pair creation. A correction can be estimated by computing the energy lost through electronic collisions as a function of the ion's kinetic energy. Neglecting this correction lead to an overestimation of charge pair creation of 4 % for U5 l.f.f in argon. Taking it into account makes the error below 0.4%. The estimation of charge pair generated along the ion tracks is performed by using W, the mean energy needed to generate a electron/ion pair. For noble gases, W is of the order of 30 eV [14].

C. Charge transport, induced signal

For each heavy ion track, charged particles are generated by using a database of W, only volume with a material referenced in the W database will contain ions and electrons. Since W is orders of magnitude smaller than the typical energy losses of heavy ion in gases, millions of charged particles are generated along the fission track. Hence, regarding the needed computing power and the approximations already done, it is vain to track each electron and ion. Instead, charged meta-particles representing thousand of electrons or ions are generated along the heavy ion track to make the simulation run in a reasonable amount of time.

Charged meta-particles are then drifted because of the polarised electrodes at a velocity given by:

$$v_p = \mu_p E \quad (3)$$

Where v_p (m/s) is the group velocity, μ_E the mobility and E the electric field. Our software relies on drift velocities from Raju [15] for the electrons and on Frost's semi-empirical model for ions[16]. It is assumed there is no charge recombination during the drift. From a practical point of view, COSICAF uses a explicit Euler scheme to move electric charges and to compute the induced signal. For each time step, meta-particles are moved by a $\vec{\Delta l}$ distance given by:

$$\vec{\Delta l} = \text{sgn}(q)\frac{\vec{E}}{E}.v_p(E).\Delta t \quad (4)$$

where dt is the time step, E the electric field and q the charge carried by the meta-particle. According to the Shockley-Ramo theorem [17], charged particles movement induces a current signal on electrode. On the k$_{th}$ electrode, it is computed by:

$$i_k = q_m v_d(E)\frac{\vec{E}}{E}.\vec{E_{0k}}. \quad (5)$$

where $\vec{E_{0k}}$ is the weighting field and $\vec{E}$ is the electric field.

All meta-particles are followed until the maximum computation time is reached or until all the charges are destroyed by hitting walls or electrodes.

D. Electrostatic computation

Electric field appears in all equations related to charged particles drift, this quantity is rather important since it is, with the weighting field, essential for signal computation. COSICAF has two different methods to describe electric field which will be used in the simulation. The first one allows to define a user function, which is useful for simple geometry since electric field derivation is straightforward. For more complex cases, it is not possible to compute the electric field analytically. COSICAF offers the possibility to solve Poisson

https://doi.org/10.1051/epjconf/202022510003

equation with Dirichlet boundary conditions using random walk technique. Let us consider the following equation in a 3 dimensional domain G:

$$\nabla.(\epsilon\nabla V(\vec{r})) = 0, \text{ for } \vec{r} \in G, \quad (6)$$

$$V(\vec{r}) = V_p, \text{on the boundary of } G, \quad (7)$$

where ϵ is the dielectric permittivity. By dividing G into a mesh with constant step h, it is possible to estimate the electric potential at $\vec{r}$ by setting particles at this position and starting random walks on the mesh until they reach boundaries. With N particles, the solution of the Dirichlet problem is given by [18], [19]:

$$V(\vec{r}) = \frac{1}{N}\sum_{i=1} NV_p(i) \quad (8)$$

with V_p the electric potential at boundaries reached with random walks. Since we are interested in the electric potential in the whole domain and not only in $\vec{r}$, the computation is speed up by updating all the potential along random walks. In COSICAF, the direction of the random walk is selected with a random number a from an uniform distribution and the probability to move in each direction which is given by:

$$P(x,y,z+h) = \frac{1}{3}\frac{\epsilon(x,y,z+h)}{\epsilon(x,y,z+h)+\epsilon(x,y,z-h)} \quad (9)$$

One of the main problem to use this numerical method is the need of a meshed geometry. COSICAF provides a routine which "voxelize" the surface defined geometry. An example of the voxelisation algorithm output is depicted in Fig.4 for a detector with a complex geometry. Once meshed geometry is available, a C++ routine computes the electric potential. The method is not computationally effective but is quite simple to implement and to understand which is perfect for an educational code. During the simulation, dedicated routine estimates the electric field directly from electric potential matrices, through numerical differentiation.

III. Advanced features of the simulation code

COSICAF is successful to simulate regular or complex fission chambers but, due to its limited database and the ways to handle sources it fails to describe other detectors like boron lined or helium 3 detector or even fission chambers were the two fission fragments are interacting with the gas. We recently extended the capability of COSICAF in order to make those simulations possible.

A. Moderation laws

New moderation laws were included in COSICAF for Tritium, Lithium and alpha particles in few materials and gases. Some are compared with TRIM computation in Fig.5. Their validity domains, between 5-0 MeV, allow to handle a wide range of reactions. Moderation laws have a fairly good agreement with TRIM computation but some discrepancies are noticeable at low energy.

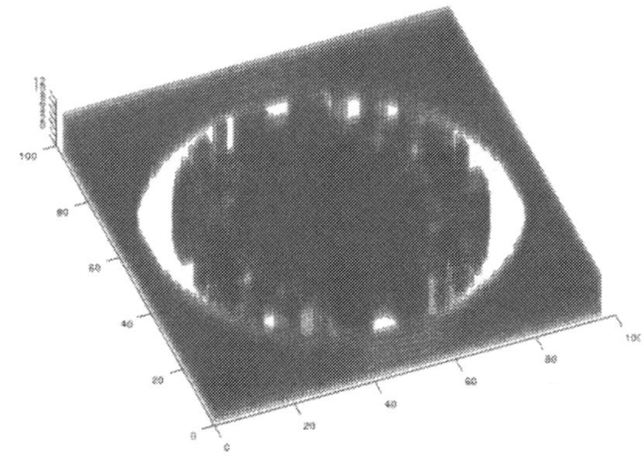

Fig. 4. Voxel representation of alumina dielectric permittivity in a complex detector geometry.

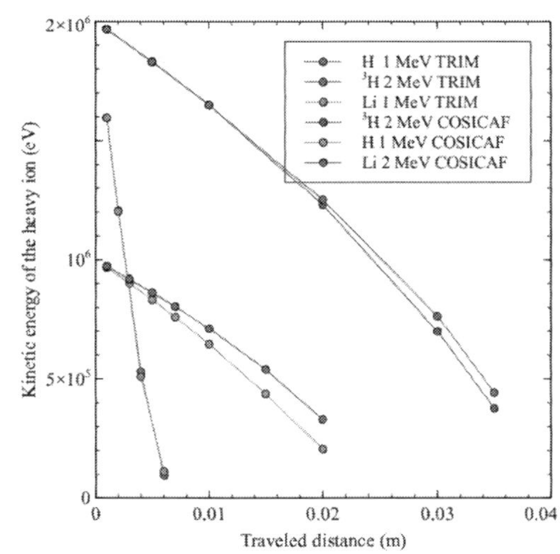

Fig. 5. Comparison of moderation laws with TRIM computations.

Hence results obtained on reactions such as neutron with ^{3}He where 0.75 MeV of kinetic energy is produced are not accurate and should be be considered as qualitative.

B. Correlated sources

To simulate the correlated emission of proton and tritium in helium 3 detector, it is necessary to modify the source definition. This was done first by uncoupling the source definition from the geometry definition. COSICAF is now able to load a source definition file with one or two heavy ions per source. The initialisation of heavy ion was modified as well to generate correlated emission of projectiles at the same location with opposite transport directions.

C. Charge multiplication

Charge multiplication is required to simulate proportional counters where the electric field is structured to make electrons energy large enough to ionise the gas and produce more electron-ion pairs. The electron multiplication is given by:

$$\frac{dn}{n} = \alpha dx \qquad (10)$$

where α is the Townsend's first ionisation coefficient and n the electron density. Ionisation coefficient are implemented in COSICAF through experimental data provided by Kruithof [20]. The Euler scheme used to move charges and compute induced signal was a bit modified to take into account the generation of new pairs. Each electronic meta-particle has its charges increased by a factor of:

$$\Delta q = q.\Delta l.\alpha \qquad (11)$$

For each time step, new charged meta-particles are generated to keep tracks of the generated ions, thus, the computation of proportional counters is order of magnitude slower than the simulation of detectors in saturation regime.

IV. EXPERIMENTAL VALIDATION

To demonstrate the interest of COSICAF, we simulated the CFTM, a special fission chamber built and designed by the CEA cadarache fission chamber workshop. It consists of a flat fission chamber with a 1.25 cm radius and a L=5 mm inter-electrode space. The detector is inflated with a mixture of Ar+20%N_2, and contains 10 μg of U235. The electric field is defined through a user function since for planar geometry, it is only given by:

$$\vec{E_{plan}} = \frac{\Delta V}{L}\vec{e_v} \qquad (12)$$

where ΔV is the polarisation voltage. Alumina spacers are not taken into account in the geometry and it is assumed that boundary effect on electric field is negligible.

A complete set of experiments was performed on the Minerve zero power reactor with the CFTM. Prior to the simulation, various charge spectra were recorded to find to different working regime of the detector. We found that experiments with a 2 bar filling pressure and a 700 V polarisation voltage was perfect since the saturation regime is well established and current pulses measurements with direct (fissile deposit holder is the cathode) and inverse (fissile deposit holder is the anode) polarisation were available. Only absolute measurements were performed thanks to carefully calibrated measurement systems in charge and in current.

Pulses were generated with Cosicaf in both inverse and direct polarisation and charge spectra were computed. Their comparisons with data recorded with a current pre-amplifier designed by CEA are available in Fig. 6 and 7.

For direct polarisation, we note that, even with the various approximations such as straight f.f. trajectories and simplified fission y ield, t he s pectra a re i n g ood a greement: t he mean

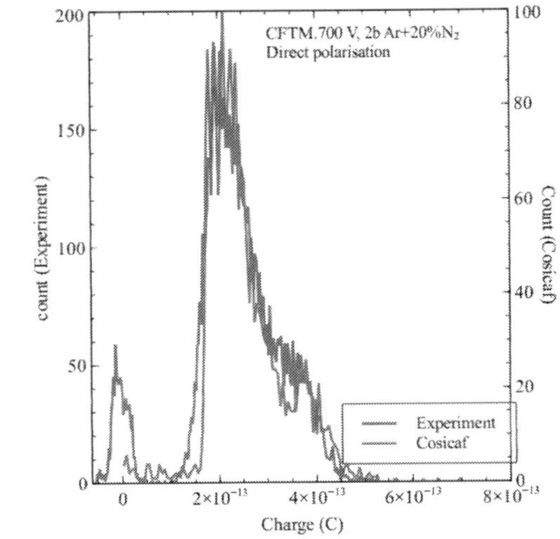

Fig. 6. Comparison between experimental and simulated CFTM charge spectra.

charge for the simulation is 2.51 10^{-13} C while it was measured at 2.53$\pm$ 0.10 10^{-13} C. The transimpedance of the pre-amplifier was measured, its uncertainty, mainly due to the non-negligible electronic noise during calibration procedure, was estimated to 4%. The slight difference between the simulation and the experimental result might be explained by the low histogram statistics and by approximations made on fission yield and on straight heavy ions trajectories.

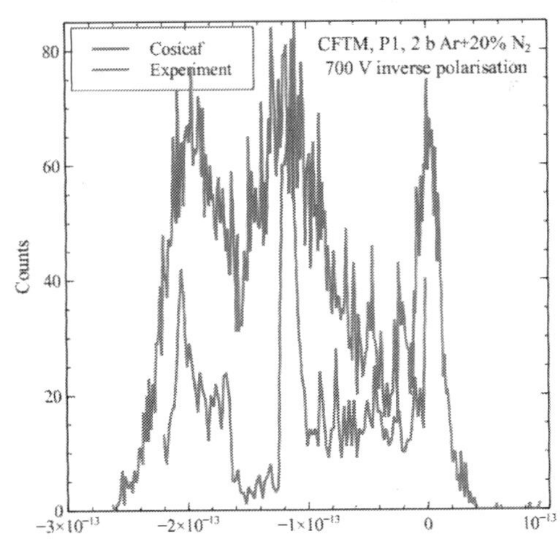

Fig. 7. Comparison between experimental and simulated CFTM charge spectra with inverse polarisation.

For inverse polarisation case, the structure of the spectrum of both simulation and experiment is similar. In this configuration, the end of fission tracks contributes more to the induced charge. After the initial energy losses, the stopping power become highly dependent to the fission yield, that is why two groups of pulses are noticeable on the experimental data. For COSICAF, two mean fission fragments are considered, which simplify the structure of each group: only the geometric effects spread the two lines corresponding to two fission fragments.

V. Conclusion

Through the paper, COSICAF, a simulation software for academic purpose was presented. It is useful to simulate current pulses at the output of ionisation chamber in saturation or in proportional regime. Since education is one of its primary goal, all the low levels function are exposed to the user as well as the physical models included in the software. The simulation helps students to understand how signal are generated in an ionisation chamber but also how to perform computational physics. Through the software, Monte-Carlo methods can be studied either as problem like electrostatic computation or particle tracking. There is still room for improvement, the COSICAF database is still reduced but additional materials and heavy ions could be easily integrated. Campbell and current mode could also be simulated from the computed pulses with additional work [1], [21]. Even if the software assumes strong approximations, it was shown that for fission chamber signal, the quantitative results are in rather good agreement with the experimental ones. That might not be the case for other ionisation chambers which rely on low energy reactions: straight trajectory is not accurate anymore and the ion transport model is too simple.

References

[1] Z. Elter, C. Jammes, I. Pázsit, L. Pál, and P. Filliatre, "Performance investigation of the pulse and campbelling modes of a fission chamber using a poisson pulse train simulation code," *NIMA*, vol. 774, pp. 60–67, 2014.
[2] S. Chabod, G. Fioni, A. Letourneau, and F. Marie, "Modelling of fission chambers in current mode–analytical approach," *Nuclear Instruments and Methods*, vol. 566, pp. 633–653, 2006.
[3] G. de Izarra, "Cosicaf software," https://github.com/gdeIzarra/cosicaf, 2019.
[4] J. W. Eaton, D. Bateman, S. Hauberg, and R. Wehbring, *GNU Octave version 4.4.1 manual: a high-level interactive language for numerical computations*, 2017. [Online]. Available: https://www.gnu.org/software/octave/doc/v4.4.1/
[5] P. Filliatre, C. Jammes, B. Geslot, and R. Veenhof, "A monte carlo simulation of the fission chambers neutron-induced pulse shape using the garfield suite," *Nuclear Instruments and Methods in Physics Research Section A*, vol. 678, pp. 139–147, 2012.
[6] Z. Elter, "pyfc: a trim-based fission chamber pulse shape simulator," CHALMERS UNIVERSITY OF TECHNOLOGY, Tech. Rep., 2015.
[7] A. K. Chung and M. A. Prelas, "The transport of heavy charged particles in a cylindrical nuclear-pumped plasma," *Nuclear Science and engineering*, vol. 86, pp. 267–274, 1984.
[8] V. Y. Mat'ev, "Energy deposition by fission fragments in nuclear-pumped lasers: 1. a general method of calculation," *Technical Physics*, vol. 46, no. 1, pp. 68–75, 2001.
[9] J. Ziegler, J. Biersack, and M. Ziegler, *SRIM - The Stopping and Range of Ions in Matter*, 2008.
[10] J. Hajnrych, P. Filliatre, and B. Geslot, "Penetration of gases by fission fragments—comparison between selected models and the data," *NIMA*, vol. 911, pp. 15–18, 2018.
[11] J. P. Biersack, "New projected range algorithm as derived from transport equations," *Z. Phys. A - Atoms and Nuclei*, vol. 305, pp. 95–101, 1982.
[12] D. Nguyen and L. Grossman, "Ionisation by fission fragments escaping from a source medium," *Nuclear science and engineering*, vol. 30, pp. 233–241, 1967.
[13] J. Ziegler, M. Ziegler, and J. Biersack, "Srim–the stopping and range of ions in matter (2010). nuclear instruments and methods," *Nuclear Instruments and Methods in Physics Research Section B: Beam Interactions with Materials and Atoms*, vol. 268, pp. 1818–1823, 2010.
[14] *ICRU report 31, average energy required to produce an ion pair*, International Commissionon Radiation Units and Measurements, 1979.
[15] G. Raju, *Gaseous Eectronics: theory and practice*. Taylor and Francis, 2006.
[16] R. Goyatina and S. Maiorov, "Approximation of the characteristics of ion drift in parent gas," *Plasma Physics Reports*, vol. 43, pp. 75–82, 2017.
[17] Z. He, "Review of the Shockley–Ramo theorem and its application in semiconductor gamma-ray detectors," *Nuclear Instruments and Methods in Physics Research A*, vol. 463, pp. 250–267, 2001.
[18] S. Kakutani, "Two-dimensional brownian motion and harmonic functions," *Proc. Imp. Acad*, vol. 20, pp. 706–714, 1944.
[19] M. Sadiky, *Numerical techniques in electromagnetics*, C. Press, Ed., 2001.
[20] A.A.Kruithof, "Townsend's ionization coefficients for neon, argon, krypton and xenon," *Physica*, vol. 7, pp. 519–540, 1940.
[21] P. Filliatre and C. Jammes, "A comprehensive numerical modelling of a fission chamber to be operated over a wide dynamics in the vessel of a sodium-cooled reactor," in *Physor 2016 proceedings*, 2016.

EPJ Web of Conferences **225**, 10004 (2020) https://doi.org/10.1051/epjconf/202022510004
ANIMMA 2019

International school for radiation measurements in Asia

Masaharu, Nomachi
Institute for Radiation Sciences
Osaka University
Toyonaka, Osaka, Japan
nomachi@rcnp.osaka-u.ac.jp

Hai, Vo Hong
University of Science
Vietnam National University, Ho Chi Minh city
Ho Chi Minh city, Vietnam
vhhai@hcmus.edu.vn

***Abstract*—Radiation measurement is a key technology for various sciences. The education of radiation science is demanding in Southeast Asian countries. We are collaborating with Universities in Southeast Asia. Hands-on exercise is important. However, it was not so easy to provide enough number of setups. Recent developments change the situation. The granularity of detectors in particle physics and medical apparatus is increasing. It means detector unit becomes smaller and less expensive. We are developing setups for radiation measurement exercises based on those new developments. Those system is portable to carry. In Osaka University, we are organizing schools for radiation measurements inviting Southeast Asian students. In addition, we are organizing schools in Southeast Asia. Compact system helps us to carry.**

***Index Terms*—International school, radiation measurements**

I. Introduction

Demands on Radiation Science increases in South East Asian countries. Interests in medical application is also glowing. We are collaborating to organizing school for radiation measurements. How to use instruments is not in our scope. We want educate the people so as to understand the instruments. Hands-on experience is very much important for students. The school provides not only lectures but also laboratory exercises. The goal of the school is,

- To train young students in the area of radiation detectors and its applications.
- To stimulate the development of real-time system in South East Asia.
- To promote the participation of young scientists in radiation measurements and related fields.

Our activites and Lessons-and-Learned are described.

II. Activities

A. JSPS school in Osaka

The school held in Osaka University with the budget of the Japan Society for the Promotion of Science (JSPS). We organized "A primer course of experimental particle and nuclear physics, Advanced Technology Training Program for Radiation Measurements in Osaka University". The school was organized three times.

- 20 February 2013 21 March 2009
- 12 July-10 August 2010

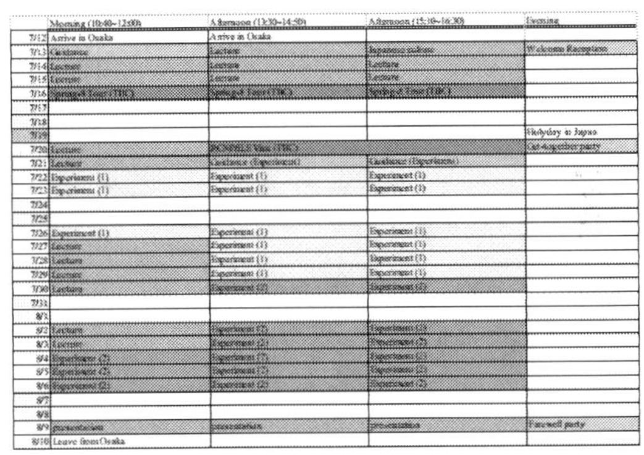

	Morning (10:40~12:00)	Afternoon (13:30~14:50)	Afternoon (15:10~16:30)	Evening
7/12	Arrive in Osaka	Arrive in Osaka		
7/13	Guidance	Lecture	Japanese culture	Welcome Reception
7/14	Lecture	Lecture	Lecture	
7/15	Lecture	Lecture	Lecture	
7/16	Spring-8 Tour (TIC)	Spring-8 Tour (TIC)	Spring-8 Tour (TIC)	
7/17				
7/18				
7/19				Holyday in Japan
7/20	Lecture	[illegible] Visit (TIC)		Get-together party
7/21	Lecture	Guidance (Experiment)	Guidance (Experiment)	
7/22	Experiment (1)	Experiment (1)	Experiment (1)	
7/23	Experiment (1)	Experiment (1)	Experiment (1)	
7/24				
7/25				
7/26	Experiment (1)	Experiment (1)	Experiment (1)	
7/27	Lecture	Experiment (1)	Experiment (1)	
7/28	Lecture	Experiment (1)	Experiment (1)	
7/29	Lecture	Experiment (1)	Experiment (1)	
7/30	Lecture	Experiment (2)	Experiment (2)	
7/31				
8/1				
8/2	Lecture	Experiment (2)	Experiment (2)	
8/3	Lecture	Experiment (2)	Experiment (2)	
8/4	Experiment (2)	Experiment (2)	Experiment (2)	
8/5	Experiment (2)	Experiment (2)	Experiment (2)	
8/6	Experiment (2)	Experiment (2)	Experiment (2)	
8/7				
8/8				
8/9	presentation	presentation	presentation	Farewell party
8/10	Leave from Osaka			

Fig. 1. Schedule in 2010.

- 8 September -3 October 2011

The school ran for one month as shown in Fig. 1. The number of students invited were 21∼24. The list of countries are in the table I. Students are graduate school students.

TABLE I
THE NUMBER OF PARTICIPANTS

Country	2009 school	2010 school	2011 school
Vietnam	13	11	8
Chine	11		7
Malaysia		5	7
Indonesia		2	
Philippine		2	
Myanmar		1	

Participants are selected by interview. The interview was done at their Universities.

The setups of the exercises are those for undergraduate exercises in physics department in Osaka University.

B. Mini-school

The number of students we can invite to Japan is limited. In order to have more students, we also did Min-schools in Asian countries. Because of limited budget, we were not able to bring large setups. However, thanks to the recent developments of electronics, we can use compact equipments. We measured speed-of-light. Laser modulated by the signal from USB function generator was measured by a photo sensor with USB oscilloscope. The school was 2-3 days as shown in Fig. 2. The schools were held as follows.

- Da Nang, 20-21 September 2010
- Can Tho 22 - 25, Sept 2010
- Ho Chi Minh city, 14-15, March 2011
- Beijing, 19-20 April 2011
- Da Nang, 1-3 December 2011
- Qui Nhon 26 July - 3 August 2012
- Kuala Lumpur 14-16 February 2013

The number of students in the exercises were about 20. The number of students in the lectures were about 20∼50. The students invited to Osaka help the school as teaching assistants.

1-Dec		2-Dec		3-Dec	
		9:00	Lecture (Tokita)	9:00	Lecture (Kishimoto)
		10:20		10:20	
		10:40	Lecture (Itahashi)	10:40	Lecture (Itahashi)
		11:30	Accelerators for applied work and their requirements	11:30	Accelerators for applied work and their requirements
13:30	Lecture (Ogawa) Physics of Excitons	13:30	Experiments	13:50	Experiments
14:50			1) speed of light		1) speed of light
			2) Radioactivity in our environment		2) Radioactivity in our environment
15:10	Lecture (Kuno) Introduction to Osaka University				
16:30	Muon Science and MuSIC@Osaka	16:30		16:30	

Fig. 2. Schedule of DaNang school in 2011

C. *JST SAKURA Science program*

The school held in Osaka with the budget of Japan Science Technology Agency (JST), Sakura Science Plan. We organized "International school for Radiation Measurements in Osaka University". The school was organized five times and will be held also in 2019.

- 18 January - 27 January 2015 10 participants
- 9 November -18 November 2015 10 participants
- 24 November - 3 December 2016 18 participants
- 8 November - 17 November 2017 18 participants
- 31 October -9 November 2018 18 participants

The school ran for 10 days as shown in the schedule. Students invited were from Hanoi, DaNang, Dalat, Ho Chi Minh City, Kuala Lumpur, Jakarta, Bandung. Students are master course students and last year in undergraduate.

Since the there is not enough time to do full course of exercise, just introductory ones were provided. Activation experiments by AVF cyclotron beam were provided. In addition, we did pulse shape analysis and time of flight measurements.

D. *IEEE/NPSS school*

We collaborate with IEEE/NPSS for "Real time school". Thanks to distinguish lectures of IEEE/NPSS, we had lectures

	Morning	Afternoon	Evening
10/31 (Wed)	Arrive in Osaka	Guidance RCNP visit	
11/1 (Thu)	Lecture	Experiment (preparation)	
11/2 (Fri)	Experiment (preparation)	Experiment (Measurement)	
11/3 (Sat)	Experiment (Measurement)	Experiment (Measurement)	
11/4 (Sun)	Free		
11/5 (Mon)	Visit (Osaka – Shirakawa Gou – Toyama)		
11/6 (Tue)	Visit (Kamioka - Osaka)		
11/7 (Wed)	Lecture	Analysis	
11/8 (Thu)	Analysis	Presentation	
11/9 (Fri)	Leave from Osaka		

Fig. 3. Schedule in 2018 SAKURA Science plan.

and exercises by top level researchers from international laboratories. So far, we organized three times. Osaka University hosted the school in Osaka and co-hosted the school in Vietnam.

- 2-7 June, 2014, Osaka
- 18-26 July, 2016, Ho Chi Minh city
- 7-17 July, 2018, Cape Town

The schedule of the school in Ho Chi Minh city is shown in Fig. 4

July 2016	18 Mon.	19 Tue.	20 Wed.	21 Thu.	22 Fri.	23 Sat.	24 Sun.	25 Mon.	26 Tue.
Morning 9:00 ~10:20		Orientation	Lecture A-2	Exercise-1	Lecture C-2	Exercise-2	Excursion	Lecture D-2	Students Presentation
Morning 10:40 ~12:00		Lecture A-1	Lecture B-2	Exercise-1	Lecture D-1	Exercise-2		Lecture E	
		Lunch	Lunch	Lunch	Lunch	Lunch		Lunch	
Afternoon 13:30 ~14:50		Lecture B-1	Exercise-1	Exercise-1	Exercise-2	Exercise-2		Analysis	
Afternoon 15:10 ~16:30	Registration	Lecture C-1	Exercise-1	Exercise-1	Exercise-2	Exercise-2			Farewell Party
Evening		Welcome Party				Move to downtown			
	University Guest house					Hotel in down town			

Fig. 4. Schedule of Ho Chi Minh city school in 2016.

The exercises use Raspberry-Pi [1] and DRS-4 [2] evaluation module. Data acquisition on Raspberry Pi is studied. FPGA programing is also studied.

III. Lessons and Learned

A. *Setups of the exercises*

We used rather cheap setups for the exercises. It saves our budget or makes possible to have more setups. In addition, affordable price means the Universities of participants can buy to repeat after the school. Cheap does not mean poor performance. Thanks to recent developments in HEP, Medical etc., Price-per-Channel is dramatically decreasing, High

 https://doi.org/10.1051/epjconf/202022510004

performance evaluation boards are available such as DRS-4 [2], Raspberry Pi [1] etc. Recent developments also make downsizing. Our setup fits in our luggages. We don't need to carry or send heavy and legacy CAMAC crates for exercises.

B. Hands on exercises

So as all students can do hands-on experiences, we use one setup for 2∼3 students. Cheaper setup allows us to prepare enough number of the setups.

At the end of the school, students make presentation. Not only follow the procedure written in the text, students prepare and discuss the presentation to explain what they did. It is very useful to understand what they did.

We asked young scientists in South East Asian countries to participate as a teaching assistant. We can show them how to do exercises. We expent they repeat the exercises at home institute. It may motivate the future collaboration. We also hire PhD students as a teaching assistant. It should be a good experience in their carrier.

C. Interview

Selection of participants are important. We selected students with interview. Interview was useful to find motivated students We also were able to know the level of students and what they want. The level of school (lectures / exercises) must fit to participants. We need to adjust the level to the students. Interview with local professors was useful to show what kind of students were selected. In addition, Inviting professors, we show the school to them. It is important to continue the collaboration.

The interview were done at the University which has candidates. SKYPE connection was also applyed. Face to face interview at their University was valuable for the collaboration involving young scientist.

REFERENCES

[1] https://www.raspberrypi.org
[2] https://www.psi.ch/en/drs/evaluation-board

AUTHOR INDEX

AUTHOR INDEX

AUTHOR INDEX

AUTHOR INDEX

AUTHOR INDEX

EDP Sciences – Web of Conferences
17, Avenue du Hoggar
Parc d'Activité de Courtabouf
BP 112
F-91944 Les Ulis Cedex A
France

ISBN 978-1-7138-0288-4

CURRAN ASSOCIATES INC.
proceedings.com